Modern Control Engineering

Modern Control Engineering

Second Edition

Katsuhiko Ogata

University of Minnesota

PRENTICE HALL, Englewood Cliffs, New Jersey 07632

Library of Congress Cataloging-in-Publication Data

Ogata, Katsuhiko.
    Modern control engineering / Katsuhiko Ogata. -- 2nd ed.
        p.   cm.
    ISBN 0-13-589128-0
    1. Automatic control. 2. Control theory.   I. Title.
TJ213.O28 1990
629.8--dc20   89-26499
                                                        CIP

Editorial/production supervision: Christina Burghard
Manufacturing buyer: Lori Bulwin

 © 1990, 1970 by Prentice-Hall, Inc.
A Division of Simon & Schuster
Englewood Cliffs, New Jersey 07632

Printed in the United States of America
10 9 8 7 6 5 4 3 2

ISBN 0-13-589128-0

Prentice-Hall International (UK) Limited, *London*
Prentice-Hall of Australia Pty. Limited, *Sydney*
Prentice-Hall Canada Inc., *Toronto*
Prentice-Hall Hispanoamericana, S.A., *Mexico*
Prentice-Hall of India Private Limited, *New Delhi*
Prentice-Hall of Japan, Inc., *Tokyo*
Simon & Schuster Asia Pte. Ltd., *Singapore*
Editora Prentice-Hall do Brasil, Ltda., *Rio de Janeiro*

Contents

2 Mathematical Modeling of Dynamic Systems

3 Basic Control Actions and Industrial Automatic Controllers

4 Transient-Response Analysis and Steady-State Error Analysis

PART II
Control Systems Analysis and Design by Conventional Methods

5 Root-Locus Analysis

6 Frequency-Response Analysis

7 Design and Compensation Techniques

8 Describing-Function Analysis of Nonlinear Control Systems

PART III
Control Systems Analysis and Design by State-Space Methods

9 Analysis of Control Systems in State Space

10

Design of Control Systems by State-Space Methods   772

Appendix   Vector-Matrix Analysis   890

References   946

Index   951

Preface

This book presents a comprehensive treatment of the analysis and design of continuous-time control systems. It is intended to be used as a text for a first course in control systems. It is written at the level of the senior engineering (electrical, mechanical, aerospace, or chemical) student. The prerequisite on the part of the reader is that he or she has had courses on introductory differential equations, introductory vector-matrix analysis, introductory circuit analysis, and mechanics.

This second edition includes many new subjects that were not discussed in the first edition. These subjects include the pole placement approach to the design of control systems, design of observers, computer simulation of control systems, among others.

The material of this book is organized into three parts. Part I, which consists of the first four chapters, deals with basic materials for analyzing control systems by both conventional and state-space approaches. Part II, which consists of the next four chapters, treats the root-locus method, frequency-response methods, design of control systems based on these methods, and the describing-function analysis of nonlinear control systems. Part III, which consists of the last two chapters, presents the analysis and design of control systems by state-space methods.

The outline of each chapter is as follows: Chapter 1 gives the fundamental concepts of feedback control systems and a basic mathematical background necessary for the understanding of the book. Chapter 2 treats mathematical modeling of physical components and systems and develops transfer function models and state-space models of such components and systems. Chapter 3 discusses various control actions and gives details of pneumatic and hydraulic controllers. Chapter 4 presents transient response analysis, steady-state error analysis, and computer simulation of control systems.

Chapter 5 treats the root-locus analysis of control systems. Chapter 6 deals with the frequency-response analysis of control systems. Chapter 7 presents design and compensation techniques using the root-locus and frequency-response approaches. Chapter 8 gives the describing-function analysis of nonlinear control systems.

Chapter 9 discusses basic materials of modern control theory, including concepts of controllability and observability and Liapunov stability analysis. Chapter 10 deals with the design of control systems based on state-space approaches. The topics include pole-placement design techniques, state observers, design of servo systems, quadratic optimal control systems, model-reference control systems, and adaptive control systems. Since a good background of vector-matrix analysis is needed for state-space analysis, a summary of vector-matrix analysis is provided in the appendix for convenient reference.

The book is written from the engineer's point of view. The basic concepts involved are emphasized and highly mathematical arguments are carefully avoided in the presentation of the materials. (Mathematical proofs are provided when they contribute to the understanding of the subjects presented.) All the material has been organized toward a gradual development of control theory.

Throughout the book, carefully chosen examples are presented at strategic points so that the reader will have a clear understanding of the subject matter discussed. Also, numerous solved problems are provided at the end of each chapter and the appendix. Since these problems are an integral part of the text, it is suggested that the reader study all such problems carefully. Also, many problems (without solutions) of various degrees of difficulty are included to test the reader's ability to apply the theory involved.

Most of the materials including solved and unsolved problems presented in the book have been class-tested in senior and first-year graduate level courses in the field of control systems at the University of Minnesota. In the examples and problems, physical quantities are introduced in the International System of Units (SI), the Metric Engineering System of Units, or the British Engineering System of Units so that the reader will be able to work with any of these systems of units.

This book may be used in the following ways: For a four-hour one-quarter course (or three-hour one-semester course) if the objective of the course is to study basic materials of control systems including an introductory state-space approach, then Parts I and II (with possible omission of Chapter 8) may be covered, and if the objective of the course is to emphasize the state-space approach to the analysis and design of control systems, then Parts I and III (with possible omission of Chapter 3) may be covered. In a four-hour one-semester course most of Parts I, II, and III may be covered with flexibility in omitting certain subjects. In a four-hour two-quarter sequence of courses the entire book may be covered. This book can also serve as a reference book or self-study book for practicing engineers who wish to study control theory.

I would like to express my appreciation to many former students who solved numerous problems included in this book, to those students who made many constructive comments, and to Scott Hanowski and Tim Berg for improving the presentation of certain materials in the book. Appreciation is also due to anonymous reviewers who made valuable suggestions at the midstage of the revising process. Finally, I would like to thank Tim Bozik, Executive Editor, for his enthusiasm in publishing this second edition.

Katsuhiko Ogata

CHAPTER 1

Introduction to Control Systems Analysis

1–1 INTRODUCTION

Automatic control has played a vital role in the advance of engineering and science. In addition to its extreme importance in space-vehicle systems, missile-guidance systems, aircraft-autopiloting systems, robotic systems, and the like, automatic control has become an important and integral part of modern manufacturing and industrial processes. For example, automatic control is essential in the numerical control of machine tools in the manufacturing industries. It is also essential in such industrial operations as controlling pressure, temperature, humidity, viscosity, and flow in process industries.

Since advances in the theory and practice of automatic control provide the means for attaining optimal performance of dynamic systems, improving productivity, relieving the drudgery of many routine repetitive manual operations, and more, most engineers and scientists must now have a good understanding of this field.

Historical review. The first significant work in automatic control was James Watt's centrifugal governor for the speed control of a steam engine in the eighteenth century. Other significant works in the early stages of development of control theory were due to Minorsky, Hazen, and Nyquist, among many others. In 1922, Minorsky worked on automatic controllers for steering ships and showed how stability could be determined from the differential equations describing the system. In 1932, Nyquist developed a relatively simple procedure for determining the stability of closed-loop systems on the basis of open-loop response to steady-state sinusoidal inputs. In 1934 Hazen, who introduced the term servomechanisms for position control systems, discussed the design of relay servomechanisms capable of closely following a changing input.

During the decade of the 1940s, frequency-response methods made it possible for engineers to design linear closed-loop control systems that satisfied performance requirements. From the end of the 1940s to early 1950s, the root-locus method due to Evans was fully developed.

The frequency-response and root-locus methods, which are the core of classical control theory, lead to systems that are stable and satisfy a set of more or less arbitrary performance requirements. Such systems are, in general, acceptable but not optimal in any meaningful sense. Since the late 1950s, the emphasis in control design problems has been shifted from the design of one of many systems that work to the design of one optimal system in some meaningful sense.

As modern plants with many inputs and outputs become more and more complex, the description of a modern control system requires a large number of equations. Classical control theory, which deals only with single-input, single-output systems, becomes powerless for multiple-input, multiple-output systems. Since about 1960, because the availability of digital computers made possible time-domain analysis of complex systems, modern control theory, based on time-domain analysis and synthesis using state variables, has been developed to cope with the increased complexity of modern plants and the stringent requirements on accuracy, weight, and cost in military, space, and industrial applications.

Recent developments in modern control theory are in the field of optimal control of both deterministic and stochastic systems, as well as the adaptive and learning control of complex systems. Now that digital computers have become cheaper and more compact, they are used as integral parts of these control systems. Recent applications of modern control theory include such nonengineering systems as biological, biomedical, economic, and socioeconomic systems.

Definitions.   The *controlled* variable is the quantity or condition that is measured and controlled. The *manipulated* variable is the quantity or condition that is varied by the controller so as to affect the value of the controlled variable. Normally, the controlled variable is the output of the system. *Control* means measuring the value of the controlled variable of the system and applying the manipulated variable to the system to correct or limit deviation of the measured value from a desired value.

In studying control engineering, we need to define additional terms that are necessary to describe control systems, such as plants, disturbances, feedback control, and feedback control systems. In what follows, definitions of these terms are given. Then a description of closed-loop and open-loop control systems follows and the advantages and disadvantages of closed-loop and open-loop control systems are compared. Finally, definitions of adaptive and learning control systems are given.

Plants.   A plant is a piece of equipment, perhaps just a set of machine parts functioning together, the purpose of which is to perform a particular operation. In this book, we shall call any physical object to be controlled (such as a heating furnace, a chemical reactor, or a spacecraft) a *plant*.

Processes.   The *Merriam–Webster Dictionary* defines a process to be a natural, progressively continuing operation or development marked by a series of gradual changes that succeed one another in a relatively fixed way and lead toward a particular result or end; or an artifical or voluntary, progressively continuing operation that consists of a series of controlled actions or movements systematically directed toward a particular result or end.

In this book we shall call any operation to be controlled a *process*. Examples are chemical, economic, and biological processes.

Systems. A system is a combination of components that act together and perform a certain objective. A system is not limited to physical ones. The concept of the system can be applied to abstract, dynamic phenomena such as those encountered in economics. The word system should, therefore, be interpreted to imply physical, biological, economic, and the like, systems.

Disturbances. A disturbance is a signal that tends to adversely affect the value of the output of a system. If a disturbance is generated within the system, it is called *internal*, while an *external* disturbance is generated outside the system and is an input.

Feedback Control. Feedback control refers to an operation that, in the presence of disturbances, tends to reduce the difference between the output of a system and some reference input and that does so on the basis of this difference. Here only unpredictable disturbances are so specified, since predictable or known disturbances can always be compensated for within the system.

Feedback Control Systems. A system that maintains a prescribed relationship between the output and some reference input by comparing them and using the difference as a means of control is called a *feedback control system*. An example would be a room-temperature control system. By measuring the actual room temperature and comparing it with the reference temperature (desired temperature), the thermostat turns the heating or cooling equipment on or off in such a way as to ensure that the room temperature remains at a comfortable level regardless of outside conditions.

Feedback control systems are not limited to engineering but can be found in various nonengineering fields as well. The human body, for instance, is a highly advanced feedback control system. Both body temperature and blood pressure are kept constant by means of physiological feedback. In fact, feedback performs a vital function: It makes the human body relatively insensitive to external disturbances, thus enabling it to function properly in a changing environment.

As another example, consider the control of automobile speed by a human operator. The driver decides on an appropriate speed for the situation, which may be the posted speed limit on the road or highway involved. This speed acts as the reference speed. The driver observes the actual speed by looking at the speedometer. If he is traveling too slowly, he depresses the accelerator and the car speeds up. If the actual speed is too high, he releases the pressure on the accelerator and the car slows down. This is a feedback control system with a human operator. The human operator here can easily be replaced by a mechanical, electrical, or similar device. Instead of the driver observing the speedometer, an electric generator can be used to produce a voltage that is proportional to the speed. This voltage can be compared with a reference voltage that corresponds to the desired speed. The difference in the voltages can then be used as the error signal to position the throttle to increase or decrease the speed as needed.

Servo Systems. A servo system (or servomechanism) is a feedback control system in which the output is some mechanical position, velocity, or acceleration. Therefore, the terms servo system and position- (or velocity- or acceleration-) control system are synonymous. Servo systems are extensively used in modern industry. For example, the completely automatic operation of machine tools, together with programmed instruction, may be accomplished by the use of servo systems. It is noted that a control system, whose output (such as the position of an aircraft in space in an automatic landing system) is required to follow a prescribed path in space, is sometimes called a servo system, also. Examples include the robot-hand control system, where the robot hand must follow a prescribed path in space, and the aircraft automatic landing system, where the aircraft must follow a prescribed path in space.

Automatic Regulating Systems. An automatic regulating system is a feedback control system in which the reference input or the desired output is either constant or slowly varying with time and in which the primary task is to maintain the actual output at the desired value in the presence of disturbances. There are many examples of automatic regulating systems, some of which are the Watt's flyball governor (for details, see Section 1–2), automatic regulation of voltage at an electric power plant in the presence of a varying electrical power load, and automatic control of the pressure and temperature of a chemical process.

Process Control Systems. An automatic regulating system in which the output is a variable, such as temperature, pressure, flow, liquid level, or pH, is called a *process control system*. Process control is widely applied in industry. Programmed controls such as the temperature control of heating furnaces in which the furnace temperature is controlled according to a preset program are often used in such systems. For example, a preset program may be such that the furnace temperature is raised to a given temperature in a given time interval and then lowered to another given temperature in some other given time interval. In such a program control the set point is varied according to the preset time schedule. The controller then functions to maintain the furnace temperature close to the varying set point.

Closed-loop Control Systems. Feedback control systems are often referred to as *closed-loop control systems*. In practice, the terms feedback control and closed-loop control are used interchangeably. In a closed-loop control system the actuating error signal, which is the difference between the input signal and the feedback signal (which may be the output signal itself or a function of the output signal and its derivatives), is fed to the controller so as to reduce the error and bring the output of the system to a desired value. The term closed-loop control always implies the use of feedback control action in order to reduce system error.

Open-loop Control Systems. Those systems in which the output has no effect on the control action are called *open-loop control systems*. In other words, in an open-loop control system the output is neither measured nor fed back for comparison with the input. One practical example is a washing machine. Soaking, washing, and rinsing in the washer operate on a time basis. The machine does not measure the output signal, that is, the cleanliness of the clothes.

In any open-loop control system the output is not compared with the reference input. Thus, to each reference input there corresponds a fixed operating condition; as a result, the accuracy of the system depends on calibration. In the presence of disturbances, an open-loop control system will not perform the desired task. Open-loop control can be used, in practice, only if the relationship between the input and output is known and if there are neither internal nor external disturbances. Clearly, such systems are not feedback control systems. Note that any control system that operates on a time basis is open loop. For example, traffic control by means of signals operated on a time basis is another example of open-loop control.

Closed-loop versus Open-loop Control Systems. An advantage of the closed-loop control system is the fact that the use of feedback makes the system response relatively insensitive to external disturbances and internal variations in system parameters. It is thus possible to use relatively inaccurate and inexpensive components to obtain the accurate control of a given plant, whereas doing so is impossible in the open-loop case.

From the point of view of stability, the open-loop control system is easier to build because system stability is not a major problem. On the other hand, stability *is* a major problem in the closed-loop control system, which may tend to overcorrect errors that can cause oscillations of constant or changing amplitude.

It should be emphasized that for systems in which the inputs are known ahead of time and in which there are no disturbances it is advisable to use open-loop control. Closed-loop control systems have advantages only when unpredictable disturbances and/or unpredictable variations in system components are present. Note that the output power rating partially determines the cost, weight, and size of a control system. The number of components used in a closed-loop control system is more than that for a corresponding open-loop control system. Thus, the closed-loop control system is generally higher in cost and power. To decrease the required power of a system, open-loop control may be used where applicable. A proper combination of open-loop and closed-loop controls is usually less expensive and will give satisfactory overall system performance.

Adaptive Control Systems. The dynamic characteristics of most control systems are not constant for several reasons, such as the deterioration of components as time elapses or the changes in parameters and environment. Although the effects of small changes on the dynamic characteristics are attenuated in a feedback control system, if changes in the system parameters and environment are significant, a satisfactory system must have the ability of adaptation. Adaptation implies the ability to self-adjust or self-modify in accordance with unpredictable changes in conditions of environment or structure. The control system having a candid ability of adaptation (that is, the control system itself detects changes in the plant parameters and makes necessary adjustments to the controller parameters in order to maintain an optimal performance) is called the *adaptive control system*.

In an adaptive control system, the dynamic characteristics must be identified at all times so that the controller parameters can be adjusted in order to maintain optimal performance. (Thus, an adaptive control system is a nonstationary system.) The concept of adaptive control has a great deal of appeal to the system designer since an adaptive control system, besides accommodating environmental changes, will also accommodate moderate engineering design errors or uncertainties and will compensate for the failure of minor system components, thereby increasing overall system reliability.

Learning Control Systems. Many apparently open-loop control systems can be converted into closed-loop control systems if a human operator is considered a controller, comparing the input and output and making the corrective action based on the resulting difference or error.

If we attempt to analyze such human-operated closed-loop control systems, we encounter the difficult problem of writing equations that describe the behavior of a human being. One of the many complicating factors in this case is the learning ability of the human operator. As the operator gains more experience, he or she will become a better controller, and this must be taken into consideration in analyzing such a system. Control systems having an ability to learn are called *learning control systems*. Recent advances in adaptive and learning control applications are available in the literature.

Classification of control systems.   Control systems may be classified in many different ways. Some of them are given next.

Linear versus Nonlinear Control Systems.   Strictly speaking, most physical systems are nonlinear to various extents. However, if the range of variations of the system variables is not wide, then the system may be linearized within a relatively small range of variation of variables. For linear systems, the principle of superposition applies. Those systems for which this principle does not apply are nonlinear systems. (More details on linear and nonlinear systems are given in Chapter 2.)

It is noted that, in some cases, nonlinear elements are intentionally introduced to the control system to optimize the performance. For example, time-optimal control systems use on–off types of controls. Many missile and spacecraft control systems also use on–off controls.

Time-invariant versus Time-varying Control Systems.   A time-invariant control system (constant coefficient control system) is one whose parameters do not vary with time. The response of such a system is independent of the time at which an input is applied. A time-varying control system is a system in which one or more parameters vary with time. The response depends on the time at which an input is applied. An example of time-varying control systems is a space-vehicle control system, where the mass decreases with time as the fuel is consumed during flight.

Continuous-time versus Discrete-time Control Systems.   In a continuous-time control system, all system variables are functions of a continuous time t. A discrete-time control system involves one or more variables that are known only at discrete instants of time.

Single-input, Single-output versus Multiple-input, Multiple-output Control Systems.   A system may have one input and one output. An example is a position control system, where there is one command input (desired position) and one controlled output (output position). Such a system is called a single-input, single-output control system. Some systems may have multiple inputs and multiple outputs. An example of such multiple-input, multiple-output systems is a process control system that has two inputs (pressure input and temperature input) and two outputs (pressure output and temperature output).

Lumped-parameter versus Distributed-parameter Control Systems.   Control systems that can be described by ordinary differential equations are lumped-parameter control systems, whereas distributed-parameter control systems are those that may be described by partial differential equations.

Deterministic versus Stochastic Control Systems.   A control system is deterministic if the response to input is predictable and repeatable. If not, the control system is a stochastic control system.

1–2 EXAMPLES OF CONTROL SYSTEMS

In this section we shall present several examples of closed-loop control systems.

Speed control system.   The basic principle of a Watt's speed governor for an engine is illustrated in the schematic diagram of Figure 1–1. The amount of fuel admitted to the engine is adjusted according to the difference between the desired and the actual engine speeds.

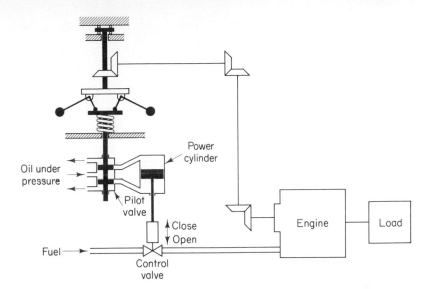

Figure 1–1
Speed control system.

The sequence of actions may be stated as follows: The speed governor is adjusted such that, at the desired speed, no pressured oil will flow into either side of the power cylinder. If the actual speed drops below the desired value due to disturbance, then the decrease in the centrifugal force of the speed governor causes the control valve to move downward, supplying more fuel, and the speed of the engine increases until the desired value is reached. On the other hand, if the speed of the engine increases above the desired value, then the increase in the centrifugal force of the governor causes the control valve to move upward. This decreases the supply of fuel, and the speed of the engine decreases until the desired value is reached.

In this speed control system, the plant (controlled system) is the engine and the controlled variable is the speed of the engine. The difference between the desired speed and the actual speed is the error signal. The control signal (the amount of fuel) to be applied to the plant (engine) is the actuating signal. The external input to disturb the controlled variable is the disturbance. An unexpected change in the load is a disturbance.

Robot control system. Industrial robots are frequently used in industry to improve productivity. The robot can handle monotonous jobs as well as complex jobs without errors in operation. The robot can work in an environment intolerable to human operators. For example, it can work in extreme temperatures (both high and low) or in a high- or low-pressure environment or under water or in space. There are special robots for fire fighting, underwater exploration, and space exploration, among many others.

The industrial robot must handle mechanical parts that have particular shapes and weights. Hence, it must have at least an arm, a wrist, and a hand. It must have sufficient power to perform the task and the capability for at least limited mobility. In fact, some robots of today are able to move freely by themselves in a limited space in a factory.

The industrial robot must have some sensory devices. In low-level robots, microswitches are installed in the arms as sensory devices. The robot first touches an object and then, through the microswitches, confirms the existence of the object in space and proceeds in the next step to grasp it.

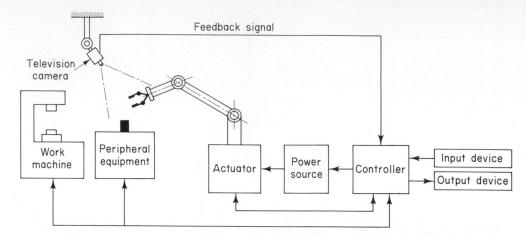

Figure 1–2
Robot using a pattern recognition process.

In a high-level robot, an optical means (such as a television system) is used to scan the background of the object. It recognizes the pattern and determines the presence and orientation of the object. A computer is necessary to process signals in the pattern-recognition process (see Figure 1–2). In some applications, the computerized robot recognizes the presence and orientation of each mechanical part by a pattern recognition process that consists of reading the code numbers attached to it. Then the robot picks up the part and moves it to an appropriate place for assembling, and there it assembles several parts into a component. A well-programmed digital computer acts as a controller.

Robot arm control system. Figure 1–3 shows a schematic diagram for a simplified version of the robot arm control system. The diagram shows a straight-line motion control of the arm. A straight-line motion is a 1-degree-of-freedom motion. The actual robot arm has 3 degrees of freedom (up-and-down motion, forward-and-backward motion, and left-

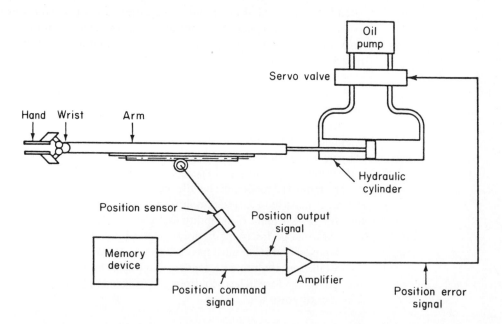

Figure 1–3
Robot arm control system.

Chapter 1 / Introduction to Control Systems Analysis

and-right motion). The wrist attached to the end of the arm also has 3 degrees of freedom, and the hand has 1 degree of freedom (grasp motion). Altogether the robot arm system has 7 degrees of freedom. Additional degrees of freedom are required if the robot body must move on a plane. In general, robot hands may be interchangeable parts: a different type of grasping device can be attached to the wrist to serve as a hand to handle each different type of mechanical object.

A servo system is used to position the arm and wrist. Since the robot arm motion frequently requires speed and power, hydraulic pressure or pneumatic pressure is used as the source of power. For medium power requirements, dc motors may be used. And for small power requirements, step motors may be used.

For control of sequential motions, command signals are stored on magnetic disks. In high-level robot systems, the playback mode of control is frequently used. In this mode, a human operator first "teaches" the robot the desired sequence of movements by working some mechanism attached to the arm; the computer in the robot memorizes the desired sequential movements. Then, from the second time on, the robot repeats faithfully the sequence of movements.

Robot hand grasping force control system.   Figure 1–4 shows a schematic diagram for a grasping force control system using a force-sensing device and a slip-sensing device. If the grasping force is too small, the robot hand will drop the mechanical object, and if it is too large, the hand may damage or crush the object. In the system shown, the grasping force is preset at a moderate level before the hand touches the mechanical object. The hand picks up and raises the object with the preset grasping force. If there is a slip in the raising motion, it will be observed by the slip-sensing device and a signal will be sent back to the controller, which will then increase the grasping force. In this way, a reasonable grasping force can be realized that prevents slipping but does not damage the mechanical object.

Numerical control systems.   Numerical control is a method of controlling the motions of machine components by use of numbers. In numerical control the motion of a workhead may be controlled by the binary information contained on a disk.

The system shown in Figure 1–5 works as follows: A magnetic disk is prepared in binary form representing the desired part P. To start the system, the disk is fed through the reader. The frequency-modulated input pulse signal is compared with the feedback pulse signal. The controller carries out mathematical operations on the difference in the pulse signals. The digital-to-analog converter converts the controller output pulse into an analog signal that

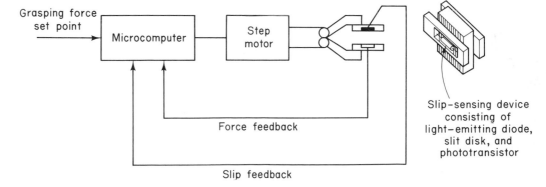

Figure 1–4
Robot hand grasping
force control system.

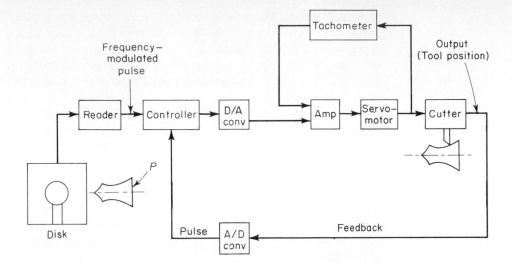

Figure 1–5
Numerical control of a machine.

represents a certain magnitude of voltage, which, in turn, causes the servomotor to rotate. The position of the cutterhead is controlled according to the input to the servomotor. The transducer attached to the cutterhead converts the motion into an electrical signal, which is converted to the pulse signal by the analog-to-digital converter. Then this signal is compared with the input pulse signal. If there is any difference between these two, the controller sends a signal to the servomotor to reduce it, as stated earlier.

An advantage of numerical control is that complex parts can be produced with uniform tolerances at the maximum milling speed.

Temperature control system.   Figure 1–6 shows a schematic diagram of temperature control of an electric furnace. The temperature in the electric furnace is measured by a thermometer, which is an analog device. The analog temperature is converted to a digital temperature by an A/D converter. The digital temperature is fed to a controller through an interface. This digital temperature is compared with the programmed input temperature, and if there is any discrepancy (error), the controller sends out a signal to the heater, through an interface, amplifier, and relay, to bring the furnace temperature to a desired value.

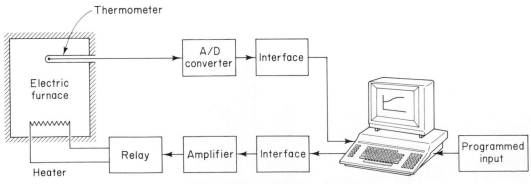

Figure 1–6   Temperature control system.

Temperature control of the passenger compartment of a car. Figure 1–7 shows a functional diagram of temperature control of the passenger compartment of a car. The desired temperature, converted to a voltage, is the input to the controller. The actual temperature of the passenger compartment is converted to a voltage through a sensor and is fed back to the controller for comparison with the input. The ambient temperature and radiation heat transfer from the sun, which are not constant while the car is driven, act as disturbances. This system employs both feedback control and feedforward control. (Feedforward control gives corrective action before the disturbances affect the output.)

The temperature of the passenger car compartment differs considerably depending on the place where it is measured. Instead of using multiple sensors for temperature measurement and averaging the measured values, it is economical to install a small suction blower at the place where passengers normally sense the temperature. The temperature of the air from the suction blower is an indication of the passenger compartment temperature and is considered the output of the system.

The controller receives the input signal, output signal, and signals from sensors from disturbance sources. The controller sends out an optimal control signal to the air conditioner to control the amount of cooling air so that the passenger compartment temperature is equal to the desired temperature.

Traffic control systems. As stated in Section 1–1, traffic control by means of traffic signals operated on a time basis constitutes an open-loop control system. If, however, the number of cars waiting at each traffic signal in a congested area of a city is continuously measured and the information fed to the central control computer that controls the traffic signals, then such a system becomes closed loop.

Traffic movement in networks is quite complex because the variation in traffic volume depends heavily on the hour and day of the week, as well as on many other factors. In some cases the Poisson distribution may be assumed for the arrivals at intersections, but this is not necessarily valid for all traffic problems. In fact, minimizing the average waiting time is a very complex control problem.

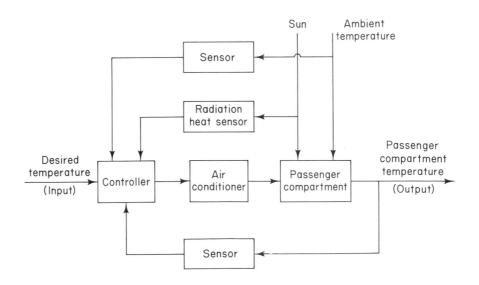

Figure 1–7
Temperature control
of passenger
compartment of a car.

Biological systems. Consider the competition of two species of bacteria whose populations are x_1 and x_2. The two are competing in the sense that they consume the same food supply. Under certain conditions, the populations x_1 and x_2 change with time according to

$$\dot{x}_1 = a_{11}x_1 - a_{12}x_1x_2$$

$$\dot{x}_2 = a_{21}x_2 - a_{22}x_1x_2$$

where a_{11}, a_{12}, a_{21}, and a_{22} are positive constants and x_1 and x_2 are nonnegative. These equations are called Volterra's competition equations.

If a certain chemical is given to the species, the populations change according to the following equations:

$$\dot{x}_1 = a_{11}x_1 - a_{12}x_1x_2 - b_1u$$

$$\dot{x}_2 = a_{21}x_2 - a_{22}x_1x_2 - b_2u$$

where b_1 and b_2 are positive constants and u is the controlling input (the amount of chemical in this example). An interesting problem is to minimize within a given time period the population x_1 while keeping the population x_2 as large as possible. This is an example of a biological system to which control theory can be applied.

Inventory control systems. The industrial programming of production rate and inventory level is another example of a closed-loop control system. The actual inventory level, which is the output of the system, is compared with the desired inventory level, which may change from time to time according to the market. If there is any difference between the actual inventory level and the desired inventory level, then the production rate is adjusted so that the output is always at or near the desired level, which is chosen to maximize profit.

Business systems. A business system may consist of many groups. Each task assigned to a group will represent a dynamic element of the system. Feedback methods of reporting the accomplishments of each group must be established in such a system for proper operation. The cross-coupling between functional groups must be made a minimum in order to reduce undesirable delay times in the system. The smaller this cross-coupling, the smoother the flow of work signals and materials will be.

A business system is a closed-loop system. A good design will reduce the managerial control required. Note that disturbances in this system are the lack of personnel or materials, interruption of communication, human errors, and the like.

The establishment of a well-founded estimating system based on statistics is mandatory to proper management. (Note that it is a well-known fact that the performance of such a system can be improved by the use of lead time or "anticipation.")

To apply control theory to improve the performance of such a system, we must represent the dynamic characteristic of the component groups of the system by a relatively simple set of equations.

Although it is certainly a difficult problem to derive mathematical representations of the component groups, the application of optimization techniques to business systems significantly improves the performance of the business system.

The Laplace transform method is an operational method that can be used advantageously for solving linear differential equations. By use of Laplace transforms, one can convert many common functions, such as sinusoidal fuctions, damped sinusoidal functions, and exponential functions, into algebraic functions of a complex variable s. Operations such as differentiation and integration can be replaced by algebraic operations in the complex plane. Thus, a linear differential equation can be transformed into an algebraic equation in a complex variable s. If the algebraic equation in s is solved for the dependent variable, then the solution of the differential equation (the inverse Laplace transform of the dependent variable) may be found by use of a Laplace transform table or by use of the partial fraction expansion technique, which is presented in Section 1–4.

An advantage of the Laplace transform method is that it allows the use of graphical techniques for predicting the system performance without actually solving system differential equations. Another advantage of the Laplace transform method is that, when one solves the differential equation, both the transient component and steady-state component of the solution can be obtained simultaneously.

Review of complex variables and complex functions. Before we present the Laplace transformation, we shall review the complex variable and complex function. We shall also review Euler's theorem, which relates the sinusoidal functions to exponential functions.

Complex Variable. A complex number has a real part and an imaginary part, both of which are constant. If the real part and/or imaginary part are variables, a complex number is called a *complex variable*. In the Laplace transformation we use the notation s as a complex variable; that is,

$$s = \sigma + j\omega$$

where σ is the real part and ω is the imaginary part.

Complex Function. A complex function $F(s)$, a function of s, has a real part and an imaginary part or

$$F(s) = F_x + jF_y$$

where F_x and F_y are real quantities. The magnitude of $F(s)$ is $\sqrt{F_x^2 + F_y^2}$, and the angle θ of $F(s)$ is $\tan^{-1}(F_y/F_x)$. The angle is measured counterclockwise from the positive real axis. The complex conjugate of $F(s)$ is $\bar{F}(s) = F_x - jF_y$.

Complex functions commonly encountered in linear control systems analysis are single-valued functions of s and are uniquely determined for a given value of s.

A complex function $G(s)$ is said to be *analytic* in a region if $G(s)$ and all its derivatives exist in that region. The derivative of an analytic function $G(s)$ is given by

$$\frac{d}{ds}G(s) = \lim_{\Delta s \to 0} \frac{G(s + \Delta s) - G(s)}{\Delta s} = \lim_{\Delta s \to 0} \frac{\Delta G}{\Delta s}$$

Since $\Delta s = \Delta\sigma + j\Delta\omega$, Δs can approach zero along an infinite number of different paths. It can be shown, but is stated without a proof here, that if the derivatives taken along two particular paths, that is, $\Delta s = \Delta\sigma$ and $\Delta s = j\Delta\omega$, are equal, then the derivative is unique for any other path $\Delta s = \Delta\sigma + j\Delta\omega$ and so the derivative exists.

For a particular path $\Delta s = \Delta\sigma$ (which means that the path is parallel to the real axis),

$$\frac{d}{ds} G(s) = \lim_{\Delta\sigma \to 0} \left(\frac{\Delta G_x}{\Delta\sigma} + j \frac{\Delta G_y}{\Delta\sigma} \right) = \frac{\partial G_x}{\partial\sigma} + j \frac{\partial G_y}{\partial\sigma}$$

For another particular path $\Delta s = j\,\Delta\omega$ (which means that the path is parallel to the imaginary axis),

$$\frac{d}{ds} G(s) = \lim_{j\Delta\omega \to 0} \left(\frac{\Delta G_x}{j\Delta\omega} + j \frac{\Delta G_y}{j\Delta\omega} \right) = -j \frac{\partial G_x}{\partial\omega} + \frac{\partial G_y}{\partial\omega}$$

If these two values of the derivative are equal

$$\frac{\partial G_x}{\partial\sigma} + j \frac{\partial G_y}{\partial\sigma} = \frac{\partial G_y}{\partial\omega} - j \frac{\partial G_x}{\partial\omega}$$

or if the following two conditions

$$\frac{\partial G_x}{\partial\sigma} = \frac{\partial G_y}{\partial\omega} \qquad \text{and} \qquad \frac{\partial G_y}{\partial\sigma} = -\frac{\partial G_x}{\partial\omega}$$

are satisfied, then the derivative $dG(s)/ds$ is uniquely determined. These two conditions are known as the Cauchy–Riemann conditions. If these conditions are satisifed, the function $G(s)$ is analytic.

As an example, consider the following $G(s)$:

$$G(s) = \frac{1}{s + 1}$$

Then

$$G(\sigma + j\omega) = \frac{1}{\sigma + j\omega + 1} = G_x + jG_y$$

where

$$G_x = \frac{\sigma + 1}{(\sigma + 1)^2 + \omega^2} \qquad \text{and} \qquad G_y = \frac{-\omega}{(\sigma + 1)^2 + \omega^2}$$

It can be seen that, except at $s = -1$ (that is, $\sigma = -1$, $\omega = 0$), $G(s)$ satisfies the Cauchy–Riemann conditions:

$$\frac{\partial G_x}{\partial\sigma} = \frac{\partial G_y}{\partial\omega} = \frac{\omega^2 - (\sigma + 1)^2}{[(\sigma + 1)^2 + \omega^2]^2}$$

$$\frac{\partial G_y}{\partial\sigma} = -\frac{\partial G_x}{\partial\omega} = \frac{2\omega(\sigma + 1)}{[(\sigma + 1)^2 + \omega^2]^2}$$

Hence $G(s) = 1/(s + 1)$ is analytic in the entire s plane except at $s = -1$. The derivative

$dG(s)/ds$ except at $s = -1$ is found to be

$$\frac{d}{ds} G(s) = \frac{\partial G_x}{\partial \sigma} + j \frac{\partial G_y}{\partial \sigma} = \frac{\partial G_y}{\partial \omega} - j \frac{\partial G_x}{\partial \omega}$$

$$= -\frac{1}{(\sigma + j\omega + 1)^2} = -\frac{1}{(s + 1)^2}$$

Note that the derivative of an analytic function can be obtained simply by differentiating $G(s)$ with respect to s. In this example,

$$\frac{d}{ds}\left(\frac{1}{s + 1}\right) = -\frac{1}{(s + 1)^2}$$

Points in the s plane at which the function $G(s)$ is analytic are called *ordinary* points, while points in the s plane at which the function $G(s)$ is not analytic are called *singular* points. Singular points at which the function $G(s)$ or its derivatives approach infinity are called *poles*. In the previous example, $s = -1$ is a singular point and is a pole of the function $G(s)$.

If $G(s)$ approaches infinity as s approaches $-p$ and if the function

$$G(s)(s + p)^n \qquad (n = 1, 2, 3, \ldots)$$

has a finite, nonzero value at $s = -p$, then $s = -p$ is called a pole of order n. If $n = 1$, the pole is called a simple pole. If $n = 2, 3, \ldots$, the pole is called a second-order pole, a third-order pole, and so on. Points at which the function $G(s)$ equals zero are called zeros.

To illustrate, consider the complex function

$$G(s) = \frac{K(s + 2)(s + 10)}{s(s + 1)(s + 5)(s + 15)^2}$$

$G(s)$ has zeros at $s = -2$, $s = -10$, simple poles at $s = 0$, $s = -1$, $s = -5$, and a double pole (multiple pole of order 2) at $s = -15$. Note that $G(s)$ becomes zero at $s = \infty$. Since for large values of s

$$G(s) \doteq \frac{K}{s^3}$$

$G(s)$ possesses a triple zero (multiple zero of order 3) at $s = \infty$. If points at infinity are included, $G(s)$ has the same number of poles as zeros. To summarize, $G(s)$ has five zeros ($s = -2$, $s = -10$, $s = \infty$, $s = \infty$, $s = \infty$) and five poles ($s = 0$, $s = -1$, $s = -5$, $s = -15$, $s = -15$).

Euler's theorem. The power series expansions of $\cos \theta$ and $\sin \theta$ are, respectively,

$$\cos \theta = 1 - \frac{\theta^2}{2!} + \frac{\theta^4}{4!} - \frac{\theta^6}{6!} + \cdots$$

$$\sin \theta = \theta - \frac{\theta^3}{3!} + \frac{\theta^5}{5!} - \frac{\theta^7}{7!} + \cdots$$

And so

$$\cos \theta + j \sin \theta = 1 + (j\theta) + \frac{(j\theta)^2}{2!} + \frac{(j\theta)^3}{3!} + \frac{(j\theta)^4}{4!} + \cdots$$

Since

$$e^x = 1 + x + \frac{x^2}{2!} + \frac{x^3}{3!} + \cdots$$

we see that

$$\cos \theta + j \sin \theta = e^{j\theta} \qquad (1\text{--}1)$$

This is known as *Euler's theorem*.

By using Euler's theorem, we can express sine and cosine in terms of an exponential function. Noting that $e^{-j\theta}$ is the complex conjugate of $e^{j\theta}$ and that

$$e^{j\theta} = \cos \theta + j \sin \theta$$

$$e^{-j\theta} = \cos \theta - j \sin \theta$$

we find, after adding and subtracting these two equations,

$$\cos \theta = \frac{1}{2}(e^{j\theta} + e^{-j\theta}) \qquad (1\text{--}2)$$

$$\sin \theta = \frac{1}{2j}(e^{j\theta} - e^{-j\theta}) \qquad (1\text{--}3)$$

Laplace transformation.   We shall first present a definition of the Laplace transformation and a brief discussion of the condition for the existence of the Laplace transform and then provide examples for illustrating the derivation of Laplace transforms of several common functions.

Let us define

$f(t)$ = a function of time t such that $f(t) = 0$ for $t < 0$
s = a complex variable
$\mathscr{L}$ = an operational symbol indicating that the quantity that it prefixes
  is to be transformed by the Laplace integral $\int_0^\infty e^{-st}\, dt$
$F(s)$ = Laplace transform of $f(t)$

Then the Laplace transform of $f(t)$ is given by

$$\mathscr{L}[f(t)] = F(s) = \int_0^\infty e^{-st}\, dt[f(t)] = \int_0^\infty f(t)e^{-st}\, dt$$

The reverse process of finding the time function $f(t)$ from the Laplace transform $F(s)$ is called the *inverse Laplace transformation*. The notation for the inverse Laplace transformation is $\mathscr{L}^{-1}$. Thus

$$\mathscr{L}^{-1}[F(s)] = f(t)$$

Existence of Laplace transform.   The Laplace transform of a function $f(t)$ exists if the Laplace integral converges. The integral will converge if $f(t)$ is sectionally continuous in every finite interval in the range $t > 0$ and if it is of exponential order as t approaches infinity. A function $f(t)$ is said to be of exponential order if a real, positive costant σ exists such that the function

$$e^{-\sigma t}|f(t)|$$

approaches zero as t approaches infinity. If the limit of the function $e^{-\sigma t}|f(t)|$ approaches zero for σ greater than σ_c and the limit approaches infinity for σ less than σ_c, the value σ_c is called the *abscissa of convergence*.

For the function $f(t) = Ae^{-\alpha t}$

$$\lim_{t \to \infty} e^{-\sigma t}|Ae^{-\alpha t}|$$

approaches zero if $\sigma > -\alpha$. The abscissa of convergence in this case is $\sigma_c = -\alpha$. The integral $\int_0^\infty f(t)e^{-st}\, dt$ converges only if σ, the real part of s, is greater than the abscissa of convergence σ_c. Thus the operator s must be chosen as a constant such that this integral coverges.

In terms of the poles of the function $F(s)$, the abscissa of convergence σ_c corresponds to the real part of the pole located farthest to the right in the s plane. For example, for the following function $F(s)$,

$$F(s) = \frac{K(s + 3)}{(s + 1)(s + 2)}$$

the abscissa of convergence σ_c is equal to -1. It can be seen that for such functions as t, $\sin \omega t$, and $t \sin \omega t$ the abscissa of convergence is equal to zero. For functions like e^{-ct}, te^{-ct}, $e^{-ct} \sin \omega t$, and so on, the abscissa of convergence is equal to $-c$. For functions that increase faster than the exponential function, however, it is impossible to find suitable values of the abcissa of convergence. Therefore, such functions as e^{t^2} and te^{t^2} do not possess Laplace transforms.

The reader should be cautioned that although e^{t^2} (for $0 \le t \le \infty$) does not possess a Laplace transform, the time function defined by

$$
\begin{aligned}
f(t) &= e^{t^2} \quad && \text{for } 0 \le t \le T < \infty \\
&= 0 && \text{for } t < 0, T < t
\end{aligned}
$$

does possess a Laplace transform since $f(t) = e^{t^2}$ for only a limited time interval $0 \le t \le T$ and not for $0 \le t \le \infty$. Such a signal can be physically generated. Note that the signals that we can physically generate always have corresponding Laplace transforms.

If a function $f(t)$ has a Laplace transform, then the Laplace transform of $Af(t)$, where A is a constant, is given by

$$\mathscr{L}[Af(t)] = A\mathscr{L}[f(t)]$$

This is obvious from the definition of the Laplace transform. Similarly, if functions $f_1(t)$ and $f_2(t)$ have Laplace transforms, then the Laplace transform of the function $f_1(t) + f_2(t)$ is given by

$$\mathscr{L}[f_1(t) + f_2(t)] = \mathscr{L}[f_1(t)] + \mathscr{L}[f_2(t)]$$

Again the proof of this relationship is evident from the definition of the Laplace transform.

In what follows, we derive Laplace transforms of a few commonly encountered functions.

Exponential function.   Consider the exponential function

$$f(t) = 0 \qquad \text{for } t < 0$$

$$= Ae^{-\alpha t} \qquad \text{for } t \geq 0$$

where A and α are constants. The Laplace transform of this exponential function can be obtained as follows:

$$\mathcal{L}[Ae^{-\alpha t}] = \int_0^\infty Ae^{-\alpha t}e^{-st}\, dt = A\int_0^\infty e^{-(\alpha+s)t}\, dt = \frac{A}{s+\alpha}$$

It is seen that the exponential function produces a pole in the complex plane.

In deriving the Laplace transform of $f(t) = Ae^{-\alpha t}$, we required that the real part of s was greater than $-\alpha$ (the abscissa of convergence). A question may immediately arise as to whether or not the Laplace transform thus obtained is valid in the range where $\sigma < -\alpha$ in the s plane. To answer this question, we must resort to the theory of complex variables. In the theory of complex variables, there is a theorem known as the analytic extension theorem. It states that, if two analytic functions are equal for a finite length along any arc in a region in which both are analytic, then they are equal everywhere in the region. The arc of equality is usually the real axis or a portion of it. By using this theorem the form of $F(s)$ determined by an integration in which s is allowed to have any real positive value greater than the abscissa of convergence holds for any complex values of s at which $F(s)$ is analytic. Thus, although we require the real part of s to be greater than the abscissa of convergence to make the integral $\int_0^\infty f(t)e^{-st}\, dt$ absolutely convergent, once the Laplace transform $F(s)$ is obtained, $F(s)$ can be considered valid throughout the entire s plane except at the poles of $F(s)$.

Step function.   Consider the step function

$$f(t) = 0 \qquad \text{for } t < 0$$

$$= A \qquad \text{for } t > 0$$

where A is a constant. Note that it is a special case of the exponential function $Ae^{-\alpha t}$, where $\alpha = 0$. The step function is undefined at $t = 0$. Its Laplace transform is given by

$$\mathcal{L}[A] = \int_0^\infty Ae^{-st}\, dt = \frac{A}{s}$$

In performing this integration, we assumed that the real part of s was greater than zero (the abscissa of convergence) and therefore that $\lim_{t\to\infty} e^{-st}$ was zero. As stated earlier, the Laplace transform thus obtained is valid in the entire s plane except at the pole $s = 0$.

The step function whose height is unity is called a *unit-step* function. The unit-step function that occurs at $t = t_0$ is frequently written as $u(t - t_0)$ or $1(t - t_0)$. In this book we shall use the latter notation unless otherwise stated. The step function of height A can then be written as $f(t) = A\,1(t)$. The Laplace transform of the unit-step function, which is defined by

$$1(t) = 0 \qquad \text{for } t < 0$$

$$= 1 \qquad \text{for } t > 0$$

is 1/s, or

$$\mathscr{L}[1(t)] = \frac{1}{s}$$

Physically, a step function occurring at $t = 0$ corresponds to a constant signal suddenly applied to the system at time t equals zero.

Ramp function.   Consider the ramp function

$$f(t) = 0 \qquad \text{for } t < 0$$
$$= At \qquad \text{for } t \geq 0$$

where A is a constant. The Laplace transform of this ramp function is obtained as

$$\mathscr{L}[At] = \int_0^\infty At e^{-st} \, dt = At \frac{e^{-st}}{-s} \bigg|_0^\infty - \int_0^\infty \frac{A e^{-st}}{-s} \, dt$$

$$= \frac{A}{s} \int_0^\infty e^{-st} \, dt = \frac{A}{s^2}$$

Sinusoidal function.   The Laplace transform of the sinusoidal function

$$f(t) = 0 \qquad\qquad \text{for } t < 0$$
$$= A \sin \omega t \qquad \text{for } t \geq 0$$

where A and ω are constants, is obtained as follows. Referring to Equation (1–3) $\sin \omega t$ can be written

$$\sin \omega t = \frac{1}{2j} (e^{j\omega t} - e^{-j\omega t})$$

Hence

$$\mathscr{L}[A \sin \omega t] = \frac{A}{2j} \int_0^\infty (e^{j\omega t} - e^{-j\omega t}) e^{-st} \, dt$$

$$= \frac{A}{2j} \frac{1}{s - j\omega} - \frac{A}{2j} \frac{1}{s + j\omega} = \frac{A\omega}{s^2 + \omega^2}$$

Similarly, the Laplace transform of $A \cos \omega t$ can be derived as follows:

$$\mathscr{L}[A \cos \omega t] = \frac{As}{s^2 + \omega^2}$$

Comments.   The Laplace transform of any Laplace transformable function $f(t)$ can be found by multiplying $f(t)$ by e^{-st} and then integrating the product from $t = 0$ to $t = \infty$. Once we know the method of obtaining the Laplace transform, however, it is not necessary to derive the Laplace transform of $f(t)$ each time. Laplace transform tables can conveniently be used to find the transform of a given function $f(t)$. Table 1–1 shows Laplace transforms of time functions that will frequently appear in linear control systems analysis.

In the following discussion we present Laplace transforms of functions as well as theorems on the Laplace transformation that are useful in the study of linear control systems.

Table 1–1 Laplace Transform Pairs

	$f(t)$	$F(s)$
1	Unit impulse $\delta(t)$	1
2	Unit step $1(t)$	$\dfrac{1}{s}$
3	t	$\dfrac{1}{s^2}$
4	$\dfrac{t^{n-1}}{(n-1)!}\quad (n = 1,2,3,\ \dots)$	$\dfrac{1}{s^n}$
5	$t^n\quad (n = 1,2,3,\ \dots)$	$\dfrac{n!}{s^{n+1}}$
6	e^{-at}	$\dfrac{1}{s+a}$
7	te^{-at}	$\dfrac{1}{(s+a)^2}$
8	$\dfrac{1}{(n-1)!}t^{n-1}e^{-at}\quad (n = 1,2,3,\ \dots)$	$\dfrac{1}{(s+a)^n}$
9	$t^n e^{-at}\quad (n = 1,2,3,\ \dots)$	$\dfrac{n!}{(s+a)^{n+1}}$
10	$\sin \omega t$	$\dfrac{\omega}{s^2+\omega^2}$
11	$\cos \omega t$	$\dfrac{s}{s^2+\omega^2}$
12	$\sinh \omega t$	$\dfrac{\omega}{s^2-\omega^2}$
13	$\cosh \omega t$	$\dfrac{s}{s^2-\omega^2}$
14	$\dfrac{1}{a}(1 - e^{-at})$	$\dfrac{1}{s(s+a)}$
15	$\dfrac{1}{b-a}(e^{-at} - e^{-bt})$	$\dfrac{1}{(s+a)(s+b)}$
16	$\dfrac{1}{b-a}(be^{-bt} - ae^{-at})$	$\dfrac{s}{(s+a)(s+b)}$
17	$\dfrac{1}{ab}\left[1 + \dfrac{1}{a-b}(be^{-at} - ae^{-bt})\right]$	$\dfrac{1}{s(s+a)(s+b)}$

Table 1–1 Cont'd

18	$\dfrac{1}{a^2}(1 - e^{-at} - ate^{-at})$	$\dfrac{1}{s(s + a)^2}$
19	$\dfrac{1}{a^2}(at - 1 + e^{-at})$	$\dfrac{1}{s^2(s + a)}$
20	$e^{-at}\sin \omega t$	$\dfrac{\omega}{(s + a)^2 + \omega^2}$
21	$e^{-at}\cos \omega t$	$\dfrac{s + a}{(s + a)^2 + \omega^2}$
22	$\dfrac{\omega_n}{\sqrt{1 - \zeta^2}}e^{-\zeta\omega_n t}\sin \omega_n \sqrt{1 - \zeta^2}\,t$	$\dfrac{\omega_n^2}{s^2 + 2\zeta\omega_n s + \omega_n^2}$
23	$-\dfrac{1}{\sqrt{1 - \zeta^2}}e^{-\zeta\omega_n t}\sin (\omega_n \sqrt{1 - \zeta^2}\,t - \phi)$ $\phi = \tan^{-1}\dfrac{\sqrt{1 - \zeta^2}}{\zeta}$	$\dfrac{s}{s^2 + 2\zeta\omega_n s + \omega_n^2}$
24	$1 - \dfrac{1}{\sqrt{1 - \zeta^2}}e^{-\zeta\omega_n t}\sin (\omega_n \sqrt{1 - \zeta^2}\,t + \phi)$ $\phi = \tan^{-1}\dfrac{\sqrt{1 - \zeta^2}}{\zeta}$	$\dfrac{\omega_n^2}{s(s^2 + 2\zeta\omega_n s + \omega_n^2)}$
25	$1 - \cos \omega t$	$\dfrac{\omega^2}{s(s^2 + \omega^2)}$
26	$\omega t - \sin \omega t$	$\dfrac{\omega^3}{s^2(s^2 + \omega^2)}$
27	$\sin \omega t - \omega t \cos \omega t$	$\dfrac{2\omega^3}{(s^2 + \omega^2)^2}$
28	$\dfrac{1}{2\omega}t \sin \omega t$	$\dfrac{s}{(s^2 + \omega^2)^2}$
29	$t \cos \omega t$	$\dfrac{s^2 - \omega^2}{(s^2 + \omega^2)^2}$
30	$\dfrac{1}{\omega_2^2 - \omega_1^2}(\cos \omega_1 t - \cos \omega_2 t) \quad (\omega_1^2 \neq \omega_2^2)$	$\dfrac{s}{(s^2 + \omega_1^2)(s^2 + \omega_2^2)}$
31	$\dfrac{1}{2\omega}(\sin \omega t + \omega t \cos \omega t)$	$\dfrac{s^2}{(s^2 + \omega^2)^2}$

Translated function.   Let us obtain the Laplace transform of the translated function $f(t - \alpha)1(t - \alpha)$, where $\alpha \geq 0$. This function is zero for $t < \alpha$. The functions $f(t)1(t)$ and $f(t - \alpha)1(t - \alpha)$ are shown in Figure 1–8.

By definition, the Laplace transform of $f(t - \alpha)1(t - \alpha)$ is

$$\mathcal{L}[f(t - \alpha)1(t - \alpha)] = \int_0^\infty f(t - \alpha)1(t - \alpha)e^{-st}\,dt$$

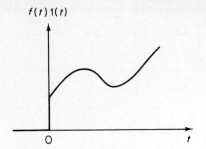

f(t) 1(t)

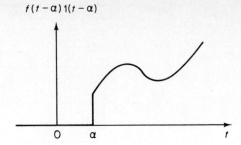

f(t − α) 1(t − α)

Figure 1–8
Function $f(t)1(t)$ and
translated function
$f(t - \alpha)1(t - \alpha)$.

By changing the independent variable from t to τ, where $\tau = t - \alpha$, we obtain

$$\int_0^\infty f(t - \alpha)1(t - \alpha)e^{-st}\,dt = \int_{-\alpha}^\infty f(\tau)1(\tau)e^{-s(\tau + \alpha)}\,d\tau$$

Noting that $f(\tau)1(\tau) = 0$ for $\tau < 0$, we can change the lower limit of itegration from $-\alpha$ to 0. Thus

$$\int_{-\alpha}^\infty f(\tau)1(\tau)e^{-s(\tau + \alpha)}\,d\tau = \int_0^\infty f(\tau)1(\tau)e^{-s(\tau + \alpha)}\,d\tau$$

$$= \int_0^\infty f(\tau)e^{-s\tau}e^{-\alpha s}\,d\tau$$

$$= e^{-\alpha s}\int_0^\infty f(\tau)e^{-s\tau}\,d\tau = e^{-\alpha s}F(s)$$

where

$$F(s) = \mathcal{L}[f(t)] = \int_0^\infty f(t)e^{-st}\,dt$$

And so

$$\mathcal{L}[f(t - \alpha)1(t - \alpha)] = e^{-\alpha s}F(s), \qquad \alpha \geq 0$$

This last equation states that the translation of the time function $f(t)1(t)$ by α (where $\alpha \geq 0$) corresponds to the multiplication of the transform $F(s)$ by $e^{-\alpha s}$.

Pulse function. Consider the pulse function

$$f(t) = \frac{A}{t_0} \qquad \text{for } 0 < t < t_0$$

$$= 0 \qquad \text{for } t < 0,\, t_0 < t$$

where A and t_0 are constants.

The pulse function here may be considered a step function of height A/t_0 that begins at $t = 0$ and that is superimposed by a negative step function of height A/t_0 beginning at $t = t_0$; that is,

$$f(t) = \frac{A}{t_0}1(t) - \frac{A}{t_0}1(t - t_0)$$

Then the Laplace transform of $f(t)$ is obtained as

$$\mathcal{L}[f(t)] = \mathcal{L}\left[\frac{A}{t_0}1(t)\right] - \mathcal{L}\left[\frac{A}{t_0}1(t-t_0)\right]$$

$$= \frac{A}{t_0 s} - \frac{A}{t_0 s}e^{-st_0}$$

$$= \frac{A}{t_0 s}(1 - e^{-st_0}) \tag{1-4}$$

Impulse function. The impulse function is a special limiting case of the pulse function. Consider the impulse function

$$f(t) = \lim_{t_0 \to 0}\frac{A}{t_0} \qquad \text{for } 0 < t < t_0$$

$$= 0 \qquad \text{for } t < 0, t_0 < t$$

Since the height of the impulse function is A/t_0 and the duration is t_0, the area under the impulse is equal to A. As the duration t_0 approaches zero, the height A/t_0 approaches infinity, but the area under the impulse remains equal to A. Note that the magnitude of the impulse is measured by its area.

Referring to Equation (1-4), the Laplace transform of this impulse function is shown to be

$$\mathcal{L}[f(t)] = \lim_{t_0 \to 0}\left[\frac{A}{t_0 s}(1 - e^{-st_0})\right]$$

$$= \lim_{t_0 \to 0}\frac{\dfrac{d}{dt_0}[A(1 - e^{-st_0})]}{\dfrac{d}{dt_0}(t_0 s)} = \frac{As}{s} = A$$

Thus the Laplace transform of the impulse function is equal to the area under the impulse.

The impulse function whose area is equal to unity is called the *unit-impulse function* or the *Dirac delta function*. The unit-impulse function occurring at $t = t_0$ is usually denoted by $\delta(t - t_0)$. $\delta(t - t_0)$ satisfies the following:

$$\delta(t - t_0) = 0 \qquad \text{for } t \neq t_0$$

$$\delta(t - t_0) = \infty \qquad \text{for } t = t_0$$

$$\int_{-\infty}^{\infty}\delta(t - t_0)\,dt = 1$$

It should be mentioned that an impulse that has an infinite magnitude and zero duration is mathematical fiction and does not occur in physical systems. If, however, the magnitude of a pulse input to a system is very large and its duration is very short compared to the system time constants, then we can approximate the pulse input by an impulse function. For instance, if a force or torque input $f(t)$ is applied to a system for a very short time duration, $0 < t < t_0$, where the magnitude of $f(t)$ is sufficiently large so that the integral $\int_0^{t_0} f(t)\,dt$ is not negligible, then this input can be considered an impulse input. (Note that when we describe the impulse input the area or magnitude of the impulse is most important, but the exact shape of the impulse is usually immaterial.) The impulse input supplies energy to the system in an infinitesimal time.

The concept of the impulse function is quite useful in differentiating discontinuous functions. The unit-impulse function $\delta(t - t_0)$ can be considered the derivative of the unit-step function $1(t - t_0)$ at the point of discontinuity $t = t_0$ or

$$\delta(t - t_0) = \frac{d}{dt} 1(t - t_0)$$

Conversely, if the unit-impulse function $\delta(t - t_0)$ is integrated, the result is the unit-step function $1(t - t_0)$. With the concept of the impulse function we can differentiate a function containing discontinuities, giving impulses, the magnitudes of which are equal to the magnitude of each corresponding discontinuity.

Multiplication of $f(t)$ by $e^{-\alpha t}$.   If $f(t)$ is Laplace transformable, its Laplace transform being $F(s)$, then the Laplace transform of $e^{-\alpha t}f(t)$ is obtained as

$$\mathscr{L}[e^{-\alpha t}f(t)] = \int_0^\infty e^{-\alpha t}f(t)e^{-st}\,dt = F(s + \alpha) \tag{1--5}$$

We see that the multiplication of $f(t)$ by $e^{-\alpha t}$ has the effect of replacing s by $(s + \alpha)$ in the Laplace transform. Conversely, changing s to $(s + \alpha)$ is equivalent to multiplying $f(t)$ by $e^{-\alpha t}$. (Note that α may be real or complex.)

The relationship given by Equation (1–5) is useful in finding the Laplace transforms of such functions as $e^{-\alpha t}\sin \omega t$ and $e^{-\alpha t}\cos \omega t$. For instance, since

$$\mathscr{L}[\sin \omega t] = \frac{\omega}{s^2 + \omega^2} = F(s), \qquad \mathscr{L}[\cos \omega] = \frac{s}{s^2 + \omega^2} = G(s)$$

it follows from Equation (1–5) that the Laplace transforms of $e^{-\alpha t}\sin \omega t$ and $e^{-\alpha t}\cos \omega t$ are given, respectively, by

$$\mathscr{L}[e^{-\alpha t}\sin \omega t] = F(s + \alpha) = \frac{\omega}{(s + \alpha)^2 + \omega^2}$$

$$\mathscr{L}[e^{-\alpha t}\cos \omega t] = G(s + \alpha) = \frac{s + \alpha}{(s + \alpha)^2 + \omega^2}$$

Change of time scale.   In analyzing physical systems, it is sometimes desirable to change the time scale or normalize a given time function. The result obtained in terms of normalized time is useful because it can be applied directly to different systems having similar mathematical equations.

If t is changed into t/α, where α is a positive constant, then the function $f(t)$ is changed into $f(t/\alpha)$. If we denote the Laplace transform of $f(t)$ by $F(s)$, then the Laplace transform of $f(t/\alpha)$ may be obtained as follows:

$$\mathscr{L}\left[f\left(\frac{t}{\alpha}\right)\right] = \int_0^\infty f\left(\frac{t}{\alpha}\right) e^{-st}\,dt$$

Letting $t/\alpha = t_1$ and $\alpha s = s_1$, we obtain

$$\mathscr{L}\left[f\left(\frac{t}{\alpha}\right)\right] = \int_0^\infty f(t_1)e^{-s_1 t_1}\, d(\alpha t_1)$$

$$= \alpha \int_0^\infty f(t_1)e^{-s_1 t_1}\, dt_1$$

$$= \alpha F(s_1)$$

or

$$\mathscr{L}\left[f\left(\frac{t}{\alpha}\right)\right] = \alpha F(\alpha s)$$

As an example, consider $f(t) = e^{-t}$ and $f(t/5) = e^{-0.2t}$. We obtain

$$\mathscr{L}[f(t)] = \mathscr{L}[e^{-t}] = F(s) = \frac{1}{s+1}$$

Hence

$$\mathscr{L}\left[f\left(\frac{t}{5}\right)\right] = \mathscr{L}[e^{-0.2t}] = 5F(5s) = \frac{5}{5s+1}$$

This result can be verified easily by taking the Laplace transform of $e^{-0.2t}$ directly as follows:

$$\mathscr{L}[e^{-0.2t}] = \frac{1}{s+0.2} = \frac{5}{5s+1}$$

Comments on the lower limit of the Laplace integral.   In some cases, $f(t)$ possesses an impulse function at $t = 0$. Then the lower limit of the Laplace integral must be clearly specified as to whether it is $0-$ or $0+$, since the Laplace transforms of $f(t)$ differ for these two lower limits. If such a distinction of the lower limit of the Laplace integral is necessary, we use the notations

$$\mathscr{L}_+[f(t)] = \int_{0+}^\infty f(t)e^{-st}\, dt$$

$$\mathscr{L}_-[f(t)] = \int_{0-}^\infty f(t)e^{-st}\, dt = \mathscr{L}_+[f(t)] + \int_{0-}^{0+} f(t)e^{-st}\, dt$$

If $f(t)$ involves an impulse function at $t = 0$, then

$$\mathscr{L}_+[f(t)] \neq \mathscr{L}_-[f(t)]$$

since

$$\int_{0-}^{0+} f(t)e^{-st}\, dt \neq 0$$

for such a case. Obviously, if $f(t)$ does not possess an impulse function at $t = 0$ (that is, if the function to be transformed is finite between $t = 0-$ and $t = 0+$), then

$$\mathscr{L}_+[f(t)] = \mathscr{L}_-[f(t)]$$

Real differentiation theorem. The Laplace transform of the derivative of a function $f(t)$ is given by

$$\mathscr{L}\left[\frac{d}{dt}f(t)\right] = sF(s) - f(0) \qquad (1\text{-}6)$$

where $f(0)$ is the initial value of $f(t)$ evaluated at $t = 0$.

For a given function $f(t)$, the values of $f(0+)$ and $f(0-)$ may be the same or different, as illustrated in Figure 1–9. The distinction between $f(0+)$ and $f(0-)$ is important when $f(t)$ has a discontinuity at $t = 0$ because in such a case $df(t)/dt$ will involve an impulse function at $t = 0$. If $f(0+) \neq f(0-)$, Equation (1–6) must be modified to

$$\mathscr{L}_+\left[\frac{d}{dt}f(t)\right] = sF(s) - f(0+)$$

$$\mathscr{L}_-\left[\frac{d}{dt}f(t)\right] = sF(s) - f(0-)$$

To prove the real differentiation theorem, Equation (1–6), we proceed as follows. Integrating the Laplace integral by parts gives

$$\int_0^\infty f(t)e^{-st}\,dt = f(t)\frac{e^{-st}}{-s}\bigg|_0^\infty - \int_0^\infty \left[\frac{d}{dt}f(t)\right]\frac{e^{-st}}{-s}\,dt$$

Hence

$$F(s) = \frac{f(0)}{s} + \frac{1}{s}\mathscr{L}\left[\frac{d}{dt}f(t)\right]$$

It follows that

$$\mathscr{L}\left[\frac{d}{dt}f(t)\right] = sF(s) - f(0)$$

Similarly, we obtain the following relationship for the second derivative of $f(t)$:

$$\mathscr{L}\left[\frac{d^2}{dt^2}f(t)\right] = s^2F(s) - sf(0) - \dot{f}(0)$$

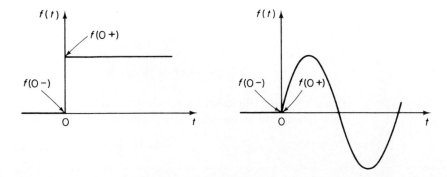

Figure 1–9
Step function and sine function indicating initial values at $t = 0-$ and $t = 0+$.

where $\dot{f}(0)$ is the value of $df(t)/dt$ evaluated at $t = 0$. To derive this equation, define

$$\frac{d}{dt} f(t) = g(t)$$

Then

$$\mathscr{L}\left[\frac{d^2}{dt^2} f(t)\right] = \mathscr{L}\left[\frac{d}{dt} g(t)\right] = s\mathscr{L}[g(t)] - g(0)$$

$$= s\mathscr{L}\left[\frac{d}{dt} f(t)\right] - \dot{f}(0)$$

$$= s^2 F(s) - sf(0) - \dot{f}(0)$$

Similarly, for the nth derivative of $f(t)$, we obtain

$$\mathscr{L}\left[\frac{d^n}{dt^n} f(t)\right] = s^n F(s) - s^{n-1} f(0) - s^{n-2} \dot{f}(0) - \cdots - s\overset{(n-2)}{f}(0) - \overset{(n-1)}{f}(0)$$

where $f(0)$, $\dot{f}(0)$, . . . , $\overset{(n-1)}{f}(0)$ represent the values of $f(t)$, $df(t)/dt$, . . . , $d^{n-1}f(t)/dt^{n-1}$, respectively, evaluated at $t = 0$. If the distinction between $\mathscr{L}_+$ and $\mathscr{L}_-$ is necessary, we substitute $t = 0+$ or $t = 0-$ into $f(t)$, $df(t)/dt$, . . . , $d^{n-1}f(t)/dt^{n-1}$, depending on whether we take $\mathscr{L}_+$ or $\mathscr{L}_-$.

Note that, in order for Laplace transforms of derivatives of $f(t)$ to exist, $d^n f(t)/dt^n$ ($n = 1, 2, 3, . . .$) must be Laplace transformable.

Note also that if all the initial values of $f(t)$ and its derivatives are equal to zero, then the Laplace transform of the nth derivative of $f(t)$ is given by $s^n F(s)$.

EXAMPLE 1–1   Consider the cosine function.

$$g(t) = 0 \qquad \text{for } t < 0$$
$$= \cos \omega t \qquad \text{for } t \geq 0$$

The Laplace transform of this cosine function can be obtained directly as in the case of the sinusoidal function. The use of the real differentiation theorem, however, will be demonstrated here by deriving the Laplace transform of the cosine function from the Laplace transform of the sine function. If we define

$$f(t) = 0 \qquad \text{for } t < 0$$
$$= \sin \omega t \qquad \text{for } t \geq 0$$

then

$$\mathscr{L}[\sin \omega t] = F(s) = \frac{\omega}{s^2 + \omega^2}$$

The Laplace transform of the cosine function is obtained as

$$\mathscr{L}[\cos \omega t] = \mathscr{L}\left[\frac{d}{dt}\left(\frac{1}{\omega} \sin \omega t\right)\right] = \frac{1}{\omega}[sF(s) - f(0)]$$

$$= \frac{1}{\omega}\left[\frac{s\omega}{s^2 + \omega^2} - 0\right] = \frac{s}{s^2 + \omega^2}$$

Final value theorem. The final value theorem relates the steady-state behavior of $f(t)$ to the behavior of $sF(s)$ in the neighborhood of $s = 0$. This theorem, however, applies if and only if $\lim_{t\to\infty} f(t)$ exists [which means that $f(t)$ settles down to a definite value for $t \to \infty$]. If all poles of $sF(s)$ lie in the left half s plane, $\lim_{t\to\infty} f(t)$ exists. But if $sF(s)$ has poles on the imaginary axis or in the right half s plane, $f(t)$ will contain oscillating or exponentially increasing time functions, respectively, and $\lim_{t\to\infty} f(t)$ will not exist. The final value theorem does not apply in such cases. For instance, if $f(t)$ is the sinusoidal function $\sin \omega t$, $sF(s)$ has poles at $s = \pm j\omega$ and $\lim_{t\to\infty} f(t)$ does not exist. Therefore, this theorem is not applicable to such a function.

The final value theorem may be stated as follows. If $f(t)$ and $df(t)/dt$ are Laplace transformable, if $F(s)$ is the Laplace transform of $f(t)$, and if $\lim_{t\to\infty} f(t)$ exists, then

$$\lim_{t\to\infty} f(t) = \lim_{s\to0} sF(s)$$

To prove the theorem, we let s approach zero in the equation for the Laplace transform of the derivative of $f(t)$ or

$$\lim_{s\to0} \int_0^\infty \left[\frac{d}{dt}f(t)\right] e^{-st}\, dt = \lim_{s\to0} [sF(s) - f(0)]$$

Since $\lim_{s\to0} e^{-st} = 1$, we obtain

$$\int_0^\infty \left[\frac{d}{dt}f(t)\right] dt = f(t)\Big|_0^\infty = f(\infty) - f(0)$$
$$= \lim_{s\to0} sF(s) - f(0)$$

from which

$$f(\infty) = \lim_{t\to\infty} f(t) = \lim_{s\to0} sF(s)$$

The final value theorem states that the steady-state behavior of $f(t)$ is the same as the behavior of $sF(s)$ in the neighborhood of $s = 0$. Thus, it is possible to obtain the value of $f(t)$ at $t = \infty$ directly from $F(s)$.

EXAMPLE 1–2.   Given

$$\mathscr{L}[f(t)] = F(s) = \frac{1}{s(s + 1)}$$

what is $\lim_{t\to\infty} f(t)$?

Since the pole of $sF(s) = 1/(s + 1)$ lies in the left half s plane, $\lim_{t\to\infty} f(t)$ exists. So the final value theorem is applicable in this case.

$$\lim_{t\to\infty} f(t) = f(\infty) = \lim_{s\to0} sF(s) = \lim_{s\to0} \frac{s}{s(s + 1)} = \lim_{s\to0} \frac{1}{s + 1} = 1$$

In fact, this result can easily be verified, since

$$f(t) = 1 - e^{-t} \qquad \text{for } t \geq 0$$

Initial value theorem. The initial value theorem is the counterpart of the final value theorem. By using this theorem, we are able to find the value of $f(t)$ at $t = 0+$ directly from the Laplace transform of $f(t)$. The initial value theorem does not give the value of $f(t)$ at exactly $t = 0$ but at a time slightly greater than zero.

The initial value theorem may be stated as follows. If $f(t)$ and $df(t)/dt$ are both Laplace transformable and if $\lim_{s \to \infty} sF(s)$ exists, then

$$f(0+) = \lim_{s \to \infty} sF(s)$$

To prove this theorem, we use the equation for the $\mathcal{L}_+$ transform of $df(t)/dt$:

$$\mathcal{L}_+ \left[\frac{d}{dt} f(t) \right] = sF(s) - f(0+)$$

For the time interval $0+ \le t \le \infty$, as s approaches infinity, e^{-st} approaches zero. (Note that we must use $\mathcal{L}_+$ rather than $\mathcal{L}_-$ for this condition.) And so

$$\lim_{s \to \infty} \int_{0+}^{\infty} \left[\frac{d}{dt} f(t) \right] e^{-st} \, dt = \lim_{s \to \infty} [sF(s) - f(0+)] = 0$$

or

$$f(0+) = \lim_{s \to \infty} sF(s)$$

In applying the initial value theorem, we are not limited as to the locations of the poles of $sF(s)$. Thus the initial value theorem is valid for the sinusoidal function.

It should be noted that the initial value theorem and the final value theorem provide a convenient check on the solution, since they enable us to predict the system behavior in the time domain without actually transforming functions in s back to time functions.

Real integration theorem. If $f(t)$ is of exponential order, then the Laplace transform of $\int f(t) \, dt$ exists and is given by

$$\mathcal{L} \left[\int f(t) \, dt \right] = \frac{F(s)}{s} + \frac{f^{-1}(0)}{s} \qquad (1\text{--}7)$$

where $F(s) = \mathcal{L}[f(t)]$ and $f^{-1}(0) = \int f(t) \, dt$, evaluated at $t = 0$.

Note that if $f(t)$ involves an impulse function at $t = 0$, then $f^{-1}(0+) \ne f^{-1}(0-)$. So if $f(t)$ involves an impulse function at $t = 0$, we must modify Equation (1–7) as follows:

$$\mathcal{L}_+ \left[\int f(t) \, dt \right] = \frac{F(s)}{s} + \frac{f^{-1}(0+)}{s}$$

$$\mathcal{L}_- \left[\int f(t) \, dt \right] = \frac{F(s)}{s} + \frac{f^{-1}(0-)}{s}$$

The real integration theorem given by Equation (1–7) can be proved in the following way. Integration by parts yields

$$\mathcal{L}\left[\int f(t)\, dt\right] = \int_0^\infty \left[\int f(t)\, dt\right] e^{-st}\, dt$$

$$= \left[\int f(t)\, dt\right] \frac{e^{-st}}{-s}\bigg|_0^\infty - \int_0^\infty f(t)\frac{e^{-st}}{-s}\, dt$$

$$= \frac{1}{s}\int f(t)\, dt\bigg|_{t=0} + \frac{1}{s}\int_0^\infty f(t)e^{-st}\, dt$$

$$= \frac{f^{-1}(0)}{s} + \frac{F(s)}{s}$$

and the theorem is proved.

We see that integration in the time domain is converted into division in the s domain. If the initial value of the integral is zero, the Laplace transform of the integral of $f(t)$ is given by $F(s)/s$.

The preceding real integration theorem given by Equation (1–7) can be modified slightly to deal with the definite integral of $f(t)$. If $f(t)$ is of exponential order, the Laplace transform of the definite integral $\int_0^t f(t)\, dt$ is given by

$$\mathcal{L}\left[\int_0^t f(t)\, dt\right] = \frac{F(s)}{s} \tag{1–8}$$

where $F(s) = \mathcal{L}[f(t)]$. This is also referred to as the real integration theorem. Note that if $f(t)$ involves an impulse function at $t = 0$, then $\int_{0+}^t f(t)\, dt \neq \int_{0-}^t f(t)\, dt$, and the following distinction must be observed:

$$\mathcal{L}_+\left[\int_{0+}^t f(t)\, dt\right] = \frac{\mathcal{L}_+[f(t)]}{s}$$

$$\mathcal{L}_-\left[\int_{0-}^t f(t)\, dt\right] = \frac{\mathcal{L}_-[f(t)]}{s}$$

To prove Equation (1–8), first note that

$$\int_0^t f(t)\, dt = \int f(t)\, dt - f^{-1}(0)$$

where $f^{-1}(0)$ is equal to $\int f(t)\, dt$ evaluated at $t = 0$ and is a constant. Hence

$$\mathcal{L}\left[\int_0^t f(t)\, dt\right] = \mathcal{L}\left[\int f(t)\, dt\right] - \mathcal{L}[f^{-1}(0)]$$

Noting that $f^{-1}(0)$ is a constant so that

$$\mathcal{L}[f^{-1}(0)] = \frac{f^{-1}(0)}{s}$$

we obtain

$$\mathcal{L}\left[\int_0^t f(t)\, dt\right] = \frac{F(s)}{s} + \frac{f^{-1}(0)}{s} - \frac{f^{-1}(0)}{s} = \frac{F(s)}{s}$$

Complex differentiation theorem. If $f(t)$ is Laplace transformable, then, except at poles of $F(s)$,

$$\mathcal{L}[tf(t)] = -\frac{d}{ds} F(s)$$

where $F(s) = \mathcal{L}[f(t)]$. This is known as the complex differentiation theorem. Also,

$$\mathcal{L}[t^2 f(t)] = \frac{d^2}{ds^2} F(s)$$

In general,

$$\mathcal{L}[t^n f(t)] = (-1)^n \frac{d^n}{ds^n} F(s) \qquad (n = 1,2,3, \ldots)$$

To prove the complex differentiation theorem, we proceed as follows:

$$\mathcal{L}[tf(t)] = \int_0^\infty tf(t)e^{-st}\, dt = -\int_0^\infty f(t) \frac{d}{ds}(e^{-st})\, dt$$

$$= -\frac{d}{ds}\int_0^\infty f(t)e^{-st}\, dt = -\frac{d}{ds} F(s)$$

Hence the theorem. Similarly, by defining $tf(t) = g(t)$, the result is

$$\mathcal{L}[t^2 f(t)] = \mathcal{L}[tg(t)] = -\frac{d}{ds} G(s) = -\frac{d}{ds}\left[-\frac{d}{ds} F(s)\right]$$

$$= (-1)^2 \frac{d^2}{ds^2} F(s) = \frac{d^2}{ds^2} F(s)$$

Repeating the same process, we obtain

$$\mathcal{L}[t^n f(t)] = (-1)^n \frac{d^n}{ds^n} F(s) \qquad (n = 1,2,3, \ldots)$$

Convolution integral. Consider the Laplace transform of

$$\int_0^t f_1(t - \tau)f_2(\tau)\, d\tau$$

This integral is often written as

$$f_1(t)*f_2(t)$$

The mathematical operation $f_1(t)*f_2(t)$ is called *convolution*. Note that if we put $t - \tau = \xi$, then

$$\int_0^t f_1(t - \tau)f_2(\tau)\, d\tau = -\int_t^0 f_1(\xi)f_2(t - \xi)\, d\xi$$

$$= \int_0^t f_1(\tau)f_2(t - \tau)\, d\tau$$

Hence

$$f_1(t)*f_2(t) = \int_0^t f_1(t - \tau)f_2(\tau)\,d\tau$$

$$= \int_0^t f_1(\tau)f_2(t - \tau)\,d\tau$$

$$= f_2(t)*f_1(t)$$

Figure 1–10(a) shows sample curves of $f_1(t)$, $f_1(t - \tau)$, and $f_2(\tau)$. Figure 1–10(b) shows the product of $f_1(t - \tau)$ and $f_2(\tau)$. The shape of the curve $f_1(t - \tau)f_2(\tau)$ depends on t.

If $f_1(t)$ and $f_2(t)$ are piecewise continuous and of exponential order, then the Laplace transform of

$$\int_0^t f_1(t - \tau)f_2(\tau)\,d\tau$$

can be obtained as follows:

$$\mathscr{L}\left[\int_0^t f_1(t - \tau)f_2(\tau)\,d\tau\right] = F_1(s)F_2(s) \tag{1-9}$$

where

$$F_1(s) = \int_0^\infty f_1(t)e^{-st}\,dt = \mathscr{L}[f_1(t)]$$

$$F_2(s) = \int_0^\infty f_2(t)e^{-st}\,dt = \mathscr{L}[f_2(t)]$$

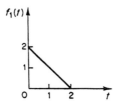
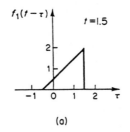
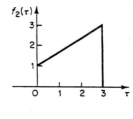

(a)

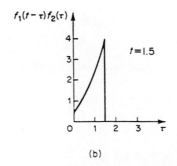

(b)

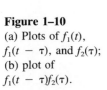

Figure 1–10
(a) Plots of $f_1(t)$, $f_1(t - \tau)$, and $f_2(\tau)$; (b) plot of $f_1(t - \tau)f_2(\tau)$.

To prove Equation (1–9), note that $f_1(t - \tau)1(t - \tau) = 0$ for $\tau > t$. Hence

$$\int_0^t f_1(t - \tau)f_2(\tau)\, d\tau = \int_0^\infty f_1(t - \tau)1(t - \tau)f_2(\tau)\, d\tau$$

Then

$$\mathscr{L}\left[\int_0^t f_1(t - \tau)f_2(\tau)\, d\tau\right] = \mathscr{L}\left[\int_0^\infty f_1(t - \tau)1(t - \tau)f_2(\tau)\, d\tau\right]$$

$$= \int_0^\infty e^{-st}\left[\int_0^\infty f_1(t - \tau)1(t - \tau)f_2(\tau)\, d\tau\right] dt$$

Substituting $t - \tau = \lambda$ in this last equation and changing the order of integration, which is valid in this case because $f_1(t)$ and $f_2(t)$ are Laplace transformable, we obtain

$$\mathscr{L}\left[\int_0^t f_1(t - \tau)f_2(\tau)\, d\tau\right] = \int_0^\infty f_1(t - \tau)1(t - \tau)e^{-st}\, dt \int_0^\infty f_2(\tau)d\tau$$

$$= \int_0^\infty f_1(\lambda)e^{-s(\lambda + \tau)}\, d\lambda \int_0^\infty f_2(\tau)\, d\tau$$

$$= \int_0^\infty f_1(\lambda)e^{-s\lambda}\, d\lambda \int_0^\infty f_2(\tau)e^{-s\tau}\, d\tau$$

$$= F_1(s)F_2(s)$$

This last equation gives the Laplace transform of the convolution integral. Conversely, if the Laplace transform of a function is given by a product of two Laplace transform functions, $F_1(s)F_2(s)$, then the corresponding time function (the inverse Laplace transform) is given by the convolution integral $f_1(t)*f_2(t)$.

Summary. Table 1–2 summarizes the properties and theorems of the Laplace transforms. Most of them have been derived or proved in this section.

Table 1–2 Properties of Laplace Transforms

1	$\mathscr{L}[Af(t)] = AF(s)$
2	$\mathscr{L}[f_1(t) \pm f_2(t)] = F_1(s) \pm F_2(s)$
3	$\mathscr{L}_\pm\left[\dfrac{d}{dt}f(t)\right] = sF(s) - f(0\pm)$
4	$\mathscr{L}_\pm\left[\dfrac{d^2}{dt^2}f(t)\right] = s^2F(s) - sf(0\pm) - \dot{f}(0\pm)$
5	$\mathscr{L}_\pm\left[\dfrac{d^n}{dt^n}f(t)\right] = s^nF(s) - \displaystyle\sum_{k=1}^{n} s^{n-k}\overset{(k-1)}{f}(0\pm)$ $\text{where } \overset{(k-1)}{f}(t) = \dfrac{d^{k-1}}{dt^{k-1}}f(t)$

Table 1–2 Cont'd

6	$$\mathcal{L}_{\pm}\left[\int f(t)\,dt\right] = \frac{F(s)}{s} + \frac{\left[\int f(t)\,dt\right]_{t=0\pm}}{s}$$
7	$$\mathcal{L}_{\pm}\left[\int\int f(t)\,dt\,dt\right] = \frac{F(s)}{s^2} + \frac{\left[\int f(t)\,dt\right]_{t=0\pm}}{s^2} + \frac{\left[\int\int f(t)\,dt\,dt\right]_{t=0\pm}}{s}$$
8	$$\mathcal{L}_{\pm}\left[\int\cdots\int f(t)(dt)^n\right] = \frac{F(s)}{s^n} + \sum_{k=1}^{n}\frac{1}{s^{n-k+1}}\left[\int\cdots\int f(t)(dt)^k\right]_{t=0\pm}$$
9	$$\mathcal{L}\left[\int_0^t f(t)\,dt\right] = \frac{F(s)}{s}$$
10	$$\int_0^\infty f(t)\,dt = \lim_{s\to 0} F(s) \qquad \text{if } \int_0^\infty f(t)\,dt \text{ exists}$$
11	$$\mathcal{L}[e^{-at}f(t)] = F(s + a)$$
12	$$\mathcal{L}[f(t - \alpha)1(t - \alpha)] = e^{-\alpha s}F(s) \qquad \alpha \geq 0$$
13	$$\mathcal{L}[tf(t)] = -\frac{dF(s)}{ds}$$
14	$$\mathcal{L}[t^2f(t)] = \frac{d^2}{ds^2}F(s)$$
15	$$\mathcal{L}[t^nf(t)] = (-1)^n\frac{d^n}{ds^n}F(s) \qquad n = 1,2,3,\ldots$$
16	$$\mathcal{L}\left[\frac{1}{t}f(t)\right] = \int_s^\infty F(s)\,ds$$
17	$$\mathcal{L}\left[f\left(\frac{t}{a}\right)\right] = aF(as)$$

1–4 INVERSE LAPLACE TRANSFORMATION

The mathematical process of passing from the complex variable expression to the time expression is called an *inverse transformation*. The notation for the inverse Laplace transformation is $\mathcal{L}^{-1}$, so that

$$\mathcal{L}^{-1}[F(s)] = f(t)$$

In solving problems by use of the Laplace transform method, we are confronted with the question of how to find $f(t)$ from $F(s)$. Mathematically, $f(t)$ is found from $F(s)$ by the following inversion integral:

$$f(t) = \frac{1}{2\pi j} \int_{c-j\infty}^{c+j\infty} F(s)e^{st}\,ds \qquad (t > 0)$$

where c, the abscissa of convergence, is a real constant and is chosen larger than the real parts of all singular points of $F(s)$. Thus, the path of integration is parallel to the $j\omega$ axis and is displaced by the amount c from it. This path of integration is to the right of all singular points.

Evaluating the inversion integral appears complicated. Fortunately, there are simpler methods by which $f(t)$ can be found from $F(s)$ than by performing this integration directly. A convenient method for obtaining inverse Laplace transforms is to use a table of Laplace transforms. In this case, the Laplace transform must be in a form immediately recognizable in such a table. Quite often the function in question may not appear in tables of Laplace transforms available to the engineer. If a particular transform $F(s)$ cannot be found in a table, then we may expand it into partial fractions and write $F(s)$ in terms of simple functions of s for which the inverse Laplace transforms are already known.

Note that these simpler methods for finding inverse Laplace transforms are based on the fact that the unique correspondence of a time function and its inverse Laplace transform holds for any continuous time function.

Partial-fraction expansion method for finding inverse Laplace transforms. For problems in control systems analysis, $F(s)$, the Laplace transform of $f(t)$, frequently occurs in the form

$$F(s) = \frac{B(s)}{A(s)}$$

where $A(s)$ and $B(s)$ are polynomials in s, and the degree of $B(s)$ is less than that of $A(s)$. If $F(s)$ is broken up into components,

$$F(s) = F_1(s) + F_2(s) + \cdots + F_n(s)$$

and if the inverse Laplace transforms of $F_1(s)$, $F_2(s)$, . . . , $F_n(s)$ are readily available, then

$$\mathcal{L}^{-1}[F(s)] = \mathcal{L}^{-1}[F_1(s)] + \mathcal{L}^{-1}[F_2(s)] + \cdots + \mathcal{L}^{-1}[F_n(s)]$$
$$= f_1(t) + f_2(t) + \cdots + f_n(t)$$

where $f_1(t), f_2(t), . . . , f_n(t)$ are the inverse Laplace transforms of $F_1(s), F_2(s), . . . , F_n(s)$, respectively. The inverse Laplace transform of $F(s)$ thus obtained is unique except possibly at points where the time function is discontinuous. Whenever the time function is continuous, the time function $f(t)$ and its Laplace transform $F(s)$ have a one-to-one correspondence.

The advantage of the partial-fraction expansion approach is that the individual terms of $F(s)$, resulting from the expansion into partial-fraction form, are very simple functions of s; consequently, it is not necessary to refer to a Laplace transform table if we memorize several simple Laplace transform pairs. It should be noted, however, that in applying the partial-fraction expansion technique in the search for the inverse Laplace transform of $F(s) = B(s)/A(s)$ the roots of the denominator polynomial $A(s)$ must be known in advance. That is this method does not apply until the denominator polynomial has been factored.

In the expansion of $F(s) = B(s)/A(s)$ into partial-fraction form, it is important that the highest power of s in $A(s)$ be greater than the highest power of s in $B(s)$. If such is not the case, the numerator $B(s)$ must be divided by the denominator $A(s)$ in order to produce a polynomial in s plus a remainder (a ratio of polynomials in s whose numerator is of lower degree than the denominator.) (For details, see Example 1–4.)

Partial-fraction expansion when $F(s)$ involves distinct poles only. Consider $F(s)$ written in the factored form

$$F(s) = \frac{B(s)}{A(s)} = \frac{K(s + z_1)(s + z_2) \cdots (s + z_m)}{(s + p_1)(s + p_2) \cdots (s + p_n)} \quad (m < n)$$

where $p_1, p_2, \ldots, p_n$ and $z_1, z_2, \ldots, z_m$ are either real or complex quantities, but for each complex p_i or z_i there will occur the complex conjugate of p_i or z_i, respectively. If $F(s)$ involves distinct poles only, then it can be expanded into a sum of simple partial fractions as follows:

$$F(s) = \frac{B(s)}{A(s)} = \frac{a_1}{s + p_1} + \frac{a_2}{s + p_2} + \cdots + \frac{a_n}{s + p_n} \qquad (1\text{--}10)$$

where a_k ($k = 1, 2, \ldots, n$) are constants. The coefficient a_k is called the *residue* at the pole at $s = -p_k$. The value of a_k can be found by multiplying both sides of Equation (1–10) by $(s + p_k)$ and letting $s = -p_k$, which gives

$$\left[(s + p_k) \frac{B(s)}{A(s)} \right]_{s = -p_k} = \left[\frac{a_1}{s + p_1}(s + p_k) + \frac{a_2}{s + p_2}(s + p_k) \right.$$

$$\left. + \cdots + \frac{a_k}{s + p_k}(s + p_k) + \cdots + \frac{a_n}{s + p_n}(s + p_k) \right]_{s = -p_k}$$

$$= a_k$$

We see that all the expanded terms drop out with the exception of a_k. Thus the residue a_k is found from

$$a_k = \left[(s + p_k) \frac{B(s)}{A(s)} \right]_{s = -p_k} \qquad (1\text{--}11)$$

Note that, since $f(t)$ is a real function of time, if p_1 and p_2 are complex conjugates, then the residues a_1 and a_2 are also complex conjugates. Only one of the conjugates, a_1 or a_2, needs to be evaluated because the other is known automatically.

Since

$$\mathcal{L}^{-1}\left[\frac{a_k}{s + p_k} \right] = a_k e^{-p_k t}$$

$f(t)$ is obtained as

$$f(t) = \mathcal{L}^{-1}[F(s)] = a_1 e^{-p_1 t} + a_2 e^{-p_2 t} + \cdots + a_n e^{-p_n t} \qquad (t \geq 0)$$

EXAMPLE 1–3   Find the inverse Laplace transform of

$$F(s) = \frac{s + 3}{(s + 1)(s + 2)}$$

The partial-fraction expansion of $F(s)$ is

$$F(s) = \frac{s + 3}{(s + 1)(s + 2)} = \frac{a_1}{s + 1} + \frac{a_2}{s + 2}$$

where a_1 and a_2 are found by using Equation (1–11).

$$a_1 = \left[(s + 1) \frac{s + 3}{(s + 1)(s + 2)} \right]_{s = -1} = \left[\frac{s + 3}{s + 2} \right]_{s = -1} = 2$$

$$a_2 = \left[(s + 2) \frac{s + 3}{(s + 1)(s + 2)} \right]_{s = -2} = \left[\frac{s + 3}{s + 1} \right]_{s = -2} = -1$$

Thus

$$f(t) = \mathscr{L}^{-1}[F(s)]$$

$$= \mathscr{L}^{-1} \left[\frac{2}{s + 1} \right] + \mathscr{L}^{-1} \left[\frac{-1}{s + 2} \right]$$

$$= 2e^{-t} - e^{-2t} \qquad (t \geq 0)$$

EXAMPLE 1–4   Obtain the inverse Laplace transform of

$$G(s) = \frac{s^3 + 5s^2 + 9s + 7}{(s + 1)(s + 2)}$$

Here, since the degree of the numerator polynomial is higher than that of the denominator polynomial, we must divide the numerator by the denominator.

$$G(s) = s + 2 + \frac{s + 3}{(s + 1)(s + 2)}$$

Note that the Laplace transform of the unit impulse function $\delta(t)$ is 1 and that the Laplace transform of $d\delta(t)/dt$ is s. The third term on the right-hand side of this last equation is $F(s)$ in Example 1–3. So the inverse Laplace transform of $G(s)$ is given as

$$g(t) = \frac{d}{dt}\delta(t) + 2\delta(t) + 2e^{-t} - e^{-2t} \qquad (t \geq 0-)$$

EXAMPLE 1–5   Find the inverse Laplace transform of

$$F(s) = \frac{2s + 12}{s^2 + 2s + 5}$$

Notice that the denominator polynomial can be factored as

$$s^2 + 2s + 5 = (s + 1 + j2)(s + 1 - j2)$$

If the function $F(s)$ involves a pair of complex conjugate poles, it it convenient not to expand $F(s)$ into the usual partial fractions but to expand it into the sum of a damped sine and a damped cosine function.

Noting that $s^2 + 2s + 5 = (s + 1)^2 + 2^2$ and referring to the Laplace transforms of $e^{-\alpha t} \sin \omega t$ and $e^{-\alpha t} \cos \omega t$, rewritten thus,

$$\mathcal{L}[e^{-\alpha t} \sin \omega t] = \frac{\omega}{(s + \alpha)^2 + \omega^2}$$

$$\mathcal{L}[e^{-\alpha t} \cos \omega t] = \frac{s + \alpha}{(s + \alpha)^2 + \omega^2}$$

the given $F(s)$ can be written as a sum of a damped sine and a damped cosine function.

$$F(s) = \frac{2s + 12}{s^2 + 2s + 5} = \frac{10 + 2(s + 1)}{(s + 1)^2 + 2^2}$$

$$= 5 \frac{2}{(s + 1)^2 + 2^2} + 2 \frac{s + 1}{(s + 1)^2 + 2^2}$$

It follows that

$$f(t) = \mathcal{L}^{-1}[F(s)]$$

$$= 5\mathcal{L}^{-1}\left[\frac{2}{(s + 1)^2 + 2^2}\right] + 2\mathcal{L}^{-1}\left[\frac{s + 1}{(s + 1)^2 + 2^2}\right]$$

$$= 5e^{-t} \sin 2t + 2e^{-t} \cos 2t \qquad (t \geq 0)$$

Partial-fraction expansion when $F(s)$ involves multiple poles. Instead of discussing the general case, we shall use an example to show how to obtain the partial-fraction expansion of $F(s)$. (See also Problem A–1–19.)

Consider the following $F(s)$:

$$F(s) = \frac{s^2 + 2s + 3}{(s + 1)^3}$$

The partial-fraction expansion of this $F(s)$ involves three terms,

$$F(s) = \frac{B(s)}{A(s)} = \frac{b_3}{(s + 1)^3} + \frac{b_2}{(s + 1)^2} + \frac{b_1}{s + 1}$$

where b_3, b_2, and b_1 are determined as follows. By multiplying both sides of this last equation by $(s + 1)^3$, we have

$$(s + 1)^3 \frac{B(s)}{A(s)} = b_3 + b_2(s + 1) + b_1(s + 1)^2 \qquad (1\text{–}12)$$

Then letting $s = -1$, Equation (1–12) gives

$$\left[(s + 1)^3 \frac{B(s)}{A(s)}\right]_{s=-1} = b_3$$

Also, differentiation of both sides of Equation (1–12) with respect to s yields

$$\frac{d}{ds}\left[(s+1)^3 \frac{B(s)}{A(s)}\right] = b_2 + 2b_1(s+1) \tag{1–13}$$

If we let $s = -1$ in Equation (1–13), then

$$\frac{d}{ds}\left[(s+1)^3 \frac{B(s)}{A(s)}\right]_{s=-1} = b_2$$

By differentiating both sides of Equation (1–13) with respect to s, the result is

$$\frac{d^2}{ds^2}\left[(s+1)^3 \frac{B(s)}{A(s)}\right] = 2b_1$$

From the preceding analysis it can be seen that the values of b_1, b_2, and b_3 are found systematically as follows:

$$b_3 = \left[(s+1)^3 \frac{B(s)}{A(s)}\right]_{s=-1}$$
$$= (s^2 + 2s + 3)_{s=-1}$$
$$= 2$$

$$b_2 = \left\{\frac{d}{ds}\left[(s+1)^3 \frac{B(s)}{A(s)}\right]\right\}_{s=-1}$$
$$= \left[\frac{d}{ds}(s^2 + 2s + 3)\right]_{s=-1}$$
$$= (2s + 2)_{s=-1}$$
$$= 0$$

$$b_1 = \frac{1}{2!}\left\{\frac{d^2}{ds^2}\left[(s+1)^3 \frac{B(s)}{A(s)}\right]\right\}_{s=-1}$$
$$= \frac{1}{2!}\left[\frac{d^2}{ds^2}(s^2 + 2s + 3)\right]_{s=-1}$$
$$= \frac{1}{2}(2) = 1$$

We thus obtain

$$f(t) = \mathcal{L}^{-1}[F(s)]$$
$$= \mathcal{L}^{-1}\left[\frac{2}{(s+1)^3}\right] + \mathcal{L}^{-1}\left[\frac{0}{(s+1)^2}\right] + \mathcal{L}^{-1}\left[\frac{1}{s+1}\right]$$
$$= t^2 e^{-t} + 0 + e^{-t}$$
$$= (t^2 + 1)e^{-t} \qquad (t \geq 0)$$

1–5 SOLVING LINEAR, TIME-INVARIANT, DIFFERENTIAL EQUATIONS

In this section we are concerned with the use of the Laplace transform method in solving linear, time-invariant, differential equations.

The Laplace transform method yields the complete solution (complementary solution and particular solution) of linear, time-invariant, differential equations. Classical methods for finding the complete solution of a differential equation require the evaluation of the integration constants from the initial conditions. In the case of the Laplace transform method, however, this requirement is unnecessary because the initial conditions are automatically included in the Laplace transform of the differential equation.

If all initial conditions are zero, then the Laplace transform of the differential equation is obtained simply by replacing d/dt with s, d^2/dt^2 with s^2, and so on.

In solving linear, time-invariant, differential equations by the Laplace transform method, two steps are needed.

1. By taking the Laplace transform of each term in the given differential equation, convert the differential equation into an algebraic equation in s and obtain the expression for the Laplace transform of the dependent variable by rearranging the algebraic equation.
2. The time solution of the differential equation is obtained by finding the inverse Laplace transform of the dependent variable.

In the following discussion, two examples are used to demonstrate the solution of linear, time-invariant, differential equations by the Laplace transform method.

EXAMPLE 1–6

Find the solution $x(t)$ of the differential equation

$$\ddot{x} + 3\dot{x} + 2x = 0, \qquad x(0) = a, \qquad \dot{x}(0) = b$$

where a and b are constants.

By writing the Laplace transform of $x(t)$ as $X(s)$ or

$$\mathcal{L}[x(t)] = X(s)$$

we obtain

$$\mathcal{L}[\dot{x}] = sX(s) - x(0)$$

$$\mathcal{L}[\ddot{x}] = s^2X(s) - sx(0) - \dot{x}(0)$$

And so the given differential equation becomes

$$[s^2X(s) - sx(0) - \dot{x}(0)] + 3[sX(s) - x(0)] + 2X(s) = 0$$

By substituting the given initial conditions into this last equation,

$$[s^2X(s) - as - b] + 3[sX(s) - a] + 2X(s) = 0$$

or

$$(s^2 + 3s + 2)X(s) = as + b + 3a$$

Solving for $X(s)$, we have

$$X(s) = \frac{as + b + 3a}{s^2 + 3s + 2} = \frac{as + b + 3a}{(s + 1)(s + 2)} = \frac{2a + b}{s + 1} - \frac{a + b}{s + 2}$$

The inverse Laplace transform of $X(s)$ gives

$$x(t) = \mathcal{L}^{-1}[X(s)] = \mathcal{L}^{-1}\left[\frac{2a + b}{s + 1}\right] - \mathcal{L}^{-1}\left[\frac{a + b}{s + 2}\right]$$

$$= (2a + b)e^{-t} - (a + b)e^{-2t} \qquad (t \geq 0)$$

which is the solution of the given differential equation. Notice that the initial conditions a and b appear in the solution. Thus $x(t)$ has no undetermined constants.

EXAMPLE 1–7   Find the solution $x(t)$ of the differential equation

$$\ddot{x} + 2\dot{x} + 5x = 3, \qquad x(0) = 0, \qquad \dot{x}(0) = 0$$

Noting that $\mathcal{L}[3] = 3/s$, $x(0) = 0$, and $\dot{x}(0) = 0$, the Laplace transform of the differential equation becomes

$$s^2 X(s) + 2sX(s) + 5X(s) = \frac{3}{s}$$

Solving for $X(s)$, we find

$$X(s) = \frac{3}{s(s^2 + 2s + 5)} = \frac{3}{5}\frac{1}{s} - \frac{3}{5}\frac{s + 2}{s^2 + 2s + 5}$$

$$= \frac{3}{5}\frac{1}{s} - \frac{3}{10}\frac{2}{(s + 1)^2 + 2^2} - \frac{3}{5}\frac{s + 1}{(s + 1)^2 + 2^2}$$

Hence the inverse Laplace transform becomes

$$x(t) = \mathcal{L}^{-1}[X(s)]$$

$$= \frac{3}{5}\mathcal{L}^{-1}\left[\frac{1}{s}\right] - \frac{3}{10}\mathcal{L}^{-1}\left[\frac{2}{(s + 1)^2 + 2^2}\right] - \frac{3}{5}\mathcal{L}^{-1}\left[\frac{s + 1}{(s + 1)^2 + 2^2}\right]$$

$$= \frac{3}{5} - \frac{3}{10}e^{-t}\sin 2t - \frac{3}{5}e^{-t}\cos 2t \qquad (t \geq 0)$$

which is the solution of the given differential equation.

1–6   TRANSFER FUNCTION

In control theory, functions called transfer functions are commonly used to characterize the input–output relationships of components or systems that can be described by linear, time-invariant, differential equations. We begin by defining the transfer function and follow with a derivation of the transfer function of a mechanical system. (Additional examples appear in Chapter 2.)

Transfer function.   The *transfer function* of a linear, time-invariant, differential equation system is defined as the ratio of the Laplace transform of the output (response function) to the Laplace transform of the input (driving function) under the assumption that all initial conditions are zero.

Consider the linear time-invariant system defined by the following differential equation:

$$\overset{(n)}{a_0 y} + \overset{(n-1)}{a_1 y} + \cdots + a_{n-1}\dot{y} + a_n y$$

$$= \overset{(m)}{b_0 x} + \overset{(m-1)}{b_1 x} + \cdots + b_{m-1}\dot{x} + b_m x \qquad (n \geq m) \qquad (1\text{–}14)$$

where y is the output of the system and x is the input. The transfer function of this system is obtained by taking the Laplace transforms of both sides of Equation (1–14), under the assumption that all initial conditions are zero, or

$$\text{Transfer function} = G(s) = \left.\frac{\mathscr{L}[\text{output}]}{\mathscr{L}[\text{input}]}\right|_{\text{zero initial conditions}}$$

$$= \frac{Y(s)}{X(s)} = \frac{b_0 s^m + b_1 s^{m-1} + \cdots + b_{m-1}s + b_m}{a_0 s^n + a_1 s^{n-1} + \cdots + a_{n-1}s + a_n}$$

By using the concept of transfer function, it is possible to represent system dynamics by algebraic equations in s. If the highest power of s in the denominator of the transfer function is equal to n, the system is called an *nth-order system.*

Comments on transfer function. The applicability of the concept of the transfer function is limited to linear, time-invariant, differential equation systems. The transfer function approach, however, is extensively used in the analysis and design of such systems. In what follows, we shall list important comments concerning the transfer function. (Note that in the list a system referred to is one described by a linear, time-invariant, differential equation.)

1. The transfer function of a system is a mathematical model in that it is an operational method of expressing the differential equation that relates the output variable to the input variable.
2. The transfer function is a property of a system itself, independent of the magnitude and nature of the input or driving function.
3. The transfer function includes the units necessary to relate the input to the output; however, it does not provide any information concerning the physical structure of the system. (The transfer functions of many physically different systems can be identical.)
4. If the transfer function of a system is known, the output or response can be studied for various forms of inputs with a view toward understanding the nature of the system.
5. If the transfer function of a system in unknown, it may be established experimentally by introducing known inputs and studying the output of the system. Once established, a transfer function gives a full description of the dynamic characteristics of the system, as distinct from its physical description.

Mechanical system. Consider the satellite attitude control system shown in Figure 1–11. The diagram shows the control of only the yaw angle θ. (In the actual system there are controls about three axes.) Small jets apply reaction forces to rotate the satellite body into the desired attitude. The two skew symmetrically placed jets denoted by A or B operate in pairs. Assume that each jet thrust is $F/2$ and a torque $T = Fl$ is applied to the system. The jets are applied for a certain time duration and thus the torque can be written as $T(t)$. The moment of inertia about the axis of rotation at the center of mass is J.

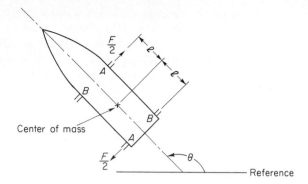

Figure 1–11
Schematic diagram of a satellite attitude control system.

Let us obtain the transfer function of this system by assuming that torque $T(t)$ is the input, and the angular displacement $\theta(t)$ of the satellite is the output.

To derive the transfer function, we proceed according to the following steps.

1. Write the differential equation for the system.
2. Take the Laplace transform of the differential equation, assuming all initial conditions are zero.
3. Take the ratio of the output $\Theta(s)$ to the input $T(s)$. This ratio is the transfer function.

Applying Newton's second law to the present system and noting that there is no friction in the environment of the satellite, we have

$$J\frac{d^2\theta}{dt^2} = T$$

Taking the Laplace transform of both sides of this last equation and assuming all initial conditions to be zero yields

$$Js^2\Theta(s) = T(s)$$

where $\Theta(s) = \mathscr{L}[\theta(t)]$ and $T(s) = \mathscr{L}[T(t)]$. The transfer function of the system is thus obtained as

$$\text{Transfer function} = \frac{\Theta(s)}{T(s)} = \frac{1}{Js^2}$$

1–7 BLOCK DIAGRAMS

A control system may consist of a number of components. To show the functions performed by each component, in control engineering, we commonly use a diagram called the *block diagram*. This section explains what a block diagram is, presents a method for obtaining block diagrams for physical systems, and, finally, discusses techniques to simplify such diagrams.

Block diagrams.   A *block diagram* of a system is a pictorial representation of the functions performed by each component and of the flow of signals. Such a diagram depicts the interrelationships that exist among the various components. Differing from a purely abstract mathematical representation, a block diagram has the advantage of indicating more realistically the signal flows of the actual system.

In a block diagram all system variables are linked to each other through functional blocks. The *functional block* or simply *block* is a symbol for the mathematical operation on the input signal to the block that produces the output. The transfer functions of the components are usually entered in the corresponding blocks, which are connected by arrows to indicate the direction of the flow of signals. Note that the signal can pass only in the direction of the arrows. Thus a block diagram of a control system explicitly shows a unilateral property.

Figure 1–12 shows an element of the block diagram. The arrowhead pointing toward the block indicates the input, and the arrowhead leading away from the block represents the output. Such arrows are referred to as *signals*.

Note that the dimensions of the output signal from the block are the dimensions of the input signal multiplied by the dimensions of the transfer function in the block.

The advantages of the block diagram representation of a system lie in the fact that it is easy to form the overall block diagram for the entire system by merely connecting the blocks of the components according to the signal flow and that it is possible to evaluate the contribution of each component to the overall performance of the system.

In general, the functional operation of the system can be visualized more readily by examining the block diagram than by examining the physical system itself. A block diagram contains information concerning dynamic behavior, but it does not include any information on the physical construction of the system. Consequently, many dissimilar and unrelated systems can be represented by the same block diagram.

It should be noted that in a block diagram the main source of energy is not explicitly shown and that the block diagram of a given system is not unique. A number of different block diagrams can be drawn for a system, depending on the point of view of the analysis.

Summing Point. Referring to Figure 1–13, a circle with a cross is the symbol that indicates a summing operation. The plus or minus sign at each arrowhead indicates whether that signal is to be added or subtracted. It is important that the quantities being added or substracted have the same dimensions and the same units.

Branch Point. A *branch point* is a point from which the signal from a block goes concurrently to other blocks or summing points.

Block diagram of a closed-loop system.

Figure 1–14 shows an example of a block diagram of a closed-loop system. The output $C(s)$ is fed back to the summing point, where it is compared with the reference input $R(s)$. The closed-loop nature of the system is clearly indicated by the figure. The output of the block, $C(s)$ in this case, is obtained by multiplying the transfer function $G(s)$ by the input to the block, $E(s)$. Any linear control system may be represented by a block diagram consisting of blocks, summing points, and branch points.

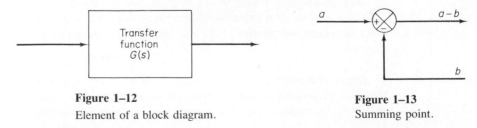

Figure 1–12
Element of a block diagram.

Figure 1–13
Summing point.

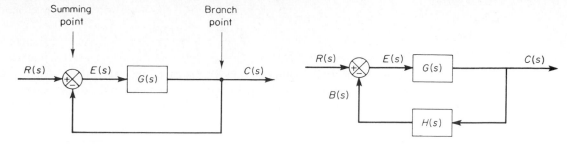

Figure 1–14
Block diagram of a closed-loop system.

Figure 1–15
Closed-loop system.

When the output is fed back to the summing point for comparison with the input, it is necessary to convert the form of the output signal to that of the input signal. For example, in a temperature control system, the output signal is usually the controlled temperature. The output signal, which has the dimension of temperature, must be converted to a force or position or voltage before it can be compared with the input signal. This conversion is accomplished by the feedback element whose transfer function is $H(s)$, as shown in Figure 1–15. The role of the feedback element is to modify the output before it is compared with the input. (In most cases the feedback element is a sensor that measures the output of the plant. The output of the sensor is compared with the input, and the actuating error signal is generated.) In the present example, the feedback signal that is fed back to the summing point for comparison with the input is $B(s) = H(s)C(s)$.

Open-loop transfer function and feedforward transfer function.   Referring to Figure 1–15, the ratio of the feedback signal $B(s)$ to the actuating error signal $E(s)$ is called the *open-loop transfer function*. That is,

$$\text{Open-loop transfer function} = \frac{B(s)}{E(s)} = G(s)H(s)$$

The ratio of the output $C(s)$ to the actuating error signal $E(s)$ is called the *feedforward transfer function*, so that

$$\text{Feedforward transfer function} = \frac{C(s)}{E(s)} = G(s)$$

If the feedback transfer function $H(s)$ is unity, then the open-loop transfer function and the feedforward transfer function are the same.

Closed-loop transfer function.   For the system shown in Figure 1–15, the output $C(s)$ and input $R(s)$ are related as follows:

$$C(s) = G(s)E(s)$$
$$E(s) = R(s) - B(s)$$
$$= R(s) - H(s)C(s)$$

Eliminating $E(s)$ from these equations gives

$$C(s) = G(s)[R(s) - H(s)C(s)]$$

or

$$\frac{C(s)}{R(s)} = \frac{G(s)}{1 + G(s)H(s)} \tag{1-15}$$

The transfer function relating $C(s)$ to $R(s)$ is called the *closed-loop transfer function*. This transfer function relates the closed-loop system dynamics to the dynamics of the feedforward elements and feedback elements.

From Equation (1–15), $C(s)$ is given by

$$C(s) = \frac{G(s)}{1 + G(s)H(s)} R(s)$$

Thus the output of the closed-loop system clearly depends on both the closed-loop transfer function and the nature of the input.

Closed-loop system subjected to a disturbance.   Figure 1–16 shows a closed-loop system subjected to a disturbance. When two inputs (the reference input and disturbance) are present in a linear system, each input can be treated independently of the other; and the outputs corresponding to each input alone can be added to give the complete output. The way each input is introduced into the system is shown at the summing point by either a plus or minus sign.

Consider the system shown in Figure 1–16. In examining the effect of the disturbance $N(s)$, we may assume that the system is at rest initially with zero error; we may then calculate the response $C_N(s)$ to the disturbance only. This response can be found from

$$\frac{C_N(s)}{N(s)} = \frac{G_2(s)}{1 + G_1(s)G_2(s)H(s)}$$

On the other hand, in considering the response to the reference input $R(s)$, we may assume that the disturbance is zero. Then the response $C_R(s)$ to the reference input $R(s)$ can be obtained from

$$\frac{C_R(s)}{R(s)} = \frac{G_1(s)G_2(s)}{1 + G_1(s)G_2(s)H(s)}$$

The response to the simultaneous application of the reference input and disturbance can be obtained by adding the two individual responses. In other words, the response $C(s)$ due to

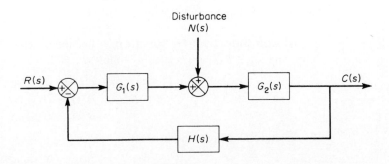

Figure 1–16
Closed-loop system subjected to a disturbance.

the simultaneous application of the reference input $R(s)$ and disturbance $N(s)$ is given by

$$C(s) = C_R(s) + C_N(s)$$

$$= \frac{G_2(s)}{1 + G_1(s)G_2(s)H(s)}[G_1(s)R(s) + N(s)]$$

Consider now the case where $|G_1(s)H(s)| \gg 1$ and $|G_1(s)G_2(s)H(s)| \gg 1$. In this case, the closed-loop transfer function $C_N(s)/N(s)$ becomes almost zero, and the effect of the disturbance is suppressed. This is an advantage of the closed-loop system.

On the other hand, the closed-loop transfer function $C_R(s)/R(s)$ approaches $1/H(s)$ as the gain of $G_1(s)G_2(s)H(s)$ increases. This means that if $|G_1(s)G_2(s)H(s)| \gg 1$ then the closed-loop transfer function $C_R(s)/R(s)$ becomes independent of $G_1(s)$ and $G_2(s)$ and becomes inversely proportional to $H(s)$ so that the variations of $G_1(s)$ and $G_2(s)$ do not affect the closed-loop transfer function $C_R(s)/R(s)$. This is another advantage of the closed-loop system. It can easily be seen that any closed-loop system with unity feedback, $H(s) = 1$, tends to equalize the input and output.

Procedures for drawing a block diagram. To draw a block diagram for a system, first write the equations that describe the dynamic behavior of each component. Then take the Laplace transforms of these equations, assuming zero initial conditions, and represent each Laplace-transformed equation individually in block form. Finally, assemble the elements into a complete block diagram.

As an example, consider the RC circuit shown in Figure 1–17(a). The equations for this circuit are

$$i = \frac{e_i - e_o}{R} \tag{1–16}$$

$$e_o = \frac{\int i \, dt}{C} \tag{1–17}$$

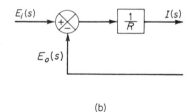

Figure 1–17
(a) RC circuit; (b) block diagram representing Eq. (1–18); (c) block diagram representing Eq. (1–19); (d) block diagram of the RC circuit.

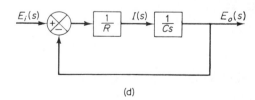

(a)   (b)   (c)   (d)

The Laplace transforms of Equations (1–16) and (1–17), with zero initial condition, become

$$I(s) = \frac{E_i(s) - E_o(s)}{R} \qquad (1\text{--}18)$$

$$E_o(s) = \frac{I(s)}{Cs} \qquad (1\text{--}19)$$

Equation (1–18) represents a summing operation, and the corresponding diagram is shown in Figure 1–17(b). Equation (1–19) represents the block as shown in Figure 1–17(c). Assembling these two elements, we obtain the overall block diagram for the system as shown in Figure 1–17(d).

Block diagram reduction. It is important to note that blocks can be connected in series only if the output of one block is not affected by the next following block. If there are any loading effects between the components, it is necessary to combine these components into a single block.

Any number of cascaded blocks representing nonloading components can be replaced by a single block, the transfer function of which is simply the product of the individual transfer functions.

A complicated block diagram involving many feedback loops can be simplified by a step-by-step rearrangement, using rules of block diagram algebra. Some of these important rules are given in Table 1–3. They are obtained by writing the same equation in a different way. Simplification of the block diagram by rearrangements and substitutions considerably reduces the labor needed for subsequent mathematical analysis. It should be noted, however, that as the block diagram is simplified the transfer functions in new blocks become more complex because new poles and new zeros are generated.

In simplifying a block diagram, remember the following.

1. The product of the transfer functions in the feedforward direction must remain the same.
2. The product of the transfer functions around the loop must remain the same.

EXAMPLE 1–8 Consider the system shown in Figure 1–18(a). Simplify this diagram by using the rules given in Table 1–3.

By moving the summing point of the negative feedback loop containing H_2 outside the positive feedback loop containing H_1, we obtain Figure 1–18(b). Eliminating the positive feedback loop, we have Figure 1–18(c). Then elimination of the loop containing H_2/G_1 gives Figure 1–18(d). Finally, eliminating the feedback loop results in Figure 1–18(e).

Notice that the numerator of the closed-loop transfer function $C(s)/R(s)$ is the product of the transfer functions of the feedforward path. The denominator of $C(s)/R(s)$ is equal to

$$1 - \sum (\text{product of the transfer functions around each loop})$$
$$= 1 - (G_1G_2H_1 - G_2G_3H_2 - G_1G_2G_3)$$
$$= 1 - G_1G_2H_1 + G_2G_3H_2 + G_1G_2G_3$$

(The positive feedback loop yields a negative term in the denominator.)

EXAMPLE 1–9 Draw a block diagram for the mechanical system shown in Figure 1–19. Then simplify the block diagram and obtain the transfer function between $X_o(s)$ and $X_i(s)$. Assume that displacement x_o is measured from the equilibrium position when $x_i = 0$.

Chapter 1 / Introduction to Control Systems Analysis

Table 1–3 Rules of Block Diagram Algebra

	Original Block Diagrams	Equivalent Block Diagrams
1	$A \rightarrow \otimes \xrightarrow{A-B} \otimes \xrightarrow{A-B+C}$, $B\uparrow$, $C\uparrow$	$A \rightarrow \otimes \xrightarrow{A+C} \otimes \xrightarrow{A-B+C}$, $C\uparrow$, $B\uparrow$
2	$A \rightarrow \otimes \xrightarrow{A-B+C}$, $C\downarrow$, $B\uparrow$	$A \rightarrow \otimes \xrightarrow{A-B} \otimes \xrightarrow{A-B+C}$, $B\uparrow$, $C\downarrow$
3	$A \rightarrow \boxed{G_1} \xrightarrow{AG_1} \boxed{G_2} \xrightarrow{AG_1G_2}$	$A \rightarrow \boxed{G_2} \xrightarrow{AG_2} \boxed{G_1} \xrightarrow{AG_1G_2}$
4	$A \rightarrow \boxed{G_1} \xrightarrow{AG_1} \boxed{G_2} \xrightarrow{AG_1G_2}$	$A \rightarrow \boxed{G_1G_2} \xrightarrow{AG_1G_2}$
5	$A \rightarrow \boxed{G_1} \xrightarrow{AG_1} \otimes \xrightarrow{AG_1+AG_2}$, $\boxed{G_2} \xrightarrow{AG_2}$	$A \rightarrow \boxed{G_1+G_2} \xrightarrow{AG_1+AG_2}$
6	$A \rightarrow \boxed{G} \xrightarrow{AG} \otimes \xrightarrow{AG-B}$, $B\uparrow$	$A \rightarrow \otimes \xrightarrow{A-\frac{B}{G}} \boxed{G} \xrightarrow{AG-B}$, $\frac{B}{G}\uparrow$, $\boxed{\frac{1}{G}} \leftarrow B$
7	$A \rightarrow \otimes \xrightarrow{A-B} \boxed{G} \xrightarrow{AG-BG}$, $B\uparrow$	$A \rightarrow \boxed{G} \xrightarrow{AG} \otimes \xrightarrow{AG-BG}$, $B \rightarrow \boxed{G} \xrightarrow{BG}$
8	$A \rightarrow \boxed{G} \xrightarrow{AG}$, $\xrightarrow{AG}$	$A \rightarrow \boxed{G} \xrightarrow{AG}$, $\boxed{G} \xrightarrow{AG}$
9	$A \rightarrow \boxed{G} \xrightarrow{AG}$, $\xrightarrow{A}$	$A \rightarrow \boxed{G} \xrightarrow{AG}$, $AG \rightarrow \boxed{\frac{1}{G}} \xrightarrow{A}$
10	$A \rightarrow \otimes \xrightarrow{A-B}$, $\xrightarrow{A-B}$, $B\uparrow$	$B\downarrow \otimes \xrightarrow{A-B}$, $A \rightarrow \otimes \xrightarrow{A-B}$, $B\uparrow$
11	$A \rightarrow \boxed{G_1} \xrightarrow{AG_1} \otimes \xrightarrow{AG_1+AG_2}$, $\boxed{G_2} \xrightarrow{AG_2}$	$A \rightarrow \boxed{G_1} \xrightarrow{AG_1} \otimes \xrightarrow{AG_1+AG_2}$, $\boxed{\frac{G_2}{G_1}}\uparrow$
12	$A \rightarrow \otimes \rightarrow \boxed{G_1} \xrightarrow{B}$, $\boxed{G_2} \leftarrow$	$A \rightarrow \boxed{\frac{1}{G_2}} \rightarrow \otimes \rightarrow \boxed{G_2} \rightarrow \boxed{G_1} \xrightarrow{B}$
13	$A \rightarrow \otimes \rightarrow \boxed{G_1} \xrightarrow{B}$, $\boxed{G_2} \leftarrow$	$A \rightarrow \boxed{\dfrac{G_1}{1+G_1G_2}} \xrightarrow{B}$

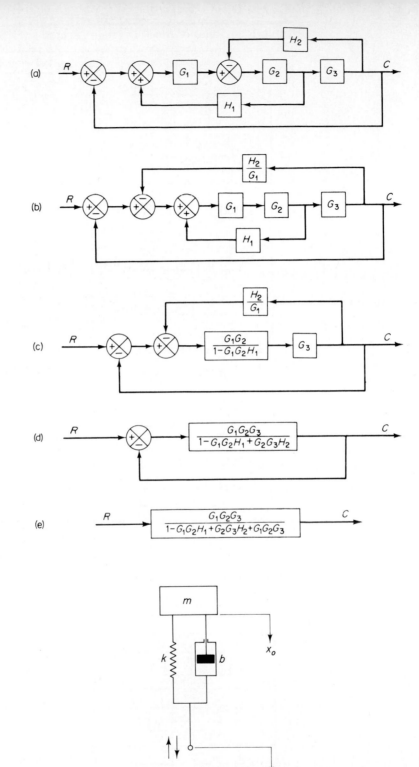

Figure 1–18
(a) Multiple-loop
system; (b)–(e)
successive reduction
of the block diagram
shown in (a).

Figure 1–19
Mechanical system.

Defining the sum of the forces acting on mass m as F, we obtain equations for the system as follows:

$$m\ddot{x}_o = F \tag{1–20}$$

$$F = -b(\dot{x}_o - \dot{x}_i) - k(x_o - x_i) \tag{1–21}$$

Rewriting Equations (1–20) and (1–21) in Laplace-transformed forms, assuming the zero initial conditions, we have

$$ms^2 X_o(s) = F(s) \tag{1–22}$$

$$F(s) = -b[sX_o(s) - sX_i(s)] - k[X_o(s) - X_i(s)]$$

$$= (bs + k)[X_i(s) - X_o(s)] \tag{1–23}$$

From Equations (1–22) and (1–23), we obtain elements of the block diagram as shown in Figure 1–20(a). By connecting signals properly, we can construct a block diagram for the system, as shown in Figure 1–20(b). Referring to the rules of block diagram algebra given in Table 1–3, this diagram can be simplified into the one shown in Figure 1–20(c). Elimination of the feedback loop results in Figure 1–20(d). The transfer function between $X_o(s)$ and $X_i(s)$ is thus

$$\frac{X_o(s)}{X_i(s)} = \frac{bs + k}{ms^2 + bs + k}$$

Figure 1–20
(a) Elements of a block diagram; (b) block diagram as a result of combining elements; (c) and (d) simplified block diagrams.

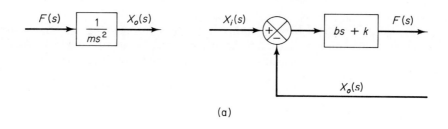

(a)

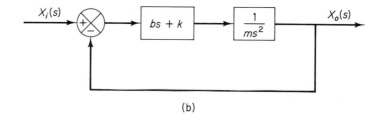

(b)

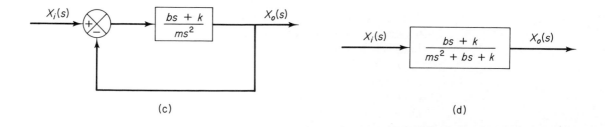

(c)   (d)

We have presented examples showing how to simplify block diagrams. More on simplification of block diagrams will be given in Chapters 2 and 3, where the reader will have opportunities to practice simplification or reduction of complex block diagrams.

Comments. It is important to note that, in the transfer function concept, we assume that the transfer function of any block is not affected by the succeeding block. That is, the succeeding block does not load the first block. If any loading effects are present, the two blocks must be considered as one and we derive a single transfer function for the two blocks combined. (More details will be given in Chapter 2.) It is also noted that the sequence of individual transfer functions may be changed. The product of two transfer functions $G_1 G_2$ may be written as $G_2 G_1$.

1-8 SIGNAL FLOW GRAPHS

The block diagram is useful for graphically representing control system dynamics and is used extensively in the analysis and design of control systems. An alternate approach for graphically representing control system dynamics is the signal flow graph approach, due to S. J. Mason. It is noted that the signal flow graph approach and the block diagram approach yield the same information and one is in no sense superior to the other.

Signal flow graphs. A signal flow graph is a diagram that represents a set of simultaneous linear algebraic equations. When applying the signal flow graph method to analyses of control systems, we must first transform linear differential equations into algebraic equations in s.

A signal flow graph consists of a network in which nodes are connected by directed branches. Each node represents a system variable, and each branch connected between two nodes acts as a signal multiplier. Note that the signal flows in only one direction. The direction of signal flow is indicated by an arrow placed on the branch, and the multiplication factor is indicated along the branch. The signal flow graph depicts the flow of signals from one point of a system to another and gives the relationships among the signals.

As mentioned earlier, a signal flow graph contains essentially the same information as a block diagram. If a signal flow graph is used to represent a control system, then a gain formula, called Mason's gain formula, may be used to obtain the relationships among system variables without carrying out reduction of the graph.

Definitions. Before we discuss signal flow graphs, we must define certain terms.

Node. A node is a point representing a variable or signal.

Transmittance. The transmittance is a real gain or complex gain between two nodes. Such gains can be expressed in terms of the transfer function between two nodes.

Branch. A branch is a directed line segment joining two nodes. The gain of a branch is a transmittance.

Input node or source. An input node or source is a node that has only outgoing branches. This corresponds to an independent variable.

Output node or sink. An output node or sink is a node that has only incoming branches. This corresponds to a dependent variable.

Chapter 1 / Introduction to Control Systems Analysis

Mixed node. A mixed node is a node that has both incoming and outgoing branches.

Path. A path is a traversal of connected branches in the direction of the branch arrows. If no node is crossed more than once, the path is open. If the path ends at the same node from which it began and does not cross any other node more than once, it is closed. If a path crosses some node more than once but ends at a different node from which it began, it is neither open nor closed.

Loop. A loop is a closed path.

Loop gain. The loop gain is the product of the branch transmittances of a loop.

Nontouching loops. Loops are nontouching if they do not possess any common nodes.

Forward path. A forward path is a path from an input node (source) to an output node (sink) that does not cross any nodes more than once.

Forward path gain. A forward path gain is the product of the branch transmittances of a forward path.

Figure 1–21 shows nodes and branches, together with transmittances.

Properties of signal flow graphs. A few important properties of signal flow graphs are as follows:

1. A branch indicates the functional dependence of one signal on another. A signal passes through only in the direction specified by the arrow of the branch.
2. A node adds the signals of all incoming branches and transmits this sum to all outgoing branches.
3. A mixed node, which has both incoming and outgoing branches, may be treated as an output node (sink) by adding an outgoing branch of unity transmittance. (See Figure 1–21. Notice that a branch with unity transmittance is directed from x_3 to another node, also denoted by x_3.) Note, however, that we cannot change a mixed node to a source by this method.
4. For a given system, a signal flow graph is not unique. Many different signal flow graphs can be drawn for a given system by writing the system equations differently.

Signal flow graph algebra. A signal flow graph of a linear system can be drawn using the foregoing definitions. In doing so, we usually bring the input nodes (sources) to the left and the output nodes (sinks) to the right. The independent and dependent variables of the equations become the input nodes (sources) and output nodes (sinks), respectively. The branch transmittances can be obtained from the coefficients of the equations.

To determine the input–output relationship, we may use Mason's formula, which will

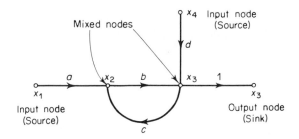

Figure 1–21
Signal flow graph.

be given later, or we may reduce the signal flow graph to a graph containing only input and output nodes. To accomplish this, we use the following rules:

1. The value of a node with one incoming branch, as shown in Figure 1–22(a), is $x_2 = ax_1$.
2. The total transmittance of cascaded branches is equal to the product of all the branch transmittances. Cascaded branches can thus be combined into a single branch by multiplying the transmittances, as shown in Figure 1–22(b).
3. Parallel branches may be combined by adding the transmittances, as shown in Figure 1–22(c).
4. A mixed node may be eliminated, as shown in Figure 1–22(d).
5. A loop may be eliminated, as shown in Figure 1–22(e). Note that

$$x_3 = bx_2, \qquad x_2 = ax_1 + cx_3$$

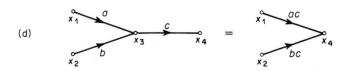

Figure 1–22
Signal flow graphs
and simplifications.

Hence

$$x_3 = abx_1 + bcx_3 \qquad (1\text{--}24)$$

or

$$x_3 = \frac{ab}{1 - bc} x_1 \qquad (1\text{--}25)$$

Equation (1–24) corresponds to a diagram having a self-loop of transmittance bc. Elimination of the self-loop yields Equation (1–25), which clearly shows that the overall transmittance is $ab/(1 - bc)$.

Signal flow graph representation of linear systems.   Signal flow graphs are widely applied to linear-system analysis. Here the graph can be drawn from the system equations or, with practice, can be drawn by inspection of the physical system. Routine reduction by use of the foregoing rules gives the relation between an input and output variable.

Consider a system defined by the following set of equations:

$$x_1 = a_{11}x_1 + a_{12}x_2 + a_{13}x_3 + b_1u_1 \qquad (1\text{--}26)$$

$$x_2 = a_{21}x_1 + a_{22}x_2 + a_{23}x_3 + b_2u_2 \qquad (1\text{--}27)$$

$$x_3 = a_{31}x_1 + a_{32}x_2 + a_{33}x_3 \qquad (1\text{--}28)$$

where u_1 and u_2 are input variables and x_1, x_2, and x_3 are output variables. A signal flow graph for this system, a graphical representation of these three simultaneous equations, indicating the interdependence of the variables, can be obtained as follows: First locate the nodes x_1, x_2, and x_3 as shown in Figure 1–23(a). Note that a_{ij} is the transmittance between

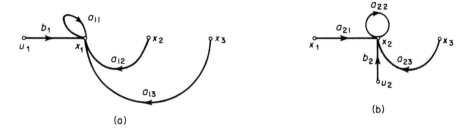

Figure 1–23   Signal flow graphs representing (a) Eq. (1–26), (b) Eq. (1–27), and (c) Eq. (1–28); (d) complete signal flow graph for the system described by Eqs. (1–26)–(1–28).

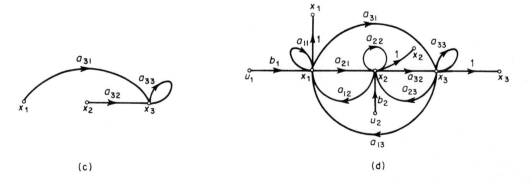

x_j and x_i. Equation (1–26) states that x_1 is equal to the sum of the four signals $a_{11}x_1$, $a_{12}x_2$, $a_{13}x_3$, and b_1u_1. The signal flow graph representing Equation (1–26) is shown in Figure 1–23(a). Equation (1–27) states that x_2 is equal to the sum of $a_{21}x_1$, $a_{22}x_2$, $a_{23}x_3$, and b_2u_2. The corresponding signal flow graph is shown in Figure 1–23(b). The signal flow graph representing Equation (1–28) is shown in Figure 1–23(c).

The signal flow graph representing Equations (1–26), (1–27), and (1–28) is then obtained by combining Figures 1–23(a), (b), and (c). Finally, the complete signal flow graph for the given simultaneous equations is shown in Figure 1–23(d).

In dealing with a signal flow graph, the input nodes (sources) may be considered one at a time. The output signal is then equal to the sum of the individual contributions of each input.

The overall gain from an input to an output may be obtained directly from the signal flow graph by inspection, by use of Mason's formula, or by a reduction of the graph to a simpler form.

Signal flow graphs of control systems. Some signal flow graphs of simple control systems are shown in Figure 1–24. For such simple graphs, the closed-loop transfer function $C(s)/R(s)$ [or $C(s)/N(s)$] can be obtained easily by inspection. For more complicated signal flow graphs, Mason's gain formula is quite useful.

Mason's gain formula for signal flow graphs. In many practical cases, we wish to determine the relationship between an input variable and an output variable of the signal flow graph. The transmittance between an input node and an output node is the overall gain, or overall transmittance, between these two nodes.

Mason's gain formula, which is applicable to the overall gain, is given by

$$P = \frac{1}{\Delta} \sum_k P_k \Delta_k$$

where

P_k = path gain or transmittance of kth forward path

Δ = determinant of graph

= 1 − (sum of all individual loop gains) + (sum of gain products of all possible combinations of two nontouching loops) − (sum of gain products of all possible combinations of three nontouching loops) + $\cdots$

$$= 1 - \sum_a L_a + \sum_{b,c} L_b L_c - \sum_{d,e,f} L_d L_e L_f + \cdots$$

$\displaystyle\sum_a L_a$ = sum of all individual loop gains

$\displaystyle\sum_{b,c} L_b L_c$ = sum of gain products of all possible combinations of two non-touching loops

$\displaystyle\sum_{d,e,f} L_d L_e L_f$ = sum of gain products of all possible combinations of three non-touching loops

Δ_k = cofactor of the kth forward path determinant of the graph with the loops touching the kth forward path removed, that is, the cofactor Δ_k is obtained from Δ by removing the loops that touch path P_k

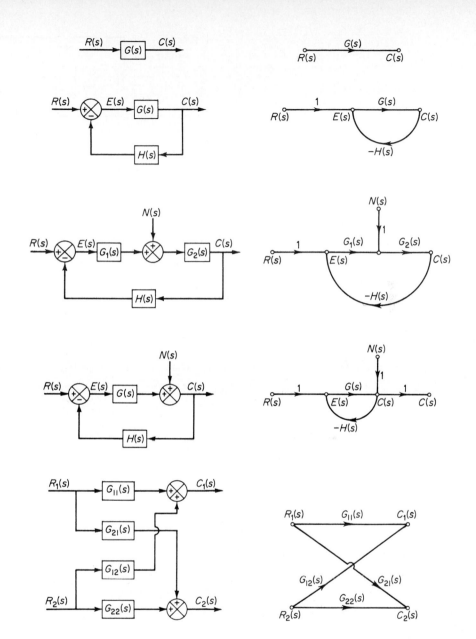

Figure 1–24
Block diagrams and corresponding signal flow graphs.

(Note that the summations are taken over all possible paths from input to output.)

In the following, we shall illustrate the use of Mason's gain formula by means of two examples.

EXAMPLE 1–10   Consider the system shown in Figure 1–25. A signal flow graph for this system is shown in Figure 1–26. Let us obtain the closed-loop transfer function $C(s)/R(s)$ by use of Mason's gain formula.

In this system there is only one forward path between the input $R(s)$ and the output $C(s)$. The forward path gain is

$$P_1 = G_1 G_2 G_3$$

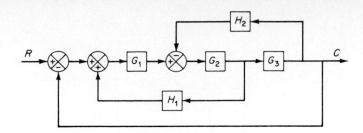

Figure 1–25
Multiple-loop system.

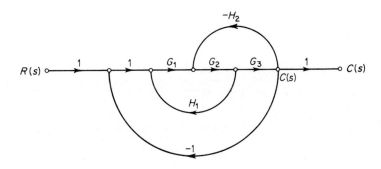

Figure 1–26
Signal flow graph for
the system shown in
Fig. 1–25.

From Figure 1–26, we see that there are three individual loops. The gains of these loops are

$$L_1 = G_1 G_2 H_1$$

$$L_2 = -G_2 G_3 H_2$$

$$L_3 = -G_1 G_2 G_3$$

Note that since all three loops have a common branch, there are no nontouching loops. Hence, the determinant Δ is given by

$$\begin{aligned}\Delta &= 1 - (L_1 + L_2 + L_3)\\&= 1 - G_1 G_2 H_1 + G_2 G_3 H_2 + G_1 G_2 G_3\end{aligned}$$

The cofactor Δ_1 of the determinant along the forward path connecting the input node and output node is obtained from Δ by removing the loops that touch this path. Since path P_1 touches all three loops, we obtain

$$\Delta_1 = 1$$

Therefore, the overall gain between the input $R(s)$ and the output $C(s)$, or the closed-loop transfer function, is given by

$$\begin{aligned}\frac{C(s)}{R(s)} = P &= \frac{P_1 \Delta_1}{\Delta}\\&= \frac{G_1 G_2 G_3}{1 - G_1 G_2 H_1 + G_2 G_3 H_2 + G_1 G_2 G_3}\end{aligned}$$

which is the same as the closed-loop transfer function obtained by block diagram reduction. Mason's gain formula thus gives the overall gain $C(s)/R(s)$ without a reduction of the graph.

EXAMPLE 1–11 Consider the system shown in Figure 1–27. Obtain the closed-loop transfer function $C(s)/R(s)$ by use of Mason's gain formula.

Chapter 1 / Introduction to Control Systems Analysis

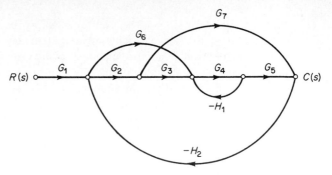

Figure 1–27
Signal flow graph for a system.

In this system, there are three forward paths between the input $R(s)$ and the output $C(s)$. The forward path gains are

$$P_1 = G_1 G_2 G_3 G_4 G_5$$

$$P_2 = G_1 G_6 G_4 G_5$$

$$P_3 = G_1 G_2 G_7$$

There are four individual loops. The gains of these loops are

$$L_1 = -G_4 H_1$$

$$L_2 = -G_2 G_7 H_2$$

$$L_3 = -G_6 G_4 G_5 H_2$$

$$L_4 = -G_2 G_3 G_4 G_5 H_2$$

Loop L_1 does not touch loop L_2. Hence, the determinant Δ is given by

$$\Delta = 1 - (L_1 + L_2 + L_3 + L_4) + L_1 L_2 \qquad (1\text{–}29)$$

The cofactor Δ_1 is obtained from Δ by removing the loops that touch path P_1. Therefore, by removing L_1, L_2, L_3, L_4, and $L_1 L_2$ from Equation (1–29), we obtain

$$\Delta_1 = 1$$

Similarly, the cofactor Δ_2 is

$$\Delta_2 = 1$$

The cofactor Δ_3 is obtained by removing L_2, L_3, L_4, and $L_1 L_2$ from Equation (1–29), giving

$$\Delta_3 = 1 - L_1$$

The closed-loop transfer function $C(s)/R(s)$ is then

$$\frac{C(s)}{R(s)} = P = \frac{1}{\Delta}(P_1 \Delta_1 + P_2 \Delta_2 + P_3 \Delta_3)$$

$$= \frac{G_1 G_2 G_3 G_4 G_5 + G_1 G_6 G_4 G_5 + G_1 G_2 G_7 (1 + G_4 H_1)}{1 + G_4 H_1 + G_2 G_7 H_2 + G_6 G_4 G_5 H_2 + G_2 G_3 G_4 G_5 H_2 + G_4 H_1 G_2 G_7 H_2}$$

Comments. The usual application of signal flow graphs is in system diagramming. The set of equations describing a linear system is represented by a signal flow graph by establishing nodes that represent the system variables and by interconnecting the nodes with weighted, directed, transmittances, which represent the relationships among the variables.

Mason's gain formula may be used to establish the relationship between an input and an output. (Alternatively, the variables in the system may be eliminated one by one with reduction techniques.) Mason's gain formula is especially useful in reducing large and complex system diagrams in one step, without requiring step-by-step reductions.

Finally, it is noted that in applying the Mason's gain formula to a given system, one must be careful not to make mistakes in calculating the cofactors of the forward paths, Δ_k, since any errors, if they exist, may not easily be detected.

1–9 STATE-SPACE APPROACH TO CONTROL SYSTEM ANALYSIS

In this section we shall present introductory material on state-space analysis of control systems.

Modern control theory. The modern trend in engineering systems is toward greater complexity, due mainly to the requirements of complex tasks and good accuracy. Complex systems may have multiple inputs and multiple outputs and may be time varying. Because of the necessity of meeting increasingly stringent requirements on the performance of control systems, the increase in system complexity, and easy access to large-scale computers, modern control theory, which is a new approach to the analysis and design of complex control systems, has been developed since around 1960. This new approach is based on the concept of state. The concept of state by itself is not new since it has been in existence for a long time in the field of classical dynamics and other fields.

Modern control theory versus conventional control theory. Modern control theory is contrasted with conventional control theory in that the former is applicable to multiple-input, multiple-output systems, which may be linear or nonlinear, time-invariant or time-varying, while the latter is applicable only to linear time-invariant single-input, single-output systems. Also, modern control theory is essentially a time-domain approach, while conventional control theory is a complex frequency-domain approach.

Before we proceed further, we must define state, state variables, state vector, and state space.

State. The state of a dynamic system is the smallest set of variables (called *state variables*) such that the knowledge of these variables at $t = t_0$, together with the knowledge of the input for $t \geq t_0$, completely determines the behavior of the system for any time $t \geq t_0$.

Thus, the state of a dynamic system at time t is uniquely determined by the state at time t_0 and the input for $t \geq t_0$, and it is independent of the state and input before t_0. Note that, in dealing with linear time-invariant systems, we usually choose the reference time t_0 to be zero.

Note that the concept of state is by no means limited to physical systems. It is applicable to biological systems, economic systems, social systems, and others.

State variables. The state variables of a dynamic system are the variables making up the smallest set of variables that determine the state of the dynamic system. If at least n variables $x_1, x_2, \ldots, x_n$ are needed to completely describe the behavior of a dynamic system (so that once the input is given for $t \geq t_0$ and the initial state at $t = t_0$ is specified,

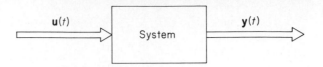

Figure 1–28
Dynamic system.

the future state of the system is completely determined), then such n variables are a set of state variables.

Note that state variables need not be physically measurable or observable quantities. Variables that do not represent physical quantities and those that are neither measurable nor observable can be chosen as state variables. Such freedom in choosing state variables is an advantage of the state space methods. Practically speaking, however, it is convenient to choose easily measurable quantities for the state variables, if this is possible at all, because optimal control laws will require the feedback of all state variables with suitable weighting.

State vector.   If n state variables are needed to completely describe the behavior of a given system, then these n state variables can be considered the n components of a vector **x**. Such a vector is called a *state vector*. A state vector is thus a vector that determines uniquely the system state $\mathbf{x}(t)$ for any time $t \geq t_0$, once the state at $t = t_0$ is given and the input $\mathbf{u}(t)$ for $t \geq t_0$ is specified.

State space.   The n-dimensional space whose coordinate axes consist of the x_1 axis, x_2 axis, . . . , x_n axis is called a *state space*. Any state can be represented by a point in the state space.

State-space equations.   In state-space analysis we are concerned with three types of variables that are involved in the modeling of dynamic systems: input variables, output variables, and state variables. As we shall see in Section 2–2, the state-space representation for a given system is not unique, except that the number of state variables is the same for any of the different state-space representations of the same system.

Consider the dynamic system shown in Figure 1–28. In the diagram the heavy arrows imply that the signals are vector quantities. In this system the output $\mathbf{y}(t)$ for $t \geq t_1$ depends on the value $\mathbf{y}(t_1)$ and the input $\mathbf{u}(t)$ for $t \geq t_1$. The dynamic system must involve elements that memorize the values of the input for $t \geq t_1$. Since integrators in a continuous-time control system serve as memory devices, the outputs of such integrators can be considered as the variables that define the internal state of the dynamic system. Thus the outputs of integrators serve as state variables. The number of state variables to completely define the dynamics of the system is equal to the number of integrators involved in the system.

Assume that a multiple-input, multiple-output system involves n integrators. Assume also that there are r inputs $u_1(t), u_2(t), \ldots, u_r(t)$ and m outputs $y_1(t), y_2(t), \ldots, y_m(t)$. Define n outputs of the integrators as state variables: $x_1(t), x_2(t), \ldots, x_n(t)$. Then the system may be described by

$$\dot{x}_1(t) = f_1(x_1, x_2, \ldots, x_n; u_1, u_2, \ldots, u_r; t)$$
$$\dot{x}_2(t) = f_2(x_1, x_2, \ldots, x_n; u_1, u_2, \ldots, u_r; t)$$
$$\cdot$$
$$\cdot \qquad\qquad\qquad\qquad\qquad\qquad\qquad\qquad (1\text{–}30)$$
$$\cdot$$
$$\dot{x}_n(t) = f_n(x_1, x_2, \ldots, x_n; u_1, u_2, \ldots, u_r; t)$$

The outputs $y_1(t)$, $y_2(t)$, . . ., $y_m(t)$ of the system may be given by

$$y_1(t) = g_1(x_1, x_2, \ldots, x_n; u_1, u_2, \ldots, u_r; t)$$
$$y_2(t) = g_2(x_1, x_2, \ldots, x_n; u_1, u_2, \ldots, u_r; t)$$
$$\cdot$$
$$\cdot$$
$$\cdot$$
$$y_m(t) = g_m(x_1, x_2, \ldots, x_n; u_1, u_2, \ldots, u_r; t)$$

(1–31)

If we define

$$\mathbf{x}(t) = \begin{bmatrix} x_1(t) \\ x_2(t) \\ \cdot \\ \cdot \\ \cdot \\ x_n(t) \end{bmatrix}, \quad \mathbf{f}(\mathbf{x}, \mathbf{u}, t) = \begin{bmatrix} f_1(x_1, x_2, \ldots, x_n; u_1, u_2, \ldots, u_r; t) \\ f_2(x_1, x_2, \ldots, x_n; u_1, u_2, \ldots, u_r; t) \\ \cdot \\ \cdot \\ \cdot \\ f_n(x_1, x_2, \ldots, x_n; u_1, u_2, \ldots, u_r; t) \end{bmatrix}$$

$$\mathbf{y}(t) = \begin{bmatrix} y_1(t) \\ y_2(t) \\ \cdot \\ \cdot \\ \cdot \\ y_m(t) \end{bmatrix}, \quad \mathbf{g}(\mathbf{x}, \mathbf{u}, t) = \begin{bmatrix} g_1(x_1, x_2, \ldots, x_n; u_1, u_2, \ldots, u_r; t) \\ g_2(x_1, x_2, \ldots, x_n; u_1, u_2, \ldots, u_r; t) \\ \cdot \\ \cdot \\ \cdot \\ g_m(x_1, x_2, \ldots, x_n; u_1, u_2, \ldots, u_r; t) \end{bmatrix}, \quad \mathbf{u}(t) = \begin{bmatrix} u_1(t) \\ u_2(t) \\ \cdot \\ \cdot \\ \cdot \\ u_r(t) \end{bmatrix}$$

then Equations (1–30) and (1–31) become

$$\dot{\mathbf{x}}(t) = \mathbf{f}(\mathbf{x}, \mathbf{u}, t) \tag{1–32}$$

$$\mathbf{y}(t) = \mathbf{g}(\mathbf{x}, \mathbf{u}, t) \tag{1–33}$$

where Equation (1–32) is the state equation and Equation (1–33) is the output equation. If vector functions $\mathbf{f}$ and/or $\mathbf{g}$ involve time t explicitly, then the system is called a time-varying system.

If Equations (1–32) and (1–33) are linearized about the operating state, then we have the following linearized state equation and output equation:

$$\dot{\mathbf{x}}(t) = \mathbf{A}(t)\mathbf{x}(t) + \mathbf{B}(t)\mathbf{u}(t) \tag{1–34}$$

$$\mathbf{y}(t) = \mathbf{C}(t)\mathbf{x}(t) + \mathbf{D}(t)\mathbf{u}(t) \tag{1–35}$$

where $\mathbf{A}(t)$ is called the state matrix, $\mathbf{B}(t)$ the input matrix, $\mathbf{C}(t)$ the output matrix, and $\mathbf{D}(t)$ the direct transmission matrix. A block diagram representation of Equations (1–34) and (1–35) is shown in Figure 1–29.

If vector functions $\mathbf{f}$ and $\mathbf{g}$ do not involve time t explicitly, then the system is called a time-invariant system. In this case, Equations (1–32) and (1–33) can be simplified to

$$\dot{\mathbf{x}}(t) = \mathbf{f}(\mathbf{x}, \mathbf{u}) \tag{1–36}$$

$$\mathbf{y}(t) = \mathbf{g}(\mathbf{x}, \mathbf{u}) \tag{1–37}$$

Figure 1–29
Block diagram of the
linear continuous-time
control system
represented in state
space.

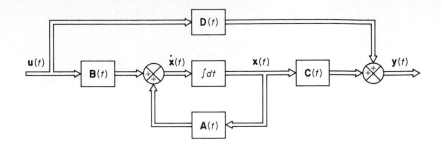

Equations (1–36) and (1–37) can be linearized about the operating state as follows:

$$\dot{\mathbf{x}}(t) = \mathbf{A}\mathbf{x}(t) + \mathbf{B}\mathbf{u}(t) \qquad (1\text{–}38)$$

$$\mathbf{y}(t) = \mathbf{C}\mathbf{x}(t) + \mathbf{D}\mathbf{u}(t) \qquad (1\text{–}39)$$

Equation (1–38) is the state equation of the linear, time-invariant system. Equation (1–39) is the output equation for the same system. In this book we shall be concerned mostly with systems described by Equations (1–38) and (1–39).

   In what follows we shall present an example for deriving a state equation and output equation.

EXAMPLE 1–12

Consider the mechanical system shown in Figure 1–30. We assume that the system is linear. The external force $u(t)$ is the input to the system, and the displacement $y(t)$ of the mass is the output. The displacement $y(t)$ is measured from the equilibrium position in the absence of the external force. This system is a single-input, single-output system.

   From the diagram, the system equation is

$$m\ddot{y} + b\dot{y} + ky = u \qquad (1\text{–}40)$$

This system is of second order. This means that the system involves two integrators. Let us define state variables $x_1(t)$ and $x_2(t)$ as

$$x_1(t) = y(t)$$

$$x_2(t) = \dot{y}(t)$$

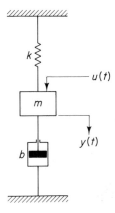

Figure 1–30
Mechanical system.

Then we obtain

$$\dot{x}_1 = x_2$$

$$\dot{x}_2 = \frac{1}{m}(-ky - b\dot{y}) + \frac{1}{m}u$$

or

$$\dot{x}_1 = x_2 \qquad\qquad (1\text{–}41)$$

$$\dot{x}_2 = -\frac{k}{m}x_1 - \frac{b}{m}x_2 + \frac{1}{m}u \qquad\qquad (1\text{–}42)$$

The output equation is

$$y = x_1 \qquad\qquad (1\text{–}43)$$

In a vector-matrix form, Equations (1–41) and (1–42) can be written as

$$\begin{bmatrix} \dot{x}_1 \\ \dot{x}_2 \end{bmatrix} = \begin{bmatrix} 0 & 1 \\ -\dfrac{k}{m} & -\dfrac{b}{m} \end{bmatrix} \begin{bmatrix} x_1 \\ x_2 \end{bmatrix} + \begin{bmatrix} 0 \\ \dfrac{1}{m} \end{bmatrix} u \qquad\qquad (1\text{–}44)$$

The output equation, Equation (1–43), can be written as

$$y = \begin{bmatrix} 1 & 0 \end{bmatrix} \begin{bmatrix} x_1 \\ x_2 \end{bmatrix} \qquad\qquad (1\text{–}45)$$

Equation (1–44) is a state equation and Equation (1–45) is an output equation for the system. Equations (1–44) and (1–45) are in the standard form:

$$\dot{\mathbf{x}} = \mathbf{Ax} + \mathbf{B}u$$

$$y = \mathbf{Cx} + \mathbf{D}u$$

where

$$\mathbf{A} = \begin{bmatrix} 0 & 1 \\ -\dfrac{k}{m} & -\dfrac{b}{m} \end{bmatrix}, \qquad \mathbf{B} = \begin{bmatrix} 0 \\ \dfrac{1}{m} \end{bmatrix}, \qquad \mathbf{C} = \begin{bmatrix} 1 & 0 \end{bmatrix}, \qquad D = 0$$

Figure 1–31 is a block diagram for the system. Notice that the outputs of the integrators are state variables.

Figure 1–31
Block diagram of the mechanical system shown in Fig. 1–30.

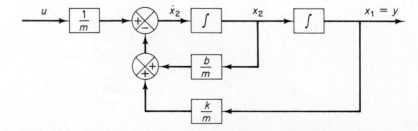

Correlation between transfer functions and state-space equations. In what follows we shall show how to derive the transfer function of a single-input, single-output system from the state-space equations.

Let us consider the system whose transfer function is given by

$$\frac{Y(s)}{U(s)} = G(s) \tag{1-46}$$

This system may be represented in state space by the following equations:

$$\dot{\mathbf{x}} = \mathbf{A}\mathbf{x} + \mathbf{B}u \tag{1-47}$$

$$y = \mathbf{C}\mathbf{x} + Du \tag{1-48}$$

where $\mathbf{x}$ is the state vector, u is the input, and y is the output. The Laplace transforms of Equations (1–47) and (1–48) are given by

$$s\mathbf{X}(s) - \mathbf{x}(0) = \mathbf{A}\mathbf{X}(s) + \mathbf{B}U(s) \tag{1-49}$$

$$Y(s) = \mathbf{C}\mathbf{X}(s) + DU(s) \tag{1-50}$$

Since the transfer function was previously defined as the ratio of the Laplace transform of the output to the Laplace transform of the input when the initial conditions were zero, we assume that $\mathbf{x}(0)$ in Equation (1–49) is zero. Then we have

$$s\mathbf{X}(s) - \mathbf{A}\mathbf{X}(s) = \mathbf{B}U(s)$$

or

$$(s\mathbf{I} - \mathbf{A})\mathbf{X}(s) = \mathbf{B}U(s)$$

By premultiplying $(s\mathbf{I} - \mathbf{A})^{-1}$ to both sides of this last equation, we obtain

$$\mathbf{X}(s) = (s\mathbf{I} - \mathbf{A})^{-1}\mathbf{B}U(s) \tag{1-51}$$

By substituting Equation (1–51) into Equation (1–50), we get

$$Y(s) = [\mathbf{C}(s\mathbf{I} - \mathbf{A})^{-1}\mathbf{B} + D]U(s) \tag{1-52}$$

Upon comparing Equation (1–52) with Equation (1–46), we see that

$$G(s) = \mathbf{C}(s\mathbf{I} - \mathbf{A})^{-1}\mathbf{B} + D \tag{1-53}$$

This is the transfer-function expression in terms of $\mathbf{A}$, $\mathbf{B}$, $\mathbf{C}$, and D.

Note that the right-hand side of Equation (1–53) involves $(s\mathbf{I} - \mathbf{A})^{-1}$. Hence $G(s)$ can be written as

$$G(s) = \frac{Q(s)}{|s\mathbf{I} - \mathbf{A}|}$$

where $Q(s)$ is a polynomial in s. Therefore, $|s\mathbf{I} - \mathbf{A}|$ is equal to the characteristic polynomial of $G(s)$. In other words, the eigenvalues of $\mathbf{A}$ are identical to the poles of $G(s)$.

EXAMPLE 1–13 Consider again the mechanical system shown in Figure 1–30. State-space equations for the system are given by Equations (1–44) and (1–45). We shall obtain the transfer function for the system from the state-space equations.

By substituting $\mathbf{A}$, $\mathbf{B}$, $\mathbf{C}$, and D into Equation (1–53), we obtain

$$G(s) = \mathbf{C}(s\mathbf{I} - \mathbf{A})^{-1}\mathbf{B} + D$$

$$= \begin{bmatrix} 1 & 0 \end{bmatrix} \left\{ \begin{bmatrix} s & 0 \\ 0 & s \end{bmatrix} - \begin{bmatrix} 0 & 1 \\ -\dfrac{k}{m} & -\dfrac{b}{m} \end{bmatrix} \right\}^{-1} \begin{bmatrix} 0 \\ \dfrac{1}{m} \end{bmatrix} + 0$$

$$= \begin{bmatrix} 1 & 0 \end{bmatrix} \begin{bmatrix} s & -1 \\ \dfrac{k}{m} & s + \dfrac{b}{m} \end{bmatrix}^{-1} \begin{bmatrix} 0 \\ \dfrac{1}{m} \end{bmatrix}$$

Since

$$\begin{bmatrix} s & -1 \\ \dfrac{k}{m} & s + \dfrac{b}{m} \end{bmatrix}^{-1} = \cfrac{1}{s^2 + \dfrac{b}{m}s + \dfrac{k}{m}} \begin{bmatrix} s + \dfrac{b}{m} & 1 \\ -\dfrac{k}{m} & s \end{bmatrix}$$

we have

$$G(s) = \begin{bmatrix} 1 & 0 \end{bmatrix} \cfrac{1}{s^2 + \dfrac{b}{m}s + \dfrac{k}{m}} \begin{bmatrix} s + \dfrac{b}{m} & 1 \\ -\dfrac{k}{m} & s \end{bmatrix} \begin{bmatrix} 0 \\ \dfrac{1}{m} \end{bmatrix}$$

$$= \frac{1}{ms^2 + bs + k}$$

which is the transfer function of the system. The same transfer function can be obtained from Equation (1–40).

Transfer matrix. Next, consider a multiple-input, multiple-output system. Assume that there are r inputs $u_1, u_2, \ldots, u_r$ and m outputs $y_1, y_2, \ldots, y_m$. Define

$$\mathbf{y} = \begin{bmatrix} y_1 \\ y_2 \\ \cdot \\ \cdot \\ \cdot \\ y_m \end{bmatrix}, \qquad \mathbf{u} = \begin{bmatrix} u_1 \\ u_2 \\ \cdot \\ \cdot \\ \cdot \\ u_r \end{bmatrix}$$

The transfer matrix $\mathbf{G}(s)$ relates the output $\mathbf{Y}(s)$ to the input $\mathbf{U}(s)$, or

$$\mathbf{Y}(s) = \mathbf{G}(s)\mathbf{U}(s) \tag{1–54}$$

Since the input vector $\mathbf{u}$ is r dimensional and the output vector $\mathbf{y}$ is m dimensional, the transfer matrix is an $m \times r$ matrix. We shall discuss details of the transfer matrix in Chapter 9.

General requirements of control systems.   Any control system must be stable. This is a primary requirement. In addition to absolute stability, a control system must have a reasonable relative stability; that is, the response must show reasonable damping. Also, the speed of response must be reasonably fast and the control system must be capable of reducing errors to zero or to some small tolerable value. Any useful control system must satisfy these requirements.

The requirement of reasonable relative stability and that of steady-state accuracy tend to be incompatible. In designing control systems, we therefore find it necessary to make the most effective compromise between these two requirements.

Classical versus modern control theory.   Classical control theory utilizes extensively the transfer function concept. Analysis and design are done in the *s* domain and/or frequency domain. Modern control theory, which is based on the state space concept, utilizes extensively vector-matrix analysis. Analysis and design are done in the time domain.

Classical control theory generally yields satisfactory results for single-input, single-output control systems. However, classical control theory cannot handle multiple-input, multiple-output control systems.

In this book we present both classical control approaches, often called conventional control approaches, and modern control approaches to the analysis and design of closed-loop control systems. Note that classical or conventional control approaches place emphasis on physical understanding and use less mathematics than modern control approaches. Thus, classical or conventional control approaches are easier to understand.

Mathematical modeling.   The components involved in control systems are widely different. They may be electromechanical, hydraulic, pneumatic, electronic, and so on. In control engineering, rather than dealing with hardware devices, we replace such devices or components by their mathematical models.

To obtain a reasonably accurate mathematical model of a physical component is one of the most important problems in control engineering. Note that to be useful a mathematical model must be neither too complicated nor too simplified. A mathematical model must represent the essential aspects of a physical component. The predictions of the system behavior based on the mathematical model must be reasonably accurate. Note also that seemingly different systems may be represented by the same mathematical model. Use of such mathematical models enables control engineers to develop a unified control theory. In control engineering, linear, time-invariant differential equations, transfer functions, and state variable equations are commonly used for mathematical models of linear, time-invariant, continuous-time systems.

Although the input–output relationships of many components are nonlinear, we normally linearize such relationships about the operating points by limiting the range of variables to be small. Obviously, such linear models are much easier to deal with analytically and computationally. Details of mathematical modeling techniques of physical systems are presented in Chapter 2.

Analysis and design of control systems.   At this point it is desirable to define what is meant by analysis, design, transient response analysis, and the like. By the *analysis* of a control system, we mean the investigation, under specified conditions, of the performance of the system whose mathematical model is known. Since any system is made up of components, analysis must start with a mathematical description of each component. Once a mathematical model of the complete system has been derived, the manner in which analysis is carried out is independent of whether the physical system is pneumatic, electrical, mechanical, and so on. By *transient response analysis* we generally mean the determination of the responses of a plant to command inputs and disturbance inputs. By *steady-state analysis* we mean the determination of the response after the transient response has disappeared.

To *design* a system means to find one that accomplishes a given task. If the dynamic response characteristics and/or steady-state characteristics are not satisfactory, we must add a compensator to the system. In general, the design of a suitable compensator is not straightforward but will require trial-and-error methods.

By *synthesis*, we mean finding by a direct procedure a control system that will perform in a specified way. Usually, such a procedure is entirely mathematical from the start to the end of the design process. Synthesis procedures are available for linear networks and for optimal linear control systems.

In recent years, digital computers have been playing an important role in the analysis, design, and operation of control systems. The computer may be used to carry out necessary computations, to simulate a plant or system components, or to control a system. Computer control has become increasingly common, and many industrial control systems, airborne systems, and robot control systems utilize digital controllers.

Basic approach to control system design. The basic approach to the design of any practical control system will necessarily involve trial-and-error procedures. The synthesis of linear control systems is theoretically possible, and the control engineer can systematically determine the components necessary to perform the given objective. In practice, however, the system may be subjected to many constraints or may be nonlinear, and for such cases no synthesis methods are available at present. In addition, the characteristics of components may not be precisely known. Thus, trial-and-error procedures are always necessary.

Situations are often encountered in practice where a plant is unalterable (that is, we are not free to change the dynamics of the plant), and the control engineer has to design the rest of the system so that the whole will meet the given specifications in accomplishing the given task.

The specifications may include such factors as the speed of response, reasonable damping, steady-state accuracy, reliability, and cost. In certain cases the requirements or specifications may be given explicitly and in other cases they may not be. All requirements or specifications must be interpreted in mathematical terms. In the conventional design, we must make sure that the closed-loop system is stable and has acceptable transient response characteristics (that is, reasonable speed and reasonable damping) and acceptable steady-state accuracy.

It is important to remember that some of the specifications may not be realistic. In such a case, the specifications must be revised in early stages of the design. Also, the given specifications may include conflicting requirements. Then the designer must successfully resolve the conflicts among many given requirements.

Design via modern control theory requires that the designer have a reasonable performance index to guide the design of a control system. A *performance index* is a quantitative measure of the performance, measuring deviation from ideal performance. The selection of a particular performance index is determined by the objectives of the control system.

The performance index may be an integral of a function of the error variable that must be minimized. Such performance indexes based on the minimization of the error integral may be used with both conventional and modern control approaches. In general, however, minimization of a performance index can be achieved much easier by using modern control approaches.

The specification of the control signal over the operating time interval is called the *control law*. Mathematically, the basic control problem is to determine the optimal control law, subject to various engineering and economic constraints, that minimizes (or maximizes as the case may be) a given performance index. For relatively simple systems, the control law may be obtained analytically. For complex systems, it may be necessary to generate the optimal control law with an on-line digital computer.

For spacecraft control systems, the performance index may be minimum time or minimum fuel. For industrial control systems, the performance index may be minimum cost, maximum reliability, and so on. It is important to point out that the selection of the performance index is very important, since the nature of the optimal control system designed depends on the particular performance index selected. We need to choose the most appropriate performance index for the situation.

Design steps. Given a plant (in most cases its dynamics are unalterable), we must first choose an appropriate sensor(s) and actuator(s). Then we must obtain mathematical models of the plant, sensor(s), and actuator(s). Then, using the mathematical models obtained, we design a controller such that the closed-loop system will satisfy the given specifications. The controller designed is the solution to the mathematical version of the design problem. (Note that optimal control theory is very useful in this stage of the design because it gives the upper boundary of system performance for a given performance index.)

After the mathematical design has been completed, the control engineer simulates the model on a computer to test the behavior of the resulting system in response to various signals and disturbances. Usually, the initial system configuration is not satisfactory. Then the system must be redesigned and corresponding analysis completed. This process of design and analysis is repeated until a satisfactory system is obtained. Then a prototype physical system can be constructed.

Note that this process of constructing a prototype is the reverse of that of modeling. The prototype is a physical system that represents the mathematical model with reasonable accuracy. Once the prototype has been built, the engineer tests it to see whether or not it is satisfactory. If it is, the design is complete. If not, the prototype must be modified and tested. This process continues until the prototype is completely satisfactory.

It is noted that, in the case of some process control systems, standard forms of controllers (such as PD, PI, and PID controllers) may be used and the controller parameters are determined experimentally by following a standard established procedure. In this case, no mathematical models are needed. However, this is rather a special case.

Comments

1. The controller produces control signals based on the difference between the reference input and the output. In practical situations there are always some disturbances acting on the plant. These may be of external or internal origin, and they may be random or predictable. The controller must take into consideration any disturbances that will affect the output variables. In general, a good control system must follow the command input closely, but must not be sensitive to external noises and parameter variations.

2. It is desirable to measure and control directly the variable that indicates the state of the system or quality of the product. In the case of process control systems, we may want to measure and control directly the quality of the product. This may present a difficult problem, however, since the quality may be difficult to measure. If this is the case, it becomes necessary to control a secondary variable. For example, variables (such as temperature and pressure) that are directly related to the quality may be controlled. Because other variables may affect the relationship between the quality and the measured variable, the indirect control of a system is usually not as effective as direct control. Although it may be difficult, we should always try to control the primary variable as directly as possible.

1–11 OUTLINE OF THE TEXT

This book deals with continuous-time control systems only. In what follows, we shall briefly present the arrangements and contents of the book.

Part I, which consists of the first four chapters, presents basic analysis of control systems by conventional and state-space methods. Chapter 1 has given introductory materials on control systems analysis. Chapter 2 treats mathematical modeling of physical components and systems. Here we derive differential equation models, transfer function models, and state space models for a variety of physical systems. Chapter 3 presents various control actions. Detailed analysis of pneumatic and hydraulic controllers is given in this chapter. Chapter 4 treats the time-domain analysis of control systems. The transient response of control systems is investigated in detail. Routh's stability criterion is discussed for the stability analysis of higher-order systems. The chapter includes the steady-state error analysis and an introduction to system optimization. Finally, the chapter discusses the state space approach to transient response analysis of control systems. A brief discussion of computer solution of state equations is included.

Part II consists of the next four chapters and presents control systems analysis and design by conventional methods. Specifically, Chapter 5 presents root-locus analysis, while Chapter 6 treats frequency-response analysis. The Nyquist stability criterion is derived and applied to the stability analysis of control systems. Chapter 7 presents design and compensation techniques. Examples of system compensation by use of the root-locus and frequency-response techniques are presented. Chapter 8 discusses the describing-function analysis of nonlinear control systems.

Part III consists of the last two chapters. It presents control systems analysis and design by state space methods. Chapter 9 gives basic materials of modern control theory, including concepts of controllability and observability. Chapter 10 treats the design of control systems via pole placement, design of state observers, design of servo systems, design of optimal control systems based on quadratic performance indexes, and discussions on model-reference control systems and adaptive control systems. Although it is assumed that the reader has a necessary background on vector-matrix analysis, a reasonably detailed account of vector-matrix analysis is provided in the appendix for convenient reference.

The root-locus and frequency-response methods, the core of classical or conventional control theory, are widely used in industry. They are useful in treating linear, time-invariant, single-input, single-output control systems. Modern control theory is applicable to both single-input, single-output systems and multiple-input, multiple-output systems. Also, modern control theory enables the designer to take into account arbitrary initial conditions in the synthesis of optimal control systems. In such synthesis we may need to deal only with the analytical aspects of the problem. A digital computer may be programmed to handle all the necessary numerical computations. This is one of the basic advantages of modern control theory.

It is important to note that modern control theory does not completely replace conventional (or classical) control theory. The two approaches supplement each other. The aim of this book is to present useful aspects of both conventional control theory and modern control theory and give the reader a good background in using various tools for the analysis and design of modern control systems.

Example Problems and Solutions

A–1–1. List the major advantages and disadvantages of open-loop control systems.

Solution. The advantages of open-loop control systems are as follows:

1. Simple construction and ease of maintenance.
2. Less expensive than a corresponding closed-loop system.
3. There is no stability problem.
4. Convenient when output is hard to measure or economically not feasible. (For example, in the washer system, it would be quite expensive to provide a device to measure the quality of the output, cleanliness of the clothes, of the washer.)

The disadvantages of open-loop control systems are as follows:

1. Disturbances and changes in calibration cause errors, and the output may be different from what is desired.
2. To maintain the required quality in the output, recalibration is necessary from time to time.

A–1–2. Figure 1–32(a) is a schematic diagram of a liquid-level control system. Here the automatic controller maintains the liquid level by comparing the actual level with a desired level and correcting any error by adjusting the opening of the pneumatic valve. Figure 1–32(b) is a block diagram of the control system. Draw the corresponding block diagram for a human-operated liquid-level control system.

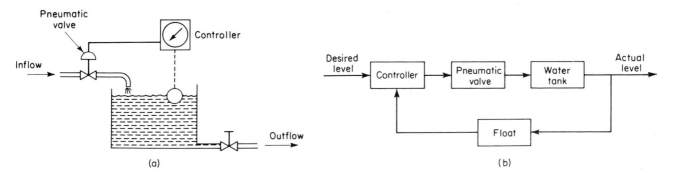

Figure 1–32 (a) Liquid-level control system; (b) block diagram.

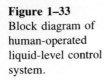

Figure 1–33
Block diagram of
human-operated
liquid-level control
system.

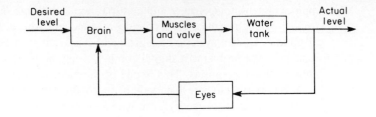

Solution. In the human-operated system, the eyes, brain, and muscles correspond to the sensor, controller, and pneumatic valve, respectively. A block diagram is shown in Figure 1–33.

A–1–3. An engineering organizational system is composed of major groups, such as management, research and development, preliminary design, experiments, product design and drafting, fabrication and assembling, and testing. These groups are interconnected to make up the whole operation.

   The system may be analyzed by reducing it to the most elementary set of components necessary that can provide the analytical detail required and by representing the dynamic characteristics of each component by a set of simple equations. (The dynamic performance of such a system may be determined from the relation between progressive accomplishment and time.)

   Draw a functional block diagram showing an engineering organizational system.

Solution. A functional block diagram can be drawn by using blocks to represent the functional activities and interconnecting signal lines to represent the information or product output of the system operation. A possible block diagram is shown in Figure 1–34.

A–1–4. Find the poles of the following $F(s)$:

$$F(s) = \frac{1}{1 - e^{-s}}$$

Solution. The poles are found from

$$e^{-s} = 1$$

or

$$e^{-(\sigma + j\omega)} = e^{-\sigma}(\cos \omega - j \sin \omega) = 1$$

From this it follows that $\sigma = 0$, $\omega = \pm 2n\pi$ ($n = 0,1,2, \ldots$). Thus, the poles are located at

$$s = \pm j2n\pi \qquad (n = 0,1,2, \ldots)$$

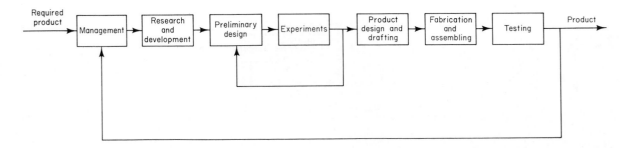

Figure 1–34 Block diagram of an engineering organizational system.

A–1–5. Find the Laplace transform of $f(t)$ defined by

$$f(t) = 0 \qquad (t < 0)$$
$$= te^{-3t} \qquad (t \geq 0)$$

Solution. Since

$$\mathscr{L}[t] = G(s) = \frac{1}{s^2}$$

referring to Equation (1–5), we obtain

$$F(s) = \mathscr{L}[te^{-3t}] = G(s + 3) = \frac{1}{(s + 3)^2}$$

A–1–6. What is the Laplace transform of

$$f(t) = 0 \qquad (t < 0)$$
$$= \sin (\omega t + \theta) \qquad (t \geq 0)$$

where θ is a constant?

Solution. Noting that

$$\sin (\omega t + \theta) = \sin \omega t \cos \theta + \cos \omega t \sin \theta$$

we have

$$\mathscr{L}[\sin (\omega t + \theta)] = \cos \theta \, \mathscr{L}[\sin \omega t] + \sin \theta \, \mathscr{L}[\cos \omega t]$$
$$= \cos \theta \frac{\omega}{s^2 + \omega^2} + \sin \theta \frac{s}{s^2 + \omega^2}$$
$$= \frac{\omega \cos \theta + s \sin \theta}{s^2 + \omega^2}$$

A–1–7. Find the Laplace transform $F(s)$ of the function $f(t)$ shown in Figure 1–35. Also find the limiting value of $F(s)$ as a approaches zero.

Solution. The function $f(t)$ can be written

$$f(t) = \frac{1}{a^2} \, 1(t) - \frac{2}{a^2} \, 1(t - a) + \frac{1}{a^2} \, 1(t - 2a)$$

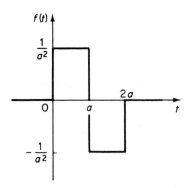

Figure 1–35
Function $f(t)$.

Then

$$F(s) = \mathcal{L}[f(t)]$$

$$= \frac{1}{a^2}\mathcal{L}[1(t)] - \frac{2}{a^2}\mathcal{L}[1(t-a)] + \frac{1}{a^2}\mathcal{L}[1(t-2a)]$$

$$= \frac{1}{a^2}\frac{1}{s} - \frac{2}{a^2}\frac{1}{s}e^{-as} + \frac{1}{a^2}\frac{1}{s}e^{-2as}$$

$$= \frac{1}{a^2 s}(1 - 2e^{-as} + e^{-2as})$$

As a approaches zero we have

$$\lim_{a \to 0} F(s) = \lim_{a \to 0} \frac{1 - 2e^{-as} + e^{-2as}}{a^2 s} = \lim_{a \to 0} \frac{\dfrac{d}{da}(1 - 2e^{-as} + e^{-2as})}{\dfrac{d}{da}(a^2 s)}$$

$$= \lim_{a \to 0} \frac{2se^{-as} - 2se^{-2as}}{2as} = \lim_{a \to 0} \frac{e^{-as} - e^{-2as}}{a}$$

$$= \lim_{a \to 0} \frac{\dfrac{d}{da}(e^{-as} - e^{-2as})}{\dfrac{d}{da}(a)} = \lim_{a \to 0} \frac{-se^{-as} + 2se^{-2as}}{1}$$

$$= -s + 2s = s$$

A–1–8. Find the initial value of $df(t)/dt$ when the Laplace transform of $f(t)$ is given by

$$F(s) = \mathcal{L}[f(t)] = \frac{2s + 1}{s^2 + s + 1}$$

Solution. Using the initial value theorem,

$$\lim_{t \to 0+} f(t) = f(0+) = \lim_{s \to \infty} sF(s) = \lim_{s \to \infty} \frac{s(2s + 1)}{s^2 + s + 1} = 2$$

Since the $\mathcal{L}_+$ transform of $df(t)/dt = g(t)$ is given by

$$\mathcal{L}_+[g(t)] = sF(s) - f(0+)$$

$$= \frac{s(2s + 1)}{s^2 + s + 1} - 2 = \frac{-s - 2}{s^2 + s + 1}$$

the initial value of $df(t)/dt$ is obtained as

$$\lim_{t \to 0+} \frac{df(t)}{dt} = g(0+) = \lim_{s \to \infty} s[sF(s) - f(0+)]$$

$$= \lim_{s \to \infty} \frac{-s^2 - 2s}{s^2 + s + 1} = -1$$

A–1–9. The derivative of the unit-impulse function $\delta(t)$ is called a *unit-doublet* function. (Thus, the integral of the unit-doublet function is the unit-impulse function.) Mathematically, an example of the unit-doublet function, which is usually denoted by $u_2(t)$, may be given by

$$u_2(t) = \lim_{t_0 \to 0} \frac{1(t) - 2[1(t - t_0)] + 1(t - 2t_0)}{t_0^2}$$

Obtain the Laplace transform of $u_2(t)$.

Solution. The Laplace transform of $u_2(t)$ is given by

$$\mathscr{L}[u_2(t)] = \lim_{t_0 \to 0} \frac{1}{t_0^2} \left(\frac{1}{s} - \frac{2}{s} e^{-t_0 s} + \frac{1}{s} e^{-2t_0 s} \right)$$

$$= \lim_{t_0 \to 0} \frac{1}{t_0^2 s} \left[1 - 2 \left(1 - t_0 s + \frac{t_0^2 s^2}{2} + \cdots \right) + \left(1 - 2t_0 s + \frac{4 t_0^2 s^2}{2} + \cdots \right) \right]$$

$$= \lim_{t_0 \to 0} \frac{1}{t_0^2 s} \left[t_0^2 s^2 + \text{(higher-order terms in } t_0 s) \right] = s$$

A–1–10. Find the Laplace transform of $f(t)$ defined by

$$f(t) = 0 \qquad (t < 0)$$
$$= t^2 \sin \omega t \qquad (t \geq 0)$$

Solution. Since

$$\mathscr{L}[\sin \omega t] = \frac{\omega}{s^2 + \omega^2}$$

applying the complex differentiation theorem

$$\mathscr{L}[t^2 f(t)] = \frac{d^2}{ds^2} F(s)$$

to this problem, we have

$$\mathscr{L}[f(t)] = \mathscr{L}[t^2 \sin \omega t] = \frac{d^2}{ds^2} \left[\frac{\omega}{s^2 + \omega^2} \right] = \frac{-2\omega^3 + 6\omega s^2}{(s^2 + \omega^2)^3}$$

A–1–11. Prove that if $f(t)$ is of exponential order and if $\int_0^\infty f(t)\, dt$ exists [which means that $\int_0^\infty f(t)\, dt$ assumes a definite value], then

$$\int_0^\infty f(t)\, dt = \lim_{s \to 0} F(s)$$

where $F(s) = \mathscr{L}[f(t)]$.

Solution. Note that

$$\int_0^\infty f(t)\, dt = \lim_{t \to \infty} \int_0^t f(t)\, dt$$

Referring to Equation (1–8),

$$\mathscr{L}\left[\int_0^t f(t)\, dt \right] = \frac{F(s)}{s}$$

Since $\int_0^\infty f(t)\, dt$ exists, by applying the final value theorem to this case,

$$\lim_{t \to \infty} \int_0^t f(t)\, dt = \lim_{s \to 0} s \frac{F(s)}{s}$$

or

$$\int_0^\infty f(t)\, dt = \lim_{s \to 0} F(s)$$

A–1–12. Determine the Laplace transform of the convolution integral

$$f_1(t)*f_2(t) = \int_0^t \tau[1 - e^{-(t-\tau)}] \, d\tau = \int_0^t (t - \tau)(1 - e^{-\tau}) \, d\tau$$

where

$$f_1(t) = f_2(t) = 0 \qquad \text{for } t < 0$$

$$f_1(t) = t \qquad \text{for } t \geq 0$$

$$f_2(t) = 1 - e^{-t} \qquad \text{for } t \geq 0$$

Solution. Note that

$$\mathcal{L}[t] = F_1(s) = \frac{1}{s^2}$$

$$\mathcal{L}[1 - e^{-t}] = F_2(s) = \frac{1}{s} - \frac{1}{s + 1}$$

The Laplace transform of the convolution integral is given by

$$\mathcal{L}[f_1(t)*f_2(t)] = F_1(s)F_2(s) = \frac{1}{s^2}\left(\frac{1}{s} - \frac{1}{s + 1}\right)$$

$$= \frac{1}{s^3} - \frac{1}{s^2(s + 1)} = \frac{1}{s^3} - \frac{1}{s^2} + \frac{1}{s} - \frac{1}{s + 1}$$

To verify that it is indeed the Laplace transform of the convolution integral, let us first perform integration of the convolution integral and then take its Laplace transform.

$$f_1(t)*f_2(t) = \int_0^t \tau[1 - e^{-(t-\tau)}] \, d\tau = \int_0^t (t - \tau)(1 - e^{-\tau}) \, d\tau$$

$$= \frac{t^2}{2} - t + 1 - e^{-t}$$

And so

$$\mathcal{L}\left[\frac{t^2}{2} - t + 1 - e^{-t}\right] = \frac{1}{s^3} - \frac{1}{s^2} + \frac{1}{s} - \frac{1}{s + 1}$$

A–1–13. Prove that if $f(t)$ is a periodic function with period T, then

$$\mathcal{L}[f(t)] = \frac{\int_0^T f(t)e^{-st} \, dt}{1 - e^{-Ts}}$$

Solution.

$$\mathcal{L}[f(t)] = \int_0^\infty f(t)e^{-st} \, dt = \sum_{n=0}^\infty \int_{nT}^{(n+1)T} f(t) \, e^{-st} \, dt$$

By changing the independent variable from t to τ, where $\tau = t - nT$,

$$\mathcal{L}[f(t)] = \sum_{n=0}^\infty e^{-nTs} \int_0^T f(\tau)e^{-s\tau} \, d\tau$$

Noting that

$$\sum_{n=0}^{\infty} e^{-nTs} = 1 + e^{-Ts} + e^{-2Ts} + \cdots$$

$$= 1 + e^{-Ts}(1 + e^{-Ts} + e^{-2Ts} + \cdots)$$

$$= 1 + e^{-Ts}\left(\sum_{n=0}^{\infty} e^{-nTs}\right)$$

we obtain

$$\sum_{n=0}^{\infty} e^{-nTs} = \frac{1}{1 - e^{-Ts}}$$

It follows that

$$\mathcal{L}[f(t)] = \frac{\int_{0}^{T} f(t)e^{-st}\,dt}{1 - e^{-Ts}}$$

A–1–14. What is the Laplace transform of the periodic function shown in Figure 1–36?

Solution. Note that

$$\int_{0}^{T} f(t)e^{-st}\,dt = \int_{0}^{T/2} e^{-st}\,dt + \int_{T/2}^{T} (-1)e^{-st}\,dt$$

$$= \frac{e^{-st}}{-s}\bigg|_{0}^{T/2} - \frac{e^{-st}}{-s}\bigg|_{T/2}^{T}$$

$$= \frac{e^{-(1/2)Ts} - 1}{-s} + \frac{e^{-Ts} - e^{-(1/2)Ts}}{s}$$

$$= \frac{1}{s}[e^{-Ts} - 2e^{-(1/2)Ts} + 1]$$

$$= \frac{1}{s}[1 - e^{-(1/2)Ts}]^2$$

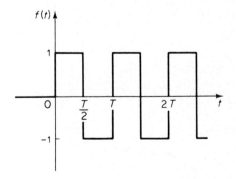

Figure 1–36
Periodic function
(square wave).

Referring to Problem A–1–13, we have

$$F(s) = \frac{\int_0^T f(t)e^{-st}\,dt}{1 - e^{-Ts}} = \frac{(1/s)[1 - e^{-(1/2)Ts}]^2}{1 - e^{-Ts}}$$

$$= \frac{1 - e^{-(1/2)Ts}}{s[1 + e^{-(1/2)Ts}]} = \frac{1}{s}\tanh\frac{Ts}{4}$$

A–1–15. Find the inverse Laplace transform of $F(s)$, where

$$F(s) = \frac{1}{s(s^2 + 2s + 2)}$$

Solution. Since

$$s^2 + 2s + 2 = (s + 1 + j1)(s + 1 - j1)$$

we notice that $F(s)$ involves a pair of complex conjugate poles, and so we expand $F(s)$ into the form

$$F(s) = \frac{1}{s(s^2 + 2s + 2)} = \frac{a_1}{s} + \frac{a_2 s + a_3}{s^2 + 2s + 2}$$

where a_1, a_2, and a_3 are determined from

$$1 = a_1(s^2 + 2s + 2) + (a_2 s + a_3)s$$

By comparing coefficients of s^2, s, and s^0 terms on both sides of this last equation, respectively, we obtain

$$a_1 + a_2 = 0, \qquad 2a_1 + a_3 = 0, \qquad 2a_1 = 1$$

from which

$$a_1 = \frac{1}{2}, \qquad a_2 = -\frac{1}{2}, \qquad a_3 = -1$$

Therefore,

$$F(s) = \frac{1}{2}\frac{1}{s} - \frac{1}{2}\frac{s + 2}{s^2 + 2s + 2}$$

$$= \frac{1}{2}\frac{1}{s} - \frac{1}{2}\frac{1}{(s + 1)^2 + 1^2} - \frac{1}{2}\frac{s + 1}{(s + 1)^2 + 1^2}$$

The inverse Laplace transform of $F(s)$ gives

$$f(t) = \frac{1}{2} - \frac{1}{2}e^{-t}\sin t - \frac{1}{2}e^{-t}\cos t \qquad (t \geq 0)$$

A–1–16. Obtain the inverse Laplace transform of

$$F(s) = \frac{5(s + 2)}{s^2(s + 1)(s + 3)}$$

Solution.

$$F(s) = \frac{5(s + 2)}{s^2(s + 1)(s + 3)} = \frac{b_2}{s^2} + \frac{b_1}{s} + \frac{a_1}{s + 1} + \frac{a_2}{s + 3}$$

where

$$a_1 = \left. \frac{5(s + 2)}{s^2(s + 3)} \right|_{s=-1} = \frac{5}{2}$$

$$a_2 = \left. \frac{5(s + 2)}{s^2(s + 1)} \right|_{s=-3} = \frac{5}{18}$$

$$b_2 = \left. \frac{5(s + 2)}{(s + 1)(s + 3)} \right|_{s=0} = \frac{10}{3}$$

$$b_2 = \frac{d}{ds} \left[\frac{5(s + 2)}{(s + 1)(s + 3)} \right]_{s=0}$$

$$= \left. \frac{5(s + 1)(s + 3) - 5(s + 2)(2s + 4)}{(s + 1)^2(s + 3)^2} \right|_{s=0} = -\frac{25}{9}$$

Thus

$$F(s) = \frac{10}{3} \frac{1}{s^2} - \frac{25}{9} \frac{1}{s} + \frac{5}{2} \frac{1}{s + 1} + \frac{5}{18} \frac{1}{s + 3}$$

The inverse Laplace transform of $F(s)$ is

$$f(t) = \frac{10}{3} t - \frac{25}{9} + \frac{5}{2} e^{-t} + \frac{5}{18} e^{-3t} \qquad (t \geq 0)$$

A–1–17.   Find the inverse Laplace transform of

$$F(s) = \frac{s^4 + 2s^3 + 3s^2 + 4s + 5}{s(s + 1)}$$

Solution. Since the numerator polynomial is of higher degree than the denominator polynomial, by dividing the numerator by the denominator until the remainder is a fraction, we have

$$F(s) = s^2 + s + 2 + \frac{2s + 5}{s(s + 1)} = s^2 + s + 2 + \frac{a_1}{s} + \frac{a_2}{s + 1}$$

where

$$a_1 = \left. \frac{2s + 5}{s + 1} \right|_{s=0} = 5$$

$$a_2 = \left. \frac{2s + 5}{s} \right|_{s=-1} = -3$$

It follows that

$$F(s) = s^2 + s + 2 + \frac{5}{s} - \frac{3}{s + 1}$$

The inverse Laplace transform of $F(s)$ is

$$f(t) = \mathcal{L}^{-1}[F(s)] = \frac{d^2}{dt^2} \delta(t) + \frac{d}{dt} \delta(t) + 2 \delta(t) + 5 - 3e^{-t} \qquad (t \geq 0-)$$

A–1–18. Derive the inverse Laplace transform of

$$F(s) = \frac{1}{s(s^2 + \omega^2)}$$

Solution.

$$F(s) = \frac{1}{s(s^2 + \omega^2)} = \frac{1}{\omega^2}\frac{1}{s} - \frac{1}{\omega^2}\frac{s}{s^2 + \omega^2}$$

Hence the inverse Laplace transform of $F(s)$ is obtained as

$$f(t) = \mathcal{L}^{-1}[F(s)] = \frac{1}{\omega^2}(1 - \cos \omega t) \qquad (t \geq 0)$$

A–1–19. Obtain the inverse Laplace transform of the following $F(s)$:

$$F(s) = \frac{B(s)}{A(s)} = \frac{B(s)}{(s + p_1)^r(s + p_{r+1})(s + p_{r+2}) \cdots (s + p_n)}$$

where the degree of polynomial $B(s)$ is lower than that of polynomial $A(s)$.

Solution. The partial-fraction expansion of $F(s)$ is

$$F(s) = \frac{B(s)}{A(s)} = \frac{b_r}{(s + p_1)^r} + \frac{b_{r-1}}{(s + p_1)^{r-1}} + \cdots + \frac{b_1}{s + p_1}$$

$$+ \frac{a_{r+1}}{s + p_{r+1}} + \frac{a_{r+2}}{s + p_{r+2}} + \cdots + \frac{a_n}{s + p_n} \qquad (1\text{--}55)$$

where $b_r, b_{r-1}, \ldots, b_1$ are given by

$$b_r = \left[(s + p_1)^r \frac{B(s)}{A(s)} \right]_{s = -p_1}$$

$$b_{r-1} = \left\{ \frac{d}{ds} \left[(s + p_1)^r \frac{B(s)}{A(s)} \right] \right\}_{s = -p_1}$$

$$\vdots$$

$$b_{r-j} = \frac{1}{j!} \left\{ \frac{d^j}{ds^j} \left[(s + p_1)^r \frac{B(s)}{A(s)} \right] \right\}_{s = -p_1}$$

$$\vdots$$

$$b_1 = \frac{1}{(r-1)!} \left\{ \frac{d^{r-1}}{ds^{r-1}} \left[(s + p_1)^r \frac{B(s)}{A(s)} \right] \right\}_{s = -p_1}$$

The foregoing relationships for the b's may be obtained as follows: By multiplying both sides of Equation (1–55) by $(s + p_1)^r$ and letting s approach $-p_1$, we obtain

$$b_r = \left[(s + p_1)^r \frac{B(s)}{A(s)} \right]_{s = -p_1}$$

If we multiply both sides of Equation (1–55) by $(s + p_1)^r$ and then differentiate with respect to s,

$$\frac{d}{ds}\left[(s + p_1)^r \frac{B(s)}{A(s)}\right] = b_r \frac{d}{ds}\left[\frac{(s + p_1)^r}{(s + p_1)^r}\right] + b_{r-1} \frac{d}{ds}\left[\frac{(s + p_1)^r}{(s + p_1)^{r-1}}\right]$$

$$+ \cdots + b_1 \frac{d}{ds}\left[\frac{(s + p_1)^r}{s + p_1}\right] + a_{r+1} \frac{d}{ds}\left[\frac{(s + p_1)^r}{s + p_{r+1}}\right]$$

$$+ \cdots + a_n \frac{d}{ds}\left[\frac{(s + p_1)^r}{s + p_n}\right]$$

The first term on the right-hand side of this last equation vanishes. The second term becomes b_{r-1}. Each of the other terms contains some power of $(s + p_1)$ as a factor, with the result that, when s is allowed to approach $-p_1$, these terms drop out. Hence

$$b_{r-1} = \lim_{s \to -p_1} \frac{d}{ds}\left[(s + p_1)^r \frac{B(s)}{A(s)}\right]$$

$$= \left\{\frac{d}{ds}\left[(s + p_1)^r \frac{B(s)}{A(s)}\right]\right\}_{s = -p_1}$$

Similarly, by performing successive differentiations with respect to s and by letting s approach $-p_1$, we obtain equations for the b_{r-j}, where $j = 2, 3, \ldots r - 1$.

Note that the inverse Laplace transform of $1/(s + p_1)^n$ is given by

$$\mathcal{L}^{-1}\left[\frac{1}{(s + p_1)^n}\right] = \frac{t^{n-1}}{(n - 1)!} e^{-p_1 t}$$

The constants $a_{r+1}, a_{r+2}, \ldots, a_n$ in Equation (1–55) are determined from

$$a_k = \left[(s + p_k) \frac{B(s)}{A(s)}\right]_{s = -p_k} \qquad (k = r + 1, r + 2, \ldots, n)$$

The inverse Laplace transform of $F(s)$ is then obtained as follows:

$$f(t) = \mathcal{L}^{-1}[F(s)] = \left[\frac{b_r}{(r - 1)!} t^{r-1} + \frac{b_{r-1}}{(r - 2)!} t^{r-2} + \cdots + b_2 t + b_1\right] e^{-p_1 t}$$

$$+ a_{r+1}e^{-p_{r+1}t} + a_{r+2}e^{-p_{r+2}t} + \cdots + a_n e^{-p_n t} \qquad (t \geq 0)$$

A–1–20. Find the Laplace transform of the following differential equation:

$$\ddot{x} + 3\dot{x} + 6x = 0, \qquad x(0) = 0, \qquad \dot{x}(0) = 3$$

Taking the inverse Laplace transform of $X(s)$, obtain the time solution $x(t)$.

Solution. The Laplace transform of the differential equation is

$$s^2 X(s) - sx(0) - \dot{x}(0) + 3sX(s) - 3x(0) + 6X(s) = 0$$

Substituting the initial conditions and solving for $X(s)$,

$$X(s) = \frac{3}{s^2 + 3s + 6} = \frac{2\sqrt{3}}{\sqrt{5}} \frac{\frac{\sqrt{15}}{2}}{(s + 1.5)^2 + \left(\frac{\sqrt{15}}{2}\right)^2}$$

The inverse Laplace transform of $X(s)$ is

$$x(t) = \frac{2\sqrt{3}}{\sqrt{5}} e^{-1.5t} \sin\left(\frac{\sqrt{15}}{2} t\right)$$

A–1–21. Consider the mechanical system shown in Figure 1–37. Suppose that the system is set into motion by a unit-impulse force. Find the resulting oscillation. Assume that the system is at rest initially.

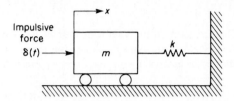

Figure 1–37
Mechanical system.

Solution. The system is excited by an impulse input. Hence

$$m\ddot{x} + kx = \delta(t)$$

Taking the Laplace transform of both sides of this equation gives

$$m[s^2 X(s) - sx(0) - \dot{x}(0)] + kX(s) = 1$$

By substituting the initial conditions $x(0) = 0$ and $\dot{x}(0) = 0$ into this last equation and solving for $X(s)$, we obtain

$$X(s) = \frac{1}{ms^2 + k}$$

The inverse Laplace transform of $X(s)$ becomes

$$x(t) = \frac{1}{\sqrt{mk}} \sin\sqrt{\frac{k}{m}}\, t$$

The oscillation is a simple harmonic motion. The amplitude of the oscillation is $1/\sqrt{mk}$.

A–1–22. Simplify the block diagram shown in Figure 1–38.

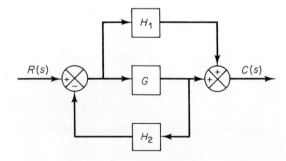

Figure 1–38
Block diagram of a system.

Solution. First, move the branch point of the path involving H_1 outside the loop involving H_2, as shown in Figure 1–39(a). Then eliminating two loops results in Figure 1–39(b). Combining two blocks into one gives Figure 1–39(c).

Chapter 1 / Introduction to Control Systems Analysis

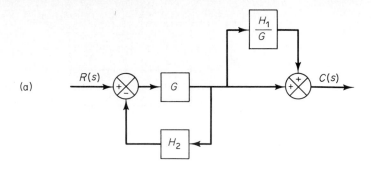

(a)

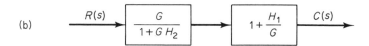

(b)

Figure 1–39
Simplified block
diagrams for the
system shown in Fig.
1–38.

(c)

A–1–23. Simplify the block diagram shown in Figure 1–40. Obtain the transfer function relating $C(s)$ and $R(s)$.

Solution. The block diagram of Figure 1–40 can be modified to that shown in Figure 1–41(a). Eliminating the minor feedforward path, we obtain Figure 1–41(b), which can be simplified to that shown in Figure 1–41(c). The transfer function $C(s)/R(s)$ is thus given by

$$\frac{C(s)}{R(s)} = G_1 G_2 + G_2 + 1$$

The same result can also be obtained by proceeding as follows: Since signal $X(s)$ is the sum of two signals $G_1 R(s)$ and $R(s)$, we have

$$X(s) = G_1 R(s) + R(s)$$

The output signal $C(s)$ is the sum of $G_2 X(s)$ and $R(s)$. Hence

$$C(s) = G_2 X(s) + R(s) = G_2 [G_1 R(s) + R(s)] + R(s)$$

And so we have the same result as before:

$$\frac{C(s)}{R(s)} = G_1 G_2 + G_2 + 1$$

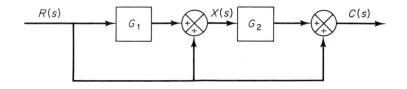

Figure 1–40
Block diagram of a
system.

Example Problems and Solutions

83

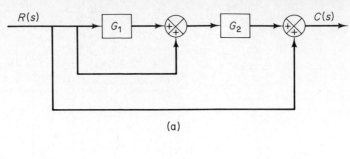

(a)

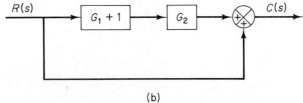

(b)

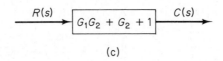

(c)

Figure 1–41
Reduction of the
block diagram shown
in Fig. 1–40.

A–1–24. Consider the system shown in Figure 1–42. Obtain the closed-loop transfer function $H(s)/Q(s)$.

Solution. In this system there is only one forward path that connects the input $Q(s)$ and the output $H(s)$. Thus,

$$P_1 = \frac{1}{C_1 s} \frac{1}{R_1} \frac{1}{C_2 s}$$

There are three individual loops. Thus,

$$L_1 = -\frac{1}{C_1 s} \frac{1}{R_1}$$

$$L_2 = -\frac{1}{C_2 s} \frac{1}{R_2}$$

$$L_3 = -\frac{1}{R_1} \frac{1}{C_2 s}$$

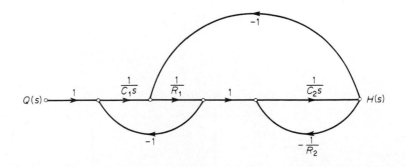

Figure 1–42
Signal flow graph of a
control system.

Chapter 1 / Introduction to Control Systems Analysis

Loop L_1 does not touch loop L_2. (Loop L_1 touches loop L_3, and loop L_2 touches loop L_3). Hence the determinant Δ is given by

$$\Delta = 1 - (L_1 + L_2 + L_3) + (L_1 L_2)$$

$$= 1 + \frac{1}{R_1 C_1 s} + \frac{1}{R_2 C_2 s} + \frac{1}{R_1 C_2 s} + \frac{1}{R_1 C_1 R_2 C_2 s^2}$$

Since all three loops touch the forward path P_1, we remove L_1, L_2, L_3, and $L_1 L_2$ from Δ and evaluate the cofactor Δ_1 as follows:

$$\Delta_1 = 1$$

Thus we obtain the closed-loop transfer function as shown:

$$\frac{H(s)}{Q(s)} = \frac{P_1 \Delta_1}{\Delta}$$

$$= \frac{\dfrac{1}{R_1 C_1 C_2 s^2}}{1 + \dfrac{1}{R_1 C_1 s} + \dfrac{1}{R_2 C_2 s} + \dfrac{1}{R_1 C_2 s} + \dfrac{1}{R_1 C_1 R_2 C_2 s^2}}$$

$$= \frac{R_2}{R_1 C_1 R_2 C_2 s^2 + (R_1 C_1 + R_2 C_2 + R_2 C_1)s + 1}$$

A–1–25. Figure 1–43 is the block diagram of an engine-speed control system. The speed is measured by a set of flyweights. Draw a signal flow graph for this system.

Solution. Referring to Figure 1–22(e), a signal flow graph for

$$\frac{Y(s)}{X(s)} = \frac{1}{s + 140}$$

may be drawn as shown in Figure 1–44(a). Similarly, a signal flow graph for

$$\frac{Z(s)}{X(s)} = \frac{1}{s^2 + 140s + 100^2} = \frac{\dfrac{1}{s + 140}\dfrac{1}{s}}{1 + \dfrac{100^2}{s + 140}\dfrac{1}{s}}$$

may be drawn as shown in Figure 1–44(b).

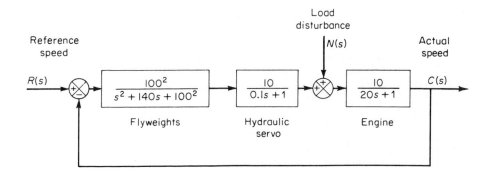

Figure 1–43
Block diagram of an engine-speed control system.

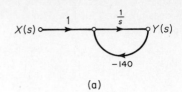

(a)

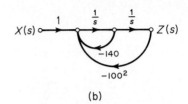

(b)

Figure 1–44
(a) Signal flow graph
for $Y(s)/X(s) = 1/(s$
$+ 140)$; (b) signal
flow graph for $Z(s)/$
$X(s) = 1/(s^2 + 140s$
$+ 100^2)$; (c) signal
flow graph for the
system shown in Fig.
1–43.

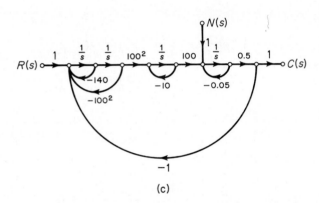

(c)

Drawing a signal flow graph for each of the system components and combining them together, a signal flow graph for the complete system may be obtained as shown in Figure 1–44(c).

A–1–26. Obtain a state-space representation of the system shown in Figure 1–45.

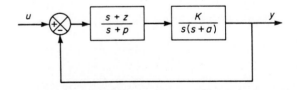

Figure 1–45
Control system.

Solution. In this problem, we shall illustrate a method for deriving a state-space representation of systems given in block-diagram form.

In the present problem, first expand $(s + z)/(s + p)$ into partial fractions.

$$\frac{s + z}{s + p} = 1 + \frac{z - p}{s + p}$$

Next convert $K/[s(s + a)]$ into the product of K/s and $1/(s + a)$. Then redraw the block diagram, as shown in Figure 1–46. Defining a set of state variables, as shown in Figure 1–46, we obtain the following equations:

Figure 1–46
Block diagram defining the state variables for the system shown in Fig. 1–45.

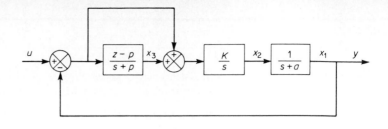

$$\dot{x}_1 = -ax_1 + x_2$$

$$\dot{x}_2 = -Kx_1 + Kx_3 + Ku$$

$$\dot{x}_3 = -(z - p)x_1 - px_3 + (z - p)u$$

$$y = x_1$$

Rewriting gives

$$\begin{bmatrix} \dot{x}_1 \\ \dot{x}_2 \\ \dot{x}_3 \end{bmatrix} = \begin{bmatrix} -a & 1 & 0 \\ -K & 0 & K \\ -(z-p) & 0 & -p \end{bmatrix} \begin{bmatrix} x_1 \\ x_2 \\ x_3 \end{bmatrix} + \begin{bmatrix} 0 \\ K \\ z-p \end{bmatrix} u$$

$$y = \begin{bmatrix} 1 & 0 & 0 \end{bmatrix} \begin{bmatrix} x_1 \\ x_2 \\ x_3 \end{bmatrix}$$

Notice that the output of the integrator and the outputs of the first-order delayed integrators $[1/(s + a)$ and $(z - p)/(s + p)]$ are chosen as state variables. It is important to remember that the output of the block $(s + z)/(s + p)$ in Figure 1–45 cannot be a state variable, because this block involves a derivative term, $s + z$.

A–1–27. Consider a system defined by the following state-space equations:

$$\begin{bmatrix} \dot{x}_1 \\ \dot{x}_2 \end{bmatrix} = \begin{bmatrix} -5 & -1 \\ 3 & -1 \end{bmatrix} \begin{bmatrix} x_1 \\ x_2 \end{bmatrix} + \begin{bmatrix} 2 \\ 5 \end{bmatrix} u$$

$$y = \begin{bmatrix} 1 & 2 \end{bmatrix} \begin{bmatrix} x_1 \\ x_2 \end{bmatrix}$$

Obtain the transfer function $G(s)$ of the system.

Solution. Referring to Equation (1–53), the transfer function of the system can be obtained as follows (note that $D = 0$ in this case):

$$G(s) = \mathbf{C}(s\mathbf{I} - \mathbf{A})^{-1}\mathbf{B}$$

$$= \begin{bmatrix} 1 & 2 \end{bmatrix} \begin{bmatrix} s+5 & 1 \\ -3 & s+1 \end{bmatrix}^{-1} \begin{bmatrix} 2 \\ 5 \end{bmatrix}$$

$$= \begin{bmatrix} 1 & 2 \end{bmatrix} \begin{bmatrix} \dfrac{s+1}{(s+2)(s+4)} & \dfrac{-1}{(s+2)(s+4)} \\ \dfrac{3}{(s+2)(s+4)} & \dfrac{s+5}{(s+2)(s+4)} \end{bmatrix} \begin{bmatrix} 2 \\ 5 \end{bmatrix}$$

$$= \frac{12s + 59}{(s+2)(s+4)}$$

Example Problems and Solutions

B–1–1. Many closed-loop and open-loop control systems may be found in homes. List several examples and describe them.

B–1–2. Figure 1–47 shows a tension control system. Explain the sequence of control actions when the feed speed is suddenly changed for a short time.

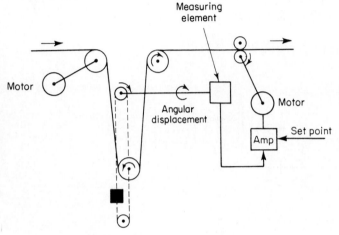

Figure 1–47 Tension control system.

B–1–3. Many machines, such as lathes, milling machines, and grinders, are provided with tracers to reproduce the contour of templates. Figure 1–48 shows a schematic diagram of a tracing

system in which the tool duplicates the shape of the template on the workpiece. Explain the operation of this system.

B–1–4. Find the Laplace transforms of the following functions:

(a)
$$f_1(t) = 0 \qquad\qquad (t < 0)$$
$$= e^{-0.4t} \cos 12t \qquad (t \geq 0)$$

(b)
$$f_2(t) = 0 \qquad\qquad (t < 0)$$
$$= \sin\left(4t + \frac{\pi}{3}\right) \qquad (t \geq 0)$$

B–1–5. Find the Laplace transforms of the following functions:

(a)
$$f_1(t) = 0 \qquad\qquad (t < 0)$$
$$= 3 \sin (5t + 45°) \qquad (t \geq 0)$$

(b)
$$f_2(t) = 0 \qquad\qquad (t < 0)$$
$$= 0.03(1 - \cos 2t) \qquad (t \geq 0)$$

B–1–6. Obtain the Laplace transform of the function defined by

$$f(t) = 0 \qquad\qquad (t < 0)$$
$$= t^2 e^{-at} \qquad (t \geq 0)$$

B–1–7. Obtain the Laplace transform of the function defined by

$$f(t) = 0 \qquad\qquad (t < 0)$$
$$= \cos 2\omega t \cdot \cos 3\omega t \qquad (t \geq 0)$$

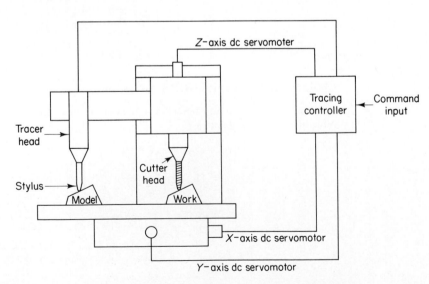

Figure 1–48 Schematic diagram of a tracing system.

B–1–8. What is the Laplace transform of the function $f(t)$ shown in Figure 1–49?

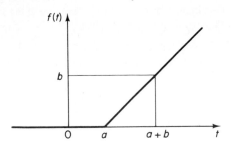

Figure 1–49 Function $f(t)$.

B–1–9. Obtain the Laplace transform of the function $f(t)$ shown in Figure 1–50.

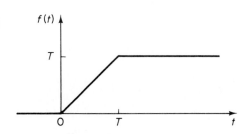

Figure 1–50 Function $f(t)$.

B–1–10. Find the Laplace transform of the function $f(t)$ shown in Figure 1–51. Also, find the limiting value of $\mathcal{L}[f(t)]$ as a approaches zero.

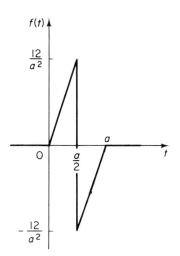

Figure 1–51 Function $f(t)$.

B–1–11. By applying the final value theorem, find the final value of $f(t)$ whose Laplace transform is given by

$$F(s) = \frac{10}{s(s+1)}$$

Verify this result by taking the inverse Laplace transform of $F(s)$ and letting $t \to \infty$.

B–1–12. Given

$$F(s) = \frac{1}{(s+2)^2}$$

determine the values of $f(0+)$ and $\dot{f}(0+)$. (Use the initial value theorem.)

B–1–13. Find the inverse Laplace transform of

$$F(s) = \frac{s+1}{s(s^2 + s + 1)}$$

B–1–14. Find the inverse Laplace transforms of the following functions.

(a) $$F_1(s) = \frac{6s + 3}{s^2}$$

(b) $$F_2(s) = \frac{5s + 2}{(s+1)(s+2)^2}$$

B–1–15. Find the inverse Laplace transform of

$$F(s) = \frac{1}{s^2(s^2 + \omega^2)}$$

B–1–16. What is the solution of the following differential equation?

$$2\ddot{x} + 7\dot{x} + 3x = 0, \qquad x(0) = 3, \qquad \dot{x}(0) = 0$$

B–1–17. Solve the differential equation

$$\dot{x} + 2x = \delta(t), \qquad x(0-) = 0$$

B–1–18. Solve the following differential equation:

$$\ddot{x} + 2\zeta\omega_n\dot{x} + \omega_n^2 x = 0, \qquad x(0) = a, \qquad \dot{x}(0) = b$$

where a and b are constants.

B–1–19. Obtain the solution of the differential equation

$$\dot{x} + ax = A \sin \omega t, \qquad x(0) = b$$

B–1–20. Consider the system shown in Figure 1–37. The system is initially at rest. Suppose that the cart is set into motion by an impulsive force whose strength is unity. Can it be stopped by another such impulsive force?

B–1–21. Simplify the block diagram shown in Figure 1–52 and obtain the closed-loop transfer function $C(s)/R(s)$.

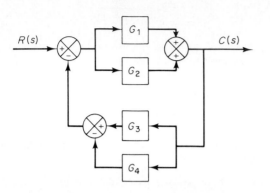

Figure 1–52 Block diagram of a system.

B–1–22. Simplify the block diagram shown in Figure 1–53 and obtain the closed-loop transfer function $C(s)/R(s)$.

B–1–23. Simplify the block diagram shown in Figure 1–54 and obtain the transfer function $C(s)/R(s)$.

B–1–24. Obtain the transfer functions $Y(s)/X(s)$ of the systems shown in Figures 1–55(a), (b), and (c).

B–1–25. Consider the network system shown in Figure 1–56. Choosing v_c and i_L as the state variables, obtain the state equation of the system.

B–1–26. Consider the system described by

$$\dddot{y} + 3\ddot{y} + 2\dot{y} = u$$

Derive a state-space representation of the system.

B–1–27. Consider the system described by

$$\begin{bmatrix} \dot{x}_1 \\ \dot{x}_2 \end{bmatrix} = \begin{bmatrix} -4 & -1 \\ 3 & -1 \end{bmatrix} \begin{bmatrix} x_1 \\ x_2 \end{bmatrix} + \begin{bmatrix} 1 \\ 1 \end{bmatrix} u$$

$$y = \begin{bmatrix} 1 & 0 \end{bmatrix} \begin{bmatrix} x_1 \\ x_2 \end{bmatrix}$$

Obtain the transfer function of the system.

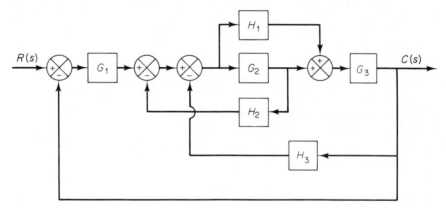

Figure 1–53 Block diagram of a system.

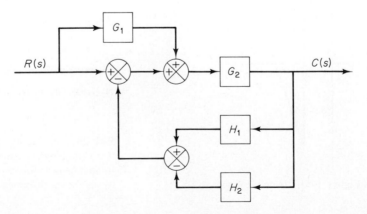

Figure 1–54 Block diagram of a system.

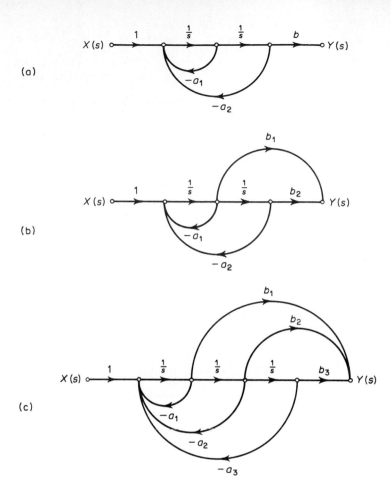

(a)

(b)

(c)

Figure 1–55 Signal flow graphs of systems.

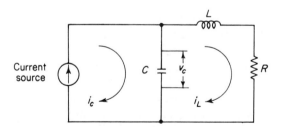

Figure 1–56 Network system.

CHAPTER 2
Mathematical Modeling of Dynamic Systems

2-1 INTRODUCTION

This chapter discusses mathematical modeling and computer simulation of dynamic systems. In studying control systems the reader must be able to model dynamic systems and analyze dynamic characteristics. A mathematical model of a dynamic system is defined as a set of equations that represents the dynamics of the system accurately or, at least, fairly well. Note that a mathematical model is not unique to a given system. A system may be represented in many different ways and, therefore, may have many mathematical models, depending on one's perspective. In this chapter we obtain mathematical models that relate the output of a system to its input.

The dynamics of many systems, whether they are mechanical, electrical, thermal, economic, biological, and so on, may be described in terms of differential equations. Such differential equations may be obtained by utilizing physical laws governing a particular system, for example, Newton's laws for mechanical systems and Kirchhoff's laws for electrical systems. The response of a dynamic system to an input (or forcing function) may be obtained if the equations involved are solved.

Mathematical models. The first step in the analysis of a dynamic system is to derive its mathematical model. We must always keep in mind that deriving a reasonable mathematical model is the most important part of the entire analysis.

Mathematical models may assume many different forms. Depending on the particular system and the particular circumstances, one mathematical model may be better suited than other models. For example, in optimal control problems, it is advantageous to use state-space representations. On the other hand, for the transient-response or frequency-response analysis of single-input, single-output, linear, time-invariant systems, the transfer function representation may be more convenient than any other. Once a mathematical model of a system is obtained, various analytical and computer tools can be used for analysis and synthesis purposes.

It is noted that the state-space representation is best suited in dealing with multiple-input, multiple-output systems. Also, the state-space representation is convenient in implementing computer-aided design algorithms.

Simplicity versus accuracy.   It is possible to improve the accuracy of a mathematical model by increasing its complexity. In some cases, we include hundreds of equations to describe a complete system. In obtaining a mathematical model, however, we must make a compromise between the simplicity of the model and the accuracy of the results of the analysis. If extreme accuracy is not needed, however, it is preferable to obtain only a reasonably simplified model. In fact, we are generally satisfied if we can obtain a mathematical model that is adequate for the problem under consideration. It is important to note, however, that the results obtained from the analysis are valid only to the extent that the model approximates a given dynamic system.

In deriving a reasonably simplified mathematical model, we frequently find it necessary to ignore certain inherent physical properties of the system. In particular, if a linear lumped-parameter mathematical model (that is, one employing ordinary differential equations) is desired, it is always necessary to ignore certain nolinearities and distributed parameters (that is, ones giving rise to partial differential equations) that may be present in the physical system. If the effects that these ignored properties have on the response are small, good agreement will be obtained between the results of the analysis of a mathematical model and the results of the experimental study of the physical system.

In general, in solving a new problem, we find it desirable first to build a simplified model so that we can get a general feeling for the solution. A more complete mathematical model may then be built and used for a more complete analysis.

We must be well aware of the fact that a linear lumped-parameter model, which may be valid in low-frequency operations, may not be valid at sufficiently high frequencies since the neglected property of distributed parameters may become an important factor in the dynamic behavior of the system. For example, the mass of a spring may be neglected in low-frequency operations but it becomes an important property of the system at high frequencies.

Linear systems.   A system is called linear if the principle of superposition applies. The principle of superposition states that the response produced by the simultaneous application of two different forcing functions is the sum of the two individual responses. Hence, for the linear system, the response to several inputs can be calculated by treating one input at a time and adding the results. It is this principle that allows one to build up complicated solutions to the linear differential equation from simple solutions.

In an experimental investigation of a dynamic system, if cause and effect are proportional, thus implying that the principle of superposition holds, then the system can be considered linear.

Linear time-invariant systems and linear time-varying systems.   A differential equation is linear if the coefficients are constants or functions only of the independent variable.

Dynamic systems that are composed of linear time-invariant lumped-parameter components may be described by linear time-invariant (constant-coefficient) differential equations. Such systems are called *linear time-invariant* (or *linear constant-coefficient*) systems. Systems that are represented by differential equations whose coefficients are functions of time are called linear time-varying systems. An example of a time-varying control system is a spacecraft control system. (The mass of a spacecraft changes due to fuel consumption.)

Nonlinear systems. A system is nonlinear if the principle of superposition does not apply. Thus, for a nonlinear system the response to two inputs cannot be calculated by treating one input at a time and adding the results. Examples of nonlinear differential equations are

$$\frac{d^2x}{dt^2} + \left(\frac{dx}{dt}\right)^2 + x = A \sin \omega t$$

$$\frac{d^2x}{dt^2} + (x^2 - 1)\frac{dx}{dt} + x = 0$$

$$\frac{d^2x}{dt^2} + \frac{dx}{dt} + x + x^3 = 0$$

Although many physical relationships are often represented by linear equations, in most cases actual relationships are not quite linear. In fact, a careful study of physical systems reveals that even so-called "linear systems" are really linear only in limited operating ranges. In practice, many electromechanical systems, hydraulic systems, pneumatic systems, and so on, involve nonlinear relationships among the variables. For example, the output of a component may saturate for large input signals. There may be a dead space that affects small signals. (The dead space of a component is a small range of input variations to which the component is insensitive.) Square-law nonlinearity may occur in some components. For instance, dampers used in physical systems may be linear for low-velocity operations but may become nonlinear at high velocities, and the damping force may become proportional to the square of the operating velocity. Examples of characteristic curves for these nonlinearities are shown in Figure 2–1.

Note that some important control systems are nonlinear for signals of any size. For example, in on–off control systems, the control action is either on or off, and there is no linear relationship between the input and output of the controller.

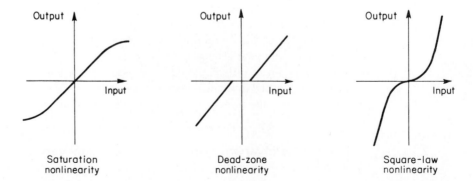

Figure 2–1
Characteristic curves for various nonlinearities.

Saturation nonlinearity

Dead-zone nonlinearity

Square-law nonlinearity

Procedures for finding the solutions of problems involving such nonlinear systems, in general, are extremely complicated. Because of this mathematical difficulty attached to nonlinear systems, one often finds it necessary to introduce "equivalent" linear systems in place of nonlinear ones. Such equivalent linear systems are valid for only a limited range of operation. Once a nonlinear system is approximated by a linear mathematical model, a number of linear tools may be applied for analysis and design purposes.

Linearization of nonlinear systems. In control engineering a normal operation of the system may be around an equilibrium point, and the signals may be considered small signals around the equilibrium. (It should be pointed out that there are many exceptions to such a case.) However, if the system operates around an equilibrium point and if the signals involved are small signals, then it is possible to approximate the nonlinear system by a linear system. Such a linear system is equivalent to the nonlinear system considered within a limited operating range. Such a linearized model (linear, time-invariant model) is very important in control engineering. We shall discuss a linearization technique in Section 2–10.

Outline of the chapter. Section 2–1 has presented an introduction to the mathematical modeling of dynamic systems, including discussions of linear and nonlinear systems. Section 2–2 discusses state-space representation of control systems. Section 2–3 treats mathematical modeling of mechanical systems. We shall discuss Newton's approach to modeling mechanical systems.

Section 2–4 deals with mathematical modeling of various electrical circuits and electronic circuits involving operational amplifiers. Section 2–5 presents analogies existing between one type of dynamic system and other types of dynamic systems. Here we focus on analogies between mechanical systems and electrical systems. Section 2–6 treats electric servomotors, which are very commonly used in control systems. Section 2–7 deals with liquid-level systems. We model such systems in terms of resistance and capacitance. Section 2–8 presents mathematical modeling of thermal systems in terms of thermal resistance and thermal capacitance. (Mathematical modeling of pneumatic systems and hydraulic systems is given in Chapter 3.) Section 2–9 discusses robot-arm systems and their simulators. Finally, Section 2–10 treats the linearization of nonlinear mathematical models.

2–2   STATE-SPACE REPRESENTATION OF DYNAMIC SYSTEMS

A dynamic system consisting of a finite number of lumped elements may be described by ordinary differential equations in which time is the independent variable. By use of vector-matrix notation, an nth-order differential equation may be expressed by a first-order vector-matrix differential equation. If n elements of the vector are a set of state variables, then the vector-matrix differential equation is called a *state* equation. In this section we shall present methods for obtaining state-space representations of continuous-time systems.

State-space representation of nth-order systems of linear differential equations in which the forcing function does not involve derivative terms. Consider the following nth-order system:

$$\overset{(n)}{y} + a_1 \overset{(n-1)}{y} + \ldots + a_{n-1}\dot{y} + a_n y = u \qquad (2\text{–}1)$$

Noting that the knowledge of $y(0)$, $\dot{y}(0)$, . . . $\overset{(n-1)}{y}(0)$, together with the input $u(t)$ for $t \geq 0$, determines completely the future behavior of the system, we may take $y(t)$, $\dot{y}(t)$, . . . , $\overset{(n-1)}{y}(t)$ as a set of n state variables. (Mathematically, such a choice of state variables is quite convenient. Practically, however, because higher-order derivative terms are inaccurate, due to the noise effects inherent in any practical situations, such a choice of the state variables may not be desirable.)

Let us define

$$x_1 = y$$
$$x_2 = \dot{y}$$
$$\cdot$$
$$\cdot$$
$$\cdot$$
$$x_n = \overset{(n-1)}{y}$$

Then Equation (2–1) can be written as

$$\dot{x}_1 = x_2$$
$$\dot{x}_2 = x_3$$
$$\cdot$$
$$\cdot$$
$$\cdot$$
$$\dot{x}_{n-1} = x_n$$
$$\dot{x}_n = -a_n x_1 - \cdots - a_1 x_n + u$$

or

$$\dot{\mathbf{x}} = \mathbf{A}\mathbf{x} + \mathbf{B}u \qquad (2\text{–}2)$$

where

$$\mathbf{x} = \begin{bmatrix} x_1 \\ x_2 \\ \cdot \\ \cdot \\ \cdot \\ x_n \end{bmatrix}, \quad \mathbf{A} = \begin{bmatrix} 0 & 1 & 0 & \ldots & 0 \\ 0 & 0 & 1 & \ldots & 0 \\ \cdot & \cdot & \cdot & & \cdot \\ \cdot & \cdot & \cdot & & \cdot \\ \cdot & \cdot & \cdot & & \cdot \\ 0 & 0 & 0 & \ldots & 1 \\ -a_n & -a_{n-1} & -a_{n-2} & \ldots & -a_1 \end{bmatrix}, \quad \mathbf{B} = \begin{bmatrix} 0 \\ 0 \\ \cdot \\ \cdot \\ \cdot \\ 0 \\ 1 \end{bmatrix}$$

The output can be given by

$$y = \begin{bmatrix} 1 & 0 & \ldots & 0 \end{bmatrix} \begin{bmatrix} x_1 \\ x_2 \\ \cdot \\ \cdot \\ \cdot \\ x_n \end{bmatrix}$$

or

$$y = \mathbf{C}\mathbf{x} \qquad (2\text{–}3)$$

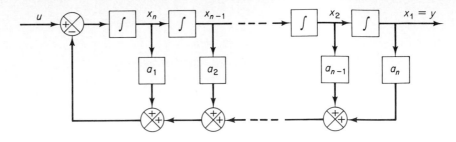

Figure 2–2
Block diagram
realization of state
equation and output
equation given by
Eqs. (2–2) and (2–3),
respectively.

where

$$\mathbf{C} = [1 \quad 0 \quad \ldots \quad 0]$$

The first-order differential equation, Equation (2–2), is the state equation, and the algebraic equation, Equation (2–3), is the output equation. A block diagram realization of the state equation and output equation given by Equations (2–2) and (2–3), respectively, is shown in Figure 2–2.

State-space representation of *n*th-order systems of linear differential equations in which the forcing function involves derivative terms.

If the differential equation of the system involves derivatives of the forcing function, such as

$$\overset{(n)}{y} + a_1 \overset{(n-1)}{y} + \cdots + a_{n-1} \dot{y} + a_n y = b_0 \overset{(n)}{u} + b_1 \overset{(n-1)}{u} + \cdots + b_{n-1} \dot{u} + b_n u \qquad (2\text{–}4)$$

then the set of *n* variables $y, \dot{y}, \ddot{y}, \ldots, \overset{(n-1)}{y}$ does not qualify as a set of state variables, and the straightforward method previously employed cannot be used. This is because *n* first-order differential equations

$$\dot{x}_1 = x_2$$

$$\dot{x}_2 = x_3$$

$$\cdot$$

$$\cdot$$

$$\cdot$$

$$\dot{x}_n = -a_n x_1 - a_{n-1} x_2 - \cdots - a_1 x_n + b_0 \overset{(n)}{u} + b_1 \overset{(n-1)}{u} + \cdots + b_n u$$

where $x_1 = y$, may not yield a unique solution.

The main problem in defining the state variables for this case lies in the derivative terms on the right side of the last of the preceding *n* equations. The state variables must be such that they will eliminate the derivatives of *u* in the state equation.

One way to obtain a state equation and output equation is to define the following *n* variables as a set of *n* state variables:

$$x_1 = y - \beta_0 u$$

$$x_2 = \dot{y} - \beta_0 \dot{u} - \beta_1 u = \dot{x}_1 - \beta_1 u$$

$$x_3 = \ddot{y} - \beta_0 \ddot{u} - \beta_1 \dot{u} - \beta_2 u = \dot{x}_2 - \beta_2 u$$

$$\cdot$$

$$\cdot \qquad\qquad\qquad\qquad\qquad\qquad\qquad\qquad (2\text{–}5)$$

$$\cdot$$

$$x_n = \overset{(n-1)}{y} - \beta_0 \overset{(n-1)}{u} - \beta_1 \overset{(n-2)}{u} - \cdots - \beta_{n-2} \dot{u} - \beta_{n-1} u = \dot{x}_{n-1} - \beta_{n-1} u$$

where β_0, β_1, β_2, . . . , β_n are determined from

$$\beta_0 = b_0$$
$$\beta_1 = b_1 - a_1\beta_0$$
$$\beta_2 = b_2 - a_1\beta_1 - a_2\beta_0$$
$$\beta_3 = b_3 - a_1\beta_2 - a_2\beta_1 - a_3\beta_0 \qquad (2\text{–}6)$$

.

.

.

$$\beta_n = b_n - a_1\beta_{n-1} - \cdots - a_{n-1}\beta_1 - a_n\beta_0$$

With this choice of state variables the existence and uniqueness of the solution of the state equation is guaranteed. (Note that this is not the only choice of a set of state variables.) With the present choice of state variables, we obtain

$$\dot{x}_1 = x_2 + \beta_1 u$$
$$\dot{x}_2 = x_3 + \beta_2 u$$

.

. $\qquad (2\text{–}7)$

.

$$\dot{x}_{n-1} = x_n + \beta_{n-1} u$$
$$\dot{x}_n = -a_n x_1 - a_{n-1} x_2 - \cdots - a_1 x_n + \beta_n u$$

[To derive Equation (2–7), see Problem A–2–1.] In terms of vector-matrix equations, Equation (2–7) and the output equation can be written as

$$
\begin{bmatrix} \dot{x}_1 \\ \dot{x}_2 \\ \cdot \\ \cdot \\ \cdot \\ \dot{x}_{n-1} \\ \dot{x}_n \end{bmatrix}
=
\begin{bmatrix}
0 & 1 & 0 & \cdots & 0 \\
0 & 0 & 1 & \cdots & 0 \\
\cdot & \cdot & \cdot & & \cdot \\
\cdot & \cdot & \cdot & & \cdot \\
\cdot & \cdot & \cdot & & \cdot \\
0 & 0 & 0 & \cdots & 1 \\
-a_n & -a_{n-1} & -a_{n-2} & \cdots & -a_1
\end{bmatrix}
\begin{bmatrix} x_1 \\ x_2 \\ \cdot \\ \cdot \\ \cdot \\ x_{n-1} \\ x_n \end{bmatrix}
+
\begin{bmatrix} \beta_1 \\ \beta_2 \\ \cdot \\ \cdot \\ \cdot \\ \beta_{n-1} \\ \beta_n \end{bmatrix} u
$$

$$
y = [1 \quad 0 \quad \cdots \quad 0]
\begin{bmatrix} x_1 \\ x_2 \\ \cdot \\ \cdot \\ \cdot \\ x_n \end{bmatrix}
+ \beta_0 u
$$

or

$$\dot{\mathbf{x}} = \mathbf{A}\mathbf{x} + \mathbf{B}u \qquad (2\text{–}8)$$
$$y = \mathbf{C}\mathbf{x} + Du \qquad (2\text{–}9)$$

where

$$
\mathbf{x} = \begin{bmatrix} x_1 \\ x_2 \\ \cdot \\ \cdot \\ \cdot \\ x_{n-1} \\ x_n \end{bmatrix}, \qquad
\mathbf{A} = \begin{bmatrix} 0 & 1 & 0 & \cdots & 0 \\ 0 & 0 & 1 & \cdots & 0 \\ \cdot & & \cdot & & \cdot \\ \cdot & & \cdot & & \cdot \\ \cdot & & \cdot & & \cdot \\ 0 & 0 & 0 & \cdots & 1 \\ -a_n & -a_{n-1} & -a_{n-2} & \cdots & -a_1 \end{bmatrix}
$$

$$
\mathbf{B} = \begin{bmatrix} \beta_1 \\ \beta_2 \\ \cdot \\ \cdot \\ \cdot \\ \beta_{n-1} \\ \beta_n \end{bmatrix}, \qquad
\mathbf{C} = [1 \quad 0 \quad \cdots \quad 0], \qquad D = \beta_0 = b_0
$$

The initial condition $\mathbf{x}(0)$ may be determined by use of Equation (2–5).

In this state-space representation, the matrix $\mathbf{A}$ is exactly the same as that for the system of Equation (2–1). The derivatives on the right side of Equation (2–4) affect only the elements of the $\mathbf{B}$ matrix.

Note that the state-space representation for the transfer function

$$
\frac{Y(s)}{U(s)} = \frac{b_0 s^n + b_1 s^{n-1} + \cdots + b_{n-1} s + b_n}{s^n + a_1 s^{n-1} + \cdots + a_{n-1} s + a_n} \tag{2–10}
$$

is given also by Equations (2–8) and (2–9). Figure 2–3 is a block diagram realization of the state equation and output equation, given by Equations (2–8) and (2–9), respectively.

It is noted that there are many other representations of the system in state space (such as controllable canonical form, observable canonical form, diagonal canonical form, and Jordan canonical form). They will be presented in Chapter 9.

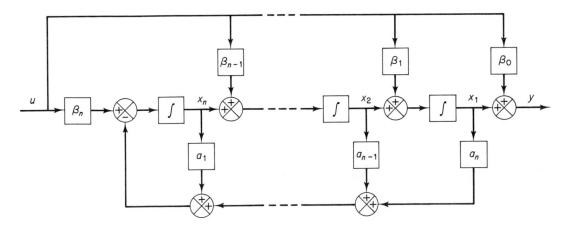

Figure 2–3 Block diagram realization of state equation and output equation given by Eqs. (2–8) and (2–9), respectively.

EXAMPLE 2–1   Consider the control system shown in Figure 2–4. The closed-loop transfer function is

$$\frac{Y(s)}{U(s)} = \frac{160(s + 4)}{s^3 + 18s^2 + 192s + 640}$$

The corresponding differential equation is

$$\dddot{y} + 18\ddot{y} + 192\dot{y} + 640y = 160\dot{u} + 640u$$

Obtain a state-space representation of the system.

Referring to Equation (2–5), let us define

$$x_1 = y - \beta_0 u$$

$$x_2 = \dot{y} - \beta_0 \dot{u} - \beta_1 u = \dot{x}_1 - \beta_1 u$$

$$x_3 = \ddot{y} - \beta_0 \ddot{u} - \beta_1 \dot{u} - \beta_2 u = \dot{x}_2 - \beta_2 u$$

where β_0, β_1, and β_2 are determined from Equation (2–6) as follows:

$$\beta_0 = b_0 = 0$$

$$\beta_1 = b_1 - a_1\beta_0 = 0$$

$$\beta_2 = b_2 - a_1\beta_1 - a_2\beta_0 = 160$$

$$\beta_3 = b_3 - a_1\beta_2 - a_2\beta_1 - a_3\beta_0 = -2240$$

Then the state equation for the system becomes

$$\begin{bmatrix} \dot{x}_1 \\ \dot{x}_2 \\ \dot{x}_3 \end{bmatrix} = \begin{bmatrix} 0 & 1 & 0 \\ 0 & 0 & 1 \\ -640 & -192 & -18 \end{bmatrix} \begin{bmatrix} x_1 \\ x_2 \\ x_3 \end{bmatrix} + \begin{bmatrix} 0 \\ 160 \\ -2240 \end{bmatrix} u$$

The output equation becomes

$$y = \begin{bmatrix} 1 & 0 & 0 \end{bmatrix} \begin{bmatrix} x_1 \\ x_2 \\ x_3 \end{bmatrix}$$

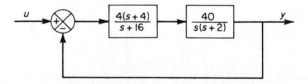

Figure 2–4
Control system.

2–3   MECHANICAL SYSTEMS

In this section we shall discuss mathematical modeling of mechanical systems. The fundamental law governing mechanical systems is Newton's second law. It can be applied to any mechanical systems. In what follows we shall derive mathematical models of a few mechanical systems.

Chapter 2 / Mathematical Modeling of Dynamic Systems

Mechanical translational systems.   Consider the spring–mass–dashpot system mounted on a cart as shown in Figure 2–5. A dashpot is a device that provides viscous friction, or damping. It consists of a piston and oil-filled cylinder. Any relative motion between the piston rod and the cylinder is resisted by the oil because the oil must flow around the piston (or through orifices provided in the piston) from one side of the piston to the other. The dashpot essentially absorbs energy. This absorbed energy is dissipated as heat, and the dashpot does not store any kinetic or potential energy. The dashpot is also called a damper.

Let us obtain a mathematical model of this spring–mass–dashpot system mounted on a cart by assuming that the cart is standing still for $t < 0$. In this system, $u(t)$ is the displacement of the cart and is the input to the system. At $t = 0$, the cart is moved at a constant speed, or $\dot{u}$ = constant. The displacement $y(t)$ of the mass is the output. (The displacement is relative to the ground.) In this system, m denotes the mass, b denotes the viscous friction coefficient, and k denotes the spring constant. We assume that the friction force of the dashpot is proportional to $\dot{y} - \dot{u}$ and that the spring is a linear spring; that is, the spring force is proportional to $y - u$.

For translational systems, Newton's second law states that

$$ma = \sum F$$

where m = mass, kg
$\quad\quad\ a$ = acceleration, m/sec^2
$\quad\quad\ F$ = force, N

Applying Newton's second law to the present system, we obtain

$$m\frac{d^2y}{dt^2} = -b\left(\frac{dy}{dt} - \frac{du}{dt}\right) - k(y - u)$$

or

$$m\frac{d^2y}{dt^2} + b\frac{dy}{dt} + ky = b\frac{du}{dt} + ku \qquad (2\text{–}11)$$

Equation (2–11) gives a mathematical model of the system considered.

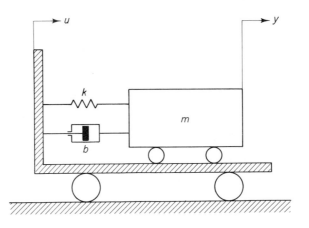

Figure 2–5
Spring–mass–dashpot system mounted on a cart.

A transfer function model is another way of representing a mathematical model of a linear, time-invariant system. For the present mechanical system, the transfer function model can be obtained as follows: Taking the Laplace transform of each term of Equation (2–11) gives

$$\mathscr{L}\left[m\frac{d^2y}{dt^2}\right] = m[s^2Y(s) - sy(0) - \dot{y}(0)]$$

$$\mathscr{L}\left[b\frac{dy}{dt}\right] = b[sY(s) - y(0)]$$

$$\mathscr{L}[ky] = kY(s)$$

$$\mathscr{L}\left[b\frac{du}{dt}\right] = b\left[sU(s) - u(0)\right]$$

$$\mathscr{L}[ku] = kU(s)$$

If we set the initial conditions equal to zero, or set $y(0) = 0$, $\dot{y}(0) = 0$, and $u(0) = 0$, the Laplace transform of Equation (2–11) can be written as

$$(ms^2 + bs + k)Y(s) = (bs + k)U(s)$$

Taking the ratio of $Y(s)$ to $U(s)$, we find the transfer function of the system to be

$$\text{Transfer function} = G(s) = \frac{Y(s)}{U(s)} = \frac{bs + k}{ms^2 + bs + k}$$

Such a transfer-function representation of a mathematical model is used very frequently in control engineering. It should be noted, however, that transfer-function models apply only to linear, time-invariant systems, since the transfer functions are defined only for such systems.

Next we shall obtain a state-space model of this system. We shall first compare the differential equation for this system

$$\ddot{y} + \frac{b}{m}\dot{y} + \frac{k}{m}y = \frac{b}{m}\dot{u} + \frac{k}{m}u$$

with the standard form

$$\ddot{y} + a_1\dot{y} + a_2y = b_0\ddot{u} + b_1\dot{u} + b_2u$$

and identify a_1, a_2, b_0, b_1, and b_2 as follows:

$$a_1 = \frac{b}{m}, \qquad a_2 = \frac{k}{m}, \qquad b_0 = 0, \qquad b_1 = \frac{b}{m}, \qquad b_2 = \frac{k}{m}$$

Referring to Equation (2–6), we have

$$\beta_0 = b_0 = 0$$

$$\beta_1 = b_1 - a_1\beta_0 = \frac{b}{m}$$

$$\beta_2 = b_2 - a_1\beta_1 - a_2\beta_0 = \frac{k}{m} - \left(\frac{b}{m}\right)^2$$

Then, referring to Equation (2–5), define

$$x_1 = y - \beta_0 u = y$$

$$x_2 = \dot{x}_1 - \beta_1 u = \dot{x}_1 - \frac{b}{m} u$$

From Equation (2–7) we have

$$\dot{x}_1 = x_2 + \beta_1 u = x_2 + \frac{b}{m} u$$

$$\dot{x}_2 = -a_2 x_1 - a_1 x_2 + \beta_2 u = -\frac{k}{m} x_1 - \frac{b}{m} x_2 + \left[\frac{k}{m} - \left(\frac{b}{m} \right)^2 \right] u$$

and the output equation becomes

$$y = x_1$$

or

$$
\begin{bmatrix} \dot{x}_1 \\ \dot{x}_2 \end{bmatrix} = \begin{bmatrix} 0 & 1 \\ -\dfrac{k}{m} & -\dfrac{b}{m} \end{bmatrix} \begin{bmatrix} x_1 \\ x_2 \end{bmatrix} + \begin{bmatrix} \dfrac{b}{m} \\ \dfrac{k}{m} - \left(\dfrac{b}{m} \right)^2 \end{bmatrix} u \tag{2–12}
$$

and

$$
y = \begin{bmatrix} 1 & 0 \end{bmatrix} \begin{bmatrix} x_1 \\ x_2 \end{bmatrix} \tag{2–13}
$$

Equations (2–12) and (2–13) give a state-space representation of the system. (Note that this is not the only state-space representation. There are infinitely many state-space representations for the system.)

Mechanical rotational system.   Consider the system shown in Figure 2–6. The system consists of a load inertia and a viscous-friction damper. For such a mechanical rotational system, Newton's second law states that

$$J\alpha = \sum T$$

where J = moment of inertia of the load, kg-m^2
α = angular acceleration of the load, rad/sec^2
T = torque applied to the system, N-m

For sets of consistent units for mass, moment of inertia, and torque, see Table 2–1. Also, for the conversion of systems of unit, see the footnote.*

---

* 14.594 kg = 1 slug. A slug is a unit of mass (slug = lb-sec^2/ft). When acted upon by 1 lb of force, a 1-slug mass accelerates at 1 ft/sec^2. To obtain mass in pounds, multiply the number of slugs by 32.174, and to obtain mass in slugs, multiple the number of pounds by 3.108×10^{-2}.
The relationship between kg-m^2 and slug-ft^2 is given by 1.356 kg-m^2 = 1 slug-ft^2. To obtain the moment of inertia in lb-ft^2, multiply the number of slug-ft^2 by 32.174. Conversely, to obtain the moment of inertia in slug-ft^2, multiply the number of lb-ft^2 by 3.108×10^{-2}.

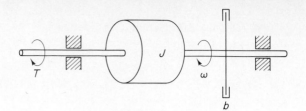

Figure 2–6
Mechanical rotational
system.

Table 2–1 Sets of Consistent Units for Mass, Moment of
Inertia, and Torque

Mass	Moment of Inertia	Torque
kilogram	$kg\text{-}m^2$	newton-m
gram	$g\text{-}cm^2$	dyne-cm
slug	$slug\text{-}ft^2$	lb-ft

Applying Newton's second law to the present system, we obtain

$$J\dot{\omega} = -b\omega + T$$

where J = moment of inertia of the load, $kg\text{-}m^2$
$\quad b$ = viscous-friction coefficient, N-m/rad/sec
$\quad \omega$ = angular velocity, rad/sec
$\quad T$ = torque, N-m

This last equation may be written as

$$J\dot{\omega} + b\omega = T$$

which is a mathematical model of the mechanical rotational system considered.

The transfer function model for the system can be obtained by taking the Laplace transform of the differential equation, assuming the zero initial condition, and writing the ratio of the output (angular velocity ω) and the input (applied torque T) as follows:

$$\frac{\Omega(s)}{T(s)} = \frac{1}{Js + b}$$

where $\Omega(s) = \mathscr{L}[\omega(t)]$ and $T(s) = \mathscr{L}[T(t)]$.

EXAMPLE 2–2   An inverted pendulum mounted on a motor-driven cart is shown in Figure 2–7. This is a model of the attitude control of a space booster on takeoff. (The objective of the attitude control problem is to keep the space booster in a vertical position.) The inverted pendulum is unstable in that it may fall over any time in any direction unless a suitable control force is applied. Here we consider only a two-dimensional problem that the pendulum moves only in the plane of the page. Assume that the pendulum mass is concentrated at the end of the rod, as shown in the figure. (The rod is massless.) The control force u is applied to the cart. Obtain a mathematical model for the system.

Define the angle of the rod from the vertical line as θ. (Since we want to keep the inverted pendulum vertical, angle θ is assumed to be small.) Define also the (x, y) coordinates of the center of gravity of the mass as (x_G, y_G). Then

$$x_G = x + l \sin \theta$$

$$y_G = l \cos \theta$$

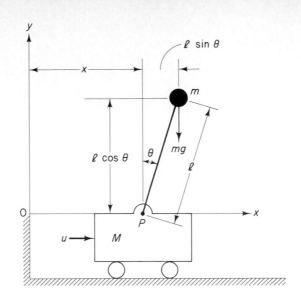

Figure 2–7
Inverted pendulum
system.

Applying Newton's second law to the x direction of motion yields

$$M \frac{d^2x}{dt^2} + m \frac{d^2x_G}{dt^2} = u$$

or

$$M \frac{d^2x}{dt^2} + m \frac{d^2}{dt^2} (x + l \sin \theta) = u \qquad (2\text{--}14)$$

Noting that

$$\frac{d}{dt} \sin \theta = (\cos \theta)\dot{\theta}$$

$$\frac{d^2}{dt^2} \sin \theta = -(\sin \theta)\dot{\theta}^2 + (\cos \theta)\ddot{\theta}$$

$$\frac{d}{dt} \cos \theta = -(\sin \theta)\dot{\theta}$$

$$\frac{d^2}{dt^2} \cos \theta = -(\cos \theta)\dot{\theta}^2 - (\sin \theta)\ddot{\theta}$$

Equation (2–14) can be rewritten as

$$(M + m)\ddot{x} - ml(\sin \theta)\dot{\theta}^2 + ml(\cos \theta)\ddot{\theta} = u \qquad (2\text{--}15)$$

The equation of motion of the mass m in the y direction cannot be written without considering the motion of the mass m in the x direction. Therefore, instead of considering the motion of the mass m in the y direction, we consider the rotational motion of the mass m around point P. Applying Newton's second law to the rotational motion, we obtain

$$m \frac{d^2x_G}{dt^2} l \cos \theta - m \frac{d^2y_G}{dt^2} l \sin \theta = mgl \sin \theta$$

or

$$\left[m \frac{d^2}{dt^2} (x + l \sin \theta) \right] l \cos \theta - \left[m \frac{d^2}{dt^2} (l \cos \theta) \right] l \sin \theta = mgl \sin \theta$$

which can be simplified as follows:

$$m[\ddot{x} - l(\sin\theta)\dot{\theta}^2 + l(\cos\theta)\ddot{\theta}]l\cos\theta - m[-l(\cos\theta)\dot{\theta}^2 - l(\sin\theta)\ddot{\theta}]l\sin\theta = mgl\sin\theta$$

Further simplification results in

$$m\ddot{x}\cos\theta + ml\ddot{\theta} = mg\sin\theta \qquad (2\text{--}16)$$

Clearly, Equations (2–15) and (2–16) are nonlinear differential equations. Since we must keep the inverted pendulum vertical, we can assume that $\theta(t)$ and $\dot{\theta}(t)$ are small quantities such that $\sin\theta \doteq \theta$, $\cos\theta \doteq 1$, and $\theta\dot{\theta}^2 \doteq 0$. Then Equations (2–15) and (2–16) can be linearized as follows:

$$(M + m)\ddot{x} + ml\ddot{\theta} = u \qquad (2\text{--}17)$$

$$m\ddot{x} + ml\ddot{\theta} = mg\theta \qquad (2\text{--}18)$$

These linearized equations are valid as long as θ and $\dot{\theta}$ are small. Equations (2–17) and (2–18) define a mathematical model of the inverted pendulum system.

EXAMPLE 2–3 Referring to the inverted pendulum system considered in Example 2–2, obtain a state-space representation of the linearized system.

The linearized system equations, Equations (2–17) and (2–18), can be modified to

$$Ml\ddot{\theta} = (M + m)g\theta - u \qquad (2\text{--}19)$$

$$M\ddot{x} = u - mg\theta \qquad (2\text{--}20)$$

Equation (2–19) was obtained by eliminating $\ddot{x}$ from Equations (2–17) and (2–18). Equation (2–20) was obtained by eliminating $\ddot{\theta}$ from Equations (2–17) and (2–18). Define state variables x_1, x_2, x_3, and x_4 by

$$x_1 = \theta$$

$$x_2 = \dot{\theta}$$

$$x_3 = x$$

$$x_4 = \dot{x}$$

Note that angle θ indicates the rotation of the pendulum rod about point P, and x is the location of the cart. We consider θ and x as the outputs of the system, or

$$\mathbf{y} = \begin{bmatrix} y_1 \\ y_2 \end{bmatrix} = \begin{bmatrix} \theta \\ x \end{bmatrix} = \begin{bmatrix} x_1 \\ x_3 \end{bmatrix}$$

(Notice that both θ and x are easily measurable quantities.) Then, from the definition of the state variables and Equations (2–19) and (2–20), we obtain

$$\dot{x}_1 = x_2$$

$$\dot{x}_2 = \frac{M + m}{Ml}gx_1 - \frac{1}{Ml}u$$

$$\dot{x}_3 = x_4$$

$$\dot{x}_4 = -\frac{m}{M}gx_1 + \frac{1}{M}u$$

In terms of vector-matrix equations, we have

$$
\begin{bmatrix} \dot{x}_1 \\ \dot{x}_2 \\ \dot{x}_3 \\ \dot{x}_4 \end{bmatrix} = \begin{bmatrix} 0 & 1 & 0 & 0 \\ \dfrac{M+m}{Ml}g & 0 & 0 & 0 \\ 0 & 0 & 0 & 1 \\ -\dfrac{m}{M}g & 0 & 0 & 0 \end{bmatrix} \begin{bmatrix} x_1 \\ x_2 \\ x_3 \\ x_4 \end{bmatrix} + \begin{bmatrix} 0 \\ -\dfrac{1}{Ml} \\ 0 \\ \dfrac{1}{M} \end{bmatrix} u \qquad (2\text{--}21)
$$

$$
\begin{bmatrix} y_1 \\ y_2 \end{bmatrix} = \begin{bmatrix} 1 & 0 & 0 & 0 \\ 0 & 0 & 1 & 0 \end{bmatrix} \begin{bmatrix} x_1 \\ x_2 \\ x_3 \\ x_4 \end{bmatrix} \qquad (2\text{--}22)
$$

Equations (2–21) and (2–22) give a state-space representation of the inverted pendulum system. (Note that state-space representation of the system is not unique. There are infinitely many such representations.)

2–4 ELECTRICAL SYSTEMS

In this section we shall deal with electrical circuits involving resistors, capacitors, inductors, and operational amplifiers.

Basic laws governing electrical circuits are Kirchhoff's current law and voltage law. Kirchhoff's current law (node law) states that the algebraic sum of all currents entering and leaving a node is zero. (This law can also be stated as follows: The sum of currents entering a node is equal to the sum of currents leaving the same node.) Kirchhoff's voltage law (loop law) states that at any given instant the algebraic sum of the voltages around any loop in an electrical circuit is zero. (This law can also be stated as follows: The sum of the voltage drops is equal to the sum of the voltage rises around a loop.) A mathematical model of an electrical circuit can be obtained by applying one or both of Kirchhoff's laws to it.

L-R-C circuit.   Consider the electrical circuit shown in Figure 2–8. The circuit consists of an inductance L (henry), a resistance R (ohm), and a capacitance C (farad). Applying Kirchhoff's voltage law to the system, we obtain the following equations:

$$
L\frac{di}{dt} + Ri + \frac{1}{C}\int i\, dt = e_i \qquad (2\text{--}23)
$$

$$
\frac{1}{C}\int i\, dt = e_o \qquad (2\text{--}24)
$$

Equations (2–23) and (2–24) give a mathematical model of the circuit.

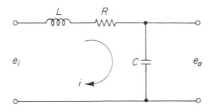

Figure 2–8
Electrical circuit.

A transfer function model of the circuit can also be obtained as follows: Taking the Laplace transforms of Equations (2–23) and (2–24), assuming zero initial conditions, we obtain

$$LsI(s) + RI(s) + \frac{1}{C}\frac{1}{s}I(s) = E_i(s)$$

$$\frac{1}{C}\frac{1}{s}I(s) = E_o(s)$$

If e_i is assumed to be the input and e_o the output, then the transfer function of this system is found to be

$$\frac{E_o(s)}{E_i(s)} = \frac{1}{LCs^2 + RCs + 1} \qquad (2\text{–}25)$$

Complex impedances.   In deriving transfer functions for electrical circuits, we frequently find it convenient to write the Laplace-transformed equations directly, without writing the differential equations. Consider the system shown in Figure 2–9(a). In this system, Z_1 and Z_2 represent complex impedances. The complex impedance $Z(s)$ of a two-terminal circuit is the ratio of $E(s)$, the Laplace transform of the voltage across the terminals, to $I(s)$, the Laplace transform of the current through the element, under the assumption that the initial conditions are zero, so that $Z(s) = E(s)/I(s)$. If the two-terminal element is a resistance R, capacitance C, or inductance L, then the complex impedance is given by R, $1/Cs$, or Ls, respectively. If complex impedances are connected in series, the total impedance is the sum of the individual complex impedances.

Remember that the impedance approach is valid only if the initial conditions involved are all zeros. Since the transfer function requires zero initial conditions, the impedance approach can be applied to obtain the transfer function of the electrical circuit. This approach greatly simplifies the derivation of transfer functions of electrical circuits.

Consider the circuit shown in Figure 2–9(b). Assume that the voltages e_i and e_o are the input and output of the circuit, respectively. Then the transfer function of this circuit is

$$\frac{E_o(s)}{E_i(s)} = \frac{Z_2(s)}{Z_1(s) + Z_2(s)}$$

For the system shown in Figure 2–8,

$$Z_1 = Ls + R, \qquad Z_2 = \frac{1}{Cs}$$

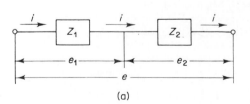

(a)

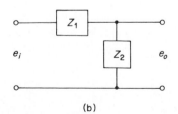

(b)

Figure 2–9
Electrical circuits.

Hence the transfer function $E_o(s)/E_i(s)$ can be found as follows:

$$\frac{E_o(s)}{E_i(s)} = \frac{\dfrac{1}{Cs}}{Ls + R + \dfrac{1}{Cs}} = \frac{1}{LCs^2 + RCs + 1}$$

which is, of course, identical to Equation (2–25).

State-space representation.   A state-space model of the system shown in Figure 2–8 may be obtained as follows: First, note that the differential equation for the system can be obtained from Equation (2–25) as

$$\ddot{e}_o + \frac{R}{L}\dot{e}_o + \frac{1}{LC}e_o = \frac{1}{LC}e_i$$

Then by defining state variables by

$$x_1 = e_o$$

$$x_2 = \dot{e}_o$$

and the input and output variables by

$$u = e_i$$

$$y = e_o = x_1$$

we obtain

$$\begin{bmatrix} \dot{x}_1 \\ \dot{x}_2 \end{bmatrix} = \begin{bmatrix} 0 & 1 \\ -\dfrac{1}{LC} & -\dfrac{R}{L} \end{bmatrix} \begin{bmatrix} x_1 \\ x_2 \end{bmatrix} + \begin{bmatrix} 0 \\ \dfrac{1}{LC} \end{bmatrix} u$$

and

$$y = \begin{bmatrix} 1 & 0 \end{bmatrix} \begin{bmatrix} x_1 \\ x_2 \end{bmatrix}$$

These two equations give a mathematical model of the system in state space.

Transfer functions of cascaded elements.   Many feedback systems have components that load each other. Consider the system shown in Figure 2–10. Assume that e_i is the

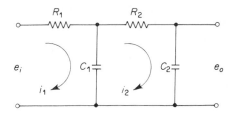

Figure 2–10
Electrical system.

input and e_o is the output. In this system the second stage of the circuit (R_2C_2 portion) produces a loading effect on the first stage (R_1C_1 portion). The equations for this system are

$$\frac{1}{C_1} \int (i_1 - i_2) \, dt + R_1 i_1 = e_i \qquad (2\text{–}26)$$

and

$$\frac{1}{C_1} \int (i_2 - i_1) \, dt + R_2 i_2 = -\frac{1}{C_2} \int i_2 \, dt = -e_o \qquad (2\text{–}27)$$

Taking the Laplace transforms of Equations (2–26) and (2–27), respectively, assuming zero initial conditions, we obtain

$$\frac{1}{C_1 s} [I_1(s) - I_2(s)] + R_1 I_1(s) = E_i(s) \qquad (2\text{–}28)$$

$$\frac{1}{C_1 s} [I_2(s) - I_1(s)] + R_2 I_2(s) = -\frac{1}{C_2 s} I_2(s) = -E_o(s) \qquad (2\text{–}29)$$

Eliminating $I_1(s)$ and $I_2(s)$ from Equations (2–28) and (2–29), we find the transfer function between $E_o(s)$ and $E_i(s)$ to be

$$\begin{aligned}\frac{E_o(s)}{E_i(s)} &= \frac{1}{(R_1 C_1 s + 1)(R_2 C_2 s + 1) + R_1 C_2 s} \\ &= \frac{1}{R_1 C_1 R_2 C_2 s^2 + (R_1 C_1 + R_2 C_2 + R_1 C_2)s + 1} \end{aligned} \qquad (2\text{–}30)$$

The term $R_1 C_2 s$ in the denominator of the transfer function represents the interaction of two simple *RC* circuits. Since $(R_1 C_1 + R_2 C_2 + R_1 C_2)^2 > 4R_1 C_1 R_2 C_2$, the two roots of the denominator of Equation (2–30) are real.

The present analysis shows that, if two *RC* circuits are connected in cascade so that the output from the first circuit is the input to the second, the overall transfer function is not the product of $1/(R_1 C_1 s + 1)$ and $1/(R_2 C_2 s + 1)$. The reason for this is that, when we derive the tranfer function for an isolated circuit, we implicitly assume that the output is unloaded. In other words, the load impedance is assumed to be infinite, which means that no power is being withdrawn at the output. When the second circuit is connected to the output of the first, however, a certain amount of power is withdrawn, and thus the assumption of no loading is violated. Therefore, if the transfer function of this system is obtained under the assumption of no loading, then it is not valid. The degree of the loading effect determines the amount of modification of the transfer function.

Transfer functions of nonloading cascaded elements.   The transfer function of a system consisting of two nonloading cascaded elements can be obtained by eliminating the intermediate input and output. For example, consider the system shown in Figure 2–11(a). The transfer functions of the elements are

$$G_1(s) = \frac{X_2(s)}{X_1(s)} \qquad \text{and} \qquad G_2(s) = \frac{X_3(s)}{X_2(s)}$$

If the input impedance of the second element is infinite, the output of the first element is

(a)   (b)

Figure 2–11 (a) System consisting of two nonloading cascaded elements; (b) an equivalent system.

not affected by connecting it to the second element. Then the transfer function of the whole system becomes

$$G(s) = \frac{X_3(s)}{X_1(s)} = \frac{X_2(s)\,X_3(s)}{X_1(s)\,X_2(s)} = G_1(s)G_2(s)$$

The transfer function of the whole system is thus the product of the transfer functions of the individual elements. This is shown in Figure 2–11(b).

As an example, consider the system shown in Figure 2–12. The insertion of an isolating amplifier between the circuits to obtain nonloading characteristics is frequently used in combining circuits. Since amplifiers have very high input impedances, an isolation amplifier inserted between the two circuits justifies the nonloading assumption.

The two simple *RC* circuits, isolated by an amplifier as shown in Figure 2–12, have negligible loading effects, and the transfer function of the entire circuit equals the product of the individual transfer functions. Thus, in this case,

$$\frac{E_o(s)}{E_i(s)} = \left(\frac{1}{R_1C_1s + 1}\right)(K)\left(\frac{1}{R_2C_2s + 1}\right)$$
$$= \frac{K}{(R_1C_1s + 1)(R_2C_2s + 1)}$$

Passive elements and active elements.   Some elements in a system, such as capacitors and inductors, store energy. This energy can later be introduced into the system. The amount of energy that can be introduced cannot exceed the amount the element has stored; and unless such an element stored energy beforehand, it cannot deliver any energy to the system. Because of this, such an element is called a *passive* element. A system containing only passive elements is called a *passive* system. Examples of passive elements are capacitors, resistors, and inductors in electrical systems and masses, inertias, dampers, and springs in mechanical systems. For passive systems, every term in the homogeneous system differential equation has the same sign.

A physical element that can deliver external energy into a system is called an *active* element. For example, an amplifier is an active element since it has a power source and

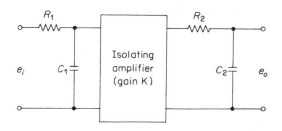

Figure 2–12
Electrical system.

supplies power to the system. External force, torque, or velocity sources and current or voltage sources are also active elements.

Operational amplifiers. Operational amplifiers, often called op amps, are frequently used to amplify signals in sensor circuits. Op amps are also frequently used in filters used for compensation purposes. Figure 2–13 shows an op amp. It is a common practice to choose the ground as 0 volt and measure the input voltages e_1 and e_2 relative to the ground. The input e_1 to the minus terminal of the amplifier is inverted, and the input e_2 to the plus terminal is not inverted. The total input to the amplifier thus becomes $e_2 - e_1$. Hence, for the circuit shown in Figure 2–13, we have

$$e_o = K(e_2 - e_1) = -K(e_1 - e_2)$$

where the inputs e_1 and e_2 may be dc or ac signals and K is the differential gain or voltage gain. The magnitude of K is approximately $10^5 \sim 10^6$ for dc signals and ac signals with frequencies less than approximately 10 Hz. (The differential gain K decreases with the signal frequency and becomes about unity for frequencies of 1 MHz $\sim$ 50 MHz.) Note that the op amp amplifies the difference in voltages e_1 and e_2. Such an amplifier is commonly called a differential amplifier. Since the gain of the op amp is very high, it is necessary to have a negative feedback from the output to the input to make the amplifier stable. (The feedback is made from the output to the inverted input so that the feedback is a negative feedback.)

In the ideal op amp, no current flows into the input terminals, and the output voltage is not affected by the load connected to the output terminal. In other words, the input impedance is infinity and the output impedance is zero. In an actual op amp, a very small (almost negligible) current flows into an input terminal and the output cannot be loaded too much. In our analysis here, we make the assumption that the op amps are ideal.

Inverting amplifier. Consider the operational amplifier circuit shown in Figure 2–14. Let us obtain the output voltage e_0.

The equation for this circuit can be obtained as follows: Define

$$i_1 = \frac{e_i - e'}{R_1}, \qquad i_2 = \frac{e' - e_o}{R_2}$$

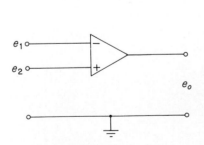

Figure 2–13
Operational amplifier.

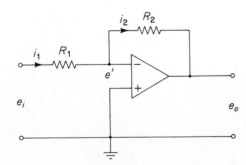

Figure 2–14
Inverting amplifier.

Since only a negligible current flows into the amplifier, the current i_1 must be equal to current i_2. Thus

$$\frac{e_i - e'}{R_1} = \frac{e' - e_o}{R_2}$$

Since $K(0 - e') = e_o$ and $K \gg 1$, e' must be almost zero, or $e' \doteq 0$. Hence we have

$$\frac{e_i}{R_1} = \frac{-e_o}{R_2}$$

or

$$e_o = -\frac{R_2}{R_1} e_i$$

Thus the circuit shown is an inverting amplifier. If $R_1 = R_2$, then the op-amp circuit shown acts as a sign inverter.

Noninverting amplifier.   Figure 2–15(a) shows a noninverting amplifier. A circuit equivalent to this one is shown in Figure 2–15(b). For the circuit of Figure 2–15(b), we have

$$e_o = K \left(e_i - \frac{R_1}{R_1 + R_2} e_o \right)$$

where K is the differential gain of the amplifier. From this last equation, we get

$$e_i = \left(\frac{R_1}{R_1 + R_2} + \frac{1}{K} \right) e_o$$

Since $K \gg 1$, if $R_1/(R_1 + R_2) \gg 1/K$, then

$$e_o = \left(1 + \frac{R_2}{R_1} \right) e_i$$

This equation gives the output voltage e_o. Since e_o and e_i have the same signs, the op-amp circuit shown in Figure 2–15(a) is noninverting.

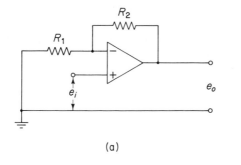

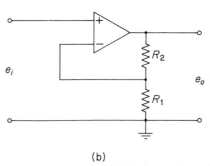

Figure 2–15
(a) Noninverting operational amplifier; (b) equivalent circuit.

(a)   (b)

EXAMPLE 2–4   Figure 2–16 shows an electrical circuit involving an operational amplifier. Obtain the output e_o.
Let us define

$$i_1 = \frac{e_i - e'}{R_1}, \qquad i_2 = C\frac{d(e' - e_o)}{dt}, \qquad i_3 = \frac{e' - e_o}{R_2}$$

Noting that the current flowing into the amplifier is negligible, we have

$$i_1 = i_2 + i_3$$

Hence

$$\frac{e_i - e'}{R_1} = C\frac{d(e' - e_o)}{dt} + \frac{e' - e_o}{R_2}$$

Since $e' \doteq 0$, we have

$$\frac{e_i}{R_1} = -C\frac{de_o}{dt} - \frac{e_o}{R_2}$$

Taking the Laplace transform of this last equation, assuming the zero initial condition, we have

$$\frac{E_i(s)}{R_1} = -\frac{R_2Cs + 1}{R_2}E_o(s)$$

which can be written as

$$\frac{E_o(s)}{E_i(s)} = -\frac{R_2}{R_1}\frac{1}{R_2Cs + 1}$$

The op-amp circuit shown in Figure 2–16 is a first-order lag circuit. (Several other circuits involving op amps are shown in Table 7–1 together with their transfer functions.)

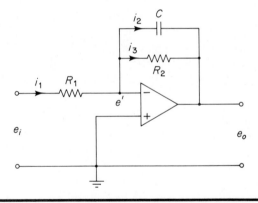

Figure 2–16
First-order lag circuit
using operational
amplifier.

Impedance approach for obtaining transfer functions.   Consider the op-amp circuit shown in Figure 2–17. Similar to the case of electrical circuits we discussed earlier, the impedance approach can be applied to op-amp circuits to obtain their transfer functions. For the circuit shown in Figure 2–17, we have

$$E_i(s) = Z_1(s)I(s), \qquad E_o(s) = -Z_2(s)I(s)$$

Hence, the transfer function for the circuit is obtained as

$$\frac{E_o(s)}{E_i(s)} = -\frac{Z_2(s)}{Z_1(s)}$$

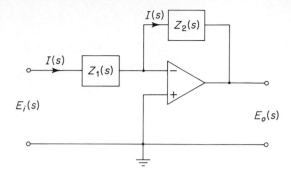

Figure 2–17
Operational amplifier circuit.

EXAMPLE 2–5

Referring to the op amp circuit shown in Figure 2–16, obtain the transfer function $E_o(s)/E_i(s)$ by use of the impedance approach.

The complex impedances $Z_1(s)$ and $Z_2(s)$ for this circuit are

$$Z_1(s) = R_1 \quad \text{and} \quad Z_2(s) = \frac{1}{Cs + \dfrac{1}{R_2}} = \frac{R_2}{R_2Cs + 1}$$

Hence, $E_i(s)$ and $E_o(s)$ are obtained as

$$E_i(s) = R_1 I(s), \qquad E_o(s) = -\frac{R_2}{R_2Cs + 1}I(s)$$

The transfer function $E_o(s)/E_i(s)$ is, therefore, obtained as

$$\frac{E_o(s)}{E_i(s)} = -\frac{R_2}{R_1}\frac{1}{R_2Cs + 1}$$

which is, of course, the same as that obtained in Example 2–4.

2–5 ANALOGOUS SYSTEMS

Systems that can be represented by the same mathematical model but that are different physically are called *analogous* systems. Thus analogous systems are described by the same differential or integrodifferential equations or set of equations.

The concept of analogous systems is very useful in practice for the following reasons.

1. The solution of the equation describing one physical system can be directly applied to analogous systems in any other field.
2. Since one type of system may be easier to handle experimentally than another, instead of building and studying a mechanical system (or hydraulic system or pneumatic system), we can build and study its electrical analog, for electrical or electronic systems are, in general, much easier to deal with experimentally.

This section presents analogies between mechanical and electrical systems. The concept of analogous systems, however, is applicable to other kinds of systems, and analogies among mechanical, electrical, hydraulic, pneumatic, thermal, and other systems may be established.

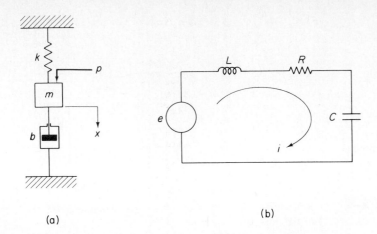

Figure 2–18
Analogous mechanical
and electrical systems.

(a)　　　　　　　　　　　　　　　(b)

Mechanical–electrical analogies. Mechanical systems can be studied through the use of their electrical analogs, which may be more easily constructed than models of the corresponding mechanical systems. There are two electrical analogies for mechanical systems: the force–voltage analogy and the force–current analogy.

Force–voltage analogy. Consider the mechanical system of Figure 2–18(a) and the electrical system of Figure 2–18(b). The system equation for the former is

$$m\frac{d^2x}{dt^2} + b\frac{dx}{dt} + kx = p \tag{2–31}$$

whereas the system equation for the latter is

$$L\frac{di}{dt} + Ri + \frac{1}{C}\int i\,dt = e$$

In terms of electric charge q, this last equation becomes

$$L\frac{d^2q}{dt^2} + R\frac{dq}{dt} + \frac{1}{C}q = e \tag{2–32}$$

Comparing Equations (2–31) and (2–32), we see that the differential equations for the two systems are of identical form. Thus these two systems are analogous systems. The terms that occupy corresponding positions in the differential equations are called *analogous quantities*, a list of which appears in Table 2–2. The analogy here is called the *force–voltage analogy* (or *mass–inductance analogy*).

Table 2–2 Force–Voltage Analogy

Mechanical Systems	Electrical Systems
Force p (torque T)	Voltage e
Mass m (moment of inertia J)	Inductance L
Viscous-friction coefficient b	Resistance R
Spring constant k	Reciprocal of capacitance, $1/C$
Displacement x (angular displacement θ)	Charge q
Velocity $\dot{x}$ (angular velocity $\dot{\theta}$)	Current i

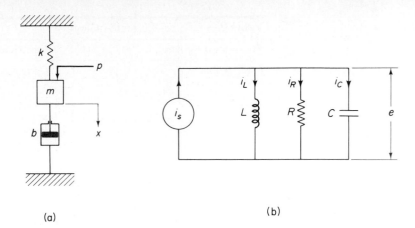

Figure 2–19
Analogous mechanical
and electrical systems.

(a)

(b)

Force–current analogy. Another analogy between electrical and mechanical systems is based on the force–current analogy. Consider the mechanical system shown in Figure 2–19(a). The system equation can be obtained as

$$m \frac{dx^2}{dt^2} + b \frac{dx}{dt} + kx = p \tag{2-33}$$

Consider next the electrical system shown in Figure 2–19(b). Application of Kirchhoff's current law gives

$$i_L + i_R + i_C = i_s \tag{2-34}$$

where

$$i_L = \frac{1}{L} \int e \, dt, \qquad i_R = \frac{e}{R}, \qquad i_C = C \frac{de}{dt}$$

Equation (2–34) can be written as

$$\frac{1}{L} \int e \, dt + \frac{e}{R} + C \frac{de}{dt} = i_s \tag{2-35}$$

Since the magnetic flux linkage ψ is related to voltage e by the equation

$$\frac{d\psi}{dt} = e$$

in terms of ψ, Equation (2–35) can be written as

$$C \frac{d^2\psi}{dt^2} + \frac{1}{R} \frac{d\psi}{dt} + \frac{1}{L} \psi = i_s \tag{2-36}$$

Comparing Equations (2–33) and (2–36), we find that the two systems are analogous. The analogous quantities are listed in Table 2–3. The analogy here is called the *force–current analogy* (or *mass–capacitance analogy*).

It should be remembered that analogies between two systems may break down if the regions of operation are extended too far. In other words, since the differential equations

Table 2–3 Force–Current Analogy

Mechanical Systems	Electrical Systems
Force p (torque T)	Current i
Mass m (moment of inertia J)	Capacitance C
Viscous-friction coefficient b	Reciprocal of resistance, $1/R$
Spring constant k	Reciprocal of inductance, $1/L$
Displacement x (angular displacement θ)	Magnetic flux linkage ψ
Velocity $\dot{x}$ (angular velocity $\dot{\theta}$)	Voltage e

upon which the analogies are based are only approximations to the dynamic characteristics of physical systems in a certain operating region, the analogy may break down if the operating region of one system is very wide. If the operating region of a given mechanical system is wide, however, it may be divided into two or more subregions, and analogous electrical systems may be built for each subregion. As a matter of fact, analogies are not limited to electrical systems and mechanical systems; they are applicable to any systems as long as their differential equations, or transfer functions, are of identical form.

2–6 ELECTROMECHANICAL SYSTEMS

The electromechanical systems that we shall discuss here are dc servomotors and two-phase servomotors. Conventional dc motors use mechanical brushes and commutators that require regular maintenance. Due to improvements that have been made in the brushes and commutators, however, many dc motors used in servo systems can be operated almost maintenance free. Some dc motors use electronic commutation. They are called brushless dc motors.

DC servomotors. There are many types of dc motors in use in industries. DC motors that are used in servo systems are called dc servomotors. In dc servomotors, the rotor inertias have been made very small, with the result that motors with very high torque-to-inertia ratios are commercially available. Some dc servomotors have extremely small time constants. DC servomotors with relatively small power ratings are used in instruments and computer-related equipment such as disk drives, tape drives, printers, and word processors. DC servomotors with medium and large power ratings are used in robot systems, numerically controlled milling machines, and so on.

In dc servomotors, the field windings may be connected in series with the armature or the field windings may be separate from the armature. (That is, the magnetic field is produced by a separate circuit.) In the latter case, where the field is excited separately, the magnetic flux is independent of the armature current. In some dc servomotors, the magnetic field is produced by a permanent magnet and, therefore, the magnetic flux is constant. Such dc servomotors are called permanent magnet dc servomotors. DC servomotors with separately excited fields, as well as permanent magnet dc servomotors, can be controlled by the armature current. Such a scheme to control the output of the dc servomotor by the armature current is called armature control of dc servomotors.

In the case where the armature current is maintained constant and the speed is controlled by the field voltage, the dc motor is called a field-controlled dc motor. (Some speed control systems use field-controlled dc motors.) The requirement of constant armature current is a serious disadvantage. (Providing a constant current source is much more difficult than providing a constant voltage source.) The time constants of the field-controlled dc motor are generally large compared with the time constants of a comparable armature-controlled dc motor.

A dc servomotor may also be driven by an electronic motion controller, frequently called a servodriver, as a motor–driver combination. The servodriver controls the motion of a dc servomotor and operates in various modes. Some of the features are point-to-point positioning, velocity profiling, and programmable acceleration. The use of an electronic motion controller using a pulse-width-modulated driver to control a dc servomotor is frequently used in robot control systems, numerical control systems, and other position and/or speed control systems.

In what follows we shall discuss armature control of dc servomotors and electronic motion control of dc servomotors.

Armature control of dc servomotors. Consider the armature-controlled dc servomotors shown in Figure 2–20(a), where the field current is held constant. In this system,

R_a = armature resistance, ohm
L_a = armature inductance, henry
i_a = armature current, ampere
i_f = field current, ampere
e_a = applied armature voltage, volt
e_b = back emf, volt
θ = angular displacement of the motor shaft, radian
T = torque developed by the motor, N-m
J = equivalent moment of inertia of the motor and load referred to the motor shaft, kg-m^2
b = equivalent viscous-friction coefficient of the motor and load referred to the motor shaft, N-m/rad/sec

The torque T developed by the motor is proportional to the product of the armature current

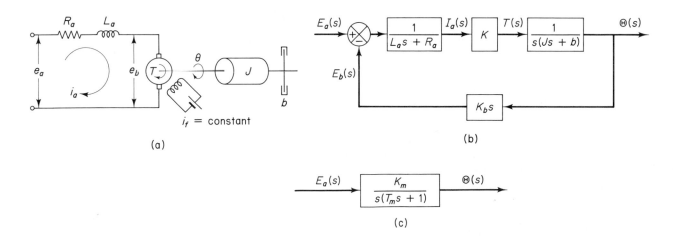

Figure 2–20 (a) Schematic diagram of armature-controlled dc motor; (b) block diagram obtained from Eqs. (2–40)–(2–42); (c) simplified block diagram.

i_a and the air gap flux ψ, which in turn is proportional to the field current, or

$$\psi = K_f i_f$$

where K_f is a constant. The torque T can therefore be written as

$$T = K_f i_f K_1 i_a$$

where K_1 is a constant.

Note that for a constant field current the flux becomes constant, and the torque becomes directly proportional to the armature current so that

$$T = K i_a$$

where K is a motor-torque constant. Notice that if the sign of the current i_a is reversed the sign of the torque T will be reversed, which will result in the reversion of the direction of rotor rotation.

When the armature is rotating, a voltage proportional to the product of the flux and angular velocity is induced in the armature. For a constant flux, the induced voltage e_b is directly proportional to the angular velocity $d\theta/dt$, or

$$e_b = K_b \frac{d\theta}{dt} \tag{2-37}$$

where e_b is the back emf and K_b is a back emf constant.

The speed of an armature-controlled dc servomotor is controlled by the armature voltage e_a. (The armature voltage e_a is the output of a power amplifier, which is not shown in the diagram.) The differential equation for the armature circuit is

$$L_a \frac{di_a}{dt} + R_a i_a + e_b = e_a \tag{2-38}$$

The armature current produces the torque that is applied to the inertia and friction; hence

$$J \frac{d^2\theta}{dt^2} + b \frac{d\theta}{dt} = T = K i_a \tag{2-39}$$

Assuming that all initial conditions are zero, and taking the Laplace transforms of Equations (2–37), (2–38), and (2–39), we obtain the following equations:

$$K_b s\Theta(s) = E_b(s) \tag{2-40}$$

$$(L_a s + R_a)I_a(s) + E_b(s) = E_a(s) \tag{2-41}$$

$$(Js^2 + bs)\Theta(s) = T(s) = KI_a(s) \tag{2-42}$$

Considering $E_a(s)$ as the input and $\Theta(s)$ as the output, it is possible to construct a block diagram from Equations (2–40), (2–41), and (2–42). [See Figure 2–20(b).] The armature-controlled dc servomotor is, itself, a feedback system. The effect of the back emf is seen to be the feedback signal proportional to the speed of the motor. This back emf thus increases the effective damping of the system. The transfer function for the dc servomotor considered here is obtained as

$$\frac{\Theta(s)}{E_a(s)} = \frac{K}{s[L_a Js^2 + (L_a b + R_a J)s + R_a b + KK_b]} \tag{2-43}$$

The inductance L_a in the armature circuit is usually small and may be neglected. If L_a is neglected, then the transfer function given by Equation (2–43) reduces to

$$\frac{\Theta(s)}{E_a(s)} = \frac{K_m}{s(T_m s + 1)} \tag{2–44}$$

where $K_m = K/(R_a b + KK_b)$ = motor gain constant
$T_m = R_a J/(R_a b + KK_b)$ = motor time constant

Figure 2–20(c) shows a simplified block diagram.

From Equations (2–43) and (2–44), it can be seen that the transfer functions involve the term $1/s$. Thus, this system possesses an integrating property. In Equation (2–44), notice that the time constant of the motor is smaller for a smaller R_a and smaller J. With small J, as the resistance R_a is reduced, the motor time constant approaches zero, and the motor acts as an ideal integrator.

State-space representation. A state-space model for the armature-controlled dc motor system just considered may be obtained as follows: First, notice that from Equation (2-44) the differential equation for the system is

$$\ddot{\theta} + \frac{1}{T_m} \dot{\theta} = \frac{K_m}{T_m} e_a$$

Define state variables x_1 and x_2 by

$$x_1 = \theta$$

$$x_2 = \dot{\theta}$$

the input variable u by

$$u = e_a$$

and the output variable y by

$$y = \theta = x_1$$

Then the state-space representation of the dc motor system is given by

$$\begin{bmatrix} \dot{x}_1 \\ \dot{x}_2 \end{bmatrix} = \begin{bmatrix} 0 & 1 \\ 0 & -\dfrac{1}{T_m} \end{bmatrix} \begin{bmatrix} x_1 \\ x_2 \end{bmatrix} + \begin{bmatrix} 0 \\ \dfrac{K_m}{T_m} \end{bmatrix} u$$

$$y = \begin{bmatrix} 1 & 0 \end{bmatrix} \begin{bmatrix} x_1 \\ x_2 \end{bmatrix}$$

Electronic motion control of dc servomotors. There are many different types of electronic motion controllers or servodrivers for servomotors. Most servodrivers are designed for speed control of dc servomotors. They improve the efficiency of operating servomotors. Figure 2–21(a) shows a block diagram of a high-speed, high-precision positional servo with speed control using a servodriver and servomotor combination. This servodriver is designed

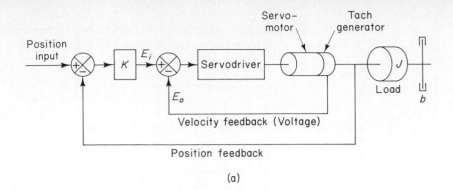

(a)

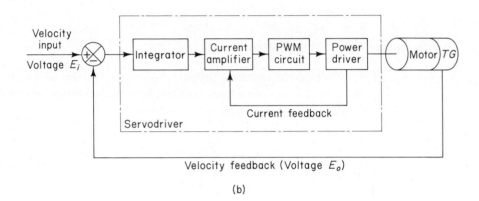

Figure 2–21
(a) High-speed, high-precision positional servo system with speed control using a servodriver, servomotor combination; (b) functional diagram of a servodriver.

(b)

to obtain motor speed proportional to the voltage E_i. Figure 2–21(b) shows a functional block diagram for the servodriver.

Comments. Figure 2–22(a) shows a basic configuration of a positional servo system. This configuration has been used in industry for many years. It represents a simple, low-cost positional servo system. Figure 2–22(b) shows a positional servo system with a servodriver. The system involves a velocity feedback loop. The integrator and gain shown as the transfer function of the servodriver are a very simplified version of the mathematical model of the servodriver. This system represents a high-speed, high-precision positional servo system. The positional servo systems of this type are used frequently in today's position control systems.

Two-phase servomotors. A two-phase servomotor, commonly used for instrument servos, is similar to a conventional two-phase induction motor except for the fact that the rotor has a small diameter-to-length ratio to minimize the moment of inertia and to obtain a good accelerating characteristic. The two-phase servomotor is very rugged and reliable. In many practical applications, the power range for which two-phase servomotors are used is between a fraction of a watt and a few hundred watts.

A schematic diagram of a two-phase servomotor is shown in Figure 2–23(a). Here one phase (fixed field) of the motor is continuously excited from the reference voltage, the frequency of which is usually 60, 400, or 1000 Hz; and the other phase (control field) is driven with the control voltage (a suppressed carrier signal), which is 90° phase-shifted in

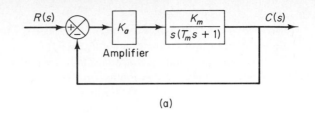

(a)

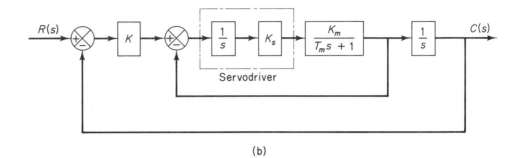

(b)

Figure 2–22
(a) Simple, low-cost positional servo system; (b) high-speed, high-precision positional servo system.

time with respect to the reference voltage. (The control voltage is of variable magnitude and polarity.)

Note that the voltage of the control phase is made 90° out of phase with respect to the voltage of the fixed phase. The stator windings for the fixed and control phases are placed 90° apart in space. These considerations are based on the fact that torque is produced most

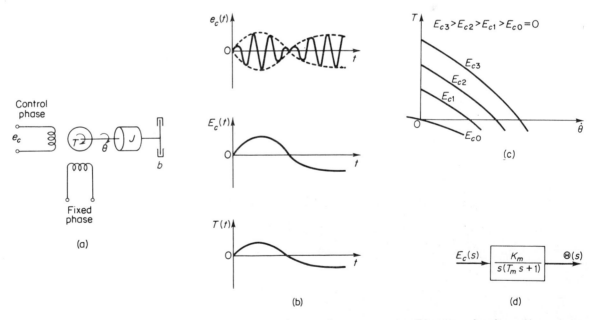

Figure 2–23 (a) Schematic diagram of a two-phase servomotor; (b) curves showing $e_c(t)$ versus t, $E_c(t)$ versus t, and $T(t)$ versus t; (c) torque–speed curves; (d) block diagram of a two-phase servomotor.

efficiently on a shaft when the phase-winding axes are in space quadrature and voltages in the two phases are in time quadrature.

The two stator windings are normally excited by a two-phase power supply. If a two-phase power supply is not available, however, then the fixed phase winding may be connected to a single-phase power supply through a capacitor, which will provide the 90° phase shift. The amplifier to which the control phase winding is connected is supplied from the same single-phase power supply.

In the two-phase servomotor, the polarity of the control voltage determines the direction of rotation. The instantaneous control voltage $e_c(t)$ is of the form

$$e_c(t) = E_c(t) \sin \omega t \qquad \text{for} \quad E_c(t) > 0$$

$$= |E_c(t)| \sin (\omega t + \pi) \qquad \text{for} \quad E_c(t) < 0$$

This means that a change in the sign of $E_c(t)$ shifts the phase by π radians. Thus, a change in the sign of the control voltage $E_c(t)$ reverses the direction of rotation of the motor. Since the reference voltage is constant, the torque T and angular speed $\dot{\theta}$ are also functions of the control voltage $E_c(t)$. If variations in $E_c(t)$ are slow compared with the ac supply frequency, the torque developed by the motor is proportional to $E_c(t)$. Figure 2–23(b) shows the curves $e_c(t)$ versus t, $E_c(t)$ versus t, and torque $T(t)$ versus t. The angular speed at steady state is proportional to the control voltage $E_c(t)$.

A family of torque-speed curves, when the rated voltage is applied to the fixed phase winding and various voltages are applied to the control phase winding, gives the steady-state characteristics of the two-phase servomotor. The transfer function of a two-phase servomotor may be obtained from such torque–speed curves if they are parallel and equidistant straight lines. Generally, the torque–speed curves are parallel for a relatively wide speed range but they may not be equidistant; that is, for a given speed, the torque may not vary linearly with respect to the control voltage. In a low-speed region, however, the torque–speed curves are usually straight lines and equidistant in a region of low control voltages. Since the two-phase servomotor seldom operates at high speeds, the linear portions of the torque–speed curves may be extended to the high-speed region. If the assumption is made that they are equidistant for all control voltages, then the servomotor may be considered linear.

Figure 2–23(c) shows a set of torque–speed curves for various values of control voltages. The torque–speed curve corresponding to zero control voltage passes through the origin. Since the slope of this curve is normally negative, if the control phase voltage becomes equal to zero, the motor develops that torque necessary to stop the rotation.

The servomotor provides a large torque at zero speed. This torque is necessary for rapid acceleration. From Figure 2–23(c), we see that the torque T generated is a function of the motor-shaft angular speed $\dot{\theta}$ and the control voltage E_c. The equation for any linearized torque–speed line is

$$T = -K_n\dot{\theta} + K_cE_c \tag{2-45}$$

where K_n and K_c are positive constants. The torque-balance equation for the two-phase servomotor is

$$T = J\ddot{\theta} + b\dot{\theta} \tag{2-46}$$

Where J is the moment of inertia of the motor and load referred to the motor shaft and b is the viscous-friction coefficient of the motor and load referred to the motor shaft. From Equations (2–45) and (2–46), we obtain the following equation:

$$J\ddot{\theta} + (b + K_n)\dot{\theta} = K_c E_c \qquad (2\text{–}47)$$

Noting that the control voltage E_c is the input and the displacement of the motor shaft is the output, we see from Equation (2–47) that the transfer function of the system is given by

$$\frac{\Theta(s)}{E_c(s)} = \frac{K_c}{Js^2 + (b + K_n)s} = \frac{K_m}{s(T_m s + 1)} \qquad (2\text{–}48)$$

where $k_m = K_c/(b + K_n) =$ motor gain constant

$\quad T_m = J/(b + K_n) =$ motor time constant

Figure 2–23(d) shows a block diagram for this system. From the transfer function of this system, we can see that $(b + K_n)s$ is a viscous-friction term produced by the motor and load. Thus, K_n, the negative of the slope of the torque–speed curve, together with b, defines the equivalent viscous friction of the motor and load combination. For steeper torque–speed curves, the damping of the motor is higher. If the rotor inertia is sufficiently low, then for most of the frequency range we have $|T_m s| \ll 1$ and the servomotor acts as an integrator.

The transfer function given by Equation (2–48) is based on the assumption that the servomotor is linear. In practice, however, it is not quite so. For torque–speed curves not quite parallel and equidistant, the value of K_n is not constant and, therefore, the values of K_m and T_m are also not constant; they vary with the control voltage.

Torque-to-inertia ratio. The maximum acceleration that the servomotor can achieve may be indicated by the torque-to-inertia ratio, which is the ratio of the maximum torque at standstill to the rotor inertia. The higher this ratio, the better the acceleration characteristic is.

Effect of load on servomotor dynamics. Most important among the characteristics of the servomotor is the maximum acceleration obtainable. For a given available torque, the rotor moment of inertia must be a minimum. Since the servomotor operates under continuously varying conditions, acceleration and deceleration of the rotor occur from time to time. The servomotor must be able to absorb mechanical energy as well as to generate it. The performance of the servomotor when used as a brake should be satisfactory.

Let J_m and b_m be, respectively, the moment of inertia and viscous-friction coefficient of the rotor, and let J_L and b_L be, respectively, the moment of inertia and viscous-friction coefficient of the load on the output shaft. Assume that the moment of inertia and viscous-friction coefficient of the gear train are either negligible or included in J_L and b_L, respectively. Then, the equivalent moment of inertia J_{eq} referred to the motor shaft and equivalent viscous–friction coefficient b_{eq} referred to the motor shaft can be written as (for details, see Problem A–2–6)

$$J_{eq} = J_m + n^2 J_L \qquad (n < 1)$$

$$b_{eq} = b_m + n^2 f_L \qquad (n < 1)$$

where n is the gear ratio between the motor and load. If the gear ratio n is small and $J_m \gg n^2 J_L$, then the moment of inertia of the load referred to the motor shaft is negligible with respect to the rotor moment of inertia. A similar argument applies to the load friction. In general, when the gear ratio n is small, the transfer function of the electric servomotor may be obtained without taking into account the load moment of inertia and friction. If neither J_m nor $n^2 J_L$ is negligibly small compared with the other, however, then the equivalent moment of inertia J_{eq} must be used for evaluating the transfer function of the motor–load combination.

2–7 LIQUID-LEVEL SYSTEMS

In analyzing systems involving fluid flow, we find it necessary to divide flow regimes into laminar flow and turbulent flow, according to the magnitude of the Reynolds number. If the Reynolds number is greater than about 3000 ~ 4000, then the flow is turbulent. The flow is laminar if the Reynolds number is less than about 2000. In the laminar case, fluid flow occurs in streamlines with no turbulence. Systems involving turbulent flow often have to be represented by nonlinear differential equations, while systems involving laminar flow may be represented by linear differential equations. (Industrial processes often involve flow of liquids through connecting pipes and tanks. The flow in such processes is often turbulent and not laminar.)

In this section we shall derive mathematical models of liquid-level systems. By introducing the concept of resistance and capacitance for such liquid-level systems, it is possible to describe the dynamic characteristics of such systems in simple forms.

Resistance and capacitance of liquid-level systems. Consider the flow through a short pipe connecting two tanks. The resistance for liquid flow in such a pipe or restriction is defined as the change in the level difference (the difference of the liquid levels of the two tanks) necessary to cause a unit change in flow rate; that is,

$$R = \frac{\text{change in level difference, m}}{\text{change in flow rate, m}^3/\text{sec}}$$

Since the relationship between the flow rate and level difference differs for the laminar flow and turbulent flow, we shall consider both cases in the following.

Consider the liquid-level system shown in Figure 2–24(a). In this system the liquid spouts through the load valve in the side of the tank. If the flow through this restriction is laminar, the relationship between the steady-state flow rate and steady-state head at the level of the restriction is given by

$$Q = KH$$

where Q = steady-state liquid flow rate, m³/sec
$\quad K$ = coefficient, m²/sec
$\quad H$ = steady-state head, m

Notice that the law governing laminar flow is analogous to Coulomb's law, which states that the current is directly proportional to the potential difference.

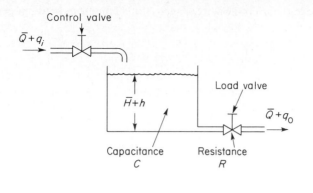

Figure 2–24
(a) Liquid-level system; (b) head versus flow rate curve.

(a)

(b)

For laminar flow, the resistance R_l is obtained as

$$R_l = \frac{dH}{dQ} = \frac{H}{Q} \qquad (2\text{–}49)$$

The laminar-flow resistance is constant and is analogous to the electrical resistance.

If the flow through the restriction is turbulent, the steady-state flow rate is given by

$$Q = K\sqrt{H} \qquad (2\text{–}50)$$

where Q = steady-state liquid flow rate, m³/sec
$\quad\quad K$ = coefficient, m^{2.5}/sec
$\quad\quad H$ = steady-state head, m

The resistance R_t for turbulent flow is obtained from

$$R_t = \frac{dH}{dQ}$$

Since from Equation (2–50) we obtain

$$dQ = \frac{K}{2\sqrt{H}}\, dH$$

we have

$$\frac{dH}{dQ} = \frac{2\sqrt{H}}{K} = \frac{2\sqrt{H}\,\sqrt{H}}{Q} = \frac{2H}{Q}$$

Thus,

$$R_t = \frac{2H}{Q} \qquad (2\text{–}51)$$

The value of the turbulent-flow resistance R_t depends upon the flow rate and the head. The value of R_t, however, may be considered constant if the changes in head and flow rate are small.

By use of the turbulent-flow resistance, the relationship between Q and H can be given by

$$Q = \frac{2H}{R_t}$$

Such linearization is valid, provided that changes in the head and flow rate from their respective steady-state values are small.

In many practical cases, the value of the coefficient K in Equation (2–50), which depends on the flow coefficient and the area of restriction, is not known. Then the resistance may be determined by plotting the head versus flow rate curve based on experimental data and measuring the slope of the curve at the operating condition. An example of such a plot is shown in Figure 2–24(b). In the figure, point P is the steady-state operating point. The tangent line to the curve at point P intersects the ordinate at point $(-\bar{H}, 0)$. Thus, the slope of this tangent line is $2\bar{H}/\bar{Q}$. Since the resistance R_t at the operating point P is given by $2\bar{H}/\bar{Q}$, the resistance R_t is the slope of the curve at the operating point.

Consider the operating condition in the neighborhood of point P. Define a small deviation of the head from the steady-state value as h and the corresponding small change of the flow rate as q. Then the slope of the curve at point P can be given by

$$\text{Slope of curve at point } P = \frac{h}{q} = \frac{2\bar{H}}{\bar{Q}} = R_t$$

The linear approximation is based on the fact that the actual curve does not differ much from its tangent line if the operating condition does not vary too much.

The capacitance C of a tank is defined to be the change in quantity of stored liquid necessary to cause a unit change in the potential (head). (The potential is the quantity that indicates the energy level of the system.)

$$C = \frac{\text{change in liquid stored, m}^3}{\text{change in head m}}$$

It should be noted that the capacity (m³) and the capacitance (m²) are different. The capacitance of the tank is equal to its cross-sectional area. If this is constant, the capacitance is constant for any head.

Liquid-level systems.   Consider the system shown in Figure 2-24(a). The variables are defined as follows:

$\bar{Q}$ = steady-state flow rate (before any change has occurred), m³/sec
q_i = small deviation of inflow rate from its steady-state value, m³/sec
q_o = small deviation of outflow rate from its steady-state value, m³/sec
$\bar{H}$ = steady-state head (before any change has occurred), m
h = small deviation of head from its steady-state value, m

As stated previously, a system can be considered linear if the flow is laminar. Even if the flow is turbulent, the system can be linearized if changes in the variables are kept small. Based on the assumption that the system is either linear or linearized, the differential equation of this system can be obtained as follows: Since the inflow minus outflow during the small time interval dt is equal to the additional amount stored in the tank, we see that

$$C \, dh = (q_i - q_o) \, dt$$

From the definition of resistance, the relationship between q_o and h is given by

$$q_o = \frac{h}{R}$$

The differential equation for this system for a constant value of R becomes

$$RC \frac{dh}{dt} + h = Rq_i \tag{2-52}$$

Note that RC is the time constant of the system. Taking the Laplace transforms of both sides of Equation (2–52), assuming the zero initial condition, we obtain

$$(RCs + 1)H(s) = RQ_i(s)$$

where

$$H(s) = \mathcal{L}[h] \qquad \text{and} \qquad Q_i(s) = \mathcal{L}[q_i]$$

If q_i is considered the input and h the output, the transfer function of the system is

$$\frac{H(s)}{Q_i(s)} = \frac{R}{RCs + 1}$$

If, however, q_o is taken as the output, the input being the same, then the transfer function is

$$\frac{Q_o(s)}{Q_i(s)} = \frac{1}{RCs + 1}$$

where we have used the relationship

$$Q_o(s) = \frac{1}{R} H(s)$$

Liquid-level systems with interaction. Consider the system shown in Figure 2–25. In this system, the two tanks interact. Thus the transfer function of the system is not the product of two first-order transfer functions.

In the following, we shall assume only small variations of the variables from the steady-state values. Using the symbols as defined in Figure 2–25, we can obtain the following equations for this system:

$$\frac{h_1 - h_2}{R_1} = q_1 \tag{2-53}$$

$$C_1 \frac{dh_1}{dt} = q - q_1 \tag{2-54}$$

$$\frac{h_2}{R_2} = q_2 \tag{2-55}$$

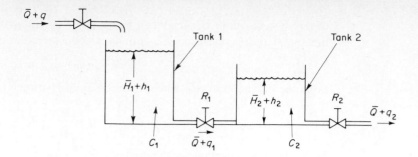

Figure 2–25
Liquid-level system
with interaction.

$\bar{Q}$: Steady-state flow rate
$\bar{H}_1$: Steady-state liquid level of tank 1
$\bar{H}_2$: Steady-state liquid level of tank 2

$$C_2 \frac{dh_2}{dt} = q_1 - q_2 \qquad (2\text{--}56)$$

If q is considered the input and q_2 the output, the transfer function of the system is

$$\frac{Q_2(s)}{Q(s)} = \frac{1}{R_1 C_1 R_2 C_2 s^2 + (R_1 C_1 + R_2 C_2 + R_2 C_1)s + 1} \qquad (2\text{--}57)$$

It is instructive to obtain Equation (2–57), the transfer function of the interacted system, by block diagram reduction. From Equations (2–53) through (2–56), we obtain elements of the block diagram, as shown in Figure 2–26(a). By connecting signals properly, we can construct a block diagram, as shown in Figure 2–26(b). By use of the rules of block diagram algebra given in Table 1–3, this block diagram can be simplified, as shown in Figure 2–26(c). Further simplification results in Figures 2–26(d) and (e). Figure 2–26(e) is equivalent to Equation (2–57).

Notice the similarity and difference between the transfer function given by Equation (2–57) and that given by Equation (2–30). The term $R_2 C_1 s$ that appears in the denominator of Equation (2–57) exemplifies the interaction between the two tanks. Similarly, the term $R_1 C_2 s$ in the denominator of Equation (2–30) represents the interaction between the two RC circuits shown in Figure 2–10.

State-space representation. A state-space representation for this system may be obtained as follows: Noting that the differential equation for this system is

$$R_1 C_1 R_2 C_2 \ddot{q}_2 + (R_1 C_1 + R_2 C_2 + R_2 C_1)\dot{q}_2 + q_2 = q$$

or

$$\ddot{q}_2 + \left(\frac{1}{R_2 C_2} + \frac{1}{R_1 C_1} + \frac{1}{R_1 C_2} \right)\dot{q}_2 + \frac{1}{R_1 C_1 R_2 C_2} q_2 = q$$

define state variables by

$$x_1 = q_2$$

$$x_2 = \dot{q}_2$$

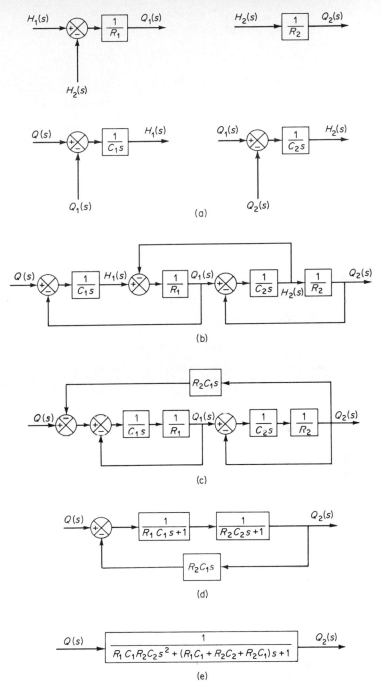

Figure 2–26 (a) Elements of the block diagram of the system shown in Fig. 2–25; (b) block diagram of the system; (c)–(e) successive reduction of the block diagram.

the input variable by

$$u = q$$

and the output variable by

$$y = q_2 = x_1$$

Then we obtain

$$\begin{bmatrix} \dot{x}_1 \\ \dot{x}_2 \end{bmatrix} = \begin{bmatrix} 0 & 1 \\ -\dfrac{1}{R_1C_1R_2C_2} & -\left(\dfrac{1}{R_2C_2} + \dfrac{1}{R_1C_1} + \dfrac{1}{R_1C_2}\right) \end{bmatrix} \begin{bmatrix} x_1 \\ x_2 \end{bmatrix} + \begin{bmatrix} 0 \\ 1 \end{bmatrix} u$$

$$y = \begin{bmatrix} 1 & 0 \end{bmatrix} \begin{bmatrix} x_1 \\ x_2 \end{bmatrix}$$

These two equations give a state-space representation of the system when q is considered as the input and q_2 is considered as the output.

If q is considered as the input and h_2 is considered as the output, then we get a different state-space representation. Let us define state variables by

$$x_1 = h_2$$

$$x_2 = h_1$$

the input variable u by

$$u = q$$

and the output variable y by

$$y = h_2$$

Then the corresponding state-space representation can be obtained as follows. From Equations (2–53) through (2–56), we obtain

$$C_2 \frac{dh_2}{dt} = \frac{h_1 - h_2}{R_1} - \frac{h_2}{R_2}$$

$$C_1 \frac{dh_1}{dt} = q - \frac{h_1 - h_2}{R_1}$$

or

$$\frac{dh_2}{dt} = -\left(\frac{1}{R_1C_2} + \frac{1}{R_2C_2}\right) h_2 + \frac{1}{R_1C_2} h_1$$

$$\frac{dh_1}{dt} = \frac{1}{R_1C_1} h_2 - \frac{1}{R_1C_1} h_1 + \frac{1}{C_1} q$$

The state-space representation now becomes

$$
\begin{bmatrix} \dot{x}_1 \\ \dot{x}_2 \end{bmatrix} = \begin{bmatrix} -\left(\dfrac{1}{R_1 C_2} + \dfrac{1}{R_2 C_2} \right) & \dfrac{1}{R_1 C_2} \\ \dfrac{1}{R_1 C_1} & -\dfrac{1}{R_1 C_1} \end{bmatrix} \begin{bmatrix} x_1 \\ x_2 \end{bmatrix} + \begin{bmatrix} 0 \\ \dfrac{1}{C_1} \end{bmatrix} u
$$

$$
y = \begin{bmatrix} 1 & 0 \end{bmatrix} \begin{bmatrix} x_1 \\ x_2 \end{bmatrix}
$$

Note that many different state-space representations for this system are possible.

2–8 THERMAL SYSTEMS

Thermal systems are those that involve the transfer of heat from one substance to another. Thermal systems may be analyzed in terms of resistance and capacitance, although the thermal capacitance and thermal resistance may not be represented accurately as lumped parameters since they are usually distributed throughout the substance. For precise analysis, distributed-parameter models must be used. Here, however, to simplify the analysis we shall assume that a thermal system can be represented by a lumped-parameter model, that substances that are characterized by resistance to heat flow have negligible heat capacitance, and that substances that are characterized by heat capacitance have negligible resistance to heat flow.

There are three different ways heat can flow from one substance to another: conduction, convection, and radiation.

For conduction or convection heat transfer,

$$
q = K \, \Delta\theta
$$

where q = heat flow rate, kcal/sec
$\Delta\theta$ = temperature difference, °C
K = coefficient, kcal/sec °C

The coefficient K is given by

$$
K = \frac{kA}{\Delta X} \qquad \text{for conduction}
$$

$$
= HA \qquad \text{for convection}
$$

where k = thermal conductivity, kcal/m sec °C
A = area normal to heat flow, m^2
ΔX = thickness of conductor, m
H = convection coefficient, kcal/m^2 sec °C

For radiation heat transfer, the heat flow is given by

$$
q = K_r(\theta_1^4 - \theta_2^4) \tag{2–58}
$$

where q = heat flow rate, kcal/sec
$\quad K_r$ = coefficient that depends on the emissivity, size, and configuration of the emanating surface and those of the receiving surface
$\quad \theta_1$ = absolute temperature of emitter, K
$\quad \theta_2$ = absolute temperature of receiver, K

Since the constant K_r is a very small number, radiation heat transfer is appreciable only if the temperature of the emitter is very high compared to that of the receiver, or $\theta_1 \gg \theta_2$. For such a case, Equation (2–58) may be approximated by

$$q = K_r \bar{\theta}^4 \qquad (2-59)$$

where $\bar{\theta}$ is an effective temperature difference of the emitter and receiver. The effective temperature difference $\bar{\theta}$ is given by

$$\bar{\theta} = \sqrt[4]{\theta_1^4 - \theta_2^4}$$

where $\theta_1 \gg \theta_2$.

Thermal resistance and thermal capacitance. The thermal resistance R for heat transfer between two substances may be defined as follows:

$$R = \frac{\text{change in temperature difference, °C}}{\text{change in heat flow rate, kcal/sec}}$$

The thermal resistance for conduction or convection heat transfer is given by

$$R = \frac{d(\Delta\theta)}{dq} = \frac{1}{K}$$

Since the thermal conductivity and convection coefficient are almost constant, the thermal resistance for either conduction or convection is constant. Referring to Equation (2–59), the thermal resistance for radiation heat transfer may be given by

$$R = \frac{d\bar{\theta}}{dq} = \frac{1}{4K_r\bar{\theta}^3}$$

where $\bar{\theta}$ is an effective temperature difference of the emitter and receiver. Note that the radiation resistance may be considered constant only for a small range of the operating condition.

The thermal capacitance C is defined by

$$C = \frac{\text{change in heat stored, kcal}}{\text{change in temperature, °C}}$$

or

$$C = Wc_p$$

where W = weight of substance considered, kg
$\quad c_p$ = specific heat of substance, kcal/kg °C

Thermal systems.   Consider the system shown in Figure. 2–27(a). It is assumed that the tank is insulated to eliminate heat loss to the surrounding air. It is also assumed that there is no heat storage in the insulation and that the liquid in the tank is perfectly mixed so that it is at a uniform temperature. Thus, a single temperature is used to describe the temperature of the liquid in the tank and of the outflowing liquid.

Let us define

$\bar{\Theta}_i$ = steady-state temperature of inflowing liquid, °C
$\bar{\Theta}_o$ = steady-state temperature of outflowing liquid, °C
G = steady-state liquid flow rate, kg/sec
M = mass of liquid in tank, kg
c = specific heat of liquid, kcal/kg °C
R = thermal resistance, °C sec/kcal
C = thermal capacitance, kcal/°C
$\bar{H}$ = steady-state heat input rate, kcal/sec

Assume that the temperature of the inflowing liquid is kept constant and that the heat input rate to the system (heat supplied by the heater) is suddenly changed from $\bar{H}$ to $\bar{H}$ + h_i, where h_i represents a small change in the heat input rate. The heat outflow rate will then change gradually from $\bar{H}$ to $\bar{H}$ + h_o. The temperature of the outflowing liquid will also be changed from $\bar{\Theta}_o$ to $\bar{\Theta}_o$ + θ. For this case, h_o, C, and R are obtained, respectively, as

$$h_o = Gc\theta$$

$$C = Mc$$

$$R = \frac{\theta}{h_o} = \frac{1}{Gc}$$

The differential equation for this system is

$$C\frac{d\theta}{dt} = h_i - h_o$$

which may be rewritten as

$$RC\frac{d\theta}{dt} + \theta = Rh_i$$

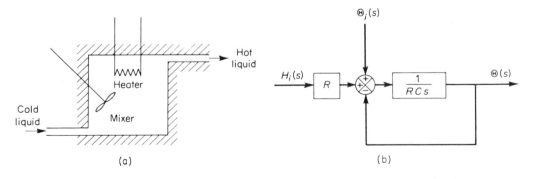

Figure 2–27
(a) Thermal system;
(b) block diagram of
the system.

Note that the time constant of the system is equal to RC or M/G seconds. The transfer function relating θ and h_i is given by

$$\frac{\Theta(s)}{H_i(s)} = \frac{R}{RCs + 1}$$

where $\Theta(s) = \mathcal{L}[\theta(t)]$ and $H_i(s) = \mathcal{L}[h_i(t)]$.

In practice, the temperature of the inflowing liquid may fluctuate and may act as a load disturbance. (If a constant outflow temperature is desired, an automatic controller may be installed to adjust the heat inflow rate to compensate for the fluctuations in the temperature of the inflowing liquid.) If the temperature of the inflowing liquid is suddenly changed from $\bar{\Theta}_i$ to $\bar{\Theta}_i + \theta_i$ while the heat input rate H and the liquid flow rate G are kept constant, then the heat outflow rate will be changed from $\bar{H}$ to $\bar{H} + h_o$, and the temperature of the outflowing liquid will be changed from $\bar{\Theta}_o$ to $\bar{\Theta}_o + \bar{\theta}$. The differential equation for this case is

$$C\frac{d\theta}{dt} = Gc\theta_i - h_o$$

which may be rewritten

$$RC\frac{d\theta}{dt} + \theta = \theta_i$$

The transfer function relating θ and θ_i is given by

$$\frac{\Theta(s)}{\Theta_i(s)} = \frac{1}{RCs + 1}$$

where $\Theta(s) = \mathcal{L}[\theta(t)]$ and $\Theta_i(s) = \mathcal{L}[\theta_i(t)]$.

If the present thermal system is subjected to changes in both the temperature of the inflowing liquid and the heat input rate, while the liquid flow rate is kept constant, the change θ in the temperature of the outflowing liquid can be given by the following equation:

$$RC\frac{d\theta}{dt} + \theta = \theta_i + Rh_i$$

A block diagram corresponding to this case is shown in Figure 2-27(b). (Notice that the system involves two inputs.)

State-space representation.   A state-space representation of this system can be obtained in the following way. Since

$$\frac{d\theta}{dt} = -\frac{1}{RC}\theta + \frac{1}{RC}\theta_i + \frac{1}{C}h_i$$

define the state variable x by

$$x = \theta$$

the input variables u_1 and u_2 by

$$u_1 = \theta_i$$

$$u_2 = h_i$$

and the output variable y by

$$y = \theta = x$$

Then we obtain

$$\dot{x} = -\frac{1}{RC}x + \begin{bmatrix} \dfrac{1}{RC} & \dfrac{1}{C} \end{bmatrix} \begin{bmatrix} u_1 \\ u_2 \end{bmatrix}$$

$$y = x$$

The last two equations represent a state-space model of the system.

2–9 ROBOT-ARM SYSTEMS

In this section we shall discuss robot-arm systems and their simulators. Robot-arm systems may have a few to several or more degrees of freedom depending on their configurations. A schematic diagram of a simple robot-arm system is shown in Figure 2–28. In a robot-arm system, if the grasping force of the robot hand is too small, the robot hand will drop the mechanical object, and if it is too great, the hand may damage or crush the object. Thus, robot hands must have touch-sensing devices and slip-sensing devices. In addition, the robot hands must have force-sensing devices. Semiconductor strain gauges are frequently used for force-sensing purposes. Such a strain gauge transforms force into a voltage that is proportional to deflection. For the slip-sensing device, one approach is to attach a roller device to the contact surface. Slipping is measured by the rotation angle of the roller.

Robot-arm systems are controlled by digital controllers. For satisfactory computer control, we must have an accurate mathematical model of the robot-arm system. Since most robot-arm systems involve many joints and extendable links, it is not a simple matter to determine which joints must be rotated and which links must be extended to bring the hand to a desired location and orientation. To compute the location and orientation of the hand, it is necessary to know the amounts of rotations and linear movements involved in the various components of the robot-arm system.

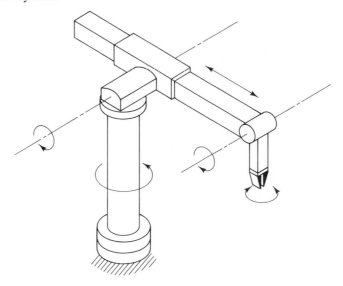

Figure 2–28
Robot-arm system.

Robot systems may be classified in many different ways. They may be classified according to the coordinate systems associated with them. That is, they may be classified into the rectangular coordinate robot systems, circular coordinate robot systems, polar coordinate robot systems, and multiple-joint-type robot systems. Schematic diagrams for each of these robot systems are shown in Table 2–4, together with the coordinate representation for point P, the center of grasp in the hand.

Table 2–4 Types of Robot Systems

Type	Configuration	Coordinates of Point P
Rectangular coordinate robot system		x, y, z
Circular coordinate robot system		$r \cos \theta, r \sin \theta, z$
Polar coordinate robot system		$r \cos \theta \cos \phi, r \sin \theta \cos \phi, r \sin \phi$
Multiple-joint-type robot system		$(r_1 \cos \phi_1 + r_2 \cos \phi_2) \cos \theta,$ $(r_1 \cos \phi_1 + r_2 \cos \phi_2) \sin \theta,$ $(r_1 \sin \phi_1 + r_2 \sin \phi_2)$

Robot systems may also be classified according to the nature of the actuator as the electric type, hydraulic type, or pneumatic type. The electric type, using dc motors and step motors, is easy to handle. The hydraulic type is used where large power is required. The pneumatic type may be used for linear motion, but because of the difficulty associated with pneumatic systems for precise position control, stoppers are generally required.

Robot systems may also be classified according to their functions, such as playback robots, numerical control robots, and intelligence robots.

Computing the location and orientation of various points of the robot-arm system. For computing the location and orientation of various points of the robot-arm system, it is a common practice to use coordinate transformation techniques. In what follows, we shall consider only a simple robot-arm simulator involving coordinate transformations.

Robot-arm simulator. The robot-arm simulator is a computer simulator of the motion of a robot-arm system, given the rotational angles of joints and the linear displacements of extendable links.

Consider the simple robot-arm system shown in Figure 2–29. Assume that the robot arms move only in the plane of the page. The role of the robot-arm simulator is to find the motion of each point of the arm from the knowledge of rotational angles θ and θ'. (In the present robot-arm system, the lengths of links are constant.)

From the geometrical consideration, the x, y, z coordinates of point P can be found as

$$x = a \cos \theta + b \cos(\theta + \theta')$$

$$y = a \sin \theta + b \sin(\theta + \theta')$$

$$z = 0$$

Writing and calculating such coordinates of many different points in a robot-arm system are time consuming. In the robot-arm simulator, instead of geometically determining the coordinates of many points, it uses transformation matrices to simplify the determination of the coordinates of many points.

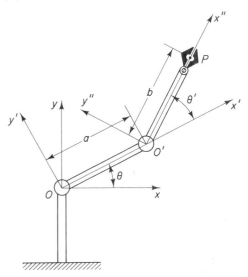

Figure 2–29
Simple robot-arm system.

Coordinate transformation. Consider the coordinate systems shown in Figure 2–30. Assume that the O–xyz coordinate system is fixed in space. The origin of this coordinate system is point O. The O–$x'y'z'$ coordinate system is a rotating coordinate system having its origin at point O.

Define angles between the axes of the fixed coordinate system and the axes of the rotating coordinate system as follows: Define the angle between the x axis and x' axis as $\theta_{x'x}$, the angle between the y axis and y' axis as $\theta_{y'y}$, the angle between the z axis and z' axis as $\theta_{z'z}$; the angle between the x axis and y' axis as $\theta_{y'x}$, the angle between the x axis and z' axis as $\theta_{z'x}$; the angle between the y axis and x' axis as $\theta_{x'y}$, the angle between the y axis and z' axis as $\theta_{z'y}$; the angle between the z axis and x' axis as $\theta_{x'z}$, and the angle between the z axis and y' axis as $\theta_{y'z}$.

Consider point P, whose coordinates in the rotating coordinate system is (x', y', z'). Define the coordinates of point P in the fixed coordinate system as (x, y, z). Once the coordinates (x, y, z) of point P are found, the equation of motion of point P in the fixed coordinate system can be derived. (To determine the motion of a point in the fixed space, the equations of motion must be written in terms of coordinates in the fixed coordinate system.)

Now define unit vectors along the x, y, and z axes as $\mathbf{i}$, $\mathbf{j}$, and $\mathbf{k}$; and unit vectors along the x', y', and z' axes as $\mathbf{i'}$, $\mathbf{j'}$, and $\mathbf{k'}$. Notice that point P is the tip of vector $\overrightarrow{OP}$. The vector $\overrightarrow{OP}$ can be written in two different ways:

$$\overrightarrow{OP} = x\mathbf{i} + y\mathbf{j} + z\mathbf{k} = [\mathbf{i} \quad \mathbf{j} \quad \mathbf{k}] \begin{bmatrix} x \\ y \\ z \end{bmatrix} \tag{2–60}$$

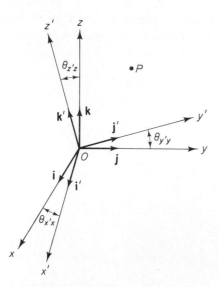

Figure 2–30
Fixed coordinate system and rotating coordinate system.

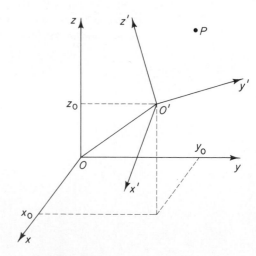

Figure 2–31
Fixed coordinate system and moving coordinate system.

and

$$\overrightarrow{OP} = x'\mathbf{i}' + y'\mathbf{j}' + z'\mathbf{k}' = [\mathbf{i}' \quad \mathbf{j}' \quad \mathbf{k}'] \begin{bmatrix} x' \\ y' \\ z' \end{bmatrix} \qquad (2\text{-}61)$$

Since the unit vectors $(\mathbf{i}, \mathbf{j}, \mathbf{k})$ and $(\mathbf{i}', \mathbf{j}', \mathbf{k}')$ are related by

$$\begin{bmatrix} \mathbf{i}' \\ \mathbf{j}' \\ \mathbf{k}' \end{bmatrix} = \begin{bmatrix} \cos\theta_{x'x} & \cos\theta_{x'y} & \cos\theta_{x'z} \\ \cos\theta_{y'x} & \cos\theta_{y'y} & \cos\theta_{y'z} \\ \cos\theta_{z'x} & \cos\theta_{z'y} & \cos\theta_{z'z} \end{bmatrix} \begin{bmatrix} \mathbf{i} \\ \mathbf{j} \\ \mathbf{k} \end{bmatrix} \qquad (2\text{-}62)$$

[see Problem A–2–20 for the derivation of Equation (2–62)], by taking the transpose of Equation (2–62) and substituting it into Equation (2–61), we obtain

$$\overrightarrow{OP} = [\mathbf{i}' \quad \mathbf{j}' \quad \mathbf{k}'] \begin{bmatrix} x' \\ y' \\ z' \end{bmatrix} = [\mathbf{i} \quad \mathbf{j} \quad \mathbf{k}] \begin{bmatrix} \cos\theta_{x'x} & \cos\theta_{y'x} & \cos\theta_{z'x} \\ \cos\theta_{x'y} & \cos\theta_{y'y} & \cos\theta_{z'y} \\ \cos\theta_{x'z} & \cos\theta_{y'z} & \cos\theta_{z'z} \end{bmatrix} \begin{bmatrix} x' \\ y' \\ z' \end{bmatrix} \qquad (2\text{-}63)$$

Equating Equations (2–60) and (2–63), we obtain

$$\begin{bmatrix} x \\ y \\ z \end{bmatrix} = \begin{bmatrix} \cos\theta_{x'x} & \cos\theta_{y'x} & \cos\theta_{z'x} \\ \cos\theta_{x'y} & \cos\theta_{y'y} & \cos\theta_{z'y} \\ \cos\theta_{x'z} & \cos\theta_{y'z} & \cos\theta_{z'z} \end{bmatrix} \begin{bmatrix} x' \\ y' \\ z' \end{bmatrix} \qquad (2\text{-}64)$$

Equation (2–64) enables us to find the (x, y, z) coordinates of any point (x', y', z') in the rotating coordinate system.

Thus far we did not consider the motion of the origin of the rotating coordinate system. Next we consider the case where the origin of the rotating coordinate system moves in the fixed coordinate system. Referring to Figure 2–31, assume that the origin of the rotating coordinate system is (x_0, y_0, z_0). Assume that point P is a fixed point in the $O'\text{-}x'y'z'$ coordinate system. The coordinates of point P in the $O'\text{-}x'y'z'$ coordinate system is (x', y', z'). Assume also that the coordinates of point P in the fixed $O\text{-}xyz$ coordinate system is (x, y, z). Then the (x, y, z) coordinates of point P can be obtained by adding to the right side of Equation (2–64) the coordinates (x_0, y_0, z_0) of the origin of the rotating coordinate system, or

$$\begin{bmatrix} x \\ y \\ z \end{bmatrix} = \begin{bmatrix} \cos\theta_{x'x} & \cos\theta_{y'x} & \cos\theta_{z'x} \\ \cos\theta_{x'y} & \cos\theta_{y'y} & \cos\theta_{z'y} \\ \cos\theta_{x'z} & \cos\theta_{y'z} & \cos\theta_{z'z} \end{bmatrix} \begin{bmatrix} x' \\ y' \\ z' \end{bmatrix} + \begin{bmatrix} x_0 \\ y_0 \\ z_0 \end{bmatrix} \qquad (2\text{-}65)$$

Equation (2–65) can be modified to

$$\begin{bmatrix} x \\ y \\ z \\ 1 \end{bmatrix} = \begin{bmatrix} \cos\theta_{x'x} & \cos\theta_{y'x} & \cos\theta_{z'x} & x_0 \\ \cos\theta_{x'y} & \cos\theta_{y'y} & \cos\theta_{z'y} & y_0 \\ \cos\theta_{x'z} & \cos\theta_{y'z} & \cos\theta_{z'z} & z_0 \\ 0 & 0 & 0 & 1 \end{bmatrix} \begin{bmatrix} x' \\ y' \\ x' \\ 1 \end{bmatrix} \qquad (2\text{-}66)$$

Noting that point O' is a movable point; the $O'\text{-}x'y'z'$ coordinate system can translate and rotate in the fixed $O\text{-}xyz$ coordinate system. We shall call the $O'\text{-}x'y'z'$ coordinate system as a moving coordinate system. Equation (2–66) gives the coordinates (x,y,z) in the fixed coordinate system of any point in the moving coordinate system. The 4×4 matrix in Equation (2–66) is called the transformation matrix.

In what follows, we shall obtain the transformation matrices for the robot-arm simulator for the robot-arm system shown in Figure 2–29. Referring to Figure 2–29, point P has the coordinates $(b, 0, 0)$ in the $O'-x''y''z''$ coordinate system. (The z'' axis is perpendicular to the plane of page and is not shown in Figure 2–29.) The $O-x'y'z'$ coordinate system and the $O'-x''y''z''$ coordinate system are related by

$$\begin{bmatrix} x' \\ y' \\ z' \\ 1 \end{bmatrix} = \begin{bmatrix} \cos\theta' & -\sin\theta' & 0 & a \\ \sin\theta' & \cos\theta' & 0 & 0 \\ 0 & 0 & 1 & 0 \\ 0 & 0 & 0 & 1 \end{bmatrix} \begin{bmatrix} x'' \\ y'' \\ z'' \\ 1 \end{bmatrix} \qquad (2\text{–}67)$$

(For the derivation of this transformation matrix, see Problem A–2–21.) Similarly, the $O-xyz$ coordinate system and $O-x'y'z'$ coordinate system are related by

$$\begin{bmatrix} x \\ y \\ z \\ 1 \end{bmatrix} = \begin{bmatrix} \cos\theta & -\sin\theta & 0 & 0 \\ \sin\theta & \cos\theta & 0 & 0 \\ 0 & 0 & 1 & 0 \\ 0 & 0 & 0 & 1 \end{bmatrix} \begin{bmatrix} x' \\ y' \\ z' \\ 1 \end{bmatrix} \qquad (2\text{–}68)$$

Hence, the (x, y, z) coordinates of any point in the $O'-x''y''z''$ coordinate system can be obtained from

$$\begin{bmatrix} x \\ y \\ z \\ 1 \end{bmatrix} = \begin{bmatrix} \cos\theta & -\sin\theta & 0 & 0 \\ \sin\theta & \cos\theta & 0 & 0 \\ 0 & 0 & 1 & 0 \\ 0 & 0 & 0 & 1 \end{bmatrix} \begin{bmatrix} \cos\theta' & -\sin\theta' & 0 & a \\ \sin\theta' & \cos\theta' & 0 & 0 \\ 0 & 0 & 1 & 0 \\ 0 & 0 & 0 & 1 \end{bmatrix} \begin{bmatrix} x'' \\ y'' \\ z'' \\ 1 \end{bmatrix}$$

$$= \begin{bmatrix} \cos(\theta + \theta') & -\sin(\theta + \theta') & 0 & a\cos\theta \\ \sin(\theta + \theta') & \cos(\theta + \theta') & 0 & a\sin\theta \\ 0 & 0 & 1 & 0 \\ 0 & 0 & 0 & 1 \end{bmatrix} \begin{bmatrix} x'' \\ y'' \\ z'' \\ 1 \end{bmatrix} \qquad (2\text{–}69)$$

Now we shall find the (x, y, z) coordinates of point P. The coordinates of point P in the $O'-x''z''z''$ coordinate system are $(b, 0, 0)$. By substituting these coordinates into Equation (2–69), we obtain

$$\begin{bmatrix} x \\ y \\ z \\ 1 \end{bmatrix} = \begin{bmatrix} \cos(\theta + \theta') & -\sin(\theta + \theta') & 0 & a\cos\theta \\ \sin(\theta + \theta') & \cos(\theta + \theta') & 0 & a\sin\theta \\ 0 & 0 & 1 & 0 \\ 0 & 0 & 0 & 1 \end{bmatrix} \begin{bmatrix} b \\ 0 \\ 0 \\ 1 \end{bmatrix}$$

$$= \begin{bmatrix} b\cos(\theta + \theta') + a\cos\theta \\ b\sin(\theta + \theta') + a\sin\theta \\ 0 \\ 1 \end{bmatrix}$$

The result here agrees with what we obtained earlier from geometrical considerations.

The robot-arm simulator for the two-dimensional robot-arm system shown in Figure 2–29 consists of two transformation matrices. By programming transformation matrices given by Equations (2–67) and (2–68) on the computer, the motion of point P, as input angles θ and θ' are varied, can be realized on the computer screen.

If a robot-arm system involves several joints and extendable links, the coordinates (x, y, z) of any point in any moving coordinate system can be obtained by multiplying a series of transformation matrices. Note that each transformation matrix will involve one angle, which relates the rotation between the two coordinate systems, and one linear translation, which relates the translation or displacement of the origin of the one coordinate system relative to the origin of the other coordinate system. (If the origins of the two coordinate systems coincide, then the translation term is zero.) If two transformation matrices are multiplied, then the resulting transformation matrix will involve two angles and two translations. (One or both translation terms may be zero.)

2–10 LINEARIZATION OF NONLINEAR MATHEMATICAL MODELS

In this section we present a linearization technique that is applicable to many nonlinear systems. The process of linearizing nonlinear systems is important, for by linearizing nonlinear equations, it is possible to apply numerous linear analysis methods that will produce information on the behavior of nonlinear systems. The linearization procedure presented here is based on the expansion of the nonlinear function into a Taylor series about the operating point and the retention of only the linear term. Because we neglect higher-order terms of Taylor series expansion, these neglected terms must be small enough; that is, the variables deviate only slightly from the operating condition.

In what follows we shall first present mathematical aspects of the linearization technique and then apply the technique to a liquid-level system and a hydraulic servo system to obtain linear models for both systems. (Linearization of nonlinear elements in the frequency domain is treated in Chapter 8.)

Linear approximation of nonlinear mathematical models. To obtain a linear mathematical model for a nonlinear system, we assume that the variables deviate only slightly from some operating condition. Consider a system whose input is $x(t)$ and output is $y(t)$. The relationship between $y(t)$ and $x(t)$ is given by

$$y = f(x) \qquad (2\text{–}70)$$

If the normal operating condition corresponds to $\bar{x}, \bar{y}$, then Equation (2–70) may be expanded into a Taylor series about this point as follows:

$$
\begin{aligned}
y &= f(x) \\
&= f(\bar{x}) + \frac{df}{dx}(x - \bar{x}) + \frac{1}{2!}\frac{d^2f}{dx^2}(x - \bar{x})^2 + \cdots
\end{aligned}
\qquad (2\text{–}71)
$$

where the derivatives df/dx, d^2f/dx^2, . . . are evaluated at $x = \bar{x}$. If the variation $x - \bar{x}$ is small, we may neglect the higher-order terms in $x - \bar{x}$. Then Equation (2–71) may be written

$$y = \bar{y} + K(x - \bar{x}) \qquad (2\text{–}72)$$

where

$$\bar{y} = f(\bar{x})$$

$$K = \left. \frac{df}{dx} \right|_{x = \bar{x}}$$

Equation (2–72) may be rewritten as

$$y - \bar{y} = K(x - \bar{x}) \qquad (2\text{–}73)$$

which indicates that $y - \bar{y}$ is proportional to $x - \bar{x}$. Equation (2–73) gives a linear mathematical model for the nonlinear system given by Eq. (2–70) near the operating point $x = \bar{x}$, $y = \bar{y}$.

Next, consider a nonlinear system whose output y is a function of two inputs x_1 and x_2, so that

$$y = f(x_1, x_2) \qquad (2\text{–}74)$$

To obtain a linear approximation to this nonlinear system, we may expand Equation (2–74) into a Taylor series about the normal operating point $\bar{x}_1$, $\bar{x}_2$. Then Equation (2–74) becomes

$$
\begin{aligned}
y = f(\bar{x}_1, \bar{x}_2) &+ \left[\frac{\partial f}{\partial x_1}(x_1 - \bar{x}_1) + \frac{\partial f}{\partial x_2}(x_2 - \bar{x}_2) \right] \\
&+ \frac{1}{2!} \left[\frac{\partial^2 f}{\partial x_1^2}(x_1 - \bar{x}_1)^2 + 2 \frac{\partial^2 f}{\partial x_1 \, \partial x_2}(x_1 - \bar{x}_1)(x_2 - \bar{x}_2) \right. \\
&+ \left. \frac{\partial^2 f}{\partial x_2^2}(x_2 - \bar{x}_2)^2 \right] + \cdots
\end{aligned}
$$

where the partial derivatives are evaluated at $x_1 = \bar{x}_1$, $x_2 = \bar{x}_2$. Near the normal operating point, the higher-order terms may be neglected. The linear mathematical model of this nonlinear system in the neighborhood of the normal operating condition is then given by

$$y - \bar{y} = K_1(x_1 - \bar{x}_1) + K_2(x_2 - \bar{x}_2)$$

where

$$\bar{y} = f(\bar{x}_1, \bar{x}_2)$$

$$K_1 = \left. \frac{\partial f}{\partial x_1} \right|_{x_1 = \bar{x}_1, x_2 = \bar{x}_2}$$

$$K_2 = \left. \frac{\partial f}{\partial x_2} \right|_{x_1 = \bar{x}_1, x_2 = \bar{x}_2}$$

The linearization technique presented here is valid in the vicinity of the operating condition. If the operating conditions vary widely, however, such linearized equations are not adequate, and nonlinear equations must be dealt with. It is important to remember that a particular mathematical model used in analysis and design may accurately represent the dynamics of an actual system for certain operating conditions, but may not be accurate for other operating conditions.

Linearization of liquid-level system. Consider the liquid-level system shown in Figure 2–32. At steady state the inflow rate is $Q_i = \bar{Q}$, the outflow rate is $Q_o = \bar{Q}$, and head is $H = \bar{H}$. If the flow is turbulent, then we have

$$\bar{Q} = K\sqrt{\bar{H}}$$

Assume that at $t = 0$ the inflow rate is changed from $Q_i = \bar{Q}$ to $Q_i = \bar{Q} + q_i$. This change causes the head to change from $H = \bar{H}$ to $H = \bar{H} + h$, which, in turn, causes the outflow rate to change from $Q_o = \bar{Q}$ to $Q_o = \bar{Q} + q_o$. For this system we have

$$C\frac{dH}{dt} = Q_i - Q_o = Q_i - K\sqrt{H}$$

where C is the capacitance of the tank. Let us define

$$\frac{dH}{dt} = f(H, Q_i) = \frac{1}{C}Q_i - \frac{K\sqrt{H}}{C} \qquad (2\text{--}75)$$

Note that the steady-state operating condition is $(\bar{H}, \bar{Q})$ and $H = \bar{H} + h$, $Q_i = \bar{Q} + q_i$. Since at steady-state operation $dH/dt = 0$, we have $f(\bar{H}, \bar{Q}) = 0$.

Using the linearization technique just presented, a linearized equation for Equation (2–75) is

$$\frac{dH}{dt} - f(\bar{H}, \bar{Q}_i) = \frac{\partial f}{\partial H}(H - \bar{H}) + \frac{\partial f}{\partial Q_i}(Q_i - \bar{Q}_i) \qquad (2\text{--}76)$$

where

$$f(\bar{H}, \bar{Q}_i) = 0$$

$$\left. \frac{\partial f}{\partial H} \right|_{H=\bar{H}, Q_i=\bar{Q}} = -\frac{K}{2C\sqrt{\bar{H}}} = -\frac{\bar{Q}}{\sqrt{\bar{H}}}\frac{1}{2C\sqrt{\bar{H}}} = -\frac{\bar{Q}}{2C\bar{H}} = -\frac{1}{RC}$$

where we used the resistance R defined by

$$R = \frac{2\bar{H}}{\bar{Q}}$$

Also,

$$\left. \frac{\partial f}{\partial Q_i} \right|_{H=\bar{H}, Q_i=\bar{Q}} = \frac{1}{C}$$

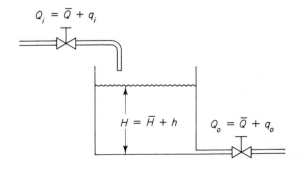

Figure 2–32
Liquid-level system.

Then Equation (2–76) can be written as

$$\frac{dH}{dt} = -\frac{1}{RC}(H - \bar{H}) + \frac{1}{C}(Q_i - \bar{Q}_i) \qquad (2\text{–}77)$$

Since $H - \bar{H} = h$ and $Q_i - \bar{Q}_i = q_i$, Equation (2–77) can be written as

$$\frac{dh}{dt} = -\frac{1}{RC}h + \frac{1}{C}q_i$$

or

$$RC\frac{dh}{dt} + h = Rq_i$$

which is the linearized equation for the liquid-level system and is the same as Equation (2–52) we obtained in Section 2–7.

Linearization of a hydraulic servo system. Figure 2–33 shows a hydraulic servomotor. It is essentially a pilot-valve-controlled hydraulic power amplifier and actuator. The pilot valve is a balanced valve, in the sense that the pressure forces acting on it are all balanced. A very large power output can be controlled by a pilot valve, which can be positioned with very little power.

In practice, ports a and b shown in Figure 2–33 are often made wider than the corresponding valves A and B. In this case, there is always leakage through the valve. This improves both the sensitivity and the linearity of the hydraulic servomotor. We shall make this assumption in the following analysis. [Note that sometimes a dither signal, a high-frequency signal of very small amplitude (with respect to the maximum displacement of the valve), is superimposed on the motion of the pilot valve. This also improves the sensitivity and linearity. In this case also there is leakage through the valve.]

Let us define

$$Q = \text{rate of flow of oil to the power cylinder}$$
$$\Delta P = P_1 - P_2 = \text{pressure difference across the power piston}$$
$$x = \text{displacement of pilot valve}$$

In Figure 2–33, we can see that Q is a function of x and ΔP. In general, the relationship among the variables Q, x, and ΔP is given by a nonlinear equation:

$$Q = f(x, \Delta P)$$

Linearizing this nonlinear equation near the normal operating point $\bar{Q}$, $\bar{x}$, $\Delta \bar{P}$, we obtain

$$Q - \bar{Q} = \frac{\partial f}{\partial x}(x - \bar{x}) + \frac{\partial f}{\partial \Delta P}(\Delta P - \Delta \bar{P}) \qquad (2\text{–}78)$$

where the partial derivatives are evaluated at $x = \bar{x}$, $\Delta P = \Delta \bar{P}$, and $\bar{Q} = f(\bar{x}, \Delta \bar{P})$. Define

$$K_1 = \left.\frac{\partial f}{\partial x}\right|_{x=\bar{x},\Delta P=\Delta \bar{P}} > 0$$

$$K_2 = -\left.\frac{\partial f}{\partial \Delta P}\right|_{x=\bar{x},\Delta P=\Delta \bar{P}} > 0$$

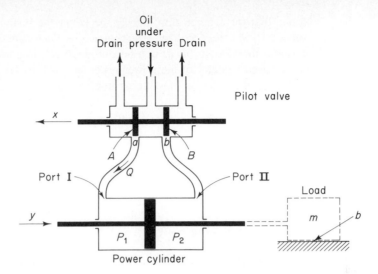

Figure 2–33
Schematic diagram of
a hydraulic
servomotor.

Since the normal operating condition corresponds to $\bar{Q} = 0$, $\bar{x} = 0$, and $\Delta\bar{P} = 0$, we obtain from Equation (2–78)

$$Q = K_1 x - K_2 \Delta P \qquad (2\text{–}79)$$

Figure 2–34 shows this linearized relationship among Q, x, and ΔP. The straight lines shown are the characteristic curves of the linearized hydraulic servomotor. This family of curves consists of equidistant parallel straight lines, parameterized by x. Note that the region near the origin is most important because system operation usually occurs near this point. Such a linearized mathematical model is useful in analyzing the performance of hydraulic control valves.

Referring to Figure 2–33, we see that the rate of flow of oil Q times dt is equal to the power piston displacement dy times the piston area A times the density of oil ρ. Thus, we obtain

$$A\,\rho\,dy = Q\,dt$$

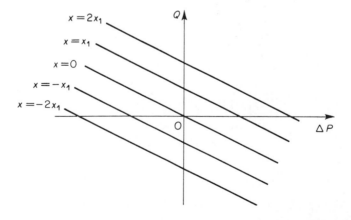

Figure 2–34
Characteristic curves
of the linearized
hydraulic servomotor.

Notice that for a given flow rate Q the larger the piston area A is, the lower will be the velocity dy/dt. Hence, if the piston area A is made smaller, the other variables remaining constant, the velocity dy/dt will become higher. Also, an increased flow rate Q will cause an increased velocity of the power piston and will make the response time shorter.

Equation (2–79) can now be written as

$$\Delta P = \frac{1}{K_2}\left(K_1 x - A\rho\frac{dy}{dt}\right)$$

The force developed by the power piston is equal to the pressure difference ΔP times the piston area A or

$$\text{Force developed by the power piston} = A\,\Delta P$$
$$= \frac{A}{K_2}\left(K_1 x - A\,\rho\frac{dy}{dt}\right)$$

For a given maximum force, if the pressure difference is sufficiently high, the piston area, or the volume of oil in the cylinder, can be made small. Consequently, to minimize the weight of the controller, we must make the supply pressure sufficiently high.

Assume that the power piston moves a load consisting of a mass and viscous friction. Then the force developed by the power piston is applied to the load mass and friction, and we obtain

$$m\ddot{y} + b\dot{y} = \frac{A}{K_2}(K_1 x - A\rho\dot{y})$$

or

$$m\ddot{y} + \left(b + \frac{A^2\rho}{K_2}\right)\dot{y} = \frac{AK_1}{K_2}x \tag{2–80}$$

where m is the mass of the load and b is the viscous-friction coefficient.

Assuming that the pilot valve displacement x is the input and the power piston displacement y is the output, we find that the transfer function for the hydraulic servomotor is, from Eq. (2–80),

$$\frac{Y(s)}{X(s)} = \frac{1}{s\left[\left(\dfrac{mK_2}{AK_1}\right)s + \dfrac{bK_2}{AK_1} + \dfrac{A\rho}{K_1}\right]}$$
$$= \frac{K}{s(Ts + 1)} \tag{2–81}$$

where

$$K = \frac{1}{\dfrac{bK_2}{AK_1} + \dfrac{A\rho}{K_1}} \qquad \text{and} \qquad T = \frac{mK_2}{bK_2 + A^2\rho}$$

From Equation (2–81), we see that this transfer function is of the second order. If the ratio $mK_2/(bK_2 + A^2\rho)$ is negligibly small or the time constant T is negligible, the transfer function can be simplified to give

$$\frac{Y(s)}{X(s)} = \frac{K}{s}$$

It is noted that a more detailed analysis shows that if oil leakage, compressibility (including the effects of dissolved air), expansion of pipelines, and the likes are taken into consideration, the transfer function becomes

$$\frac{Y(s)}{X(s)} = \frac{K}{s(T_1 s + 1)(T_2 s + 1)}$$

where T_1 and T_2 are time constants. As a matter of fact, these time constants depend on the volume of oil in the operating circuit. The smaller the volume, the smaller the time constants.

Example Problems and Solutions

A–2–1   Show that for the differential equation system

$$\dddot{y} + a_1 \ddot{y} + a_2 \dot{y} + a_3 y = b_0 \dddot{u} + b_1 \ddot{u} + b_2 \dot{u} + b_3 u \tag{2–82}$$

state and output equations can be given, respectively, by

$$\begin{bmatrix} \dot{x}_1 \\ \dot{x}_2 \\ \dot{x}_3 \end{bmatrix} = \begin{bmatrix} 0 & 1 & 0 \\ 0 & 0 & 1 \\ -a_3 & -a_2 & -a_1 \end{bmatrix} \begin{bmatrix} x_1 \\ x_2 \\ x_3 \end{bmatrix} + \begin{bmatrix} \beta_1 \\ \beta_2 \\ \beta_3 \end{bmatrix} u \tag{2–83}$$

and

$$y = \begin{bmatrix} 1 & 0 & 0 \end{bmatrix} \begin{bmatrix} x_1 \\ x_2 \\ x_3 \end{bmatrix} + \beta_0 u \tag{2–84}$$

where state variables are defined by

$$x_1 = y - \beta_0 u$$

$$x_2 = \dot{y} - \beta_0 \dot{u} - \beta_1 u = \dot{x}_1 - \beta_1 u$$

$$x_3 = \ddot{y} - \beta_0 \ddot{u} - \beta_1 \dot{u} - \beta_2 u = \dot{x}_2 - \beta_2 u$$

and

$$\beta_0 = b_0$$

$$\beta_1 = b_1 - a_1 \beta_0$$

$$\beta_2 = b_2 - a_1 \beta_1 - a_2 \beta_0$$

$$\beta_3 = b_3 - a_1 \beta_2 - a_2 \beta_1 - a_3 \beta_0$$

Solution. From the definition of state variables x_2 and x_3, we have

$$\dot{x}_1 = x_2 + \beta_1 u \tag{2–85}$$

$$\dot{x}_2 = x_3 + \beta_2 u \tag{2–86}$$

To derive the equation for $\dot{x}_3$, we first note that

$$\dddot{y} = -a_1\ddot{y} - a_2\dot{y} - a_3y + b_0\dddot{u} + b_1\ddot{u} + b_2\dot{u} + b_3u$$

Since

$$x_3 = \ddot{y} - \beta_0\ddot{u} - \beta_1\dot{u} - \beta_2u$$

we have

$$\dot{x}_3 = \dddot{y} - \beta_0\dddot{u} - \beta_1\ddot{u} - \beta_2\dot{u}$$

$$= (-a_1\ddot{y} - a_2\dot{y} - a_3y) + b_0\dddot{u} + b_1\ddot{u} + b_2\dot{u} + b_3u - \beta_0\dddot{u} - \beta_1\ddot{u} - \beta_2\dot{u}$$

$$= -a_1(\ddot{y} - \beta_0\ddot{u} - \beta_1\dot{u} - \beta_2u) - a_1\beta_0\ddot{u} - a_1\beta_1\dot{u} - a_1\beta_2u$$

$$\quad -a_2(\dot{y} - \beta_0\dot{u} - \beta_1u) - a_2\beta_0\dot{u} - a_2\beta_1u - a_3(y - \beta_0u) - a_3\beta_0u$$

$$\quad + b_0\dddot{u} + b_1\ddot{u} + b_2\dot{u} + b_3u - \beta_0\dddot{u} - \beta_1\ddot{u} - \beta_2\dot{u}$$

$$= -a_1x_3 - a_2x_2 - a_3x_1 + (b_0 - \beta_0)\dddot{u} + (b_1 - \beta_1 - a_1\beta_0)\ddot{u}$$

$$\quad +(b_2 - \beta_2 - a_1\beta_1 - a_2\beta_0)\dot{u} + (b_3 - a_1\beta_2 - a_2\beta_1 - a_3\beta_0)u$$

$$= -a_1x_3 - a_2x_2 - a_3x_1 + (b_3 - a_1\beta_2 - a_2\beta_1 - a_3\beta_0)u$$

$$= -a_1x_3 - a_2x_2 - a_3x_1 + \beta_3u$$

Hence, we get

$$\dot{x}_3 = -a_3x_1 - a_2x_2 - a_1x_3 + \beta_3u \tag{2-87}$$

Combining Equations (2–85), (2–86), and (2–87) into a vector-matrix differential equation, we obtain Equation (2–83). Also, from the definition of state variable x_1, we get the output equation given by Equation (2–84).

A–2–2. Obtain a state-space model of the system shown in Figure 2–35.

Solution. The system involves one integrator and two delayed integrators. The output of each integrator or delayed integrator can be a state variable. Let us define the output of the plant as x_1, the output of the controller as x_2, and the output of the sensor as x_3. Then we obtain

$$\frac{X_1(s)}{X_2(s)} = \frac{10}{s + 5}$$

$$\frac{X_2(s)}{U(s) - X_3(s)} = \frac{1}{s}$$

$$\frac{X_3(s)}{X_1(s)} = \frac{1}{s + 1}$$

$$Y(s) = X_1(s)$$

which can be rewritten as

$$sX_1(s) = -5X_1(s) + 10X_2(s)$$

$$sX_2(s) = -X_3(s) + U(s)$$

$$sX_3(s) = X_1(s) - X_3(s)$$

$$Y(s) = X_1(s)$$

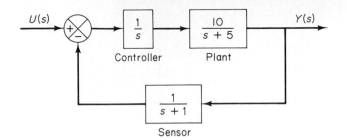

Figure 2–35
Control system.

By taking the inverse Laplace transforms of the preceding four equations, we obtain

$$\dot{x}_1 = -5x_1 + 10x_2$$

$$\dot{x}_2 = -x_3 + u$$

$$\dot{x}_3 = x_1 - x_3$$

$$y = x_1$$

Thus, a state-space model of the system in the standard form is given by

$$\begin{bmatrix} \dot{x}_1 \\ \dot{x}_2 \\ \dot{x}_1 \end{bmatrix} = \begin{bmatrix} -5 & 10 & 0 \\ 0 & 0 & -1 \\ 1 & 0 & -1 \end{bmatrix} \begin{bmatrix} x_1 \\ x_2 \\ x_3 \end{bmatrix} + \begin{bmatrix} 0 \\ 1 \\ 0 \end{bmatrix} u$$

$$y = \begin{bmatrix} 1 & 0 & 0 \end{bmatrix} \begin{bmatrix} x_1 \\ x_2 \\ x_3 \end{bmatrix}$$

It is important to note that this is not the only state-space representation of the system. Many other state-space representations are possible. However, the number of state variables is the same in any state-space representation of the same system. In the present system, the number of state variables is three, regardless of what variables are chosen as state variables.

A–2–3.   Obtain a state-space model for the system shown in Figure 2–36(a).

Solution. First, notice that $(as + b)/s^2$ involves a derivative term. Such a derivative term may be avoided if we modify $(as + b)/s^2$ as

$$\frac{as + b}{s^2} = \left(a + \frac{b}{s} \right) \frac{1}{s}$$

Using this modification, the block diagram of Figure 2–36(a) can be modified to that shown in Figure 2–36(b).

Define the outputs of the integrators as state variables, as shown in Figure 2–36(b). Then from Figure 2–36(b) we obtain

$$\frac{X_1(s)}{X_2(s) + a[U(s) - X_1(s)]} = \frac{1}{s}$$

$$\frac{X_2(s)}{U(s) - X_1(s)} = \frac{b}{s}$$

$$Y(s) = X_1(s)$$

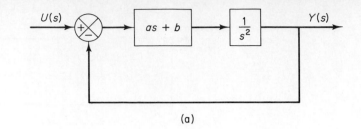

(a)

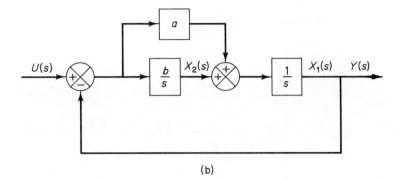

Figure 2–36
(a) Control system;
(b) modified block
diagram.

(b)

which may be modified to

$$sX_1(s) = X_2(s) + a[U(s) - X_1(s)]$$

$$sX_2(s) = -bX_1(s) + bU(s)$$

$$Y(s) = X_1(s)$$

Taking the inverse Laplace transforms of the preceding three equations, we obtain

$$\dot{x}_1 = -ax_1 + x_2 + au$$

$$\dot{x}_2 = -bx_1 + bu$$

$$y = x_1$$

Rewriting the state and output equations in the standard vector-matrix form, we obtain

$$\begin{bmatrix} \dot{x}_1 \\ \dot{x}_2 \end{bmatrix} = \begin{bmatrix} -a & 1 \\ -b & 0 \end{bmatrix} \begin{bmatrix} x_1 \\ x_2 \end{bmatrix} + \begin{bmatrix} a \\ b \end{bmatrix} u$$

$$y = \begin{bmatrix} 1 & 0 \end{bmatrix} \begin{bmatrix} x_1 \\ x_2 \end{bmatrix}$$

A–2–4. Figure 2–37(a) shows a schematic diagram of an automobile suspension system. As the car moves along the road, the vertical displacements at the tires act as the motion excitation to the automobile suspension system. The motion of this system consists of a translational motion of the center of mass and a rotational motion about the center of mass. Mathematical modeling of the complete system is quite complicated.

A very simplified version of the suspension system is shown in Fig. 2–27(b). Assuming that the motion x_i at point P is the input to the system and the vertical motion x_o of the body is the output,

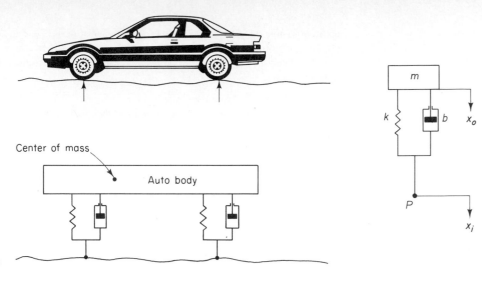

Figure 2–37
(a) Automobile suspension system; (b) simplified suspension system.

(a)

(b)

obtain the transfer function $X_o(s)/X_i(s)$. (Consider the motion of the body only in the vertical direction.) Displacement x_o is measured from the equilibrium position in the absence of input x_i.

Solution. The equation of motion for the system shown in Fig. 2–37(b) is

$$m\ddot{x}_o + b(\dot{x}_o - \dot{x}_i) + k(x_o - x_i) = 0$$

or

$$m\ddot{x}_o + b\dot{x}_o + kx_o = b\dot{x}_i + kx_i$$

Taking the Laplace transform of this last equation, assuming zero initial conditions, we obtain

$$(ms^2 + bs + k)X_o(s) = (bs + k)X_i(s)$$

Hence the transfer function $X_o(s)/X_i(s)$ is given by

$$\frac{X_o(s)}{X_i(s)} = \frac{bs + k}{ms^2 + bs + k}$$

A–2–5. In real-life situations, the motion of a mechanical system may be simultaneously translational and rotational in three-dimensional space, and parts of the system may have path constraints on where they can move. The geometrical description of such motions can become complicated, but the fundamental physical laws still apply.

For a complex system, more than one coordinate may be necessary in describing its motion. The term used to describe the minimum number of independent coordinates required to specify this motion is *degrees of freedom*. The number of degrees of freedom that a mechanical system possesses is the minimum number of independent coordinates required to specify the positions of all its elements. For instance, if only one independent coordinate is needed to specify the geometric location of the mass of a system in space completely, it is a one-degree-of-freedom system. That is, a rigid body rotating on an axis has one degree of freedom, whereas a rigid body in space has six degrees of freedom— three translational and three rotational.

Example Problems and Solutions

It is important to note that, in general, neither the number of masses nor any other obvious quantity will always lead to a correct assessment of the number of degrees of freedom.

In terms of the number of equations of motion and the number of constraints, the degrees of freedom may be written as

Number of degrees of freedom
    = (Number of equations of motion) − (Number of equations of constraint)

Now consider the systems shown in Figure 2–38. Find the degrees of freedom for each system.

Solution. (a) For the system shown in Fig. 2–38(a), if mass m is constrained to move vertically, only one coordinate x is required to define the location of the mass at any time. Thus the system shown in Fig. 2–38(a) has one degree of freedom.

We can verify this statement by counting the number of equations of motion and the number of equations of constraint. This system has one equation of motion

$$m\ddot{x} + b\dot{x} + (k_1 + k_2)x = 0$$

and no equation of constraint. Consequently,

$$\text{Degree of freedom} = 1 - 0 = 1$$

(b) Next, consider the system shown in Fig. 2–38(b). The equations of motion here are

$$m\ddot{x}_1 + k_1(x_1 - x_2) + k_2 x_1 = 0$$
$$k_1(x_1 - x_2) = b_1\dot{x}_2$$

So the number of equations of motion is two. There is no equation of constraint. Therefore,

$$\text{Degrees of freedom} = 2 - 0 = 2$$

(c) Finally, consider the pendulum system shown in Figure 2–38(c). Define the coordinates of the pendulum mass as (x, y). Then the equations of motion are

$$m\ddot{x} = -T\sin\theta$$
$$m\ddot{y} = mg - T\cos\theta$$

Thus the number of equations of motion is two. The constraint equation for this system is

$$x^2 + y^2 = l^2$$

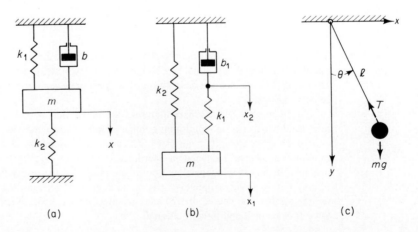

Figure 2–38
Mechanical systems.

(a)   (b)   (c)

The number of equation of constraint is one. And so

$$\text{Degree of freedom} = 2 - 1 = 1$$

Note that when physical constraints are present the most convenient coordinate system may not be the rectangular one. In the pendulum system of Figure 2–38(c), the pendulum is constrained to move in a circular path. The most convenient coordinate system here would be a polar coordinate system. Then the only coordinate that is needed is the angle θ through which the pendulum has swung. The rectangular coordinates x, y and polar coordinates θ, l (where l is a constant) are related by

$$x = l \sin \theta, \qquad y = l \cos \theta$$

In terms of the polar coordinate system, the equation of motion becomes

$$ml^2\ddot{\theta} = -mgl \sin \theta$$

or

$$\ddot{\theta} + \frac{g}{l} \sin \theta = 0$$

Note that since l is a constant the configuration of the system can be specified by one coordinate, θ. Consequently, it is a *one-degree-of-freedom system*, as we found earlier.

A–2–6. Gear trains are often used in servo systems to reduce speed, to magnify torque, or to obtain the most efficient power transfer by matching the driving member to the given load.

Consider the gear train system shown in Fig. 2–39. In this system, a load is driven by a motor through the gear train. Assuming that the stiffness of the shafts of the gear train is infinite (there is neither backlash nor elastic deformation) and that the number of teeth on each gear is proportional to the radius of the gear, obtain the equivalent moment of inertia and equivalent viscous-friction coefficient referred to the motor shaft and referred to the load shaft.

In Fig. 2–39, the numbers of teeth on gears 1, 2, 3, and 4 are N_1, N_2, N_3, and N_4, respectively. The angular displacements of shafts 1, 2, and 3 are θ_1, θ_2, and θ_3, respectively. Thus, $\theta_2/\theta_1 = N_1/N_2$ and $\theta_3/\theta_2 = N_3/N_4$. The moment of inertia and viscous-friction coefficient of each gear train component are denoted by J_1, b_1; J_2, b_2; and J_3, b_3; respectively. (J_3 and b_3 include the moment of inertia and friction of the load.)

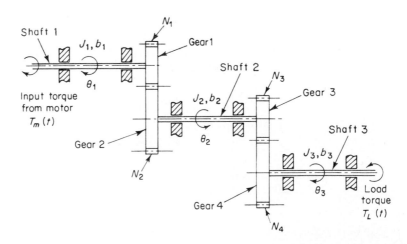

Figure 2–39
Gear train system.

Solution. For this gear train system, we can obtain the following three equations: For shaft 1,

$$J_1\ddot{\theta}_1 + b_1\dot{\theta}_1 + T_1 = T_m \tag{2-88}$$

where T_m is the torque developed by the motor and T_1 is the load torque on gear 1 due to the rest of the gear train. For shaft 2,

$$J_2\ddot{\theta}_2 + b_2\dot{\theta}_2 + T_3 = T_2 \tag{2-89}$$

where T_2 is the torque transmitted to gear 2 and T_3 is the load torque on gear 3 due to the rest of the gear train. Since the work done by gear 1 is equal to that of gear 2,

$$T_1\theta_1 = T_2\theta_2 \quad \text{or} \quad T_2 = T_1\frac{N_2}{N_1}$$

If $N_1/N_2 < 1$, the gear ratio reduces the speed as well as magnifies the torque. For the third shaft,

$$J_3\ddot{\theta}_3 + b_3\dot{\theta}_3 + T_L = T_4 \tag{2-90}$$

where T_L is the load torque and T_4 is the torque transmitted to gear 4. T_3 and T_4 are related by

$$T_4 = T_3\frac{N_4}{N_3}$$

and θ_3 and θ_1 are related by

$$\theta_3 = \theta_2\frac{N_3}{N_4} = \theta_1\frac{N_1}{N_2}\frac{N_3}{N_4}$$

Elimination of T_1, T_2, T_3, and T_4 from Equations (2–88), (2–89), and (2–90) yields

$$J_1\ddot{\theta}_1 + b_1\dot{\theta}_1 + \frac{N_1}{N_2}(J_2\ddot{\theta}_2 + b_2\dot{\theta}_2) + \frac{N_1 N_3}{N_2 N_4}(J_3\ddot{\theta}_3 + b_3\dot{\theta}_3 + T_L) = T_m$$

Eliminating θ_2 and θ_3 from this last equation and writing the resulting equation in terms of θ_1 and its time derivatives, we obtain

$$\left[J_1 + \left(\frac{N_1}{N_2}\right)^2 J_2 + \left(\frac{N_1}{N_2}\right)^2\left(\frac{N_3}{N_4}\right)^2 J_3\right]\ddot{\theta}_1$$

$$+ \left[b_1 + \left(\frac{N_1}{N_2}\right)^2 b_2 + \left(\frac{N_1}{N_2}\right)^2\left(\frac{N_3}{N_4}\right)^2 b_3\right]\dot{\theta}_1 + \left(\frac{N_1}{N_2}\right)\left(\frac{N_3}{N_4}\right)T_L = T_m \tag{2-91}$$

Thus, the equivalent moment of inertia and viscous-friction coefficient of the gear train referred to shaft 1 are given, respectively, by

$$J_{1eq} = J_1 + \left(\frac{N_1}{N_2}\right)^2 J_2 + \left(\frac{N_1}{N_2}\right)^2\left(\frac{N_3}{N_4}\right)^2 J_3$$

$$b_{1eq} = b_1 + \left(\frac{N_1}{N_2}\right)^2 b_2 + \left(\frac{N_1}{N_2}\right)^2\left(\frac{N_3}{N_4}\right)^2 b_3$$

Similarly, the equivalent moment of inertia and viscous friction coefficient of the gear train referred to the load shaft (shaft 3) are given, respectively, by

$$J_{3eq} = J_3 + \left(\frac{N_4}{N_3}\right)^2 J_2 + \left(\frac{N_2}{N_1}\right)^2\left(\frac{N_4}{N_3}\right)^2 J_1$$

$$b_{3eq} = b_3 + \left(\frac{N_4}{N_3}\right)^2 b_2 + \left(\frac{N_2}{N_1}\right)^2 \left(\frac{N_4}{N_3}\right)^2 b_1$$

The relationship between J_{1eq} and J_{3eq} is thus

$$J_{1eq} = \left(\frac{N_1}{N_2}\right)^2 \left(\frac{N_3}{N_4}\right)^2 J_{3eq}$$

and that between b_{1eq} and b_{3eq} is

$$b_{1eq} = \left(\frac{N_1}{N_2}\right)^2 \left(\frac{N_3}{N_4}\right)^2 b_{3eq}$$

The effect of J_2 and J_3 on an equivalent moment of inertia is determined by the gear ratios N_1/N_2 and N_3/N_4. For speed-reducing gear trains, the ratios N_1/N_2 and N_3/N_4 are usually less than unity. If $N_1/N_2 \ll 1$ and $N_3/N_4 \ll 1$, then the effect of J_2 and J_3 on the equivalent moment of inertia J_{1eq} is negligible. Similar comments apply to the equivalent viscous-friction coefficient b_{1eq} of the gear train. In terms of the equivalent moment of inertia J_{1eq} and equivalent viscous-friction coefficient b_{1eq}, Equation (2–91) can be simplified to give

$$J_{1eq}\ddot{\theta}_1 + b_{1eq}\dot{\theta}_1 + nT_L = T_m$$

where

$$n = \frac{N_1}{N_2}\frac{N_3}{N_4}$$

A–2–7. Obtain the transfer function $E_o(s)/E_i(s)$ of the op-amp circuit shown in Figure 2–40.

Solution. Define the voltage at point A as e_A. Then

$$\frac{E_A(s)}{E_i(s)} = \frac{R_1}{\dfrac{1}{Cs} + R_1} = \frac{R_1 Cs}{R_1 Cs + 1}$$

Define the voltage at point B as e_B. Then

$$E_B(s) = \frac{R_3}{R_2 + R_3} E_o(s)$$

Noting that

$$[E_A(s) - E_B(s)]K = E_o(s)$$

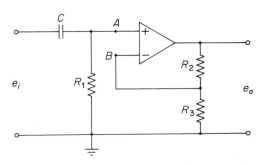

Figure 2–40
Operational amplifier
circuit.

and $K \gg 1$, we must have

$$E_A(s) = E_B(s)$$

Hence

$$E_A(s) = \frac{R_1 Cs}{R_1 Cs + 1} E_i(s) = E_B(s) = \frac{R_3}{R_2 + R_3} E_o(s)$$

from which we obtain

$$\frac{E_o(s)}{E_i(s)} = \frac{R_2 + R_3}{R_3} \frac{R_1 Cs}{R_1 Cs + 1} = \frac{\left(1 + \dfrac{R_2}{R_3}\right)s}{s + \dfrac{1}{R_1 C}}$$

A-2-8. Obtain the transfer function $E_o(s)/E_i(s)$ of the op-amp circuit shown in Figure 2–41.

Solution. The voltage at point A is

$$e_A = \frac{1}{2}(e_i - e_o) + e_o$$

The Laplace transformed version of this last equation is

$$E_A(s) = \frac{1}{2}\left[E_i(s) + E_o(s)\right]$$

The voltage at point B is

$$E_B(s) = \frac{\dfrac{1}{Cs}}{R_2 + \dfrac{1}{Cs}} E_i(s) = \frac{1}{R_2 Cs + 1} E_i(s)$$

Since $[E_B(s) - E_A(s)]K = E_o(s)$ and $K \gg 1$, we must have $E_A(s) = E_B(s)$. Thus

$$\frac{1}{2}[E_i(s) + E_o(s)] = \frac{1}{R_2 Cs + 1} E_i(s)$$

Hence

$$\frac{E_o(s)}{E_i(s)} = -\frac{R_2 Cs - 1}{R_2 Cs + 1} = -\frac{s - \dfrac{1}{R_2 C}}{s + \dfrac{1}{R_2 C}}$$

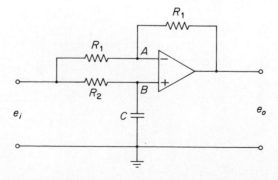

Figure 2–41
Operational amplifier
circuit.

A–2–9. Show that the systems in Figures 2–42(a) and (b) are analogous systems. (Show that the transfer functions of the two systems are of identical form.)

Solution. The equations of motion for the mechanical system shown in Figure 2–42(a) are

$$b_1(\dot{x}_i - \dot{x}_o) + k_1(x_i - x_o) = b_2(\dot{x}_o - \dot{y})$$

$$b_2(\dot{x}_o - \dot{y}) = k_2 y$$

By taking the Laplace transforms of these two equations, assuming zero initial conditions, we have

$$b_1[sX_i(s) - sX_o(s)] + k_1[X_i(s) - X_o(s)] = b_2[sX_o(s) - sY(s)]$$

$$b_2[sX_o(s) - sY(s)] = k_2 Y(s)$$

If we eliminate $Y(s)$ from the last two equations, then we obtain

$$b_1[sX_i(s) - sX_o(s)] + k_1[X_i(s) - X_o(s)] = b_2 s X_o(s) - b_2 s \frac{b_2 s X_o(s)}{b_2 s + k_2}$$

or

$$(b_1 s + k_1)X_i(s) = \left(b_1 s + k_1 + b_2 s - b_2 s \frac{b_2 s}{b_2 s + k_2} \right) X_o(s)$$

Hence the transfer function $X_o(s)/X_i(s)$ can be obtained as

$$\frac{X_o(s)}{X_i(s)} = \frac{\left(\dfrac{b_1}{k_1} s + 1 \right)\left(\dfrac{b_2}{k_2} s + 1 \right)}{\left(\dfrac{b_1}{k_1} s + 1 \right)\left(\dfrac{b_2}{k_2} s + 1 \right) + \dfrac{b_2}{k_1} s}$$

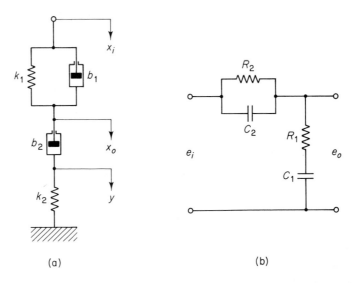

Figure 2–42
(a) Mechanical system; (b) analogous electrical system.

(a)

(b)

Example Problems and Solutions

159

For the electrical system shown in Figure 2–42(b), the transfer function $E_o(s)/E_i(s)$ is found to be

$$\frac{E_o(s)}{E_i(s)} = \frac{R_1 + \dfrac{1}{C_1 s}}{\dfrac{1}{(1/R_2) + C_2 s} + R_1 + \dfrac{1}{C_1 s}}$$

$$= \frac{(R_1 C_1 s + 1)(R_2 C_2 s + 1)}{(R_1 C_1 s + 1)(R_2 C_2 s + 1) + R_2 C_1 s}$$

A comparison of the transfer functions shows that the systems shown in Figures 2–42(a) and (b) are analogous.

A–2–10.   Using the force–voltage analogy, obtain an electrical analog of the mechanical system shown in Figure 2–43.

Solution. The equations of motion for the mechanical system are

$$m_1 \ddot{x}_1 + b_1 \dot{x}_1 + k_1 x_1 + b_2(\dot{x}_1 - \dot{x}_2) + k_2(x_1 - x_2) = 0$$

$$m_2 \ddot{x}_2 + b_2(\dot{x}_2 - \dot{x}_1) + k_2(x_2 - x_1) = 0$$

By use of the force–voltage analogy, the equations for an analogous electrical system may be written as

$$L_1 \ddot{q}_1 + R_1 \dot{q}_1 + \frac{1}{C_1} q_1 + R_2(\dot{q}_1 - \dot{q}_2) + \frac{1}{C_2}(q_1 - q_2) = 0$$

$$L_2 \ddot{q}_2 + R_2(\dot{q}_2 - \dot{q}_1) + \frac{1}{C_2}(q_2 - q_1) = 0$$

Then substituting $\dot{q}_1 = i_1$ and $\dot{q}_2 = i_2$ into the last two equations gives

$$L_1 \frac{di_1}{dt} + R_1 i_1 + \frac{1}{C_1} \int i_1 \, dt + R_2(i_1 - i_2) + \frac{1}{C_2} \int (i_1 - i_2) \, dt = 0 \qquad (2\text{–}92)$$

$$L_2 \frac{di_2}{dt} + R_2(i_2 - i_1) + \frac{1}{C_2} \int (i_2 - i_1) \, dt = 0 \qquad (2\text{–}93)$$

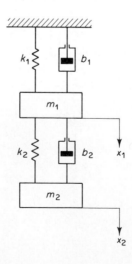

Figure 2–43
Mechanical system.

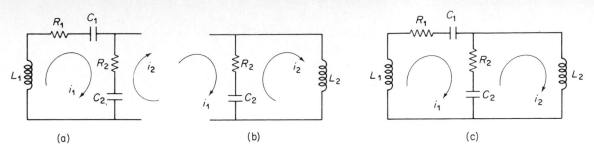

Figure 2–44 (a) Electrical circuit corresponding to Eq. (2–92); (b) electrical circuit corresponding to Eq. (2–93); (c) electrical system analogous to the mechanical system shown in Fig. 2–43 (force–voltage analogy).

These two equations are loop-voltage equations. From Equation (2–92) we obtain the diagram shown in Figure 2–44(a). Similarly, from Equation (2–93) we obtain the one given in Figure 2–44(b). Combining these two diagrams produces the desired analogous electrical system as shown in Figure 2–44(c).

A–2–11. Obtain the closed-loop transfer function for the positional servo system shown in Figure 2–45. Assume that the input and output of the system are the input shaft position and the output shaft position, respectively. Assume the following numerical values for system constants:

r = angular displacement of the reference input shaft, radians
c = angular displacement of the output shaft, radians
θ = angular displacement of the motor shaft, radians
K_1 = gain of the potentiometric error detector = $24/\pi$ volts/rad
K_p = amplifier gain = 10 volts/volt
e_a = applied armature voltage, volts
e_b = back emf, volts
R_a = armature-winding resistance = 0.2 ohms
L_a = armature-winding inductance = negligible
i_a = armature-winding current, amperes
K_b = back emf constant = 5.5×10^{-2} volts-sec/rad
K = motor torque constant = 6×10^{-5} N-m/ampere
J_m = moment of inertia of the motor = 1×10^{-5} kg-m^2
b_m = viscous-friction coefficient of the motor = negligible
J_L = moment of inertia of the load = 4.4×10^{-3} kg-m^2
b_L = viscous-friction coefficient of the load = 4×10^{-2} N-m/rad/sec
n = gear ratio N_1/N_2 = 1/10

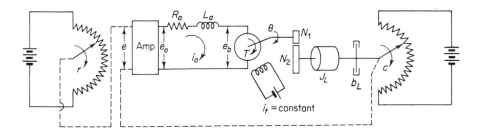

Figure 2–45
Positional servo system.

Example Problems and Solutions

161

Solution. The equations describing the system dynamics are as follows:

For the potentiometric error detector;

$$E(s) = K_1[R(s) - C(s)] = 7.64[R(s) - C(s)] \tag{2–94}$$

For the amplifier,

$$E_a(s) = K_p E(s) = 10E(s) \tag{2–95}$$

For the armature-controlled dc motor, the equivalent moment of inertia J and equivalent viscous-friction coefficient b referred to the motor shaft are, respectively,

$$J = J_m + n^2 J_L$$
$$= 1 \times 10^{-5} + 4.4 \times 10^{-5} = 5.4 \times 10^{-5}$$

$$b = b_m + n^2 b_L$$
$$= 4 \times 10^{-4}$$

Referring to Equation (2–44), we obtain

$$\frac{\Theta(s)}{E_a(s)} = \frac{K}{s(T_m s + 1)}$$

where

$$K_m = \frac{K}{R_a b + KK_b} = \frac{6 \times 10^{-5}}{(0.2)(4 \times 10^{-4}) + (6 \times 10^{-5})(5.5 \times 10^{-2})} = 0.72$$

$$T_m = \frac{R_a J}{R_a b + KK_b} = \frac{(0.2)(5.4 \times 10^{-5})}{(0.2)(4 \times 10^{-4}) + (6 \times 10^{-5})(5.5 \times 10^{-2})} = 0.13$$

Thus

$$\frac{\Theta(s)}{E_a(s)} = \frac{10C(s)}{E_a(s)} = \frac{0.72}{s(0.13s + 1)} \tag{2–96}$$

Using Equations (2–94), (2–95), and (2–96), we can draw the block diagram of the system as shown in Figure 2–46(a). Simplifying this block diagram, we obtain Figure 2–46(b). The closed-loop transfer function of this system is

$$\frac{C(s)}{R(s)} = \frac{5.5}{0.13s^2 + s + 5.5} = \frac{42.3}{s^2 + 7.69s + 42.3}$$

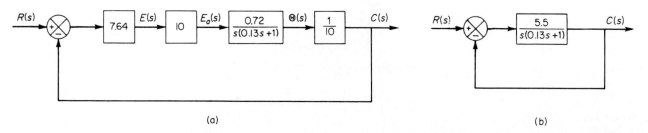

(a)   (b)

Figure 2–46   (a) Block diagram of the system shown in Fig. 2–45; (b) simplified block diagram.

A–2–12. Show that the torque-to-inertia ratios referred to the motor shaft and to the load shaft differ from each other by a factor of n. Show also that the torque squared-to-inertia ratios referred to the motor shaft and to the load shaft are the same.

Solution. Suppose that T_{max} is the maximum torque that can be produced on the motor shaft. Then the torque-to-inertia ratio referred to the motor shaft is

$$\frac{T_{max}}{J_m + n^2 J_L}$$

where J_m = moment of inertia of the rotor
J_L = moment of inertia of the load
n = gear ratio

The torque-to-inertia ratio referred to the load shaft is

$$\frac{\dfrac{T_{max}}{n}}{J_L + \dfrac{J_m}{n^2}} = \frac{n T_{max}}{J_m + n^2 J_L}$$

Clearly, they differ from each other by a factor of n. Hence, in comparing torque-to-inertia ratios of motors, we find it necessary to specify which shaft is the reference.

Note that the ratio of torque squared to inertia referred to the motor shaft is

$$\frac{T_{max}^2}{J_m + n^2 J_L}$$

and that referred to the load shaft is

$$\frac{\dfrac{T_{max}^2}{n^2}}{J_L + \dfrac{J_m}{n^2}} = \frac{T_{max}^2}{J_m + n^2 J_L}$$

These two ratios are clearly the same.

A–2–13. Assuming that a two-phase servomotor has a linear torque–speed curve such that the no-load speed is ω_0 rad/sec and the stall torque is T_s N-m, find the maximum motor-shaft output power P_{max} watts. If $T_s = 0.0353$ N-m and $\omega_0 = 418.9$ rad/sec (4000 rpm), what is the maximum power output P_{max} and at what frequency does it occur?

Solution. The torque–speed curve is

$$T = T_s - \left(\frac{\omega}{\omega_0}\right) T_s$$

The shaft output power P is given by

$$P = T\omega$$
$$= \left(T_s - \frac{\omega}{\omega_0} T_s\right)\omega$$

To find P_{max}, let us differentiate P with respect to ω:

$$\frac{dP}{d\omega} = T_s - 2\frac{\omega}{\omega_0} T_s$$

By setting $dP/d\omega = 0$, we obtain

$$\omega = \frac{\omega_0}{2}$$

Clearly, $d^2P/d\omega^2 = -2T_s/\omega_0 < 0$. Hence P is maximum at $\omega = \omega_0/2$. P_{max} is

$$P_{max} = T\omega|_{\omega = \omega_0/2}$$

$$= \left(T_s - \frac{T_s}{2}\right)\frac{\omega_0}{2}$$

$$= \frac{T_s\omega_0}{4}$$

For $T_s = 0.0353$ N-m and $\omega_0 = 418.9$ rad/sec, the maximum power output P_{max} is

$$P_{max} = \frac{0.0353 \times 418.9}{4} = 3.697 \text{ watts}$$

The maximum power occurs at $\omega = 209.4$ rad/sec or 2000 rpm.

A–2–14.   Referring to the robot-arm system shown in Figure 2–47, consider a problem of controlling the angular motion about the z axis. (In this problem we consider only the motion of the base joint. We do not consider angular motions about other joints.) The motor is an armature-controlled dc servomotor. The gear ratio involved is $n = N_1/N_2 \ll 1$. The moment of inertia of the robot-arm system about the z axis is not constant. The value of moment of inertia J_z about the z axis depends on the configuration of the arms.

The moment of inertia J about the z axis of the entire system referred to the motor shaft is

$$J = J_m + n^2J_z \qquad (n < 1)$$

where J_m is the moment of inertia of the motor plus gear 1. Because n^2 will be small, the effect of the variation of J_z due to various robot-arm configurations can be made small through the gear train.

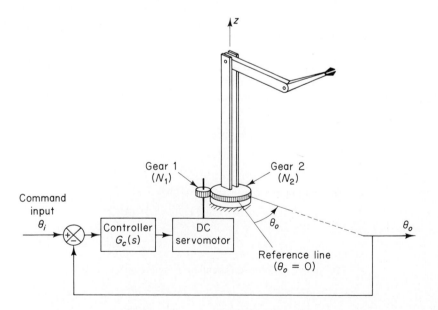

Figure 2–47
Robot-arm system.

Chapter 2 / Mathematical Modeling of Dynamic Systems

Define the effective viscous-friction coefficient of the rotational motion about the z axis as b_z. Then the viscous-friction coefficient referred to the motor shaft is

$$b = b_m + n^2 b_z \qquad (n < 1)$$

where b_m is the viscous-friction coefficient of the motor plus gear 1.

The angular displacement of the motor shaft is θ_m and the angular displacement of the robot-arm system about the z axis is θ_o. Angle θ_o is the output of the system under consideration. The command angular displacement is θ_i. The transfer function of the controller is $G_c(s)$.

Draw a block diagram for this system. Then obtain the transfer function $\Theta_o(s)/\Theta_i(s)$.

Solution. Referring to Figure 2–20, we have

$$J\ddot{\theta}_m + b\dot{\theta}_m = T = Ki_a \tag{2–97}$$

$$L_a\dot{i}_a + R_a i_a + e_b = e_a \tag{2–98}$$

where

$$e_b = K_b\dot{\theta}_m$$

Taking the Laplace transforms of Equations (2–97) and (2–98), assuming zero initial conditions, we obtain

$$(Js^2 + bs)\,\Theta_m(s) = T(s) = KI_a(s)$$

$$(L_a s + R_a)I_a(s) + K_b s\Theta_m(s) = E_a(s)$$

which can be modified to

$$\Theta_m(s) = \frac{K}{s(Js + b)}\,I_a(s) \tag{2–99}$$

$$I_a(s) = \frac{1}{L_a s + R_a}\left[E_a(s) - K_b s\Theta_m(s)\right] \tag{2–100}$$

Based on Equations (2–99) and (2–100) and referring to Figure 2–47, we obtain a block diagram as shown in Figure 2–48.

From the block diagram, we have

$$\frac{\Theta_m(s)}{E_a(s)} = \frac{K}{(L_a s + R_a)s(Js + b) + KK_b s}$$

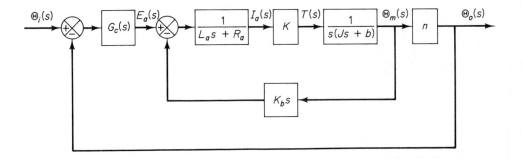

Figure 2–48
Block diagram of
robot-arm system.

Hence

$$\frac{\Theta_o(s)}{\Theta_i(s)} = \frac{G_c(s)\dfrac{K}{(L_as + R_a)s(Js + b) + KK_bs}n}{1 + G_c(s)\dfrac{K}{(L_as + R_a)s(Js + b) + KK_bs}n}$$

$$= \frac{G_c(s)Kn}{(L_as + R_a)s(Js + b) + KK_bs + G_c(s)Kn}$$

This is the transfer function of the system.

A–2–15. In the liquid-level system of Figure 2–49, assume that the outflow rate Q m³/s through the outflow valve is related to the head H m by

$$Q = K\sqrt{H} = 0.01\sqrt{H}$$

Assume also that when the inflow rate Q_i is 0.015 m³/s, the head stays constant. At $t = 0$ the inflow valve is closed and so there is no inflow for $t \geq 0$. Find the time necessary to empty the tank to half the original head. The capacitance C of the tank is 2 m².

Solution. When the head is stationary, the inflow rate equals the outflow rate. Thus head H_o at $t = 0$ is obtained from

$$0.015 = 0.01\sqrt{H_o}$$

or

$$H_o = 2.25 \text{ m}$$

The equation for the system for $t > 0$ is

$$-C \, dH = Q \, dt$$

or

$$\frac{dH}{dt} = -\frac{Q}{C} = \frac{-0.01\sqrt{H}}{2}$$

Hence

$$\frac{dH}{\sqrt{H}} = -0.005 \, dt$$

Assume that, at $t = t_1$, $H = 1.125$ m. Integrating both sides of this last equation, we obtain

$$\int_{2.25}^{1.125} \frac{dH}{\sqrt{H}} = \int_0^{t_1} (-0.005) \, dt = -0.005t_1$$

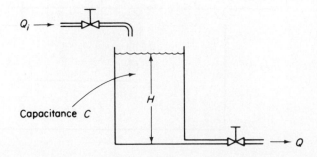

Figure 2–49
Liquid-level system.

Chapter 2 / Mathematical Modeling of Dynamic Systems

It follows that

$$2\sqrt{H}\bigg|_{2.25}^{1.125} = 2\sqrt{1.125} - 2\sqrt{2.25} = -0.005t_1$$

or

$$t_1 = 176 \ s$$

Thus, the head becomes half the original value (2.25 m) in 176 sec.

A–2–16. Consider the liquid-level system shown in Figure 2–50. At steady state, the inflow rate and outflow rate are both $\bar{Q}$ and the flow rate between the tanks is zero. The heads of tanks 1 and 2 are both $\bar{H}$. At $t = 0$, the inflow rate is changed from $\bar{Q}$ to $\bar{Q} + q$, where q is a small change in the inflow rate. The resulting changes in the heads (h_1 and h_2) and flow rates (q_1 and q_2) are assumed to be small. The capacitances of tanks 1 and 2 are C_1 and C_2, respectively. The resistance of the valve between the tanks is R_1 and that of the outflow valve is R_2.

Derive mathematical models for the system when (a) q is the input and h_2 the output, (b) q is the input and q_2 the output, and (c) q is the input and h_1 the output.

Solution. (a) For tank 1, we have

$$C_1 \, dh_1 = q_1 \, dt$$

where

$$q_1 = \frac{h_2 - h_1}{R_1}$$

Consequently,

$$R_1 C_1 \frac{dh_1}{dt} + h_1 = h_2 \tag{2–101}$$

For tank 2, we get

$$C_2 \, dh_2 = (q - q_1 - q_2) \, dt$$

where

$$q_1 = \frac{h_2 - h_1}{R_1}, \qquad q_2 = \frac{h_2}{R_2}$$

It follows that

$$R_2 C_2 \frac{dh_2}{dt} + \frac{R_2}{R_1} h_2 + h_2 = R_2 q + \frac{R_2}{R_1} h_1 \tag{2–102}$$

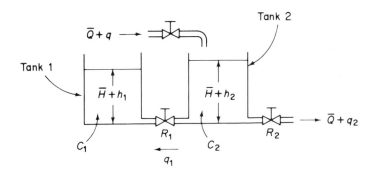

Figure 2–50
Liquid-level system.

By eliminating h_1 from Equations (2–101) and (2–102), we have

$$R_1C_1R_2C_2\frac{d^2h_2}{dt^2} + (R_1C_1 + R_2C_2 + R_2C_1)\frac{dh_2}{dt} + h_2 = R_1R_2C_1\frac{dq}{dt} + R_2q \qquad (2\text{–}103)$$

In terms of the transfer function, we have

$$\frac{H_2(s)}{Q(s)} = \frac{R_2(R_1C_1s + 1)}{R_1C_1R_2C_2s^2 + (R_1C_1 + R_2C_2 + R_2C_1)s + 1}$$

This is the desired mathematical model in which q is considered the input and h_2 is the output.

(b) Substitution of $h_2 = R_2q_2$ into Equation (2–103) gives

$$R_1C_1R_2C_2\frac{d^2q_2}{dt^2} + (R_1C_1 + R_2C_2 + R_2C_1)\frac{dq_2}{dt} + q_2 = R_1C_1\frac{dq}{dt} + q$$

This equation is a mathematical model of the system when q is considered the input and q_2 is the output. In terms of the transfer function, we obtain

$$\frac{Q_2(s)}{Q(s)} = \frac{R_1C_1s + 1}{R_1C_1R_2C_2s^2 + (R_1C_1 + R_2C_2 + R_2C_1)s + 1}$$

(c) Elimination of h_2 from Equations (2–101) and (2–102) yields

$$R_1C_1R_2C_2\frac{d^2h_1}{dt^2} + (R_1C_1 + R_2C_2 + R_2C_1)\frac{dh_1}{dt} + h_1 = R_2q$$

which is a mathematical model of the system in which q is considered the input and h_1 is the output. In terms of the transfer function, we get

$$\frac{H_1(s)}{Q(s)} = \frac{R_2}{R_1C_1R_2C_2s^2 + (R_1C_1 + R_2C_2 + R_2C_1)s + 1}$$

A–2–17. Consider the liquid-level system shown in Figure 2–51. In the system, $\bar{Q}_1$ and $\bar{Q}_2$ are steady-state inflow rates and $\bar{H}_1$ and $\bar{H}_2$ are steady-state heads. The quantities q_{i1}, q_{i2}, h_1, h_2, q_1, and q_o are considered small. Obtain a state-space representation for the system when h_1 and h_2 are the outputs and q_{i1} and q_{i2} are the inputs.

Solution. The equations for the system are

$$C_1dh_1 = (q_{i1} - q_1)\,dt \qquad (2\text{–}104)$$

$$\frac{h_1 - h_2}{R_1} = q_1 \qquad (2\text{–}105)$$

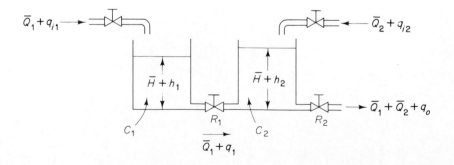

Figure 2–51
Liquid-level system.

$$C_2 dh_2 = (q_1 + q_{i2} - q_o) \, dt \qquad (2\text{–}106)$$

$$\frac{h_2}{R_2} = q_o \qquad (2\text{–}107)$$

Elimination of q_1 from Equation (2–104) using Equation (2–105) results in

$$\frac{dh_1}{dt} = \frac{1}{C_1}\left(q_{i1} - \frac{h_1 - h_2}{R_1}\right) \qquad (2\text{–}108)$$

Eliminating q_1 and q_o from Equation (2–106) by using Equations (2–105) and (2–107) gives

$$\frac{dh_2}{dt} = \frac{1}{C_2}\left(\frac{h_1 - h_2}{R_1} + q_{i2} - \frac{h_2}{R_2}\right) \qquad (2\text{–}109)$$

Define state variables x_1 and x_2 by

$$x_1 = h_1$$

$$x_2 = h_2$$

the input variables u_1 and u_2 by

$$u_1 = q_{i1}$$

$$u_2 = q_{i2}$$

and the output variables y_1 and y_2 by

$$y_1 = h_1 = x_1$$

$$y_2 = h_2 = x_2$$

Then Equations (2–108) and (2–109) can be written as

$$\dot{x}_1 = -\frac{1}{R_1 C_1} x_1 + \frac{1}{R_1 C_1} x_2 + \frac{1}{C_1} u_1$$

$$\dot{x}_2 = \frac{1}{R_1 C_2} x_1 - \left(\frac{1}{R_1 C_2} + \frac{1}{R_2 C_2}\right) x_2 + \frac{1}{C_2} u_2$$

In the form of the standard vector-matrix representation, we have

$$\begin{bmatrix} \dot{x}_1 \\ \dot{x}_2 \end{bmatrix} = \begin{bmatrix} -\dfrac{1}{R_1 C_1} & \dfrac{1}{R_1 C_1} \\[2ex] \dfrac{1}{R_1 C_2} & -\left(\dfrac{1}{R_1 C_2} + \dfrac{1}{R_2 C_2}\right) \end{bmatrix} \begin{bmatrix} x_1 \\ x_2 \end{bmatrix} + \begin{bmatrix} \dfrac{1}{C_1} & 0 \\[2ex] 0 & \dfrac{1}{C_2} \end{bmatrix} \begin{bmatrix} u_1 \\ u_2 \end{bmatrix}$$

which is the state equation, and

$$\begin{bmatrix} y_1 \\ y_2 \end{bmatrix} = \begin{bmatrix} 1 & 0 \\ 0 & 1 \end{bmatrix} \begin{bmatrix} x_1 \\ x_2 \end{bmatrix}$$

which is the output equation.

A–2–18. Considering small deviations from steady-state operation, draw a block diagram of the air heating system shown in Figure 2–52. Assume that the heat loss to the surroundings and the heat capacitance of the metal parts of the heater are negligible.

Example Problems and Solutions

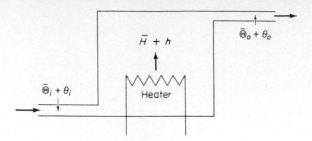

Figure 2–52
Air heating system.

Solution. Let us define

$\bar{\Theta}_i$ = steady-state temperature of inlet air, °C
$\bar{\Theta}_o$ = steady-state temperature of outlet air, °C
G = flow rate of air through the heating chamber, kg/sec
M = air contained in the heating chamber, kg
c = specific heat of air, kcal/kg °C
R = thermal resistance, °C sec/kcal
C = thermal capacitance of air contained in the heating chamber = Mc, kcal/°C
$\bar{H}$ = steady-state heat input, kcal/sec

Let us assume that the heat input is suddenly changed from $\bar{H}$ to $\bar{H} + h$ and the inlet air temperature is suddenly changed from $\bar{\Theta}_i$ to $\bar{\Theta}_i + \theta_i$. Then the outlet air temperature will be changed from $\bar{\Theta}_o$ to $\bar{\Theta}_o + \theta_o$.

The equation describing the system behavior is

$$C \, d\theta_o = [h + Gc(\theta_i - \theta_o)] \, dt$$

or

$$C \frac{d\theta_o}{dt} = h + Gc(\theta_i - \theta_o)$$

Noting that

$$Gc = \frac{1}{R}$$

we obtain

$$C \frac{d\theta_o}{dt} = h + \frac{1}{R}(\theta_i - \theta_o)$$

or

$$RC \frac{d\theta_o}{dt} + \theta_o = Rh + \theta_i$$

Taking the Laplace transforms of both sides of this last equation and substituting the initial condition that $\theta_o(0) = 0$, we obtain

$$\Theta_o(s) = \frac{R}{RCs + 1} H(s) + \frac{1}{RCs + 1} \Theta_i(s)$$

The block diagram of the system corresponding to this equation is shown in Figure 2–53.

A–2–19. Consider the thin, glass-wall, mercury thermometer system shown in Fig. 2–54. Assume that the thermometer is at a uniform temperature $\bar{\Theta}$ °C (ambient temperature) and that at $t = 0$ it is immersed

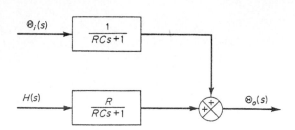

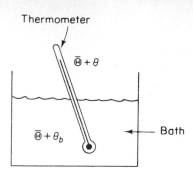

Figure 2–53
Block diagram of the air heating
system shown in Fig. 2–52.

Figure 2–54
Thin, glass-wall, mercury
thermometer system.

in a bath of temperature $\bar{\Theta} + \theta_b$ °C, where θ_b is the bath temperature (which may be constant or changing) measured from the ambient temperature $\bar{\Theta}$. Define the instantaneous thermometer temperature by $\bar{\Theta} + \theta$ °C so that θ is the change in the thermometer temperature satisfying the condition that $\theta(0) = 0$. Obtain the transfer function $\Theta(s)/\Theta_b(s)$. Also obtain an electrical analog of the thermometer system.

Solution. A mathematical model for the system can be derived by considering heat balance as follows: The heat entering the thermometer during dt sec is $q\,dt$, where q is the heat flow rate to the thermometer. This heat is stored in the thermal capacitance C of the thermometer, thereby raising its temperature by $d\theta$. Thus the heat-balance equation is

$$C\,d\theta = q\,dt \qquad (2\text{--}110)$$

Since thermal resistance R may be written as

$$R = \frac{d(\Delta\theta)}{dq} = \frac{\Delta\theta}{q}$$

heat flow rate q may be given, in terms of thermal resistance R, as

$$q = \frac{(\bar{\Theta} + \theta_b) - (\bar{\Theta} + \theta)}{R} = \frac{\theta_b - \theta}{R}$$

where $\bar{\Theta} + \theta_b$ is the bath temperature and $\bar{\Theta} + \theta$ is the thermometer temperature. Hence, we can rewrite Equation (2–110) as

$$C\frac{d\theta}{dt} = \frac{\theta_b - \theta}{R}$$

or

$$RC\frac{d\theta}{dt} + \theta = \theta_b \qquad (2\text{--}111)$$

Equation (2–111) is a mathematical model of the thermometer system.

Referring to Equation (2–111), an electrical analog for the thermometer system can be written as

$$RC\frac{de_o}{dt} + e_o = e_i$$

An electrical circuit represented by this last equation is shown in Figure 2–55.

Example Problems and Solutions

171

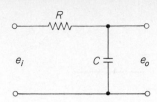

A–2–20. Consider the xyz coordinate system and $x'y'x'$ coordinate system having the origin at the same point, point O. Define unit vectors along the x, y, and z axes as $\mathbf{i}$, $\mathbf{j}$, and $\mathbf{k}$, and unit vectors along the x', y', and z' axes as $\mathbf{i'}$, $\mathbf{j'}$, and $\mathbf{k'}$, respectively. Show that

$$\begin{bmatrix} \mathbf{i'} \\ \mathbf{j'} \\ \mathbf{k'} \end{bmatrix} = \begin{bmatrix} \cos\theta_{x'x} & \cos\theta_{x'y} & \cos\theta_{x'z} \\ \cos\theta_{y'x} & \cos\theta_{y'y} & \cos\theta_{y'z} \\ \cos\theta_{z'x} & \cos\theta_{z'y} & \cos\theta_{z'z} \end{bmatrix} \begin{bmatrix} \mathbf{i} \\ \mathbf{j} \\ \mathbf{k} \end{bmatrix} \qquad (2\text{--}112)$$

where $\theta_{\xi'\eta}$, is the angle between the axis η and axis ξ' ($\eta = x, y, z$; $\xi' = x', y', z'$).

Solution. First, consider the two-dimensional case. Referring to Figure 2–56, notice that vectors $\mathbf{i'}$ and $\mathbf{j'}$ can be written, respectively, as

$$\mathbf{i'} = \cos\theta\,\mathbf{i} + \sin\theta\,\mathbf{j}$$

$$\mathbf{j'} = -\sin\theta\,\mathbf{i} + \cos\theta\,\mathbf{j}$$

Hence, we have

$$\begin{bmatrix} \mathbf{i'} \\ \mathbf{j'} \end{bmatrix} = \begin{bmatrix} \cos\theta & \sin\theta \\ -\sin\theta & \cos\theta \end{bmatrix} \begin{bmatrix} \mathbf{i} \\ \mathbf{j} \end{bmatrix} \qquad (2\text{--}113)$$

Referring to Figure 2–56, it can be seen that

$$\cos\theta_{x'x} = \cos\theta, \qquad \cos\theta_{x'y} = \sin\theta$$

$$\cos\theta_{y'x} = -\sin\theta, \qquad \cos\theta_{y'y} = \cos\theta$$

Therefore, we may rewrite Equation (2–113) as

$$\begin{bmatrix} \mathbf{i'} \\ \mathbf{j'} \end{bmatrix} = \begin{bmatrix} \cos\theta_{x'x} & \cos\theta_{x'y} \\ \cos\theta_{y'x} & \cos\theta_{y'y} \end{bmatrix} \begin{bmatrix} \mathbf{i} \\ \mathbf{j} \end{bmatrix}$$

Next, consider the three-dimensional case. Figure 2–57(a) shows unit vectors $\mathbf{i}$, $\mathbf{j}$, $\mathbf{k}$ and $\mathbf{i'}$, $\mathbf{j'}$, $\mathbf{k'}$. Consider the unit vector $\mathbf{i'}$. Referring to Figure 2–57(b), the projection of unit vector $\mathbf{i'}$ on the x axis is $\cos\theta_{x'x}\mathbf{i}$. Similarly, the projections of unit vector $\mathbf{i'}$ on the y and z axes are $\cos\theta_{x'y}\mathbf{j}$ and $\cos\theta_{x'z}\mathbf{k}$,

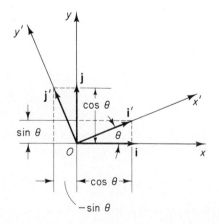

Figure 2–56
O–xy coordinate
system and rotated
coordinate system O–
$x'y'$.

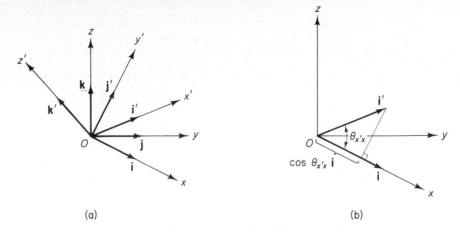

Figure 2–57
(a) Coordinate systems and associated unit vectors; (b) diagram showing projection of unit vector **i′** on the *x* axis.

(a)   (b)

respectively. Consequently, we can write unit vector **i′** as

$$\mathbf{i'} = \cos\theta_{x'x}\mathbf{i} + \cos\theta_{x'y}\mathbf{j} + \cos\theta_{x'z}\mathbf{k}$$

In the same way, we can write vectors **j′** and **k′** as follows:

$$\mathbf{j'} = \cos\theta_{y'x}\mathbf{i} + \cos\theta_{y'y}\mathbf{j} + \cos\theta_{y'z}\mathbf{k}$$

$$\mathbf{k'} = \cos\theta_{z'x}\mathbf{i} + \cos\theta_{z'y}\mathbf{j} + \cos\theta_{z'z}\mathbf{k}$$

If we combine the last three equations into one vector-matrix equation, we obtain Equation (2–112).

A–2–21. Consider the O–$x'y'z'$ and O–$x''y''z''$ coordinate systems as shown in Figure 2–58. (These coordinate systems are the same as those shown in Figure 2–29.) The z' and z'' axes are parallel to each other in space. Show that

$$\begin{bmatrix} x' \\ y' \\ z' \\ 1 \end{bmatrix} = \begin{bmatrix} \cos\theta' & -\sin\theta' & 0 & a \\ \sin\theta' & \cos\theta' & 0 & 0 \\ 0 & 0 & 1 & 0 \\ 0 & 0 & 0 & 1 \end{bmatrix} \begin{bmatrix} x'' \\ y'' \\ z'' \\ 1 \end{bmatrix}$$

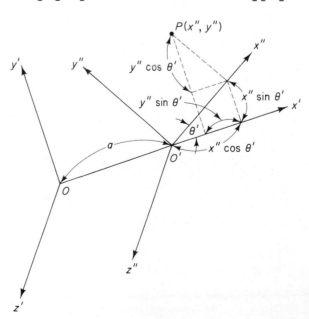

Figure 2–58
O–$x'y'z'$ coordinate system and O'–$x''y''z''$ coordinate system.

Solution. Consider a point P in the $x'y'$ plane (which is the same as the $x''y''$ plane). The coordinates of point P in the $O{-}x''y''z''$ coordinate system are (x'', y'', z''). The coordinates of point P in the $O{-}x'y'z'$ coordinate system are (x', y', z'). The coordinates $P(x', y', z')$ and $P(x'', y'', z'')$ are related as follows:

$$x' = x'' \cos \theta' - y'' \sin \theta' + a$$

$$y' = x'' \sin \theta' + y'' \cos \theta'$$

$$z' = z'' = 0$$

Hence

$$\begin{bmatrix} x' \\ y' \end{bmatrix} = \begin{bmatrix} \cos \theta' & -\sin \theta' \\ \sin \theta' & \cos \theta' \end{bmatrix} \begin{bmatrix} x'' \\ y'' \end{bmatrix} + \begin{bmatrix} a \\ 0 \end{bmatrix}$$

$$z' = z''$$

The last two equations can be combined into a vector-matrix equation as follows:

$$\begin{bmatrix} x' \\ y' \\ z' \\ 1 \end{bmatrix} = \begin{bmatrix} \cos \theta' & -\sin \theta' & 0 & a \\ \sin \theta' & \cos \theta' & 0 & 0 \\ 0 & 0 & 1 & 0 \\ 0 & 0 & 0 & 1 \end{bmatrix} \begin{bmatrix} x'' \\ y'' \\ z'' \\ 1 \end{bmatrix}$$

A–2–22. Gyros for sensing angular motion are commonly used in inertial guidance systems, autopilot systems, and the like. Figure 2–59(a) shows a single-degree-of-freedom gyro. The spinning wheel is mounted in a movable gimbal, which is, in turn, mounted in a gyro case. The gimbal is free to move relative to the case about the output axis OB. Note that the output axis is perpendicular to the wheel spin axis. The input axis around which a turning rate, or angle, is measured is perpendicular to both the output and spin axes. Information on the input signal (the turning rate or angle around the input axis) is obtained from the resulting motion of the gimbal about the output axis, relative to the case.

Figure 2–59(b) shows a functional diagram of the gyro system. The equation of motion about the output axis can be obtained by equating the rate of change of angular momentum to the sum of the external torques.

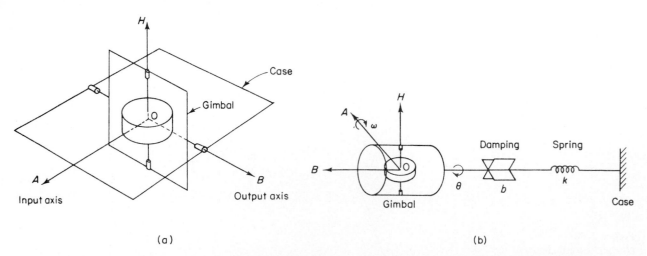

(a)   (b)

Figure 2–59 (a) Schematic diagram of a single-degree-of-freedom gyro; (b) functional diagram of the gyro shown in part (a).

The change in angular momentum about axis OB consists of two parts: $I\ddot{\theta}$, the change due to acceleration of the gimbal around axis OB, and $-H\omega\cos\theta$, the change due to the turning of the wheel angular-momentum vector around axis OA. The external torques consist of $-b\dot{\theta}$, the damping torque, and $-k\theta$, the spring torque. Thus the equation of the gyro system is

$$I\ddot{\theta} - H\omega\cos\theta = -b\dot{\theta} - k\theta$$

or

$$I\ddot{\theta} + b\dot{\theta} + k\theta = H\omega\cos\theta \tag{2–114}$$

In practice, θ is a very small angle, usually not more than $\pm 2.5°$.

Obtain a state-space representation of the gyro system.

Solution. In this system, θ and $\dot{\theta}$ may be chosen as state variables. The input variable is ω and the output variable is θ. Let us define

$$\mathbf{x} = \begin{bmatrix} x_1 \\ x_2 \end{bmatrix} = \begin{bmatrix} \theta \\ \dot{\theta} \end{bmatrix}, \qquad u = \omega, \qquad y = \theta$$

Then Equation (2–114) can be written as follows:

$$\dot{x}_1 = x_2$$

$$\dot{x}_2 = -\frac{k}{I}x_1 - \frac{b}{I}x_2 + \frac{H}{I}u\cos x_1$$

or

$$\dot{\mathbf{x}} = \mathbf{f}(\mathbf{x}, u)$$

where

$$\mathbf{x} = \begin{bmatrix} x_1 \\ x_2 \end{bmatrix}, \qquad \mathbf{f}(\mathbf{x}, u) = \begin{bmatrix} f_1(\mathbf{x}, u) \\ f_2(\mathbf{x}, u) \end{bmatrix} = \begin{bmatrix} x_2 \\ -\dfrac{k}{I}x_1 - \dfrac{b}{I}x_2 + \dfrac{H}{I}u\cos x_1 \end{bmatrix}$$

Clearly, $f_2(\mathbf{x}, u)$ involves a nonlinear term in x_1 and u. By expanding $\cos x_1$ into its series representation,

$$\cos x_1 = 1 - \frac{1}{2}x_1^2 + \cdots$$

and noting that x_1 is a very small angle, we may approximate $\cos x_1$ by unity to obtain the following linearized state equation:

$$\begin{bmatrix} \dot{x}_1 \\ \dot{x}_2 \end{bmatrix} = \begin{bmatrix} 0 & 1 \\ -\dfrac{k}{I} & -\dfrac{b}{I} \end{bmatrix} \begin{bmatrix} x_1 \\ x_2 \end{bmatrix} + \begin{bmatrix} 0 \\ \dfrac{H}{I} \end{bmatrix} u$$

The output equation is

$$y = \begin{bmatrix} 1 & 0 \end{bmatrix} \begin{bmatrix} x_1 \\ x_2 \end{bmatrix}$$

A–2–23. Linearize the nonlinear equation

$$z = xy$$

in the region $5 \le x \le 7$, $10 \le y \le 12$. Find the error if the linearized equation is used to calculate the value of z when $x = 5$, $y = 10$.

Solution. Since the region considered is given by $5 \le x \le 7$, $10 \le y \le 12$, choose $\bar{x} = 6$, $\bar{y} = 11$. Then $\bar{z} = \bar{x}\bar{y} = 66$. Let us obtain a linearized equation for the nonlinear equation near a point $\bar{x} = 6$, $\bar{y} = 11$.

Expanding the nonlinear equation into a Taylor series about point $x = \bar{x}$, $y = \bar{y}$ and neglecting the higher-order terms, we have

$$z - \bar{z} = a(x - \bar{x}) + b(y - \bar{y})$$

where

$$a = \left. \frac{\partial(xy)}{\partial x} \right|_{x=\bar{x}, y=\bar{y}} = \bar{y} = 11$$

$$b = \left. \frac{\partial(xy)}{\partial y} \right|_{x=\bar{x}, y=\bar{y}} = \bar{x} = 6$$

Hence the linearized equation is

$$z - 66 = 11(x - 6) + 6(y - 11)$$

or

$$z = 11x + 6y - 66$$

When $x = 5$, $y = 10$, the value of z given by the linearized equation is

$$z = 11x + 6y - 66 = 55 + 60 - 66 = 49$$

The exact value of z is $z = xy = 50$. The error is thus $50 - 49 = 1$. In terms of percentage, the error is 2%.

PROBLEMS

B–2–1. Obtain a state-space representation of the system shown in Fig. 2–60 using the method presented in Section 2–2.

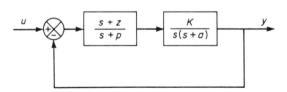

Figure 2–60 Control system.

B–2–2. Obtain a state-space representation of the mechanical system shown in Figure 2–61, where u_1 and u_2 are the inputs and y_1 and y_2 are the outputs.

B–2–3. Figure 2–62 is a schematic diagram of an accelerometer. Assume that the case of the accelerometer is attached to an aircraft frame. In the diagram, x is the displacement of mass m relative to inertial space and is measured from the position where the spring is neither compressed nor stretched,

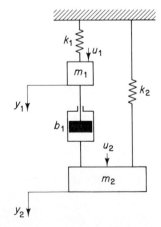

Figure 2–61 Mechanical system.

and y is the displacement of the case relative to inertial space. The accelerometer indicates the acceleration of its case with respect to inertia space. The tilt angle θ measured from the

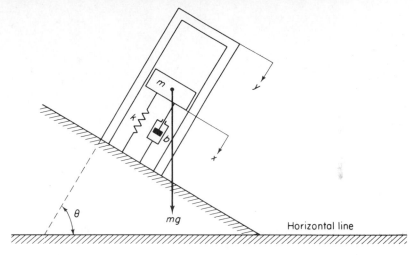

Figure 2–62 Schematic diagram of an accelerometer system.

horizontal line is assumed to be constant during the measurement period. The equation of motion for the system is

$$m\ddot{x} + b(\dot{x} - \dot{y}) + k(x - y) = mg \sin \theta$$

Considering the acceleration $\ddot{y}$ (acceleration of the case relative to the inertial space) as the input to this system and z, where

$$z = x - y - \frac{mg}{k} \sin \theta$$

as the output, obtain a mathematical model (transfer function) relating z and $\ddot{y}$.

B–2–4. Obtain mathematical models of the mechanical systems shown in Figures 2–63(a) and (b).

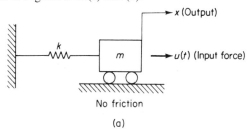

(a)

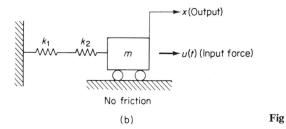

(b)

Fig

Figure 2–63 Mechanical systems.

B–2–5. Consider the inverted pendulum system shown in Figure 2–64. Assume that the mass of the inverted pendulum is m and is evenly distributed along the length of the rod. (The center of gravity of the pendulum is located at the center of the rod.) Apply Newton's second law to the system and derive the equations of motion when the angle θ is small. Then obtain a mathematical model of the system in a state space.

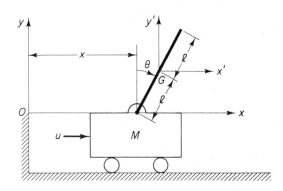

Figure 2–64 Inverted pendulum system.

B–2–6. Consider the spring-loaded pendulum system shown in Figure 2–65. Assume that the spring force acting on the pendulum is zero when the pendulum is vertical, or $\theta = 0$. Assume also that the friction involved is negligible and the angle of oscillation θ is small. Obtain a mathematical model of the system.

B–2–7. Obtain the transfer function $X_o(s)/X_i(s)$ of each of the three mechanical systems shown in Figure 2–66. In the dia-

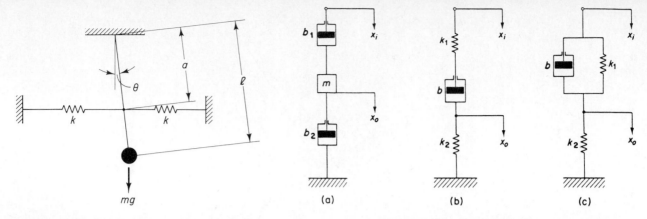

Figure 2–65 Spring-loaded pendulum system.

Figure 2–66 Mechanical systems.

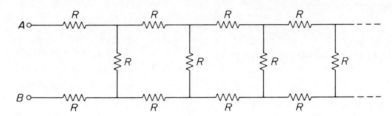

Figure 2–67 Electrical circuit.

grams, x_i denotes the input displacement and x_o denotes the output displacement. (Each displacement is measured from its equilibrium position.)

B–2–8. Obtain the resistance between points A and B of the circuit shown in Figure 2–67, which consists of an infinite number of resistors connected in the form of a ladder.

B–2–9. Obtain the transfer function $E_o(s)/E_i(s)$ of the op-amp circuit shown in Figure 2–68.

B–2–10. Obtain the transfer function $E_o(s)/E_i(s)$ of the op-amp circuit shown in Figure 2–69.

B–2–11. Derive the transfer function of the electrical system shown in Figure 2–70. Draw a schematic diagram of an equivalent mechanical system.

B–2–12. Derive the transfer function of the mechanical system shown in Figure 2–71. Draw a schematic diagram of an equivalent electrical system.

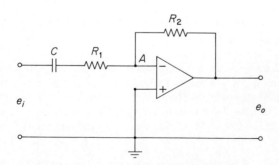

Figure 2–68 Operational amplifier circuit.

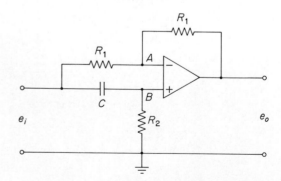

Figure 2–69 Operational amplifier cicuit.

B–2–13. Figure 2–72 shows a block diagram of a speed control system. Obtain the transfer function between $\Omega(s)$ and $E_i(s)$.

B–2–14. Obtain the transfer function of the two-phase servo-motor whose torque–speed curve is shown in Figure 2–73. The maximum-rated fixed-phase and control-phase voltages are 115 volts. The moment of inertia J of the rotor (including the effect

of load) is 7.77×10^{-4} oz-in.-sec^2 and the viscous-friction coefficient of the motor (including the effect of load) is 0.005 oz-in./rad/sec.

B–2–15. Consider the system shown in Figure 2–74. An armature-controlled dc servomotor drives a load consisting of moment of inertia J_L. The torque developed by the motor is T.

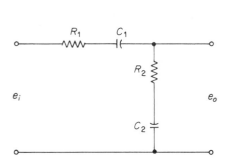

Figure 2–70   Electrical system.

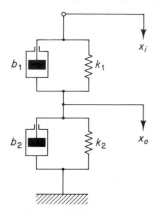

Figure 2–71   Mechanical system.

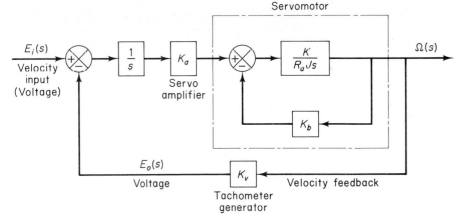

Figure 2–72   Speed control system.

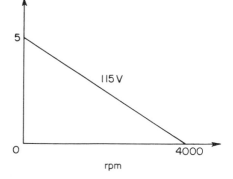

Figure 2–73   Torque–speed curve.

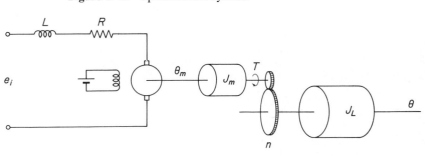

Figure 2–74   Armature-controlled dc servomotor system.

The angular displacements of the motor rotor and the load element are θ_m and θ, respectively. The gear ratio is $n = \theta/\theta_m$. Obtain the transfer function $\Theta(s)/E_i(s)$.

B–2–16. Consider the liquid-level system shown in Figure 2–75. Assuming that $\bar{H} = 3$ m, $\bar{Q} = 0.02$ m³/sec, and the cross-sectional area of the tank is equal to 5 m², obtain the time constant of the system at the operating point $(\bar{H}, \bar{Q})$. Assume that the flow through the valve is turbulent.

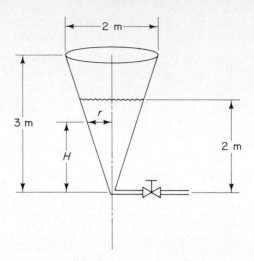

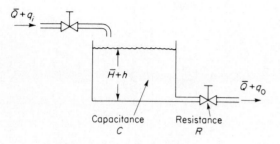

Figure 2–75 Liquid-level system.

B–2–17. Consider the conical water tank system shown in Figure 2–76. The flow through the valve is turbulent and is related to the head H by

$$Q = 0.005\sqrt{H}$$

where Q is the flow rate measured in m³/sec and H is in meters.

Suppose that the head is 2 m at $t = 0$. What will be the head at $t = 60$ sec?

B–2–18. Consider the liquid-level system shown in Figure 2–77. At steady state the inflow rate is $\bar{Q}$ and the outflow rate is also $\bar{Q}$. Assume that at $t = 0$ the inflow rate is changed from

Figure 2–76 Conical water tank system.

$\bar{Q}$ to $\bar{Q} + q_i$, where q_i is a small quantity. The disturbance input is q_d, which is also a small quantity. Draw a block diagram of the system and simplify it to obtain $H_2(s)$ as a function of $Q_i(s)$ and $Q_d(s)$, where $H_2(s) = \mathcal{L}[h_2(t)]$, $Q_i(s) = \mathcal{L}[q_i(t)]$, and $Q_d(s) = \mathcal{L}[q_d(t)]$. The capacitances of tanks 1 and 2 are C_1 and C_2, respectively.

B–2–19. A thermocouple has a time constant of 2 sec. A thermal well has a time constant of 30 sec. When the thermocouple is inserted into the well, this temperature-measuring device can be considered a two-capacitance system.

Determine the time constants of the combined thermocouple–thermal well system. Assume that the weight of the thermocouple is 8 g and the weight of the thermal well is 40 g. Assume also that the specific heats of the thermocouple and thermal well are the same.

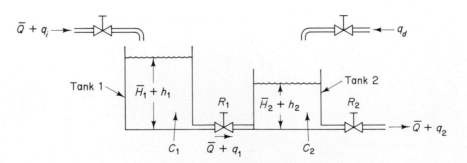

Figure 2–77 Liquid-level system.

B–2–20. Consider the O–xyz coordinate system shown in Figure 2–78(a). Assume that this coordinate system is rotated about the x axis by an angle θ. Define the resulting coordinate system as O–$x'y'z'$ coordinate system. Then the O–$x'y'z'$ coordinate system is rotated about the y' axis by an angle ϕ, as shown in Figure 2–78(b). Define the resulting coordinate system as the O–$x''y''z''$ coordinate system. Obtain the transformation matrix that relates (x, y, z) and (x'', y'', z'').

B–2–21. Suppose that the flow rate Q and head H in a liquid-level system are related by

$$Q = 0.002 \sqrt{H}$$

Obtain a linearized mathematical model relating the flow rate

and head near the steady-state operating point $(\bar{H}, \bar{Q})$, where $\bar{H} = 2.25$ m and $\bar{Q} = 0.003$ m³/s.

B–2–22. Find a linearized equation for

$$y = 0.2x^3$$

about a point $\bar{x} = 2$.

B–2–23. Linearize the nonlinear equation

$$z = x^2 + 4xy + 6y^2$$

in the region defined by $8 \le x \le 10$, $2 \le y \le 4$.

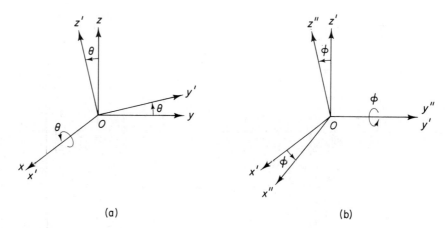

(a)   (b)

Figure 2–78   (a) O–xyz coordinate system and O–$x'y'z'$ coordinate system; (b) O–$x'y'z'$ coordinate system and O–$x''y''z''$ coordinate system.

CHAPTER 3

Basic Control Actions and Industrial Automatic Controllers

3–1 INTRODUCTION

An automatic controller compares the actual value of the plant output with the reference input (desired value), determines the deviation, and produces a control signal that will reduce the deviation to zero or to a small value. The manner in which the automatic controller produces the control signal is called the *control action*.

In this chapter we shall first discuss the basic control actions used in industrial control systems. Next, we shall present principles of pneumatic controllers and hydraulic controllers. Then we shall discuss effects of integral and derivative control actions on system performance. Finally, we shall present methods for reducing parameter variations by use of feedback.

The outline of this chapter is as follows: Section 3–1 gives introductory material. Section 3–2 presents the basic control actions commonly used in industrial automatic controllers. Sections 3–3 and 3–4 discuss pneumatic controllers and hydraulic controllers, respectively. Here we introduce the principle of operation of pneumatic and hydraulic controllers and methods for generating various control actions. Section 3–5 discusses the effects of particular control actions on the system performance. Finally, Section 3–6 presents a brief discussion of the reduction of parameter variations by the use of feedback.

3–2 BASIC CONTROL ACTIONS

In this section we shall discuss the details of basic control actions used in industrial analog controllers. We shall begin with classifications of industrial analog controllers.

Classifications of industrial analog controllers.   Industrial analog controllers may be classified according to their control actions as:

1. Two-position or on–off controllers
2. Proportional controllers
3. Integral controllers
4. Proportional-plus-integral controllers
5. Proportional-plus-derivative controllers
6. Proportional-plus-integral-plus-derivative controllers

Most industrial analog controllers use electricity or pressurized fluid such as oil or air as power sources. Analog controllers may also be classified according to the kind of power employed in the operation, such as pneumatic controllers, hydraulic controllers, or electronic controllers. The kind of controller to use must be decided based on the nature of the plant and the operating conditions, including such considerations as safety, cost, availability, reliability, accuracy, weight, and size.

Automatic controller, actuator, and sensor (measuring element).   Figure 3–1 is a block diagram of an industrial control system, which consists of an automatic controller, an actuator, a plant, and a sensor (measuring element). The controller detects the actuating error signal, which is usually at a very low power level, and amplifies it to a sufficiently high level. (Thus the automatic controller comprises an error detector and amplifier. Quite often a suitable feedback circuit, together with an amplifier, is used to alter the actuating error signal by amplifying and sometimes by differentiating and/or integrating it to produce a better control signal.) The actuator is a power device that produces the input to the plant according to the control signal so that the feedback signal will correspond to the reference input signal. The output of an automatic controller is fed to an actuator, such as a pneumatic motor or valve, a hydraulic motor, or an electric motor.

The sensor or measuring element is a device that converts the output variable into another suitable variable, such as a displacement, pressure, or voltage, that can be used to compare the output to the reference input signal. This element is in the feedback path of the closed-loop system. The set point of the controller must be converted to a reference input with the same units as the feedback signal from the sensor or measuring element.

Figure 3–1
Block diagram of an industrial control system, which consists of an automatic controller, an actuator, a plant, and a sensor (measuring element).

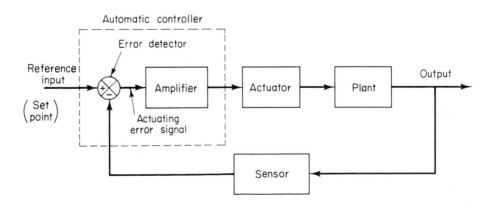

Self-operated controllers.   In most industrial automatic controllers, separate units are used for the measuring element and for the actuator. In a very simple one, however, such as a self-operated controller, these elements are assembled in one unit. Self-operated controllers utilize power developed by the measuring element and are very simple and inexpensive. An example of such a self-operated controller is shown in Figure 3–2. The set point is determined by the adjustment of the spring force. The controlled pressure is measured by the diaphragm. The actuating error signal is the net force acting on the diaphragm. Its position determines the valve opening.

The operation of the self-operated controller is as follows: Suppose that the output pressure is lower than the reference pressure, as determined by the set point. Then the downward spring force is greater than the upward pressure force, resulting in a downward movement of the diaphragm. This increases the flow rate and raises the output pressure. When the upward pressure force equals the downward spring force, the valve plug stays stationary and the flow rate is constant. Conversely, if the output pressure is higher than the reference pressure, the valve opening becomes small and reduces the flow rate through the valve opening. Such a self-operated controller is widely used for water and gas pressure control.

Control actions.   The following six basic control actions are very common among industrial analog controllers: two-position or on–off, proportional, integral, proportional-plus-integral, proportional-plus-derivative, and proportional-plus-integral-plus-derivative control action. These six will be discussed in this chapter. Note that an understanding of the basic characteristics of the various control actions is necessary in order for the control engineer to select the one best suited to his particular application.

Two-position or on–off control action.   In a two-position control system, the actuating element has only two fixed positions, which are, in many cases, simply on and off. Two-position or on–off control is relatively simple and inexpensive and, for this reason, is very widely used in both industrial and domestic control systems.

Let the output signal from the controller be $u(t)$ and the actuating error signal be $e(t)$. In two-position control, the signal $u(t)$ remains at either a maximum or minimum value, depending on whether the actuating error signal is positive or negative, so that

$$u(t) = U_1 \qquad \text{for } e(t) > 0$$
$$= U_2 \qquad \text{for } e(t) < 0$$

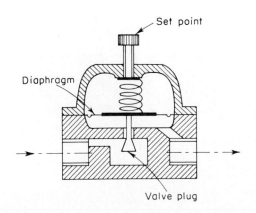

Figure 3–2
Self-operated
controller.

Figure 3–3
(a) Block diagram of
an on–off controller;
(b) block diagram of
an on–off controller
with differential gap.

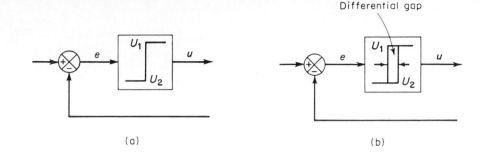

(a) (b)

where U_1 and U_2 are constants. The minimum value U_2 is usually either zero or $-U_1$. Two-position controllers are generally electrical devices, and an electric solenoid-operated valve is widely used in such controllers. Pneumatic proportional controllers with very high gains act as two-position controllers and are sometimes called pneumatic two-position controllers.

Figures 3–3(a) and (b) show the block diagrams for two-position controllers. The range through which the actuating error signal must move before the switching occurs is called the differential gap. A differential gap is indicated in Figure 3–3(b). Such a differential gap causes the controller output $u(t)$ to maintain its present value until the actuating error signal has moved slightly beyond the zero value. In some cases, the differential gap is a result of unintentional friction and lost motion; however, quite often it is intentionally provided in order to prevent too frequent operation of the on–off mechanism.

Consider the liquid-level control system shown in Figure 3–4(a), where the electromagnetic valve shown in Figure 3–4(b) is used for controlling the inflow rate. This valve is either open or closed. With this two-position control, the water inflow rate is either a positive constant or zero. As shown in Figure 3–5, the output signal continuously moves between the two limits required to cause the actuating element to move from one fixed position to the other. Notice that the output curve follows one of two exponential curves, one corresponding to the filling curve and the other to the emptying curve. Such output oscillation between two limits is a typical response characteristic of a system under two-position control.

From Figure 3–5, we notice that the amplitude of the output oscillation can be reduced by decreasing the differential gap. The decrease in the differential gap, however, increases the number of on–off switchings per minute and reduces the useful life of the component.

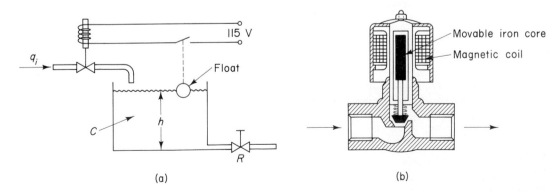

(a) (b)

Figure 3–4 (a) Liquid-level control system; (b) electromagnetic valve.

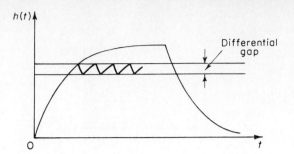

Figure 3–5
Level $h(t)$ versus t
curve for the system
shown in Fig. 3–4(a).

The magnitude of the differential gap must be determined from such considerations as the accuracy required and the life of the component.

Proportional control action. For a controller with proportional control action, the relationship between the output of the controller $u(t)$ and the actuating error signal $e(t)$ is

$$u(t) = K_p e(t)$$

or, in Laplace-transformed quantities,

$$\frac{U(s)}{E(s)} = K_p$$

where K_p is termed the proportional gain.

Whatever the actual mechanism may be and whatever the form of the operating power, the proportional controller is essentially an amplifier with an adjustable gain. A block diagram of such a controller is shown in Figure 3–6.

Integral control action. In a controller with integral control action, the value of the controller output $u(t)$ is changed at a rate proportional to the actuating error signal $e(t)$. That is,

$$\frac{du(t)}{dt} = K_i e(t)$$

or

$$u(t) = K_i \int_0^t e(t)\, dt$$

where K_i is an adjustable constant. The transfer function of the integral controller is

$$\frac{U(s)}{E(s)} = \frac{K_i}{s}$$

If the value of $e(t)$ is doubled, then the value of $u(t)$ varies twice as fast. For zero actuating error, the value of $u(t)$ remains stationary. The integral control action is sometimes called reset control. Figure 3–7 shows a block diagram of such a controller.

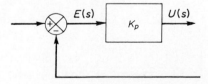

Figure 3–6
Block diagram of a
proportional
controller.

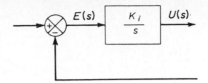

Figure 3–7
Block diagram of an
integral controller.

Proportional-plus-integral control action. The control action of a proportional-plus-integral controller is defined by the following equation:

$$u(t) = K_p e(t) + \frac{K_p}{T_i} \int_0^t e(t)\, dt$$

or the transfer function of the controller is

$$\frac{U(s)}{E(s)} = K_p \left(1 + \frac{1}{T_i s}\right)$$

where K_p is the proportional gain, and T_i is called the integral time. Both K_p and T_i are adjustable. The integral time adjusts the integral control action, while a change in the value of K_p affects both the proportional and integral parts of the control action. The inverse of the integral time T_i is called the reset rate. The reset rate is the number of times per minute that the proportional part of the control action is duplicated. Reset rate is measured in terms of repeats per minute. Figure 3–8(a) shows a block diagram of a proportional-plus-integral controller. If the actuating error signal $e(t)$ is a unit-step function as shown in Figure 3–8(b), then the controller output $u(t)$ becomes as shown in Figure 3–8(c).

Proportional-plus-derivative control action. The control action of a proportional-plus-derivative controller is defined by the following equation:

$$u(t) = K_p e(t) + K_p T_d \frac{de(t)}{dt}$$

and the transfer function is

$$\frac{U(s)}{E(s)} = K_p(1 + T_d s)$$

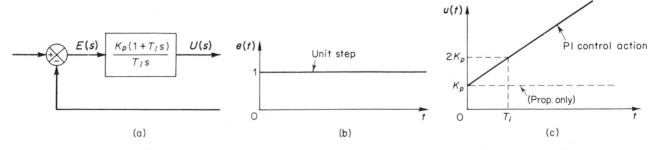

(a) (b) (c)

Figure 3–8 (a) Block diagram of a proportional-plus-integral controller; (b) and (c) diagrams depicting a unit-step input and the controller output.

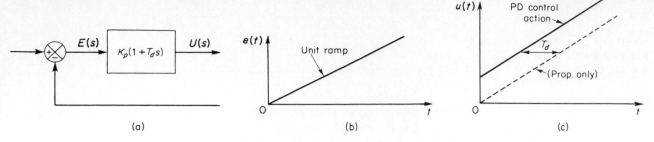

$E(s)$ $K_p(1 + T_d s)$ $U(s)$

$e(t)$

Unit ramp

O t

(b)

$u(t)$ PD control action

T_d

(Prop. only)

O t

(c)

Figure 3–9 (a) Block diagram of a proportional-plus-derivative controller; (b) and (c) diagrams depicting a unit-ramp input and the controller output.

where K_p is the proportional gain and T_d is a constant called the *derivative time*. Both K_p and T_d are adjustable. The derivative control action, sometimes called rate control, is where the magnitude of the controller output is proportional to the rate of change of the actuating error signal. The derivative time T_d is the time interval by which the rate action advances the effect of the proportional control action. Figure 3–9(a) shows a block diagram of a proportional-plus-derivative controller. If the actuating error signal $e(t)$ is a unit-ramp function as shown in Figure 3–9(b), then the controller output $u(t)$ becomes as shown in Figure 3–9(c). As may be seen from Figure 3–9(c), the derivative control action has an anticipatory character. As a matter of course, however, derivative control action can never anticipate any action that has not yet taken place.

While derivative control action has an advantage of being anticipatory, it has the disadvantages that it amplifies noise signals and may cause a saturation effect in the actuator.

Note that derivative control action can never be used alone because this control action is effective only during transient periods.

Proportional-plus-integral-plus-derivative control action. The combination of proportional control action, integral control action, and derivative control action is termed proportional-plus-integral-plus-derivative control action. This combined action has the advantages of each of the three individual control actions. The equation of a controller with this combined action is given by

$$u(t) = K_p e(t) + \frac{K_p}{T_i} \int_0^t e(t) \, dt + K_p T_d \frac{de(t)}{dt}$$

or the transfer function is

$$\frac{U(s)}{E(s)} = K_p \left(1 + \frac{1}{T_i s} + T_d s \right)$$

where K_p is the proportional gain, T_i is the integral time, and T_d is the derivative time. The block diagram of a proportional-plus-integral-plus-derivative controller is shown in Figure 3–10(a). If $e(t)$ is a unit-ramp function as shown in Figure 3–10(b), then the controller output $u(t)$ becomes as shown in Figure 3–10(c).

Effects of the sensor (measuring element) on system performance. Since the dynamic and static characteristics of the sensor or measuring element affect the indication

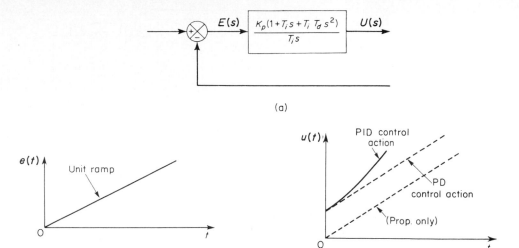

(a)

Figure 3–10

(a) Block diagram of a proportional-plus-integral-plus-derivative controller; (b) and (c) diagrams depicting a unit-ramp input and the controller output.

(b)

(c)

of the actual value of the output variable, the sensor plays an important role in determining the overall performance of the control system. The sensor usually determines the transfer function in the feedback path. If the time constants of a sensor are negligibly small compared with other time constants of the control system, the transfer function of the sensor simply becomes a constant. Figures 3–11(a), (b), and (c) show block diagrams of automatic controllers having a first-order sensor, an overdamped second-order sensor, and an underdamped second-order sensor, respectively. The response of a thermal sensor is often of the overdamped second-order type.

Block diagrams of automatic control systems. A block diagram of a simple automatic control system may be obtained by connecting the plant to the automatic controller, as shown in Figure 3–12. Feedback of the output signal is accomplished by the sensor. The

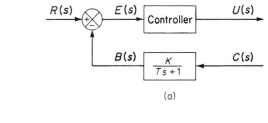

(a)

Figure 3–11

Block diagrams of automatic controllers with (a) first-order sensor; (b) overdamped second-order sensor; (c) underdamped second-order sensor.

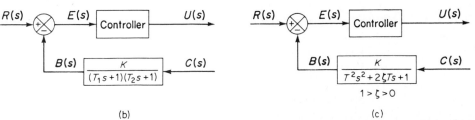

(b)

(c)

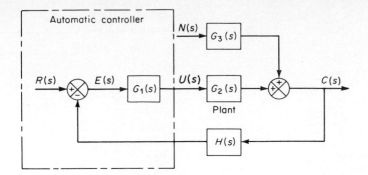

Figure 3–12
Block diagram of a
control system.

equation relating the output variable $C(s)$ to the reference input $R(s)$ and disturbance variable $N(s)$ may be obtained as follows:

$$C(s) = \frac{G_1(s)G_2(s)}{1 + G_1(s)G_2(s)H(s)} R(s) + \frac{G_3(s)}{1 + G_1(s)G_2(s)H(s)} N(s)$$

In process control systems, we are usually interested in the response to the load disturbance $N(s)$. In servo systems, however, the response to a varying input $R(s)$ is of most interest. We shall postpone the analysis of the system response to changes in load disturbances to Section 3–5. The system response to changes in the reference input will be studied in detail in Chapter 4.

3–3 PNEUMATIC CONTROLLERS

As the most versatile medium for transmitting signals and power, fluids, either as liquids or gases, have wide usage in industry. Liquids and gases can be distinguished basically by their relative incompressibilities and the fact that a liquid may have a free surface, whereas a gas expands to fill its vessel. In the engineering field the term *pneumatic* describes fluid systems that use air or gases and *hydraulic* applies to those using oil.

Pneumatic systems are extensively used in the automation of production machinery and in the field of automatic controllers. For instance, pneumatic circuits that convert the energy of compressed air into mechanical energy enjoy wide usage, and various types of pneumatic controllers are found in industry.

Since pneumatic systems and hydraulic systems are often compared, in what follows we shall give a brief comparison of these two kinds of systems.

Comparison between pneumatic systems and hydraulic systems. The fluid generally found in pneumatic systems is air; in hydraulic systems it is oil. And it is primarily the different properties of the fluids involved that characterize the differences between the two systems. These differences can be listed as follows.

1. Air and gases are compressible, whereas oil is incompressible.
2. Air lacks lubricating property and always contains water vapor. Oil functions as a hydraulic fluid as well as a lubricator.
3. The normal operating pressure of pneumatic systems is very much lower than that of hydraulic systems.

4. Output powers of pneumatic systems are considerably less than those of hydraulic systems.

5. Accuracy of pneumatic actuators is poor at low velocities, whereas accuracy of hydraulic actuators may be made satisfactory at all velocities.

6. In pneumatic systems external leakage is permissible to a certain extent, but internal leakage must be avoided because the effective pressure difference is rather small. In hydraulic systems internal leakage is permissible to a certain extent, but external leakage must be avoided.

7. No return pipes are required in pneumatic systems when air is used, whereas they are always needed in hydraulic systems.

8. Normal operating temperature for pneumatic systems is 5° to 60°C (41° to 140°F). The pneumatic system, however, can be operated in the 0° to 200°C (32° to 392°F) range. Pneumatic systems are insensitive to temperature changes, in contrast to hydraulic systems, where fluid friction due to viscosity depends greatly on temperature. Normal operating temperature for hydraulic systems is 20° to 70°C (68° to 158°F).

9. Pneumatic systems are fire- and explosion-proof, whereas hydraulic systems are not.

In what follows we begin with a mathematical modeling of pneumatic systems. Then we shall present pneumatic proportional controllers. We shall illustrate the fact that proportional controllers utilize the principle of negative feedback in themselves. We shall give a detailed discussion of the principle by which proportional controllers operate. Finally, we shall treat methods for obtaining derivative and integral control actions. Throughout the discussions, we shall place emphasis on the fundamental principles, rather than on the details of the operation of the actual mechanisms.

Pneumatic systems. The past decades have seen a great development in low-pressure pneumatic controllers for industrial control systems, and today they are used extensively in industrial processes. Reasons for their broad appeal include an explosion-proof character, simplicity, and ease of maintenance.

Resistance and capacitance of pressure systems. Many industrial processes and pneumatic controllers involve the flow of a gas or air through connected pipelines and pressure vessels.

Consider the pressure system shown in Figure 3–13(a). The gas flow through the restriction is a function of the gas pressure difference $p_i - p_o$. Such a pressure system may be characterized in terms of a resistance and a capacitance.

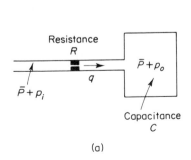

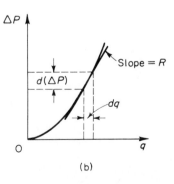

Figure 3–13
(a) Schematic diagram of a pressure system;
(b) pressure difference versus flow rate curve.

(a)

(b)

The gas flow resistance R may be defined as follows:

$$R = \frac{\text{change in gas pressure difference, lb}_f/\text{ft}^2}{\text{change in gas flow rate, lb/sec}}$$

or

$$R = \frac{d\,(\Delta P)}{dq} \tag{3-1}$$

where $d\,(\Delta P)$ is a small change in the gas pressure difference and dq is a small change in the gas flow rate. Computation of the value of the gas flow resistance R may be quite time consuming. Experimentally, however, it can be easily determined from a plot of the pressure difference versus flow curve by calculating the slope of the curve at a given operating condition, as shown in Figure 3–13(b).

The capacitance of the pressure vessel may be defined by

$$C = \frac{\text{change in gas stored, lb}}{\text{change in gas pressure, lb}_f/\text{ft}^2}$$

or

$$C = \frac{dm}{dp} = V\frac{d\rho}{dp} \tag{3-2}$$

where C = capacitance, lb-ft^2/lb$_f$

m = mass of gas in vessel, lb

p = gas pressure, lb$_f$/ft^2

V = volume of vessel, ft^3

ρ = density, lb/ft^3

The capacitance of the pressure system depends on the type of expansion process involved. The capacitance can be calculated by use of the ideal gas law. (See Problem A–3–3.) If the gas expansion process is polytropic and the change of state of the gas is between isothermal and adiabatic, then

$$p\left(\frac{V}{m}\right)^n = \frac{p}{\rho^n} = \text{constant} \tag{3-3}$$

where n = polytropic exponent.

For ideal gases,

$$p\bar{v} = \bar{R}T \quad \text{or} \quad pv = \frac{\bar{R}}{M}T$$

where p = absolute pressure, lb$_f$/ft^2

$\bar{v}$ = the volume occupied by 1 mole of a gas, ft^3/lb-mole

$\bar{R}$ = universal gas constant, ft-lb$_f$/lb-mole °R

T = absolute temperature, °R

v = specific volume of gas, ft^3/lb

M = molecular weight of gas per mole, lb/lb-mole

Thus

$$pv = \frac{p}{\rho} = \frac{\bar{R}}{M} T = R_{\text{gas}}T \qquad (3\text{--}4)$$

where R_{gas} = gas constant, ft-lb$_f$/lb °R.

The polytropic exponent n is unity for isothermal expansion. For adiabatic expansion, n is equal to the ratio of specific heats c_p/c_v, where c_p is the specific heat at constant pressure and c_v is the specific heat at constant volume. In many practical cases, the value of n is approximately constant, and thus the capacitance may be considered constant. The value of $d\rho/dp$ is obtained from Equations (3–3) and (3–4) as

$$\frac{d\rho}{dp} = \frac{1}{nR_{\text{gas}}T}$$

The capacitance is then obtained as

$$C = \frac{V}{nR_{\text{gas}}T} \qquad (3\text{--}5)$$

The capacitance of a given vessel is constant if the temperature stays constant. (In many practical cases, the polytropic exponent n is approximately $1.0 \sim 1.2$ for gases in uninsulated metal vessels.)

Pressure systems. Consider the system shown in Fig. 3–13(a). If we assume only small deviations in the variables from their respective steady-state values, then this system may be considered linear.

Let us define

$\bar{P}$ = gas pressure in the vessel at steady state (before changes in pressure have occurred), lb$_f$/ft^2

p_i = small change in inflow gas pressure, lb$_f$/ft^2

p_o = small change in gas pressure in the vessel, lb$_f$/ft^2

V = volume of the vessel, ft^3

m = mass of gas in vessel, lb

q = gas flow rate, lb/sec

ρ = density of gas, lb/ft^3

For small values of p_i and p_o, the resistance R given by Equation (3–1) becomes constant and may be written as

$$R = \frac{p_i - p_o}{q}$$

The capacitance C is given by Equation (3–2), rewritten

$$C = \frac{dm}{dp} = V\frac{d\rho}{dp}$$

Since the pressure change dp_o times the capacitance C is equal to the gas added to the vessel during dt seconds, we obtain

$$C\,dp_o = q\,dt$$

or

$$C \frac{dp_o}{dt} = \frac{p_i - p_o}{R}$$

which can be written as

$$RC \frac{dp_o}{dt} + p_o = p_i$$

If p_i and p_o are considered the input and output, respectively, then the transfer function of the system is

$$\frac{P_o(s)}{P_i(s)} = \frac{1}{RCs + 1}$$

where RC has the dimension of time and is the time constant of the system.

Pneumatic nozzle–flapper amplifiers. A schematic diagram of a pneumatic nozzle–flapper amplifier is shown in Figure 3–14(a). The power source for this amplifer is a supply of air at constant pressure. The nozzle–flapper amplifier converts small changes in the position of the flapper into large changes in the back pressure in the nozzle. Thus a large power output can be controlled by the very little power that is needed to position the flapper.

In Figure 3–14(a), pressurized air is fed through the orifice, and the air is ejected from the nozzle toward the flapper. Generally, the supply pressure P_s for such a controller is 20 psig (1.4 kg$_f$/cm^2 gage). The diameter of the orifice is on the order of 0.01 in. (0.25 mm) and that of the nozzle is on the order of 0.016 in. (0.4 mm). To ensure proper functioning of the amplifier, the nozzle diameter must be larger than the orifice diameter.

In operating this system, the flapper is positioned against the nozzle opening. The nozzle back pressure P_b is controlled by the nozzle–flapper distance X. As the flapper approaches the nozzle, the opposition to the flow of air through the nozzle increases, with the result that the nozzle back pressure P_b increases. If the nozzle is completely closed by the flapper,

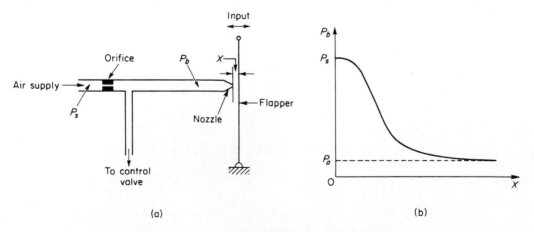

(a)   (b)

Figure 3–14 (a) Schematic diagram of a pneumatic nozzle-flapper amplifier; (b) characteristic curve relating nozzle back pressure and nozzle-flapper distance.

the nozzle back pressure P_b becomes equal to the supply pressure P_s. If the flapper is moved away from the nozzle, so that the nozzle–flapper distance is wide (on the order of 0.01 in.), then there is practically no restriction to flow, and the nozzle back pressure P_b takes on a minimum value that depends on the nozzle–flapper device. (The lowest possible pressure will be the ambient pressure P_a.)

Note that, because the air jet puts a force against the flapper, it is necessary to make the nozzle diameter as small as possible.

A typical curve relating the nozzle back pressure P_b to the nozzle–flapper distance X is shown in Figure 3–14(b). The steep and almost linear part of the curve is utilized in the actual operation of the nozzle–flapper amplifier. Because the range of flapper displacements is restricted to a small value, the change in output pressure is also small, unless the curve happens to be very steep.

The nozzle–flapper amplifier converts displacement into a pressure signal. Since industrial process control systems require large output power to operate large pneumatic actuating valves, the power amplification of the nozzle–flapper amplifier is usually insufficient. Consequently, a pneumatic relay often serves as a power amplifier in connection with the nozzle–flapper amplifier.

Pneumatic relays. In practice, in a pneumatic controller, a nozzle–flapper amplifier acts as the first-stage amplifier and a pneumatic relay as the second-stage amplifier. The pneumatic relay is capable of handling a large quantity of airflow.

A schematic diagram of a pneumatic relay is shown in Figure 3–15(a). As the nozzle back pressure P_b increases, the diaphragm valve moves downward. The opening to the atmosphere decreases and the opening to the pneumatic valve increases, thereby increasing the control pressure P_c. When the diaphragm valve closes the opening to the atmosphere, the control pressure P_c becomes equal to the supply pressure P_s. When the nozzle back pressure P_b decreases and the diaphragm valve moves upward and shuts off the air supply, the control pressure P_c drops to the ambient pressure P_a. The control pressure P_c can thus be made to vary from 0 psig to full supply pressure, usually 20 psig.

The total movement of the diaphragm valve is very small. In all positions of the valve, except at the position to shut off the air supply, air continues to bleed into the atmosphere,

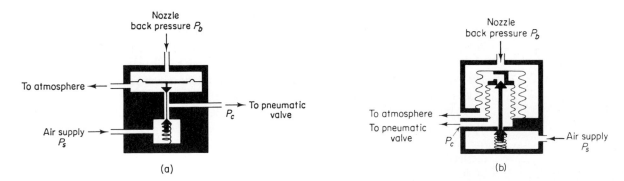

Figure 3–15 (a) Schematic diagram of a bleed-type relay; (b) schematic diagram of a nonbleed-type relay.

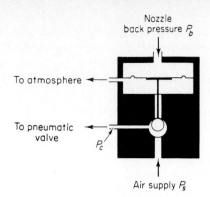

Figure 3–16
Reverse acting relay.

even after the equilibrium condition is attained between the nozzle back pressure and the control pressure. Thus the relay shown in Figure 3–15(a) is called a bleed-type relay.

There is another type of relay, the nonbleed type. In this one the air bleed stops when the equilibrium condition is obtained and, therefore, there is no loss of pressurized air at steady-state operation. Note, however, that the nonbleed-type relay must have an atmospheric relief to release the control pressure P_c from the pneumatic actuating valve. A schematic diagram of a nonbleed-type relay is shown in Figure 3–15(b).

In either type of relay, the air supply is controlled by a valve, which is in turn controlled by the nozzle back pressure. Thus, the nozzle back pressure is converted into the control pressure with power amplification.

Since the control pressure P_c changes almost instantaneously with changes in the nozzle back pressure P_b, the time constant of the pneumatic relay is negligible compared with the other larger time constants of the pneumatic controller and the plant.

It is noted that some pneumatic relays are reverse acting. For example, the relay shown in Figure 3–16 is a reverse-acting relay. Here, as the nozzle back pressure P_b increases, the ball valve is forced toward the lower seat, thereby decreasing the control pressure P_c. Thus, this relay is a reverse acting relay.

Pneumatic proportional controllers (force–distance type).   Two types of pneumatic controllers, one called the force–distance type and the other the force–balance type, are used extensively in industry. Regardless of how differently industrial pneumatic controllers may appear, careful study will show the close similarity in the functions of the pneumatic circuit. Here we shall consider the force–distance type of pneumtic controllers.

Figure 3–17(a) shows a schematic diagram of such a proportional controller. The nozzle–flapper amplifier constitutes the first-stage amplifier, and the nozzle back pressure is controlled by the nozzle–flapper distance. The relay-type amplifier constitutes the second-stage amplifier. The nozzle back pressure determines the position of the diaphragm valve for the second-stage amplifier, which is capable of handling a large quantity of air flow.

In most pneumatic controllers, some type of pneumatic feedback is employed. Feedback of the pneumatic output reduces the amount of actual movement of the flapper. Instead of mounting the flapper on a fixed point, as shown in Figure 3–17(b), it is often pivoted on the feedback bellows, as shown in Figure 3–17(c). The amount of feedback can be regulated by introducing a variable linkage between the feedback bellows and the flapper connecting point. The flapper then becomes a floating link. It can be moved by both the error signal and the feedback signal.

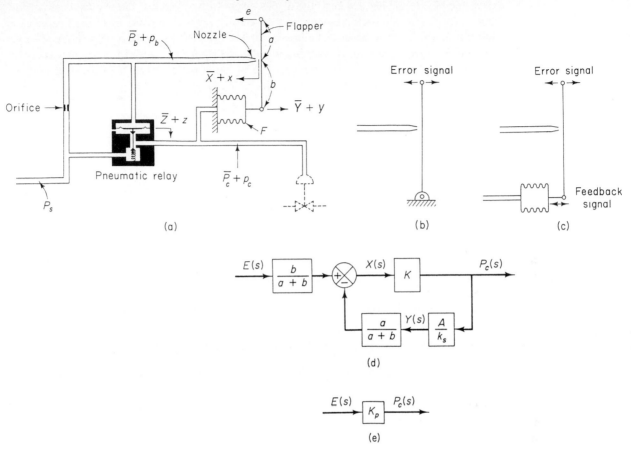

Figure 3–17 (a) Schematic diagram of a force–distance type pneumatic proportional controller; (b) flapper mounted on a fixed point; (c) flapper mounted on a feedback bellows; (d) block diagram for the controller; (e) simplified block diagram for the controller.

The operation of the controller shown in Figure 3–17(a) is as follows. The input signal to the two-stage pneumatic amplifier is the actuating error signal. Increasing the actuating error signal moves the flapper to the left. This move will, in turn, increase the nozzle back pressure, and the diaphragm valve moves downward. This results in an increase of the control pressure. This increase will cause bellows F to expand and move the flapper to the right, thus opening the nozzle. Because of this feedback, the nozzle–flapper displacement is very small, but the change in the control pressure can be large.

It should be noted that proper operation of the controller requires that the feedback bellows move the flapper less than that movement caused by the error signal alone. (If these two movements were equal, no control action would result.)

Equations for this controller can be derived as follows. When the actuating error is zero, or $e = 0$, an equilibrium state exists with the nozzle–flapper distance equal to $\bar{X}$, the displacement of bellows equal to $\bar{Y}$, the displacement of the diaphragm equal to $\bar{Z}$, the nozzle back pressure equal to $\bar{P}_b$, and the control pressure equal to $\bar{P}_c$. When an actuating error exists, the nozzle-flapper distance, the displacement of the bellows, the displacement of the diaphragm, the nozzle back pressure, and the control pressure deviate from their respective equilibrium values. Let these deviations be x, y, z, p_b, and p_c, respectively. (The positive direction for each displacement variable is indicated by an arrowhead in the diagram.)

Assuming that the relationship between the variation in the nozzle back pressure and the variation in the nozzle–flapper distance is linear, we have

$$p_b = K_1 x \tag{3-6}$$

where K_1 is a positive constant. For the diaphragm valve

$$p_b = K_2 z \tag{3-7}$$

where K_2 is a positive constant. The position of the diaphragm valve determines the control pressure. If the diaphragm valve is such that the relationship between p_c and z is linear, then

$$p_c = K_3 z \tag{3-8}$$

where K_3 is a positive constant. From Equations (3–6), (3–7), and (3–8), we obtain

$$p_c = \frac{K_3}{K_2} p_b = Kx \tag{3-9}$$

where $K = K_1 K_3 / K_2$ is a positive constant. For the flapper movement, we have

$$x = \frac{b}{a+b} e - \frac{a}{a+b} y \tag{3-10}$$

The bellows acts like a spring, and the following equation holds true:

$$A p_c = k_s y \tag{3-11}$$

where A is the effective area of the bellows and k_s is the equivalent spring constant, that is, the stiffness due to the action of the corrugated side of the bellows.

Assuming that all variations in the variables are within a linear range, we can obtain a block diagram for this system from Equations (3–9), (3–10), and (3–11) as shown in Figure 3–17(d). From Figure 3–17(d), it can be clearly seen that the pneumatic controller shown in Figure 3–17(a) itself is a feedback system. The transfer function between p_c and e is given by

$$\frac{P_c(s)}{E(s)} = \frac{\dfrac{b}{a+b} K}{1 + K \dfrac{a}{a+b} \dfrac{A}{k_s}} = K_p \tag{3-12}$$

A simplified block diagram is shown in Figure 3–17(e). Since p_c and e are proportional, the pneumatic controller shown in Figure 3–17(a) is called a *pneumatic proportional controller*. As seen from Equation (3–12) the gain of the pneumatic proportional controller can be widely varied by adjusting the effective value of k_s. This can be accomplished easily by adjusting the flapper connecting linkage. [The flapper connecting linkage is not shown in Figure 3–17(a).] In most commercial proportional controllers an adjusting knob or other mechanism is provided for varying the gain by adjusting this linkage.

As noted earlier, the actuating error signal moved the flapper in one direction, and the feedback bellows moved the flapper in the opposite direction, but to a smaller degree. The

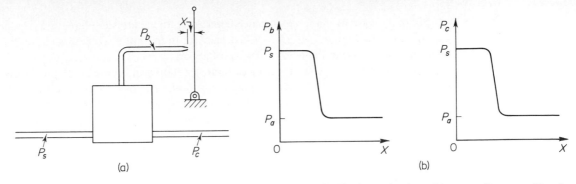

(a)   (b)

Figure 3–18 (a) Pneumatic controller without a feedback mechanism; (b) curves P_b versus X and P_c versus X.

effect of the feedback bellows is thus to reduce the sensitivity of the controller. The principle of feedback is commonly used to obtain wide proportional-band controllers.

Pneumatic controllers that do not have feedback mechanisms [which means that one end of the flapper is fixed, as shown in Figure 3–18(a)] have high sensitivity and are called *pneumatic two-position controllers* or *pneumatic on–off controllers*. In such a controller, only a small motion between the nozzle and the flapper is required to give a complete change from the maximum to the minimum control pressure. The curves relating P_b to X and P_c to X are shown in Figure 3–18(b). Notice that a small change in X can cause a large change in P_b, which causes the diaphragm valve to be completely open or completely closed.

Pneumatic proportional controllers (force–balance type).   Figure 3–19 shows a schematic diagram of a force–balance pneumatic proportional controller. Force–balance controllers are in extensive use in industry. Such controllers are called stack controllers. The basic principle of operation does not differ from that of the force–distance controller. The main advantage of the force–balance controller is that it eliminates many mechanical linkages and pivot joints, thereby reducing the effects of friction.

In what follows, we shall consider the principle of the force–balance controller. In the controller shown in Figure 3–19, the reference input pressure P_r and the output pressure P_o are fed to large diaphragm chambers. Note that a force–balance pneumatic controller operates only on pressure signals. Therefore, it is necessary to convert the reference input and system output to corresponding pressure signals.

Figure 3–19

Schematic diagram of a force–balance type pneumatic proportional controller.

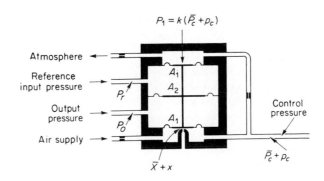

As in the case of the force–distance controller, this controller employs a flapper, nozzle, and orifices. In Figure 3–19, the drilled opening in the bottom chamber is the nozzle. The diaphragm just above the nozzle acts as a flapper.

The operation of the force–balance controller shown in Figure 3–19 may be summarized as follows: 20-psig air from an air supply flows through an orifice, causing a reduced pressure in the bottom chamber. Air in this chamber escapes to the atmosphere through the nozzle. The flow through the nozzle depends on the gap and the pressure drop across it. An increase in the reference input pressure P_r, while the output pressure P_o remains the same, causes the valve stem to move down, decreasing the gap between the nozzle and the flapper diaphragm. This causes the control pressure P_c to increase. Let

$$p_e = P_r - P_o \tag{3–13}$$

If $p_e = 0$, there is an equilibrium state with the nozzle–flapper distance equal to $\bar{X}$ and the control pressure equal to $\bar{P}_c$. At this equilibrium state, $P_1 = \bar{P}_c k$ (where $k < 1$) and

$$\bar{X} = \alpha(\bar{P}_c A_1 - \bar{P}_c k A_1) \tag{3–14}$$

where α is a constant.

Let us assume that $p_e \neq 0$ and define small variations in the nozzle–flapper distance and control pressure as x and p_c, respectively. Then we obtain the following equation:

$$\bar{X} + x = \alpha[(\bar{P}_c + p_c)A_1 - (\bar{P}c + p_c)k A_1 - p_e(A_2 - A_1)] \tag{3–15}$$

From Equations (3–14) and (3–15), we obtain

$$x = \alpha[p_c(1 - k)A_1 - p_e(A_2 - A_1)] \tag{3–16}$$

At this point, we must examine the quantity x. In the design of pneumatic controllers, the nozzle–flapper distance is quite small. In view of the fact that x/α is a higher-order term than $p_c(1 - k)A_1$ or $p_e(A_2 - A_1)$, that is, for $p_e \neq 0$,

$$\frac{x}{\alpha} \ll p_c(1 - k)A_1$$

$$\frac{x}{\alpha} \ll p_e(A_2 - A_1)$$

we may neglect the term x in our analysis. Equation (3–16) can then be rewritten to reflect this assumption as follows:

$$p_c(1 - k)A_1 = p_e(A_2 - A_1)$$

and the transfer function between p_c and p_e becomes

$$\frac{P_c(s)}{P_e(s)} = \frac{A_2 - A_1}{A_1} \frac{1}{1 - k} = K_p$$

where p_e is defined by Equation (3–13). The controller shown in Figure 3–19 is a proportional controller. The value of gain K_p increases as k approaches unity. Note that the value of k depends on the diameters of the orifices in the inlet and outlet pipes of the feedback chamber. (The value of k approaches unity as the resistance to flow in the orifice of the inlet pipe is made smaller.)

Pneumatic actuating valves. One characteristic of pneumatic controls is that they almost exclusively employ pneumatic actuating valves. A pneumatic actuating valve can provide a large power output. (Since a pneumatic actuator requires a large power input to produce a large power output, it is necessary that a sufficient quantity of pressurized air be available.) In practical pneumatic actuating valves, the valve characteristics may not be linear; that is, the flow may not be directly proportional to the valve stem position, and also there may be other nonlinear effects, such as hysteresis.

Consider the schematic diagram of a pneumatic actuating valve shown in Figure 3–20. Assume that the area of the diaphragm is A. Assume also that when the actuating error is zero, the control pressure is equal to $\bar{P}_c$ and the valve displacement is equal to $\bar{X}$.

In the following analysis, we shall consider small variations in the variables and linearize the pneumatic actuating valve. Let us define the small variation in the control pressure and the corresponding valve displacement to be p_c and x, respectively. Since a small change in the pneumatic pressure force applied to the diaphragm repositions the load, consisting of the spring, viscous friction, and mass, the force balance equation becomes

$$Ap_c = m\ddot{x} + b\dot{x} + kx \qquad (3\text{--}17)$$

where m = mass of the valve and valve stem
$\quad b$ = viscous-friction coefficient
$\quad k$ = spring constant

If the force due to the mass and viscous friction are negligibly small, then Equation (3–17) can be simplified to

$$Ap_c = kx$$

The transfer function between x and p_c thus becomes

$$\frac{X(s)}{P_c(s)} = \frac{A}{k} = K_c$$

where $X(s) = \mathscr{L}[x]$ and $P_c(s) = \mathscr{L}[p_c]$. If q_i, the change in flow through the pneumatic

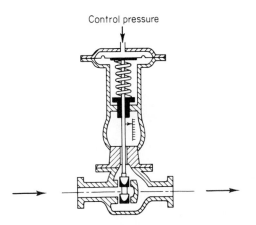

Figure 3–20
Schematic diagram of a pneumatic actuating valve.

actuating valve, is proportional to x, the change in the valve-stem displacement, then

$$\frac{Q_i(s)}{X(s)} = K_q$$

where $Q_i(s) = \mathcal{L}[q_i]$ and K_q is a constant. The transfer function between q_i and p_c becomes

$$\frac{Q_i(s)}{P_c(s)} = K_c K_q = K_v$$

where K_v is a constant.

The standard control pressure for this kind of a pneumatic actuating valve is between 3 and 15 psig. The valve-stem displacement is limited by the allowable stroke of the diaphragm and is only a few inches. If a longer stroke is needed, a piston–spring combination may be employed.

In pneumatic actuating valves, the static-friction force must be limited to a low value so that excessive hysteresis does not result. Because of the compressibility of air, the control action may not be positive; that is, an error may exist in the valve-stem position. The use of a valve positioner results in improvements in the performance of a pneumatic actuating valve.

Proportional control of a first-order system. Consider the liquid-level control system shown in Figure 3–21(a). [The controller is assumed to be the proportional controller shown in Figure 3–17(a).] We assume that all the variables r, q_i, h_1, and q_o are measured from their respective steady-state values $\bar{R}$, $\bar{Q}$, $\bar{H}$, and $\bar{Q}$. We also assume that the magnitudes

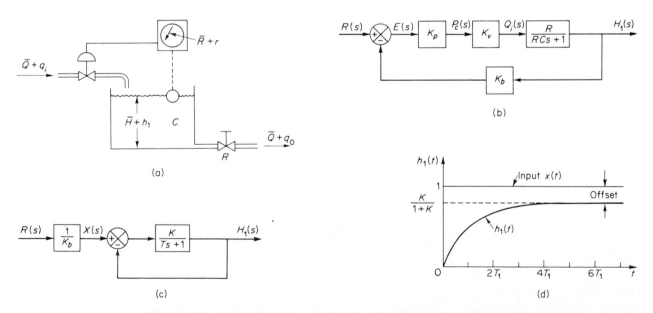

Figure 3–21 (a) Liquid-level control system; (b) block diagram; (c) simplified block diagram; (d) curve $h_1(t)$ versus t.

of the variables r, q_i, h_1, and q_o are sufficiently small so that the system can be approximated by a linear mathematical model, that is, a transfer function.

Referring to Section 2–7, we can obtain the transfer function of the liquid-level system as

$$\frac{H_1(s)}{Q_i(s)} = \frac{R}{RCs + 1}$$

Since the controller is a proportional controller, the change in inflow q_i is proportional to the actuating error e so that $q_i = K_p K_v e$, where K_p is the gain of the controller and K_v is the gain of the control valve. In terms of Laplace-transformed quantities,

$$Q_i(s) = K_p K_v E(s)$$

A block diagram of this system is shown in Figure 3–21(b). A simplified block diagram is given in Figure 3–21(c), where $X(s) = (1/K_b)R(s)$, $K = K_p K_v RK_b$, and $T = RC$.

In what follows we shall investigate the response $h_1(t)$ to a change in the reference input. We shall assume a unit-step change in $x(t)$, where $x(t) = (1/K_b)r(t)$. The closed-loop transfer function between $H_1(s)$ and $X(s)$ is given by

$$\frac{H_1(s)}{X(s)} = \frac{K}{Ts + 1 + K} \qquad\qquad (3\text{–}18)$$

Since the Laplace transform of the unit-step function is $1/s$, substituting $X(s) = 1/s$ into Equation (3–18) gives

$$H_1(s) = \frac{K}{Ts + 1 + K} \frac{1}{s}$$

Expanding $H_1(s)$ into partial fractions gives

$$H_1(s) = \frac{K}{1 + K} \frac{1}{s} - \frac{TK}{1 + K} \frac{1}{Ts + 1 + K}$$

Taking the inverse Laplace transforms of both sides of this last equation, we obtain the following time solution $h_1(t)$:

$$h_1(t) = \frac{K}{1 + K} (1 - e^{-t/T_1}) \qquad (t > 0) \qquad\qquad (3\text{–}19)$$

where

$$T_1 = \frac{T}{1 + K}$$

The response curve $h_1(t)$ is plotted in Figure 3–21(d). From Equation (3–19), notice that the time constant T_1 of the closed-loop system is different from the time constant T of the feedforward block.

From Equation (3–19), we see that as t approaches infinity the value of $h_1(t)$ approaches $K/(1 + K)$, or

$$h_1(\infty) = \frac{K}{1 + K}$$

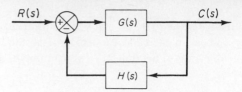

Figure 3–22
Control system.

Since $x(\infty) = 1$, there is a steady-state error of $1/(1 + K)$. Such an error is called *offset*. The value of the offset becomes smaller as the gain K becomes larger.

Offset is a characteristic of the proportional control of a plant whose transfer function does not possess an integrating element. (In such a case we need a nonzero error to provide a nonzero output.) To eliminate such offset, we must add integral control action. (Refer to Section 3–5.)

Basic principle for obtaining derivative and integral control actions. We shall now present methods for obtaining derivative and integral control actions. We shall again place the emphasis on the principle and not on the details of the actual mechanisms.

The basic principle for generating a desired control action is to insert the inverse of the desired transfer function in the feedback path. For the system shown in Figure 3–22, the closed-loop transfer function is

$$\frac{C(s)}{R(s)} = \frac{G(s)}{1 + G(s)H(s)}$$

If $|G(s)H(s)| \gg 1$, then $C(s)/R(s)$ can be modified to

$$\frac{C(s)}{R(s)} = \frac{1}{H(s)}$$

Thus, if proportional-plus-derivative control action is desired, we insert an element having the transfer function $1/(Ts + 1)$ in the feedback path.

Consider the pneumatic controller shown in Figure 3–23(a). Considering small changes in the variables, we can draw a block diagram of this controller as shown in Figure 3–23(b). From the block diagram we see that the controller is of proportional type.

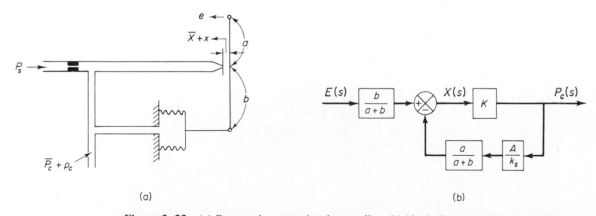

(a)   (b)

Figure 3–23 (a) Pneumatic proportional controller; (b) block diagram of the controller.

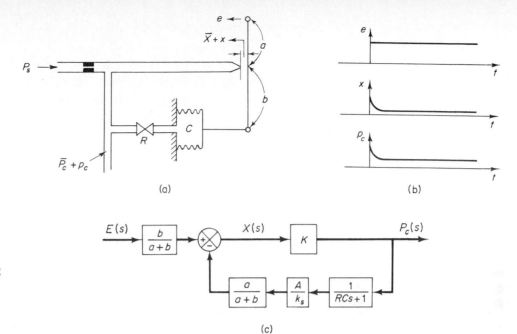

Figure 3–24
(a) Pneumatic proportional-plus-derivative controller; (b) step change in e and the corresponding changes in x and p_c plotted versus t; (c) block diagram of the controller.

We shall now show that the addition of a restriction in the negative feedback path will modify the proportional controller to a proportional-plus-derivative controller.

Consider the pneumatic controller shown in Figure 3–24(a). Assuming again small changes in the actuating error, nozzle–flapper distance, and control pressure, we can summarize the operation of this controller as follows: Let us first assume a small step change in e. Then the change in the control pressure p_c will be instantaneous. The restriction R will momentarily prevent the feedback bellows from sensing the pressure change p_c. Thus the feedback bellows will not respond momentarily, and the pneumatic actuating valve will feel the full effect of the movement of the flapper. As time goes on, the feedback bellows will expand or contract. The change in the nozzle–flapper distance x and the change in the control pressure p_c can be plotted against time t, as shown in Figure 3–24(b). At steady state, the feedback bellows acts like an ordinary feedback mechanism. The curve p_c versus t clearly shows that this controller is of the proportional-plus-derivative type.

A block diagram corresponding to this pneumatic controller is shown in Figure 3–24(c). In the block diagram, K is a constant, A is the area of the bellows, and k_s is the equivalent spring constant of the bellows. The transfer function between p_c and e can be obtained from the block diagram as follows:

$$\frac{P_c(s)}{E(s)} = \frac{\dfrac{b}{a+b}K}{1 + \dfrac{Ka}{a+b}\dfrac{A}{k_s}\dfrac{1}{RCs+1}}$$

In such a controller the loop gain $|KaA/[(a+b)k_s(RCs+1)]|$ is normally very much greater than unity. Thus the transfer function $P_c(s)/E(s)$ can be simplified to give

$$\frac{P_c(s)}{E(s)} = K_p(1 + T_d s)$$

where

$$K_p = \frac{bk_s}{aA}, \qquad T_d = RC$$

Thus, delayed negative feedback, or the transfer function $1/(RCs + 1)$ in the feedback path, modifies the proportional controller to a proportional-plus-derivative controller.

Note that if the feedback valve is fully opened, the control action becomes proportional. If the feedback valve is fully closed, the control action becomes narrow-band proportional (on–off).

Obtaining pneumatic proportional-plus-integral control action. Consider the proportional controller shown in Figure 3–23(a). Considering small changes in the variables, we can show that the addition of delayed positive feedback will modify this proportional controller to a proportional-plus-integral controller.

Consider the pneumatic controller shown in Figure 3–25(a). The operation of this controller is as follows: The bellows denoted by I is connected to the control pressure source without any restriction. The bellows denoted by II is connected to the control pressure source through a restriction. Let us assume a small step change in the actuating error. This will cause the back pressure in the nozzle to change instantaneously. Thus a change in the control pressure p_c also occurs instantaneously. Due to the restriction of the valve in the path to bellows II, there will be a pressure drop across the valve. As time goes on, air will flow across the valve in such a way that the change in pressure in bellows II attains the value p_c. Thus bellows II will expand or contract as time elapses in such a way as to move the flapper an additional amount in the direction of the original displacement e. This will cause the back pressure p_c in the nozzle to change continuously, as shown in Figure 3–25(b).

Note that the integral control action in the controller takes the form of slowly canceling the feedback that the proportional control originally provided.

A block diagram of this controller under the assumption of small variations in the variables is shown in Figure 3–25(c). A simplification of this block diagram yields Figure 3–25(d). The transfer function of this controller is

$$\frac{P_c(s)}{E(s)} = \frac{\dfrac{b}{a+b}K}{1 + \dfrac{Ka}{a+b}\dfrac{A}{k_s}\left(1 - \dfrac{1}{RCs+1}\right)}$$

where K is a constant, A is the area of the bellows, and k_s is the equivalent spring constant of the combined bellows. If $|KaARCs/[(a+b)k_s(RCs+1)]| \gg 1$, which is usually the case, the transfer function can be simplified to

$$\frac{P_c(s)}{E(s)} = K_p\left(1 + \frac{1}{T_i s}\right)$$

where

$$K_p = \frac{bk_s}{aA}, \qquad T_i = RC$$

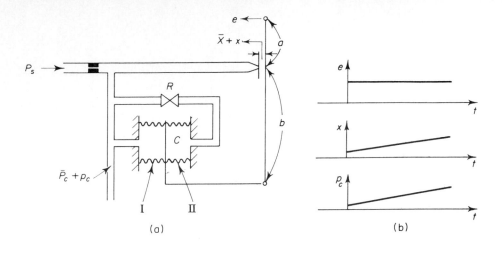

(a)

(b)

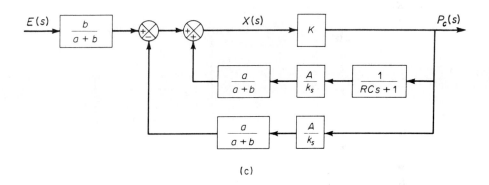

(c)

Figure 3–25
(a) Pneumatic
proportional-plus-
integral controller; (b)
step change in e and
the corresponding
changes in x and p_c
plotted versus t; (c)
block diagram of the
controller; (d)
simplified block
diagram.

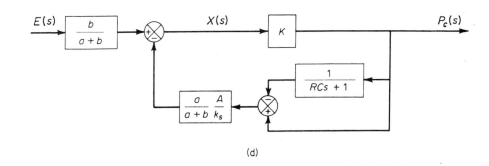

(d)

Obtaining pneumatic proportional-plus-integral-plus-derivative control action. A combination of the pneumatic controllers shown in Figures 3–24(a) and 3–25(a) yields a proportional-plus-integral-plus-derivative controller. Figure 3–26(a) shows a schematic diagram of such a controller. Figure 3–26(b) shows a block diagram of this controller under the assumption of small variations in the variables.

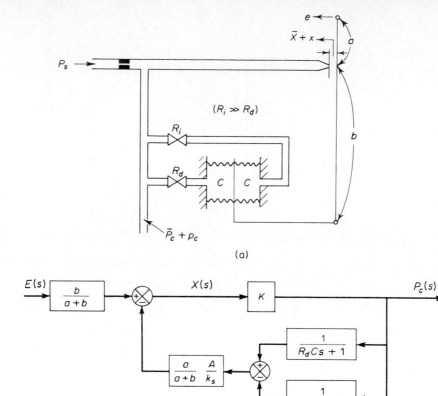

Figure 3–26
(a) Pneumatic proportional-plus-integral-plus-derivative controller; (b) block diagram of the controller.

The transfer function of this controller is

$$\frac{P_c(s)}{E(s)} = \frac{\dfrac{bK}{a+b}}{1 + \dfrac{Ka}{a+b}\dfrac{A}{k_s}\dfrac{(R_iC - R_dC)s}{(R_dCs + 1)(R_iCs + 1)}}$$

By defining

$$T_i = R_iC, \qquad T_d = R_dC$$

and noting that under normal operation $|KaA(T_i - T_d)s/[(a + b)k_s(T_ds + 1)(T_is + 1)]| \gg 1$ and $T_i \gg T_d$, we obtain

$$\frac{P_c(s)}{E(s)} \doteq \frac{bk_s}{aA}\frac{(T_ds + 1)(T_is + 1)}{(T_i - T_d)s}$$

$$\doteq \frac{bk_s}{aA}\frac{T_dT_is^2 + T_is + 1}{T_is}$$

$$= K_p\left(1 + \frac{1}{T_is} + T_ds\right) \qquad (3\text{--}20)$$

where

$$K_p = \frac{bk_s}{aA}$$

Equation (3–20) indicates that the controller shown in Figure 3–26(a) is a proportional-plus-integral-plus-derivative controller.

3–4 HYDRAULIC CONTROLLERS

Except for low-pressure pneumatic controllers, compressed air has seldom been used for the continuous control of the motion of devices having significant mass under external load forces. For such a case, hydraulic controllers are generally preferred.

Hydraulic systems.   The widespread use of hydraulic circuitry in machine tool applications, aircraft control systems, and similar operations occurs because of such factors as positiveness, accuracy, flexibility, high horsepower-to-weight ratio, fast starting, stopping, and reversal with smoothness and precision, and simplicity of operations.

The operating pressure in hydraulic systems is somewhere between 145 and 5000 lb_F/in.^{2} (between 1 and 35 MPa). In some special applications, the operating pressure may go up to 10,000 lb_F/in.^{2} (70 MPa). For the same power requirement, the weight and size of the hydraulic unit can be made smaller by increasing the supply pressure. With high-pressure hydraulic systems, very large force can be obtained. Rapid-acting accurate positioning of heavy loads is possible with hydraulic systems. A combination of electronic and hydraulic systems is widely used because it combines the advantages of both electronic control and hydraulic power.

Advantages and disadvantages of hydraulic systems.   There are certain advantages and disadvantages in using hydraulic systems rather than other systems. Some of the advantages are:

1. Hydraulic fluid acts as a lubricant, in addition to carrying away heat generated in the system to a convenient heat exchanger.
2. Comparatively small sized hydraulic actuators can develop large forces or torques.
3. Hydraulic actuators have a higher speed of response with fast starts, stops, and speed reversals.
4. Hydraulic actuators can be operated under continuous, intermittent, reversing, and stalled conditions without damage.
5. Availability of both linear and rotary actuators gives flexibility in design.
6. Because of low leakages in hydraulic actuators, speed drop when loads are applied is small.

On the other hand, several disadvantages tend to limit their use.

1. Hydraulic power is not readily available compared to electric power.
2. Cost of a hydraulic system may be higher than a comparable electrical system performing a similar function.
3. Fire and explosion hazards exist unless fire-resistant fluids are used.

4. Because it is difficult to maintain a hydraulic system that is free from leaks, the system tends to be messy.
5. Contaminated oil may cause failure in the proper functioning of a hydraulic system.
6. As a result of the nonlinear and other complex characteristics involved, the design of sophisticated hydraulic systems is quite involved.
7. Hydraulic circuits have generally poor damping characteristics. If a hydraulic circuit is not designed properly, some unstable phenomena may occur or disappear, depending on the operating condition.

Comments.   Particular attention is necessary to ensure that the hydraulic system is stable and satisfactory under all operating conditions. Since the viscosity of hydraulic fluid can greatly affect damping and friction effects of the hydraulic circuits, stability tests must be carried out at the highest possible operating temperature.

Note that most hydraulic systems are nonlinear. Sometimes, however, it is possible to linearize nonlinear systems so as to reduce their complexity and permit solutions that are sufficiently accurate for most purposes. A useful linearization technique for dealing with nonlinear systems was presented in Section 2–10.

Hydraulic integral controllers.   The hydraulic servomotor shown in Figure 3–27 is essentially a pilot-valve-controlled hydraulic power amplifier and actuator. The pilot valve is a balanced valve in the sense that the pressure forces acting on it are all balanced. A very large power output can be controlled by a pilot valve, which can be positioned with very little power.

It will be shown in the following that for negligibly small load mass the servomotor shown in Figure 3–27 acts as an integrator or an integral controller. Such a servomotor constitutes the basis of the hydraulic control circuit.

In the hydraulic servomotor shown in Figure 3–27, the pilot valve (a four-way valve) has two lands on the spool. If the width of the land is smaller than the port in the valve sleeve, the valve is said to be *underlapped*. *Overlapped* valves have a land width greater than the port width. A *zero-lapped* valve has a land width that is identical to the port width. (If the pilot valve is not a zero-lapped valve, analyses of hydraulic servomotors become very complicated.)

In the present analysis, we assume that hydraulic fluid is incompressible and that the inertia force of the power piston and load is negligible compared to the hydraulic force at

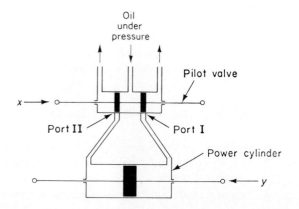

Figure 3–27
Hydraulic servomotor.

the power piston. We also assume that the pilot valve is a zero-lapped valve, and the oil flow rate is proportional to the pilot valve displacement.

Operation of this hydraulic servomotor is as follows. If input x moves the pilot valve to the right, port I is uncovered, and so high-pressure oil enters the right side of the power piston. Since port II is connected to the drain port, the oil in the left side of the power piston is returned to the drain. The oil flowing into the power cylinder is at high pressure; the oil flowing out from the power cylinder into the drain is at low pressure. The resulting difference in pressure on both sides of the power piston will cause it to move to the left.

Note that the rate of flow of oil q (kg/sec) times dt (sec) is equal to the power piston displacement dy (m) times the piston area A (m²) times the density of oil ρ (kg/m³). Therefore,

$$A\rho \, dy = q \, dt \qquad (3\text{–}21)$$

Because of the assumption that the oil flow rate q is proportional to the pilot valve displacement x, we have

$$q = K_1 x \qquad (3\text{–}22)$$

where K_1 is a positive constant. From Equations (3–21) and (3–22) we obtain

$$A\rho \, \frac{dy}{dt} = K_1 x$$

The Laplace transform of this last equation, assuming a zero initial condition, gives

$$A\rho sY(s) = K_1 X(s)$$

or

$$\frac{Y(s)}{X(s)} = \frac{K_1}{A\rho s} = \frac{K}{s}$$

where $K = K_1/(A\rho)$. Thus the hydraulic servomotor shown in Figure 3–27 acts as an integral controller.

Hydraulic proportional controllers.   It has been shown that the servomotor in Figure 3–27 acts as an integral controller. This servomotor can be modified to a proportional controller by means of a feedback link. Consider the hydraulic controller shown in Figure 3–28(a).

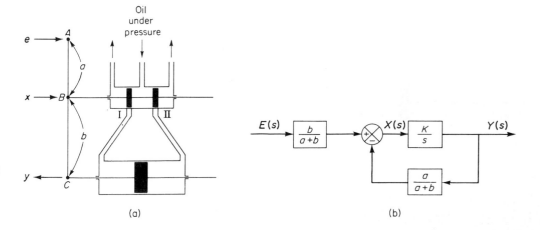

Figure 3–28
(a) Servomotor that acts as a proportional controller; (b) block diagram of the servomotor.

(a)

(b)

The left side of the pilot valve is joined to the left side of the power piston by a link ABC. This link is a floating link rather than one moving about a fixed pivot.

The controller here operates in the following way. If input x moves the pilot valve to the right, port II will be uncovered and high-pressure oil will flow through port II into the right side of the power piston and force this piston to the left. The power piston, in moving to the left, will carry the feedback link ABC with it, thereby moving the pilot valve to the left. This action continues until the pilot piston again covers ports I and II. A block diagram of the system can be drawn as in Figure 3–28(b). The transfer function between $Y(s)$ and $E(s)$ is given by

$$\frac{Y(s)}{E(s)} = \frac{\dfrac{b}{a+b}\dfrac{K}{s}}{1 + \dfrac{K}{s}\dfrac{a}{a+b}}$$

$$= \frac{bK}{s(a+b) + Ka}$$

Noting that under the normal operating conditions we have $|Ka/[s(a+b)]| \gg 1$, this last equation can be simplified to

$$\frac{Y(s)}{E(s)} = \frac{b}{a} = K_p$$

The transfer function between y and x becomes a constant. Thus, the hydraulic controller shown in Figure 3–28(a) acts as a proportional controller, the gain of which is K_p. This gain can be adjusted by effectively changing the lever ratio b/a. (The adjusting mechanism is not shown in the diagram.)

We have thus seen that the addition of a feedback lever will cause the hydraulic servomotor to act as a proportional controller.

Dashpots. The dashpot shown in Figure 3–29(a) acts as a differentiating element. Suppose we introduce a step displacement to the piston position x. Then the displacement y becomes equal to x momentarily. Because of the spring force, however, the oil will flow through the resistance R and the cylinder will come back to the original position. The curves x versus t and y versus t are shown in Figure 3–29(b).

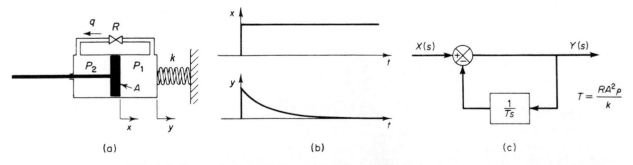

Figure 3–29 (a) Dashpot; (b) step change in x and the corresponding change in y plotted versus t; (c) block diagram of the dashpot.

Chapter 3 / Basic Control Actions and Industrial Automatic Controllers

Let us derive the transfer function between the displacement y and displacement x. Define the pressures existing on the right and left sides of the piston as P_1 (lb/in.^{2}) and P_2 (lb/in.^{2}), respectively. Suppose that the inertia force involved is negligible. Then the force acting on the piston must balance the spring force. Thus

$$A(P_1 - P_2) = ky$$

where A = piston area, in.^{2}

k = spring constant, lb$_f$/in.

The flow rate q is given by

$$q = \frac{P_1 - P_2}{R}$$

where q = flow rate through the restriction, lb/sec

R = resistance to flow at the restriction, lb$_f$-sec/in.^{2}-lb

Since the flow through the restriction during dt seconds must equal the change in the mass of oil to the left of the piston during the same dt seconds, we obtain

$$q \, dt = A\rho(dx - dy)$$

where ρ = density, lb/in.^{3}. (We assume that the fluid is incompressible or ρ = constant.) This last equation can be rewritten as

$$\frac{dx}{dt} - \frac{dy}{dt} = \frac{q}{A\rho} = \frac{P_1 - P_2}{RA\rho} = \frac{ky}{RA^2\rho}$$

or

$$\frac{dx}{dt} = \frac{dy}{dt} + \frac{ky}{RA^2\rho}$$

Taking the Laplace transforms of both sides of this last equation, assuming zero initial conditions, we obtain

$$sX(s) = sY(s) + \frac{k}{RA^2\rho} Y(s)$$

The transfer function of this system thus becomes

$$\frac{Y(s)}{X(s)} = \frac{s}{s + \dfrac{k}{RA^2\rho}}$$

Let us define $RA^2\rho/k = T$. Then

$$\frac{Y(s)}{X(s)} = \frac{Ts}{Ts + 1} = \frac{1}{1 + \dfrac{1}{Ts}}$$

Figure 3–29(c) shows a block diagram representation for this system.

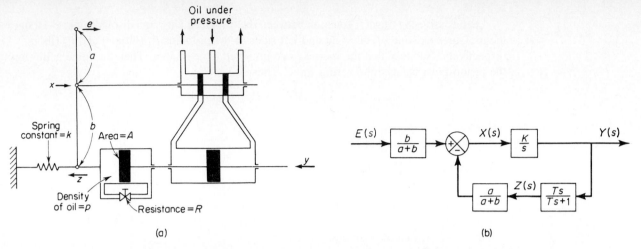

Figure 3–30 (a) Schematic diagram of a hydraulic proportional-plus-integral controller; (b) block diagram of the controller.

Obtaining hydraulic proportional-plus-integral control action. Figure 3–30(a) shows a schematic diagram of a hydraulic proportional-plus-integral controller. A block diagram of this controller is shown in Figure 3–30(b). The transfer function $Y(s)/E(s)$ is given by

$$\frac{Y(s)}{E(s)} = \frac{\dfrac{b}{a+b}\dfrac{K}{s}}{1 + \dfrac{Ka}{a+b}\dfrac{T}{Ts+1}}$$

In such a controller, under normal operation $|KaT/[(a+b)(Ts+1)]| \gg 1$, with the result that

$$\frac{Y(s)}{E(s)} = K_p\left(1 + \frac{1}{T_i s}\right)$$

where

$$K_p = \frac{b}{a}, \qquad T_i = T = \frac{RA^2\rho}{k}$$

Thus the controller shown in Figure 3–30(a) is a proportional-plus-integral controller.

Obtaining hydraulic proportional-plus-derivative control action. Figure 3–31(a) shows a schematic diagram of a hydraulic proportional-plus-derivtive controller. The cylinders are fixed in space and the pistons can move. For this system, notice that

$$k(y - z) = A(P_2 - P_1)$$

$$q = \frac{P_2 - P_1}{R}$$

$$q\,dt = \rho A\,dz$$

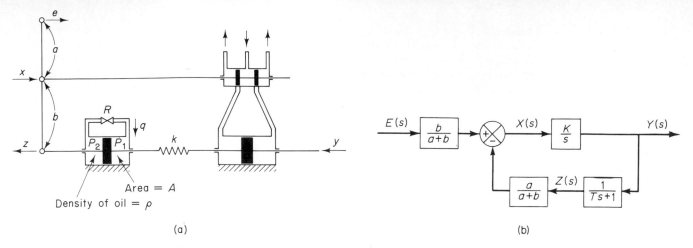

(a)

(b)

Figure 3–31 (a) Schematic diagram of a hydraulic proportional-plus-derivative controller; (b) block diagram of the controller.

Hence

$$y = z + \frac{A}{k} qR = z + \frac{RA^2\rho}{k} \frac{dz}{dt}$$

or

$$\frac{Z(s)}{Y(s)} = \frac{1}{Ts + 1}$$

where

$$T = \frac{RA^2\rho}{k}$$

A block diagram for this system is shown in Figure 3–31(b). From the block diagram the transfer function $Y(s)/E(s)$ can be obtained as

$$\frac{Y(s)}{E(s)} = \frac{\dfrac{b}{a + b}\dfrac{K}{s}}{1 + \dfrac{a}{a + b}\dfrac{K}{s}\dfrac{1}{Ts + 1}}$$

Under normal operation we have $|aK/[(a + b)s(Ts + 1)]| \gg 1$. Hence

$$\frac{Y(s)}{E(s)} = K_p(1 + Ts)$$

where

$$K_p = \frac{b}{a}, \qquad T = \frac{RA^2\rho}{k}$$

Thus, the controller shown in Figure 3–31(a) is a proportional-plus-derivative controller.

EXAMPLE 3–1

Consider the liquid-level control system shown in Figure 3–32. The inlet valve is controlled by a hydraulic integral controller. Assume that the steady-state inflow rate is $\bar{Q}$ and steady-state outflow rate is also $\bar{Q}$, the steady-state head is $\bar{H}$, steady-state pilot valve displacement is $\bar{X} = 0$, and steady-state valve position is $\bar{Y}$. We assume that the set point $\bar{R}$ corresponds to the steady-state head $\bar{H}$. The set point is fixed. Assume also that the disturbance inflow rate q_d, which is a small quantity, is applied to the water tank at $t = 0$. This disturbance causes the head to change from $\bar{H}$ to $\bar{H} + h$. This change results in a change in the outflow rate by q_o. Through the hydraulic controller the change in head causes a change in the inflow rate from $\bar{Q}$ to $\bar{Q} + q_i$. (The integral controller tends to keep the head constant as much as possible in the presence of disturbances.) We assume that all changes are of small quantities.

Assuming that h is the output and q_d is the input, derive a mathematical model of the system in state space.

Since the increase of water in the tank during dt seconds is equal to the net inflow to the tank during the same dt seconds, we have

$$C\,dh = (q_i - q_o + q_d)\,dt \tag{3–23}$$

where

$$q_o = \frac{h}{R} \tag{3–24}$$

For the feedback lever mechanism, we have

$$x = \frac{a}{a + b}h \tag{3–25}$$

We assume that the velocity of the power piston (valve) is proportional to pilot valve displacement x,

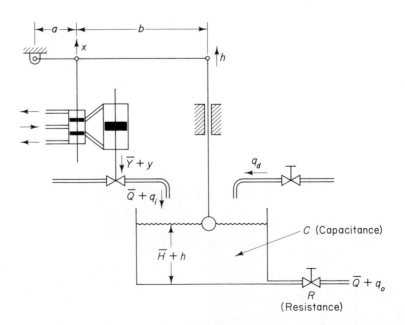

Figure 3–32
Liquid-level control system.

or

$$\frac{dy}{dt} = K_1 x \qquad (3\text{--}26)$$

where K_1 is a positive constant. We also assume that the change in the inflow rate q_i is negatively proportional to the change in the valve opening y, or

$$q_i = -K_v y \qquad (3\text{--}27)$$

where K_v is a positive constant.

Now we obtain the equations for the system as follows: From Equations (3–23), (3–24), and (3–27), we get

$$C\frac{dh}{dt} = -K_v y - \frac{h}{R} + q_d \qquad (3\text{--}28)$$

From Equations (3–25) and (3–26), we have

$$\frac{dy}{dt} = \frac{K_1 a}{a + b} h \qquad (3\text{--}29)$$

Let us define state variables x_1 and x_2 and the input variable u as follows:

$$x_1 = h$$

$$x_2 = y$$

$$u = q_d$$

Then Equations (3–28) and (3–29) become

$$\dot{x}_1 = -\frac{1}{RC}x_1 - \frac{K_v}{C}x_2 + \frac{1}{C}u$$

$$\dot{x}_2 = \frac{K_1 a}{a + b}x_1$$

Combining the last two equations into one vector-matrix equation, we obtain the state equation for the system.

$$\begin{bmatrix} \dot{x}_1 \\ \dot{x}_2 \end{bmatrix} = \begin{bmatrix} -\dfrac{1}{RC} & -\dfrac{K_v}{C} \\ \dfrac{K_1 a}{a + b} & 0 \end{bmatrix} \begin{bmatrix} x_1 \\ x_2 \end{bmatrix} + \begin{bmatrix} \dfrac{1}{C} \\ 0 \end{bmatrix} u \qquad (3\text{--}30)$$

The output equation for the system when h is the output is given by

$$h = \begin{bmatrix} 1 & 0 \end{bmatrix} \begin{bmatrix} x_1 \\ x_2 \end{bmatrix} \qquad (3\text{--}31)$$

Equations (3–30) and (3–31) give a mathematical model of the system in state space.

For convenience in visualizing the system dynamics, a block diagram for the system is shown in Figure 3–33. Notice that state variable x_1 is the output of the first-order delay element (delayed integrator), and state variable x_2 is the output of the integrator.

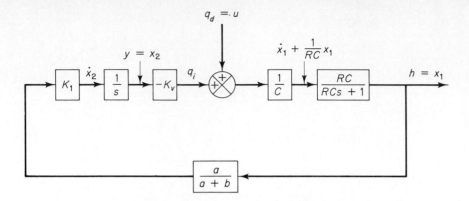

Figure 3–33
Block diagram of the liquid-level control system considered in Example 3–1.

EXAMPLE 3–2

Referring to the system of Example 3–1, obtain the response $h(t)$ when the disturbance input q_d is a unit-step function. Assume the following numerical values for the system:

$$C = 2 \text{ m}^2, \qquad R = 0.5 \text{ sec/m}^2, \qquad K_v = 1 \text{ m}^2/\text{sec}$$

$$a = 0.25 \text{ m}, \qquad b = 0.75 \text{ m}, \qquad K_1 = 4 \text{ sec}^{-1}$$

By substituting the given numerical values into Equations (3–28) and (3–29), we obtain

$$2\frac{dh}{dt} = -y - 2h + q_d$$

$$\frac{dy}{dt} = h$$

Taking the Laplace transforms of the preceding two equations, assuming zero initial conditions, we obtain

$$2sH(s) = -Y(s) - 2H(s) + Q_d(s)$$

$$sY(s) = H(s)$$

By eliminating $Y(s)$ from the last two equations and noting that the disturbance input is a unit-step

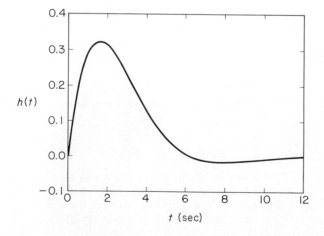

Figure 3–34
Response curve $h(t)$ versus t when disturbance input is a unit step.

function, or $Q_d = 1/s$, we get

$$H(s) = \frac{s}{2s^2 + 2s + 1} \frac{1}{s} = \frac{0.5}{(s + 0.5)^2 + 0.5^2}$$

The inverse Laplace transform of $H(s)$ gives the time response $h(t)$.

$$h(t) = e^{-0.5t} \sin 0.5t$$

Notice that the unit-step disturbance input q_d caused a transient error in the head as shown in Figure 3–34. However, the error becomes zero at steady state. The integral controller thus eliminated the error caused by the disturbance input q_d.

3–5 EFFECTS OF INTEGRAL AND DERIVATIVE CONTROL ACTIONS ON SYSTEM PERFORMANCE

In this section, we shall investigate the effects of integral and derivative control actions on the system performance. Here we shall consider only simple systems so that the effects of integral and derivative control actions on system performance can be clearly seen.

Integral control action.   In the proportional control of a plant whose transfer function does not possess an integrator $1/s$, there is a steady-state error, or offset, in the response to a step input. Such an offset can be eliminated if the integral control action is included in the controller.

In the integral control of a plant, the control signal, the output signal from the controller, at any instant is the area under the actuating error signal curve up to that instant. The control signal $u(t)$ can have a nonzero value when the actuating error signal $e(t)$ is zero, as shown in Figure 3–35(a). This is impossible in the case of the proportional controller since a nonzero control signal requires a nonzero actuating error signal. (A nonzero actuating error signal at steady state means that there is an offset.) Figure 3–35(b) shows the curve $e(t)$ versus t and the corresponding curve $u(t)$ versus t when the controller is of the proportional type.

Note that integral control action, while removing offset or steady-state error, may lead to oscillatory response of slowly decreasing amplitude or even increasing amplitude, both of which are usually undesirable.

Figure 3–35
(a) Plots of $e(t)$ and $u(t)$ curves showing nonzero control signal when the actuating error signal is zero (integral control); (b) plots of $e(t)$ and $u(t)$ curves showing zero control signal when the actuating error signal is zero (proportional control).

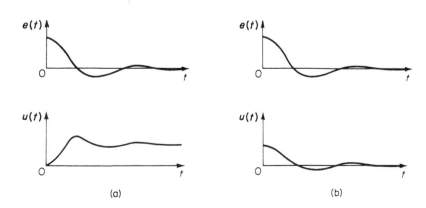

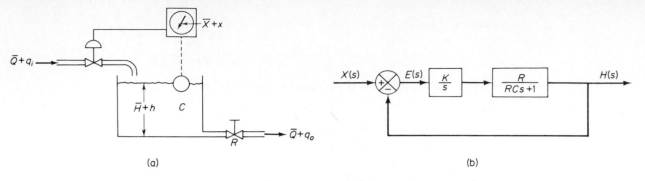

(a)   (b)

Figure 3–36 (a) Liquid-level control system; (b) block diagram of the system.

Integral control of liquid-level control systems. In Section 3–3, we found that the proportional control of a liquid-level system will result in a steady-state error with a step input. We shall now show that such an error can be eliminated if integral control action is included in the controller.

Figure 3–36(a) shows a liquid-level control system. We assume that the controller is an integral controller. We also assume that the variables x, q_i, h, and q_o, which are measured from their respective steady-state values $\bar{X}$, $\bar{Q}$, $\bar{H}$, and $\bar{Q}$, are small quantities so that the system can be considered linear. Under these assumptions, the block diagram of the system can be obtained as shown in Figure 3–36(b). From Figure 3–36(b), the closed-loop transfer function between $H(s)$ and $X(s)$ is

$$\frac{H(s)}{X(s)} = \frac{KR}{RCs^2 + s + KR}$$

Hence

$$\frac{E(s)}{X(s)} = \frac{X(s) - H(s)}{X(s)}$$

$$= \frac{RCs^2 + s}{RCs^2 + s + KR}$$

Since the system is stable, the steady-state error for the unit-step response is obtained by applying the final value theorem as follows:

$$e_{ss} = \lim_{s \to 0} sE(s)$$

$$= \lim_{s \to 0} \frac{s(RCs^2 + s)}{RCs^2 + s + KR} \frac{1}{s}$$

$$= 0$$

Integral control of the liquid-level system thus eliminates the steady-state error in the response to the step input. This is an important improvement over the proportional control alone, which gives offset.

Response to torque disturbances (proportional control). Let us investigate the effect of a torque disturbance occurring at the load element. Consider the system shown in

Chapter 3 / Basic Control Actions and Industrial Automatic Controllers

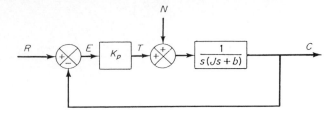

Figure 3–37
Control system with a torque disturbance.

Figure 3–37. The proportional controller delivers torque T to position the load element, which consists of moment of inertia and viscous friction. Torque disturbance is denoted by N.

Assuming that the reference input is zero or $R(s) = 0$, the transfer function between $C(s)$ and $N(s)$ is given by

$$\frac{C(s)}{N(s)} = \frac{1}{Js^2 + bs + K_p}$$

Hence

$$\frac{E(s)}{N(s)} = -\frac{C(s)}{N(s)} = -\frac{1}{Js^2 + bs + K_p}$$

The steady-state error due to a step disturbance torque of magnitude T_n is given by

$$e_{ss} = \lim_{s \to 0} sE(s)$$

$$= \lim_{s \to 0} \frac{-s}{Js^2 + bs + K_p} \frac{T_n}{s}$$

$$= -\frac{T_n}{K_p}$$

At steady state, the proportional controller provides the torque $-T_n$, which is equal in magnitude but opposite in sign to the disturbance torque T_n. The steady-state output due to the step disturbance torque is

$$c_{ss} = -e_{ss} = \frac{T_n}{K_p}$$

The steady-state error can be reduced by increasing the value of the gain K_p. Increasing this value, however, will cause the system response to be more oscillatory. Typical response curves for a small value of K_p and a large value of K_p are shown in Figure 3–38.

Since the value of the gain K_p cannot be increased too much, it is desirable to modify the proportional controller to a proportional-plus-integral controller.

Response to torque disturbances (proportional-plus-integral control). To eliminate offset due to torque disturbance, the proportional controller may be replaced by a proportional-plus-integral controller.

If integral control action is added to the controller, then as long as there is an error signal, a torque is developed by the controller to reduce this error, provided the control system is a stable one.

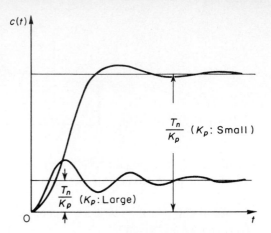

Figure 3–38
Typical response
curves to a step torque
disturbance.

Figure 3–39 shows the proportional-plus-integral control of the load element, consisting of moment of inertia and viscous friction.

The closed-loop transfer function between $C(s)$ and $N(s)$ is

$$\frac{C(s)}{N(s)} = \frac{s}{Js^3 + bs^2 + K_p s + \dfrac{K_p}{T_i}}$$

In the absence of the reference input, or $r(t) = 0$, the error signal is obtained from

$$E(s) = -\frac{s}{Js^3 + bs^2 + K_p s + \dfrac{K_p}{T_i}} N(s)$$

If this control system is stable, that is, if the roots of the characteristic equation

$$Js^3 + bs^2 + K_p s + \frac{K_p}{T_i} = 0$$

have negative real parts, then the steady-state error in the response to a step disturbance torque of magnitude T_n can be obtained by applying the final value theorem as follows:

$$e_{ss} = \lim_{s \to 0} s\, E(s)$$

$$= \lim_{s \to 0} \frac{-s^2}{Js^3 + bs^2 + K_p s + \dfrac{K_p}{T_i}} \frac{T_n}{s}$$

$$= 0$$

Figure 3–39
Proportional-plus-
integral control of a
load element
consisting of moment
of inertia and viscous
friction.

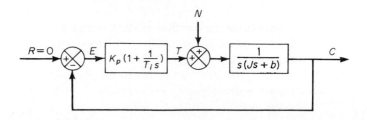

Figure 3–40
Integral control of a load element consisting of moment of inertia and viscous friction.

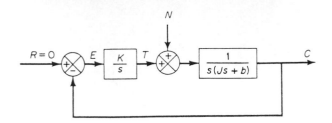

Thus steady-state error to the step disturbance torque can be eliminated if the controller is of the proportional-plus-integral type.

Note that the integral control action added to the proportional controller has converted the originally second-order system to a third-order one. Hence the control system may become unstable for a large value of K_p since the roots of the characteristic equation may have positive real parts. (The second-order system is always stable if the coefficients in the system differential equation are all positive.)

It is important to point out that if the controller were an integral controller, as in Figure 3–40, then the system always becomes unstable because the characteristic equation

$$Js^3 + bs^2 + K = 0$$

will have roots with positive real parts. Such an unstable system cannot be used in practice.

Note that in the system of Figure 3–39 the proportional control action tends to stabilize the system, while the integral control action tends to eliminate or reduce steady-state error in response to various inputs.

Derivative control action.   Derivative control action, when added to a proportional controller, provides a means of obtaining a controller with high sensitivity. An advantage of using derivative control action is that it responds to the rate of change of the actuating error and can produce a significant correction before the magnitude of the actuating error becomes too large. Derivative control thus anticipates the actuating error, initiates an early corrective action, and tends to increase the stability of the system.

Although derivative control does not affect the steady-state error directly, it adds damping to the system and thus permits the use of a larger value of the gain K, which will result in an improvement in the steady-state accuracy.

Because derivative control operates on the rate of change of the actuating error and not the actuating error itself, this mode is never used alone. It is always used in combination with proportional or proportional-plus-integral control action.

Proportional control of systems with inertia load.   Before we discuss the effect of derivative control action on system performance, we shall consider the proportional control of an inertia load.

Consider the system shown in Figure 3–41(a). The closed-loop transfer function is obtained as

$$\frac{C(s)}{R(s)} = \frac{K_p}{Js^2 + K_p}$$

Since the roots of the characteristic equation

$$Js^2 + K_p = 0$$

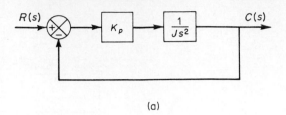

(a)

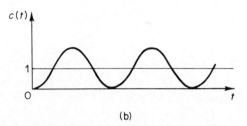

(b)

Figure 3–41
(a) Proportional
control of a system
with inertia load; (b)
response to a unit-step
input.

are imaginary, the response to a unit-step input continues to oscillate indefinitely, as shown in Figure 3–41(b).

Control systems exhibiting such response characteristics are not desirable. We shall see that the addition of derivative control will stabilize the system.

Proportional-plus-derivative control of a system with inertia load.   Let us modify the proportional controller to a proportional-plus-derivative controller whose transfer function is $K_p(1 + T_d s)$. The torque developed by the controller is proportional to $K_p(e + T_d \dot{e})$. Derivative control is essentially anticipatory, measures the instantaneous error velocity, and predicts the large overshoot ahead of time and produces an appropriate counteraction before too large an overshoot occurs.

Consider the system shown in Figure 3–42(a). The closed-loop transfer function is given by

$$\frac{C(s)}{R(s)} = \frac{K_p(1 + T_d s)}{Js^2 + K_p T_d s + K_p}$$

The characteristic equation

$$Js^2 + K_p T_d s + K_p = 0$$

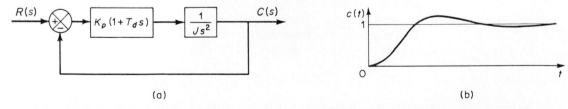

(a)   (b)

Figure 3–42   (a) Proportional-plus-derivative control of a system with inertia load; (b) response to a unit-step input.

now has two roots with negative real parts for positive values of J, K_p, and T_d. Thus derivative control introduces a damping effect. A typical response curve $c(t)$ to a unit-step input is shown in Figure 3–42(b). Clearly, the response curve shows a marked improvement over the original response curve shown in Figure 3–41(b).

3–6 REDUCTION OF PARAMETER VARIATIONS BY USE OF FEEDBACK

The primary purpose of using feedback in control systems is to reduce the sensitivity of the system to parameter variations and unwanted disturbances.

If we are to construct a suitable open-loop control system, we must select all the components of the open-loop transfer function $G(s)$ very carefully so that they respond accurately. In the case of constructing a closed-loop control system, however, the components can be less accurate since the sensitivity to parameter variations in $G(s)$ is reduced by a factor of $1 + G(s)$.

To illustrate this, consider the open-loop system and the closed-loop system shown in Figures 3–43(a) and (b), respectively. Suppose that, due to parameter variations, $G(s)$ is changed to $G(s) + \Delta G(s)$, where $|G(s)| \gg |\Delta G(s)|$. Then, in the open-loop system shown in Figure 3–43(a), the output is given by

$$C(s) + \Delta C(s) = [G(s) + \Delta G(s)]R(s)$$

Hence the change in the output is given by

$$\Delta C(s) = \Delta G(s)R(s)$$

In the closed-loop system shown in Figure 3–43(b),

$$C(s) + \Delta C(s) = \frac{G(s) + \Delta G(s)}{1 + G(s) + \Delta G(s)} R(s)$$

or

$$\Delta C(s) \doteq \frac{\Delta G(s)}{1 + G(s)} R(s)$$

Thus, the change in the output of the closed-loop system, due to the parameter variations in $G(s)$, is reduced by a factor of $1 + G(s)$. In many practical cases, the magnitude of $1 + G(s)$ is generally much greater than 1.

Note that in reducing the effects of the parameter variations of the components we very often bridge the offending component with a feedback loop.

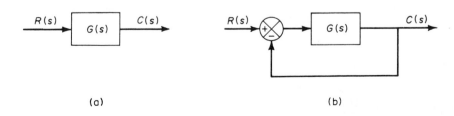

Figure 3–43
(a) Open-loop system;
(b) closed-loop system.

(a)   (b)

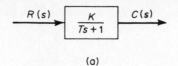

(a)

Figure 3-44
(a) Open-loop system;
(b) closed-loop system
with time constant $T/(1 + Ka)$; (c) closed-loop system with time constant $T - bK$.

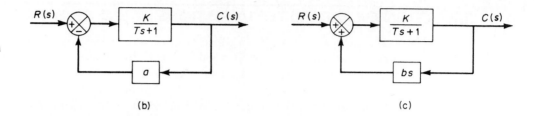

(b)   (c)

Changing time constants by use of feedback.   Consider the system shown in Fig. 3-44(a). The time constant of the system is T. The addition of a negative feedback loop around this element reduces the time constant. Figure 3-44(b) shows the system with the same feedforward transfer function as that shown in Figure 3-44(a), with the exception that a negative feedback loop has been added. The time constant of this system has been reduced to $T/(1 + Ka)$. Note also that the gain constant for this system has also been reduced from K to $K/(1 + Ka)$.

If, instead of a negative feedback loop, a positive feedback loop is added around the transfer function $K/(Ts + 1)$ and if the feedback transfer function is properly chosen, then the time constant can be made zero or a very small value. Consider the system shown in Figure 3-44(c). Since the closed-loop transfer function is

$$\frac{C(s)}{R(s)} = \frac{K}{(T - bK)s + 1}$$

the time cosntant can be reduced by properly choosing the value of b. If b is set equal to T/K, then the time constant becomes zero. Note, however, that if disturbances cause $T - bK$ to be negative instead of zero, the system becomes unstable. Hence if positive feedback is employed to reduce the time constant to a small value, we must be very careful so that $T - bK$ never becomes negative. (Some safety device must be provided.)

Increasing loop gains by use of positive feedback.   The system shown in Figure 3-45(a) has the transfer function $C(s)/R(s) = G(s)$. Consider now the system shown in Figure 3-45(b). The closed-loop transfer function for this system is

$$\frac{C(s)}{R(s)} = \frac{G(s)}{1 - G_f(s) + G(s)}$$

If $G_f(s)$ is chosen nearly unity, or $G_f(s) \doteq 1$, then

$$\frac{C(s)}{R(s)} \doteq 1$$

Essentially, this means that the inner loop, using positive feedback, has increased the feedforward gain to a very large value. As we stated earlier, when the loop gain is very large,

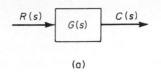

(a)

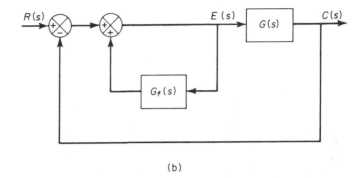

Figure 3–45
(a) Open-loop system;
(b) closed-loop system
whose transfer
function is nearly
unity if $G_f(s) \doteq 1$.

(b)

the closed-loop transfer function $C(s)/R(s)$ becomes equal to the inverse of the transfer function of the feedback element. Since the system shown in Figure 3–45(b) has unity feedback, $C(s)/R(s)$ becomes almost equal to unity. [Thus $C(s)/R(s)$ is not sensitive to the parameter variations in $G(s)$.]

Elimination of integration. Addition of a minor loop around an integrator modifies it to a first-order delay element. Consider the system shown in Figure 3 46(a). Negative feedback of the output, as shown in Figure 3–46(b), modifies the integrator K/s to a first-order delay element $K/(s + K)$.

Comments on the use of feedback loops. As we have seen in the previous discussion, feedback control, or closed-loop control, reduces the sensitivity of a system to parameter variations and therefore decreases the effects of gain variations in the feedforward path in response to variations of supply pressure, supply voltage, temperature, and the like. In the study of controllers made in Sections 3–3 and 3–4, we have also seen that the elements that perform the various control actions are in the feedback part of the controller mechanisms and that the feedback elements in a controller essentially increase the linearity of the amplifier and increase the range of the proportional sensitivity.

The use of feedback loops in control systems, however, will increase the number of components of the systems, will thereby increase the complexity, and will also introduce the possibility of instability.

Figure 3–46
(a) Integrating
element; (b) first-order
delay element.

(a)   (b)

Example Problems and Solutions

A-3-1. The term commonly used to define the gain or sensitivity of a proportional controller is the *proportional band*. This is the percentage of change in the input to the controller (error signal) required to cause 100% change in the output of the actuator. Thus a small proportional band corresponds to high gain or high proportional sensitivity.

What is the proportional band if the controller and actuator have an overall gain of 4%/%? (Note that the total changes in the input to the controller and the output of the actuator are given as 100%. Thus a gain of 4%/% means that there is a change of 4% in the output if the change in input is 1%.)

Solution

$$\text{Proportional band} = \frac{100\%}{\text{gain in }\%/\%} = \frac{100\%}{4\%/\%} = 25\%$$

A-3-2. Consider an ideal gas changing from a state represented by (p_1, v_1, T_1) to a state represented by (p_2, v_2, T_2). If we keep the temperature constant at T but change the pressure from p_1 to p_2 then the volume of gas will change from v_1 to v' such that

$$p_1 v_1 = p_2 v' \qquad (3\text{-}32)$$

Now keep the pressure constant but change the temperature to T_2. Then the volume of gas reaches v_2. Thus

$$\frac{v'}{T_1} = \frac{v_2}{T_2} \qquad (3\text{-}33)$$

By eliminating v' between Equations (3-32) and (3-33), we obtain

$$\frac{p_1 v_1}{T_1} = \frac{p_2 v_2}{T_2}$$

This means that, for a fixed quantity of gas, no matter what physical changes occur, pv/T will be constant. We may therefore write

$$pv = kT$$

where the value of constant k depends on the quantity and nature of the gas considered.

In dealing with gas systems, we find it convenient to work in molar quantities since 1 mole of any gas contains the same number of molecules. Thus 1 mole occupies the same volume if measured under the same conditions of temperature and pressure.

If we consider 1 mole of gas, then

$$p\bar{v} = \bar{R}T \qquad (3\text{-}34)$$

The value of $\bar{R}$ is the same for all gases under all conditions. The constant $\bar{R}$ is called the universal gas constant. At standrd temperature and pressure (that is, at 492°R and 14.7 psia), 1 lb-mole of any gas is found to occupy 359 ft³. [For example, at 492°R (= 32°F) and 14.7 psia, the volume occupied by 2 lb of hydrogen, 32 lb of oxygen, or 28 lb of nitrogen is the same, 359 ft³.] This volume is called the molal volume and is denoted by $\bar{v}$.

Obtain the value of the universal gas constant.

Solution. By substituting $p = 14.7$ lb$_f$/in.², $\bar{v} = 359$ ft³/lb-mole, and $T = 492°R$ into Equation (3-34), we obtain

$$\bar{R} = \frac{14.7 \times 144 \times 359}{492}$$
$$= 1545 \text{ ft-lb}_f/\text{lb-mole °R}$$
$$= 1.985 \text{ Btu/lb-mole °R}$$

A–3–3. The value of the gas constant for any gas may be determined from accurate experimental observations of simultaneous values of p, v, and T.

Obtain the gas constant R_{air} for air. Note that at 32°F and 14.7 psia the specific volume of air is 12.39 ft³/lb. Then obtain the capacitance of a 20-ft³ pressure vessel that contains air at 160°F. Assume that the expansion process is isothermal.

Solution

$$R_{\text{air}} = \frac{pv}{T} = \frac{14.7 \times 144 \times 12.39}{460 + 32} = 53.3 \text{ ft-lb}_\text{f}/\text{lb °R}$$

Referring to Equation (3–5), the capacitance of a 20-ft³ pressure vessel is

$$C = \frac{V}{nR_{\text{air}}T} = \frac{20}{1 \times 53.3 \times 620} = 6.05 \times 10^{-4} \frac{\text{lb}}{\text{lb}_\text{f}/\text{ft}^2}$$

A–3–4. Consider the liquid-level control system shown in Figure 3–47. Assume that the set point of the controller is fixed. Assuming a step disturbance of magnitude n_o, determine the error. Assume that n_o is small and the variations in the variables from their respective steady-state values are also small. The controller is proportional.

If the controller is not proportional, but integral, what is the steady-state error?

Solution. Figure 3–48 is a block diagram of the system when the controller is proportional with gain K_p. (We assume the transfer function of the pneumatic valve to be unity.) Since the set point is fixed, the variation in the set point is zero, or $X(s) = 0$. The Laplace transform of $h(t)$ is

$$H(s) = \frac{K_p R}{RCs + 1} E(s) + \frac{R}{RCs + 1} N(s)$$

Then

$$E(s) = -H(s) = -\frac{K_p R}{RCs + 1} E(s) - \frac{R}{RCs + 1} N(s)$$

Hence

$$E(s) = -\frac{R}{RCs + 1 + K_p R} N(s)$$

Since

$$N(s) = \frac{n_0}{s}$$

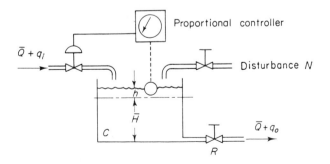

Figure 3–47 Liquid-level control system.

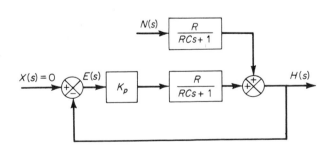

Figure 3–48 Block diagram of the liquid-level control system shown in Fig. 3–47.

Example Problems and Solutions

229

we obtain

$$E(s) = -\frac{R}{RCs + 1 + K_pR}\frac{n_0}{s}$$

$$= \frac{Rn_0}{1 + K_pR}\left(\frac{1}{s + \dfrac{1 + K_pR}{RC}}\right) - \frac{Rn_0}{1 + K_pR}\frac{1}{s}$$

The time solution for $t > 0$ is

$$e(t) = \frac{Rn_0}{1 + K_pR}\left[\exp\left(-\frac{1 + K_pR}{RC}t\right) - 1\right]$$

Thus, the time constant is $RC/(1 + K_pR)$. (In the absence of the controller, the time constant is equal to RC.) As the gain of the controller is increased, the time constant is decreased. The steady-state error is

$$e(\infty) = -\frac{Rn_0}{1 + K_pR}$$

As the gain K_p of the controller is increased, the steady-state error, or offset, is reduced. Thus, mathematically, the larger the gain K_p is, the smaller the offset and time constant are. In practical systems, however, if the gain K_p of the proportional controller is increased to a very large value, oscillation may result in the output since in our analysis all the small lags and small time constants that may exist in the actual control system are neglected. (If these small lags and time constants are included in the analysis, the transfer function becomes higher order, and for very large values of K_p the possibility of oscillation or even instability may occur.)

If the controller is integral, then assuming the transfer function of the controller to be

$$G_c = \frac{K}{s}$$

we obtain

$$E(s) = -\frac{Rs}{RCs^2 + s + KR}N(s)$$

The steady-state error for a step disturbance $N(s) = n_o/s$ is

$$e(\infty) = \lim_{s \to 0} sE(s)$$

$$= \lim_{s \to 0}\frac{-Rs^2}{RCs^2 + s + KR}\frac{n_o}{s}$$

$$= 0$$

Thus, an integral controller eliminates steady-state error or offset due to the step disturbance. (The value of K must be chosen so that the transient response due to the command input and/or disturbance damps out with a reasonable speed. See Chapter 4 for transient-response analysis.)

A–3–5. In the pneumatic pressure system of Figure 3–49(a), assume that, for $t < 0$, the system is at steady state and that the pressure of the entire system is $\bar{P}$. Also, assume that the two bellows are identical. At $t = 0$ the input pressure is changed from $\bar{P}$ to $\bar{P} + p_i$. Then the pressures in bellows 1 and 2 will change from $\bar{P}$ to $\bar{P} + p_1$ and from $\bar{P}$ to $\bar{P} + p_2$, respectively. The capacity (volume) of each bellows is 5×10^{-4} m^3, and the operating pressure difference Δp (difference between p_i and p_1 or difference

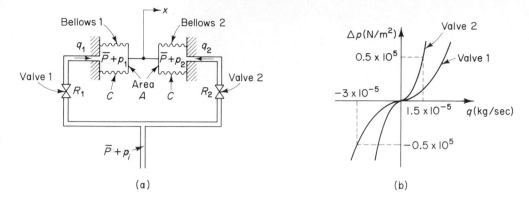

Figure 3–49
(a) Pneumatic pressure system; (b) pressure difference versus mass flow rate curves.

(a)

(b)

between p_i and p_2) is between -0.5×10^5 N/m² and 0.5×10^5 N/m². The corresponding mass flow rates (kg/sec) through the valves are shown in Figure 3–49(b). Assume that the bellows expand or contract linearly with the air pressures applied to them, that the equivalent spring constant of the bellows system is $k = 1 \times 10^5$ N/m, and that each bellows has area $A = 15 \times 10^{-4}$ m².

Defining the displacement of the midpoint of the rod that connects two bellows as x, find the transfer function $X(s)/P_i(s)$. Assume that the expansion process is isothermal and that the temperature of the entire system stays at 30°C.

Solution. Referring to Section 3–3, transfer function $P_1(s)/P_i(s)$ can be obtained as

$$\frac{P_1(s)}{P_i(s)} = \frac{1}{R_1 Cs + 1} \tag{3–35}$$

Similarly, transfer function $P_2(s)/P_i(s)$ is

$$\frac{P_2(s)}{P_i(s)} = \frac{1}{R_2 Cs + 1} \tag{3–36}$$

The force acting on bellows 1 in the x direction is $A(\bar{P} + p_1)$, and the force acting on bellows 2 in the negative x direction is $A(\bar{P} + p_2)$. The resultant force balances with kx, the equivalent spring force of the corrugated side of the bellows.

$$A(p_1 - p_2) = kx$$

or

$$A[P_1(s) - P_2(s)] = kX(s) \tag{3–37}$$

Referring to Equations (3–35) and (3–36), we see that

$$P_1(s) - P_2(s) = \left(\frac{1}{R_1 Cs + 1} - \frac{1}{R_2 Cs + 1} \right) P_i(s)$$

$$= \frac{R_2 Cs - R_1 Cs}{(R_1 Cs + 1)(R_2 Cs + 1)} P_i(s)$$

By substituting this last equation into Equation (3–37) and rewriting, the transfer function $X(s)/P_i(s)$ is obtained as

$$\frac{X(s)}{P_i(s)} = \frac{A}{k} \frac{(R_2 C - R_1 C)s}{(R_1 Cs + 1)(R_2 Cs + 1)} \tag{3–38}$$

Example Problems and Solutions

231

The numerical values of average resistances R_1 and R_2 are

$$R_1 = \frac{d\,\Delta p}{dq_1} = \frac{0.5 \times 10^5}{3 \times 10^{-5}} = 0.167 \times 10^{10}\,\frac{\text{N/m}^2}{\text{kg/sec}}$$

$$R_2 = \frac{d\,\Delta p}{dq_2} = \frac{0.5 \times 10^5}{1.5 \times 10^{-5}} = 0.333 \times 10^{10}\,\frac{\text{N/m}^2}{\text{kg/sec}}$$

The numerical value of capacitance C of each bellows is

$$C = \frac{V}{nR_{\text{air}}T} = \frac{5 \times 10^{-4}}{1 \times 287 \times (273 + 30)} = 5.75 \times 10^{-9}\,\frac{\text{kg}}{\text{N/m}^2}$$

Consequently,

$$R_1 C = 0.167 \times 10^{10} \times 5.75 \times 10^{-9} = 9.60\ \text{sec}$$

$$R_2 C = 0.333 \times 10^{10} \times 5.75 \times 10^{-9} = 19.2\ \text{sec}$$

By substituting the numerical values for A, k, R_1C, and R_2C into Equation (3–38), we have

$$\frac{X(s)}{P_i(s)} = \frac{1.44 \times 10^{-7}\,s}{(9.6s + 1)(19.2s + 1)}$$

A–3–6. Draw a block diagram of the pneumatic controller shown in Figure 3–50. Then derive the transfer function of this controller.

    If the resistance R_d is removed (replaced by the line-sized tubing), what control action do we get? If the resistance R_i is removed (replaced by the line-sized tubing), what control action do we get?

Solution. Let us assume that when $e = 0$ the nozzle–flapper distance is equal to $\bar{X}$ and the control pressure is equal to $\bar{P}_c$. In the present analysis, we shall assume small deviations from the respective reference values as follows:

e = small error signal

x = small change in the nozzle–flapper distance

p_c = small change in the control pressure

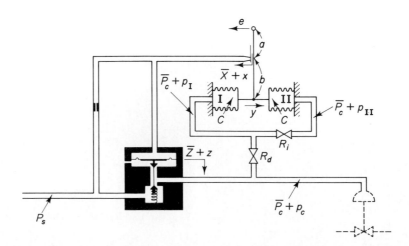

Figure 3–50
Schematic diagram of a pneumatic controller.

p_I = small pressure change in bellows I due to small change in the control pressure

p_II = small pressure change in bellows II due to small change in the control pressure

y = small displacement at the lower end of the flapper

In this controller, p_c is transmitted to bellows I through the resistance R_d. Similarly, p_c is transmitted to bellows II through the series of resistances R_d and R_i. An approximate relationship between p_I and p_c is

$$\frac{P_\mathrm{I}(s)}{P_c(s)} = \frac{1}{R_d Cs + 1} = \frac{1}{T_d s + 1}$$

where $T_d = R_d C$ = derivative time. Similarly, p_II and p_I are related by the transfer function

$$\frac{P_\mathrm{II}(s)}{P_\mathrm{I}(s)} = \frac{1}{R_i Cs + 1} = \frac{1}{T_i s + 1}$$

where $T_i = R_i C$ = integral time. The force–balance equation for the two bellows is

$$(p_\mathrm{I} - p_\mathrm{II})A = k_s y$$

where k_s is the stiffness of the two connected bellows and A is the cross-sectional area of the bellows. The relationship among the variables e, x, and y is

$$x = \frac{b}{a + b} e - \frac{a}{a + b} y$$

The relationship between p_c and x is

$$p_c = Kx \qquad (K > 0)$$

From the equations just derived, a block diagram of the controller can be drawn, as shown in Figure 3–51(a). Simplification of this block diagram results in Figure 3–51(b).

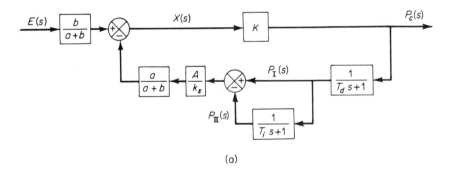

(a)

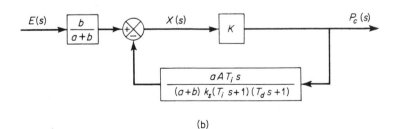

(b)

Figure 3–51
(a) Block diagram of the pneumatic controller shown in Fig. 3–50; (b) simplified block diagram.

The transfer function between $P_c(s)$ and $E(s)$ is

$$\frac{P_c(s)}{E(s)} = \frac{\dfrac{b}{a+b}K}{1 + K\dfrac{a}{a+b}\dfrac{A}{k_s}\left(\dfrac{T_i s}{T_i s + 1}\right)\left(\dfrac{1}{T_d s + 1}\right)}$$

For a practical controller, under normal operation $|KaAT_i s/[(a+b)k_s(T_i s + 1)(T_d s + 1)]|$ is very much greater than unity and $T_i \gg T_d$. Therefore, the transfer function can be simplified as follows:

$$\frac{P_c(s)}{E(s)} \doteq \frac{bk_s(T_i s + 1)(T_d s + 1)}{aAT_i s}$$

$$= \frac{bk_s}{aA}\left(\frac{T_i + T_d}{T_i} + \frac{1}{T_i s} + T_d s\right)$$

$$\doteq K_p\left(1 + \frac{1}{T_i s} + T_d s\right)$$

where

$$K_p = \frac{bk_s}{aA}$$

Thus the controller shown in Figure 3–50 is a proportional-plus-integral-plus-derivative one.

If the resistance R_d is removed, or $R_d = 0$, the action becomes that of a proportional-plus-integral controller.

If the resistance R_i is removed, or $R_i = 0$, the action becomes that of a narrow-band proportional, or two-position, controller. (Note that the actions of two feedback bellows cancel each other, and there is no feedback.)

A–3–7. Figure 3–52 is a schematic diagram of a pneumatic diaphragm valve. At steady state the control pressure from a controller is $\bar{P}_c$, the pressure in the valve is also $\bar{P}_c$, and the valve-stem diaplacement is $\bar{X}$. Assume that at $t = 0$ the control pressure is changed from $\bar{P}_c$ to $\bar{P}_c + p_c$. Then the valve pressure will be changed from $\bar{P}_c$ to $\bar{P}_c + p_v$. The change in valve pressure p_v will cause the valve-stem

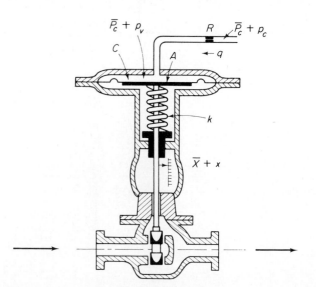

Figure 3–52
Pneumatic diaphragm
valve.

displacement to change from $\bar{X}$ to $\bar{X} + x$. Find the transfer function between the change in valve-stem displacement x and the change in control pressure p_c.

Solution. Let us define the air flow rate to the diaphragm valve through resistance R as q. Then

$$q = \frac{p_c - p_v}{R}$$

For the air chamber in the diaphragm valve, we have

$$C \, dp_v = q \, dt$$

Consequently,

$$C \frac{dp_v}{dt} = q = \frac{p_c - p_v}{R}$$

from which

$$RC \frac{dp_v}{dt} + p_v = p_c$$

Noting that

$$Ap_v = kx$$

we have

$$\frac{k}{A}\left(RC \frac{dx}{dt} + x\right) = p_c$$

The transfer function between x and p_c is

$$\frac{X(s)}{P_c(s)} = \frac{A/k}{RCs + 1}$$

A–3–8. Figure 3–53 shows a hydraulic jet-pipe controller. Hydraulic fluid is ejected from the jet pipe. If the jet pipe is shifted to the right from the neutral position, the power piston moves to the left, and vice versa. The jet pipe valve is not used as much as the flapper valve because of large null flow, slower response, and rather unpredictable characteristics. Its main advantage lies in its insensitivity to dirty fluids.

Suppose that the power piston is connected to a light load so that the inertia force of the load element is negligible compared to the hydraulic force developed by the power piston. What type of control action does this controller produce?

Solution. Define the displacement of the jet nozzle from the neutral position as x and the displacement of the power piston as y. If the jet nozzle is moved to the right by a small displacement x, the oil flows to the right side of the power piston, and the oil in the left side of the power piston is returned to the drain. The oil flowing into the power cylinder is at high pressure; the oil flowing out from the power cylinder into the drain is at low pressure. The resulting pressure difference causes the power piston to move to the left.

For a small jet nozzle displacement x, the flow rate q to the power cylinder is proportional to x; that is,

$$q = K_1 x$$

For the power cylinder,

$$A\rho \, dy = q \, dt$$

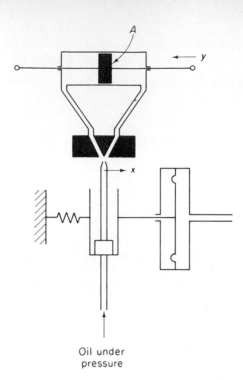

Figure 3–53
Hydraulic jet-pipe
controller.

where A is the power piston area and ρ is the density of oil. Hence

$$\frac{dy}{dt} = \frac{q}{A\rho} = \frac{K_1}{A\rho}x = Kx$$

where $K = K_1/(A\rho) =$ constant. The transfer function $Y(s)/X(s)$ is thus

$$\frac{Y(s)}{X(s)} = \frac{K}{s}$$

The controller produces the integral control action.

A–3–9 Figure 3–54 shows a hydraulic jet pipe controller applied to a flow control system. The jet pipe controller governs the position of the butterfly valve. Discuss the operation of this system. Plot a possible curve relating the displacement x of the nozzle to the total force F acting on the power piston.

Solution. The operation of this system is as follows: The flow rate is measured by the orifice, and the pressure difference produced by this orifice is transmitted to the diaphragm of the pressure-measuring device. The diaphragm is connected to the free swinging nozzle, or jet pipe, through a linkage. High-pressure oil ejects from the nozzle all the time. When the nozzle is at a neutral position, no oil flows through either of the pipes to move the power piston. If the nozzle is displaced by the motion of the balance arm to one side, the high-pressure oil flows through the corresponding pipe, and the oil in the power cylinder flows back to the sump through the other pipe.

Assume that the system is initially at rest. If the reference input is changed suddenly to a higher flow rate, then the nozzle is moved in such a direction as to move the power piston and open the butterfly valve. Then the flow rate will increase, the pressure difference across the orifice becomes larger, and the nozzle will move back to the neutral position. The movement of the power piston stops when x, the displacement of the nozzle, comes back to and stays at the neutral position. (The jet pipe controller thus possesses an integrating property.)

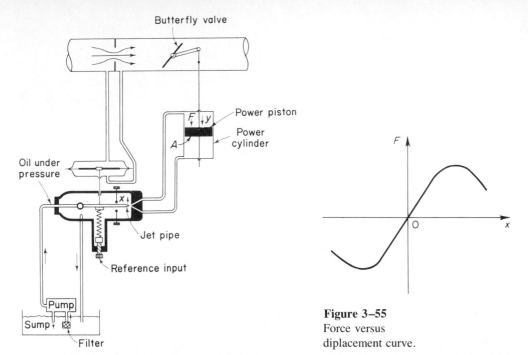

Figure 3–54
Schematic diagram of a flow control system using a hydraulic jet-pipe controller.

Figure 3–55
Force versus diplacement curve.

The relationship between the total force F acting on the power piston and the displacement x of the nozzle is shown in Figure 3–55. The total force is equal to the pressure difference ΔP across the piston times the area A of the power piston. For a small displacement x of the nozzle, the total force F and displacement x may be considered proportional.

A–3–10. Consider the hydraulic servo system shown in Figure 3–56. Assuming that signal $e(t)$ is the input and power piston displacement $y(t)$ the output, find the transfer function $Y(s)/E(s)$.

Solution. A block diagram for the system can be drawn as shown in Figure 3–57. Assuming that $|K_1 a_1/[s(a_1 + a_2)]| \gg 1$ and $|K_2 b_1/[s(b_1 + b_2)]| \gg 1$, we obtain

$$\frac{Z(s)}{E(s)} = \frac{\dfrac{a_2}{a_1 + a_2} \cdot \dfrac{K_1}{s}}{1 + \dfrac{K_1}{s} \cdot \dfrac{a_1}{a_1 + a_2}} \doteq \frac{a_2}{a_1 + a_2} \cdot \frac{a_1 + a_2}{a_1} = \frac{a_2}{a_1}$$

$$\frac{W(s)}{E(s)} = \frac{a_1 + a_2 + a_3}{a_1 + a_2} \cdot \frac{Z(s)}{E(s)} + \frac{a_3}{a_1 + a_2} = \frac{a_2 + a_3}{a_1}$$

$$\frac{Y(s)}{W(s)} = \frac{\dfrac{K_2}{s}}{1 + \dfrac{b_1}{b_1 + b_2} \cdot \dfrac{K_2}{s}} \doteq \frac{b_1 + b_2}{b_1}$$

Hence

$$\frac{Y(s)}{E(s)} = \frac{Y(s)}{W(s)} \cdot \frac{W(s)}{E(s)} = \frac{(a_2 + a_3)(b_1 + b_2)}{a_1 b_1}$$

This servo system is a proportional controller.

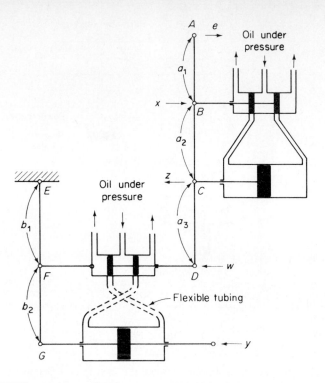

Figure 3–56
Hydraulic servo system.

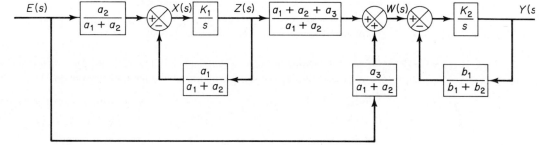

Figure 3–57
Block diagram for the system shown in Fig. 3–56.

A–3–11. Explain the operation of the speed control system shown in Figure 3–58.

Solution. If the engine speed increases, the sleeve of the fly-ball governor moves upward. This movement acts as the input to the hydraulic controller. A positive error signal (upward motion of the sleeve) causes the power piston to move downward, reduces the fuel-valve opening, and decreases the engine speed. A block diagram for the system is shown in Figure 3–59.

From the block diagram the transfer function $Y(s)/E(s)$ can be obtained as

$$\frac{Y(s)}{E(s)} = \frac{a_2}{a_1 + a_2} \frac{\dfrac{K}{s}}{1 + \dfrac{a_1}{a_1 + a_2}\dfrac{bs}{bs + k}\dfrac{K}{s}}$$

If the following condition applies,

$$\left| \frac{a_1}{a_1 + a_2}\frac{bs}{bs + k}\frac{K}{s} \right| \gg 1$$

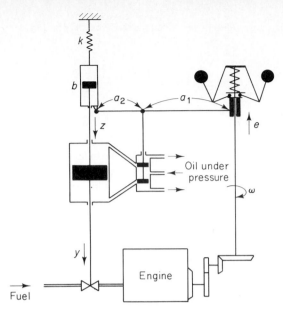

Figure 3–58 Speed control system.

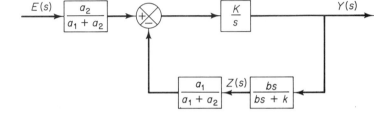

Figure 3–59 Block diagram for the speed control system shown in Fig. 3–58.

the transfer function $Y(s)/E(s)$ becomes

$$\frac{Y(s)}{E(s)} \doteq \frac{a_2}{a_1 + a_2} \frac{a_1 + a_2}{a_1} \frac{bs + k}{bs} = \frac{a_2}{a_1}\left(1 + \frac{k}{bs}\right)$$

The speed controller has proportional-plus-integral control action.

A–3–12. The block diagram of Figure 3–60 shows a speed control system in which the output member of the system is subject to a torque disturbance. In the diagram, $\Omega_r(s)$, $\Omega(s)$, $T(s)$, and $N(s)$ are the Laplace transforms of the reference speed, output speed, driving torque, and disturbance torque, respectively. In the absence of a disturbance torque, the output speed is equal to the reference speed.

Investigate the response of this system to a unit step disturbance torque. Assume that the reference input is zero, or $\Omega_r(s) = 0$.

Solution. Figure 3–61 is a modified block diagram convenient for the present analysis. The closed-loop transfer function is

$$\frac{\Omega_N(s)}{N(s)} = \frac{1}{Js + K}$$

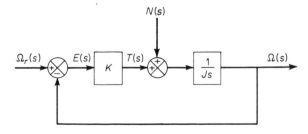

Figure 3–60 Block diagram of a speed control system.

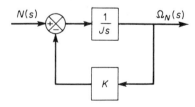

Figure 3–61 Block diagram of the speed control system of Fig. 3–60 when $\Omega_r(s) = 0$.

where $\Omega_N(s)$ is the Laplace transform of the output speed due to the disturbance torque. For a unit-step disturbance torque, the steady-state output velocity is

$$\omega_N(\infty) = \lim_{s \to 0} s\Omega_N(s)$$

$$= \lim_{s \to 0} \frac{s}{Js + K} \frac{1}{s}$$

$$= \frac{1}{K}$$

From this analysis, we conclude that, if a step disturbance torque is applied to the output member of the system, an error speed will result so that the ensuing motor torque will exactly cancel the disturbance torque. To develop this motor torque, it is necessary that there be an error in speed so that nonzero torque will result.

A–3–13. In the system considered in Problem A–3–12, it is desired to eliminate as much as possible the speed errors due to torque disturbances.

Is it possible to cancel the effect of a disturbance torque at steady state so that a constant disturbance torque applied to the output member will cause no speed change at steady state?

Solution. Suppose we choose a suitable controller whose transfer function is $G_c(s)$, as shown in Figure 3–62. Then in the absence of the reference input the closed-loop transfer function between the output velocity $\Omega_N(s)$ and the disturbance torque $N(s)$ is

$$\frac{\Omega_N(s)}{N(s)} = \frac{\dfrac{1}{Js}}{1 + \dfrac{1}{Js} G_c(s)}$$

$$= \frac{1}{Js + G_c(s)}$$

The steady-state output speed due to a unit-step disturbance torque is

$$\omega_N(\infty) = \lim_{s \to 0} s\Omega_N(s)$$

$$= \lim_{s \to 0} \frac{s}{Js + G_c(s)} \frac{1}{s}$$

$$= \frac{1}{G_c(0)}$$

To satisfy the requirement that

$$\omega_N(\infty) = 0$$

we must choose $G_c(0) = \infty$. This can be realized if we choose

$$G_c(s) = \frac{K}{s}$$

Integral control action will continue to correct until the error is zero. This controller, however, presents a stability problem because the characteristic equation will have two imaginary roots.

One method of stabilizing such a system is to add a proportional mode to the controller or choose

$$G_c(s) = K_p + \frac{K}{s}$$

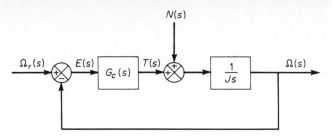

Figure 3–62 Block diagram of a speed control system.

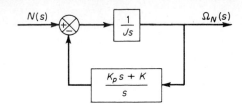

Figure 3–63 Block diagram of the speed control system of Fig. 3–62 when $G_c(s) = K_p + (K/s)$ and $\Omega_r(s) = 0$.

With this controller, the block diagram of Figure 3–62 in the absence of the reference input can be modified to that of Figure 3–63. The closed-loop transfer function $\Omega_N(s)/N(s)$ becomes

$$\frac{\Omega_N(s)}{N(s)} = \frac{s}{Js^2 + K_p s + K}$$

For a unit-step disturbance torque, the steady state output speed is

$$\omega_n(\infty) = \lim_{s \to 0} s\Omega_N(s) = \lim_{s \to 0} \frac{s^2}{Js^2 + K_p s + K} \frac{1}{s} = 0$$

Thus, we see that the proportional-plus-integral controller eliminates speed error at steady state.

The use of integral control action has increased the order of the system by 1. (This tends to produce an oscillatory response.)

In the present system, a step disturbance torque will cause a transient error in the output speed, but the error will become zero at steady state. The integrator provides a nonzero output with zero error. (The nonzero output of the integrator produces a motor torque that exactly cancels the disturbance torque.)

Note that the integrator in the transfer function of the plant does not eliminate the steady-state error due to a step disturbance torque. To eliminate this, we must have an integrator before the point where the disturbance torque enters.

A–3–14. Consider the open-loop control system and closed-loop control system shown in Figure 3–64. In the open loop one, gain K_c is calibrated so that $K_c = 1/K$. Thus, the transfer function of the open-loop

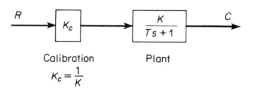

Calibration   Plant

$$K_c = \frac{1}{K}$$

Figure 3–64
Block diagrams of an open-loop control system and a closed-loop control system.

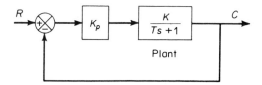

Plant

control system is

$$G_0(s) = \frac{1}{K} \frac{K}{Ts + 1} = \frac{1}{Ts + 1}$$

In the closed-loop control system, gain K_p of the controller is set so that $K_p K \gg 1$.

Assuming a unit-step input, compare the steady-state errors for these control systems.

Solution. For the open-loop control system, the error signal is

$$e(t) = r(t) - c(t)$$

or

$$
\begin{aligned}
E(s) &= R(s) - C(s) \\
&= [1 - G_0(s)]R(s)
\end{aligned}
$$

The steady-state error in the unit-step response is

$$
\begin{aligned}
e_{ss} &= \lim_{s \to 0} sE(s) \\
&= \lim_{s \to 0} s[1 - G_0(s)] \frac{1}{s} \\
&= 1 - G_0(0)
\end{aligned}
$$

If $G_0(0)$, the dc gain of the open-loop control system, is equal to unity, then the steady-state error is zero. Due to environmental changes and aging of components, however, the dc gain $G_0(0)$ will drift from unity as time elapses, and the steady-state error will no longer be equal to zero. Such steady-state error in an open-loop control system will remain until the system is recalibrated.

For the closed-loop control system, the error signal is

$$
\begin{aligned}
E(s) &= R(s) - C(s) \\
&= \frac{1}{1 + G(s)} R(s)
\end{aligned}
$$

where

$$G(s) = \frac{K_p K}{Ts + 1}$$

The steady-state error in the unit-step response is

$$
\begin{aligned}
e_{ss} &= \lim_{s \to 0} s \left[\frac{1}{1 + G(s)} \right] \frac{1}{s} \\
&= \frac{1}{1 + G(0)} \\
&= \frac{1}{1 + K_p K}
\end{aligned}
$$

In the closed-loop control system, gain K_p is set at a large value compared with $1/K$. Thus the steady-state error can be made small, although not exactly zero.

Let us assume the following variation in the transfer function of the plant, assuming K_c and K_p constant:

$$\frac{K + \Delta K}{Ts + 1}$$

For simplicity, let us assume $K = 10$, $\Delta K = 1$, or $\Delta K / K = 0.1$. Then the steady-state error in the unit-step response for the open-loop control system becomes

$$e_{ss} = 1 - \frac{1}{K}(K + \Delta K)$$
$$= 1 - 1.1 = -0.1$$

For the closed-loop control system, if K_p is set at $100/K$, then the steady-state error in the unit-step response becomes

$$e_{ss} = \frac{1}{1 + G(0)}$$
$$= \frac{1}{1 + \dfrac{100}{K}(K + \Delta K)}$$
$$= \frac{1}{1 + 110} = 0.009$$

Thus, the closed-loop control system is superior to the open-loop control system in the presence of environmental changes, aging of components, and the like, which definitely affect the steady-state performance.

A–3–15. Actual spool valves are either overlapped or underlapped because of manufacturing tolerances. Consider the overlapped and underlapped spool valves shown in Figure 3–65(a) and (b). Sketch curves relating the uncovered port area A versus displacement x.

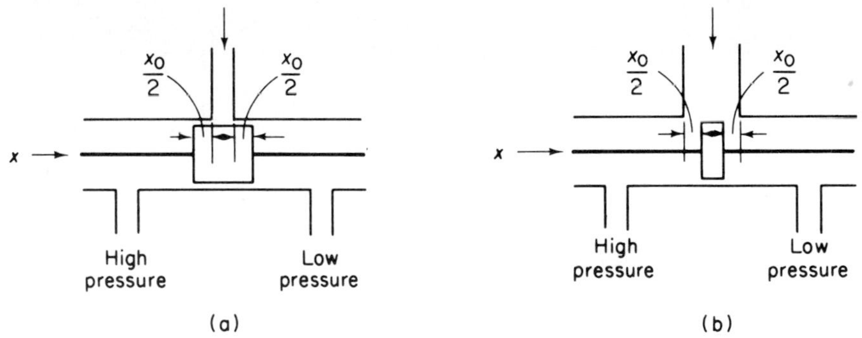

Figure 3–65
(a) Overlapped spool valve; (b) underlapped spool valve.

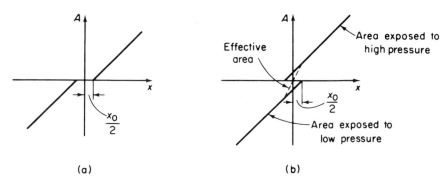

(a)

(b)

Solution. For the overlapped valve, a dead zone exists between $-\frac{1}{2}x_0$ and $\frac{1}{2}x_0$, or $-\frac{1}{2}x_0 < x < \frac{1}{2}x_0$. The uncovered port area A versus displacement x curve is shown in Figure 3–66(a). Such an overlapped valve is unfit as a control valve.

Figure 3–66
(a) Uncovered port area A versus displacement x curve for the overlapped valve; (b) uncovered port area A versus displacement x curve for the underlapped valve.

(a)

(b)

For the underlapped valve, the port area A versus displacement x curve is shown in Figure 3–66(b). The effective curve for the underlapped region has a higher slope, meaning a higher sensitivity. Valves used for controls are usually underlapped.

PROBLEMS

B–3–1. Consider industrial automatic controllers whose control actions are proportional, integral, proportional-plus-integral, proportional-plus-derivative, and proportional-plus-integral-plus-derivative. The transfer functions of these controllers can be given, respectively, by

$$\frac{U(s)}{E(s)} = K_p$$

$$\frac{U(s)}{E(s)} = \frac{K_i}{s}$$

$$\frac{U(s)}{E(s)} = K_p\left(1 + \frac{1}{T_i s}\right)$$

$$\frac{U(s)}{E(s)} = K_p(1 + T_d s)$$

$$\frac{U(s)}{E(s)} = K_p\left(1 + \frac{1}{T_i s} + T_d s\right)$$

where $U(s)$ is the Laplace transform of $u(t)$, the controller output, and $E(s)$ the Laplace transform of $e(t)$, the actuating error signal. Sketch $u(t)$ versus t curves for each of the five types of controllers when the actuating error signal is

(a) $e(t) =$ unit-step function

(b) $e(t) =$ unit-ramp function

In sketching curves, assume that the numerical values of K_p, K_i, T_i, and T_d are given as

$$K_p = \text{proportional gain} = 4$$

$$K_i = \text{integral gain} = 2$$

$$T_i = \text{integral time} = 2 \text{ sec}$$

$$T_d = \text{derivative time} = 0.8 \text{ sec}$$

B–3–2. Consider the pneumatic system shown in Figure 3–67. Obtain the transfer function $X(s)/P_i(s)$.

B–3–3. Figure 3–68 shows a pneumatic controller. What kind of control action does this controller produce? Derive the transfer function $P_c(s)/E(s)$.

B–3–4. Consider the pneumatic controller shown in Figure

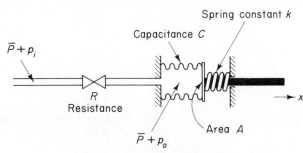

Figure 3–67 Pneumatic system.

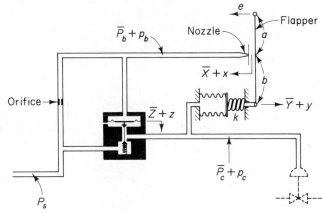

Figure 3–68 Pneumatic controller.

3–69. Assuming that the pneumatic relay has the characteristics that $p_c = Kp_b$ (where $K > 0$), determine the control action of this controller. The input to the controller is e and the output is p_c.

B–3–5. Figure 3–70 shows a pneumatic controller. The signal e is the input and the change in the control pressure p_c is the output. Obtain the transfer function $P_c(s)/E(s)$. Assume that the pneumatic relay has the characteristics that $p_c = Kp_b$, where $K > 0$.

B–3–6. Consider the pneumatic controller shown in Figure 3–71. What control action does this controller produce? Assume that the pneumatic relay has the characteristics that $p_c = Kp_b$, where $K > 0$.

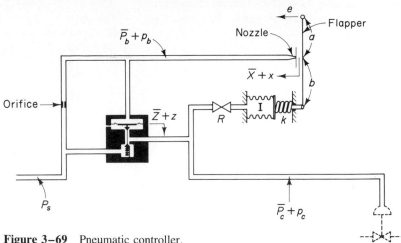

Figure 3–69 Pneumatic controller.

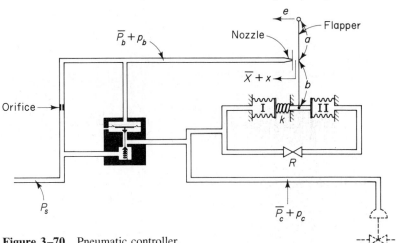

Figure 3–70 Pneumatic controller.

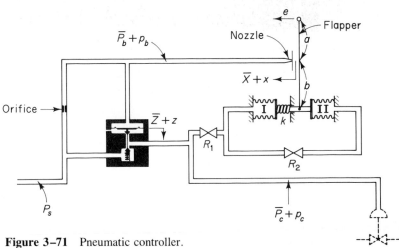

Figure 3–71 Pneumatic controller.

B–3–7. Figure 3–72 shows an electric-pneumatic transducer. Show that the change in the output pressure is proportional to the change in the input current.

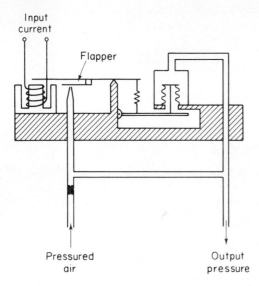

Figure 3–72 Electric-pneumatic transducer.

B–3–8. Figure 3–73 shows a flapper valve. It is placed between two opposing nozzles. If the flapper is moved slightly to the right, the pressure unbalance occurs in the nozzles and the power piston moves to the left, and vice versa. Such a device is frequently used in hydraulic servos as the first-stage valve in two-stage servovalves. This usage occurs because considerable force may be needed to stroke larger spool valves that result from the steady-state flow force. To reduce or compensate this force, two-stage valve configuration is often employed; a flapper valve or jet pipe is used as the first-stage valve to provide a necessary force to stroke the second-stage spool valve.

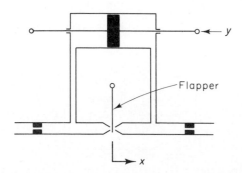

Figure 3–73 Flapper valve connected to a power cylinder.

Figure 3–74 shows a schematic diagram of a hydraulic servomotor in which the error signal is amplified in two stages using a jet pipe and a pilot valve. Draw a block diagram of the system of Figure 3–74 and then find the transfer function between y and x.

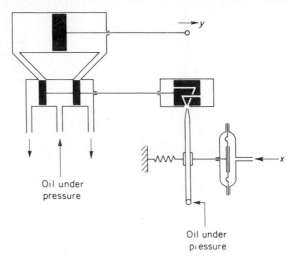

Figure 3–74 Schematic diagram of a hydraulic servomotor.

B–3–9. Figure 3–75 is a schematic diagram of an aircraft elevator control system. The input to the system is the deflection angle θ of the control lever, and the output is the elevator angle ϕ. Assume that angles θ and ϕ are relatively small. Show that for each angle θ of the control lever there is a corresponding (steady-state) elevator angle ϕ.

B–3–10. Consider the controller shown in Figure 3–76. The input is the air pressure p_i and the output is the displacement y of the power piston. Obtain the transfer function $Y(s)/P_i(s)$.

B–3–11. Figure 3–77 shows a schematic diagram of the edge-position control system. It is desired to keep the edge of the running paper at the desired position. The edge of the running paper is detected by the detecting head D. The hydraulic controller repositions the roller so as to keep the edge of the running paper at the desired position. Draw a block diagram of the system.

B–3–12. Referring to Examples 3–1 and 3–2, obtain the response $h(t)$ when the disturbance flow q_d is given by

$$q_d(t) = 1(t) - 1(t - 50)$$

That is, q_d is unity for $0 < t < 50$ and zero elsewhere. Also obtain $q_i(t)$.

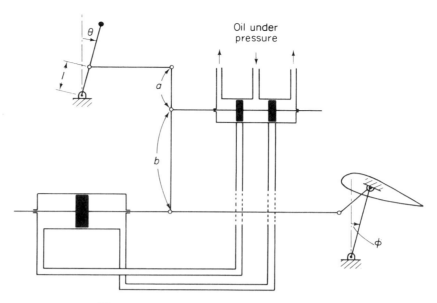

Figure 3–75 Aircraft elevator control system.

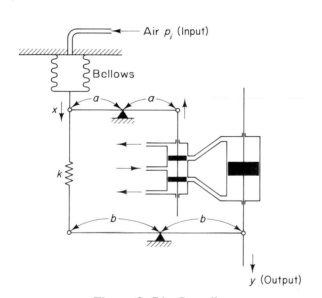

Figure 3–76 Controller.

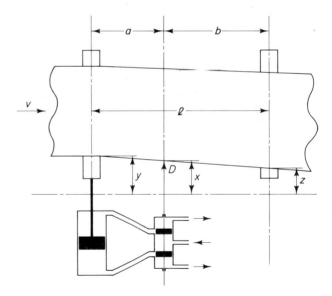

Figure 3–77 Edge-position control system.

B–3–13. Consider the multiple-loop system shown in Figure 3–78. Obtain the closed-loop transfer function between $C(s)$ and $N(s)$.

B–3–14. If the feedforward path of a control system contains at least one integrating element, then the output continues to change as long as an error is present. The output stops when the error is precisely zero. If an external disturbance enters the system, it is desirable to have an integrating element between the error-measuring element and the point where the disturbance enters so that the effect of the external disturbance may be made zero at steady state.

where

$$\beta = \tan^{-1} \frac{\omega_d}{\sigma} = \tan^{-1} 1.95 = 1.10$$

Thus, t_r is

$$t_r = 0.65 \text{ sec}$$

Settling time t_s: For the 2% criterion,

$$t_s = \frac{4}{\sigma} = 2.48 \text{ sec}$$

For the 5% criterion,

$$t_s = \frac{3}{\sigma} = 1.86 \text{ sec}$$

4–5 HIGHER-ORDER SYSTEMS

In this section, we shall first discuss the unit-step response of a particular type of third-order system. We shall then present a transient response analysis of higher-order systems in general terms. Finally, we shall present a discussion of stability analysis in the complex plane.

Unit-step response of third-order systems. We shall discuss the unit-step response of a commonly encountered third-order system whose closed-loop transfer function is

$$\frac{C(s)}{R(s)} = \frac{\omega_n^2 p}{(s^2 + 2\zeta\omega_n s + \omega_n^2)(s + p)} \qquad (0 < \zeta < 1)$$

The unit-step response of this system can be obtained as follows:

$$c(t) = 1 - \frac{e^{-\zeta\omega_n t}}{\beta\zeta^2(\beta - 2) + 1} \left\{ \beta\zeta^2(\beta - 2) \cos \sqrt{1 - \zeta^2}\, \omega_n t \right.$$

$$\left. + \frac{\beta\zeta[\zeta^2(\beta - 2) + 1]}{\sqrt{1 - \zeta^2}} \sin \sqrt{1 - \zeta^2}\, \omega_n t \right\} - \frac{e^{-pt}}{\beta\zeta^2(\beta - 2) + 1} \qquad (t \geq 0)$$

where

$$\beta = \frac{p}{\zeta\omega_n}$$

Note that since

$$\beta\zeta^2(\beta - 2) + 1 = \zeta^2(\beta - 1)^2 + (1 - \zeta^2) > 0$$

the coefficient of the term e^{-pt} is always negative.

The effect of the real pole $s = -p$ on the unit-step response is that of reducing the maximum overshoot and increasing the settling time. Figure 4–28 shows unit-step response curves of this third-order system with $\zeta = 0.5$. The ratio $\beta = p/(\zeta\omega_n)$ is a parameter in the family of curves.

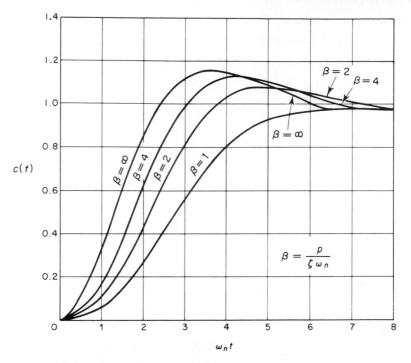

Figure 4–28 Unit-step response curves of the third-order system

$$\frac{C(s)}{R(s)} = \frac{\omega_n^2 p}{(s^2 + 2\zeta\omega_n s + \omega_n^2)(s + p)}, \qquad \zeta = 0.5$$

If the real pole is located to the right of the complex-conjugate poles, then there is a tendency for sluggish response. The system will behave like an overdamped system. The complex-conjugate poles will add ripple to the response curve.

Transient response of higher-order systems.   Consider the system shown in Figure 4–29. The closed-loop transfer function is

$$\frac{C(s)}{R(s)} = \frac{G(s)}{1 + G(s)H(s)} \qquad (4\text{--}37)$$

In general, $G(s)$ and $H(s)$ are given as ratios of polynomials in s, or

$$G(s) = \frac{p(s)}{q(s)} \qquad \text{or} \qquad H(s) = \frac{n(s)}{d(s)}$$

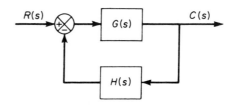

Figure 4–29
Control system.

where $p(s)$, $q(s)$, $n(s)$, and $d(s)$ are polynomials in s. The closed-loop transfer function given by Equation (4–37) may then be written

$$\frac{C(s)}{R(s)} = \frac{p(s)d(s)}{q(s)d(s) + p(s)n(s)}$$

$$= \frac{b_0 s^m + b_1 s^{m-1} + \cdots + b_{m-1}s + b_m}{a_0 s^n + a_1 s^{n-1} + \cdots + a_{n-1}s + a_n} \qquad (m \leq n)$$

The transient response of this system to any given input can be obtained by a computer simulation (see Section 4–10). If an analytical expression for the transient response is desired, then it is necessary to factor the denominator polynomial. (Various methods are available for factoring polynomials.) Once the denominator polynomial has been factored, $C(s)/R(s)$ can be written as

$$\frac{C(s)}{R(s)} = \frac{K(s + z_1)(s + z_2) \cdots (s + z_m)}{(s + p_1)(s + p_2) \cdots (s + p_n)} \qquad (4\text{–}38)$$

Note that the numerator can be factored easily because it is a product of $p(s)$ and $d(s)$. [Normally, $p(s)$ and $d(s)$ are lower-order polynomials and can be factored easily.]

Let us examine the response behavior of this system to a unit-step input. We assume that the closed-loop poles are all distinct. (This is usually the case in practical systems.) For a unit-step input, Equation (4–38) can be written

$$C(s) = \frac{a}{s} + \sum_{i=1}^{n} \frac{a_i}{s + p_i} \qquad (4\text{–}39)$$

where a_i is the residue of the pole at $s = -p_i$.

If all closed-loop poles lie in the left-half s plane, the relative magnitudes of the residues determine the relative importance of the components in the expanded form of $C(s)$. If there is a closed-loop zero close to a closed-loop pole, then the residue at this pole is small and the coefficient of the transient-response term corresponding to this pole becomes small. A pair of closely located poles and zeros will effectively cancel each other. If a pole is located very far from the origin, the residue at this pole may be small. The transients corresponding to such a remote pole are small and last a short time. Terms in the expanded form of $C(s)$ having very small residues contribute little to the transient response, and these terms may be neglected. If this is done, the higher-order system may be approximated by a lower-order one. (Such an approximation often enables us to estimate the response characteristics of a higher-order system from those of a simplified one.)

The poles of $C(s)$ consist of real poles and pairs of complex-conjugate poles. A pair of complex-conjugate poles yields a second-order term in s. Since the factored form of the higher-order characteristic equation consists of first- and second-order terms, Equation (4–39) can be rewritten

$$C(s) = \frac{K \prod_{i=1}^{m} (s + z_i)}{s \prod_{j=1}^{q} (s + p_j) \prod_{k=1}^{r} (s^2 + 2\zeta_k \omega_k s + \omega_k^2)} \qquad (4\text{–}40)$$

where $q + 2r = n$. If the closed-loop poles are distinct, Equation (4–40) can be expanded into partial fractions as follows:

$$C(s) = \frac{a}{s} + \sum_{j=1}^{q} \frac{a_j}{s + p_j} + \sum_{k=1}^{r} \frac{b_k(s + \zeta_k\omega_k) + c_k\omega_k\sqrt{1 - \zeta_k^2}}{s^2 + 2\zeta_k\omega_k s + \omega_k^2}$$

From this last equation, we see that the response of a higher-order system is composed of a number of terms involving the simple functions found in the responses of first- and second-order systems. The unit-step response $c(t)$, the inverse Laplace transform of $C(s)$, is then

$$c(t) = a + \sum_{j=1}^{q} a_j e^{-p_j t} + \sum_{k=1}^{r} b_k e^{-\zeta_k\omega_k t} \cos \omega_k\sqrt{1 - \zeta_k^2}\, t$$

$$+ \sum_{k=1}^{r} c_k e^{-\zeta_k\omega_k t} \sin \omega_k\sqrt{1 - \zeta_k^2}\, t \qquad (t \geq 0) \qquad (4\text{--}41)$$

If all closed-loop poles lie in the left-half s plane, then the exponential terms and the damped exponential terms in Equation (4–41) will approach zero as time t increases. The steady-state output is then $c(\infty) = a$.

Let us assume that the system considered is a stable one. Then the closed-loop poles that are located far from the $j\omega$ axis have large negative real parts. The exponential terms that correspond to these poles decay very rapidly to zero. (Note that the horizontal distance from a closed-loop pole to the $j\omega$ axis determines the settling time of transients due to that pole. The smaller the distance, the longer the settling time.)

The response curve of a stable higher-order system is the sum of a number of exponential curves and damped sinusoidal curves. Examples of step-response curves of higher-order systems are shown in Figure 4–30. A particular characteristic of such response curves is that small oscillations are superimposed on larger oscillations or on exponential curves. Fast-decaying components have significance only in the initial part of the transient response.

Remember that the type of transient response is determined by the closed-loop poles, while the shape of the transient response is primarily determined by the closed-loop zeros. As we have seen earlier, the poles of the input $R(s)$ yield the steady-state response terms in

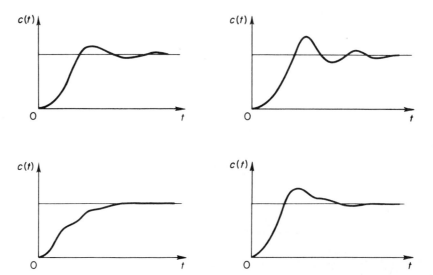

Figure 4–30
Step-response curves
of higher-order
systems.

the solution, while the poles of $C(s)/R(s)$ enter into the exponential transient-response terms and/or damped sinusoidal transient-response terms. The zeros of $C(s)/R(s)$ do not affect the exponents in the exponential terms, but they do affect the magnitudes and signs of the residues.

Dominant closed-loop poles. The relative dominance of closed-loop poles is determined by the ratio of the real parts of the closed-loop poles, as well as by the relative magnitudes of the residues evaluated at the closed-loop poles. The magnitudes of the residues depend on both the closed-loop poles and zeros.

If the ratios of the real parts exceed 5, and there are no zeros nearby, then the closed-loop poles nearest the $j\omega$ axis will dominate in the transient-response behavior because these poles correspond to transient-response terms that decay slowly. Those closed-loop poles that have dominant effects on the transient-response behavior are called *dominant closed-loop* poles. Quite often the dominant closed-loop poles occur in the form of a complex-conjugate pair. The dominant closed-loop poles are most important among all closed-loop poles.

The gain of a higher-order system is often adjusted so that there will exist a pair of dominant complex-conjugate closed-loop poles. The presence of such poles in a stable system reduces the effect of such nonlinearities as dead zone, backlash, and coulomb friction.

Remember that, although the concept of dominant closed-loop poles is useful for estimating the dynamic behavior of a closed-loop system, we must be careful to see that the underlying assumptions are met before using it.

Stability analysis in the complex plane. The stability of a linear closed-loop system can be determined from the location of the closed-loop poles in the s plane. If any of these poles lie in the right-half s plane, then with increasing time they give rise to the dominant mode, and the transient response increases monotonically or oscillates with increasing amplitude. This represents an unstable system. For such a system, as soon as the power is turned on, the output may increase with time. If no saturation takes place in the system and no mechanical stop is provided, then the system may eventually be subjected to damage and fail since the response of a real physical system cannot increase indefinitely. Therefore, closed-loop poles in the right-half s plane are not permissible in the usual linear control system. If all closed-loop poles lie to the left of the $j\omega$ axis, any transient response eventually reaches equilibrium. This represents a stable system.

Whether a linear system is stable or unstable is a property of the system itself and does not depend on the input or driving function of the system. The poles of the input, or driving function, do not affect the property of stability of the system but they contribute only to steady-state response terms in the solution. Thus, the problem of absolute stability can be solved readily by choosing no closed-loop poles in the right-half s plane, including the $j\omega$ axis. (Mathematically, closed-loop poles on the $j\omega$ axis will yield oscillations, the amplitude of which is neither decaying nor growing with time. In practical cases, where noise is present, however, the amplitude of oscillations may increase at a rate determined by the noise power level. Therefore, a control system should not have closed-loop poles on the $j\omega$ axis.)

Note that the mere fact that all closed-loop poles lie in the left-half s plane does not guarantee satisfactory transient-response characteristics. If dominant complex-conjugate closed-loop poles lie close to the $j\omega$ axis, the transient response may exhibit excessive oscillations or may be very slow. Therefore, to gurantee fast, yet well-damped, transient-

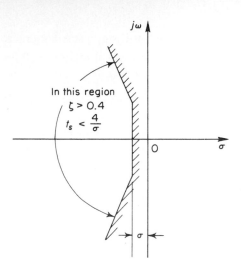

Figure 4–31
Region in the complex plane satisfying the conditions $\zeta > 0.4$ and $t_s < 4/\sigma$.

In this region
$\zeta > 0.4$
$t_s < \dfrac{4}{\sigma}$

response characteristics, it is necessary that the closed-loop poles of the system lie in a particular region in the complex plane, such as the region bounded by the shaded area in Figure 4–31.

Since the relative stability and transient performance of a closed-loop control system are directly related to the closed-loop pole-zero configuration in the s plane, it is frequently necessary to adjust one or more system parameters in order to obtain suitable configurations. The effects of varying system parameters on the closed-loop poles will be discussed in detail in Chapter 5.

4–6 ROUTH'S STABILITY CRITERION

The most important problem in linear control systems concerns stability. That is, under what conditions will a system become unstable? If it is unstable, how should we stabilize the system? In Section 4–5 it was stated that a control system is stable if and only if all closed-loop poles lie in the left-half s plane. Since most linear closed-loop systems have closed-loop transfer functions of the form

$$\frac{C(s)}{R(s)} = \frac{b_0 s^m + b_1 s^{m-1} + \cdots + b_{m-1}s + b_m}{a_0 s^n + a_1 s^{n-1} + \cdots + a_{n-1}s + a_n} = \frac{B(s)}{A(s)}$$

where the a's and b's are constants and $m \le n$, we must first factor the polynomial $A(s)$ in order to find the closed-loop poles. This process is very time consuming for a polynomial of degree greater than second. A simple criterion, known as Routh's stability criterion, enables us to determine the number of closed-loop poles that lie in the right-half s plane without having to factor the polynomial.

Routh's stability criterion.   Routh's stability criterion tells us whether or not there are positive roots in a polynomial equation without actually solving for them. This stability criterion applies to polynomials with only a finite number of terms. When the criterion is applied to a control system, information about absolute stability can be obtained directly from the coefficients of the characteristic equation.

The procedure in Routh's stability criterion is as follows:

1. Write the polynomial in s in the following form:

$$a_0 s^n + a_1 s^{n-1} + \cdots + a_{n-1} s + a_n = 0 \qquad (4\text{--}42)$$

where the coefficients are real quantities. We assume that $a_n \neq 0$; that is, any zero root has been removed.

2. If any of the coefficients are zero or negative in the presence of at least one positive coefficient, there is a root or roots that are imaginary or that have positive real parts. Therefore, in such a case, the system is not stable. If we are interested in only the absolute stability, there is no need to follow the procedure further. Note that all the coefficients must be positive. This is a necessary condition, as may be seen from the following argument: A polynomial in s having real coefficients can always be factored into linear and quadratic factors, such as $(s + a)$ and $(s^2 + bs + c)$, where a, b, and c are real. The linear factors yield real roots and the quadratic factors yield complex roots of the polynomial. The factor $(s^2 + bs + c)$ yields roots having negative real parts only if b and c are both positive. For all roots to have negative real parts, the constants a, b, c, and so on, in all factors must be positive. The product of any number of linear and quadratic factors containing only positive coefficients always yields a polynomial with positive coefficients. It is important to note that the condition that all the coefficients are positive is not sufficient to assure stability. The necessary but not sufficient condition for stability is that the coefficients of Equation (4–42) all be present and all have a positive sign. (If all a's are negative, they can be made positive by multiplying both sides of the equation by -1.)

3. If all coefficients are positive, arrange the coefficients of the polynomial in rows and columns according to the following pattern:

$$
\begin{array}{c|cccccc}
s^n   & a_0 & a_2 & a_4 & a_6 & \cdot & \cdot & \cdot \\
s^{n-1} & a_1 & a_3 & a_5 & a_7 & \cdot & \cdot & \cdot \\
s^{n-2} & b_1 & b_2 & b_3 & b_4 & \cdot & \cdot & \cdot \\
s^{n-3} & c_1 & c_2 & c_3 & c_4 & \cdot & \cdot & \cdot \\
s^{n-4} & d_1 & d_2 & d_3 & d_4 & \cdot & \cdot & \cdot \\
\cdot   & \cdot & \cdot &   &   &   &   &   \\
\cdot   & \cdot & \cdot &   &   &   &   &   \\
\cdot   & \cdot & \cdot &   &   &   &   &   \\
s^2   & e_1 & e_2 &   &   &   &   &   \\
s^1   & f_1 &   &   &   &   &   &   \\
s^0   & g_1 &   &   &   &   &   &   \\
\end{array}
$$

The coefficients b_1, b_2, b_3, and so on, are evaluated as follows:

$$b_1 = \frac{a_1 a_2 - a_0 a_3}{a_1}$$

$$b_2 = \frac{a_1 a_4 - a_0 a_5}{a_1}$$

$$b_3 = \frac{a_1 a_6 - a_0 a_7}{a_1}$$

$$\cdot$$
$$\cdot$$
$$\cdot$$

The evaluation of the b's is continued until the remaining ones are all zero. The same pattern of cross-multiplying the coefficients of the two previous rows is followed in evaluating the c's, d's, e's, and so on. That is,

$$c_1 = \frac{b_1 a_3 - a_1 b_2}{b_1}$$

$$c_2 = \frac{b_1 a_5 - a_1 b_3}{b_1}$$

$$c_3 = \frac{b_1 a_7 - a_1 b_4}{b_1}$$

$$\vdots$$

and

$$d_1 = \frac{c_1 b_2 - b_1 c_2}{c_1}$$

$$d_2 = \frac{c_1 b_3 - b_1 c_3}{c_1}$$

$$\vdots$$

This process is continued until the nth row has been completed. The complete array of coefficients is triangular. Note that in developing the array an entire row may be divided or multiplied by a positive number in order to simplify the subsequent numerical calculation without altering the stability conclusion.

Routh's stability criterion states that the number of roots of Equation (4–42) with positive real parts is equal to the number of changes in sign of the coefficients of the first column of the array. It should be noted that the exact values of the terms in the first column need not be known; instead only the signs are needed. The necessary and sufficient condition that all roots of Equation (4–42) lie in the left-half s plane is that all the coefficients of Equation (4–42) be positive and all terms in the first column of the array have positive signs.

EXAMPLE 4–3   Let us apply Routh's stability criterion to the following third-order polynomial:

$$a_0 s^3 + a_1 s^2 + a_2 s + a_3 = 0$$

where all the coefficients are positive numbers. The array of coefficients becomes

s^3	a_0	a_2
s^2	a_1	a_3
s^1	$\dfrac{a_1 a_2 - a_0 a_3}{a_1}$	
s^0	a_3	

The condition that all roots have negative real parts is given by

$$a_1 a_2 > a_0 a_3$$

EXAMPLE 4–4   Consider the following polynomial:

$$s^4 + 2s^3 + 3s^2 + 4s + 5 = 0$$

Let us follow the procedure just presented and construct the array of coefficients. (The first two rows can be obtained directly from the given polynomial. The remaining terms are obtained from these. If any coefficients are missing, they may be replaced by zeros in the array.)

s^4	1	3	5		s^4	1	3	5	
s^3	2	4	0		s^3	$\not{2}$	$\not{4}$	$\not{0}$	The second row is divided
						1	2	0	by 2.
s^2	1	5			s^2	1	5		
s^1	-6				s^1	-3			
s_0	5				s^0	5			

In this example, the number of changes in sign of the coefficients in the first column is two. This means that there are two roots with positive real parts. Note that the result is unchanged when the coefficients of any row are multiplied or divided by a positive number in order to simplify the computation.

Special cases.   If a first-column term in any row is zero, but the remaining terms are not zero or there is no remaining term, then the zero term is replaced by a very small positive number ϵ and the rest of the array is evaluated. For example, consider the following equation:

$$s^3 + 2s^2 + s + 2 = 0 \tag{4–43}$$

The array of coefficients is

$$
\begin{array}{c|cc}
s^3 & 1 & 1 \\
s^2 & 2 & 2 \\
s^1 & 0 \approx \epsilon \\
s^0 & 2
\end{array}
$$

If the sign of the coefficient above the zero (ϵ) is the same as that below it, it indicates that there are a pair of imaginary roots. Actually, Equation (4–43) has two roots at $s = \pm j$.

If, however, the sign of the coefficient above the zero (ϵ) is opposite that below it, it indicates that there is one sign change. For example, for the following equation,

$$s^3 - 3s + 2 = (s - 1)^2(s + 2) = 0$$

the array of coefficients is

One sign change: $\left(\begin{array}{c|ccc} s^3 & 1 & -3 \\ s^2 & 0 \approx \epsilon & 2 \\ \end{array}\right.$

One sign change: $\left.\begin{array}{c|cc} s^1 & -3 - \dfrac{2}{\epsilon} \\ s^0 & 2 \end{array}\right.$

There are two sign changes of the coefficients in the first column. This agrees with the correct result indicated by the factored form of the polynomial equation.

If all the coefficients in any derived row are zero, it indicates that there are roots of equal magnitude lying radially opposite in the s plane, that is, two real roots with equal magnitudes and opposite signs and/or two conjugate imaginary roots. In such a case, the evaluation of the rest of the array can be continued by forming an auxiliary polynomial with the coefficients of the last row and by using the coefficients of the derivative of this polynomial in the next row. Such roots with equal magnitudes and lying radially opposite in the s plane can be found by solving the auxiliary polynomial, which is always even. For a $2n$-degree auxiliary polynomial, there are n pairs of equal and opposite roots. For example, consider the following equation:

$$s^5 + 2s^4 + 24s^3 + 48s^2 - 25s - 50 = 0$$

The array of coefficients is

$$
\begin{array}{llll}
s^5 & 1 & 24 & -25 \\
s^4 & 2 & 48 & -50 \quad \leftarrow\text{Auxiliary polynomial } P(s) \\
s^3 & 0 & 0 \\
\end{array}
$$

The terms in the s^3 row are all zero. The auxiliary polynomial is then formed from the coefficients of the s^4 row. The auxiliary polynomial $P(s)$ is

$$P(s) = 2s^4 + 48s^2 - 50$$

which indicates that there are two pairs of roots of equal magnitude and opposite sign. These pairs are obtained by solving the auxiliary polynomial equation $P(s) = 0$. The derivative of $P(s)$ with respect to s is

$$\frac{dP(s)}{ds} = 8s^3 + 96s$$

The terms in the s^3 row are replaced by the coefficients of the last equation, that is, 8 and 96. The array of coefficients then becomes

$$
\begin{array}{llll}
s^5 & 1 & 24 & -25 \\
s^4 & 2 & 48 & -50 \\
s^3 & 8 & 96 & \quad \leftarrow\text{Coefficients of } dP(s)/ds \\
s^2 & 24 & -50 \\
s^1 & 112.7 & 0 \\
s^0 & -50 \\
\end{array}
$$

We see that there is one change in sign in the first column of the new array. Thus, the original equation has one root with a positive real part. By solving for roots of the auxiliary polynomial equation,

$$2s^4 + 48s^2 - 50 = 0$$

we obtain

$$s^2 = 1, \qquad s^2 = -25$$

or

$$s = \pm 1, \qquad s = \pm j5$$

These two pairs of roots are a part of the roots of the original equation. As a matter of fact, the original equation can be written in factored form as follows:

$$(s + 1)(s - 1)(s + j5)(s - j5)(s + 2) = 0$$

Clearly, the original equation has one root with a positive real part.

Relative stability analysis.   Routh's stability criterion provides the answer to the question of absolute stability. This, in many practical cases, is not sufficient. We usually require information about the relative stability of the system. A useful approach for examining relative stability is to shift the s-plane axis and apply Routh's stability criterion. That is, we substitute

$$s = \hat{s} - \sigma \qquad (\sigma = \text{constant})$$

into the characteristic equation of the system, write the polynomial in terms of $\hat{s}$, and apply Routh's stability criterion to the new polynomial in $\hat{s}$. The number of changes of sign in the first column of the array developed for the polynomial in $\hat{s}$ is equal to the number of roots that are located to the right of the vertical line $s = -\sigma$. Thus, this test reveals the number of roots that lie to the right of the vertical line $s = -\sigma$.

Application of Routh's stability criterion to control system analysis.   Routh's stability criterion is of limited usefulness in linear control system analysis mainly because it does not suggest how to improve relative stability or how to stabilize an unstable system. It is possible, however, to determine the effects of changing one or two parameters of a system by examining the values that cause instability. In the following, we shall consider the problem of determining the stability range of a parameter value.

Consider the system shown in Figure 4–32. Let us determine the range of K for stability. The closed-loop transfer function is

$$\frac{C(s)}{R(s)} = \frac{K}{s(s^2 + s + 1)(s + 2) + K}$$

The characteristic equation is

$$s^4 + 3s^3 + 3s^2 + 2s + K = 0$$

The array of coefficients becomes

$$
\begin{array}{c|ccc}
s^4 & 1 & 3 & K \\
s^3 & 3 & 2 & 0 \\
s^2 & \frac{7}{3} & K & \\
s^1 & 2 - \frac{9}{7}K & & \\
s^0 & K & &
\end{array}
$$

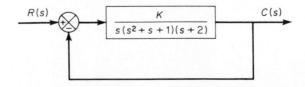

Figure 4–32
Control system.

Chapter 4 / Transient-Response Analysis and Steady-State Error Analysis

For stability, K must be positive, and all coefficients in the first column must be positive. Therefore,

$$\tfrac{14}{9} > K > 0$$

When $K = \tfrac{14}{9}$, the system becomes oscillatory and, mathematically, the oscillation is sustained at constant amplitude.

4–7 STEADY-STATE ERROR ANALYSIS

Any physical control system inherently suffers steady-state error in response to certain types of inputs. A system may have no steady-state error to a step input, but the same system may exhibit nonzero steady-state error to a ramp input. (The only way we may be able to eliminate this error is to modify the system structure.) Whether or not a given system will exhibit steady-state error for a given type of input depends on the type of open-loop transfer function of the system, to be discussed in what follows.

Classification of control systems. Control systems may be classified according to their ability to follow step inputs, ramp inputs, parabolic inputs, and so on. This is a reasonable classification scheme because actual inputs may frequently be considered combinations of such inputs. The magnitudes of the steady-state errors due to these individual inputs are indicative of the "goodness" of the system.

Consider the following open-loop transfer function $G(s)H(s)$:

$$G(s)H(s) = \frac{K(T_a s + 1)(T_b s + 1) \cdots (T_m s + 1)}{s^N (T_1 s + 1)(T_2 s + 1) \cdots (T_p s + 1)}$$

It involves the term s^N in the denominator, representing a pole of multiplicity N at the origin. The present classification scheme is based on the number of integrations indicated by the open-loop transfer function. A system is called type 0, type 1, type 2, . . . , if $N = 0$, $N = 1, N = 2, . . .$, respectively. Note that this classification is different from that of the order of a system. As the type number is increased, accuracy is improved; however, increasing the type number aggravates the stability problem. A compromise between steady-state accuracy and relative stability is always necessary. In practice, it is rather exceptional to have type 3 or higher systems because we find it generally difficult to design stable systems having more than two integrations in the feedforward path.

We shall see later that if $G(s)H(s)$ is written so that each term in the numerator and denominator, except the term s^N, approaches unity as s approaches zero, then the open-loop gain K is directly related to the steady-state error.

Steady-state errors. Consider the system shown in Figure 4–33. The closed-loop transfer function is

$$\frac{C(s)}{R(s)} = \frac{G(s)}{1 + G(s)H(s)}$$

The transfer function between the actuating error signal $e(t)$ and the input signal $r(t)$ is

$$\frac{E(s)}{R(s)} = 1 - \frac{C(s)H(s)}{R(s)} = \frac{1}{1 + G(s)H(s)}$$

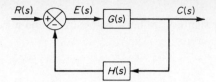

Figure 4–33
Control system.

where the actuating error $e(t)$ is the difference between the input signal and the feedback signal.

The final value theorem provides a convenient way to find the steady-state performance of a stable system. Since $E(s)$ is

$$E(s) = \frac{1}{1 + G(s)H(s)} R(s)$$

the steady-state actuating error is

$$e_{ss} = \lim_{t \to \infty} e(t) = \lim_{s \to 0} sE(s) = \lim_{s \to 0} \frac{sR(s)}{1 + G(s)H(s)}$$

The static error constants defined in the following are figures of merit of control systems. The higher the constants, the smaller the steady-state error. In a given system, the output may be the position, velocity, pressure, temperature, or the like. The physical form of the output, however, is immaterial to the present analysis. Therefore, in what follows, we shall call the output "position," the rate of change of the output "velocity," and so on. This means that in a temperature control system "position" represents the output temperature, "velocity" represents the rate of change of the output temperature, and so on.

Static position error constant K_p.   The steady-state actuating error of the system for a unit-step input is

$$e_{ss} = \lim_{s \to 0} \frac{s}{1 + G(s)H(s)} \frac{1}{s}$$

$$= \frac{1}{1 + G(0)H(0)}$$

The static position error constant K_p is defined by

$$K_p = \lim_{s \to 0} G(s)H(s) = G(0)H(0)$$

Thus, the steady-state actuating error in terms of the static position error constant K_p is given by

$$e_{ss} = \frac{1}{1 + K_p}$$

For a type 0 system,

$$K_p = \lim_{s \to 0} \frac{K(T_a s + 1)(T_b s + 1) \cdots}{(T_1 s + 1)(T_2 s + 1) \cdots} = K$$

For a type 1 or higher system,

$$K_p = \lim_{s \to 0} \frac{K(T_a s + 1)(T_b s + 1) \cdots}{s^N(T_1 s + 1)(T_2 s + 1) \cdots} = \infty \qquad (N \geq 1)$$

Hence, for a type 0 system, the static position error constant K_p is finite, while for a type 1 or higher system K_p is infinite.

For a unit-step input, the steady-state actuating error e_{ss} may be summarized as follows:

$$e_{ss} = \frac{1}{1 + K} \qquad \text{for type 0 systems}$$

$$e_{ss} = 0 \qquad \text{for type 1 or higher systems}$$

From the foregoing analysis, it is seen that the response of a feedback control system to a step input involves a steady-state error if there is no integration in the feedforward path. (If small errors for step inputs can be tolerated, then a type 0 system may be permissible, provided that the gain K is sufficiently large. If the gain K is too large, however, it is difficult to obtain reasonable relative stability.) If zero steady-state error for a step input is desired, the type of the system must be one or higher.

Static velocity error constant K_v. The steady-state actuating error of the system with a unit-ramp input is given by

$$e_{ss} = \lim_{s \to 0} \frac{s}{1 + G(s)H(s)} \frac{1}{s^2}$$

$$= \lim_{s \to 0} \frac{1}{sG(s)H(s)}$$

The static velocity error constant K_v is defined by

$$K_v = \lim_{s \to 0} sG(s)H(s)$$

Thus, the steady-state actuating error in terms of the static velocity error constant K_v is given by

$$e_{ss} = \frac{1}{K_v}$$

The term *velocity error* is used here to express the steady-state error for a ramp input. The dimension of the velocity error is the same as the system error. That is, velocity error is not an error in velocity, but it is an error in position due to a ramp input.

For a type 0 system,

$$K_v = \lim_{s \to 0} \frac{sK(T_a s + 1)(T_b s + 1) \cdots}{(T_1 s + 1)(T_2 s + 1) \cdots} = 0$$

For a type 1 system,

$$K_v = \lim_{s \to 0} \frac{sK(T_a s + 1)(T_b s + 1) \cdots}{s(T_1 s + 1)(T_2 s + 1) \cdots} = K$$

For a type 2 or higher system,

$$K_v = \lim_{s \to 0} \frac{sK(T_a s + 1)(T_b s + 1) \cdots}{s^N(T_1 s + 1)(T_2 s + 1) \cdots} = \infty \qquad (N \geq 2)$$

The steady-state actuating error e_{ss} for the unit-ramp input can be summarized as follows:

$$e_{ss} = \frac{1}{K_v} = \infty \qquad \text{for type 0 systems}$$

$$e_{ss} = \frac{1}{K_v} = \frac{1}{K} \qquad \text{for type 1 systems}$$

$$e_{ss} = \frac{1}{K_v} = 0 \qquad \text{for type 2 or higher systems}$$

The foregoing analysis indicates that a type 0 system is incapable of following a ramp input in the steady state. The type 1 system with unity feedback can follow the ramp input with a finite error. In steady-state operation, the output velocity is exactly the same as the input velocity, but there is a positional error. This error is proportional to the velocity of the input and is inversely proportional to the gain K. Figure 4–34 shows an example of the response of a type 1 system with unity feedback to a ramp input. The type 2 or higher system can follow a ramp input with zero actuating error at steady state.

Static acceleration error constant K_a.   The steady-state actuating error of the system with a unit-parabolic input (acceleration input), which is defined by

$$r(t) = \frac{t^2}{2} \qquad \text{for } t \geq 0$$

$$= 0 \qquad \text{for } t < 0$$

is given by

$$e_{ss} = \lim_{s \to 0} \frac{s}{1 + G(s)H(s)} \frac{1}{s^3}$$

$$= \frac{1}{\lim_{s \to 0} s^2 G(s)H(s)}$$

The static acceleration error constant K_a is defined by the equation

$$K_a = \lim_{s \to 0} s^2 G(s)H(s)$$

The steady-state actuating error is then

$$e_{ss} = \frac{1}{K_a}$$

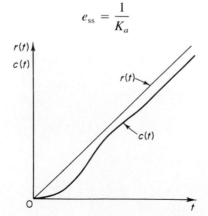

Figure 4–34
Response of a type 1
unity-feedback system
to a ramp input.

Note that the acceleration error, the steady-state error due to a parabolic input, is an error in position.

The values of K_a are obtained as follows:

For a type 0 system,

$$K_a = \lim_{s \to 0} \frac{s^2 K(T_a s + 1)(T_b s + 1) \cdots}{(T_1 s + 1)(T_2 s + 1) \cdots} = 0$$

For a type 1 system,

$$K_a = \lim_{s \to 0} \frac{s^2 K(T_a s + 1)(T_b s + 1) \cdots}{s(T_1 s + 1)(T_2 s + 1) \cdots} = 0$$

For a type 2 system,

$$K_a = \lim_{s \to 0} \frac{s^2 K(T_a s + 1)(T_b s + 1) \cdots}{s^2(T_1 s + 1)(T_2 s + 1) \cdots} = K$$

For a type 3 or higher system,

$$K_a = \lim_{s \to 0} \frac{s^2 K(T_a s + 1)(T_b s + 1) \cdots}{s^N(T_1 s + 1)(T_2 s + 1) \cdots} = \infty \qquad (N \geq 3)$$

Thus, the steady-state actuating error for the unit parabolic input is

$$e_{ss} = \infty \qquad \text{for type 0 and type 1 systems}$$

$$e_{ss} = \frac{1}{K} \qquad \text{for type 2 systems}$$

$$e_{ss} = 0 \qquad \text{for type 3 or higher systems}$$

Note that both type 0 and type 1 systems are incapable of following a parabolic input in the steady state. The type 2 system with unity feedback can follow a parabolic input with a finite actuating error signal. Figure 4–35 shows an example of the response of a type 2 system with unity feedback to a parabolic input. The type 3 or higher system with unity feedback follows a parabolic input with zero actuating error at steady state.

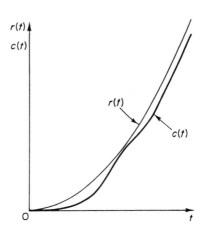

Figure 4–35
Response of a type 2
unity-feedback system
to a parabolic input.

Summary. Table 4–1 summarizes the steady-state errors for type 0, type 1, and type 2 systems when they are subjected to various inputs. The finite values for steady-state errors appear on the diagonal line. Above the diagonal, the steady-state errors are infinity; below the diagonal, they are zero.

Remember that the terms *position error*, *velocity error*, and *acceleration error* mean steady-state deviations in the output position. A finite velocity error implies that after transients have died out the input and output move at the same velocity but have a finite position difference.

The error constants K_p, K_v, and K_a describe the ability of a system to reduce or eliminate steady-state error. Therefore, they are indicative of the steady-state performance. It is generally desirable to increase the error constants, while maintaining the transient response within an acceptable range. If there is any conflict between the static velocity error constant and the static acceleration error constant, then the latter may be considered less important than the former. It is noted that to improve the steady-state performance we can increase the type of the system by adding an integrator or integrators to the feedforward path. This, however, introduces an additional stability problem. The design of a satisfactory system with more than two integrators in series in the feedforward path is generally difficult.

Table 4–1 Steady-State Error in Terms of Gain K

	Step Input $r(t) = 1$	Ramp Input $r(t) = t$	Acceleration Input $r(t) = \frac{1}{2}t^2$
Type 0 system	$\dfrac{1}{1 + K}$	∞	∞
Type 1 system	0	$\dfrac{1}{K}$	∞
Type 2 system	0	0	$\dfrac{1}{K}$

Correlation between integral of actuating error in step response and steady-state error in ramp response. We shall next discuss the correlation between the integral of the actuating error in the unit-step response of the system shown in Figure 4–33 and the steady-state actuating error in the unit-ramp response of the same system.

It will be seen that the total area under the actuating error curve $\displaystyle\int_0^\infty e(t)\, dt$ as a result of unit-step response gives the steady-state actuating error in the response to the unit-ramp input. To prove this, let us define

$$\mathscr{L}[e(t)] = \int_0^\infty \epsilon^{-st} e(t)\, dt = E(s)$$

Then

$$\lim_{s \to 0} \int_0^\infty \epsilon^{-st} e(t)\, dt = \int_0^\infty e(t)\, dt = \lim_{s \to 0} E(s)$$

Note that

$$\frac{E(s)}{R(s)} = 1 - \frac{H(s)C(s)}{R(s)} = \frac{1}{1 + G(s)H(s)}$$

Hence

$$\int_0^\infty e(t)\, dt = \lim_{s \to 0} \left[\frac{R(s)}{1 + G(s)H(s)} \right]$$

For a unit-step input,

$$\int_0^\infty e(t)\, dt = \lim_{s \to 0} \left[\frac{1}{1 + G(s)H(s)} \frac{1}{s} \right]$$

$$= \lim_{s \to 0} \frac{1}{sG(s)H(s)}$$

$$= \frac{1}{K_v}$$

$$= \text{steady-state actuating error in unit-ramp response}$$

Hence

$$\int_0^\infty e(t)\, dt = e_{\text{ssr}} \qquad (4\text{-}44)$$

where $e(t)$ = actuating error in the unit-step response

e_{ssr} = steady-state actuating error in the unit-ramp response

If e_{ssr} is zero, then $e(t)$ must change its sign at least once. This means that a zero-velocity-error system (a system having $K_v = \infty$) will exhibit at least one overshoot when the system is subjected to a step input.

4-8 INTRODUCTION TO SYSTEM OPTIMIZATION

In this section, we shall solve a simple optimization problem in which we minimize certain error performance indexes. We shall introduce the direct approach and the Laplace transform approach to the solution of this problem. We shall present a state-space approach to the solution of similar optimization problems in Chapter 10.

In the design of a control system, it is important that the system meet given performance specifications. Since control systems are dynamic, the performance specifications may be given in terms of the transient-response behavior to specific inputs, such as step inputs, ramp inputs, and so on, or the specifications may be given in terms of a performance index.

Performance indexes.   A performance index is a number that indicates the "good-ness" of system performance. A control system is considered optimal if the values of the parameters are chosen so that the selected performance index is minimum or maximum depending on the situation. The optimal values of the parameters depend directly on the performance index selected.

Requirements of performance indexes.   A performance index must offer selectivity: that is, an optimal adjustment of the parameters must clearly distinguish nonoptimal adjustments of the parameters. In addition, a performance index must yield a single positive number or zero, the latter being obtained if and only if the measure of the deviation is identically zero. To be useful, a performance index must be a function of the parameters of the system, and it must exhibit a maximum or minimum. Finally, to be practical, a performance index must be easily computed, analytically or computationally.

Error performance indexes. In what follows, we shall discuss several error criteria in which the corresponding performance indexes are integrals of some function or weighted function of the deviation of the system output from the input. Since the values of the integrals can be obtained as functions of the system parameters, once a performance index is specified, the optimal system can be designed by adjusting the parameters to yield, say, the smallest value of the integral.

Various error performance indexes have been proposed in the literature. We shall discuss the following four in this section.

$$\int_0^\infty e^2(t)\, dt, \qquad \int_0^\infty te^2(t)\, dt, \qquad \int_0^\infty |e(t)|\, dt, \qquad \int_0^\infty t\,|e(t)|\, dt$$

Consider a control system whose input and output are $r(t)$ and $c(t)$, respectively. We shall define the error $e(t)$ as

$$e(t) = r(t) - c(t)$$

Note that unless

$$\lim_{t \to \infty} e(t) = 0$$

the performance indexes approach infinity. If $\lim_{t \to \infty} e(t)$ does not approach zero, we may define

$$e(t) = c(\infty) - c(t)$$

With this definition of the error, the performance indexes will yield finite numbers, if the system is a stable one.

Integral square-error criterion. According to the integral square-error (ISE) criterion, the quality of system performance is evaluated by the following integral:

$$\int_0^\infty e^2(t)\, dt$$

where the upper limit ∞ may be replaced by T, which is chosen sufficiently large so that $e(t)$ for $T < t$ is negligible. The optimal system is the one that minimizes this integral. This performance index has been used extensively for both deterministic inputs (such as step inputs) and statistical inputs because of the ease of computing the integral both analytically and experimentally. Figure 4–36 shows $r(t)$, $c(t)$, $e(t)$, $e^2(t)$, and $\int e^2(t)\, dt$ when the input $r(t)$ is a unit step. The integral of $e^2(t)$ from 0 to T is the total area under the curve $e^2(t)$.

A characteristic of this performance index is that it weighs large errors heavily and small errors lightly. This criterion is not very selective since, for the following second-order system,

$$\frac{C(s)}{R(s)} = \frac{1}{s^2 + 2\zeta s + 1}$$

a change in ζ from 0.5 to 0.7 does not yield much change in the value of the integral.

A system designed by this criterion tends to show a rapid decrease in a large initial error. Hence the response is fast and ocillatory. Thus the system has poor relative stability.

Note, however, that the integral square-error criterion is often of practical significance because the minimization of the performance index results in the minimization of power consumption for some systems, such as spacecraft systems.

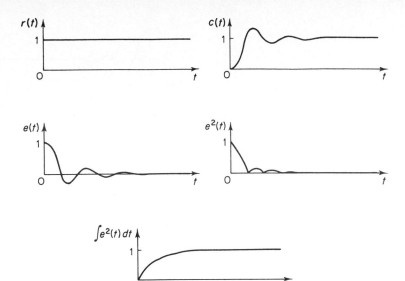

Figure 4–36
Curves showing input $r(t)$, output $c(t)$, error $e(t)$, squared error $e^2(t)$, and integral squared error $\int e^2(t)\,dt$ as functions of t.

Integral-of-time-multiplied square-error criterion.   The performance index based on the integral-of-time-multiplied square-error (ITSE) criterion is

$$\int_0^\infty te^2(t)\,dt$$

The optimal system is the one that minimizes this integral.

This criterion has a characteristic that in the unit-step response of the system a large initial error is weighed lightly, while errors occurring late in the transient response are penalized heavily. This criterion has a better selectivity than the integral square-error criterion.

Integral absolute-error criterion.   The performance index defined by the integral absolute-error (IAE) criterion is

$$\int_0^\infty |e(t)|\,dt$$

This is one of the most easily applied performance indexes. If this criterion is used, both highly underdamped and highly overdamped systems cannot be made optimum. An optimum system based on this criterion is a system that has reasonable damping and a satisfactory transient-response characteristic. However, this performance index cannot easily be evaluated by analytical means. Figure 4–37 shows the $e(t)$ versus t curve and the $|e(t)|$ versus t curve.

Note that minimization of the integral absolute error is directly related to the minimization of fuel consumption of spacecraft systems.

Integral-of-time-multiplied absolute-error criterion.   According to this criterion, the ITAE criterion, the optimum system is the one that minimizes the following performance index:

$$\int_0^\infty t|e(t)|\,dt$$

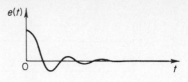

Figure 4–37
Curves showing $e(t)$
versus t and $|e(t)|$
versus t.

As in the preceding criteria, a large initial error in a unit-step response is weighed lightly, and errors occurring late in the transient response are penalized heavily.

A system designed by use of this criterion has a characteristic that the overshoot in the transient response is small and oscillations are well damped. This criterion possesses good selectivity and is an improvement over the integral absolute-error criterion. It is, however, very difficult to evaluate analytically, although it can be easily measured experimentally.

Comparison of various error criteria. Figure 4–38 shows several error performance curves. The system considered is

$$\frac{C(s)}{R(s)} = \frac{1}{s^2 + 2\zeta s + 1}$$

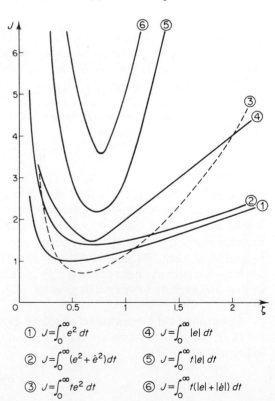

$$\text{①} \quad J = \int_0^\infty e^2 \, dt \qquad\qquad \text{④} \quad J = \int_0^\infty |e| \, dt$$

$$\text{②} \quad J = \int_0^\infty (e^2 + \dot{e}^2) \, dt \qquad \text{⑤} \quad J = \int_0^\infty t|e| \, dt$$

$$\text{③} \quad J = \int_0^\infty t e^2 \, dt \qquad\qquad \text{⑥} \quad J = \int_0^\infty t(|e| + |\dot{e}|) \, dt$$

Figure 4–38
Error performance
curves.

Inspection of these curves reveals that the integral square-error criterion does not have good selectivity because the curve is rather flat near the point where the performance index is minimum. Addition of the square of error rate to the integrand of the performance index, or

$$\int_0^\infty [e^2(t) + \dot{e}^2(t)]\, dt$$

makes the optimal value of ζ to be 0.707, but the selectivity is again poor. From Figure 4–38, it is clearly seen that other performance indexes have good selectivity.

The curves in Figure 4–38 indicate that in the second-order system considered $\zeta = 0.7$ corresponds to the optimal, or near optimal, value with respect to each of the performance indexes employed. When the damping ratio ζ is 0.7, the second-order system results in swift response to a step input with approximately 5% overshoot.

Application of the ITAE criterion to nth-order systems.* The ITAE criterion has been applied to the following nth-order system:

$$\frac{C(s)}{R(s)} = \frac{a_n}{s^n + a_1 s^{n-1} + \cdots + a_{n-1}s + a_n}$$

and the coefficients that will minimize

$$\int_0^\infty t|e|\, dt$$

have been determined. Clearly, this transfer function yields zero steady-state error to step inputs.

Table 4–2 shows the optimal form of the closed-loop transfer function based on this ITAE criterion. Figure 4–39 shows the step-response curves of the optimal systems.

Table 4–2 Optimal Form of the Closed-Loop Transfer Function Based on the ITAE Criterion (Zero Step Error Systems)

$\dfrac{C(s)}{R(s)} = \dfrac{a_n}{s^n + a_1 s^{n-1} + \cdots + a_{n-1}s + a_n},\qquad a_n = \omega_n^n$
$s + \omega_n$
$s^2 + 1.4\omega_n s + \omega_n^2$
$s^3 + 1.75\omega_n s^2 + 2.15\omega_n^2 s + \omega_n^3$
$s^4 + 2.1\omega_n s^3 + 3.4\omega_n^2 s^2 + 2.7\omega_n^3 s + \omega_n^4$
$s^5 + 2.8\omega_n s^4 + 5.0\omega_n^2 s^3 + 5.5\omega_n^3 s^2 + 3.4\omega_n^4 s + \omega_n^5$
$s^6 + 3.25\omega_n s^5 + 6.60\omega_n^2 s^4 + 8.60\omega_n^3 s^3 + 7.45\omega_n^4 s^2 + 3.95\omega_n^5 s + \omega_n^6$

* Reference G-4.

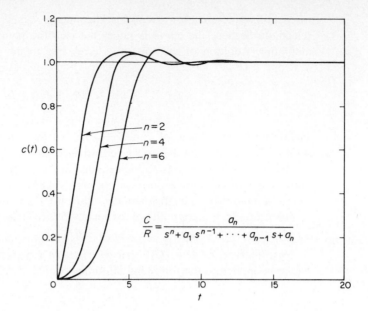

Figure 4–39
Step-response curves
of systems with
optimal transfer
functions based on the
ITAE criterion.

$$\frac{C}{R} = \frac{a_n}{s^n + a_1 s^{n-1} + \cdots + a_{n-1} s + a_n}$$

Laplace transform approach to the computation of ISE performance indexes. Consider the control system shown in Figure 4–40. Suppose that a unit-step input $r(t)$ is applied to the system at $t = 0$. We assume that the system is initially at rest. The problem is to determine the optimal value of ζ $(0 < \zeta < 1)$ such that

$$J = \int_0^\infty x^2(t)\, dt \tag{4-45}$$

is minimum, where $x(t)$ is the error signal. [We use the notation $x(t)$ for the error signal to avoid confusion with the exponential function $e(t)$.]

We shall compute $\int_0^\infty f(t)\, dt$, where $f(t) = x^2(t)$, by use of the Laplace transform method. In doing this, we shall use the real integration theorem, which was presented in Section 1–3 [see Equation (1–8)].

$$\mathcal{L}\left[\int_0^t f(t)\, dt \right] = \frac{F(s)}{s}$$

where $F(s) = \mathcal{L}[f(t)]$.

By use of the real integration theorem and the final value theorem, we obtain

$$\int_0^\infty f(t)\, dt = \lim_{t \to \infty} \int_0^t f(t)\, dt$$

$$= \lim_{s \to 0} s \frac{F(s)}{s}$$

$$= \lim_{s \to 0} F(s)$$

Figure 4–40
Control system where
$0 < \zeta < 1$.

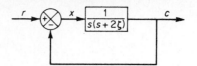

This is the basic equation for computing the ISE performance index. In the present problem

$$f(t) = x^2(t)$$

Hence

$$\lim_{t \to \infty} \int_0^t x^2(t)\, dt = \lim_{s \to 0} F(s)$$

where $F(s) = \mathcal{L}[x^2(t)]$.

Consider again the control system shown in Figure 4–40. For this system

$$\frac{X(s)}{R(s)} = \frac{s^2 + 2\zeta s}{s^2 + 2\zeta s + 1} \qquad (0 < \zeta < 1)$$

Hence, for the unit-step input,

$$X(s) = \frac{s^2 + 2\zeta s}{s^2 + 2\zeta s + 1}\,\frac{1}{s}$$

$$= \frac{s + \zeta}{s^2 + 2\zeta s + 1} + \frac{\zeta}{s^2 + 2\zeta s + 1}$$

Since $0 < \zeta < 1$, the inverse Laplace transform is

$$x(t) = e^{-\zeta t}\left(\cos\sqrt{1 - \zeta^2}\,t + \frac{\zeta}{\sqrt{1 - \zeta^2}}\sin\sqrt{1 - \zeta^2}\,t\right) \qquad (t \geq 0)$$

Hence

$$x^2(t) = e^{-2\zeta t}\left[\frac{1}{2(1 - \zeta^2)} + \frac{1}{2}\frac{1 - 2\zeta^2}{1 - \zeta^2}\cos 2\sqrt{1 - \zeta^2}\,t + \frac{\zeta}{\sqrt{1 - \zeta^2}}\sin 2\sqrt{1 - \zeta^2}\,t\right]$$

Then

$$F(s) = \mathcal{L}[x^2(t)] = \frac{1}{2(1 - \zeta^2)}\frac{1}{s + 2\zeta} + \frac{1 - 2\zeta^2}{2(1 - \zeta^2)}\frac{s + 2\zeta}{(s + 2\zeta)^2 + 4(1 - \zeta^2)}$$

$$+ \frac{\zeta}{\sqrt{1 - \zeta^2}}\frac{2\sqrt{1 - \zeta^2}}{(s + 2\zeta)^2 + 4(1 - \zeta^2)} \qquad (4\text{–}46)$$

Taking the limit as s approaches zero gives

$$\lim_{s \to 0} F(t) = \frac{1}{2(1 - \zeta^2)}\frac{1}{2\zeta} + \frac{1 - 2\zeta^2}{2(1 - \zeta^2)}\frac{\zeta}{2} + \frac{\zeta}{2}$$

$$= \zeta + \frac{1}{4\zeta}$$

Hence

$$J = \int_0^\infty x^2(t)\, dt$$

$$= \zeta + \frac{1}{4\zeta} \qquad (4\text{--}47)$$

The optimal value of ζ can be obtained by differentiating J with respect to ζ, equating $dJ/d\zeta$ to zero, and solving for ζ. Since

$$\frac{dJ}{d\zeta} = 1 - \frac{1}{4\zeta^2} = 0$$

the optimal value of ζ ($0 < \zeta < 1$) is

$$\zeta = 0.5$$

(Clearly, $\zeta = 0.5$ corresponds to a minimum since $d^2J/d\zeta^2 > 0$.) The minimum value of J is

$$\min J = 0.5 + 0.5 = 1$$

Laplace transform approach to the computation of ITSE performance indexes. In evaluating the performance index $\int_0^\infty tf(t)\, dt$, where $f(t) = x^2(t)$, we find it convenient to use the complex differentiation theorem, which was presented in Section 1–3. This theorem states that if $f(t)$ is Laplace transformable, then, except at poles of $F(s)$,

$$\mathscr{L}[tf(t)] = -\frac{d}{ds} F(s)$$

where $F(s) = \mathscr{L}[f(t)]$.

Consider the system shown in Figure 4–40, with the following performance index:

$$J = \int_0^\infty tx^2(t)\, dt$$

We again assume that the system is at rest before a unit-step input is applied. Let us find the optimal value of ζ that minimizes this performance index.

Define $x^2(t) = f(t)$. By using the complex differentiation theorem, we obtain

$$\mathscr{L}[tx^2(t)] = -\frac{d}{ds} F(s)$$

where $F(s) = \mathscr{L}[f(t)] = \mathscr{L}[x^2(t)]$. In the present problem, $F(s)$ is given by Equation (4–46).

$$\mathscr{L}[tx^2(t)] = -\frac{d}{ds} F(s)$$

$$= \frac{1}{2(1 - \zeta^2)} \frac{1}{(s + 2\zeta)^2} + \frac{1 - 2\zeta^2}{2(1 - \zeta^2)} \frac{(s + 2\zeta)^2 - 4(1 - \zeta^2)}{[(s + 2\zeta)^2 + 4(1 - \zeta^2)]^2}$$

$$+ \frac{4\zeta(s + 2\zeta)}{[(s + 2\zeta)^2 + 4(1 - \zeta^2)]^2}$$

Since

$$\lim_{s \to 0} \int_0^\infty t x^2(t) e^{-st} \, dt = \lim_{s \to 0} \left[-\frac{d}{ds} F(s) \right]$$

we have

$$\int_0^\infty t x^2(t) \, dt = \lim_{s \to 0} \left[-\frac{d}{ds} F(s) \right]$$

Taking the limit of $-dF(s)/ds$ as s approaches zero, we obtain

$$\lim_{s \to 0} \left[-\frac{d}{ds} F(s) \right] = \frac{1}{2(1 - \zeta^2)} \frac{1}{4\zeta^2} + \frac{1 - 2\zeta^2}{2(1 - \zeta^2)} \frac{2\zeta^2 - 1}{4} + \frac{\zeta^2}{2}$$

$$= \zeta^2 + \frac{1}{8\zeta^2}$$

Hence

$$J = \int_0^\infty t x^2(t) \, dt$$

$$= \zeta^2 + \frac{1}{8\zeta^2} \tag{4-48}$$

To obtain the optimal value of ζ that minimizes J, we differentiate J with respect to ζ and set the result equal to zero:

$$\frac{dJ}{d\zeta} = 2\zeta - \frac{1}{4\zeta^3} = 0$$

which yields

$$\zeta = 0.595$$

Clearly, $d^2J/d\zeta^2 > 0$. The optimal value of ζ is 0.595 and the minimum value of J is

$$\min J = 0.595^2 + \frac{1}{8 \times 0.595^2} = 0.707$$

Comments.   In this section we have presented an introduction to system optimization. More in-depth treatment of optimization problems is given later in Chapter 10. Specifically, in Chapter 10 we shall solve optimization problems based on quadratic performance indexes by use of the Liapunov approach in state space.

4-9 SOLVING THE TIME-INVARIANT STATE EQUATION

In this section, we shall obtain the general solution of the linear time-invariant state equation. We shall first consider the homogeneous case and then consider the nonhomogeneous case.

Solution of homogeneous state equations.   Before we solve vector-matrix differential equations, let us review the solution of the scalar differential equation

$$\dot{x} = ax \tag{4-49}$$

In solving this equation, we may assume a solution $x(t)$ of the form

$$x(t) = b_0 + b_1 t + b_2 t^2 + \cdots + b_k t^k + \cdots \qquad (4\text{–}50)$$

By substituting this assumed solution into Equation (4–49), we obtain

$$b_1 + 2b_2 t + 3b_3 t^2 + \cdots + kb_k t^{k-1} + \cdots$$
$$= a(b_0 + b_1 t + b_2 t^2 + \cdots + b_k t^k + \cdots) \qquad (4\text{–}51)$$

If the assumed solution is to be the true solution, Equation (4–51) must hold for any t. Hence, equating the coefficients of the equal powers of t, we obtain

$$b_1 = ab_0$$

$$b_2 = \frac{1}{2} ab_1 = \frac{1}{2} a^2 b_0$$

$$b_3 = \frac{1}{3} ab_2 = \frac{1}{3 \times 2} a^3 b_0$$

$$\cdot$$
$$\cdot$$
$$\cdot$$

$$b_k = \frac{1}{k!} a^k b_0$$

The value of b_0 is determined by substituting $t = 0$ into Equation (4–50), or

$$x(0) = b_0$$

Hence the solution $x(t)$ can be written as

$$x(t) = \left(1 + at + \frac{1}{2!} a^2 t^2 + \cdots + \frac{1}{k!} a^k t^k + \cdots \right) x(0)$$
$$= e^{at} x(0)$$

We shall now solve the vector-matrix differential equation

$$\dot{\mathbf{x}} = \mathbf{A}\mathbf{x} \qquad (4\text{–}52)$$

where $\mathbf{x} = n$-vector
$\mathbf{A} = n \times n$ constant matrix

By analogy with the scalar case we assume that the solution is in the form of a vector power series in t, or

$$\mathbf{x}(t) = \mathbf{b}_0 + \mathbf{b}_1 t + \mathbf{b}_2 t^2 + \cdots + \mathbf{b}_k t^k + \cdots \qquad (4\text{–}53)$$

By substituting this assumed solution into Equation (4–52), we obtain

$$\mathbf{b}_1 + 2\mathbf{b}_2 t + 3\mathbf{b}_3 t^2 + \cdots + k\mathbf{b}_k t^{k-1} + \cdots$$
$$= \mathbf{A}(\mathbf{b}_0 + \mathbf{b}_1 t + \mathbf{b}_2 t^2 + \cdots + k\mathbf{b}_k t^k + \cdots) \qquad (4\text{–}54)$$

If the assumed solution is to be the true solution, Equation (4–54) must hold for all t. Thus, by equating the coefficients of like powers of t on both sides of Equation (4–54), we obtain

$$\mathbf{b}_1 = \mathbf{A}\mathbf{b}_0$$

$$\mathbf{b}_2 = \frac{1}{2}\mathbf{A}\mathbf{b}_1 = \frac{1}{2}\mathbf{A}^2\mathbf{b}_0$$

$$\mathbf{b}_3 = \frac{1}{3}\mathbf{A}\mathbf{b}_2 = \frac{1}{3 \times 2}\mathbf{A}^3\mathbf{b}_0$$

$$\vdots$$

$$\mathbf{b}_k = \frac{1}{k!}\mathbf{A}^k\mathbf{b}_0$$

By substituting $t = 0$ into Equation (4–53), we obtain

$$\mathbf{x}(0) = \mathbf{b}_0$$

Thus the solution $\mathbf{x}(t)$ can be written as

$$\mathbf{x}(t) = \left(\mathbf{I} + \mathbf{A}t + \frac{1}{2!}\mathbf{A}^2 t^2 + \cdots + \frac{1}{k!}\mathbf{A}^k t^k + \cdots\right)\mathbf{x}(0)$$

The expression in the parentheses in the right side of this last equation is an $n \times n$ matrix. Because of its similarity to the infinite power series for a scaler exponential, we call it the matrix exponential and write

$$\mathbf{I} + \mathbf{A}t + \frac{1}{2!}\mathbf{A}^2 t^2 + \cdots + \frac{1}{k!}\mathbf{A}^k t^k + \cdots = e^{\mathbf{A}t}$$

In terms of the matrix exponential, the solution to Equation (4–52) can be written as

$$\mathbf{x}(t) = e^{\mathbf{A}t}\mathbf{x}(0) \tag{4–55}$$

Since the matrix exponential is very important in the state-space analysis of linear systems, we shall next examine the properties of the matrix exponential.

Matrix exponential.   It can be proved that the matrix exponential of an $n \times n$ matrix $\mathbf{A}$

$$e^{\mathbf{A}t} = \sum_{k=0}^{\infty} \frac{\mathbf{A}^k t^k}{k!}$$

converges absolutely for all finite t. (Hence computer calculations for evaluating the elements of $e^{\mathbf{A}t}$ by using the series expansion can be easily carried out.)

Because of the convergence of the infinite series $\sum_{k=0}^{\infty} \mathbf{A}^k t^k / k!$, the series can be differentiated term by term to give

$$\frac{d}{dt} e^{\mathbf{A}t} = \mathbf{A} + \mathbf{A}^2 t + \frac{\mathbf{A}^3 t^2}{2!} + \cdots + \frac{\mathbf{A}^k t^{k-1}}{(k-1)!} + \cdots$$

$$= \mathbf{A} \left[\mathbf{I} + \mathbf{A}t + \frac{\mathbf{A}^2 t^2}{2!} + \cdots + \frac{\mathbf{A}^{k-1} t^{k-1}}{(k-1)!} + \cdots \right] = \mathbf{A} e^{\mathbf{A}t}$$

$$= \left[\mathbf{I} + \mathbf{A}t + \frac{\mathbf{A}^2 t^2}{2!} + \cdots + \frac{\mathbf{A}^{k-1} t^{k-1}}{(k-1)!} + \cdots \right] \mathbf{A} = e^{\mathbf{A}t} \mathbf{A}$$

The matrix exponential has the property that

$$e^{\mathbf{A}(t+s)} = e^{\mathbf{A}t} e^{\mathbf{A}s}$$

This can be proved as follows:

$$e^{\mathbf{A}t} e^{\mathbf{A}s} = \left(\sum_{k=0}^{\infty} \frac{\mathbf{A}^k t^k}{k!} \right) \left(\sum_{k=0}^{\infty} \frac{\mathbf{A}^k s^k}{k!} \right)$$

$$= \sum_{k=0}^{\infty} \mathbf{A}^k \left(\sum_{i=0}^{k} \frac{t^i s^{k-i}}{i!(k-i)!} \right)$$

$$= \sum_{k=0}^{\infty} \mathbf{A}^k \frac{(t+s)^k}{k!}$$

$$= e^{\mathbf{A}(t+s)}$$

In particular, if $s = -t$, then

$$e^{\mathbf{A}t} e^{-\mathbf{A}t} = e^{-\mathbf{A}t} e^{\mathbf{A}t} = e^{\mathbf{A}(t-t)} = \mathbf{I}$$

Thus the inverse of $e^{\mathbf{A}t}$ is $e^{-\mathbf{A}t}$. Since the inverse of $e^{\mathbf{A}t}$ always exists, $e^{\mathbf{A}t}$ is nonsingular.

It is very important to remember that

$$e^{(\mathbf{A}+\mathbf{B})t} = e^{\mathbf{A}t} e^{\mathbf{B}t} \qquad \text{if } \mathbf{AB} = \mathbf{BA}$$

$$e^{(\mathbf{A}+\mathbf{B})t} \neq e^{\mathbf{A}t} e^{\mathbf{B}t} \qquad \text{if } \mathbf{AB} \neq \mathbf{BA}$$

To prove this, note that

$$e^{(\mathbf{A}+\mathbf{B})t} = \mathbf{I} + (\mathbf{A}+\mathbf{B})t + \frac{(\mathbf{A}+\mathbf{B})^2}{2!} t^2 + \frac{(\mathbf{A}+\mathbf{B})^3}{3!} t^3 + \cdots$$

$$e^{\mathbf{A}t} e^{\mathbf{B}t} = \left(\mathbf{I} + \mathbf{A}t + \frac{\mathbf{A}^2 t^2}{2!} + \frac{\mathbf{A}^3 t^3}{3!} + \cdots \right) \left(\mathbf{I} + \mathbf{B}t + \frac{\mathbf{B}^2 t^2}{2!} + \frac{\mathbf{B}^3 t^3}{3!} + \cdots \right)$$

$$= \mathbf{I} + (\mathbf{A}+\mathbf{B})t + \frac{\mathbf{A}^2 t^2}{2!} + \mathbf{AB}t^2 + \frac{\mathbf{B}^2 t^2}{2!} + \frac{\mathbf{A}^3 t^3}{3!}$$

$$+ \frac{\mathbf{A}^2 \mathbf{B} t^3}{2!} + \frac{\mathbf{AB}^2 t^3}{2!} + \frac{\mathbf{B}^3 t^3}{3!} + \cdots$$

Hence

$$e^{(\mathbf{A}+\mathbf{B})t} - e^{\mathbf{A}t} e^{\mathbf{B}t} = \frac{\mathbf{BA} - \mathbf{AB}}{2!} t^2$$

$$+ \frac{\mathbf{BA}^2 + \mathbf{ABA} + \mathbf{B}^2 \mathbf{A} + \mathbf{BAB} - 2\mathbf{A}^2 \mathbf{B} - 2\mathbf{AB}^2}{3!} t^3 + \cdots$$

The difference between $e^{(A+B)t}$ and $e^{At}e^{Bt}$ vanishes if $\mathbf{A}$ and $\mathbf{B}$ commute.

Laplace transform approach to the solution of homogeneous state equations.
Let us first consider the scalar case:

$$\dot{x} = ax \qquad (4\text{--}56)$$

Taking the Laplace transform of Equation (4–56), we obtain

$$sX(s) - x(0) = aX(s) \qquad (4\text{--}57)$$

where $X(s) = \mathcal{L}[x]$. Solving Equation (4–57) for $X(s)$ gives

$$X(s) = \frac{x(0)}{s - a} = (s - a)^{-1}x(0)$$

The inverse Laplace transform of this last equation gives the solution

$$x(t) = e^{at}x(0)$$

The foregoing approach to the solution of the homogeneous scalar differential equation can be extended to the homogeneous state equation:

$$\dot{\mathbf{x}}(t) = \mathbf{A}\mathbf{x}(t) \qquad (4\text{--}58)$$

Taking the Laplace transform of both sides of Equation (4–58), we obtain

$$s\mathbf{X}(s) - \mathbf{x}(0) = \mathbf{A}\mathbf{X}(s)$$

where $\mathbf{X}(s) = \mathcal{L}[\mathbf{x}]$. Hence

$$(s\mathbf{I} - \mathbf{A})\mathbf{X}(s) = \mathbf{x}(0)$$

Premultiplying both sides of this last equation by $(s\mathbf{I} - \mathbf{A})^{-1}$, we obtain

$$\mathbf{X}(s) = (s\mathbf{I} - \mathbf{A})^{-1}\mathbf{x}(0)$$

The inverse Laplace transform of $\mathbf{X}(s)$ gives the solution $\mathbf{x}(t)$. Thus

$$\mathbf{x}(t) = \mathcal{L}^{-1}[(s\mathbf{I} - \mathbf{A})^{-1}]\mathbf{x}(0) \qquad (4\text{--}59)$$

Note that

$$(s\mathbf{I} - \mathbf{A})^{-1} = \frac{\mathbf{I}}{s} + \frac{\mathbf{A}}{s^2} + \frac{\mathbf{A}^2}{s^3} + \cdots$$

Hence, the inverse Laplace transform of $(s\mathbf{I} - \mathbf{A})^{-1}$ gives

$$\mathcal{L}^{-1}[(s\mathbf{I} - \mathbf{A})^{-1}] = \mathbf{I} + \mathbf{A}t + \frac{\mathbf{A}^2 t^2}{2!} + \frac{\mathbf{A}^3 t^3}{3!} + \cdots = e^{\mathbf{A}t} \qquad (4\text{--}60)$$

(The inverse Laplace transform of a matrix is the matrix consisting of the inverse Laplace transforms of all elements.) From Equations (4–59) and (4–60), the solution of Equation (4–58) is obtained as

$$\mathbf{x}(t) = e^{\mathbf{A}t}\mathbf{x}(0)$$

The importance of Equation (4–60) lies in the fact that it provides a convenient means for finding the closed solution for the matrix exponential.

State transition matrix.   We can write the solution of the homogeneous state equation

$$\dot{\mathbf{x}} = \mathbf{A}\mathbf{x} \qquad\qquad (4\text{–}61)$$

as

$$\mathbf{x}(t) = \mathbf{\Phi}(t)\mathbf{x}(0) \qquad\qquad (4\text{–}62)$$

where $\mathbf{\Phi}(t)$ is an $n \times n$ matrix and is the unique solution of

$$\dot{\mathbf{\Phi}}(t) = \mathbf{A}\mathbf{\Phi}(t), \qquad \mathbf{\Phi}(0) = \mathbf{I}$$

To verify this, note that

$$\mathbf{x}(0) = \mathbf{\Phi}(0)\mathbf{x}(0) = \mathbf{x}(0)$$

and

$$\dot{\mathbf{x}}(t) = \dot{\mathbf{\Phi}}(t)\mathbf{x}(0) = \mathbf{A}\mathbf{\Phi}(t)\mathbf{x}(0) = \mathbf{A}\mathbf{x}(t)$$

We thus confirm that Equation (4–62) is the solution of Equation (4–61).

From Equations (4–55), (4–59), and (4–62), we obtain

$$\mathbf{\Phi}(t) = e^{\mathbf{A}t} = \mathscr{L}^{-1}[(s\mathbf{I} - \mathbf{A})^{-1}]$$

Note that

$$\mathbf{\Phi}^{-1}(t) = e^{-\mathbf{A}t} = \mathbf{\Phi}(-t)$$

From Equation (4–62), we see that the solution of Equation (4–61) is simply a transformation of the initial condition. Hence the unique matrix $\mathbf{\Phi}(t)$ is called the state-transition matrix. The state-transition matrix contains all the information about the free motions of the system defined by Equation (4–61).

If the eigenvalues $\lambda_1, \lambda_2, \ldots, \lambda_n$ of the matrix $\mathbf{A}$ are distinct, then $\mathbf{\Phi}(t)$ will contain the n exponentials

$$e^{\lambda_1 t}, e^{\lambda_2 t}, \ldots, e^{\lambda_n t}$$

In particular, if the matrix $\mathbf{A}$ is diagonal, then

$$\mathbf{\Phi}(t) = e^{\mathbf{A}t} = \begin{bmatrix} e^{\lambda_1 t} & & & & 0 \\ & e^{\lambda_2 t} & & & \\ & & \cdot & & \\ & & & \cdot & \\ 0 & & & & e^{\lambda_n t} \end{bmatrix} \qquad (\mathbf{A}: \text{diagonal})$$

If there is a multiplicity in the eigenvalues, for example, if the eigenvalues of $\mathbf{A}$ are

$$\lambda_1, \lambda_1, \lambda_1, \lambda_4, \lambda_5, \ldots, \lambda_n$$

then $\mathbf{\Phi}(t)$ will contain, in addition to the exponentials $e^{\lambda_1 t}, e^{\lambda_4 t}, e^{\lambda_5 t}, \ldots, e^{\lambda_n t}$, terms like $te^{\lambda_1 t}$ and $t^2 e^{\lambda_1 t}$.

Properties of state-transition matrices. We shall now summarize the important properties of the state-transition matrix $\mathbf{\Phi}(t)$. For the time-invariant system

$$\dot{\mathbf{x}} = \mathbf{A}\mathbf{x}$$

for which

$$\mathbf{\Phi}(t) = e^{\mathbf{A}t}$$

we have

1. $\mathbf{\Phi}(0) = e^{\mathbf{A}0} = \mathbf{I}$
2. $\mathbf{\Phi}(t) = e^{\mathbf{A}t} = (e^{-\mathbf{A}t})^{-1} = [\mathbf{\Phi}(-t)]^{-1}$ or $\mathbf{\Phi}^{-1}(t) = \mathbf{\Phi}(-t)$
3. $\mathbf{\Phi}(t_1 + t_2) = e^{\mathbf{A}(t_1 + t_2)} = e^{\mathbf{A}t_1}e^{\mathbf{A}t_2} = \mathbf{\Phi}(t_1)\mathbf{\Phi}(t_2) = \mathbf{\Phi}(t_2)\mathbf{\Phi}(t_1)$
4. $[\mathbf{\Phi}(t)]^n = \mathbf{\Phi}(nt)$
5. $\mathbf{\Phi}(t_2 - t_1)\mathbf{\Phi}(t_1 - t_0) = \mathbf{\Phi}(t_2 - t_0) = \mathbf{\Phi}(t_1 - t_0)\mathbf{\Phi}(t_2 - t_1)$

EXAMPLE 4–5 Obtain the state-transition matrix $\mathbf{\Phi}(t)$ of the following system:

$$\begin{bmatrix} \dot{x}_1 \\ \dot{x}_2 \end{bmatrix} = \begin{bmatrix} 0 & 1 \\ -2 & -3 \end{bmatrix}\begin{bmatrix} x_1 \\ x_2 \end{bmatrix}$$

Obtain also the inverse of the state-transition matrix, $\mathbf{\Phi}^{-1}(t)$.
For this system,

$$\mathbf{A} = \begin{bmatrix} 0 & 1 \\ -2 & -3 \end{bmatrix}$$

The state-transition matrix $\mathbf{\Phi}(t)$ is given by

$$\mathbf{\Phi}(t) = e^{\mathbf{A}t} = \mathscr{L}^{-1}[(s\mathbf{I} - \mathbf{A})^{-1}]$$

Since

$$s\mathbf{I} - \mathbf{A} = \begin{bmatrix} s & 0 \\ 0 & s \end{bmatrix} - \begin{bmatrix} 0 & 1 \\ -2 & -3 \end{bmatrix} = \begin{bmatrix} s & -1 \\ 2 & s+3 \end{bmatrix}$$

the inverse of $(s\mathbf{I} - \mathbf{A})$ is given by

$$(s\mathbf{I} - \mathbf{A})^{-1} = \frac{1}{(s+1)(s+2)}\begin{bmatrix} s+3 & 1 \\ -2 & s \end{bmatrix}$$

$$= \begin{bmatrix} \dfrac{s+3}{(s+1)(s+2)} & \dfrac{1}{(s+1)(s+2)} \\ \dfrac{-2}{(s+1)(s+2)} & \dfrac{s}{(s+1)(s+2)} \end{bmatrix}$$

Hence

$$\mathbf{\Phi}(t) = e^{\mathbf{A}t} = \mathscr{L}^{-1}[(s\mathbf{I} - \mathbf{A})^{-1}]$$

$$= \begin{bmatrix} 2e^{-t} - e^{-2t} & e^{-t} - e^{-2t} \\ -2e^{-t} + 2e^{-2t} & -e^{-t} + 2e^{-2t} \end{bmatrix}$$

Section 4–9 / Solving the Time-Invariant State Equation

Noting that $\mathbf{\Phi}^{-1}(t) = \mathbf{\Phi}(-t)$, we obtain the inverse of the state-transition matrix as follows:

$$\mathbf{\Phi}^{-1}(t) = e^{-\mathbf{A}t} = \begin{bmatrix} 2e^t - e^{2t} & e^t - e^{2t} \\ -2e^t + 2e^{2t} & -e^t + 2e^{2t} \end{bmatrix}$$

Solution of nonhomogeneous state equations. We shall begin by considering the scalar case

$$\dot{x} = ax + bu \tag{4–63}$$

Let us rewrite Equation (4–63) as

$$\dot{x} - ax = bu$$

Multiplying both sides of this equation by e^{-at}, we obtain

$$e^{-at}[\dot{x}(t) - ax(t)] = \frac{d}{dt}[e^{-at}x(t)] = e^{-at}bu(t)$$

Integrating this equation between 0 and t gives

$$e^{-at}x(t) = x(0) + \int_0^t e^{-a\tau}bu(\tau)\,d\tau$$

or

$$x(t) = e^{at}x(0) + e^{at}\int_0^t e^{-a\tau}bu(\tau)\,d\tau$$

The first term on the right side is the response to the initial condition and the second term is the response to the input $u(t)$.

Let us now consider the nonhomogeneous state equation described by

$$\dot{\mathbf{x}} = \mathbf{A}\mathbf{x} + \mathbf{B}\mathbf{u} \tag{4–64}$$

where $\mathbf{x} = n$-vector
$\quad \mathbf{u} = r$-vector
$\quad \mathbf{A} = n \times n$ constant matrix
$\quad \mathbf{B} = n \times r$ constant matrix

By writing Equation (4–64) as

$$\dot{\mathbf{x}}(t) - \mathbf{A}\mathbf{x}(t) = \mathbf{B}\mathbf{u}(t)$$

and premultiplying both sides of this equation by $e^{-\mathbf{A}t}$, we obtain

$$e^{-\mathbf{A}t}[\dot{\mathbf{x}}(t) - \mathbf{A}\mathbf{x}(t)] = \frac{d}{dt}[e^{-\mathbf{A}t}\mathbf{x}(t)] = e^{-\mathbf{A}t}\mathbf{B}\mathbf{u}(t)$$

Integrating the preceding equation between 0 and t gives

$$e^{-\mathbf{A}t}\mathbf{x}(t) = \mathbf{x}(0) + \int_0^t e^{-\mathbf{A}\tau}\mathbf{B}\mathbf{u}(\tau)\,d\tau$$

or

$$\mathbf{x}(t) = e^{\mathbf{A}t}\mathbf{x}(0) + \int_0^t e^{\mathbf{A}(t-\tau)}\mathbf{B}\mathbf{u}(\tau)\, d\tau \qquad (4\text{--}65)$$

Equation (4–65) can also be written as

$$\mathbf{x}(t) = \mathbf{\Phi}(t)\mathbf{x}(0) + \int_0^t \mathbf{\Phi}(t-\tau)\mathbf{B}\mathbf{u}(\tau)\, d\tau \qquad (4\text{--}66)$$

where $\mathbf{\Phi}(t) = e^{\mathbf{A}t}$. Equation (4–65) or (4–66) is the solution of Equation (4–64). The solution $\mathbf{x}(t)$ is clearly the sum of a term consisting of the transition of the initial state and a term arising from the input vector.

Laplace transform approach to the solution of nonhomogeneous state equations. The solution of the nonhomogeneous state equation

$$\dot{\mathbf{x}} = \mathbf{A}\mathbf{x} + \mathbf{B}\mathbf{u}$$

can also be obtained by the Laplace transform approach. The Laplace transform of this last equation yields

$$s\mathbf{X}(s) - \mathbf{x}(0) = \mathbf{A}\mathbf{X}(s) + \mathbf{B}\mathbf{U}(s)$$

or

$$(s\mathbf{I} - \mathbf{A})\mathbf{X}(s) = \mathbf{x}(0) + \mathbf{B}\mathbf{U}(s)$$

Premultiplying both sides of this last equation by $(s\mathbf{I} - \mathbf{A})^{-1}$, we obtain

$$\mathbf{X}(s) = (s\mathbf{I} - \mathbf{A})^{-1}\mathbf{x}(0) + (s\mathbf{I} - \mathbf{A})^{-1}\mathbf{B}\mathbf{U}(s)$$

Using the relationship given by Equation (4–60) gives

$$\mathbf{X}(s) = \mathscr{L}[e^{\mathbf{A}t}]\mathbf{x}(0) + \mathscr{L}[e^{\mathbf{A}t}]\mathbf{B}\mathbf{U}(s)$$

The inverse Laplace transform of this last equation can be obtained by use of the convolution integral as follows:

$$\mathbf{x}(t) = e^{\mathbf{A}t}\mathbf{x}(0) + \int_0^t e^{\mathbf{A}(t-\tau)}\mathbf{B}\mathbf{u}(\tau)\, d\tau$$

Solution in terms of x(t₀). Thus far we have assumed the initial time to be zero. If, however, the initial time is given by t_0 instead of 0, then the solution to Equation (4–64) must be modified to

$$\mathbf{x}(t) = e^{\mathbf{A}(t-t_0)}\mathbf{x}(t_0) + \int_{t_0}^t e^{\mathbf{A}(t-\tau)}\mathbf{B}\mathbf{u}(\tau)\, d\tau \qquad (4\text{--}67)$$

EXAMPLE 4–6   Obtain the time response of the following system:

$$\begin{bmatrix} \dot{x}_1 \\ \dot{x}_2 \end{bmatrix} = \begin{bmatrix} 0 & 1 \\ -2 & -3 \end{bmatrix}\begin{bmatrix} x_1 \\ x_2 \end{bmatrix} + \begin{bmatrix} 0 \\ 1 \end{bmatrix} u$$

where $u(t)$ is the unit-step function occurring at $t = 0$, or

$$u(t) = 1(t)$$

For this system

$$\mathbf{A} = \begin{bmatrix} 0 & 1 \\ -2 & -3 \end{bmatrix}, \qquad \mathbf{B} = \begin{bmatrix} 0 \\ 1 \end{bmatrix}$$

The state-transition matrix $\mathbf{\Phi}(t) = e^{\mathbf{A}t}$ was obtained in Example 4–5 as

$$\mathbf{\Phi}(t) = e^{\mathbf{A}t} = \begin{bmatrix} 2e^{-t} - e^{-2t} & e^{-t} - e^{-2t} \\ -2e^{-t} + 2e^{-2t} & -e^{-t} + 2e^{-2t} \end{bmatrix}$$

The response to the unit-step input is then obtained as

$$\mathbf{x}(t) = e^{\mathbf{A}t}\mathbf{x}(0) + \int_0^t \begin{bmatrix} 2e^{-(t-\tau)} - e^{-2(t-\tau)} & e^{-(t-\tau)} - e^{-2(t-\tau)} \\ -2e^{-(t-\tau)} + 2e^{-2(t-\tau)} & -e^{-(t-\tau)} + 2e^{-2(t-\tau)} \end{bmatrix}\begin{bmatrix} 0 \\ 1 \end{bmatrix}[1]\, d\tau$$

or

$$\begin{bmatrix} x_1(t) \\ x_2(t) \end{bmatrix} = \begin{bmatrix} 2e^{-t} - e^{-2t} & e^{-t} - e^{-2t} \\ -2e^{-t} + 2e^{-2t} & -e^{-t} + 2e^{-2t} \end{bmatrix}\begin{bmatrix} x_1(0) \\ x_2(0) \end{bmatrix} + \begin{bmatrix} \dfrac{1}{2} - e^{-t} + \dfrac{1}{2}e^{-2t} \\ e^{-t} - e^{-2t} \end{bmatrix}$$

If the initial state is zero, or $\mathbf{x}(0) = \mathbf{0}$, then $\mathbf{x}(t)$ can be simplified to

$$\begin{bmatrix} x_1(t) \\ x_2(t) \end{bmatrix} = \begin{bmatrix} \dfrac{1}{2} - e^{-t} + \dfrac{1}{2}e^{-2t} \\ e^{-t} - e^{-2t} \end{bmatrix}$$

4–10 COMPUTER SOLUTION OF STATE EQUATIONS

Section 4–9 presented analytical methods for solving state equations. In this section we shall discuss computer solutions of state equations. One of the most important computations involved in the control system analysis and design is computations of the transient response to step inputs. Therefore, we shall present a general method for computing the response to such inputs. Several methods (such as the Runge–Kutta method, Runge–Kutta–Gill method, Euler's method, and modified Euler's method) are available for such purposes. In what follows, we shall present the Runge–Kutta method for computing the step response.

Runge–Kutta method.　Supose that we wish to find the solution of a scalar differential equation

$$\frac{dx}{dt} = f(t, x), \qquad x(0) = x_0 \tag{4–68}$$

Assume that, in the t–x plane, point (t_i, x_i) is known and we wish to find point (t_{i+1}, x_{i+1}), as shown in Figure 4–41(a). The incremental time $t_{i+1} - t_i = h$ is the time interval for computation, or the sampling period.

According to the Runge–Kutta method, given x_1 for $t = t_i$, we determine x_{i+1} for $t = t_{i+1}$ as follows:

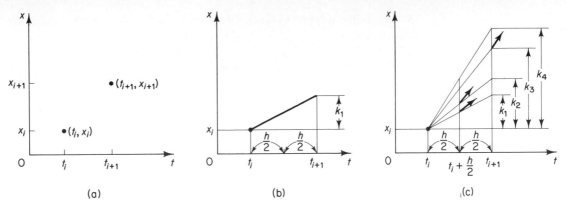

Figure 4–41 (a) Points (t_i, x_i) and $(t_i + 1, x_i + 1)$; (b) slope of the solution curve at point (t_i, x_i) and an increase of x_i value at $t = t_{i+1}$ if the slope is extended; (c) lines drawn from point (t_i, x_i) parallel to the slopes of the curve at points $(t_i + \frac{1}{2}h, x_i + \frac{1}{2}k_1)$ and $(t_i + \frac{1}{2}h, x_i + \frac{1}{2}k_2)$ and a line drawn from point (t_i, x_i) parallel to the slope of the curve at point $(t_i + h, x_i + k_3)$.

1. At point (t_i, x_i) determine the slope of the curve:

$$\text{Slope of curve at point } (t_i, x_i) = f(t_i, x_i)$$

Then determine k_1, the change of the x value at $t = t_{i+1}$ if this slope continued for the entire time interval h, or

$$k_1 = hf(t_i, x_i)$$

as shown in Figure 4–41(b).

2. At point $(t_i + \frac{1}{2}h, x_i + \frac{1}{2}k_1)$, where t is increased by $\frac{1}{2}h$ and x is increased by $\frac{1}{2}k_1$, determine the slope of the curve. The slope is

$$\text{Slope at point } (t_i + \tfrac{1}{2}h, x_i + \tfrac{1}{2}k_1) = f(t_i + \tfrac{1}{2}h, x_i + \tfrac{1}{2}k_1)$$

Using this slope, draw a line from point (t_i, x_i) and determine k_2, the change of the x value at $t = t_{i+1}$, or

$$k_2 = hf(t_1 + \tfrac{1}{2}h, x_i + \tfrac{1}{2}k_1)$$

as shown in Figure 4–41(c).

3. Similarly, at point $(t_i + \frac{1}{2}h, x_i + \frac{1}{2}k_2)$, determine the slope of the curve. The slope is $f(t_i + \frac{1}{2}h, x_i + \frac{1}{2}k_2)$. Using this slope, draw a line from point (t_i, x_i) and determine k_3, the change of the x value at $t = t_{i+1}$, or

$$k_3 = hf(t_i + \tfrac{1}{2}h, x_i + \tfrac{1}{2}k_2)$$

as shown in Figure 4–41(c).

4. At point $(t_i + h, x_i + k_3)$, determine the slope of the curve. Using this slope, draw a line from point (t_i, x_i) and determine k_4, the change of the x value at $t = t_{i+1}$, or

$$k_4 = hf(t_i + h, x_i + k_3)$$

as shown in Figure 4–41(c).

5. Obtain the weighted average of k_1, k_2, k_3, and k_4, where the weights are 1, 2, 2, and 1, respectively, or

$$\Delta x_i = x_{i+1} - x_i = \frac{1}{6}(k_1 + 2k_2 + 2k_3 + k_4)$$

Then the value of x_{i+1} can be given by

$$x_{i+1} = x_i + \Delta x_i = x_i + \frac{1}{6}(k_1 + 2k_2 + 2k_3 + k_4) \qquad (4\text{--}69)$$

The reason why we got Equation (4–69) for x_{i+1} is explained in Problem A–4–23.

It is noted that Equation (4–69) is called the fourth-order Runge–Kutta equation, because it involves four values of k's. The third-order Runge–Kutta equation involves three values of k's, and x_{i+1} may be given by

$$x_{i+1} = x_i + \frac{1}{6}(k_1 + 4k_2 + k_3)$$

The fourth-order Runge–Kutta equation is commonly used in the numerical solution of the differential equation.

In obtaining a computational solution of Equation (4–68), we use Equation (4–69) repeatedly, shifting the value of x_{i+1} to x_i. In Example 4–7 we shall show a BASIC computer program for obtaining a step response of a differential equation system. (To obtain a computer solution of the ramp response, refer to Problem A–4–25.)

EXAMPLE 4–7

Consider the system shown in Figure 4–42. Obtain the response to a unit-step input.

To obtain a computer solution for this example problem, we need to obtain a state-space representation for the system. To obtain such a representation, we first obtain the closed-loop transfer function.

$$\frac{Y(s)}{U(s)} = \frac{25.04(s + 0.2)}{25.04(s + 0.2) + s(s + 5.02)(s + 0.01247)}$$

$$= \frac{25.04s + 5.008}{s^3 + 5.03247s^2 + 25.1026s + 5.008}$$

Then, by taking the inverse Laplace transforms, we obtain the differential equation for the system as follows:

$$\dddot{y} + 5.03247\ddot{y} + 25.1026\dot{y} + 5.008y = 25.04\dot{u} + 5.008u \qquad (4\text{--}70)$$

which can be written as

$$\dddot{y} + a_1\ddot{y} + a_2\dot{y} + a_3y = b_0\dddot{u} + b_1\ddot{u} + b_2\dot{u} + b_3u$$

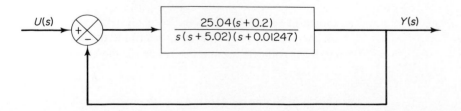

Figure 4–42
Control system.

where

$$a_1 = 5.03247, \qquad a_2 = 25.1026, \qquad a_3 = 5.008$$

$$b_0 = 0, \qquad b_1 = 0, \qquad b_2 = 25.04, \qquad b_3 = 5.008$$

Referring to Equations (2–5) and (2–6), we define state variables x_1, x_2, and x_3 as follows:

$$x_1 = y - \beta_0 u$$

$$x_2 = \dot{x}_1 - \beta_1 u$$

$$x_3 = \dot{x}_2 - \beta_2 u$$

where

$$\beta_0 = b_0 = 0$$

$$\beta_1 = b_1 - a_1\beta_0 = 0$$

$$\beta_2 = b_2 - a_1\beta_1 - a_2\beta_0 = 25.04$$

Then

$$\dot{x}_3 = -a_3 x_1 - a_2 x_2 - a_1 x_3 + \beta_3 u$$

where

$$\beta_3 = b_3 - a_1\beta_2 - a_2\beta_1 - a_3\beta_0 = 5.008 - 5.03247 \times 25.04$$
$$= -121.005$$

Thus, the state variables are given by

$$x_1 = y$$

$$x_2 = \dot{x}_1$$

$$x_3 = \dot{x}_2 - 25.04u$$

Referring to Equations (2–8) and (2–9), the state equation and output equation can be obtained as

$$\begin{bmatrix} \dot{x}_1 \\ \dot{x}_2 \\ \dot{x}_3 \end{bmatrix} = \begin{bmatrix} 0 & 1 & 0 \\ 0 & 0 & 1 \\ -5.008 & -25.1026 & -5.03247 \end{bmatrix} \begin{bmatrix} x_1 \\ x_2 \\ x_3 \end{bmatrix} + \begin{bmatrix} 0 \\ 25.04 \\ -121.005 \end{bmatrix} u \qquad (4\text{–}71)$$

$$y = \begin{bmatrix} 1 & 0 & 0 \end{bmatrix} \begin{bmatrix} x_1 \\ x_2 \\ x_3 \end{bmatrix} \qquad (4\text{–}72)$$

To write a BASIC computer program for finding the response of the system to a unit-step input ($u = 1$), define x_1 = X1, x_2 = X2, x_3 = X3 and denote such functions $\dot{x}_1$, $\dot{x}_2$, $\dot{x}_3$, as FNX1D(T, X1, X2, X3), FNX2D(T, X1, X2, X3), and FNX3D(T, X1, X2, X3), respectively. Then, from Equation (4–71), these functions can be given as follows:

FNX1D(T, X1, X2, X3) = X2

FNX2D(T, X1, X2, X3) = X3 + 25.04

FNX3D(T, X1, X2, X3) = −5.008*X1 − 25.1026*X2 − 5.03247*X3 − 121.005

A BASIC computer program for obtaining the unit-step response may be written as shown in Table 4–3. This computer program clearly shows the steps involved in the Runge–Kutta method. The unit-step response curve obtained in the computer simulation is shown in Figure 4–43. It is noted that many different computer programs can be written for such a problem. (A different computer program for solving a similar problem is given in Problem A–4–24.)

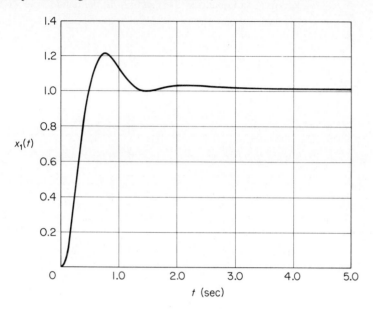

Figure 4–43
Unit-step response curve for the system shown in Fig. 4–42.

Table 4–3 BASIC Computer Program for Solving Equation (4–71) with Zero Initial Conditions and $u = 1$ (Unit-Step Response)

```
 10 OPEN "O",#1,"BUSH.BAS"
 20 DEF FNX1D(T,X1,X2,X3) = X2
 30 DEF FNX2D(T,X1,X2,X3) = X3 + 25.04
 40 DEF FNX3D(T,X1,X2,X3) = -5.008*X1 - 25.1026*X2 - 5.03247*X3 - 121.005
 50 H = .05
 60 T = 0 : X1 = 0 : X2 = 0 : X3 = 0
 70 TT = 5
 80 PRINT "         TIME          X1        X2        X3      "
 90 PRINT "   -------------------------------------------------"
100 IF T > TT THEN GOTO 1000
110 PRINT #1, USING "#####.#####"; X1
120 PRINT USING "#####.#####"; T, X1, X2, X3
130 K1 = H*FNX1D(T, X1, X2, X3)
140 L1 = H*FNX2D(T, X1, X2, X3)
150 M1 = H*FNX3D(T, X1, X2, X3)
160 K2 = H*FNX1D(T+.5*H, X1+.5*K1, X2+.5*L1, X3+.5*M1)
170 L2 = H*FNX2D(T+.5*H, X1+.5*K1, X2+.5*L1, X3+.5*M1)
180 M2 = H*FNX3D(T+.5*H, X1+.5*K1, X2+.5*L1, X3+.5*M1)
190 K3 = H*FNX1D(T+.5*H, X1+.5*K2, X2+.5*L2, X3+.5*M2)
200 L3 = H*FNX2D(T+.5*H, X1+.5*K2, X2+.5*L2, X3+.5*M2)
```

Table 4–3 Cont'd

```
210  M3  =  H*FNX3D(T+.5*H,  X1+.5*K2,  X2+.5*L2,  X3+.5*M2)
220  K4  =  H*FNX1D(T+H,  X1+K3,  X2+L3,  X3+M3)
230  L4  =  H*FNX2D(T+H,  X1+K3,  X2+L3,  X3+M3)
240  M4  =  H*FNX3D(T+H,  X1+K3,  X2+L3,  X3+M3)
250  X1  =  X1  +  (K1  +  2*K2  +  2*K3  +  K4)/6
260  X2  =  X2  +  (L1  +  2*L2  +  2*L3  +  L4)/6
270  X3  =  X3  +  (M1  +  2*M2  +  2*M3  +  M4)/6
280  T  =  T  +  H
290  GOTO  100
1000 CLOSE  #1
1010 END
```

Example Problems and Solutions

A–4–1.   In the system of Figure 4–44, $x(t)$ is the input displacement and $\theta(t)$ is the output angular displacement. Assume that the masses involved are negligibly small and that all motions are restricted to be small; therefore, the system can be considered linear. The initial conditions for x and θ are zeros, or $x(0-)$ = 0 and $\theta(0-) = 0$. Show that this system is a differentiating element. Then obtain the response $\theta(t)$ when $x(t)$ is a unit-step input.

Solution.   The equation for the system is

$$b(\dot{x} - L\dot{\theta}) = kL\theta$$

or

$$L\dot{\theta} + \frac{k}{b}L\theta = \dot{x}$$

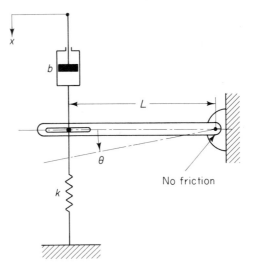

Figure 4–44
Mechanical system.

The Laplace transform of this last equation, using zero initial conditions, gives

$$\left(Ls + \frac{k}{b}L\right)\Theta(s) = sX(s)$$

And so

$$\frac{\Theta(s)}{X(s)} = \frac{1}{L}\frac{s}{s + (k/b)}$$

Thus the system is a differentiating system.

For the unit-step input $X(s) = 1/s$, the output $\Theta(s)$ becomes

$$\Theta(s) = \frac{1}{L}\frac{1}{s + (k/b)}$$

The inverse Laplace transform of $\Theta(s)$ gives

$$\theta(t) = \frac{1}{L}e^{-(k/b)t}$$

Note that if the value of k/b is large the response $\theta(t)$ approaches a pulse signal as shown in Figure 4–45.

A–4–2. When the system shown in Figure 4–46(a) is subjected to a unit-step input, the system output responds as shown in Figure 4–46(b). Determine the values of K and T from the response curve.

Solution. The maximum overshoot of 25.4% corresponds to $\zeta = 0.4$. From the response curve we have

$$t_p = 3$$

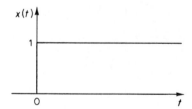

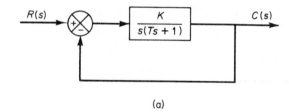

(a)

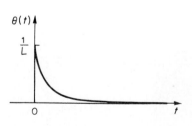

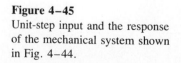

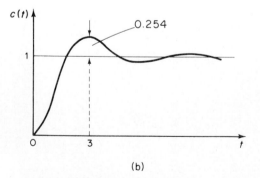

(b)

Figure 4–45
Unit-step input and the response
of the mechanical system shown
in Fig. 4–44.

Figure 4–46
(a) Closed-loop system; (b) unit-step
response curve.

Chapter 4 / Transient-Response Analysis and Steady-State Error Analysis

Consequently,

$$t_p = \frac{\pi}{\omega_d} = \frac{\pi}{\omega_n\sqrt{1 - \zeta^2}} = \frac{\pi}{\omega_n\sqrt{1 - 0.4^2}} = 3$$

It follows that

$$\omega_n = 1.14$$

From the block diagram we have

$$\frac{C(s)}{R(s)} = \frac{K}{Ts^2 + s + K}$$

from which

$$\omega_n = \sqrt{\frac{K}{T}}, \qquad 2\zeta\omega_n = \frac{1}{T}$$

Therefore, the values of T and K are determined as

$$T = \frac{1}{2\zeta\omega_n} = \frac{1}{2 \times 0.4 \times 1.14} = 1.09$$

$$K = \omega_n^2 T = 1.14^2 \times 1.09 = 1.42$$

A-4-3. Determine the values of K and k of the closed-loop system shown in Figure 4–47 so that the maximum overshoot in unit-step response is 25% and the peak time is 2 sec. Assume that $J = 1$ kg-m².

Solution. The closed-loop transfer function is

$$\frac{C(s)}{R(s)} = \frac{K}{Js^2 + Kks + K}$$

By sububstituting $J = 1$ kg-m² into this last equation, we have

$$\frac{C(s)}{R(s)} = \frac{K}{s^2 + Kks + K}$$

Note that

$$\omega_n = \sqrt{K}, \qquad 2\zeta\omega_n = Kk$$

The maximum overshoot M_p is

$$M_p = e^{-\zeta\pi/\sqrt{1 - \zeta^2}}$$

which is specified as 25%. Hence

$$e^{-\zeta\pi/\sqrt{1 - \zeta^2}} = 0.25$$

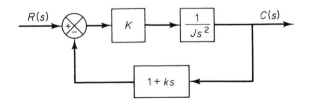

Figure 4–47
Closed-loop system.

from which

$$\frac{\zeta\pi}{\sqrt{1 - \zeta^2}} = 1.386$$

or

$$\zeta = 0.404$$

The peak time t_p is specified as 2 sec. And so

$$t_p = \frac{\pi}{\omega_d} = 2$$

or

$$\omega_d = 1.57$$

Then the undamped natural frequency ω_n is

$$\omega_n = \frac{\omega_d}{\sqrt{1 - \zeta^2}} = \frac{1.57}{\sqrt{1 - 0.404^2}} = 1.72$$

Therefore, we obtain

$$K = \omega_n^2 = 1.72^2 = 2.95 \text{ N-m}$$

$$k = \frac{2\zeta\omega_n}{K} = \frac{2 \times 0.404 \times 1.72}{2.95} = 0.471 \text{ sec}$$

A–4–4. What is the unit-step response of the system shown in Figure 4–48?

Solution. The closed-loop transfer function is

$$\frac{C(s)}{R(s)} = \frac{10s + 10}{s^2 + 10s + 10}$$

For the unit-step input $[R(s) = 1/s]$, we have

$$
\begin{aligned}
C(s) &= \frac{10s + 10}{s^2 + 10s + 10} \frac{1}{s} \\
&= \frac{10s + 10}{(s + 5 + \sqrt{15})(s + 5 - \sqrt{15})s} \\
&= \frac{-4 - \sqrt{15}}{3 + \sqrt{15}} \frac{1}{s + 5 + \sqrt{15}} + \frac{-4 + \sqrt{15}}{3 - \sqrt{15}} \frac{1}{s + 5 - \sqrt{15}} + \frac{1}{s}
\end{aligned}
$$

The inverse Laplace transform of $C(s)$ gives

$$
\begin{aligned}
c(t) &= -\frac{4 + \sqrt{15}}{3 + \sqrt{15}} e^{-(5 + \sqrt{15})t} + \frac{4 - \sqrt{15}}{-3 + \sqrt{15}} e^{-(5 - \sqrt{15})t} + 1 \\
&= -1.1455e^{-8.87t} + 0.1455e^{-1.13t} + 1
\end{aligned}
$$

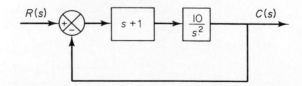

Figure 4–48
Closed-loop system.

Clearly, the output will not exhibit any oscillation. The response curve exponentially approaches the final value $c(\infty) = 1$.

A–4–5. Figure 4–49(a) shows a mechanical vibratory system. When 2 lb of force (step input) is applied to the system, the mass oscillates, as shown in Figure 4–49(b). Determine m, b, and k of the system from this response curve. The displacement x is measured from the equilibrium position.

Solution. The transfer function of this system is

$$\frac{X(s)}{P(s)} = \frac{1}{ms^2 + bs + k}$$

Since

$$P(s) = \frac{2}{s}$$

we obtain

$$X(s) = \frac{2}{s(ms^2 + bs + k)}$$

It follows that the steady-state value of x is

$$x(\infty) = \lim_{s \to 0} sX(s) = \frac{2}{k} = 0.1 \text{ ft}$$

Hence

$$k = 20 \text{ lb}_f/\text{ft}$$

Note that $M_p = 9.5\%$ corresponds to $\zeta = 0.6$. The peak time t_p is given by

$$t_p = \frac{\pi}{\omega_d} = \frac{\pi}{\omega_n\sqrt{1 - \zeta^2}} = \frac{\pi}{0.8\omega_n}$$

The experimental curve shows that $t_p = 2$ sec. Therefore,

$$\omega_n = \frac{3.14}{2 \times 0.8} = 1.96 \text{ rad/sec}$$

Since $\omega_n^2 = k/m = 20/m$, we obtain

$$m = \frac{20}{\omega_n^2} = \frac{20}{1.96^2} = 5.2 \text{ slugs} = 166 \text{ lb}$$

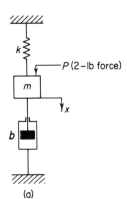

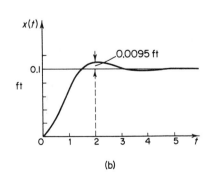

Figure 4–49
(a) Mechanical vibratory system; (b) step-response curve.

(a)

(b)

(Note that 1 slug $=$ 1 lb$_f$-sec^2/ft.) Then b is determined from

$$2\zeta\omega_n = \frac{b}{m}$$

or

$$b = 2\zeta\omega_n m = 2 \times 0.6 \times 1.96 \times 5.2 = 12.2 \text{ lb}_f\text{/ft/sec}$$

A–4–6. Assuming that the mechanical system shown in Figure 4–50 is at rest before excitation force $P \sin \omega t$ is given, derive the complete solution $x(t)$ and the steady-state solution $x_{ss}(t)$. The displacement x is measured from the equilibrium position. Assume that the system is underdamped.

Solution. The equation of motion for the system is

$$m\ddot{x} + b\dot{x} + kx = P \sin \omega t$$

Noting that $x(0) = 0$ and $\dot{x}(0) = 0$, the Laplace transform of this equation is

$$(ms^2 + bs + k)X(s) = P\frac{\omega}{s^2 + \omega^2}$$

or

$$X(s) = \frac{P\omega}{(s^2 + \omega^2)\,(ms^2 + bs + k)}$$

Since the system is underdamped, $X(s)$ can be written as follows:

$$X(s) = \frac{P\omega}{m}\,\frac{1}{s^2 + \omega^2}\,\frac{1}{s^2 + 2\zeta\omega_n s + \omega_n^2} \qquad (0 < \zeta < 1)$$

where $\omega_n = \sqrt{k/m}$ and $\zeta = b/(2\sqrt{mk})$. $X(s)$ can be expanded as

$$X(s) = \frac{P\omega}{m}\left(\frac{as + c}{s^2 + \omega^2} + \frac{-as + d}{s^2 + 2\zeta\omega_n s + \omega_n^2}\right)$$

By simple calculations it can be found that

$$a = \frac{-2\zeta\omega_n}{(\omega_n^2 - \omega^2)^2 + 4\zeta^2\omega_n^2\omega^2}, \qquad c = \frac{(\omega_n^2 - \omega^2)}{(\omega_n^2 - \omega^2)^2 + 4\zeta^2\omega_n^2\omega^2}, \qquad d = \frac{4\zeta^2\omega_n^2 - (\omega_n^2 - \omega^2)}{(\omega_n^2 - \omega^2)^2 + 4\zeta^2\omega_n^2\omega^2}$$

Hence

$$X(s) = \frac{P\omega}{m}\,\frac{1}{(\omega_n^2 - \omega^2)^2 + 4\zeta^2\omega_n^2\omega^2}\left[\frac{-2\zeta\omega_n s + (\omega_n^2 - \omega^2)}{s^2 + \omega^2} + \frac{2\zeta\omega_n(s + \zeta\omega_n) + 2\zeta^2\omega_n^2 - (\omega_n^2 - \omega^2)}{s^2 + 2\zeta\omega_n s + \omega_n^2}\right]$$

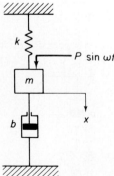

Figure 4–50
Mechanical system.

The inverse Laplace transform of $X(s)$ gives

$$x(t) = \frac{P\omega}{m[(\omega_n^2 - \omega^2)^2 + 4\zeta^2\omega_n^2\omega^2]}\left[-2\zeta\omega_n \cos \omega t + \frac{(\omega_n^2 - \omega^2)}{\omega}\sin \omega t \right.$$
$$\left. + 2\zeta\omega_n e^{-\zeta\omega_n t} \cos \omega_n\sqrt{1-\zeta^2}\,t + \frac{2\zeta^2\omega_n^2 - (\omega_n^2 - \omega^2)}{\omega_n\sqrt{1-\zeta^2}} e^{-\zeta\omega_n t} \sin \omega_n\sqrt{1-\zeta^2}\,t \right]$$

At steady state $(t \to \infty)$ the terms involving $e^{-\zeta\omega_n t}$ approach zero. Thus at steady state

$$x_{ss}(t) = \frac{P\omega}{m[(\omega_n^2 - \omega^2)^2 + 4\zeta^2\omega_n^2\omega^2]}\left(-2\zeta\omega_n \cos \omega t + \frac{\omega_n^2 - \omega^2}{\omega}\sin \omega t \right)$$

$$= \frac{P\omega}{(k - m\omega^2)^2 + b^2\omega^2}\left(-b\cos \omega t + \frac{k - m\omega^2}{\omega}\sin \omega t \right)$$

$$= \frac{P}{\sqrt{(k - m\omega^2)^2 + b^2\omega^2}}\sin\left(\omega t - \tan^{-1}\frac{b\omega}{k - m\omega^2} \right)$$

A–4–7. Consider the unit-step response of the second-order system

$$\frac{C(s)}{R(s)} = \frac{\omega_n^2}{s^2 + 2\zeta\omega_n s + \omega_n^2}$$

The amplitude of the exponentially damped sinusoid changes as a geometric series. At time $t = t_p = \pi/\omega_d$, the amplitude is equal to $e^{-(\sigma/\omega_d)\pi}$. After one oscillation, or at $t = t_p + 2\pi/\omega_d = 3\pi/\omega_d$, the amplitude is equal to $e^{-(\sigma/\omega_d)3\pi}$; after another cycle of oscillation, the amplitude is $e^{-(\sigma/\omega_d)5\pi}$. The logarithm of the ratio of successive amplitudes is called the *logarithmic decrement*. Determine the logarithmic decrement for this second-order system. Describe a method for experimental determination of the damping ratio from the rate of decay of the oscillation.

Solution. Let us define the amplitude of the output oscillation at $t = t_i$ to be x_i, where $t_i = t_p + (i - 1)T$ (T = period of oscillation). The amplitude ratio per one period of damped oscillation is

$$\frac{x_1}{x_2} = \frac{e^{-(\sigma/\omega_d)\pi}}{e^{-(\sigma/\omega_d)3\pi}} = e^{2(\sigma/\omega_d)\pi} = e^{2\zeta\pi/\sqrt{1-\zeta^2}}$$

Thus, the logarithmic decrement δ is

$$\delta = \ln\frac{x_1}{x_2} = \frac{2\zeta\pi}{\sqrt{1-\zeta^2}}$$

It is a function only of the damping ratio ζ. Thus, the damping ratio ζ can be determined by use of the logarithmic decrement.

In the experimental determination of the damping ratio ζ from the rate of decay of the oscillation, we measure the amplitude x_1 at $t = t_p$ and amplitude x_n at $t = t_p + (n - 1)T$. Note that it is necessary to choose n large enough so that the ratio x_1/x_n is not near unity. Then

$$\frac{x_1}{x_n} = e^{(n-1)2\zeta\pi/\sqrt{1-\zeta^2}}$$

or

$$\ln\frac{x_1}{x_n} = (n - 1)\frac{2\zeta\pi}{\sqrt{1-\zeta^2}}$$

Hence

$$\zeta = \frac{\dfrac{1}{n-1}\left(\ln\dfrac{x_1}{x_n}\right)}{\sqrt{4\pi^2 + \left[\dfrac{1}{n-1}\left(\ln\dfrac{x_1}{x_n}\right)\right]^2}}$$

A–4–8. In the system shown in Figure 4–51, the numerical values of m, b, and k are given as $m = 1$ kg, $b = 2$ N-sec/m, and $k = 100$ N/m. The mass is displaced 0.05 m and released without initial velocity. Find the frequency observed in the vibration. In addition, find the amplitude four cycles later. The displacement x is measured from the equilibrium position.

Solution. The equation of motion for the system is

$$m\ddot{x} + b\dot{x} + kx = 0$$

Substituting the numerical values for m, b, and k into this equation gives

$$\ddot{x} + 2\dot{x} + 100x = 0$$

where the initial conditions are $x(0) = 0.05$ and $\dot{x}(0) = 0$. From this last equation the undamped natural frequency ω_n and the damping ratio ζ are found to be

$$\omega_n = 10, \qquad \zeta = 0.1$$

The frequency actually observed in the vibration is the damped natural frequency ω_d.

$$\omega_d = \omega_n\sqrt{1 - \zeta^2} = 10\sqrt{1 - 0.01} = 9.95 \text{ rad/sec}$$

In the present analysis, $\dot{x}(0)$ is given as zero. Thus, solution $x(t)$ can be written as

$$x(t) = x(0)e^{-\zeta\omega_n t}\left(\cos \omega_d t + \frac{\zeta}{\sqrt{1 - \zeta^2}} \sin \omega_d t\right)$$

It follows that at $t = nT$, where $T = 2\pi/\omega_d$,

$$x(nT) = x(0)e^{-\zeta\omega_n nT}$$

Consequently, the amplitude four cycles later becomes

$$x(4T) = x(0)e^{-\zeta\omega_n 4T} = x(0)e^{-(0.1)(10)(4)(0.6315)}$$
$$= 0.05e^{-2.526} = 0.05 \times 0.07998 = 0.004 \text{ m}$$

A–4–9. Consider a system whose closed-loop poles and closed-loop zero are located in the s plane on a line parallel to the $j\omega$ axis, as shown in Figure 4–52. Show that the impulse response of such a system is a damped cosine function.

Solution. The closed-loop transfer function is

$$\frac{C(s)}{R(s)} = \frac{K(s+\sigma)}{(s + \sigma + j\omega_d)(s + \sigma - j\omega_d)}$$

For a unit-impulse input, $R(s) = 1$ and

$$C(s) = \frac{K(s+\sigma)}{(s+\sigma)^2 + \omega_d^2}$$

The inverse Laplace transform of $C(s)$ is

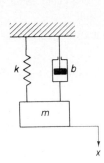

Figure 4–51
Spring–mass–damper
system.

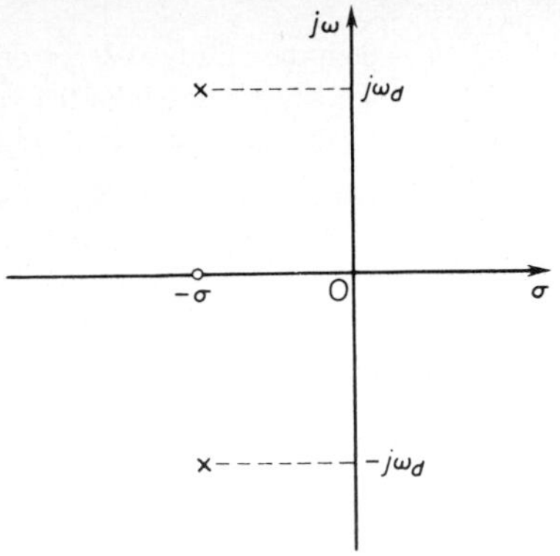

Figure 4–52
Closed-loop pole–zero configuration of system
whose impulse response is a damped cosine
function.

$$c(t) = Ke^{-\sigma t} \cos \omega_d t \qquad (t \geq 0)$$

which is a damped cosine function.

A–4–10. Consider the system shown in Figure 4–53(a). The steady-state error to a unit-ramp input is $e_{ss} = 2\zeta/\omega_n$. Show that the steady-state error for following a ramp input may be eliminated if the input is introduced to the system through a proportional-plus-derivative filter, as shown in Figure 4–53(b), and the value of k is properly set. Note that the error $e(t)$ is given by $r(t) - c(t)$.

Solution. The closed-loop transfer function of the system shown in Figure 4–53(b) is

$$\frac{C(s)}{R(s)} = \frac{(1 + ks)\omega_n^2}{s^2 + 2\zeta\omega_n s + \omega_n^2}$$

Then

$$R(s) - C(s) = \left(\frac{s^2 + 2\zeta\omega_n s - \omega_n^2 ks}{s^2 + 2\zeta\omega_n s + \omega_n^2} \right) R(s)$$

If the input is a unit ramp, then the steady-state error is

$$\begin{aligned} e(\infty) &= r(\infty) - c(\infty) \\ &= \lim_{s \to 0} s \left(\frac{s^2 + 2\zeta\omega_n s - \omega_n^2 ks}{s^2 + 2\zeta\omega_n s + \omega_n^2} \right) \frac{1}{s^2} \\ &= \frac{2\zeta\omega_n - \omega_n^2 k}{\omega_n^2} \end{aligned}$$

Therefore, if k is chosen as

$$k = \frac{2\zeta}{\omega_n}$$

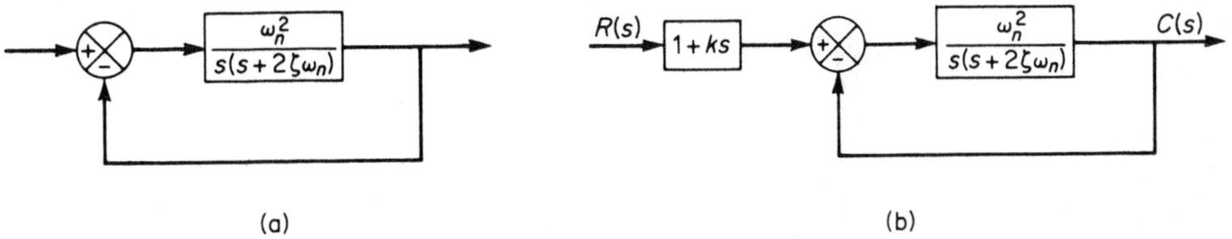

Figure 4–53
(a) Control system;
(b) control system
with input filter.

(a)

(b)

then the steady-state error for following a ramp input can be made equal to zero. Note that if there are any variations in the values of ζ and/or ω_n due to environmental changes or aging, then a nonzero steady-state error for a ramp response may result.

A–4–11. Obtain the unit-step response of a unity-feedback system whose open-loop transfer function is

$$G(s) = \frac{5(s + 20)}{s(s + 4.59)(s^2 + 3.41s + 16.35)}$$

Solution. The closed-loop transfer function is

$$\frac{C(s)}{R(s)} = \frac{5(s + 20)}{s(s + 4.59)(s^2 + 3.41s + 16.35) + 5(s + 20)}$$

$$= \frac{5s + 100}{s^4 + 8s^3 + 32s^2 + 80s + 100}$$

$$= \frac{5(s + 20)}{(s^2 + 2s + 10)(s^2 + 6s + 10)}$$

The unit-step response of this system is then

$$C(s) = \frac{5(s + 20)}{s(s^2 + 2s + 10)(s^2 + 6s + 10)}$$

$$= \frac{1}{s} + \frac{\frac{3}{8}(s + 1) - \frac{17}{8}}{(s + 1)^2 + 3^2} + \frac{-\frac{11}{8}(s + 3) - \frac{13}{8}}{(s + 3)^2 + 1^2}$$

The time response $c(t)$ can be found by taking the inverse Laplace transform of $C(s)$ as follows:

$$c(t) = 1 + \tfrac{3}{8}e^{-t}\cos 3t - \tfrac{17}{24}e^{-t}\sin 3t - \tfrac{11}{8}e^{-3t}\cos t - \tfrac{13}{8}e^{-3t}\sin t \qquad (t \geq 0)$$

(A computer solution of this problem is presented in Problem A–4–24.)

A–4–12. Consider the following characteristic equation:

$$s^4 + Ks^3 + s^2 + s + 1 = 0$$

Determine the range of K for stability.

Solution. The Routh array of coefficients is

s^4	1	1	1
s^3	K	1	0
s^2	$\dfrac{K-1}{K}$	1	0
s^1	$1 - \dfrac{K^2}{K-1}$	0	
s^0	1		

For stability, we require that

$$K > 0$$

$$\frac{K-1}{K} > 0$$

$$1 - \frac{K^2}{K-1} > 0$$

From the first and second conditions K must be greater than 1. For $K > 1$ notice that the term $1 - [K^2/(K - 1)]$ is always negative, since

$$\frac{K - 1 - K^2}{K - 1} = \frac{-1 + K(1 - K)}{K - 1} < 0$$

Thus, the three conditions cannot be fulfilled simultaneously. Therefore, there is no value of K that allows stability of the system.

A–4–13. Consider the characteristic equation given by

$$a_0 s^n + a_1 s^{n-1} + a_2 s^{n-2} + \cdots + a_{n-1} s + a_n = 0 \qquad (4\text{–}73)$$

The Hurwitz stability criterion, given below, gives conditions for all the roots to have negative real parts in terms of the coefficients of the polynomial. As stated in the discussions of Routh's stability criterion in Section 4–6, for all the roots to have negative real parts, all the coefficients a's must be positive. This is a necessary condition but not a sufficient condition. If this condition is not satisfied, it indicates that some of the roots have positive real parts or are imaginary or zero. A sufficient condition for all the roots to have negative real parts is given in the following Hurwitz stability criterion: If all the coefficients of the polynomial are positive, arrange these coefficients in the following determinant:

$$\Delta_n = \begin{vmatrix} a_1 & a_3 & a_5 & \cdots & 0 & 0 & 0 \\ a_0 & a_2 & a_4 & \cdots & \cdot & \cdot & \cdot \\ 0 & a_1 & a_3 & \cdots & a_n & 0 & 0 \\ 0 & a_0 & a_2 & \cdots & a_{n-1} & 0 & 0 \\ \cdot & \cdot & \cdot & & a_{n-2} & a_n & 0 \\ \cdot & \cdot & \cdot & & a_{n-3} & a_{n-1} & 0 \\ 0 & 0 & 0 & \cdots & a_{n-4} & a_{n-2} & a_n \end{vmatrix}$$

where we substituted zero for a_s if $s > n$. For all the roots to have negative real parts, it is necessary and sufficient that successive principal minors of Δ_n be positive. The successive principal minors are the following determinants:

$$\Delta_i = \begin{vmatrix} a_1 & a_3 & \cdots & a_{2i-1} \\ a_0 & a_2 & \cdots & a_{2i-2} \\ 0 & a_1 & \cdots & a_{2i-3} \\ \cdot & \cdot & & \cdot \\ 0 & 0 & \cdots & a_i \end{vmatrix} \qquad (i = 1, 2, \ldots, n)$$

where $a_s = 0$ if $s > n$. (It is noted that some of the conditions for the lower-order determinants are included in the conditions for the higher-order determinants.) If all these determinants are positive, and $a_0 > 0$ as already assumed, the equilibrium state of the system whose characteristic equation is given by Equation (4–73) is asymptotically stable. Note that exact values of determinants are not needed; instead, only signs of these determinants are needed for the stability criterion.

Now consider the following characteristic equation:

$$a_0 s^4 + a_1 s^3 + a_2 s^2 + a_3 s + a_4 = 0$$

Obtain the condition for stability using the Hurwitz stability criterion.

Solution. The conditions for stability are that all the a's be positive and that

$$\Delta_2 = \begin{vmatrix} a_1 & a_3 \\ a_0 & a_2 \end{vmatrix} = a_1 a_2 - a_0 a_3 > 0$$

$$\Delta_3 = \begin{vmatrix} a_1 & a_3 & 0 \\ a_0 & a_2 & a_4 \\ 0 & a_1 & a_3 \end{vmatrix}$$

$$= a_1(a_2 a_3 - a_1 a_4) - a_0 a_3^2$$

$$= a_3(a_1 a_2 - a_0 a_3) - a_1^2 a_4 > 0$$

It is clear that if all the a's are positive and if the condition $\Delta_3 > 0$ is satisfied, the condition $\Delta_2 > 0$ is also satisfied. Therefore, for all the roots of the given characteristic equation to have negative real parts, it is necessary and sufficient that all the coefficients a's are positive and $\Delta_3 > 0$.

A–4–14. Explain why the proportional control of a plant that does not possess an integrating property (which means that the plant transfer function does not include the factor $1/s$) suffers offset in response to step inputs.

Solution. Consider, for example, the system shown in Figure 4–54. At steady state, if c were equal to a nonzero constant r, then $e = 0$ and $u = Ke = 0$, resulting in $c = 0$, which contradicts the assumption that $c = r = $ nonzero constant.

A nonzero offset must exist for proper operation of such a control system. In other words, at steady state, if e were equal to $r/(1 + K)$, then $u = Kr/(1 + K)$ and $c = Kr/(1 + K)$, which results in the assumed error signal $e = r/(1 + K)$. Thus the offset of $r/(1 + K)$ must exist in such a system.

A–4–15. Consider a unity-feedback control system whose open-loop transfer function is

$$G(s) = \frac{K}{s(Js + B)}$$

Discuss the effects that varying the values of K and B has on the steady-state error in unit-ramp response. Sketch typical unit-ramp response curves for a small value, medium value, and large value of K.

Solution. The closed-loop transfer function is

$$\frac{C(s)}{R(s)} = \frac{K}{Js^2 + Bs + K}$$

For a unit-ramp input, $R(s) = 1/s^2$. Thus

$$\frac{E(s)}{R(s)} = \frac{R(s) - C(s)}{R(s)} = \frac{Js^2 + Bs}{Js^2 + Bs + K}$$

or

$$E(s) = \frac{Js^2 + Bs}{Js^2 + Bs + K} \frac{1}{s^2}$$

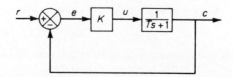

Figure 4–54
Control system.

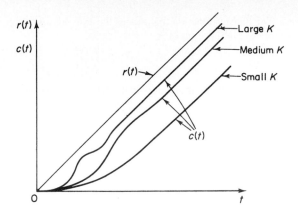

Figure 4–55
Unit-ramp response curves of the system considered in Problem A–4–15.

The steady-state error is

$$e_{ss} = e(\infty) = \lim_{s \to 0} sE(s) = \frac{B}{K}$$

We see that we can reduce the steady-state error e_{ss} by increasing the gain K or decreasing the viscous-friction coefficient B. Increasing the gain or decreasing the viscous-friction coefficient, however, causes the damping ratio to decrease, with the result that the transient response of the system will become more oscillatory. Doubling K decreases e_{ss} to half of its original value, while ζ is decreased to 0.707 of its original value since ζ is inversely proportional to the square root of K. On the other hand, decreasing B to half of its original value decreases both e_{ss} and ζ to the halves of their original values, respectively. Therefore, it is advisable to increase the value of K rather than to decrease the value of B. After the transient response has died out and a steady state is reached, the output velocity becomes the same as the input velocity. However, there is a steady-state positional error between the input and the output. Examples of the unit-ramp response of the system for three different values of K are illustrated in Figure 4–55.

A–4–16. Consider the system shown in Figure 4–56. This system is subjected to two signals, one the reference input and the other the external disturbance. Show that the characteristic equation of this system is the same regardless of which signal is chosen as input.

Solution. The transfer function that relates the reference input and the corresponding output, without considering the external disturbance, is

$$\frac{C(s)}{R(s)} = \frac{G_1(s)G_2(s)}{1 + G_1(s)G_2(s)H(s)} \tag{4–74}$$

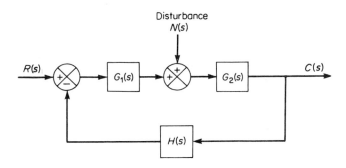

Figure 4–56
Control system.

Example Problems and Solutions

The transfer function that relates the external disturbance and the corresponding output in the absence of the reference input is

$$\frac{C(s)}{N(s)} = \frac{G_2(s)}{1 + G_1(s)G_2(s)H(s)} \qquad (4\text{--}75)$$

Note that the denominators of Equations (4–74) and (4–75) are the same. The characteristic equation is

$$1 + G_1(s)G_2(s)H(s) = 0$$

It contains the information necessary to determine the basic characteristics of the system response. (Remember that for a given system there is only one characteristic equation. This means that the characteristic equation of a given transfer function is the same regardless of which signal is chosen as input.)

A–4–17. Consider the liquid-level control system shown in Figure 4–57. The controller is of the proportional type. The set point of the controller is fixed.

Draw a block diagram of the system, assuming that changes in the variables are small. Investigate the response of the level of the second tank subjected to a step disturbance u.

Solution. Figure 4–58(a) is a block diagram of this system when changes in the variables are small. Since the set point of the controller is fixed, $r = 0$. (Note that r is the change in set point.)

To investigate the response of the level of the second tank subjected to a step disturbance u, we find it convenient to modify the block diagram of Figure 4–58(a) to the one shown in Figure 4–58(b).

The transfer function between $H_2(s)$ and $U(s)$ can be obtained as

$$\frac{H_2(s)}{U(s)} = \frac{R_2(R_1C_1s + 1)}{(R_1C_1s + 1)(R_2C_2s + 1) + KR_2}$$

From this equation, the response $H_2(s)$ to a disturbance $U(s)$ can be found. The effect of the controller is seen by the presence of K in the denominator of this last equation.

For a step disturbance of magnitude U_0, we obtain

$$h_2(\infty) = \frac{R_2}{1 + KR_2} U_0$$

or

$$\text{Steady-state error} = -\frac{R_2}{1 + KR_2} U_0$$

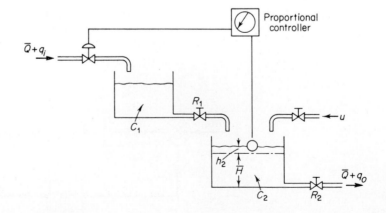

Figure 4–57
Liquid-level control system.

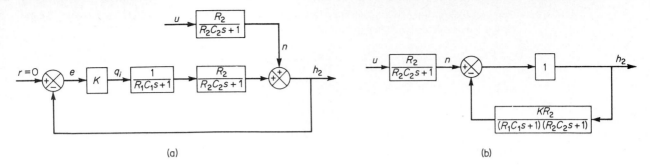

(a)   (b)

Figure 4–58 (a) Block diagram of the system shown in Fig. 4–57; (b) modified block diagram.

The system exhibits offset in the response to a step disturbance.

Note that the characteristic equation for the disturbance input and that for the reference input are the same. (As stated in Problem A–4–16, the characteristic equation is unique for any given system.) The characteristic equation for this system is

$$(R_1C_1s + 1)(R_2C_2s + 1) + KR_2 = 0$$

which can be modified to

$$s^2 + \left(\frac{R_1C_1 + R_2C_2}{R_1C_1R_2C_2}\right)s + \frac{1 + KR_2}{R_1C_1R_2C_2} = 0$$

The undamped natural frequency ω_n and the damping ratio ζ are given by

$$\omega_n = \sqrt{\frac{1 + KR_2}{R_1C_1R_2C_2}}, \qquad \zeta = \frac{R_1C_1 + R_2C_2}{2\sqrt{R_1C_1R_2C_2}\,\sqrt{1 + KR_2}}$$

Both the undamped natural frequency and the damping ratio depend on the value of the gain K. This gain must be adjusted so that the transient responses to both the reference input and disturbance input show reasonable damping and reasonable speed.

A–4–18. Are the open-loop zeros and closed-loop zeros identical in a closed-loop system?

Solution. No. Consider a closed-loop system whose feedforward transfer function is $G(s) = p(s)/q(s)$ and feedback transfer function is $H(s) = n(s)/d(s)$, where $p(s)$, $q(s)$, $n(s)$, and $d(s)$ are polynomials in s. Then

$$\frac{C(s)}{R(s)} = \frac{G(s)}{1 + G(s)H(s)}$$

$$= \frac{p(s)d(s)}{q(s)d(s) + p(s)n(s)}$$

We see that the zeros of the closed-loop function are those values of s that make $p(s)d(s) = 0$ and those of the open-loop transfer function are those values of s that make $p(s)n(s) = 0$. (Thus, some of the zeros of the closed-loop transfer function are the same as those of the open-loop transfer function.)

A–4–19. Consider the unity-feedback control system with feedforward transfer function $G(s)$. Suppose that the closed-loop transfer function can be written

$$\frac{C(s)}{R(s)} = \frac{G(s)}{1 + G(s)} = \frac{(T_a s + 1)(T_b s + 1) \cdots (T_m s + 1)}{(T_1 s + 1)(T_2 s + 1) \cdots (T_n s + 1)} \qquad (m \le n)$$

Show that

$$\int_0^\infty e(t)dt = (T_1 + T_2 + \cdots + T_n) - (T_a + T_b + \cdots + T_m)$$

where $e(t)$ is the error in the unit-step response. Show also that

$$\frac{1}{\lim\limits_{s \to 0} sG(s)} = (T_1 + T_2 + \cdots + T_n) - (T_a + T_b + \cdots + T_m)$$

Solution. Let us define

$$(T_a s + 1)(T_b s + 1) \cdots (T_m s + 1) = P(s)$$

and

$$(T_1 s + 1)(T_2 s + 1) \cdots (T_n s + 1) = Q(s)$$

Then

$$\frac{C(s)}{R(s)} = \frac{P(s)}{Q(s)}$$

and

$$E(s) = \frac{Q(s) - P(s)}{Q(s)} R(s)$$

For a unit-step input, $R(s) = 1/s$ and

$$E(s) = \frac{Q(s) - P(s)}{sQ(s)}$$

Using the fact that

$$\int_0^\infty e(t)\, dt = \lim_{s \to 0} s\, \frac{E(s)}{s} = \lim_{s \to 0} E(s)$$

we obtain

$$\int_0^\infty e(t)\, dt = \lim_{s \to 0} \frac{Q(s) - P(s)}{sQ(s)}$$

$$= \lim_{s \to 0} \frac{Q'(s) - P'(s)}{Q(s) + sQ'(s)}$$

$$= \lim_{s \to 0} [Q'(s) - P'(s)]$$

Since

$$\lim_{s \to 0} P'(s) = T_a + T_b + \cdots + T_m$$

$$\lim_{s \to 0} Q'(s) = T_1 + T_2 + \cdots + T_n$$

we have

$$\int_0^\infty e(t)\, dt = (T_1 + T_2 + \cdots + T_n) - (T_a + T_b + \cdots + T_m)$$

Note that since

$$\int_0^\infty e(t)\, dt = \frac{1}{K_v} = \frac{1}{\lim_{s \to 0} sG(s)}$$

we immediately obtain the following relationship:

$$\frac{1}{\lim_{s \to 0} sG(s)} = (T_1 + T_2 + \cdots + T_n) - (T_a + T_b + \cdots + T_m)$$

Note that zeros will improve K_v. Poles close to the origin cause low velocity-error constants unless there are zeros nearby.

A–4–20. Consider the system whose closed-loop transfer function is

$$\frac{C(s)}{R(s)} = \frac{1}{s^2 + 2\zeta s + 1} \qquad (0 < \zeta < 1)$$

Compute $\int_0^\infty c^2(t)\, dt$ for a unit-impulse input.

Solution. For a unit-impulse input, $R(s) = 1$. Hence

$$
\begin{aligned}
C(s) &= \frac{1}{s^2 + 2\zeta s + 1} \\
&= \frac{1}{\sqrt{1 - \zeta^2}} \frac{\sqrt{1 - \zeta^2}}{(s + \zeta)^2 + 1 - \zeta^2}
\end{aligned}
$$

The inverse Laplace transform of $C(s)$ is

$$c(t) = \frac{1}{\sqrt{1 - \zeta^2}} e^{-\zeta t} \sin \sqrt{1 - \zeta^2}\, t \qquad (t \ge 0)$$

Since $c^2(t)$ is

$$
\begin{aligned}
c^2(t) &= \frac{1}{1 - \zeta^2} e^{-2\zeta t} \sin^2 \sqrt{1 - \zeta^2}\, t \\
&= \frac{1}{1 - \zeta^2} e^{-2\zeta t} \frac{1}{2} (1 - \cos 2\sqrt{1 - \zeta^2}\, t)
\end{aligned}
$$

we get

$$\mathcal{L}[c^2(t)] = \frac{1}{2(1 - \zeta^2)} \left[\frac{1}{s + 2\zeta} - \frac{s + 2\zeta}{(s + 2\zeta)^2 + 4(1 - \zeta^2)} \right]$$

Then

$$
\begin{aligned}
\int_0^\infty c^2(t)\, dt &= \lim_{s \to 0} \mathcal{L}[c^2(t)] \\
&= \lim_{s \to 0} \frac{1}{2(1 - \zeta^2)} \left[\frac{1}{s + 2\zeta} - \frac{s + 2\zeta}{(s + 2\zeta)^2 + 4(1 - \zeta^2)} \right] \\
&= \frac{1}{4\zeta}
\end{aligned}
$$

A–4–21. Consider the following system:

$$\ddot{x} + 2\zeta\dot{x} + x = 0, \qquad x(0) = 1, \qquad \dot{x}(0) = 0 \qquad (4\text{–}76)$$

where $0 < \zeta < 1$. Find the value of ζ that minimizes the following performance index J:

$$J = \int_0^\infty [x^2(t) + \dot{x}^2(t)]\, dt$$

Solution. The solution of Equation (4–76) is

$$x(t) = k_1 e^{\lambda_1 t} + k_2 e^{\lambda_2 t}$$

where

$$\lambda_1 = -\zeta + \sqrt{\zeta^2 - 1}, \qquad \lambda_2 = -\zeta - \sqrt{\zeta^2 - 1}$$

and k_1 and k_2 are determined from the initial conditions. Hence

$$\dot{x}(t) = k_1 \lambda_1 e^{\lambda_1 t} + k_2 \lambda_2 e^{\lambda_2 t}$$

Then $x^2(t)$ and $\dot{x}^2(t)$ are obtained as

$$x^2(t) = k_1^2 e^{2\lambda_1 t} + 2k_1 k_2 e^{(\lambda_1 + \lambda_2)t} + k_2^2 e^{2\lambda_2 t}$$

and

$$\dot{x}^2(t) = k_1^2 \lambda_1^2 e^{2\lambda_1 t} + 2k_1 k_2 \lambda_1 \lambda_2 e^{(\lambda_1 + \lambda_2)t} + k_2^2 \lambda_2^2 e^{2\lambda_2 t}$$

The performance index becomes

$$J = \int_0^\infty [x^2(t) + \dot{x}^2(t)]\, dt$$

$$= \int_0^\infty [k_1^2(1 + \lambda_1^2)e^{2\lambda_1 t} + 2k_1 k_2(1 + \lambda_1 \lambda_2)e^{(\lambda_1 + \lambda_2)t} + k_2^2(1 + \lambda_2^2)e^{2\lambda_2 t}]\, dt$$

Performing the integration and evaluating at $t = \infty$ and $t = 0$, and noting that the real parts of λ_1 and λ_2 are negative, we obtain

$$J = -\left[\frac{k_1^2(1 + \lambda_1^2)}{2\lambda_1} + \frac{2k_1 k_2(1 + \lambda_1 \lambda_2)}{\lambda_1 + \lambda_2} + \frac{k_2^2(1 + \lambda_2^2)}{2\lambda_2}\right]$$

Since

$$\lambda_1 + \lambda_2 = -2\zeta, \qquad \lambda_1 \lambda_2 = 1$$

we obtain

$$J = \zeta(k_1^2 + k_2^2) + \frac{2k_1 k_2}{\zeta}$$

$$= \zeta(k_1 + k_2)^2 - 2k_1 k_2\left(\zeta - \frac{1}{\zeta}\right) \qquad (4\text{–}77)$$

From the initial conditions,

$$x(0) = 1 = k_1 + k_2, \qquad \dot{x}(0) = 0 = k_1 \lambda_1 + k_2 \lambda_2$$

Since $0 < \zeta < 1$, we have $\lambda_1 \neq \lambda_2$ and k_1 and k_2 are obtained as

$$k_1 = \frac{\lambda_2}{\lambda_2 - \lambda_1}, \qquad k_2 = \frac{-\lambda_1}{\lambda_2 - \lambda_1}$$

Substitution of these values of k_1 and k_2 into Equation (4–77) yields

$$J = \zeta + \frac{1}{2\zeta} \tag{4–78}$$

To find the minimum value of J, first differentiate J with respect to ζ and then set the result to zero.

$$\frac{\partial J}{\partial \zeta} = 1 - \frac{1}{2\zeta^2} = 0$$

Solving this equation

$$\zeta = \sqrt{\frac{1}{2}} = 0.707 \tag{4–79}$$

Since $\partial^2 J/\partial \zeta^2 > 0$, Equation (4–79) gives the optimal value of ζ. The minimum value of J is

$$\min J = \zeta + \frac{1}{2\zeta} = 1.414$$

A–4–22. Compute the following performance indexes:

$$\int_0^\infty |e(t)| dt, \qquad \int_0^\infty t|e(t)| dt, \qquad \int_0^\infty t[|e(t)| + |\dot{e}(t)|] \, dt$$

for the system

$$\frac{C(s)}{R(s)} = \frac{1}{s^2 + 2\zeta s + 1} \qquad (\zeta \geq 1)$$

$$E(s) = \mathcal{L}[e(t)] = R(s) - C(s)$$

Assume that the system is initially at rest and is subjected to a unit-step input.

Solution. For a unit-step input, $R(s) = 1/s$ and

$$E(s) = \frac{s^2 + 2\zeta s}{s^2 + 2\zeta s + 1} \frac{1}{s}$$

$$= \frac{-\zeta + \sqrt{\zeta^2 - 1}}{2\sqrt{\zeta^2 - 1}} \frac{1}{s + \zeta + \sqrt{\zeta^2 - 1}} + \frac{\zeta + \sqrt{\zeta^2 - 1}}{2\sqrt{\zeta^2 - 1}} \frac{1}{s + \zeta - \sqrt{\zeta^2 - 1}}$$

For $\zeta \geq 1$, the system does not overshoot. Hence $|e(t)| = e(t)$ for all $t \geq 0$. The given performance indexes then can be computed as follows:

1.

$$\int_0^\infty |e(t)| \, dt = \int_0^\infty e(t) \, dt$$

$$= \lim_{s \to 0} E(s)$$

$$= \frac{-\zeta + \sqrt{\zeta^2 - 1}}{2\sqrt{\zeta^2 - 1}} \frac{1}{\zeta + \sqrt{\zeta^2 - 1}} + \frac{\zeta + \sqrt{\zeta^2 - 1}}{2\sqrt{\zeta^2 - 1}} \frac{1}{\zeta - \sqrt{\zeta^2 - 1}}$$

$$= 2\zeta \qquad (\zeta \geq 1)$$

2.
$$\int_0^\infty t|e(t)|\,dt = \int_0^\infty te(t)\,dt$$

$$= \lim_{s \to 0}\left[-\frac{d}{ds}E(s)\right]$$

$$= \lim_{s \to 0}\left[\frac{-\zeta + \sqrt{\zeta^2 - 1}}{2\sqrt{\zeta^2 - 1}}\frac{1}{(s + \zeta + \sqrt{\zeta^2 - 1})^2}\right.$$

$$\left. + \frac{\zeta + \sqrt{\zeta^2 - 1}}{2\sqrt{\zeta^2 - 1}}\frac{1}{(s + \zeta - \sqrt{\zeta^2 - 1})^2}\right]$$

$$= 4\zeta^2 - 1 \qquad (\zeta \geq 1)$$

3. Noting that $\dot{e}(t) < 0$ and $|\dot{e}(t)| = -\dot{e}(t)$ for $t \geq 0$, we obtain

$$\int_0^\infty t(|e(t)| + |\dot{e}(t)|)\,dt = \int_0^\infty [te(t) - t\dot{e}(t)]\,dt$$

$$= \lim_{s \to 0}\left[-\frac{d}{ds}E(s) + \frac{d}{ds}sE(s)\right]$$

$$= \lim_{s \to 0}\left[(-1 + s)\frac{d}{ds}E(s) + E(s)\right]$$

$$= 4\zeta^2 - 1 + 2\zeta \qquad (\zeta \geq 1)$$

For purposes of comparison, we shall list a few other performance indexes computed for the same system [refer to Equations (4–47), (4–48), and (4–78)].

$$\int_0^\infty e^2(t)\,dt = \zeta + \frac{1}{4\zeta} \qquad (\zeta > 0)$$

$$\int_0^\infty te^2(t)\,dt = \zeta^2 + \frac{1}{8\zeta^2} \qquad (\zeta > 0)$$

$$\int_0^\infty [e^2(t) + \dot{e}^2(t)]\,dt = \zeta + \frac{1}{2\zeta} \qquad (\zeta > 0)$$

Plots of these six performance indexes are shown in Figure 4–38 as functions of ζ. (Note that the curves shown in Figure 4–38 are valid for $\zeta > 0$. The three performance indexes computed in this problem are valid only for $\zeta \geq 1$.)

A–4–23. Consider the differential equation

$$\frac{dx}{dt} = f(t, x)$$

Derive the following fourth-order Runge–Kutta equation:

$$x_{i+1} = x_i + \Delta x_i = x_i + \frac{1}{6}(K_1 + 2K_2 + 2K_3 + K_4)$$

where

$$K_1 = hf(t_i, x_i)$$

$$K_2 = hf(t_i + \tfrac{1}{2}h, x_i + \tfrac{1}{2}K_1)$$

$$K_3 = hf(t_i + \tfrac{1}{2}h, x_i + \tfrac{1}{2}K_2)$$

$$K_4 = hf(t_i + h, x_i + K_3)$$

and

$$h = t_{i+1} - t_i$$

Solution. Note that the values of x at $t = t_i$ and $t = t_i + h$ can be written as $x_i = x(t_i)$ and $x_{i+1} = x(t_{i+1})$, respectively. If we expand x_{i+1} into a Taylor series about point x_i, we obtain

$$x_{i+1} = x_i + hx_i' + \frac{h^2}{2!}x_i'' + \frac{h^3}{3!}x_i''' + \frac{h^4}{4!}x_i^{(IV)} + R_5 \qquad (4\text{--}80)$$

where

$$R_5 = \frac{h^5}{5!}x_i^{(V)}(t_i + \theta h) \qquad (0 < \theta < 1)$$

and

$$x' = f(t, x)$$

$$x'' = \frac{d^2x}{dt^2} = \frac{\partial f}{\partial t} + \frac{\partial f}{\partial x}f$$

$$x''' = \frac{d^3x}{dt^3} = \frac{\partial^2 f}{\partial t^2} + 2f\frac{\partial^2 f}{\partial t\,\partial x} + f^2\frac{\partial^2 f}{\partial x^2} + \frac{\partial f}{\partial x}\frac{\partial f}{\partial t} + f\left(\frac{\partial f}{\partial x}\right)^2$$

$$x^{(IV)} = \frac{d^4x}{dt^4} = \frac{\partial^3 f}{\partial t^3} + 3f\frac{\partial^3 f}{\partial t^2\,\partial x} + 3f^2\frac{\partial^3 f}{\partial t\,\partial x^2} + f^3\frac{\partial^3 f}{\partial x^3} + \frac{\partial f}{\partial x}\frac{\partial^2 f}{\partial t^2}$$

$$+ 2f\frac{\partial f}{\partial x}\frac{\partial^2 f}{\partial t\,\partial x} + f^2\frac{\partial f}{\partial x}\frac{\partial^2 f}{\partial x^2} + 3\frac{\partial f}{\partial t}\frac{\partial^2 f}{\partial t\,\partial x} + 3f\frac{\partial f}{\partial t}\frac{\partial^2 f}{\partial x^2}$$

$$+ 3f\frac{\partial f}{\partial x}\frac{\partial^2 f}{\partial t\,\partial x} + 3f^2\frac{\partial f}{\partial x}\frac{\partial^2 f}{\partial x^2} + \left(\frac{\partial f}{\partial x}\right)^2\frac{\partial f}{\partial t} + f\left(\frac{\partial f}{\partial x}\right)^2\frac{\partial f}{\partial x}$$

Define

$$D = \frac{\partial}{\partial t_i} + f\frac{\partial}{\partial x_i}$$

Then Equation (4–80) can be written as

$$\Delta x_i = hf(t_i, x_i) + \frac{h^2}{2!}Df(t_i, x_i)$$

$$+ \frac{h^3}{3!}\left\{D^2f(t_i, x_i) + \frac{\partial f(t_i, x_i)}{\partial x_i}Df(t_i, x_i)\right\}$$

$$+ \frac{h^4}{4!}\left[D^3f(t_i, x_i) + \frac{\partial f(t_i, x_i)}{\partial x_i}D^2f(t_i, x_i)\right.$$

$$+ \left.\left\{\frac{\partial f(t_i, x_i)}{\partial x_i}\right\}^2 Df(t_i, x_i) + 3D\left\{\frac{\partial f(t_i, x_i)}{\partial x_i}\right\}Df(t_i, x_i)\right]$$

The error in the above equation is on the order $0(h^5)$. Define

$$K_1 = hf(t_i, x_i)$$

$$K_2 = hf(t_i + \alpha_1 h, x_i + \beta_1 K_1)$$

$$K_3 = hf(t_i + \alpha_2 h, x_i + \beta_2 K_1 + \gamma_2 K_2)$$

$$K_4 = hf(t_i + \alpha_3 h, x_i + \beta_3 K_1 + \gamma_3 K_2 + \delta_3 K_3)$$

and

$$\Delta \tilde{x}_i = \lambda_1 K_1 + \lambda_2 K_2 + \lambda_3 K_3 + \lambda_4 K_4$$

By expanding K_1, K_2, K_3, and K_4 into Taylor series about point (t_i, x_i) and choosing thirteen undetermined constants $\alpha_1, \alpha_2, \alpha_3, \beta_1, \beta_2, \beta_3, \gamma_2, \gamma_3, \delta_3, \lambda_1, \lambda_2, \lambda_3$, and λ_4 such that

$$\Delta \tilde{x}_i = \Delta x_i$$

we obtain

$$\lambda_1 + \lambda_2 + \lambda_3 + \lambda_4 = 1$$

$$\lambda_2 \alpha_1 + \lambda_3 \alpha_2 + \lambda_4 \alpha_3 = \frac{1}{2}$$

$$\lambda_2 \alpha_1^2 + \lambda_3 \alpha_2^2 + \lambda_4 \alpha_3^2 = \frac{1}{3}$$

$$\lambda_2 \alpha_1^3 + \lambda_3 \alpha_2^3 + \lambda_4 \alpha_3^3 = \frac{1}{4}$$

$$\lambda_3 \alpha_1 \gamma_2 + \lambda_4 (\alpha_1 \gamma_3 + \alpha_2 \delta_3) = \frac{1}{6}$$

$$\lambda_3 \alpha_1^2 \gamma_2 + \lambda_4 (\alpha_1^2 \gamma_3 + \alpha_2^2 \delta_3) = \frac{1}{12}$$

$$\lambda_3 \alpha_1 \alpha_2 \gamma_2 + \lambda_4 (\alpha_1 \alpha_3 \gamma_3 + \alpha_2 \alpha_3 \delta_3) = \frac{1}{8}$$

$$\lambda_4 \alpha_1 \gamma_2 \delta_3 = \frac{1}{24}$$

$$\alpha_1 = \beta_1$$

$$\alpha_2 = \beta_2 + \gamma_2$$

$$\alpha_3 = \beta_3 + \gamma_3 + \delta_3$$

Here we have 11 equations and 13 unknowns. Consequently, we may choose two unknown variables arbitrarily. Let us choose

$$\alpha_1 = \frac{1}{2}, \qquad \delta_3 = 1$$

Then the remaining 11 unknowns can be determined as follows:

$$\alpha_2 = \frac{1}{2}, \qquad \alpha_3 = 1$$

$$\beta_1 = \frac{1}{2}, \qquad \beta_2 = 0, \qquad \beta_3 = 0$$

$$\gamma_2 = \frac{1}{2}, \qquad \gamma_3 = 0,$$

$$\lambda_1 = \frac{1}{6}, \qquad \lambda_2 = \frac{1}{3}, \qquad \lambda_3 = \frac{1}{3}, \qquad \lambda_4 = \frac{1}{6}$$

Therefore, we get

$$K_1 = hf(t_i, x_i)$$

$$K_2 = hf(t_i + \tfrac{1}{2}h, x_i + \tfrac{1}{2}K_1)$$

$$K_3 = hf(t_i + \tfrac{1}{2}h, x_i + \tfrac{1}{2}K_2)$$

$$K_4 = hf(t_i + h, x_i + K_3)$$

and

$$\Delta x_i = \frac{1}{6}K_1 + \frac{1}{3}K_2 + \frac{1}{3}K_3 + \frac{1}{6}K_4$$

$$= \frac{1}{6}(K_1 + 2K_2 + 2K_3 + K_4)$$

or

$$x_{i+1} = x_1 + \Delta x_i = x_i + \frac{1}{6}(K_1 + 2K_2 + 2K_3 + K_4)$$

This is the fourth-order Runge–Kutta equation.

A–4–24. Obtain a computer solution of the unit-step response of the following system:

$$\frac{Y(s)}{U(s)} = \frac{5s + 100}{s^4 + 8s^3 + 32s^2 + 80s + 100}$$

Write a computer program. Plot a unit-step response curve, $y(t)$ versus t. (For an analytical solution of this problem, see Problem A–4–11.)

Solution. From the given transfer function, the corresponding differential equation can be obtained as

$$\ddddot{y} + 8\dddot{y} + 32\ddot{y} + 80\dot{y} + 100y = 5\dot{u} + 100u$$

Comparing this differential equation with the following differential equation,

$$\ddddot{y} + a_1\dddot{y} + a_2\ddot{y} + a_3\dot{y} + a_4y = b_0\dddot{u} + b_1\ddot{u} + b_2\ddot{u} + b_3\dot{u} + b_4u$$

we obtain

$$a_1 = 8, \qquad a_2 = 32, \qquad a_3 = 80, \qquad a_4 = 100$$

$$b_0 = 0, \qquad b_1 = 0, \qquad b_2 = 0, \qquad b_3 = 5, \qquad b_4 = 100$$

Referring to Equations (2–5) and (2–6), we define state variables x_1, x_2, x_3, and x_4 as follows:

$$x_1 = y - \beta_0 u$$

$$x_2 = \dot{x}_1 - \beta_1 u$$

$$x_3 = \dot{x}_2 - \beta_2 u$$

$$x_4 = \dot{x}_3 - \beta_3 u$$

Example Problems and Solutions

339

Table 4–4 BASIC Computer Program for Solving the State Equation Given in Problem A–4–24, where All Initial Conditions are Zero and $u = 1$ (Unit-Step Response)

```
  10  ORDER  =  4
  20  X(1)  =  0
  30  X(2)  =  0
  40  X(3)  =  0
  50  X(4)  =  0
  60  H  =  .01
  70  T  =  0
  80  TK  =  0
  90  TF  =  4
 100  OPEN  "O",#1,  "ANS.BAS"
 110  PRINT  "        TIME        X(1)        X(2)        X(3)        X(4)        "
 120  PRINT  "------------------------------------------------------------"
 130  PRINT  #1,  USING  "####.######";  X(1)
 140  PRINT  USING  "####.#####";  T,  X(1),  X(2),  X(3),  X(4)
 150  IF  T  >  TF  THEN  GOTO  5000
 160  GOSUB  1000
 170  GOTO  130
1000  TK  =  T
1010  GOSUB  2000
1020  FOR  I  =  1  TO  ORDER
1030  XK(I)  =  X(I)
1040  K(1,I) =  DX(I)
1050  T  =  TK  +  H/2
1060  X(I)  =  XK(I)  +  (H/2)*K(1,I)
1070  NEXT  I
1080  GOSUB  2000
1090  FOR  I  =  1  TO  ORDER
1100  K(2,I)  =  DX(I)
1110  X(I)  =  XK(I)  +  (H/2)*K(2,I)
1120  NEXT  I
1130  GOSUB  2000
1140  FOR  I  =  1  TO  ORDER
1150  K(3,I)  =  DX(I)
1160  T  =  TK  +  H
1170  X(I)  =  XK(I)  +  H*K(3,I)
1180  NEXT  I
1190  GOSUB  2000
1200  FOR  I  =  1  TO  ORDER
1210  K(4,I)  =  DX(I)
1220  X(I)  =  XK(I)  +  (H/6)*(K(1,I)  +  2*K(2,I)  +  2*K(3,I)  +  K(4,I))
1230  NEXT  I
1240  RETURN
2000  DX(1)  =  X(2)
2010  DX(2)  =  X(3)
2020  DX(3)  =  X(4)  +  5
2030  DX(4)  =  - 100*X(1)  -  80*X(2)  -  32*X(3)  -  8*X(4)  +  60
2040  RETURN
4900  CLOSE  #1
5000  END
```

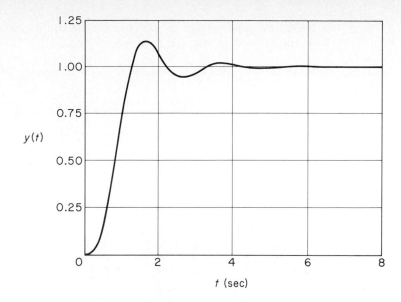

Figure 4–59
Unit-step response curve $y(t)$ versus t for the system considered in Problem A–4–24.

where

$$\beta_0 = b_0 = 0$$

$$\beta_1 = b_1 - a_1\beta_0 = 0$$

$$\beta_2 = b_2 - a_1\beta_1 - a_2\beta_0 = 0$$

$$\beta_3 = b_3 - a_1\beta_2 - a_2\beta_1 - a_3\beta_0 = 5$$

$$\beta_4 = b_4 - a_1\beta_3 - a_2\beta_2 - a_3\beta_1 - a_4\beta_0 = 100 - 8 \times 5 = 60$$

Thus, the state variables are given by

$$x_1 = y$$

$$x_2 = \dot{x}_1$$

$$x_3 = \dot{x}_2$$

$$x_4 = \dot{x}_3 - 5u$$

Note that

$$\dot{x}_4 = -a_4x_1 - a_3x_2 - a_2x_3 - a_1x_4 + \beta_4u$$
$$= -100x_1 - 80x_2 - 32x_3 - 8x_4 + 60u$$

The state equation and output equation can be obtained as

$$
\begin{bmatrix} \dot{x}_1 \\ \dot{x}_2 \\ \dot{x}_3 \\ \dot{x}_4 \end{bmatrix} =
\begin{bmatrix} 0 & 1 & 0 & 0 \\ 0 & 0 & 1 & 0 \\ 0 & 0 & 0 & 1 \\ -100 & -80 & -32 & -8 \end{bmatrix}
\begin{bmatrix} x_1 \\ x_2 \\ x_3 \\ x_4 \end{bmatrix} +
\begin{bmatrix} 0 \\ 0 \\ 5 \\ 60 \end{bmatrix} u
$$

$$
y = \begin{bmatrix} 1 & 0 & 0 & 0 \end{bmatrix}
\begin{bmatrix} x_1 \\ x_2 \\ x_3 \\ x_4 \end{bmatrix}
$$

Example Problems and Solutions

A BASIC computer program for obtaining the unit-step response of the system is given in Table 4–4. The unit-step response curve $y(t)$ versus t [where $y(t) = x_1(t)$] is given in Figure 4–59.

A–4–25. Consider the system shown in Figure 4–42. (Refer to Example 4–7.) The system is defined by

$$\frac{Y(s)}{U(s)} = \frac{25.04s + 5.008}{s^3 + 5.03247s^2 + 25.1026s + 5.008}$$

Write a computer program to obtain the response $y(t)$ of the system to a unit-ramp input, $u(t) = t$.

Solution. In Example 4–7, we derived state and output equations for the system, rewritten as

$$\begin{bmatrix} \dot{x}_1 \\ \dot{x}_2 \\ \dot{x}_3 \end{bmatrix} = \begin{bmatrix} 0 & 1 & 0 \\ 0 & 0 & 1 \\ -5.008 & -25.1026 & -5.03247 \end{bmatrix} \begin{bmatrix} x_1 \\ x_2 \\ x_3 \end{bmatrix} + \begin{bmatrix} 0 \\ 25.04 \\ -121.005 \end{bmatrix} u$$

$$y = \begin{bmatrix} 1 & 0 & 0 \end{bmatrix} \begin{bmatrix} x_1 \\ x_2 \\ x_3 \end{bmatrix}$$

In Example 4–7, the input $u(t)$ was a unit-step function. In this problem the input $u(t)$ is a unit-ramp function. We can use the computer program given in Table 4–3 with a minor modification in the state equations. The state equations for the unit-ramp response are

FNX1D(T, X1, X2, X3) = X2

FNX2D(T, X1, X2, X3) = X3 + 25.04*T   (4–81)

FNX3D(T, X1, X2, X3) = −5.008*X1 − 25.1026*X2 − 5.03247*X3 − 121.005*T   (4–82)

Thus, a BASIC computer program for obtaining the unit-ramp response is the same as Table 4–3, except that lines 30 and 40 must be replaced by Equations (4–81) and (4–82), respectively. The unit-ramp response curve $y(t)$ versus t [where $y(t) = x_1(t)$] is shown in Figure 4–60.

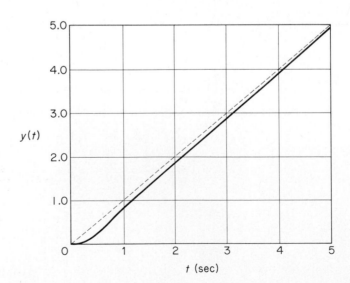

Figure 4–60
Unit-ramp response
for the system shown
in Fig. 4–42.

PROBLEMS

B–4–1. A thermometer requires 1 min to indicate 98% of the response to a step input. Assuming the thermometer to be a first-order system, find the time constant.

If the thermometer is placed in a bath, the temperature of which is changing linearly at a rate 10°/min, how much error does the thermometer show?

B–4–2. Obtain the unit-step response of a unity-feedback system whose open-loop transfer function is

$$G(s) = \frac{4}{s(s + 5)}$$

B–4–3. Consider the unit-step response of a unity-feedback control system whose open-loop transfer function is

$$G(s) = \frac{1}{s(s + 1)}$$

Obtain the rise time, peak time, maximum overshoot, and settling time.

B–4–4. Consider the closed-loop system given by

$$\frac{C(s)}{R(s)} = \frac{\omega_n^2}{s^2 + 2\zeta\omega_n s + \omega_n^2}$$

Determine the values of ζ and ω_n so that the system responds to a step input with approximately 5% overshoot and with a settling time of 2 sec. (Use the 2% criterion.)

B–4–5. Figure 4–61 is a block diagram of a space-vehicle attitude-control system. Assuming the time constant T of the controller to be 3 sec and the ratio of torque to inertia K/J to be $\frac{2}{9}$ rad^2/sec^2, find the damping ratio of the system.

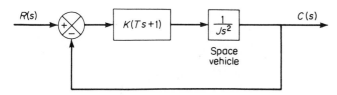

Figure 4–61 Space-vehicle attitude-control system.

B–4–6. Consider a unity-feedback control system whose open-loop transfer function is

$$G(s) = \frac{0.4s + 1}{s(s + 0.6)}$$

Obtain the response to a unit-step input. What is the rise time for this system? What is the maximum overshoot?

B–4–7. Obtain the unit-impulse response and the unit-step response of a unity-feedback system whose open-loop transfer function is

$$G(s) = \frac{2s + 1}{s^2}$$

B–4–8. Consider the system shown in Figure 4–62. Show that the transfer function $Y(s)/X(s)$ has a zero in the right-half s plane. Then obtain $y(t)$ when $x(t)$ is a unit step. Plot $y(t)$ versus t.

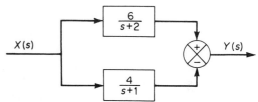

Figure 4–62 System with a zero in the right-half s plane (nonminimum phase system).

B–4–9. An oscillatory system is known to have a transfer function of the following form:

$$G(s) = \frac{\omega_n^2}{s^2 + 2\zeta\omega_n s + \omega_n^2}$$

Assume that a record of a damped oscillation is available as shown in Figure 4–63. Determine the damping ratio ζ of the system from the graph.

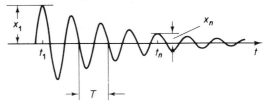

Figure 4–63 Decaying oscillation.

B–4–10. Referring to the system in Figure 4–64, determine the values of K and k such that the system has a damping ratio ζ of 0.7 and an undamped natural frequency ω_n of 4 rad/sec.

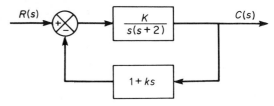

Figure 4–64 Closed-loop system.

B–4–11. Figure 4–65 shows a position control system with velocity feedback. What is the response $c(t)$ to the unit-step input?

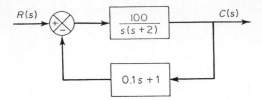

Figure 4–65 Block diagram of a position control system with velocity feedback.

B–4–12. Consider the system shown in Figure 4–66. Determine the value of k such that the damping ratio ζ is 0.5. Then obtain the rise time t_r, peak time t_p, maximum overshoot M_p, and settling time t_s in the unit-step response.

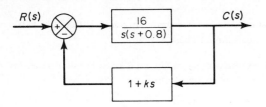

Figure 4–66 Block diagram of a system.

B–4–13. Obtain the unit-ramp response of the system given by

$$\frac{C(s)}{R(s)} = \frac{\omega_n^2}{s^2 + 2\zeta\omega_n s + \omega_n^2}$$

Obtain also the steady-state error.

B–4–14. Consider the system shown in Figure 4–67(a). The damping ratio of this system is 0.158 and the undamped natural frequency is 3.16 rad/sec. To improve the relative stability, we employ tachometer feedback. Figure 4–67(b) shows such a tachometer-feedback system.

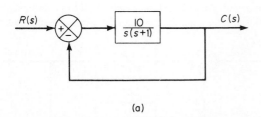

(a)

Figure 4–67 (a) Control system; (b) control system with tachometer feedback.

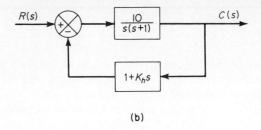

(b)

Figure 4–67 Cont'd.

Determine the value of K_h so that the damping ratio of the system is 0.5. Draw unit-step response curves of both the original and tachometer-feedback systems. Also draw the error-versus-time curves for the unit-ramp response of both systems.

B–4–15. Apply Routh's stability criterion to the following characteristic equation:

$$s^4 + s^3 + Ks^2 + s + 1 = 0$$

Determine the range of K for stability.

B–4–16. Determine the range of K for stability of a unity-feedback control system whose open-loop transfer function is

$$G(s) = \frac{K}{s(s + 1)(s + 2)}$$

B–4–17. Consider the unity-feedback control system with the following open-loop transfer function:

$$G(s) = \frac{10}{s(s - 1)(2s + 3)}$$

Is this system stable?

B–4–18. Consider a unity-feedback control system with the closed-loop transfer function

$$\frac{C(s)}{R(s)} = \frac{Ks + b}{s^2 + as + b}$$

Determine the open-loop transfer function $G(s)$.

Show that the steady-state error in the unit-ramp response is given by

$$e_{ss} = \frac{1}{K_v} = \frac{a - K}{b}$$

(Thus, if $K = a$, then there will be no steady-state error in the response to the ramp input.)

B–4–19. Show that the steady-state error in the response to ramp inputs can be made zero if the closed-loop transfer function is given by

$$\frac{C(s)}{R(s)} = \frac{a_{n-1}s + a_n}{s^n + a_1 s^{n-1} + \cdots + a_{n-1}s + a_n}$$

B–4–20. Consider a system described by

$$\ddot{x} + 2\zeta\dot{x} + x = 0, \qquad x(0) = 1, \qquad \dot{x}(0) = 0$$

where $0 < \zeta < 1$. Determine the value of the damping ratio ζ such that

$$\int_0^\infty (x^2 + \mu\dot{x}^2)\, dt \qquad (\mu \geq 0)$$

is minimum.

B–4–21. The closed-loop transfer function for a disturbance input $n(t)$ is given by

$$\frac{C(s)}{N(s)} = \frac{s(s + a)}{s^2 + as + 10}$$

Determine the value of the constant a such that the integral square error due to a step disturbance is minimized.

B–4–22. Consider a unity-feedback control system with the open-loop transfer function

$$G(s) = \frac{(a + b)s + ab}{s^2}$$

Define the error signal $e(t)$ by

$$e(t) = r(t) - c(t)$$

where $r(t)$ is the input and $c(t)$ is the output. For a unit-step input, compute the following performance indexes:

$$J_1 = \int_0^\infty e^2(t)\, dt$$

$$J_2 = \int_0^\infty te^2(t)\, dt$$

If $a = b$, what are the values of J_1 and J_2?

B–4–23. Write a computer program to obtain a unit-step response of the following system:

$$\frac{Y(s)}{U(s)} = \frac{13}{s^2 + 4s + 13}$$

where $U(s)$ is the input and $Y(s)$ is the output.

B–4–24. Consider the position control system shown in Figure 4–68. Write a computer program to obtain a unit-step response

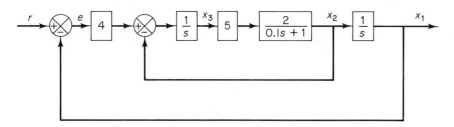

Figure 4–68 Position control system.

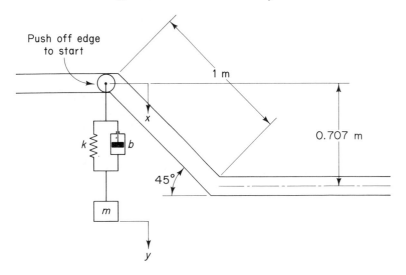

Figure 4–69 Wheel with hanging spring–mass–damper system.

and a unit-ramp response of the system. Plot curves $x_1(t)$ versus t, $x_2(t)$ versus t, $x_3(t)$ versus t, and $e(t)$ versus t [where $e(t) = r(t) - x_1(t)$] for both the unit-step response and the unit-ramp response.

B–4–25. Consider the system shown in Figure 4–69. A wheel has a spring–mass–damper system hanging from it. The wheel is in a track that contains a flat portion, a slanted (downward at 45°) portion, and another flat portion. We start the motion of the system by nudging the wheel over the edge of the ramp. As the wheel drops down the ramp for a total of 0.707 m (vertically measured), the mass hanging from the spring and damper drops with it. The mass gains momentum that dissipates

gradually when the wheel rolls on the second flat portion of the track.

Obtain an analytical solution to this problem and find $y(t)$. Assume that the initial conditions $y(0)$ and $\dot{y}(0)$ are zeros. Obtain also a computer solution by simulating this system on the computer. Plot curves $x(t)$ versus t and $y(t)$ versus t. Assume the following numerical values for m, b, and k:

$$m = 4\ \text{kg}, \qquad b = 40\ \text{N-sec/m}, \qquad k = 400\ \text{N/m}$$

Assume no friction in the system, except viscous friction in the damper.

PART II Control Systems Analysis and Design by Conventional Methods

CHAPTER 5
Root-Locus Analysis

5–1 INTRODUCTION

The basic characteristic of the transient response of a closed-loop system is closely related to the location of the closed-loop poles. If the system has a variable loop gain, then the location of the closed-loop poles depends on the value of the loop gain chosen. It is important, therefore, that the designer know how the closed-loop poles move in the s plane as the loop gain is varied.

From the design viewpoint, in some systems simple gain adjustment may move the closed-loop poles to desired locations. Then the design problem may become the selection of an appropriate gain value. In this chapter we discuss design problems that involve the selection of a particular parameter value (usually the open-loop gain value) such that the transient response characteristics are satisfactory. If the gain adjustment alone does not yield a desired result, addition of a compensator to the system will become necessary. (This subject is discussed in detail in Chapter 7.)

The closed-loop poles are the roots of the characteristic equation. Finding the roots of the characteristic equation of degree higher than three is laborious and will need computer solution. However, just finding the roots of the characteristic equation may be of limited value, because as the gain of the open-loop transfer function varies, the characteristic equation changes and the computations must be repeated.

A simple method for finding the roots of the characteristic equation has been developed by W. R. Evans and used extensively in control engineering. This method, called the *root-locus method*, is one in which the roots of the characteristic equation are plotted for all values of a system parameter. The roots corresponding to a particular value of this parameter can then be located on the resulting graph. Note that the parameter is usually the gain, but any other variable of the open-loop transfer function may be used. Unless otherwise stated, we shall assume that the gain of the open-loop transfer function is the parameter to be varied through all values, from zero to infinity.

By using the root-locus method the designer can predict the effects on the location of the closed-loop poles when varying the gain value or adding open-loop poles and/or open-loop zeros. Therefore, it is desired that the designer have a good understanding of the method for sketching the root loci of the closed-loop system.

Root-locus method. The basic idea behind the root-locus method is that the values of s that make the transfer function around the loop equal -1 must satisfy the characteristic equation of the system.

The locus of roots of the characteristic equation of the closed-loop system as the gain is varied from zero to infinity gives the method its name. Such a plot clearly shows the contributions of each open-loop pole or zero to the locations of the closed-loop poles.

The root-locus method enables us to find the closed-loop poles from the open-loop poles and zeros with the gain as parameter. It avoids inherent difficulties existing in classical techniques by providing a graphical display of all closed-loop poles for all values of the gain of the open-loop transfer function.

In designing a linear control system, we find that the root-locus method proves quite useful since it indicates the manner in which the open-loop poles and zeros should be modified so that the response meets system performance specifications. This method is particularly suited to obtaining approximate results very quickly.

Since the method is a graphical one for finding the roots of the characteristic equation, it provides an effective graphical procedure for finding the roots of any polynomial equation arising in the study of physical systems.

Some control systems may involve more than one parameter to be adjusted. The root-locus diagram for a system having multiple parameters may be constructed by varying one parameter at a time. In this chapter we include the discussion of the root loci for a system having two parameters. The root loci for such a case is called the *root contour*.

Outline of the chapter. This chapter introduces the basic concept of the root-locus method and presents useful rules for graphically constructing the root loci.

The outline of the chapter is as follows: Section 5–1 has presented an introduction to the root-locus method. Section 5–2 details the concepts underlying the root-locus method. Section 5–3 presents the general procedure for sketching root loci using illustrative examples. Section 5–4 summarizes general rules for constructing root loci. Section 5–5 analyzes closed-loop systems by use of the root-locus method. Section 5–6 extends the root-locus method to treat closed-loop systems with transport lag. Section 5–7 discusses root-contour plots. Finally, Section 5–8 gives the conclusions of the chapter.

5–2 ROOT-LOCUS PLOTS

Angle and magnitude conditions. Consider the system shown in Figure 5–1. The closed-loop transfer function is

$$\frac{C(s)}{R(s)} = \frac{G(s)}{1 + G(s)H(s)} \tag{5–1}$$

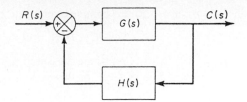

Figure 5–1
Control system.

The characteristic equation for this closed-loop system is obtained by setting the denominator of the right side of Equation (5–1) equal to zero. That is,

$$1 + G(s)H(s) = 0$$

or

$$G(s)H(s) = -1 \qquad (5\text{–}2)$$

Here we assume that $G(s)H(s)$ is a ratio of polynomials in s. [Later in Section 5–6 we extend the analysis to the case where $G(s)H(s)$ involves the transport lag e^{-Ts}.] Since $G(s)H(s)$ is a complex quantity, Equation (5–2) can be split into two equations by equating the angles and magnitudes of both sides, respectively, to obtain

Angle condition:

$$\underline{/G(s)H(s)} = \pm 180°(2k + 1) \qquad (k = 0, 1, 2, \ldots) \qquad (5\text{–}3)$$

Magnitude condition:

$$|G(s)H(s)| = 1 \qquad (5\text{–}4)$$

The values of s that fulfill the angle and magnitude conditions are the roots of the characteristic equation, or the closed-loop poles. A plot of the points of the complex plane satisfying the angle condition alone is the root locus. The roots of the characteristic equation (the closed-loop poles) corresponding to a given value of the gain can be determined from the magnitude condition. The details of applying the angle and magnitude conditions to obtain the closed-loop poles are presented in Section 5–3.

In many cases, $G(s)H(s)$ involves a gain parameter K and the characteristic equation may be written as

$$1 + \frac{K(s + z_1)(s + z_2) \cdots (s + z_m)}{(s + p_1)(s + p_2) \cdots (s + p_n)} = 0 \qquad (5\text{–}5)$$

Then the root loci for the system are the loci of the closed-loop poles as the gain K is varied from zero to infinity.

Note that to begin sketching the root loci of a system by the root-locus method we must know the location of the poles and zeros of $G(s)H(s)$. Remember that the angles of the complex quantities originating from the open-loop poles and open-loop zeros to the test point s are measured in the counterclockwise direction. For example, if $G(s)H(s)$ is given by

$$G(s)H(s) = \frac{K(s + z_1)}{(s + p_1)(s + p_2)(s + p_3)(s + p_4)}$$

where $-p_2$ and $-p_3$ are complex-conjugate poles, then the angle of $G(s)H(s)$ is

$$\angle G(s)H(s) = \phi_1 - \theta_1 - \theta_2 - \theta_3 - \theta_4$$

where ϕ_1, θ_1, θ_2, θ_3, and θ_4 are measured counterclockwise as shown in Figures 5–2(a) and (b). The magnitude of $G(s)$ for this system is

$$|G(s)H(s)| = \frac{KB_1}{A_1 A_2 A_3 A_4}$$

where A_1, A_2, A_3, A_4, and B_1 are the magnitudes of the complex quantities $s + p_1$, $s + p_2$, $s + p_3$, $s + p_4$, and $s + z_1$, respectively, as shown in Figure 5–2(a).

Note that because the open-loop complex-conjugate poles and complex-conjugate zeros, if any, are always located symmetrically about the real axis, the root loci are always symmetrical with respect to this axis. Therefore, we only need to construct the upper half of the root loci and draw the mirror image of the upper half in the lower-half s plane.

Root-locus plots of second-order systems.

Before presenting a method for constructing such plots in detail, a root-locus plot of a simple second-order system is illustrated.

Consider the system shown in Figure 5–3. The open-loop transfer function $G(s)H(s)$, where $H(s) = 1$ in this system, is

$$G(s)H(s) = \frac{K}{s(s + 1)}$$

The closed-loop transfer function is then

$$\frac{C(s)}{R(s)} = \frac{K}{s^2 + s + K}$$

The characteristic equation is

$$s^2 + s + K = 0 \tag{5–6}$$

We wish to find the locus of roots of this equation as K is varied from zero to infinity.

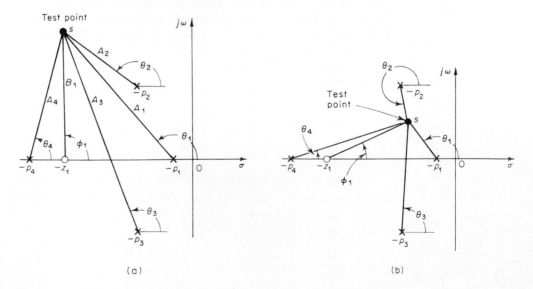

Figure 5–2
(a) and (b) Diagrams showing angle measurements from open-loop poles and open-loop zero to test point s.

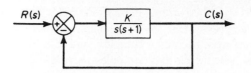

Figure 5–3
Control system.

To give a clear idea of what a root-locus plot looks like for this system, we shall first obtain the roots of the characteristic equation analytically in terms of K and then vary K from zero to infinity. It should be noted that this is not the proper way to construct the root-locus plot. The proper way is through a graphical trial-and-error approach. The graphical work can be simplified greatly by applying the general rules to be presented in Section 5–3. (Obviously, if an analytical solution for the characteristic roots can be found easily, there is no need for the root-locus method.) The roots of the characteristic equation, Equation (5–6), are

$$s_1 = -\tfrac{1}{2} + \tfrac{1}{2}\sqrt{1 - 4K}, \qquad s_2 = -\tfrac{1}{2} - \tfrac{1}{2}\sqrt{1 - 4K}$$

The roots are real for $K \le \tfrac{1}{4}$ and are complex for $K > \tfrac{1}{4}$.

The loci of the roots corresponding to all values of K are plotted in Figure 5–4(a). The root loci are graduated with K as parameter. (The motion of the roots with increasing K is shown by arrows.) Once we draw such a plot, we can immediately determine the value of K that will yield a root, or a closed-loop pole, at a desired point. From this analysis, it is clear that the closed-loop poles corresponding to $K = 0$ are the same as the poles of $G(s)H(s)$. As the value of K is increased from zero to $\tfrac{1}{4}$, the closed-loop poles move toward point $(-\tfrac{1}{2}, 0)$. For values of K between zero and $\tfrac{1}{4}$, all the closed-loop poles are on the real axis. This corresponds to an overdamped system, and the impulse response is nonoscillatory. At

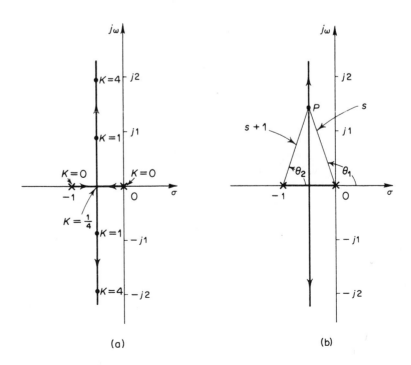

Figure 5–4
Root-locus diagrams for the system shown in Fig. 5–3.

(a)   (b)

$K = \frac{1}{4}$, the two real closed-loop poles coalesce. This corresponds to the case of a critically damped system. As K is increased from $\frac{1}{4}$, the closed-loop poles break away from the real axis, becoming complex, and since the real part of the closed-loop pole is constant for $K > \frac{1}{4}$, the closed-loop poles move along the line $s = -\frac{1}{2}$. Hence, for $K > \frac{1}{4}$, the system becomes underdamped. For a given value of K, one of the conjugate closed-loop poles moves toward $s = -\frac{1}{2} + j\infty$.

We shall show that any point on the root locus satisfies the angle condition. The angle condition given by Equation (5–3) is

$$\Big/ \frac{K}{s(s+1)} = -\angle s - \angle s + 1 = \pm 180°(2k+1) \qquad (k = 0, 1, 2 \ldots)$$

Consider point P on the root locus shown in Figure 5–4(b). The complex quantities s and $s + 1$ have angles θ_1 and θ_2, respectively, and magnitudes $|s|$ and $|s + 1|$, respectively. (Note that all angles are considered positive when they are measured in the counterclockwise direction.) The sum of the angles θ_1 and θ_2 is clearly 180°.

If point P is located on the real axis between 0 and -1, then $\theta_1 = 180°$ and $\theta_2 = 0°$. Thus, we see that any point on the root locus satisfies the angle condition. We also see that if point P is not a point on the root locus, then the sum of θ_1 and θ_2 is not equal to $\pm 180°(2k + 1)$, where $k = 0, 1, 2, \ldots$. Thus, points that are not on the root locus do not satisfy the angle condition and, therefore, cannot be closed-loop poles for any value of K.

If the closed-loop poles are specified on the root locus, then the corresponding value of K is determined from the magnitude condition, Equation (5–4). If, for example, the closed-loop poles selected are $s = -\frac{1}{2} \pm j2$, then the corresponding value of K is found from

$$|G(s)H(s)| = \left| \frac{K}{s(s+1)} \right|_{s = -(1/2)+j2} = 1$$

or

$$K = |s(s+1)|_{s=-(1/2)+j2} = \frac{17}{4}$$

Since complex poles are conjugates, if one of them, for example $s = -\frac{1}{2} + j2$, is specified, then the other is automatically fixed. In evaluating the value of K, either pole can be used.

From the root-locus diagram of Figure 5–4(a), we clearly see the effects of changes in the value of K on the transient-response behavior of the second-order system. An increase in the value of K will decrease the damping ratio ζ, resulting in an increase in the overshoot of the response. An increase in the value of K will also result in increases in the damped and undamped natural frequencies. (If K is greater than the critical value, that which corresponds to a critically damped system, increasing K will have no effect on the value of the real part of the closed-loop poles.) From the root-locus plot, it is clear that the closed-loop poles are always in the left half of the s plane; thus no matter how much K is increased, the system always remains stable. Hence, the second-order system is always stable. (Note, however, that if the gain is set at a very high value, the effects of some of the neglected time constants may become important, and the system, which is supposedly of second order, but actually of higher order, may become unstable.)

Table 5–1 shows a collection of simple root-locus plots.

Table 5–1 Collection of Simple Root-Locus Plots

$G(s)H(s)$	Open-loop Pole-zero Locations and Root Loci	$G(s)H(s)$	Open-loop Pole-zero Locations and Root Loci
$\dfrac{K}{s}$		$\dfrac{K}{s^2}$	
$\dfrac{K}{s+p}$	$-p$	$\dfrac{K}{s^2+\omega_1^2}$	$j\omega_1$, $-j\omega_1$
$\dfrac{K(s+z)}{s+p}$ $(z>p)$	$-z$ $-p$	$\dfrac{K}{(s+\sigma)^2+\omega_1^2}$	$j\omega_1$, $-\sigma$, $-j\omega_1$
$\dfrac{K(s+z)}{s+p}$ $(z<p)$	$-p$ $-z$	$\dfrac{K}{(s+p_1)(s+p_2)}$	$-p_1$ $-p_2$

5–3 ILLUSTRATIVE EXAMPLES

In this section we shall present three illustrative examples for constructing root-locus plots. Although computer approaches to the construction of the root loci are available, here we shall use graphical computation, combined with inspection, to determine the root loci upon which the roots of the characteristic equation of the closed-loop system must lie. Such a graphical approach will enhance understanding of how the closed-loop poles move in the complex plane as the open-loop poles and zeros are moved. Although we employ only simple systems for illustrative purposes, the procedure for finding the root loci is no more complicated for higher-order systems.

The first step in the procedure for constructing a root-locus plot is to seek out the loci of possible roots using the angle condition. Then the loci are scaled, or graduated, in gain using the magnitude condition.

Because graphical measurements of angles and magnitudes are involved in the analysis, we find it necessary to use the same divisions on the abscissa as on the ordinate axis when sketching the root locus on graph paper.

EXAMPLE 5–1

Consider the system shown in Figure 5–5. (We assume that the value of gain K is nonnegative.) For this system,

$$G(s) = \frac{K}{s(s + 1)(s + 2)}, \qquad H(s) = 1$$

Let us sketch the root-locus plot and then determine the value of K so that the damping ratio ζ of a pair of dominant complex-conjugate closed-loop poles is 0.5.

For the given system, the angle condition becomes

$$\angle G(s) = \Big/ \frac{K}{s(s + 1)(s + 2)}$$

$$= -\angle s - \angle s + 1 - \angle s + 2$$

$$= \pm 180°(2k + 1) \qquad (k = 0, 1, 2, \dots)$$

The magnitude condition is

$$|G(s)| = \left| \frac{K}{s(s + 1)(s + 2)} \right| = 1$$

A typical procedure for sketching the root-locus plot is as follows:

1. Determine the root loci on the real axis. The first step in constructing a root-locus plot is to locate the open-loop poles, $s = 0$, $s = -1$, and $s = -2$, in the complex plane. (There are no open-loop zeros in this system.) The locations of the open-loop poles are indicated by crosses. (The locations of the open-loop zeros in this book will be indicated by small circles.) Note that the starting points of the root loci (the points corresponding to $K = 0$) are open-loop poles. The number of individual root loci for this system is three, which is the same as the number of open-loop poles.

To determine the root loci on the real axis, we select a test point, s. If the test point is on the positive real axis, then

$$\angle s = \angle s + 1 = \angle s + 2 = 0°$$

This shows that the angle condition cannot be satisfied. Hence, there is no root locus on the positive real axis. Next, select a test point on the negative real axis between 0 and -1. Then

$$\angle s = 180°, \qquad \angle s + 1 = \angle s + 2 = 0°$$

Thus

$$-\angle s - \angle s + 1 - \angle s + 2 = -180°$$

and the angle condition is satisfied. Therefore, the portion of the negative real axis between 0 and -1 forms a portion of the root locus. If a test point is selected between -1 and -2, then

$$\angle s = \angle s + 1 = 180°, \qquad \angle s + 2 = 0°$$

and

$$-\angle s - \angle s + 1 - \angle s + 2 = -360°$$

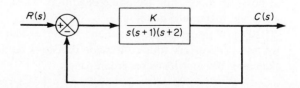

Figure 5–5
Control system.

It can be seen that the angle condition is not satisfied. Therefore, the negative real axis from -1 to -2 is not a part of the root locus. Similarly, if a test point is located on the negative real axis from -2 to $-\infty$, the angle condition is satisfied. Thus, root loci exist on the negative real axis between 0 and -1 and between -2 and $-\infty$.

2. Determine the asymptotes of the root loci. The asymptotes of the root loci as s approaches infinity can be determined as follows: If a test point s is selected very far from the origin, then

$$\lim_{s \to \infty} G(s) = \lim_{s \to \infty} \frac{K}{s(s+1)(s+2)} = \lim_{s \to \infty} \frac{K}{s^3}$$

and the angle condition becomes

$$-3\underline{/s} = \pm 180°(2k+1) \qquad (k = 0, 1, 2, \ldots)$$

or

$$\text{Angles of asymptotes} = \frac{\pm 180°(2k+1)}{3} \qquad (k = 0, 1, 2, \ldots)$$

Since the angle repeats itself as k is varied, the distinct angles for the asymptotes are determined as $60°$, $-60°$, and $180°$. Thus, there are three asymptotes. The one having the angle of $180°$ is the negative real axis.

Before we can draw these asymptotes in the complex plane, we must find where they intersect the real axis. Since the characteristic equation of this system is

$$\frac{K}{s(s+1)(s+2)} = -1$$

or

$$s^3 + 3s^2 + 2s = -K \tag{5-7}$$

if s is assumed to be very large, then we can approximate this characteristic equation by the following equation:

$$(s+1)^3 = 0$$

The abscissa of the intersection of the asymptotes and the real axis can be obtained by letting $s = \sigma_a$ and solving for σ_a. Thus

$$\sigma_a = -1$$

and the point of origin of the asymptotes is $(-1, 0)$. The asymptotes are almost part of the root loci in regions very far from the origin.

3. Determine the breakaway point. To plot root loci accurately, we must find the breakaway point, where the root-locus branches originating from the poles at 0 and -1 break away (as K is increased) from the real axis and move into the complex plane. The breakaway point corresponds to a point in the s plane where multiple roots of the characteristic equation occur.

A simple method for finding the breakaway point is available. We shall present this method in the following: Let us write the characteristic equation as

$$f(s) = B(s) + KA(s) = 0 \tag{5-8}$$

where $A(s)$ and $B(s)$ do not contain K. Note that $f(s) = 0$ has multiple roots at points where

$$\frac{df(s)}{ds} = 0$$

This can be seen as follows: Suppose that $f(s)$ has a multiple roots of order r. Then $f(s)$ may be written as

$$f(s) = (s - s_1)^r(s - s_2) \cdots (s - s_n)$$

If we differentiate this equation with respect to s and set $s = s_1$, then we get

$$\left.\frac{df(s)}{ds}\right|_{s=s_1} = 0 \qquad (5-9)$$

This means that multiple roots of $f(s)$ will satisfy Equation (5–9). From Equation (5–8), we obtain

$$\frac{df(s)}{ds} = B'(s) + KA'(s) = 0 \qquad (5-10)$$

where

$$A'(s) = \frac{dA(s)}{ds}, \qquad B'(s) = \frac{dB(s)}{ds}$$

The particular value of K that will yield multiple roots of the characteristic equation is obtained from Equation (5–10) as

$$K = -\frac{B'(s)}{A'(s)}$$

If we substitute this value of K into Equation (5–8), we get

$$f(s) = B(s) - \frac{B'(s)}{A'(s)} A(s) = 0$$

or

$$B(s)A'(s) - B'(s)A(s) = 0 \qquad (5-11)$$

If Equation (5–11) is solved for s, the points where multiple roots occur can be obtained. On the other hand, from Equation (5–8) we obtain

$$K = -\frac{B(s)}{A(s)}$$

and

$$\frac{dK}{ds} = -\frac{B'(s)A(s) - B(s)A'(s)}{A^2(s)}$$

If dK/ds is set equal to zero, we get the same equation as Equation (5–11). Therefore, the breakaway points can be simply determined from the roots of

$$\frac{dK}{ds} = 0$$

It should be noted that not all the solutions of Equation (5–11) or of $dK/ds = 0$ correspond to actual breakaway points. We shall demonstrate this using the present example where

$$f(s) = s^3 + 3s^2 + 2s + K = 0 \qquad (5-12)$$

Referring to the plot of $f(\sigma)$ versus σ shown in Figure 5–6, notice that points P and Q correspond to $df(\sigma)/d\sigma = 0$. Clearly, point Q corresponds to $K < 0$, which means that point Q cannot be a breakaway point of the system under consideration, since in the present system the gain constant K

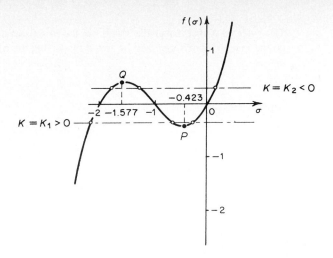

Figure 5–6
Plot of $f(\sigma)$ versus σ.

must be nonnegative. [If a point at which $df(s)/ds = 0$ is on a root locus, it is an actual breakaway or break-in point. Stated differently, if at a point at which $df(s)/ds = 0$ the value of K takes a real positive value, then that point is an actual breakaway or break-in point.]

For the present example, we have from Equation (5–7).

$$K = -(s^3 + 3s^2 + 2s)$$

By setting $dK/ds = 0$, we obtain

$$\frac{dK}{ds} = -(3s^2 + 6s + 2) = 0$$

or

$$s = -0.4226, \qquad s = -1.5774$$

Since the breakaway point must lie on a root locus between 0 and -1, it is clear that $s = -0.4226$ corresponds to the actual breakaway point. Point $s = -1.5774$ is not on the root locus. Hence, this point is not an actual breakaway or break-in point. In fact, evaluation of the values of K corresponding to $s = -0.4226$ and $s = -1.5774$ yields

$$K = 0.3849 \qquad \text{for } s = -0.4226$$

$$K = -0.3849 \qquad \text{for } s = -1.5774$$

4. Determine the points where the root loci cross the imaginary axis. These points can be found by use of Routh's stability criterion as follows: Since the characteristic equation for the present system is

$$s^3 + 3s^2 + 2s + K = 0$$

the Routh array becomes

$$
\begin{array}{c c c}
s^3 & 1 & 2 \\
s^2 & 3 & K \\
s^1 & \dfrac{6 - K}{3} & \\
s^0 & K &
\end{array}
$$

The value of K that makes the s^1 term in the first column equal zero is $K = 6$. The crossing points on the imaginary axis can then be found by solving the auxiliary equation obtained from the s^2 row; that is,

$$3s^2 + K = 3s^2 + 6 = 0$$

which yields

$$s = \pm j\sqrt{2}$$

The frequencies at the crossing points on the imaginary axis are thus $\omega = \pm\sqrt{2}$. The gain value corresponding to the crossing points is $K = 6$.

An alternate approach is to let $s = j\omega$ in the characteristic equation, equate both the real part and the imaginary part to zero, and then solve for ω and K. For the present system, the characteristic equation, with $s = j\omega$, is

$$(j\omega)^3 + 3(j\omega)^2 + 2(j\omega) + K = 0$$

or

$$(K - 3\omega^2) + j(2\omega - \omega^3) = 0$$

Equating both the real and imaginary parts of this last equation to zero, we obtain

$$K - 3\omega^2 = 0, \qquad 2\omega - \omega^3 = 0$$

from which

$$\omega = \pm\sqrt{2}, \qquad K = 6 \qquad \text{or} \qquad \omega = 0, \qquad K = 0$$

Thus, root loci cross the imaginary axis at $\omega = \pm\sqrt{2}$, and the value of K at the crossing points is 6. Also, a root-locus branch on the real axis touches the imaginary axis at $\omega = 0$.

5. Choose a test point in the broad neighborhood of the $j\omega$ axis and the origin, as shown in Figure 5–7, and apply the angle condition. If a test point is on the root loci, then the sum of the three angles, $\theta_1 + \theta_2 + \theta_3$, must be 180°. If the test point does not satisfy the angle condition, select another test point until it satisfies the condition. (The sum of the angles at the test point will indicate which direction the test point should be moved.) Continue this process and locate a sufficient number of points satisfying the angle condition.

6. Based on the information obtained in the foregoing steps, draw the root loci as shown in Figure 5–8.

7. Determine a pair of dominant complex-conjugate closed-loop poles such that the damping ratio ζ is 0.5. Closed-loop poles with $\zeta = 0.5$ lie on lines passing through the origin and making the angles

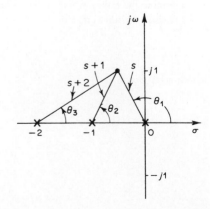

Figure 5–7
Construction of root locus.

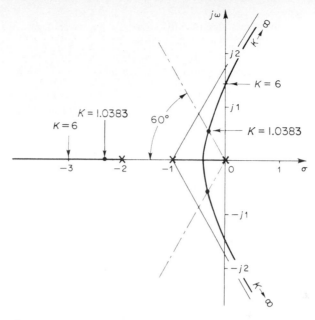

Figure 5–8
Root-locus diagram.

$\pm \cos^{-1} \zeta = \pm \cos^{-1} 0.5 = \pm 60°$ with the negative real axis. From Figure 5–8, such closed-loop poles having $\zeta = 0.5$ are obtained as follows:

$$s_1 = -0.3337 + j0.5780, \qquad s_2 = -0.3337 - j0.5780$$

The value of K that yields such poles is found from the magnitude condition as follows:

$$K = |s(s + 1)(s + 2)|_{s = -0.3337 + j0.5780}$$
$$= 1.0383$$

Using this value of K, the third pole is found at $s = -2.3326$.

Note that, from step 4, it can be seen that for $K = 6$ the dominant closed-loop poles lie on the imaginary axis at $s = \pm j\sqrt{2}$. With this value of K, the system will exhibit sustained oscillations. For $K > 6$ the dominant closed-loop poles lie in the right-half s plane, resulting in an unstable system.

Finally, note that, if necessary, the root loci can be easily graduated in terms of K by use of the magnitude condition. We simply pick out a point on a root locus, measure the magnitudes of the three complex quantities s, $s + 1$, and $s + 2$, multiply these three magnitudes, and the product is equal to the gain value K at that point, or

$$|s| \cdot |s + 1| \cdot |s + 2| = K$$

EXAMPLE 5–2

In this example, we shall sketch the root-locus plot of a system with complex-conjugate open-loop poles. Consider the system shown in Figure 5–9. For this system,

$$G(s) = \frac{K(s + 2)}{s^2 + 2s + 3}, \qquad H(s) = 1$$

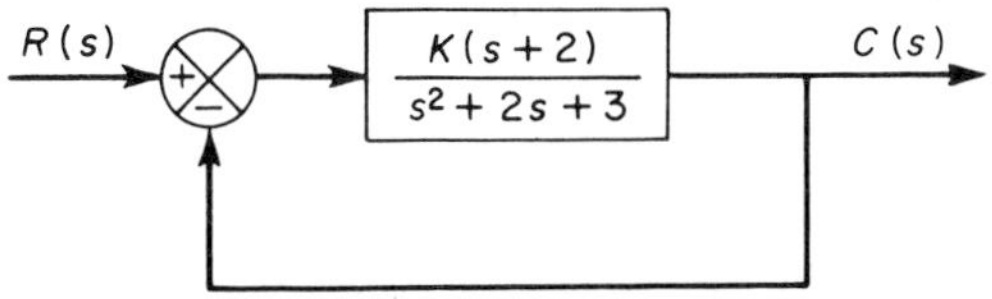

$R(s)$ $+$ $-$ $\dfrac{K(s+2)}{s^2 + 2s + 3}$ $C(s)$

Figure 5–9
Control system.

It is seen that $G(s)$ has a pair of complex-conjugate poles at

$$s = -1 + j\sqrt{2}, \qquad s = -1 - j\sqrt{2}$$

A typical procedure for sketching the root-locus plot is as follows:

1. Determine the root loci on the real axis. For any test point s on the real axis, the sum of the angular contributions of the complex-conjugate poles is 360°, as shown in Figure 5–10. Thus, the net effect of the complex-conjugate poles is zero on the real axis. The location of the root locus on the real axis is determined from the open-loop zero on the negative real axis. A simple test reveals that a section of the negative real axis, that between -2 and $-\infty$, is a part of the root locus. It is noted that since this locus lies between two zeros (at $s = -2$ and $s = -\infty$), it is actually a part of two root loci, each of which starts from one of the two complex-conjugate poles. In other words, two root loci break in the part of the negative real axis between -2 and $-\infty$.

Since there are two open-loop poles and one zero, there is one asymptote, which coincides with the negative real axis.

2. Determine the angle of departure from the complex-conjugate open-loop poles. The presence of a pair of complex-conjugate open-loop poles requires determination of the angle of departure from these poles. Knowledge of this angle is important since the root locus near a complex pole yields information as to whether the locus originating from the complex pole migrates toward the real axis or extends toward the asymptote.

Referring to Figure 5–11, if we choose a test point and move it in the very vicinity of the complex open-loop pole at $s = -p_1$, we find that the sum of the angular contributions from the pole at $s = -p_2$ and zero at $s = -z_1$ to the test point can be considered remaining the same. If the test point is to be on the root locus, then the sum of ϕ_1', $-\theta_1$, and $-\theta_2'$ must be $\pm 180°(2k + 1)$, where $k = 0$, $1, 2, \ldots$. Thus, in this example,

$$\phi_1' = (\theta_1 + \theta_2') = \pm 180°(2k + 1)$$

or

$$\theta_1 = 180° - \theta_2' + \phi_1' = 180° - \theta_2 + \phi_1$$

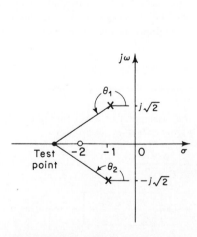

Figure 5–10 Determination of the root locus on the real axis.

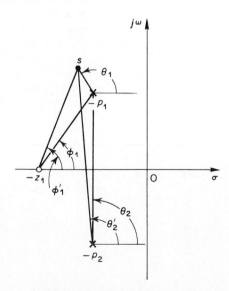

Figure 5–11 Determination of the angle of departure.

The angle of departure is then

$$\theta_1 = 180° - \theta_2 + \phi_1 = 180° - 90° + 55° = 145°$$

Since the root locus is symmetric about the real axis, the angle of departure from the pole at $s = -p_2$ is $-145°$.

3. Determine the break-in point. A break-in point exists where a pair of root-locus branches coalesce as K is increased. For this problem, the break-in point can be found as follows: Since

$$K = -\frac{s^2 + 2s + 3}{s + 2}$$

we have

$$\frac{dK}{ds} = -\frac{(2s + 2)(s + 2) - (s^2 + 2s + 3)}{(s + 2)^2} = 0$$

which gives

$$s^2 + 4s + 1 = 0$$

or

$$s = -3.7320 \quad \text{or} \quad s = -0.2680$$

Notice that point $s = -3.7320$ is on the root locus. Hence this point is an actual break-in point. (Note that at point $s = -3.7320$ the corresponding gain value is $K = 5.4641$.) Since point $s = -0.2680$ is not on the root locus, it cannot be a break-in point. (For point $s = -0.2680$, the corresponding gain value is $K = -1.4641$.)

4. Based on the information obtained in the foregoing steps, sketch a root-locus plot. To determine accurate root loci, several points must be found by trial and error between the break-in point and the complex open-loop poles. (To facilitate sketching the root-locus plot, we should find the direction in which the test point should be moved by mentally summing up the changes on the angles of the poles and zeros.) Figure 5–12 shows a complete root-locus diagram for the system considered.

The value of the gain K at any point on the root locus can be found by applying the magnitude condition. For example, the value of K at which the complex-conjugate closed-loop poles have the damping ratio $\zeta = 0.7$ can be found by locating the roots, as shown in Figure 5–12, and computing the value of K as follows:

$$K = \left| \frac{(s + 1 - j\sqrt{2})(s + 1 + j\sqrt{2})}{s + 2} \right|_{s = -1.6659 + j1.6995} = 1.3318$$

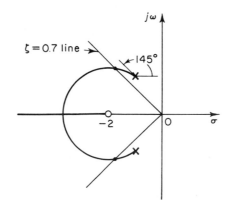

Figure 5–12
Root-locus diagram.

It is noted that in this system the root locus in the complex plane is a part of a circle. Such a circular root locus will not occur in most systems. Circular root loci may occur in systems that involve two poles and one zero, two poles and two zeros, or one pole and two zeros. Even in such systems, whether or not circular root loci occur depends on the locations of poles and zeros involved.

To show the occurrence of a circular root locus in the present system, we need to derive the equation for the root locus. For the present system, the angle condition is

$$\underline{/s + 2} - \underline{/s + 1 - j\sqrt{2}} - \underline{/s + 1 + j\sqrt{2}} = \pm 180°(2k + 1)$$

If $s = \sigma + j\omega$ is substituted into this last equation, we obtain

$$\underline{/\sigma + 2 + j\omega} - \underline{/\sigma + 1 + j\omega - j\sqrt{2}} - \underline{/\sigma + 1 + j\omega + j\sqrt{2}} = \pm 180°(2k + 1)$$

which can be written as

$$\tan^{-1}\left(\frac{\omega}{\sigma + 2}\right) - \tan^{-1}\left(\frac{\omega - \sqrt{2}}{\sigma + 1}\right) - \tan^{-1}\left(\frac{\omega + \sqrt{2}}{\sigma + 1}\right) = \pm 180°(2k + 1)$$

or

$$\tan^{-1}\left(\frac{\omega - \sqrt{2}}{\sigma + 1}\right) + \tan^{-1}\left(\frac{\omega + \sqrt{2}}{\sigma + 1}\right) = \tan^{-1}\left(\frac{\omega}{\sigma + 2}\right) \pm 180°(2k + 1)$$

Taking tangents of both sides of this last equation using the relationship

$$\tan(x \pm y) = \frac{\tan x \pm \tan y}{1 \mp \tan x \tan y} \qquad (5\text{–}13)$$

we obtain

$$\tan\left[\tan^{-1}\left(\frac{\omega - \sqrt{2}}{\sigma + 1}\right) + \tan^{-1}\left(\frac{\omega + \sqrt{2}}{\sigma + 1}\right)\right] = \tan\left[\tan^{-1}\left(\frac{\omega}{\sigma + 2}\right) \pm 180°(2k + 1)\right]$$

or

$$\frac{\dfrac{\omega - \sqrt{2}}{\sigma + 1} + \dfrac{\omega + \sqrt{2}}{\sigma + 1}}{1 - \left(\dfrac{\omega - \sqrt{2}}{\sigma + 1}\right)\left(\dfrac{\omega + \sqrt{2}}{\sigma + 1}\right)} = \frac{\dfrac{\omega}{\sigma + 2} \pm 0}{1 \mp \dfrac{\omega}{\sigma + 2} \times 0}$$

which can be simplified to

$$\frac{2\omega(\sigma + 1)}{(\sigma + 1)^2 - (\omega^2 - 2)} = \frac{\omega}{\sigma + 2}$$

or

$$\omega[(\sigma + 2)^2 + \omega^2 - 3] = 0$$

This last equation is equivalent to

$$\omega = 0 \qquad \text{or} \qquad (\sigma + 2)^2 + \omega^2 = (\sqrt{3})^2$$

These two equations are the equations for the root loci for the present system. Notice that the first equation, $\omega = 0$, is the equation for the real axis. The real axis from $s = -2$ to $s = -\infty$ corresponds to a root locus for $K \geq 0$. The remaining part of the real axis corresponds to a root locus when K is negative. (In the present system, K is nonnegative.) The second equation for the root locus is an equation of circle with center at $\sigma = -2$, $\omega = 0$ and the radius equal to $\sqrt{3}$. That part of the circle to the left of the complex-conjugate poles corresponds to a root locus for $K \geq 0$. The remaining part of the circle corresponds to a root locus when K is negative.

It is important to note that easily interpretable equations for the root locus can be derived for simple systems only. For complicated systems having many poles and zeros, any attempt to derive equations for the root loci is discouraged. Such derived equations are very complicated and their configuration in the complex plane is difficult to visualize.

Constructing root loci when a variable parameter does not appear as a multiplying factor. In some cases the variable parameter K may not appear as a multiplying factor of $G(s)H(s)$. In such cases it may be possible to rewrite the characteristic equation such that the variable parameter K appears as a multiplying factor of $G(s)H(s)$. Example 5–3 illustrates how to proceed in such a case.

EXAMPLE 5–3   Consider the system shown in Figure 5–13. Draw a root-locus diagram. Then determine the value of k such that the damping ratio of the dominant closed-loop poles is 0.4

Here the system involves velocity feedback. The open-loop transfer function is

$$G(s)H(s) = \frac{20(1 + ks)}{s(s + 1)(s + 4)}$$

Notice that the adjustable variable k does not appear as a multiplying factor. The characteristic equation for the system is

$$1 + \frac{20(1 + ks)}{s(s + 1)(s + 4)} = 0 \tag{5–14}$$

or

$$s^3 + 5s^2 + 4s + 20 + 20ks = 0 \tag{5–15}$$

By defining

$$20k = K$$

and dividing both sides of Equation (5–15) by the sum of the terms that do not contain k, we get

$$1 + \frac{Ks}{s^3 + 5s^2 + 4s + 20} = 0$$

or

$$1 + \frac{Ks}{(s + j2)(s - j2)(s + 5)} = 0 \tag{5–16}$$

Equation (5–16) is now of the form of Equation (5–5).

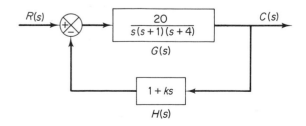

Figure 5–13
Control system.

We shall now sketch the root loci of the system given by Equation (5–16). Notice that the open-loop poles are located at $s = j2$, $s = -j2$, $s = -5$, and the open-loop zero is located at $s = 0$. The root locus exists on the real axis between 0 and -5. Since

$$\lim_{s \to \infty} \frac{Ks}{(s + j2)(s - j2)(s + 5)} = \lim_{s \to \infty} \frac{K}{s^2}$$

we have

$$\text{Angle of asymptote} = \frac{\pm 180°(2k + 1)}{2} = \pm 90°$$

The intersection of the asymptotes with the real axis can be found from

$$\lim_{s \to \infty} \frac{Ks}{s^3 + 5s^2 + 4s + 20} = \lim_{s \to \infty} \frac{K}{s^2 + 5s + \cdots} = \lim_{s \to \infty} \frac{K}{(s + 2.5)^2}$$

as

$$\sigma_a = -2.5$$

The angle of departure (angle θ) from the pole at $s = j2$ is obtained as follows:

$$\theta = 180° - 90° - 21.8° + 90° = 158.2°$$

The angle of departure from the pole at $s = j2$ is 158.2°. Figure 5–14 shows the root-locus diagram for the system.

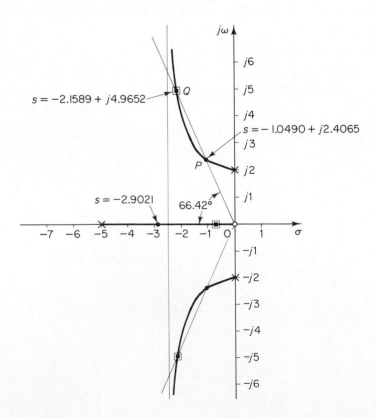

Figure 5–14
Root-locus diagram for the system shown in Fig. 5–13.

Chapter 5 / Root-Locus Analysis

Note that the closed-loop poles with $\zeta = 0.4$ must lie on straight lines passing through the origin and making the angles $\pm 66.42°$ with the negative real axis. In the present case, there are two intersections of the root-locus branch in the upper half s plane and the straight line of angle $66.42°$. Thus, two values of K will give the damping ration ζ of the closed-loop poles equal to 0.4. At point P the value of K is

$$K = \left| \frac{(s + j2)(s - j2)(s + 5)}{s} \right|_{s = -1.0490 + j2.4065} = 8.9801$$

Hence

$$k = \frac{K}{20} = 0.4490 \qquad \text{at point } P$$

At point Q, the value of K is

$$K = \left| \frac{(s + j2)(s - j2)(s + 5)}{s} \right|_{s = -2.1589 + j4.9652} = 28.260$$

Hence

$$k = \frac{K}{20} = 1.4130 \qquad \text{at point } Q$$

Thus, we have two solutions for this problem. For $k = 0.4490$, the three closed-loop poles are located at

$$s = -1.0490 + j2.4065, \qquad s = -1.0490 - j2.4065, \qquad s = -2.9021$$

For $k = 1.4130$, the three closed-loop poles are located at

$$s = -2.1589 + j4.9652, \qquad s = -2.1589 - j4.9652, \qquad s = -0.6823$$

It is important to point out that the zero at the origin is the open-loop zero, but not the closed-loop zero. This is evident, because the original system shown in Figure 5–13 does not have a closed-loop zero, since

$$\frac{C(s)}{R(s)} = \frac{20}{s(s + 1)(s + 4) + 20(1 + ks)}$$

The open-loop zero at $s = 0$ was introduced in the process of modifying the characteristic equation such that the adjustable variable $K = 20k$ was to appear as a multiplying factor.

We have obtained two different values of k to satisfy the requirement that the damping ratio of the dominant closed-loop poles be equal to 0.4. The closed-loop transfer function with $k = 0.4490$ is given by

$$\frac{C(s)}{R(s)} = \frac{20}{s^3 + 5s^2 + 12.98s + 20}$$

$$= \frac{20}{(s + 1.0490 + j2.4065)(s + 1.0490 - j2.4065)(s + 2.9021)}$$

The closed loop transfer function with $k = 1.4130$ is given by

$$\frac{C(s)}{R(s)} = \frac{20}{s^3 + 5s^2 + 32.26s + 20}$$

$$= \frac{20}{(s + 2.1589 + j4.9652)(s + 2.1589 - j4.9652)(s + 0.6823)}$$

Notice that the system with $k = 0.4490$ has a pair of dominant complex-conjugate closed-loop poles, while in the system with $k = 1.4130$ the real closed-loop pole at $s = -0.6823$ is dominant, and the complex-conjugate closed-loop poles are not dominant. In this case, the response characteristic is primarily determined by the real closed-loop pole.

For the unit-step input, the response of the system with $k = 0.4490$ becomes

$$C(s) = \frac{1}{s} - \frac{0.747}{s + 2.9021} - \frac{0.2530(s + 1.0490)}{(s + 1.0490)^2 + (2.4065)^2}$$

$$- \frac{1.0113(2.4065)}{(s + 1.0490)^2 + (2.4065)^2}$$

or

$$c(t) = 1 - 0.747e^{-2.9021t} - 0.2530e^{-1.0490t}\cos(2.4065t)$$
$$- 1.0113e^{-1.0490t}\sin(2.4065t)$$

The second term in the right side of this equation damps out quickly and the response becomes oscillatory. The unit-step response curve is shown in Figure 5–15.

For the system with $k = 1.4130$, we have for the unit-step input

$$C(s) = \frac{1}{s} - \frac{1.0924}{s + 0.6823} + \frac{0.0924(s + 2.1589)}{(s + 2.1589)^2 + (4.9652)^2}$$

$$- \frac{0.1102(4.9652)}{(s + 2.1589)^2 + (4.9652)^2}$$

or

$$c(t) = 1 - 1.0924e^{-0.6823t} + 0.0924e^{-2.1589t}\cos(4.9652t)$$
$$- 0.1102e^{-2.1589t}\sin(4.9652t)$$

Clearly, the oscillatory terms damp out much faster than the purely exponential term. The response is dominated by this exponential term. The unit-step response curve is also shown in Figure 5–15.

The system with $k = 0.4490$ (which exhibits a faster response with relatively small overshoot) has a much better response characteristic than the system with $k = 1.4130$ (which exhibits a slow overdamped response). Therefore, we should choose $k = 0.4490$ for the present system.

Figure 5–15
Unit-step response curves for the system shown in Fig. 5–13 when the damping ratio ζ of the dominant closed-loop poles is set equal to 0.4. (Two possible values of k give the damping ratio ζ equal to 0.4.)

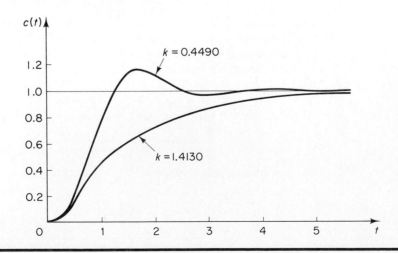

Summary. From the preceding examples, it may be clear that it is possible to sketch a reasonably accurate root-locus diagram for a given system by following simple rules. (The reader is suggested to study various root-locus diagrams shown in the solved problems at the end of the chapter.) At preliminary design stages, we may not need the precise locations of the closed-loop poles. Often their approximate locations are all that is needed to make an estimate of system performance. Thus, it is important that the designer have the capability of quickly sketching the root loci for a given system. In this regard the designer must be familiar with general rules for constructing the root loci. Such rules are summarized in the next section.

5-4 SUMMARY OF GENERAL RULES FOR CONSTRUCTING ROOT LOCI

For a complicated system with many open-loop poles and zeros, constructing a root-locus plot may seem complicated, but actually it is not difficult if the rules for constructing the root loci are applied. By locating particular points and asymptotes, and by computing angles of departure from complex poles and angles of arrival at complex zeros, we can construct the general form of the root loci without difficulty. As a matter of fact, the full advantage of the root-locus method can be realized in the case of higher-order systems for which other methods of finding closed-loop poles are extremely laborious.

Some of the rules for constructing root loci were given in Section 5–3. The purpose of this section is to summarize the general rules for constructing root loci of the system shown in Figure 5–16. While the root-locus method is essentially based on a trial-and-error technique, the number of trials required can be greatly reduced if we use these rules.

General rules for constructing root loci. We shall now summarize the general rules and procedure for constructing the root loci of the system shown in Figure 5–16.

1. First obtain the characteristic equation

$$1 + G(s)H(s) = 0$$

and rearrange this equation so that the parameter of interest appears as the multiplying factor in the form

$$1 + \frac{K(s + z_1)(s + z_2) \cdots (s + z_m)}{(s + p_1)(s + p_2) \cdots (s + p_n)} = 0$$

In the present discussions, we assume that the parameter of interest is the gain K, where $K > 0$. (If $K < 0$, which corresponds to the positive feedback case, the angle condition must be modified. See Problem A–5–4.) Note, however, that the method is still applicable to systems with parameters of interest other than gain.

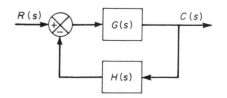

Figure 5–16
Control system.

From the factored form of the open-loop transfer function, locate the open-loop poles and zeros in the s plane. [Note that the open-loop zeros are the zeros of $G(s)H(s)$, while the closed-loop zeros consist of the zeros of $G(s)$ and the poles of $H(s)$.]

Note that the root loci are symmetrical about the real axis of the s plane, because the complex poles and complex zeros occur only in conjugate pairs.

2. Find the starting points and terminating points of the root loci and find also the number of separate root loci. The points on the root loci corresponding to $K = 0$ are open-loop poles. This can be seen from the magnitude condition by letting K approach zero, or

$$\lim_{K \to 0} \left| \frac{(s + z_1)(s + z_2) \cdots (s + z_m)}{(s + p_1)(s + p_2) \cdots (s + p_n)} \right| = \lim_{K \to 0} \frac{1}{K} = \infty$$

This last equation implies that as K is decreased the value of s must approach one of the open-loop poles. Each root locus thus originates at a pole of the open-loop transfer function $G(s)H(s)$. As K is increased to infinity, each root-locus approaches either a zero of the open-loop transfer function or infinity in the complex plane. This can be seen as follows: If we let K approach infinity in the magnitude condition, then

$$\lim_{K \to 0} \left| \frac{(s + z_1)(s + z_2) \cdots (s + z_m)}{(s + p_1)(s + p_2) \cdots (s + p_n)} \right| = \lim_{K \to 0} \frac{1}{K} = 0$$

Hence the value of s must approach one of the open-loop zeros or an open-loop zero at infinity. [If the zeros at infinity are included in the count, $G(s)H(s)$ has the same number of zeros as poles.]

A root-locus plot will have just as many branches as there are roots of the characteristic equation. Since the number of open-loop poles generally exceeds that of zeros, the number of branches equals that of poles. If the number of closed-loop poles is the same as the number of open-loop poles, then the number of individual root-locus branches terminating at finite open-loop zeros is equal to the number m of the open-loop zeros. The remaining $n - m$ branches terminate at infinity ($n - m$ implicit zeros at infinity) along asymptotes.

It is important to note, however, that if we consider a purely mathematical problem, then the number of the closed-loop poles can be made the same as the number of open-loop zeros rather than the number of open-loop poles. In such a case, the number of root-locus branches is equal to the number of open-loop zeros. For example, consider the following polynomial equation:

$$s^2 + s + 1 = 0$$

This equation may be rewritten

$$1 + \frac{s^2}{s + 1} = 0$$

Then the transfer function $s^2/(s + 1)$ may be considered the open-loop transfer function. It has two zeros and one pole. Hence the number of finite zeros is greater than that of finite poles. The number of root-locus branches is equal to that of open-loop zeros.

If we include poles and zeros at infinity, the number of open-loop poles is equal to that of open-loop zeros. Hence we can always state that the root loci start at the poles of $G(s)H(s)$ and end at the zeros of $G(s)H(s)$, as K increases from zero to infinity, where the poles and zeros include both those in the finite s plane and those at infinity.

3. Determine the root loci on the real axis. Root loci on the real axis are determined by open-loop poles and zeros lying on it. The complex-conjugate poles and zeros of the open-loop transfer function have no effect on the location of the root loci on the real axis because the angle contribution of a pair of complex-conjugate poles or zeros is 360° on the real axis. Each portion of the root locus on the real axis extends over a range from a pole or zero to another pole or zero. In constructing the root loci on the real axis, choose a test point on it. If the total number of real poles and real zeros to the right of this test point is odd, then this point lies on a root locus. The root locus and its complement form alternate segments along the real axis.

4. Determine the asymptotes of the root loci. If the test point s is located far from the origin, then the angle of each complex quantity may be considered the same. One open-loop zero and one open-loop pole then cancel the effects of the other. Therefore, the root loci for very large values of s must be asymptotic to straight lines whose angles (slopes) are given by

$$\text{Angles of asymptotes} = \frac{\pm 180°(2k + 1)}{n - m} \qquad (k = 0, 1, 2, \ldots)$$

where n = number of finite poles of $G(s)H(s)$
m = number of finite zeros of $G(s)H(s)$

Here, $k = 0$ corresponds to the asymptotes with the smallest angle with the real axis. Although k assumes an infinite number of values, as k is increased, the angle repeats itself, and the number of distinct asymptotes is $n - m$.

All the asymptotes intersect on the real axis. The point at which they do so is obtained as follows: If both the numerator and denominator of the open-loop transfer function are expanded, the result is

$$G(s)H(s) = \frac{K[s^m + (z_1 + z_2 + \cdots + z_m)s^{m-1} + \cdots + z_1 z_2 \cdots z_m]}{s^n + (p_1 + p_2 + \cdots + p_n)s^{n-1} + \cdots + p_1 p_2 \cdots p_n}$$

If a test point is located very far from the origin, then $G(s)H(s)$ may be written

$$G(s)H(s) = \frac{K}{s^{n-m} + [(p_1 + p_2 + \cdots + p_n) - (z_1 + z_2 + \cdots + z_m)]s^{n-m-1} + \cdots}$$

Since the characteristic equation is

$$G(s)H(s) = -1$$

it may be written

$$s^{n-m} + [(p_1 + p_2 + \cdots + p_n) - (z_1 + z_2 + \cdots + z_m)]s^{n-m-1} + \cdots = -K \qquad (5\text{--}17)$$

For a large value of s, Equation (5–17) may be approximated by

$$\left[s + \frac{(p_1 + p_2 + \cdots + p_n) - (z_1 + z_2 + \cdots + z_m)}{n - m} \right]^{n-m} = 0$$

If the abscissa of the intersection of the asymptotes and the real axis is denoted by $s = \sigma_a$, then

$$\sigma_a = -\frac{(p_1 + p_2 + \cdots + p_n) - (z_1 + z_2 + \cdots + z_m)}{n - m} \qquad (5\text{--}18)$$

or

$$\sigma_a = \frac{(\text{sum of poles}) - (\text{sum of zeros})}{n - m} \qquad (5\text{–}19)$$

Because all the complex poles and zeros occur in conjugate pairs, σ_a is always a real quantity. Once the intersection of the asymptotes and the real axis is found, the asymptotes can be readily drawn in the complex plane.

It is important to note that the asymptotes show the behavior of the root loci for $|s| \gg 1$. A root locus branch may lie on one side of the corresponding asymptote or may cross the corresponding asymptote from one side to the other side.

5. Find the breakaway and break-in points. Because of the conjugate symmetry of the root loci, the breakaway points and break-in points either lie on the real axis or occur in complex-conjugate pairs.

If a root locus lies between two adjacent open-loop poles on the real axis, then there exists at least one breakaway point between the two poles. Similarly, if the root locus lies between two adjacent zeros (one zero may be located at $-\infty$) on the real axis, then there always exists at least one break-in point between the two zeros. If the root locus lies between an open-loop pole and a zero (finite or infinite) on the real axis, then there may exist no breakaway or break-in points or there may exist both breakaway and break-in points.

Suppose that the characteristic equation is given by

$$B(s) + KA(s) = 0$$

The breakaway points and break-in points correspond to multiple roots of the characteristic equation. Hence, the breakaway and break-in points can be determined from the roots of

$$\frac{dK}{ds} = -\frac{B'(s)A(s) - B(s)A'(s)}{A^2(s)} = 0 \qquad (5\text{–}20)$$

where the prime indicates differentiation with respect to s. It is important to note that the breakaway points and break-in points must be the roots of Equation (5–20), but not all roots of Equation (5–20) are breakaway or break-in points. If a real root of Equation (5–20) lies on the root locus portion of the real axis, then it is an actual breakaway or break-in point. If a real root of Equation (5–20) is not on the root locus portion of the real axis, then this root corresponds to neither a breakaway point nor a break-in point. If two roots $s = s_1$ and $s = -s_1$ of Equation (5–20) are a complex-conjugate pair and if it is not certain whether they are on root loci, then it is necessary to check the corresponding K value. If the value of K corresponding to a root $s = s_1$ of $dK/ds = 0$ is positive, point $s = s_1$ is an actual breakaway or break-in point. (Since K is assumed to be nonnegative, if the value of K thus obtained is negative, then point $s = s_1$ is neither a breakaway nor break-in point.)

6. Determine the angles of departure (or angles of arrival) of root loci from the complex poles (or at complex zeros). To sketch the root loci with reasonable accuracy, we must find the directions of the root loci near the complex poles and zeros. If a test point is chosen and moved in the very vicinity of a complex pole (or complex zero), the sum of the angular contributions from all other poles and zeros can be considered remaining the same. Therefore, the angle of departure (or angle of arrival) of the root locus from a complex pole (or at a complex zero) can be found by substracting from 180° the sum of all the angles of vectors from all other poles and zeros to the complex pole (or complex zero) in question, with appropriate signs included.

Angle of departure from a complex pole = 180°
- (sum of the angles of vectors to a complex pole in question from other poles)
+ (sum of the angles of vectors to a complex pole in question from zeros)

Angle of arrival at a complex zero = 180°
- (sum of the angles of vectors to a complex zero in question from other zeros)
+ (sum of the angles of vectors to a complex zero in question from poles)

The angle of departure is shown in Figure 5–17.

7. Find the points where the root loci cross the imaginary axis. The points where the root loci intersect the $j\omega$ axis can be found easily by (1) use of Routh's stability criterion, (2) a trial-and-error approach, or (3) letting $s = j\omega$ in the characteristic equation, equating both the real part and the imaginary part to zero, and solving for ω and K. The values of ω thus found give the frequencies at which root loci cross the imaginary axis. The K value corresponding to each crossing frequency gives the gain at the crossing point.

8. Any point on the root loci is a possible closed-loop pole. A particular point will be a closed-loop pole when the value of K satisfies the magnitude condition. Thus, the magnitude condition enables us to determine the value of the gain K at any specific root location on the locus. (If necessary, the root loci may be graduated in terms of K. The root loci are continuous with K.)

The value of K corresponding to any point s on a root locus can be obtained using the magnitude condition, or

$$K = \frac{\text{product of lengths between point } s \text{ to poles}}{\text{product of lengths between point } s \text{ to zeros}}$$

This value can be evaluated either graphically or analytically.

If the gain K of the open-loop transfer function is given in the problem, then by applying the magnitude condition we can find the correct locations of the closed-loop poles for a given K on each branch of the root loci by a trial-and-error approach.

Once the dominant closed-loop poles (or the closed-loop poles closest to the $j\omega$ axis) are found by the root-locus method, then the remaining closed-loop poles may be found by dividing the characteristic equation by the factors corresponding to the dominant closed-loop poles. The remainder corresponds to the remaining closed-loop poles. Often this division may not be done exactly. This is unavoidable because of inaccuracies introduced in graphical analysis.

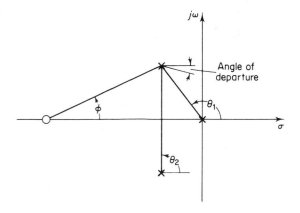

Figure 5–17
Construction of the root locus. [Angle of departure = 180° − (θ_1 + θ_2) + ϕ.]

It is noted that the characteristic equation of the system whose open-loop transfer function is

$$G(s)H(s) = \frac{K(s^m + b_1 s^{m-1} + \cdots + b_m)}{s^n + a_1 s^{n-1} + \cdots + a_n} \qquad (n \geq m)$$

is an nth-degree algebraic equation in s. If the order of the numerator of $G(s)H(s)$ is lower than that of the denominator by two or more (which means that there are two or more zeros at infinity), then the coefficient a_1 is the negative sum of the roots of the equation and is independent of K. In such a case, if some of the roots move on the locus toward the left as K is increased, then the other roots must move toward the right as K is increased. This information is helpful in finding the general shape of the root loci.

9. Determine the root loci in the broad neighborhood of the $j\omega$ axis and the origin. The most important part of the root loci is on neither the real axis nor the asymptotes but the part in the broad neighborhood of the $j\omega$ axis and the origin. The shape of the root loci in this important region in the s plane must be obtained with sufficient accuracy.

A comment on the root-locus plots. It is noted that a slight change in the pole–zero configuration may cause significant changes in the root-locus configurations. Figure 5–18 demonstrates the fact that a slight change in the location of a zero or pole will make the root-locus configuration look quite different.

Summary of the steps for constructing root loci

1. Locate the poles and zeros of $G(s)H(s)$ on the s plane. The root-locus branches start from open-loop poles and terminate at zeros (finite zeros or zeros at infinity).
2. Determine the root-locus on the real axis.
3. Determine the asymptotes of root-locus branches.
4. Find the breakaway and break-in points.
5. Determine the angle of departure (angle of arrival) of the root locus from a complex pole (at a complex zero).
6. Find the points where the root loci may cross the imaginary axis.
7. Taking a series of test points in the broad neighborhood of the origin of the s plane, sketch the root loci.

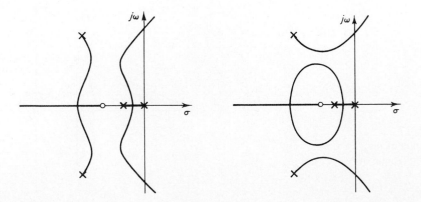

Figure 5–18
Root-locus diagrams.

Chapter 5 / Root-Locus Analysis

8. Locate the closed-loop poles on the root-locus branches and determine the corresponding value of gain K by use of the magnitude condition. Or, using the magnitude condition, determine the location of the closed-loop poles for a given value of gain K.

Cancellation of poles of $G(s)$ with zeros of $H(s)$.   It is important to note that if the denominator of $G(s)$ and the numerator of $H(s)$ involve common factors, then the corresponding open-loop poles and zeros will cancel each other, reducing the degree of the characteristic equation by one or more. For example, consider the system shown in Fig. 5–19(a). (This system has velocity feedback.) The closed-loop transfer function $C(s)/R(s)$ is

$$\frac{C(s)}{R(s)} = \frac{K}{s(s+1)(s+2) + K(s+1)}$$

The characteristic equation is

$$[s(s+2) + K](s+1) = 0$$

Because of the cancellation of the terms $(s+1)$ appearing in $G(s)$ and $H(s)$, however, we have

$$\begin{aligned} 1 + G(s)H(s) &= 1 + \frac{K(s+1)}{s(s+1)(s+2)} \\ &= \frac{s(s+2) + K}{s(s+2)} \end{aligned}$$

The reduced characteristic equation is

$$s(s+2) + K = 0$$

The root-locus plot of $G(s)H(s)$ does not show all the roots of the characteristic equation, only the roots of the reduced equation.

To obtain the complete set of closed-loop poles, we must add the canceled pole of $G(s)H(s)$ to those closed-loop poles obtained from the root-locus plot of $G(s)H(s)$. The important thing to remember is that the canceled pole of $G(s)H(s)$ is a closed-loop pole of the system, as seen from Fig. 5–19(b).

Typical pole-zero configurations and corresponding root loci.   In concluding this section, we show several open-loop pole–zero configurations and their corresponding root loci in Table 5–2. The pattern of the root loci depends only on the relative separation of the open-loop poles and zeros. If the number of open-loop poles exceeds the number of finite zeros by three or more, there is a value of the gain K beyond which root loci enter

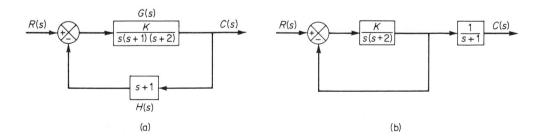

Figure 5–19
Control systems.

Table 5–2 Open-Loop Pole–Zero Configurations and
the Corresponding Root Loci

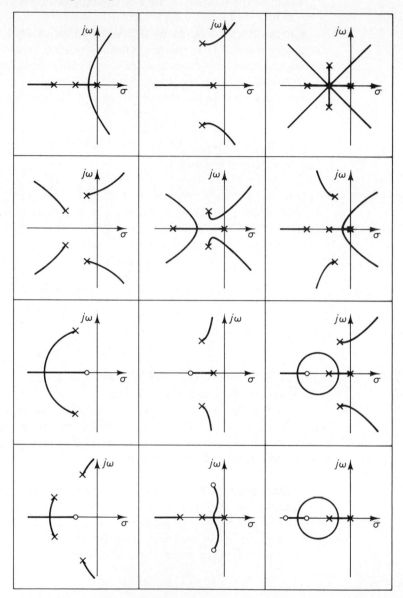

the right-half s plane, and thus the system can become unstable. A stable system must have all its closed-loop poles in the left-half s plane.

Note that once we have some experience with the method we can easily evaluate the changes in the root loci due to the changes in the number and location of the open-loop poles and zeros by visualizing the root-locus plots resulting from various pole–zero configurations.

5-5 ROOT-LOCUS ANALYSIS OF CONTROL SYSTEMS

In this section we shall first discuss orthogonality of the root loci and constant gain loci for the closed-loop systems. Then we compare the effects of the derivative control and velocity feedback on the performance of a servo system. Next, we discuss conditionally stable systems. Finally, we analyze nonminimum-phase systems.

Orthogonality of root loci and constant gain loci. Consider the system whose open-loop transfer function is $G(s)H(s)$. In the $G(s)H(s)$ plane, the loci of $|G(s)H(s)| =$ constant are circles centered at the origin, and the loci corresponding to $\angle G(s)H(s) = \pm 180°(2k + 1)(k = 0, 1, 2, \ldots)$ lie on the negative real axis of the $G(s)H(s)$ plane, as shown in Figure 5–20. [Note that the complex plane employed here is not the s plane but the $G(s)H(s)$ plane.]

The root loci and constant gain loci in the s plane are conformal mappings of the loci of $\angle G(s)H(s) = \pm 180°(2k + 1)$ and of $|G(s)H(s)| =$ constant in the $G(s)H(s)$ plane.

Since the constant phase and constant gain loci in the $G(s)H(s)$ plane are orthogonal, the root loci and constant gain in the s plane are orthogonal. Figure 5–21(a) shows the root loci and constant gain loci for the following system:

$$G(s) = \frac{K(s + 2)}{s^2 + 2s + 3}, \qquad H(s) = 1$$

Notice that since the pole–zero configuration is symmetrical about the real axis, the constant gain loci are also symmetrical about the real axis.

Figure 5–21(b) shows the root loci and constant gain loci for the system:

$$G(s) = \frac{K}{s(s + 1)(s + 2)}, \qquad H(s) = 1$$

Notice that since the configuration of the poles in the s plane is symmetrical about the real axis and the line parallel to the imaginary axis passing through point ($\sigma = -1$, $\omega = 0$), the constant gain loci are symmetrical about the $\omega = 0$ line (real axis) and the $\sigma = -1$ line.

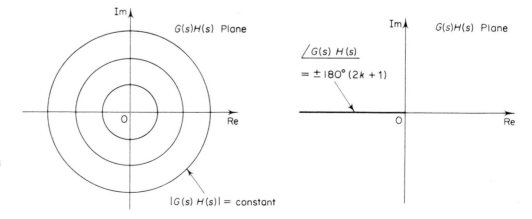

Figure 5–20
Plots of constant gain and constant phase loci in the $G(s)H(s)$ plane.

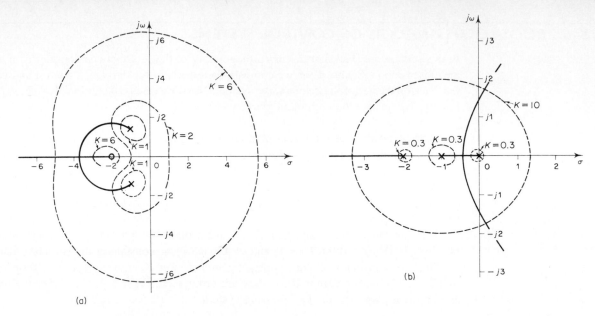

Figure 5–21 Plots of root loci and constant gain loci. (a) System with $G(s) = K(s + 2)/(s^2 + 2s + 3)$, $H(s) = 1$; (b) system with $G(s) = K/[s(s + 1)(s + 2)]$, $H(s) = 1$.

Comparison of the effects of derivative control and velocity feedback (tachometer feedback) on the performance of positional servo systems. System I shown in Figure 5–22 is a positional servo system. (The output is position.) System II shown in Figure 5–22 is positional servo system utilizing proportional-plus-derivative control action. System III shown in Figure 5–22 is a positional servo system utilizing velocity feedback or tachometer feedback. Let us compare the relative merits of derivative control and velocity feedback.

The root-locus diagram for System I is shown in Figure 5–23(a). The closed-loop poles are located at $s = -0.1 \pm j0.995$.

The open-loop transfer function of System II is

$$G_{\text{II}}(s)H_{\text{II}}(s) = \frac{5(1 + 0.8s)}{s(5s + 1)}$$

The open-loop transfer function of System III is

$$G_{\text{III}}(s)H_{\text{III}}(s) = \frac{5(1 + 0.8s)}{s(5s + 1)}$$

Thus Systems II and III have identical open-loop transfer functions. (Both systems have the same open-loop poles and zero.) The root-locus diagrams for Systems II and III are therefore identical and are shown in Figures 5–23(b) and (c), respectively.

Note, however, that the closed-loop transfer functions of System II and System III are clearly different. System II has two closed-loop poles and one finite closed-loop zero, while System III has two closed-loop poles and no finite closed-loop zero. (Velocity feedback, or tachometer feedback, possesses an open-loop zero but not a closed-loop zero.) The closed-loop poles of Systems II and III are $s = -0.5 \pm j0.866$.

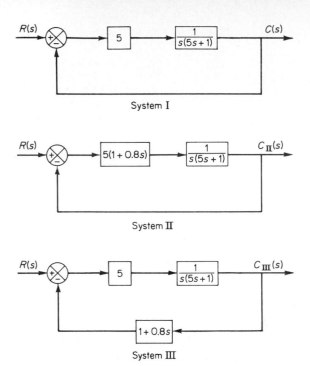

System I

System II

System III

Figure 5–22
Positional servo
systems.

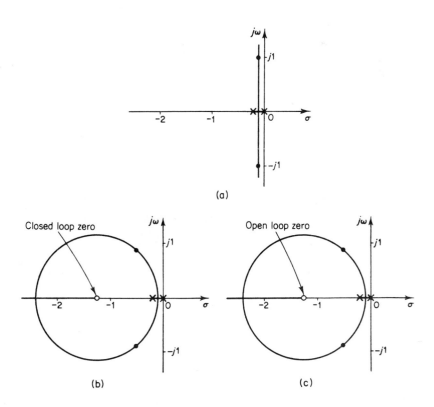

(a)

Closed loop zero

Open loop zero

Figure 5–23
Root-locus diagrams
for the systems shown
in Fig. 5–22: (a)
System I; (b) System
II; (c) System III.

(b)   (c)

The closed-loop transfer function of System II is

$$\frac{C_{II}(s)}{R(s)} = \frac{1 + 0.8s}{(s + 0.5 + j0.866)(s + 0.5 - j0.866)}$$

For a unit-impulse input,

$$C_{II}(s) = \frac{0.4 + j0.346}{s + 0.5 + j0.866} + \frac{0.4 - j0.346}{s + 0.5 - j0.866}$$

The residue at the closed-loop pole $s = -0.5 - j0.866$ is $0.4 + j0.346$ and that at the closed-loop pole $s = -0.5 + j0.866$ is $0.4 - j0.346$. The inverse Laplace transform of $C_{II}(s)$ yields

$$c_{II}(t)_{impulse} = e^{-0.5t}(0.8 \cos 0.866t + 0.693 \sin 0.866t) \qquad (t \geq 0)$$

The closed-loop transfer function of System III is

$$\frac{C_{III}(s)}{R(s)} = \frac{1}{(s + 0.5 + j0.866)(s + 0.5 - j0.866)}$$

For a unit-impulse input,

$$C_{III}(s) = \frac{j0.577}{s + 0.5 + j0.866} + \frac{-j0.577}{s + 0.5 - j0.866}$$

The residue at the closed-loop pole $s = -0.5 - j0.866$ is $j0.577$ and that at the closed-loop pole $s = -0.5 + j0.866$ is $-j0.577$. The inverse Laplace transform of $C_{III}(s)$ yields

$$c_{III}(t)_{impulse} = 1.155e^{-0.5t} \sin 0.866t \qquad (t \geq 0)$$

The unit-impulse responses of Systems II and III are clearly different because the residues at the same pole are different for the two systems. (Remember that the residues depend on both the closed-loop poles and zeros.) Figure 5–24 shows the unit-impulse response curves for the three systems.

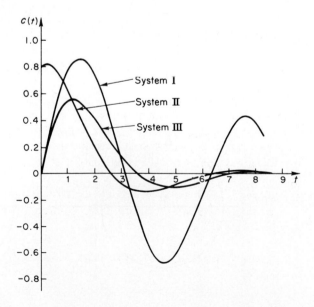

Figure 5–24
Unit-impulse response curves for Systems I, II, and III shown in Fig. 5–22.

Chapter 5 / Root-Locus Analysis

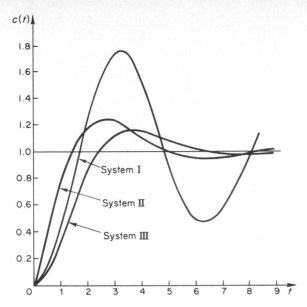

Figure 5–25 Unit-step response curves for Systems I, II, and III shown in Fig. 5–22.

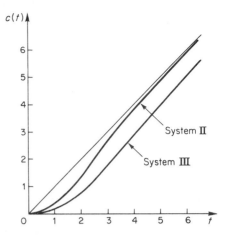

Figure 5–26 Unit-ramp response curves for Systems II and III shown in Fig. 5–22.

Note that the unit-step response can be obtained either directly or by integrating the unit-impulse response. For example, for System III, the unit-step response is obtained as follows:

$$c_{III}(t)_{step} = \int_0^t c_{III}(t)_{impulse} \, dt$$

$$= \int_0^t 1.155 e^{-0.5t} \sin 0.866t \, dt$$

$$= 1 - e^{-0.5t}(\cos 0.866t + 0.577 \sin 0.866t)$$

Figure 5–25 shows unit-step response curves for the three systems. The system utilizing proportional-plus-derivative control action exhibits the shortest rise time. The system with velocity feedback has the least maximum overshoot, or the best relative stability, of the three systems.

Figure 5–26 shows the unit-ramp response curves for Systems II and III. System II has the advantage of quicker response and less steady-state error to a ramp input.

The main reason why the system utilizing proportional-plus-derivative control action has superior response characteristics is that derivative control responds to the rate of change of the error signal and can produce early corrective action before the magnitude of the error becomes large.

Notice that the output of System III is the output of System II delayed by a first-order lag term $1/(1 + 0.8s)$. Figure 5–27 shows the relationship between the outputs of System II and System III.

Conditionally stable systems. Consider the system shown in Figure 5–28(a). The root loci for this system can be plotted by applying the general rules and procedure for constructing root loci. A root locus diagram for this system is shown in Figure 5–28(b). It

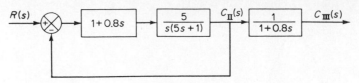

Figure 5–27 Block diagram showing the relationship between the output of System II and that of System III shown in Fig. 5–22.

can be seen that this system is stable only for limited ranges of the value of K; that is, $0 < K < 14$ and $64 < K < 195$. The system becomes unstable for $14 < K < 64$ and $195 < K$. If K assumes a value corresponding to unstable operation, the system may break down or may become nonlinear due to a saturation nonlinearity that may exist. Such a system is called *conditionally stable*.

In practice, conditionally stable systems are not desirable. Conditional stability is dangerous but does occur in certain systems, in particular, a system that has an unstable feedforward path. Such a feedforward path may occur if the system has a minor loop. It is advisable to avoid such conditional stability since, if the gain drops beyond the critical value for some reason or other, the system becomes unstable. Note that the addition of a proper compensating network will eliminate conditional stability. [An addition of a zero will cause the root loci to bend to the left. (See Section 7–3.) Hence conditional stability may be eliminated by adding proper compensation.]

Nonminimum phase systems. If all the poles and zeros of a system lie in the left-half s plane, then the system is called minimum phase. If a system has at least one pole or zero in the right-half s plane, then the system is called *nonminimum phase*. The term non-

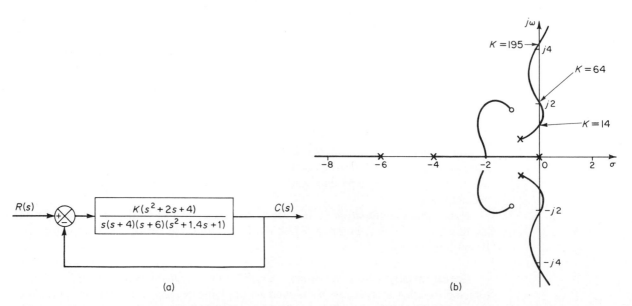

Figure 5–28 (a) Control system; (b) root-locus diagram.

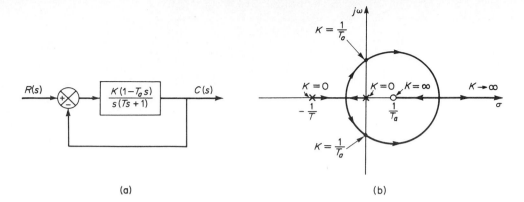

Figure 5–29
(a) Nonminimum phase system; (b) root-locus diagram.

(a)

(b)

minimum phase comes from the phase shift characteristics of such a system when subjected to sinusoidal inputs. (See Section 6–2.)

Consider the system shown in Figure 5–29(a). In this system

$$G(s) = \frac{K(1 - T_a s)}{s(Ts + 1)} \qquad (T_a > 0), \qquad H(s) = 1$$

This is a nonminimum phase system since there is one zero in the right-half s plane. For this system, the angle condition becomes

$$\angle G(s) = \left/ -\frac{K(T_a s - 1)}{s(Ts + 1)} \right.$$

$$= \left/ \frac{K(T_a s - 1)}{s(Ts + 1)} + 180° \right.$$

$$= \pm 180°(2k + 1) \qquad (k = 0, 1, 2, \ldots)$$

or

$$\left/ \frac{K(T_a s - 1)}{s(Ts + 1)} \right. = 0° \tag{5–21}$$

The root loci can be obtained from Equation (5–21). Figure 5–29(b) shows a root-locus diagram for this system. From the diagram, we see that the system is stable if the gain K is less than $1/T_a$.

5–6 ROOT LOCI FOR SYSTEMS WITH TRANSPORT LAG

Figure 5–30 shows a thermal system in which hot air is circulated to keep the temperature of a chamber constant. In this system, the measuring element is placed downstream a distance L ft from the furnace, the air velocity is v ft/sec, and $T = L/v$ sec would elapse before any change in the furnace temperature was sensed by the thermometer. Such a delay in measuring, delay in controller action, or delay in actuator operation, and the like, is called *transport lag* or *dead time*. Dead time is present in most process control systems.

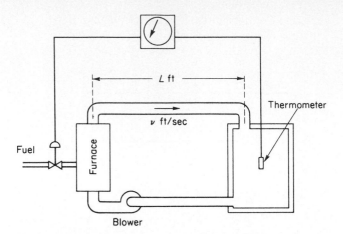

Figure 5–30
Thermal system.

The input $x(t)$ and the output $y(t)$ of a transport lag or dead time element are related by

$$y(t) = x(t - T)$$

where T is dead time. The transfer function of transport lag or dead time is given by

$$\text{Transfer function of transport lag or dead time} = \frac{\mathscr{L}[x(t - T)1(t - T)]}{\mathscr{L}[x(t)1(t)]}$$

$$= \frac{X(s)e^{-Ts}}{X(s)} = e^{-Ts}$$

Suppose that the feedforward transfer function of this thermal system can be approximated by

$$G(s) = \frac{Ke^{-Ts}}{s + 1}$$

as shown in Figure 5–31. Let us construct a root-locus plot for this system. The characteristic equation for this closed-loop system is

$$1 + \frac{Ke^{-Ts}}{s + 1} = 0 \tag{5–22}$$

It is noted that for systems with transport lag, the rules of construction presented earlier need to be modified. For example, the number of the root-locus branches is infinite, since the characteristic equation has an infinite number of roots. The number of asymptotes is infinite. They are all parallel to the real axis of the s plane.

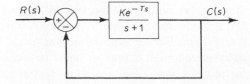

Figure 5–31
Block diagram of the system shown in Fig. 5–30.

From Equation (5–22), we obtain

$$\frac{Ke^{-Ts}}{s + 1} = -1$$

Thus, the angle condition becomes

$$\left/\frac{Ke^{-Ts}}{s + 1}\right. = \underline{/e^{-Ts}} - \underline{/s + 1} = \pm 180°(2k + 1) \qquad (k = 0, 1, 2, \ldots) \qquad (5\text{–}23)$$

To find the angle of e^{-Ts}, we write $s = \sigma + j\omega$. Then we obtain

$$e^{-Ts} = e^{-T\sigma - j\omega T}$$

Since $e^{-T\sigma}$ is a real quantity, the angle of $e^{-T\sigma}$ is zero. Hence

$$\underline{/e^{-Ts}} = \underline{/e^{-j\omega T}} = \underline{/\cos \omega T - j \sin \omega T}$$

$$= -\omega T \qquad \text{(radians)}$$

$$= -57.3\omega T \qquad \text{(degrees)}$$

The angle condition, Equation (5–23), then becomes

$$-57.3\omega T - \underline{/s + 1} = \pm 180°(2k + 1)$$

Since T is a given constant, the angle of e^{-Ts} is a function of ω only.

We shall next determine the angle contribution due to e^{-Ts}. For $k = 0$, the angle condition may be written

$$\underline{/s + 1} = \pm 180° - 57.3°\omega T \qquad (5\text{–}24)$$

Since the angle contribution of e^{-Ts} is zero for $\omega = 0$, the real axis from -1 to $-\infty$ forms a part of the root loci. Now assume a value ω_1 for ω and compute $57.3°\omega_1 T$. At point -1 on the negative real axis, draw a line that makes an angle of $180° - 57.3°\omega_1 T$ with the real axis. Find the intersection of this line and the horizontal line $\omega = \omega_1$. This intersection, point P in Figure 5–32(a), is a point satisfying Equation (5–24) and hence is on a root locus. Continuing the same process, we obtain the root-locus plot as shown in Figure 5–32(b).

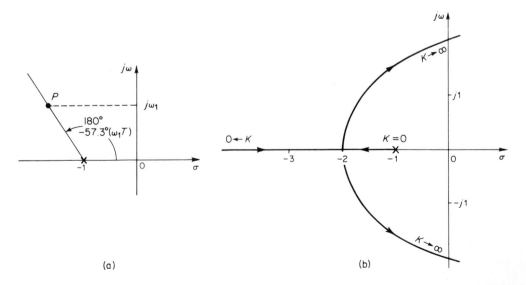

Figure 5–32
(a) Construction of the root locus; (b) root-locus diagram.

(a)   (b)

Note that as s approaches minus infinity, the open-loop transfer function

$$\frac{Ke^{-Ts}}{s + 1}$$

approaches minus infinity since

$$\lim_{s = -\infty} \frac{Ke^{-Ts}}{s + 1} = \frac{\dfrac{d}{ds}(Ke^{-Ts})}{\dfrac{d}{ds}(s + 1)}\Bigg|_{s = -\infty}$$

$$= -KTe^{-Ts}\big|_{s = -\infty}$$

$$= -\infty$$

Therefore, $s = -\infty$ is a pole of the open-loop transfer function. Thus, root loci start from $s = -1$ or $s = -\infty$ and terminate at $s = \infty$, as K increases from zero to infinity. Since the right side of the angle condition given by Equation (5–23) has an infinite number of values, there are an infinite number of root loci, as the value of k ($k = 0, 1, 2, \ldots$) goes from zero to infinity. For example, if $k = 1$, the angle condition becomes

$$\angle s + 1 = \pm 540° - 57.3°\omega T \quad \text{(degrees)}$$

$$= \pm 3\pi - \omega T \quad \text{(radians)}$$

The construction of the root loci for $k = 1$ is the same as that for $k = 0$. A plot of root loci for $k = 0, 1,$ and 2 when $T = 1$ is shown in Figure 5–33.

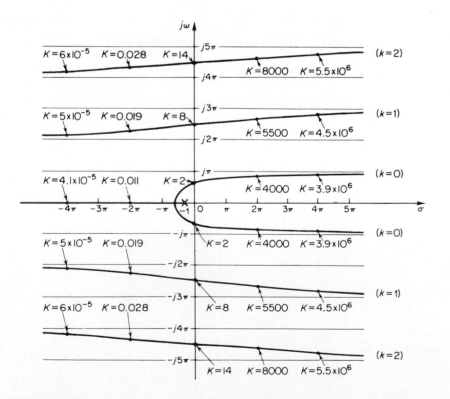

Figure 5–33
Root-locus diagram
for the system shown
in Fig. 5–31 ($T = 1$).

384   Chapter 5 / Root-Locus Analysis

The magnitude condition states that

$$\left|\frac{Ke^{-Ts}}{s+1}\right| = 1$$

Since the magnitude of e^{-Ts} is equal to that of $e^{-T\sigma}$ or

$$\left|e^{-Ts}\right| = \left|e^{-T\sigma}\right|\cdot\left|e^{-j\omega T}\right| = e^{-T\sigma}$$

the magnitude condition becomes

$$|s+1| = Ke^{-T\sigma}$$

The root loci shown in Figure 5–33 are graduated in terms of K when $T = 1$.

Although there are an infinite number of root-locus branches, the primary branch that lies between $-j\pi$ and $j\pi$ is most important. Referring to Figure 5–33, the critical value of K at the primary branch is equal to 2, while the critical values of K at other branches are much higher (8, 14, . . .). Therefore, the critical value $K = 2$ on the primary branch is most significant from the stability viewpoint. The transient response of the system is determined by the roots located closest to the $j\omega$ axis and lie on the primary branch. In summary, the root-locus branch corresponding to $k = 0$ is the dominant one; other branches corresponding to $k = 1, 2, 3, . . .$ are not so important and may be neglected.

This example illustrates the fact that dead time can cause instability even in the first-order system because the root loci enter the right-half s plane for large values of K. Therefore, although the gain K of the first-order system can be set at a high value in the absence of dead time, it cannot be set too high if dead time is present. (For the system considered here, the value of the gain K must be considerably less than 2 for a satisfactory operation.)

Approximation of transport lag or dead time.   If the dead time T is very small, then e^{-Ts} may be approximated by

$$e^{-Ts} \doteq 1 - Ts$$

or

$$e^{-Ts} \doteq \frac{1}{Ts+1}$$

Such approximations are good if the dead time is very small and, in addition, the input time function $f(t)$ to the dead-time element is smooth and continuous. [This means that the second- and higher-order derivatives of $f(t)$ are small.]

5–7   ROOT-CONTOUR PLOTS

Effects of parameter variations on closed-loop poles.   In many design problems, the effects on the closed-loop poles of the variations of parameters other than the gain K need to be investigated. Such effects can be easily investigated by the root-locus method. When two (or more) parameters are varied, the corresponding root loci are called *root contours*.

We shall use an example to illustrate the construction of the root contours when two parameters are varied, respectively, from zero to infinity.

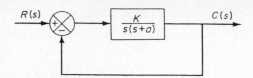

Figure 5–34
Control system.

Consider the system shown in Figure 5–34. We wish to investigate the effect of varying the parameter a as well as the gain K. The closed-loop transfer function of this system is

$$\frac{C(s)}{R(s)} = \frac{K}{s^2 + as + K}$$

The characteristic equation is

$$s^2 + as + K = 0 \tag{5–25}$$

which may be rewritten

$$1 + \frac{as}{s^2 + K} = 0$$

or

$$\frac{as}{s^2 + K} = -1 \tag{5–26}$$

In Equation (5–26), the parameter a is written as a multiplying factor. For a given value of K, the effect of a on the closed-loop poles can be investigated from Equation (5–26). The root contours for this system can be constructed by following the usual procedure for constructing root loci.

We shall now construct the root contours as K and a vary, respectively, from zero to infinity. The root contours start and terminate at the poles and zeros of $as/(s^2 + K)$.

We shall first construct the locus of roots when $a = 0$. This can be done easily as follows: Substitute $a = 0$ into Equation (5–25). Then

$$s^2 + K = 0$$

or

$$\frac{K}{s^2} = -1 \tag{5–27}$$

The open-loop poles are thus a double pole at the origin. The root-locus plot of Equation (5–27) is shown in Figure 5–35(a).

To construct the root contours, let us assume that K is a constant; for example, $K = 4$. Then Equation (5–26) becomes

$$\frac{as}{s^2 + 4} = -1 \tag{5–28}$$

The open-loop poles are $s \pm j2$. The finite open-loop zero is at the origin. The root-locus plot corresponding to Equation (5–28) is shown in Figure 5–35(b). For different values of K, Equation (5–28) yields similar root loci.

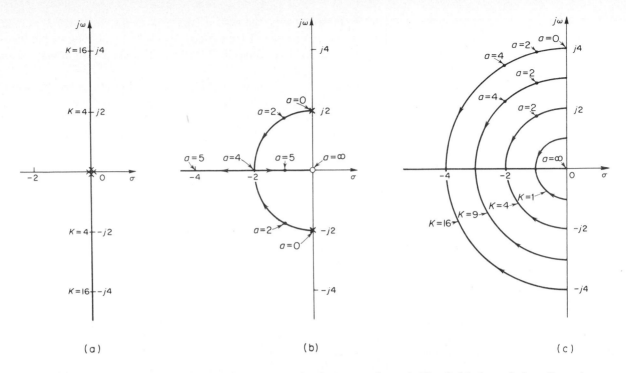

Figure 5–35 (a) Root-locus diagram for the system shown in Fig. 5–34. ($a = 0$, $0 \leq K \leq \infty$); (b) root locus diagram ($0 \leq a \leq \infty$, $K = 4$); (c) root-contour diagram.

The root contour, the diagram showing the root loci corresponding to $0 \leq K \leq \infty$, $0 \leq a \leq \infty$, can be sketched as in Figure 5–35(c). Clearly, the root contours start at the poles of and end at the zeros of the transfer function $as/(s^2 + K)$. The arrowheads on the root contours indicate the direction of increase in the value of a.

The root contours show the effects of the variations of system parameters on the closed-loop poles. From the root-contour plot shown in Figure 5–35(c), we see that, for $0 < K < \infty$, $0 < a < \infty$, the closed-loop poles lie in the left-half s plane and the system is stable.

Note that if the value of K is fixed, say $K = 4$, then the root contours become simply the root loci, as shown in Figure 5–35(b).

We have illustrated a method for constructing root contours when the gain K and parameter a are varied, respectively, from zero to infinity. Basically, we assign one parameter a constant value at a time and vary the other parameter from 0 to ∞ and sketch the root loci. Then we change the value of the first parameter and repeat sketching the root loci. By repeating this process we can sketch the root contour.

5–8 CONCLUSIONS

The transient response of a closed-loop system depends on the location of the closed-loop poles. In this chapter we have presented the root-locus method, a very powerful graphical technique, for investigating the effects of the variation of a system parameter on the location of the closed-loop poles. In most cases, the system parameter is the loop gain K, although the parameter can be any other variable of the system. If the designer follows the general rules for constructing the root loci, sketching the root loci of a given system may become a simple matter.

The general procedure for constructing the root loci presented in this chapter enables us to obtain a reasonable sketch of the root loci. If an accurate plot of the root loci is needed, it is a relatively simple matter to get a plot of the root loci for a given system by using a computer. Note that experience in sketching the root loci by hand is invaluable for interpreting computer-generated root loci.

By using the root-locus method, it is possible to determine the value of the loop gain K that will make the damping ratio of the dominant closed-loop poles as prescribed. If the location of an open-loop pole or zero is a system variable, then the root-locus method suggests the way to choose the location of an open-loop pole or zero. (See Example 5–3 and Problems A–5–14 through A–5–17.) More on the control system design based on the root-locus method will be given in Chapter 7.

Example Problems and Solutions

A–5–1. Sketch the root loci for the system shown in Figure 5–36(a). (The gain K is assumed to be positive.) Observe that for small or large values of K the system is overdamped and for medium values of K it is underdamped.

Solution. The procedure for plotting the root loci is as follows:

1. Locate the open-loop poles and zeros on the complex plane. Root loci exist on the negative real axis between 0 and -1 and between -2 and -3.

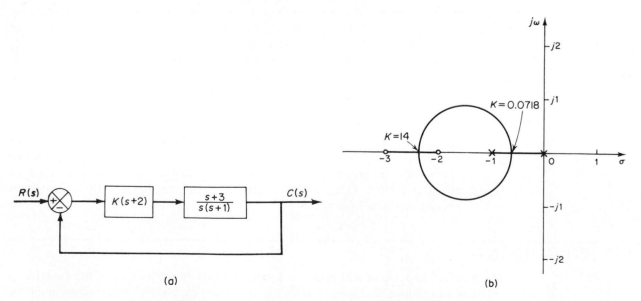

(a)

(b)

Figure 5–36 (a) Control system; (b) root-locus diagram.

2. The number of open-loop poles and that of finite zeros are the same. This means that there are no asymptotes in the complex region of the s plane.

3. Determine the breakaway and break-in points. The characteristic equation for the system is

$$1 + \frac{K(s + 2)(s + 3)}{s(s + 1)} = 0$$

or

$$K = - \frac{s(s + 1)}{(s + 2)(s + 3)}$$

The breakaway and break-in points are determined from

$$\frac{dK}{ds} = - \frac{(2s + 1)(s + 2)(s + 3) - s(s + 1)(2s + 5)}{[(s + 2)(s + 3)]^2}$$

$$= - \frac{4(s + 0.634)(s + 2.366)}{[(s + 2)(s + 3)]^2}$$

$$= 0$$

as follows:

$$s = -0.634, \qquad s = -2.366$$

Notice that both points are on root loci. Therefore, they are actual breakaway or break-in points. At point $s = -0.634$, the value of K is

$$K = - \frac{(-0.634)(0.366)}{(1.366)(2.366)} = 0.0718$$

Similarly, as $s = -2.366$,

$$K = - \frac{(-2.366)(-1.366)}{(-0.366)(0.634)} = 14$$

(Because point $s = -0.634$ lies between two poles, it is a breakaway point and because point $s = -2.366$ lies between two zeros, it is a break-in point.)

4. Determine a sufficient number of points that satisfy the angle condition. (It can be found that the root locus is a circle with center at -1.5 that passes through the breakaway and break-in points.) The root-locus plot for this system is shown in Figure 5–36(b).

Note that this system is stable for any positive value of K since all the root loci lie in the left-half s plane.

Small values of K ($0 < K < 0.0718$) correspond to an overdamped system. Medium values of K ($0.718 < K < 14$) correspond to an underdamped system. Finally, large values of K ($14 < K$) correspond to an overdamped system. With a large value of K, the steady state can be reached in much shorter time than with a small value of K.

The value of K should be adjusted so that system performance is optimum according to a given performance index.

A–5–2. Find the roots of the following polynomial by use of the root-locus method:

$$3s^4 + 10s^3 + 21s^2 + 24s - 16 = 0 \qquad (5\text{–}29)$$

Solution. First rearrange the polynomial and put it into the form

$$\frac{P(s)}{Q(s)} = -1$$

where $P(s)$ and $Q(s)$ are factored polynomials. Then apply the general rules presented in Section 5–4 to locate the roots of the polynomial.

Equation (5–29) may be rearranged in a convenient way as follows:

$$3s^4 + 10s^3 + 21s^2 = -24s + 16$$

In this case the polynomial can be rewritten

$$\frac{8(s - \frac{2}{3})}{s^2(s^2 + \frac{10}{3}s + 7)} = -1 \tag{5–30}$$

This form has two poles at the origin, two complex poles, and one zero on the positive real axis. Since Equation (5–30) has the form $G(s)H(s) = -1$, the root-locus method can be applied to find the roots of the polynomial.

Equation (5–29) can, of course, be rearranged in different ways. For example, it can be rewritten as

$$3s^4 + 10s^3 = -21s^2 - 24s + 16$$

or

$$\frac{7(s^2 + \frac{8}{7}s - \frac{16}{21})}{s^3(s + \frac{10}{3})} = -1 \tag{5–31}$$

In this case, however, the system involves three poles at the origin, one pole and one zero on the negative real axis, and one zero on the positive real axis. The amount of graphical work needed to sketch the root-locus plot of Equation (5–31) is almost the same as that of Equation (5–30).

Note that if two root-locus plots, one corresponding to Equation (5–30) and the other corresponding to Equation (5–31), are sketched on the same diagram, the intersections of the two give the roots for the polynomial. (If the magnitude condition is used, only one root-locus plot need be sketched.)

In this problem, we shall sketch only one root-locus plot based on Equation (5–30) and utilize the magnitude condition to determine the roots of the polynomial. Equation (5–30) can be rewritten

$$\frac{8(s - \frac{2}{3})}{s^2(s + 1.67 + j2.06)(s + 1.67 - j2.06)} = -1 \tag{5–32}$$

To determine the root loci, replace the constant 8 in the numerator of Equation (5–32) by K and write

$$\frac{K(s - \frac{2}{3})}{s^2(s + 1.67 + j2.06)(s + 1.67 - j2.06)} = -1$$

To sketch the root loci, we follow this procedure:

1. Locate the poles and zero in the complex plane. Root loci exist on the real axis between $\frac{2}{3}$ and 0 and between 0 and $-\infty$.
2. Determine the asymptotes of the root loci. There are three asymptotes, which make angles of

$$\frac{\pm 180°(2k + 1)}{4 - 1} = 60°, -60°, 180°$$

with the positive real axis. Referring to Equation (5–18) the abscissa of the intersection of the asymptotes with the real axis is given by

$$\sigma_a = -\frac{(0 + 0 + \frac{5}{3} + j2.06 + \frac{5}{3} - j2.06) + \frac{2}{3}}{4 - 1} = -\frac{4}{3}$$

3. Using Routh's stability criterion, determine the value of K at which root loci cross the imaginary axis. The characteristic equation is

$$s^2(s^2 + \tfrac{10}{3}s + 7) = -K(s - \tfrac{2}{3})$$

or

$$s^4 + \tfrac{10}{3}s^3 + 7s^2 + Ks - \tfrac{2}{3}K = 0$$

The Routh array becomes

s^4	1	7	$-\tfrac{2}{3}K$
s^3	$\tfrac{10}{3}$	K	0
s^2	$7 - \tfrac{3}{10}K$	$-\tfrac{2}{3}K$	
s^1	$\dfrac{-\tfrac{3}{10}K^2 + \tfrac{83}{9}K}{7 - \tfrac{3}{10}K}$	0	
s^0	$-\tfrac{2}{3}K$		

To have roots on the imaginary axis we must have a row with zero elements. Notice that the elements on the s^1 row can be made zero by a proper choice of K. The value of K (other than $K = 0$) that makes the s^1 term in the first column equal zero is $K = 30.7$. The crossing points on the imaginary axis can be found by solving the auxiliary equation obtained from the s^2 row, or

$$(7 - \tfrac{3}{10}K)s^2 - \tfrac{2}{3}K = 0$$

where $K = 30.7$. The result is

$$s = \pm j3.04$$

The crossing points on the imaginary axis are, therefore, $s = \pm j3.04$.

4. See if there are any breakaway or break-in points. From step 3 we know that there are only two crossing points on the $j\omega$ axis. Hence, in this particular pole–zero configuration, there cannot exit any breakaway or break-in points.

5. Find the angles of departure of the root loci from the complex poles. At the pole $s = -1.67 + j2.06$, the angle of departure θ is found from

$$\theta = 180° - 129.03° - 129.03° - 90° + 138.60°$$

as follows:

$$\theta = -29.46°$$

The angle of departure from the pole $s = -1.67 - j2.06$ is $29.46°$.

6. In the broad neighborhood of the $j\omega$ axis and the origin, locate a sufficient number of points that satisfy the angle condition. Based on the information obtained thus far, the root loci for this system can be sketched as shown in Figure 5–37.

7. Use the magnitude condition

$$K = \left| \frac{s^2(s + 1.67 + j2.06)(s + 1.67 - j2.06)}{s - \tfrac{2}{3}} \right|$$

to determine the points on the root locus at which $K = 8$. By use of a trial-and-error procedure, we find

$$s = -0.79 \pm j2.16$$

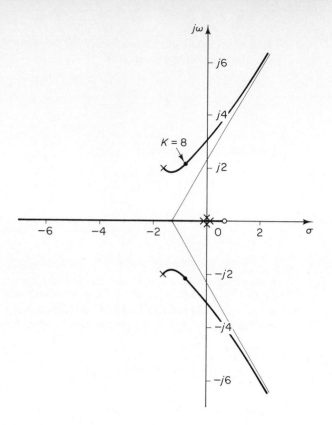

Figure 5–37
Root-locus diagram
for the system given
by Eq. (5–30).

A trial-and-error procedure can be used to locate the two remaining roots. It may be simpler, however, to factor the known roots from the given polynomial.

$$3s^4 + 10s^3 + 21s^2 + 24s - 16 = (s + 0.79 + j2.16)(s + 0.79 - j2.16)(3s^2 + 5.28s - 3.06)$$
$$= 3(s + 0.79 + j2.16)(s + 0.79 - j2.16)(s + 2.22)(s - 0.46)$$

Thus, the roots of the given polynomial are

$$s_1 = -0.79 - j2.16, \qquad s_2 = -0.79 + j2.16, \qquad s_3 = -2.22, \qquad s_4 = 0.46$$

A–5–3. A simplified form of the open-loop transfer function of an airplane with an autopilot in the longitudinal mode is

$$G(s)H(s) = \frac{K(s + a)}{s(s - b)(s^2 + 2\zeta\omega_n s + \omega_n^2)}, \qquad a > 0, \qquad b > 0$$

Such a system involving an open-loop pole in the right-half s plane may be conditionally stable. Sketch the root loci when $a = b = 1$, $\zeta = 0.5$, and $\omega_n = 4$. Find the range of gain K for stability.

Solution. The open-loop transfer function for the system is

$$G(s)H(s) = \frac{K(s + 1)}{s(s - 1)(s^2 + 4s + 16)}$$

To sketch the root loci, we follow this procedure:

1. Locate the open-loop poles and zero in the complex plane. Root loci exist on the real axis between 1 and 0 and between -1 and $-\infty$.

2. Determine the asymptotes of the root loci. There are three asymptotes whose angles can be determined as

$$\text{Angles of asymptotes} = \frac{180°(2k + 1)}{4 - 1} = 60°, \ -60°, \ 180°$$

The abscissa of the intersection of the asymptotes and the real axis is

$$\sigma_a = -\frac{(0 - 1 + 2 + j2\sqrt{3} + 2 - j2\sqrt{3}) - 1}{4 - 1} = -\frac{2}{3}$$

3. Determine the breakaway and break-in points. Since the characteristic equation is

$$1 + \frac{K(s + 1)}{s(s - 1)(s^2 + 4s + 16)} = 0$$

we obtain

$$K = -\frac{s(s - 1)(s^2 + 4s + 16)}{s + 1}$$

By differentiating K with respect to s, we get

$$\frac{dK}{ds} = -\frac{3s^4 + 10s^3 + 21s^2 + 24s - 16}{(s + 1)^2}$$

In Problem A–5–2, it was shown that

$$3s^4 + 10s^3 + 21s^2 + 24s - 16$$
$$= 3(s + 0.79 + j2.16)(s + 0.79 - j2.16)(s + 2.22)(s - 0.46)$$

Points $s = 0.46$ and $s = -2.22$ are on root loci on the real axis. Hence, these points are actual breakaway and break-in points, respectively. Points $s = -0.79 \pm j2.16$ do not satisfy the angle condition. Hence, they are neither breakaway nor break-in points.

4. Using Routh's stability criterion, determine the value of K at which the root loci cross the imaginary axis. Since the characteristic equation is

$$s^4 + 3s^3 + 12s^2 + (K - 16)s + K = 0$$

the Routh array becomes

s^4	1	12	K
s^3	3	$K - 16$	0
s^2	$\dfrac{52 - K}{3}$	K	0
s^1	$\dfrac{-K^2 + 59K - 832}{3(52 - K)}$	0	
s^0	K		

The values of K that make the s^1 term in the first column equal zero are $K = 35.7$ and $K = 23.3$.

The crossing points on the imaginary axis can be found by solving the auxiliary equation obtained from the s^2 row, that is, by solving the following equation for s:

$$\frac{52 - K}{3}s^2 + K = 0$$

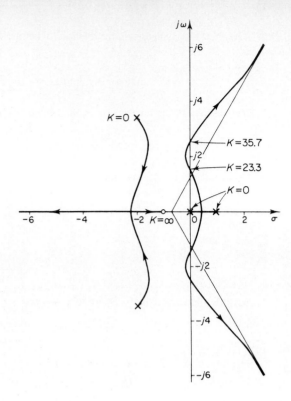

Figure 5–38
Root-locus diagram.

The results are

$$s = \pm j2.56 \qquad \text{for } K = 35.7$$
$$s = \pm j1.56 \qquad \text{for } K = 23.3$$

The crossing points on the imaginary axis are thus $s = \pm j2.56$ and $s = \pm j1.56$.

5. Find the angles of departure of the root loci from the complex poles. For the open-loop pole at $s = -2 + j2\sqrt{3}$, the angle of departure θ is

$$\theta = 180° - 120° - 130.5° - 90° + 106°$$

or

$$\theta = -54.5°$$

(The angle of departure from the open-loop pole at $s = -2 - j2\sqrt{3}$ is 54.5°.)

6. Choose a test point in the broad neighborhood of the $j\omega$ axis and the origin, and apply the angle condition. If the test point does not satisfy the angle condition, select another test point until it does. Continue the same process and locate a sufficient number of points that satisfy the angle condition.

Figure 5–38 shows the root loci for this system. From step 4 the system is stable for $23.3 < K < 35.7$. Otherwise it is unstable.

A–5–4.* In a complex control system, there may be a positive feedback inner loop as shown in Figure 5–39. Such a loop is usually stabilized by the outer loop. In this problem we shall be concerned only with

* Reference W–5.

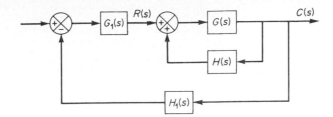

Figure 5–39
Control system.

the positive feedback inner loop. The closed-loop transfer function of the inner loop is

$$\frac{C(s)}{R(s)} = \frac{G(s)}{1 - G(s)H(s)}$$

The characteristic equation is

$$1 - G(s)H(s) = 0 \qquad\qquad (5\text{–}33)$$

This equation can be solved in a manner similar to the development of the root-locus method in Sections 5–3 and 5–4. The angle condition, however, must be altered.

Equation (5–33) can be rewritten

$$G(s)H(s) = 1$$

which is equivalent to the following two equations:

$$\angle G(s)H(s) = 0° \pm k360° \qquad (k = 0, 1, 2, \ldots)$$
$$|G(s)H(s)| = 1$$

The total sum of all angles from the open-loop poles and zeros must be equal to $0° \pm k360°$. Thus, the root locus follows a $0°$ locus in contrast to the $180°$ locus considered previously. The magnitude condition remains unaltered.

Sketch the root-locus plot for the positive feedback system with the following transfer functions:

$$G(s) = \frac{K(s + 2)}{(s + 3)(s^2 + 2s + 2)}, \qquad H(s) = 1$$

The gain K is assumed to be positive.

Solution. The general rules for constructing root loci given in Section 5–4 must be modified in the following way:

Rule 3 is modified as follows: If the total number of real poles and real zeros to the right of a test point on the real axis is even, then this test point lies on the root locus.

Rule 4 is modified as follows:

$$\text{Angles of asymptotes} = \frac{\pm k360°}{n - m} \qquad (k = 0, 1, 2, \ldots)$$

where n = number of finite poles of $G(s)H(s)$
m = number of finite zeros of $G(s)H(s)$

Rule 6 is modified as follows: When calculating the angle of departure (or angle of arrival) from a complex open-loop pole (or at a complex zero), subtract from $0°$ the sum of all the angles of the vectors from all the other poles and zeros to the complex pole (or complex zero) in question, with appropriate signs included.

Example Problems and Solutions

Other rules for constructing the root-locus plot remain the same. We shall now apply the modified rules to construct the root-locus plot. The closed-loop transfer function for the positive feedback system is given by

$$\frac{C(s)}{R(s)} = \frac{G(s)}{1 - G(s)H(s)}$$

$$= \frac{K(s + 2)}{(s + 3)(s^2 + 2s + 2) - K(s + 2)}$$

1. Plot the open-loop poles ($s = -1 + j$, $s = -1 - j$, $s = -3$) and zero ($s = -2$) in the complex plane. As K is increased from 0 to ∞, the closed-loop poles start at the open-loop poles and terminate at the open-loop zeros (finite or infinite), just as in the case of negative feedback systems.
2. Determine the root loci on the real axis. Root loci exist on the real axis between -2 and $+\infty$ and between -3 and $-\infty$.
3. Determine the asymptotes of the root loci. For the present system,

$$\text{Angle of asymptote} = \frac{\pm k360°}{3 - 1} = \pm 180°$$

This simply means that root-locus branches are on the real axis.
4. Determine the breakaway and break-in points. Since the characteristic equation is

$$(s + 3)(s^2 + 2s + 2) - K(s + 2) = 0$$

we obtain

$$K = \frac{(s + 3)(s^2 + 2s + 2)}{s + 2}$$

By differentiating K with respect to s, we obtain

$$\frac{dK}{ds} = \frac{2s^3 + 11s^2 + 20s + 10}{(s + 2)^2}$$

Note that

$$2s^3 + 11s^2 + 20s + 10 = 2(s + 0.8)(s^2 + 4.7s + 6.24)$$
$$= 2(s + 0.8)(s + 2.35 + j0.77)(s + 2.35 - j0.77)$$

Point $s = -0.8$ is on the root locus. Since this point lies between two zeros (a finite zero and an infinite zero), it is an actual break-in point. Points $s = -2.35 \pm j0.77$ do not satisfy the angle condition and, therefore, they are neither breakaway nor break-in points.
5. Find the angle of departure of the root locus from a complex pole. For the complex pole at $s = -1 + j$, the angle of departure θ is

$$\theta = 0° - 27° - 90° + 45°$$

or

$$\theta = -72°$$

(The angle of departure from the complex pole at $s = -1 - j$ is 72°.)
6. Choose a test point in the broad neighborhood of the $j\omega$ axis and the origin and apply the angle condition. Locate a sufficient number of points that satisfy the angle condition.

Figure 5–40 shows the root loci for the given positive feedback system. The root loci are shown with dotted lines and curve.

Figure 5–40

Root-locus diagram for the positive feedback system with $G(s) = K(s + 2)/[(s + 3)(s^2 + 2s + 2)]$, $H(s) = 1$.

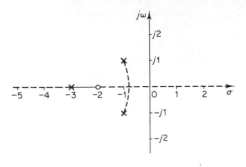

Note that if

$$K > \left. \frac{(s + 3)(s^2 + 2s + 2)}{s + 2} \right|_{s=0} = 3$$

one real root enters the right-half s plane. Hence, for values of K greater than 3, the system becomes unstable. (For $K > 3$, the system must be stabilized with an outer loop.)

To compare this root-locus plot with that of the corresponding negative feedback system, we show in Figure 5–41 the root loci for the negative feedback system whose closed-loop transfer function is

$$\frac{C(s)}{R(s)} = \frac{K(s + 2)}{(s + 3)(s^2 + 2s + 2) + K(s + 2)}$$

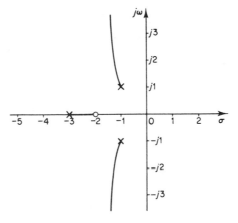

Figure 5–41

Root-locus diagram for the negative feedback system with $G(s) = K(s + 2)/[(s + 3)(s^2 + 2s + 2)]$, $H(s) = 1$.

Table 5–3 shows various root-locus plots of negative feedback and positive feedback systems. The closed-loop transfer functions are given by

$$\frac{C}{R} = \frac{G}{1 + GH} \qquad \text{for negative feedback systems}$$

$$\frac{C}{R} = \frac{G}{1 - GH} \qquad \text{for positive feedback systems}$$

where GH is the open-loop transfer function. In Table 5–3, the root loci for negative feedback systems are drawn with heavy lines and curves and those for positive feedback systems are drawn with dotted lines and curves.

Example Problems and Solutions

Table 5–3 Root-Locus Diagrams of Negative Feedback and Positive Feedback Systems

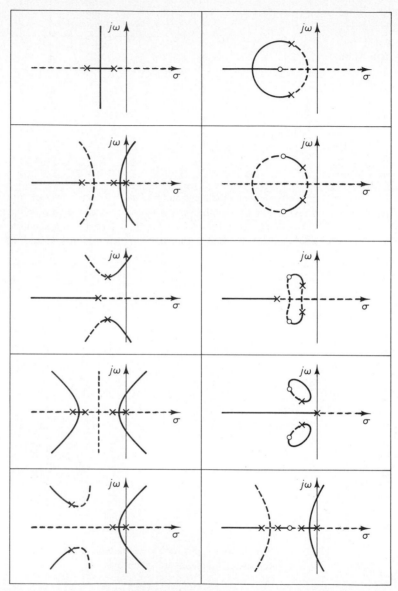

Heavy lines and curves correspond to negative feedback systems; dotted lines and curves correspond to positive feedback systems.

A–5–5. Sketch the root loci of the control system shown in Figure 5–42(a).

Solution. The open-loop poles are located at $s = 0$, $s = -3 + j4$, and $s = -3 - j4$. A root locus branch exists on the real axis between the origin and $-\infty$. There are three asymptotes for the root loci. The angles of asymptotes are

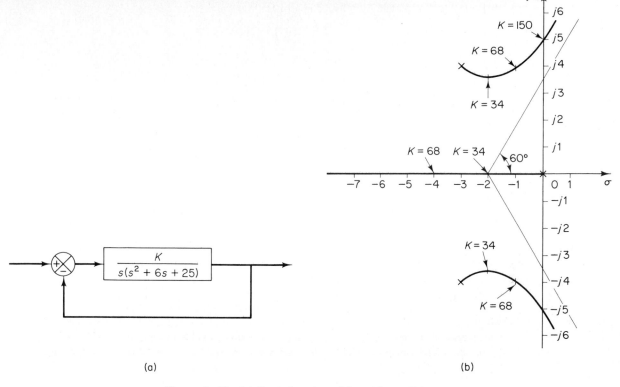

Figure 5–42 (a) Control system; (b) root-locus diagram.

$$\text{Angles of asymptotes} = \frac{\pm 180°(2k + 1)}{3} = 60°, -60°, 180°$$

The intersection of the asymptotes and the real axis is obtained as

$$\sigma_a = -\frac{0 + 3 + 3}{3} = -2$$

Next we check the breakaway and break-in points. For this system we have

$$K = -s(s^2 + 6s + 25)$$

Now we set

$$\frac{dK}{ds} = -(3s^2 + 12s + 25) = 0$$

which yields

$$s = -2 + j2.0817, \qquad s = -2 - j2.0817$$

Notice that at points $s = -2 \pm j2.0817$ the angle condition is not satisfied. Hence they are neither breakaway nor break-in points. In fact, if we calculate the value of K, we obtain

$$K = -s(s^2 + 6s + 25)\Big|_{s = -2 \pm j2.0817} = 34 \pm j18.04$$

(To be an actual breakaway or break-in point, the corresponding value of K must be real and positive.)

The angle of departure from the complex pole in the upper half s plane is

$$\theta = 180° - 126.87° - 90°$$

or

$$\theta = -36.87°$$

The points where root-locus branches cross the imaginary axis may be found by substituting $s = j\omega$ into the characteristic equation and solving the equation for ω and K as follows: Noting that the characteristic equation is

$$s^3 + 6s^2 + 25s + K = 0$$

we have

$$(j\omega)^3 + 6(j\omega)^2 + 25(j\omega) + K = (-6\omega^2 + K) + j\omega(25 - \omega^2) = 0$$

which yields

$$\omega = \pm 5, \qquad K = 150 \qquad \text{or} \qquad \omega = 0, \qquad K = 0$$

Root-locus branches cross the imaginary axis at $\omega = 5$ and $\omega = -5$. The value of gain K at the crossing points is 150. Also, the root-locus branch on the real axis touches the imaginary axis at $\omega = 0$. Figure 5–42(b) shows a root-locus diagram for the system.

It is noted that if the order of the numerator of $G(s)H(s)$ is lower than that of the denominator by two or more and if some of the closed-loop poles move on the root locus toward the right as gain K is increased, then other closed-loop poles must move toward the left as gain K is increased. This fact can be seen clearly in this problem. If the gain K is increased from $K = 34$ to $K = 68$, the complex-conjugate closed-loop poles are moved from $s = -2 + j3.65$ to $s = -1 + j4$; the third pole is moved from $s = -2$ (which corresponds to $K = 34$) to $s = -4$ (which corresponds to $K = 68$). Thus, the movements of two complex-conjugate closed-loop poles to the right by one unit cause the remaining closed-loop pole (real pole in this case) to move to the left by two units.

A–5–6. Consider the system shown in Figure 5–43(a). Sketch the root loci for the system. Observe that for small or large values of K the system is underdamped and for medium values of K it is overdamped.

Solution. A root locus exists on the real axis between the origin and $-\infty$. The angles of asymptotes of the root-locus branches are obtained as

$$\text{Angles of asymptotes} = \frac{\pm 180°(2k + 1)}{3} = 60°, -60°, -180°$$

The intersection of the asymptotes and the real axis is located on the real axis at

$$\sigma_a = -\frac{0 + 2 + 2}{3} = -1.3333$$

The breakaway and break-in points are found from $dK/ds = 0$. Since the characteristic equation is

$$s^3 + 4s^2 + 5s + K = 0$$

we have

$$K = -(s^3 + 4s^2 + 5s)$$

Now we set

$$\frac{dK}{ds} = -(3s^2 + 8s + 5) = 0$$

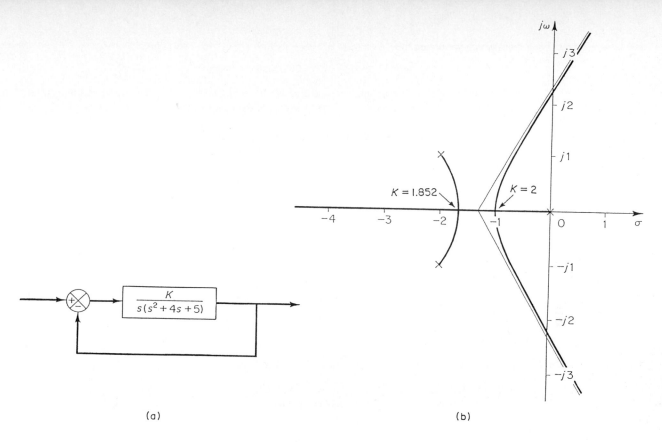

(a)   (b)

Figure 5–43   (a) Control system; (b) root-locus diagram.

which yields

$$s = -1, \qquad s = -1.6667$$

Since these points are on the root locus, they are actual breakaway or break-in points. (At point $s = -1$, the value of K is 2, and at point $s = -1.6667$, the value of K is 1.852.)

The angle of departure from a complex pole in the upper half s plane is obtained from

$$\theta = 180° - 153.43° - 90°$$

or

$$\theta = -63.43°$$

The root-locus branch from the complex pole in the upper half s plane breaks into the real axis at $s = -1.6667$.

Next we determine the points where root-locus branches cross the imaginary axis. By substituting $s = j\omega$ into the characteristic equation, we have

$$(j\omega)^3 + 4(j\omega)^2 + 5(j\omega) + K = 0$$

or

$$(K - 4\omega^2) + j\omega(5 - \omega^2) = 0$$

from which we obtain

$$\omega = \pm\sqrt{5}, \qquad K = 20 \qquad \text{or} \qquad \omega = 0, \qquad K = 0$$

Root-locus branches cross the imaginary axis at $\omega = \sqrt{5}$ and $\omega = -\sqrt{5}$, The root-locus branch on the real axis touches the $j\omega$ axis at $\omega = 0$. A sketch of the root loci for the system is shown in Figure 5–43(b).

Note that since this system is of third order, there are three closed-loop poles. The nature of the system response to a given input depends on the locations of the closed-loop poles.

For $0 < K < 1.852$ there are a set of complex-conjugate closed-loop poles and a real closed-loop pole. For $1.852 \le K \le 2$ there are three real closed-loop poles. For example, the closed-loop poles are located at

$$s = -1.667, \qquad s = -1.667, \qquad s = -0.667 \qquad \text{for } K = 1.852$$

$$s = -1, \qquad s = -1, \qquad s = -2 \qquad \text{for } K = 2$$

For $2 < K$, there are a set of complex-conjugate closed-loop poles and a real closed-loop pole. Thus, small values of K ($0 < K < 1.852$) correspond to an underdamped system. (Since the real closed-loop pole dominates, only a small ripple may show up in the transient response.) Medium values of K ($1.852 \le K \le 2$) correspond to an overdamped system. Large values of K ($2 < K$) correspond to an underdamped system. With a large value of K, the system responds much faster than with a smaller value of K.

A–5–7. Sketch the root loci for the system shown in Figure 5–44(a).

Solution. The open-loop poles are located at $s = 0$, $s = -1$, $s = -2 + j3$, and $s = -2 - j3$. A root locus exists on the real axis between points $s = 0$ and $s = -1$. The asymptotes are found as follows:

$$\text{Angles of asymptotes} = \frac{\pm 180°(2k + 1)}{4} = 45°, -45°, 135°, -135°$$

The intersection of the asymptotes and the real axis is found from

$$\sigma_a = -\frac{0 + 1 + 2 + 2}{4} = -1.25$$

The breakaway and break-in points are found from $dK/ds = 0$. Noting that

$$K = -s(s + 1)(s^2 + 4s + 13) = -(s^4 + 5s^3 + 17s^2 + 13s)$$

we have

$$\frac{dK}{ds} = -(4s^3 + 15s^2 + 34s + 13) = 0$$

from which we get

$$s = -0.467, \qquad s = -1.642 + j2.067, \qquad s = -1.642 - j2.067$$

The point $s = -0.467$ is on a root locus. Therefore, it is an actual breakaway point. The gain values K corresponding to points $s = -1.642 \pm j2.067$ are complex quantities. Since the gain values are not real positive, these points are neither breakaway nor break-in points.

The angle of departure from the complex pole in the upper half s plane is

$$\theta = 180° - 123.69° - 108.44° - 90°$$

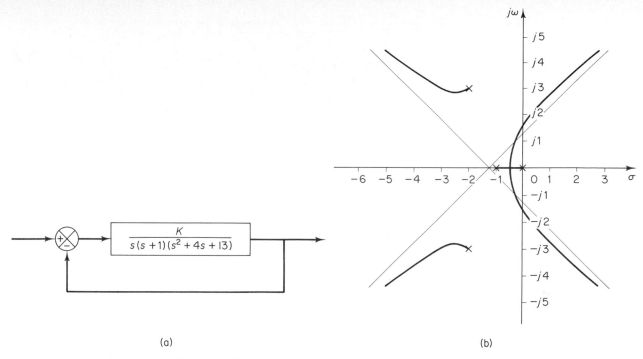

$$\frac{K}{s(s+1)(s^2+4s+13)}$$

(a)   (b)

Figure 5–44   (a) Control system; (b) root-locus diagram.

or

$$\theta = -142.13°$$

Next we shall find the points where root loci may cross the $j\omega$ axis. Since the characteristic equation is

$$s^4 + 5s^3 + 17s^2 + 13s + K = 0$$

by substituting $s = j\omega$ into it we obtain

$$(j\omega)^4 + 5(j\omega)^3 + 17(j\omega)^2 + 13(j\omega) + K = 0$$

or

$$(K + \omega^4 - 17\omega^2) + j\omega(13 - 5\omega^2) = 0$$

from which we obtain

$$\omega = \pm 1.6125, \quad K = 37.44 \quad \text{or} \quad \omega = 0, \quad K = 0$$

The root-locus branches that extend to the right-half s plane cross the imaginary axis at $\omega = \pm 1.6125$. Also, the root-locus branch on the real axis touches the imaginary axis at $\omega = 0$. Figure 5–44(b) shows a sketch of the root loci for the system. Notice that each root-locus branch that extends to the right-half s plane crosses its own asymptote.

A–5–8.   Sketch the root loci for the system shown in Figure 5–45(a).

Solution. A root locus exists on the real axis between points $s = -1$ and $s = -3.6$. The asymptotes can be determined as follows:

Example Problems and Solutions   **403**

$$\text{Angles of asymptotes} = \frac{\pm 180°(2k + 1)}{3 - 1} = 90°, -90°$$

The intersection of the asymptotes and the real exis is found from

$$\sigma_a = -\frac{0 + 0 + 3.6 - 1}{3 - 1} = -1.3$$

Since the characteristic equation is

$$s^3 + 3.6s^2 + K(s + 1) = 0$$

we have

$$K = -\frac{s^3 + 3.6s^2}{s + 1}$$

The breakaway and break-in points are found from

$$\frac{dK}{ds} = -\frac{(3s^2 + 7.2s)(s + 1) - (s^3 + 3.6s^2)}{(s + 1)^2} = 0$$

or

$$s^3 + 3.3s^2 + 3.6s = 0$$

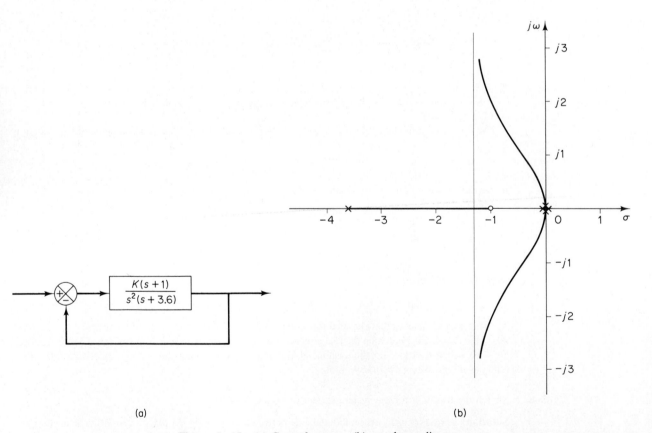

(a) (b)

Figure 5–45 (a) Control system; (b) root-locus diagram.

from which we get

$$s = 0, \qquad s = -1.65 + j0.9367, \qquad s = -1.65 - j0.9367$$

Point $s = 0$ corresponds to the actual breakaway point. But points $s = -1.65 \pm j0.9367$ are neither breakaway nor break-in points, because the corresponding gain values K become complex quantities.

To check the points where root-locus branches may cross the imaginary axis, substitute $s = j\omega$ into the characteristic equation.

$$(j\omega)^3 + 3.6(j\omega)^2 + Kj\omega + K = 0$$

or

$$(K - 3.6\omega^2) + j\omega(K - \omega^2) = 0$$

Notice that this equation can be satisfied only if $\omega = 0$, $K = 0$. Because of the presence of a double pole at the origin, the root locus is tangent to the $j\omega$ axis at $\omega = 0$. The root locus branches do not cross the $j\omega$ axis. Figure 5–45(b) shows a sketch of the root loci for this system.

A–5–9. Sketch the root loci for the system shown in Figure 5–46(a).

Solution. A root locus exists on the real axis between points $s = -0.4$ and $s = -3.6$. The asymptotes can be found as follows:

$$\text{Angles of asymptotes} = \frac{\pm 180°(2k + 1)}{3 - 1} = 90°, -90°$$

The intersection of the asymptotes and the real axis is obtained from

$$\sigma_a = -\frac{0 + 0 + 3.6 - 0.4}{3 - 1} = -1.6$$

Next we shall find the breakaway points. Since the characteristic equation is

$$s^3 + 3.6s^2 + Ks + 0.4K = 0$$

we have

$$K = -\frac{s^3 + 3.6s^2}{s + 0.4}$$

The breakaway and break-in points are found from

$$\frac{dK}{ds} = -\frac{(3s^2 + 7.2s)(s + 0.4) - (s^3 + 3.6s^2)}{(s + 0.4)^2} = 0$$

from which we get

$$s^3 + 2.4s^2 + 1.44s = 0$$

or

$$s(s + 1.2)^2 = 0$$

Thus, the breakaway or break-in points are at $s = 0$ and $s = -1.2$. Note that $s = -1.2$ is a double root. When a double root occurs in $dK/ds = 0$ at point $s = -1.2$, $d^2K/(ds^2) = 0$ at this point. The value of gain K at point $s = -1.2$ is

$$K = -\frac{s^3 + 3.6s^2}{s + 0.4}\bigg|_{s = -1.2} = 4.32$$

This means that with $K = 4.32$ the characteristic equation has a triple root at point $s = -1.2$. This can be easily verified as follows:

$$s^3 + 3.6s^2 + 4.32s + 1.728 = (s + 1.2)^3 = 0$$

Hence three root-locus branches meet at point $s = -1.2$. The angles of departures at point $s = -1.2$ of the root locus branches that approach the asymptotes are $\pm 180°/3$, that is, $60°$ and $-60°$. (See Problem A–5–10.)

Finally, we shall examine if root-locus branches cross the imaginary axis. By substituting $s = j\omega$ into the characteristic equation, we have

$$(j\omega)^3 + 3.6(j\omega)^2 + K(j\omega) + 0.4K = 0$$

or

$$(0.4K - 3.6\omega^2) + j\omega(K - \omega^2) = 0$$

This equation can be satisfied only if $\omega = 0$, $K = 0$. At point $\omega = 0$, the root locus is tangent to the $j\omega$ axis because of the presence of a double pole at the origin. There are no points that root-locus branches cross the imaginary axis.

A sketch of the root loci for this system is shown in Figure 5–46(b).

A–5–10. Referring to Problem A–5–9, obtain the equations for the root-locus branches for the system shown in Figure 5–46(a). Show that the root-locus branches cross the real axis at the breakaway point at angles $\pm 60°$.

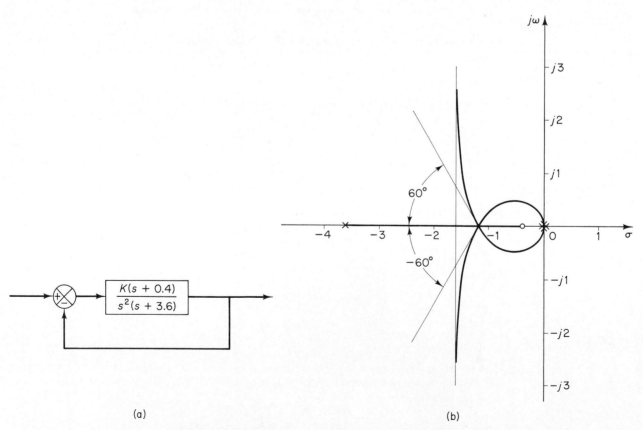

Figure 5–46   (a) Control system; (b) root-locus diagram.

Solution. The equations for the root-locus branches can be obtained from the angle condition

$$\left/ \frac{K(s + 0.4)}{s^2(s + 3.6)} \right. = \pm 180°(2k + 1)$$

which can be rewritten as

$$\angle s + 0.4 - 2 \angle s - \angle s + 3.6 = \pm 180°(2k + 1)$$

By substituting $s = \sigma + j\omega$, we obtain

$$\angle \sigma + j\omega + 0.4 - 2 \angle \sigma + j\omega - \angle \sigma + j\omega + 3.6 = \pm 180°(2k + 1)$$

or

$$\tan^{-1}\left(\frac{\omega}{\sigma + 0.4}\right) - 2 \tan^{-1}\left(\frac{\omega}{\sigma}\right) - \tan^{-1}\left(\frac{\omega}{\sigma + 3.6}\right) = \pm 180°(2k + 1)$$

By rearranging, we have

$$\tan^{-1}\left(\frac{\omega}{\sigma + 0.4}\right) - \tan^{-1}\left(\frac{\omega}{\sigma}\right) = \tan^{-1}\left(\frac{\omega}{\sigma}\right) + \tan^{-1}\left(\frac{\omega}{\sigma + 3.6}\right) \pm 180°(2k + 1)$$

Taking tangents of both sides of this last equation, and noting that

$$\tan\left[\tan^{-1}\left(\frac{\omega}{\sigma + 3.6}\right) \pm 180°(2k + 1)\right] = \frac{\omega}{\sigma + 3.6}$$

we obtain

$$\frac{\dfrac{\omega}{\sigma + 0.4} - \dfrac{\omega}{\sigma}}{1 + \dfrac{\omega}{\sigma + 0.4}\dfrac{\omega}{\sigma}} = \frac{\dfrac{\omega}{\sigma} + \dfrac{\omega}{\sigma + 3.6}}{1 - \dfrac{\omega}{\sigma}\dfrac{\omega}{\sigma + 3.6}}$$

which can be simplified to

$$\frac{\omega\sigma - \omega(\sigma + 0.4)}{(\sigma + 0.4)\sigma + \omega^2} = \frac{\omega(\sigma + 3.6) + \omega\sigma}{\sigma(\sigma + 3.6) - \omega^2}$$

or

$$\omega(\sigma^3 + 2.4\sigma^2 + 1.44\sigma + 1.6\omega^2 + \sigma\omega^2) = 0$$

which can be further simplified to

$$\omega[\sigma(\sigma + 1.2)^2 + (\sigma + 1.6)\omega^2] = 0$$

For $\sigma \neq -1.6$, we may write

$$\omega\left[\omega - (\sigma + 1.2)\sqrt{\frac{-\sigma}{\sigma + 1.6}}\right]\left[\omega + (\sigma + 1.2)\sqrt{\frac{-\sigma}{\sigma + 1.6}}\right] = 0$$

which gives the equations for the root-locus branches as follows:

$$\omega = 0$$

$$\omega = (\sigma + 1.2)\sqrt{\frac{-\sigma}{\sigma + 1.6}}$$

$$\omega = -(\sigma + 1.2)\sqrt{\frac{-\sigma}{\sigma + 1.6}}$$

Example Problems and Solutions

The equation $\omega = 0$ represents the real axis. The root locus for $0 \le K \le \infty$ is between points $s = -0.4$ and $s = -3.6$. (The real axis other than this line segment and the origin $s = 0$ corresponds to the root locus for $-\infty \le K < 0$.)

The equations

$$\omega = \pm(\sigma + 1.2) \sqrt{\dfrac{-\sigma}{\sigma + 1.6}} \qquad (5\text{–}34)$$

represent the complex branches for $0 \le K \le \infty$. These two branches lie between $\sigma = -1.6$ and $\sigma = 0$. [See Figure 5–46(b).] The slopes of the complex root-locus branches at the breakaway point ($\sigma = -1.2$) can be found by evaluating $d\omega/d\sigma$ of Equation (5–34) at point $\sigma = -1.2$.

$$\left.\dfrac{d\omega}{d\sigma}\right|_{\sigma = -1.2} = \pm \left.\sqrt{\dfrac{-\sigma}{\sigma + 1.6}}\right|_{\sigma = -1.2} = \pm \sqrt{\dfrac{1.2}{0.4}} = \pm\sqrt{3}$$

Since $\tan^{-1}\sqrt{3} = 60°$, the root-locus branches intersect the real axis with angles $\pm60°$.

A–5–11. Consider the system shown in Figure 5–47, which has an unstable feedforward transfer function. Sketch the root-locus plot and locate the closed-loop poles. Show that, although the closed-loop poles lie on the negative real axis and the system is not oscillatory, the unit-step response curve will exhibit overshoot.

Solution. The root-locus plot for this system is shown in Figure 5–48. The closed-loop poles are located at $s = -2$ and $s = -5$.

The closed-loop transfer function becomes

$$\dfrac{C(s)}{R(s)} = \dfrac{10(s + 1)}{s^2 + 7s + 10}$$

The unit-step response of this system is

$$C(s) = \dfrac{10(s + 1)}{s(s + 2)(s + 5)}$$

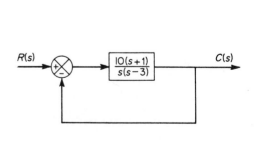

Figure 5–47 Control system.

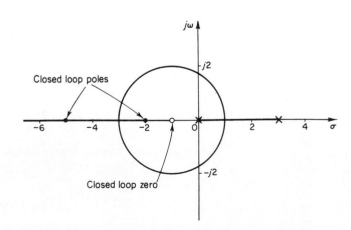

Figure 5–48 Root-locus diagram for the system shown in Fig. 5–47.

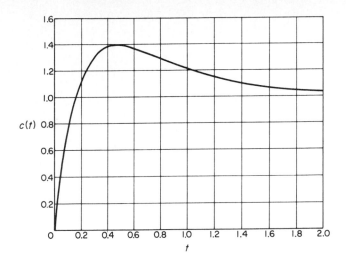

Figure 5-49
Unit-step response
curve for the system
shown in Fig. 5-47.

The inverse Laplace transform of $C(s)$ gives

$$c(t) = 1 + 1.666e^{-2t} - 2.666e^{-5t} \qquad (t \geq 0)$$

The unit-step response curve is shown in Figure 5–49. Although the system is not oscillatory, the unit-step response curve exhibits overshoot. (This is due to the presence of a zero at $s = -1$.)

A–5–12. Sketch the root loci of the control system shown in Figure 5–50(a). Determine the range of gain K for stability.

Solution. Open-loop poles are located at $s = 1$, $s = -2 + j\sqrt{3}$, and $s = -2 - j\sqrt{3}$. A root locus exists on the real axis between points $s = 1$ and $s = -\infty$. The asymptotes of the root-locus branches are found as follows:

$$\text{Angles of asymptotes} = \frac{\pm 180°(2k + 1)}{3} = 60°, -60°, 180°$$

The intersection of the asymptotes and the real axis is obtained from

$$\sigma_a = -\frac{-1 + 2 + 2}{3} = -1$$

The breakaway and break-in points can be located from $dK/ds = 0$. Since

$$K = -(s - 1)(s^2 + 4s + 7) = -(s^3 + 3s^2 + 3s - 7)$$

we have

$$\frac{dK}{ds} = -(3s^2 + 6s + 3) = 0$$

which yields

$$(s + 1)^2 = 0$$

Thus the equation $dK/ds = 0$ has a double root at $s = -1$. The breakaway point is located at $s = -1$. Three root locus branches meet at this breakaway point. The angles of departure of the branches at the breakaway point are $\pm 180°/3$, that is, $60°$ and $-60°$.

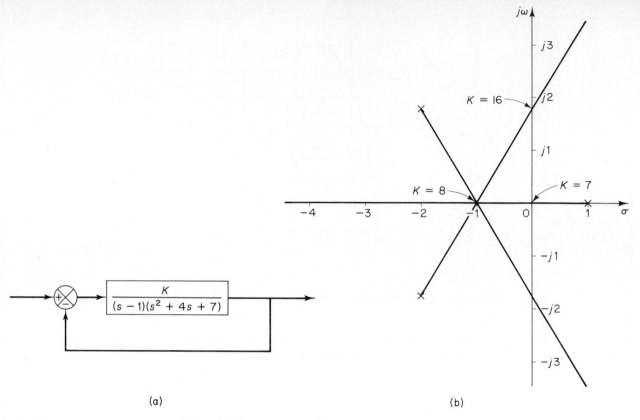

Figure 5–50   (a) Control system; (b) root-locus diagram.

We shall next determine the points where root-locus branches may cross the imaginary axis. Noting that the characteristic equation is

$$(s - 1)(s^2 + 4s + 7) + K = 0$$

or

$$s^3 + 3s^2 + 3s - 7 + K = 0$$

we substitute $s = j\omega$ into it and obtain

$$(j\omega)^3 + 3(j\omega)^2 + 3(j\omega) - 7 + K = 0$$

By rewriting this last equation, we have

$$(K - 7 - 3\omega^2) + j\omega(3 - \omega^2) = 0$$

This equation is satisfied when

$$\omega = \pm \sqrt{3}, \qquad K = 7 + 3\omega^2 = 16 \qquad \text{or} \qquad \omega = 0, \qquad K = 7$$

The root-locus branches cross the imaginary axis at $\omega = \pm \sqrt{3}$ (where $K = 16$) and $\omega = 0$ (where $K = 7$). Since the value of gain K at the origin is 7, the range of gain value K for stability is

$$7 < K < 16$$

Figure 5–50(b) shows a sketch of the root loci for the system. Notice that all branches consist of parts of straight lines.

The fact that the root-locus branches consist of straight lines can be verified as follows: Since the angle condition is

$$\left/ \frac{K}{(s-1)(s+2+j\sqrt{3})(s+2-j\sqrt{3})} \right. = \pm 180°(2k+1)$$

we have

$$-\underline{/s-1} - \underline{/s+2+j\sqrt{3}} - \underline{/s+2-j\sqrt{3}} = \pm 180°(2k+1)$$

By substituting $s = \sigma + j\omega$ into this last equation,

$$\underline{/\sigma - 1 + j\omega} + \underline{/\sigma + 2 + j\omega + j\sqrt{3}} + \underline{/\sigma + 2 + j\omega - j\sqrt{3}} = \pm 180°(2k+1)$$

or

$$\underline{/\sigma + 2 + j(\omega + \sqrt{3})} + \underline{/\sigma + 2 + j(\omega - \sqrt{3})} = -\underline{/\sigma - 1 + j\omega} \pm 180°(2k+1)$$

which can be rewritten as

$$\tan^{-1}\left(\frac{\omega + \sqrt{3}}{\sigma + 2}\right) + \tan^{-1}\left(\frac{\omega - \sqrt{3}}{\sigma + 2}\right) = -\tan^{-1}\left(\frac{\omega}{\sigma - 1}\right) \pm 180°(2k+1)$$

Taking tangents of both sides of this last equation, we obtain

$$\frac{\dfrac{\omega + \sqrt{3}}{\sigma + 2} + \dfrac{\omega - \sqrt{3}}{\sigma + 2}}{1 - \left(\dfrac{\omega + \sqrt{3}}{\sigma + 2}\right)\left(\dfrac{\omega - \sqrt{3}}{\sigma + 2}\right)} = -\frac{\omega}{\sigma - 1}$$

or

$$\frac{2\omega(\sigma + 2)}{\sigma^2 + 4\sigma + 4 - \omega^2 + 3} = -\frac{\omega}{\sigma - 1}$$

which can be simplified to

$$2\omega(\sigma + 2)(\sigma - 1) = -\omega(\sigma^2 + 4\sigma + 7 - \omega^2)$$

or

$$\omega(3\sigma^2 + 6\sigma + 3 - \omega^2) = 0$$

Further simplification of this last equation yields

$$\omega\left(\sigma + 1 + \frac{1}{\sqrt{3}}\omega\right)\left(\sigma + 1 - \frac{1}{\sqrt{3}}\omega\right) = 0$$

which defines three lines:

$$\omega = 0, \qquad \sigma + 1 + \frac{1}{\sqrt{3}}\omega = 0, \qquad \sigma + 1 - \frac{1}{\sqrt{3}}\omega = 0$$

Thus, the root-locus branches consist of three lines. Note that the root loci for $K > 0$ consist of portions of the straight lines as shown in Figure 5–50(b). (Note that each straight line extends to infinity in the direction of 180°, 60°, or −60° measured from the real axis.) The remaining portion of each straight line corresponds to $K < 0$.

Example Problems and Solutions

411

A–5–13. Consider the system shown in Figure 5–51(a). Sketch the root loci.

Solution. The open-loop zeros of the system are located at $s = \pm j$. The open-loop poles are located at $s = 0$ and $s = -2$. This system involves two poles and two zeroes. Hence, there is a possibility that a circular root locus branch exists. In fact, such a circular root locus exists in this case, as shown in the following. The angle condition is

$$\left/ \frac{K(s + j)(s - j)}{s(s + 2)} \right. = \pm 180°(2k + 1)$$

or

$$\underline{/s + j} + \underline{/s - j} - \underline{/s} - \underline{/s + 2} = \pm 180°(2k + 1)$$

By substituting $s = \sigma + j\omega$ into this last equation, we obtain

$$\underline{/\sigma + j\omega + j} + \underline{/\sigma + j\omega - j} = \underline{/\sigma + j\omega} + \underline{/\sigma + 2 + j\omega} \pm 180°(2k + 1)$$

or

$$\tan^{-1}\left(\frac{\omega + 1}{\sigma}\right) + \tan^{-1}\left(\frac{\omega - 1}{\sigma}\right) = \tan^{-1}\left(\frac{\omega}{\sigma}\right) + \tan^{-1}\left(\frac{\omega}{\sigma + 2}\right) \pm 180°(2k + 1)$$

Taking tangents of both sides of this equation and noting that

$$\tan\left[\tan^{-1}\left(\frac{\omega}{\sigma + 2}\right) \pm 180°\right] = \frac{\omega}{\sigma + 2}$$

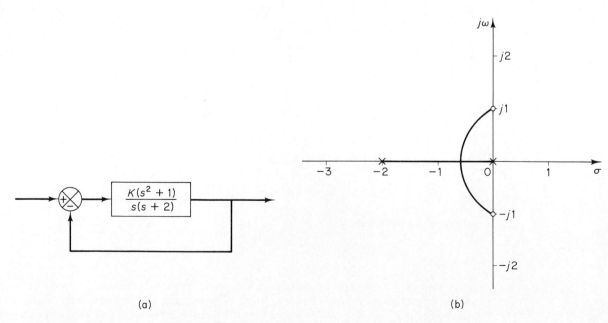

(a)   (b)

Figure 5–51   (a) Control system; (b) root-locus diagram.

we obtain

$$\frac{\dfrac{\omega+1}{\sigma}+\dfrac{\omega-1}{\sigma}}{1-\dfrac{\omega+1}{\sigma}\dfrac{\omega-1}{\sigma}}=\frac{\dfrac{\omega}{\sigma}+\dfrac{\omega}{\sigma+2}}{1-\dfrac{\omega}{\sigma}\dfrac{\omega}{\sigma+2}}$$

or

$$\omega\left[\left(\sigma-\frac{1}{2}\right)^2+\omega^2-\frac{5}{4}\right]=0$$

which is equivalent to

$$\omega=0 \quad\text{or}\quad \left(\sigma-\frac{1}{2}\right)^2+\omega^2=\frac{5}{4}$$

These two equations are equations for the root loci. The first equation corresponds to the root locus on the real axis. (The segment between $s=0$ and $s=-2$ corresponds to the root locus for $0\le K<\infty$. The remaining parts of the real axis correspond to the root locus for $K<0$.) The second equation is an equation for a circle. Thus, there exists a circular root locus with center at $\sigma=\frac{1}{2}$, $\omega=0$ and the radius equal to $\sqrt{5}/2$. The root loci are sketched in Figure 5–51(b). [That part of the circular locus to the left of the imaginary zeros corresponds to $K>0$. The portion of the circular locus not shown in Figure 5–51(b) corresponds to $K<0$.]

A–5–14. Consider the system shown in Figure 5–52. Determine the value of α such that the damping ratio ζ of the dominant closed-loop poles is 0.5.

Solution. In this system the characteristic equation is

$$1+\frac{2(s+\alpha)}{s(s+1)(s+3)}=0$$

Notice that the variable α is not a multiplying factor. Hence, we need to rewrite the characteristic equation

$$s(s+1)(s+3)+2s+2\alpha=0$$

as follows:

$$1+\frac{2\alpha}{s^3+4s^2+5s}=0$$

Define

$$2\alpha=K$$

Then we get the characteristic equation in the form

$$1+\frac{K}{s(s^2+4s+5)}=0 \tag{5–35}$$

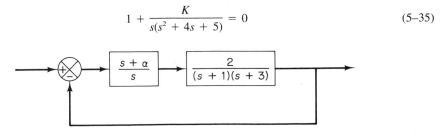

Figure 5–52
Control system.

In Problem A–5–6 we constructed the root-locus diagram for the system defined by Equation (5–35). Hence, the solution to this problem is available in Problem A–5–6. In that problem we obtained the closed-loop poles having the damping ratio $\zeta = 0.5$ located at $s = -0.63 \pm j1.09$. The value of K at point $s = -0.63 + j1.09$ was found in Problem A–5–6 as 4.32. Hence the value of α in this problem is obtained as follows:

$$\alpha = \frac{K}{2} = 2.16$$

A–5–15. Consider the system shown in Figure 5–53(a). Determine the value of α such that the damping ratio ζ of the dominant closed poles is 0.5.

Solution. The characteristic equation is

$$1 + \frac{10(s + \alpha)}{s(s + 1)(s + 8)} = 0$$

The variable α is not a multiplying factor. Hence we need to modify the characteristic equation. Since the characteristic equation can be written as

$$s^3 + 9s^2 + 18s + 10\alpha = 0$$

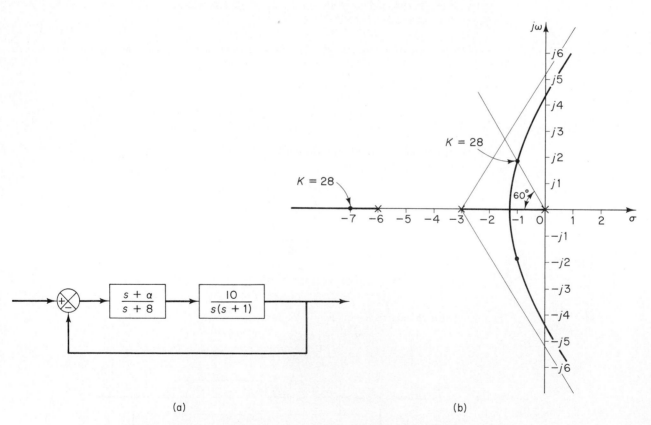

(a)   (b)

Figure 5–53 (a) Control system; (b) root-locus diagram, where $K = 10\alpha$.

we rewrite this equation such that α appears as a multiplying factor as follows:

$$1 + \frac{10\alpha}{s(s^2 + 9s + 18)} = 0$$

Define

$$10\alpha = K$$

Then the characteristic equation becomes

$$1 + \frac{K}{s(s^2 + 9s + 18)} = 0$$

Notice that the characteristic equation is in a suitable form for the construction of the root loci.

This system involves three poles and no zero. The three poles are at $s = 0$, $s = -3$, and $s = -6$. A root locus branch exists on the real axis between points $s = 0$ and $s = -3$. Also, another branch exists between points $s = -6$ and $s = -\infty$.

The asymptotes for the root loci are found as follows:

$$\text{Angles of asymptotes} = \frac{\pm 180°(2k + 1)}{3} = 60°, -60°, 180°$$

The intersection of the asymptotes and the real axis is obtained from

$$\sigma_a = -\frac{0 + 3 + 6}{3} = -3$$

The breakaway and break-in points can be determined from $dK/ds = 0$, where

$$K = -(s^3 + 9s^2 + 18s)$$

Now we set

$$\frac{dK}{ds} = -(3s^2 + 18s + 18) = 0$$

which yields

$$s^2 + 6s + 6 = 0$$

or

$$s = -1.268, \qquad s = -4.732$$

Point $s = -1.268$ is on a root-locus branch. Hence, point $s = -1.268$ is an actual breakaway point. But point $s = -4.732$ is not on the root locus and therefore is neither a breakaway nor break-in point.

Next we shall find points where root-locus branches cross the imaginary axis. We substitute $s = j\omega$ in the characteristic equation, which is

$$s^3 + 9s^2 + 18s + K = 0$$

as follows:

$$(j\omega)^3 + 9(j\omega)^2 + 18(j\omega) + K = 0$$

or

$$(K - 9\omega^2) + j\omega(18 - \omega^2) = 0$$

Example Problems and Solutions

from which we get

$$\omega = \pm 3\sqrt{2}, \qquad K = 9\omega^2 = 162 \qquad \text{or} \qquad \omega = 0, \qquad K = 0$$

The crossing points are at $\omega = \pm 3\sqrt{2}$ and the corresponding value of gain K is 162. Also, a root-locus branch touches the imaginary axis at $\omega = 0$. Figure 5–53(b) shows a sketch of the root loci for the system.

Since the damping ratio of the dominant closed-loop poles is specified as 0.5, the desired closed-loop pole in the upper-half s plane is located at the intersection of the root-locus branch in the upper-half s plane and a straight line having an angle of 60° with the negative real axis. The desired dominant closed-loop poles are located at

$$s = -1 + j1.732, \qquad s = -1 - j1.732$$

At these points, the value of gain K is 28. Hence

$$\alpha = \frac{K}{10} = 2.8$$

Since the system involves two or more poles than zeros (in fact, three poles and no zero), the third pole can be located on the negtive real axis from the fact that the sum of the three closed-loop poles is -9. Hence, the third pole is found to be at

$$s = -9 - (-1 + j1.732) - (-1 - j1.732)$$

or

$$s = -7$$

A–5–16. Consider the system shown in Figure 5–54(a). Sketch the root loci of the system as the velocity feedback gain k varies from zero to infinity. Determine the value of k such that the closed-loop poles have the damping ratio ζ of 0.7.

Solution. The characteristic equation for the system is

$$1 + \frac{10(1 + ks)}{s(s + 1)} = 0$$

Since k is not a multiplying factor, we modify the equation such that k appears as a multiplying factor. Since the characteristic equation is

$$s^2 + s + 10ks + 10 = 0$$

we rewrite this equation as follows:

$$1 + \frac{10ks}{s^2 + s + 10} = 0 \qquad\qquad (5\text{–}36)$$

Define

$$10k = K$$

Then Equation (5–36) becomes

$$1 + \frac{Ks}{s^2 + s + 10} = 0$$

Notice that a zero is located at the origin of the s plane, and poles are located at $s = -0.5 \pm j3.1225$. Since this system involves two poles and one zero, there is a possibility that a circular root locus exists.

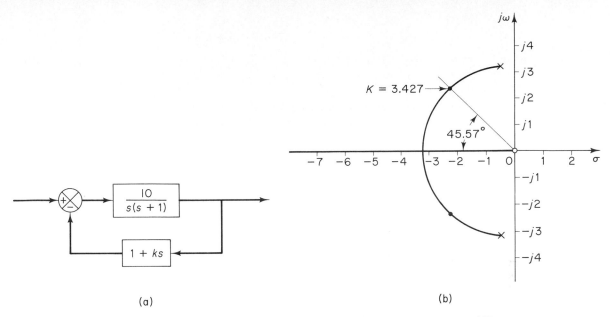

(a)   (b)

Figure 5–54 (a) Control system; (b) root-locus diagram, where $K = 10k$.

In fact, this system has a circular root locus, as will be shown. Since the angle condition is

$$\Big/ \frac{Ks}{s^2 + s + 10} = \pm 180°(2k + 1)$$

we have

$$\Big/ s - \Big/ s + 0.5 + j3.1225 - \Big/ s + 0.5 - j3.1225 = \pm 180°(2k + 1)$$

By substituting $s = \sigma + j\omega$ into this last equation and rearranging, we obtain

$$\Big/ \sigma + 0.5 + j(\omega + 3.1225) + \Big/ \sigma + 0.5 + j(\omega - 3.1225) = \Big/ \sigma + j\omega \pm 180°(2k + 1)$$

which can be rewritten as

$$\tan^{-1}\left(\frac{\omega + 3.1225}{\sigma + 0.5}\right) + \tan^{-1}\left(\frac{\omega - 3.1225}{\sigma + 0.5}\right) = \tan^{-1}\left(\frac{\omega}{\sigma}\right) \pm 180°(2k + 1)$$

Taking tangents of both sides of this last equation, we obtain

$$\frac{\dfrac{\omega + 3.1225}{\sigma + 0.5} + \dfrac{\omega - 3.1225}{\sigma + 0.5}}{1 - \left(\dfrac{\omega + 3.1225}{\sigma + 0.5}\right)\left(\dfrac{\omega - 3.1225}{\sigma + 0.5}\right)} = \frac{\omega}{\sigma}$$

which can be simplified to

$$\frac{2\omega(\sigma + 0.5)}{(\sigma + 0.5)^2 - (\omega^2 - 3.1225^2)} = \frac{\omega}{\sigma}$$

Example Problems and Solutions

or

$$\omega(\sigma^2 - 10 + \omega^2) = 0$$

which yields

$$\omega = 0 \quad \text{or} \quad \sigma^2 + \omega^2 = 10$$

Notice that $\omega = 0$ corresponds to the real axis. The negative real axis (between $s = 0$ and $s = -\infty$) corresponds to $K \geq 0$, and the positive real axis corresponds to $K < 0$. The equation

$$\sigma^2 + \omega^2 = 10$$

is an equation of a circle with center at $\sigma = 0$, $\omega = 0$ with the radius equal to $\sqrt{10}$. A portion of this circle that lies to the left of the complex poles corresponds to the root locus for $K > 0$. The portion of the circle which lies to the right of the complex poles corresponds to the root locus for $K < 0$. Hence, this portion is not a root locus for the present system, where $K > 0$. Figure 5–54(b) shows a sketch of the root loci.

Since we require $\zeta = 0.7$ for the closed-loop poles, we find the intersection of the circular root locus and a line having an angle of $45.57°$ (note that $\cos 45.57° = 0.7$) with the negative real axis. The intersection is at $s = -2.214 + j2.258$. The gain K corresponding to this point is 3.427. Hence, the desired value of the velocity feedback gain k is

$$k = \frac{K}{10} = 0.3427$$

A–5–17. Consider the system shown in Figure 5–55(a). It is a part of a control system. Sketch the root loci for the system. Determine the value of k such that the damping ratio ζ of the dominant closed-loop poles is 0.5.

Solution. The characteristic equation is

$$1 + \frac{s + 1}{s + 7} \frac{10}{s(s + 1) + 10k} = 0$$

Since the variable k is not a multiplying factor, we need to rewrite this last equation as

$$s^3 + 8s^2 + 17s + 10 + 10k(s + 7) = 0$$

which can be modified to

$$1 + \frac{10k(s + 7)}{(s + 1)(s + 2)(s + 5)} = 0$$

Define

$$10k = K$$

Then the characteristic equation can be rewritten in a form convenient for constructing the root loci as follows:

$$1 + \frac{K(s + 7)}{(s + 1)(s + 2)(s + 5)} = 0$$

Root-locus branches exist on the real axis between points $s = -1$ and $s = -2$ and between points $s = -5$ and $s = -7$. The asymptotes of the root locus branches are found as follows:

$$\text{Angles of asymptotes} = \frac{\pm 180°(2k + 1)}{3 - 1} = 90°, -90°$$

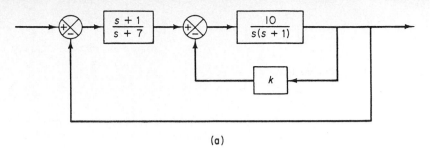

(a)

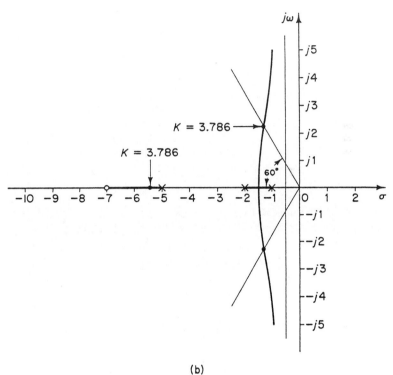

Figure 5–55
(a) System; (b) root-locus diagram, where $K = 10k$.

(b)

The intersection of the asymptotes and the real axis is obtained as

$$\sigma_a = -\frac{1 + 2 + 5 - 7}{3 - 1} = -0.5$$

The breakaway or break-in points can be found from $dK/ds = 0$. Since

$$K = -\frac{(s + 1)(s + 2)(s + 5)}{s + 7}$$

we have

$$\frac{dK}{ds} = -\frac{(3s^2 + 16s + 17)(s + 7) - (s^3 + 8s^2 + 17s + 10)}{(s + 7)^2} = 0$$

or

$$s^3 + 14.5s^2 + 56s + 54.5 = 0$$

which can be factored as

$$(s + 1.49)(s + 4.1175)(s + 8.8925) = 0$$

Thus we have

$$s = -1.49, \qquad s = -4.1175, \qquad s = -8.8925$$

Point $s = -1.49$ is on a root locus. Thus this point is an actual breakaway point. Points $s = -4.1175$ and $s = -8.8925$ are not on root-locus branches, and therefore these points are neither breakaway nor break-in points.

Next we shall examine if root-locus branches cross the $j\omega$ axis. By substituting $s = j\omega$ into the characteristic equation, which is

$$s^3 + 8s^2 + 17s + 10 + K(s + 7) = 0$$

we have

$$(j\omega)^3 + 8(j\omega)^2 + 17(j\omega) + 10 + K(j\omega) + 7K = 0$$

or

$$(10 + 7K - 8\omega^2) + j\omega(17 + K - \omega^2) = 0$$

From this last equation, we get

$$10 + 7K - 8\omega^2 = 0, \qquad 17 + K - \omega^2 = 0$$

which yields

$$\omega^2 = -109, \qquad K = -126$$

This implies that no root-locus branches cross the imaginary axis. Figure 5–55(b) shows a sketch of the root loci.

To locate the closed-loop poles with damping ratio $\zeta = 0.5$, we determine the point of intersection of the root-locus branch in the upper-half s plane and a line having an angle of 60° with the negative real axis. The intersection is located at $s = -1.30 + j2.252$. The value of K at this point is

$$K = -\left.\frac{(s + 1)(s + 2)(s + 5)}{s + 7}\right|_{s = -1.30 + j2.252} = 3.786$$

The desired value of k is then determined as

$$k = \frac{K}{10} = 0.3786$$

Note that the three closed-loop poles are located at

$$s = -1.30 + j2.252, \qquad s = -1.30 - j2.252, \qquad s = -5.4$$

A–5–18. Consider the system with transport lag shown in Figure 5–56(a). Sketch the root loci and find the two pairs of closed-loop poles nearest the $j\omega$ axis.

Using only the dominant closed-loop poles, obtain the unit-step response and sketch the response curve.

Solution. The characteristic equation is

$$\frac{2e^{-0.3s}}{s + 1} + 1 = 0$$

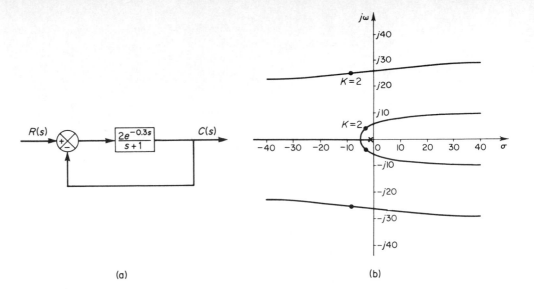

Figure 5–56
(a) Control system with transport lag; (b) root-locus diagram.

(a)

(b)

which is equivalent to the following angle and magnitude conditions:

$$\left/ \frac{2e^{-0.3s}}{s+1} \right. = \pm 180°(2k+1)$$

$$\left| \frac{2e^{-0.3s}}{s+1} \right| = 1$$

The angle condition reduces to

$$\angle s + 1 = \pm \pi(2k+1) - 0.3\omega \qquad \text{(radians)}$$

For $k = 0$,

$$\angle s + 1 = \pm \pi - 0.3\omega \qquad \text{(radians)}$$

$$= \pm 180° - 17.2°\omega \qquad \text{(degrees)}$$

For $k = 1$,

$$\angle s + 1 = \pm 3\pi - 0.3\omega \qquad \text{(radians)}$$

$$= \pm 540° - 17.2°\omega \qquad \text{(degrees)}$$

The root-locus plot for this system is shown in Figure 5–56(b).

Let us set $s = \sigma + j\omega$ in the magnitude condition and replace 2 by K. Then we obtain

$$\frac{\sqrt{(1+\sigma)^2 + \omega^2}}{e^{-0.3\sigma}} = K$$

By evaluating K at different points on the root loci, the points may be found for which $K = 2$. These points are closed-loop poles. The dominant pair of closed-loop poles is

$$s = -2.5 \pm j3.9$$

The next pair of closed-loop poles is

$$s = -8.6 \pm j25.1$$

Example Problems and Solutions

Using only the pair of dominant closed-loop poles, the closed-loop transfer function may be approximated as follows: Noting that

$$\frac{C(s)}{R(s)} = \frac{2e^{-0.3s}}{1 + s + 2e^{-0.3s}}$$

$$= \frac{2e^{-0.3s}}{1 + s + 2\left(1 - 0.3s + \dfrac{0.09s^2}{2} + \cdots\right)}$$

$$= \frac{2e^{-0.3s}}{3 + 0.4s + 0.09s^2 + \cdots}$$

and

$$(s + 2.5 + j3.9)(s + 2.5 - j3.9) = s^2 + 5s + 21.46$$

we may approximate $C(s)/R(s)$ by

$$\frac{C(s)}{R(s)} = \frac{\frac{2}{3}(21.46)e^{-0.3s}}{s^2 + 5s + 21.46}$$

or

$$\frac{C(s)}{R(s)} = \frac{14.31e^{-0.3s}}{(s + 2.5)^2 + 3.9^2}$$

For a unit-step input,

$$C(s) = \frac{14.31e^{-0.3s}}{[(s + 2.5)^2 + 3.9^2]s}$$

Note that

$$\frac{14.31}{[(s + 2.5)^2 + 3.9^2]s} = \frac{\frac{2}{3}}{s} + \frac{-\frac{2}{3}s - \frac{10}{3}}{(s + 2.5)^2 + 3.9^2}$$

Hence

$$C(s) = \left(\frac{\frac{2}{3}}{s}\right)e^{-0.3s} + \left[\frac{-\frac{2}{3}s - \frac{10}{3}}{(s + 2.5)^2 + 3.9^2}\right]e^{-0.3s}$$

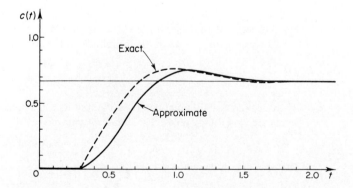

Figure 5–57
Unit-step response curves for the system shown in Fig. 5–56(a).

The inverse Laplace transform of $C(s)$ gives

$$c(t) = \tfrac{2}{3}[1 - e^{-2.5(t-0.3)}\cos 3.9(t - 0.3) - 0.641e^{-2.5(t-0.3)}\sin 3.9(t - 0.3)]1(t - 0.3)$$

where $1(t - 0.3)$ is the unit-step function occurring at $t = 0.3$.

Figure 5–57 shows the approximate response curve thus obtained, together with the exact unit-step response curve obtained by a computer simulation. Note that in this system a fairly good approximation can be obtained by use of only the dominant closed-loop poles.

PROBLEMS

B–5–1. Consider the unity-feedback system whose feedforward transfer function is

$$G(s) = \frac{K}{s(s + 1)}$$

The constant gain locus for the system for a given value of K is defined by the following equation:

$$\left| \frac{K}{s(s + 1)} \right| = 1$$

Show that the constant gain loci for $0 \leq K < \infty$ may be given by

$$[\sigma(\sigma + 1) + \omega^2]^2 + \omega^2 = K^2$$

Sketch the constant gain loci for $K = 1, 2, 5, 10,$ and 20 on the s plane.

B–5–2. Show that the root loci for a control system with

$$G(s) = \frac{K(s^2 + 6s + 10)}{s^2 + 2s + 10}, \qquad H(s) = 1$$

are arcs of the circle centered at the origin with radius equal to $\sqrt{10}$.

B–5–3. Sketch the root loci for a system with

$$G(s) = \frac{K}{(s^2 + 2s + 2)(s^2 + 2s + 5)}, \qquad H(s) = 1$$

Determine the exact points where the root loci cross the $j\omega$ axis.

B–5–4. Sketch the root loci for the system with

$$G(s) = \frac{K}{s(s + 0.5)(s^2 + 0.6s + 10)}, \qquad H(s) = 1$$

B–5–5. Consider the system shown in Figure 5–58. Determine the values of the gain K and the velocity feedback coefficient K_h so that the closed-loop poles are at $s = -1 \pm j\sqrt{3}$. Then, using the determined value of K_h, sketch the root loci.

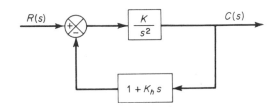

Figure 5–58 Control system.

B–5–6. Sketch the root loci for the system shown in Figure 5–59. If the value of gain K is set equal to 2, where are the closed-loop poles located?

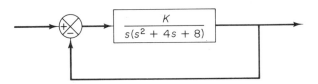

Figure 5–59 Control system.

B–5–7. Consider the system shown in Figure. 5–60. The system involves velocity feedback. Determine the value of gain K such that the dominant closed-loop poles have a damping ratio of 0.5. Using the gain K thus determined, obtain the unit-step response of the system.

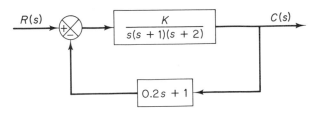

Figure 5–60 Control system.

B–5–8. Sketch the root loci for the closed-loop control system with

$$G(s) = \frac{K}{s(s + 1)(s^2 + 4s + 5)}, \qquad H(s) = 1$$

B–5–9. Sketch the root loci for a closed-loop control system with

$$G(s) = \frac{K(s + 9)}{s(s^2 + 4s + 11)}, \qquad H(s) = 1$$

Locate the closed-loop poles on the root loci such that the dominant closed-loop poles have a damping ratio equal to 0.5.

B–5–10. Sketch the root loci for a closed-loop control system with

$$G(s) = \frac{K(s + 0.2)}{s^2(s + 3.6)}, \qquad H(s) = 1$$

B–5–11. Sketch the root loci for a closed-loop control system with

$$G(s) = \frac{K(s + 0.5)}{s^3 + s^2 + 1}, \qquad H(s) = 1$$

B–5–12. Sketch the root loci for the system shown in Figure 5–61. Determine the range of gain K for stability.

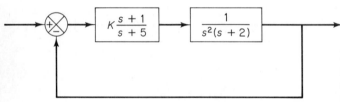

Figure 5–61 Control system.

B–5–13. Consider the system shown in Figure 5–62. Sketch the root loci as α varies from 0 to ∞. Determine the value of

α such that the damping ratio of the dominant closed-loop poles is 0.5.

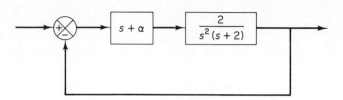

Figure 5–62 Control system.

B–5–14. Consider the system shown in Fig. 5–63. Sketch the root loci. Locate the closed-loop poles when the gain K is set equal to 2.

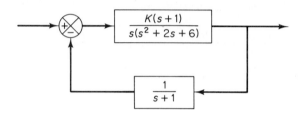

Figure 5–63 Control system.

B–5–15. Consider the system shown in Figure 5–64. Sketch the root loci as the value of k varies from 0 to ∞. What value of k will give the damping ratio of the dominant closed-loop poles equal to 0.5? Find the static velocity error constant with this value of k.

B–5–16. Sketch the root loci for the system shown in Figure 5–65. Show that the system may become unstable for large values of K.

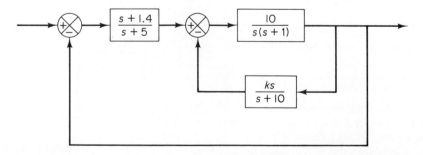

Figure 5–64 Control system.

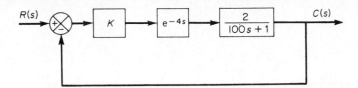

Figure 5–65 Control system.

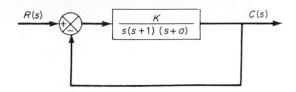

Figure 5–66 Control system.

B–5–17. Sketch the root contours for the system shown in Figure 5–66 when the gain K and parameter a vary, respectively, from zero to infinity.

B–5–18. Consider the system shown in Figure 5–67. Assuming that the value of gain K varies from 0 to ∞, sketch the root loci when $K_h = 0.5$. Then sketch the root contours for $0 \le K < \infty$ and $0 \le K_h < \infty$. Locate the closed-loop poles on the root contour when $K = 10$ and $K_h = 0.5$.

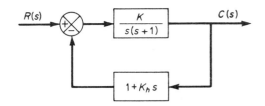

Figure 5–67 Control system.

CHAPTER 6

Frequency-Response Analysis

6-1 INTRODUCTION

Frequency-response methods. By the term frequency response, we mean the steady-state response of a system to a sinusoidal input. In the frequency-response methods, the most conventional methods available to control engineers for analysis and design of control systems, we vary the frequency of the input signal over a range of interest and study the resulting response.

Industrial control systems are often designed by use of frequency-response methods. Many techniques are available in the frequency-response methods for the analysis and design of control systems. They will be discussed in this and next chapters.

The Nyquist stability criterion, to be given in Section 6–5, enables us to investigate both the absolute and relative stabilities of linear closed-loop systems from a knowledge of their open-loop frequency-response characteristics. In using this stability criterion, we do not have to determine the roots of the characteristic equation. This is one advantage of the frequency-response approach. Another advantage of this approach is that frequency-response tests are, in general, simple and can be made accurately by use of readily available sinusoidal signal generators and precise measurement equipments. Often the transfer functions of complicated components can be determined experimentally by frequency-response tests. Such experimentally obtained transfer functions can be easily incorporated in the frequency-response approach. Also, the frequency-response methods can be applied to systems that do not have rational functions, such as those with transport lags. In addition, plants with uncertainties or plants that are poorly known can be handled by the frequency-response methods. A system may be designed by use of the frequency-response approach such that the effects of undesirable noises are negligible. Finally, frequency-response analyses and designs can be extended to certain nonlinear control systems.

Although the frequency response of a control system presents a qualitative picture of the transient response, the correlation between frequency and transient responses is indirect, except for the case of second-order systems. In designing a closed-loop system, we may adjust the frequency-response characteristic by using several design criteria (to be given later in this chapter) in order to obtain acceptable transient-response characteristics.

In what follows, we shall obtain the relationship between the transfer function and the frequency response of the linear stable system.

Obtaining steady-state outputs to sinusoidal inputs.   We shall first prove the basic fact that the frequency-response characteristics of a system can be obtained directly from the sinusoidal transfer function, that is, the transfer function in which s is replaced by $j\omega$, where ω is frequency.

Consider the stable linear time-invariant system shown in Figure 6–1. The input and output of the system, whose transfer function is $G(s)$, are denoted by $x(t)$ and $y(t)$, respectively. If the input $x(t)$ is a sinusoidal signal, the steady-state output will also be a sinusoidal signal of the same frequency but with possibly different magnitude and phase angle.

Let us assume that the input signal is given by

$$x(t) = X \sin \omega t$$

Suppose that the transfer function $G(s)$ can be written as a ratio of two polynomials in s; that is,

$$G(s) = \frac{p(s)}{q(s)} = \frac{p(s)}{(s + s_1)(s + s_2) \cdots (s + s_n)}$$

The Laplace-transformed output $Y(s)$ is then

$$Y(s) = G(s)X(s) = \frac{p(s)}{q(s)} X(s) \qquad (6\text{–}1)$$

where $X(s)$ is the Laplace transform of the input $x(t)$.

It will be shown that after waiting until steady-state conditions are reached the frequency response can be calculated by replacing s in the transfer function by $j\omega$. It will also be shown that the steady-state response can be given by

$$G(j\omega) = Me^{j\phi} = M\angle\phi$$

where M is the amplitude ratio of the output and input sinusoids and ϕ is the phase shift between the input sinusoid and the output sinusoid. In the frequency-response test, the input frequency ω is varied until the entire frequency range of interest is covered.

The steady-state response of a stable linear time-invariant system to a sinusoidal input does not depend on the initial conditions. (Thus, we can assume the zero initial condition.)

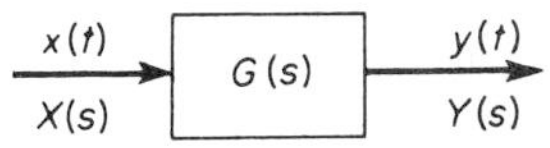

Figure 6–1
Stable linear time-invariant system.

$$\xrightarrow[X(s)]{x(t)} \boxed{G(s)} \xrightarrow[Y(s)]{y(t)}$$

If $Y(s)$ has only distinct poles, then the partial fraction expansion of Equation (6–1) yields

$$Y(s) = G(s)X(s) = G(s)\frac{\omega X}{s^2 + \omega^2}$$

$$= \frac{a}{s + j\omega} + \frac{\bar{a}}{s - j\omega} + \frac{b_1}{s + s_1} + \frac{b_2}{s + s_2} + \cdots + \frac{b_n}{s + s_n} \qquad (6\text{–}2)$$

where a and the b_i (where $i = 1, 2, \ldots, n$) are constants and $\bar{a}$ is the complex conjugate of a. The inverse Laplace transform of Equation (6–2) gives

$$y(t) = ae^{-j\omega t} + \bar{a}e^{j\omega t} + b_1 e^{-s_1 t} + b_2 e^{-s_2 t} + \cdots + b_n e^{-s_n t} \qquad (t \geq 0) \qquad (6\text{–}3)$$

For a stable system, $-s_1, -s_2, \ldots, -s_n$ have negative real parts. Therefore, as t approaches infinity, the terms $e^{-s_1 t}, e^{-s_2 t}, \ldots,$ and $e^{-s_n t}$ approach zero. Thus, all the terms on the right side of Equation (6–3), except the first two, drop out at steady state.

If $Y(s)$ involves multiple poles s_j of multiplicity m_j, then $y(t)$ will involve terms such as $t^{h_j} e^{-s_j t}$ ($h_j = 0, 1, 2, \ldots, m_j - 1$). For a stable system, the terms $t^{h_j} e^{-s_j t}$ approach zero as t approaches infinity.

Thus, regardless of whether or not the system is of the distinct-pole type, the steady-state response becomes

$$y_{ss}(t) = ae^{-j\omega t} + \bar{a}e^{j\omega t} \qquad (6\text{–}4)$$

where the constant a can be evaluated from Equation (6–2) as follows:

$$a = G(s)\frac{\omega X}{s^2 + \omega^2}(s + j\omega)\bigg|_{s = -j\omega} = -\frac{XG(-j\omega)}{2j}$$

Note that

$$\bar{a} = G(s)\frac{\omega X}{s^2 + \omega^2}(s - j\omega)\bigg|_{s = j\omega} = \frac{XG(j\omega)}{2j}$$

Since $G(j\omega)$ is a complex quantity, it can be written in the following form:

$$G(j\omega) = |G(j\omega)|e^{j\phi}$$

where $|(Gj\omega)|$ represents the magnitude and ϕ represents the angle of $G(j\omega)$; that is,

$$\phi = \angle G(j\omega) = \tan^{-1}\left[\frac{\text{imaginary part of } G(j\omega)}{\text{real part of } G(j\omega)}\right]$$

The angle ϕ may be negative, positive, or zero. Similarly, we obtain the following expression for $G(-j\omega)$:

$$G(-j\omega) = |G(-j\omega)|e^{-j\phi} = |G(j\omega)|e^{-j\phi}$$

Then, Equation (6–4) can be written

$$y_{ss}(t) = X|G(j\omega)|\frac{e^{j(\omega t + \phi)} - e^{-j(\omega t + \phi)}}{2j}$$

$$= X|G(j\omega)|\sin(\omega t + \phi)$$

$$= Y\sin(\omega t + \phi) \qquad (6\text{–}5)$$

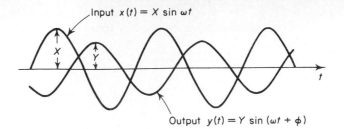

Figure 6–2
Input and output
sinusoidal signals.

Input $x(t) = X \sin \omega t$

Output $y(t) = Y \sin (\omega t + \phi)$

where $Y = X|G(j\omega)|$. We see that a stable linear time-invariant system subjected to a sinusoidal input will, at steady state, have a sinusoidal output of the same frequency as the input. But the amplitude and phase of the output will, in general, be different from those of the input. In fact, the amplitude of the output is given by the product of that of the input and $|G(j\omega)|$, while the phase angle differs from that of the input by the amount $\phi = \angle G(j\omega)$. An example of input and output sinusoidal signals is shown in Figure 6–2.

On the basis of this, we obtain this important result: For sinusoidal inputs,

$$|G(j\omega)| = \left|\frac{Y(j\omega)}{X(j\omega)}\right| = \text{amplitude ratio of the output sinusoid to the input sinusoid}$$

$$\angle G(j\omega) = \angle \frac{Y(j\omega)}{X(j\omega)} = \text{phase shift of the output sinusoid with respect to the input sinusoid}$$

Hence, the response characteristics of a system to a sinusoidal input can be obtained directly from

$$\frac{Y(j\omega)}{X(j\omega)} = G(j\omega)$$

The function $G(j\omega)$ is called the *sinusoidal transfer function*. It is the ratio of $Y(j\omega)$ to $X(j\omega)$, is a complex quantity, and can be represented by the magnitude and phase angle with frequency as a parameter. (A negative phase angle is called *phase lag*, and a positive phase angle is called *phase lead*.) The sinusoidal transfer function of any linear system is obtained by substituting $j\omega$ for s in the transfer function of the system. To completely characterize a linear system in the frequency domain, we must specify both the amplitude ratio and the phase angle as functions of the frequency ω.

EXAMPLE 6–1

Consider the system shown in Figure 6–3. The transfer function $G(s)$ is

$$G(s) = \frac{K}{Ts + 1}$$

For the sinusoidal input $x(t) = X \sin \omega t$, the steady-state output $y_{ss}(t)$ can be found as follows: Substituting $j\omega$ for s in $G(s)$ yields

$$G(j\omega) = \frac{K}{jT\omega + 1}$$

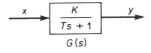

Figure 6–3
First-order system.

$G(s)$

The amplitude ratio of the output to input is

$$|G(j\omega)| = \frac{K}{\sqrt{1 + T^2\omega^2}}$$

while the phase angle ϕ is

$$\phi = \underline{/G(j\omega)} = -\tan^{-1} T\omega$$

Thus, for the input $x(t) = X \sin \omega t$, the steady-state output $y_{ss}(t)$ can be obtained from Equation (6–5) as follows:

$$y_{ss}(t) = \frac{XK}{\sqrt{1 + T^2\omega^2}} \sin (\omega t - \tan^{-1} T\omega) \qquad (6\text{--}6)$$

From Equation (6–6), it can be seen that for small ω the amplitude of the steady-state output $y_{ss}(t)$ is almost equal to K times the amplitude of the input. The phase shift of the output is small for small ω. For large ω, the amplitude of the output is small and almost inversely proportional to ω. The phase shift approaches $-90°$ as ω approaches infinity.

Frequency response from pole–zero plots. The frequency response can be determined graphically from the pole–zero plot of the transfer function. Consider the following transfer function:

$$G(s) = \frac{K(s + z)}{s(s + p)}$$

where p and z are real. The frequency response of this transfer function can be obtained from

$$G(j\omega) = \frac{K(j\omega + z)}{j\omega(j\omega + p)}$$

The factors $j\omega + z$, $j\omega$, and $j\omega + p$ are complex quantities, as shown in Figure 6–4. The magnitude of $G(j\omega)$ is

$$|G(j\omega)| = \frac{K|j\omega + z|}{|j\omega||j\omega + p|}$$

$$= \frac{K|\overline{AP}|}{|\overline{OP}| \cdot |\overline{BP}|}$$

and the phase angle of $G(j\omega)$ is

$$\underline{/G(j\omega)} = \underline{/j\omega + z} - \underline{/j\omega} - \underline{/j\omega + p}$$

$$= \tan^{-1} \frac{\omega}{z} - 90° - \tan^{-1} \frac{\omega}{p}$$

$$= \phi - \theta_1 - \theta_2$$

where the angles ϕ, θ_1, and θ_2 are defined in Figure 6–4. Note that a counterclockwise rotation is defined as the positive direction for the measurement of angles.

From the transient-response analysis of closed-loop systems, we know that a complex-conjugate pair of poles near the $j\omega$ axis will produce a highly oscillatory mode of transient

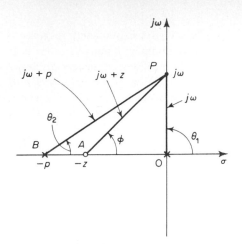

Figure 6–4
Determination of the
frequency response in
the complex plane.

response. In the case of the frequency response, such a pair of poles will produce a highly peaked response.

Consider, for example, the following transfer function:

$$G(s) = \frac{K}{(s + p_1)(s + p_2)}$$

where p_1 and p_2 are complex conjugates, as shown in Figure 6–5. The frequency response of this transfer function can be found from

$$|G(j\omega)| = \frac{K}{|j\omega + p_1||j\omega + p_2|}$$

$$= \frac{K}{|\overline{AP}||\overline{BP}|}$$

$$\angle G(j\omega) = -\theta_1 - \theta_2$$

where the angles θ_1 and θ_2 are defined in Figure 6–5. Since $|\overline{AP}||\overline{BP}|$ is very small near $\omega = \omega_1$, $|G(j\omega_1)|$ is very large. Thus a pair of complex-conjugate poles near the $j\omega$ axis will cause a highly peaked frequency response.

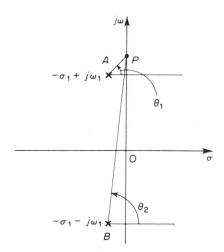

Figure 6–5
Determination of the
frequency response in
the complex plane.

Conversely, if the frequency response is not highly peaked, then the transfer function will not have complex-conjugate poles near the $j\omega$ axis. Such a transfer function will not exhibit highly oscillatory transient response either. Since the frequency response indirectly describes the location of the poles and zeros of the transfer function, we can estimate the transient-response characteristics of a system from frequency-response characteristics. We shall present a detailed discussion of this subject in Section 6–7.

Once we understand the indirect correlation between several measures of the transient response and the frequency response, the frequency-response approach may be used to advantage. The design of a control system through this approach is based on the interpretation of the desired dynamic characteristics in terms of the frequency-response characteristics. Such analysis of a control system indicates graphically what changes have to be made in the open-loop transfer function in order to obtain the desired transient-response characteristics.

Outline of the chapter. Section 6–1 has presented an introductory material on the frequency response and has introduced the sinusoidal transfer function. Section 6–2 presents Bode diagrams and discusses methods for plotting Bode diagrams. Section 6–3 treats polar plots of sinusoidal transfer functions. Section 6–4 briefly discusses log-magnitude versus phase plots. Section 6–5 presents detailed account of Nyquist stability criterion. Section 6–6 discusses stability analysis of closed-loop systems using the Nyquist stability criterion. Section 6–7 treats relative stability analysis of closed-loop systems. Measures of relative stability such as phase margin and gain margin are introduced. The correlation between the transient response and frequency response is also discussed. Section 6–8 presents a method for obtaining closed-loop frequency response from the open-loop frequency response by use of M and N circles. Use of the Nichols chart is discussed for obtaining the closed-loop frequency response. Finally, Section 6–9 deals with the determination of the transfer function based on an experimental Bode diagram.

6–2 BODE DIAGRAMS

This section and the following two are concerned primarily with presenting frequency-response characteristics of linear control systems.

The sinusoidal transfer function, a complex function of the frequency ω, is characterized by its magnitude and phase angle, with frequency as the parameter. There are three commonly used representations of sinusoidal transfer functions. They are

1. Bode diagram or logarithmic plot
2. Polar plot
3. Log-magnitude versus phase plot

This section presents Bode diagrams of sinusoidal transfer functions. Polar plots and log-magnitude versus phase plots are presented, respectively, in Sections 6–3 and 6–4.

Bode diagrams or logarithmic plots. A sinusoidal transfer function may be represented by two separate plots, one giving the magnitude versus frequency and the other the phase angle (in degrees) versus frequency. A Bode diagram consists of two graphs: One is a plot of the logarithm of the magnitude of a sinusoidal transfer function; the other is a plot of the phase angle; both are plotted against the frequency in logarithmic scale.

The standard representation of the logarithmic magnitude of $G(j\omega)$ is $20 \log |G(j\omega)|$, where the base of the logarithm is 10. The unit used in this representation of the magnitude is the decibel, usually abbreviated db. In the logarithmic representation, the curves are drawn on semilog paper, using the log scale for frequency and the linear scale for either magnitude (but in decibels) or phase angle (in degrees). (The frequency range of interest determines the number of logarithmic cycles required on the abscissa.)

The main advantage of using the logarithmic plot is that multiplication of magnitudes can be converted into addition. Furthermore, a simple method for sketching an approximate log-magnitude curve is available. It is based on asymptotic approximations. Such approximation by straight-line asymptotes is sufficient if only rough information on the frequency-response characteristics is needed. Should exact curves be desired, corrections can be made easily to these basic asymptotic ones. The phase-angle curves can be drawn easily if a template for the phase-angle curve of $1 + j\omega$ is available.

Note that the experimental determination of a transfer function can be made simple if frequency-response data are presented in the form of a Bode diagram.

The logarithmic representation is useful in that it shows both the low- and high-frequency characteristics of the transfer function in one diagram. Expanding the low-frequency range by use of a logarithmic scale for the frequency is very advantageous since characteristics at low frequencies are most important in practical systems. Although it is not possible to plot the curves right down to zero frequency because of the logarithmic frequency ($\log 0 = -\infty$), this does not create a serious problem.

Basic factors of $G(j\omega)H(j\omega)$.   As stated earlier, the main advantage in using the logarithmic plot is in the relative ease of plotting frequency-response curves. The basic factors that very frequently occur in an arbitrary transfer function $G(j\omega)H(j\omega)$ are

1. Gain K
2. Integral and derivative factors $(j\omega)^{\mp 1}$
3. First-order factors $(1 + j\omega T)^{\pm 1}$
4. Quadratic factors $[1 + 2\zeta(j\omega/\omega_n) + (j\omega/\omega_n)^2]^{\mp 1}$

Once we become familiar with the logarithmic plots of these basic factors, it is possible to utilize them in constructing a composite logarithmic plot for any general form of $G(j\omega)H(j\omega)$ by sketching the curves for each factor and adding individual curves graphically because adding the logarithms of the gains corresponds to multiplying them together.

The process of obtaining the logarithmic plot can be further simplified by using asymptotic approximations to the curves for each factor. (If necessary, corrections can be made easily to an approximate plot to obtain an accurate one.)

The gain K.   A number greater than unity has a positive value in decibels, while a number smaller than unity has a negative value. The log-magnitude curve for a constant gain K is a horizontal straight line at the magnitude of $20 \log K$ decibels. The phase angle of the gain K is zero. The effect of varying the gain K in the transfer function is that it raises or lowers the log-magnitude curve of the transfer function by the corresponding constant amount, but it has no effect on the phase angle.

A number–decibel conversion line is given in Figure 6–6. The decibel value of any number can be obtained from this line. As a number increases by a factor of 10, the

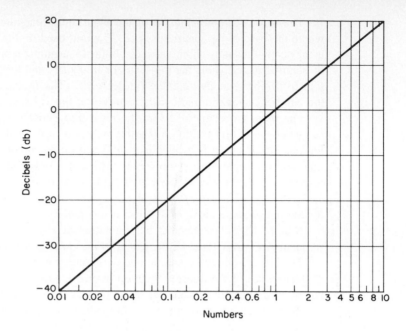

Figure 6–6
Number–decibel
conversion line.

corresponding decibel value increases by a factor of 20. This may be seen from the following:

$$20 \log (K \times 10^n) = 20 \log K + 20n$$

Note that, when expressed in decibels, the reciprocal of a number differs from its value only in sign; that is, for the number K,

$$20 \log K = -20 \log \frac{1}{K}$$

Integral and derivative factors $(j\omega)^{\mp 1}$.   The logarithmic magnitude of $1/j\omega$ in decibels is

$$20 \log \left| \frac{1}{j\omega} \right| = -20 \log \omega \ \text{db}$$

The phase angle of $1/j\omega$ is constant and equal to $-90°$.

In Bode diagrams, frequency ratios are expressed in terms of octaves or decades. An octave is a frequency band from ω_1 to $2\omega_1$, where ω_1 is any frequency value. A decade is a frequency band from ω_1 to $10\omega_1$ where again ω_1 is any frequency. (On the logarithmic scale of semilog paper, any given frequency ratio can be represented by the same horizontal distance. For example, the horizontal distance from $\omega = 1$ to $\omega = 10$ is equal to that from $\omega = 3$ to $\omega = 30$.)

If the log magnitude $-20 \log \omega$ db is plotted against ω on a logarithmic scale, it is a straight line. To draw this straight line, we need to locate one point (0 db, $\omega = 1$) on it. Since

$$(-20 \log 10\omega) \ \text{db} = (-20 \log \omega - 20) \ \text{db}$$

the slope of the line is -20 db/decade (or -6 db/octave).

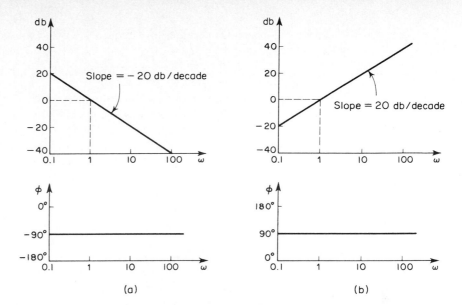

Figure 6–7
(a) Frequency-response curves of $1/j\omega$; (b) frequency-response curves of $j\omega$.

(a)   (b)

Similarly, the log magnitude of $j\omega$ in decibels is

$$20 \log |j\omega| = 20 \log \omega \text{ db}$$

The phase angle of $j\omega$ is constant and equal to 90°. The log-magnitude curve is a straight line with a slope of 20 db/decade. Figures 6–7(a) and (b) show frequency-response curves for $1/j\omega$ and $j\omega$, respectively. We can clearly see that the differences in the frequency responses of the factors $1/j\omega$ and $j\omega$ lie in the signs of the slopes of the log-magnitude curves and in the signs of the phase angles. Both log magnitudes become equal in 0 db at $\omega = 1$.

If the transfer function contains the factor $(1/j\omega)^n$ or $(j\omega)^n$, the log magnitude becomes, respectively,

$$20 \log \left| \frac{1}{(j\omega)^n} \right| = -n \times 20 \log |j\omega| = -20n \log \omega \text{ db}$$

or

$$20 \log |(j\omega)^n| = n \times 20 \log |j\omega| = 20n \log \omega \text{ db}$$

The slopes of the log-magnitude curves for the factors $(1/j\omega)^n$ and $(j\omega)^n$ are then $-20n$ db/decade and $20n$ db/decade, respectively. The phase angle of $(1/j\omega)^n$ is equal to $-90° \times n$ over the entire frequency range, while that of $(j\omega)^n$ is equal to $90° \times n$ over the entire frequency range. These magnitude curves will pass through the point (0 db, $\omega = 1$).

First-order factors $(1 + j\omega T)^{\mp 1}$. The log magnitude of the first-order factor $1/(1 + j\omega T)$ is

$$20 \log \left| \frac{1}{1 + j\omega T} \right| = -20 \log \sqrt{1 + \omega^2 T^2} \text{ db}$$

For low frequencies, such that $\omega \ll 1/T$, the log magnitude may be approximated by

$$-20 \log \sqrt{1 + \omega^2 T^2} \doteq -20 \log 1 = 0 \text{ db}$$

Thus, the log-magnitude curve at low frequencies is the constant 0-db line. For high frequencies, such that $\omega \gg 1/T$,

$$-20 \log \sqrt{1 + \omega^2 T^2} \doteq -20 \log \omega T \text{ db}$$

This is an approximate expression for the high-frequency range. At $\omega = 1/T$, the log magnitude equals 0 db; at $\omega = 10/T$, the log magnitude is -20 db. Thus, the value of $-20 \log \omega T$ db decreases by 20 db for every decade of ω. For $\omega \gg 1/T$, the log-magnitude curve is thus a straight line with a slope of -20 db/decade (or -6 db/octave.)

The analysis above shows that the logarithmic representation of the frequency-response curve of the factor $1/(1 + j\omega T)$ can be approximated by two straight-line asymptotes, one a straight line at 0 db for the frequency range $0 < \omega < 1/T$ and the other a straight line with slope -20 db/decade (or -6 db/octave) for the frequency range $1/T < \omega < \infty$. The exact log-magnitude curve, the asymptotes, and the exact phase-angle curve are shown in Figure 6–8.

The frequency at which the two asymptotes meet is called the *corner* frequency or *break* frequency. For the factor $1/(1 + j\omega T)$, the frequency $\omega = 1/T$ is the corner frequency since at $\omega = 1/T$ the two asymptotes have the same value. (The low-frequency asymptotic expression at $\omega = 1/T$ is 20 log 1 db = 0 db and the high-frequency asymptotic expression at $\omega = 1/T$ is also 20 log 1 db = 0 db.) The corner frequency divides the frequency-response curve into two regions, a curve for the low-frequency region and a curve for the high-frequency region. The corner frequency is very important in sketching logarithmic frequency-response curves.

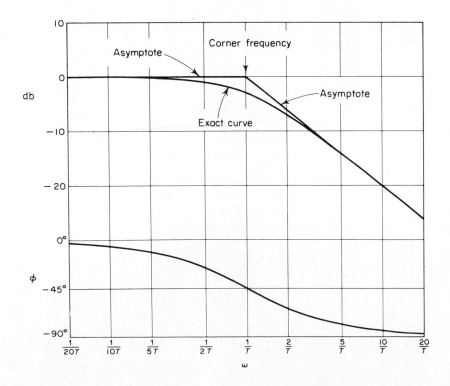

Figure 6–8
Log-magnitude curve together with the asymptotes and phase angle curve of $1/(1 + j\omega T)$.

The exact phase angle ϕ of the factor $1/(1 + j\omega T)$ is

$$\phi = -\tan^{-1} \omega T$$

At zero frequency, the phase angle is $0°$. At the corner frequency, the phase angle is

$$\phi = -\tan^{-1} \frac{T}{T} = -\tan^{-1} 1 = -45°$$

At infinity, the phase angle becomes $-90°$. Since the phase angle is given by an inverse-tangent function, the phase angle is skew symmetric about the inflection point at $\phi = -45°$.

The error in the magnitude curve caused by the use of asymptotes can be calculated. The maximum error occurs at the corner frequency and is approximately equal to -3 db since

$$-20 \log \sqrt{1 + 1} + 20 \log 1 = -10 \log 2 = -3.03 \text{ db}$$

The error at the frequency one octave below the corner frequency, that is, at $\omega = 1/2T$, is

$$-20 \log \sqrt{\frac{1}{4} + 1} + 20 \log 1 = -20 \log \frac{\sqrt{5}}{2} = -0.97 \text{ db}$$

The error at the frequency one octave above the corner frequency, that is, at $\omega = 2/T$, is

$$-20 \log \sqrt{2^2 + 1} + 20 \log 2 = -20 \log \frac{\sqrt{5}}{2} = -0.97 \text{ db}$$

Thus, the error at one octave below or above the corner frequency is approximately equal to -1 db. Similarly, the error at one decade below or above the corner frequency is approximately -0.04 db. The error in decibels involved in using the asymptotic expression for the frequency-response curve of $1/(1 + j\omega T)$ is shown in Figure 6–9. The error is symmetric with respect to the corner frequency.

Since the asymptotes are quite easy to draw and are sufficiently close to the exact curve, the use of such approximations in drawing Bode diagrams is convenient in establishing the general nature of the frequency-response characteristics quickly with a minimum amount of calculation and may be used for most preliminary design work. If accurate frequency-response curves are desired, corrections may easily be made by referring to the curve given in Figure 6–9. In practice, an accurate frequency-response curve can be drawn by introducing a cor-

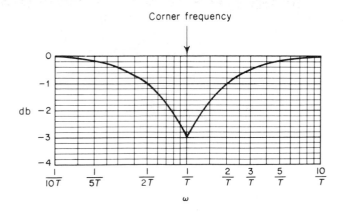

Figure 6–9
Log-magnitude error in the asymptotic expression of the frequency-response curve of $1/(1 + j\omega T)$.

rection of 3 db at the corner frequency and a correction of 1 db at points one octave below and above the corner frequency and then connecting these points by a smooth curve.

Note that varying the time constant T shifts the corner frequency to the left or to the right, but the shapes of the log-magnitude and the phase-angle curves remain the same.

The transfer function $1/(1 + j\omega T)$ has the characteristics of a low-pass filter. For frequencies above $\omega = 1/T$, the log magnitude falls off rapidly toward $-\infty$. This is essentially due to the presence of the time constant. In the low-pass filter, the output can follow a sinusoidal input faithfully at low frequencies. But as the input frequency is increased, the output cannot follow the input because a certain amount of time is required for the system to build up in magnitude. Thus, at high frequencies, the amplitude of the output approaches zero and the phase angle of the output approaches $-90°$. Therefore, if the input function contains many harmonics, then the low-frequency components are reproduced faithfully at the output, while the high-frequency components are attenuated in amplitude and shifted in phase. Thus, a first-order element yields exact, or almost exact, duplication only for constant or slowly varying phenomena.

An advantage of the Bode diagram is that for reciprocal factors, for example, the factor $1 + j\omega T$, the log-magnitude and the phase-angle curves need only be changed in sign. Since

$$20 \log |1 + j\omega T| = -20 \log \left| \frac{1}{1 + j\omega T} \right|$$

$$\angle 1 + j\omega T = \tan^{-1} \omega T = -\angle \frac{1}{1 + j\omega T}$$

the corner frequency is the same for both cases. The slope of the high-frequency asymptote of $1 + j\omega T$ is 20 db/decade, and the phase angle varies from 0 to 90° as the frequency ω is increased from zero to infinity. The log-magnitude curve together with the asymptotes and the phase-angle curve for the factor $1 + j\omega T$ are shown in Figure 6–10.

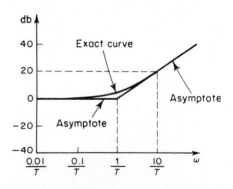

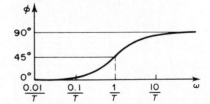

Figure 6–10
Log-magnitude curve together with the asymptotes and phase-angle curve for $1 + j\omega T$.

The shapes of phase-angle curves are the same for any factor of the form $(1 + j\omega T)^{\mp 1}$. Hence, it is convenient to have a template for the phase-angle curve on cardboard. Then such a template may be used repeatedly for constructing phase-angle curves for any function of the form $(1 + j\omega T)^{\mp 1}$. If such a template is not available, we have to locate several points on the curve. The phase angles of $(1 + j\omega T)^{\mp 1}$ are

$$\mp 45° \quad \text{at} \quad \omega = \frac{1}{T}$$

$$\mp 26.6° \quad \text{at} \quad \omega = \frac{1}{2T}$$

$$\mp 5.7° \quad \text{at} \quad \omega = \frac{1}{10T}$$

$$\mp 63.4° \quad \text{at} \quad \omega = \frac{2}{T}$$

$$\mp 84.3° \quad \text{at} \quad \omega = \frac{10}{T}$$

For the case where a given transfer function involves terms like $(1 + j\omega T)^{\mp n}$, a similar asymptotic construction may be made. The corner frequency is still at $\omega = 1/T$, and the asymptotes are straight lines. The low-frequency asymptote is a horizontal straight line at 0 db, while the high-frequency asymptote has the slope of $-20n$ db/decade or $20n$ db/decade. The error involved in the asymptotic expressions is n times that for $(1 + j\omega T)^{\mp 1}$. The phase angle is n times that of $(1 + j\omega T)^{\mp 1}$ at each frequency point.

Quadratic factors $[1 + 2\zeta(j\omega/\omega_n) + (j\omega/\omega_n)^2]^{\mp 1}$. Control systems often possess quadratic factors of the form

$$\frac{1}{1 + 2\zeta\left(j\dfrac{\omega}{\omega_n}\right) + \left(j\dfrac{\omega}{\omega_n}\right)^2} \tag{6-7}$$

If $\zeta > 1$, this quadratic factor can be expressed as a product of two first-order ones with real poles. If $0 < \zeta < 1$, this quadratic factor is the product of two complex-conjugate factors. Asymptotic approximations to the frequency-response curves are not accurate for a factor with low values of ζ. This is because the magnitude and phase of the quadratic factor depend on both the corner frequency and the damping ratio ζ.

The asymptotic frequency-response curve may be obtained as follows: Since

$$20 \log \left| \frac{1}{1 + 2\zeta\left(j\dfrac{\omega}{\omega_n}\right) + \left(j\dfrac{\omega}{\omega_n}\right)^2} \right|$$

$$= -20 \log \sqrt{\left(1 - \frac{\omega^2}{\omega_n^2}\right)^2 + \left(2\zeta\frac{\omega}{\omega_n}\right)^2}$$

This function is composed of the following factors:

$$7.5, \qquad (j\omega)^{-1}, \qquad 1 + j\frac{\omega}{3}, \qquad \left(1 + j\frac{\omega}{2}\right)^{-1}, \qquad \left[1 + j\frac{\omega}{2} + \frac{(j\omega)^2}{2}\right]^{-1}$$

The corner frequencies of the third, fourth, and fifth terms are $\omega = 3$, $\omega = 2$, and $\omega = \sqrt{2}$, respectively. Note that the last term has the damping ratio of 0.3536.

To plot the Bode diagram, the separate asymptotic curves for each of the factors are shown in Figure 6–13. The composite curve is then obtained by adding algebraically the individual curves, also shown in Figure 6–13. Note that when the individual asymptotic curves are added at each frequency, the slope of the composite curve is cumulative. Below $\omega = \sqrt{2}$, the plot has the slope of -20 db/decade. At the first corner frequency $\omega = \sqrt{2}$, the slope changes to -60 db/decade and continues to the next corner frequency $\omega = 2$, where the slope becomes -80 db/decade. At the last corner frequency $\omega = 3$, the slope changes to -60 db/decade.

Once such an approximate log-magnitude curve has been drawn, the actual curve can be obtained by adding corrections at each corner frequency and at frequencies one octave below and above the corner frequencies. For first-order factors $(1 + j\omega T)^{\mp 1}$, the corrections are ± 3 db at the corner frequency and ± 1 db at the frequencies one octave below and above the corner frequency. Corrections necessary

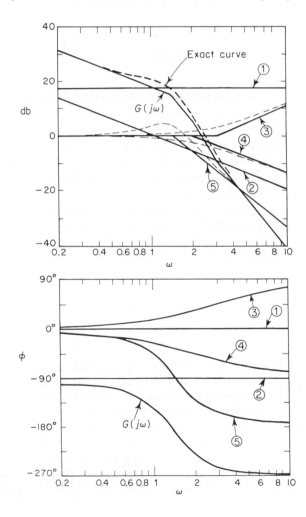

Figure 6–13
Bode diagram of the system considered in Example 6–2.

for the quadratic factor are obtained from Figure 6–11. The exact log-magnitude curve for $G(j\omega)$ is shown by a dotted curve in Figure 6–13.

Note that any change in the slope of the magnitude curve is made only at the corner frequencies of the transfer function $G(j\omega)$. Therefore, instead of drawing individual magnitude curves and adding them up, as shown above, we may sketch the magnitude curve without sketching individual curves. We may start drawing the lowest-frequency portion of the straight line (that is, the straight line with the slope -20 db/decade for $\omega < \sqrt{2}$). As the frequency is increased, we get the effect of the complex-conjugate poles (quadratic term) at the corner frequency $\omega = \sqrt{2}$. The complex-conjugate poles cause the slopes of the magnitude curve to change from -20 to -60 db/decade. At the next corner frequency, $\omega = 2$, the effect of the pole is to change the slope to -80 db/decade. Finally, at the corner frequency $\omega = 3$, the effect of the zero is to change the slope from -80 to -60 db/decade.

For plotting the complete phase-angle curve, the phase-angle curves for all factors have to be sketched. The algebraic sum of all phase-angle curves provides the complete phase-angle curve, as shown in Figure 6–13.

The simplicity in plotting the Bode diagram should now be apparent. It is noted that such a Bode diagram can be plotted by using a computer. However, it is important that the designer be familiar with a conventional graphical construction of Bode diagrams so that any erroneous computer results can be detected and identified.

Minimum phase systems and nonminimum phase systems.   Transfer functions having neither poles nor zeros in the right-half s plane are minimum-phase transfer functions, whereas those having poles and/or zeros in the right-half s plane are nonminimum phase transfer functions. Systems with minimum phase transfer functions are called *minimum phase* systems, whereas those with nonminimum phase transfer functions are called *nonminimum phase* systems.

For systems with the same magnitude characteristic, the range in phase angle of the minimum phase transfer function is minimum for all such systems, while the range in phase angle of any nonminimum phase transfer function is greater than this minimum.

It is noted that for a minimum phase system the transfer function can be uniquely determined from the magnitude curve alone. For a nonminimum phase system, this is not the case. Multiplying any transfer function by all-pass filters does not alter the magnitude curve, but the phase curve is changed.

Consider as an example the two systems whose sinusoidal transfer functions are, respectively,

$$G_1(j\omega) = \frac{1 + j\omega T}{1 + j\omega T_1}, \qquad G_2(j\omega) = \frac{1 - j\omega T}{1 + j\omega T_1} \qquad (0 < T < T_1)$$

The pole–zero configurations of these systems are shown in Figure 6–14. The two sinusoidal transfer functions have the same magnitude characteristics, but they have different phase-angle characteristics, as shown in Figure 6–15. These two systems differ from each other by the factor

$$G(j\omega) = \frac{1 - j\omega T}{1 + j\omega T}$$

The magnitude of the factor $(1 - j\omega T)/(1 + j\omega T)$ is always unity. But the phase angle equals $-2 \tan^{-1} \omega T$ and varies from $0°$ to $-180°$ as ω is increased from zero to infinity.

As stated earlier, for a minimum phase system, the magnitude and phase-angle characteristics are uniquely related. This means that if the magnitude curve of a system is specified

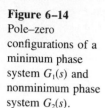

Figure 6–14
Pole–zero
configurations of a
minimum phase
system $G_1(s)$ and
nonminimum phase
system $G_2(s)$.

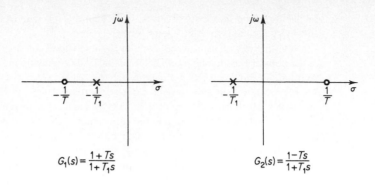

Figure 6–15
Phase angle
characteristics of the
systems $G_1(s)$ and
$G_2(s)$ shown in Fig.
6–14.

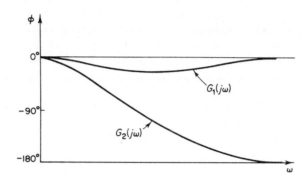

over the entire frequency range from zero to infinity, then the phase-angle curve is uniquely determined, and vice versa. This, however, does not hold for a nonminimum phase system.

Nonminimum phase situations may arise in two different ways. One is simply when a system includes a nonminimum phase element or elements. The other situation may arise in the case where a minor loop is unstable.

For a minimum phase system, the phase angle at $\omega = \infty$ becomes $-90°(q - p)$, where p and q are the degrees of the numerator and denominator polynomials of the transfer function, respectively. For a nonminimum phase system, the phase angle at $\omega = \infty$ differs from $-90°(q - p)$. In either system, the slope of the log-magnitude curve at $\omega = \infty$ is equal to $-20(q - p)$ db/decade. It is therefore possible to detect whether or not the system is minimum phase by examining both the slope of the high-frequency asymptote of the log-magnitude curve and the phase angle at $\omega = \infty$. If the slope of the log-magnitude curve as ω approaches infinity is $-20(q - p)$ db/decade and the phase angle at $\omega = \infty$ is equal to $-90°(q - p)$, then the system is minimum phase.

Nonminimum phase systems are slow in response because of their faulty behavior at the start of response. In most practical control systems, excessive phase lag should be carefully avoided. In designing a system, if fast speed of response is of primary importance, we should not use nonminimum phase components. (A common example of nonminimum phase elements that may be present in control system is transport lag.)

It is noted that the techniques of frequency-response analysis and design to be presented in this and the next chapter are valid for both minimum phase and nonminimum phase systems.

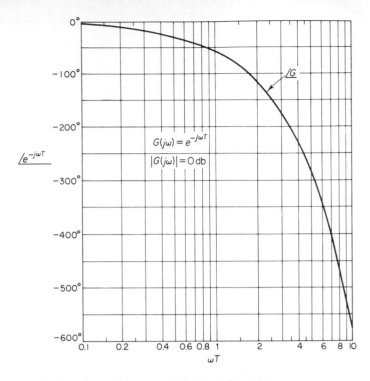

Figure 6–16
Phase angle characteristic of transport lag.

Transport lag.   Transport lag is of nonminimum phase behavior and has an excessive phase lag with no attentuation at high frequencies. Such transport lags normally exist in thermal, hydraulic, and pneumatic systems.

Consider the transport lag given by

$$G(j\omega) = e^{-j\omega T}$$

The magnitude is always equal to unity since

$$|G(j\omega)| = |\cos \omega T - j \sin \omega T| = 1$$

Therefore, the log magnitude of the transport lag $e^{-j\omega T}$ is equal to 0 db. The phase angle of the transport lag is

$$\angle G(j\omega) = -\omega T \quad \text{(radians)}$$
$$= -57.3 \, \omega T \quad \text{(degrees)}$$

The phase angle varies linearly with the frequency ω. The phase-angle characteristic of transport lag is shown in Figure 6–16.

EXAMPLE 6–3.   Draw the Bode diagram of the following transfer function:

$$G(j\omega) = \frac{e^{-j\omega L}}{1 + j\omega T}$$

The log magnitude is

$$20 \log|G(j\omega)| = 20 \log|e^{-j\omega L}| + 20 \log \left| \frac{1}{1 + j\omega T} \right|$$
$$= 0 + 20 \log \left| \frac{1}{1 + j\omega T} \right|$$

Section 6–2   /   Bode Diagrams

447

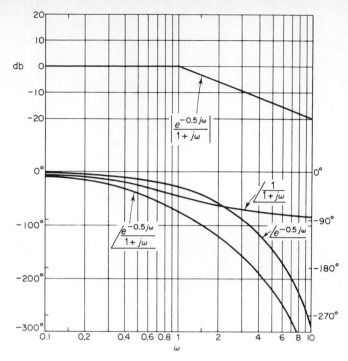

Figure 6–17
Bode diagram for the system $e^{-j\omega L}/(1 + j\omega T)$ with $L = 0.5$ and $T = 1$.

The phase angle of $G(j\omega)$ is

$$\angle G(j\omega) = \angle e^{-j\omega L} + \left\angle \frac{1}{1 + j\omega T}\right.$$

$$= -\omega L - \tan^{-1}\omega T$$

The log-magnitude and phase-angle curves for this transfer function with $L = 0.5$ and $T = 1$ are shown in Figure 6–17.

Relationship between system type and log-magnitude curve. The static position, velocity, and acceleration error constants describe the low-frequency behavior of type 0, type 1, and type 2 systems, respectively. For a given system, only one of the static error constants is finite and significant. (The larger the value of the finite static error constant, the higher the loop gain is as ω approaches zero.)

The type of the system determines the slope of the log-magnitude curve at low frequencies. Thus, information concerning the existence and magnitude of the steady-state error of a control system to a given input can be determined from the observation of the low-frequency region of the log-magnitude curve.

Determination of static position error constants. Figure 6–18 shows an example of the log-magnitude plot of a type 0 system. In such a system, the magnitude of $G(j\omega)H(j\omega)$ equals K_p at low frequencies, or

$$\lim_{\omega \to 0} G(j\omega)H(j\omega) = K_p$$

It follows that the low-frequency asymptote is a horizontal line at $20 \log K_p$ db.

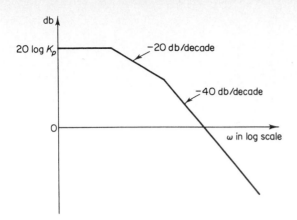

Figure 6–18 Log-magnitude curve of a type 0 system.

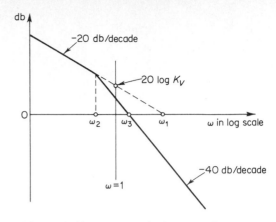

Figure 6–19 Log-magnitude curve of a type 1 system.

Determination of static velocity error constants. Figure 6–19 shows an example of the log-magnitude plot of a type 1 system. The intersection of the initial -20-db/decade segment (or its extension) with the line $\omega = 1$ has the magnitude $20 \log K_v$. This may be seen as follows: In a type 1 system,

$$G(j\omega)H(j\omega) = \frac{K_v}{j\omega} \qquad \text{for } \omega \ll 1$$

Thus,

$$20 \log \left| \frac{K_v}{j\omega} \right|_{\omega=1} = 20 \log K_v$$

The intersection of the initial -20-db/decade segment (or its extension) with the 0-db line has a frequency numerically equal to K_v. To see this, define the frequency at this intersection to be ω_1; then

$$\left| \frac{K_v}{j\omega_1} \right| = 1$$

or

$$K_v = \omega_1$$

As an example, consider the type 1 system with unity feedback whose open-loop transfer function is

$$G(s) = \frac{K}{s(Js + F)}$$

If we define the corner frequency to be ω_2 and the frequency at the intersection of the -40-db/decade segment (or its extension) with 0-db line to be ω_3, then

$$\omega_2 = \frac{F}{J}, \qquad \omega_3^2 = \frac{K}{J}$$

Since

$$\omega_1 = K_v = \frac{K}{F}$$

it follows that

$$\omega_1 \omega_2 = \omega_3^2$$

or

$$\frac{\omega_1}{\omega_3} = \frac{\omega_3}{\omega_2}$$

On the Bode diagram

$$\log \omega_1 - \log \omega_3 = \log \omega_3 - \log \omega_2$$

Thus, the ω_3 point is just midway between the ω_2 and ω_1 points. The damping ratio ζ of the system is then

$$\zeta = \frac{F}{2\sqrt{KJ}} = \frac{\omega_2}{2\omega_3}$$

Determination of static acceleration error constants. Figure 6–20 shows an example of the log-magnitude plot of a type 2 system. The intersection of the initial -40-db/decade segment (or its extension) with the $\omega = 1$ line has the magnitude of 20 log K_a. Since at low frequencies

$$G(j\omega)H(j\omega) = \frac{K_a}{(j\omega)^2}$$

it follows that

$$20 \log \left| \frac{K_a}{(j\omega)^2} \right|_{\omega=1} = 20 \log K_a$$

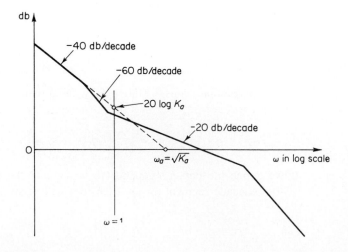

Figure 6–20
Log-magnitude curve
of a type 2 system.

The frequency ω_a at the intersection of the initial -40-db/decade segment (or its extension) with the 0-db line gives the square root of K_a numerically. This can be seen from the following:

$$20 \log \left| \frac{K_a}{(j\omega_a)^2} \right| = 20 \log 1 = 0$$

which yields

$$\omega_a = \sqrt{K_a}$$

6–3 POLAR PLOTS

The polar plot of a sinusoidal transfer function $G(j\omega)$ is a plot of the magnitude of $G(j\omega)$ versus the phase angle of $G(j\omega)$ on polar coordinates as ω is varied from zero to infinity. Thus, the polar plot is the locus of vectors $|G(j\omega)|\underline{/G(j\omega)}$ as ω is varied from zero to infinity. Note that in polar plots, a positive (negative) phase angle is measured counterclockwise (clockwise) from the positive real axis. The polar plot is often called the Nyquist plot. An example of such a plot is shown in Figure 6–21. Each point on the polar plot of $G(j\omega)$ represents the terminal point of a vector at a particular value of ω. In the polar plot, it is important to show the frequency graduation of the locus. The projections of $G(j\omega)$ on the real and imaginary axes are its real and imaginary components. Both the magnitude $|G(j\omega)|$ and phase angle $\underline{/G(j\omega)}$ must be calculated directly for each frequency ω in order to construct polar plots. Since the logarithmic plot is easy to construct, however, the data necessary for plotting the polar plot may be obtained directly from the logarithmic plot if the latter is drawn first and decibels are converted into ordinary magnitude. Or, of course, a digital computer may be used to obtain $|G(j\omega)|$ and $\underline{/G(j\omega)}$ accurately for various values of ω in the frequency range of interest.

For two systems connected in cascade, the overall transfer function of the combination in the absence of loading effects is the product of the two individual transfer functions. If multiplication of two sinusoidal transfer functions is necessary, this can be done by multiplying

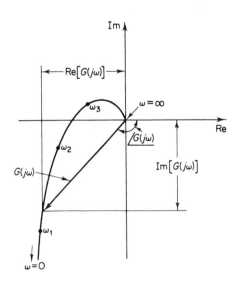

Figure 6–21
Polar plot.

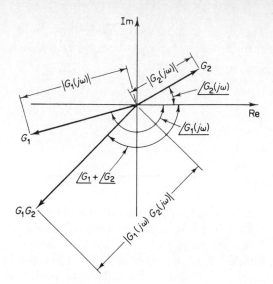

Figure 6–22
Polar plots of $G_1(j\omega)$, $G_2(j\omega)$, and $G_1(j\omega)G_2(j\omega)$.

together at each frequency the individual sinusoidal transfer functions by means of complex-algebra multiplication. That is, if $G(j\omega) = G_1(j\omega)G_2(j\omega)$, then

$$G(j\omega) = |G(j\omega)|\underline{/G(j\omega)}$$

where

$$|G(j\omega)| = |G_1(j\omega)| \cdot |G_2(j\omega)|$$

and

$$\underline{/G(j\omega)} = \underline{/G_1(j\omega)} + \underline{/G_2(j\omega)}$$

The product of $G_1(j\omega)$ and $G_2(j\omega)$ is shown in Figure 6–22.

In general, if a polar plot of $G_1(j\omega)G_2(j\omega)$ is desired, it is convenient to draw a logarithmic plot of $G_1(j\omega)G_2(j\omega)$ first and then convert this into a polar plot rather than to draw the polar plots of $G_1(j\omega)$ and $G_2(j\omega)$ and multiply these two on the complex plane to obtain a polar plot of $G_1(j\omega)G_2(j\omega)$. Again, a digital computer can also be used to obtain the product of $G_1(j\omega)$ and $G_2(j\omega)$.

An advantage in using a polar plot is that it depicts the frequency-response characteristics of a system over the entire frequency range in a single plot. One disadvantage is that the plot does not clearly indicate the contributions of each of the individual factors of the open-loop transfer function.

Integral and derivative factors $(j\omega)^{\mp 1}$. The polar plot of $G(j\omega) = 1/j\omega$ is the negative imaginary axis since

$$G(j\omega) = \frac{1}{j\omega} = -j\frac{1}{\omega} = \frac{1}{\omega}\underline{/-90°}$$

The polar plot of $G(j\omega) = j\omega$ is the positive imaginary axis.

First-order factors $(1 + j\omega T)^{\mp 1}$. For the sinusoidal transfer function

$$G(j\omega) = \frac{1}{1 + j\omega T} = \frac{1}{\sqrt{1 + \omega^2 T^2}}\underline{/-\tan^{-1}\omega T}$$

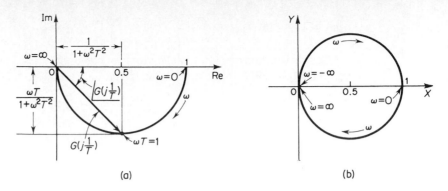

Figure 6–23
(a) Polar plot of $1/(1 + j\omega T)$; (b) plot of $G(j\omega)$ in X-Y plane.

(a)　　　　　　　(b)

the values of $G(j\omega)$ at $\omega = 0$ and $\omega = 1/T$ are, respectively,

$$G(j0) = 1\underline{/0°} \quad \text{and} \quad G\left(j\frac{1}{T}\right) = \frac{1}{\sqrt{2}}\underline{/-45°}$$

If ω approaches infinity, the magnitude of $G(j\omega)$ approaches zero and the phase angle approaches $-90°$. The polar plot of this transfer function is a semicircle as the frequency ω is varied from zero to infinity, as shown in Figure 6–23(a). The center is located at 0.5 on the real axis, and the radius is equal to 0.5.

To prove that the polar plot is a semicircle, define

$$G(j\omega) = X + jY$$

where

$$X = \frac{1}{1 + \omega^2 T^2} = \text{real part of } G(j\omega)$$

$$Y = \frac{-\omega T}{1 + \omega^2 T^2} = \text{imaginary part of } G(j\omega)$$

Then we obtain

$$\left(X - \frac{1}{2}\right)^2 + Y^2 = \left(\frac{1}{2}\frac{1 - \omega^2 T^2}{1 + \omega^2 T^2}\right)^2 + \left(\frac{-\omega T}{1 + \omega^2 T^2}\right)^2 = \left(\frac{1}{2}\right)^2$$

Thus, in the X-Y plane $G(j\omega)$ is a circle with center at $X = \frac{1}{2}$, $Y = 0$ and with radius $\frac{1}{2}$, as shown in Figure 6–23(b). The lower semicircle corresponds to $0 \leq \omega \leq \infty$ and the upper semicircle corresponds to $-\infty \leq \omega \leq 0$.

The polar plot of the transfer function $1 + j\omega T$ is simply the upper half of the straight line passing through point $(1, 0)$ in the complex plane and parallel to the imaginary axis, as shown in Figure 6–24. The polar plot of $1 + j\omega T$ has an appearance completely different

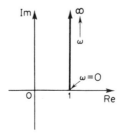

Figure 6–24
Polar plot of $1 + j\omega T$.

from that of $1/(1 + j\omega T)$.

Quadratic factors $[1 + 2\zeta(j\omega/\omega_n) + (j\omega/\omega_n)^2]^{\mp 1}$. The low- and high-frequency portions of the polar lot of the following sinusoidal transfer function

$$G(j\omega) = \frac{1}{1 + 2\zeta\left(j\dfrac{\omega}{\omega_n}\right) + \left(j\dfrac{\omega}{\omega_n}\right)^2} \qquad (\zeta > 0)$$

are given, respectively, by

$$\lim_{\omega \to 0} G(j\omega) = 1\underline{/0°} \qquad \text{and} \qquad \lim_{\omega \to \infty} G(j\omega) = 0\underline{/-180°}$$

The polar plot of this sinusoidal transfer function starts at $1\underline{/0°}$ and ends at $0\underline{/-180°}$ as ω increases from zero to infinity. Thus, the high-frequency portion of $G(j\omega)$ is tangent to the negative real axis. The values of $G(j\omega)$ in the frequency range of interest can be calculated directly or by use of the logarithmic plot.

Examples of polar plots of the transfer function just considered are shown in Figure 6–25. The exact shape of a polar plot depends on the value of the damping ratio ζ, but the general shape of the plot is the same for both the underdamped case $(1 > \zeta > 0)$ and overdamped case $(\zeta > 1)$.

For the underdamped case at $\omega = \omega_n$, we have $G(j\omega_n) = 1/(j2\zeta)$ and the phase angle at $\omega = \omega_n$ is $-90°$. Therefore, it can be seen that the frequency at which the $G(j\omega)$ locus intersects the imaginary axis is the undamped natural frequency, ω_n. In the polar plot, the frequency point whose distance from the origin is maximum corresponds to the resonant frequency ω_r. The peak value of $G(j\omega)$ is obtained as the ratio of the magnitude of the vector

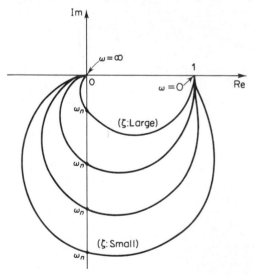

Figure 6–25 Polar plots of
$$\frac{1}{1 + 2\zeta\left(j\dfrac{\omega}{\omega_n}\right) + \left(j\dfrac{\omega}{\omega_n}\right)^2}, \qquad (\zeta > 0).$$

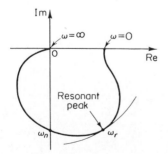

Figure 6–26
Polar plot showing the resonant peak and resonant frequency ω_r.

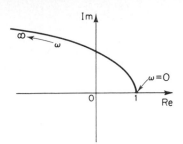

Figure 6-27 Polar plot of $1 + 2\zeta\left(j\dfrac{\omega}{\omega_n}\right) + \left(j\dfrac{\omega}{\omega_n}\right)^2$, $(\zeta > 0)$.

at the resonant frequency ω_r to the magnitude of the vector at $\omega = 0$. The resonant frequency ω_r is indicated in the polar plot shown in Figure 6-26.

For the overdamped case, as ζ increases well beyond unity, the $G(j\omega)$ locus approaches a semicircle. This may be seen from the fact that for a heavily damped system the characteristic roots are real and one is much smaller than the other. Since for sufficiently large ζ the effect of the larger root on the response becomes very small, the system behaves like a first-order one.

For the sinusoidal transfer function

$$G(j\omega) = 1 + 2\zeta\left(j\,\frac{\omega}{\omega_n}\right) + \left(j\,\frac{\omega}{\omega_n}\right)^2$$

$$= \left(1 - \frac{\omega^2}{\omega_n^2}\right) + j\left(\frac{2\zeta\omega}{\omega_n}\right)$$

the low-frequency portion of the curve is

$$\lim_{\omega \to 0} G(j\omega) = 1\underline{/0°}$$

and the high-frequency portion is

$$\lim_{\omega \to \infty} G(j\omega) = \infty\underline{/180°}$$

Since the imaginary part of $G(j\omega)$ is positive for $\omega > 0$ and is monotonically increasing and the real part of $G(j\omega)$ is monotonically decreasing from unity, the general shape of the polar plot of $G(j\omega)$ is as shown in Figure 6-27. The phase angle is between 0° and 180°.

EXAMPLE 6-4.   Consider the following second-order transfer function:

$$G(s) = \frac{1}{s(Ts + 1)}$$

Sketch a polar plot of this transfer function,
    Since the sinusoidal transfer function can be written

$$G(j\omega) = \frac{1}{j\omega(1 + j\omega T)} = -\frac{T}{1 + \omega^2 T^2} - j\frac{1}{\omega(1 + \omega^2 T^2)}$$

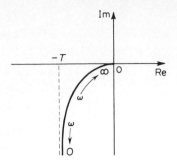

Figure 6–28
Polar plot of
$1/[j\omega (1 + j\omega T)]$.

the low-frequency portion of the polar plot becomes

$$\lim_{\omega \to 0} G(j\omega) = -T - j\infty = \infty \underline{/-90°}$$

and the high-frequency portion becomes

$$\lim_{\omega \to \infty} G(j\omega) = 0 - j0 = 0 \underline{/-180°}$$

The general shape of the polar plot of $G(j\omega)$ is shown in Figure 6–28. The $G(j\omega)$ plot is asymptotic to the vertical line passing through the point $(-T, 0)$. Since this transfer function involves an integrator $(1/s)$, the general shape of the polar plot differs substantially from those of second-order transfer functions that do not have an integrator.

Transport lag. The transport lag

$$G(j\omega) = e^{-j\omega T}$$

can be written

$$G(j\omega) = 1 \underline{/\cos \omega T - j \sin \omega T}$$

Since the magnitude of $G(j\omega)$ is always unity and the phase angle varies linearly with ω, the polar plot of the transport lag is a unit circle, as shown in Figure 6–29.

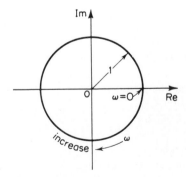

Figure 6–29 Polar plot of transport lag.

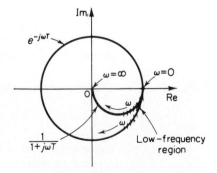

Figure 6–30 Polar plots of $e^{-j\omega T}$ and $1/(1 + j\omega T)$.

At low frequencies, the transport lag $e^{-j\omega T}$ and the first-order lag $1/(1 + j\omega T)$ behave similarly, as shown in Figure 6–30. The polar plots of $e^{-j\omega T}$ and $1/(1 + j\omega T)$ are tangent to each other at $\omega = 0$. This may be seen from the fact that, for $\omega \ll 1/T$,

$$e^{-j\omega T} \doteq 1 - j\omega T \qquad \text{and} \qquad \frac{1}{1 + j\omega T} \doteq 1 - j\omega T$$

For $\omega \gg 1/T$, however, an essential difference exists between $e^{-j\omega T}$ and $1/(1 + j\omega T)$, as may also be seen from Figure 6–30.

EXAMPLE 6–5. Obtain the polar plot of the following transfer function:

$$G(j\omega) = \frac{e^{-j\omega L}}{1 + j\omega T}$$

Since $G(j\omega)$ can be written

$$G(j\omega) = (e^{-j\omega L})\left(\frac{1}{1 + j\omega T}\right)$$

the magnitude and phase angle are, respectively,

$$|G(j\omega)| = |e^{-j\omega L}| \cdot \left|\frac{1}{1 + j\omega T}\right| = \frac{1}{\sqrt{1 + \omega^2 T^2}}$$

and

$$\underline{/G(j\omega)} = \underline{/e^{-j\omega L}} + \underline{\left/\frac{1}{1 + j\omega T}\right.} = -\omega L - \tan^{-1}\omega T$$

Since the magnitude decreases from unity monotonically and the phase angle also decreases monotonically and indefinitely, the polar plot of the given transfer function is a spiral, as shown in Figure 6–31.

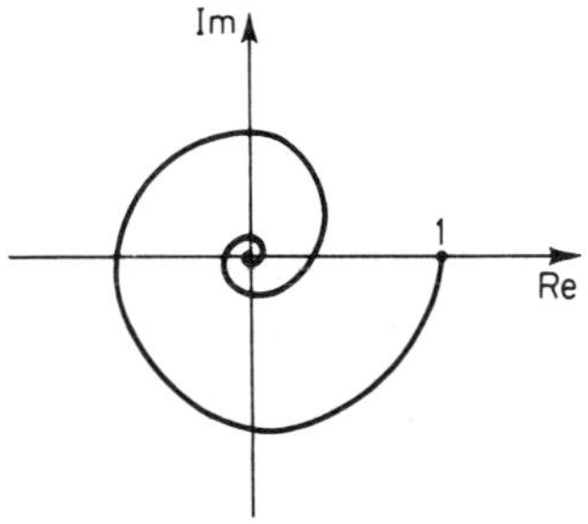

Figure 6–31
Polar plot of
$e^{-j\omega L}/(1 + j\omega T)$.

General shapes of polar plots. The polar plots of a transfer function of the form

$$G(j\omega) = \frac{K(1 + j\omega T_a)(1 + j\omega T_b) \cdots}{(j\omega)^\lambda(1 + j\omega T_1)(1 + j\omega T_2) \cdots}$$

$$= \frac{b_0(j\omega)^m + b_1(j\omega)^{m-1} + \cdots}{a_0(j\omega)^n + a_1(j\omega)^{n-1} + \cdots}$$

where $n > m$ or the degree of the denominator polynomial is greater than that of the numerator, will have the following general shapes:

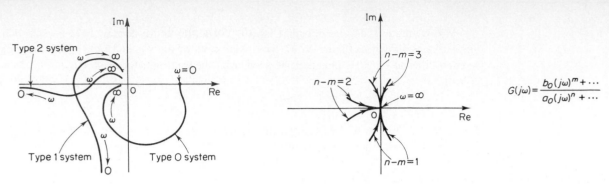

Figure 6–32 Polar plots of type 0, type 1, and type 2 systems.

Figure 6–33 Polar plots in the high-frequency range.

$$G(j\omega) = \frac{b_0(j\omega)^m + \cdots}{a_0(j\omega)^n + \cdots}$$

1. *For* $\lambda = 0$ *or type 0 systems*: The starting point of the polar plot (which corresponds to $\omega = 0$) is finite and is on the positive real axis. The tangent to the polar plot at $\omega = 0$ is perpendicular to the real axis. The terminal point, which corresponds to $\omega = \infty$, is at the origin, and the curve is tangent to one of the axes.

2. *For* $\lambda = 1$ *or type 1 systems*: The $j\omega$ term in the denominator contributes $-90°$ to the total phase angle of $G(j\omega)$ for $0 \le \omega \le \infty$. At $\omega = 0$, the magnitude of $G(j\omega)$ is infinity, and the phase angle becomes $-90°$. At low frequencies, the polar plot is asymptotic to a line parallel to the negative imaginary axis. At $\omega = \infty$, the magnitude becomes zero, and the curve converges to the origin and is tangent to one of the axes.

3. *For* $\lambda = 2$ *or type 2 systems*: The $(j\omega)^2$ term in the denominator contributes $-180°$ to the total phase angle of $G(j\omega)$ for $0 \le \omega \le \infty$. At $\omega = 0$, the magnitude of $G(j\omega)$ is infinity, and the phase angle is equal to $-180°$. At low frequencies, the polar plot is asymptotic to a line parallel to the negative real axis. At $\omega = \infty$, the magnitude becomes zero, and the curve is tangent to one of the axes.

The general shapes of the low-frequency portions of the polar plots of type 0, type 1, and type 2 systems are shown in Figure 6–32. It can be seen that if the degree of the denominator polynomial of $G(j\omega)$ is greater than that of the numerator, then the $G(j\omega)$ loci converge to the origin clockwise. At $\omega = \infty$, the loci are tangent to one or the other axes, as shown in Figure 6–33.

Figure 6–34
Polar plots of transfer functions with numerator dynamics.

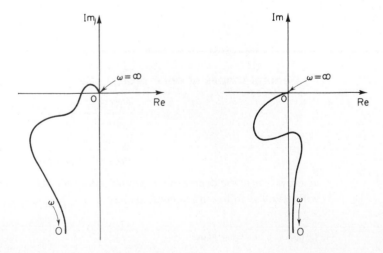

Note that any complicated shapes in the polar plot curves are caused by the numerator dynamics, that is, by the time constants in the numerator of the transfer function. Figure 6–34 shows examples of polar plots of transfer functions with numerator dynamics. In analyzing control systems, the polar plot of $G(j\omega)$ in the frequency range of interest must be accurately determined.

Table 6–1 shows sketches of polar plots of several transfer functions.

Table 6–1 Polar Plots of Simple Transfer Functions

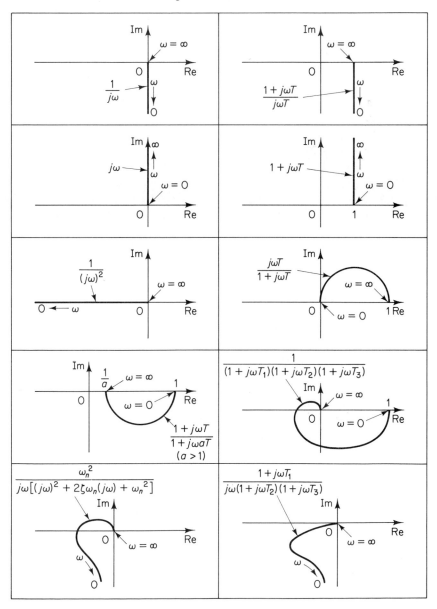

6–4 LOG-MAGNITUDE VERSUS PHASE PLOTS

Another approach to graphically portraying the frequency-response characteristics is to use the log-magnitude versus phase plot, which is a plot of the logarithmic magnitude in decibels versus the phase angle or phase margin for a frequency range of interest. [The phase margin is the difference between the actual phase angle ϕ and $-180°$; that is, $\phi - (-180°) = 180° + \phi$.] The curve is graduated in terms of the frequency ω. Such log-magnitude versus phase plots are commonly called Nichols plots.

In the Bode diagram, the frequency-response characteristics of $G(j\omega)$ are shown on semilog paper by two separate curves, the log-magnitude curve and the phase-angle curve, while in the log-magnitude versus phase plot, the two curves in the Bode diagram are combined into one. The log-magnitude versus phase plot can easily be constructed by reading values of the log magnitude and phase angle from the Bode diagram. Notice that in the log-magnitude versus phase plot, a change in the gain constant of $G(j\omega)$ merely shifts the curve up (for increasing gain) or down (for decreasing gain), but the shape of the curve remains the same.

Advantages of the log-magnitude versus phase plot are that the relative stability of the closed-loop system can be determined quickly and that compensation can be worked out easily.

The log-magnitude versus phase plots for the sinusoidal transfer functions $G(j\omega)$ and $1/G(j\omega)$ are skew symmetrical about the origin since

$$\left|\frac{1}{G(j\omega)}\right| \text{ in db} = -\left|G(j\omega)\right| \text{ in db}$$

and

$$\angle \frac{1}{G(j\omega)} = -\angle G(j\omega)$$

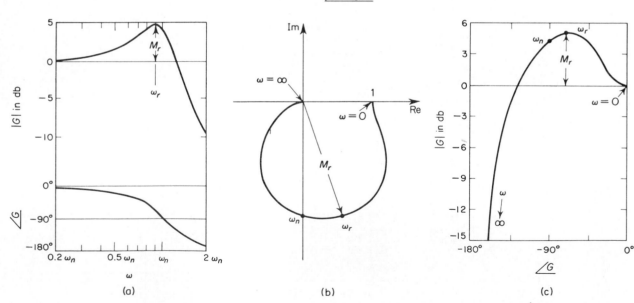

Figure 6–35 Three representations of the frequency response of $\dfrac{1}{1 + 2\zeta\left(j\dfrac{\omega}{\omega_n}\right) + \left(j\dfrac{\omega}{\omega_n}\right)^2}$,

($\zeta > 0$). (a) Bode diagram; (b) polar plot; (c) log-magnitude versus phase plot.

Table 6–2 Log-Magnitude versus Phase Plots of Simple Transfer Functions

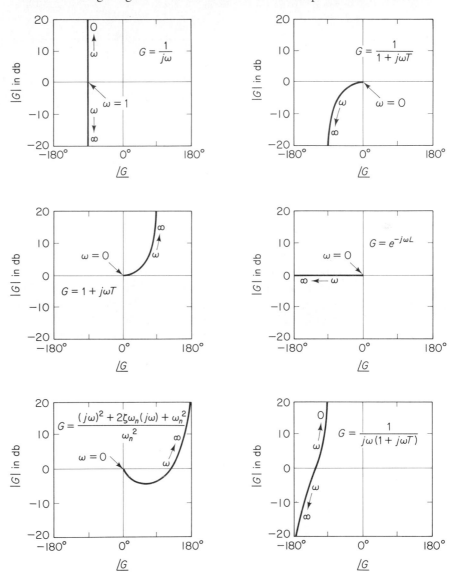

Since log-magnitude and phase-angle characteristics of basic transfer functions have been discussed in detail in Sections 6–2 and 6–3, it will be sufficient here to give examples of some log-magnitude versus phase plots. Table 6–2 shows such examples.

Figure 6–35 compares frequency-response curves of

$$G(j\omega) = \cfrac{1}{1 + 2\zeta\left(j\cfrac{\omega}{\omega_n}\right) + \left(j\cfrac{\omega}{\omega_n}\right)^2}$$

in three different representations. In the log-magnitude versus phase plot, the vertical distance between the points $\omega = 0$ and $\omega = \omega_r$, where ω_r is the resonant frequency, is the peak value of $G(j\omega)$ in decibels.

6–5   NYQUIST STABILITY CRITERION

This section presents the Nyquist stability criterion and associated mathematical backgrounds. Consider the closed-loop system shown in Figure 6–36. The closed-loop transfer function is

$$\frac{C(s)}{R(s)} = \frac{G(s)}{1 + G(s)H(s)}$$

For stability, all roots of the characteristic equation

$$1 + G(s)H(s) = 0$$

must lie in the left-half s plane. [It is noted that, although poles and zeros of the open-loop transfer function $G(s)H(s)$ may be in the right-half s plane, the system is stable if all the poles of the closed-loop transfer function (that is, the roots of the characteristic equation) are in the left-half s plane.] The Nyquist stability criterion relates the open-loop frequency response $G(j\omega)H(j\omega)$ to the number of zeros and poles of $1 + G(s)H(s)$ that lie in the right-half s plane. This criterion, derived by H. Nyquist, is useful in control engineering because the absolute stability of the closed-loop system can be determined graphically from open-loop frequency-response curves and there is no need for actually determining the closed-loop poles. Analytically obtained open-loop frequency-response curves, as well as experimentally obtained ones, can be used for the stability analysis. This is convenient because, in designing a control system, it often happens that mathematical expressions for some of the components are not known; only their frequency-response data are available.

The Nyquist stability criterion is based on a theorem from the theory of complex variables. To understand the criterion, we shall first discuss mappings of contours in the complex plane.

We shall assume that the open-loop transfer function $G(s)H(s)$ is representable as a ratio of polynomials in s. For a physically realizable system, the degree of the denominator polynomial of the closed-loop transfer function must be greater than or equal to that of the numerator polynomial. This means that the limit of $G(s)H(s)$ as s approaches infinity is zero or a constant for any physically realizable system.

Preliminary study.   The characteristic equation of the system shown in Figure 6–36 is

$$F(s) = 1 + G(s)H(s) = 0$$

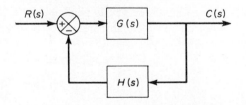

Figure 6–36
Closed-loop system.

We shall show that for a given continuous closed path in the s plane, which does not go through any singular points, there corresponds a closed curve in the $F(s)$ plane. The number and direction of encirclements of the origin of the $F(s)$ plane by the closed curve plays a particularly important role in what follows, for later we shall correlate the number and direction of encirclements with the stability of the system.

Consider, for example, the following open-loop transfer function:

$$G(s)H(s) = \frac{6}{(s + 1)(s + 2)}$$

The characteristic equation is

$$F(s) = 1 + G(s)H(s) = 1 + \frac{6}{(s + 1)(s + 2)}$$
$$= \frac{(s + 1.5 + j2.4)(s + 1.5 - j2.4)}{(s + 1)(s + 2)} = 0$$

The function $F(s)$ is analytic everywhere in the s plane except at its singular points. For each point of analyticity in the s plane, there corresponds a point in the $F(s)$ plane. For example, if $s = 1 + j2$, then $F(s)$ becomes

$$F(1 + j2) = 1 + \frac{6}{(2 + j2)(3 + j2)} = 1.115 - j0.577$$

Thus, the point $s = 1 + j2$ in the s plane maps into the point $1.115 - j0.577$ in the $F(s)$ plane.

Thus, as stated above, for a given continuous closed path in the s plane, which does not go through any singular points, there corresponds a closed curve in the $F(s)$ plane. Figure 6–37(a) shows conformal mappings of the lines $\omega = 0,1,2,3$ and the lines $\sigma = 1, 0, -1, -2, -3, -4$ in the upper-half s plane into the $F(s)$ plane. For example, the line $s = j\omega$ in the upper-half s plane ($\omega \geq 0$) maps into the curve denoted by $\sigma = 0$ in the $F(s)$ plane. Figure 6–37(b) shows conformal mappings of the lines $\omega = 0, -1, -2, -3$ and the lines $\sigma = 1, 0, -1, -2, -3, -4$ in the lower-half s plane into the $F(s)$ plane. Notice that for a given σ the curve for negative frequencies is symmetrical about the real axis with the curve for positive frequencies. Referring to Figures 6–37(a) and (b), we see that for the path $ABCD$ in the s plane traversed in the clockwise direction, the corresponding curve in the $F(s)$ plane is $A'B'C'D'$. The arrows on the curves indicate directions of traversal. Similarly, the path $DEFA$ in the s plane maps into the curve $D'E'F'A'$ in the $F(s)$ plane. Because of the property of conformal mapping, the corresponding angles in the s plane and $F(s)$ plane are equal and have the same sense. [For example, since lines AB and BC intersect at right angles to each other in the s plane, curves $A'B'$ and $B'C'$ also intersect at right angles at point B' in the $F(s)$ plane.] Referring to Figure 6–37(c), we see that on the closed contour $ABCDEFA$ in the s plane the variable s starts at point A and assumes values on this path in a clockwise direction until it returns to the starting point A. The corresponding curve in the $F(s)$ plane is denoted $A'B'C'D'E'F'A'$. If we define the area to the right of the contour when a representative point s moves in the clockwise direction to be the inside of the contour and the area to the left to be the outside, then the shaded area in Figure 6–37(c) is enclosed by the contour $ABCDEFA$ and is inside it. From Figure 6–37(c), it can be seen that when the contour in the s plane encloses two poles of $F(s)$, the locus of $F(s)$ encircles the origin of the $F(s)$ plane twice in the counterclockwise direction.

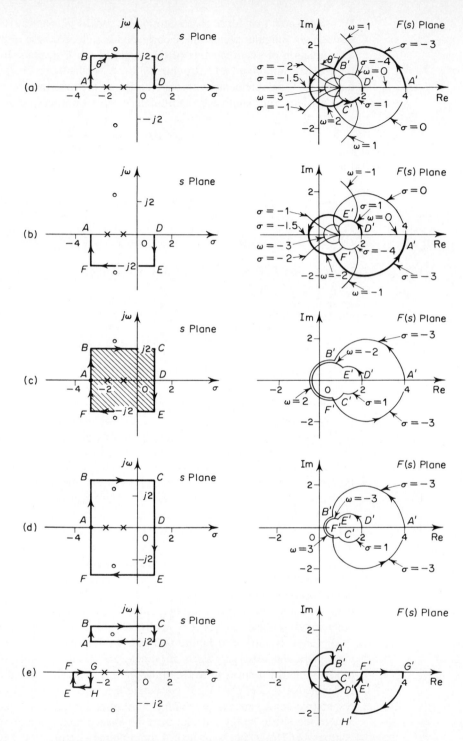

Figure 6–37 Conformal mapping of s plane grids into the $F(s)$ plane.

The number of encirclements of the origin of the $F(s)$ plane depends on the closed contour in the s plane. If this contour encloses two zeros and two poles of $F(s)$, then the corresponding $F(s)$ locus does not encircle the origin, as shown in Figure 6–37(d). If this contour encloses only one zero, the corresponding locus of $F(s)$ encircles the origin once in the clockwise direction. This is shown in Figure 6–37(e). Finally, if the closed contour in the s plane encloses neither zeros nor poles, then the locus of $F(s)$ does not encircle the origin of the $F(s)$ plane at all. This is also shown in Figure 6–37(e).

Note that for each point in the s plane, except for the singular points, there is only one corresponding point in the $F(s)$ plane; that is, the mapping from the s plane into the $F(s)$ plane is one to one. The mapping from the $F(s)$ plane into the s plane may not be one to one, however, so that a given point in the $F(s)$ plane may correspond to more than one point in the s plane. For example, point B' in the $F(s)$ plane in Figure 6–37(d) corresponds to both point $(-3, 3)$ and point $(0, -3)$ in the s plane.

From the foregoing analysis, we can see that the direction of encirclement of the origin of the $F(s)$ plane depends on whether the contour in the s plane encloses a pole or a zero. Note that the location of a pole or zero in the s plane, whether in the right-half or left-half s plane, does not make any difference, but the enclosure of a pole or zero does. If the contour in the s plane encloses k zeros and k poles ($k = 0, 1, 2, \ldots$), that is, an equal number of each, then the corresponding closed curve in the $F(s)$ plane does not encircle the origin of the $F(s)$ plane. The foregoing discussion is a graphical explanation of the mapping theorem, which is the basis for the Nyquist stability criterion.

Mapping theorem.　Let $F(s)$ be a ratio of two polynomials in s. Let P be the number of poles and Z be the number of zeros of $F(s)$ that lie inside some closed contour in the s plane, with multiplicity of poles and zeros accounted for. Let this contour be such that it does not pass through any poles or zeros of $F(s)$. This closed contour in the s plane is then mapped into the $F(s)$ plane as a closed curve. The total number N of clockwise encirclements of the origin of the $F(s)$ plane, as a representative point s traces out the entire contour in the clockwise direction, is equal to $Z - P$. (Note that by this mapping theorem the numbers of zeros and of poles cannot be found, only their difference.)

We shall not present a formal proof of this theorem here but leave the proof to Problem A–6–8. Note that a positive number N indicates an excess of zeros over poles of the function $F(s)$ and a negative N indictes an excess of poles over zeros. In control system applications, the number P can be readily determined for $F(s) = 1 + G(s)H(s)$ from the function $G(s)H(s)$. Therefore, if N is determined from the plot of $F(s)$, the number of zeros in the closed contour in the s plane can be determined readily. Note that the exact shapes of the s-plane contour and $F(s)$ locus are immaterial so far as encirclements of the origin are concerned, since encirclements depend only on the enclosure of poles and/or zeros of $F(s)$ by the s-plane contour.

Application of the mapping theorem to the stability analysis of closed-loop systems.　For analyzing the stability of linear control systems, we let the closed contour in the s plane enclose the entire right-half s plane. The contour consists of the entire $j\omega$ axis from $\omega = -\infty$ to $+\infty$ and a semicircular path of infinite radius in the right-half s plane. Such a contour is called the Nyquist path. (The direction of the path is clockwise.) The Nyquist path encloses the entire right-half s plane and encloses all the zeros and poles of $1 + G(s)H(s)$ that have positive real parts. [If there are no zeros of $1 + G(s)H(s)$ in the

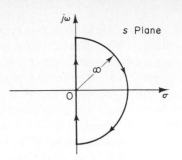

Figure 6–38
Closed contour in the
s plane.

right-half s plane, then there are no closed-loop poles there, and the system is stable.] It is necessary that the closed contour, or the Nyquist path, does not pass through any zeros and poles of $1 + G(s)H(s)$. If $G(s)H(s)$ has a pole or poles at the origin of the s plane, mapping of the point $s = 0$ becomes indeterminate. In such cases, the origin is avoided by taking a detour around it. (A detailed discussion of this special case is given later.)

If the mapping theorem is applied to the special case in which $F(s)$ is equal to $1 + G(s)H(s)$, then we can make the following statement: If the closed contour in the s plane encloses the entire right-half s plane, as shown in Figure 6–38, then the number of right-half plane zeros of the function $F(s) = 1 + G(s)H(s)$ is equal to the number of poles of the function $F(s) = 1 + G(s)H(s)$ in the right-half s plane plus the number of clockwise encirclements of the origin of the $1 + G(s)H(s)$ plane by the corresponding closed curve in this latter plane.

Because of the assumed condition that

$$\lim_{s \to \infty} [1 + G(s)H(s)] = \text{constant}$$

the function $1 + G(s)H(s)$ remains constant as s traverses the semicircle of infinite radius. Because of this, whether or not the locus of $1 + G(s)H(s)$ encircles the origin of the $1 + G(s)H(s)$ plane can be determined by considering only a part of the closed contour in the s plane, that is, the $j\omega$ axis. Encirclements of the origin, if there are any, occur only while a representative point moves from $-j\infty$ to $+j\infty$ along the $j\omega$ axis, provided that no zeros or poles lie on the $j\omega$ axis.

Note that the portion of the $1 + G(s)H(s)$ contour from $\omega = -\infty$ to $\omega = \infty$ is simply $1 + G(j\omega)H(j\omega)$. Since $1 + G j\omega)H(j\omega)$ is the vector sum of the unit vector and the vector $G j\omega)H(j\omega)$, $1 + G j\omega)H(j\omega)$ is identical to the vector drawn from the $-1 + j0$ point to the terminal point of the vector $G j\omega)H(j\omega)$, as shown in Figure 6–39. Encirclement

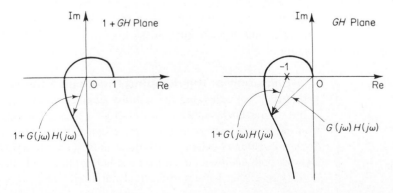

Figure 6–39
Plots of $1 + G(j\omega)H(j\omega)$ in the $1 + GH$ plane and GH plane.

of the origin by the graph of $1 + G(j\omega)H(j\omega)$ is equivalent to encirclement of the $-1 + j0$ point by just the $G j\omega)H(j\omega)$ locus. Thus, stability of a closed-loop system can be investigated by examining encirclements of the $-1 + j0$ point by the locus of $G(j\omega)H(j\omega)$. The number of clockwise encirclements of the $-1 + j0$ point can be found by drawing a vector from the $-1 + j0$ point to the $G(j\omega)H(j\omega)$ locus, starting from $\omega = -\infty$, going through $\omega = 0$, and ending at $\omega = +\infty$ and by counting the number of clockwise rotations of the vector.

Plotting $G(j\omega)H(j\omega)$ for the Nyquist path is straightforward. The map of the negative $j\omega$ axis is the mirror image about the real axis of the map of the positive $j\omega$ axis. That is, the plot of $G(j\omega)H(j\omega)$ and the plot of $G(-j\omega)H(-j\omega)$ are symmetrical with each other about the real axis. The semicircle with infinite radius maps into either the origin of the GH plane or a point on the real axis of the GH plane.

In the preceding discussion, $G(s)H(s)$ has been assumed to be the ratio of two polynomials in s. Thus, the transport lag e^{-Ts} has been excluded from the discussion. Note, however, that a similar discussion applies to systems with transport lag, although a proof of this is not given here. The stability of a system with transport lag can be determined from the open-loop frequency-response curves by examining the number of encirclements of the $-1 + j0$ point, just as in the case of a system whose open-loop transfer function is a ratio of two polynomials in s.

Nyquist stability criterion.   The foregoing analysis, utilizing the encirclement of the $-1 + j0$ point by the $G(j\omega)H(j\omega)$ locus, is summarized in the following Nyquist stability criterion:

Nyquist stability criterion [*For a special case where $G(s)H(s)$ has neither poles nor zeros on the $j\omega$ axis.*]: In the system shown in Figure 6–36, if the open-loop transfer function $G(s)H(s)$ has k poles in the right-half s plane and $\lim_{s \to \infty} G(s)H(s) = $ constant, then for stability the $G(j\omega)H(j\omega)$ locus, as ω varies from $-\infty$ to ∞, must encircle the $-1 + j0$ point k times in the counterclockwise direction.

Remarks on the Nyquist stability criterion

1. This criterion can be expressed as

$$Z = N + P$$

where $Z = $ number of zeros of $1 + G(s)H(s)$ in the right-half s plane
  $N = $ number of clockwise encirclements of the $-1 + j0$ point
  $P = $ number of poles of $G(s)H(s)$ in the right-half s plane

If P is not zero, for a stable control system, we must have $Z = 0$, or $N = -P$, which means that we must have P counterclockwise encirclements of the $-1 + j0$ point.

If $G(s)H(s)$ does not have any poles in the right-half s plane, then $Z = N$. Thus for stability there must be no encirclement of the $-1 + j0$ point by the $G(j\omega)H(j\omega)$ locus. In this case it is not necessary to consider the locus for the entire $j\omega$ axis, only for the positive frequency portion. The stability of such a system can be determined by seeing if the $-1 + j0$ point is enclosed by the Nyquist plot of $G(j\omega)H(j\omega)$. The region enclosed by the Nyquist plot is shown in Figure 6–40. For stability, the $-1 + j0$ point must lie outside the shaded region.

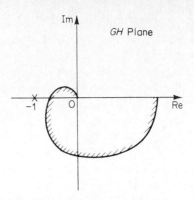

Figure 6–40
Region enclosed by a
Nyquist plot.

2. We must be careful when testing the stability of multiple-loop systems since they may include poles in the right-half s plane. (Note that although an inner loop may be unstable the entire closed-loop system can be made stable by proper design.) Simple inspection of encirclements of the $-1 + j0$ point by the $G(j\omega)H(j\omega)$ locus is not sufficient to detect instability in multiple-loop systems. In such cases, however, whether or not any pole of $1 + G(s)H(s)$ is in the right-half s plane can be determined easily by applying the Routh stability criterion to the denominator of $G(s)H(s)$.

If transcendental functions, such as transport lag e^{-Ts}, are included in $G(s)H(s)$, they must be approximated by a series expansion before the Routh stability criterion can be applied. One form of a series expansion of e^{-Ts} is given below:

$$e^{-Ts} = \frac{1 - \dfrac{Ts}{2} + \dfrac{(Ts)^2}{8} - \dfrac{(Ts)^3}{48} + \cdots}{1 + \dfrac{Ts}{2} + \dfrac{(Ts)^2}{8} + \dfrac{(Ts)^3}{48} + \cdots}$$

As a first approximation, we may take only the first two terms in the numerator and denominator, respectively, or

$$e^{-Ts} \doteq \frac{1 - \dfrac{Ts}{2}}{1 + \dfrac{Ts}{2}} = \frac{2 - Ts}{2 + Ts}$$

This gives a good approximation to transport lag for the frequency range $0 \le \omega \le (0.5/T)$. [Note that the magnitude of $(2 - j\omega T)/(2 + j\omega T)$ is always unity and the phase lag of $(2 - j\omega T)/(2 + j\omega T)$ closely approximates that of transport lag within the stated frequency range.]

3. If the locus of $G(j\omega)H(j\omega)$ passes through the $-1 + j0$ point, then zeros of the characteristic equation, or closed-loop poles, are located on the $j\omega$ axis. This is not desirable for practical control systems. For a well-designed closed-loop system, none of the roots of the characteristic equation should lie on the $j\omega$ axis.

Special case where $G(s)H(s)$ involves poles and/or zeros on the $j\omega$ axis. In the previous discussion, we assumed that the open-loop transfer function $G(s)H(s)$ has neither

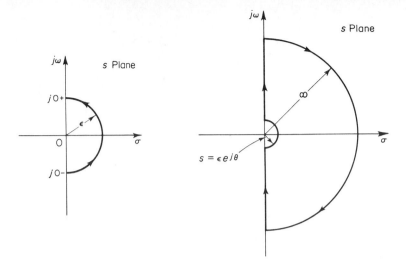

Figure 6–41
Closed contours in the
s plane avoiding poles
and zeros at the
origin.

poles nor zeros at the origin. We now consider the case where $G(s)H(s)$ involves poles and/or zeros on the $j\omega$ axis.

Since the Nyquist path must not pass through poles or zeros of $G(s)H(s)$, if the function $G(s)H(s)$ has poles or zeros at the origin (or on the $j\omega$ axis at points other than the origin), the contour in the s plane must be modified. The usual way of modifying the contour near the origin is to use a semicircle with infinitesimal radius ϵ, as shown in Figure 6–41. A representative point s moves along the negative $j\omega$ axis from $-j\infty$ to $j0-$. From $s = j0-$ to $s = j0+$, the point moves along the semicircle of radius ϵ (where $\epsilon \ll 1$) and then moves along the positive $j\omega$ axis from $j0+$ to $j\infty$. From $s = j\infty$, the contour follows a semicircle with infinite radius, and the representative point moves back to the starting point. The area that the modified closed contour avoids is very small and approaches zero as the radius ϵ approaches zero. Therefore, all the poles and zeros, if any, in the right-half s plane are enclosed by this contour.

Consider, for example, a closed-loop system whose open-loop transfer function is given by

$$G(s)H(s) = \frac{K}{s(Ts + 1)}$$

The points corresponding to $s = j0+$ and $s = j0-$ on the locus of $G(s)H(s)$ in the $G(s)H(s)$ plane are $-j\infty$ and $j\infty$, respectively. On the semicircular path with radius ϵ (where $\epsilon \ll 1$), the complex variable s can be written

$$s = \epsilon e^{j\theta}$$

where θ varies from $-90°$ to $+90°$. Then $G(s)H(s)$ becomes

$$G(\epsilon e^{j\theta})H(\epsilon e^{j\theta}) = \frac{K}{\epsilon e^{j\theta}} = \frac{K}{\epsilon} e^{-j\theta}$$

The value K/ϵ approaches infinity as ϵ approaches zero, and $-\theta$ varies from $90°$ to $-90°$ as a representative point s moves along the semicircle. Thus, the points $G(j0-)H(j0-) = j\infty$ and $G(j0+)H(j0+) = -j\infty$ are joined by a semicircle of infinite radius in the right-

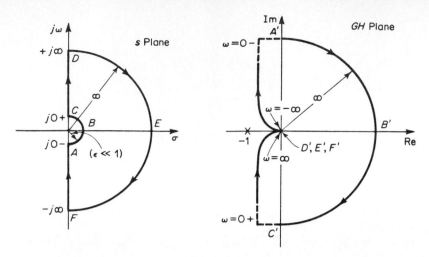

Figure 6–42 s-plane contour and the $G(s)H(s)$ locus in the GH plane where $G(s)H(s) = K/[s(Ts + 1)]$.

half *GH* plane. The infinitesimal semicircular detour around the origin maps into the *GH* plane as a semicircle of infinite radius. Figure 6–42 shows the s-plane contour and the $G(s)H(s)$ locus in the *GH* plane. Points A, B, and C on the s-plane contour map into the respective points A', B', and C' on the $G(s)H(s)$ locus. As seen from Figure 6–42, points D, E, and F on the semicircle of infinite radius in the s plane map into the origin of the *GH* plane. Since there is no pole in the right-half s plane and the $G(s)H(s)$ locus does not encircle the $-1 + j0$ point, there are no zeros of the function $1 + G(s)H(s)$ in the right-half s plane. Therefore, the system is stable.

For an open-loop transfer function $G(s)H(s)$ involving a $1/s^n$ factor (where $n = 2, 3, \ldots$), the plot of $G(s)H(s)$ has n clockwise semicircles of infinite radius about the origin as a representative point s moves along the semicircle of radius ϵ (where $\epsilon \ll 1$). For example, consider the following open-loop transfer function:

$$G(s)H(s) = \frac{K}{s^2(Ts + 1)}$$

Then

$$\lim_{s \to \epsilon e^{j\theta}} G(s)H(s) = \frac{K}{\epsilon^2 e^{2j\theta}} = \frac{K}{\epsilon^2} e^{-2j\theta}$$

As θ varies from $-90°$ to $90°$ in the s plane, the angle of $G(s)H(s)$ varies from $180°$ to $-180°$, as shown in Figure 6–43. Since there is no pole in the right-half s plane and the locus encircles the $-1 + j0$ point twice clockwise for any positive value of K, there are two zeros of $1 + G(s)H(s)$ in the right-half s plane. Therefore, this system is always unstable.

Note that a similar analysis can be made if $G(s)H(s)$ involves poles and/or zeros on the $j\omega$ axis. The Nyquist stability criterion can now be generalized as follows:

Nyquist stability criterion [For a general case where $G(s)H(s)$ has poles and/or zeros on the $j\omega$ axis.]: In the system shown in Figure 6–36, if the open-loop transfer function

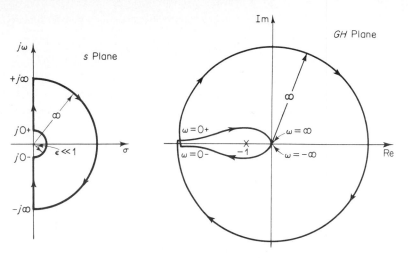

Figure 6-43 s-plane contour and the $G(s)H(s)$ locus in the GH plane where $G(s)H(s) = K/[s^2(Ts + 1)]$.

$G(s)H(s)$ has k poles in the right-half s plane, then for stability the $G(s)H(s)$ locus, as a representative point s traces out the modified Nyquist path in the clockwise direction, must encircle the $-1 + j0$ point k times in the counterclockwise direction.

6-6 STABILITY ANALYSIS

In this section, we shall present several illustrative examples of the stability analysis of control systems using the Nyquist stability criterion.

If the Nyquist path in the s plane encircles Z zeros and P poles of $1 + G(s)H(s)$ and does not pass through any poles or zeros of $1 + G(s)H(s)$ as a representative point s moves in the clockwise direction along the Nyquist path, then the corresponding contour in the $G(s)H(s)$ plane encircles the $-1 + j0$ point $N = Z - P$ times in the clockwise direction. (Negtive values of N imply counterclockwise encirclements.)

In examining the stability of liner control systems using the Nyquist stability criterion, we see that three possibilities can occur.

1. There is no encirclement of the $-1 + j0$ point. This implies that the system is stable if there are no poles of $G(s)H(s)$ in the right-half s plane; otherwise, the system is unstable.
2. There is a counterclockwise encirclement or encirclements of the $-1 + j0$ point. In this case the system is stable if the number of counterclockwise encirclements is the same as the number of poles of $G(s)H(s)$ in the right-half s plane; otherwise, the system is unstable.
3. There is a clockwise encirclement or encirclements of the $-1 + j0$ point. In this case the system is unstable.

In the following examples, we assume that the values of the gain K and the time constants (such as T, T_1, and T_2) are all positive.

EXAMPLE 6–6   Consider a closed-loop system whose open-loop transfer function is given by

$$G(s)H(s) = \frac{K}{(T_1 s + 1)(T_2 s + 1)}$$

Examine the stability of the system.

A plot of $G(j\omega)H(j\omega)$ is shown in Figure 6–44. Since $G(s)H(s)$ does not have any poles in the right-half s plane and the $-1 + j0$ point is not encircled by the $G(j\omega)H(j\omega)$ locus, this system is stable for any positive values of K, T_1, and T_2.

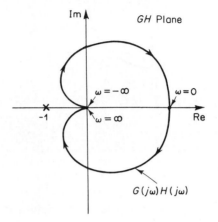

Figure 6–44
Polar plot of
$G(j\omega)H(j\omega)$
considered in Example
6–6.

EXAMPLE 6–7   Consider the system with the following open-loop transfer function:

$$G(s) = \frac{K}{s(T_1 s + 1)(T_2 s + 1)}$$

Determine the stability of the system for two cases: (1) the gain K is small, (2) K is large.

The Nyquist plots of the open-loop transfer function with a small value of K and a large value of K are shown in Figure 6–45. The number of poles of $G(s)H(s)$ in the right-half s plane is zero. Therefore,

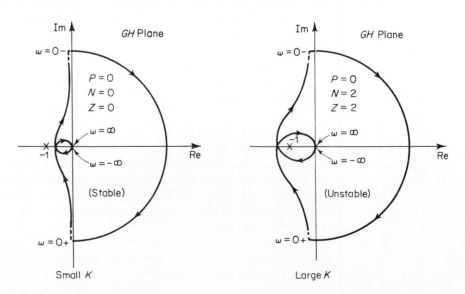

Figure 6–45
Polar plots of the
system considered in
Example 6–7.

for this system to be stable, it is necessary that $N = Z = 0$ or that the $G(s)H(s)$ locus does not encircle the $-1 + j0$ point.

For small values of K, there is no encirclement of the $-1 + j0$ point. Hence the system is stable for small values of K. For large values of K, the locus of $G(s)H(s)$ encircles the $-1 + j0$ point twice in the clockwise direction, indicating two closed-loop poles in the right-half s plane and the system is unstable. (For good accuracy, K should be large. From the stability viewpoint, however, a large value of K causes poor stability or even instability. To compromise between accuracy and stability, it is necessary to insert a compensation network into the system. Compensating techniques in the frequency domain are discussed in Chapter 7.)

EXAMPLE 6–8 The stability of a closed-loop system with the following open-loop transfer function

$$G(s)H(s) = \frac{K(T_2 s + 1)}{s^2(T_1 s + 1)}$$

depends on the relative magnitudes of T_1 and T_2. Draw Nyquist plots and determine the stability of the system.

Plots of the locus of $G(s)H(s)$ for three cases, $T_1 < T_2$, $T_1 = T_2$, and $T_1 > T_2$, are shown in Figure 6–46. For $T_1 < T_2$, the locus of $G(s)H(s)$ does not encircle the $-1 + j0$ point, and the closed-loop system is stable. For $T_1 = T_2$, the $G(s)H(s)$ locus passes through the $-1 + j0$ point, which indicates that the closed-loop poles are located on the $j\omega$ axis. For $T_1 > T_2$, the locus of $G(s)H(s)$ encircles the $-1 + j0$ point twice in the clockwise direction. Thus, the closed-loop system has two closed-loop poles in the right-half s plane, and the system is unstable.

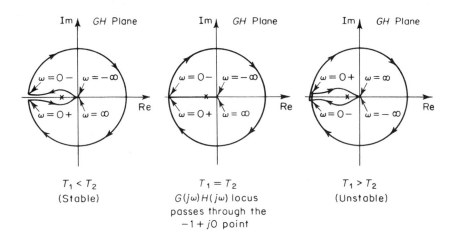

Figure 6–46
Polar plots of the system considered in Example 6–8.

$T_1 < T_2$
(Stable)

$T_1 = T_2$
$G(j\omega)H(j\omega)$ locus passes through the $-1 + j0$ point

$T_1 > T_2$
(Unstable)

EXAMPLE 6–9 Consider the closed-loop system having the following open-loop transfer function:

$$G(s)H(s) = \frac{K}{s(Ts - 1)}$$

Determine the stability of the system.

The function $G(s)H(s)$ has one pole ($s = 1/T$) in the right-half s plane. Therefore, $P = 1$. The Nyquist plot shown in Figure 6–47 indicates that the $G(s)H(s)$ plot encircles the $-1 + j0$ point once clockwise. Thus, $N = 1$. Since $Z = N + P$, we find that $Z = 2$. This means that the closed-loop system has two closed-loop poles in the right-half s plane and is unstable.

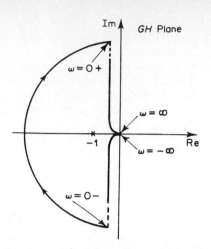

Figure 6–47
Polar plot of the system considered in Example 6–9.

EXAMPLE 6–10   Investigate the stability of a closed-loop system with the following open-loop transfer function:

$$G(s)H(s) = \frac{K(s + 3)}{s(s - 1)} \qquad (K > 1)$$

The open-loop transfer function has one pole ($s = 1$) in the right-half s plane, or $P = 1$. The open-loop system is unstable. The Nyquist plot shown in Figure 6–48 indicates that the $-1 + j0$ point is encircled by the $G(s)H(s)$ locus once in the counterclockwise direction. Therefore, $N = -1$. Thus, Z is found from $Z = N + P$ to be zero, which indicates that there is no zero of $1 + G(s)H(s)$ in the right-half s plane, and the closed-loop system is stable. This is one of the examples where an unstable open-loop system becomes stable when the loop is closed.

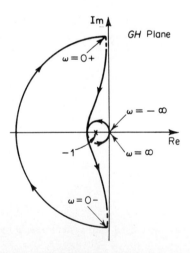

Figure 6–48
Polar plot of the system considered in Example 6–10.

Conditionally stable systems.   Figure 6–49 shows an example of a $G(j\omega)H(j\omega)$ locus for which the closed-loop system can be made unstable by varying the open-loop gain. If the open-loop gain is increased sufficiently, the $G(j\omega)H(j\omega)$ locus encloses the $-1 + j0$ point twice, and the system becomes unstable. If the open-loop gain is decreased sufficiently, again the $G(j\omega)H(j\omega)$ locus encloses the $-1 + j0$ point twice. For stable operation

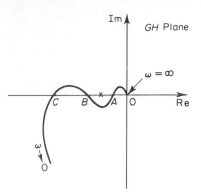

Figure 6–49
Polar plot of a conditionally stable system.

of the system considered here, the critical point $-1 + j0$ must not be located in the regions between OA and BC shown in Figure 6–49. Such a system that is stable only for limited ranges of values of the open-loop gain for which the $-1 + j0$ point is completely outside the $G(j\omega)H(j\omega)$ locus is a conditionally stable system.

A conditionally stable system is stable for the value of the open-loop gain lying between critical values, but it is unstable if the open-loop gain is either increased or decreased sufficiently. Such a system becomes unstable when large input signals are applied since a large signal may cause saturation, which in turn reduces the open-loop gain of the system. It is advisable to avoid such a situation.

Multiple-loop systems. Consider the system shown in Figure 6–50. This is a multiple-loop system. The inner loop has the transfer function

$$G(s) = \frac{G_2(s)}{1 + G_2(s)H_2(s)}$$

If $G(s)$ is unstable, the effects of instability are to produce a pole or poles in the right-half s plane. Then the characteristic equation of the inner loop, $1 + G_2(s)H_2(s) = 0$, has a zero or zeros in this portion of the plane. If $G_2(s)$ and $H_2(s)$ have P_1 poles here, then the number Z_1 of right-half plane zeros of $1 + G_2(s)H_2(s)$ can be found from $Z_1 = N_1 + P_1$, where N_1 is the number of clockwise encirclements of the $-1 + j0$ point by the $G_2(s)H_2(s)$ locus. Since the open-loop transfer function of the entire system is given by $G_1(s)G(s)H_1(s)$, the

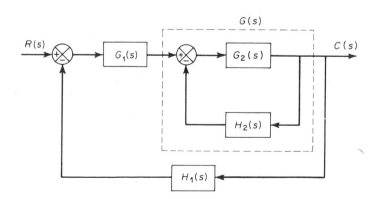

Figure 6–50
Multiple-loop system.

stability of this closed-loop system can be found from the Nyquist plot of $G_1(s)G(s)H_1(s)$ and knowledge of the right-half plane poles of $G_1(s)G(s)H_1(s)$.

Notice that if a feedback loop is eliminated by means of block diagram reductions, there is a possibility that unstable poles are introduced; if the feedforward branch is eliminated by means of block diagram reductions, there is a possibility that right-half plane zeros are introduced. Therefore, we must note all right-half plane poles and zeros as they appear from subsidiary loop reductions. This knowledge is necessary in determining the stability of multiple-loop systems.

EXAMPLE 6–11   Consider the control system shown in Figure 6–51. The system involves two loops. Determine the range of gain K for stability of the system by use of the Nyquist stability criterion. (The gain K is positive.)

To examine the stability of the control system, we need to sketch the Nyquist locus of $G(s)$, where

$$G(s) = G_1(s)G_2(s)$$

However, the poles of $G(s)$ are not known at this point. Therefore, we need to examine the minor loop if there are right-half s plane poles. This can be done easily by use of the Routh stability criterion. Since

$$G_2(s) = \frac{1}{s^3 + s^2 + 1}$$

the Routh array becomes as follows:

$$
\begin{array}{ccc}
s^3 & 1 & 0 \\
s^2 & 1 & 1 \\
s^1 & -1 & 0 \\
s^0 & 1 &
\end{array}
$$

Notice that there are two sign changes in the first column. Hence, there are two poles of $G_2(s)$ in the right-half s plane.

Once we find the number of right-half s plane poles of $G_2(s)$, we proceed to sketch the Nyquist locus of $G(s)$, where

$$G(s) = G_1(s)G_2(s) = \frac{K(s + 0.5)}{s^3 + s^2 + 1}$$

Our problem is to determine the range of gain K for stability. Hence, instead of plotting Nyquist loci of $G(j\omega)$ for various values of K, we plot the Nyquist locus of $G(j\omega)/K$. Figure 6–52 shows the Nyquist plot or polar plot of $G(j\omega)/K$.

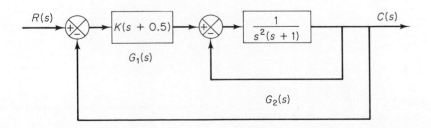

Figure 6–51
Control system.

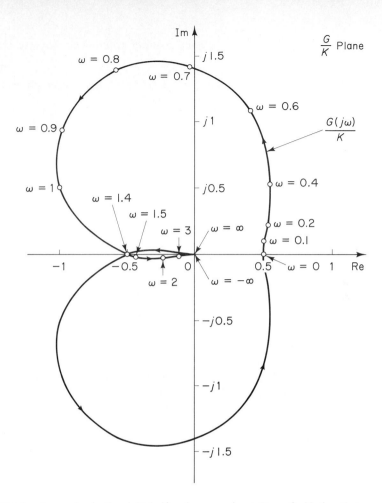

Figure 6–52
Polar plot of $G(j\omega)/K$.

Since $G(s)$ has two poles in the right-half s plane, we have $P_1 = 2$. Noting that

$$Z_1 = N_1 + P_1$$

for stability, we require $Z_1 = 0$ or $N_1 = -2$. That is, the Nyquist locus of $G(j\omega)$ must encircle the $-1 + j0$ point twice counterclockwise. From Figure 6–52 we see that, if the critical point lies between 0 and -0.5, then the $G(j\omega)/K$ locus encircles the critical point twice counterclockwise. Therefore, we require

$$-0.5K < -1$$

The range of gain K for stability is

$$2 < K$$

Nyquist stability criterion applied to inverse polar plots. In the previous analyses, the Nyquist stability criterion was applied to polar plots of the open-loop transfer function $G(s)H(s)$.

In analyzing multiple-loop systems, the inverse transfer function may sometimes be used in order to permit graphical analysis; this avoids much of the numerical calculation. (The Nyquist stability criterion can be applied equally well to inverse polar plots. The mathematical derivation of the Nyquist stability criterion for inverse polar plots is the same as that for direct polar plots.)

The inverse polar plot of $G(j\omega)H(j\omega)$ is a graph of $1/[G(j\omega)H(j\omega)]$ as a function of ω. For example, if $G(j\omega)H(j\omega)$ is

$$G(j\omega)H(j\omega) = \frac{j\omega T}{1 + j\omega T}$$

then

$$\frac{1}{G(j\omega)H(j\omega)} = \frac{1}{j\omega T} + 1$$

The inverse polar plot for $\omega \geq 0$ is the lower half of the vertical line starting at the point $(1, 0)$ on the real axis.

The Nyquist stability criterion applied to inverse polar plots may be stated as follows: For a closed-loop system to be stable, the encirclement, if any, of the $-1 + j0$ point by the $1/[G(s)H(s)]$ locus (as s moves along the Nyquist path) must be counterclockwise, and the number of such encirclements must be equal to the number of poles of $1/[G(s)G(s)]$ [that is, the zeros of $G(s)H(s)$] that lie in the right-half s plane. [The number of zeros of $G(s)H(s)$ in the right-half s plane may be determined by use of the Routh stability criterion.] If the open-loop transfer function $G(s)H(s)$ has no zeros in the right-half s plane, then for a closed-loop system to be stable the number of encirclements of the $-1 + j0$ point by the $1/[G(s)H(s)]$ locus must be zero.

Note that although the Nyquist stability criterion can be applied to inverse polar plots, if experimental frequency-response data are incorporated, counting the number of encirclements of the $1/[G(s)H(s)]$ locus may be difficult because the phase shift corresponding to the infinite semicircular path in the s plane is difficult to measure. For example, if the open-loop transfer function $G(s)H(s)$ involves transport lag such that

$$G(s)H(s) = \frac{Ke^{-j\omega L}}{s(Ts + 1)}$$

then the number of encirclements of the $-1 + j0$ point by the $1/[G(s)H(s)]$ locus becomes infinite, and the Nyquist stability criterion cannot be applied to the inverse polar plot of such an open-loop transfer function.

In general, if experimental frequency-response data cannot be put into analytical form, both the $G(j\omega)H(j\omega)$ and $1/[G(j\omega)H(j\omega)]$ loci must be plotted. In addition, the number of right-half plane zeros of $G(s)H(s)$ must be determined. It is more difficult to determine the right-half plane zeros of $G(s)H(s)$ (in other words, to determine whether or not a given component is minimum phase) than it is to determine the right-half plane poles of $G(s)H(s)$ (in other words, to determine whether or not the component is stable).

Depending on whether the data are graphical or analytical and whether or not nonminimum phase components are included, an appropriate stability test must be used for multiple-loop systems. If the data are given in analytical form or if mathematical expressions for all the components are known, the application of the Nyquist stability criterion to inverse polar plots causes no difficulty and multiple-loop systems may be analyzed and designed in the inverse GH plane.

EXAMPLE 6–12 Consider the control system shown in Figure 6–51. (Refer to Example 6–11.) Using the inverse polar plot, determine the range of gain K for stability.

Since

$$G_2(s) = \frac{1}{s^3 + s^2 + 1}$$

we have

$$G(s) = G_1(s)G_2(s) = \frac{K(s + 0.5)}{s^3 + s^2 + 1}$$

Hence

$$\frac{1}{G(s)} = \frac{s^3 + s^2 + 1}{K(s + 0.5)}$$

Notice that $1/G(s)$ has a pole at $s = -0.5$. It does not have any pole in the right-half s plane. Therefore, the Nyquist stability equation

$$Z = N + P$$

reduces to $Z = N$ since $P = 0$. The reduced equation states that the number Z of the zeros of $1 + [1/G(s)]$ in the right-half s plane is equal to N, the number of clockwise encirclements of the $-1 + j0$ point. For stability, N must be equal to zero, or there should be no encirclement. Figure 6–53 shows the Nyquist plot or polar plot of $K/G(j\omega)$.

Notice that since

$$\frac{K}{G(j\omega)} = \left[\frac{(j\omega)^3 + (j\omega)^2 + 1}{j\omega + 0.5} \right] \left(\frac{0.5 - j\omega}{0.5 - j\omega} \right)$$

$$= \frac{0.5 - 0.5\omega^2 - \omega^4 + j\omega(-1 + 0.5\omega^2)}{0.25 + \omega^2}$$

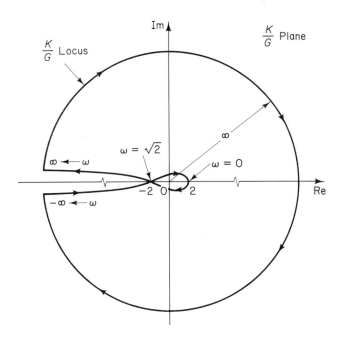

Figure 6–53
Polar plot of $K/G(j\omega)$.

the $K/G(j\omega)$ locus crosses the negative real axis at $\omega = \sqrt{2}$ and the crossing point at the negative real axis is -2.

From Fig. 6–53 we see that if the critical point lies in the region between -2 and $-\infty$, then the critical point is not encircled. Hence, for stability we require

$$-1 < \frac{-2}{K}$$

Thus, the range of gain K for stability is

$$2 < K$$

which is the same result as we obtained in Example 6–11.

Relative stability analysis through modified Nyquist paths. The Nyquist path for stability tests can be modified in order that we may investigate the relative stability of closed-loop systems. For the following second-order characteristic equation,

$$s^2 + 2\zeta\omega_n s + \omega_n^2 = 0 \qquad (0 < \zeta < 1)$$

the roots are complex conjugates and are

$$s_1 = -\zeta\omega_n + j\omega_n\sqrt{1 - \zeta^2}, \qquad s_2 = -\zeta\omega_n - j\omega_n\sqrt{1 - \zeta^2}$$

If these roots are plotted in the s plane, as shown in Figure 6–54, then we see that $\sin\theta = \zeta$ or the angle θ is indicative of the damping ratio ζ. As θ becomes smaller, so does the value of ζ.

If we modify the Nyquist path and use radial lines with angle θ_x, instead of the $j\omega$ axis, as shown in Figure 6–55, then it can be said, following the same reasoning as in the case of the Nyquist stability criterion, that if the $G(s)H(s)$ locus corresponding to the modified s-plane contour does not encircle the $-1 + j0$ point and none of the poles of $G(s)H(s)$ lie within the closed s-plane contour, then this contour does not enclose any zeros of $1 + G(s)H(s)$. The characteristic equation, $1 + G(s)H(s) = 0$, then does not have any roots within the modified s-plane contour. If no closed loop poles of a higher-order system are

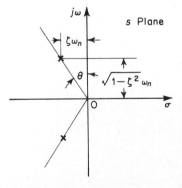

Figure 6–54 Plot of complex-conjugate roots in the s plane.

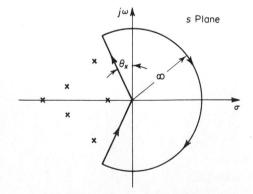

Figure 6–55 Modified Nyquist path.

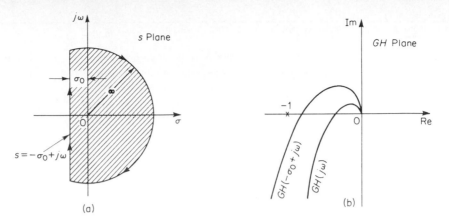

Figure 6-56
(a) Modified Nyquist path; (b) polar plots of $G(-\sigma_0 + j\omega)H(-\sigma_0 + j\omega)$ locus and $G(j\omega)H(j\omega)$ locus in the GH plane.

enclosed by this contour, we can say that the damping ratio of each pair of complex-conjugate closed-loop poles of the system is greater than $\sin \theta_x$.

Suppose that the s-plane contour consists of a line to the left of and parallel to the $j\omega$ axis at a distance $-\sigma_0$ (or the line $s = -\sigma_0 + j\omega$) and the semicircle of infinite radius enclosing the entire right-half s plane and that part of the left-half s plane between the lines $s = -\sigma_0 + j\omega$ and $s = j\omega$, as shown in Figure 6-56(a). If the $G(s)H(s)$ locus corresponding to this s-plane contour does not encircle the $-1 + j0$ point and $G(s)H(s)$ has no poles within the enclosed s-plane contour, then the characteristic equation does not have any zeros in the region enclosed by the modified s-plane contour. All roots of the characteristic equation lie to the left of the line $s = -\sigma_0 + j\omega$. An example of a $G(-\sigma_0 + j\omega)H(-\sigma_0 + j\omega)$ locus, together with a $G(j\omega)H(j\omega)$ locus, is shown in Figure 6-56(b). The magnitude $1/\sigma_0$ is indicative of the time constant of the dominant closed-loop poles. If all roots lie outside the s-plane contour, all time constants of the closed-loop transfer function are less than $1/\sigma_0$. If the s-plane contour is chosen as shown in Fig. 6-57, then the test of encirclements of the $-1 + j0$ point reveals the existence or nonexistence of the roots of the characteristic equation of the closed-loop system within this s-plane contour. If the test reveals that no roots lie in the s-plane contour, then it is clear that all the closed-loop poles have damping ratios greater than ζ_x and time constants less than $1/\sigma_0$. Thus, by taking an appropriate s-plane contour, we can investigate time constants and damping ratios of closed-loop poles from given open-loop transfer functions.

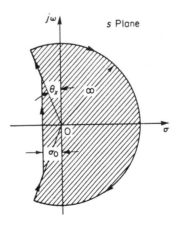

Figure 6-57
Modified Nyquist path.

In designing a control system, we require that the system be stable. Furthermore, it is necessary that the system have adequate relative stability.

In this section, we shall show that the Nyquist plot indicates not only whether or not a system is stable but also the degree of stability of a stable system. The Nyquist plot also gives information as to how stability may be improved, if this is necessary. (For details, see Chapter 7.)

In the following discussion, we shall assume that the systems considered have unity feedback. Note that it is always possible to reduce a system with feedback elements to a unity-feedback system, as shown in Figure 6–58. Hence the extension of relative stability analysis for the unity-feedback system to nonunity-feedback systems is possible.

We shall also assume that, unless otherwise stated, the systems are minimum phase systems; that is, the open-loop transfer function $G(s)$ has neither poles nor zeros in the right-half s plane.

Relative stability analysis via conformal mapping. One of the important problems in analyzing a control system is to find all closed-loop poles or at least those closest to the $j\omega$ axis (or the dominant pair of closed-loop poles). If the open-loop frequency-response characteristics of a system are known, it may be possible to estimate the closed-loop poles closest to the $j\omega$ axis. It is noted that the Nyquist locus $G(j\omega)$ need not be an analytically known function of ω. The entire Nyquist locus may be an experimentally obtained one. The technique to be presented here is essentially a graphical one and is based on a conformal mapping of the s plane into the $G(s)$ plane.

Consider the conformal mapping of constant-σ lines (lines $s = \sigma + j\omega$ where σ is constant and ω varies) and constant-ω lines (lines $s = \sigma + j\omega$, where ω is constant and σ varies) in the s plane. The $\sigma = 0$ line (the $j\omega$ axis) in the s plane maps into the Nyquist plot in the $G(s)$ plane. The constant-σ lines in the s plane map into curves that are similar to the Nyquist plot and are in a sense parallel to the Nyquist plot, as shown in Fig. 6–59. The constant-ω lines in the s plane map into curves, also shown in Fig. 6–59.

Although the shapes of constant-σ and constant-ω loci in the $G(s)$ plane and the closeness of approach of the $G(j\omega)$ locus to the $-1 + j0$ point depend on a particular $G(s)$, the

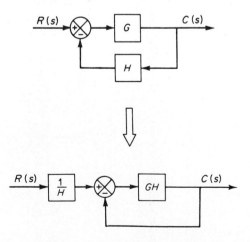

Figure 6–58
Modification of a system with feedback elements to a unity-feedback system.

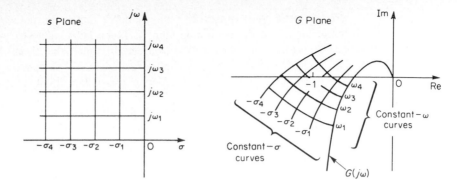

Figure 6–59
Conformal mapping of
s-plane girds into the
G(s) plane.

closeness of approach of the $G(j\omega)$ locus to the $-1 + j0$ point is an indication of the relative stability of a stable system. In general, we may expect that the closer the $G(j\omega)$ locus is to the $-1 + j0$ point, the larger the maximum overshoot is in the step transient response and the longer it takes to damp out.

Consider the two systems shown in Figures 6–60(a) and (b). (In Figure 6–60, the $\times$'s indicate closed-loop poles.) System (a) is obviously more stable than system (b) because the closed-loop poles of system (a) are located farther left of those of system (b). Figures

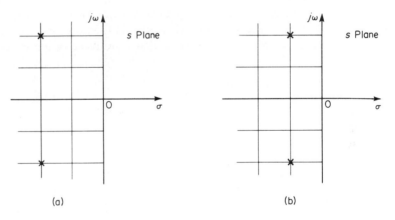

Figure 6–60
Two systems with two
closed-loop poles.

6–61(a) and (b) show the conformal mapping of s-plane grids into the $G(s)$ plane. The closer the closed-loop poles are located to the $j\omega$ axis, the closer the $G(j\omega)$ locus is to the $-1 + j0$ point.

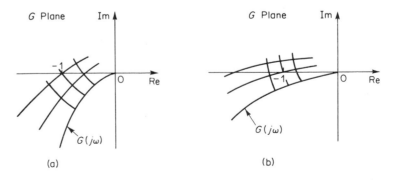

Figure 6–61
Conformal mappings
of s-plane grids for
the systems shown in
Fig. 6–60 into the
G(s) plane.

Phase and gain margins. Figure 6–62 shows the polar plots of $G(j\omega)$ for three different values of the open-loop gain K. For a large value of the gain K, the system is unstable. As the gain is decreased to a certain value, the $G(j\omega)$ locus passes through the $-1 + j0$ point. This means that with this gain value the system is on the verge of instability, and the system will exhibit sustained oscillations. For a small value of the gain K, the system is stable.

In general, the closer the $G(j\omega)$ locus comes to encircling the $-1 + j0$ point, the more oscillatory is the system response. The closeness of the $G(j\omega)$ locus to the $-1 + j0$ point can be used as a measure of the margin of stability. (This does not apply, however, to conditionally stable systems.) It is common practice to represent the closeness in terms of phase margin and gain margin.

Phase margin: The phase margin is that amount of additional phase lag at the gain crossover frequency required to bring the system to the verge of instability. The gain crossover frequency is the frequency at which $|G(j\omega)|$, the magnitude of the open-loop transfer function, is unity. The phase margin γ is $180°$ plus the phase angle ϕ of the open-loop transfer function at the gain crossover frequency, or

$$\gamma = 180° + \phi$$

Figures 6–63(a), (b), and (c) illustrate the phase margin of both a stable system and an unstable system in Bode diagrams, polar plots, and log-magnitude versus phase plots. In the polar plot, a line may be drawn from the origin to the point at which the unit circle crosses the $G(j\omega)$ locus. The angle from the negative real axis to this line is the phase margin. The phase margin is positive for $\gamma > 0$ and negative for $\gamma < 0$. For a minimum phase system to be stable, the phase margin must be positive. In the logarithmic plots, the critical point in the complex plane corresponds to the 0 db and $-180°$ lines.

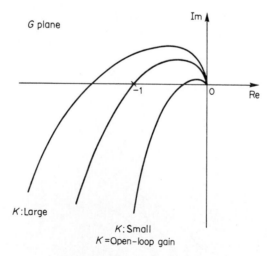

Figure 6–62 Polar plots of
$$\frac{K(1 + j\omega T_a)(1 + j\omega T_b)\cdots}{(j\omega)^{\lambda}(1 + j\omega T_1)(1 + j\omega T_2)\cdots}$$

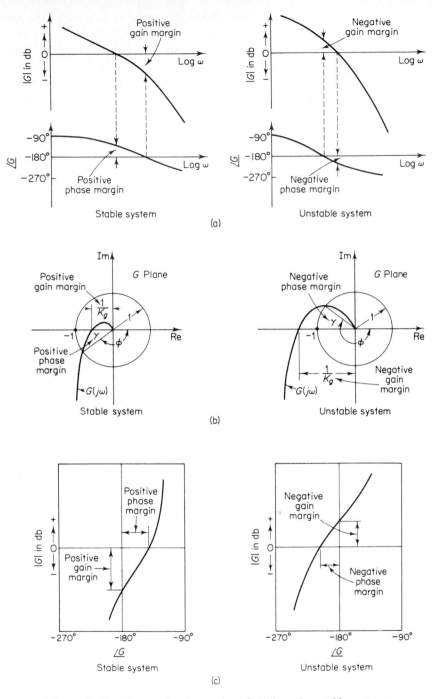

Figure 6–63 Phase and gain margins of stable and unstable systems. (a) Bode diagrams; (b) polar plots; (c) log-magnitude versus phase plots.

Gain margin: The gain margin is the reciprocal of the magnitude $|G(j\omega)|$ at the frequency where the phase angle is $-180°$. Defining the phase crossover frequency ω_1 to be the frequency at which the phase angle of the open-loop transfer function equals $-180°$ gives the gain margin K_g:

$$K_g = \frac{1}{|G(j\omega_1)|}$$

In terms of decibels,

$$K_g \text{ db} = 20 \log K_g = -20 \log |G(j\omega_1)|$$

The gain margin expressed in decibels is positive if K_g is greater than unity and negative if K_g is smaller than unity. Thus, a positive gain margin (in decibels) means that the system is stable, and a negative gain margin (in decibels) means that the system is unstable. The gain margin is shown in Figures 6–63(a), (b), and (c).

For a stable minimum phase system, the gain margin indicates how much the gain can be increased before the system becomes unstable. For an unstable system, the gain margin is indicative of how much the gain must be decreased to make the system stable.

The gain margin of a first- or second-order system is infinite since the polar plots for such systems do not cross the negative real axis. Thus, theoretically, first- or second-order systems cannot be unstable. (Note, however, that so-called first- or second-order systems are only approximations in the sense that small time lags are neglected in deriving the system equations and are thus not truly first- or second-order systems. If these small lags are accounted for, the so-called first- or second-order systems may become unstable.)

It is important to point out that for a nonminimum phase system the stability condition will not be satisfied unless the $G(j\omega)$ plot encircles the $-1 + j0$ point. Hence, a stable nonminimum phase system will have negative phase and gain margins.

It is also important to point out that conditionally stable systems will have two or more phase crossover frequencies, and some higher-order systems with complicated numerator

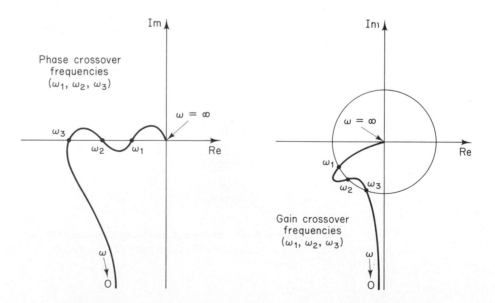

Figure 6–64
Polar plots showing more than two phase or gain crossover frequencies.

Chapter 6 / Frequency-Response Analysis

dynamics may also have two or more gain crossover frequencies, as shown in Figure 6–64. For stable systems having two or more gain crossover frequencies, the phase margin is measured at the highest gain crossover frequency.

A few comments on phase and gain margins. The phase and gain margins of a control system are a measure of the closeness of the polar plot to the $-1 + j0$ point. Therefore, these margins may be used as design criteria.

It should be noted that either the gain margin alone or the phase margin alone does not give a sufficient indication of the relative stability. Both should be given in the determination of relative stability.

For a minimum phase system, both the phase and gain margins must be positive for the system to be stable. Negative margins indicate instability.

Proper phase and gain margins ensure us against variations in the system components and are specified for definite values of frequency. The two values bound the behavior of the closed-loop system near the resonant frequency. For satisfactory performance, the phase margin should be between 30° and 60°, and the gain margin should be greater than 6 db. With these values, a minimum phase system has guaranteed stability, even if the open-loop gain and time constants of the components vary to a certain extent. Although the phase and gain margins give only rough estimates of the effective damping ratio of the closed-loop system, they do offer a convenient means for designing control systems or adjusting the gain constants of systems.

For minimum phase systems, the magnitude and phase characteristics of the open-loop transfer function are definitely related. The requirement that the phase margin be between 30° and 60° means that in a Bode diagram the slope of the log-magnitude curve at the gain crossover frequency should be more gradual than -40 db/decade. In most practical cases, a slope of -20 db/decade is desirable at the gain crossover frequency for stability. If it is -40 db/decade, the system could be either stable or unstable. (Even if the system is stable, however, the phase margin is small.) If the slope at the gain crossover frequency is -60 db/decade or steeper, the system is most likely unstable.

EXAMPLE 6–13

Obtain the phase and gain margins of the system shown in Figure 6–65 for the two cases where $K = 10$ and $K = 100$.

The phase and gain margins can easily be obtained from the Bode diagram. A Bode diagram of the given open-loop transfer function with $K = 10$ is shown in Figure 6–66(a). The phase and gain margins for $K = 10$ are

$$\text{Phase margin} = 21°, \qquad \text{Gain margin} = 8 \text{ db}$$

Therefore, the system gain may be increased by 8 db before the instability occurs.

Increasing the gain from $K = 10$ to $K = 100$ shifts the 0-db axis down by 20 db, as shown in Figure 6–66(b). The phase and gain margins are

$$\text{Phase margin} = -30°, \qquad \text{Gain margin} = -12 \text{ db}$$

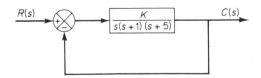

Figure 6–65
Control system.

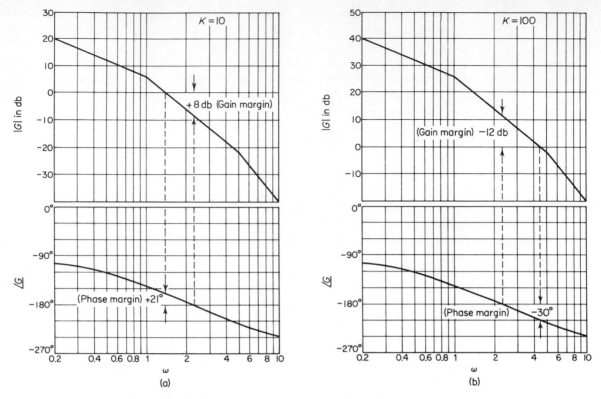

Figure 6–66 Bode diagrams of the system shown in Fig. 6–65 with $K = 10$ and $K = 100$.

Thus, the system is stable for $K = 10$ but unstable for $K = 100$.

Notice that one of the very convenient aspects of the Bode diagram approach is the ease with which the effects of gain changes can be evaluated.

Note that to obtain satisfactory performance, we must increase the phase margin to $30° \sim 60°$. This can be done by decreasing the gain K. Decreasing K is not desirable, however, since a small value of K will yield a large error for the ramp input. This suggests that reshaping of the open-loop frequency-response curve by adding compensation may be necessary. Such techniques are discussed in detail in Chapter 7.

Resonant peak magnitude M_r and resonant peak frequency ω_r. Consider the system shown in Figure 6–67. The closed-loop transfer function is

$$\frac{C(s)}{R(s)} = \frac{\omega_n^2}{s^2 + 2\zeta\omega_n s + \omega_n^2} \qquad (6\text{–}15)$$

where ζ and ω_n are the damping ratio and the undamped natural frequency, respectively. The closed-loop frequency response is

$$\frac{C(j\omega)}{R(j\omega)} = \frac{1}{\left(1 - \dfrac{\omega^2}{\omega_n^2}\right) + j2\zeta\dfrac{\omega}{\omega_n}} = Me^{j\alpha}$$

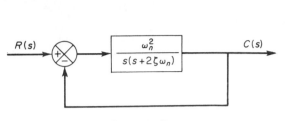

$R(s)$ $\dfrac{\omega_n^2}{s(s+2\zeta\omega_n)}$ $C(s)$

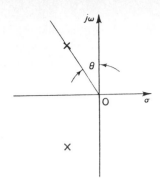

Figure 6–67
Control system.

Figure 6–68
Definition of the angle θ.

where

$$M = \frac{1}{\sqrt{\left(1 - \dfrac{\omega^2}{\omega_n^2}\right)^2 + \left(2\zeta\dfrac{\omega}{\omega_n}\right)^2}}, \qquad \alpha = -\tan^{-1}\frac{2\zeta\dfrac{\omega}{\omega_n}}{1 - \dfrac{\omega^2}{\omega_n^2}}$$

As given by Equation (6–12), for $0 \le \zeta \le 0.707$ the maximum value of M occurs at the frequency ω_r, where

$$\omega_r = \omega_n\sqrt{1 - 2\zeta^2} = \omega_n\sqrt{\cos 2\theta} \qquad (6\text{–}16)$$

The angle θ is defined in Figure 6–68. The frequency ω_r is the resonant frequency. At the resonant frequency, the value of M is maximum and is given by Equation (6–13), rewritten thus:

$$M_r = \frac{1}{2\zeta\sqrt{1 - \zeta^2}} = \frac{1}{\sin 2\theta} \qquad (6\text{–}17)$$

where M_r is defined as the *resonant peak magnitude*. The resonant peak magnitude is related to the damping of the system.

The magnitude of the resonant peak gives an indication of the relative stability of the system. A large resonant peak magnitude indicates the presence of a pair of dominant closed-loop poles with small damping ratio, which will yield an undesirable transient response. A smaller resonant peak magnitude, on the other hand, indicates the absence of a pair of dominant closed-loop poles with small damping ratio, meaning that the system is well damped.

Remember that ω_r is real only if $\zeta < 0.707$. Thus, there is no closed-loop resonance if $\zeta > 0.707$. [The value of M_r is unity for $\zeta > 0.707$. See Equation (6–14).] Since the values of M_r and ω_r can be easily measured in a physical system, they are quite useful for checking agreement between theoretical and experimental analyses.

It is noted, however, that in prctical design problems the phase margin and gain margin are more frequently specified than the resonant peak magnitude to indicate the degree of damping in a system.

Correlation between step transient response and frequency response in the standard second-order system. The maximum overshoot in the unit-step response of the standard second-order system, as shown in Figure 6–67, can be exactly correlated to the resonant peak magnitude in the frequency response. Hence, essentially the same information of the system dynamics is contained in the frequency response as is in the transient response.

For a unit-step input, the output of the system shown in Figure 6–67 is given by Equation (4–19), or

$$c(t) = 1 - e^{-\zeta\omega_n t}\left(\cos\omega_d t + \frac{\zeta}{\sqrt{1-\zeta^2}}\sin\omega_d t\right) \qquad (t \geq 0)$$

where

$$\omega_d = \omega_n\sqrt{1-\zeta^2} = \omega_n\cos\theta \qquad (6\text{–}18)$$

On the other hand, the maximum overshoot M_p for the unit-step response is given by Equation (4–28), or

$$M_p = e^{-(\zeta/\sqrt{1-\zeta^2})\pi} \qquad (6\text{–}19)$$

This maximum overshoot occurs in the transient response that has the damped natural frequency $\omega_d = \omega_n\sqrt{1-\zeta^2}$. The maximum overshoot becomes excessive for values of $\zeta < 0.4$.

Since the second-order system shown in Figure 6–67 has the following open-loop transfer function

$$G(s) = \frac{\omega_n^2}{s(s + 2\zeta\omega_n)}$$

for sinusoidal operation, the magnitude of $G(j\omega)$ becomes unity when

$$\omega = \omega_n\sqrt{\sqrt{1 + 4\zeta^4} - 2\zeta^2}$$

which can be obtained by equating $|G(j\omega)|$ to unity and solving for ω. At this frequency, the phase angle of $G(j\omega)$ is

$$\angle G(j\omega) = -\angle j\omega - \angle j\omega + 2\zeta\omega_n = -90° - \tan^{-1}\frac{\sqrt{\sqrt{1 + 4\zeta^4} - 2\zeta^2}}{2\zeta}$$

Thus, the phase margin γ is

$$\gamma = 180° + \angle G(j\omega)$$

$$= 90° - \tan^{-1}\frac{\sqrt{\sqrt{1 + 4\zeta^4} - 2\zeta^2}}{2\zeta}$$

$$= \tan^{-1}\frac{2\zeta}{\sqrt{\sqrt{1 + 4\zeta^4} - 2\zeta^2}} \qquad (6\text{–}20)$$

Equation (6–20) gives the relationship between the damping ratio ζ and the phase margin γ. (Notice that the phase margin γ is a function only of the damping ratio ζ.)

In the following, we shall summarize the correlation between the step transient response and frequency response of the second-order system given by Equation (6–15):

1. The phase margin and the damping ratio are directly related. Figure 6–69 shows a plot of the phase margin γ as a function of the damping ratio ζ. It is noted that for the

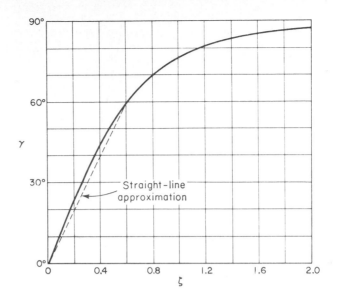

Figure 6–69
Curve γ (phase margin) versus ζ for the system shown in Fig. 6–67.

standard second-order system shown in Figure 6–67 the phase margin γ and the damping ratio ζ are related approximately by a straight line for $0 \le \zeta \le 0.6$, as follows:

$$\zeta = \frac{\gamma}{100}$$

Thus a phase margin of 60° corresponds to a damping ratio of 0.6. For higher-order systems having a dominant pair of closed-loop poles, this relationship may be used as a rule of thumb in estimating the relative stability in transient response (that is, the damping ratio) from the frequency response.

2. Referring to Equations (6–16) and (6–18), we see that the values of ω_r and ω_d are almost the same for small values of ζ. Thus, for small values of ζ, the value of ω_r is indicative of the speed of the transient response of the system.

3. From Equations (6–17) and (6–19), we note that the smaller the value of ζ is, the larger the values of M_r and M_p are. The correlation between M_r and M_p as a function of ζ is shown in Figure 6–70. A close relationship between M_r and M_p can be seen for $\zeta > 0.4$. For very small values of ζ, M_r becomes very large ($M_r \gg 1$), while the value of M_p does not exceed 1.

Correlation between step transient response and frequency response in general systems. The design of control systems is very often carried out on the basis of frequency response. The main reason for this is the relative simplicity of this approach as compared to others. Since in many applications it is the transient response of the system to aperiodic inputs rather than the steady-state response to sinusoidal inputs that is of primary concern, the question of correlation between transient response and frequency response arises.

For the second-order system shown in Figure 6–67, mathematical relationships correlating the step transient response and frequency response can be obtained easily. The time response of a second-order system can be exactly predicted from a knowledge of the M_r and ω_r of its closed-loop frequency response.

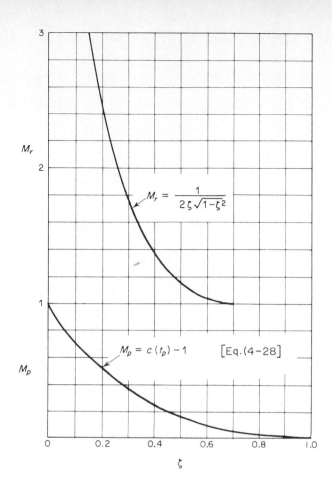

Figure 6–70
Curves M_r versus ζ and M_p versus ζ for the system shown in Fig. 6–67.

For higher-order systems, the correlation is more complex, and transient response may not be predicted easily from frequency response because additional poles may change the correlation between the step transient response and frequency response existing for a second-order system. Mathematical techniques for obtaining the exact correlation are available, but they are very laborious and of little practical value.

The applicability of the transient-response–frequency-response correlation existing for the second-order system shown in Figure 6–67 to higher-order systems depends on the presence of a dominant pair of complex-conjugate closed-loop poles in the latter systems. Clearly, if the frequency response of a higher-order system is dominated by a pair of complex-conjugate closed-loop poles, the transient-response–frequency-response correlation existing for the second-order system can be extended to the higher-order system.

For linear time-invariant higher-order systems having a dominant pair of complex-conjugate closed-loop poles, the following relationships generally exist between the step transient response and frequency response:

1. The value of M_r is indicative of the relative stability. Satisfactory transient performance is usually obtained if the value of M_r is in the range $1.0 < M_r < 1.4$ (0 db $< M_r <$ 3 db), which corresponds to an effective damping ratio of $0.4 < \zeta < 0.7$. For values of M_r greater than 1.5, the step transient response may exhibit several overshoots. (Note that in general a large value of M_r corresponds to a large overshoot in the step transient response. If the system is subjected to noise signals whose frequencies are near the resonant frequency ω_r, the noise will be amplified in the output and will present serious problems.)

2. The magnitude of the resonant frequency ω_r is indicative of the speed of the transient response. The larger the value of ω_r, the faster the time response is. In other words, the rise time varies inversely with ω_r. In terms of the open-loop frequency response, the damped natural frequency in the transient response is somewhere between the gain crossover frequency and phase crossover frequency.

3. The resonant peak frequency ω_r and the damped natural frequency ω_d for step transient response are very close to each other for lightly damped systems.

The three relationships just listed are useful for correlating the step transient response with the frequency response of higher-order systems, provided they can be approximated by a second-order system or a pair of complex-conjugate closed-loop poles. If a higher-order system satisfies this condition, a set of time-domain specifications may be translated into frequency-domain specifications. This simplifies greatly the design work or compensation work of higher-order systems.

In addition to the phase margin, gain margin, resonant peak M_r, and resonant peak frequency ω_r, there are other frequency-domain quantities commonly used in performance specifications. They are the cutoff frequency, bandwidth, and the cutoff rate. These will be defined in what follows.

Cutoff frequency and bandwidth.　Referring to Figure 6–71, the frequency ω_b at which the magnitude of the closed-loop frequency is 3 db below its zero-frequency value is called the *cutoff frequency*. Thus

$$\left|\frac{C(j\omega)}{R(j\omega)}\right| < \left|\frac{C(j0)}{R(j0)}\right| - 3\text{ db} \qquad (\omega > \omega_b)$$

For systems where $|C(j0)/R(j0)| = 0$ db,

$$\left|\frac{C(j\omega)}{R(j\omega)}\right| < -3\text{ db} \qquad (\omega > \omega_b)$$

The closed-loop system filters out the signal components whose frequencies are greater than the cutoff frequency and transmits those signal components with frequencies lower than the cutoff frequency.

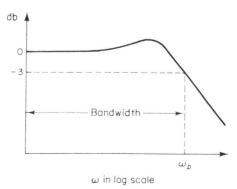

Figure 6–71
Logarithmic plot showing cutoff frequency ω_b and bandwidth.

The frequency range $0 \le \omega \le \omega_b$, in which the magnitude of the closed loop does not drop -3 db is called the *bandwidth* of the system. The bandwidth indicates the frequency where the gain starts to fall off from its low-frequency value. Thus, the bandwidth indicates how well the system will track an input sinusoid. Note that for a given ω_n, the rise time increases with increasing damping ratio ζ. On the other hand, the bandwidth decreases with the increase of ζ. Therefore, the rise time and the bandwidth are inversely proportional to each other.

The specification of the bandwidth may be determined by the following factors:

1. The ability to reproduce the input signal. A large bandwidth corresponds to a small rise time, or fast response. Roughly speaking, we can say that the bandwidth is proportional to the speed of response.
2. The necessary filtering characteristics for high-frequency noise.

For the system to follow arbitrary inputs accurately, it is necessary that the system have a large bandwidth. From the viewpoint of noise, however, the bandwidth should not be too large. Thus, there are conflicting requirements on the bandwidth, and a compromise is usually necessary for good design. Note that a system with large bandwidth requires high performance components. So cost of components usually increases with the bandwidth.

Cutoff rate.   The cutoff rate is the slope of the log-magnitude curve near the cutoff frequency. The cutoff rate indicates the ability of a system to distinguish the signal from noise.

It is noted that a closed-loop frequency response curve with a steep cutoff characteristic may have a large resonant peak magnitude, which implies that the system has relatively small stability margin.

EXAMPLE 6–14   Consider the following two systems:

$$\text{System I:} \quad \frac{C(s)}{R(s)} = \frac{1}{s+1}, \qquad \text{System II:} \quad \frac{C(s)}{R(s)} = \frac{1}{3s+1}$$

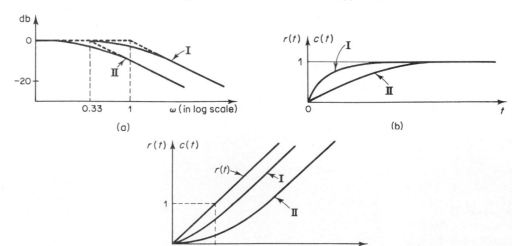

Figure 6–72
Comparison of dynamic characteristics of the two systems considered in Example 6–14. (a) Closed-loop frequency response curves; (b) unit-step response curves; (c) unit-ramp response curves.

Compare the bandwidths of these two systems. Show that the system with the larger bandwidth has a faster speed of response and can follow the input much better than the one with a smaller bandwidth.

Figure 6–72(a) shows the closed-loop frequency-response curves for the two systems. (Asymptotic curves are shown by dotted lines.) We find that the bandwidth of system I is $0 \leq \omega \leq 1$ rad/sec and that of system II is $0 \leq \omega \leq 0.33$ rad/sec. Figures 6–72(b) and (c) show, respectively, the unit-step response and unit-ramp response curves for the two systems. Clearly, system I, whose bandwidth is three times wider than that of system II, has a faster speed of response and can follow the input much better.

6–8 CLOSED-LOOP FREQUENCY RESPONSE

Closed-loop frequency response of unity-feedback systems.   For a stable closed-loop system, the frequency response can be obtained easily from that of the open loop. Consider the unity-feedback system shown in Figure 6–73(a). The closed-loop transfer function is

$$\frac{C(s)}{R(s)} = \frac{G(s)}{1 + G(s)}$$

In the Nyquist or polar plot shown in Figure 6–73(b), the vector $\overrightarrow{OA}$ represents $G(j\omega_1)$, where ω_1 is the frequency at point A. The length of the vector $\overrightarrow{OA}$ is $|G(j\omega_1)|$ and the angle of the vector $\overrightarrow{OA}$ is $\underline{/G(j\omega_1)}$. The vector $\overrightarrow{PA}$, the vector from the $-1 + j0$ point to the Nyquist locus, represents $1 + G(j\omega_1)$. Therefore, the ratio of $\overrightarrow{OA}$ to $\overrightarrow{PA}$ represents the closed-loop frequency response, or

$$\frac{\overrightarrow{OA}}{\overrightarrow{PA}} = \frac{G(j\omega_1)}{1 + G(j\omega_1)} = \frac{C(j\omega_1)}{R(j\omega_1)}$$

The magnitude of the closed-loop transfer function at $\omega = \omega_1$ is the ratio of the magnitudes of $\overrightarrow{OA}$ to $\overrightarrow{PA}$. The phase angle of the closed-loop transfer function at $\omega = \omega_1$ is the angle formed by the vectors $\overrightarrow{OA}$ to $\overrightarrow{PA}$, that is, $\phi - \theta$, shown in Figure 6–73(b). By measuring the magnitude and phase angle at different frequency points, the closed-loop frequency-response curve can be obtained.

Figure 6–73
(a) Unity-feedback system; (b) Determination of closed-loop frequency response from open-loop frequency response.

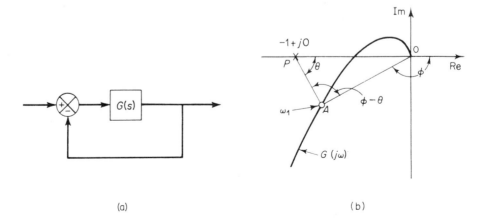

(a)

(b)

Let us define the magnitude of the closed-loop frequency response as M and the phase angle as α, or

$$\frac{C(j\omega)}{R(j\omega)} = Me^{j\alpha}$$

In the following, we shall find the constant magnitude loci and constant phase-angle loci. Such loci are convenient in determining the closed-loop frequency response from the polar plot or Nyquist plot.

Constant magnitude loci (M circles). To obtain the constant magnitude loci, let us first note that $G(j\omega)$ is a complex quantity and can be written as follows:

$$G(j\omega) = X + jY$$

where X and Y are real quantities. Then M is given by

$$M = \frac{|X + jY|}{|1 + X + jY|}$$

and M^2 is

$$M^2 = \frac{X^2 + Y^2}{(1 + X)^2 + Y^2}$$

Hence

$$X^2(1 - M^2) - 2M^2 X - M^2 + (1 - M^2)Y^2 = 0 \qquad (6\text{--}21)$$

If $M = 1$, then from Equation (6–21) we obtain $X = -\frac{1}{2}$. This is the equation of a straight line parallel to the Y axis and passing through the point $(-\frac{1}{2}, 0)$.

If $M \neq 1$, Equation (6–21) can be written

$$X^2 + \frac{2M^2}{M^2 - 1}X + \frac{M^2}{M^2 - 1} + Y^2 = 0$$

If the term $M^2/(M^2 - 1)^2$ is added to both sides of this last equation, we obtain

$$\left(X + \frac{M^2}{M^2 - 1}\right)^2 + Y^2 = \frac{M^2}{(M^2 - 1)^2} \qquad (6\text{--}22)$$

Equation (6–22) is the equation of a circle with center at $X = -M^2/(M^2 - 1)$, $Y = 0$ and with radius $|M/(M^2 - 1)|$.

The constant M loci on the $G(s)$ plane are thus a family of circles. The center and radius of the circle for a given value of M can be easily calculated. For example, for $M = 1.3$, the center is at $(-2.45, 0)$ and the radius is 1.88. A family of constant M circles is shown in Figure 6–74. It is seen that as M becomes larger compared with 1, the M circles become smaller and converge to the $-1 + j0$ point. For $M > 1$, the centers of the M circles lie to the left of the $-1 + j0$ point. Similarly, as M becomes smaller compared with 1, the M circle becomes smaller and converges to the origin. For $0 < M < 1$, the centers of the M circles lie to the right of the origin. $M = 1$ corresponds to the locus of points equidistant from the origin and from the $-1 + j0$ point. As stated earlier, it is a straight line passing through the point $(-\frac{1}{2}, 0)$ and parallel to the imaginary axis. (The constant M circles cor-

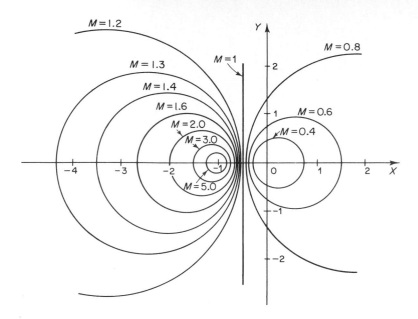

Figure 6–74
A family of constant
M circles.

responding to $M > 1$ lie to the left of the $M = 1$ line and those corresponding to $0 <$
$M < 1$ lie to the right of the $M = 1$ line.) The M circles are symmetrical with respect to
the straight line corresponding to $M = 1$ and with respect to the real axis.

Constant phase-angle loci (N circles). We shall obtain the phase angle α in terms
of X and Y. Since

$$\angle e^{j\alpha} = \left/ \frac{X + jY}{1 + X + jY} \right.$$

the phase angle α is

$$\alpha = \tan^{-1}\left(\frac{Y}{X}\right) - \tan^{-1}\left(\frac{Y}{1 + X}\right)$$

If we define

$$\tan \alpha = N$$

then

$$N = \tan\left[\tan^{-1}\left(\frac{Y}{X}\right) - \tan^{-1}\left(\frac{Y}{1 + X}\right)\right]$$

Since

$$\tan(A - B) = \frac{\tan A - \tan B}{1 + \tan A \tan B}$$

we obtain

$$N = \frac{\dfrac{Y}{X} - \dfrac{Y}{1+X}}{1 + \dfrac{Y}{X}\left(\dfrac{Y}{1+X}\right)} = \frac{Y}{X^2 + X + Y^2}$$

or

$$X^2 + X + Y^2 - \frac{1}{N}Y = 0$$

The addition of $(\frac{1}{4}) + 1/(2N)^2$ to both sides of this last equation yields

$$\left(X + \frac{1}{2}\right)^2 + \left(Y - \frac{1}{2N}\right)^2 = \frac{1}{4} + \left(\frac{1}{2N}\right)^2 \tag{6–23}$$

This is an equation of a circle with center at $X = -\frac{1}{2}$, $Y = 1/(2N)$ and with radius $\sqrt{(\frac{1}{4}) + 1/(2N)^2}$. For example, if $\alpha = 30°$, then $N = \tan\alpha = 0.577$, and the center and the radius of the circle corresponding to $\alpha = 30°$ are found to be $(-0.5, 0.866)$ and unity, respectively. Since Equation (6–23) is satisfied when $X = Y = 0$ and $X = -1$, $Y = 0$ regardless of the value of N, each circle passes through the origin and the $-1 + j0$ point. The constant α loci can be drawn easily once the value of N is given. A family of constant N circles are shown in Figure 6–75 with α as a parameter.

It should be noted that the constant N locus for a given value of α is actually not the entire circle but only an arc. In other words, the $\alpha = 30°$ and $\alpha = -150°$ arcs are parts

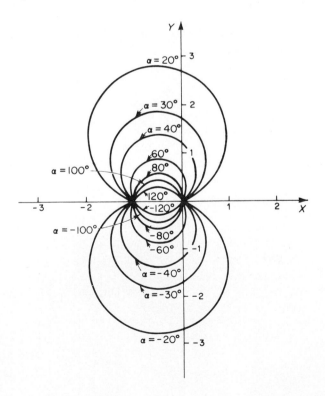

Figure 6–75
A family of constant
N circles.

of the same circle. This is so because the tangent of an angle remains the same if $\pm 180°$ (or multiplies thereof) is added to the angle.

The use of the M and N circles enables us to find the entire closed-loop frequency response from the open-loop frequency response $G(j\omega)$ without calculating the magnitude and phase of the closed-loop transfer function at each frequency. The intersections of the $G(j\omega)$ locus and the M circles and N circles give the values of M and N at frequency points on the $G(j\omega)$ locus.

The N circles are multivalued in the sense that the circle for $\alpha = \alpha_1$ and that for $\alpha = \alpha_1 \pm 180°n$ ($n = 1, 2, \ldots$) are the same. In using the N circles for the determination of the phase angle of closed-loop systems, we must interpret the proper value of α. To avoid any error, start at zero frequency, which corresponds to $\alpha = 0°$, and proceed to higher frequencies. The phase-angle curve must be continuous.

Graphically, the intersections of the $G(j\omega)$ locus and M circles give the values of M at the frequencies denoted on the $G(j\omega)$ locus. Thus, the constant M circle with the smallest radius that is tangent to the $G(j\omega)$ locus gives the value of the resonant peak magnitude M_r. If it is desired to keep the resonant peak value less than a certain value, then the system should not enclose the critical point ($-1 + j0$ point) and at the same time there should be no intersections with the particular M circle and the $G(j\omega)$ locus.

Figure 6–76(a) shows the $G(j\omega)$ locus superimposed on a family of M circles. Figure 6–76(b) shows the $G(j\omega)$ locus superimposed on a family of N circles. From these plots, it is possible to obtain the closed-loop frequency response by inspection. It is seen that the $M = 1.1$ circle intersects the $G(j\omega)$ locus at frequency point $\omega = \omega_1$. This means that at this frequency the magnitude of the closed-loop transfer function is 1.1. In Figure 6–76(a), the $M = 2$ circle is just tangent to the $G(j\omega)$ locus. Thus, there is only one point on the $G(j\omega)$ locus for which $|C(j\omega))/R((j\omega)|$ is equal to 2. Figure 6–76(c) shows the closed-loop frequency-response curve for the system. The upper curve is the M versus frequency ω curve and the lower curve is the phase angle α versus frequency ω curve.

The resonant peak value is the value of M corresponding to the M circle of smallest radius that is tangent to the $G(j\omega)$ locus. Thus, in the Nyquist diagram, the resonant peak value M_r and the resonant frequency ω_r can be found from the M-circle tangency to the $G(j\omega)$ locus. (In the present example, $M_r = 2$ and $\omega_r = \omega_4$.)

Nichols chart.   In dealing with design problems, we find it convenient to construct the M and N loci in the log-magnitude versus phase plane. The chart consisting of the M and N loci in the log-magnitude versus phase diagram is called the Nichols chart. This chart is shown in Fig. 6–77 for phase angles between $0°$ and $-240°$.

Note that the critical point ($-1 + j0$ point) is mapped to the Nichols chart as the point (0 db, $-180°$). The Nichols chart contains curves of constant closed-loop magnitude and phase angle. The designer can graphically determine the phase margin, gain margin, resonant peak magnitude, resonant peak frequency, and bandwidth of the clsoed-loop system from the plot of the open-loop locus, $G(j\omega)$.

The Nichols chart is symmetric about the $-180°$ axis. The M and N loci repeat for every $360°$, and there is symmetry at every $180°$ interval. The M loci are centered about the critical point (0 db, $-180°$). The Nichols chart is quite useful for determining the closed-loop frequency response from that of the open loop. If the open-loop frequency-response curve is superimposed on the Nichols chart, the intersections of the open-loop frequency-response curve $G(j\omega)$ and the M and N loci give the values of the magnitude M and phase angle α

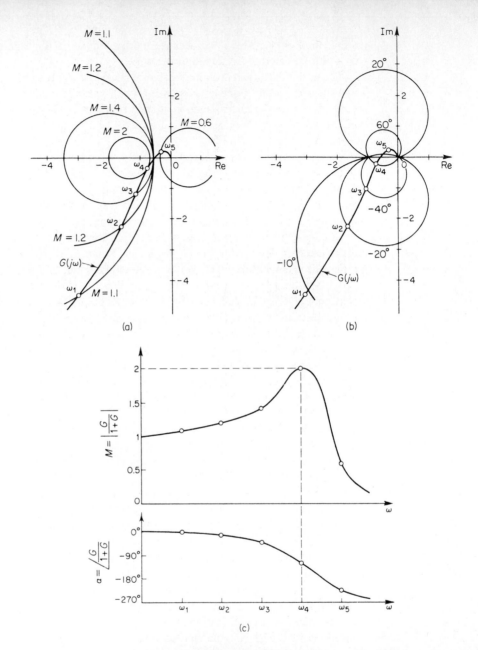

Figure 6–76
(a) $G(j\omega)$ locus
superimposed on a
family of M circles;
(b) $G(j\omega)$ locus
superimposed on a
family of N circles;
(c) closed-loop
frequency-response
curves.

of the closed-loop frequency response at each frequency point. If the $G(j\omega)$ locus does not intersect the $M = M_r$ locus but is tangent to it, then the resonant peak value of M of the closed-loop frequency response is given by M_r. The resonant peak frequency is given by the frequency at the point of tangency.

As an example, consider the unity-feedback system with the following open-loop transfer function:

$$G(j\omega) = \frac{K}{s(s + 1)(0.5s + 1)}, \qquad K = 1$$

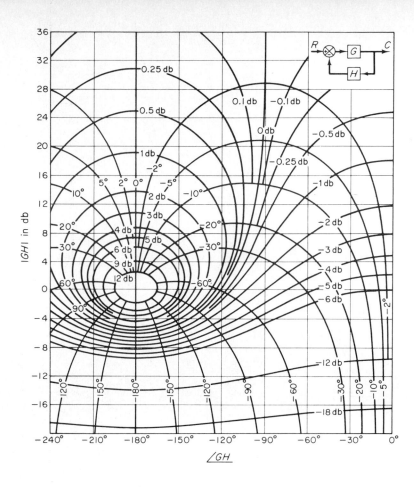

Figure 6–77
Nichols chart.

To find the closed-loop frequency response by use of the Nichols chart, the $G(j\omega)$ locus is constructed in the log-magnitude versus phase plane from the Bode diagram. The use of the Bode diagram eliminates the lengthy numerical calculation of $G(j\omega)$. Figure 6–78(a) shows the $G(j\omega)$ locus together with the M and N loci. The closed-loop frequency-response curves may be constructed by reading the magnitudes and phase angles at various frequency points on the $G(j\omega)$ locus from the M and N loci, as shown in Figure 6–78(b). Since the largest magnitude contour touched by the $G(j\omega)$ locus is 5 db, the resonant peak magnitude M_r is 5 db. The corresponding resonant peak frequency is 0.8 rad/sec.

Notice that the phase crossover point is the point where the $G(j\omega)$ locus intersects the $-180°$ axis (for the present system, $\omega = 1.4$ rad/sec), and the gain crossover point is the point where the locus intersect the 0-db axis (for the present system, $\omega = 0.76$ rad/sec). The phase margin is the horizontal distance (measured in degrees) between the gain crossover point and the critical point (0 db, $-180°$). The gain margin is the distance (in decibels) between the phase crossover point and the critical point.

The bandwidth of the closed-loop system can easily be found from the $G(j\omega)$ locus in the Nichols diagram. The frequency at the intersection of the $G(j\omega)$ locus and the $M = -3$ db locus gives the bandwidth.

If the open-loop gain K is varied, the shape of the $G(j\omega)$ locus in the log-magnitude versus phase diagram remains the same, but it is shifted up (for increasing K) or down (for

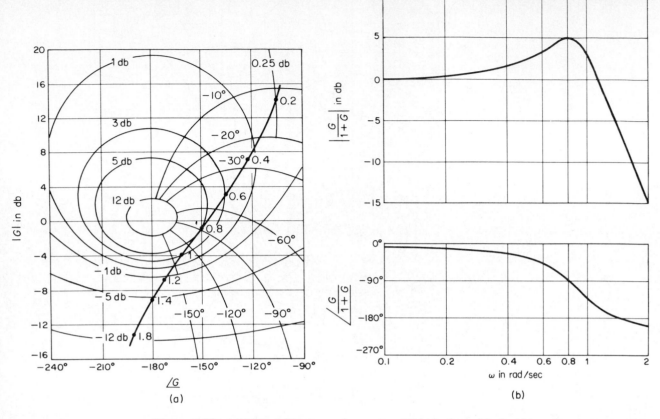

Figure 6–78 (a) Plot of $G(j\omega)$ superimposed on Nichols chart; (b) closed-loop frequency-response curves.

decreasing K) along the vertical axis. Therefore, the $G(j\omega)$ locus intersects the M and N loci differently, resulting in a different closed-loop frequency-response curve. For a small value of the gain K, the $G(j\omega)$ locus will not be tangent to any of the M loci, which means that there is no resonance in the closed-loop frequency response.

Closed-loop frequency response for nonunity-feedback systems. In the preceding sections, our discussions were limited to closed-loop systems with unity feedback. The constant M and N loci and the Nichols chart cannot be directly applied to control systems with nonunity feedback but rather require a slight modification.

If the closed-loop system involves a nonunity-feedback transfer function, then the closed-loop transfer function may be written

$$\frac{C(s)}{R(s)} = \frac{G(s)}{1 + G(s)H(s)}$$

where $G(s)$ is the feedforward transfer function and $H(s)$ is the feedback transfer function. Then $C(j\omega)/R(j\omega)$ can be written

$$\frac{C(j\omega)}{R(j\omega)} = \frac{1}{H(j\omega)} \frac{G(j\omega)H(j\omega)}{1 + G(j\omega)H(j\omega)}$$

The magnitude and phase angle of

$$\frac{G_1(j\omega)}{1 + G_1(j\omega)}$$

where $G_1(j\omega) = G(j\omega)H(j\omega)$, may be obtained easily by plotting the $G_1(j\omega)$ locus on the Nichols chart and reading the values of M and N at various frequency points. The closed-loop frequency response $C(j\omega)/R(j\omega)$ may then be obtained by multiplying $G_1(j\omega)/[1 + G_1(j\omega)]$ by $1/H(j\omega)$. This multiplication can be made without difficulty if we draw Bode diagrams for $G_1(j\omega)/[1 + G_1(j\omega)]$ and $H(j\omega)$ and then graphically subtract the magnitude of $H(j\omega)$ from that of $G_1(j\omega)/[1 + G_1(j\omega)]$ and also graphically subtract the phase angle of $H(j\omega)$ from that of $G_1(j\omega)/[1 + G_1(j\omega)]$. Then the resulting log-magnitude curve and phase-angle curve give the closed-loop frequency response $C(j\omega)/R(j\omega)$.

To obtain acceptable values of M_r, ω_r, and ω_b for $|C(j\omega)/R(j\omega)|$, a trial-and-error process may be necessary. In each trial, the $G_1(j\omega)$ locus is varied in shape. Then Bode diagrams for $G_1(j\omega)/[1 + G_1(j\omega)]$ and $H(j\omega)$ are drawn, and the closed-loop frequency response $C(j\omega)/R(j\omega)$ is obtained. The values of M_r, ω_r, and ω_b are checked until they are acceptable.

Gain adjustments. The concept of M circles will now be applied to the design of control systems. In obtaining suitable performance, the adjustment of gain is usually the first consideration. The adjustment may be based on a desirable value for the resonant peak.

In the following, we shall demonstrate a method for determining the gain K so that the system will have some maximum value M_r, not exceeded over the entire frequency range.

Referring to Figure 6–79, we see that the tangent line drawn from the origin to the desired M_r circle has an angle of ψ, as shown, if M_r is greater than unity. The value of $\sin \psi$ is

$$\sin \psi = \left| \frac{\dfrac{M_r}{M_r^2 - 1}}{\dfrac{M_r^2}{M_r^2 - 1}} \right| = \frac{1}{M_r} \qquad (6\text{–}24)$$

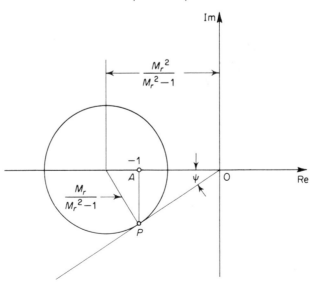

Figure 6–79
M circle.

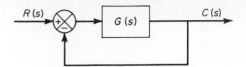

Figure 6–80
Control system.

It can easily be proved that the line drawn from point P, perpendicular to the negative real axis, intersects this axis at the $-1 + j0$ point.

Consider the system shown in Figure 6–80. The procedure for determining the gain K so that $G(j\omega) = KG_1(j\omega)$ will have a desired value of M_r (where $M_r > 1$) can be summarized as follows:

1. Draw the polar plot of the normalized open-loop transfer function $G_1(j\omega) = G(j\omega)/K$.
2. Draw from the origin the line that makes an angle of $\psi = \sin^{-1}(1/M_r)$ with the negative real axis.
3. Fit a circle with center on the negative real axis tangent to both the $G_1(j\omega)$ locus and the line PO.
4. Draw a perpendicular line to the negative real axis from point P, the point of tangency of this circle with the line PO. The perpendicular line PA intersects the negative real axis at point A.
5. For the circle just drawn to correspond to the desired M_r circle, point A should be the $-1 + j0$ point.
6. The desired value of the gain K is that value which changes the scale so that point A becomes the $-1 + j0$ point. Thus, $K = 1/\overline{OA}$.

Note that the resonant frequency ω_r is the frequency of the point at which the circle is tangent to the $G_1(j\omega)$ locus. The present procedure may not yield a satisfactory value for ω_r. If this is the case, the system must be compensated in order to increase the value of ω_r without changing the value of M_r. (For the compensation of control systems, see Chapter 7.)

Note also that if the system has nonunity feedback, then the method requires some cut-and-try steps.

EXAMPLE 6–15   Consider the unity-feedback control system whose open-loop transfer function is

$$G(j\omega) = \frac{K}{j\omega(1 + j\omega)}$$

Determine the value of the gain K so that $M_r = 1.4$.

The first step in the determination of the gain K is to sketch the polar plot of

$$\frac{G(j\omega)}{K} = \frac{1}{j\omega(1 + j\omega)}$$

as shown in Figure 6–81. The value of ψ corresponding to $M_r = 1.4$ is obtained from

$$\psi = \sin^{-1}\frac{1}{M_r} = \sin^{-1}\frac{1}{1.4} = 45.6°$$

The next step is to draw the line OP that makes an angle $\psi = 45.6°$ with the negative real axis. Then draw the circle that is tangent to both the $G(j\omega)/K$ locus and the line OP. Define the point where the

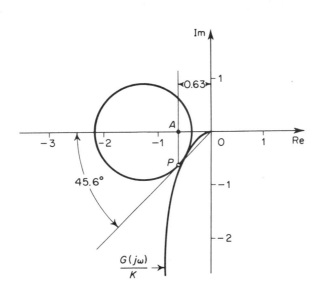

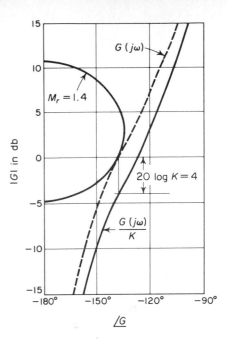

Figure 6–81 Determination of the gain K using an M circle.

Figure 6–82 Determination of the gain K using the Nichols chart.

circle is tangent to the 45.6° line as point P. The perpendicular line drawn from point P intersects the negative real axis at $(-0.63, 0)$. Then the gain K of the system is determined as follows:

$$K = \frac{1}{0.63} = 1.58$$

It should be noted that such determination of the gain can also be done easily on the log-magnitude versus phase plot. In what follows, we shall demonstrate how the log-magnitude versus phase diagram can be used to determine the gain K so that the system will have a desired value of M_r.

Figure 6–82 shows the $M_r = 1.4$ locus and the $G(j\omega)/K$ locus. Changing the gain has no effect on the phase angle but merely moves the curve vertically up for $K > 1$ and down for $K < 1$. In Figure 6–82, the $G(j\omega)/K$ locus must be raised by 4 db in order that it be tangent to the desired M_r locus and that the entire $G(j\omega)/K$ locus be outside the $M_r = 1.4$ locus. The amount of vertical shift of the $G(j\omega)/K$ locus determines the gain necessary to yield the desired value of M_r. Thus, by solving

$$20 \log K = 4$$

we obtain

$$K = 1.58$$

Thus, we have the same result as we obtained earlier.

6–9 EXPERIMENTAL DETERMINATION OF TRANSFER FUNCTIONS

The first step in the analysis and design of a control system is to derive a mathematical model of the plant under consideration. Obtaining a model analytically may be quite difficult. We may have to obtain it by means of experimental analysis. The importance of the frequency-response methods is that the transfer function of the plant, or any other component of a system, may be determined by simple frequency-response measurements.

If the amplitude ratio and phase shift have been measured at a sufficient number of frequencies within the frequency range of interest, they may be plotted on the Bode diagram. Then the transfer function can be determined by asymptotic approximations. We build up asymptotic log-magnitude curves consisting of several segments. With some trial-and-error juggling of the corner frequencies, it is usually possible to find a very close fit to the curve. (Note that if the frequency is plotted in cycles per second rather than radians per second, the corner frequencies must be converted to radians per second before computing the time constants.)

Sinusoidal-signal generators.   In performing a frequency-response test, suitable sinusoidal-signal generators must be available. The signal may have to be in mechanical, electrical, or pneumatic form. The frequency ranges needed for the test are approximately 0.001 to 10 Hz for large time-constant systems and 0.1 to 1000 Hz for small time-constant systems. The sinusoidal signal must be reasonably free from harmonics or distortion.

For very low frequency ranges (below 0.01 Hz), a mechanical signal generator (together with a suitable pneumatic or electrical transducer if necessary) may be used. For the frequency range from 0.01 to 1000 Hz, a suitable electrical-signal generator (together with a suitable transducer if necessary) may be used.

Determination of minimum phase transfer functions from Bode diagrams. As stated previously, whether or not a system is minimum phase can be determined from the frequency-response curves by examining the high-frequency characteristics.

To determine the transfer function, we first draw asymptotes to the experimentally obtained log-magnitude curve. The asymptotes must have slopes of multiples of ± 20 db/decade. If the slope of the experimentally obtained log-magnitude curve changes from -20 to -40 db/decade at $\omega = \omega_1$, it is clear that a factor $1/[1 + j(\omega/\omega_1)]$ exists in the transfer function. If the slope changes by -40 db/decade at $\omega = \omega_2$, there must be a quadratic factor of the form

$$\frac{1}{1 + 2\zeta\left(j\dfrac{\omega}{\omega_2}\right) + \left(j\dfrac{\omega}{\omega_2}\right)^2}$$

in the transfer function. The undamped natural frequency of this quadratic factor is equal to the corner frequency ω_2. The damping ratio ζ can be determined from the experimentally obtained log-magnitude curve by measuring the amount of resonant peak near the corner frequency ω_2 and comparing this with the curves shown in Figure 6–11.

Once the factors of the transfer function $G(j\omega)$ have been determined, the gain can be determined from the low-frequency portion of the log-magnitude curve. Since such terms as $1 + j(\omega/\omega_1)$ and $1 + 2\zeta(j\omega/\omega_2) + (j\omega/\omega_2)^2$ become unity as ω approaches zero, at very low frequencies, the sinusoidal transfer function $G(j\omega)$ can be written

$$\lim_{\omega \to 0} G(j\omega) = \frac{K}{(j\omega)^\lambda}$$

In many practical systems, λ equals 0, 1, or 2.

1. For $\lambda = 0$, or type 0 systems,

$$G(j\omega) = K \qquad \text{for } \omega \ll 1$$

or

$$20 \log |G(j\omega)| = 20 \log K \qquad \text{for } \omega \ll 1$$

The low-frequency asymptote is a horizontal line at $20 \log K$ db. The value of K can thus be found from this horizontal asymptote.

2. For $\lambda = 1$, or type 1 systems,

$$G(j\omega) = \frac{K}{j\omega} \qquad \text{for } \omega \ll 1$$

or

$$20 \log|G(j\omega)| = 20 \log K - 20 \log \omega \qquad \text{for } \omega \ll 1$$

which indicates that the low-frequency asymptote has the slope -20 db/decade. The frequency at which the low-frequency asymptote (or its extension) intersects the 0-db line is numerically equal to K.

3. For $\lambda = 2$, or type 2 systems,

$$G(j\omega) = \frac{K}{(j\omega)^2} \qquad \text{for } \omega \ll 1$$

or

$$20 \log|G(j\omega)| = 20 \log K - 40 \log \omega \qquad \text{for } \omega \ll 1$$

The slope of the low-frequency asymptote is -40 db/decade. The frequency at which this asymptote (or its extension) intersects the 0-db line is numerically equal to $\sqrt{K}$.

Examples of log-magnitude curves for type 0, type 1, and type 2 systems are shown in Figure 6–83, together with the frequency to which the gain K is related.

The experimentally obtained phase-angle curve provides a means of checking the transfer function obtained from the log-magnitude curve. For a minimum phase system, the experimental phase-angle curve should agree reasonably well with the theoretical phase-angle curve obtained from the transfer function just determined. These two phase-angle curves should agree exactly in both the very low frequency range and the very high frequency range. If the experimentally obtained phase angle at very high frequencies (compared with the corner frequencies) is not equal to $-90°(q - p)$, where p and q are the degrees of the numerator and denominator polynomials of the transfer function, respectively, then the transfer function must be a nonminimum phase transfer function.

Nonminimum phase transfer functions.   If, at the high-frequency end, the computed phase lag is $180°$ less than the experimentally obtained phase lag, then one of the zeros of the transfer function should have been in the right-half s plane instead of the left-half s plane.

If the computed phase lag differed from the experimentally obtained phase lag by a constant rate of change of phase, then transport lag, or dead time, is present. If we assume the transfer function to be of the form

$$G(s)e^{-Ts}$$

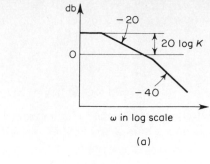

(a)

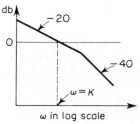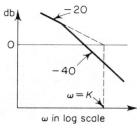

(b)

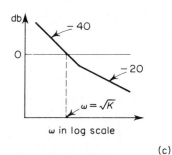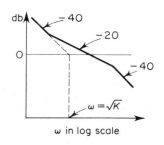

(c)

Figure 6–83
(a) Log-magnitude curve of a type 0 system; (b) log-magnitude curves of type 1 systems; (c) log-magnitude curves of type 2 systems. (The slopes shown are in db/decade.)

where $G(s)$ is a ratio of two polynomials in s, then

$$\lim_{\omega \to \infty} \frac{d}{d\omega} \angle G(j\omega)e^{-j\omega T} = \lim_{\omega \to \infty} \frac{d}{d\omega} \left[\angle G(j\omega) + \angle e^{-j\omega T} \right]$$

$$= \lim_{\omega \to \infty} \frac{d}{d\omega} \left[\angle G(j\omega) - \omega T \right]$$

$$= 0 - T = -T$$

Thus, from this last equation, we can evaluate the magnitude of the transport lag T.

A few remarks on the experimental determination of transfer functions

1. It is usually easier to make accurate amplitude measurements than accurate phase-shift measurements. Phase-shift measurements may involve errors that may be caused by instrumentation or by misinterpretation of the experimental records.

2. The frequency response of measuring equipment used to measure the system output must have nearly flat magnitude versus frequency curves. In addition, the phase angle must be nearly proportional to the frequency.

3. Physical systems may have several kinds of nonlinearities. Therefore, it is necessary to consider carefully the amplitude of input sinusoidal signals. If the amplitude of the input signal is too large, the system will saturate, and the frequency-response test will yield inaccurate results. On the other hand, a small signal will cause errors due to dead zone. Hence, a careful choice of the amplitude of the input sinusoidal signal must be made. It is necessary to sample the waveform of the system output to make sure that the waveform is sinusoidal and that the system is operating in the linear region during the test period. (The waveform of the system output is not sinusoidal when the system is operating in its nonlinear region.)

4. If the system under consideration is operating continuously for days and weeks, then normal operation need not be stopped for frequency-response tests. The sinusoidal test signal may be superimposed on the normal inputs. Then, for linear systems, the output due to the test signal is superimposed on the normal output. For the determination of the transfer function while the system is in normal operation, stochastic signals (white noise signals) also are often used. By use of correlation functions, the transfer function of the system can be determined without interrupting normal operation.

EXAMPLE 6–16   Determine the transfer function of the system whose experimental frequency-response curves are as shown in Figure 6–84.

The first step in determining the transfer function is to approximate the log-magnitude curve by asymptotes with slopes ± 20 db/decade and multiplies thereof, as shown in Figure 6–84. We then estimate the corner frequencies. For the system shown in Figure 6–84, the following form of the transfer function is estimated.

$$G(j\omega) = \frac{K(1 + 0.5j\omega)}{j\omega(1 + j\omega)\left[1 + 2\zeta\left(j\dfrac{\omega}{8}\right) + \left(j\dfrac{\omega}{8}\right)^2\right]}$$

The value of the damping ratio ζ is estimated by examining the peak resonance near $\omega = 6$ rad/sec. Referring to Figure 6–11, ζ is determined to be 0.5. The gain K is numerically equal to the frequency at the intersection of the extension of the low-frequency asymptote and the 0-db line. The value of K is thus found to be 10. Therefore, $G(j\omega)$ is tentatively determined as

$$G(j\omega) = \frac{10(1 + 0.5j\omega)}{j\omega(1 + j\omega)\left[1 + \left(j\dfrac{\omega}{8}\right) + \left(j\dfrac{\omega}{8}\right)^2\right]}$$

or

$$G(s) = \frac{320(s + 2)}{s(s + 1)(s^2 + 8s + 64)}$$

This transfer function is tentative because we have not examined the phase-angle curve yet.

Once the corner frequencies are noted on the log-magnitude curve, the corresponding phase-angle curve for each component factor of the transfer function can easily be drawn. The sum of these component phase-angle curves is that of the assumed transfer function. The phase-angle curve for $G(j\omega)$ is denoted by $\angle G$ in Figure 6–84. From Figure 6–84 we clearly notice a discrepancy between the computed phase-angle curve and the experimentally obtained phase-angle curve. The difference between the two

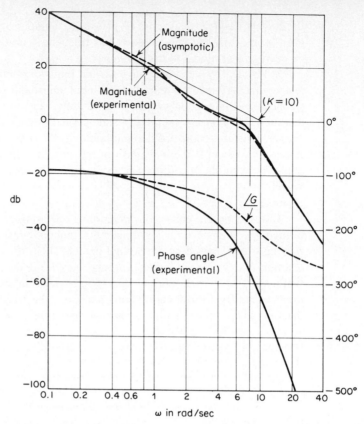

Figure 6–84
Bode diagram of a
system. (Solid curves
are experimentally
obtained curves.)

curves at very high frequencies appears to be a constant rate of change. Thus the discrepancy in the phase-angle curves must be caused by transport lag.

Hence we assume the complete transfer function to be $G(s)e^{-Ts}$. Since the discrepancy between the computed and experimental phase angles is -0.2ω rad for very high frequencies, we can determine the value of T as follows:

$$\lim_{\omega \to \infty} \frac{d}{d\omega} \underline{/G(j\omega)e^{-j\omega T}} = -T = -0.2$$

or

$$T = 0.2 \text{ sec}$$

The presence of transport lag can thus be determined, and the complete transfer function determined from the experimental curves is

$$G(s)e^{-Ts} = \frac{320(s + 2)e^{-0.2s}}{s(s + 1)(s^2 + 8s + 64)}$$

Example Problems and Solutions

A–6–1. Consider the circuit shown in Figure 6–85. Assume that the voltage e_i is applied to the input terminals and the voltage e_o appears at the output terminals. Assume that the input to the system is

$$e_i(t) = E_i \sin \omega t$$

Figure 6–85
Electrical circuit.

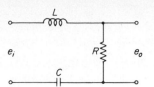

Obtain the steady-state current $i(t)$ flowing through the resistance R.

Solution. Applying Kirchhoff's voltage law to the circuit, we obtain

$$L\frac{di}{dt} + Ri + \frac{1}{C}\int i\, dt = e_i$$

The transfer function between $I(s)$ and $E_i(s)$ is

$$\frac{I(s)}{E_i(s)} = \frac{1}{Ls + R + \dfrac{1}{Cs}} = G(s)$$

For the given input $e_i(t) = E_i \sin \omega t$, the steady-state current $i_{ss}(t)$ is

$$i_{ss}(t) = E_i|G(j\omega)|\sin[\omega t + \underline{/G(j\omega)}]$$

where

$$G(j\omega) = \frac{1}{Lj\omega + R - j\dfrac{1}{C\omega}}$$

Since

$$|G(j\omega)| = \frac{1}{\sqrt{R^2 + \left(L\omega - \dfrac{1}{C\omega}\right)^2}}$$

and

$$\underline{/G(j\omega)} = -\tan^{-1}\left(\frac{L\omega - \dfrac{1}{C\omega}}{R}\right)$$

we have

$$i_{ss}(t) = \frac{E_i}{\sqrt{R^2 + \left(L\omega - \dfrac{1}{C\omega}\right)^2}}\sin\left[\omega t - \tan^{-1}\left(\frac{L\omega - \dfrac{1}{C\omega}}{R}\right)\right]$$

A–6–2. Consider a system whose closed-loop transfer function is

$$\frac{C(s)}{R(s)} = \frac{10(s + 1)}{(s + 2)(s + 5)}$$

(This is the same system considered in Problem A–5–11.) Clearly, the closed-loop poles are located at $s = -2$ and $s = -5$, and the system is not oscillatory. (The unit-step response, however, exhibits overshoot due to the presence of a zero at $s = -1$. See Figure 5–49.)

Show that the closed-loop frequency response of this system will exhibit a resonant peak, although the damping ratio of the closed-loop poles are greater than unity.

Solution. Figure 6–86 shows the Bode diagram for the system. The resonant peak value is approximately 3.5 db. (Note that, in the absence of a zero, the second-order system with $\zeta > 0.7$ will not exhibit a resonant peak; however, the presence of a closed-loop zero will cause such a peak.)

A–6–3. Prove that the polar plot of the sinusoidal transfer function

$$G(j\omega) = \frac{j\omega T}{1 + j\omega T} \qquad (0 \le \omega \le \infty)$$

is a semicircle. Find the center and radius of the circle.

Solution. The given sinusoidal transfer function $G(j\omega)$ can be written as follows:

$$G(j\omega) = X + jY$$

where

$$X = \frac{\omega^2 T^2}{1 + \omega^2 T^2}, \qquad Y = \frac{\omega T}{1 + \omega^2 T^2}$$

Then

$$\left(X - \frac{1}{2}\right)^2 + Y^2 = \frac{(\omega^2 T^2 - 1)^2}{4(1 + \omega^2 T^2)^2} + \frac{\omega^2 T^2}{(1 + \omega^2 T^2)^2} = \frac{1}{4}$$

Hence we see that the plot of $G(j\omega)$ is a circle centered at $(0.5, 0)$ with radius equal to 0.5. The upper semicircle corresponds to $0 \le \omega \le \infty$, and the lower semicircle corresponds to $-\infty \le \omega \le 0$.

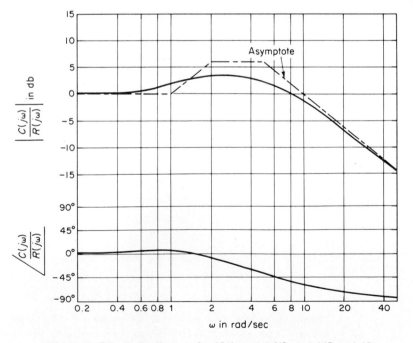

Figure 6–86 Bode diagram for $10(1 + j\omega)/[(2 + j\omega)(5 + j\omega)]$.

A–6–4. Plot a Bode diagram for the following open-loop transfer function $G(s)$:

$$G(s) = \frac{20(s^2 + s + 0.5)}{s(s + 1)(s + 10)}$$

Solution. By substituting $s = j\omega$ into $G(s)$, we have

$$G(j\omega) = \frac{20\,[(j\omega)^2 + (j\omega) + 0.5]}{j\omega(j\omega + 1)(j\omega + 10)}$$

Notice that ω_n and ζ of the quadratic term in the numerator are

$$\omega_n = \sqrt{0.5} \qquad \text{and} \qquad \zeta = 0.707$$

This quadratic term can be written as

$$\omega_n^2\left[\left(\frac{j\omega}{\omega_n}\right)^2 + 2\,\zeta\left(j\,\frac{\omega}{\omega_n}\right) + 1\right] = (\sqrt{0.5})^2\left[\left(\frac{j\omega}{\sqrt{0.5}}\right)^2 + 2 \times 0.707\left(j\,\frac{\omega}{\sqrt{0.5}}\right) + 1\right]$$

Notice that the corner frequency is at $\omega = \sqrt{0.5} = 0.707$ rad/sec. Now $G(j\omega)$ can be written as

$$G(j\omega) = \frac{\left(j\,\dfrac{\omega}{\sqrt{0.5}}\right)^2 + 1.414\left(j\,\dfrac{\omega}{\sqrt{0.5}}\right) + 1}{j\omega(j\omega + 1)(0.1\,j\omega + 1)}$$

The Bode diagram for $G(j\omega)$ is shown in Figure 6–87.

A–6–5. Draw the Bode diagram of the following nonminimum phase system:

$$\frac{C(s)}{R(s)} = 1 - Ts$$

Obtain the unit-ramp response of the system and plot $c(t)$ versus t.

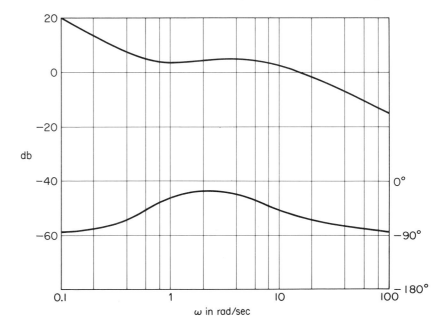

Figure 6–87
Bode diagram for
$G(j\omega)$ of Problem A–
6–4.

A–6–7. Consider the function

$$F(s) = \frac{s + 1}{s - 1}$$

The conformal mapping of the lines $\omega = 0, \pm 1, \pm 2$ and the lines $\sigma = 0, \pm 1, \pm 2$ yields circles in the $F(s)$ plane, as shown in Figure 6–91. Show that if the contour in the s plane encloses the pole of $F(s)$, there is one encirclement of the origin of the $F(s)$ plane in the counterclockwise direction. If the contour in the s plane encloses the zero of $F(s)$, there is one encirclement of the origin of the $F(s)$ plane in the clockwise direction. If the contour in the s plane encloses both the zero and pole or if the contour encloses neither the zero nor pole, then there is no encirclement of the origin of the $F(s)$ plane by the locus of $F(s)$. (Note that in the s plane a representative point s traces out a contour in the clockwise direction.)

Solution. A graphical solution is given in Figure 6–92; this shows closed contours in the s plane and their corresponding closed curves in the $F(s)$ plane.

A–6–8. Prove the following mapping theorem: Let $F(s)$ be a ratio of polynomials in s. Let P be the number of poles and Z be the number of zeros of $F(s)$ that lie inside a closed contour in the s plane, multiplicity accounted for. Let the closed contour be such that it does not pass through any poles or zeros of $F(s)$. The closed contour in the s plane then maps into the $F(s)$ plane as a closed curve. The number N of clockwise encirclements of the origin of the $F(s)$ plane, as a representative point s traces out the entire contour in the s plane in the clockwise direction, is equal to $Z - P$.

Solution. To prove this theorem, we use Cauchy's theorem and the residue theorem. Cauchy's theorem states that the integral of $F(s)$ around a closed contour in the s plane is zero if $F(s)$ is analytic within and on the closed contour, or

$$\oint F(s)\, ds = 0$$

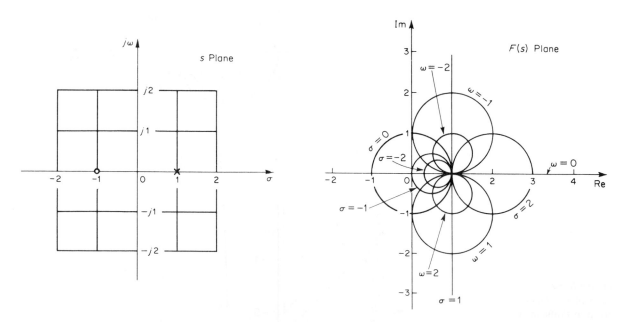

Figure 6–91 Conformal mapping of the s-plane grids into the $F(s)$ plane where $F(s) = (s + 1)/(s - 1)$.

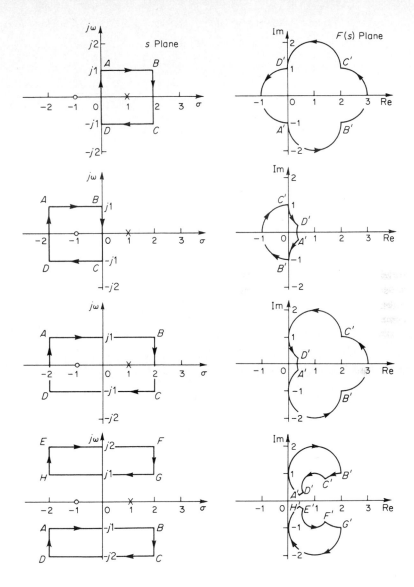

Figure 6–92
Conformal mapping of the s-plane contours into the $F(s)$ plane where $F(s) = (s + 1)/(s - 1)$.

Suppose that $F(s)$ is given by

$$F(s) = \frac{(s + z_1)^{k_1}(s + z_2)^{k_2} \cdots}{(s + p_1)^{m_1}(s + p_2)^{m_2} \cdots} X(s)$$

where $X(s)$ is analytic in the closed contour in the s plane and all the poles and zeros are located in the contour. Then the ratio $F'(s)/F(s)$ can be written

$$\frac{F'(s)}{F(s)} = \left(\frac{k_1}{s + z_1} + \frac{k_2}{s + z_2} + \cdots \right) - \left(\frac{m_1}{s + p_1} + \frac{m_2}{s + p_2} + \cdots \right) + \frac{X'(s)}{X(s)} \quad (6\text{–}25)$$

This may be seen from the following consideration: If $F(s)$ is given by

$$F(s) = (s + z_1)^k X(s)$$

then $F(s)$ has a zero of kth order at $s = -z_1$. Differentiating $F(s)$ with respect to s yields

$$F'(s) = k(s + z_1)^{k-1}X(s) + (s + z_1)^k X'(s)$$

Hence,

$$\frac{F'(s)}{F(s)} = \frac{k}{s + z_1} + \frac{X'(s)}{X(s)} \tag{6-26}$$

We see that by taking the ratio $F'(s)/F(s)$ the kth-order zero of $F(s)$ becomes a simple pole of $F'(s)/F(s)$.

If the last term on the right side of Equation (6–26) does not contain any poles or zeros in the closed contour in the s plane, $F'(s)/F(s)$ is analytic in this contour except at the zero $s = -z_1$. Then, referring to Equation (6–25) and using the residue theorem, which states that the integral of $F'(s)/F(s)$ taken in the clockwise direction around a closed contour in the s plane is equal to $-2\pi j$ times the residues at the simple poles of $F'(s)/F(s)$, or

$$\oint \frac{F'(s)}{F(s)} ds = -2\pi j (\Sigma \text{ residues})$$

we have

$$\oint \frac{F'(s)}{F(s)} ds = -2\pi j[(k_1 + k_2 + \cdots) - (m_1 + m_2 + \cdots)] = -2\pi j(Z - P)$$

where $Z = k_1 + k_2 + \cdots$ = total number of zeros of $F(s)$ enclosed in the closed contour in the s plane

$\quad\quad P = m_1 + m_2 + \cdots$ = total number of poles of $F(s)$ enclosed in the closed contour in the s plane

[The k multiple zeros (or poles) are considered k zeros (or poles) located at the same point.] Since $F(s)$ is a complex number, $F(s)$ can be written

$$F(s) = |F|e^{j\theta}$$

and

$$\ln F(s) = \ln|F| + j\theta$$

Noting that $F'(s)/F(s)$ can be written

$$\frac{F'(s)}{F(s)} = \frac{d \ln F(s)}{ds}$$

we obtain

$$\frac{F'(s)}{F(s)} = \frac{d \ln|F|}{ds} + j\frac{d\theta}{ds}$$

If the closed contour in the s plane is mapped into the closed contour Γ in the $F(s)$ plane, then

$$\oint \frac{F'(s)}{F(s)} ds = \oint_\Gamma d \ln|F| + j\oint_\Gamma d\theta = j\int d\theta = 2\pi j(P - Z)$$

The integral $\oint_\Gamma d \ln|F|$ is zero since the magnitude $\ln|F|$ is the same at the initial point and the final point of the contour Γ. Thus, we obtain

$$\frac{\theta_2 - \theta_1}{2\pi} = P - Z$$

The angular difference between the final and initial values of θ is equal to the total change in the phase angle of $F'(s)/F(s)$ as a representative point in the s plane moves along the closed contour. Noting that N is the number of clockwise encirclements of the origin of the $F(s)$ plane and $\theta_2 - \theta_1$ is zero or a multiple of 2π rad, we obtain

$$\frac{\theta_2 - \theta_1}{2\pi} = -N$$

Thus, we have the relationship

$$N = Z - P$$

This proves the theorem.

Note that by this mapping theorem, the exact numbers of zeros and of poles cannot be found, only their difference. Note also that from Figures 6–93(a) and (b) we see that, if θ does not change through 2π rad, then the origin of the $F(s)$ plane cannot be encircled.

A–6–9. The polar plot of the open-loop frequency response of a unity-feedback control system is shown in Figure 6–94. Assuming that the Nyquist path in the s plane encloses the entire right-half s plane, draw a complete Nyquist plot in the G plane. Then answer the following questions:

(a) If the open-loop transfer function has no poles in the right-half s plane, is the closed-loop system stable?
(b) If the open-loop transfer function has one pole and no zeros in the right-half s plane, is the closed-loop system stable?
(c) If the open-loop transfer function has one zero and no poles in the right-half s plane, is the closed-loop system stable?

Solution. Figure 6–95 shows a complete Nyquist plot in the G plane. Answers to the three questions are as follows:

(a) The closed-loop system is stable, because the critical point $(-1 + j0)$ is not encircled by the Nyquist plot. That is, since $P = 0$ and $N = 0$, we have $Z = N + P = 0$.
(b) The open-loop transfer function has one pole in the right-half s plane. Hence, $P = 1$. (The open-loop system is unstable.) For the closed-loop system to be stable, the Nyquist plot must encircle

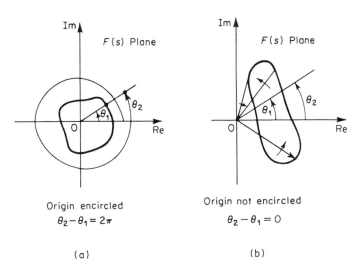

Figure 6–93
Determination of encirclement of the origin of $F(s)$ plane.

Origin encircled
$\theta_2 - \theta_1 = 2\pi$

(a)

Origin not encircled
$\theta_2 - \theta_1 = 0$

(b)

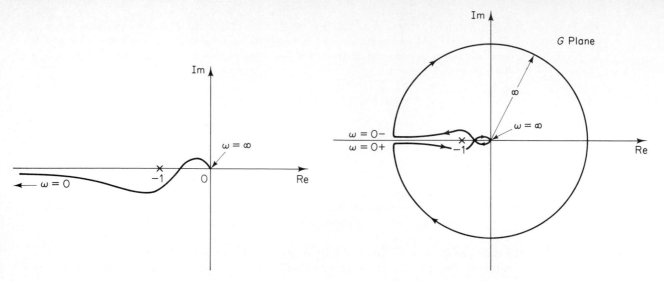

Figure 6–94 Polar plot.   **Figure 6–95** Complete Nyquist plot in the G plane.

the critical point $(-1 + j0)$ once counterclockwise. However, the Nyquist plot does not encircle the critical point. Hence $N = 0$. Therefore, $Z = N + P = 1$. The closed-loop system is unstable.

(c) Since the open-loop transfer function has one zero but no poles in the right-half s plane, we have $Z = N + P = 0$. Thus, the closed-loop system is stable. (Note that the zeros of the open-loop transfer function do not affect the stability of the closed-loop system.)

A–6–10.   Is the system with the following open-loop transfer function and with $K = 2$ stable?

$$G(s)H(s) = \frac{K}{s(s + 1)(2s + 1)}$$

Find the critical value of the gain K for stability.

Solution. The open-loop transfer function is

$$G(j\omega)H(j\omega) = \frac{K}{j\omega(j\omega + 1)(2j\omega + 1)}$$

$$= \frac{K}{-3\omega^2 + j\omega(1 - 2\omega^2)}$$

This open-loop transfer function has no poles in the right-half s plane. Thus, for stability, the $-1 + j0$ point should not be encircled by the Nyquist plot. Let us find the point where the Nyquist plot crosses the negative real axis. Let the imaginary part of $G(j\omega)H(j\omega)$ be zero, or

$$1 - 2\omega^2 = 0$$

from which

$$\omega = \pm \frac{1}{\sqrt{2}}$$

Substituting $\omega = 1/\sqrt{2}$ into $G(j\omega)H(j\omega)$, we obtain

$$G\left(j\frac{1}{\sqrt{2}}\right) H\left(j\frac{1}{\sqrt{2}}\right) = -\frac{2K}{3}$$

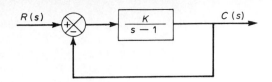

Figure 6–96
Closed-loop system.

The critical value of the gain K is obtained by equating $-2K/3$ to -1, or

$$-\frac{2}{3}K = -1$$

Hence

$$K = \frac{3}{2}$$

The system is stable if $0 < K < \frac{3}{2}$. Hence the system with $K = 2$ is unstable.

A–6–11. Consider the closed-loop system shown in Figure 6–96. Determine the critical value of K for stability by use of the Nyquist stability criterion.

Solution. The polar plot of

$$G(j\omega) = \frac{K}{j\omega - 1}$$

is a circle with center at $-K/2$ on the negative real axis and radius $K/2$, as shown in Figure 6–97(a). As ω is increased from $-\infty$ to ∞, the $G(j\omega)$ locus makes a counterclockwise rotation. In this system,

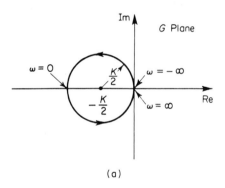

(a)

Figure 6–97
(a) Polar plot of $K/(j\omega - 1)$; (b) polar plots of $K/(j\omega - 1)$ for stable and unstable cases.

$P = 1$
$N = -1$ (Stable)
$Z = 0$

$K > 1$

$P = 1$
$N = 0$ (Unstable)
$Z = 1$

$K < 1$

(b)

$P = 1$ because there is one pole of $G(s)$ in the right-half s plane. For the closed-loop system to be stable, Z must be equal to zero. Therefore, $N = Z - P$ must be equal to -1, or there must be one counterclockwise encirclement of the $-1 + j0$ point for stability. (If there is no encirclement of the $-1 + j0$ point, the system is unstable.) Thus, for stability, K must be greater than unity, and $K = 1$ gives the stability limit. Figure 6–97(b) shows both stable and unstable cases of $G(j\omega)$ plots.

A–6–12. Consider a unity-feedback system whose open-loop transfer function is

$$G(s) = \frac{Ke^{-0.8s}}{s + 1}$$

Using the Nyquist plot, determine the critical value of K for stability.

Solution. For this system,

$$G(j\omega) = \frac{Ke^{-0.8j\omega}}{j\omega + 1}$$

$$= \frac{K(\cos 0.8\omega - j \sin 0.8\omega)(1 - j\omega)}{1 + \omega^2}$$

$$= \frac{K}{1 + \omega^2}[(\cos 0.8\omega - \omega \sin 0.8\omega) - j(\sin 0.8\omega + \omega \cos 0.8\omega)]$$

The imaginary part of $G(j\omega)$ is equal to zero if

$$\sin 0.8\omega + \omega \cos 0.8\omega = 0$$

Hence

$$\omega = -\tan 0.8\omega$$

Solving this equation for the smallest positive value of ω, we obtain

$$\omega = 2.4482$$

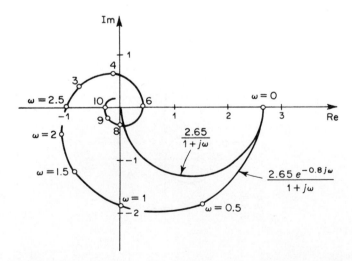

Figure 6–98 Polar plots of $2.65e^{-0.8j\omega}/(1 + j\omega)$ and $2.65/(1 + j\omega)$.

Substituting $\omega = 2.4482$ into $G(j\omega)$, we obtain

$$G(j2.4482) = \frac{K}{1 + 2.4482^2}(\cos 1.9586 - 2.4482 \sin 1.9586) = -0.378\,K$$

The critical value of K for stability is obtained by letting $G(j2.4482)$ equal -1. Hence

$$0.378K = 1$$

or

$$K = 2.65$$

Figure 6–98 shows the Nyquist or polar plots of $2.65\,e^{-0.8j\omega}/(1 + j\omega)$ and $2.65/(1 + j\omega)$. The first-order system without transport lag is stable for all values of K, but the one with transport lag of 0.8 sec becomes unstable for $K > 2.65$.

A–6–13. Given

$$G(s) = \frac{\omega_n^2}{s^2 + 2\zeta\omega_n s + \omega_n^2}$$

show that

$$|G(j\omega_n)| = \frac{1}{2\zeta}$$

Solution. Noting that

$$G(j\omega) = \frac{\omega_n^2}{(j\omega)^2 + 2\zeta\omega_n(j\omega) + \omega_n^2}$$

$$= \frac{1}{\left(j\dfrac{\omega}{\omega_n}\right)^2 + 2\zeta\left(j\dfrac{\omega}{\omega_n}\right) + 1}$$

we have

$$|G(j\omega_n)| = \left|\frac{1}{-1 + 2\zeta j + 1}\right| = \frac{1}{2\zeta}$$

A–6–14. Suppose that a system possesses at least one pair of complex-conjugate closed-loop poles. If the $-1 + j0$ point is found at the intersection of a constant-σ curve and a constant-ω curve in the $G(s)$ plane, then those particular values of σ and ω, which we define as $-\sigma_c$ and ω_c, respectively, characterize the closed-loop pole closest to the $j\omega$ axis in the upper-half s plane. (Note that $-\sigma_c$ represents the exponential decay and ω_c represents the damped natural frequency of the step transient-response term due to the pair of the closed-loop poles closest to the $j\omega$ axis.) Probable values of $-\sigma_c$ and ω_c may be estimated from the plot, as shown in Figure 6–99. Thus, the pair of complex-conjugate closed-loop poles that lies closest to the $j\omega$ axis can be determined graphically. It should be noted that all closed-loop poles are mapped into the $-1 + j0$ point in the $G(s)$ plane. Although the complex conjugate closed-loop poles closest to the $j\omega$ axis can be found easily by the technique above, finding other closed-loop poles, if any, by this technique is practically impossible.

If the data on $G(j\omega)$ are experimental, then a curvilinear square near the $-1 + j0$ point can be constructed by extrapolation. Referring to Figure 6–100, we can find the location of the dominant closed-loop poles in the s plane, or the damping ratio ζ and the damped natural frequency ω_d, by drawing the line AB that connects the $-1 + j0$ point (point A) and point B, the nearest approach to

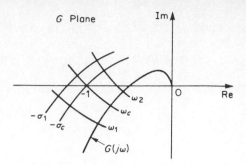

Figure 6–99
Estimation of $-\sigma_c$ and ω_c.

the $-1 + j0$ point, and then constructing a curvilinear square $CDEF$, as shown in Figure 6–100. This curvilinear square $CDEF$ may be constructed by drawing the most likely curve PQ (where PQ is the conformal mapping of a line parallel to the $j\omega$ axis in the s plane) passing through the $-1 + j0$ point and "parallel" to the $G(j\omega)$ locus, and adjusting the points C, D, E, and F such that $\widehat{FB} = \widehat{BE}$, $\widehat{CA} = \widehat{AD}$, and $\widehat{FE} + \widehat{CD} = \widehat{FC} + \widehat{ED}$. The corresponding s-plane contour $C'D'E'F'$, together with point A', the closed-loop pole closest to the $j\omega$ axis, is shown in Figure 6–100. The value of the frequency interval $\Delta\omega_1$ between points E and F is approximately equal to the value of σ_1 shown in Figure 6–100. The frequency at point B is approximately equal to the damped natural frequency ω_d. The closed-loop poles closest to the $j\omega$ axis are then estimated as

$$s = -\sigma_1 \pm j\omega_d$$

Then the damping ratio ζ of these closed-loop poles can be obtained from

$$\frac{\zeta}{\sqrt{1 - \zeta^2}} = \frac{\sigma_1}{\omega_d} = \frac{\Delta\omega_1}{\omega_d}$$

It should be noted that the damped natural frequency ω_d of the step transient response actually is on the frequency contour that passes through the $-1 + j0$ point and is not necessarily the point of nearest approach to the $G(j\omega)$ locus. Therefore, the value ω_d obtained by the technique above is somewhat in error.

From the analysis above, we may conclude that it is possible to estimate the closed-loop poles closest to the $j\omega$ axis from the closeness of approach of the $G(j\omega)$ locus to the $-1 + j0$ point, the frequency at the point of nearest approach, and the frequency graduation near this point.

Referring to the frequency-response plot of $G(j\omega)$ of a unity-feedback system, as shown in Figure 6–101, find the closed-loop poles closest to the $j\omega$ axis.

Solution. The line connecting the $-1 + j0$ point and the point of nearest approach of the $G(j\omega)$ locus to the $-1 + j0$ point is drawn first. Then the curvilinear square $ABCD$ is constructed. Since the

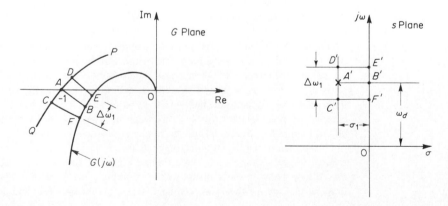

Figure 6–100
Conformal mapping of a curvilinear square near the $-1 + j0$ point in the $G(s)$ plane into the s plane.

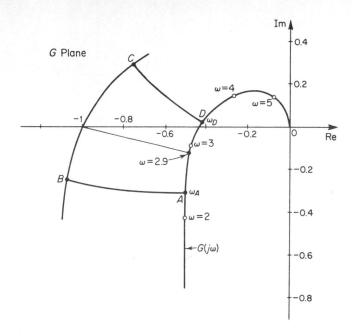

Figure 6–101
Polar plot and a
curvilinear square.

frequency at the point of nearest approach is $\omega = 2.9$, the damped natural frequency is approximately 2.9, or $\omega_d = 2.9$. From the curvilinear square $ABCD$, it is found that

$$\Delta\omega = \omega_D - \omega_A = 3.4 - 2.4 = 1.0$$

The closed-loop poles closest to the $j\omega$ axis are then estimated as

$$s = -1 \pm j2.9$$

The $G(j\omega)$ locus shown in Figure 6–101 is actually a plot of the following open-loop transfer function:

$$G(s) = \frac{5(s + 20)}{s(s + 4.59)(s^2 + 3.41s + 16.35)}$$

The exact closed-loop poles of this system are $s = -1 \pm j3$ and $s = -3 \pm j1$. The closed-loop poles closest to the $j\omega$ axis are $s = -1 \pm j3$. In this particular example, we see that the error involved is fairly small. In general, this error depends on a particular $G(j\omega)$ curve. The nearer the $G(j\omega)$ locus is to the $-1 + j0$ point, the smaller the error.

A–6–15. Figure 6–102 shows a block diagram of a space vehicle control system. Determine the gain K such that the phase margin is 50°. What is the gain margin in this case?

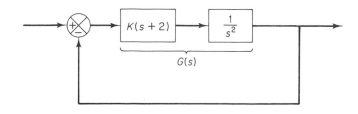

Figure 6–102
Space vehicle control
system.

Solution. Since

$$G(j\omega) = \frac{K(j\omega + 2)}{(j\omega)^2}$$

we have

$$\underline{/G(j\omega)} = \underline{/j\omega + 2} - 2\,\underline{/j\omega} = \tan^{-1}\frac{\omega}{2} - 180°$$

The requirement that the phase margin be 50° means that $\underline{/G(j\omega_c)}$ must be equal to $-130°$, where ω_c is the gain crossover frequency, or

$$\underline{/G(j\omega_c)} = -130°$$

Hence we set

$$\tan^{-1}\frac{\omega_c}{2} = 50°$$

from which we obtain

$$\omega_c = 2.3835 \text{ rad/sec}$$

Since the phase curve never crosses the $-180°$ line, the gain margin is $+\infty$ db.

Noting that the magnitude of $G(j\omega)$ must be equal to 0 db at $\omega = 2.3835$, we have

$$\left|\frac{K(j\omega + 2)}{(j\omega)^2}\right|_{\omega = 2.3835} = 1$$

from which we get

$$K = \frac{2.3835^2}{\sqrt{2^2 + 2.3835^2}} = 1.8259$$

This K value will give the phase margin of 50°.

A–6–16. Draw a Bode diagram of the open-loop transfer function $G(s)$ of the closed-loop system shown in Figure 6–103. Determine the phase margin and gain margin.

Solution. Note that

$$G(j\omega) = \frac{20(j\omega + 1)}{j\omega(j\omega + 5)[(j\omega)^2 + 2j\omega + 10]}$$

$$= \frac{0.4(j\omega + 1)}{j\omega(0.2j\omega + 1)\left[\left(\dfrac{j\omega}{\sqrt{10}}\right)^2 + \dfrac{2}{10}j\omega + 1\right]}$$

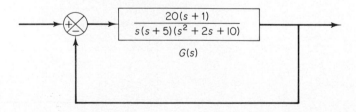

$$G(s)$$

The quadratic term in the denominator has the corner frequency of $\sqrt{10}$ rad/sec and the damping ratio ζ of 0.3162, or

$$\omega_n = \sqrt{10}, \qquad \zeta = 0.3162$$

The Bode diagram of $G(j\omega)$ is shown in Figure 6–104. From this diagram we find the phase margin to be $100°$ and the gain margin to be $+13.3$ db.

A–6–17. For the standard second-order system

$$\frac{C(s)}{R(s)} = \frac{\omega_n^2}{s^2 + 2\zeta\omega_n s + \omega_n^2}$$

show that the bandwidth ω_b is given by

$$\omega_b = \omega_n \left(1 - 2\zeta^2 + \sqrt{4\zeta^4 - 4\zeta^2 + 2} \right)^{1/2}$$

Note that ω_b/ω_n is a function only of ζ. Plot a curve ω_b/ω_n versus ζ.

Solution. The bandwidth ω_b is determined from $|C(j\omega_b)/R(j\omega_b)| = -3$ db. Quite often, instead of -3 db, we use -3.01 db, which is equal to 0.707. Thus,

$$\left| \frac{C(j\omega_b)}{R(j\omega_b)} \right| = \left| \frac{\omega_n^2}{(j\omega_b)^2 + 2\zeta\omega_n(j\omega_b) + \omega_n^2} \right| = 0.707$$

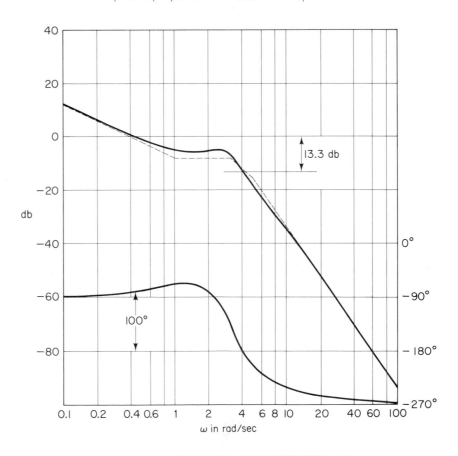

Figure 6–104
Bode diagram of
$G(j\omega)$ of system
shown in Figure 6–
103.

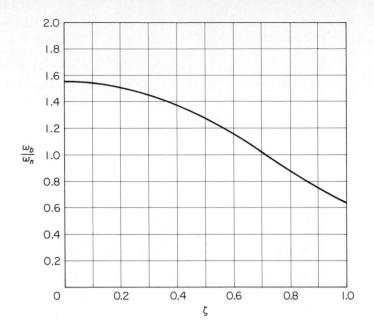

Figure 6–105
Curve relating ω_b/ω_n versus ζ, where ω_b is the bandwidth.

then

$$\frac{\omega_n^2}{\sqrt{(\omega_n^2 - \omega_b^2)^2 + (2\zeta\omega_n\omega_b)^2}} = 0.707$$

from which we get

$$\omega_n^4 = 0.5[(\omega_n^2 - \omega_b^2)^2 + 4\zeta^2\omega_n^2\omega_b^2]$$

By dividing both sides of this last equation by ω_n^4, we obtain

$$1 = 0.5\left\{\left[1 - \left(\frac{\omega_b}{\omega_n}\right)^2\right]^2 + 4\zeta^2\left(\frac{\omega_b}{\omega_n}\right)^2\right\}$$

Solving this last equation for $(\omega_b/\omega_n)^2$ yields

$$\left(\frac{\omega_b}{\omega_n}\right)^2 = -2\zeta^2 + 1 \pm \sqrt{4\zeta^4 - 4\zeta^2 + 2}$$

Since $(\omega_b/\omega_n)^2 > 0$, we take the plus sign in this last equation. Then

$$\omega_b^2 = \omega_n^2\left(1 - 2\zeta^2 + \sqrt{4\zeta^4 - 4\zeta^2 + 2}\right)$$

or

$$\omega_b = \omega_n\left(1 - 2\zeta^2 + \sqrt{4\zeta^4 - 4\zeta^2 + 2}\right)^{1/2}$$

Figure 6–105 shows a curve relating ω_b/ω_n versus ζ.

A–6–18. A unity-feedback control system has the following open-loop transfer function:

$$G(s) = \frac{K}{s(s + 1)(s + 2)}$$

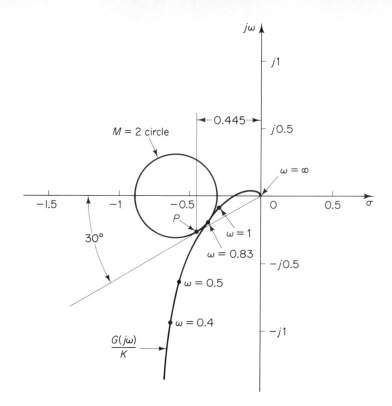

Figure 6–106
Plot of $G(j\omega)/K$ of the system considered in Problem A–6–18.

Consider the frequency response of this system. Plot a polar locus of $G(j\omega)/K$. Then determine the value of gain K such that the resonant peak magnitude M_r of the closed-loop frequency response is 2.

Solution. A plot of $G(j\omega)/K$ is shown in Figure 6–106. The value of angle ψ corresponding to $M_r = 2$ obtained from Equation (6–24) is

$$\psi = \sin^{-1}\frac{1}{2} = 30°$$

Hence, we draw the line $\overline{OP}$ that passes through the origin and makes an angle of 30° with the negative real axis as shown in Figure 6–106. We then draw the circle that is tangent to both the $G(j\omega)/K$ locus and line $\overline{OP}$. Define the point where the circle and line OP are tangent as point P. The perpendicular line drawn from point P intersects the negative real axis at $(-0.445, 0)$. Then, the gain K is determined as

$$K = \frac{1}{0.445} = 2.247$$

From Figure 6–106 we notice that the resonant frequency is approximately $\omega = 0.83$ rad/sec.

A–6–19. Figure 6–107 shows a block diagram of a chemical reactor system. Draw a Bode diagram of $G(j\omega)$. Also, draw the $G(j\omega)$ locus on the Nichols chart. From the Nichols diagram, read magnitudes and phase angles of the closed-loop frequency response and then plot the Bode diagram of the closed-loop system, $G(j\omega)/[1 + G(j\omega)]$.

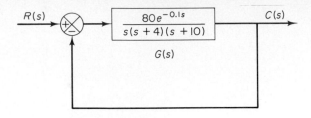

Figure 6–107
Block diagram of a
chemical reactor
system.

Solution. Noting that

$$G(s) = \frac{80e^{-0.1s}}{s(s + 4)(s + 10)} = \frac{2e^{-0.1s}}{s(0.25s + 1)(0.1s + 1)}$$

we have

$$G(j\omega) = \frac{2e^{-0.1j\omega}}{j\omega(0.25j\omega + 1)(0.1j\omega + 1)}$$

The phase angle of the transport lag $e^{-0.1j\omega}$ is

$$\underline{/e^{-0.1j\omega}} = \underline{/\cos(0.1\omega) - j\sin(0.1\omega)} = -0.1\omega \qquad \text{(rad)}$$

$$= -5.73\omega \qquad \text{(degrees)}$$

The Bode diagram of $G(j\omega)$ is shown in Figure 6–108.

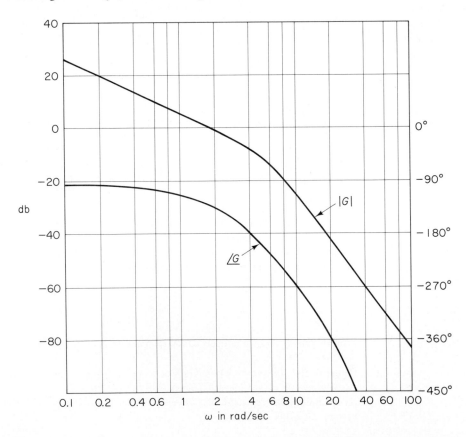

Figure 6–108
Bode diagram of
$G(j\omega)$ of the system
shown in Fig. 6–107.

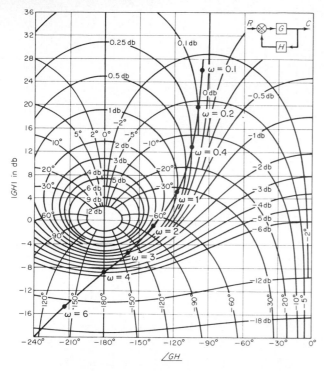

Figure 6–109 $G(j\omega)$ locus superimposed on Nichols chart. (Problem A–6–19.)

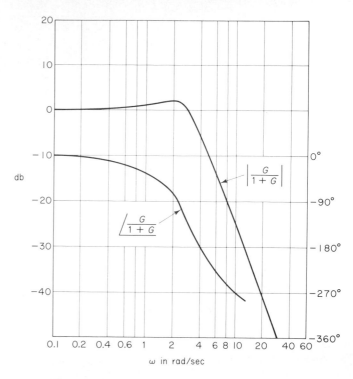

Figure 6–110 Bode diagram of the closed-loop frequency response. (Problem A–6–19.)

Next, by reading magnitudes and phase angles of $G(j\omega)$ for various values of ω, it is possible to plot the gain versus phase plot on a Nichols chart. Figure 6–109 shows such a $G(j\omega)$ locus superimposed on the Nichols chart. From this diagram, magnitudes and phase angles of the closed-loop system at various frequency points can be read. Figure 6–110 depicts the Bode diagram of the closed-loop frequency response $G(j\omega)/[1 + G(j\omega)]$.

A–6–20.   A Bode diagram of the open-loop transfer function $G(s)$ of a unity feedback control system is shown in Figure 6–111. It is known that the open-loop transfer function is minimum phase. From the diagram it can be seen that there is a pair of complex-conjugate poles at $\omega = 2$ rad/sec. Determine the damping ratio of the quadratic term involving these complex-conjugate poles. Also, determine the transfer function $G(s)$.

Solution.   Referring to Figure 6–11 and examining the Bode diagram of Figure 6–111, we find the damping ratio ζ and undamped natural frequency ω_n of the quadratic term to be

$$\zeta = 0.1, \qquad \omega_n = 2 \text{ rad/sec}$$

Noting that there is another corner frequency at $\omega = 0.5$ rad/sec and the slope of the magnitude curve

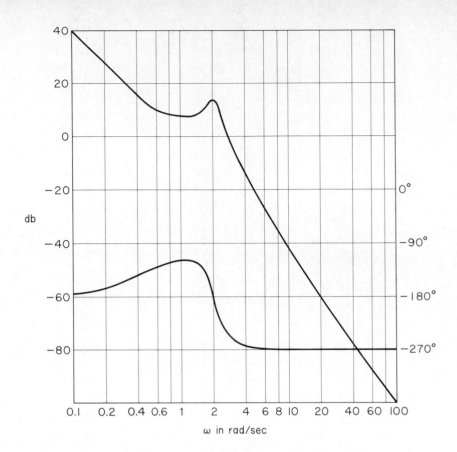

Figure 6–111
Bode diagram of the open-loop transfer function of a unity-feedback control system.

in the low-frequency region is -40 db/decade, $G(j\omega)$ can be tentatively determined as follows:

$$G(j\omega) = \frac{K\left(\dfrac{j\omega}{0.5} + 1\right)}{(j\omega)^2\left[\left(\dfrac{j\omega}{2}\right)^2 + 0.1(j\omega) + 1\right]}$$

Since from Figure 6–111 we find $|G(j0.1)| = 40$ db, the gain value K can be determined as unity. Also, the calculated phase curve, $\angle G(j\omega)$ versus ω, agrees with the given phase curve. Hence the transfer function $G(s)$ can be determined as

$$G(s) = \frac{4(2s + 1)}{s^2(s^2 + 0.4s + 4)}$$

A–6–21. A closed-loop control system may include an unstable element within the loop. When the Nyquist stability criterion is to be applied to such a system, the frequency-response curves for the unstable element must be obtained.

   How can we obtain experimentally the frequency-response curves for such an unstable element? Suggest a possible approach to the experimental determination of the frequency response of an unstable linear element.

Solution. One possible approach is to measure the frequency-response characteristics of the unstable element by using it as a part of a stable system.

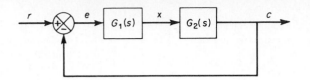

Figure 6–112
Control system.

Consider the system shown in Figure 6–112. Suppose that the element $G_1(s)$ is unstable. The complete system may be made stable by choosing a suitable linear element $G_2(s)$. We apply a sinusoidal signal at the input. At steady state, all signals in the loop will be sinusoidal. We measure the signals $e(t)$, the input to the unstable element, and $x(t)$, the output of the unstable element. By changing the frequency [and possibly the amplitude for the convenience of measuring $e(t)$ and $x(t)$] of the input sinusoid and repeating this process, it is possible to obtain the frequency response of the unstable linear element.

PROBLEMS

B–6–1. Consider the unity-feedback system with the open-loop transfer function

$$G(s) = \frac{10}{s + 1}$$

Obtain the steady-state output of the system when it is subjected to each of the following inputs:

(a) $r(t) = \sin (t + 30°)$

(b) $r(t) = 2 \cos (2t - 45°)$

(c) $r(t) = \sin (t + 30°) - 2 \cos (2t - 45°)$

B–6–2. Consider the system whose closed-loop transfer function is

$$\frac{C(s)}{R(s)} = \frac{K(T_2 s + 1)}{T_1 s + 1}$$

Obtain the steady-state output of the system when it is subjected to the input $r(t) = R \sin \omega t$.

B–6–3. Sketch the Bode diagrams of the following three transfer functions:

(a) $G(s) = \dfrac{T_1 s + 1}{T_2 s + 1}$ $(T_1 > T_2 > 0)$

(b) $G(s) = \dfrac{T_1 s - 1}{T_2 s + 1}$ $(T_1 > T_2 > 0)$

(c) $G(s) = \dfrac{-T_1 s + 1}{T_2 s + 1}$ $(T_1 > T_2 > 0)$

B–6–4. Plot the Bode diagram of

$$G(s) = \frac{9(s^2 + 0.2s + 1)}{s(s^2 + 1.2s + 9)}$$

B–6–5. Sketch the polar plots of the open-loop transfer function

$$G(s)H(s) = \frac{K(T_a s + 1)(T_b s + 1)}{s^2(T s + 1)}$$

for the following two cases:

(a) $T_a > T > 0,$ $T_b > T > 0$

(b) $T > T_a > 0,$ $T > T_b > 0$

B–6–6. The pole–zero configurations of complex functions $F_1(s)$ and $F_2(s)$ are shown in Figures 6–113(a) and (b), respectively. Assume that the closed contours in the s plane are those shown in Figures 6–113(a) and (b). Sketch qualitatively the corresponding closed contours in the $F_1(s)$ plane and $F_2(s)$ plane.

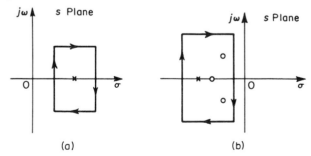

(a)   (b)

Figure 6–113   (a) s plane representation of complex function $F_1(s)$ and a closed contour; (b) s plane representation of complex function $F_2(s)$ and a closed contour.

B–6–7. Draw a Nyquist locus for the unity feedback control system with the open-loop transfer function

$$G(s) = \frac{K(1-s)}{s+1}$$

Using the Nyquist stability criterion, determine the stability of the closed-loop system.

B–6–8. A system with the open-loop transfer function

$$G(s)H(s) = \frac{K}{s^2(T_1 s + 1)}$$

is inherently unstable. This system can be stabilized by adding derivative control. Sketch the polar plots for the open-loop transfer function with and without derivative control.

B–6–9. Consider the closed-loop system with the following open-loop transfer function:

$$G(s)H(s) = \frac{10K(s+0.5)}{s^2(s+2)(s+10)}$$

Plot both the direct and inverse polar plots of $G(s)H(s)$ with $K = 1$ and $K = 10$. Apply the Nyquist stability criterion to the plots and determine the stability of the system with these values of K.

B–6–10. Consider the closed-loop system whose open-loop transfer function is

$$G(s)H(s) = \frac{Ke^{-2s}}{s}$$

Find the maximum value of K for which the system is stable.

B–6–11. Consider the unity-feedback control system whose open-loop transfer function is

$$G(s) = \frac{as+1}{s^2}$$

Determine the value of a so that the phase margin is 45°.

B–6–12. Consider the system shown in Figure 6–114. Draw a Bode diagram of the open-loop transfer function $G(s)$. Determine the phase margin and gain margin.

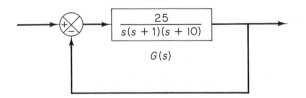

Figure 6–114 Control system.

B–6–13. Consider the system shown in Figure 6–115. Draw a Bode diagram of the open-loop transfer function $G(s)$. Determine the phase margin and gain margin.

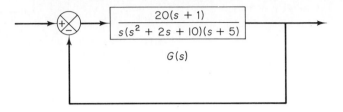

Figure 6–115 Control system.

B–6–14. Consider a unity-feedback control system with the following open-loop transfer function:

$$G(s) = \frac{K}{s(s^2+s+4)}$$

Determine the value of the gain K such that the phase margin is 50°. What is the gain margin for this case?

B–6–15. Consider the system shown in Figure 6–116. Draw a Bode diagram of the open-loop transfer functin and determine the value of the gain K such that the phase margin is 50°. What is the gain margin of this system with this gain K?

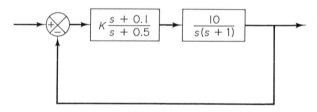

Figure 6–116 Control system.

B–6–16. Consider a unity-feedback control system whose open-loop transfer function is

$$G(s) = \frac{K}{s(s^2+s+0.5)}$$

Determine the value of the gain K such that the resonant peak magnitude in the frequency response is 2 db, or $M_r = 2$ db.

B–6–17. Figure 6–117 shows a block diagram of a process

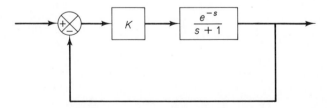

Figure 6–117 Process control system.

control system. Determine the range of gain K for stability.

B–6–18. Consider a closed-loop system whose open-loop transfer function is

$$G(s)H(s) = \frac{Ke^{-Ts}}{s(s+1)}$$

Determine the maximum value of the gain K for stability as a function of dead time T.

B–6–19. Sketch the polar plot of

$$G(s) = \frac{(Ts)^2 - 6(Ts) + 12}{(Ts)^2 + 6(Ts) + 12}$$

Show that for the frequency range $0 < \omega T < 2\sqrt{3}$, this equation gives a good approximation to the transfer function of transport lag, e^{-Ts}.

B–6–20. Figure 6–118 shows a Bode diagram of a transfer function $G(s)$. Determine this transfer function.

B–6–21. The experimentally determined Bode diagram of a system $G(j\omega)$ is shown in Figure 6–119. Determine the transfer function $G(s)$.

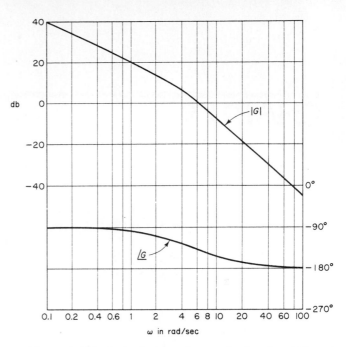

Figure 6–118 Bode diagram of a transfer function $G(s)$.

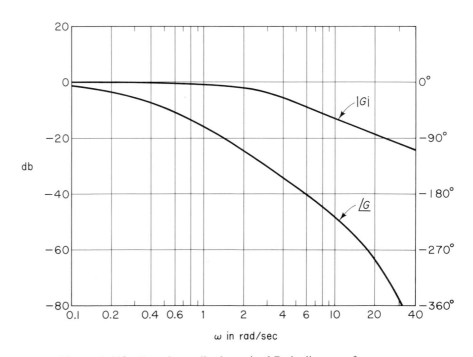

Figure 6–119 Experimentally determined Bode diagram of a system.

CHAPTER 7
Design and Compensation Techniques

7–1 INTRODUCTION

The primary objective of this chapter is to present procedures for the design and compensation of single-input, single-output linear time-invariant control systems. Compensation is the modification of the system dynamics to satisfy the given specifications. The approaches to the control system design and compensation used in this chapter are the root-locus approach and the frequency-response approach. (System design based on state-space approaches will be presented in Chapter 10.)

Performance specifications. Control systems are designed to perform specific tasks. The requirements imposed on the control system are usually spelled out as performance specifications. They generally relate to accuracy, relative stability, and speed of response.

For routine design problems, the performance specifications may be given in terms of precise numerical values. In other cases, they may be given partially in terms of precise numerical values and partially in terms of qualitative statements. In the latter case, the specifications may have to be modified during the course of design since the given specifications may never be satisfied (because of conflicting requirements) or may lead to a very expensive system.

Generally, the performance specifications should not be more stringent than necessary to perform the given task. If the accuracy at steady-state operation is of prime importance in a given control system, then we should not require unnecessarily rigid performance specifications on the transient response since such specifications will require expensive components. Remember that the most important part of control system design is to state the performance specifications precisely so that they will yield an optimal control system for the given purpose.

Conventional approach to system design.   In most practical cases, the design method to be used may be determined by the performance specifications applicable to the particular case. In designing control systems, if the performance specifications are given in terms of time-domain performance measures, such as rise time, maximum overshoot, or settling time, or frequency-domain performance measures, such as phase margin, gain margin, resonant peak value, or bandwidth, then we have no choice but to use a conventional approach based on the root-locus and/or frequency-response methods.

The systems that may be designed by a conventional approach are usually limited to single-input, single-output linear time-invariant systems. The designer seeks to satisfy all performance specifications by means of educated trial-and-error repetition. After a system is designed, the designer checks to see if the designed system satisfies all the performance specifications. If it does not, then he repeats the design process by adjusting parameter settings or by changing the system configuration until the given specifications are met. Although the design is based on a trial-and-error procedure, the ingenuity and know-how of the designer will play an important role in a successful design. An experienced designer may be able to design an acceptable system without using many trials.

In building a control system, we know that proper modification of the plant dynamics may be a simple way to meet the performance specifications. This, however, may not be possible in many practical situations because the plant may be fixed and may not be modified. Then we must adjust parameters other than those in the fixed plant. In this chapter, we assume that the plant is given and unalterable.

System compensation.   Setting the gain is the first step in adjusting the system for satisfactory performance. In many practical cases, however, the adjustment of the gain alone may not provide sufficient alteration of the system behavior to meet the given specifications. As is frequently the case, increasing the gain value will improve the steady-state behavior but will result in poor stability or even instability. It is then necessary to redesign the system (by modifying the structure or by incorporating additional devices or components) to alter the overall behavior so that the system will behave as desired. Such a redesign or addition of a suitable device is called *compensation*. A device inserted into the system for the purpose of satisfying the specifications is called a *compensator*. The compensator compensates for deficit performance of the original system.

Design of complex systems.   The root-locus and frequency-response approaches to designs that essentially consist of gain adjustment and of the design of compensators are quite useful but are limited to relatively simple control systems, such as single-input, single-output linear time-invariant ones.

While control system design via the root-locus and frequency-response approaches is an engineering endeavor, system design in the context of modern control theory (state-space methods) employs mathematical formulations of the problem and applies mathematical theory to design problems in which the system can have multiple inputs and multiple outputs and can be time varying. By applying modern control theory, the designer is able to start from a performance index, together with constraints imposed on the system, and to proceed to design a stable system by a completely analytical procedure. The advantage of design based on such modern control theory is that it enables the designer to produce a control system that is optimal with respect to the performance index considered.

It is important to note, however, that such a design technique cannot be applied if the performance specifications are given in terms of time-domain or frequency-domain quantities, in which case the root-locus or frequency-response techniques prove to be quite useful.

Series compensation and feedback (or parallel) compensation. Figures 7–1(a) and (b) show compensation schemes commonly used for feedback control systems. Figure 7–1(a) shows the configuration where the compensator $G_c(s)$ is placed in series with the plant. This scheme is called *series compensation*.

An alternative to series compensation is to feed back the signal(s) from some element(s) and place a compensator in the resulting inner feedback path, as shown in Figure 7–1(b). Such compensation is called *feedback compensation* or *parallel compensation*.

In compensating control systems, we see that the problem usually boils down to a suitable design of a series or feedback compensator. The choice between series compensation and feedback compensation depends on the nature of the signals in the system, the power levels at various points, available components, the designer's experience, economic considerations, and so on.

In general, series compensation may be simpler than feedback compensation; however, series compensation frequently requires additional amplifiers to increase the gain and/or to provide isolation. (To avoid power dissipation, the series compensator is inserted at the lowest energy point in the feedforward path.) Note that, in general, the number of components required in feedback compensation will be less than the number of components in series compensation, provided a suitable signal is available, because the energy transfer is from a higher power level to a lower level. (This means that additional amplifiers may not be necessary.)

In discussing compensators, we frequently use such terminologies as *lead network*, *lag network*, and *lag–lead network*. If a sinusoidal input e_i is applied to the input of a network

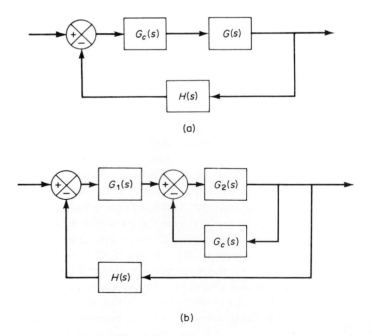

Figure 7–1
(a) Series compensation; (b) feedback or parallel compensation.

and the steady-state output e_o (which is also sinusoidal) has a phase lead, then the network is called a lead network. (The amount of phase lead angle is a function of the input frequency.) If the steady-state output e_o has a phase lag, then the network is called a lag network. In a lag–lead network, both phase lag and phase lead occur in the output but in different frequency regions; phase lag occurs in the low-frequency region and phase lead occurs in the high frequency region. A compensator having a characteristic of a lead network, lag network, or lag–lead network is called a lead compensator, lag compensator, or lag–lead compensator.

Compensators. If a compensator is needed to meet the performance specifications, the designer must realize a physical device that has the prescribed transfer function of the compensator.

Numerous physical devices have been used for such purposes. In fact, many noble and useful ideas for physically constructing compensators may be found in the literature.

Among the many kinds of compensators, widely employed compensators are the lead compensators, lag compensators, lag–lead compensators, and velocity-feedback (tachometer) compensators. In this chapter we shall limit our discussions mostly to these types. Lead, lag, and lag–lead compensators may be electronic devices (such as circuits using operational amplifiers) or RC networks (electrical, mechanical, pneumatic, hydraulic, or combinations thereof) and amplifiers.

In the actual design of a control system, whether or not to use an electronic, pneumatic, hydraulic, or electrical compensator is a matter that must be decided partially based on the nature of the controlled plant. For example, if the controlled plant involves flammable fluid, then we have to choose pneumatic components (both a compensator and an actuator) to avoid the possibility of sparks. If, however, no fire hazard exists, then electronic compensators are most commonly used. (In fact, we often transform nonelectrical signals into electrical signals because of the simplicity of transmission, increased accuracy, increased reliability, ease of compensation, and the like.)

Design procedures. In the trial-and-error approach to system design, we set up a mathematical model of the control system and adjust the parameters of a compensator. The most time-consuming part of such work is the checking of the performance specifications by analysis with each adjustment of the parameters. The designer should make use of a digital computer to avoid much of the numerical drudgery necessary for this checking.

Once a satisfactory mathematical model has been obtained, the designer must construct a prototype and test the open-loop system. If absolute stability of the closed loop is assured, the designer closes the loop and tests the performance of the resulting closed-loop system. Because of the neglected loading effects among the components, nonlinearities, distributed parameters, and so on, which were not taken into consideration in the original design work, the actual performance of the prototype system will probably differ from the theoretical predictions. Thus, the first design may not satisfy all the requirements on performance. By trial and error, the designer must make changes in the prototype until the system meets the specifications. In doing this, he must analyze each trial, and the results of the analysis must be incorporated into the next trial. The designer must see that the final system meets the performance specifications and, at the same time, is reliable and economical.

It is noted that in designing control systems by the root-locus or frequency-response methods, the final result is not unique, because the best or optimal solution may not be

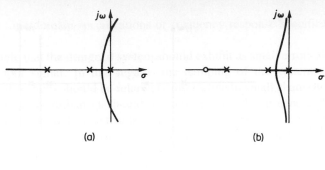

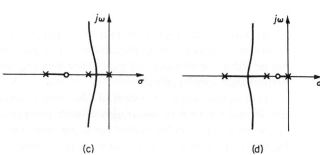

Figure 7–3
(a) Root-locus plot of a three-pole system; (b), (c), and (d) root-locus plots showing effects of addition of a zero to the three-pole system.

estimate of the speed of transient response); and static error constants (they give the steady-state accuracy). Although the correlation between the transient response and frequency response is indirect, the frequency domain specifications can be conveniently met in the Bode diagram approach.

After the open loop has been designed by the frequency-response method, the closed-loop poles and zeros can be determined. Then transient-response characteristics must be checked to see whether or not the designed system satisfies the requirements in the time domain. If it does not, then the compensator must be modified and the analysis repeated until a satisfactory result is obtained.

Design in the frequency domain is simple and straightforward. The frequency-response plot indicates clearly the manner in which the system should be modified, although the exact quantitative prediction of the transient-response characteristics cannot be made. The frequency-response approach can be applied to systems or components whose dynamic characteristics are given in the form of frequency-response data. Note that because of difficulty in deriving the equations governing certain components, such as pneumatic and hydraulic components, the dynamic characteristics of such components are usually determined experimentally through frequency-response tests. The experimentally obtained frequency-response plots can be combined easily with other such plots when the Bode diagram approach is used. Note also that in dealing with high-frequency noises we find the frequency-response approach is more convenient than other approaches.

There are basically two approaches in the frequency-domain design. One is the polar plot approach and the other is the Bode diagram approach. When a compensator is added, the polar plot does not retain the original shape, and, therefore, we need to draw a new polar plot, which will take time and is thus inconvenient. On the other hand, a Bode diagram of the compensator can be simply added to the original Bode diagram, and thus plotting the complete Bode diagram is a simple matter. Also, if the open-plot gain is varied, the magnitude curve is shifted up or down without changing the shape of the curve, and the phase curve remains the same. For design purposes, therefore, it is best to work in the Bode diagram.

A common approach to the Bode diagram design is that we first adjust the open-loop gain so that the requirement on the steady-state accuracy is met. Then the magnitude and phase curves of the uncompensated open loop (with the open-loop gain just adjusted) is plotted. If the specifications on the phase margin and gain margin are not satisfied, then a suitable compensator that will reshape the open-loop transfer function is determined. Finally, if there are any other requirements to be met, we try to satisfy them, unless some of them are contradictory to each other.

Information obtainable from open-loop frequency response. The low-frequency region (the region far below the gain crossover frequency) of the locus indicates the steady-state behavior of the closed-loop system. The medium-frequency region (the region near the $-1 + j0$ point) of the locus indicates relative stability. The high-frequency region (the region far above the gain crossover frequency) indicates the complexity of the system.

Requirements on open-loop frequency response. We might say that, in many practical cases, compensation is essentially a compromise between steady-state accuracy and relative stability.

To have a high value of the velocity error constant and yet satisfactory relative stability, we find it necessary to reshape the open-loop frequency-response curve.

The gain in the low-frequency region should be large enough, and also, near the gain crossover frequency, the slope of the log-magnitude curve in the Bode diagram should be -20 db/decade. This slope should extend over a sufficiently wide frequency band to assure a proper phase margin. For the high-frequency region, the gain should be attenuated as rapidly as possible to minimize the effects of noise.

Examples of generally desirable and undesirable open-loop and closed-loop frequency-response curves are shown in Figure 7–4.

Referring to Figure 7–5, we see that the reshaping of the open-loop frequency-response curve may be done if the high-frequency portion of the locus follows the $G_1(j\omega)$ locus, while

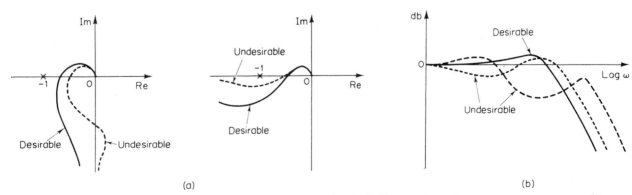

Figure 7–4 (a) Examples of desirable and undesirable open-loop frequency-response curves; (b) examples of desirable and undesirable closed-loop frequency-response curves.

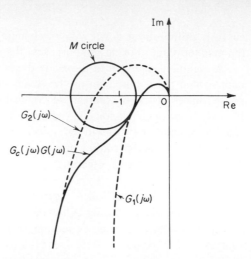

Figure 7–5
Reshaping of the
open-loop frequency-
response curve.

the low-frequency portion of the locus follows the $G_2(j\omega)$ locus. The reshaped locus $G_c(j\omega)G(j\omega)$ should have reasonable phase and gain margins or should be tangent to a proper M circle, as shown.

Basic characteristics of lead, lag, and lag–lead compensation. Lead compensation essentially yields an appreciable improvement in transient response and a small change in steady-state accuracy. It may accentuate high-frequency noise effects. Lag compensation, on the other hand, yields an appreciable improvement in steady-state accuracy at the expense of increasing the transient-response time. Lag compensation will suppress the effects of high-frequency noise effects. Lag–lead compensation combines the characteristics of both lead compensation and lag compensation. The use of a lead or lag compensator raises the order of the system by one (unless cancellation occurs between the zero of the compensator and a pole of the uncompensated open-loop transfer function). The use of a lag–lead compensator raises the order of the system by two [unless cancellation occurs between zero(s) of the lag–lead compensator and pole(s) of the uncompensated open-loop transfer function], which means that the system becomes more complex and it is more difficult to control the transient response behavior. The particular situation determines the type of compensation to be used.

7–3 LEAD COMPENSATION

In this section we first discuss the physical realization of continuous-time (or analog) lead compensators. We then present procedures for designing lead compensators based on the root-locus and frequency-response approaches. Finally, we comment on the PD control.

Lead compensators. There are many ways to realize continuous-time (or analog) lead compensators. Here we shall present three types of lead compensators: electronic networks using operational amplifiers, an electrical RC network, and a mechanical spring–dashpot network. Compensators using operational amplifiers are commonly used in practice. We shall discuss these three types of networks in the following.

Electronic lead networks using operational amplifiers. Figure 7–6(a) shows an electronic circuit using an operational amplifier. The transfer function for this circuit can be

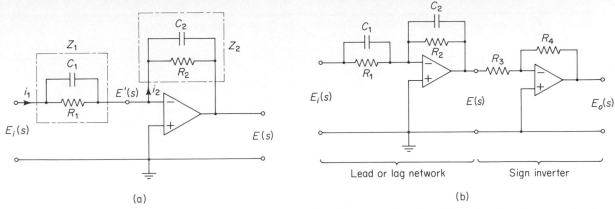

Figure 7–6 (a) Operational amplifier circuit; (b) operational amplifier circuit used as a lead or lag compensator.

obtained as follows: Define the input impedance and feedback impedance as Z_1 and Z_2, respectively. Then

$$Z_1 = \frac{R_1}{R_1 C_1 s + 1}, \qquad Z_2 = \frac{R_2}{R_2 C_2 s + 1}$$

Since the current flowing into the amplifier is negligible, current i_1 is equal to current i_2. Thus $i_1 = i_2$, or

$$\frac{E_i(s) - E'(s)}{Z_1} = \frac{E'(s) - E(s)}{Z_2}$$

Since $E'(s) \doteq 0$, we have

$$\frac{E(s)}{E_i(s)} = -\frac{Z_2}{Z_1} = -\frac{R_2}{R_1} \frac{R_1 C_1 s + 1}{R_2 C_2 s + 1} = -\frac{C_1}{C_2} \frac{s + \dfrac{1}{R_1 C_1}}{s + \dfrac{1}{R_2 C_2}} \tag{7-1}$$

Notice that the transfer function in Equation (7–1) contains a minus sign. Thus, this circuit is sign inverting. If such a sign inversion is not convenient in the actual application, a sign inverter may be connected to either the input or the output of the circuit of Figure 7–6(a). An example is shown in Figure 7–6(b). The sign inverter has the transfer function of

$$\frac{E_o(s)}{E(s)} = -\frac{R_4}{R_3}$$

The sign inverter has the gain of $-R_4/R_3$. Hence the network shown in Figure 7–6(b) has the following transfer function:

$$\frac{E_o(s)}{E_i(s)} = \frac{R_2 R_4}{R_1 R_3} \frac{R_1 C_1 s + 1}{R_2 C_2 s + 1} = \frac{R_4 C_1}{R_3 C_2} \frac{s + \dfrac{1}{R_1 C_1}}{s + \dfrac{1}{R_2 C_2}}$$

Table 7–1 Operational Amplifier Circuits That May Be Used as Compensators

	Control Action	$G(s) = \dfrac{E_o(s)}{E_i(s)}$	Operational Amplifier Circuits
1	P	$-\dfrac{R_2}{R_1}$	
2	I	$\dfrac{R_4}{R_3}\,\dfrac{1}{R_1 C_2 s}$	
3	PD	$\dfrac{R_4}{R_3}\,\dfrac{R_2}{R_1}\,(R_1 C_1 s + 1)$	
4	PI	$\dfrac{R_4}{R_3}\,\dfrac{R_2}{R_1}\,\dfrac{R_2 C_2 s + 1}{R_2 C_2 s}$	
5	PID	$\dfrac{R_4}{R_3}\,\dfrac{R_2}{R_1}\,\dfrac{(R_1 C_1 s + 1)(R_2 C_2 s + 1)}{R_2 C_2 s}$	
6	Lead or lag	$\dfrac{R_4}{R_3}\,\dfrac{R_2}{R_1}\,\dfrac{R_1 C_1 s + 1}{R_2 C_2 s + 1}$	
7	Lag–lead	$\dfrac{R_6}{R_5}\,\dfrac{R_4}{R_3}\,\dfrac{[(R_1 + R_3)\,C_1 s + 1]\,(R_2 C_2 s + 1)}{(R_1 C_1 s + 1)\,[(R_2 + R_4)\,C_2 s + 1]}$	

$$= K_c\alpha \frac{Ts + 1}{\alpha Ts + 1} = K_c \frac{s + \dfrac{1}{T}}{s + \dfrac{1}{\alpha T}} \qquad (7\text{-}2)$$

where

$$T = R_1 C_1, \qquad \alpha T = R_2 C_2, \qquad K_c = \frac{R_4 C_1}{R_3 C_2}$$

Notice that

$$K_c\alpha = \frac{R_4 C_1}{R_3 C_2} \frac{R_2 C_2}{R_1 C_1} = \frac{R_2 R_4}{R_1 R_3}, \qquad \alpha = \frac{R_2 C_2}{R_1 C_1}$$

This network has a dc gain of $K_c\alpha = R_2 R_4/(R_1 R_3)$.

From Equation (7-2), we see that this network is a lead network if $R_1 C_1 > R_2 C_2$, or $\alpha < 1$. It is a lag network if $R_1 C_1 < R_2 C_2$. Table 7-1 shows several circuits involving operational amplifiers that can be used as compensators.

Electrical lead network. A schematic diagram of an electrical lead network is shown in Figure 7-7. Let us derive the transfer function for this network. As usual in the derivation of the transfer function of any four-terminal network, we assume that the source impedance that the network sees is zero and that the output load impedance is infinite.

Using the symbols defined in Figure 7-7, we find that the complex impedances Z_1 and Z_2 are

$$Z_1 = \frac{R_1}{R_1 Cs + 1}, \qquad Z_2 = R_2$$

The transfer function between the output $E_o(s)$ and the input $E_i(s)$ is

$$\frac{E_o(s)}{E_i(s)} = \frac{Z_2}{Z_1 + Z_2} = \frac{R_2}{R_1 + R_2} \frac{R_1 Cs + 1}{\dfrac{R_1 R_2}{R_1 + R_2} Cs + 1}$$

Define

$$R_1 C = T, \qquad \frac{R_2}{R_1 + R_2} = \alpha < 1$$

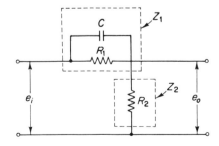

Figure 7-7
Electrical lead
network.

Then the transfer function becomes

$$\frac{E_o(s)}{E_i(s)} = \alpha \frac{Ts + 1}{\alpha Ts + 1} = \frac{s + \dfrac{1}{T}}{s + \dfrac{1}{\alpha T}}$$

If this *RC* circuit is used as a lead compensator, it is usually necessary to add an amplifier with an adjustable gain K_c so that the transfer function of the compensator is

$$G_c(s) = K_c \alpha \frac{Ts + 1}{\alpha Ts + 1} = K_c \frac{s + \dfrac{1}{T}}{s + \dfrac{1}{\alpha T}}$$

Mechanical lead network.   Figure 7–8 shows a schematic diagram of a mechanical lead network. From the diagram, we obtain the following equations:

$$b_2(\dot{x}_i - \dot{x}_o) = b_1(\dot{x}_o - \dot{y})$$

$$b_1(\dot{x}_o - \dot{y}) = ky$$

Taking the Laplace transforms of these two equations, assuming zero initial conditions, and then eliminating $Y(s)$, we obtain

$$\frac{X_o(s)}{X_i(s)} = \frac{b_2}{b_1 + b_2} \frac{\dfrac{b_1}{k} s + 1}{\dfrac{b_2}{b_1 + b_2} \dfrac{b_1}{k} s + 1}$$

This is the transfer function between $X_o(s)$ and $X_i(s)$. By defining

$$\frac{b_1}{k} = T, \qquad \frac{b_2}{b_1 + b_2} = \alpha < 1$$

we obtain

$$\frac{X_o(s)}{X_i(s)} = \alpha \frac{Ts + 1}{\alpha Ts + 1} = \frac{s + \dfrac{1}{T}}{s + \dfrac{1}{\alpha T}}$$

Just as in the case of the electrical *RC* circuit discussed above, if this mechanical network is used as a lead compensator, it is necessary to add a linkage device with an adjustable gain K_c so that the transfer function of the compensator is

$$G_c(s) = K_c \alpha \frac{Ts + 1}{\alpha Ts + 1} = K_c \frac{s + \dfrac{1}{T}}{s + \dfrac{1}{\alpha T}}$$

Characteristics of lead compensators. Consider a lead compensator having the following transfer function:

$$K_c \alpha \frac{Ts + 1}{\alpha Ts + 1} = K_c \frac{s + \dfrac{1}{T}}{s + \dfrac{1}{\alpha T}} \qquad (0 < \alpha < 1)$$

It has a zero at $s = -1/T$ and a pole at $s = -1(\alpha T)$. Since $0 < \alpha < 1$, we see that the zero is always located to the right of the pole in the complex plane. Note that for a small value of α the pole is located far to the left. The minimum value of α is limited by the physical construction of the lead compensator. The minimum value of α is usually taken to be about 0.07. (This means that the maximum phase lead that may be produced by a lead compensator is about 60°.)

Figure 7–9 shows the polar plot of

$$K_c \alpha \frac{j\omega T + 1}{j\omega \alpha T + 1} \qquad (0 < \alpha < 1)$$

with $K_c = 1$. For a given value of α, the angle between the positive real axis and the tangent line drawn from the origin to the semicircle gives the maximum phase lead angle, ϕ_m. We shall call the frequency at the tangent point as ω_m. From Figure 7–9 the phase angle at $\omega = \omega_m$ is ϕ_m, where

$$\sin \phi_m = \frac{\dfrac{1 - \alpha}{2}}{\dfrac{1 + \alpha}{2}} = \frac{1 - \alpha}{1 + \alpha} \qquad (7\text{–}3)$$

Equation (7–3) relates the maximum phase lead angle and the value of α.

Figure 7–10 shows the Bode diagram of a lead compensator when $K_c = 1$ and $\alpha = 0.1$. The corner frequencies for the lead compensator are $\omega = 1/T$ and $\omega = 1/(\alpha T) = 10/T$. By

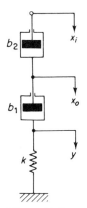

Figure 7–8
Mechanical lead
network.

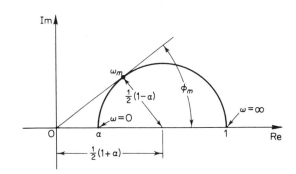

Figure 7–9
Polar plot of a lead compensator
$\alpha(j\omega T + 1)/(j\omega \alpha T + 1)$, where
$0 < \alpha < 1$.

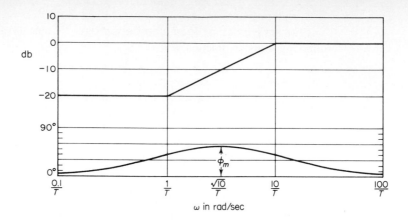

Figure 7–10
Bode diagram of a
lead compensator
$\alpha(j\omega T + 1)/(j\omega\alpha T + 1)$, where $\alpha = 0.1$.

ω in rad/sec

examining Figure 7–10, we see that ω_m is the geometric mean of the two corner frequencies, or

$$\log \omega_m = \frac{1}{2}\left(\log\frac{1}{T} + \log\frac{1}{\alpha T}\right)$$

Hence

$$\omega_m = \frac{1}{\sqrt{\alpha}\, T} \tag{7–4}$$

As seen from Figure 7–10, the lead compensator is basically a high-pass filter. (The high frequencies are passed but low frequencies are attenuated.)

Lead compensation techniques based on the root-locus approach.   The root-locus approach to design is very powerful when the specifications are given in terms of time-domain quantities, such as the damping ratio and undamped natural frequency of the desired dominant closed-loop poles, maximum overshoot, rise time, and settling time.

Consider a design problem in which the original system either is unstable for all values of gain or is stable but has undesirable transient-response characteristics. In such a case, the reshaping of the root locus is necessary in the broad neighborhood of the $j\omega$ axis and the origin in order that the dominant closed-loop poles be at desired locations in the complex plane. This problem may be solved by inserting an appropriate lead compensator in cascade with the feedforward transfer function.

The procedures for designing a lead compensator for the system shown in Figure 7–11 by the root-locus method may be stated as follows:

1. From the performance specifications, determine the desired location for the dominant closed-loop poles.

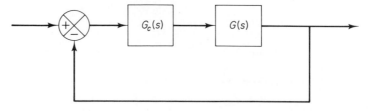

Figure 7–11
Control system.

2. By drawing the root-locus plot, ascertain whether or not the gain adjustment alone can yield the desired closed-loop poles. If not, calculate the angle deficiency ϕ. This angle must be contributed by the lead compensator if the new root locus is to pass through the desired locations for the dominant closed-loop poles.

3. Assume the lead compensator $G_c(s)$ to be

$$G_c(s) = K_c\alpha \frac{Ts + 1}{\alpha Ts + 1} = K_c \frac{s + \dfrac{1}{T}}{s + \dfrac{1}{\alpha T}} \qquad (0 < \alpha < 1)$$

where α and T are determined from the angle deficiency. K_c is determined from the requirement of the open-loop gain.

4. If static error constants are not specified, determine the location of the pole and zero of the lead compensator so that the lead compensator will contribute the necessary angle ϕ. If no other requirements are imposed on the system, try to make the value of α as large as possible. A larger value of α generally results in a larger value of K_v, which is desirable. (If a particular static error constant is specified, it is generally simpler to use the frequency-response approach.)

5. Determine the open-loop gain of the compensated system from the magnitude condition.

Once a compensator has been designed, check to see whether or not all performance specifications have been met. If the compensated system does not meet the performance specifications, then repeat the design procedure by adjusting the compensator pole and zero until all such specifications are met. If a large static error constant is required, cascade a lag network or alter the lead compensator to a lag–lead compensator.

Note that if the selected dominant closed-loop poles are not really dominant, it will be necessary to modify the location of the pair of closed-loop poles. (The closed-loop poles other than dominant ones modify the response obtained from the dominant closed-loop poles alone. The amount of modification depends on the location of these remaining closed-loop poles.)

EXAMPLE 7–1 Consider the system shown in Figure 7–12(a). The feedforward transfer function is

$$G(s) = \frac{4}{s(s + 2)}$$

The root-locus plot for this system is shown in Figure 7–12(b). The closed-loop transfer function becomes

$$\frac{C(s)}{R(s)} = \frac{4}{s^2 + 2s + 4}$$

$$= \frac{4}{(s + 1 + j\sqrt{3})(s + 1 - j\sqrt{3})}$$

The closed-loop poles are located at

$$s = -1 \pm j\sqrt{3}$$

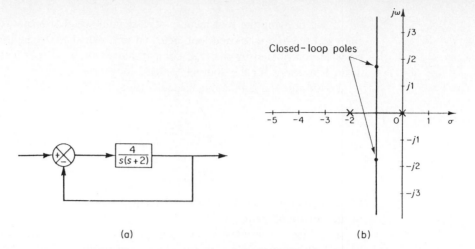

Figure 7–12
(a) Control system;
(b) root-locus plot.

(a)

(b)

The damping ratio of the closed-loop poles is 0.5. The undamped natural frequency of the closed-loop poles is 2 rad/sec. The static velocity error constant is 2 sec^{-1}.

It is desired to modify the closed-loop poles so that an undamped natural frequency $\omega_n = 4$ rad/sec is obtained, without changing the value of the damping ratio, $\zeta = 0.5$.

Recall that in the complex plane the damping ratio ζ of a pair of complex poles can be expressed in terms of the angle θ, which is measured from the $j\omega$ axis, as shown in Figure 7–13(a), with

$$\zeta = \sin \theta$$

In other words, lines of constant damping ratio ζ are radial lines passing through the origin as shown in Figure 7–13(b). For example, a damping ratio of 0.5 requires that the complex poles lie on the lines drawn through the origin making angles of $\pm 60°$ with the negative real axis. (If the real part of a pair of complex poles is positive, which means that the system is unstable, the corresponding ζ is negative.) The damping ratio determines the angular location of the poles, while the distance of the pole from the origin is determined by the undamped natural frequency ω_n.

In the present example, the desired locations of the closed-loop poles are

$$s = -2 \pm j2\sqrt{3}$$

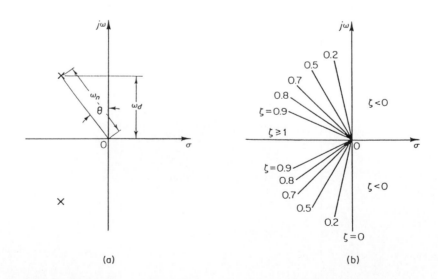

Figure 7–13
(a) Complex poles; (b) lines of constant damping ratio ζ.

(a)

(b)

In some cases, after the root loci of the original system have been obtained, the dominant closed-loop poles may be moved to the desired location by simple gain adjustment. This is, however, not the case for the present system. Therefore, we shall insert a lead compensator in the feedforward path.

A general procedure for determining the lead compensator is as follows: First, find the sum of the angles at the desired location of one of the dominant closed-loop poles with the open-loop poles and zeros of the original system, and determine the necessary angle ϕ to be added so that the total sum of the angles is equal to $\pm 180°(2k + 1)$. The lead compensator must contribute this angle ϕ. (If the angle ϕ is quite large, then two or more lead networks may be needed rather than a single one.)

If the original system has the open-loop transfer function $G(s)$, then the compensated system will have the open-loop transfer function

$$G_c(s)G(s) = \left(K_c \frac{s + \dfrac{1}{T}}{s + \dfrac{1}{\alpha T}} \right) G(s)$$

where

$$G_c(s) = K_c\alpha \frac{Ts + 1}{\alpha Ts + 1} = K_c \frac{s + \dfrac{1}{T}}{s + \dfrac{1}{\alpha T}} \qquad (0 < \alpha < 1)$$

Notice that there are many possible values for T and α that will yield the necessary angle contribution at the desired closed-loop poles.

The next step is to determine the locations of the zero and pole of the lead compensator. In choosing the value of T, we shall introduce a procedure to obtain the largest possible value for α. First, draw a horizontal line passing through point P, the desired location for one of the dominant closed-loop poles. This is shown as line PA in Figure 7–14. Draw also a line connecting point P and the origin. Bisect the angle between the lines PA and PO, as shown in Figure 7–14. Draw two lines PC and PD that make angles $\pm\phi/2$ with the bisector PB. The intersections of PC and PD with the negative real axis give the necessary location for the pole and zero of the lead network. The compensator thus designed will make point P a point on the root locus of the compensated system. The open-loop gain is determined by use of the magnitude condition.

In the present system, the angle of $G(s)$ at the desired closed-loop pole is

$$\left/ \frac{4}{s(s + 2)} \right|_{s = -2 + j2\sqrt{3}} = -210°$$

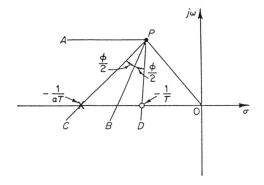

Figure 7–14
Determination of the
pole and zero of a
lead network.

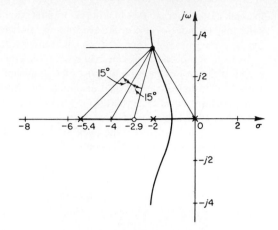

Figure 7–15
Root-locus plot of the
compensated system.

Thus, if we need to force the root locus to go through the desired closed-loop pole, the lead compensator must contribute $\phi = 30°$ at this point. By following the foregoing design procedure, we determine the zero and pole of the lead compensator, as shown in Figure 7–15, to be

$$\text{Zero at } s = -2.9, \qquad \text{Pole at } s = -5.4$$

or

$$T = \frac{1}{2.9} = 0.345, \qquad \alpha T = \frac{1}{5.4} = 0.185$$

Thus, $\alpha = 0.537$. The open-loop transfer function of the compensated system becomes

$$G_c(s)G(s) = K_c \frac{s + 2.9}{s + 5.4} \frac{4}{s(s + 2)} = \frac{K(s + 2.9)}{s(s + 2)(s + 5.4)}$$

where $K = 4K_c$. The root-locus plot for the compensated system is shown in Figure 7–15. The gain K is evaluated from the magnitude condition as follows: Referring to the root-locus plot for the compensated system shown in Figure 7–15, the gain K is evaluated from the magnitude condition as

$$\left| \frac{K(s + 2.9)}{s(s + 2)(s + 5.4)} \right|_{s = -2 + j2\sqrt{3}} = 1$$

or

$$K = 18.7$$

It follows that

$$G_c(s)G(s) = \frac{18.7(s + 2.9)}{s(s + 2)(s + 5.4)}$$

The constant K_c of the lead compensator is

$$K_c = \frac{18.7}{4} = 4.68$$

Hence, $K_c \alpha = 2.51$. The lead compensator, therefore, has the transfer function

$$G_c(s) = 2.51 \frac{0.345s + 1}{0.185s + 1} = 4.68 \frac{s + 2.9}{s + 5.4}$$

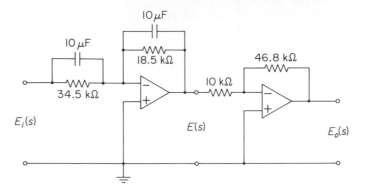

Figure 7–16
Lead compensator.

If the electronic circuit using operational amplifiers as shown in Figure 7–6(b) is used as the lead compensator just designed, then the parameter values of the lead compensator are determined from

$$\frac{E_o(s)}{E_i(s)} = \frac{R_2 R_4}{R_1 R_3} \frac{R_1 C_1 s + 1}{R_2 C_2 s + 1} = 2.51 \frac{0.345s + 1}{0.185s + 1}$$

as shown in Figure 7–16, where we have arbitrarily chosen as $C_1 = C_2 = 10\ \mu\text{F}$ and $R_3 = 10\ \text{k}\Omega$. The static velocity error constant K_v is obtained from the expression

$$K_v = \lim_{s \to 0} sG_c(s)G(s)$$
$$= \lim_{s \to 0} \frac{s18.7(s + 2.9)}{s(s + 2)(s + 5.4)}$$
$$= 5.02\ \text{sec}^{-1}$$

Note that the third closed-loop pole of the designed system is found by dividing the characteristic equation by the known factors as follows:

$$s(s + 2)(s + 5.4) + 18.7(s + 2.9) = (s + 2 + j2\sqrt{3})(s + 2 - j2\sqrt{3})(s + 3.4)$$

The foregoing compensation method enables us to place the dominant closed-loop poles at the desired points in the complex plane. The third pole at $s = -3.4$ is close to the added zero at $s = -2.9$. Therefore, the effect of this pole on the transient response is relatively small. Since no restriction has been imposed on the nondominant pole and no specification has been given concerning the value of the static velocity error coefficient, we conclude that the present design is satisfactory.

It is noted that we may place the zero of the compensator at $s = -2$ and pole at $s = -4$ so that the angle contribution of the lead compensator is 30°. (In this case the zero of the lead compensator will cancel a pole of the plant, resulting in the second-order system, rather than the third-order system as we designed.) It can be seen that the K_v value in this case is 4 sec^{-1}. Other combinations can be selected that will yield 30° phase lead. (For different combinations of a zero and pole of the compensator that contribute 30°, the value of α will be different and the value of K_v will also be different.) Although a certain change in the value of K_v can be made by altering the pole–zero location of the lead compensator, if a large increase in the value of K_v is desired, then we must alter the lead compensator to a lag–lead compensator. (See Section 7–5 for lag–lead compensation.)

Lead compensation techniques based on the frequency-response approach.
The primary function of the lead compensator is to reshape the frequency-response curve to provide sufficient phase lead angle to offset the excessive phase lag associated with the components of the fixed system.

Consider again the system shown in Figure 7–11. Assume that the performance specifications are given in terms of phase margin, gain margin, static velocity error constants, and so on. The procedure for designing a lead compensator by the frequency-response approach may be stated as follows:

1. Assume the following lead compensator:

$$G_c(s) = K_c \alpha \frac{Ts + 1}{\alpha Ts + 1} = K_c \frac{s + \dfrac{1}{T}}{s + \dfrac{1}{\alpha T}} \qquad (0 < \alpha < 1)$$

Define

$$K_c \alpha = K$$

Then

$$G_c(s) = K \frac{Ts + 1}{\alpha Ts + 1}$$

The open-loop transfer function of the compensated system is

$$G_c(s)G(s) = K \frac{Ts + 1}{\alpha Ts + 1} G(s) = \frac{Ts + 1}{\alpha Ts + 1} KG(s) = \frac{Ts + 1}{\alpha Ts + 1} G_1(s)$$

where

$$G_1(s) = KG(s)$$

Determine gain K to satisfy the requirement on the given static error constant.

2. Using the gain K thus determined, draw a Bode diagram of $G_1(j\omega)$, the gain adjusted but uncompensated system. Evaluate the phase margin.

3. Determine the necessary phase lead angle ϕ to be added to the system.

4. Determine the attenuation factor α by use of Equation (7–3). Determine the frequency where the magnitude of the uncompensated system $G_1(j\omega)$ is equal to $-20 \log (1/\sqrt{\alpha})$. Select this frequency as the new gain crossover frequency. This frequency corresponds to $\omega_m = 1/(\sqrt{\alpha}T)$, and the maximum phase shift ϕ_m occurs at this frequency.

5. Determine the corner frequencies of the lead compensator as follows:

$$\text{Zero of lead compensator:} \qquad \omega = \frac{1}{T}$$

$$\text{Pole of lead compensator:} \qquad \omega = \frac{1}{\alpha T}$$

6. Using the value of K determined in step 1 and that of α determined in step 4, calculate constant K_c from

$$K_c = \frac{K}{\alpha}$$

7. Check the gain margin to be sure it is satisfactory. If not, repeat the design process by modifying the pole–zero location of the compensator until a satisfactory result is obtained.

EXAMPLE 7–2   Consider the system shown in Figure 7–17. The open-loop transfer function is

$$G(s) = \frac{4}{s(s + 2)}$$

It is desired to design a compensator for the system so that the static velocity error constant K_v is 20 sec^{-1}, the phase margin is at least 50°, and the gain margin is at least 10 db.

We shall use a lead compensator of the form

$$G_c(s) = K_c \alpha \frac{Ts + 1}{\alpha Ts + 1} = K_c \frac{s + \dfrac{1}{T}}{s + \dfrac{1}{\alpha T}}$$

The compensated system will have the open-loop transfer function $G_c(s)G(s)$.

Define

$$G_1(s) = KG(s) = \frac{4K}{s(s + 2)}$$

where $K = K_c \alpha$.

The first step in the design is to adjust the gain K to meet the steady-state performance specification or to provide the required static velocity error constant. Since this constant is given as 20 sec^{-1}, we obtain

$$K_v = \lim_{s \to 0} sG_c(s)G(s) = \lim_{s \to 0} s \frac{Ts + 1}{\alpha Ts + 1} G_1(s) = \lim_{s \to 0} \frac{s4K}{s(s + 2)} = 2K = 20$$

or

$$K = 10$$

With $K = 10$, the compensated system will satisfy the steady-state requirement.

We shall next plot the Bode diagram of

$$G_1(j\omega) = \frac{40}{j\omega(j\omega + 2)} = \frac{20}{j\omega(0.5 j\omega + 1)}$$

Figure 7–18 shows the magnitude and phase angle curves of $G_1(j\omega)$. From this plot, the phase and gain margins of the system are found to be 17° and $+\infty$ db, respectively. (A phase margin of 17° implies that the system is quite oscillatory. Thus, satisfying the specification on the steady state yields a poor transient-response performance.) The specification calls for a phase margin of at least 50°. We thus find that the additional phase lead necessary to satisfy the relative stability requirement is 33°. To achieve a phase margin of 50° without decreasing the value of K, the lead compensator must contribute the required phase angle.

Noting that the addition of a lead compensator modifies the magnitude curve in the Bode diagram, we realize that the gain crossover frequency will be shifted to the right. We must offset the increased phase lag of $G_1(j\omega)$ due to this increase in the gain crossover frequency. Considering the shift of the

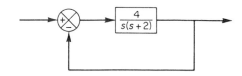

Figure 7–17
Control system.

system has the following open-loop transfer function:

$$G_c(s)G(s) = 41.7 \frac{s + 4.41}{s + 18.4} \frac{4}{s(s + 2)}$$

The solid curves in Figure 7–19 show the magnitude curve and phase-angle curve for the compensated system. The lead compensator causes the gain crossover frequency to increase from 6.3 to 9 rad/sec. The increase in this frequency means an increase in bandwidth. This implies an increase in the speed of response. The phase and gain margins are seen to be approximately 50° and $+\infty$ db, respectively. The compensated system shown in Figure 7–20 therefore meets both the steady-state and the relative-stability requirements.

Note that for type 1 systems, such as the system just considered, the value of the static velocity error constant K_v is merely the value of the frequency corresponding to the intersection of the extension of the initial -20-db/decade slope line and the 0-db line, as shown in Figure 7–19.

Figure 7–21 shows the polar plots of the uncompensated system $G_1(j\omega) = 10G(j\omega)$ and the compensated system $G_c(j\omega)G(j\omega)$. From Figure 7–21, we see that the resonant frequency of the uncompensated system is about 6 rad/sec and that of the compensated system is about 7 rad/sec. (This also indicates that the bandwidth has been increased.)

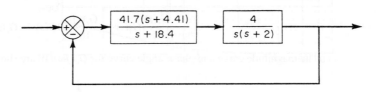

Figure 7–20
Compensated system.

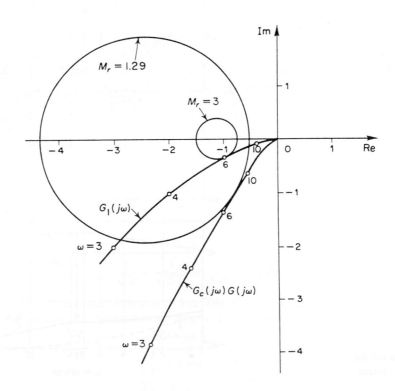

Figure 7–21
Polar plots of the uncompensated and compensated open-loop transfer function. (G_1: uncompensated system; G_cG: compensated system.)

From Figure 7–21, we find that the value of the resonant peak M_r for the uncompensated system with $K = 10$ is 3. The value of M_r for the compensated system is found to be 1.29. This clearly shows that the compensated system has improved relative stability. (Note that the value of M_r may be obtained easily by transferring the data from the Bode diagram to the Nichol's chart. See Example 7–4.)

Note that if the phase angle of $G_1(j\omega)$ decreases rapidly near the gain crossover frequency, lead compensation becomes ineffective because the shift in the gain crossover frequency to the right makes it difficult to provide enough phase lead at the new gain crossover frequency. This means that, to provide the desired phase margin, we must use a very small value for α. The value of α, however, should not be smaller than 0.07 nor should the maximum phase lead ϕ_m be more than 60° because such values will require an additional gain of excessive value. [If more than 60° is needed, two (or more) lead networks may be used in series with an isolating amplifier.]

PD control.   The proportional-plus-derivative controller is a simplified version of the lead compensator. The PD controller has the transfer function $G_c(s)$, where

$$G_c(s) = K_p(1 + T_d s)$$

The value of K_p is usually determined to satisfy the steady-state requirement. The corner frequency $1/T_d$ is chosen such that the phase lead occurs in the neighborhood of the gain crossover frequency. Although the phase margin can be increased, the magnitude of the compensator continues to increase for the frequency region $1/T_d < \omega$. (Thus the PD controller is a high-pass filter.) Such a continued increase of the magnitude is undesirable, since it amplifies high-frequency noises that may be present in the system. Therefore, lead compensation is preferred over the PD control. Lead compensation can provide a sufficient phase lead, but the increase of the magnitude for the high-frequency region is very much smaller than that for the PD control.

Because the transfer function of the PD controller involves one zero but no pole, it is not possible to electrically realize it by passive *RLC* elements only. Realization of the PD controller using op amps, resistors, and capacitors is possible, but because the PD controller is a high-pass filter, as mentioned earlier, the differentiation process involved may cause serious noise problems in some cases. There is, however, no problem if the PD controller is realized by use of hydraulic or pneumatic elements.

Finally, the PD control, as in the case of the lead compensation, improves the transient response (that is, fast rise time with small overshoot).

7–4   LAG COMPENSATION

In this section we shall first present realization of lag compensators; an electronic lag compensator using operational amplifiers, an electrical lag compensator using an *RC* network, and a mechanical lag compensator using a spring and dashpots. Then we discuss design procedures using lag compensators. Finally, we compare the PI controller with a lag compensator.

Electronic lag compensator using operational amplifiers.   The configuration of the electronic lag compensator using operational amplifiers is the same as that for the lead compensator shown in Figure 7–6(b). If we choose $R_2 C_2 > R_1 C_1$ in the circuit shown in Figure 7–6(b), it becomes a lag compensator. Referring to Figure 7–6(b), the transfer function of the lag compensator is given by

$$\frac{E_o(s)}{E_i(s)} = K_c\beta \frac{Ts + 1}{\beta Ts + 1} = K_c \frac{s + \dfrac{1}{T}}{s + \dfrac{1}{\beta T}}$$

where

$$T = R_1 C_1, \qquad \beta T = R_2 C_2, \qquad \beta = \frac{R_2 C_2}{R_1 C_1} > 1$$

Note that we use β instead of α in the above expressions. In the lead compensator we used α to indicate the ratio $R_2 C_2/(R_1 C_1)$, which was less than 1, or $0 < \alpha < 1$. In this chapter we always assume that $0 < \alpha < 1$ and $\beta > 1$.

Lag compensator using electrical *RC* network.   Figure 7–22 shows an electrical lag network. The complex impedances Z_1 and Z_2 are

$$Z_1 = R_1, \qquad Z_2 = R_2 + \frac{1}{Cs}$$

The transfer function between the output voltage $E_o(s)$ and the input voltage $E_i(s)$ is given by

$$\frac{E_o(s)}{E_i(s)} = \frac{Z_2}{Z_1 + Z_2} = \frac{R_2 Cs + 1}{(R_1 + R_2)Cs + 1}$$

Define

$$R_2 C = T, \qquad \frac{R_1 + R_2}{R_2} = \beta > 1$$

Then the transfer function becomes

$$\frac{E_o(s)}{E_i(s)} = \frac{Ts + 1}{\beta Ts + 1} = \frac{1}{\beta}\left(\frac{s + \dfrac{1}{T}}{s + \dfrac{1}{\beta T}} \right)$$

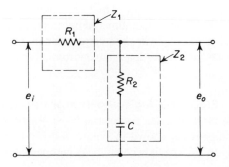

Figure 7–22
Electrical lag network.

Chapter 7 / Design and Compensation Techniques

If this *RC* circuit is used as a lag compensator, then it is usually necessary to add an amplifier with an adjustable gain $K_c\beta$ so that the transfer function of the compensator is

$$G_c(s) = K_c\beta\,\frac{Ts + 1}{\beta Ts + 1} = K_c\,\frac{s + \dfrac{1}{T}}{s + \dfrac{1}{\beta T}}$$

Mechanical lag network. Figure 7–23 shows a mechanical lag network. It consists of a spring and two dashpots. The differential equation for this mechanical network is

$$b_2(\dot{x}_i - \dot{x}_o) + k(x_i - x_o) = b_1\dot{x}_o$$

Taking the Laplace transforms of both sides of this equation, assuming zero initial conditions and then rewriting, we obtain

$$\frac{X_o(s)}{X_i(s)} = \frac{b_2 s + k}{(b_1 + b_2)s + k} = \frac{\dfrac{b_2}{k}s + 1}{\dfrac{b_1 + b_2}{k}s + 1}$$

If we define

$$\frac{b_2}{k} = T, \qquad \frac{b_1 + b_2}{b_2} = \beta > 1$$

then the transfer function $X_o(s)/X_i(s)$ becomes

$$\frac{X_o(s)}{X_i(s)} = \frac{Ts + 1}{\beta Ts + 1} = \frac{1}{\beta}\left(\frac{s + \dfrac{1}{T}}{s + \dfrac{1}{\beta T}}\right)$$

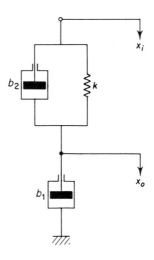

Figure 7–23
Mechanical lag network.

Just as in the case of the electrical lag network, if this mechanical network is used as a lag compensator, then it is necessary to add a linkage device with an adjustable gain $K_c\beta$ so that the transfer function of the compensator is

$$G_c(s) = K_c\beta\,\frac{Ts + 1}{\beta Ts + 1} = K_c\,\frac{s + \dfrac{1}{T}}{s + \dfrac{1}{\beta T}}$$

Characteristics of lag compensators. Consider a lag compensator having the following transfer function:

$$G_c(s) = K_c\beta\,\frac{Ts + 1}{\beta Ts + 1} = K_c\,\frac{s + \dfrac{1}{T}}{s + \dfrac{1}{\beta T}} \qquad (\beta > 1)$$

In the complex plane, a lag compensator has a zero at $s = -1/T$ and a pole at $s = -1/(\beta T)$. The pole is located to the right of the zero.

Figure 7–24 shows a polar plot of the lag compensator. Figure 7–25 shows a Bode diagram of the compensator, where $K_c = 1$ and $\beta = 10$. The corner frequencies of the lag compensator are at $\omega = 1/T$ and $\omega = 1/(\beta T)$. As seen from Figure 7–25, where the values of K_c and β are set equal to 1 and 10, respectively, the magnitude of the lag compensator becomes 10 (or 20 db) at low frequencies and unity (or 0 db) at high frequencies. Thus, the lag compensator is essentially a low-pass filter.

Lag compensation techniques based on the root-locus approach. Consider the problem of finding a suitable compensation network for the case where the system exhibits satisfactory transient-response characteristics but unsatisfactory steady-state characteristics. Compensation in this case essentially consists of increasing the open-loop gain without appreciably changing the transient-response characteristics. This means that the root locus in the neighborhood of the dominant closed-loop poles should not be changed appreciably, but the open-loop gain should be increased as much as needed. This can be accomplished if a lag compensator is put in cascade with the given feedforward transfer function.

To avoid an appreciable change in the root loci, the angle contribution of the lag network should be limited to a small amount, say 5°. To assure this, we place the pole and zero of

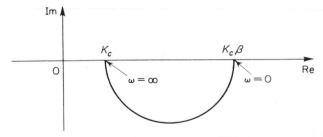

Figure 7–24 Polar plot of a lag compensator $K_c\beta(j\omega T + 1)/(j\omega\beta T + 1)$.

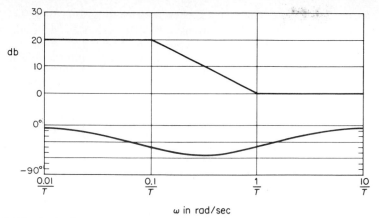

Figure 7–25 Bode diagram of a lag compensator $\beta(j\omega T + 1)/(j\omega\beta T + 1)$, with $\beta = 10$.

the lag network relatively close together and near the origin of the s plane. Then the closed-loop poles of the compensated system will be shifted only slightly from their original locations. Hence, the transient-response characteristics will be essentially unchanged.

Consider a lag compensator $G_c(s)$, where

$$G_c(s) = K_c\beta \frac{Ts + 1}{\beta Ts + 1} = K_c \frac{s + \dfrac{1}{T}}{s + \dfrac{1}{\beta T}} \tag{7-5}$$

If we place the zero and pole of the lag compensator very close to each other, then at $s = s_1$, where s_1 is one of the dominant closed-loop poles, the magnitudes $s_1 + (1/T)$ and $s_1 + [1/(\beta T)]$ are almost equal, or

$$|G_c(s_1)| = \left| K_c \frac{s_1 + \dfrac{1}{T}}{s_1 + \dfrac{1}{\beta T}} \right| \doteq K_c$$

This implies that if gain K_c of the lag compensator is set equal to 1, then the transient response characteristics will not be altered. (This means that the overall gain of the open-loop transfer function can be increased by a factor of β, where $\beta > 1$.) If the pole and zero are placed very close to the origin, then the value of β can be made large. Note that usually $1 < \beta < 15$, and $\beta = 10$ is a good choice. (It is noted that the value of T must be large, but its exact value is not critical. However, it should not be too large in order to avoid difficulties in realizing the phase lag compensator by physical components.)

An increase in the gain means an increase in the static error constants. If the open-loop transfer function of the uncompensated system is $G(s)$, then the static velocity error constant K_v is

$$K_v = \lim_{s \to 0} sG(s)$$

If the compensator is chosen as given by Equation (7–5), then for the compensated system with the open-loop transfer function $G_c(s)G(s)$ the static velocity error constant $\hat{K}_v$ becomes

$$\hat{K}_v = \lim_{s \to 0} sG_c(s)G(s)$$
$$= \lim_{s \to 0} G_c(s)K_v$$
$$= K_c\beta K_v$$

Thus if the compensator is given by Equation (7–5), then the static velocity error constant is increased by a factor of $K_c\beta$, where K_c is approximately unity.

Design procedures for lag compensation by the root-locus method. The procedure for designing lag compensators for the system shown in Figure 7–26 by the root-locus method may be stated as follows: (We assume that the uncompensated system meets the transient-response specifications by simple gain adjustment. If this is not the case, refer to Section 7–5.)

1. Draw the root-locus plot for the uncompensated system whose open-loop transfer function is $G(s)$. Based on the transient-response specifications, locate the dominant closed-loop poles on the root locus.

2. Assume the transfer function of the lag compensator to be

$$G_c(s) = K_c\beta \frac{Ts + 1}{\beta Ts + 1} = K_c \frac{s + \dfrac{1}{T}}{s + \dfrac{1}{\beta T}}$$

Then the open-loop transfer function of the compensated system becomes $G_c(s)G(s)$.

3. Evaluate the particular static error constant specified in the problem.

4. Determine the amount of increase in the static error constant necessary to satisfy the specifications.

5. Determine the pole and zero of the lag compensator that produce the necessary increase in the particular static error constant without appreciably altering the original root loci. (Note that the ratio of the value of gain required in the specifications and the gain found in the uncompensated system is the required ratio between the distance of the zero from the origin and that of the pole from the origin.)

6. Draw a new root-locus plot for the compensated system. Locate the desired dominant closed-loop poles on the root locus. (If the angle contribution of the lag network is very

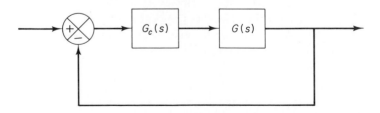

Figure 7–26
Control system.

Chapter 7 / Design and Compensation Techniques

small, that is, a few degrees, then the original and new root loci are almost identical. Otherwise, there will be a slight discrepancy between them. Then locate, on the new root locus, the desired dominant closed-loop poles based on the transient-response specifications.)

7. Adjust gain K_c of the compensator from the magnitude condition that the dominant closed-loop poles lie at the desired location.

EXAMPLE 7–3   Consider the system shown in Figure 7–27(a). The feedforward transfer function is

$$G(s) = \frac{1.06}{s(s + 1)(s + 2)}$$

The root-locus plot for the system is shown in Figure 7–27(b). The closed-loop transfer function becomes

$$\frac{C(s)}{R(s)} = \frac{1.06}{s(s + 1)(s + 2) + 1.06}$$
$$= \frac{1.06}{(s + 0.33 - j0.58)(s + 0.33 + j0.58)(s + 2.33)}$$

The dominant closed-loop poles are

$$s = -0.33 \pm j0.58$$

The damping ratio of the dominant closed-loop poles is $\zeta = 0.5$. The undamped natural frequency of the dominant closed-loop poles is 0.67 rad/sec. The static velocity error constant is 0.53 sec^{-1}.

It is desired to increase the static velocity error constant K_v to about 5 sec^{-1} without appreciably changing the location of the dominant closed-loop poles.

To meet this specification, let us insert a lag compensator as given by Equation (7–5) in cascade with the given feedforward transfer function. To increase the static velocity error constant by a factor

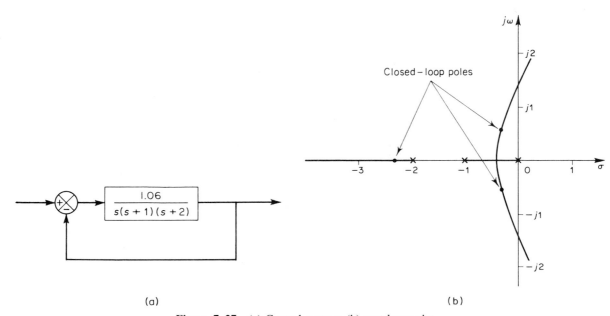

(a)   (b)

Figure 7–27   (a) Control system; (b) root-locus plot.

of about 10, let us choose $\beta = 10$ and place the zero and pole of the lag compensator at $s = -0.1$ and $s = -0.01$, respectively. The transfer function of the lag compensator becomes

$$G_c(s) = K_c \frac{s + 0.1}{s + 0.01}$$

The angle contribution of this lag network near a dominant closed-loop pole is around seven degrees. (This is about the maximum we can allow.) Because the angle contribution of this lag network is not very small, there is a small change in the new root locus near the desired dominant closed-loop poles.

The open-loop transfer function of the compensated system then becomes

$$G_c(s)G(s) = K_c \frac{s + 0.1}{s + 0.01} \frac{1.06}{s(s + 1)(s + 2)}$$

$$= \frac{K(s + 0.1)}{s(s + 0.01)(s + 1)(s + 2)}$$

where

$$K = 1.06K_c$$

The block diagram of the compensated system is shown in Figure 7–28(a). The root-locus plot for the compensated system near the dominant closed-loop poles is shown in Figure 7–28(b), together with the original root locus.

If the damping ratio of the new dominant closed-loop poles is kept the same, then the poles are obtained from the new root-locus plot as follows:

$$s_1 = -0.28 + j0.51, \qquad s_2 = -0.28 - j0.51$$

The open-loop gain K is

$$K = \left| \frac{s(s + 0.01)(s + 1)(s + 2)}{s + 0.1} \right|_{s = -0.28 + j0.51} = 0.98$$

Then the gain K_c is determined as

$$K_c = \frac{K}{1.06} = \frac{0.98}{1.06} = 0.925$$

Thus the transfer function of the lag compensator designed is

$$G_c(s) = 0.925 \frac{s + 0.1}{s + 0.01} = 9.25 \frac{10s + 1}{100s + 1}$$

Then the compensated system has the following open-loop transfer function:

$$G_1(s) = \frac{0.98(s + 0.1)}{s(s + 0.01)(s + 1)(s + 2)} = \frac{4.9(10s + 1)}{s(100s + 1)(s + 1)(0.5s + 1)}$$

The static velocity error constant K_v is

$$K_v = \lim_{s \to 0} sG_1(s) = 4.9 \text{ sec}^{-1}$$

In the compensated system, the static velocity error constant has increased to 4.9 sec^{-1}, or 4.9/0.53 = 9.25 times the original value. (The steady-state error with ramp inputs has decreased to about 11% of that of the original system.) We have essentially accomplished the design objective of increasing the static velocity error constant to about 5 sec^{-1}. (If we wish to increase the static velocity error constant to exactly 5 sec^{-1}, we can either modify the locations of the pole and zero of the lag compensator or use the present lag compensator, but choose $K_c = 0.944$. In the latter case, however, the damping

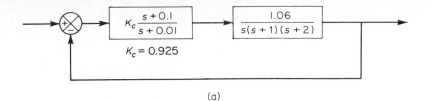

(a)

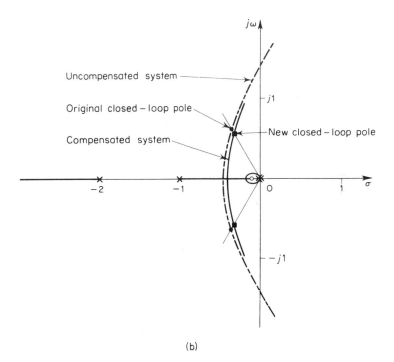

(b)

Figure 7–28
(a) Compensated system; (b) root-locus plots for the compensated system and the uncompensated system.

ratio of the dominant closed-loop poles will be slightly decreased. In the present problem, we may consider the present design as quite acceptable.)

Note that, since the pole and zero of the lag compensator are placed close together and are located very near the origin, their effect on the shape of the original root loci has been small. Except for the presence of a small closed root locus near the origin, the root loci of the compensated and the uncompensated systems are very similar to each other. However, the value of the static velocity error constant of the compensated system is 9.25 times greater than that of the uncompensated system.

The two other closed-loop poles for the compensated system are found as follows:

$$s_3 = -2.31, \qquad s_4 = -0.137$$

The addition of the lag compensator increases the order of the system from three to four, adding one additional closed-loop pole close to the zero of the lag compensator. Because the added closed-loop pole at $s = -0.137$ is close to the zero at $s = -0.1$, the effect of this pole on the transient response is small. Since the pole at $s = -2.31$ is very far from the $j\omega$ axis compared with the dominant closed-loop poles, the effect of this pole on the transient response is also small. We may therefore discard, with little error, the closed-loop poles s_3 and s_4. The conclusion is that the two closed-loop poles s_1 and s_2 are truly dominant. We can predict a fairly accurate response by considering only the dominant closed-loop poles.

The undamped natural frequency of the compensated system is 0.58 rad/sec. This value is about 13% less than the original value, 0.67 rad/sec. This implies that the transient response of the compensated system is slower than that of the original. It will take a longer time to settle down. If this can be tolerated, the lag compensation as discussed here presents a satisfactory solution to the given design problem.

Lag compensation techniques based on the frequency-response approach. The primary function of a lag compensator is to provide attenuation in the high-frequency range to give a system sufficient phase margin. The phase lag characteristic is of no consequence in lag compensation.

The procedure for designing lag compensators for the system shown in Figure 7–26 by the frequency-response approach may be stated as follows:

1. Assume the following lag compensator:

$$G_c(s) = K_c \beta \frac{Ts + 1}{\beta Ts + 1} = K_c \frac{s + \dfrac{1}{T}}{s + \dfrac{1}{\beta T}} \qquad (\beta > 1)$$

Define

$$K_c \beta = K$$

Then

$$G_c(s) = K \frac{Ts + 1}{\beta Ts + 1}$$

The open-loop transfer function of the compensated system is

$$G_c(s)G(s) = K \frac{Ts + 1}{\beta Ts + 1} G(s) = \frac{Ts + 1}{\beta Ts + 1} KG(s) = \frac{Ts + 1}{\beta Ts + 1} G_1(s)$$

where

$$G_1(s) = KG(s)$$

Determine gain K to satisfy the requirement on the given static error constant.

2. If the uncompensated system $G_1(j\omega) = KG(j\omega)$ does not satisfy the specifications on the phase and gain margins, then find the frequency point where the phase angle of the open-loop transfer function is equal $-180°$ plus the required phase margin. The required phase margin is the specified phase margin plus $5°$ to $12°$. (The addition of $5°$ to $12°$ compensates for the phase lag of the lag compensator.) Choose this frequency as the new gain crossover frequency.

3. To prevent detrimental effects of phase lag due to the lag compensator, the pole and zero of the lag compensator must be located substantially lower than the new gain crossover frequency. Therefore, choose the corner frequency $\omega = 1/T$ (corresponding to the zero of the lag compensator) one octave to one decade below the new gain crossover frequency. (If the time constants of the lag compensator do not become too large, the corner frequency $\omega = 1/T$ may be chosen one decade below the new gain crossover frequency.)

4. Determine the attenuation necessary to bring the magnitude curve down to 0 db at the new gain crossover frequency. Noting that this attenuation is $-20 \log \beta$, determine the value of β. Then the other corner frequency (corresponding to the pole of the lag compensator) is determined from $\omega = 1/(\beta T)$.

5. Using the value of K determined in step 1 and that of β determined in step 5, calculate constant K_c from

$$K_c = \frac{K}{\beta}$$

EXAMPLE 7–4

Consider the system shown in Figure 7–29. The open-loop transfer function is given by

$$G(s) = \frac{1}{s(s + 1)(0.5s + 1)}$$

It is desired to compensate the system so that the static velocity error constant K_v is 5 sec^{-1}, the phase margin is at least 40°, and the gain margin is at least 10 db.

We shall use a lag compensator of the form

$$G_c(s) = K_c\beta\frac{Ts + 1}{\beta Ts + 1} = K_c\frac{s + \dfrac{1}{T}}{s + \dfrac{1}{\beta T}} \qquad (\beta > 1)$$

Define

$$K_c\beta = K$$

Define also

$$G_1(s) = KG(s) = \frac{K}{s(s + 1)(0.5s + 1)}$$

The first step in the design is to adjust the gain K to meet the required static velocity error constant. Thus

$$K_v = \lim_{s \to 0} sG_c(s)G(s) = \lim_{s \to 0} s\frac{Ts + 1}{\beta Ts + 1}G_1(s) = \lim_{s \to 0} sG_1(s)$$

$$= \lim_{s \to 0} \frac{sK}{s(s + 1)(0.5s + 1)} = K = 5$$

or

$$K = 5$$

With $K = 5$, the compensated system satisfies the steady-state performance requirement.

We shall next plot the Bode diagram of

$$G_1(j\omega) = \frac{5}{j\omega(j\omega + 1)(0.5j\omega + 1)}$$

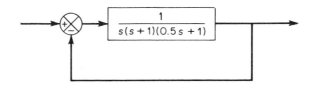

Figure 7–29
Control system.

The magnitude curve and phase-angle curve of $G_1(j\omega)$ are shown in Figure 7–30. From this plot, the phase margin is found to be $-20°$, which means that the system is unstable.

Noting that the addition of a lag compensator modifies the phase curve of the Bode diagram, we must allow 5° to 12° to the specified phase margin to compensate for the modification of the phase curve. Since the frequency corresponding to a phase margin of 40° is 0.7 rad/sec, the new gain crossover frequency (of the compensated system) must be chosen near this value. To avoid overly large time constants for the lag compensator, we shall choose the corner frequency $\omega = 1/T$ (which corresponds to the zero of the lag compensator) to be 0.1 rad/sec. Since this corner frequency is not too far below the new gain crossover frequency, the modification in the phase curve may not be small. Hence we add about 12° to the given phase margin as an allowance to account for the lag angle introduced by the lag compensator. The required phase margin is now 52°. The phase angle of the uncompensated open-loop transfer function is $-128°$ at about $\omega = 0.5$ rad/sec. So we choose the new gain crossover frequency to be 0.5 rad/sec. To bring the magnitude curve down to 0 db at this new gain crossover frequency, the lag compensator must give the necessary attenuation, which, in this case, is -20 db. Hence

$$20 \log \frac{1}{\beta} = -20$$

or

$$\beta = 10$$

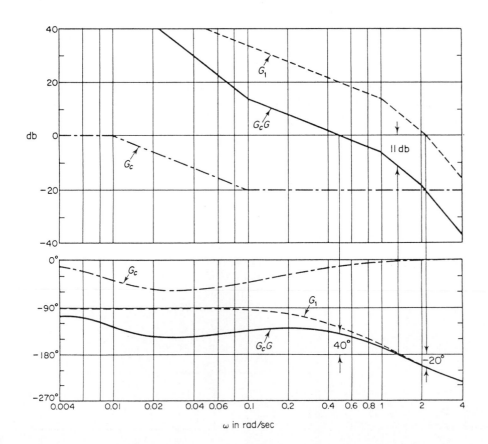

Figure 7–30
Bode diagrams for the uncompensated system, the compensator, and the compensated system. (G_1: uncompensated system, G_c: compensator, G_cG: compensated system.)

The other corner frequency $\omega = 1/(\beta T)$, which corresponds to the pole of the lag compensator, is then determined as

$$\frac{1}{\beta T} = 0.01 \text{ rad/sec}$$

Thus, the transfer function of the lag compensator is

$$G_c(s) = K_c(10)\frac{10s + 1}{100s + 1} = K_c \frac{s + \dfrac{1}{10}}{s + \dfrac{1}{100}}$$

Since the gain K was determined to be 5 and β was determined to be 10, we have

$$K_c = \frac{K}{\beta} = \frac{5}{10} = 0.5$$

The open-loop transfer function of the compensated system is

$$G_c(s)G(s) = \frac{5(10s + 1)}{s(100s + 1)(s + 1)(0.5s + 1)}$$

The magnitude and phase-angle curves of $G_c(j\omega)G(j\omega)$ are also shown in Figure 7–30.

The phase margin of the compensated system is about 40°, which is the required value. The gain margin is about 11 db, which is quite acceptable. The static velocity error constant is 5 sec^{-1}, as required. The compensated system, therefore, satisfies the requirements on both the steady state and the relative stability.

Note that the new gain crossover frequency is decreased from 2.1 to 0.5 rad/sec. This means that the bandwidth of the system is reduced.

To further show the effects of lag compensation, the log-magnitude versus phase plots of the uncompensated system $G_1(j\omega)$ and of the compensated system $G_c(j\omega)G(j\omega)$ are shown in Figure 7–31. The plot of $G_1(j\omega)$ clearly shows that the uncompensated system is unstable. The addition of

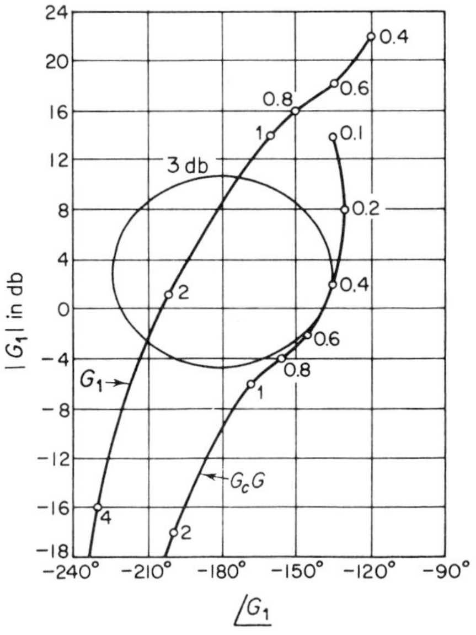

Figure 7–31
Log magnitude versus phase plots of the uncompensated system and the compensated system. (G_1: uncompensated system, G_cG: compensated system.)

the lag compensator stabilizes the system. The plot of $G_c(j\omega)G(j\omega)$ is tangent to the $M = 3$ db locus. Thus the resonant peak value is 3 db, or 1.4, and this peak occurs at $\omega = 0.5$ rad/sec.

Compensators designed by different methods or by different designers (even using the same approach) may look sufficiently different. Any of the well-designed systems, however, will give similar transient and steady-state performance. The best among many alternatives may be chosen from the economic consideration that the time constants of the lag compensator should not be too large.

A few comments on lag compensation

1. Lag compensators are essentially low-pass filters. Therefore, lag compensation permits a high gain at low frequencies (which improves the steady-state performance) and reduces gain in the higher critical range of frequencies so as to improve the phase margin. Note that in lag compensation we utilize the attenuation characteristic of the lag compensator at high frequencies rather than the phase lag characteristic. (The phase lag characteristic is of no use for compensation purposes.)

2. Suppose that the zero and pole of a lag compensator are located at $s = -z$ and $s = -p$, respectively. Then the exact location of the zero and pole is not critical provided that they are close to the origin and the ratio z/p is equal to the required multiplication factor of the static velocity error constant.

It should be noted, however, that the zero and pole of the lag compensator should not be located unnecessarily close to the origin, because the lag compensator will create an additional closed-loop pole in the same region as the zero and pole of the lag compensator.

The closed-loop pole located near the origin gives a very slowly decaying transient response, although its magnitude will become very small because the zero of the lag compensator will almost cancel the effect of this pole. However, the transient response (decay) due to this pole is so slow that the settling time will be adversely affected.

It is also noted that in the system compensated by a lag compensator the transfer function between the plant disturbance and the system error may not involve a zero that is near this pole. Therefore, the transient response to the disturbance input may last very long.

3. The attenuation due to the lag compensator will shift the gain crossover frequency to a lower frequency point where the phase margin is acceptable. Thus, the lag compensator will reduce the bandwidth of the system and will result in slower transient response. [The phase angle curve of $G_c(j\omega)G(j\omega)$ is relatively unchanged near and above the new gain crossover frequency.]

4. Since the lag compensator tends to integrate the input signal, it acts approximately as a proportional-plus-integral controller. Because of this, a lag-compensated system tends to become less stable. To avoid this undesirable feature, the time constant T should be made sufficiently larger than the largest time constant of the system.

5. Conditional stability may occur when a system having saturation or limiting is compensated by use of a lag compensator. When the saturation or limiting takes place in the system, it reduces the effective loop gain. Then the system becomes less stable, and even unstable operation may result, as shown in Figure 7–32. To avoid this, the system must be designed so that the effect of lag compensation becomes significant only when the amplitude of the input to the saturating element is small. (This can be done by means of minor feedback-loop compensation.)

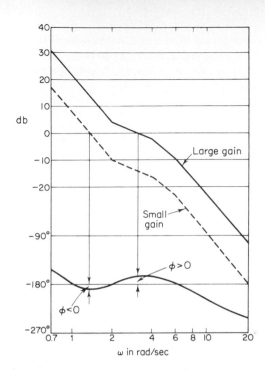

Figure 7–32
Bode diagram of a conditionally stable system.

PI control. The proportional-plus-integral controller whose control action is characterized by the transfer function

$$G_c(s) = K_p \left(1 + \frac{1}{T_i s} \right)$$

is a lag compensator. The PI controller possesses a zero at $s = -1/T_i$ and a pole at $s = 0$. Thus the characteristic of the PI controller is infinite gain at zero frequency. This improves the steady-state characteristics. However, inclusion of the PI control action in the system increases the type number of the compensated system by 1, and this causes the compensated system to be less stable or even make the system unstable. Therefore, the values of K_p and T_i must be chosen carefully to ensure a proper transient response. By properly designing the PI controller it is possible to make the transient response to a step input exhibit relatively small or no overshoot. The speed of response, however, becomes much slower. This is because the PI controller, being a low-pass filter, attenuates the high-frequency components of the signal.

7–5 LAG–LEAD COMPENSATION

Lead compensation basically increases bandwidth and speeds up the response and decreases the maximum overshoot in the step response. Lag compensation increases the low-frequency gain and thus improves the steady-state accuracy of the system, but reduces the speed of responses due to the reduced bandwidth.

If improvements in both transient response and steady-state response are desired, then both a lead compensator and a lag compensator may be used simultaneously. Rather than introducing both a lead compensator and a lag compensator as separate elements, however, it is economical to use a single lag–lead compensator.

Lag–lead compensation combines the advantages of lag and lead compensations. Since the lag–lead compensator possesses two poles and two zeros, such a compensation increases the order of the system by two, unless cancellation of pole(s) and zero(s) occurs in the compensated system.

Note that for a sinusoidal input, the steady-state output of a lag–lead compensator is sinusoidal with a phase shift that is a function of the input frequency. This phase angle varies from lag to lead as the frequency is increased from zero to infinity. (Thus, phase lag and phase lead occur in different frequency bands.)

Electronic lag–lead compensator using operational amplifiers. Figure 7–33 shows an electronic lag–lead compensator using operational amplifiers. The transfer function for this compensator may be obtained as follows: The complex impedance Z_1 is given by

$$\frac{1}{Z_1} = \frac{1}{R_1 + \dfrac{1}{C_1 s}} + \frac{1}{R_3}$$

or

$$Z_1 = \frac{(R_1 C_1 s + 1)R_3}{(R_1 + R_3)C_1 s + 1}$$

Similarly, complex impedance Z_2 is given by

$$Z_2 = \frac{(R_2 C_2 s + 1)R_4}{(R_2 + R_4)C_2 s + 1}$$

Hence, we have

$$\frac{E(s)}{E_i(s)} = -\frac{Z_2}{Z_1} = -\frac{R_4}{R_3} \frac{(R_1 + R_3)C_1 s + 1}{R_1 C_1 s + 1} \cdot \frac{R_2 C_2 s + 1}{(R_2 + R_4)C_2 s + 1}$$

The sign inverter has the transfer function

$$\frac{E_o(s)}{E(s)} = -\frac{R_6}{R_5}$$

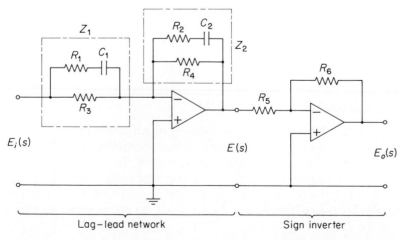

Figure 7–33
Lag–lead
compensator.

Lag–lead network   Sign inverter

Thus, the transfer function of the compensator shown in Figure 7–33 is

$$\frac{E_o(s)}{E_i(s)} = \frac{E_o(s)}{E(s)}\frac{E(s)}{E_i(s)} = \frac{R_4 R_6}{R_3 R_5}\left[\frac{(R_1 + R_3)C_1 s + 1}{R_1 C_1 s + 1}\right]\left[\frac{R_2 C_2 s + 1}{(R_2 + R_4)C_2 s + 1}\right] \qquad (7\text{--}6)$$

Let us define

$$T_1 = (R_1 + R_3)C_1, \qquad \frac{T_1}{\gamma} = R_1 C_1, \qquad T_2 = R_2 C_2, \qquad \beta T_2 = (R_2 + R_4)C_2$$

Then Equation (7–6) becomes

$$\frac{E_o(s)}{E_i(s)} = K_c\frac{\beta}{\gamma}\left(\frac{T_1 s + 1}{\dfrac{T_1}{\gamma}s + 1}\right)\left(\frac{T_2 s + 1}{\beta T_2 s + 1}\right) = K_c\frac{\left(s + \dfrac{1}{T_1}\right)\left(s + \dfrac{1}{T_2}\right)}{\left(s + \dfrac{\gamma}{T_1}\right)\left(s + \dfrac{1}{\beta T_2}\right)} \qquad (7\text{--}7)$$

where

$$\gamma = \frac{R_1 + R_3}{R_1} > 1, \qquad \beta = \frac{R_2 + R_4}{R_2} > 1, \qquad K_c = \frac{R_2 R_4 R_6}{R_1 R_3 R_5}\frac{R_1 + R_3}{R_2 + R_4}$$

Note that β is often chosen to be equal to γ.

Electrical lag–lead network. Figure 7–34 shows an electrical lag–lead network. Let us obtain the transfer function of the lag–lead network.

The complex impedances Z_1 and Z_2 are

$$Z_1 = \frac{R_1}{R_1 C_1 s + 1}, \qquad Z_2 = R_2 + \frac{1}{C_2 s}$$

The transfer function between $E_o(s)$ and $E_i(s)$ is

$$\frac{E_o(s)}{E_i(s)} = \frac{Z_2}{Z_1 + Z_2} = \frac{(R_1 C_1 s + 1)(R_2 C_2 s + 1)}{(R_1 C_1 s + 1)(R_2 C_2 s + 1) + R_1 C_2 s}$$

The denominator of this transfer function can be factored into two real terms. Let us define

$$R_1 C_1 = T_1, \qquad R_2 C_2 = T_2, \qquad R_1 C_1 + R_2 C_2 + R_1 C_2 = \frac{T_1}{\beta} + \beta T_2 \qquad (\beta > 1)$$

Then $E_o(s)/E_i(s)$ can be simplified to

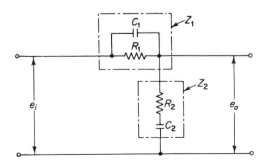

Figure 7–34
Electrical lag–lead
network.

Figure 7-39
Control system.

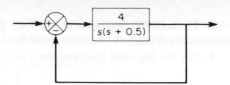

It is desired to make the damping ratio of the dominant closed-loop poles equal to 0.5 and to increase the undamped natural frequency to 5 rad/sec and the static velocity error constant to 80 sec^{-1}. Design an appropriate compensator to meet all the performance specifications.

Let us assume that we use a lag–lead compensator having the transfer function

$$G_c(s) = K_c \left(\frac{s + \dfrac{1}{T_1}}{s + \dfrac{\gamma}{T_1}} \right) \left(\frac{s + \dfrac{1}{T_2}}{s + \dfrac{1}{\beta T_2}} \right) \qquad (\gamma > 1, \beta > 1)$$

where γ need not be equal to β. Then the compensated system will have the transfer function

$$G_c(s)G(s) = K_c \left(\frac{s + \dfrac{1}{T_1}}{s + \dfrac{\gamma}{T_1}} \right) \left(\frac{s + \dfrac{1}{T_2}}{s + \dfrac{1}{\beta T_2}} \right) G(s)$$

From the performance specifications, the dominant closed-loop poles must be at

$$s = -2.50 \pm j4.33$$

Since

$$\left. \angle \frac{4}{s(s + 0.5)} \right|_{s = -2.50 + j4.33} = -235°$$

the phase lead portion of the lag–lead compensator must contribute 55° so that the root locus passes through the desired location of the dominant closed-loop poles.

To design the phase lead portion of the compensator, we first determine the location of the zero and pole that will give 55° contribution. There are many possible choices, but we shall here choose the zero at $s = -0.5$ so that this zero will cancel the pole at $s = -0.5$ of the plant. Once the zero is chosen, the pole can be located such that the angle contribution is 55°. By simple calculation or graphical analysis, the pole must be located at $s = -5.021$. Thus, the phase lead portion of the lag–lead compensator becomes

$$K_c \frac{s + \dfrac{1}{T_1}}{s + \dfrac{\gamma}{T_1}} = K_c \frac{s + 0.5}{s + 5.021}$$

Thus

$$T_1 = 2, \qquad \gamma = \frac{5.021}{0.5} = 10.04$$

Next we determine the value of K_c from the magnitude condition:

$$\left| K_c \frac{s + 0.5}{s + 5.021} \frac{4}{s(s + 0.5)} \right|_{s = -2.5 + j4.33} = 1$$

Hence

$$K_c = \left| \frac{(s + 5.021)s}{4} \right|_{s = -2.5 + j4.33} = 6.26$$

The phase lag portion of the compensator can be designed as follows: First, the value of β is determined to satisfy the requirement on the static velocity error constant:

$$K_v = \lim_{s \to 0} sG_c(s)G(s) = \lim_{s \to 0} sK_c \frac{\beta}{\gamma} G(s)$$

$$= \lim_{s \to 0} s(6.26) \frac{\beta}{10.04} \frac{4}{s(s + 0.5)} = 4.988\beta = 80$$

Hence, β is determined as

$$\beta = 16.04$$

Finally, we choose the value of T_2 large enough so that

$$\left| \frac{s + \dfrac{1}{T_2}}{s + \dfrac{1}{16.04T_2}} \right|_{s = -2.5 + j4.33} \doteq 1$$

and

$$-5° < \left/ \frac{s + \dfrac{1}{T_2}}{s + \dfrac{1}{16.04T_2}} \right|_{s = -2.5 + j4.33} < 0°$$

Since $T_2 \doteq 5$ (or any number greater than 5) satisfies the above two requirements, we may choose

$$T_2 = 5$$

Now the transfer function of the designed lag–lead compensator is given by

$$G_c(s) = (6.26) \left(\frac{s + \dfrac{1}{2}}{s + \dfrac{10.04}{2}} \right) \left(\frac{s + \dfrac{1}{5}}{s + \dfrac{1}{16.04 \times 5}} \right)$$

$$= 6.26 \left(\frac{s + 0.5}{s + 5.02} \right) \left(\frac{s + 0.2}{s + 0.01247} \right)$$

$$= \frac{10(2s + 1)(5s + 1)}{(0.1992s + 1)(80.19s + 1)}$$

The compensated system will have the open-loop transfer function

$$G_c(s)G(s) = \frac{25.04(s + 0.2)}{s(s + 5.02)(s + 0.01247)}$$

Because of the cancellation of the $(s + 0.5)$ terms, the compensated system is a third-order system. (Mathematically, this cancellation is exact, but practically such cancellation will not be exact because some approximations are usually involved in deriving the mathematical model of the system and, as

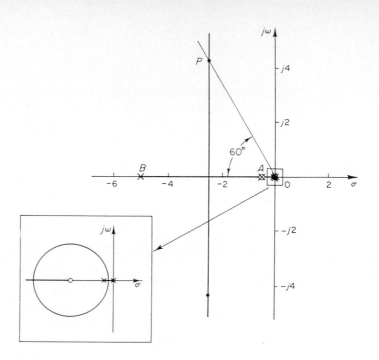

Figure 7–40
Root-locus plot of the compensated system.

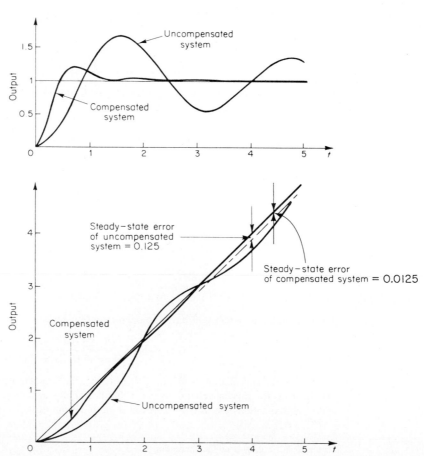

Figure 7–41
Transient response curves for the uncompensated system and the compensated system. (a) Unit-step response curves; (b) unit-ramp response curves.

a result, the time constants are not precise.) The root-locus plot of the compensated system is shown in Figure 7–40. Because the angle contribution of the phase lag portion of the lag–lead compensator is quite small, there is only a small change in the location of the dominant closed-loop poles from the desired location, $s = -2.5 \pm j4.33$. In fact, the new closed-loop poles are located at $s = -2.412 \pm j4.275$. (The new damping ratio is $\zeta = 0.49$.) Therefore, the compensated system meets all the required performance specifications. The third closed-loop pole of the compensated system is located at $s = -0.2085$. Since this closed-loop pole is very close to the zero at $s = -0.2$, the effect of this pole on the response is small. (Note that, in general, if a pole and a zero lie close to each other on the negative real axis near the origin, then such a pole–zero combination will yield a long tail of small amplitude in the transient response.)

The unit-step response curves and unit-ramp response curves before and after compensation are shown in Figure 7–41. (For the computer simulation of the compensated system, see Example 4–7.)

EXAMPLE 7–6   Consider the control system of Example 7–5. Suppose that we use an electrical lag–lead compensator of the form given by Equation (7–10), or

$$G_c(s) = K_c \frac{\left(s + \dfrac{1}{T_1}\right)\left(s + \dfrac{1}{T_2}\right)}{\left(s + \dfrac{\beta}{T_1}\right)\left(s + \dfrac{1}{\beta T_2}\right)} \quad (\beta > 1)$$

Assuming the specifications are the same as those given in Example 7–5, design a compensator $G_c(s)$.

The desired locations for the dominant closed-loop poles are at

$$s = -2.50 \pm j4.33$$

The open-loop transfer function of the compensated system is

$$G_c(s)G(s) = K_c \frac{\left(s + \dfrac{1}{T_1}\right)\left(s + \dfrac{1}{T_2}\right)}{\left(s + \dfrac{\beta}{T_1}\right)\left(s + \dfrac{1}{\beta T_2}\right)} \cdot \frac{4}{s(s + 0.5)}$$

Since the requirement on the static velocity error constant K_v is 80 sec^{-1}, we have

$$K_v = \lim_{s \to 0} sG_c(s)G(s) = \lim_{s \to 0} K_c \frac{4}{0.5} = 8K_c = 80$$

Thus

$$K_c = 10$$

The time constant T_1 and the value of β are determined from

$$\left|\frac{s + \dfrac{1}{T_1}}{s + \dfrac{\beta}{T_1}}\right|\left|\frac{40}{s(s + 0.5)}\right|_{s = -2.5 + j4.33} = \left|\frac{s + \dfrac{1}{T_1}}{s + \dfrac{\beta}{T_1}}\right|\frac{8}{4.77} = 1$$

$$\left/\!\!\underline{\frac{s + \dfrac{1}{T_1}}{s + \dfrac{\beta}{T_1}}}\right._{s = -2.5 + j4.33} = 55°$$

Referring to Figure 7–42, we can easily locate points A and B such that

$$\angle APB = 55°, \qquad \frac{\overline{PA}}{\overline{PB}} = \frac{4.77}{8}$$

The result is

$$\overline{AO} = 2.38, \qquad \overline{BO} = 8.34$$

or

$$T_1 = \frac{1}{2.38} = 0.420, \qquad \beta = 8.34T_1 = 3.503$$

The phase lead portion of the lag–lead network thus becomes

$$\frac{s + 2.38}{s + 8.34}$$

For the phase lag portion, we may choose

$$T_2 = 10$$

Then

$$\frac{1}{\beta T_2} = \frac{1}{3.503 \times 10} = 0.0285$$

Thus, the lag–lead compensator becomes

$$G_c(s) = (10) \left(\frac{s + 2.38}{s + 8.34} \right) \left(\frac{s + 0.1}{s + 0.0285} \right)$$

The compensated system will have the open-loop transfer function

$$G_c(s)G(s) = \frac{40(s + 2.38)(s + 0.1)}{(s + 8.34)(s + 0.0285)s(s + 0.5)}$$

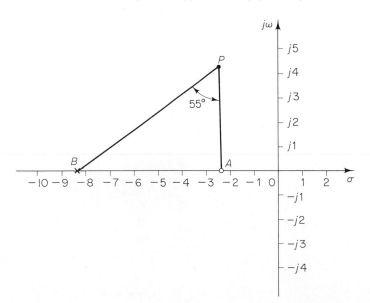

Figure 7–42
Determination of the desired pole–zero location.

Chapter 7 / Design and Compensation Techniques

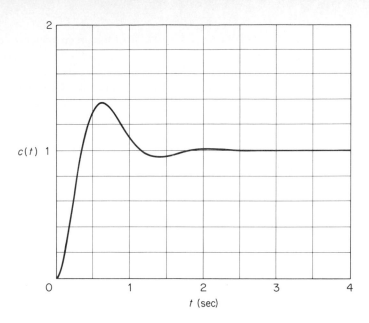

Figure 7–43
Unit-step response curve for the system designed in Example 7–6.

No cancellation occurs in this case, and the compensated system is of fourth order. Because the angle contribution of the phase lag portion of the lag–lead network is quite small, the dominant closed-loop poles are located very near the desired location.

The dominant closed-loop poles are located at $s = -2.464 \pm j4.307$. The two other closed-loop poles are located at

$$ s = -0.1003, \qquad s = -3.84 $$

Since the closed-loop pole at $s = -0.1003$ is very close to a zero at $s = -0.1$, they almost cancel each other. Thus, the effect of this closed-loop pole is very small. The remaining closed-loop pole ($s = -3.84$) does not quite cancel the zero at $s = -2.4$. The effect of this zero is to cause a larger overshoot in the step response than a similar system without such a zero. The unit-step response curve for this system is shown in Figure 7–43.

Lag–lead compensation based on the frequency-response approach. The design of a lag–lead compensator by the frequency-response approach is based on the combination of the design techniques discussed under lead compensation and lag compensation.

Let us assume that the lag–lead compensator is of the following form:

$$ G_c(s) = K_c \frac{(T_1 s + 1)(T_2 s + 1)}{\left(\dfrac{T_1}{\beta} s + 1\right)(\beta T_2 s + 1)} = K_c \frac{\left(s + \dfrac{1}{T_1}\right)\left(s + \dfrac{1}{T_2}\right)}{\left(s + \dfrac{\beta}{T_1}\right)\left(s + \dfrac{1}{\beta T_2}\right)} \qquad (7\text{--}11) $$

where $\beta > 1$. The phase lead portion of the lag–lead compensator (the portion involving T_1) alters the frequency-response curve by adding phase lead angle and increasing the phase margin at the gain crossover frequency. The phase lag portion (the portion involving T_2) provides attenuation near and above the gain crossover frequency and thereby allows an increase of gain at the low-frequency range to improve the steady-state performance.

We shall illustrate the details of the procedures for designing a lag–lead compensator by an example.

EXAMPLE 7–7   Consider the unity-feedback system whose open-loop transfer function is

$$G(s) = \frac{K}{s(s+1)(s+2)}$$

It is desired that the static velocity error constant be 10 sec^{-1}, the phase margin be 50°, and the gain margin be 10 db or more.

Assume that we use the lag–lead compensator given by Equation (7–11). The open-loop transfer function of the compensated system is $G_c(s)G(s)$. Since the gain K of the plant is adjustable, let us assume that $K_c = 1$. Then, $\lim_{s \to 0} G_c(s) = 1$.

From the requirement on the static velocity error constant, we obtain

$$K_v = \lim_{s \to 0} sG_c(s)G(s) = \lim_{s \to 0} sG_c(s)\frac{K}{s(s+1)(s+2)} = \frac{K}{2} = 10$$

Hence,

$$K = 20$$

We shall next draw the Bode diagram of the uncompensated system with $K = 20$, as shown in Figure 7–44. The phase margin of the uncompensated system is found to be $-32°$, which indicates that the uncompensated system is unstable.

The next step in the design of a lag–lead compensator is to choose a new gain crossover frequency. From the phase angle curve for $G(j\omega)$, we notice that $\underline{/G(j\omega)} = -180°$ at $\omega = 1.5$ rad/sec. It is convenient to choose the new gain crossover frequency to be 1.5 rad/sec so that the phase lead angle required at $\omega = 1.5$ rad/sec is about 50°, which is quite possible by use of a single lag–lead network.

Once we choose the gain crossover frequency to be 1.5 rad/sec, we can determine the corner frequency of the phase lag portion of the lag–lead compensator. Let us choose the corner frequency $\omega = 1/T_2$ (which corresponds to the zero of the phase lag portion of the compensator) to be one decade below the new gain crossover frequency, or at $\omega = 0.15$ rad/sec.

Recall that for the lead compensator the maximum phase lead angle ϕ_m is given by Equation (7–3), where α in Equation (7–3) is $1/\beta$ in the present case. By substituting $\alpha = 1/\beta$ in Equation (7–3), we have

$$\sin \phi_m = \frac{1 - \dfrac{1}{\beta}}{1 + \dfrac{1}{\beta}} = \frac{\beta - 1}{\beta + 1}$$

Notice that $\beta = 10$ corresponds to $\phi_m = 54.9°$. Since we need a 50° phase margin, we may choose $\beta = 10$. (Note that we will be using several degrees less than the maximum angle, 54.9°.) Thus

$$\beta = 10$$

Then the corner frequency $\omega = 1/\beta T_2$ (which corresponds to the pole of the phase lag portion of the compensator) becomes $\omega = 0.015$ rad/sec. The transfer function of the phase lag portion of the lag–lead compensator then becomes

$$\frac{s + 0.15}{s + 0.015} = 10\left(\frac{6.67s + 1}{66.7s + 1}\right)$$

The phase lead portion can be determined as follows: Since the new gain crossover frequency is $\omega = 1.5$ rad/sec, from Figure 7–44, $G(j1.5)$ is found to be 13 db. Hence if the lag–lead compensator

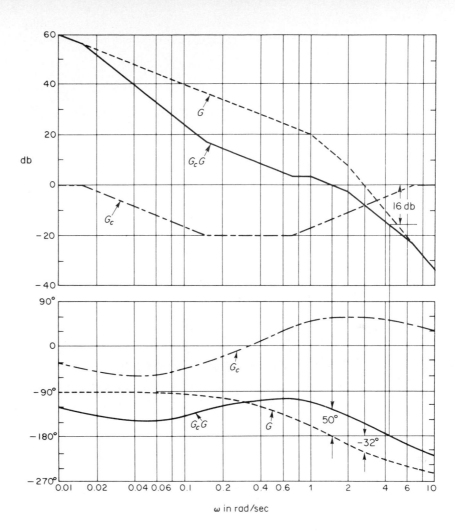

Figure 7–44
Bode diagrams for the uncompensated system, the compensator, and the compensated system. (G: uncompensated system, G_c: compensator, G_cG: compensated system.)

contributes -13 db at $\omega = 1.5$ rad/sec, then the new gain crossover frequency is as desired. From this requirement, it is possible to draw a straight line of slope 20 db/decade, passing through the point $(-13$ db, 1.5 rad/sec). The intersections of this line and the 0-db line and -20-db line determine the corner frequencies. Thus, the corner frequencies for the lead portion are $\omega = 0.7$ rad/sec and $\omega = 7$ rad/sec. Thus, the transfer function of the lead portion of the lag–lead compensator becomes

$$\frac{s + 0.7}{s + 7} = \frac{1}{10}\left(\frac{1.43s + 1}{0.143s + 1}\right)$$

Combining the transfer functions of the lag and lead portions of the compensator, we obtain the transfer function of the lag–lead compensator. Since we chose $K_c = 1$, we have

$$G_c(s) = \left(\frac{s + 0.7}{s + 7}\right)\left(\frac{s + 0.15}{s + 0.015}\right) = \left(\frac{1.43s + 1}{0.143s + 1}\right)\left(\frac{6.67s + 1}{66.7s + 1}\right)$$

The magnitude and phase-angle curves of the lag–lead compensator just designed are shown in Figure

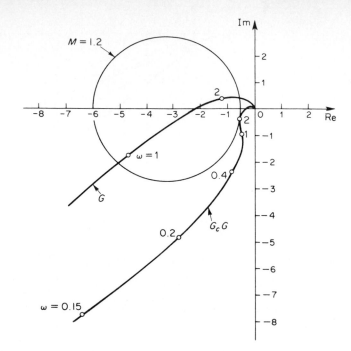

Figure 7–45
Polar plots of the uncompensated system and the compensated system. (*G*: uncompensated system, G_cG: compensated system.)

7–44. The open-loop transfer function of the compensated system is

$$G_c(s)G(s) = \frac{(s + 0.7)(s + 0.15)20}{(s + 7)(s + 0.015)s\,(s + 1)(s + 2)}$$

$$= \frac{10(1.43s + 1)(6.67s + 1)}{s(0.143s + 1)(66.7s + 1)(s + 1)(0.5s + 1)} \tag{7-12}$$

The magnitude and phase-angle curves of the system of Equation (7–12) are also shown in Figure 7–44. The phase margin of the compensated system is 50°, the gain margin is 16 db, and the static velocity error constant is 10 sec^{-1}. All the requirements are therefore met, and the design has been completed.

Figure 7–45 shows the polar plots of the uncompensated system and compensated system. The $G_c(j\omega)G(j\omega)$ locus is tangent to the $M = 1.2$ circle at about $\omega = 2$ rad/sec. Clearly, this indicates that the compensated system has satisfactory relative stability. The bandwidth of the compensated system is slightly larger than 2 rad/sec.

7–6   TUNING RULES FOR PID CONTROLLERS

PID controllers are very frequently used in industrial control systems. As presented earlier, the transfer function $G_c(s)$ of the PID controller is

$$G_c(s) = K_p\left(1 + \frac{1}{T_i s} + T_d s\right) \tag{7-13}$$

where K_p = proportional gain
T_i = integral time
T_d = derivative time

If $e(t)$ is the input to the PID controller, the output $u(t)$ from the controller is given by

$$u(t) = K_p \left[e(t) + \frac{1}{T_i} \int_{-\infty}^{t} e(t)dt + T_d \frac{de(t)}{dt} \right]$$

Constants K_p, T_i, and T_d are the controller parameters. Equation (7–13) can also be written as

$$G_c(s) = K_p + \frac{K_i}{s} + K_d s \qquad (7\text{–}14)$$

where K_p = proportional gain
$\quad\ K_i$ = integral gain
$\quad\ K_d$ = derivative gain

In this case, K_p, K_i, and K_d become controller parameters.

It is noted that in actual PID controllers, instead of adjusting the proportional gain, we adjust the proportional band. The proportional band is proportional to $1/K_p$ and is expressed in percent. (For example, 25% proportional band corresponds to $K_p = 4$.)

Figure 7–46 shows an electronic PID controller using operational amplifiers. The transfer function $E(s)/E_i(s)$ is given by

$$\frac{E(s)}{E_i(s)} = -\frac{Z_2}{Z_1}$$

where

$$Z_1 = \frac{R_1}{R_1 C_1 s + 1}, \qquad Z_2 = \frac{R_2 C_2 s + 1}{C_2 s}$$

Thus

$$\frac{E(s)}{E_i(s)} = -\left(\frac{R_2 C_2 s + 1}{C_2 s} \right)\left(\frac{R_1 C_1 s + 1}{R_1} \right)$$

Noting that

$$\frac{E_o(s)}{E(s)} = -\frac{R_4}{R_3}$$

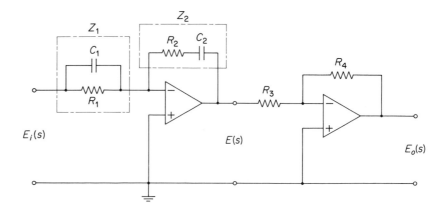

Figure 7–46
Electronic PID
controller.

we have

$$\frac{E_o(s)}{E_i(s)} = \frac{E_o(s)}{E(s)}\frac{E(s)}{E_i(s)} = \frac{R_4 R_2}{R_3 R_1}\frac{(R_1 C_1 s + 1)(R_2 C_2 s + 1)}{R_2 C_2 s}$$

$$= \frac{R_4 R_2}{R_3 R_1}\left(\frac{R_1 C_1 + R_2 C_2}{R_2 C_2} + \frac{1}{R_2 C_2 s} + R_1 C_1 s\right)$$

$$= \frac{R_4(R_1 C_1 + R_2 C_2)}{R_3 R_1 C_2}\left[1 + \frac{1}{(R_1 C_1 + R_2 C_2)s} + \frac{R_1 C_1 R_2 C_2}{R_1 C_1 + R_2 C_2}s\right] \quad (7\text{–}15)$$

Thus

$$K_p = \frac{R_4(R_1 C_1 + R_2 C_2)}{R_3 R_1 C_2}$$

$$T_i = R_1 C_1 + R_2 C_2$$

$$T_d = \frac{R_1 C_1 R_2 C_2}{R_1 C_1 + R_2 C_2}$$

In terms of the proportional gain, integral gain, and derivative gain, we have

$$K_p = \frac{R_4(R_1 C_1 + R_2 C_2)}{R_3 R_1 C_2}$$

$$K_i = \frac{R_4}{R_3 R_1 C_2}$$

$$K_d = \frac{R_4 R_2 C_1}{R_3}$$

Notice that the second operational amplifier circuit acts as a sign inverter as well as a gain adjuster.

Bode diagram of PID controller.   Consider the following PID controller:

$$G_c(s) = \frac{2(0.1s + 1)(s + 1)}{s}$$

The Bode diagram for this controller is shown in Figure 7–47. The PID controller is a lag–lead controller as clearly seen from the Bode diagram. Therefore, as is in the case of the lag–lead compensator, if the proportional gain is set high, the PID-controlled system may become conditionally stable.

PID control of plants.   Figure 7–48 shows a PID control of a plant. If a mathematical model of the plant can be derived, then it is possible to apply various design techniques for determining parameters of the controller that will meet the transient and steady-state specifications of the closed-loop system. However, if the plant is so complicated that its mathematical model cannot be easily obtained, then analytical approach to the design of a PID controller is not possible. Then we must resort to experimental approaches to the design of PID controllers.

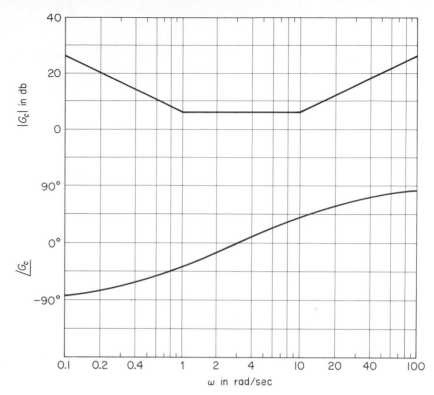

Figure 7–47
Bode diagram of PID
controller given by
$G_c(s) = 2(0.1s + 1)(s + 1)/s$.

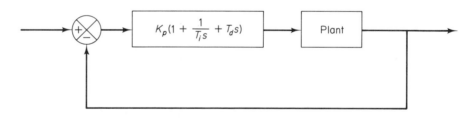

Figure 7–48
PID control of a
plant.

The process of selecting the controller parameters to meet given performance specifications is known as controller tuning. Ziegler and Nichols suggested rules for tuning PID controllers (meaning to set values of K_p, T_i, and T_d) based on experimental step responses or based on the value of K_p that results in marginal stability with only the proportional control action is used. Ziegler–Nichols rules, which are presented in the following, are very convenient when mathematical models of plants are not known. (These rules can, of course, be applied to the design of systems with known mathematical models.)

Ziegler–Nichols rules for tuning PID controllers.   Ziegler and Nichols proposed rules for determining values of the proportional gain K_p, integral time T_i, and derivative time T_d based on the transient response characteristics of a given plant. Such determination of the parameters of PID controllers or tuning of PID controllers can be made by engineers on site by experiments on the plant.

There are two methods called Ziegler–Nichols tuning rules. In both methods, they aimed at obtaining 25% maximum overshoot in step response (see Figure 7–49).

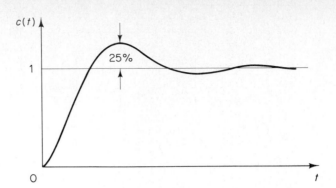

Figure 7–49
Unit-step response
curve showing 25%
maximum overshoot.

First method.   In the first method, we obtain experimentally the response of the plant
to a unit-step input, as shown in Figure 7–50. If the plant involves neither integrator(s) nor
dominant complex-conjugate poles, then such a unit-step response curve may look like an
S-shaped curve, as shown in Figure 7–51. (If the response does not exhibit an S-shaped
curve, this method does not apply.) Such step-response curves may be generated experimentally or from a dynamic simulation of the plant.

The S-shaped curve may be characterized by two constants, delay time L and time constant
T. The delay time and time constant are determined by drawing a tangent line at the inflection
point of the S-shaped curve and determining the intersections of the tangent line with the
time axis and line $c(t) = K$, as shown in Figure 7–51. The transfer function $C(s)/U(s)$ may
then be approximated by a first-order system with a transport lag.

$$\frac{C(s)}{U(s)} = \frac{Ke^{-Ls}}{Ts + 1}$$

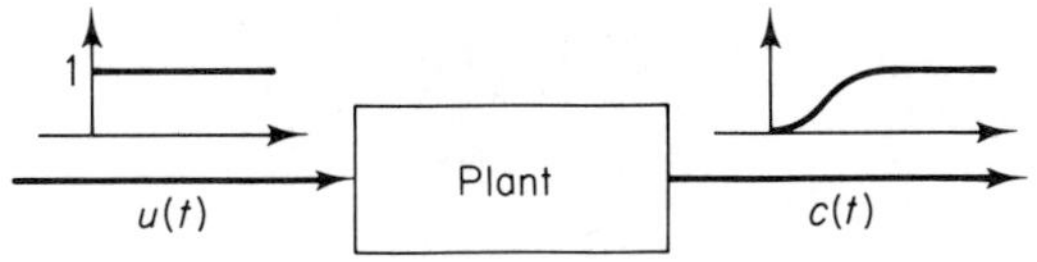

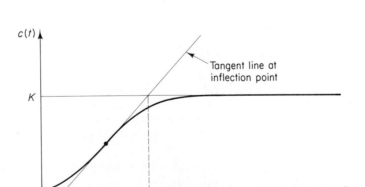

Figure 7–50
Unit-step response of
a plant.

Figure 7–51
S-shaped response
curve.

Ziegler and Nichols suggested to set the values of K_p, T_i, and T_d according to the formula shown in Table 7–2.

Notice that the PID controller tuned by the first method of Ziegler–Nichols rules gives

$$G_c(s) = K_p\left(1 + \frac{1}{T_i s} + T_d s\right)$$

$$= 1.2\frac{T}{L}\left(1 + \frac{1}{2Ls} + 0.5Ls\right)$$

$$= 0.6\,T\frac{\left(s + \dfrac{1}{L}\right)^2}{s}$$

Thus, the PID controller has a pole at the origin and double zeros at $s = -1/L$.

Table 7–2 Ziegler–Nichols Tuning Rule Based on Step Response of Plant (First Method)			
Type of Controller	$\mathbf{K}_p$	$\mathbf{T}_i$	$\mathbf{T}_d$
P	$\dfrac{T}{L}$	∞	0
PI	$0.9\dfrac{T}{L}$	$\dfrac{L}{0.3}$	0
PID	$1.2\dfrac{T}{L}$	$2L$	$0.5L$

Second method. In the second method, we first set $T_i = \infty$ and $T_d = 0$. Using the proportional control action only (see Figure 7–52), increase K_p from 0 to a critical value K_{cr} where the output first exhibits sustained oscillations. (If the output does not exhibit sustained oscillations for whatever value K_p may take, then this method does not apply.) Thus, the critical gain K_{cr} and the corresponding period P_{cr} are experimentally determined (see Figure 7–53). Ziegler and Nichols suggested to set the values of the parameters K_p, T_i, and T_d according to the formula shown in Table 7–3.

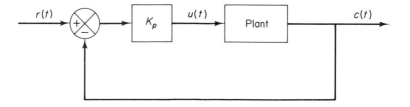

Figure 7–52
Closed-loop system with a proportional controller.

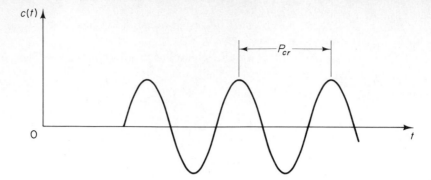

Figure 7–53
Sustained oscillation
with period P_{cr}.

Notice that the PID controller tuned by the second method of Ziegler–Nichols rules gives

$$G_c(s) = K_p\left(1 + \frac{1}{T_i s} + T_d s\right)$$

$$= 0.6K_{cr}\left(1 + \frac{1}{0.5P_{cr}s} + 0.125\,P_{cr}s\right)$$

$$= 0.075K_{cr}P_{cr}\frac{\left(s + \dfrac{4}{P_{cr}}\right)^2}{s}$$

Thus, the PID controller has a pole at the origin and double zeros at $s = -4/P_{cr}$.

Comments.   Ziegler–Nichols tuning rules have been widely used to tune PID controllers in process control systems where the plant dynamics are not precisely known. Over many years, such tuning rules proved to be very useful. Ziegler–Nichols tuning rules can, of course, be applied to plants whose dynamics are known. (If plant dynamics are known, many analytical and graphical approaches to the design of PID controllers are available, in addition to Ziegler–Nichols tuning rules.)

Table 7–3   Ziegler–Nichols Tuning Rule Based on Critical Gain K_{cr} and Critical Period P_{cr} (Second Method)

Type of Controller	$\mathbf{K}_p$	$\mathbf{T}_i$	$\mathbf{T}_d$
P	$0.5K_{cr}$	∞	0
PI	$0.45K_{cr}$	$\dfrac{1}{1.2}P_{cr}$	0
PID	$0.6K_{cr}$	$0.5P_{cr}$	$0.125P_{cr}$

If the transfer function of the plant is known, a unit-step response may be calculated or the critical gain K_{cr} and critical period P_{cr} may be calculated. Then, using those calculated values, it is possible to determine the parameters K_p, T_i, and T_d from Table 7–2 or 7–3. However, the real usefulness of Ziegler–Nichols tuning rules becomes apparent when the plant dynamics are not known so that no analytical or graphical approaches to the design of controllers are available.

Generally, for plants with complicated dynamics but no integrators, Ziegler–Nichols tuning rules can be applied. However, if the plant has an integrator, these rules may not be applied. To clearly illustrate this point, consider the following case: Suppose that a plant has the transfer function

$$G(s) = \frac{(s + 2)(s + 3)}{s(s + 1)(s + 5)}$$

Because of the presence of an integrator, the first method does not apply. Referring to Figure 7–50, the step response of this plant will not have an S-shaped response curve; rather, the response increases with time. Also, if the second method is attempted (see Figure 7–52), the closed-loop system with a proportional controller will not exhibit sustained oscillations whatever value the gain K_p may take. This can be seen from the following analysis. Since the characteristic equation is

$$s(s + 1)(s + 5) + K_p(s + 2)(s + 3) = 0$$

or

$$s^3 + (6 + K_p)s^2 + (5 + 5K_p)s + 6K_p = 0$$

the Routh array becomes

$$
\begin{array}{c c c}
s^3 & 1 & 5 + 5K_p \\
s^2 & 6 + K_p & 6K_p \\
s^1 & \dfrac{30 + 29K_p + 5K_p^2}{6 + K_p} & 0 \\
s^0 & 6K_p &
\end{array}
$$

The coefficients in the first column are positive for all values of positive K_p. Thus, the closed-loop system will not exhibit sustained oscillations and, therefore, the critical gain value K_{cr} does not exist. Hence the second method does not apply.

If the plant is such that Ziegler–Nichols rules can be applied, then the plant with a PID controller tuned by Ziegler–Nichols rules will exhibit approximately 10% ~ 60% maximum overshoot in step response. On the average (experimented on many different plants), the maximum overshoot is approximately 25%. (This is quite understandable because the values suggested in Tables 7–2 and 7–3 are based on the average.) In a given case, if the maximum overshoot is excessive, it is always possible (experimentally or otherwise) to make fine tuning so that the closed-loop system will exhibit satisfactory transient responses. In fact, Ziegler–Nichols tuning rules give an "educated guess" for the parameter values and provide a starting point for fine tuning.

EXAMPLE 7–8

Consider the control system shown in Figure 7–54 in which a PID controller is used to control the system. The PID controller has the transfer function

$$G_c(s) = K_p\left(1 + \frac{1}{T_i s} + T_d s\right)$$

Although many analytical methods are available for the design of a PID controller for the present system, let us apply a Ziegler–Nichols tuning rule for the determination of the values of parameters K_p, T_i, and T_d. Then obtain a unit-step response curve and check to see if the designed system exhibits

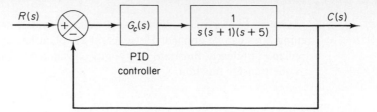

Figure 7–54
PID-controlled system

approximately 25% maximum overshoot. If the maximum overshoot is excessive (40% or more), make a fine tuning and reduce the amount of the maximum overshoot to approximately 25%.

Since the plant has an integrator, we use the second method of Ziegler–Nichols tuning rules. By setting $T_i = \infty$ and $T_d = 0$, we obtain the closed-loop transfer function as follows:

$$\frac{C(s)}{R(s)} = \frac{K_p}{s(s + 1)(s + 5) + K_p}$$

The value of K_p that makes the system marginally stable so that sustained oscillation occurs can be obtained by use of Routh's stability criterion. Since the characteristic equation for the closed-loop system is

$$s^3 + 6s^2 + 5s + K_p = 0$$

the Routh array becomes as follows:

$$
\begin{array}{ccc}
s^3 & 1 & 5 \\
s^2 & 6 & K_p \\
s^1 & \dfrac{30 - K_p}{6} & \\
s^0 & K_p &
\end{array}
$$

Examining the coefficients of the first column of the Routh table, we find that sustained oscillation will occur if $K_p = 30$. Thus, the critical gain K_{cr} is

$$K_{cr} = 30$$

With gain K_p set equal to $K_{cr}(= 30)$, the characteristic equation becomes

$$s^3 + 6s^2 + 5s + 30 = 0$$

To find the frequency of the sustained oscillation, we substitute $s = j\omega$ into this characteristic equation as follows:

$$(j\omega)^3 + 6(j\omega)^2 + 5(j\omega) + 30 = 0$$

or

$$6(5 - \omega^2) + j\omega(5 - \omega^2) = 0$$

from which we find the frequency of the sustained oscillation to be $\omega^2 = 5$ or $\omega = \sqrt{5}$. Hence, the period of sustained oscillation is

$$P_{cr} = \frac{2\pi}{\omega} = \frac{2\pi}{\sqrt{5}} = 2.81$$

Referring to Table 7–3, we determine K_p, T_i, and T_d as follows:

$$K_p = 0.6K_{cr} = 18$$

$$T_i = 0.5P_{cr} = 1.405$$

$$T_d = 0.125P_{cr} = 0.35124$$

The transfer function of the PID controller is thus

$$G_c(s) = K_p\left(1 + \frac{1}{T_i s} + T_d s\right)$$

$$= 18\left(1 + \frac{1}{1.405s} + 0.35124s\right)$$

$$= \frac{6.3223(s + 1.4235)^2}{s}$$

The PID controller has a pole at the origin and double zero at $s = -1.4235$. A block diagram of the control system with the designed PID controller is shown in Figure 7–55.

Next, let us examine the unit-step response of the system. The closed-loop transfer function $C(s)/R(s)$ is given by

$$\frac{C(s)}{R(s)} = \frac{6.3223s^2 + 18s + 12.811}{s^4 + 6s^3 + 11.3223s^2 + 18s + 12.811} \tag{7-16}$$

The unit-step response of this system can be obtained easily by using a computer simulation. The first step to a computer solution is to obtain a state-space representation of the system. Referring to Problem B–7–17, a state-space representation of Equation (7–16) can be given by

$$\begin{bmatrix} \dot{x}_1 \\ \dot{x}_2 \\ \dot{x}_3 \\ \dot{x}_4 \end{bmatrix} = \begin{bmatrix} 0 & 1 & 0 & 0 \\ 0 & 0 & 1 & 0 \\ 0 & 0 & 0 & 1 \\ -12.811 & -18 & -11.3223 & -6 \end{bmatrix} \begin{bmatrix} x_1 \\ x_2 \\ x_3 \\ x_4 \end{bmatrix} + \begin{bmatrix} 0 \\ 6.3223 \\ -19.9338 \\ 60.8308 \end{bmatrix} u$$

$$y = \begin{bmatrix} 1 & 0 & 0 & 0 \end{bmatrix} \begin{bmatrix} x_1 \\ x_2 \\ x_3 \\ x_4 \end{bmatrix}$$

where

$$x_1 = c$$

$$x_2 = \dot{x}_1$$

$$x_3 = \dot{x}_2 - 6.3223u$$

$$x_4 = \dot{x}_3 + 19.9338u$$

$$u = r$$

$$y = c = x_1$$

Figure 7–55
Block diagram of the system with PID controller designed by use of Ziegler–Nichols tuning rule (second method).

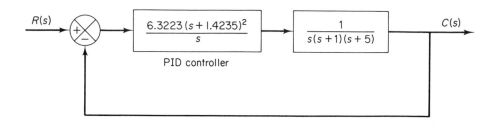

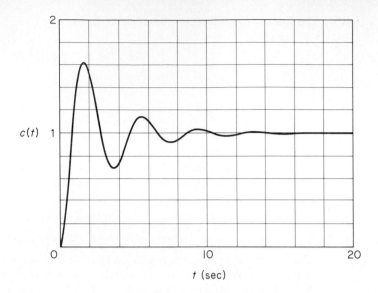

Figure 7–56
Unit-step response
curve of PID
controlled system
designed by use of
Ziegler–Nichols
tuning rule (second
method).

For a unit-step input, $u = r = 1(t)$. The state equation can be solved by a computer using a Runge–Kutta method. Figure 7–56 shows the unit-step response curve obtained by a computer simulation. [Note that $x_1(t) = c(t)$.] The maximum overshoot in the unit-step response is approximately 62%. The amount of maximum overshoot is excessive. It can be reduced by fine tuning the controller parameters. Such fine tuning can be made on the computer. We find that by keeping $K_p = 18$ and by moving the double zero of the PID controller to $s = -0.65$, that is, using the PID controller

$$G_c(s) = 18\left(1 + \frac{1}{3.077s} + 0.7692s\right) = 13.846\frac{(s + 0.65)^2}{s} \qquad (7\text{–}17)$$

the maximum overshoot in the unit-step response can be reduced to approximately 18% (see Figure 7–57). If the proportional gain K_p is increased to 39.42, without changing the location of the double

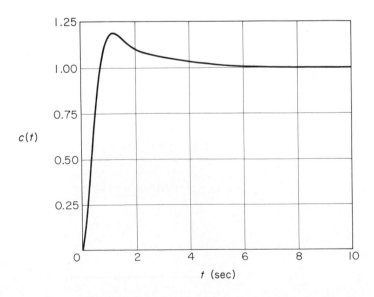

Figure 7–57
Unit-step response of
the system shown in
Fig. 7–54 with PID
controller having
parameters $K_p = 18$,
$T_i = 3.077$, and T_d
$= 0.7692$.

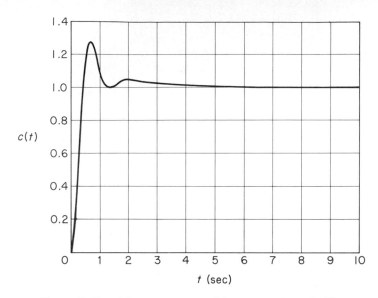

Figure 7–58 Unit-step response of the system shown in Fig. 7–54 with PID controller having parameters $K_p = 39.42$, $T_i = 3.077$, and $T_d = 0.7692$.

zero ($s = -0.65$), that is, using the PID controller

$$G_c(s) = 39.42\left(1 + \frac{1}{3.077s} + 0.7692s\right) = 30.322\frac{(s + 0.65)^2}{s} \qquad (7\text{--}18)$$

then the speed of response is increased, but the maximum overshoot is also increased to approximately 28%, as shown in Figure 7–58. Since the maximum overshoot in this case is fairly close to 25% and the response is faster than the system with $G_c(s)$ given by Equation (7–17), we may consider $G_c(s)$ as given by Equation (7–18) as acceptable. Then the tuned values of K_p, T_i, and T_d become

$$K_p = 39.42, \qquad T_i = 3.077, \qquad T_d = 0.7692$$

It is interesting to observe that these values respectively are approximately twice the values suggested by the second method of the Ziegler–Nichols tuning rule. The important thing to note here is that the Ziegler–Nichols tuning rule has provided a starting point for fine tuning.

It is instructive to note that, for the case where the double zero is located at $s = -1.4235$, increasing the value of K_p increases the speed of response, but as far as the percentage maximum overshoot is concerned, varying gain K_p has very little effect. The reason for this may be seen from the root-locus analysis. Figure 7–59 shows the root-locus diagram for the system designed by use of the second method of Ziegler–Nichols tuning rules. Since the dominant branches of root loci are along the $\zeta = 0.3$ lines for a considerable range of K, varying the value of K (from 6 to 30) will not change the damping ratio of the dominant closed-loop poles very much. However, varying the location of the double zero has a significant effect on the maximum overshoot, because the damping ratio of the dominant closed-loop poles can be changed significantly. This can also be seen from the root-locus analysis. Figure 7–60 shows the root-locus diagram for the system where the PID controller has the double zero at $s = -0.65$. Notice the change of the root-locus configuration. This change in the configuration makes it possible to change the damping ratio of the dominant closed-loop poles.

In Figure 7–60, notice that, in the case where the system has gain $K = 30.322$, the closed-loop poles at $s = -2.35 \pm j4.82$ act as dominant ones. Two additional closed-loop poles are very near

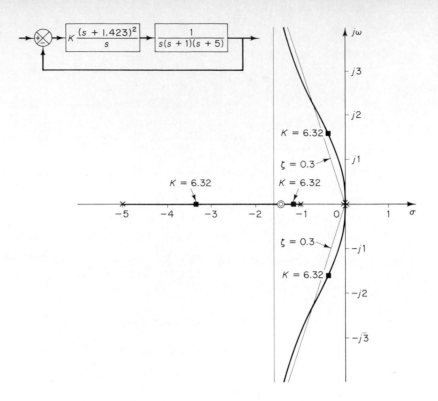

Figure 7–59
Root-locus diagram of system when PID controller has double zero at $s = -1.4235$.

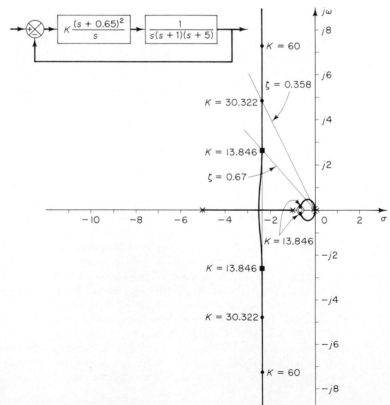

Figure 7–60
Root-locus diagram of system when PID controller has double zero at $s = -0.65$. $K = 13.846$ corresponds to $G_c(s)$ given by Eq. (7–17) and $K = 30.322$ corresponds to $G_c(s)$ given by Eq. (7–18).

the double zero at $s = -0.65$, with the result that these closed-loop poles and the double zero almost cancel each other. The dominant pair of closed-loop poles indeed determines the nature of the response. On the other hand, when the system has $K = 13.846$, the closed-loop poles at $s = -2.35 \pm j2.62$ are not quite dominant because the two other closed-loop poles near the double zero at $s = -0.65$ have considerable effect on the response. The maximum overshoot in the step response in this case (18%) is much larger than the case where the system is of second-order having only dominant closed-loop poles. (In the latter case the maximum overshoot in the step response is approximately 6%.)

7–7 SUMMARY OF CONTROL SYSTEM COMPENSATION METHODS

In Sections 7–3 through 7–6, we have presented detailed procedures for designing lead, lag, and lag–lead compensators and PID controllers by use of simple examples. A satisfactory design of a compensator or a controller for a given system will require a creative application of these basic design principles. In this section we shall first summarize what we presented in Sections 7–3 through 7–5 and then we shall discuss methods for improving PID control schemes.

Comparison of lead, lag, and lag–lead compensation.

1. Lead compensation achieves the desired result through the merits of its phase lead contribution; whereas lag compensation accomplishes the result through the merits of its attenuation property at high frequencies. (In some design problems both lag compensation and lead compensation may satisfy the specifications.)

2. In the s domain, lead compensation enables us to reshape the root locus and thus provide the desired closed-loop poles. In the frequency domain, lead compensation increases the phase margin and bandwidth. Lead compensation yields a higher gain crossover frequency than possible with lag compensation. The higher gain crossover frequency means the larger bandwidth. A large bandwidth means reduction in the settling time. The bandwidth of a system with lead compensation is always greater than that with lag compensation. Therefore, if a large bandwidth or fast response is desired, lead compensation should be employed. If, however, noise signals are present, then a large bandwidth may not be desirable since it makes the system more susceptible to noise signals because of increase in the high-frequency gain. In such a case, lag compensation should be used.

3. Lag compensation improves steady-state accuracy; however, it reduces the bandwidth. If the reduction of bandwidth is too excessive, the compensated system will exhibit sluggish response. If both fast response and good static accuracy are desired, a lag–lead compensator may be employed.

4. Lead compensation requires an additional increase in gain to offset the attenuation inherent in the lead network. This means that lead compensation will require a larger gain than that required by lag compensation. (A larger gain, in most cases, implies larger space, greater weight, and higher cost.)

5. Although a large number of practical compensation tasks can be accomplished with lead, lag, or lag–lead compensators, for complicated systems, simple compensation by use of these compensators may not yield satisfactory results. Then different compensators having different pole–zero configurations must be employed. Note that once the pole–zero configuration of a compensator has been specified the necessary compensator may be realized by use of standard network synthesis techniques.

Graphical comparison. Figure 7–61(a) shows a unit-step response curve and unit-ramp response curve of an uncompensated system. Typical unit-step response and unit-ramp response curves for the compensated system using a lead, lag, and lag–lead network, respectively, are shown in Figures 7–61(b), (c), and (d). The system with a lead compensator exhibits the fastest response, while that with a lag compensator exhibits the slowest response but with marked improvements in the unit-ramp response. The system with a lag–lead compensator will give a compromise: that is, reasonable improvements in both the transient response and steady-state response can be expected. The response curves shown depict the nature of improvements that may be expected from using different types of compensators.

Cancellation of undesirable poles. Since the transfer function of elements in cascade is the product of their individual transfer functions, it is possible to cancel some undesirable poles or zeros by placing a compensating element in cascade, with its poles and zeros being adjusted to cancel the undesirable poles or zeros of the original system. For example, a large time constant T_1 may be canceled by use of the lead network $(T_1s + 1)/(T_2s + 1)$ as follows:

$$\left(\frac{1}{T_1s + 1}\right)\left(\frac{T_1s + 1}{T_2s + 1}\right) = \frac{1}{T_2s + 1}$$

If T_2 is much smaller than T_1, we can effectively eliminate the large time constant T_1. Figure 7–62 shows the effect of canceling a large time constant in step transient response.

If an undesirable pole in the original system lies in the right-half s plane, this compensation scheme should not be used since, although mathematically it is possible to cancel the undesirable pole with an added zero, exact cancellation is physically impossible because of inaccuracies involved in the location of the poles and zeros. A pole in the right-half s plane

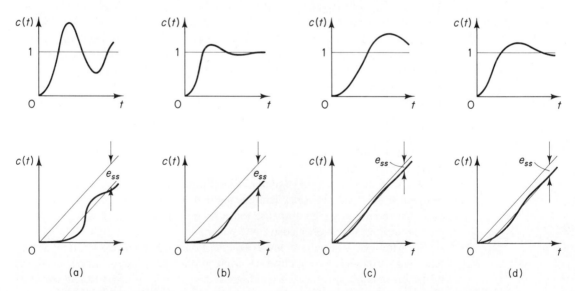

Figure 7–61 Unit-step response curves and unit-ramp response curves. (a) Uncompensated system; (b) lead-compensated system; (c) lag-compensated system; (d) lag–lead compensated system.

Figure 7–62
Step-response curves
showing the effect of
canceling a large time
constant.

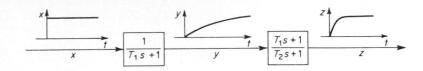

not exactly canceled by the compensator zero will eventually lead to unstable operation, because the response will involve an exponential term that increases with time.

It is noted that if a left-half plane pole is almost canceled but not exactly canceled, as is almost always the case, the uncanceled pole–zero combination will cause the response to have a small amplitude but long-lasting transient response component. If the cancellation is not exact but is reasonably good, then this component will be quite small.

It should be noted that the ideal control system is not the one that has a transfer function of unity. Physically, such a control system cannot be built since it cannot instantaneously transfer energy from the input to the output. In addition, since noise is almost always present in one form or another, a system with a unity transfer function is not desirable. A desired control system, in many practical cases, may have one set of dominant complex-conjugate closed-loop poles with a reasonable damping ratio and undamped natural frequency. The determination of the significant part of the closed-loop pole–zero configuration, such as the location of the dominant closed-loop poles, is based on the specifications that give the required system performance.

Cancellation of undesirable complex-conjugate poles.   If the transfer function of a plant contains one or more pairs of complex-conjugate poles, then a lead, lag, or lag–lead compensator may not give satisfactory results. In such a case, a network that has two zeros and two poles may prove to be useful. If the zeros are chosen so as to cancel the undesirable complex-conjugate poles of the plant, then we can essentially replace the undesirable poles by acceptable ones. That is, if the undesirable complex-conjugate poles are in the left-half s plane and are in the form

$$\frac{1}{s^2 + 2\zeta_1\omega_1 s + \omega_1^2}$$

then the insertion of a compensating network having the transfer function

$$\frac{s^2 + 2\zeta_1\omega_1 s + \omega_1^2}{s^2 + 2\zeta_2\omega_2 s + \omega_2^2}$$

will result in an effective change of the undesirable complex-conjugate poles to acceptable ones. Note that even though the cancellation may not be exact, the compensated system will exhibit better response characteristics. (As stated earlier, this approach cannot be used if the undesirable complex-conjugate poles are in the right-half s plane.)

Familiar networks consisting only of RC components whose transfer functions possess two zeros and two poles are the bridged-T networks. Examples of bridged-T networks and their transfer functions are shown in Figure 7–63.

Feedback compensation.   A tachometer is one of the rate feedback devices. Another common rate feedback device is the rate gyro. Rate gyros are commonly used in aircraft autopilot systems.

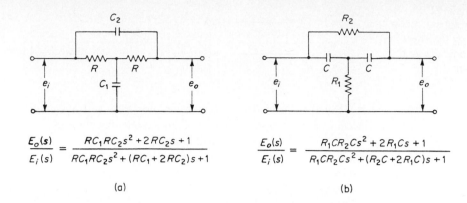

$$\frac{E_o(s)}{E_i(s)} = \frac{RC_1RC_2s^2 + 2RC_2s + 1}{RC_1RC_2s^2 + (RC_1 + 2RC_2)s + 1}$$

$$\frac{E_o(s)}{E_i(s)} = \frac{R_1CR_2Cs^2 + 2R_1Cs + 1}{R_1CR_2Cs^2 + (R_2C + 2R_1C)s + 1}$$

Figure 7–63
Bridged-*T* networks.

(a)

(b)

Velocity feedback using a tachometer is very commonly used in positional servo systems. (In Section 4–4, we discussed a simple example problem using tachometer feedback.) It is noted that, if the system is subjected to noise signals, velocity feedback may generate some difficulty if a particular velocity feedback scheme performs differentiation of the output signal. (The result is the accentuation of the noise effects.)

Eliminating the undesirable effects of disturbances by feedforward control.

If disturbances are measurable, feedforward control is a useful method of canceling their effects on the system output. By feedforward control, we mean control of undesirable effects of measurable disturbances by approximately compensating for them before they materialize. This is advantageous because, in a usual feedback control system, the corrective action starts only after the output has been affected.

As an example, consider the temperature control system shown in Figure 7–64(a). In this system, it is desired to maintain the outlet temperature at some constant value. The disturbance in this system is a change in the inflow rate, which depends on the level in the tower. The effect of a change in this rate cannot be sensed immediately at the output due to the time lags involved in the system.

The temperature controller, which controls the heat input to the heat exchanger, will not act until an error has developed. If the system involves large time lags, it will take some time before any corrective action takes place. In fact, when the error shows up after a certain delay time and the corrective action starts, it may be too late to keep the outlet temperature within the desired limits.

If feedforward control is provided in such a system, then as soon as a change in the inflow occurs a corrective measure will be taken simultaneously, by adjusting the heat input to the heat exchanger. This can be done by feeding both the signal from the flowmeter and the signal from the temperature-measuring element to the temperature controller.

Feedforward control can minimize the transient error, but since feedforward control is open-loop control, there are limitations to its functional accuracy. Feedforward control will not cancel the effects of unmeasurable disturbances under normal operating conditions. It is, therefore, necessary that a feedforward control system include a feedback loop, as shown in Figures 7–64(a) and (b).

Essentially, feedforward control minimizes the transient error caused by measurable disturbances, while feedback control compensates for any imperfections in the functioning of the feedforward control and provides for corrections for unmeasurable disturbances.

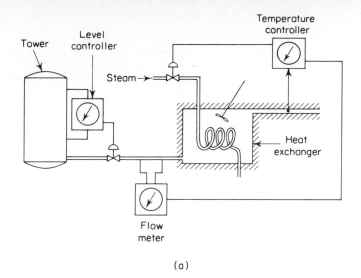

(a)

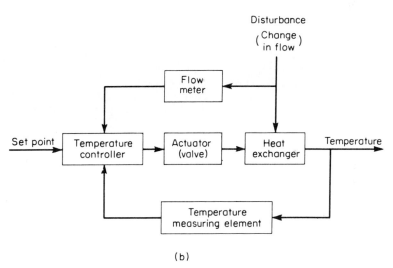

(b)

Figure 7–64
(a) Temperature
control system; (b)
block diagram.

Feedforward control of a plant. Consider the system shown in Figure 7–65. Suppose that the plant transfer function $G(s)$ and disturbance transfer function $G_n(s)$ are known. We shall illustrate a method for determining a suitable disturbance feedforward transfer function $G_1(s)$. Since the output $C(s)$ is given by

$$C(s) = G_c(s)G(s)E(s) + G_n(s)N(s)$$

where

$$E(s) = R(s) - C(s) + G_1(s)N(s)$$

we obtain

$$C(s) = G_c(s)G(s)[R(s) - C(s)] + [G_c(s)G(s)G_1(s) + G_n(s)]N(s) \qquad (7\text{–}19)$$

Equation (7–19) gives the output $C(s)$ in terms of $[R(s) - C(s)]$ and disturbance $N(s)$.

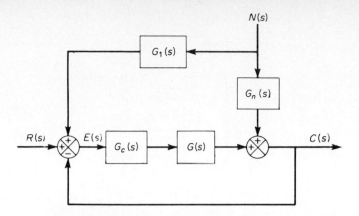

Figure 7–65
Control system.

As is true in any case, the controller transfer function $G_c(s)$ is designed to satisfy the required system specifications in the absence of disturbances. The disturbance feedforward transfer function $G_1(s)$ is determined such that the effects of $N(s)$ are eliminated in the output $C(s)$. That is, we set the coefficient term of $N(s)$ in Equation (7–19) equal to zero, or

$$G_c(s)G(s)G_1(s) + G_n(s) = 0 \qquad (7\text{–}20)$$

Since $G_c(s)$ is designed before $G_1(s)$ is to be determined, $G_c(s)$ is a known transfer function in Equation (7–20). Thus, the disturbance transfer function $G_1(s)$ can be determined by solving Equation (7–20) for $G_1(s)$, or

$$G_1(s) = -\frac{G_n(s)}{G_c(s)G(s)}$$

Two types of configurations of PID control of plants. The basic configuration of the PID-controlled system is shown in Figure 7–66. In the absence of the disturbance input $n(t)$, the closed-loop transfer function $C(s)/R(s)$ is

$$\frac{C(s)}{R(s)} = \frac{K_p\left(1 + \dfrac{1}{T_i s} + T_d s\right)G_p(s)}{1 + K_p\left(1 + \dfrac{1}{T_i s} + T_d s\right)G_p(s)} \qquad (7\text{–}21)$$

In the system shown in Figure 7–66, if the reference input $r(t)$ is changed stepwise, then due to the proportional control action and derivative control action the control signal will have a sudden change in the form of step function and impulse function. Such a change may cause saturation effects in the control signal that are not desirable.

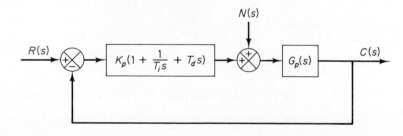

Figure 7–66
Basic configuration of
PID-controlled
system.

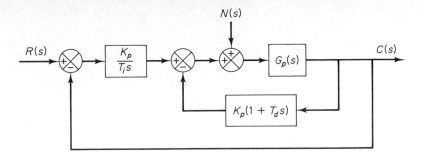

Figure 7–67
I-PD controlled
system.

Figure 7–67 shows a block diagram of a PID-controlled system in which the reference input is transmitted through the integrator path and the proportional and derivative controls act on the feedback signal. In this configuration it is possible to avoid large changes in control signals due to the proportional control and derivative control actions when there is a sudden change in the reference input. The configuration shown in Figure 7–67 is different from the previous configuration (Figure 7–66) and is called *I-PD control*.

In the absence of the disturbance input $n(t)$, the closed-loop transfer function $C(s)/R(s)$ of the I-PD controlled system is

$$\frac{C(s)}{R(s)} = \frac{\dfrac{K_p}{T_i s} G_p(s)}{1 + K_p\left(1 + \dfrac{1}{T_i s} + T_d s\right) G_p(s)} \tag{7–22}$$

Notice the difference in the numerators of $C(s)/R(s)$ in Equations (7–21) and (7–22).

It is important to point out that the PID-controlled system shown in Figure 7–66 is equivalent to the I-PD-controlled system with feedforward path from the reference input, as shown in Figure 7–68. To verify this, let us obtain the closed loop transfer function $C(s)/R(s)$ for the system shown in Figure 7–68. In the absence of $n(t)$, we have

$$U(s) = K_p(1 + T_d s)R(s) + \frac{K_p}{T_i s}[R(s) - C(s)] - K_p(1 + T_d s)C(s)$$

and

$$C(s) = G_p(s)U(s)$$

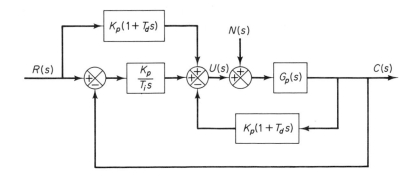

Figure 7–68
I-PD controlled
system with
feedforward control.

Hence

$$\frac{C(s)}{G_p(s)} = K_p\left(1 + \frac{1}{T_i s} + T_d s\right)R(s) - K_p\left(1 + \frac{1}{T_i s} + T_d s\right)C(s)$$

from which we obtain

$$\frac{C(s)}{R(s)} = \frac{K_p\left(1 + \dfrac{1}{T_i s} + T_d s\right) G_p(s)}{1 + K_p\left(1 + \dfrac{1}{T_i s} + T_d s\right) G_p(s)}$$

This last equation is the same as Equation (7–21), the closed-loop transfer function $C(s)/R(s)$ of the system shown in Figure 7–66.

Comments. The basic idea of the I-PD control is to avoid large control signals (which will cause saturation phenomenon) within the system. By bringing the proportional and derivative control actions to the feedback path, it is possible to choose larger values for K_p and T_d than those possible by the PID control scheme.

Consider the response of the I-PD-controlled system to the disturbance input. Since in the I-PD control of a plant it is possible to select larger values for K_p and T_d than those of the PID-controlled case, the I-PD-controlled system will attenuate the effect of disturbance faster than the PID-controlled case.

Next, consider the response of the I-PD-controlled system to a reference input. Since the PID-controlled system is equivalent to the I-PD-controlled system with feedforward control, the PID-controlled system will have faster responses than the corresponding I-PD-controlled system, provided a saturation phenomenon does not occur in the system.

Extension of I-PD control configuration to integral control with state feedback configuration. Figure 7–69(a) shows a control system involving integral control action and state feedback. The control configuration shown in Figure 7–69(a) is an extension of the I-PD control configuration shown in Figure 7–67. The control configuration shown in Figure 7–69(a) is superior to that shown in Figure 7–67 when $G_p(s)$ is of a higher-order plant. That is, in the system shown in Figure 7–67, it is possible to specify the dominant closed-loop poles (specify ζ and ω_n) but not other closed-loop poles. On the other hand, in the case of the system shown in Figure 7–69(a), it is possible to specify all closed-loop poles.

If the plant $G_p(s)$ is of nth order, then the state vector becomes an n-vector and the gain vector **K** becomes

$$\mathbf{K} = [k_1 \quad k_2 \cdots k_n]$$

so that the design parameters are k_I and $k_1, k_2, \ldots, k_n$. Thus, there are $n + 1$ parameters to be determined. This means that finer control is possible by specifying all closed-loop poles. However, the control scheme in this case is much more complex than the I-PD control.

In most practical situations, it may not be possible to measure all state variables. Then it becomes necessary to use an observer. [An observer is a device that generates estimated

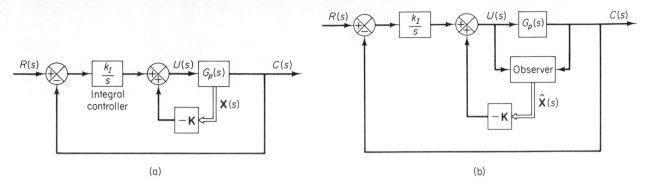

Figure 7–69 (a) Configuration of integral control with state feedback; (b) configuration of integral control with observed-state feedback.

state vector $\hat{\mathbf{x}}(t)$. For details of the observer, see Chapter 10.] Instead of the state vector $\mathbf{x}(t)$ for feedback, we use the observed state (estimated state) $\hat{\mathbf{x}}(t)$ for feedback purposes. Figure 7–69(b) shows a block diagram of the system modified to employ an observer. (Design of controllers based on the integral control with state feedback or observed-state feedback is presented in Chapter 10.)

Concluding comments. In the design examples presented in this chapter, we have been primarily concerned only with the transfer functions of compensators. In actual design problems, we must choose the hardware. Thus, we must satisfy additional design constraints such as cost, size, weight, and reliability.

The system designed may meet the specifications under normal operating conditions but may deviate considerably from the specifications when environmental changes are considerable. Since the changes in the environment affect the gain and time constants of the system, it is necessary to provide automatic or manual means to adjust the gain to compensate for such environmental changes, for nonlinear effects that were not taken into account in the design, and also to compensate for manufacturing tolerances from unit to unit in the production of system components. (The effects of manufacturing tolerances are suppressed in a closed-loop system; therefore, the effects may not be critical in closed-loop operation but critical in open-loop operation.) In addition to this, the designer must remember that any system is subject to small variations due mainly to the normal deterioration of the system.

Example Problems and Solutions

A–7–1. Show that the lead network and lag network inserted in cascade in an open loop acts as proportional-plus-derivative control (in the region of small ω) and proportional-plus-integral control (in the region of large ω), respectively.

Solution. In the region of small ω, the polar plot of the lead network is approximately the same as that of the proportional-plus-derivative controller. This is shown in Figure 7–70(a).

Similarly, in the region of large ω, the polar plot of the lag network approximates the proportional-plus-integral controller, as shown in Figure 7–70(b).

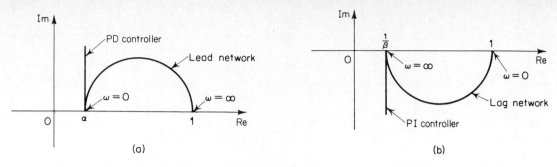

Figure 7–70 (a) Polar plots of a lead network and a proportional-plus-derivative controller; (b) polar plots of a lag network and a proportional-plus-integral controller.

A–7–2. Consider the system with the following open-loop transfer function:

$$G(s)H(s) = \frac{K(s + z)}{(s + p_1)(s + p_2)} \qquad (K \geq 0, p_1 > p_2 > 0, z > 0)$$

Examine the root loci for the system for the following three cases:

Case 1: $z > p_1 > p_2$
Case 2: $p_1 > p_2 > z$
Case 3: $p_1 > z > p_2$

Solution. For case 1, the breakpoints (breakaway point and break-in point) can be determined as follows: Since the characteristic equation for this system is

$$(s + p_1)(s + p_2) + K(s + z) = 0$$

or

$$K = -\frac{(s + p_1)(s + p_2)}{s + z}$$

the roots of $dK/ds = 0$ may be found from

$$(s + p_1)(s + p_2) - (2s + p_1 + p_2)(s + z) = 0$$

as follows:

$$s = -z \pm \sqrt{(z - p_1)(z - p_2)} \qquad (7\text{–}23)$$

Since $z > p_1 > p_2$, the quantity inside the square root sign is positive. The value of K corresponding to $s = -z + \sqrt{(z - p_1)(z - p_2)}$ is

$$K = (\sqrt{z - p_1} - \sqrt{z - p_2})^2 > 0$$

Similarly, the value of K corresponding to $s = -z - \sqrt{(z - p_1)(z - p_2)}$ is

$$K = (\sqrt{z - p_1} + \sqrt{z - p_2})^2 > 0$$

Thus, both $s = -z + \sqrt{(z - p_1)(z - p_2)}$ and $s = -z - \sqrt{(z - p_1)(z - p_2)}$ are breakpoints (one breakaway point and one break-in point).

For case 2, although the argument of the square root of Equation (7–23) is positive and thus the values of s given by Equation (7–23) are real, the corresponding values of K become negative since

$$K = -(\sqrt{p_1 - z} \pm \sqrt{p_2 - z})^2 < 0$$

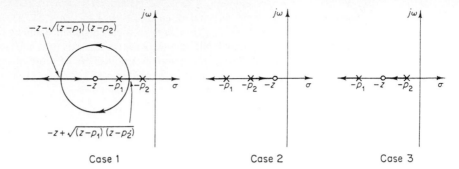

Figure 7–71
Root-locus plots of system considered in Problem A–7–2.

This means that for case 2 the points given by Equation (7–23) are not breakpoints and thus no breakpoints exist.

For case 3, the argument of the square root of Equation (7–23) becomes negative, meaning that the points given by Equation (7–23) are complex conjugates. Since breakaway and break-in points, if they exist, must be on the real axis in the present example, the points given by Equation (7–23) are not breakpoints. Therefore, there exist no breakpoints in case 3, and the root loci are simply two segments of the negative real axis.

Figure 7–71 shows the root locus plots corresponding to the three cases considered.

A–7–3. Consider the system shown in Figure 7–72(a). Draw a root-locus diagram for the system. Determine the value of K such that the damping ratio ζ of the dominant closed-loop poles is 0.5. Then determine all closed-loop poles.

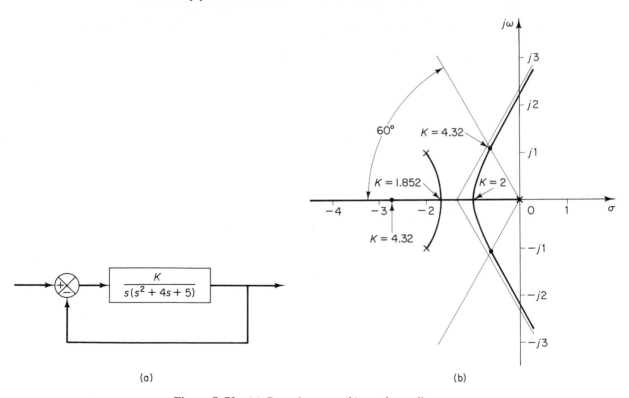

Figure 7–72 (a) Control system; (b) root-locus diagram.

Example Problems and Solutions

Solution. The open-loop poles are located at $s = 0$, $s = -2 + j1$, and $s = -2 - j1$. A root-locus diagram is shown in Figure 7–72(b). On the complex plane the damping ratio $\zeta = 0.5$ corresponds to straight lines having angles $\pm 60°$ with the negative real axis. The intersections of $\zeta = 0.5$ lines and the root-locus branches having asymptotes in the directions of $\pm 60°$ give a pair of complex-conjugate closed-loop poles. The intersections are located at $s = -0.63 \pm j1.09$. The value of the gain K can be obtained by use of the magnitude condition as follows:

$$K = |s(s + 2 + j1)(s + 2 - j1)|_{s = -0.63 + j1.09} = 4.32$$

Since the system is of third order, there are three closed-loop poles. The third pole (real pole) can be determined by dividing the characteristic equation

$$s^3 + 4s^2 + 5s + 4.32 = 0$$

by the product of the dominant closed-loop poles; that is,

$$(s + 0.63)^2 + 1.09^2 = s^2 + 1.26s + 1.585$$

The quotient is $s + 2.74$. Hence, the three closed-loop poles are located at

$$s = -0.63 + j1.09, \qquad s = -0.63 - j1.09, \qquad s = -2.74$$

A–7–4. If the open-loop transfer function $G(s)$ involves lightly damped complex-conjugate poles, then more than one M locus may be tangent to the $G(j\omega)$ locus.

Consider the unity-feedback system whose open-loop transfer function is

$$G(s) = \frac{9}{s(s + 0.5)(s^2 + 0.6s + 10)} \tag{7–24}$$

Draw the Bode diagram for this open-loop transfer function. Draw also the log-magnitude versus phase plot and show that two M loci are tangent to the $G(j\omega)$ locus. Finally, plot the Bode diagram for the closed-loop transfer function.

Solution. Figure 7–73 shows the Bode diagram of $G(j\omega)$. Figure 7–74 shows the log-magnitude versus phase plot of $G(j\omega)$. It is seen that the $G(j\omega)$ locus is tangent to the $M = 8$ db locus at $\omega = 0.97$ rad/sec and it is tangent to $M = -4$ db locus at $\omega = 2.8$ rad/sec.

Figure 7–75 shows the Bode diagram of the closed-loop transfer function. The magnitude curve of the closed-loop frequency response shows two resonant peaks. Note that such a case occurs when the closed-loop transfer function involves the product of two lightly damped second-order terms and the two corresponding resonant frequencies are sufficiently separated from each other. As a matter of fact, the closed-loop transfer function of this system can be written

$$\frac{C(s)}{R(s)} = \frac{G(s)}{1 + G(s)}$$

$$= \frac{9}{(s^2 + 0.487s + 1)(s^2 + 0.613s + 9)}$$

Clearly, the closed-loop transfer function is a product of two lightly damped second-order terms (the damping ratios are 0.243 and 0.102), and the two resonant frequencies are sufficiently separated.

A–7–5. Consider a system with an unstable plant as shown in Figure 7–76. Using the root-locus approach, design a proportional-plus-derivative controller (that is, determine the values of K_p and T_d) such that the damping ratio ζ of the closed-loop system is 0.7 and the undamped natural frequency ω_n is 0.5 rad/sec.

Solution. Note that the open-loop transfer function involves two poles at $s = 1.085$ and $s = -1.085$ and one zero at $s = -1/T_d$, which is unknown at this point.

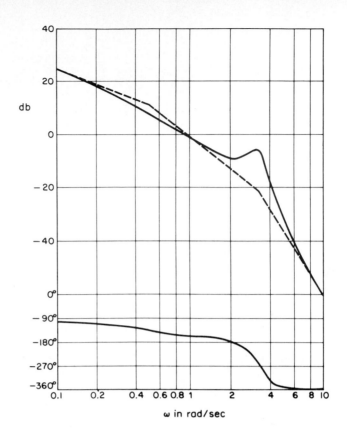

Figure 7–73
Bode diagram of
$G(j\omega)$ given by Eq.
(7–24).

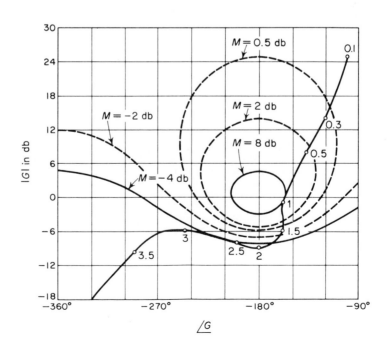

Figure 7–74
Log-magnitude versus
phase plot of $G(j\omega)$
given by Eq. (7–24).

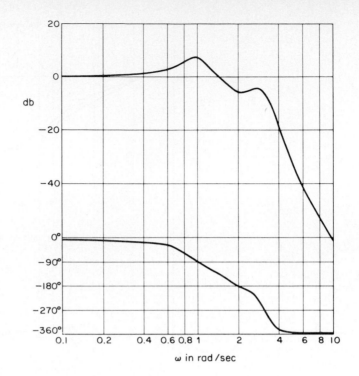

Figure 7–75
Bode diagram of
$G(j\omega)/[1 + G(j\omega)]$
where $G(j\omega)$ is given
by Eq. (7–24).

ω in rad/sec

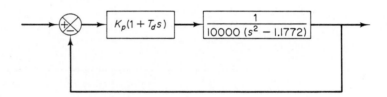

Figure 7–76
PD control of an
unstable plant.

Since the desired closed-loop poles must have $\omega_n = 0.5$ rad/sec and $\zeta = 0.7$, they must be located at

$$s = 0.5 \underline{/180° \pm 45.573°}$$

($\zeta = 0.7$ corresponds to a line having an angle of 45.573° with the negative real axis.) Hence, the desired closed-loop poles are at

$$s = 0.35 \pm j0.357$$

The open-loop poles and the desired closed-loop pole in the upper half plane are located in the diagram shown in Figure 7–77. The angle deficiency at point $s = 0.35 + j0.357$ is

$$-166.026° - 25.913° + 180° = -11.938°$$

This means that the zero at $s = -1/T_d$ must contribute 11.938°, which, in turn, determines the location of the zero as follows:

$$s = -\frac{1}{T_d} = -2.039$$

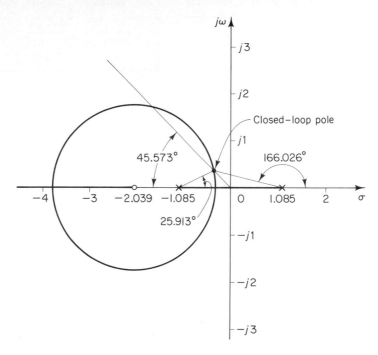

Figure 7–77
Root-locus diagram
for the system shown
in Fig. 7–76.

Hence we have

$$K_p(1 + T_d s) = K_p T_d \left(\frac{1}{T_d} + s \right) = K_p T_d (s + 2.039)$$

The value of T_d is

$$T_d = \frac{1}{2.039} = 0.4904$$

The value of gain K_p can be determined from the magnitude condition as follows:

$$\left| K_p T_d \frac{s + 2.039}{10000(s^2 - 1.1772)} \right|_{s = -0.35 + j0.357} = 1$$

or

$$K_p T_d = 6999.5$$

Hence

$$K_p = \frac{6999.5}{0.4904} = 14,273$$

Therefore,

$$K_p(1 + T_d s) = 14,273(1 + 0.4904s) = 6999.5(s + 2.039)$$

A–7–6. Consider the system shown in Figure 7–78. Design a lead compensator such that the closed-loop system
will have the phase margin of 50° and gain margin of not less than 10 db. Assume that

$$G_c(s) = K_c \alpha \left(\frac{Ts + 1}{\alpha Ts + 1} \right) \qquad (0 < \alpha < 1)$$

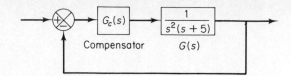

Figure 7–78
Closed-loop system.

It is desired that the bandwidth of the closed-loop system be $1 \sim 2$ rad/sec. What are the values of M_r and ω_r of the compensated system?

Solution. Notice that

$$G_c(j\omega)G(j\omega) = K_c\alpha\left(\frac{Tj\omega + 1}{\alpha Tj\omega + 1}\right)\frac{0.2}{(j\omega)^2(0.2j\omega + 1)}$$

Since the bandwidth of the closed-loop system is close to the gain crossover frequency, we choose the gain crossover frequency to be 1 rad/sec. At $\omega = 1$, the phase angle of $G(j\omega)$ is 191.31°. Hence the lead network needs to supply $50° + 11.31° = 61.31°$ at $\omega = 1$. Hence α can be determined from

$$\sin \phi_m = \sin 61.31° = \frac{1 - \alpha}{1 + \alpha} = 0.8772$$

as follows:

$$\alpha = 0.06541$$

Noting that the maximum phase lead angle ϕ_m occurs at the geometric mean of the two corner frequencies, we have

$$\omega_m = \sqrt{\frac{1}{T}\frac{1}{\alpha T}} = \frac{1}{\sqrt{\alpha}\,T} = \frac{1}{\sqrt{0.06541}\,T} = \frac{3.910}{T} = 1$$

Thus

$$\frac{1}{T} = \frac{1}{3.910} = 0.2558$$

and

$$\frac{1}{\alpha T} = \frac{0.2558}{0.06541} = 3.910$$

Hence

$$G_c(j\omega)G(j\omega) = 0.06541K_c\frac{3.910j\omega + 1}{0.2558j\omega + 1}\frac{0.2}{(j\omega)^2(0.2j\omega + 1)}$$

or

$$\frac{G_c(j\omega)G(j\omega)}{0.06541\,K_c} = \frac{3.910j\omega + 1}{0.2558j\omega + 1}\frac{0.2}{(j\omega)^2(0.2j\omega + 1)}$$

A Bode diagram for $G_c(j\omega)G(j\omega)/(0.06541K_c)$ is shown in Figure 7–79. By simple calculations (or from the Bode diagram), we find that the magnitude curve must be raised by 2.306 db so that the magnitude equals 0 db at $\omega = 1$ rad/sec. Hence we set

$$20 \log 0.06541 K_c = 2.306$$

or

$$0.06541 K_c = 1.3041$$

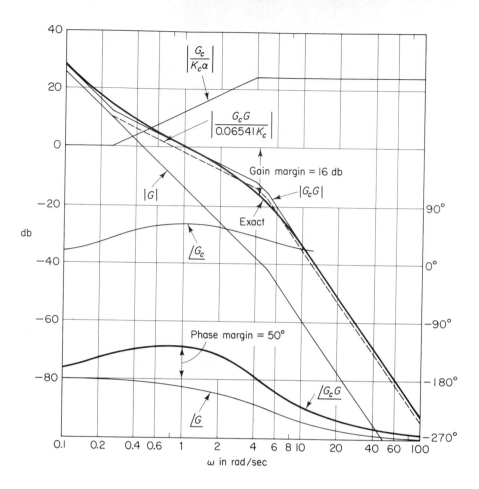

Figure 7–79
Bode diagram of the system shown in Fig. 7–78.

which yields

$$K_c = 19.94$$

The magnitude and phase curves of the compensated system show that the system has the phase margin of 50° and gain margin of 16 db. Hence the design specifications are satisfied.

Figure 7–80 shows the $G_c(j\omega)G(j\omega)$ locus superimposed on the Nichols chart. From this diagram, we find the bandwidth to be approximately 1.9 rad/sec. The values of M_r and ω_r are read from this diagram as follows:

$$M_r = 2.13 \text{ db}, \qquad \omega_r = 0.58 \text{ rad/sec}$$

A–7–7. Consider the angular positional system shown in Figure 7–81. The dominant closed-loop poles are located at $s = -3.60 \pm j4.80$. The damping ratio ζ of the dominant closed-loop poles is 0.6. The static velocity error constant K_v is 4.1 sec^{-1}, which means that for a ramp input of 360°/sec the steady-state error in following the ramp input is

$$e_v = \frac{\theta_i}{K_v} = \frac{360°/\text{sec}}{4.1 \text{ sec}^{-1}} = 87.8°$$

It is desired to decrease e_v to one-tenth of the present value, or to increase the value of the static velocity error constant K_v to 41 sec^{-1}. It is also desired to keep the damping ratio ζ of the dominant

Example Problems and Solutions

Figure 7–80
$G_c(j\omega)G(j\omega)$ locus
superimposed on
Nichols chart.
(Problem A–7–6.)

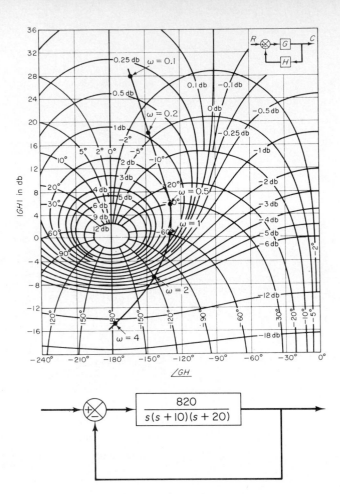

Figure 7–81
Angular-positional
system.

closed-loop poles at 0.6. A small change in the undamped natural frequency ω_n of the dominant closed-loop poles is permissible. Design a suitable lag compensator to increase the static velocity error constant as desired.

Solution. Since the characteristic equation of the uncompensated system is

$$s^3 + 30s^2 + 200s + 820 = 0$$

the uncompensated system has the closed-loop poles at

$$s = -3.60 + j4.80, \qquad s = -3.60 - j4.80, \qquad s = -22.8$$

To increase the static velocity error constant from 4.1 to 41 sec^{-1} without appreciably changing the location of the dominant closed-loop poles, we need to insert a lag compensator $G_c(s)$ whose pole and zero are located very close to the origin. For example, we may choose

$$G_c(s) = 10\,\frac{Ts + 1}{10Ts + 1}$$

where T may be chosen to be 4, or $T = 4$. Then the lag compensator becomes

$$G_c(s) = 10\,\frac{4s + 1}{40s + 1} = \frac{s + 0.25}{s + 0.025} \tag{7–25}$$

The angle contribution of this lag network at $s = -3.60 + j4.80$ is $-1.77°$, which is acceptable in the present problem.

The open-loop transfer function of the compensated system becomes

$$G_c(s)G(s) = \frac{s + 0.25}{s + 0.025} \frac{820}{s(s + 10)(s + 20)}$$
$$= \frac{820(s + 0.25)}{s(s + 0.025)(s + 10)(s + 20)}$$

Clearly, the velocity error constant K_v for the compensated system is

$$K_v = \lim_{s \to 0} sG_c(s)G(s) = 41 \text{ sec}^{-1}$$

Notice that because of the addition of the lag compensator the compensated system becomes of fourth order. The characteristic equation for the compensated system is

$$s^4 + 30.025s^3 + 200.75s^2 + 825s + 205 = 0$$

The dominant closed-loop poles for the compensated system can be found as the intersections of the root loci and the lines corresponding to $\zeta = 0.6$ in the complex plane. (Or simple trial-and-error approach may be used by noting that the damping ratio of the complex poles is 0.6.) The dominant closed-loop poles are located at

$$s = -3.4868 + j4.6697, \qquad s = 3.4868 - j4.6697$$

The other two closed-loop poles are located at

$$s = -0.2648, \qquad s = -22.787$$

The closed-loop pole at $s = -0.2648$ almost cancels the zero of the lag compensator, $s = -0.25$. Also, since the closed-loop pole at $s = -22.787$ is located very much farther to the left compared to the complex-conjugate closed-loop poles, the effect of this pole on the system response is very small. Therefore, the closed-loop poles at $s = -3.4868 \pm j4.6697$ are indeed the dominant closed-loop poles.

The undamped natural frequency ω_n of the dominant closed-loop poles is

$$\omega_n = \sqrt{3.4868^2 + 4.6697^2} = 5.828 \text{ rad/sec}$$

Since the original uncompensated system has the undamped natural frequency of 6 rad/sec, the compensated system has an approximately 3% smaller value, which would be acceptable. Hence, the lag compensator given by Equation (7–25) is satisfactory.

A–7–8. Consider the system shown in Figure 7–82. Design a lag–lead compensator such that the static velocity error constant K_v is 50 sec^{-1} and the damping ratio ζ of the dominant closed-loop poles is 0.5. (Choose the zero of the lead portion of the lag–lead compensator to cancel the pole at $s = -1$ of the plant.) Determine all closed-loop poles of the compensated system.

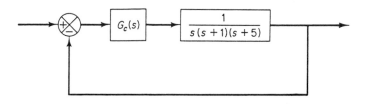

Figure 7–82
Control system.

Solution. Let us employ the lag–lead compensator given by

$$G_c(s) = K_c \frac{\left(s + \dfrac{1}{T_1}\right)\left(s + \dfrac{1}{T_2}\right)}{\left(s + \dfrac{\beta}{T_1}\right)\left(s + \dfrac{1}{\beta T_2}\right)} = K_c \frac{(T_1 s + 1)(T_2 s + 1)}{\left(\dfrac{T_1}{\beta} s + 1\right)\left(\beta T_2 s + 1\right)}$$

where $\beta > 1$. Then

$$
\begin{aligned}
K_v &= \lim_{s \to 0} s G_c(s) G(s) \\
&= \lim_{s \to 0} s \frac{K_c(T_1 s + 1)(T_2 s + 1)}{\left(\dfrac{T_1}{\beta} s + 1\right)\left(\beta T_2 s + 1\right)} \frac{1}{s(s + 1)(s + 5)} \\
&= \frac{K_c}{5}
\end{aligned}
$$

The specification that $K_v = 50 \ \text{sec}^{-1}$ determines the value of K_c, or

$$K_c = 250$$

We now choose $T_1 = 1$ so that $s + (1/T_1)$ will cancel the $(s + 1)$ term of the plant. The lead portion then becomes

$$\frac{s + 1}{s + \beta}$$

For the lag portion of the lag–lead compensator we require

$$\left| \frac{s_1 + \dfrac{1}{T_2}}{s_1 + \dfrac{1}{\beta T_2}} \right| \doteq 1, \qquad -5° < \bigg/ \frac{s_1 + \dfrac{1}{T_2}}{s_1 + \dfrac{1}{\beta T_2}} < 0°$$

where $s = s_1$ is one of the dominant closed-loop poles. For $s = s_1$, the open-loop transfer function becomes

$$G_c(s_1) G(s_1) \doteq K_c \left(\frac{s_1 + 1}{s_1 + \beta}\right) \frac{1}{s_1(s_1 + 1)(s_1 + 5)}$$

Noting that at $s = s_1$ the magnitude and angle conditions are satisfied, we have

$$\left| K_c \left(\frac{s_1 + 1}{s_1 + \beta}\right) \frac{1}{s_1(s_1 + 1)(s_1 + 5)} \right| = 1 \tag{7–26}$$

$$\bigg/ K_c \left(\frac{s_1 + 1}{s_1 + \beta}\right) \frac{1}{s_1(s_1 + 1)(s_1 + 5)} = \pm 180° (2k + 1) \tag{7–27}$$

where $k = 0, 1, 2, \ldots$. In Equations (7–26) and (7–27), β and s_1 are unknowns. Since the damping ratio ζ of the dominant closed-loop poles is specified as 0.5, the closed-loop pole $s = s_1$ can be written as

$$s_1 = -x + j\sqrt{3}\,x$$

where x is as yet undetermined.

Notice that the magnitude condition, Equation (7–26), can be rewritten as

$$\left| \frac{K_c}{(-x + j\sqrt{3}\,x)(-x + \beta + j\sqrt{3}\,x)(-x + 5 + j\sqrt{3}\,x)} \right| = 1$$

Noting that $K_c = 250$, we have

$$x\sqrt{(\beta - x)^2 + 3x^2}\,\sqrt{(5 - x)^2 + 3x^2} = 125 \qquad (7\text{–}28)$$

The angle condition, Equation (7–27), can be rewritten as

$$\bigg/\!\!\!\!\underline{\quad K_c \frac{1}{(-x + j\sqrt{3}\,x)(-x + \beta + j\sqrt{3}\,x)(-x + 5 + j\sqrt{3}\,x)}}$$

$$= -120° - \tan^{-1}\left(\frac{\sqrt{3}\,x}{-x + \beta}\right) - \tan^{-1}\left(\frac{\sqrt{3}\,x}{-x + 5}\right) = -180°$$

or

$$\tan^{-1}\left(\frac{\sqrt{3}\,x}{-x + \beta}\right) + \tan^{-1}\left(\frac{\sqrt{3}\,x}{-x + 5}\right) = 60° \qquad (7\text{–}29)$$

We need to solve Equations (7–28) and (7–29) for β and x. By several trial-and-error calculations, it can be found that

$$\beta = 16.025, \qquad x = 1.9054$$

Thus

$$s_1 = -1.9054 + j\sqrt{3}\,(1.9054) = -1.9054 + j3.3002$$

The lag portion of the lag–lead compensator can be determined as follows: Noting that the pole and zero of the lag portion of the compensator must be located near the origin, we may choose

$$\frac{1}{\beta T_2} = 0.01$$

That is,

$$\frac{1}{T_2} = 0.16025 \qquad \text{or} \qquad T_2 = 6.25$$

With the choice of $T_2 = 6.25$, we find

$$\left| \frac{s_1 + \dfrac{1}{T_2}}{s_1 + \dfrac{1}{\beta T_2}} \right| = \left| \frac{-1.9054 + j3.3002 + 0.16025}{-1.9054 + j3.3002 + 0.01} \right|$$

$$= \left| \frac{-1.74515 + j3.3002}{-1.89054 + j3.3002} \right| = 0.98 \doteq 1$$

and

$$\bigg/\!\!\!\!\underline{\quad \frac{s_1 + \dfrac{1}{T_2}}{s_1 + \dfrac{1}{\beta T_2}}} = \bigg/\!\!\!\!\underline{\quad \frac{-1.9054 + j3.3002 + 0.16025}{-1.9054 + j3.3002 + 0.01}}$$

$$= \tan^{-1}\left(\frac{3.3002}{-1.74515}\right) - \tan^{-1}\left(\frac{3.3002}{-1.89054}\right) = -1.937°$$

Since

$$-5° < -1.937° < 0°$$

our choice of $T_2 = 6.25$ is acceptable. Then the lag–lead compensator just designed can be written as

$$G_c(s) = 250 \left(\frac{s + 1}{s + 16.025} \right) \left(\frac{s + 0.16025}{s + 0.01} \right)$$

Therefore, the compensated system has the following open-loop transfer function:

$$G_c(s)G(s) = \frac{250(s + 0.16025)}{s(s + 0.01)(s + 5)(s + 16.025)}$$

Hence, the characteristic equation for the compensated system becomes

$$s(s + 0.01)(s + 5)(s + 16.025) + 250(s + 0.16025) = 0$$

or

$$s^4 + 21.035s^3 + 80.335s^2 + 250.801s + 40.0625 = 0$$

By solving this equation for the roots, we obtain the closed-loop poles as follows:

$$s = -1.8308 + j3.2359, \qquad s = -1.8308 - j3.2359$$

$$s = -17.205, \qquad s = -0.1684$$

The closed-loop pole at $s = -0.1684$ almost cancels the zero at $s = -0.16025$. Hence the effect of this closed-loop pole is very small. Since the closed-loop pole at $s = -17.205$ is located very much farther to the left compared to the closed-loop poles at $s = -1.8308 \pm j3.2359$, the effect of this real pole on the system response is also very small. Therefore, the closed-loop poles at $s = -1.8308 \pm j3.2359$ are indeed dominant closed-loop poles that determine the response characteristics of the closed-loop system.

A–7–9. Consider a lag–lead compensator $G_c(s)$ defined by

$$G_c(s) = K_c \frac{\left(s + \dfrac{1}{T_1} \right)\left(s + \dfrac{1}{T_2} \right)}{\left(s + \dfrac{\beta}{T_1} \right)\left(s + \dfrac{1}{\beta T_2} \right)}$$

Show that at frequency ω_1, where

$$\omega_1 = \frac{1}{\sqrt{T_1 T_2}}$$

the phase angle of $G_c(j\omega)$ becomes zero. (This compensator acts as a lag compensator for $0 < \omega < \omega_1$ and acts as a lead compensator for $\omega_1 < \omega < \infty$.)

Solution. The angle of $G_c(j\omega)$ is given by

$$\angle G_c(j\omega) = \Big/ j\omega + \frac{1}{T_1} + \Big/ j\omega + \frac{1}{T_2} - \Big/ j\omega + \frac{\beta}{T_1} - \Big/ j\omega + \frac{1}{\beta T_2}$$

$$= \tan^{-1} \omega T_1 + \tan^{-1} \omega T_2 - \tan^{-1} \omega T_1/\beta - \tan^{-1} \omega T_2\beta$$

At $\omega = \omega_1 = 1/\sqrt{T_1 T_2}$, we have

$$\angle G_c(j\omega_1) = \tan^{-1}\sqrt{\frac{T_1}{T_2}} + \tan^{-1}\sqrt{\frac{T_2}{T_1}} - \tan^{-1}\frac{1}{\beta}\sqrt{\frac{T_1}{T_2}} - \tan^{-1}\beta\sqrt{\frac{T_2}{T_1}}$$

Since

$$\tan\left(\tan^{-1}\sqrt{\frac{T_1}{T_2}} + \tan^{-1}\sqrt{\frac{T_2}{T_1}}\right) = \frac{\sqrt{\dfrac{T_1}{T_2}} + \sqrt{\dfrac{T_2}{T_1}}}{1 - \sqrt{\dfrac{T_1}{T_2}}\sqrt{\dfrac{T_2}{T_1}}} = \infty$$

or

$$\tan^{-1}\sqrt{\frac{T_1}{T_2}} + \tan^{-1}\sqrt{\frac{T_2}{T_1}} = 90°$$

and also

$$\tan^{-1}\frac{1}{\beta}\sqrt{\frac{T_1}{T_2}} + \tan^{-1}\beta\sqrt{\frac{T_2}{T_1}} = 90°$$

we have

$$G_c(j\omega_1) = 0°$$

Thus, the angle of $\angle G_c(j\omega_1)$ becomes $0°$ at $\omega = \omega_1 = 1/\sqrt{T_1 T_2}$.

A–7–10. Consider the system shown in Figure 7–83. It is desired to design a PDI controller $G_c(s)$ such that the dominant closed-loop poles are located at $s = -1 \pm j\sqrt{3}$. For the PDI controller, choose $a = 0.2$ and then determine the values of K and b. Sketch the root-locus diagram for the designed system.

Solution. Since

$$G_c(s)G(s) = K\frac{(s + 0.2)(s + b)}{s}\frac{1}{s^2 + 1}$$

the sum of the angles at $s = -1 + j\sqrt{3}$, one of the desired closed-loop poles, from the zero at $s = -0.2$ and poles at $s = 0$, $s = j$, and $s = -j$ is

$$114.79° - 120° - 143.80° - 110.10° = -259.11°$$

Hence the zero at $s = -b$ must contribute $79.11°$. This requires that the zero be located at

$$b = 1.3332$$

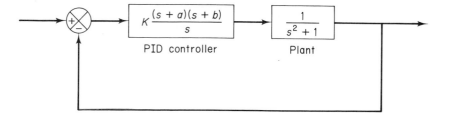

Figure 7–83
PID-controlled
system.

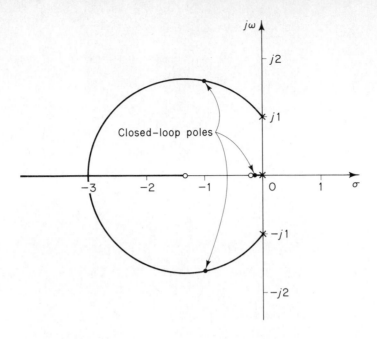

Figure 7–84
Root-locus diagram
for the system
designed in Problem
A–7–10.

The gain constant K can be determined from the magnitude condition.

$$\left| K \frac{(s + 0.2)(s + 1.3332)}{s} \frac{1}{s^2 + 1} \right|_{s = -1 + j\sqrt{3}} = 1$$

or

$$K = 2.143$$

Then the characteristic equation becomes

$$s(s^2 + 1) + 2.143(s + 0.2)(s + 1.3332) = 0$$

or

$$s^3 + 2.143s^2 + 4.2856\,s + 0.5714 = 0$$

which can be factored as follows:

$$(s^2 + 2s + 4)(s + 0.1428) = 0$$

The three closed-loop poles are therefore located at

$$s = -1 + j\sqrt{3}, \qquad s = -1 - j\sqrt{3}, \qquad s = -0.1428$$

Figure 7–84 shows the root-locus diagram for the designed system. The closed-loop pole at $s = -0.1428$ is close to a zero at $s = -0.2$. Therefore, the effect of this closed-loop pole on the response will be relatively small. Although a zero at $s = -1.3332$ has some effect on the system response, the closed-loop poles at $s = -1 \pm j\sqrt{3}$ are the dominant closed-loop poles.

A–7–11. Consider the electronic circuit involving two operational amplifiers shown in Figure 7–85. This is a modified PID controller in that the transfer function involves an integrator and a first-order lag term. Obtain the transfer function of this PID controller.

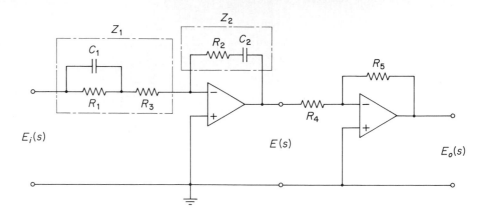

Figure 7–85
Modified PID
controller.

$E_i(s)$

$E(s)$

$E_o(s)$

Solution. Since

$$Z_1 = \frac{1}{\frac{1}{R_1} + C_1 s} + R_3 = \frac{R_1 + R_3 + R_1 R_3 C_1 s}{1 + R_1 C_1 s}$$

and

$$Z_2 = R_2 + \frac{1}{C_2 s}$$

we have

$$\frac{E(s)}{E_i(s)} = -\frac{Z_2}{Z_1} = -\frac{(R_2 C_2 s + 1)(R_1 C_1 s + 1)}{C_2 s (R_1 + R_3 + R_1 R_3 C_1 s)}$$

Also

$$\frac{E_o(s)}{E(s)} = -\frac{R_5}{R_4}$$

Consequently,

$$\frac{E_o(s)}{E_i(s)} = \frac{E_o(s)}{E(s)} \frac{E(s)}{E_i(s)} = \frac{R_5}{R_4(R_1 + R_3)C_2} \frac{(R_1 C_1 s + 1)(R_2 C_2 s + 1)}{s\left(\dfrac{R_1 R_3}{R_1 + R_3} C_1 s + 1\right)}$$

$$= \frac{R_5 R_2}{R_4 R_3} \frac{\left(s + \dfrac{1}{R_1 C_1}\right)\left(s + \dfrac{1}{R_2 C_2}\right)}{s\left(s + \dfrac{R_1 + R_3}{R_1 R_3 C_1}\right)}$$

Notice that $R_1 C_1$ and $R_2 C_2$ determine the locations of the zeros of the controller, while R_3 affects the location of the pole on the negative real axis. R_5/R_4 adjust the gain of the controller.

A–7–12. Consider the electronic PID controller shown in Figure 7–46. Determine the values of R_1, R_2, R_3, R_4, C_1, and C_2 of the controller such that the transfer function $G_c(s)$ is

$$G_c(s) = 39.42\left(1 + \frac{1}{3.077s} + 0.7692s\right) = 30.3215 \frac{(s + 0.65)^2}{s} \qquad (7\text{–}30)$$

Example Problems and Solutions

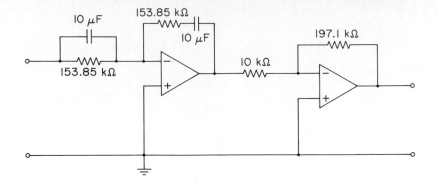

Figure 7–86
Electronic PID controller having the transfer function given by Eq. (7–30).

Solution. Referring to Equation (7–15), we have

$$K_p = \frac{R_4(R_1C_1 + R_2C_2)}{R_3R_1C_2} = 39.42$$

$$T_i = R_1C_1 + R_2C_2 = 3.077$$

$$T_d = \frac{R_1C_1R_2C_2}{R_1C_1 + R_2C_2} = 0.7692$$

First, notice that

$$(R_1C_1) + (R_2C_2) = 3.077$$

$$(R_1C_1)(R_2C_2) = 0.7692 \times 3.077 = 2.3668$$

Hence we obtain

$$R_1C_1 = 1.5385, \qquad R_2C_2 = 1.5385$$

Since we have six unknown variables and three equations, we can choose three variables arbitrarily. So we choose $C_1 = C_2 = 10\ \mu F$ and one remaining variable later. Then we get

$$R_1 = R_2 = 153.85\ k\Omega$$

From the equation for K_p, we have

$$\frac{R_4}{R_3} \frac{R_1C_1 + R_2C_2}{R_1C_2} = 39.42$$

or

$$\frac{R_4}{R_3} = 39.42 \times \frac{1}{2} = 19.71$$

We now choose arbitrarily $R_3 = 10\ k\Omega$. Then $R_4 = 197.1\ k\Omega$. The designed PID controller is shown in Figure 7–86.

A–7–13. Consider the system shown in Figure 7–87, which involves velocity feedback. Determine the values of the amplifier gain K and the velocity feedback gain K_h so that the following specifications are satisfied:

1. Damping ratio of the closed-loop poles is 0.5
2. Settling time ≤ 2 sec
3. Static velocity error constant $K_v \geq 50\ \text{sec}^{-1}$
4. $0 < K_h < 1$

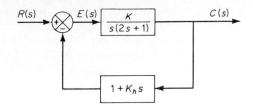

Figure 7–87
Control system.

Solution. The specification on the damping ratio requires that the closed-loop poles lie on lines that make $\pm 60°$ with the negative real axis. The specification on the settling time can be written in terms of the real part of the complex-conjugate closed-loop poles as

$$t_s = \frac{4}{\sigma} \le 2 \text{ sec}$$

or

$$\sigma \ge 2$$

Hence, the closed-loop poles must lie on the heavy lines AB and CD in the left-half s plane, as shown in Figure 7–88.

Since the velocity error constant K_v is defined by

$$K_v = \lim_{s \to 0} sG(s)H(s)$$

we obtain

$$K_v = \lim_{s \to 0} \frac{sK(1 + K_h s)}{s(2s + 1)} = K$$

From the given specification on the velocity error constant, we obtain

$$K \ge 50$$

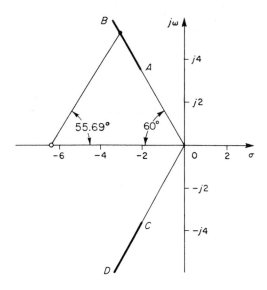

Figure 7–88
Possible location for the closed-loop poles in the s plane for the system of Problem A–7–13.

For this system, the open-loop poles are located at $s = 0$ and $s = -\frac{1}{2}$. The open-loop zero is located at $s = -1/K_h$, where K_h is an undetermined constant. First, let the closed-loop poles be at $s = -2 \pm j3.464$ (points A and C in Figure 7–88). The sum of the angles at the chosen closed-loop pole location with the open-loop poles is $120° + 113.41° = 233.41°$. Thus, we need a $53.41°$ contribution from the zero so that the total sum is $-180°$. To satisfy the angle condition, we choose the zero to be at $s = -4.572$. Then K_h is determined as

$$K_h = \frac{1}{4.572} = 0.2187$$

The magnitude condition requires that

$$\left| \frac{K(1 + 0.2187s)}{s(2s + 1)} \right|_{s = -2 + j3.464} = 1$$

Hence

$$K = 32.00$$

Since $K < 50$, the choice of the closed-loop poles at $s = -2 \pm j3.464$ is not acceptable.

As a second trial, let the closed-loop poles be at $s = -3 \pm j5.196$. The sum of the angle contributions from the open-loop poles is $235.69°$. We need a $55.69°$ contribution from the zero. This implies that the zero must be at $s = -6.546$. With this zero, the magnitude condition yields $K = 71.99$. This is quite satisfactory. Since $K_h = 1/6.546 = 0.1528$, the requirement on K_h is satisfied. Thus, all the given specifications are met. Hence a set of acceptable values of K and K_h is

$$K = 71.99, \qquad K_h = 0.1528$$

It is noted that there are infinitely many solutions for such a problem. Addition of constraints will narrow possible solutions.

A–7–14. A closed loop system has the characteristic that the closed-loop transfer function is nearly equal to the inverse of the feedback transfer function whenever the open-loop gain is much greater than unity.

The open-loop characteristic may be modified by adding an internal feedback loop with a characteristic equal to the inverse of the desired open-loop characteristic. Suppose that a unity-feedback system has the open-loop transfer function

$$G(s) = \frac{K}{(T_1 s + 1)(T_2 s + 1)}$$

Determine the transfer function $H(s)$ of the element in the internal feedback loop so that the inner loop becomes ineffective at both low and high frequencies.

Solution. Figure 7–89(a) shows the original system. Figure 7–89(b) shows the addition of the internal feedback loop around $G(s)$. Since

$$\frac{C(s)}{E(s)} = \frac{G(s)}{1 + G(s)H(s)} = \frac{1}{H(s)} \frac{G(s)H(s)}{1 + G(s)H(s)}$$

if the gain around the inner loop is large compared with unity, then $G(s)H(s)/[1 + G(s)H(s)]$ is approximately equal to unity, and the transfer function $C(s)/E(s)$ is approximately equal to $1/H(s)$.

On the other hand, if the gain $G(s)H(s)$ is much less than unity, the inner loop becomes ineffective and $C(s)/E(s)$ becomes approximately equal to $G(s)$.

To make the inner loop ineffective at both the low- and high-frequency ranges, we require that

$$G(j\omega)H(j\omega) \ll 1 \qquad \text{for } \omega \ll 1 \text{ and } \omega \gg 1$$

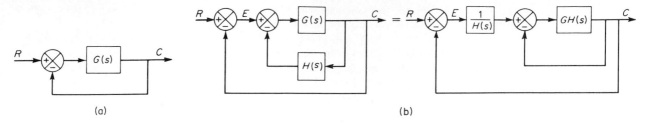

(a)   (b)

Figure 7–89 (a) Control system; (b) addition of the internal feedback loop to modify the closed-loop characteristic.

Since in this problem

$$G(j\omega) = \frac{K}{(1 + j\omega T_1)(1 + j\omega T_2)}$$

the requirement can be satisfied if $H(s)$ is chosen to be

$$H(s) = ks$$

because

$$\lim_{\omega \to 0} G(j\omega)H(j\omega) = \lim_{\omega \to 0} \frac{Kkj\omega}{(1 + j\omega T_1)(1 + j\omega T_2)} = 0$$

$$\lim_{\omega \to \infty} G(j\omega)H(j\omega) = \lim_{\omega \to \infty} \frac{Kkj\omega}{(1 + j\omega T_1)(1 + j\omega T_2)} = 0$$

Thus, with $H(s) = ks$ (velocity feedback), the inner loop becomes ineffective at both the low- and high-frequency regions. It becomes effective only in the intermediate-frequency region.

A–7–15. When a disturbance acts on a plant, it takes some time before any effect on the output can be detected. If we measure the disturbance itself (though this may not be possible or may be quite difficult) rather than the response to the disturbance, then a corrective action can be taken sooner, and we can expect a better result. Figure 7–90 is a block diagram showing feedforward compensation for the disturbance.

   Discuss the limitations of the disturbance-feedforward scheme in general. Then discuss the advantages and limitations of the scheme shown in Figure 7–90.

Figure 7–90
Control system with feedforward compensation for the disturbance.

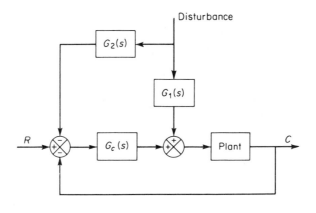

Solution. A disturbance-feedforward scheme is an open-loop scheme and thus depends on the constancy of the parameter values. Any drift in these values will result in imperfect compensation.

In the present system, both open-loop and closed-loop schemes are simultaneously in operation. Large errors due to the main disturbance source can be greatly reduced by the open-loop compensation without requiring a high loop gain. Smaller errors due to other disturbance sources can be taken care of by the closed-loop control scheme. Hence errors from all causes can be reduced without requiring a large loop gain. This is an advantage from the stability viewpoint.

Note that such a scheme cannot be used unless the main disturbance itself can be measured.

A–7–16. In some cases it is desirable to provide an input filter as shown in Figure 7–91(a). Notice that the input filter $G_f(s)$ is outside the loop. Therefore, it does not affect the stability of the closed-loop portion of the system. An advantage of having the input filter is that the zeros of the closed-loop transfer function can be modified (canceled or replaced by other zeros) so that the closed-loop response is acceptable.

Show that the configuration shown in Figure 7–91(a) can be modified to that shown in Figure 7–91(b), where $G_d(s) = [G_f(s) - 1] G_c(s)$.

Solution. For the system of Figure 7–91(a), we have

$$\frac{C(s)}{R(s)} = G_f(s) \frac{G_c(s)G_p(s)}{1 + G_c(s)G_p(s)} \tag{7-31}$$

For the system of Figure 7–91(b), we have

$$U(s) = G_d(s)R(s) + G_c(s)E(s)$$

$$E(s) = R(s) - C(s)$$

$$C(s) = G_p(s)U(s)$$

Thus

$$C(s) = G_p(s) \{G_d(s)R(s) + G_c(s) [R(s) - C(s)]\}$$

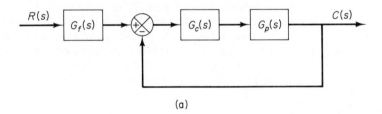

(a)

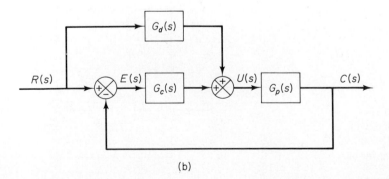

(b)

Figure 7–91
(a) Block diagram of control system with input filter; (b) modified block diagram.

or

$$\frac{C(s)}{R(s)} = \frac{[G_d(s) + G_c(s)]G_p(s)}{1 + G_c(s)G_p(s)} \tag{7-32}$$

By substituting $G_d(s) = [G_f(s) - 1]G_c(s)$ into Equation (7–32), we obtain

$$\frac{C(s)}{R(s)} = \frac{[G_f(s)G_c(s) - G_c(s) + G_c(s)]\,G_p(s)}{1 + G_c(s)G_p(s)}$$

$$= G_f(s)\frac{G_c(s)G_p(s)}{1 + G_c(s)G_p(s)}$$

which is the same as Equation (7–31). Hence, we have shown that the systems shown in Figures 7–91(a) and (b) are equivalent.

It is noted that the system shown in Figure 7–91(b) has a feedforward controller $G_d(s)$. In such a case, $G_d(s)$ does not affect the stability of the closed-loop portion of the system.

A–7–17. Consider the system shown in Figure 7–92. This is a PID control of a second-order plant $G(s)$. Assume that disturbances $U_d(s)$ enter the system as shown in the diagram. It is assumed that the reference input $r(t)$ is normally held constant, and the response characteristics to disturbances are a very important consideration in this system.

Design a control system such that the response to any step disturbance be damped out quickly (in 2 to 3 sec in terms of the 2% settling time). Choose the configuration of the closed-loop poles such that there is a pair of dominant closed-loop poles. Then obtain the response to the unit-step disturbance input. Obtain also the response to the unit-step reference input.

Solution. The PID controller has the transfer function

$$G_c(s) = \frac{K(as + 1)(bs + 1)}{s}$$

For the disturbance input in the absence of the reference input, the closed-loop transfer function becomes

$$\frac{C_d(s)}{U_d(s)} = \frac{s}{s(s^2 + 3.6s + 9) + K(as + 1)(bs + 1)}$$

$$= \frac{s}{s^3 + (3.6 + Kab)s^2 + (9 + Ka + Kb)s + K} \tag{7-33}$$

The specification requires that the response to the unit-step disturbance be such that the settling time be 2 to 3 sec and the system have a reasonable damping. We may interpret the specification as $\zeta = 0.5$ and $\omega_n = 4$ rad/sec for the dominant closed-loop poles. We may choose the third pole at $s = -10$ so that the effect of this real pole on the response is small. Then the desired characteristic equation can be written as

$$(s + 10)(s^2 + 2 \times 0.5 \times 4s + 4^2) = (s + 10)(s^2 + 4s + 16) = s^3 + 14s^2 + 56s + 160$$

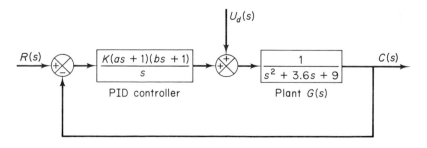

Figure 7–92
The PID-controlled system.

The characteristic equation for the system given by Equation (7–33) is

$$s^3 + (3.6 + Kab)s^2 + (9 + Ka + Kb)s + K = 0$$

Hence we require

$$3.6 + Kab = 14$$

$$9 + Ka + Kb = 56$$

$$K = 160$$

which yields

$$ab = 0.065, \quad a + b = 0.29375$$

The PID controller now becomes

$$G_c(s) = \frac{K[abs^2 + (a + b)s + 1]}{s}$$

$$= \frac{160(0.065s^2 + 0.29375s + 1)}{s}$$

$$= \frac{10.4(s^2 + 4.5192s + 15.385)}{s}$$

With this PID controller the response to the disturbance is given by

$$C_d(s) = \frac{s}{s^3 + 14s^2 + 56s + 160} U_d(s)$$

$$= \frac{s}{(s + 10)(s^2 + 4s + 16)} U_d(s)$$

Clearly, for a unit-step disturbance input, the steady-state output is zero, since

$$\lim_{t \to \infty} c_d(t) = \lim_{s \to 0} sC_d(s) = \lim_{s \to 0} \frac{s^2}{(s + 10)(s^2 + 4s + 16)} \frac{1}{s} = 0$$

The response to a unit-step disturbance is given by

$$C_d(s) = \frac{s}{(s + 10)(s^2 + 4s + 16)} \frac{1}{s}$$

$$= \frac{0.013158}{s + 10} + \frac{-0.013158s + 0.078947}{s^2 + 4s + 16}$$

$$= \frac{0.013158}{s + 10} - \frac{0.013158(s + 2)}{(s + 2)^2 + (\sqrt{12})^2} + \frac{0.03039\sqrt{12}}{(s + 2)^2 + (\sqrt{12})^2}$$

The inverse Laplace transform of this last equation gives

$$c_d(t) = 0.013158e^{-10t} - 0.013158e^{-2t} \cos \sqrt{12}\, t + 0.03039e^{-2t} \sin \sqrt{12}\, t$$

The unit-step disturbance response of this system is shown in Figure 7–93(a). (For a computer solution of the unit-step disturbance response, see Problem A–7–18.) The response curve shows that the settling time is approximately 2.7 sec. The response curve damps out quickly. Therefore, the system designed here is acceptable.

For the reference input $r(t)$, the closed-loop transfer function is

$$\frac{C_r(s)}{R(s)} = \frac{10.4(s^2 + 4.5192s + 15.385)}{s^3 + 14s^2 + 56s + 160} \tag{7–34}$$

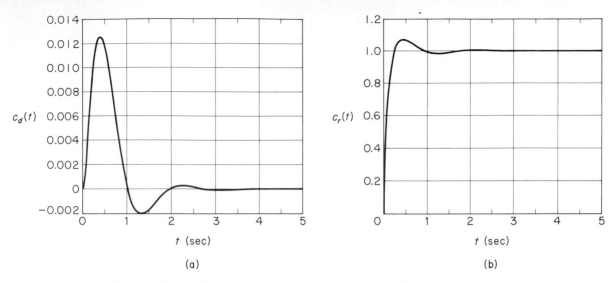

Figure 7–93 (a) Response to unit-step disturbance input; (b) response to unit-step reference input. (Problem A–7–17.)

From Equation (7–34), the response to a unit-step reference input is obtained as

$$C_r(s) = \frac{10.4(s^2 + 4.5192s + 15.385)}{s^3 + 14s^2 + 56s + 160} \frac{1}{s}$$

$$= \frac{10.4s^2 + 47s + 160}{s(s + 10)(s^2 + 4s + 16)}$$

$$= \frac{1}{s} - \frac{0.96053}{s + 10} - \frac{0.03947(s + 2)}{(s + 2)^2 + (\sqrt{12})^2} + \frac{0.20663\sqrt{12}}{(s + 2)^2 + (\sqrt{12})^2}$$

The inverse Laplace transform of this last equation is

$$c_r(t) = 1 - 0.96053e^{-10t} - 0.03947e^{-2t} \cos \sqrt{12}\, t + 0.20663e^{-2t} \sin \sqrt{12}\, t$$

The unit-step response of this system is shown in Figure 7–93(b). (For the computer solution of the response to the unit-step reference input, refer to Problem A–7–18.) This response shows that the maximum overshoot is 7.3% and the settling time is 1.2 sec. The response characteristic is quite acceptable.

A–7–18. Referring to Problem A–7–17, consider a computer simulation of the system and obtain the response to a unit-step disturbance input and that to a unit-step reference input. The response to a unit-step disturbance input is given by

$$C_d(s) = \frac{s}{s^3 + 14s^2 + 56s + 160} U_d(s) \tag{7–35}$$

The response to a unit-step reference input is given by

$$C_r(s) = \frac{10.4(s^2 + 4.5192s + 15.385)}{s^3 + 14s^2 + 56s + 160} R(s) \tag{7–36}$$

Write BASIC computer programs for the problem and obtain computer solutions for the responses.

Solution.

(1) *Response to a unit-step disturbance input*: The differential equation corresponding to Equation (7–35) is

$$\dddot{c}_d + 14\ddot{c}_d + 56\dot{c}_d + 160c_d = \dot{u}_d$$

Define $c_d = y$ and $u_d = u$. Then this last equation becomes

$$\dddot{y} + 14\ddot{y} + 56\dot{y} + 160y = \dot{u}$$

Comparing this equation with

$$\dddot{y} + a_1\ddot{y} + a_2\dot{y} + a_3y = b_0\dddot{u} + b_1\ddot{u} + b_2\dot{u} + b_3u$$

we find

$$a_1 = 14, \qquad a_2 = 56, \qquad a_3 = 160$$

$$b_0 = 0, \qquad b_1 = 0, \qquad b_2 = 1, \qquad b_3 = 0$$

Referring to Equation (2–5), we define state variables x_1, x_2, and x_3 by

$$x_1 = y - \beta_0 u$$

$$x_2 = \dot{y} - \beta_0\dot{u} - \beta_1 u = \dot{x}_1 - \beta_1 u$$

$$x_3 = \ddot{y} - \beta_0\ddot{u} - \beta_1\dot{u} - \beta_2 u = \dot{x}_2 - \beta_2 u$$

where β_0, β_1, and β_2 are determined from

$$\beta_0 = b_0 = 0$$

$$\beta_1 = b_1 - a_1\beta_0 = 0$$

$$\beta_2 = b_2 - a_1\beta_1 - a_2\beta_0 = 1$$

Thus, we have

$$x_1 = y$$

$$x_2 = \dot{x}_1$$

$$x_3 = \dot{x}_2 - u$$

Note that from Equation (2–7), we have

$$\dot{x}_3 = -a_3x_1 - a_2x_2 - a_1x_3 + \beta_3 u$$
$$= -160x_1 - 56x_2 - 14x_3 + \beta_3 u$$

where

$$\beta_3 = b_3 - a_1\beta_2 - a_2\beta_1 - a_3\beta_0 = -14$$

Thus, the state equation and output equation for the system for the case where the disturbance is the input are

$$\begin{bmatrix} \dot{x}_1 \\ \dot{x}_2 \\ \dot{x}_3 \end{bmatrix} = \begin{bmatrix} 0 & 1 & 0 \\ 0 & 0 & 1 \\ -160 & -56 & -14 \end{bmatrix} \begin{bmatrix} x_1 \\ x_2 \\ x_3 \end{bmatrix} + \begin{bmatrix} 0 \\ 1 \\ -14 \end{bmatrix} u \qquad (7\text{–}37)$$

and

$$y = \begin{bmatrix} 1 & 0 & 0 \end{bmatrix} \begin{bmatrix} x_1 \\ x_2 \\ x_3 \end{bmatrix}$$

A BASIC computer program for finding the unit-step response is given in Table 7–4. The response curve $c_d(t)$ versus t [which is $x_1(t)$ versus t where u is set equal to 1] was shown in Figure 7–93(a).

(2) *Response to a unit-step reference input*: The differential equation corresponding to Equation (7–36) is

$$\dddot{c}_r + 14\ddot{c}_r + 56\dot{c}_r + 160c_r = 10.4\ddot{r} + 47\dot{r} + 160r$$

Define $c_r = y$ and $r = u$. Then this last equation becomes

$$\dddot{y} + 14\ddot{y} + 56\dot{y} + 160y = 10.4\ddot{u} + 47\dot{u} + 160u$$

Comparing this equation with

$$\dddot{y} + a_1\ddot{y} + a_2\dot{y} + a_3y = b_0\dddot{u} + b_1\ddot{u} + b_2\dot{u} + b_3u$$

we find

$$a_1 = 14, \qquad a_2 = 56, \qquad a_3 = 160$$

$$b_0 = 0, \qquad b_1 = 10.4, \qquad b_2 = 47, \qquad b_3 = 160$$

Referring to Equation (2–5), we define state variables x_1, x_2, and x_3 by

$$x_1 = y - \beta_0 u$$

$$x_2 = \dot{y} - \beta_0\dot{u} - \beta_1 u = \dot{x}_1 - \beta_1 u$$

$$x_3 = \ddot{y} - \beta_0\ddot{u} - \beta_1\dot{u} - \beta_2 u = \dot{x}_2 - \beta_2 u$$

where β_0, β_1, and β_2 are determined from

$$\beta_0 = b_0 = 0$$

$$\beta_1 = b_1 - a_1\beta_0 = 10.4$$

$$\beta_2 = b_2 - a_1\beta_1 - a_2\beta_0 = 47 - 14 \times 10.4 = -98.6$$

Thus, we have

$$x_1 = y$$

$$x_2 = \dot{x}_1 - 10.4u$$

$$x_3 = \dot{x}_2 + 98.6u$$

Referring to Equation (2–7), we have

$$\dot{x}_3 = -a_3x_1 - a_2x_2 - a_1x_3 + \beta_3 u$$

$$= -160x_1 - 56x_2 - 14x_3 + \beta_3 u$$

where

$$\beta_3 = b_3 - a_1\beta_2 - a_2\beta_1 - a_3\beta_0 = 160 + 14 \times 98.6 - 56 \times 10.4$$

$$= 958$$

Example Problems and Solutions

Table 7–4 BASIC Computer Program for Solving Equation (7–37): Response to a Unit-Step Disturbance Input

```
  10 ORDER = 3
  20 X(1) = 0
  30 X(2) = 0
  40 X(3) = 0
  50 H = .02
  60 T = 0
  70 TK = 0
  80 TF = 5
  90 OPEN "O", #1, "ANS.BAS"
 100 PRINT "     TIME          X(1)          X(2)          X(3)     "
 110 PRINT "-----------------------------------------------------"
 120 PRINT #1, USING "####.######"; X(1)
 130 PRINT USING "####.######"; T, X(1), X(2), X(3)
 140 IF T > TF THEN GOTO 5000
 150 GOSUB 1000
 160 GOTO 120
1000 TK = T
1010 GOSUB 2000
1020 FOR I = 1 TO ORDER
1030 XK(I) = X(I)
1040 K(1,I) = DX(I)
1050 T = TK + H/2
1060 X(I) = XK(I) + (H/2)*K(1,I)
1070 NEXT I
1080 GOSUB 2000
1090 FOR I = 1 TO ORDER
1100 K(2,I) = DX(I)
1110 X(I) = XK(I) + (H/2)*K(2,I)
1120 NEXT I
1130 GOSUB 2000
1140 FOR I = 1 TO ORDER
1150 K(3,I) = DX(I)
1160 T = TK + H
1170 X(I) = XK(I) + H*K(3,I)
1180 NEXT I
1190 GOSUB 2000
1200 FOR I = 1 TO ORDER
1210 K(4,I) = DX(I)
1220 X(I) = XK(I) + H/6*(K(1,I) + 2*K(2,I) + 2*K(3,I) + K(4,I))
1230 NEXT I
1240 RETURN
2000 DX(1) = X(2)
2010 DX(2) = X(3) + 1
2020 DX(3) = - 160*X(1) - 56*X(2) - 14*X(3) - 14
2030 RETURN
4900 CLOSE #1
5000 END
```

The state equation and output equation for the system for the case where the reference input is the input to the system are

$$\begin{bmatrix} \dot{x}_1 \\ \dot{x}_2 \\ \dot{x}_3 \end{bmatrix} = \begin{bmatrix} 0 & 1 & 0 \\ 0 & 0 & 1 \\ -160 & -56 & -14 \end{bmatrix} \begin{bmatrix} x_1 \\ x_2 \\ x_3 \end{bmatrix} + \begin{bmatrix} 10.4 \\ -98.6 \\ 958 \end{bmatrix} u \qquad (7-38)$$

and

$$y = \begin{bmatrix} 1 & 0 & 0 \end{bmatrix} \begin{bmatrix} x_1 \\ x_2 \\ x_3 \end{bmatrix}$$

A BASIC computer program for solving Equation (7–38) (where $u = 1$) is the same as that given in Table 7–4, except that lines 2000, 2010, and 2020 be changed as follows:

2000 DX(1) = X(2) + 10.4

2010 DX(2) = X(3) − 98.6

2020 DX(3) = −160*X(1) − 56*X(2) − 14*X(3) + 958

The response curve $c_r(t)$ versus t [which is $x_1(t)$ versus t] was shown in Figure 7–93(b).

PROBLEMS

B–7–1. Draw Bode diagrams of the lead network and lag network shown in Figures 7–94(a) and (b), respectively.

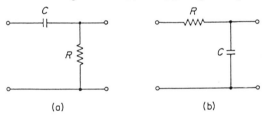

(a)   (b)

Figure 7–94 (a) Lead network; (b) lag network.

B–7–2. Consider a unity-feedback control system whose feedforward transfer function is given by

$$G(s) = \frac{K}{s(s + 1)(s + 2)(s + 3)}$$

Determine the value of K so that the dominant closed-loop poles have a damping ratio of 0.5.

B–7–3. Consider a unity-feedback system whose feedforward transfer function is given by

$$G(s) = \frac{1}{s^2}$$

It is desired to insert a series compensator so that the open-loop frequency-response curve is tangent to the $M = 3$ db circle at $\omega = 3$ rad/sec. The system is subjected to high-frequency noises and sharp cutoff is desired. Design an appropriate series compensator.

B–7–4. Determine the values of K, T_1, and T_2 of the system shown in Figure 7–95 so that the dominant closed-loop poles have $\zeta = 0.5$ and $\omega_n = 3$ rad/sec.

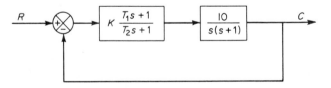

Figure 7–95 Control system.

B–7–5. A control system with

$$G(s) = \frac{K}{s^2(s + 1)}, \qquad H(s) = 1$$

is unstable for all positive values of the gain K.

Sketch the root-locus plot of the system. By using this plot, show that this system can be stabilized by adding a zero on the negative real axis or by modifying $G(s)$ to $G_1(s)$, where

$$G_1(s) = \frac{K(s + a)}{s^2(s + 1)} \qquad (0 \leq a < 1)$$

B–7–6. Referring to the closed-loop system shown in Figure 7–96, design a lead compensator $G_c(s)$ such that the phase margin is 45°, gain margin is not less than 8 db, and the velocity error constant K_v is 4.0 $\sec^{-1}$.

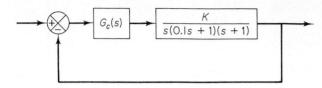

Figure 7–96 Closed-loop system

B–7–7. Figure 7–97 shows a block diagram of a space vehicle attitude control system. Determine the proportional gain constant K_p and derivative time T_d such that the bandwidth of the closed-loop system is 0.4 to 0.5 rad/sec. (Note that the closed-loop bandwidth is close to the gain crossover frequency.) The system must have an adequate phase margin. Plot both the open-loop and closed-loop frequency response curves on Bode diagrams.

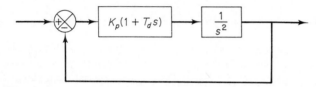

Figure 7–97 Block diagram of space vehicle attitude control system.

B–7–8. Consider the space vehicle control system shown in Figure 7–98. Design a lead compensator $G_c(s)$ such that the damping ratio ζ and the undamped natural frequency ω_n of the dominant closed-loop poles are 0.5 and 0.2 rad/sec, respectively,

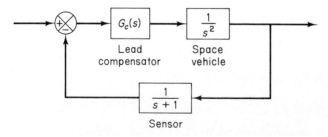

Figure 7–98 Space vehicle control system.

B–7–9. Consider the control system shown in Figure 7–99. Design a lag compensator $G_c(s)$ such that the static velocity error constant K_v is 50 $\sec^{-1}$ without appreciably changing the

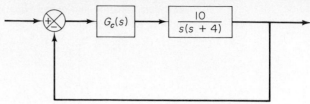

Figure 7–99 Control system.

locations of the original closed-loop poles, which are at $s = -2 \pm j\sqrt{6}$.

B–7–10. Figure 7–100 is the block diagram of an attitude-rate control system. Design a compensator $G_c(s)$ so that the dominant closed-loop poles are at $s = -2 \pm j2$.

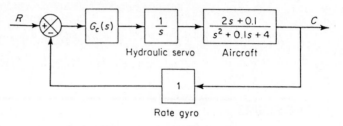

Figure 7–100 Attitude-rate control system.

B–7–11. Consider a unity-feedback control system whose feedforward transfer function is given by

$$G(s) = \frac{10}{s(s + 2)(s + 8)}$$

Design a compensator so that the static velocity error constant K_v is equal to 80 $\sec^{-1}$ and the dominant closed-loop poles are located at $s = -2 \pm j2\sqrt{3}$.

B–7–12. Draw Bode diagrams of the PI controller given by

$$G_c(s) = 5\left(1 + \frac{1}{2s}\right)$$

and the PD controller given by

$$G_c(s) = 5(1 + 0.5s)$$

B–7–13. Consider the modified PID controller shown in Figure 7–85. Show that its transfer function can be given by

$$G_c(s) = K_p + \frac{K_i}{s} + \frac{K_d s}{1 + as} \qquad (a > 0)$$

Note that the derivative control action differentiates the signal and amplifies noise effects. The pole at $s = -1/a$ added to the derivative control term smoothes rapid changes at the output of the differentiator.

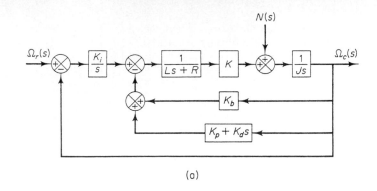

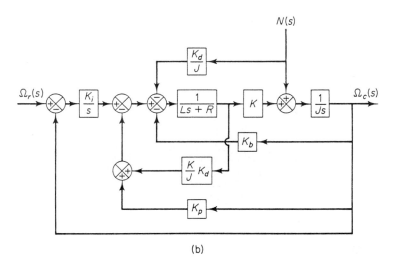

Figure 7–101 (a) I-PD control of speed of a dc motor system; (b) modified block diagram.

B–7–14. Referring to Problem A–7–12, draw a Bode diagram of

$$G_c(s) = 30.3215 \frac{(s + 0.65)^2}{s}$$

B–7–15. Figure 7–101(a) shows an I-PD control of the speed of a dc motor system. Assuming that the reference input is zero, or $\Omega_r = 0$, obtain the closed-loop transfer function for the disturbance input $N(s)$, or $\Omega_c(s)/N(s)$. Then show that the block diagram of Figure 7–101(a) can be modified to that shown in Figure 7–101(b). Notice that there is a feedforward control path for the disturbance input $N(s)$. (In the diagrams, K_b is the back emf constant.)

B–7–16. Consider the system shown in Figure 7–102. If the disturbance N can be detected, it may be passed through a transfer function G_3 and added to the feedforward path between

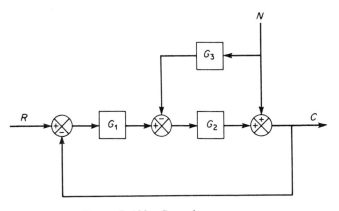

Figure 7–102 Control system.

the amplifier and the plant, as shown in Figure 7–102. To reduce the effect of this disturbance N on the steady-state error, determine an appropriate transfer function G_3. What will limit the present approach in reducing the effects of this disturbance?

B–7–17. Referring to Example 7–8, it is desired to obtain a computer solution for the unit-step response of the following system:

$$\frac{C(s)}{R(s)} = \frac{6.3223s^2 + 18s + 12.811}{s^4 + 6s^3 + 11.3223s^2 + 18s + 12.811}$$

To obtain a computer solution, we need to obtain a state-space representation for the system.

Define $c = y$ and $r = u$. Show that the state and output equations for the system may be given by

$$\dot{x}_1 = x_2 + \beta_1 u$$

$$\dot{x}_2 = x_3 + \beta_2 u$$

$$\dot{x}_3 = x_4 + \beta_3 u$$

$$\dot{x}_4 = -12.811x_1 - 18x_2 - 11.3223x_3 - 6x_4 + \beta_4 u$$

and

$$y = x_1 + \beta_0 u$$

where

$$\beta_0 = 0, \qquad \beta_1 = 0, \qquad \beta_2 = 6.3223,$$
$$\beta_3 = -19.9338, \qquad \beta_4 = 60.8308$$

Thus, in the standard vector-matrix notation, the state equation and output equation for the system can be given by

$$\begin{bmatrix} \dot{x}_1 \\ \dot{x}_2 \\ \dot{x}_3 \\ \dot{x}_4 \end{bmatrix} = \begin{bmatrix} 0 & 1 & 0 & 0 \\ 0 & 0 & 1 & 0 \\ 0 & 0 & 0 & 1 \\ -12.811 & -18 & -11.3223 & -6 \end{bmatrix} \begin{bmatrix} x_1 \\ x_2 \\ x_3 \\ x_4 \end{bmatrix}$$
$$+ \begin{bmatrix} 0 \\ 6.3223 \\ -19.9338 \\ 60.8308 \end{bmatrix} u$$

$$y = \begin{bmatrix} 1 & 0 & 0 & 0 \end{bmatrix} \begin{bmatrix} x_1 \\ x_2 \\ x_3 \\ x_4 \end{bmatrix}$$

Simulate the state equation on the computer and obtain a computer solution for the unit-step response of the system. Compare the computer solution with the unit-step response curve shown in Figure 7–56. [Note that, for a unit-step input, set $u = 1$. Then $x_1(t)$ versus t gives the unit-step response $c(t)$ versus t of the original system.]

B–7–18. In Problem A–7–8, we designed a lag–lead compensator for the system shown in Figure 7–82. The designed system has the open-loop transfer function

$$G_c(s)G(s) = \frac{250(s + 0.16025)}{s(s + 0.01)(s + 5)(s + 16.025)}$$

The system has unity feedback. Write a computer program and obtain a computer solution of the unit-step response of the system.

CHAPTER 8

Describing-Function Analysis of Nonlinear Control Systems

8–1 INTRODUCTION TO NONLINEAR SYSTEMS

It is a well-known fact that many relationships among physical quantities are not quite linear, although they are often approximated by linear equations mainly for mathematical simplicity. This simplification may be satisfactory as long as the resulting solutions are in agreement with experimental results. One of the most important characteristics of nonlinear systems is the dependence of the system response behavior on the magnitude and type of the input. For example, a nonlinear system may behave completely differently in response to step inputs of different magnitudes.

As pointed out in Chapter 2, nonlinear systems differ from linear systems greatly in that the principle of superposition does not hold for the former. Nonlinear systems exhibit many phenomena that cannot be seen in linear systems, and in investigating such systems we must be familiar with these phenomena.

In this section, we shall present a brief discussion of several of them.

Frequency-amplitude dependence. Consider the free oscillation of the mechanical system shown in Figure 8–1, which consists of a mass, viscous damper, and nonlinear spring. The differential equation describing the dynamics of this system may be written

$$m\ddot{x} + b\dot{x} + kx + k'x^3 = 0 \qquad (8\text{--}1)$$

where $kx + k'x^3$ = nonlinear spring force
$\qquad x$ = displacement of mass
$\qquad m$ = mass
$\qquad b$ = viscous-friction coefficient of damper

The parameters m, b, and k are positive constants, while k' may be either positive or negative. If k' is positive, the spring is called a hard spring; if k' is negative, a soft spring. The degree of nonlinearity of the system is characterized by the magnitude of k'. This nonlinear differential equation, Equation (8–1), is known as Duffing's equation and has often been discussed in the field of nonlinear mechanics. The solution of Equation (8–1) represents a damped oscillation if the system is subjected to a nonzero initial condition. In an experimental investigation, it is observed that as the amplitude decreases the frequency of the free oscillation either decreases or increases, depending on whether $k' > 0$ or $k' < 0$, respectively. When $k' = 0$, the frequency remains unchanged as the amplitude of the free oscillation decreases. (This corresponds to a linear system.) These characteristics are shown in Figure 8–2, which displays the waveforms of free oscillations. Figure 8–3 depicts frequency-amplitude relationships for the three cases where k' is greater than, equal to, and less than zero.

In an experimental study of nonlinear systems, the frequency-amplitude dependence can easily be detected. The frequency-amplitude dependence is one of the most fundamental characteristics of the oscillations of nonlinear systems. A graph of the form shown in Figure 8–3 reveals whether or not a nonlinearity is present and also indicates the degree of the nonlinearity.

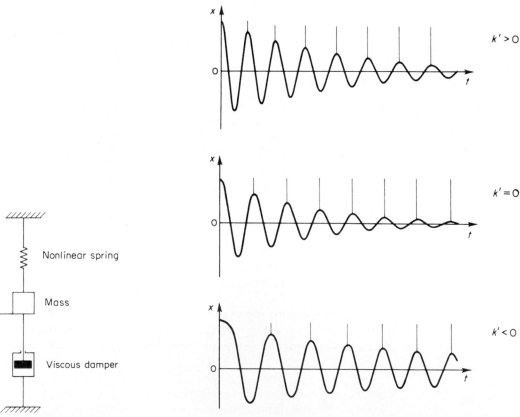

Figure 8–1
Mechanical system.

Figure 8–2 Waveforms of free oscillations in the system described by Eq. (8–1).

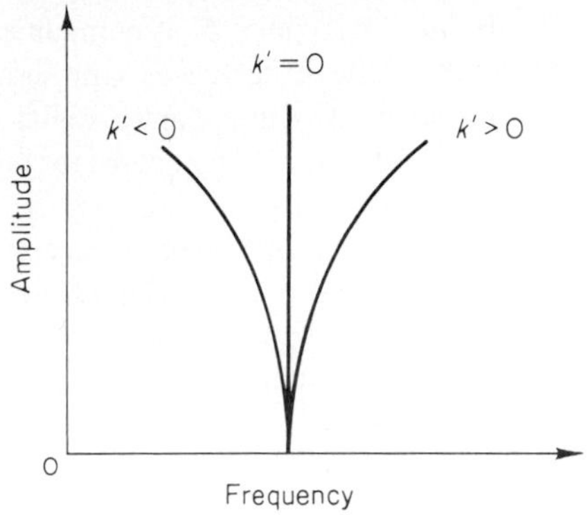

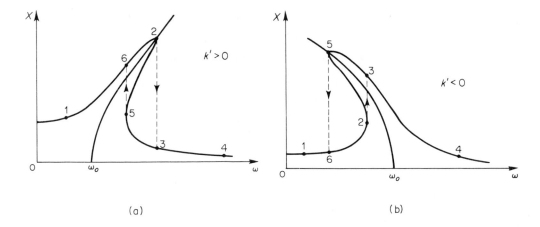

Figure 8–3
Amplitude versus frequency curves for free oscillations in the system described by Eq. (8–1).

Multivalued responses and jump resonances.   In carrying out experiments on the forced oscillations of the system shown in Figure 8–1, the differential equation of which is

$$m\ddot{x} + b\dot{x} + kx + k'x^3 = P \cos \omega t$$

where $P \cos \omega t$ is the forcing function, we may observe a number of phenomena, such as multivalued responses, jump resonances, and a variety of periodic motions (such as sub-harmonic oscillations and superharmonic oscillations). These phenomena do not occur in the responses of linear systems.

In carrying out experiments in which the amplitude P of the forcing function is held constant, while its frequency is varied slowly and the amplitude X of the response is observed, we may obtain a frequency-response curve similar to those shown in Figures 8–4(a) and (b). Suppose that $k' > 0$ and that the forcing frequency ω is low at the start at point 1 on the curve of Figure 8–4(a). As the frequency ω is increased, the amplitude X increases until point 2 is reached. A further increase in the frequency ω will cause a jump from point 2 to point 3, with accompanying changes in amplitude and phase. This phenomenon is called a *jump resonance*. As the frequency ω is increased further, the amplitude X follows the curve from point 3 toward point 4. In performing the experiment in the other direction, that is, starting from a high frequency, we observe that as ω is decreased, the amplitude X slowly increases through point 3, until point 5 is reached. A further decrease in ω will cause another

Figure 8–4
Frequency response curves showing jump resonances. (a) Mechanical system with hard spring; (b) mechanical system with soft spring.

(a)   (b)

jump from point 5 to point 6, accompanied by changes in amplitude and phase. After this jump, the amplitude X decreases with ω and follows the curve from point 6 toward point 1. Thus, the response curves are actually discontinuous, and a representative point on the response curve follows different paths for increasing and decreasing frequencies. The response oscillations corresponding to the curve between point 2 and point 5 correspond to unstable oscillations, and they cannot be observed experimentally. Similar jumps take place in the case of a system with a soft spring ($k' < 0$), as shown in Figure 8–4(b). We thus see that for a given amplitude P of the forcing function there is a range of frequencies in which either of the two stable oscillations can occur. It is noted that for jump resonance to take place, it is necessary that the damping term be small and that the amplitude of the forcing function be large enough to drive the system into a region of appreciably nonlinear operation.

Subharmonic oscillations. By the term *subharmonic oscillation*, we mean a nonlinear steady-state oscillation whsoe frequency is an integral submultiple of the forcing frequency. An example of an output waveform undergoing subharmonic oscillation is shown in Figure 8–5, together with the input waveform. The generation of subharmonic oscillations depend on system parameters and initial conditions, as well as on the amplitude and frequency of the forcing function. By the phrase *dependence on initial conditions*, we mean that subharmonic oscillations are not self-starting. It is necessary to give some sort of kick, say a sudden change in the amplitude or frequency of the forcing function, to initiate such oscillations. Once subharmonic oscillations are excited, they may be quite stable over certain frequency ranges. If the frequency of the forcing function is changed to a new value, either the subharmonic oscillation disappears or the frequency of the subharmonic oscillation changes also, to a value which is ω/n, where ω is the forcing frequency and n is the order of the subharmonic oscillation. (Note that an oscillation whose frequency is one-half that of the forcing function may occur in a linear system, if a system parameter or parameters are changed periodically with time. A linear conservative system may also exhibit oscillations that look like the subharmonic oscillations of nonlinear systems, but the oscillations in linear systems are essentially different from subharmonic oscillations.)

Self-excited oscillations or limit cycles. Another phenomenon that is observed in certain nonlinear systems is a self-excited oscillation or limit cycle. Consider a system described by the following equation:

$$m\ddot{x} - b(1 - x^2)\dot{x} + kx = 0$$

where m, b, and k are positive quantities. This equation is called the Van der Pol equation. It is nonlinear in the damping term. Upon examining this term, we notice that for small

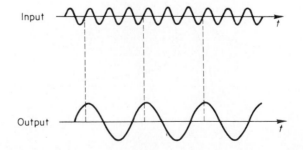

Figure 8–5
Input and output waveforms under subharmonic oscillation.

values of x the damping will be negative and will actually put energy into the system, while for large values of x it is positive and removes energy from the system. Thus, it can be expected that such a system may exhibit a sustained oscillation. Since it is not a forced system, this oscillation is called a self-excited oscillation or a limit cycle. Note that if a system possesses only one limit cycle, as in the case of the present system, the amplitude of this limit cycle does not depend on the initial condition.

Frequency entrainment. An example of an interesting phenomenon that may be observed in some nonlinear systems is frequency entrainment. If a periodic force of frequency ω is applied to a system capable of exhibiting a limit cycle of frequency ω_0, the well-known phenomenon of beats is observed. As the difference between the two frequencies decreases, the beat frequency also decreases. In a linear system, it is found, both experimentally and theoretically, that the beat frequency decreases indefinitely as ω approaches ω_0. In a self-excited nonlinear system, however, it is found experimentally that the frequency ω_0 of the limit cycle falls in synchronistically with, or is entrained by, the forcing frequency ω, within a certain band of frequencies. This phenomenon is usually called *frequency entrainment*, and the band of frequency in which entrainment occurs is called the *zone of frequency entrainment*.

Figure 8–6 shows the relationship between $|\omega - \omega_0|$ and ω. For a linear system, the relationship between $|\omega - \omega_0|$ and ω would follow the dotted lines, and $|\omega - \omega_0|$ would be zero for only one value of ω, that is, $\omega = \omega_0$. For a self-excited nonlinear system, entrainment of frequency occurs, and in the zone of frequency entrainment, which is indicated by the region $\Delta\omega$ in Figure 8–6, frequencies ω and ω_0 coalesce, and only one frequency ω exists. Such frequency entrainment is observed in the frequency response of nonlinear systems that exhibit limit cycles.

Asynchronous quenching. In a nonlinear system that exhibits a limit cycle of frequency ω_0, it is possible to quench the limit-cycle oscillation by forcing the system at a frequency ω_1, where ω_1 and ω_0 are not related to each other. This phenomenon is called *asynchronous quenching*, or *signal stabilization*.

Comment. None of the phenomena mentioned above, as well as other nonlinear phenomena not mentioned here, occur in linear systems. These phenomena cannot be explained by linear theory; to explain them, we must solve the nonlinear differential equations describing the dynamics of the system analytically or computationally.

Figure 8–6
$|\omega - \omega_0|$ versus ω curve showing zone of frequency entrainment.

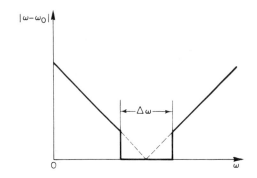

Outline of the chapter. Section 8–1 has presented an introduction to nonlinear systems and explained peculiar phenomena that may occur in many nonlinear systems. Section 8–2 gives a brief discussion of inherent and intentional nonlinearities. Section 8–3 defines the describing function and derives describing functions of a few commonly encountered nonlinearities. Section 8–4 treats the stability analysis of nonlinear control systems via the describing function approach. Section 8–5 gives a summary of describing function approach to the analysis and design of nonlinear control systems.

8–2 NONLINEAR CONTROL SYSTEMS

Many different types of nonlinearities may be found in practical control systems, and they may be divided into two classes, depending on whether they are inherent in the system or intentionally inserted into the system.

In what follows, we shall first discuss inherent nonlinearities and then intentional nonlinearities. Finally, we shall discuss approaches to the analysis and design of nonlinear control systems.

Inherent nonlinearities. Inherent nonlinearities are unavoidable in control systems. Examples of such nonlinearities are

1. Saturation
2. Dead zone
3. Hysteresis
4. Backlash
5. Static friction, coulomb friction, and other nonlinear friction
6. Nonlinear spring
7. Compressibility of fluid

Generally, the presence of such nonlinearities in the control system adversely affects system performance. For example, backlash may cause instability in the system, and dead zone will cause steady-state error.

Intentional nonlinearities. Some nonlinear elements are intentionally introduced into a system to improve system performance or to simplify the construction of the system, or both. A nonlinear system properly designed to perform a certain function is often superior from economic, weight, space, and reliability points of view to linear systems designed to perform the same task. The simplest example of such an intentionally nonlinear system is a conventional relay-operated system. Other examples may be found in optimal control systems that often employ complicated nonlinear controllers. It should be noted that, although intentional nonlinear elements may improve the system performance under certain specified operating conditions, they will, in general, degrade system performance under other operating conditions.

Effect of inherent nonlinearities on static accuracy. One characteristic of control systems is that power is transmitted through the feedforward path, while the static accuracy of the system is determined by the elements in the feedback path. Thus, the measuring device determines the uper limit on the static accuracy; the static accuracy cannot be better than the accuracy of this measuring device. Therefore, any inherent nonlinearities in the feedback elements should be kept to a minimum.

If feedback elements suffer from friction, backlash, and the like, then it is desirable to feed the error signal to an integrating device, because the system may not detect a very small error unless the small error is integrated continuously, making the magnitude large enough to be detected.

Approaches to the analysis and design of nonlinear control systems. There is no general method for dealing with all nonlinear systems because nonlinear differential equations are virtually devoid of a general method of attack. (Exact solutions can be found only for certain simple nonlinear differential equations. For many nonlinear differential equations of practical importance, only computational solutions are possible, and these solutions hold true only under the limited conditions for which they are obtained.) Because there is no general approach, we may take up each nonlinear equation, or group of similar equations, individually and attempt to develop a method of analysis that will satisfactorily apply to that particular group. (Note that although a very limited amount of generalization within the group of similar equations may be made a broad generalization of a particular solution is impossible.)

One way to analyze and design a particular group of nonlinear control systems, in which the degree of nonlinearity is small, is to use equivalent linearization techniques and to solve the resulting linearized problem. The describing-function method to be discussed in this chapter is one of the equivalent linearization methods. In many practical cases, we are primarily concerned with the stability of nonlinear control systems, and analytical solutions of nonlinear differential equations may not be necessary. (Establishing stability criteria is very much simpler than obtaining analytical solutions.) The describing-function method enables us to study the stability of many simple nonlinear control systems from a frequency-domain point of view. The describing-function method provides stability information for a system of any order, but it will not give exact information as to the time-response characteristics.

Another approach to the analysis and design of nonlinear control systems that will be given in this book is the second method of Liapunov. This method may be applied to stability analysis of any nonlinear system, but its application may be hampered because of difficulty in finding Liapunov functions for complicated nonlinear systems. (For Liapunov functions, see Chapter 9.)

Computer solutions of nonlinear problems. Modern computers have led to new methods for dealing with nonlinear problems. Comptuer simulation techniques by use of digital computers are very powerful for analyzing and designing nonlinear control systems. When the complexity of a system precludes the use of any analytical approach, computer simulations will be the only way to obtain necessary information for design purposes.

Comment. It is important to keep in mind that although the prediction of the behavior of nonlinear systems is usually difficult, in designing a control system we should not try to force the system to be as linear as possible, because the requirement of the system being linear may lead to the design of an expensive and less desirable system than a properly designed nonlinear one.

This section presents describing-function representations of commonly encountered nonlinear elements.

Describing functions.   Suppose that the input to a nonlinear element is sinusoidal. The output of the nonlinear element is, in general, not sinusoidal. Supose that the output is periodic with the same period as the input. (The output contains higher harmonics, in addition to the fundamental harmonic component.)

In the describing-function analysis, we assume that only the fundamental harmonic component of the output is significant. Such an assumption is often valid since the higher harmonics in the output of a nonlinear element are often of smaller amplitude than the amplitude of the fundamental harmonic component. In addition, most control systems are low-pass filters, with the result that the higher harmonics are very much attenuated compared with the fundamental harmonic component.

The describing function or sinusoidal describing function of a nonlinear element is defined to be the complex ratio of the fundamental harmonic component of the output to the input. That is,

$$N = \frac{Y_1}{X} \angle \phi_1$$

where N = describing function

X = amplitude of input sinusoid

Y_1 = amplitude of the fundamental harmonic component of output

ϕ_1 = phase shift of the fundamental harmonic component of output

If no energy-storage element is included in the nonlienar element, then N is a function only of the amplitude of the input to the element. On the other hand, if an energy-storage element is included, then N is a function of both the amplitude and frequency of the input.

In calculating the describing function for a given nonlinear element, we need to find the fundamental harmonic component of the output. For the sinusoidal input $x(t) = X \sin \omega t$ to the nonlinear element, the output $y(t)$ may be expressed as a Fourier series as

$$y(t) = A_0 + \sum_{n=1}^{\infty} (A_n \cos n\omega t + B_n \sin n\omega t)$$

$$= A_0 + \sum_{n=1}^{\infty} Y_n \sin (n\omega t + \phi_n)$$

where

$$A_n = \frac{1}{\pi} \int_0^{2\pi} y(t) \cos n\omega t \, d(\omega t)$$

$$B_n = \frac{1}{\pi} \int_0^{2\pi} y(t) \sin n\omega t \, d(\omega t)$$

$$Y_n = \sqrt{A_n^2 + B_n^2}$$

$$\phi_n = \tan^{-1} \left(\frac{A_n}{B_n} \right)$$

If the nonlinearity is skew symmetric, then $A_0 = 0$. The fundamental harmonic component of the output is

$$y_1(t) = A_1 \cos \omega t + B_1 \sin \omega t$$
$$= Y_1 \sin (\omega t + \phi_1)$$

The describing function is then given by

$$N = \frac{Y_1}{X} \angle \phi_1 = \frac{\sqrt{A_1^2 + B_1^2}}{X} \Big/ \tan^{-1} \left(\frac{A_1}{B_1} \right)$$

Clearly, N is a complex quantity when ϕ_1 is nonzero.

Table 8–1 shows three nonlinearities and their describing functions. (In Table 8–1, k_1, k_2, and k denote the slopes of the lines.) Illustrative calculations of describing functions for commonly encountered nonlinearities are given in what follows. (See also Problems A–8–2 through A–8–5.)

On–off nonlinearity. The on–off nonlinearity is often called a two-position nonlinearity. Consider the on–off element whose input–output characteristic curve is shown in Figure 8–7(a). The output of this element is either a positive constant or a negative constant. For a sinusoidal input, the output signal becomes a square wave. Figure 8–7(b) shows the input and output waveforms.

Table 8–1 Three Nonlinearities and Their Describing Functions

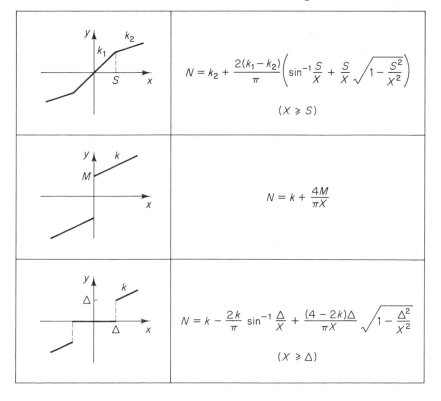

$$N = k_2 + \frac{2(k_1 - k_2)}{\pi} \left(\sin^{-1} \frac{S}{X} + \frac{S}{X} \sqrt{1 - \frac{S^2}{X^2}} \right)$$
$$(X \geqslant S)$$

$$N = k + \frac{4M}{\pi X}$$

$$N = k - \frac{2k}{\pi} \sin^{-1} \frac{\Delta}{X} + \frac{(4 - 2k)\Delta}{\pi X} \sqrt{1 - \frac{\Delta^2}{X^2}}$$
$$(X \geqslant \Delta)$$

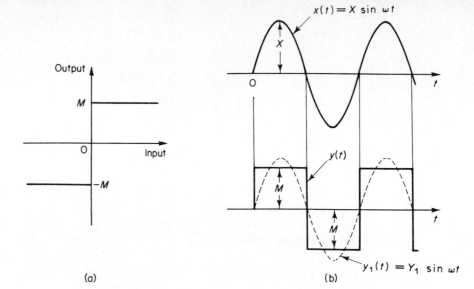

Figure 8–7
(a) Input–output characteristic curve for the on–off nonlinearity; (b) input and output waveforms for the on–off nonlinearity.

(a)   (b)

Let us obtain the Fourier series expansion of the output $y(t)$ of such an element.

$$y(t) = A_0 + \sum_{n=1}^{\infty}(A_n \cos n\omega t + B_n \sin n\omega t)$$

As shown in Figure 8–7(b), the output is an odd function. For any odd function, we have $A_n = 0$ $(n = 0, 1, 2, \ldots)$. Hence

$$y(t) = \sum_{n=1}^{\infty} B_n \sin n\omega t$$

The fundamental harmonic component of $y(t)$ is

$$y_1(t) = B_1 \sin \omega t = Y_1 \sin \omega t$$

where

$$Y_1 = \frac{1}{\pi}\int_0^{2\pi} y(t) \sin \omega t \, d(\omega t) = \frac{2}{\pi}\int_0^{\pi} y(t) \sin \omega t \, d(\omega t)$$

Substituting $y(t) = M$ into this last equation gives

$$Y_1 = \frac{2M}{\pi}\int_0^{\pi} \sin \omega t \, d(\omega t) = \frac{4M}{\pi}$$

Thus

$$y_1(t) = \frac{4M}{\pi}\sin \omega t$$

The describing function N is then given by

$$N = \frac{Y_1}{X}\angle 0° = \frac{4M}{\pi X}$$

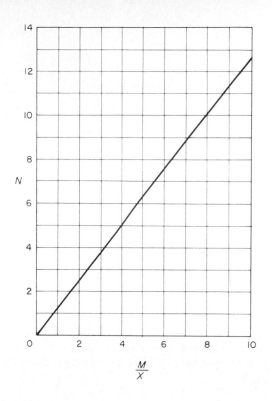

Figure 8–8
Describing function for the on–off nonlinearity.

Clearly, the describing function for the on–off element is a real quantity and is a function only of the input amplitude X. Such a nonlinearity is commonly called an amplitude-dependent nonlinearity. A plot of this describing function versus M/X is shown in Figure 8–8.

On–off nonlinearity with hysteresis. Consider the on–off element with hysteresis whose input–output characteristic curve is shown in Figure 8–9(a). For a sinusoidal input, the output signal becomes a square wave, with a phase lag of the amount $\omega t_1 = \sin^{-1}(h/X)$, as shown in Figure 8–9(b). Hence the describing function for this nonlinear element is

$$N = \frac{4M}{\pi X} \Big/ -\sin^{-1}\left(\frac{h}{X}\right)$$

It is convenient to plot

$$\frac{h}{M} N = \frac{4h}{\pi X} \Big/ -\sin^{-1}\left(\frac{h}{X}\right)$$

versus h/X rather than N versus h/X, because hN/M is a function only of h/X. A plot of hN/M versus h/X is shown in Figure 8–10.

Dead-zone nonlinearity. The dead-zone nonlinearity is sometimes referred to as the threshold nonlinearity. A typical input–output characteristic curve is shown in Figure 8–11(a). For a sinusoidal input, the output waveform becomes as shown in Figure 8–11(b). For the dead-zone element, there is no output for inputs within the dead-zone amplitude.

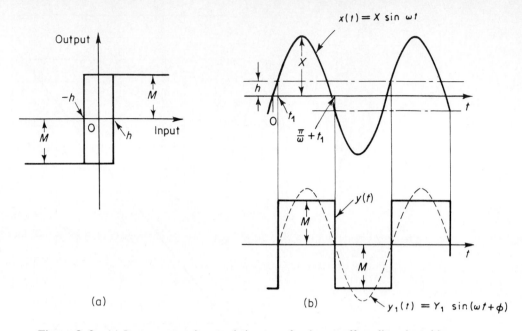

Figure 8–9 (a) Input–output characteristic curve for the on–off nonlinearity with hysteresis; (b) input and output waveforms for the on–off nonlinearity with hysteresis.

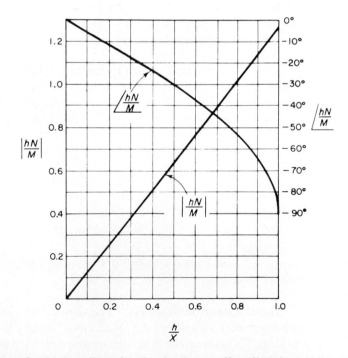

Figure 8–10 Describing function for the on–off nonlinearity with hysteresis.

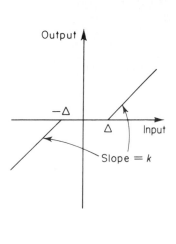

Output ▲

−Δ

Δ   Input

Slope = k

Figure 8–11
(a) Input–output characteristic curve for the dead-zone nonlinearity; (b) input and output waveforms for the dead-zone nonlinearity.

(a)

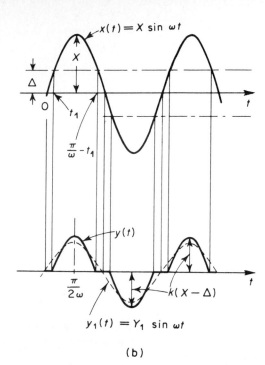

$x(t) = X \sin \omega t$

X

Δ

O t_1

$\dfrac{\pi}{\omega} - t_1$

$\dfrac{\pi}{2\omega}$

$y(t)$

$k(X - \Delta)$

$y_1(t) = Y_1 \sin \omega t$

(b)

For the element with dead-zone shown in Figure 8–11(a), the output $y(t)$ for $0 \le \omega t \le \pi$ is given by

$$y(t) = 0 \qquad\qquad \text{for } 0 < t < t_1$$

$$= k(X \sin \omega t - \Delta) \qquad \text{for } t_1 < t < \dfrac{\pi}{\omega} - t_1$$

$$= 0 \qquad\qquad \text{for } \dfrac{\pi}{\omega} - t_1 < t < \dfrac{\pi}{\omega}$$

Since the output $y(t)$ is once again an odd function, its Fourier series expansion has only sine terms. The fundamental harmonic component of the output is given by

$$y_1(t) = Y_1 \sin \omega t$$

where

$$Y_1 = \frac{1}{\pi} \int_0^{2\pi} y(t) \sin \omega t \, d(\omega t)$$

$$= \frac{4}{\pi} \int_0^{\pi/2} y(t) \sin \omega t \, d(\omega t)$$

$$= \frac{4k}{\pi} \int_{\omega t_1}^{\pi/2} (X \sin \omega t - \Delta) \sin \omega t \, d(\omega t)$$

Note that

$$\Delta = X \sin \omega t_1$$

or

$$\omega t_1 = \sin^{-1}\left(\frac{\Delta}{X}\right)$$

Hence

$$Y_1 = \frac{4Xk}{\pi}\left[\int_{\omega t_1}^{\pi/2}\sin^2\omega t \, d(\omega t) - \sin \omega t_1 \int_{\omega t_1}^{\pi/2}\sin \omega t \, d(\omega t)\right]$$

$$= \frac{2Xk}{\pi}\left[\frac{\pi}{2} - \sin^{-1}\left(\frac{\Delta}{X}\right) - \frac{\Delta}{X}\sqrt{1 - \left(\frac{\Delta}{X}\right)^2}\right]$$

The describing function for an element with dead zone can be obtained as

$$N = \frac{Y_1}{X}\angle 0°$$

$$= k - \frac{2k}{\pi}\left[\sin^{-1}\left(\frac{\Delta}{X}\right) + \frac{\Delta}{X}\sqrt{1 - \left(\frac{\Delta}{X}\right)^2}\right]$$

Figure 8–12 shows a plot of N/k as a function of Δ/X. Note that for $(\Delta/X) > 1$ the output is zero and the value of the describing function is also zero.

Saturation nonlinearity. A typical input–output characteristic curve for the saturation nonlinearity is shown in Figure 8–13(a). For small input signals, the output of a saturation element is proportional to the input. For larger input signals, the output will not increase proportionally, and for very large input signals, the output is constant at the maximum possible output value. For a sinusoidal input, the output waveform becomes as shown in Figure 8–13(b).

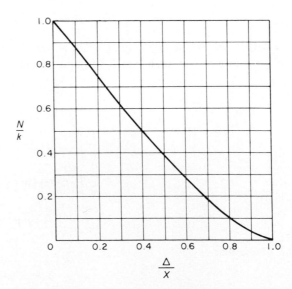

Figure 8–12
Describing function
for the dead-zone
nonlinearity.

Chapter 8 / Describing-Function Analysis of Nonlinear Control Systems

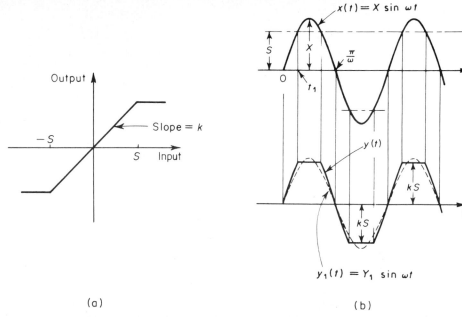

Figure 8–13
(a) Input–output
characteristic curve
for the saturation
nonlinearity; (b) input
and output waveforms
for the saturation
nonlinearity.

(a)

(b)

The describing function for the saturation nonlinearity can be obtained as

$$ N = \frac{2k}{\pi} \left[\sin^{-1} \left(\frac{S}{X} \right) + \frac{S}{X} \sqrt{1 - \left(\frac{S}{X} \right)^2} \right] $$

(See Problem A–8–4 for the derivation of this describing function.) Figure 8–14 shows a plot of N/k as a function of S/X. For $(S/X) > 1$, the value of the describing function is unity.

Comments.   If the describing function $N_{\text{saturation}}$ of the saturation nonlinearity and the describing function $N_{\text{dead zone}}$ of the dead zone nonlinearity are compared, the following

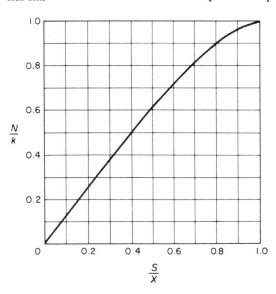

Figure 8–14
Describing function
for the saturation
nonlinearity.

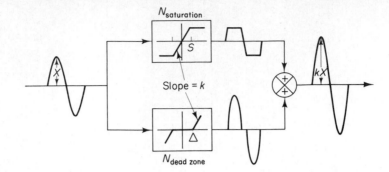

Figure 8–15
Diagram showing outputs of saturation nonlinearity and dead zone nonlinearity. ($\Delta = S$).

relationship may be found, provided that the dead-zone amplitude Δ in Figure 8–11 and the saturation amplitude S in Figure 8–13 are of the same magnitude, and the slope k of the saturation nonlinearity is equal to the slope of the dead zone nonlinearity:

$$N_{\text{dead zone}} = k - N_{\text{saturation}} \qquad \text{for } \Delta = S$$

This is not accidental, because, as may be seen from Figure 8–15, for a sinusoidal input the sum of the output of the saturation nonlinearity and that of the dead-zone nonlinearity is a sinusoidal curve with amplitude equal to kX, provided $\Delta = S$.

In general, calculations of describing functions are tedious. Therefore, time may be saved if the describing function of a nonlinearity can be obtained from a known describing function of another nonlinearity. It should be cautioned, however, that the principle of superposition does not apply to the nonlinear systems. Therefore, care must be exercised if we attempt to short-cut the derivation of the describing function of a nonlinearity by using a known describing function for another nonlinearity.

It is also noted that, if a system involves two nonlinear elements, two describing functions may be multiplied if these elements are separated by a low-pass filter as shown in Figure 8–16(a) such that the input to each nonlinear element is sinusoidal. However, if two nonlinear elements are not separated by a low-pass filter such that a sinusoidal input to the first nonlinear element does not produce a sinusoidal input to the second nonlinear element, as shown in Figure 8–16(b), then the multiplication of the two describing functions does not give a correct result. For such a case, a single describing function for the combined nonlinear elements must be obtained.

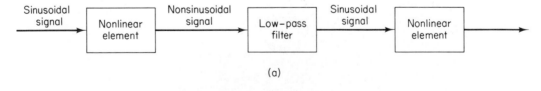

(a)

Figure 8–16
(a) Two nonlinear elements separated by a low-pass filter; (b) two nonlinear elements connected without a low-pass filter between them.

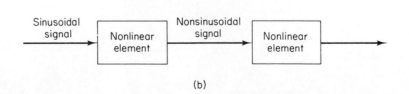

(b)

8-4 DESCRIBING-FUNCTION ANALYSIS OF NONLINEAR CONTROL SYSTEMS

Many control systems containing nonlinear elements may be represented by the block diagram shown in Figure 8–17. If the higher harmonics generated by the nonlinear element are sufficiently attenuated by the linear elements so that only the fundamental harmonic component of the output of the nonlinear element is significant, then the stability of the system can be predicted by a describing-function analysis.

Describing-function analysis. We shall first discuss how the describing functions of nonlinear elements can be used for stability analysis of nonlinear control systems. We shall show that if a sustained oscillation exists in the system output, then the amplitude and frequency of the oscillation may be determined from a graphical study in the frequency domain.

Consider the system shown in Figure 8–17, where N denotes the describing function of the nonlinear element. If the higher harmonics are sufficiently attenuated, the describing function N can be treated as a real variable or complex variable gain. Then the closed-loop frequency response becomes

$$\frac{C(j\omega)}{R(j\omega)} = \frac{NG(j\omega)}{1 + NG(j\omega)}$$

The characteristic equation is

$$1 + NG(j\omega) = 0$$

or

$$G(j\omega) = -\frac{1}{N} \tag{8-2}$$

If Equation (8–2) is satisfied, then the system output will exhibit a limit cycle. This situation corresponds to the case where the $G(j\omega)$ locus passes through the critical point. (In the conventional frequency-response analysis of linear control systems, the critical point is the $-1 + j0$ point.)

In the describing-function analysis, the conventional frequency-response analysis is modified so that the entire $-1/N$ locus becomes a locus of critical points. Thus, the relative location of the $-1/N$ locus and $G(j\omega)$ locus will provide the stability information.

To determine the stability of the system, we plot the $-1/N$ locus and the $G(j\omega)$ locus. In the present analysis, we assume that the linear part of the system is minimum phase or that all poles and zeros of $G(j\omega)$ lie in the left half of the s plane, including the $j\omega$ axis. The criterion for stability is that if the $-1/N$ locus is not enclosed by the $G(j\omega)$ locus, then the system is stable, or there is no limit cycle at steady state.

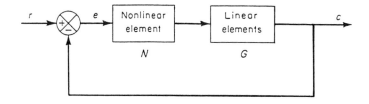

Figure 8–17
Nonlinear control system.

On the other hand, if the $-1/N$ locus is enclosed by the $G(j\omega)$ locus, then the system is unstable, and the system output when subjected to any disturbance will increase until breakdown occurs or increase to some limiting value determined by a mechanical stop or other safety device.

If the $-1/N$ locus and the $G(j\omega)$ locus intersect, then the system output may exhibit a sustained oscillation, or a *limit cycle*. Such a sustained oscillation is not sinusoidal, but it may be approximated by a sinusoidal one. The sustained oscillation is characterized by the value of X on the $-1/N$ locus and the value of ω on the $G(j\omega)$ locus at the intersection.

In general, a control system should not exhibit limit-cycle behavior, although a limit cycle of small amplitude may be acceptable in certain applications.

Stability of sustained oscillations, or limit cycles. The stability of the limit cycle can be predicted as follows: Consider the system shown in Figure 8–18. Assume that point A on the $-1/N$ locus corresponds to a small value of X, where X is the amplitude of the sinusoidal input signal to the nonlinear element, and that point B on the $-1/N$ locus corresponds to a large value of X. The value of X on the $-1/N$ locus increases in the direction from point A to point B.

Let us assume that the system is originally operated at point A. The oscillation has amplitude X_A and frequency ω_A, determined from the $-1/N$ locus and the $G(j\omega)$ locus, respectively. Assume that a slight disturbance is given to the system operating at point A so that the amplitude of the input to the nonlinear element is increased slightly. (For example, assume that the operating point moves from point A to point C on the $-1/N$ locus.) Then the operating point C corresponds to the critical point or to the $-1 + j0$ point in the complex plane for linear control systems. Therefore, as seen from Figure 8–18, the $G(j\omega)$ locus

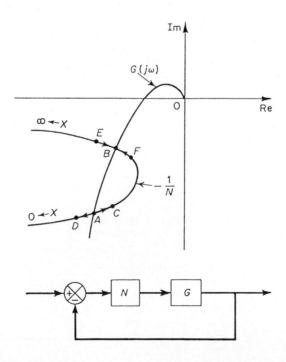

Figure 8–18
Stability analysis of limit-cycle operations of nonlinear control system.

Chapter 8 / Describing-Function Analysis of Nonlinear Control Systems encloses point C in the Nyquist sense. Since this is similar to the case where the open-loop locus of a linear system encloses the $-1 + j0$ point, the amplitude will increase and the operating point moves toward point B.

Next, suppose that a slight disturbance causes the operating point to move from point A to point D on the $-1/N$ locus. Point D then corresponds to the critical point. In this case, the $G(j\omega)$ locus does not enclose the critical point and, therefore, the amplitude of the input to the nonlinear element decreases, and the operating point moves further from point D to the left. Thus, point A possesses divergent characteristics and corresponds to an unstable limit cycle.

Consider next the case where a slight disturbance is given to the system operating at point B. Assume that the operating point is moved to point E on the $-1/N$ locus. Then the $G(j\omega)$ locus in this case does not enclose the critical point (point E). The amplitude of the sinusoidal input to the nonlinear element decreases, and the operating point moves toward point B.

Similarly, assume that a slight disturbance causes the system operating point to move from point B to point F. Then the $G(j\omega)$ locus will enclose the critical point (point F). Therefore, the amplitude of oscillation will increase, and the operating point moves from point F toward point B. Thus, point B possesses convergent characteristics, and the system operation at point B is stable; in other words, the limit cycle at this point is stable.

For the system shown in Figure 8–18, the stable limit cycle corresponding to point B can be experimentally observed, but the unstable limit cycle corresponding to point A cannot be.

Accuracy of describing-function analysis.   Note that the amplitude and frequency of the limit cycle indicated by the intersection of the $-1/N$ locus and $G(j\omega)$ locus are approximate values.

If the $-1/N$ locus and the $G(j\omega)$ locus intersect almost perpendicularly, then the accuracy of the describing-function analysis is generally good. (If the higher harmonics are all attenuated, the accuracy is excellent. Otherwise, the accuracy is good to fair.)

If the $G(j\omega)$ locus is tangent, or almost tangent, to the $-1/N$ locus, then the accuracy of the information obtained from the describing-function analysis depends on how well $G(j\omega)$ will attenuate the higher harmonics. In some cases, there is a sustained oscillation; in other cases, there is no such oscillation. It depends on the nature of $G(j\omega)$. One can say, however, that the system is almost on the verge of exhibiting a limit cycle when the $-1/N$ locus and $G(j\omega)$ locus are tangent to each other.

EXAMPLE 8–1

Figure 8–19 shows a control system with saturation nonlinearity. We assume that $G(j\omega)$ is a minimum phase transfer function. Figure 8–20(a) shows a plot of the $-1/N$ locus and the $G(j\omega)$ locus. The $-1/N$ locus starts from the -1 point on the negative real axis and extends to $-\infty$. Clearly, N is a function only of the amplitude of the input signal $x(t) = X \sin \omega t$. The $G(j\omega)$ locus is a function only

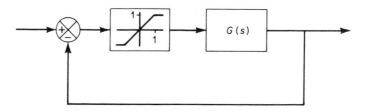

Figure 8–19
Control system with saturation nonlinearity.

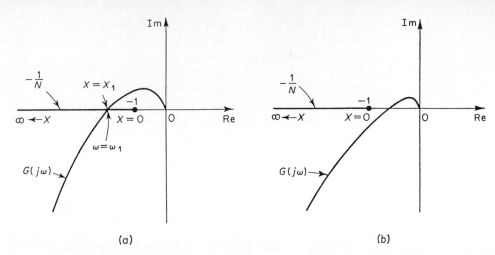

Figure 8–20
Plots of $-1/N$ and $G(j\omega)$ for stability analysis. (a) Limit cycle operation; (b) no limit cycle operation (stable operation).

of ω. The two loci may intersect as shown in Figure 8–20(a). The intersection corresponds to a stable limit cycle. The amplitude of the limit cycle is read from the $-1/N$ locus as $X = X_1$. The frequency of the limit cycle is read from the $G(j\omega)$ locus as $\omega = \omega_1$.

In the absence of any reference input, the output of this system at steady state exhibits a sustained oscillation with amplitude equal to X_1 and frequency equal to ω_1.

If the gain of the transfer function $G(s)$ is decreased so that the $-1/N$ locus and the $G(j\omega)$ locus do not intersect, as shown in Figure 8–20(b), then the system becomes stable, and any oscillations that may occur in the system output as a result of disturbances will die out and no sustained oscillation will exist at steady state. This is because the $-1/N$ locus is to the left of the $G(j\omega)$ locus, or the $G(j\omega)$ locus does not enclose the $-1/N$ locus.

EXAMPLE 8–2   Figure 8–21 shows a plot of the $-1/N$ locus for the dead-zone nonlinearity and the $G(j\omega)$ locus. In this system, the $-1/N$ locus and the $G(j\omega)$ locus intersect each other. The limit cycle in this case is unstable. The oscillation either dies out or increases in amplitude indefinitely. This indicates an undesirable situation and must be avoided.

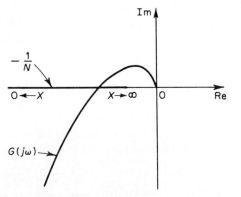

Figure 8–21
Plot of $-1/N$ and $G(j\omega)$ for stability analysis. (Dead zone nonlinearity.)

EXAMPLE 8–3   Consider the system shown in Figure 8–22. Determine the effect of hysteresis on the amplitude and frequency of the limit-cycle operation of the system.

The $-1/N$ loci for three different values of h, that is, $h = 0.1, 0.2$, and 0.3, are shown in Figure 8–23, together with the $G(j\omega)$ locus. The $-1/N$ loci are straight lines parallel to the real axis. The values of N are obtained from Figure 8–10.

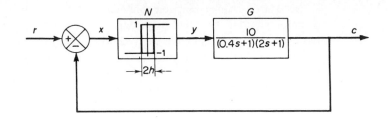

Figure 8–22
Nonlinear control system.

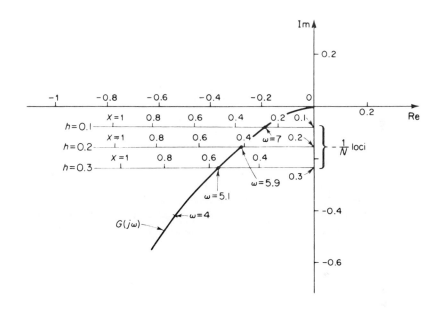

Figure 8–23
Plot of $-1/N$ and $G(j\omega)$ for the system shown in Figure 8–22.

From Figure 8–23, one can see that the amplitude and frequency of the limit cycles are

$$X = 0.27, \qquad \omega = 7 \qquad \text{if } h = 0.1$$

$$X = 0.42, \qquad \omega = 5.9 \qquad \text{if } h = 0.2$$

$$X = 0.57, \qquad \omega = 5.1 \qquad \text{if } h = 0.3$$

Inspection of these values reveals that increasing the hysteresis amplitude decreases the frequency but increases the amplitude of the limit cycle, as expected.

8–5 CONCLUDING COMMENTS

The describing-function analysis is an extension of linear techniques to the study of nonlinear systems. Therefore, typical applications are to systems with a low degree of nonlinearity. The use of describing functions in the analysis of high-degree nonlinear systems may lead to seriously erroneous results; this limits the applicability of the describing function to the analysis and design of low-degree nonlinear systems.

In concluding this chapter, we shall summarize the describing-function approach to the analysis and design of nonlinear control systems.

1. The describing-function method is an approximate method for determining the stability of unforced nonlinear control systems. In applying this method, we must keep in mind the basic assumptions and limitations. Although many practical control systems satisfy the basic assumptions of the describing function method, some systems do not. Therefore, it is always necessary to examine the validity of the method in each case.

2. In the describing-function analysis, the nature of the nonlinearity present in the system determines the complexity of the analysis. In other words, the linear elements of whatever order do not affect the complexity. It is an advantage of this method that the analysis is not materially complicated for systems with complex dynamics in their linear parts. The accuracy of the analysis is better for higher-order systems than for lower-order ones because higher-order systems generally have better low-pass filtering characteristics.

3. Although the describing-function method is quite useful in predicting the stability of unforced systems, it presents little information concerning transient-response characteristics.

4. The describing function method is convenient to apply to design problems. The use of describing functions enables us to apply frequency-response methods to the reshaping of the $G(j\omega)$ locus. The describing-function analysis is particularly useful when the designer wants a rough idea of the effects of certain nonlinearities or of the effects of modifying linear or nonlinear components within the loop. The analysis yields graphically information concerning the stability and suggests ways to improve the response characteristics, if necessary. When the $-1/N$ locus and the $G(j\omega)$ locus are plotted in the complex plane, the system performance can be estimated quickly from the plot. If any performance improvements are necessary, they may be accomplished by modifying the loci. This modification of the loci suggests the type of a suitable compensating network. [Design through describing-function analysis may suggest a particular $-1/N$ locus rather than suggesting the reshaping of the $G(j\omega)$ locus. However, the realization of a nonlinear element with a specified describing function may be difficult.] It is interesting to note that although the describing-function method may enable us to predict limit cycles with good engineering accuracy, in design problems the method is used as a negative criterion, in that system parameters are adjusted until limit-cycle conditions are eliminated and a proper relative stability is assured.

5. A physical system may possess two or more significant nonlinear elements. When only one nonlinear element becomes significant for a particular operating condition, effects of other nonlinear elements may be neglected in the analysis. For example, if the system has both small-signal nonlinearities and large-signal nonlinearities, the former may be neglected when the signal amplitude is large and vice versa. It is important to keep in mind that the describing function of two nonlinearities in series is, in general, not the same as the product of the individual describing functions. Therefore, if two or more nonlinear elements, which are not separated from each other by effective low-pass filters, become significant simultaneously under certain operating conditions, they must be combined into one block, and the equivalent describing function of this block must be obtained. In this case, the describing function may become both amplitude and frequency dependent.

6. In the usual describing-function analysis, the input to the nonlinearity is assumed to be sinusoidal, but this assumption may be extended. The input to the nonlinearity may be a sinusoidal input, plus an additional signal, although this additional complication may make the analysis very tedious. Describing functions corresponding to this case are called dual-input describing functions.

7. In some cases the stability analysis of control systems may be of primary concern, but in other cases an optimal response (in some sense) may be desired. Optimal design problems may involve the determination of a nonlinear controller (or computer) for insertion into the system. The performance of the nonlinear system depends markedly on the input signals. This means that precise descriptions of the input and desired output are necessary. Because of a weak correlation between the frequency response and time response of nonlinear systems, the describing-function approach ceases to be useful in the design of optimal control systems with aperiodic inputs.

Example Problems and Solutions

A–8–1. Figure 8–24(a) is a schematic diagram of a liquid-level control system. The movement of the float positions the electric mercury switch that energizes or deenergizes the electric solenoid-operated valve. When the valve is open, liquid is admitted to the tank. The control action is on–off with hysteresis. The inflow rate versus error curve is shown in Figure 8–24(b). (The hysteresis width $2h$ is called a *differential gap*.)

Plot the head versus time curves at steady-state operation under the following two conditions:

1. The rate of rise in head when the inflow valve is open is considerably smaller than the rate of fall in head when the valve is closed.
2. The rate of rise in head when the inflow valve is opened is considerably greater than the rate of fall in head when the valve is closed.

Solution. Assume that the liquid level is falling and that the inflow valve is closed. When the head falls to level AA' in Figure 8–24(a), the bottom of the differential gap, the mercury switch contacts will be closed and the inflow valve will be opened, admitting liquid to the tank. The head will start to rise as the inflow valve is opened. At the start there will be a high rate of rise. As the level in the tank rises, the outflow will increase due to the greater head. The result is a smaller net inflow to the tank. When the head rises to level BB' in Figure 8–24(a), the mercury switch contacts will be opened and the inflow valve will be closed. The head will then start to fall.

Figure 8–25(a) shows a head versus time curve under condition 1, when the rate of rise in head with the inflow valve open is considerably smaller than the rate of fall in head with the valve closed. Here the on time is considerably longer than the off time. Figure 8–25(b) shows a head versus time curve under condition 2. The on time is considerably shorter than the off time.

Under either condition, the level oscillates about the desired value. Thus the system exhibits limit-cycle behavior. In any case, the average inflow is equal to the average outflow. The inflow–outflow rates determine the shape of the level versus time curve. If the head rises at a rate that is equal to the fall with the valve closed, then the on time and off time will be equal.

Figure 8–24
(a) Liquid-level control system; (b) inflow-rate versus error curve of the controller.

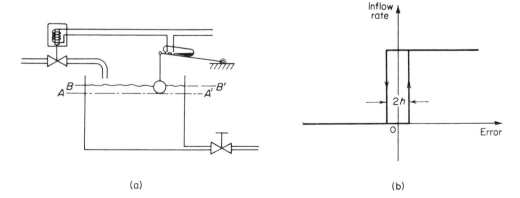

(a)

(b)

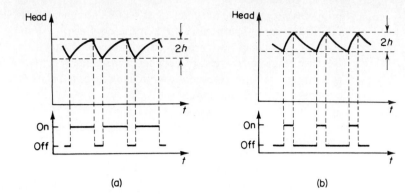

Figure 8–25
(a) Head versus time
curve under condition
1; (b) head versus
time curve under
condition 2.

(a)   (b)

Note that widening the gap will cause less frequent operation of the mercury switch and solenoid-operated valve. This normally means longer life for the equipment. The disadvantage of operating less frequently is that the variation in the head of the tank becomes larger.

Finally, it should be noted that on–off type control offers the best economy, the highest sensitivity, and ease of maintenance. Hence, if a limit cycle of small amplitude is permissible in a particular application, then the use of any other type of control would be a mistake.

A–8–2. Obtain the describing function for the on–off nonlinearity with dead zone shown in Figure 8–26.

Solution. Figure 8–27 shows the input and output waveforms for an element with the given nonlinearity. The output of the nonlinear element for $0 \leq \omega t \leq \pi$ is given by

$$y(t) = 0 \qquad \text{for } 0 < t < t_1$$

$$= M \qquad \text{for } t_1 < t < \frac{\pi}{\omega} - t_1$$

$$= 0 \qquad \text{for } \frac{\pi}{\omega} - t_1 < t < \pi$$

The output waveform is an odd function. Hence the fundamental harmonic component of the output $y(t)$ is given by

$$y_1(t) = Y_1 \sin \omega t$$

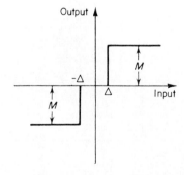

Figure 8–26
Input–output
characteristic curve
for the on–off
nonlinearity with dead
zone.

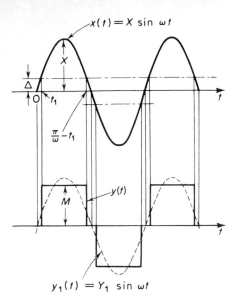

Figure 8–27 Input and output waveforms for the on–off nonlinearity with dead zone.

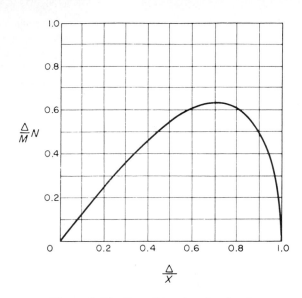

Figure 8–28 Describing function for the on–off nonlinearity with dead zone.

where

$$
\begin{aligned}
Y_1 &= \frac{1}{\pi} \int_0^{2\pi} y(t) \sin \omega t \, d(\omega t) \\
&= \frac{4}{\pi} \int_0^{\pi/2} y(t) \sin \omega t \, d(\omega t) \\
&= \frac{4}{\pi} \int_{\omega t_1}^{\pi/2} M \sin \omega t \, d(\omega t) \\
&= \frac{4M}{\pi} \cos \omega t_1
\end{aligned}
$$

Since $\sin \omega t_1 = \Delta/X$, we obtain

$$
\cos \omega t_1 = \sqrt{1 - \left(\frac{\Delta}{X}\right)^2}
$$

Thus

$$
y_1(t) = \frac{4M}{\pi} \sqrt{1 - \left(\frac{\Delta}{X}\right)^2} \sin \omega t
$$

The describing function for an on–off nonlinearity with dead zone is

$$
N = \frac{4M}{\pi X} \sqrt{1 - \left(\frac{\Delta}{X}\right)^2}
$$

The describing function in this case is a function only of the input amplitude X. Figure 8–28 shows a plot of $\Delta N/M$ versus Δ/X.

Example Problems and Solutions

A–8–3. Figure 8–29(a) shows the input–output characteristic curve for an on–off nonlinearity with dead zone and hysteresis. Figure 8–29(b) shows the input and output waveforms of a contactor with this nonlinearity.

The contactor does not close until the input exceeds the value $\Delta + h$. The contactor remains closed until the input becomes less than $\Delta - h$. In the region between $\Delta - h$ and $\Delta + h$, the output depends on the past history of the input. The contactor opens or closes in a similar fashion for a negative input.

Obtain the describing function for the contactor with this nonlinearity.

Solution. From Figure 8–29(b) we obtain the following equation:

$$y(t) = M \qquad \text{for } t_1 < t < \frac{\pi}{\omega} - t_2$$

$$= 0 \qquad \text{for } \frac{\pi}{\omega} - t_2 < t < \frac{\pi}{\omega} + t_1$$

where t_1 and t_2 are defined by

$$\sin \omega t_1 = \frac{\Delta + h}{X}$$

$$\operatorname{Sin} \omega t_2 = \frac{\Delta - h}{X}$$

As seen from Figure 8–29(b), the output lags the input. The fundamental harmonic component of the output can be obtained as follows:

$$y_1(t) = A_1 \cos \omega t + B_1 \sin \omega t$$

where

$$A_1 = \frac{2}{\pi} \int_0^{\pi} y(t) \cos \omega t \, d(\omega t)$$

$$= \frac{2}{\pi} \int_{\omega t_1}^{\pi - \omega t_2} M \cos \omega t \, d(\omega t)$$

$$= -\frac{4hM}{\pi X}$$

$$B_1 = \frac{2}{\pi} \int_0^{\pi} y(t) \sin \omega t \, d(\omega t)$$

$$= \frac{2}{\pi} \int_{\omega t_1}^{\pi - \omega t_2} M \sin \omega t \, d(\omega t)$$

$$= \frac{2M}{\pi} \left[\sqrt{1 - \left(\frac{\Delta - h}{X} \right)^2} + \sqrt{1 - \left(\frac{\Delta + h}{X} \right)^2} \right]$$

Let us define

$$h = \alpha \, \Delta, \qquad M = \beta \, \Delta$$

The values of α and β are constant for a given on–off nonlinearity with dead zone and hysteresis. Then

$$\frac{A_1}{X} = -\frac{4\alpha\beta}{\pi} \left(\frac{\Delta}{X} \right)^2$$

$$\frac{B_1}{X} = \frac{2\beta}{\pi} \frac{\Delta}{X} \left[\sqrt{1 - \left(\frac{\Delta}{X} \right)^2 (1 - \alpha)^2} + \sqrt{1 - \left(\frac{\Delta}{X} \right)^2 (1 + \alpha)^2} \right]$$

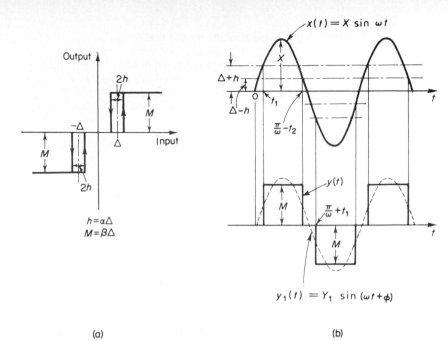

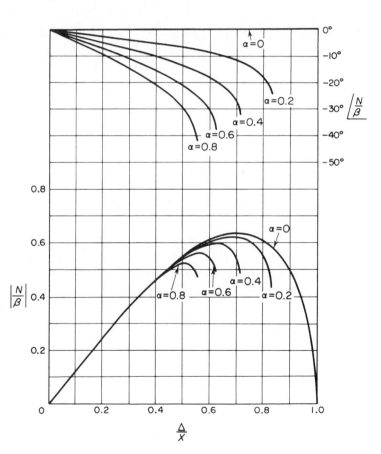

Figure 8–29
(a) Input–output characteristic curve for the on–off nonlinearity with dead zone and hysteresis; (b) input and output waveforms for the on–off nonlinearity with dead zone and hysteresis.

(a)

(b)

Figure 8–30
Describing function for the on–off nonlinearity with dead zone and hysteresis.

Thus the describing function for this nonlinearity is

$$N = \sqrt{\left(\frac{A_1}{X}\right)^2 + \left(\frac{B_1}{X}\right)^2} \bigg/ \underline{\tan^{-1}\left(\frac{A_1}{B_1}\right)}$$

This describing function is a complex quantity. Figure 8–30 shows plots of $|N/\beta|$ versus Δ/X and $\underline{/N/\beta}$ versus Δ/X.

A–8–4. Referring to the input–output characteristic curve for the saturation nonlinearity shown in Figure 8–13(a), obtain the describing function for this nonlinearity.

Solution. Referring to the characteristic curve, the gain of the linear region of the saturation nonlinearity is k. Since the characteristic curve is skew symmetric, the Fourier series expansion of $y(t)$ involves only odd harmonics. As seen from Figure 8–13(b), the input sinusoid $x(t) = X \sin \omega t$ and $y_1(t)$, the fundamental harmonic component, are in phase. Therefore, $y_1(t)$ can be written as

$$y_1(t) = Y_1 \sin \omega t$$

where

$$Y_1 = \frac{2}{\pi} \int_0^\pi y(t) \sin \omega t \, d(\omega t) \qquad (8\text{–}3)$$

From Figure 8–13(b), we have

$$y(t) = kX \sin \omega t \qquad \text{for } 0 \le t < t_1$$
$$= kS \qquad\qquad \text{for } t_1 \le t < \frac{\pi}{2\omega}$$

and

$$X \sin \omega t_1 = S$$

Hence, Equation (8–3) can be integrated to give

$$Y_1 = \frac{2Xk}{\pi}\left[\sin^{-1}\left(\frac{S}{X}\right) + \frac{S}{X}\sqrt{1 - \left(\frac{S}{X}\right)^2}\right]$$

The describing function N of the saturation nonlinearity can then be given by

$$N = \frac{Y_1}{X} = \frac{2k}{\pi}\left[\sin^{-1}\left(\frac{S}{X}\right) + \frac{S}{X}\sqrt{1 - \left(\frac{S}{X}\right)^2}\right]$$

Earlier, N/k was plotted versus S/X in Figure 8–14.

A–8–5. Figure 8–31(a) shows the input–output characteristic curve of the spring preload nonlinearity. Here the input corresponds to the displacement and the output corresponds to the force. Figure 8–31(b) shows $x(t) = X \sin \omega t$, the sinusoidal input to the nonlinearity, $y(t)$, the output from the nonlinearity, and $y_1(t)$, the fundamental harmonic component of the output $y(t)$. Obtain the describing function for this nonlinearity.

Solution. Notice that the output curve of the spring preload nonlinearity may be produced by summing up the output of a linear element with the unity gain and that of the on–off nonlinearity, as shown in Figure 8–32. (The output dead-zone amplitude M of the spring preload nonlinearity and the output amplitude M of the on–off nonlinearity are of the same magnitude.) Then the describing function of the spring preload nonlinearity may be obtained as

$$N_{\text{spring preload}} = 1 + N_{\text{on–off}}$$

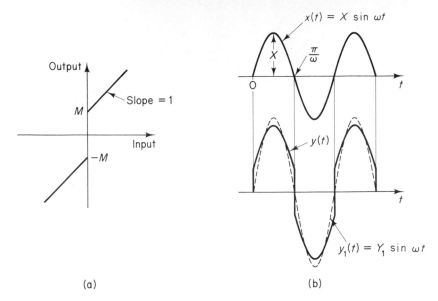

Figure 8–31
(a) Input–output characteristic curve for the spring preload nonlinearity; (b) input and output waveforms for the spring preload nonlinearity.

(a)

(b)

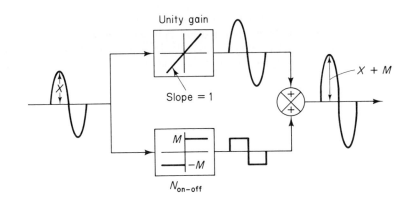

Figure 8–32
Diagram showing the output of spring preload nonlinearity equal to the sum of the outputs of linear element (unity gain) and on–off element.

The describing function $N_{\text{on–off}}$ of the on–off nonlinearity was obtained in Section 8–3 as

$$N_{\text{on–off}} = \frac{4}{\pi}\frac{M}{X}$$

Therefore, the describing function of the spring preload nonlinearity may be given as

$$N_{\text{spring preload}} = 1 + \frac{4}{\pi}\frac{M}{X}$$

A–8–6. The system shown in Figure 8–33 exhibits a limit cycle with the frequency of oscillation at 5.9 rad/ sec, as shown in Figure 8–34. It is desired to decrease the frequency of the limit cycle to 4 rad/sec. Determine the necessary change in the gain of $G(s)$, assuming that the nonlinear element is fixed.

Solution. From Figure 8–34, $\overline{OB}/\overline{OA}$ is found to be 0.36. Hence, if the value of the gain of $G(s)$ is decreased to 36% of the original value, the frequency of the new limit cycle will become 4 rad/sec. The amplitude also decreases, from 0.42 to 0.35.

Example Problems and Solutions

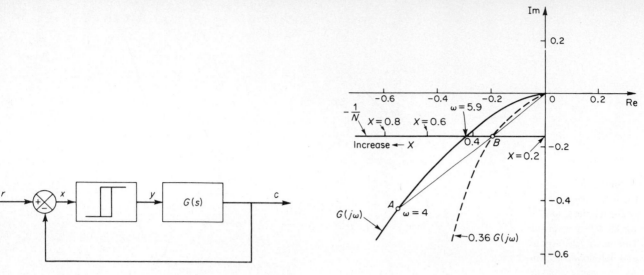

Figure 8–33 Nonlinear control system.

Figure 8–34 Plot of $-1/N$ and $G(j\omega)$.

A–8–7. Figure 8–35(a) shows a block diagram for a servo system consisting of an amplifier, a motor, a gear train, and a load. (Gear 2 shown in the diagram includes the load element.) The position of the output is fed back to the input to generate the error signal. It is assumed that the inertia of the gears and load element is negligible compared with that of the motor. It is also assumed that there is no backlash between the motor shaft and gear 1. Backlash exists between gear 1 and gear 2. The gear ratio between gear 1 and gear 2 is unity.

Because of backlash, the signals $x(t)$ and $y(t)$ are related as shown in Figure 8–35(b). Figure 8–35(c) shows the input–output characteristic curve for the backlash nonlinearity considered here. The describing function for this nonlinearity is plotted in Figure 8–36.

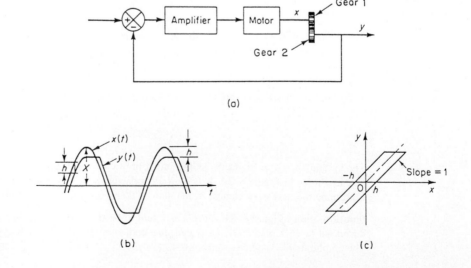

Figure 8–35
(a) Block diagram for a servo system with backlash; (b) characteristic curves for a backlash nonlinearity; (c) input–output characteristic curve for a backlash (or hysteresis) nonlinearity.

Using this describing function, determine the amplitude and frequency of the limit cycle when the transfer function of the amplifier–motor combination is given by

$$\frac{5}{s(s + 1)}$$

and the backlash amplitude is given as unity, or $h = 1$,

Solution. From the problem statement, the block diagram for the system may be drawn as shown in Figure 8–37. Information about the limit-cycle operation of the system may easily be obtained if a frequency-domain diagram is plotted.

Figure 8–38 shows a plot of the $-1/N$ locus and the $G(j\omega)$ locus on the log-magnitude versus phase diagram. As seen from the plot, there are two intersections of the two loci. Applying the stability test for the limit cycle discussed in Section 8–4 reveals that point A corresponds to a stable limit cycle and point B corresponds to an unstable limit cycle. The stable limit cycle has a frequency of 1.6 rad/sec and an amplitude of 2. (The unstable limit cycle cannot physically occur.)

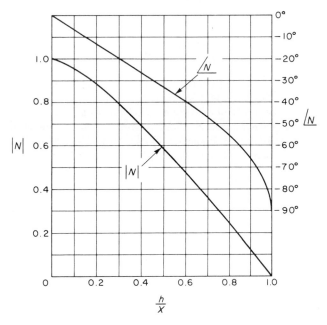

Figure 8–36 Describing function for a backlash nonlinearity, or a hysteresis nonlinearity shown in Fig. 8–35(c).

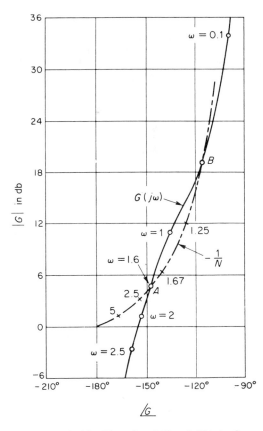

Figure 8–38 Plot of $-1/N$ and $G(j\omega)$ of the servo system shown in Fig. 8–37.

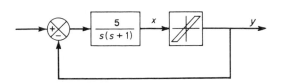

Figure 8–37 Block diagram representation for the servo system shown in Fig. 8–35(a).

To avoid limit-cycle behavior, the gain of the amplifier must be decreased sufficiently so that the $G(j\omega)$ locus lies well below the $-1/N$ locus in the log-magnitude versus phase diagram. If the two loci are tangent, or almost tangent, the accuracy of the describing-function analysis is poor, and we may expect a limit cycle or a slowly damped oscillation.

PROBLEMS

B–8–1. For the system shown in Figure 8–39, determine the amplitude and frequency of the limit cycle.

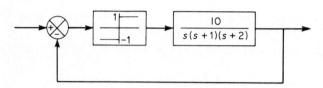

Figure 8–39 Nonlinear control system.

B–8–2. Determine the describing function for the nonlinear element described by

$$y = x^3$$

where x = input to the nonlinear element (sinusoidal signal)
y = output of the nonlinear element

B–8–3. Determine the stability of the system shown in Figure 8–40.

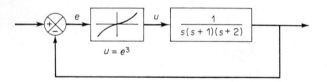

Figure 8–40 Nonlinear control system.

B–8–4. Determine the amplitude and frequency of the limit cycle of the system shown in Figure 8–41.

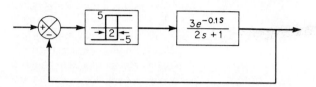

Figure 8–41 Nonlinear control system.

B–8–5. Derive the equation for the describing function N for the hysteresis nonlinearity shown in Figure 8–35(c).

B–8–6. Referring to the input and output waveforms for the saturation nonlinearity shown in Figure 8–13(b), obtain the amplitude of the third harmonic component of the output and plot Y_3/Y_1 as a function of S/X where Y_1 is the amplitude of the fundamental harmonic component and Y_3 is the amplitude of the third harmonic component.

B–8–7.* Consider a nonlinear element whose input–output characteristic is defined by

$$y = b_1 x + b_3 x^3 + b_5 x^5 + b_7 x^7 + \cdots$$

where x = input to the nonlinear element (sinusoidal signal)
y = output of the nonlinear element

Show that the describing function for this nonlinearity can be given by

$$N = b_1 + \tfrac{3}{4}b_3 X^2 + \tfrac{5}{8}b_5 X^4 + \tfrac{35}{64}b_7 X^6 + \cdots$$

where X is the amplitude of the input sinusoid $x = X \sin \omega t$.

* Reference W-1.

CHAPTER 9

Analysis of Control Systems in State Space

9–1 INTRODUCTION*

A modern complex system may have many inputs and many outputs, and these may be interrelated in a complicated manner. To analyze such a system, it is essential to reduce the complexity of the mathematical expressions, as well as to resort to computers for most of the tedious computations necessary in the analysis. The state-space approach to system analysis is best suited from this viewpoint.

While conventional control theory is based on the input–output relationship, or transfer function, modern control theory is based on the description of system equations in terms of n first-order differential equations, which may be combined into a first-order vector-matrix differential equation. The use of vector-matrix notation greatly simplifies the mathematical representation of systems of equations. The increase in the number of state variables, the number of inputs, or the number of outputs does not increase the complexity of the equations. In fact, the analysis of complicated multiple-input, multiple-output systems can be carried out by the procedures that are only slightly more complicated than those required for the analysis of systems of first-order scalar differential equations.

This chapter and the next chapter deal with state-space analysis and design of control systems. Basic materials of state-space analysis, including state-space representation of systems, controllability, observability, Liapunov stability analysis, and a brief discussion of

---

* It is noted that throughout this chapter and the next chapter an asterisk used as a superscript of matrix, such as $\mathbf{A}^*$, implies that it is a conjugate transpose of matrix $\mathbf{A}$. The conjugate transpose is the conjugate of the transpose of a matrix. For a real matrix (a matrix whose elements are all real), the conjugate transpose $\mathbf{A}^*$ is the same as the transpose $\mathbf{A}^T$.

linear time-varying systems, are presented in this chapter. Basic design methods in state space are presented in the next chapter.

Outline of the chapter.   Section 9–1 has presented an introduction to the state-space analysis of control systems. Section 9–2 deals with basic materials in state-space analysis, not covered in other sections. Section 9–3 treats the transfer matrix and discusses the design of controllers that have no cross-interaction among the inputs and outputs. Section 9–4 presents the concept of controllability and derives conditions for complete state controllability. Section 9–5 treats observability. Section 9–6 discusses state-space representation of transfer-function systems in controllable, observable, diagonal, or Jordan canonical form. Section 9–7 presents Liapunov stability analysis of linear and nonlinear systems, and Section 9–8 deals with Liapunov stability analysis of linear time-invariant systems. Finally, Section 9–9 gives the solution of state-space equations involving time-varying terms.

The materials presented in this chapter, in particular, concepts of controllability and observability and Liapunov stability analysis, serve as the basis for the design of control systems and are used extensively in Chapter 10.

9–2   BASIC MATERIALS IN STATE-SPACE ANALYSIS

In this section we shall discuss basic materials of state-space analysis, including such subjects as nonuniqueness of sets of state variables, eigenvalues of $n \times n$ matrices, invariance of eigenvalues, diagonalization of $n \times n$ matrices, state-space representation of nth-order systems with r forcing functions, Cayley–Hamilton theorem, and computation of $e^{\mathbf{A}t}$.

Nonuniqueness of set of state variables.   It has been stated that a set of state variables is not unique for a given system. Suppose that $x_1, x_2, \ldots, x_n$ are a set of state variables. Then we may take as another set of state variables any set of functions

$$\hat{x}_1 = X_1(x_1, x_2, \ldots, x_n)$$
$$\hat{x}_2 = X_2(x_1, x_2, \ldots, x_n)$$
$$\cdot$$
$$\cdot$$
$$\cdot$$
$$\hat{x}_n = X_n(x_1, x_2, \ldots, x_n)$$

provided that, for every set of values $\hat{x}_1, \hat{x}_2, \ldots, \hat{x}_n$, there corresponds a unique set of values $x_1 x_2, \ldots, x_n$, and vice versa. Thus, if $\mathbf{x}$ is a state vector, then $\hat{\mathbf{x}}$ where

$$\hat{\mathbf{x}} = \mathbf{P}\mathbf{x}$$

is also a state vector, provided the matrix $\mathbf{P}$ is nonsingular. Different state vectors convey the same information about the system behavior.

EXAMPLE 9–1   Consider the system defined by

$$\dddot{y} + 6\ddot{y} + 11\dot{y} + 6y = 6u \tag{9–1}$$

where y is the output and u is the input of the system. Obtain a state-space representation of the system.

Let us choose the state variables as

$$x_1 = y$$

$$x_2 = \dot{y}$$

$$x_3 = \ddot{y}$$

Then we obtain

$$\dot{x}_1 = x_2$$

$$\dot{x}_2 = x_3$$

$$\dot{x}_3 = -6x_1 - 11x_2 - 6x_3 + 6u$$

The last of these three equations was obtained by solving the original differential equation for the highest derivative term $\ddot{y}$ and then substituting $y = x_1$, $\dot{y} = x_2$, $\ddot{y} = x_3$ into the resulting equation. By use of vector-matrix notation, these three first-order differential equations can be combined into one, as follows:

$$\begin{bmatrix} \dot{x}_1 \\ \dot{x}_2 \\ \dot{x}_3 \end{bmatrix} = \begin{bmatrix} 0 & 1 & 0 \\ 0 & 0 & 1 \\ -6 & -11 & -6 \end{bmatrix} \begin{bmatrix} x_1 \\ x_2 \\ x_3 \end{bmatrix} + \begin{bmatrix} 0 \\ 0 \\ 6 \end{bmatrix} u \qquad (9\text{--}2)$$

The output equation is given by

$$y = \begin{bmatrix} 1 & 0 & 0 \end{bmatrix} \begin{bmatrix} x_1 \\ x_2 \\ x_3 \end{bmatrix} \qquad (9\text{--}3)$$

Equations (9–2) and (9–3) can be put in a standard form as

$$\dot{\mathbf{x}} = \mathbf{A}\mathbf{x} + \mathbf{B}u \qquad (9\text{--}4)$$

$$y = \mathbf{C}\mathbf{x} \qquad (9\text{--}5)$$

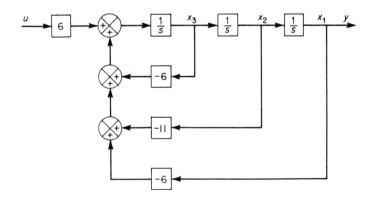

Figure 9–1
Block diagram representation of the system defined by Eqs. (9–2) and (9–3).

where

$$\mathbf{A} = \begin{bmatrix} 0 & 1 & 0 \\ 0 & 0 & 1 \\ -6 & -11 & -6 \end{bmatrix}, \quad \mathbf{B} = \begin{bmatrix} 0 \\ 0 \\ 6 \end{bmatrix}, \quad \mathbf{C} = [1 \quad 0 \quad 0]$$

Figure 9–1 shows the block-diagram representation of the present state equation and output equation. Notice that the transfer functions of the feedback blocks are identical with the negatives of the coefficients of the original differential equation, Equation (9–1).

Eigenvalues of an $n \times n$ matrix A. The eigenvalues of an $n \times n$ matrix $\mathbf{A}$ are the roots of the characteristic equation

$$|\lambda \mathbf{I} - \mathbf{A}| = 0$$

The eigenvalues are sometimes called the characteristic roots.

Consider, for example, the following matrix $\mathbf{A}$:

$$\mathbf{A} = \begin{bmatrix} 0 & 1 & 0 \\ 0 & 0 & 1 \\ -6 & -11 & -6 \end{bmatrix}$$

The characteristic equation is

$$\begin{aligned} |\lambda \mathbf{I} - \mathbf{A}| &= \begin{vmatrix} \lambda & -1 & 0 \\ 0 & \lambda & -1 \\ 6 & 11 & \lambda + 6 \end{vmatrix} \\ &= \lambda^3 + 6\lambda^2 + 11\lambda + 6 \\ &= (\lambda + 1)(\lambda + 2)(\lambda + 3) = 0 \end{aligned}$$

The eigenvalues of $\mathbf{A}$ are the roots of the characteristic equation, or -1, -2, and -3.

EXAMPLE 9–2 Consider the same system as discussed in Example 9–1. We shall show that Equation (9–2) is not the only state equation possible for this sytem. Suppose we define a set of new state variables z_1, z_2, z_3 by the transformation

$$\begin{bmatrix} x_1 \\ x_2 \\ x_3 \end{bmatrix} = \begin{bmatrix} 1 & 1 & 1 \\ -1 & -2 & -3 \\ 1 & 4 & 9 \end{bmatrix} \begin{bmatrix} z_1 \\ z_2 \\ z_3 \end{bmatrix}$$

or

$$\mathbf{x} = \mathbf{Pz} \qquad (9\text{–}6)$$

where

$$\mathbf{P} = \begin{bmatrix} 1 & 1 & 1 \\ -1 & -2 & -3 \\ 1 & 4 & 9 \end{bmatrix} \qquad (9\text{–}7)$$

Then by substituting Equation (9–6) into Equation (9–4), we obtain

$$\mathbf{P\dot{z}} = \mathbf{APz} + \mathbf{B}u$$

By premultiplying both sides of this last equation by $\mathbf{P}^{-1}$, we get

$$\dot{\mathbf{z}} = \mathbf{P}^{-1}\mathbf{APz} + \mathbf{P}^{-1}\mathbf{B}u \qquad (9\text{–}8)$$

or

$$
\begin{bmatrix} \dot{z}_1 \\ \dot{z}_2 \\ \dot{z}_3 \end{bmatrix} =
\begin{bmatrix} 3 & 2.5 & 0.5 \\ -3 & -4 & -1 \\ 1 & 1.5 & 0.5 \end{bmatrix}
\begin{bmatrix} 0 & 1 & 0 \\ 0 & 0 & 1 \\ -6 & -11 & -6 \end{bmatrix}
\begin{bmatrix} 1 & 1 & 1 \\ -1 & -2 & -3 \\ 1 & 4 & 9 \end{bmatrix}
\begin{bmatrix} z_1 \\ z_2 \\ z_3 \end{bmatrix}
$$

$$
+ \begin{bmatrix} 3 & 2.5 & 0.5 \\ -3 & -4 & -1 \\ 1 & 1.5 & 0.5 \end{bmatrix}
\begin{bmatrix} 0 \\ 0 \\ 6 \end{bmatrix} u
$$

Simplifying gives

$$
\begin{bmatrix} \dot{z}_1 \\ \dot{z}_2 \\ \dot{z}_3 \end{bmatrix} =
\begin{bmatrix} -1 & 0 & 0 \\ 0 & -2 & 0 \\ 0 & 0 & -3 \end{bmatrix}
\begin{bmatrix} z_1 \\ z_2 \\ z_3 \end{bmatrix} +
\begin{bmatrix} 3 \\ -6 \\ 3 \end{bmatrix} u \qquad (9\text{–}9)
$$

Equation (9–9) is also a state equation that describes the same system as defined by Equation (9–2). The output equation, Equation (9–5), is modified to

$$y = \mathbf{CPz}$$

or

$$
y = \begin{bmatrix} 1 & 0 & 0 \end{bmatrix}
\begin{bmatrix} 1 & 1 & 1 \\ -1 & -2 & -3 \\ 1 & 4 & 9 \end{bmatrix}
\begin{bmatrix} z_1 \\ z_2 \\ z_3 \end{bmatrix}
$$

$$
= \begin{bmatrix} 1 & 1 & 1 \end{bmatrix}
\begin{bmatrix} z_1 \\ z_2 \\ z_3 \end{bmatrix} \qquad (9\text{–}10)
$$

Notice that the transformation matrix $\mathbf{P}$, defined by Equation (9–7), modifies the coefficient matrix of $\mathbf{z}$ into the diagonal matrix. As clearly seen from Equation (9–9), the three scalar state equations are uncoupled. Notice also that the diagonal elements of the matrix $\mathbf{P}^{-1}\mathbf{AP}$ in Equation (9–8) are identical with the three eigenvalues of $\mathbf{A}$. It is very important to note that the eigenvalues of $\mathbf{A}$ and those of $\mathbf{P}^{-1}\mathbf{AP}$ are identical. We shall prove this for a general case in what follows.

Invariance of eigenvalues. To prove the invariance of the eigenvalues under a linear transformation, we must show that the characteristic polynomials $|\lambda\mathbf{I} - \mathbf{A}|$ and $|\lambda\mathbf{I} - \mathbf{P}^{-1}\mathbf{AP}|$ are identical.

Since the determinant of a product is the product of the determinants, we obtain

$$
\begin{aligned}
|\lambda\mathbf{I} - \mathbf{P}^{-1}\mathbf{AP}| &= |\lambda\mathbf{P}^{-1}\mathbf{P} - \mathbf{P}^{-1}\mathbf{AP}| \\
&= |\mathbf{P}^{-1}(\lambda\mathbf{I} - \mathbf{A})\mathbf{P}| \\
&= |\mathbf{P}^{-1}||\lambda\mathbf{I} - \mathbf{A}||\mathbf{P}| \\
&= |\mathbf{P}^{-1}||\mathbf{P}||\lambda\mathbf{I} - \mathbf{A}|
\end{aligned}
$$

Noting that the product of the determinants $|\mathbf{P}^{-1}|$ and $|\mathbf{P}|$ is the determinant of the product $|\mathbf{P}^{-1}\mathbf{P}|$, we obtain

$$|\lambda\mathbf{I} - \mathbf{P}^{-1}\mathbf{A}\mathbf{P}| = |\mathbf{P}^{-1}\mathbf{P}||\lambda\mathbf{I} - \mathbf{A}|$$
$$= |\lambda\mathbf{I} - \mathbf{A}|$$

Thus we have proved that the eigenvalues of $\mathbf{A}$ are invariant under a linear transformation.

Diagonalization of $n \times n$ matrix. Note that if an $n \times n$ matrix $\mathbf{A}$ with distinct eigenvalues is given by

$$\mathbf{A} = \begin{bmatrix} 0 & 1 & 0 & \cdots & 0 \\ 0 & 0 & 1 & \cdots & 0 \\ \cdot & \cdot & \cdot & & \cdot \\ \cdot & \cdot & \cdot & & \cdot \\ \cdot & \cdot & \cdot & & \cdot \\ 0 & 0 & 0 & \cdots & 1 \\ -a_n & -a_{n-1} & -a_{n-2} & \cdots & -a_1 \end{bmatrix} \tag{9-11}$$

the transformation $\mathbf{x} = \mathbf{P}\mathbf{z}$ where

$$\mathbf{P} = \begin{bmatrix} 1 & 1 & \cdots & 1 \\ \lambda_1 & \lambda_2 & \cdots & \lambda_n \\ \lambda_1^2 & \lambda_2^2 & \cdots & \lambda_n^2 \\ \cdot & \cdot & & \cdot \\ \cdot & \cdot & & \cdot \\ \cdot & \cdot & & \cdot \\ \lambda_1^{n-1} & \lambda_2^{n-1} & \cdots & \lambda_n^{n-1} \end{bmatrix}$$

$$\lambda_1, \lambda_2, \ldots, \lambda_n = n \text{ distinct eigenvalues of } \mathbf{A}$$

will transform $\mathbf{P}^{-1}\mathbf{A}\mathbf{P}$ into the diagonal matrix, or

$$\mathbf{P}^{-1}\mathbf{A}\mathbf{P} = \begin{bmatrix} \lambda_1 & & & & 0 \\ & \lambda_2 & & & \\ & & \cdot & & \\ & & & \cdot & \\ 0 & & & & \lambda_n \end{bmatrix}$$

If the matrix $\mathbf{A}$ defined by Equation (9-11) involves multiple eigenvalues, then diagonalization is impossible. For example, if the 3×3 matrix $\mathbf{A}$ where

$$\mathbf{A} = \begin{bmatrix} 0 & 1 & 0 \\ 0 & 0 & 1 \\ -a_3 & -a_2 & -a_1 \end{bmatrix}$$

has the eigenvalues $\lambda_1, \lambda_1, \lambda_3$, then the transformation $\mathbf{x} = \mathbf{S}\mathbf{z}$ where

$$\mathbf{S} = \begin{bmatrix} 1 & 0 & 1 \\ \lambda_1 & 1 & \lambda_3 \\ \lambda_1^2 & 2\lambda_1 & \lambda_3^2 \end{bmatrix}$$

will yield

$$\mathbf{S}^{-1}\mathbf{A}\mathbf{S} = \begin{bmatrix} \lambda_1 & 1 & 0 \\ 0 & \lambda_1 & 0 \\ 0 & 0 & \lambda_3 \end{bmatrix}$$

Such a form is called the *Jordan canonical form*.

EXAMPLE 9–3   Consider the same system as discussed in Examples 9–1 and 9–2, rewritten thus:

$$\dddot{y} + 6\ddot{y} + 11\dot{y} + 6y = 6u \tag{9–12}$$

We shall demonstrate that the state-space representation as given by Equations (9–9) and (9–10) can also be obtained by use of the partial-fraction-expansion technique.

Let us rewrite Equation (9–12) in the form of a transfer function:

$$\frac{Y(s)}{U(s)} = \frac{6}{s^3 + 6s^2 + 11s + 6} = \frac{6}{(s + 1)(s + 2)(s + 3)}$$

By expanding this transfer function into partial fractions, we obtain

$$\frac{Y(s)}{U(s)} = \frac{3}{s + 1} + \frac{-6}{s + 2} + \frac{3}{s + 3}$$

Hence

$$Y(s) = \frac{3}{s + 1} U(s) + \frac{-6}{s + 2} U(s) + \frac{3}{s + 3} U(s) \tag{9–13}$$

Let us define

$$X_1(s) = \frac{3}{s + 1} U(s) \tag{9–14}$$

$$X_2(s) = \frac{-6}{s + 2} U(s) \tag{9–15}$$

$$X_3(s) = \frac{3}{s + 3} U(s) \tag{9–16}$$

The inverse Laplace transforms of Equations (9–14), (9–15), and (9–16) give

$$\dot{x}_1 = -x_1 + 3u$$

$$\dot{x}_2 = -2x_2 - 6u$$

$$\dot{x}_3 = -3x_3 + 3u$$

In terms of vector-matrix notation, we obtain

$$\begin{bmatrix} \dot{x}_1 \\ \dot{x}_2 \\ \dot{x}_3 \end{bmatrix} = \begin{bmatrix} -1 & 0 & 0 \\ 0 & -2 & 0 \\ 0 & 0 & -3 \end{bmatrix} \begin{bmatrix} x_1 \\ x_2 \\ x_3 \end{bmatrix} + \begin{bmatrix} 3 \\ -6 \\ 3 \end{bmatrix} u \tag{9–17}$$

Since Equation (9–13) can be written as

$$Y(s) = X_1(s) + X_2(s) + X_3(s)$$

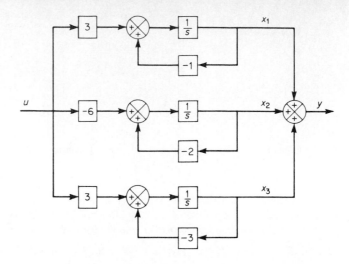

Figure 9–2
Block diagram representation of the system defined by Eqs. (9–17) and (9–18).

we obtain

$$y = x_1 + x_2 + x_3$$

or

$$y = [1 \quad 1 \quad 1] \begin{bmatrix} x_1 \\ x_2 \\ x_3 \end{bmatrix} \tag{9–18}$$

Equations (9–17) and (9–18) are of the same form as Equations (9–9) and (9–10), respectively.

Figure 9–2 shows a block-diagram representation of Equations (9–17) and (9–18). Notice that the transfer functions in the feedback blocks are identical with the eigenvalues of the system. Notice also that the residues of the poles of the transfer function, or the coefficients in the partial fraction expansions of $Y(s)/U(s)$, appear in the feedforward blocks.

State-space representation of nth-order systems of linear differential equations with r forcing functions. Consider the multiple-input, multiple-output system shown in Figure 9–3. In this system, $x_1, x_2, \ldots, x_n$ represent the state variables; $u_1, u_2, \ldots, u_r$ denote the input variables; and $y_1, y_2, \ldots, y_m$ are the output variables. For the system of Figure 9–3, we may have the system equations as follows:

$$\dot{x}_1 = a_{11}(t)x_1 + a_{12}(t)x_2 + \cdots + a_{1n}(t)x_n + b_{11}(t)u_1 + b_{12}(t)u_2 + \cdots + b_{1r}(t)u_r$$

$$\dot{x}_2 = a_{21}(t)x_1 + a_{22}(t)x_2 + \cdots + a_{2n}(t)x_n + b_{21}(t)u_1 + b_{22}(t)u_2 + \cdots + b_{2r}(t)u_r$$

$$\vdots$$

$$\dot{x}_n = a_{n1}(t)x_1 + a_{n2}(t)x_2 + \cdots + a_{nn}(t)x_n + b_{n1}(t)u_1 + b_{n2}(t)u_2 + \cdots + b_{nr}(t)u_r$$

where the $a(t)$'s and $b(t)$'s are constants or functions of t. In terms of vector-matrix notation, these n equations can be written compactly as

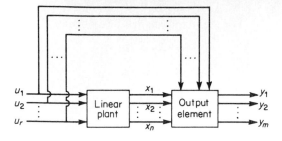

Figure 9–3
Multiple-input,
multiple-output
system.

$$\dot{\mathbf{x}} = \mathbf{A}(t)\mathbf{x} + \mathbf{B}(t)\mathbf{u} \qquad (9\text{--}19)$$

where

$$\mathbf{x} = \begin{bmatrix} x_1 \\ x_2 \\ \cdot \\ \cdot \\ \cdot \\ x_n \end{bmatrix} = \text{state vector}$$

$$\mathbf{u} = \begin{bmatrix} u_1 \\ u_2 \\ \cdot \\ \cdot \\ \cdot \\ u_r \end{bmatrix} = \text{input (or control) vector}$$

$$\mathbf{A}(t) = \begin{bmatrix} a_{11}(t) & a_{12}(t) & \cdots & a_{1n}(t) \\ a_{21}(t) & a_{22}(t) & \cdots & a_{2n}(t) \\ \cdot & \cdot & & \cdot \\ \cdot & \cdot & & \cdot \\ \cdot & \cdot & & \cdot \\ a_{n1}(t) & a_{n2}(t) & \cdots & a_{nn}(t) \end{bmatrix}$$

$$\mathbf{B}(t) = \begin{bmatrix} b_{11}(t) & b_{12}(t) & \cdots & b_{1r}(t) \\ b_{21}(t) & b_{22}(t) & \cdots & b_{2r}(t) \\ \cdot & \cdot & & \cdot \\ \cdot & \cdot & & \cdot \\ \cdot & \cdot & & \cdot \\ b_{n1}(t) & b_{n2}(t) & \cdots & b_{nr}(t) \end{bmatrix}$$

Equation (9–19) is the state equation for the system. [Note that a vector-matrix differential equation such as Equation (9–19) (or the equivalent n first-order differential equations) describing the dynamics of a system is a state equation if and only if the set of dependent variables in the vector-matrix differential equation satisfies the definition of state variables.]

For the output signals, we obtain

$$y_1 = c_{11}(t)x_1 + c_{12}(t)x_2 + \cdots + c_{1n}(t)x_n + d_{11}(t)u_1 + d_{12}(t)u_2 + \cdots + d_{1r}(t)u_r$$

$$y_2 = c_{21}(t)x_1 + c_{22}(t)x_2 + \cdots + c_{2n}(t)x_n + d_{21}(t)u_1 + d_{22}(t)u_2 + \cdots + d_{2r}(t)u_r$$

$$\begin{array}{c} \cdot \\ \cdot \\ \cdot \end{array}$$

$$y_m = c_{m1}(t)x_1 + c_{m2}(t)x_2 + \cdots + c_{mn}(t)x_n + d_{m1}(t)u_1 + d_{m2}(t)u_2 + \cdots + d_{mr}(t)u_r$$

In terms of vector-matrix notation, these m equations can be written compactly as

$$\mathbf{y} = \mathbf{C}(t)\mathbf{x} + \mathbf{D}(t)\mathbf{u} \qquad (9\text{–}20)$$

where

$$\mathbf{y} = \begin{bmatrix} y_1 \\ y_2 \\ \cdot \\ \cdot \\ \cdot \\ y_m \end{bmatrix} = \text{output vector}$$

$$\mathbf{C}(t) = \begin{bmatrix} c_{11}(t) & c_{12}(t) & \cdots & c_{1n}(t) \\ c_{21}(t) & c_{22}(t) & \cdots & c_{2n}(t) \\ \cdot & \cdot & & \cdot \\ \cdot & \cdot & & \cdot \\ \cdot & \cdot & & \cdot \\ c_{m1}(t) & c_{m2}(t) & \cdots & c_{mn}(t) \end{bmatrix}$$

$$\mathbf{D}(t) = \begin{bmatrix} d_{11}(t) & d_{12}(t) & \cdots & d_{1r}(t) \\ d_{21}(t) & d_{22}(t) & \cdots & d_{2r}(t) \\ \cdot & \cdot & & \cdot \\ \cdot & \cdot & & \cdot \\ \cdot & \cdot & & \cdot \\ d_{m1}(t) & d_{m2}(t) & \cdots & d_{mr}(t) \end{bmatrix}$$

Equation (9–20) is the output equation for the system. The matrices $\mathbf{A}(t)$, $\mathbf{B}(t)$, $\mathbf{C}(t)$, and $\mathbf{D}(t)$ completely characterize the system dynamics.

A block-diagram representation of the system defined by Equations (9–19) and (9–20) is shown in Figure 9–4(a). To indicate vector quantities, we have used double arrows in the diagram. Figure 9–4(b) shows the signal flow graph representation of the system of Figure 9–4(a).

Cayley–Hamilton theorem.   The Cayley–Hamilton theorem is very useful in proving theorems involving matrix equations or solving problems involving matrix equations.

Consider an $n \times n$ matrix $\mathbf{A}$ and its characteristic equation:

$$|\lambda\mathbf{I} - \mathbf{A}| = \lambda^n + a_1\lambda^{n-1} + \cdots + a_{n-1}\lambda + a_n = 0$$

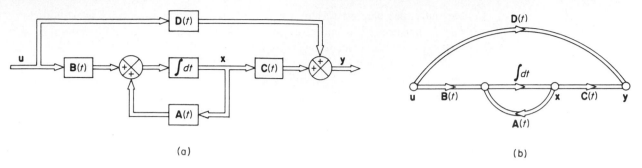

(a)

(b)

Figure 9–4

(a) Bode diagram representation of the system described by Eqs. (9–19) and (9–20); (b) signal flow graph representation of the system of (a).

The Cayley–Hamilton theorem states that the matrix $\mathbf{A}$ satisfies its characteristic equation, or that

$$\mathbf{A}^n + a_1\mathbf{A}^{n-1} + \cdots + a_{n-1}\mathbf{A} + a_n\mathbf{I} = \mathbf{0}$$

To prove this theorem, note that $\mathrm{adj}(\lambda\mathbf{I} - \mathbf{A})$ is a polynomial in λ of degree $n - 1$. That is,

$$\mathrm{adj}(\lambda\mathbf{I} - \mathbf{A}) = \mathbf{B}_1\lambda^{n-1} + \mathbf{B}_2\lambda^{n-2} + \cdots + \mathbf{B}_{n-1}\lambda + \mathbf{B}_n$$

where $\mathbf{B}_1 = \mathbf{I}$. Since

$$(\lambda\mathbf{I} - \mathbf{A})\,\mathrm{adj}\,(\lambda\mathbf{I} - \mathbf{A}) = [\mathrm{adj}\,(\lambda\mathbf{I} - \mathbf{A})](\lambda\mathbf{I} - \mathbf{A}) = |\lambda\mathbf{I} - \mathbf{A}|\mathbf{I}$$

we obtain

$$|\lambda\mathbf{I} - \mathbf{A}|\mathbf{I} = \mathbf{I}\lambda^n + a_1\mathbf{I}\lambda^{n-1} + \cdots + a_{n-1}\mathbf{I}\lambda + a_n\mathbf{I}$$

$$= (\lambda\mathbf{I} - \mathbf{A})(\mathbf{B}_1\lambda^{n-1} + \mathbf{B}_2\lambda^{n-2} + \cdots + \mathbf{B}_{n-1}\lambda + \mathbf{B}_n)$$

$$= (\mathbf{B}_1\lambda^{n-1} + \mathbf{B}_2\lambda^{n-2} + \cdots + \mathbf{B}_{n-1}\lambda + \mathbf{B}_n)(\lambda\mathbf{I} - \mathbf{A})$$

From this equation we see that $\mathbf{A}$ and $\mathbf{B}_i$ $(i = 1, 2, \ldots, n)$ commute. Hence the product of $(\lambda\mathbf{I} - \mathbf{A})$ and $\mathrm{adj}\,(\lambda\mathbf{I} - \mathbf{A})$ becomes zero if either of these is zero. If $\mathbf{A}$ is substituted for λ in this last equation, then clearly $\lambda\mathbf{I} - \mathbf{A}$ becomes zero. Hence

$$\mathbf{A}^n + a_1\mathbf{A}^{n-1} + \cdots + a_{n-1}\mathbf{A} + a_n\mathbf{I} = \mathbf{0}$$

This proves the Cayley–Hamilton theorem.

Computation of $e^{\mathbf{A}t}$. In solving control problems it often becomes necessary to compute $e^{\mathbf{A}t}$. In what follows we shall present a machine computation method and a few analytical methods for computing $e^{\mathbf{A}t}$.

Machine computation of $e^{\mathbf{A}t}$. The solution of the linear, time-invariant, continuous-time state equation involves the matrix exponential $e^{\mathbf{A}t}$. There are several methods available for computing $e^{\mathbf{A}t}$. If the number of rows of the square matrix is 4 or larger, hand computation becomes exceedingly tedious and machine computation may be necessary. (There are standard computational schemes for $e^{\mathbf{A}t}$.)

Conceptually, the simplest way to compute $e^{\mathbf{A}t}$ is to expand $e^{\mathbf{A}t}$ into a power series in t. (The matrix exponential $e^{\mathbf{A}t}$ is uniformly convergent for finite values of t.) For example, $e^{\mathbf{A}t}$ may be expanded into a power series as follows:

$$e^{\mathbf{A}t} = \mathbf{I} + (\mathbf{A}t) + \frac{\mathbf{A}t}{2}\left(\frac{\mathbf{A}t}{1!}\right) + \frac{\mathbf{A}t}{3}\left(\frac{\mathbf{A}^2 t^2}{2!}\right) + \cdots + \frac{\mathbf{A}t}{n+1}\left(\frac{\mathbf{A}^n t^n}{n!}\right) + \cdots$$

Notice that each term in parentheses is equal to the entire preceding term. This provides a convenient recursion scheme. The computation is carried out to only enough terms that additional terms are negligible by comparison with the partial sum to that point. In this approach, a norm of the matrix may be used as a check to stop computation. A norm is a scalar quantity that gives a means of determining the absolute magnitude of the n^2 elements of an $n \times n$ matrix. Several different forms of norms are commonly used. Any one of them may be used. One example is

$$\text{Norm of } \mathbf{M} = \|\mathbf{M}\| = \sum_{\substack{i=1 \\ j=1}}^{n} |m_{ij}|$$

where the m_{ij}'s are the elements of matrix $\mathbf{M}$. In machine computation, in addition to computing and adding terms in the series, the computer keeps a tab on the norm and stops the computation when the norm reaches a prescribed limit value.

The virtues of the series expansion technique are its simplicity and the ease of programming it. It is not necessary to find the eigenvalues of $\mathbf{A}$. There are, however, some computational disadvantages to the series expansion method, stemming from the convergence requirements for the series $e^{\mathbf{A}t}$. The Jordan canonical form requires considerably more programming than the series expansion method, but the program will run in a fraction of the time needed for the series solution. (A standard program is available for computing $e^{\mathbf{A}t}$ by use of the Jordan canonical form.) For details of the analytical aspects of the Jordan canonical form, see the Appendix.

There are several analytical approaches for obtaining $e^{\mathbf{A}t}$. In the following, we shall discuss three of them.

Computation of $e^{\mathbf{A}t}$: Method 1.   Transform matrix $\mathbf{A}$ into the diagonal form or into a Jordan canonical form. We shall first consider the case where matrix $\mathbf{A}$ involves only distinct eigenvectors and therefore can be transformed into the diagonal form. We shall next consider the case where matrix $\mathbf{A}$ involves multiple eigenvectors and therefore cannot be diagonalized.

Consider the state equation

$$\dot{\mathbf{x}} = \mathbf{A}\mathbf{x}$$

If a square matrix can be diagonalized, then a diagonalizing matrix (transformation matrix) exists and it can be obtained by a standard method presented in the Appendix. Let $\mathbf{P}$ be a diagonalizing matrix for $\mathbf{A}$. Let us define

$$\mathbf{x} = \mathbf{P}\hat{\mathbf{x}}$$

Then

$$\dot{\hat{\mathbf{x}}} = \mathbf{P}^{-1}\mathbf{A}\mathbf{P}\hat{\mathbf{x}} = \mathbf{D}\hat{\mathbf{x}}$$

where $\mathbf{D}$ is a diagonal matrix. The solution of this last equation is

$$\hat{\mathbf{x}}(t) = e^{\mathbf{D}t}\hat{\mathbf{x}}(0)$$

Hence

$$\mathbf{x}(t) = \mathbf{P}\hat{\mathbf{x}}(t) = \mathbf{P}e^{\mathbf{D}t}\mathbf{P}^{-1}\mathbf{x}(0)$$

Noting that $\mathbf{x}(t)$ can also be given by the equation

$$\mathbf{x}(t) = e^{\mathbf{A}t}\mathbf{x}(0)$$

we obtain $e^{\mathbf{A}t} = \mathbf{P}e^{\mathbf{D}t}\mathbf{P}^{-1}$, or

$$e^{\mathbf{A}t} = \mathbf{P}e^{\mathbf{D}t}\mathbf{P}^{-1} = \mathbf{P}\begin{bmatrix} e^{\lambda_1 t} & & & 0 \\ & e^{\lambda_2 t} & & \\ & & \ddots & \\ 0 & & & e^{\lambda_n t} \end{bmatrix}\mathbf{P}^{-1} \qquad (9\text{--}21)$$

Next, we shall consider the case where matrix $\mathbf{A}$ may be transformed into a Jordan canonical form. Consider again the state equation

$$\dot{\mathbf{x}} = \mathbf{A}\mathbf{x}$$

First obtain a transformation matrix $\mathbf{S}$ that will transform matrix $\mathbf{A}$ into a Jordan canonical form (for details, see the Appendix), so that

$$\mathbf{S}^{-1}\mathbf{A}\mathbf{S} = \mathbf{J}$$

where $\mathbf{J}$ is a matrix in a Jordan canonical form. Now define

$$\mathbf{x} = \mathbf{S}\hat{\mathbf{x}}$$

Then

$$\dot{\hat{\mathbf{x}}} = \mathbf{S}^{-1}\mathbf{A}\mathbf{S}\hat{\mathbf{x}} = \mathbf{J}\hat{\mathbf{x}}$$

The solution of this last equation is

$$\hat{\mathbf{x}}(t) = e^{\mathbf{J}t}\hat{\mathbf{x}}(0)$$

Hence

$$\mathbf{x}(t) = \mathbf{S}\hat{\mathbf{x}}(t) = \mathbf{S}e^{\mathbf{J}t}\mathbf{S}^{-1}\mathbf{x}(0)$$

Since the solution $\mathbf{x}(t)$ can also be given by the equation

$$\mathbf{x}(t) = e^{\mathbf{A}t}\mathbf{x}(0)$$

we obtain

$$e^{\mathbf{A}t} = \mathbf{S}e^{\mathbf{J}t}\mathbf{S}^{-1}$$

Note that $e^{\mathbf{J}t}$ is a triangular matrix [which means that the elements below (or above as the case may be) the principal diagonal line are zeros] whose elements are $e^{\lambda t}$, $te^{\lambda t}$, $\frac{1}{2}t^2 e^{\lambda t}$, and so forth. For example, if matrix $\mathbf{J}$ has the following Jordan canonical form:

$$\mathbf{J} = \begin{bmatrix} \lambda_1 & 1 & 0 \\ 0 & \lambda_1 & 1 \\ 0 & 0 & \lambda_1 \end{bmatrix}$$

then

$$e^{\mathbf{J}t} = \begin{bmatrix} e^{\lambda_1 t} & te^{\lambda_1 t} & \frac{1}{2}t^2 e^{\lambda_1 t} \\ 0 & e^{\lambda_1 t} & te^{\lambda_1 t} \\ 0 & 0 & e^{\lambda_1 t} \end{bmatrix}$$

Similarly, if

$$\mathbf{J} = \begin{bmatrix} \lambda_1 & 1 & 0 & & & & 0 \\ 0 & \lambda_1 & 1 & & & & \\ 0 & 0 & \lambda_1 & & & & \\ & & & \lambda_4 & 1 & & \\ & & & 0 & \lambda_4 & & \\ & & & & & \lambda_6 & \\ 0 & & & & & & \lambda_7 \end{bmatrix}$$

then

$$e^{\mathbf{J}t} = \begin{bmatrix} e^{\lambda_1 t} & te^{\lambda_1 t} & \frac{1}{2}t^2 e^{\lambda_1 t} & & & & 0 \\ 0 & e^{\lambda_1 t} & te^{\lambda_1 t} & & & & \\ 0 & 0 & e^{\lambda_1 t} & & & & \\ & & & e^{\lambda_4 t} & te^{\lambda_4 t} & & \\ & & & 0 & e^{\lambda_4 t} & & \\ & & & & & e^{\lambda_6 t} & 0 \\ 0 & & & & & 0 & e^{\lambda_7 t} \end{bmatrix}$$

As an example, consider the following matrix $\mathbf{A}$:

$$\mathbf{A} = \begin{bmatrix} 0 & 1 & 0 \\ 0 & 0 & 1 \\ 1 & -3 & 3 \end{bmatrix}$$

The characteristic equation is

$$|\lambda\mathbf{I} - \mathbf{A}| = \lambda^3 - 3\lambda^2 + 3\lambda - 1 = (\lambda - 1)^3 = 0$$

Thus matrix $\mathbf{A}$ has a multiple eigenvalue of order 3 at $\lambda = 1$. It can be shown that matrix $\mathbf{A}$ has a multiple eigenvector of order 3. The transformation matrix that will transform matrix $\mathbf{A}$ into a Jordan canonical form can be given by (for details, see the Appendix)

$$\mathbf{S} = \begin{bmatrix} 1 & 0 & 0 \\ 1 & 1 & 0 \\ 1 & 2 & 1 \end{bmatrix}$$

The inverse of matrix $\mathbf{S}$ is

$$\mathbf{S}^{-1} = \begin{bmatrix} 1 & 0 & 0 \\ -1 & 1 & 0 \\ 1 & -2 & 1 \end{bmatrix}$$

Then, it can be seen that

$$\mathbf{S}^{-1}\mathbf{A}\mathbf{S} = \begin{bmatrix} 1 & 0 & 0 \\ -1 & 1 & 0 \\ 1 & -2 & 1 \end{bmatrix}\begin{bmatrix} 0 & 1 & 0 \\ 0 & 0 & 1 \\ 1 & -3 & 3 \end{bmatrix}\begin{bmatrix} 1 & 0 & 0 \\ 1 & 1 & 0 \\ 1 & 2 & 1 \end{bmatrix}$$

$$= \begin{bmatrix} 1 & 1 & 0 \\ 0 & 1 & 1 \\ 0 & 0 & 1 \end{bmatrix} = \mathbf{J}$$

Noting that

$$e^{\mathbf{J}t} = \begin{bmatrix} e^t & te^t & \frac{1}{2}t^2e^t \\ 0 & e^t & te^t \\ 0 & 0 & e^t \end{bmatrix}$$

we find

$$e^{\mathbf{A}t} = \mathbf{S}e^{\mathbf{J}t}\mathbf{S}^{-1}$$

$$= \begin{bmatrix} 1 & 0 & 0 \\ 1 & 1 & 0 \\ 1 & 2 & 1 \end{bmatrix}\begin{bmatrix} e^t & te^t & \frac{1}{2}t^2e^t \\ 0 & e^t & te^t \\ 0 & 0 & e^t \end{bmatrix}\begin{bmatrix} 1 & 0 & 0 \\ -1 & 1 & 0 \\ 1 & -2 & 1 \end{bmatrix}$$

$$= \begin{bmatrix} e^t - te^t + \frac{1}{2}t^2e^t & te^t - t^2e^t & \frac{1}{2}t^2e^t \\ \frac{1}{2}t^2e^t & e^t - te^t - t^2e^t & te^t + \frac{1}{2}t^2e^t \\ te^t + \frac{1}{2}t^2e^t & -3te^t - t^2e^t & e^t + 2te^t + \frac{1}{2}t^2e^t \end{bmatrix}$$

Computation of $e^{\mathbf{A}t}$: Method 2. The second method of computing $e^{\mathbf{A}t}$ uses the Laplace transform approach. Referring to Equation (4–60), $e^{\mathbf{A}t}$ can be given as follows:

$$e^{\mathbf{A}t} = \mathcal{L}^{-1}[(s\mathbf{I} - \mathbf{A})^{-1}]$$

Thus, to obtain $e^{\mathbf{A}t}$, first invert the matrix $(s\mathbf{I} - \mathbf{A})$. This results in a matrix whose elements are rational functions of s. Then, take the inverse Laplace transform of each element of the matrix.

Computation of $e^{\mathbf{A}t}$: Method 3. The third method of computing $e^{\mathbf{A}t}$ uses Sylvester's interporation formula. (For the Sylvester's interporation formula, see Problem A–9–6.) We shall first consider the case where the roots of the minimal polynomial $\phi(\lambda)$ of $\mathbf{A}$ are distinct. (For the definition of the minimal polynomial, see Problem A–9–3.) Then we shall deal with the case of multiple roots.

Case 1: Minimal Polynomial of A Involves Only Distinct Roots. We shall assume that the degree of the minimal polynomial of $\mathbf{A}$ is m. By using Sylvester's interpolation formula, it can be shown that $e^{\mathbf{A}t}$ can be obtained by solving the following determinant equation:

$$
\begin{vmatrix}
1 & \lambda_1 & \lambda_1^2 & \cdots & \lambda_1^{m-1} & e^{\lambda_1 t} \\
1 & \lambda_2 & \lambda_2^2 & \cdots & \lambda_2^{m-1} & e^{\lambda_2 t} \\
\cdot & & & & \cdot & \cdot \\
\cdot & & \cdot & & \cdot & \cdot \\
\cdot & & \cdot & & \cdot & \cdot \\
1 & \lambda_m & \lambda_m^2 & \cdots & \lambda_m^{m-1} & e^{\lambda_m t} \\
\mathbf{I} & \mathbf{A} & \mathbf{A}^2 & \cdots & \mathbf{A}^{m-1} & e^{\mathbf{A}t}
\end{vmatrix} = \mathbf{0}
\tag{9–22}
$$

By solving Equation (9–22) for $e^{\mathbf{A}t}$, $e^{\mathbf{A}t}$ can be obtained in terms of the $\mathbf{A}^k$ ($k = 0, 1, 2, \ldots, m - 1$) and the $e^{\lambda_i t}$ ($i = 1, 2, 3, \ldots, m$). [Equation (9–22) may be expanded, for example, about the last column.]

Notice that solving Equation (9–22) for $e^{\mathbf{A}t}$ is the same as writing

$$
e^{\mathbf{A}t} = \alpha_0(t)\mathbf{I} + \alpha_1(t)\mathbf{A} + \alpha_2(t)\mathbf{A}^2 + \cdots + \alpha_{m-1}(t)\mathbf{A}^{m-1}
\tag{9–23}
$$

and determining the $\alpha_k(t)$ ($k = 0, 1, 2, \ldots, m - 1$) by solving the following set of m equations for the $\alpha_k(t)$:

$$
\alpha_0(t) + \alpha_1(t)\lambda_1 + \alpha_2(t)\lambda_1^2 + \cdots + \alpha_{m-1}(t)\lambda_1^{m-1} = e^{\lambda_1 t}
$$

$$
\alpha_0(t) + \alpha_1(t)\lambda_2 + \alpha_2(t)\lambda_2^2 + \cdots + \alpha_{m-1}(t)\lambda_2^{m-1} = e^{\lambda_2 t}
$$

$$
\cdot
$$
$$
\cdot
$$
$$
\cdot
$$

$$
\alpha_0(t) + \alpha_1(t)\lambda_m + \alpha_2(t)\lambda_m^2 + \cdots + \alpha_{m-1}(t)\lambda_m^{m-1} = e^{\lambda_m t}
$$

If $\mathbf{A}$ is an $n \times n$ matrix and has distinct eigenvalues, then the number of $\alpha_k(t)$'s to be determined is $m = n$. If $\mathbf{A}$ involves multiple eigenvalues but its minimal polynomial has only simple roots, however, then the number m of $\alpha_k(t)$'s to be determined is less than n.

Case 2: Minimal Polynomial of A Involves Multiple Roots. As an example, consider the case where the minimal polynomial of $\mathbf{A}$ involves three equal roots ($\lambda_1 = \lambda_2 = \lambda_3$) and has other roots ($\lambda_4, \lambda_5, \ldots, \lambda_m$) that are all distinct. By applying Sylvester's interpolation formula, it can be shown that $e^{\mathbf{A}t}$ can be obtained from the following determinant equation:

$$
\begin{vmatrix}
0 & 0 & 1 & 3\lambda_1 & \cdots & \dfrac{(m-1)(m-2)}{2}\lambda_1^{m-3} & \dfrac{t^2}{2}e^{\lambda_1 t} \\
0 & 1 & 2\lambda_1 & 3\lambda_1^2 & \cdots & (m-1)\lambda_1^{m-2} & te^{\lambda_1 t} \\
1 & \lambda_1 & \lambda_1^2 & \lambda_1^3 & \cdots & \lambda_1^{m-1} & e^{\lambda_1 t} \\
1 & \lambda_4 & \lambda_4^2 & \lambda_4^3 & \cdots & \lambda_4^{m-1} & e^{\lambda_4 t} \\
\cdot & \cdot & \cdot & \cdot & & \cdot & \cdot \\
\cdot & \cdot & \cdot & \cdot & & \cdot & \cdot \\
\cdot & \cdot & \cdot & \cdot & & \cdot & \cdot \\
1 & \lambda_m & \lambda_m^2 & \lambda_m^3 & \cdots & \lambda_m^{m-1} & e^{\lambda_m t} \\
\mathbf{I} & \mathbf{A} & \mathbf{A}^2 & \mathbf{A}^3 & \cdots & \mathbf{A}^{m-1} & e^{\mathbf{A}t}
\end{vmatrix} = \mathbf{0}
\tag{9–24}
$$

Equation (9–24) can be solved for e^{At} by expanding it about the last column.

It is noted that, just as in case 1, solving Equation (9–24) for e^{At} is the same as writing

$$e^{At} = \alpha_0(t)\mathbf{I} + \alpha_1(t)\mathbf{A} + \alpha_2(t)\mathbf{A}^2 + \cdots + \alpha_{m-1}(t)\mathbf{A}^{m-1} \qquad (9\text{–}25)$$

and determining the $\alpha_k(t)$'s $(k = 0, 1, 2, \ldots, m - 1)$ from

$$\alpha_2(t) + 3\alpha_3(t)\lambda_1 + \cdots + \frac{(m-1)(m-2)}{2}\alpha_{m-1}(t)\lambda_1^{m-3} = \frac{t^2}{2}e^{\lambda_1 t}$$

$$\alpha_1(t) + 2\alpha_2(t)\lambda_1 + 3\alpha_3(t)\lambda_1^2 + \cdots + (m-1)\alpha_{m-1}(t)\lambda_1^{m-2} = te^{\lambda_1 t}$$

$$\alpha_0(t) + \alpha_1(t)\lambda_1 + \alpha_2(t)\lambda_1^2 + \cdots + \alpha_{m-1}(t)\lambda_1^{m-1} = e^{\lambda_1 t}$$

$$\alpha_0(t) + \alpha_1(t)\lambda_4 + \alpha_2(t)\lambda_4^2 + \cdots + \alpha_{m-1}(t)\lambda_4^{m-1} = e^{\lambda_4 t}$$

$$\cdot$$
$$\cdot$$
$$\cdot$$

$$\alpha_0(t) + \alpha_1(t)\lambda_m + \alpha_2(t)\lambda_m^2 + \cdots + \alpha_{m-1}(t)\lambda_m^{m-1} = e^{\lambda_m t}$$

The extension to other cases where, for example, there are two or more sets of multiple roots will be apparent. Note that if the minimal polynomial of $\mathbf{A}$ is not found, it is possible to substitute the characteristic polynomial for the minimal polynomial. The number of computations may, of course, be increased.

EXAMPLE 9–4

Consider the matrix

$$\mathbf{A} = \begin{bmatrix} 0 & 1 \\ 0 & -2 \end{bmatrix}$$

Compute e^{At} as a sum of an infinite series.

The matrix exponential e^{At} may always be expanded into a series of matrices, which can then be added together into a closed form. In the present case,

$$e^{At} = \mathbf{I} + \sum_{k=1}^{\infty} \frac{\mathbf{A}^k t^k}{k!}$$

$$= \begin{bmatrix} 1 & 0 \\ 0 & 1 \end{bmatrix} + \begin{bmatrix} 0 & 1 \\ 0 & -2 \end{bmatrix} t + \begin{bmatrix} 0 & 1 \\ 0 & -2 \end{bmatrix}^2 \frac{t^2}{2!} + \cdots$$

$$= \begin{bmatrix} 1 & \dfrac{1}{2} - \dfrac{1}{2}\left[1 - 2t + \dfrac{(2t)^2}{2!} - \dfrac{(2t)^3}{3!} + \cdots \right] \\ 0 & 1 - 2t + \dfrac{(2t)^2}{2!} - \dfrac{(2t)^3}{3!} + \dfrac{(2t)^4}{4!} - \cdots \end{bmatrix}$$

$$= \begin{bmatrix} 1 & \frac{1}{2}(1 - e^{-2t}) \\ 0 & e^{-2t} \end{bmatrix}$$

EXAMPLE 9–5

Consider the same matrix $\mathbf{A}$ as in Example 9–4.

$$\mathbf{A} = \begin{bmatrix} 0 & 1 \\ 0 & -2 \end{bmatrix}$$

Compute e^{At} by use of the three analytical methods presented in this section.

Method 1. The eigenvalues of $\mathbf{A}$ are 0 and -2 ($\lambda_1 = 0$, $\lambda_2 = -2$). A necessary transformation matrix $\mathbf{P}$ may be obtained (see the Appendix for details) as

$$\mathbf{P} = \begin{bmatrix} 1 & 1 \\ 0 & -2 \end{bmatrix}$$

Then, from Equation (9–21), $e^{\mathbf{A}t}$ is obtained as follows:

$$e^{\mathbf{A}t} = \begin{bmatrix} 1 & 1 \\ 0 & -2 \end{bmatrix} \begin{bmatrix} e^0 & 0 \\ 0 & e^{-2t} \end{bmatrix} \begin{bmatrix} 1 & \frac{1}{2} \\ 0 & -\frac{1}{2} \end{bmatrix} = \begin{bmatrix} 1 & \frac{1}{2}(1 - e^{-2t}) \\ 0 & e^{-2t} \end{bmatrix}$$

Method 2. Since

$$s\mathbf{I} - \mathbf{A} = \begin{bmatrix} s & 0 \\ 0 & s \end{bmatrix} - \begin{bmatrix} 0 & 1 \\ 0 & -2 \end{bmatrix} = \begin{bmatrix} s & -1 \\ 0 & s + 2 \end{bmatrix}$$

we obtain

$$(s\mathbf{I} - \mathbf{A})^{-1} = \begin{bmatrix} \dfrac{1}{s} & \dfrac{1}{s(s + 2)} \\ 0 & \dfrac{1}{s + 2} \end{bmatrix}$$

Hence

$$e^{\mathbf{A}t} = \mathscr{L}^{-1}[(s\mathbf{I} - \mathbf{A})^{-1}] = \begin{bmatrix} 1 & \frac{1}{2}(1 - e^{-2t}) \\ 0 & e^{-2t} \end{bmatrix}$$

Method 3. From Equation (9–22), we get

$$\begin{vmatrix} 1 & \lambda_1 & e^{\lambda_1 t} \\ 1 & \lambda_2 & e^{\lambda_2 t} \\ \mathbf{I} & \mathbf{A} & e^{\mathbf{A}t} \end{vmatrix} = \mathbf{0}$$

Substituting 0 for λ_1 and -2 for λ_2 in this last equation, we obtain

$$\begin{vmatrix} 1 & 0 & 1 \\ 1 & -2 & e^{-2t} \\ \mathbf{I} & \mathbf{A} & e^{\mathbf{A}t} \end{vmatrix} = \mathbf{0}$$

Expanding the determinant, we obtain

$$-2e^{\mathbf{A}t} + \mathbf{A} + 2\mathbf{I} - \mathbf{A}e^{-2t} = \mathbf{0}$$

or

$$e^{\mathbf{A}t} = \tfrac{1}{2}(\mathbf{A} + 2\mathbf{I} - \mathbf{A}e^{-2t})$$

$$= \frac{1}{2}\left\{ \begin{bmatrix} 0 & 1 \\ 0 & -2 \end{bmatrix} + \begin{bmatrix} 2 & 0 \\ 0 & 2 \end{bmatrix} - \begin{bmatrix} 0 & 1 \\ 0 & -2 \end{bmatrix} e^{-2t} \right\}$$

$$= \begin{bmatrix} 1 & \frac{1}{2}(1 - e^{-2t}) \\ 0 & e^{-2t} \end{bmatrix}$$

An alternate approach is to use Equation (9–23). We first determine $\alpha_0(t)$ and $\alpha_1(t)$ from

$$\alpha_0(t) + \alpha_1(t)\lambda_1 = e^{\lambda_1 t}$$

$$\alpha_0(t) + \alpha_1(t)\lambda_2 = e^{\lambda_2 t}$$

Since $\lambda_1 = 0$ and $\lambda_2 = -2$, the last two equations become

$$\alpha_0(t) = 1$$

$$\alpha_0(t) - 2\alpha_1(t) = e^{-2t}$$

Solving for $\alpha_0(t)$ and $\alpha_1(t)$ gives

$$\alpha_0(t) = 1, \qquad \alpha_1(t) = \frac{1}{2}(1 - e^{-2t})$$

Then $e^{\mathbf{A}t}$ can be written as

$$e^{\mathbf{A}t} = \alpha_0(t)\mathbf{I} + \alpha_1(t)\mathbf{A} = \mathbf{I} + \frac{1}{2}(1 - e^{-2t})\mathbf{A} = \begin{bmatrix} 1 & \frac{1}{2}(1 - e^{-2t}) \\ 0 & e^{-2t} \end{bmatrix}$$

9–3 TRANSFER MATRIX

In this section we shall consider a problem of designing a noninteracting controller such that a change in one reference input affects only one output. Such a noninteraction property is important in process control systems.

Consider the system described by

$$\mathbf{x} = \mathbf{Ax} + \mathbf{Bu} \tag{9-26}$$

$$\mathbf{y} = \mathbf{Cx} + \mathbf{Du} \tag{9-27}$$

where $\mathbf{x} =$ state vector (n-vector)
   $\mathbf{u} =$ control vector (r-vector)
   $\mathbf{y} =$ output vector (m-vector)
   $\mathbf{A} = n \times n$ matrix
   $\mathbf{B} = n \times r$ matrix
   $\mathbf{C} = m \times n$ matrix
   $\mathbf{D} = m \times r$ matrix

The matrix $\mathbf{G}(s)$ that relates the Laplace transform of the output $\mathbf{y}(t)$ and the Laplace transform of the input (control vector) $\mathbf{u}(t)$ is called the *transfer matrix*.

$$\mathbf{Y}(s) = \mathbf{G}(s)\mathbf{U}(s) \tag{9-28}$$

In an expanded form, Equation (9–28) can be written as

$$\begin{bmatrix} Y_1(s) \\ Y_2(s) \\ \cdot \\ \cdot \\ \cdot \\ Y_m(s) \end{bmatrix} = \begin{bmatrix} G_{11}(s) & G_{12}(s) & \cdots & G_{1r}(s) \\ G_{21}(s) & G_{22}(s) & \cdots & G_{2r}(s) \\ \cdot & & & \cdot \\ \cdot & & & \cdot \\ \cdot & & & \cdot \\ G_{m1}(s) & G_{m2}(s) & \cdots & G_{mr}(s) \end{bmatrix} \begin{bmatrix} U_1(s) \\ U_2(s) \\ \cdot \\ \cdot \\ \cdot \\ U_r(s) \end{bmatrix} \tag{9-29}$$

The (i, j)th element $G_{ij}(s)$ of $\mathbf{G}(s)$ is the transfer function relating the ith output to the jth input. The transfer matrix $\mathbf{G}(s)$ is related to matrices $\mathbf{A}$, $\mathbf{B}$, $\mathbf{C}$, and $\mathbf{D}$ in the following way.

The Laplace transforms of Equations (9–26) and (9–27) are given by

$$sX(s) - x(0) = AX(s) + BU(s) \qquad (9\text{–}30)$$

$$Y(s) = CX(s) + DU(s) \qquad (9\text{–}31)$$

Just as in the case of the transfer function, we assume the initial state $x(0)$ to be zero; $x(0) = 0$. (That is, the transfer matrix is defined as the ratio of the Laplace transform of the output vector to the Laplace transform of the input vector when the initial conditions are zero.) From Equation (9–30), we obtain

$$X(s) = (sI - A)^{-1}BU(s) \qquad (9\text{–}32)$$

By substituting Equation (9–32) into Equation (9–31), we obtain

$$Y(s) = [C(sI - A)^{-1}B + D]U(s) \qquad (9\text{–}33)$$

Upon comparing Equations (9–28) and (9–33), we find that

$$G(s) = C(sI - A)^{-1}B + D \qquad (9\text{–}34)$$

Clearly, the transfer function expression given by Equation (1–53) is a special case of this transfer matrix expression.

Noninteraction in multiple-input, multiple-output systems.

Consider the system shown in Figure 9–5. The system has multiple inputs and multiple outputs. The transfer matrix of the feedforward path is $G_0(s)$, and that of the feedback path is $H(s)$. The transfer matrix between the feedback signal vector $B(s)$ and the error vector $E(s)$ is obtained as follows: Since

$$\begin{aligned} B(s) &= H(s)Y(s) \\ &= H(s)G_0(s)E(s) \end{aligned}$$

we obtain the transfer matrix between $B(s)$ and $E(s)$ to be $H(s)G_0(s)$. Thus the transfer matrix of the cascaded elements is the product of the transfer matrices of the individual elements. (Note that the order of the matrix multiplication is very important since matrix multiplication is, in general, not commutative.)

The transfer matrix of the closed-loop system is obtained as follows: Since

$$\begin{aligned} Y(s) &= G_0(s)[U(s) - B(s)] \\ &= G_0(s)[U(s) - H(s)Y(s)] \end{aligned}$$

we obtain

$$[I + G_0(s)H(s)]Y(s) = G_0(s)U(s)$$

Premultiplying both sides of this last equation by $[I + G_0(s)H(s)]^{-1}$, we obtain

$$Y(s) = [I + G_0(s)H(s)]^{-1}G_0(s)U(s)$$

The closed-loop transfer matrix $G(s)$ is then given by

$$G(s) = [I + G_0(s)H(s)]^{-1}G_0(s) \qquad (9\text{–}35)$$

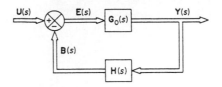

Figure 9–5
Block diagram of a
multiple-input,
multiple-output
system.

Many process control systems have multiple inputs and multiple outputs, and it is often desired that changes in one reference input affect only one output. (If such noninteraction or uncoupling can be achieved, it is easier to maintain each output value at a desired constant value in the absence of external disturbances.)

Let us assume that the transfer matrix $\mathbf{G}_p(s)$ (an $n \times n$ matrix) of a plant is given. It is desired to design a series compensator $\mathbf{G}_c(s)$ (also an $n \times n$ matrix) such that the n inputs and n outputs are uncoupled. If noninteraction, or uncoupling, between the n inputs and n outputs is desired, the closed-loop transfer matrix must be diagonal, or

$$\mathbf{G}(s) = \begin{bmatrix} G_{11}(s) & & & & 0 \\ & G_{22}(s) & & & \\ & & \cdot & & \\ & & & \cdot & \\ & & & & \cdot \\ 0 & & & & G_{nn}(s) \end{bmatrix}$$

We shall consider the case where the feedback matrix $\mathbf{H}(s)$ is the identity matrix. Then from Equation (9–35), we obtain

$$\mathbf{G}(s) = [\mathbf{I} + \mathbf{G}_0(s)]^{-1}\mathbf{G}_0(s) \tag{9–36}$$

Note that $\mathbf{G}_0(s)$ in this problem is

$$\mathbf{G}_0(s) = \mathbf{G}_p(s)\mathbf{G}_c(s)$$

and

$$\mathbf{H}(s) = \mathbf{I}$$

From Equation (9–36) we obtain

$$[\mathbf{I} + \mathbf{G}_0(s)]\mathbf{G}(s) = \mathbf{G}_0(s)$$

or

$$\mathbf{G}_0(s)[\mathbf{I} - \mathbf{G}(s)] = \mathbf{G}(s)$$

Postmultiplying both sides of this last equation by $[\mathbf{I} - \mathbf{G}(s)]^{-1}$, we obtain

$$\mathbf{G}_0(s) = \mathbf{G}(s)[\mathbf{I} - \mathbf{G}(s)]^{-1}$$

Since the desired closed-loop transfer matrix $\mathbf{G}(s)$ is a diagonal matrix, $\mathbf{I} - \mathbf{G}(s)$ is also a diagonal matrix. Then $\mathbf{G}_0(s)$, a product of two diagonal matrices, is also a diagonal matrix. This means that, to achieve noninteraction, we must make $\mathbf{G}_0(s)$ a diagonal matrix, provided the feedback matrix $\mathbf{H}(s)$ is the identity matrix.

EXAMPLE 9–6

Consider the system shown in Figure 9–6. Determine the transfer matrix of the series compensator such that the closed-loop transfer matrix is

$$\mathbf{G}(s) = \begin{bmatrix} \dfrac{1}{s+1} & 0 \\ 0 & \dfrac{1}{5s+1} \end{bmatrix}$$

Since

$$\mathbf{G}_0 = \mathbf{G}(\mathbf{I} - \mathbf{G})^{-1}$$

$$= \begin{bmatrix} \dfrac{1}{s+1} & 0 \\ 0 & \dfrac{1}{5s+1} \end{bmatrix} \begin{bmatrix} \dfrac{s+1}{s} & 0 \\ 0 & \dfrac{5s+1}{5s} \end{bmatrix} = \begin{bmatrix} \dfrac{1}{s} & 0 \\ 0 & \dfrac{1}{5s} \end{bmatrix}$$

and from Figure 9–6 the plant transfer matrix is

$$\mathbf{G}_p(s) = \begin{bmatrix} \dfrac{1}{2s+1} & 0 \\ 1 & \dfrac{1}{s+1} \end{bmatrix}$$

we have

$$\mathbf{G}_0(s) = \begin{bmatrix} \dfrac{1}{s} & 0 \\ 0 & \dfrac{1}{5s} \end{bmatrix} = \mathbf{G}_p(s)\mathbf{G}_c(s) = \begin{bmatrix} \dfrac{1}{2s+1} & 0 \\ 1 & \dfrac{1}{s+1} \end{bmatrix} \begin{bmatrix} G_{c11}(s) & G_{c12}(s) \\ G_{c21}(s) & G_{c22}(s) \end{bmatrix}$$

Hence

$$\mathbf{G}_c(s) = \begin{bmatrix} G_{c11}(s) & G_{c12}(s) \\ G_{c21}(s) & G_{c22}(s) \end{bmatrix}$$

$$= \begin{bmatrix} \dfrac{1}{2s+1} & 0 \\ 1 & \dfrac{1}{s+1} \end{bmatrix}^{-1} \begin{bmatrix} \dfrac{1}{s} & 0 \\ 0 & \dfrac{1}{5s} \end{bmatrix}$$

$$= \begin{bmatrix} 2s+1 & 0 \\ -(s+1)(2s+1) & s+1 \end{bmatrix} \begin{bmatrix} \dfrac{1}{s} & 0 \\ 0 & \dfrac{1}{5s} \end{bmatrix}$$

$$= \begin{bmatrix} \dfrac{2s+1}{s} & 0 \\ -\dfrac{(s+1)(2s+1)}{s} & \dfrac{s+1}{5s} \end{bmatrix} \tag{9-37}$$

Equation (9–37) gives the transfer matrix of the series compensator. Note that $G_{c11}(s)$ and $G_{c22}(s)$ are

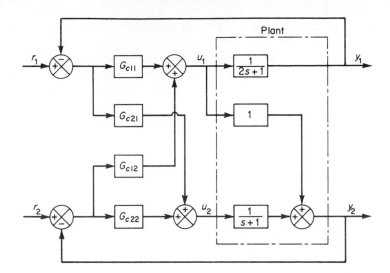

Figure 9–6
Multiple-input,
multiple-output
system.

proportional-plus-integral controllers and $G_{c21}(s)$ is a proportional-plus-integral-plus-derivative controller.

If the controller transfer matrix is chosen as given by Equation (9–37), then there is no interaction between two inputs and two outputs. Changes in r_1 affect only output y_1 and changes in r_2 affect only output y_2.

It is very important to note that in the present analysis we have not considered external disturbances. In general, in the present approach, cancellations take place in the numerator and denominator. Therefore, some of the eigenvalues will be lost in $\mathbf{G}_p(s)\mathbf{G}_c(s)$. This means that although the present approach will yield the desired result of noninteraction in the responses to reference inputs in the absence of external disturbances, if the system is disturbed by external forces, then the system may become uncontrollable because any motion caused by the canceled eigenvalue cannot be controlled. (We shall discuss details of system controllability in Section 9–4.)

9–4 CONTROLLABILITY

Controllability and observability. A system is said to be controllable at time t_0 if it is possible by means of an unconstrained control vector to transfer the system from any initial state $\mathbf{x}(t_0)$ to any other state in a finite interval of time.

A system is said to be observable at time t_0 if, with the system in state $\mathbf{x}(t_0)$, it is possible to determine this state from the observation of the output over a finite time interval.

The concepts of controllability and observability were introduced by Kalman. They play an important role in the design of control systems in state space. In fact, the conditions of controllability and observability may govern the existence of a complete solution to the control system design problem. The solution to this problem may not exist if the system considered is not controllable. Although most physical systems are controllable and observable, corresponding mathematical models may not possess the property of controllability and observability. Then it is necessary to know the conditions under which a system is controllable and observable. This section deals with controllability and the next section discusses observability.

In what follows we shall first define the linear independence of vectors and then derive the condition for complete state controllability. We also derive alternate forms of the condition for complete state controllability. Finally, we discuss complete output controllability.

Linear independence of vectors. The vectors $\mathbf{x}_1$, $\mathbf{x}_2$, . . . , $\mathbf{x}_n$ are said to be linearly independent if

$$c_1\mathbf{x}_1 + c_2\mathbf{x}_2 + \cdots + c_n\mathbf{x}_n = \mathbf{0}$$

where c_1, c_2, . . . , c_n are constants, implies

$$c_1 = c_2 = \cdots = c_n = 0$$

Conversely, the vectors $\mathbf{x}_1$, $\mathbf{x}_2$, . . . , $\mathbf{x}_n$ are said to be linearly dependent if and only if $\mathbf{x}_i$ can be expressed as a linear combination of $\mathbf{x}_j$ ($j = 1, 2, . . . , n; j \neq i$), or

$$\mathbf{x}_i = \sum_{\substack{j=1 \\ j \neq i}}^{n} c_j\mathbf{x}_j$$

for some set of constants c_j. This means that if $\mathbf{x}_i$ can be expressed as a linear combination of the other vectors in the set, it is linearly dependent on them or it is not an independent member of the set.

EXAMPLE 9–7   The vectors

$$\mathbf{x}_1 = \begin{bmatrix} 1 \\ 2 \\ 3 \end{bmatrix}, \qquad \mathbf{x}_2 = \begin{bmatrix} 1 \\ 0 \\ 1 \end{bmatrix}, \qquad \mathbf{x}_3 = \begin{bmatrix} 2 \\ 2 \\ 4 \end{bmatrix}$$

are linearly dependent since

$$\mathbf{x}_1 + \mathbf{x}_2 - \mathbf{x}_3 = \mathbf{0}$$

The vectors

$$\mathbf{y}_1 = \begin{bmatrix} 1 \\ 2 \\ 3 \end{bmatrix}, \qquad \mathbf{y}_2 = \begin{bmatrix} 1 \\ 0 \\ 1 \end{bmatrix}, \qquad \mathbf{y}_3 = \begin{bmatrix} 2 \\ 2 \\ 2 \end{bmatrix}$$

are linearly independent since

$$c_1\mathbf{y}_1 + c_2\mathbf{y}_2 + c_3\mathbf{y}_3 = \mathbf{0}$$

implies

$$c_1 = c_2 = c_3 = 0$$

Note that if an $n \times n$ matrix is nonsingular (that is, the matrix is of rank n or the determinant is nonzero), then n column (or row) vectors are linearly independent. If the $n \times n$ matrix is singular (that is, the rank of the matrix is less than n or the determinant is zero), then n column (or row) vectors are linearly dependent. To demonstrate this, notice that

$$[\mathbf{x}_1 \mid \mathbf{x}_2 \mid \mathbf{x}_3] = \begin{bmatrix} 1 & 1 & 2 \\ 2 & 0 & 2 \\ 3 & 1 & 4 \end{bmatrix} = \text{singular}$$

$$[\mathbf{y}_1 \mid \mathbf{y}_2 \mid \mathbf{y}_3] = \begin{bmatrix} 1 & 1 & 2 \\ 2 & 0 & 2 \\ 3 & 1 & 2 \end{bmatrix} = \text{nonsingular}$$

Complete state controllability of continuous-time systems.   Consider the continuous-time system

$$\dot{\mathbf{x}} = \mathbf{A}\mathbf{x} + \mathbf{B}u \tag{9-38}$$

where $\mathbf{x}$ = state vector (n-vector)
u = control signal (scalar)
$\mathbf{A}$ = $n \times n$ matrix
$\mathbf{B}$ = $n \times 1$ matrix

The system described by Equation (9–38) is said to be state controllable at $t = t_0$ if it is possible to construct an unconstrained control signal that will transfer an initial state to any final state in a finite time interval $t_0 \leq t \leq t_1$. If every state is controllable, then the system is said to be completely state controllable.

We shall now derive the condition for complete state controllability. Without loss of generality, we can assume that the final state is the origin of the state space and that the initial time is zero, or $t_0 = 0$.

The solution of Equation (9–38) is

$$\mathbf{x}(t) = e^{\mathbf{A}t}\mathbf{x}(0) + \int_0^t e^{\mathbf{A}(t-\tau)}\mathbf{B}u(\tau)d\tau$$

Applying the definition of complete state controllability just given, we have

$$\mathbf{x}(t_1) = \mathbf{0} = e^{\mathbf{A}t_1}\mathbf{x}(0) + \int_0^{t_1} e^{\mathbf{A}(t_1-\tau)}\mathbf{B}u(\tau)d\tau$$

or

$$\mathbf{x}(0) = -\int_0^{t_1} e^{-\mathbf{A}\tau}\mathbf{B}u(\tau)d\tau \tag{9-39}$$

Note that $e^{-\mathbf{A}\tau}$ can be written

$$e^{-\mathbf{A}\tau} = \sum_{k=0}^{n-1} \alpha_k(\tau)\mathbf{A}^k \tag{9-40}$$

Substituting Equation (9–40) into Equation (9–39) gives

$$\mathbf{x}(0) = -\sum_{k=0}^{n-1} \mathbf{A}^k\mathbf{B}\int_0^{t_1} \alpha_k(\tau)u(\tau)d\tau \tag{9-41}$$

Let us put

$$\int_0^{t_1} \alpha_k(\tau)u(\tau)d\tau = \beta_k$$

Then Equation (9–41) becomes

$$\mathbf{x}(0) = -\sum_{k=0}^{n-1} \mathbf{A}^k\mathbf{B}\beta_k$$

$$= -[\mathbf{B} \mid \mathbf{AB} \mid \cdots \mid \mathbf{A}^{n-1}\mathbf{B}] \begin{bmatrix} \beta_0 \\ \beta_1 \\ \vdots \\ \beta_{n-1} \end{bmatrix} \qquad (9\text{–}42)$$

If the system is completely state controllable, then, given any initial state $\mathbf{x}(0)$, Equation (9–42) must be satisfied. This requires that the rank of the $n \times n$ matrix

$$[\mathbf{B} \mid \mathbf{AB} \mid \cdots \mid \mathbf{A}^{n-1}\mathbf{B}]$$

be n.

From this analysis, we can state the condition for complete state controllability as follows: The system given by Equation (9–38) is completely state controllable if and only if the vectors $\mathbf{B}, \mathbf{AB}, \ldots, \mathbf{A}^{n-1}\mathbf{B}$ are linearly independent, or the $n \times n$ matrix

$$[\mathbf{B} \mid \mathbf{AB} \mid \cdots \mid \mathbf{A}^{n-1}\mathbf{B}]$$

is of rank n.

The result just obtained can be extended to the case where the control vector $\mathbf{u}$ is r-dimensional. If the system is described by

$$\dot{\mathbf{x}} = \mathbf{Ax} + \mathbf{Bu}$$

where $\mathbf{u}$ is an r-vector, then it can be proved that the condition for complete state controllability is that the following $n \times nr$ matrix

$$[\mathbf{B} \mid \mathbf{AB} \mid \cdots \mid \mathbf{A}^{n-1}\mathbf{B}]$$

is of rank n, or contains n linearly independent column vectors. The matrix

$$[\mathbf{B} \mid \mathbf{AB} \mid \cdots \mid \mathbf{A}^{n-1}\mathbf{B}]$$

is commonly called the *controllability matrix*.

EXAMPLE 9–8   Consider the system given by

$$\begin{bmatrix} \dot{x}_1 \\ \dot{x}_2 \end{bmatrix} = \begin{bmatrix} 1 & 1 \\ 0 & -1 \end{bmatrix} \begin{bmatrix} x_1 \\ x_2 \end{bmatrix} + \begin{bmatrix} 1 \\ 0 \end{bmatrix} u$$

Since

$$[\mathbf{B} \mid \mathbf{AB}] = \begin{bmatrix} 1 & 1 \\ 0 & 0 \end{bmatrix} = \text{singular}$$

the system is not completely state controllable.

EXAMPLE 9–9 Consider the system given by

$$\begin{bmatrix} \dot{x}_1 \\ \dot{x}_2 \end{bmatrix} = \begin{bmatrix} 1 & 1 \\ 2 & -1 \end{bmatrix} \begin{bmatrix} x_1 \\ x_2 \end{bmatrix} + \begin{bmatrix} 0 \\ 1 \end{bmatrix} [u]$$

For this case,

$$[\mathbf{B} \ \vdots \ \mathbf{AB}] = \begin{bmatrix} 0 & 1 \\ 1 & -1 \end{bmatrix} = \text{nonsingular}$$

The system is therefore completely state controllable.

Alternate form of the condition for complete state controllability. Consider the system defined by

$$\dot{\mathbf{x}} = \mathbf{Ax} + \mathbf{Bu} \qquad (9\text{–}43)$$

where $\mathbf{x}$ = state vector (n-vector)
$\quad \mathbf{u}$ = control vector (r-vector)
$\quad \mathbf{A} = n \times n$ matrix
$\quad \mathbf{B} = n \times r$ matrix

If the eigenvectors of $\mathbf{A}$ are distinct, then it is possible to find a transformation matrix $\mathbf{P}$ such that

$$\mathbf{P}^{-1}\mathbf{AP} = \mathbf{D} = \begin{bmatrix} \lambda_1 & & & & 0 \\ & \lambda_2 & & & \\ & & \cdot & & \\ & & & \cdot & \\ 0 & & & & \lambda_n \end{bmatrix}$$

Note that if the eigenvalues of $\mathbf{A}$ are distinct, then the eigenvectors of $\mathbf{A}$ are distinct; however, the converse is not true. For example, an $n \times n$ real symmetric matrix having multiple eigenvalues has n distinct eigenvectors. Note also that each column of the $\mathbf{P}$ matrix is an eigenvector of $\mathbf{A}$ associated with λ_i ($i = 1, 2, \ldots, n$).

Let us define

$$\mathbf{x} = \mathbf{Pz} \qquad (9\text{–}44)$$

Substituting Equation (9–44) into Equation (9–43), we obtain

$$\dot{\mathbf{z}} = \mathbf{P}^{-1}\mathbf{APz} + \mathbf{P}^{-1}\mathbf{Bu} \qquad (9\text{–}45)$$

By defining

$$\mathbf{P}^{-1}\mathbf{B} = \mathbf{F} = (f_{ij})$$

we can rewrite Equation (9–45) as

$$\dot{z}_1 = \lambda_1 z_1 + f_{11} u_1 + f_{12} u_2 + \cdots + f_{1r} u_r$$

$$\dot{z}_2 = \lambda_2 z_2 + f_{21} u_1 + f_{22} u_2 + \cdots + f_{2r} u_r$$

$$\vdots$$

$$\dot{z}_n = \lambda_n z_n + f_{n1} u_1 + f_{n2} u_2 + \cdots + f_{nr} u_r$$

If the elements of any one row of the $n \times r$ matrix $\mathbf{F}$ are all zero, then the corresponding state variable cannot be controlled by any of the u_i. Hence, the condition of complete state controllability is that, if the eigenvectors of $\mathbf{A}$ are distinct, then the system is completely state controllable if and only if no row of $\mathbf{P}^{-1}\mathbf{B}$ has all zero elements. It is important to note that, to apply this condition for complete state controllability, we must put the matrix $\mathbf{P}^{-1}\mathbf{A}\mathbf{P}$ in Equation (9–45) in diagonal form.

If the $\mathbf{A}$ matrix in Equation (9–43) does not possess distinct eigenvectors, then diagonalization is impossible. In such a case, we may transform $\mathbf{A}$ into the Jordan canonical form. If, for example, $\mathbf{A}$ has eigenvalues $\lambda_1, \lambda_1, \lambda_1, \lambda_4, \lambda_4, \lambda_6, \ldots, \lambda_n$ and has $n - 3$ distinct eigenvectors, then the Jordan canonical form of $\mathbf{A}$ is

$$\mathbf{J} = \begin{bmatrix} \lambda_1 & 1 & 0 & & & & & 0 \\ 0 & \lambda_1 & 1 & & & & & \\ 0 & 0 & \lambda_1 & & & & & \\ & & & \lambda_4 & 1 & & & \\ & & & 0 & \lambda_4 & & & \\ & & & & & \lambda_6 & & \\ & & & & & & \ddots & \\ 0 & & & & & & & \lambda_n \end{bmatrix}$$

The square submatrices on the main diagonal are called *Jordan blocks*.

Suppose that we can find a transformation matrix $\mathbf{S}$ such that

$$\mathbf{S}^{-1}\mathbf{A}\mathbf{S} = \mathbf{J}$$

If we define a new state vector $\mathbf{z}$ by

$$\mathbf{x} = \mathbf{S}\mathbf{z} \qquad (9\text{–}46)$$

then substitution of Equation (9–46) into Equation (9–43) yields

$$\dot{\mathbf{z}} = \mathbf{S}^{-1}\mathbf{A}\mathbf{S}\mathbf{z} + \mathbf{S}^{-1}\mathbf{B}\mathbf{u}$$
$$= \mathbf{J}\mathbf{z} + \mathbf{S}^{-1}\mathbf{B}\mathbf{u} \qquad (9\text{–}47)$$

The condition for complete state controllability of the system of Equation (9–43) may then be stated as follows: The system is completely state controllable if and only if (1) no two Jordan blocks in $\mathbf{J}$ of Equation (9–47) are associated with the same eigenvalues, (2) the elements of any row of $\mathbf{S}^{-1}\mathbf{B}$ that correspond to the last row of each Jordan block are not all zero, and (3) the elements of each row of $\mathbf{S}^{-1}\mathbf{B}$ that correspond to distinct eigenvalues are not all zero.

EXAMPLE 9–10   The following systems are completely state controllable:

$$\begin{bmatrix} \dot{x}_1 \\ \dot{x}_2 \end{bmatrix} = \begin{bmatrix} -1 & 0 \\ 0 & -2 \end{bmatrix}\begin{bmatrix} x_1 \\ x_2 \end{bmatrix} + \begin{bmatrix} 2 \\ 5 \end{bmatrix}u$$

$$\begin{bmatrix} \dot{x}_1 \\ \dot{x}_2 \\ \dot{x}_3 \end{bmatrix} = \begin{bmatrix} -1 & 1 & 0 \\ 0 & -1 & 0 \\ 0 & 0 & -2 \end{bmatrix}\begin{bmatrix} x_1 \\ x_2 \\ x_3 \end{bmatrix} + \begin{bmatrix} 0 \\ 4 \\ 3 \end{bmatrix}u$$

$$\begin{bmatrix} \dot{x}_1 \\ \dot{x}_2 \\ \dot{x}_3 \\ \dot{x}_4 \\ \dot{x}_5 \end{bmatrix} = \left[\begin{array}{ccc|cc} -2 & 1 & 0 & & 0 \\ 0 & -2 & 1 & & \\ 0 & 0 & -2 & & \\ \hline & & & -5 & 1 \\ 0 & & & 0 & -5 \end{array}\right]\begin{bmatrix} x_1 \\ x_2 \\ x_3 \\ x_4 \\ x_5 \end{bmatrix} + \begin{bmatrix} 0 & 1 \\ 0 & 0 \\ 3 & 0 \\ 0 & 0 \\ 2 & 1 \end{bmatrix}\begin{bmatrix} u_1 \\ u_2 \end{bmatrix}$$

The following systems are not completely state controllable:

$$\begin{bmatrix} \dot{x}_1 \\ \dot{x}_2 \end{bmatrix} = \begin{bmatrix} -1 & 0 \\ 0 & -2 \end{bmatrix}\begin{bmatrix} x_1 \\ x_2 \end{bmatrix} + \begin{bmatrix} 2 \\ 0 \end{bmatrix}u$$

$$\begin{bmatrix} \dot{x}_1 \\ \dot{x}_2 \\ \dot{x}_3 \end{bmatrix} = \begin{bmatrix} -1 & 1 & 0 \\ 0 & -1 & 0 \\ 0 & 0 & -2 \end{bmatrix}\begin{bmatrix} x_1 \\ x_2 \\ x_3 \end{bmatrix} + \begin{bmatrix} 4 & 2 \\ 0 & 0 \\ 3 & 0 \end{bmatrix}\begin{bmatrix} u_1 \\ u_2 \end{bmatrix}$$

$$\begin{bmatrix} \dot{x}_1 \\ \dot{x}_2 \\ \dot{x}_3 \\ \dot{x}_4 \\ \dot{x}_5 \end{bmatrix} = \left[\begin{array}{ccc|cc} -2 & 1 & 0 & & 0 \\ 0 & -2 & 1 & & \\ 0 & 0 & -2 & & \\ \hline & & & -5 & 1 \\ 0 & & & 0 & -5 \end{array}\right]\begin{bmatrix} x_1 \\ x_2 \\ x_3 \\ x_4 \\ x_5 \end{bmatrix} + \begin{bmatrix} 4 \\ 2 \\ 1 \\ 3 \\ 0 \end{bmatrix}u$$

Condition for complete state controllability in the s plane.   The condition for complete state controllability can be stated in terms of transfer functions or transfer matrices.

A necessary and sufficient condition for complete state controllability is that no cancellation occurs in the transfer function or transfer matrix. (For a proof, see Problem A–9–10.) If cancellation occurs, the system cannot be controlled in the direction of the canceled mode.

EXAMPLE 9–11   Consider the following transfer function:

$$\frac{X(s)}{U(s)} = \frac{s + 2.5}{(s + 2.5)(s - 1)}$$

Clearly, cancellation of the factor $(s + 2.5)$ occurs in the numerator and denominator of this transfer function. (Thus one degree of freedom is lost.) Because of this cancellation, this system is not completely state controllable.

The same conclusion can be obtained by writing this transfer function in the form of a state equation. A state-space representation is

$$\begin{bmatrix} \dot{x}_1 \\ \dot{x}_2 \end{bmatrix} = \begin{bmatrix} 0 & 1 \\ 2.5 & -1.5 \end{bmatrix}\begin{bmatrix} x_1 \\ x_2 \end{bmatrix} + \begin{bmatrix} 1 \\ 1 \end{bmatrix}u$$

Since

$$[\mathbf{B} \,\vdots\, \mathbf{AB}] = \begin{bmatrix} 1 & 1 \\ 1 & 1 \end{bmatrix}$$

the rank of the matrix $[\mathbf{B} \,\vdots\, \mathbf{AB}]$ is 1. Therefore we arrive at the same conclusion: The system is not completely state controllable.

Output controllability. In the practical design of a control system, we may want to control the output rather than the state of the system. Complete state controllability is neither necessary nor sufficient for controlling the output of the system. For this reason, it is desirable to define separately complete output controllability.

Consider the system described by

$$\dot{\mathbf{x}} = \mathbf{Ax} + \mathbf{Bu} \tag{9–48}$$

$$\mathbf{y} = \mathbf{Cx} + \mathbf{Du} \tag{9–49}$$

where $\mathbf{x}$ = state vector (n-vector)
  $\mathbf{u}$ = control vector (r-vector)
  $\mathbf{y}$ = output vector (m-vector)
  $\mathbf{A}$ = $n \times n$ matrix
  $\mathbf{B}$ = $n \times r$ matrix
  $\mathbf{C}$ = $m \times n$ matrix
  $\mathbf{D}$ = $m \times r$ matrix

The system described by Equations (9–48) and (9–49) is said to be completely output controllable if it is possible to construct an unconstrained control vector $\mathbf{u}(t)$ that will transfer any given initial output $\mathbf{y}(t_0)$ to any final output $\mathbf{y}(t_1)$ in a finite time interval $t_0 \le t \le t_1$.

It can be proved that the condition for complete output controllability is as follows: The system described by Equations (9–48) and (9–49) is completely output controllable if and only if the $m \times (n + 1)r$ matrix

$$[\mathbf{CB} \,\vdots\, \mathbf{CAB} \,\vdots\, \mathbf{CA^2B} \,\vdots\, \cdots \,\vdots\, \mathbf{CA^{n-1}B} \,\vdots\, \mathbf{D}]$$

is of rank m. (For a proof, see Problem A–9–11.) Note that the presence of $\mathbf{Du}$ term in Equation (9–49) always helps to establish output controllability.

9–5 OBSERVABILITY

In this section we discuss the observability of linear systems. Consider the unforced system described by the following equations:

$$\dot{\mathbf{x}} = \mathbf{Ax} \tag{9–50}$$

$$\mathbf{y} = \mathbf{Cx} \tag{9–51}$$

where $\mathbf{x}$ = state vector (n-vector)
  $\mathbf{y}$ = output vector (m-vector)
  $\mathbf{A}$ = $n \times n$ matrix
  $\mathbf{C}$ = $m \times n$ matrix

The system is said to be completely observable if every state $\mathbf{x}(t_0)$ can be determined from the observation of $\mathbf{y}(t)$ over a finite time interval, $t_0 \leq t \leq t_1$. The system is, therefore, completely observable if every transition of the state eventually affects every element of the output vector. The concept of observability is useful in solving the problem of reconstructing unmeasurable state variables from measurable ones in the minimum possible length of time. In this section we treat only linear, time-invariant systems. Therefore, without loss of generality, we can assume that $t_0 = 0$.

The concept of observability is very important because, in practice, the difficulty encountered with state feedback control is that some of the state variables are not accessible for direct measurement, with the result that it becomes necessary to estimate the ummeasurable state variables in order to construct the control signals. It will be shown in Section 10–3 that such estimates of state variables are possible if and only if the system is completely observable.

In discussing observability conditions, we consider the unforced system as given by Equations (9–50) and (9–51). The reason for this is as follows: If the system is described by

$$\dot{\mathbf{x}} = \mathbf{A}\mathbf{x} + \mathbf{B}\mathbf{u}$$

$$\mathbf{y} = \mathbf{C}\mathbf{x} + \mathbf{D}\mathbf{u}$$

then

$$\mathbf{x}(t) = e^{\mathbf{A}t}\mathbf{x}(0) + \int_0^t e^{\mathbf{A}(t-\tau)}\mathbf{B}\mathbf{u}(\tau)\,d\tau$$

and $\mathbf{y}(t)$ is

$$\mathbf{y}(t) = \mathbf{C}e^{\mathbf{A}t}\mathbf{x}(0) + \mathbf{C}\int_0^t e^{\mathbf{A}(t-\tau)}\mathbf{B}\mathbf{u}(\tau)\,d\tau + \mathbf{D}\mathbf{u}$$

Since the matrices $\mathbf{A}$, $\mathbf{B}$, $\mathbf{C}$, and $\mathbf{D}$ are known and $\mathbf{u}(t)$ is also known, the last two terms on the right side of this last equation are known quantities. Therefore, they may be subtracted from the observed value of $\mathbf{y}(t)$. Hence, for investigating a necessary and sufficient condition for complete observability, it suffices to consider the system described by Equations (9–50) and (9–51).

Complete observability of continuous-time systems. Consider the system described by Equations (9–50) and (9–51), rewritten

$$\dot{\mathbf{x}} = \mathbf{A}\mathbf{x}$$

$$\mathbf{y} = \mathbf{C}\mathbf{x}$$

The output vector $\mathbf{y}(t)$ is

$$\mathbf{y}(t) = \mathbf{C}e^{\mathbf{A}t}\mathbf{x}(0)$$

Noting that

$$e^{\mathbf{A}t} = \sum_{k=0}^{n-1} \alpha_k(t)\mathbf{A}^k$$

we obtain

$$\mathbf{y}(t) = \sum_{k=0}^{n-1} \alpha_k(t)\mathbf{C}\mathbf{A}^k\mathbf{x}(0)$$

or

$$\mathbf{y}(t) = \alpha_0(t)\mathbf{C}\mathbf{x}(0) + \alpha_1(t)\mathbf{C}\mathbf{A}\mathbf{x}(0) + \cdots + \alpha_{n-1}(t)\mathbf{C}\mathbf{A}^{n-1}\mathbf{x}(0) \qquad (9\text{--}52)$$

If the system is completely observable, then given the output $\mathbf{y}(t)$ over a time interval $0 \le t \le t_1$, $\mathbf{x}(0)$ is uniquely determined from Equation (9–52). It can be shown that this requires the rank of the $nm \times n$ matrix

$$\begin{bmatrix} \mathbf{C} \\ \hline \mathbf{C}\mathbf{A} \\ \hline \cdot \\ \cdot \\ \cdot \\ \hline \mathbf{C}\mathbf{A}^{n-1} \end{bmatrix}$$

to be n. (See Problem A–9–14 for the derivation of this condition.)

From this analysis, we can state the condition for complete observability as follows: The system described by Equations (9–50) and (9–51) is completely observable if and only if the $n \times nm$ matrix

$$[\mathbf{C}^* \ \vdots \ \mathbf{A}^*\mathbf{C}^* \ \vdots \ \cdots \ \vdots \ (\mathbf{A}^*)^{n-1}\mathbf{C}^*]$$

is of rank n or has n linearly independent column vectors. This matrix is called the *observability matrix*.

EXAMPLE 9–12   Consider the system described by

$$\begin{bmatrix} \dot{x}_1 \\ \dot{x}_2 \end{bmatrix} = \begin{bmatrix} 1 & 1 \\ -2 & -1 \end{bmatrix} \begin{bmatrix} x_1 \\ x_2 \end{bmatrix} + \begin{bmatrix} 0 \\ 1 \end{bmatrix} u$$

$$y = \begin{bmatrix} 1 & 0 \end{bmatrix} \begin{bmatrix} x_1 \\ x_2 \end{bmatrix}$$

Is this system controllable and observable?

Since the rank of the matrix

$$[\mathbf{B} \ \vdots \ \mathbf{A}\mathbf{B}] = \begin{bmatrix} 0 & 1 \\ 1 & -1 \end{bmatrix}$$

is 2, the system is completely state controllable.

For output controllability, let us find the rank of the matrix $[\mathbf{C}\mathbf{B} \ \vdots \ \mathbf{C}\mathbf{A}\mathbf{B}]$. Since

$$[\mathbf{C}\mathbf{B} \ \vdots \ \mathbf{C}\mathbf{A}\mathbf{B}] = \begin{bmatrix} 0 & 1 \end{bmatrix}$$

the rank of this matrix is 1. Hence the system is completely output controllable.

To test the observability condition, examine the rank of $[\mathbf{C}^* \ \vdots \ \mathbf{A}^*\mathbf{C}^*]$. Since

$$[\mathbf{C}^* \ \vdots \ \mathbf{A}^*\mathbf{C}^*] = \begin{bmatrix} 1 & 1 \\ 0 & 1 \end{bmatrix}$$

the rank of $[\mathbf{C}^* \ \vdots \ \mathbf{A}^*\mathbf{C}^*]$ is 2. Hence the system is completely observable.

Conditions for complete observability in the s plane. The conditions for complete observability can also be stated in terms of transfer functions or transfer matrices. The necessary and sufficient condition for complete observability is that no cancellation occurs in the transfer function or transfer matrix. If cancellation occurs, the canceled mode cannot be observed in the output.

EXAMPLE 9–13   Show that the following system is not completely observable.

$$\dot{\mathbf{x}} = \mathbf{A}\mathbf{x} + \mathbf{B}u$$

$$y = \mathbf{C}\mathbf{x}$$

where

$$\mathbf{x} = \begin{bmatrix} x_1 \\ x_2 \\ x_3 \end{bmatrix}, \quad \mathbf{A} = \begin{bmatrix} 0 & 1 & 0 \\ 0 & 0 & 1 \\ -6 & -11 & -6 \end{bmatrix}, \quad \mathbf{B} = \begin{bmatrix} 0 \\ 0 \\ 1 \end{bmatrix}, \quad \mathbf{C} = \begin{bmatrix} 4 & 5 & 1 \end{bmatrix}$$

Note that the control function u does not affect the complete observability of the system. To examine complete observability, we may simply set $u = 0$. For this system, we have

$$[\mathbf{C}^* \;\vdots\; \mathbf{A}^*\mathbf{C}^* \;\vdots\; (\mathbf{A}^*)^2\mathbf{C}^*] = \begin{bmatrix} 4 & -6 & 6 \\ 5 & -7 & 5 \\ 1 & -1 & -1 \end{bmatrix}$$

Note that

$$\begin{vmatrix} 4 & -6 & 6 \\ 5 & -7 & 5 \\ 1 & -1 & -1 \end{vmatrix} = 0$$

Hence the rank of the matrix $[\mathbf{C}^* \;\vdots\; \mathbf{A}^*\mathbf{C}^* \;\vdots\; (\mathbf{A}^*)^2\mathbf{C}^*]$ is less than 3. Therefore, the system is not completely observable.

In fact, in this system cancellation occurs in the transfer function of the system. The transfer function between $X_1(s)$ and $U(s)$ is

$$\frac{X_1(s)}{U(s)} = \frac{1}{(s + 1)(s + 2)(s + 3)}$$

and the transfer function between $Y(s)$ and $X_1(s)$ is

$$\frac{Y(s)}{X_1(s)} = (s + 1)(s + 4)$$

Therefore, the transfer function between the output $Y(s)$ and the input $U(s)$ is

$$\frac{Y(s)}{U(s)} = \frac{(s + 1)(s + 4)}{(s + 1)(s + 2)(s + 3)}$$

Clearly, the two factors $(s + 1)$ cancel each other. This means that there are nonzero initial states $\mathbf{x}(0)$, which cannot be determined from the measurement of $y(t)$.

Relationships between controllability, observability, and transfer functions. The transfer function has no cancellation if and only if the system is completely state controllable and completely observable. This means that the canceled transfer function does not carry along all the information characterizing the dynamic system.

where $\mathbf{x}$ = state vector (n-vector)
$\mathbf{u}$ = control vector (r-vector)
$\mathbf{y}$ = output vector (m-vector)
$\mathbf{A}$ = $n \times n$ matrix
$\mathbf{B}$ = $n \times r$ matrix
$\mathbf{C}$ = $m \times n$ matrix

and the dual system S_2 defined by

$$\dot{\mathbf{z}} = \mathbf{A}^*\mathbf{z} + \mathbf{C}^*\mathbf{v}$$

$$\mathbf{n} = \mathbf{B}^*\mathbf{z}$$

where $\mathbf{z}$ = state vector (n-vector)
$\mathbf{v}$ = control vector (m-vector)
$\mathbf{n}$ = output vector (r-vector)
$\mathbf{A}^*$ = conjugate transpose of $\mathbf{A}$
$\mathbf{B}^*$ = conjugate transpose of $\mathbf{B}$
$\mathbf{C}^*$ = conjugate transpose of $\mathbf{C}$

The principle of duality states that system S_1 is completely state controllable (observable) if and only if system S_2 is completely observable (state controllable).

To verify this principle, let us write down the necessary and sufficient conditions for complete state controllability and complete observability for systems S_1 and S_2.

For system S_1:

1. A necessary and sufficient condition for complete state controllability is that the rank of the $n \times nr$ matrix

$$[\mathbf{B} \ \vdots \ \mathbf{AB} \ \vdots \ \cdots \ \vdots \ \mathbf{A}^{n-1}\mathbf{B}]$$

be n.

2. A necessary and sufficient condition for complete observability is that the rank of the $n \times nm$ matrix

$$[\mathbf{C}^* \ \vdots \ \mathbf{A}^*\mathbf{C}^* \ \vdots \ \cdots \ \vdots \ (\mathbf{A}^*)^{n-1}\mathbf{C}^*]$$

be n.

For system S_2:

1. A necessary and sufficient condition for complete state controllability is that the rank of the $n \times nm$ matrix

$$[\mathbf{C}^* \ \vdots \ \mathbf{A}^*\mathbf{C}^* \ \vdots \ \cdots \ \vdots \ (\mathbf{A}^*)^{n-1}\mathbf{C}^*]$$

be n.

2. A necessary and sufficient condition for complete observability is that the rank of the $n \times nr$ matrix

$$[\mathbf{B} \ \vdots \ \mathbf{AB} \ \vdots \ \cdots \ \vdots \ \mathbf{A}^{n-1}\mathbf{B}]$$

be n.

By comparing these conditions, the truth of this principle is apparent. By use of this principle, the observability of a given system can be checked by testing the state controllability of its dual.

9–6 OBTAINING STATE-SPACE EQUATIONS IN CANONICAL FORMS

Many techniques are available for obtaining state-space representations of differential equation systems. In Chapter 4 we presented a few such techniques. This section presents methods for obtaining state-space equations in the controllable, observable, diagonal, or Jordan canonical form.

State-space representation in canonical forms. Consider a system defined by

$$\overset{(n)}{y} + a_1 \overset{(n-1)}{y} + \cdots + a_{n-1}\dot{y} + a_n y = b_0 \overset{(n)}{u} + b_1 \overset{(n-1)}{u} + \cdots + b_{n-1}\dot{u} + b_n u$$

where u is the input and y is the output. This equation can also be written as

$$\frac{Y(s)}{U(s)} = \frac{b_0 s^n + b_1 s^{n-1} + \cdots + b_{n-1}s + b_n}{s^n + a_1 s^{n-1} + \cdots + a_{n-1}s + a_n} \tag{9-55}$$

In what follows, we shall present three methods for obtaining state-space representations in canonical forms:

1. Direct programming method for obtaining controllable canonical form
2. Nested programming method for obtaining observable canonical form
3. Partial-fraction-expansion programming method for obtaining diagonal or Jordan canonical form

Direct programming method. Equation (9–55) can be written as

$$\frac{Y(s)}{U(s)} = b_0 + \frac{(b_1 - a_1 b_0)s^{n-1} + \cdots + (b_{n-1} - a_{n-1}b_0)s + (b_n - a_n b_0)}{s^n + a_1 s^{n-1} + \cdots + a_{n-1}s + a_n}$$

which can be modified to

$$Y(s) = b_0 U(s) + \hat{Y}(s) \tag{9-56}$$

where

$$\hat{Y}(s) = \frac{(b_1 - a_1 b_0)s^{n-1} + \cdots + (b_{n-1} - a_{n-1}b_0)s + (b_n - a_n b_0)}{s^n + a_1 s^{n-1} + \cdots + a_{n-1}s + a_n} U(s)$$

Let us rewrite this last equation in the following form:

$$\frac{\hat{Y}(s)}{(b_1 - a_1 b_0)s^{n-1} + \cdots + (b_{n-1} - a_{n-1}b_0)s + (b_n - a_n b_0)}$$
$$= \frac{U(s)}{s^n + a_1 s^{n-1} + \cdots + a_{n-1}s + a_n} = Q(s)$$

From this last equation, the following two equations may be obtained:

$$s^n Q(s) = -a_1 s^{n-1} Q(s) - \cdots - a_{n-1} s Q(s) - a_n Q(s) + U(s) \qquad (9\text{--}57)$$

$$\hat{Y}(s) = (b_1 - a_1 b_0) s^{n-1} Q(s) + \cdots + (b_{n-1} - a_{n-1} b_0) s Q(s)$$

$$+ (b_n - a_n b_0) Q(s) \qquad (9\text{--}58)$$

Now define state variables as follows:

$$X_1(s) = Q(s)$$
$$X_2(s) = s Q(s)$$
$$\vdots$$
$$X_{n-1}(s) = s^{n-2} Q(s)$$
$$X_n(s) = s^{n-1} Q(s)$$

Then, clearly,

$$s X_1(s) = X_2(s)$$
$$s X_2(s) = X_3(s)$$
$$\vdots$$
$$s X_{n-1}(s) = X_n(s)$$

which may be rewritten as

$$\dot{x}_1 = x_2$$
$$\dot{x}_2 = x_3$$
$$\vdots$$
$$\dot{x}_{n-1} = x_n$$

Noting that $s^n Q(s) = s X_n(s)$, Equation (9–57) can be rewritten as

$$s X_n(s) = -a_1 X_n - \cdots - a_{n-1} X_2(s) - a_n X_1(s) + U(s)$$

or

$$\dot{x}_n = -a_n x_1 - a_{n-1} x_2 - \cdots - a_1 x_n + u$$

Also, from Equations (9–56) and (9–58) we obtain

$$Y(s) = b_0 U(s) + (b_1 - a_1 b_0) s^{n-1} Q(s) + \cdots + (b_{n-1} - a_{n-1} b_0) s Q(s)$$
$$+ (b_n - a_n b_0) Q(s)$$
$$= b_0 U(s) + (b_1 - a_1 b_0) X_n(s) + \cdots + (b_{n-1} - a_{n-1} b_0) X_2(s)$$
$$+ (b_n - a_n b_0) X_1(s)$$

The inverse Laplace transform of this output equation becomes

$$y = (b_n - a_nb_0)x_1 + (b_{n-1} - a_{n-1}b_0)x_2 + \cdots + (b_1 - a_1b_0)x_n + b_0u$$

Then the state equation and output equation can be given by

$$\begin{bmatrix} \dot{x}_1 \\ \dot{x}_2 \\ \cdot \\ \cdot \\ \cdot \\ \dot{x}_{n-1} \\ \dot{x}_n \end{bmatrix} = \begin{bmatrix} 0 & 1 & 0 & \cdots & 0 \\ 0 & 0 & 1 & \cdots & 0 \\ \cdot & \cdot & \cdot & & \cdot \\ \cdot & \cdot & \cdot & & \cdot \\ \cdot & \cdot & \cdot & & \cdot \\ 0 & 0 & 0 & \cdots & 1 \\ -a_n & -a_{n-1} & -a_{n-2} & \cdots & -a_1 \end{bmatrix} \begin{bmatrix} x_1 \\ x_2 \\ \cdot \\ \cdot \\ \cdot \\ x_{n-1} \\ x_n \end{bmatrix} + \begin{bmatrix} 0 \\ 0 \\ \cdot \\ \cdot \\ \cdot \\ 0 \\ 1 \end{bmatrix} u \qquad (9\text{--}59)$$

$$y = [b_n - a_nb_0 \;\vdots\; b_{n-1} - a_{n-1}b_0 \;\vdots\; \cdots \;\vdots\; b_1 - a_1b_0] \begin{bmatrix} x_1 \\ x_2 \\ \cdot \\ \cdot \\ \cdot \\ x_n \end{bmatrix} + b_0u \qquad (9\text{--}60)$$

The state space representation given by Equations (9–59) and (9–60) is said to be in the controllable canonical form. Figure 9–7 shows a block-diagram representation of the system given by Equations (9–59) and (9–60). The controllable canonical form is important in discussing the pole placement approach to the control system design.

Nested programming method. Consider the transfer function system defined by Equation (9–55). This equation may be modified into the following form:

$$s^n[Y(s) - b_0U(s)] + s^{n-1}[a_1Y(s) - b_1U(s)] + \cdots$$
$$+ s[a_{n-1}Y(s) - b_{n-1}U(s)] + a_nY(s) - b_nU(s) = 0$$

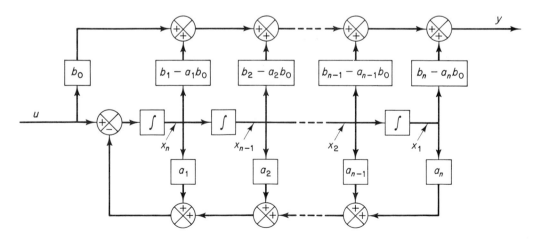

Figure 9–7
Block diagram representation of the system given by Eqs. (9–59) and (9–60). (Controllable canonical form.)

By dividing the entire equation by s^n and rearranging, we obtain

$$Y(s) = b_0 U(s) + \frac{1}{s}[b_1 U(s) - a_1 Y(s)] + \cdots$$

$$+ \frac{1}{s^{n-1}}[b_{n-1}U(s) - a_{n-1}Y(s)] + \frac{1}{s^n}[b_n U(s) - a_n Y(s)] \qquad (9\text{--}61)$$

Now define state variables as follows;

$$X_n(s) = \frac{1}{s}[b_1 U(s) - a_1 Y(s) + X_{n-1}(s)]$$

$$X_{n-1}(s) = \frac{1}{s}[b_2 U(s) - a_2 Y(s) + X_{n-2}(s)]$$

$$\cdot$$
$$\cdot \qquad\qquad\qquad\qquad\qquad\qquad (9\text{--}62)$$
$$\cdot$$

$$X_2(s) = \frac{1}{s}[b_{n-1}U(s) - a_{n-1}Y(s) + X_1(s)]$$

$$X_1(s) = \frac{1}{s}[b_n U(s) - a_n Y(s)]$$

Then Equation (9–61) can be written as

$$Y(s) = b_0 U(s) + X_n(s) \qquad (9\text{--}63)$$

By substituting Equation (9–63) into Equation (9–62) and multiplying both sides of the equations by s, we obtain

$$sX_n(s) = X_{n-1}(s) - a_1 X_n(s) + (b_1 - a_1 b_0)U(s)$$
$$sX_{n-1}(s) = X_{n-2}(s) - a_2 X_n(s) + (b_2 - a_2 b_0)U(s)$$
$$\cdot$$
$$\cdot$$
$$\cdot$$
$$sX_2(s) = X_1(s) - a_{n-1}X_n(s) + (b_{n-1} - a_{n-1}b_0)U(s)$$
$$sX_1(s) = -a_n X_n(s) + (b_n - a_n b_0)U(s)$$

Taking the inverse Laplace transforms of the preceding n equations and writing them in the reverse order, we get

$$\dot{x}_1 = -a_n x_n + (b_n - a_n b_0)u$$
$$\dot{x}_2 = x_1 - a_{n-1}x_n + (b_{n-1} - a_{n-1}b_0)u$$
$$\cdot$$
$$\cdot$$
$$\cdot$$
$$\dot{x}_{n-1} = x_{n-2} - a_2 x_n + (b_2 - a_2 b_0)u$$
$$\dot{x}_n = x_{n-1} - a_1 x_n + (b_1 - a_1 b_0)u$$

Also, the inverse Laplace transform of Equation (9–63) gives

$$y = x_n + b_0 u$$

Rewriting the state and output equations in the standard vector-matrix forms gives

$$
\begin{bmatrix} \dot{x}_1 \\ \dot{x}_2 \\ \cdot \\ \cdot \\ \cdot \\ \dot{x}_n \end{bmatrix}
=
\begin{bmatrix}
0 & 0 & \cdots & 0 & -a_n \\
1 & 0 & \cdots & 0 & -a_{n-1} \\
\cdot & \cdot & & \cdot & \cdot \\
\cdot & \cdot & & \cdot & \cdot \\
\cdot & \cdot & & \cdot & \cdot \\
0 & 0 & \cdots & 1 & -a_1
\end{bmatrix}
\begin{bmatrix} x_1 \\ x_2 \\ \cdot \\ \cdot \\ \cdot \\ x_n \end{bmatrix}
+
\begin{bmatrix} b_n - a_n b_0 \\ b_{n-1} - a_{n-1} b_0 \\ \cdot \\ \cdot \\ \cdot \\ b_1 - a_1 b_0 \end{bmatrix} u
\qquad (9\text{–}64)
$$

$$
y = \begin{bmatrix} 0 & 0 & \cdots & 0 & 1 \end{bmatrix}
\begin{bmatrix} x_1 \\ x_2 \\ \cdot \\ \cdot \\ \cdot \\ x_{n-1} \\ x_n \end{bmatrix}
+ b_0 u
\qquad (9\text{–}65)
$$

The state-space representation given by Equations (9–64) and (9–65) is said to be in the observable canonical form. Figure 9–8 shows the block diagram representation of the system given by Equations (9–64) and (9–65). Notice that the $n \times n$ state matrix of the state equation given by Equation (9–64) is the transpose of that of the state equation defined by Equation (9–59).

Partial-fraction-expansion programming method. Consider the transfer function system defined by Equation (9–55). We shall first consider the case where the denominator polynomial involves only distinct roots. Then we shall consider the case where the denominator polynomial includes multiple roots. For the distinct roots case, Equation (9–55) can be written as

$$
\frac{Y(s)}{U(s)} = \frac{b_0 s^n + b_1 s^{n-1} + \cdots + b_{n-1} s + b_n}{(s - p_1)(s - p_2) \cdots (s - p_n)}
$$

$$
= b_0 + \frac{c_1}{s - p_1} + \frac{c_2}{s - p_2} + \cdots + \frac{c_n}{s - p_n}
$$

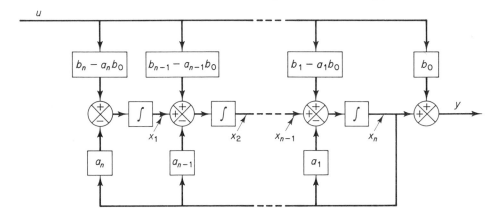

Figure 9–8
Block diagram representation of the system given by Eqs. (9–64) and (9–65). (Observable canonical form.)

where $p_i \neq p_j$. This last equation may be written as

$$Y(s) = b_0 U(s) + \frac{c_1}{s - p_1} U(s) + \frac{c_2}{s - p_2} U(s) + \cdots + \frac{c_n}{s - p_n} U(s) \qquad (9\text{--}66)$$

Define the state variables as follows:

$$X_1(s) = \frac{1}{s - p_1} U(s)$$

$$X_2(s) = \frac{1}{s - p_2} U(s)$$

$$\cdot$$
$$\cdot$$
$$\cdot$$

$$X_n(s) = \frac{1}{s - p_n} U(s)$$

which may be rewritten as

$$sX_1(s) = p_1 X_1(s) + U(s)$$
$$sX_2(s) = p_2 X_2(s) + U(s)$$
$$\cdot$$
$$\cdot$$
$$\cdot$$
$$sX_n(s) = p_n X_n(s) + U(s)$$

The inverse Laplace transforms of these equations give

$$\dot{x}_1 = p_1 x_1 + u$$
$$\dot{x}_2 = p_2 x_2 + u$$
$$\cdot$$
$$\cdot$$
$$\cdot$$
$$\dot{x}_n = p_n x_n + u$$

These n equations make up a state equation.

In terms of the state variables $X_1(s), X_2(s), \ldots, X_n(s)$, Equation (9–66) can be written as

$$Y(s) = b_0 U(s) + c_1 X_1(s) + c_2 X_2(s) + \cdots + c_n X_n(s)$$

The inverse Laplace transform of this last equation is

$$y = c_1 x_1 + c_2 x_2 + \cdots + c_n x_n + b_0 u$$

which is the output equation.

The state and output equations can now be written in the following standard form:

$$\begin{bmatrix} \dot{x}_1 \\ \dot{x}_2 \\ \cdot \\ \cdot \\ \cdot \\ \dot{x}_n \end{bmatrix} = \begin{bmatrix} p_1 & & & 0 \\ & p_2 & & \\ & & \cdot & \\ & & & \cdot \\ & & & \cdot \\ 0 & & & p_n \end{bmatrix} \begin{bmatrix} x_1 \\ x_2 \\ \cdot \\ \cdot \\ \cdot \\ x_n \end{bmatrix} + \begin{bmatrix} 1 \\ 1 \\ \cdot \\ \cdot \\ \cdot \\ 1 \end{bmatrix} u \qquad (9\text{--}67)$$

$$y = \begin{bmatrix} c_1 & c_2 & \cdots & c_n \end{bmatrix} \begin{bmatrix} x_1 \\ x_2 \\ \cdot \\ \cdot \\ \cdot \\ x_n \end{bmatrix} + b_0 u \qquad (9\text{--}68)$$

The state-space representation given by Equations (9–67) and (9–68) is said to be in the diagonal canonical form. Figure 9–9 shows a block-diagram representation of the system given by Equations (9–67) and (9–68).

Next we shall consider the case where the denominator polynomial of Equation (9–55) involves multiple roots. For this case, the preceding diagonal canonical form must be modified into the Jordan canonical form. Suppose, for example, that the p_i's are different from one another, except that the first three p_i's are equal, or $p_1 = p_2 = p_3$. Then the factored form of $Y(s)/U(s)$ becomes

$$\frac{Y(s)}{U(s)} = \frac{b_0 s^n + b_1 s^{n-1} + \cdots + b_{n-1}s + b_n}{(s - p_1)^3 (s - p_4)(s - p_5) \cdots (s - p_n)}$$

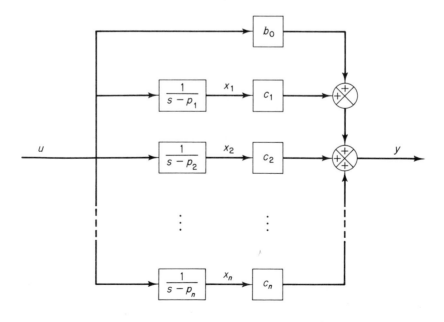

Figure 9–9
Block diagram
representation of the
system given by Eqs.
(9–67) and (9–68).
(Diagonal canonical
form.)

The partial-fraction expansion of this last equation becomes

$$\frac{Y(s)}{U(s)} = b_0 + \frac{c_1}{(s - p_1)^3} + \frac{c_2}{(s - p_1)^2} + \frac{c_3}{s - p_1} + \frac{c_4}{s - p_4} + \cdots + \frac{c_n}{s - p_n}$$

which may be written as

$$Y(s) = b_0 U(s) + \frac{c_1}{(s - p_1)^3} U(s) + \frac{c_2}{(s - p_1)^2} U(s) + \frac{c_3}{s - p_1} U(s)$$

$$+ \frac{c_4}{s - p_4} U(s) + \cdots + \frac{c_n}{s - p_n} U(s) \tag{9-69}$$

Define

$$X_1(s) = \frac{1}{(s - p_1)^3} U(s)$$

$$X_2(s) = \frac{1}{(s - p_1)^2} U(s)$$

$$X_3(s) = \frac{1}{s - p_1} U(s)$$

$$X_4(s) = \frac{1}{s - p_4} U(s)$$

$$\vdots$$

$$X_n(s) = \frac{1}{s - p_n} U(s)$$

Notice that the following relationships exist among $X_1(s)$, $X_2(s)$, and $X_3(s)$:

$$\frac{X_1(s)}{X_2(s)} = \frac{1}{s - p_1}$$

$$\frac{X_2(s)}{X_3(s)} = \frac{1}{s - p_1}$$

Then, from the definition of the state variables above and the preceding relationships, we obtain

$$sX_1(s) = p_1 X_1(s) + X_2(s)$$

$$sX_2(s) = p_1 X_2(s) + X_3(s)$$

$$sX_3(s) = p_1 X_3(s) + U(s)$$

$$sX_4(s) = p_4 X_4(s) + U(s)$$

$$\vdots$$

$$sX_n(s) = p_n X_n(s) + U(s)$$

The inverse Laplace transforms of the preceding n equations give

$$\dot{x}_1 = p_1 x_1 + x_2$$

$$\dot{x}_2 = p_1 x_2 + x_3$$

$$\dot{x}_3 = p_1 x_3 + u$$

$$\dot{x}_4 = p_4 x_4 + u$$

$$\cdot$$
$$\cdot$$
$$\cdot$$

$$\dot{x}_n = p_n x_n + u$$

The output equation, Equation (9–69), can be rewritten as

$$Y(s) = b_0 U(s) + c_1 X_1(s) + c_2 X_2(s) + c_3 X_3(s) + c_4 X_4(s) + \cdots + c_n X_n(s)$$

The inverse Laplace transform of this output equation is

$$y = c_1 x_1 + c_2 x_2 + c_3 x_3 + c_4 x_4 + \cdots + c_n x_n + b_0 u$$

Thus, the state-space representation of the system for the case where the denominator polynomial involves a triple root p_1 can be given as follows:

$$
\begin{bmatrix} \dot{x}_1 \\ \dot{x}_2 \\ \dot{x}_3 \\ \dot{x}_4 \\ \cdot \\ \cdot \\ \cdot \\ \dot{x}_n \end{bmatrix}
=
\left[\begin{array}{ccccccc}
p_1 & 1 & 0 & 0 & \cdots & & 0 \\
0 & p_1 & 1 & & & & \cdot \\
0 & 0 & p_1 & 0 & \cdots & & 0 \\
\hline
0 & \cdots & 0 & p_4 & & & 0 \\
\cdot & & \cdot & & & & \\
\cdot & & \cdot & & & & \\
\cdot & & \cdot & & & & \\
0 & \cdots & 0 & 0 & & & p_n
\end{array} \right]
\begin{bmatrix} x_1 \\ x_2 \\ x_3 \\ x_4 \\ \cdot \\ \cdot \\ \cdot \\ x_n \end{bmatrix}
+
\begin{bmatrix} 0 \\ 0 \\ 1 \\ 1 \\ \cdot \\ \cdot \\ \cdot \\ 1 \end{bmatrix} u
\qquad (9\text{–}70)
$$

$$
y = \begin{bmatrix} c_1 & c_2 & \cdots & c_n \end{bmatrix}
\begin{bmatrix} x_1 \\ x_2 \\ \cdot \\ \cdot \\ \cdot \\ x_n \end{bmatrix}
+ b_0 u
\qquad (9\text{–}71)
$$

The state-space representation in the form given by Equations (9–70) and (9–71) is said to be in the Jordan canonical form. Figure 9–10 shows a block-diagram representation of the system given by Equations (9–70) and (9–71).

EXAMPLE 9–15   Consider the system given by

$$\frac{Y(s)}{U(s)} = \frac{s + 3}{s^2 + 3s + 2}$$

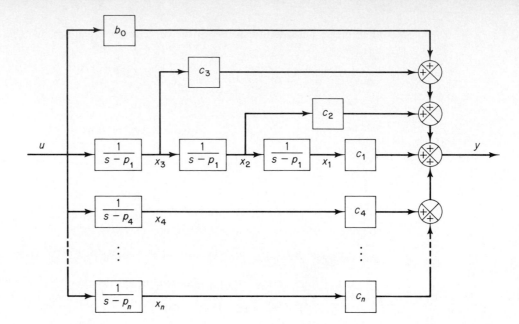

Figure 9–10
Block diagram
representation of the
system given by Eqs.
(9–70) and (9–71).
(Jordan canonical
form.)

Obtain the state-space representation in the controllable canonical form, observable canonical form, and diagonal canonical form.

Controllable canonical form:

$$\begin{bmatrix} \dot{x}_1(t) \\ \dot{x}_2(t) \end{bmatrix} = \begin{bmatrix} 0 & 1 \\ -2 & -3 \end{bmatrix} \begin{bmatrix} x_1(t) \\ x_2(t) \end{bmatrix} + \begin{bmatrix} 0 \\ 1 \end{bmatrix} u(t)$$

$$y(t) = \begin{bmatrix} 3 & 1 \end{bmatrix} \begin{bmatrix} x_1(t) \\ x_2(t) \end{bmatrix}$$

Observable canonical form:

$$\begin{bmatrix} \dot{x}_1(t) \\ \dot{x}_2(t) \end{bmatrix} = \begin{bmatrix} 0 & -2 \\ 1 & -3 \end{bmatrix} \begin{bmatrix} x_1(t) \\ x_2(t) \end{bmatrix} + \begin{bmatrix} 3 \\ 1 \end{bmatrix} u(t)$$

$$y(t) = \begin{bmatrix} 0 & 1 \end{bmatrix} \begin{bmatrix} x_1(t) \\ x_2(t) \end{bmatrix}$$

Diagonal canonical form:

$$\begin{bmatrix} \dot{x}_1(t) \\ \dot{x}_2(t) \end{bmatrix} = \begin{bmatrix} -1 & 0 \\ 0 & -2 \end{bmatrix} \begin{bmatrix} x_1(t) \\ x_2(t) \end{bmatrix} + \begin{bmatrix} 1 \\ 1 \end{bmatrix} u(t)$$

$$y(t) = \begin{bmatrix} 2 & -1 \end{bmatrix} \begin{bmatrix} x_1(t) \\ x_2(t) \end{bmatrix}$$

9–7 LIAPUNOV STABILITY ANALYSIS

For a given control system, stability is usually the most important thing to be determined. If the system is linear and time invariant, many stability criteria are available. Among them are the Nyquist stability criterion and Routh's stability criterion. If the system is nonlinear, or linear but time varying, however, then such stability criteria do not apply.

The second method of Liapunov (which is also called the direct method of Liapunov) to be presented in this section is the most general method for the determination of stability of nonlinear and/or time-varying systems. This method, of course, applies to the determination of the stability of linear, time-invariant systems. In addition, the second method is useful for solving some optimization problems. (See Sections 10–5 and 10–6.)

In what follows we first present preliminary materials, such as definitions of various types of system stability and the concepts of the definiteness of scalar functions. Then we present the main stability theorem of the second method and introduce the Liapunov function. Discussions of the application of the second method of Liapunov to the stability analysis of linear time-invariant systems are given in Section 9–8.

Second method of Liapunov. In 1892, A. M. Liapunov presented two methods (called the first and the second methods) for determining the stability of dynamic systems described by ordinary differential equations.

The first method consists of all procedures in which the explicit form of the solutions of the differential equations are used for the analysis.

The second method, on the other hand, does not require the solutions of the differential equations. That is, by using the second method of Liapunov, we can determine the stability of a system without solving the state equations. This is quite advantageous because solving nonlinear and/or time-varying state equations is usually very difficult.

Although the second method of Liapunov, when applied to the stability analysis of nonlinear systems, requires considerable experience and ingenuity, it can answer the question of stability of nonlinear systems when other methods fail.

System. The system we consider here is defined by

$$\dot{\mathbf{x}} = \mathbf{f}(\mathbf{x}, t) \tag{9–72}$$

where $\mathbf{x}$ is a state vector (n-vector) and $\mathbf{f}(\mathbf{x}, t)$ is an n-vector whose elements are functions of $x_1, x_2, \ldots, x_n$, and t. We assume that the system of Equation (9–72) has a unique solution starting at the given initial condition. We shall denote the solution of Equation (9–72) as $\boldsymbol{\phi}(t; \mathbf{x}_0, t_0)$, where $\mathbf{x} = \mathbf{x}_0$ at $t = t_0$ and t is the observed time. Thus,

$$\boldsymbol{\phi}(t_0; \mathbf{x}_0, t_0) = \mathbf{x}_0$$

Equilibrium state. In the system of Equation (9–72), a state $\mathbf{x}_e$ where

$$\mathbf{f}(\mathbf{x}_e, t) = \mathbf{0} \qquad \text{for all } t \tag{9–73}$$

is called an equilibrium state of the system. If the system is linear time invariant, that is, if $\mathbf{f}(\mathbf{x}, t) = \mathbf{A}\mathbf{x}$, then there exists only one equilibrium state if $\mathbf{A}$ is nonsingular, and there exist infinitely many equilibrium states if $\mathbf{A}$ is singular. For nonlinear systems, there may be one or more equilibrium states. These states correspond to the constant solutions of the system ($\mathbf{x} = \mathbf{x}_e$ for all t). Determination of the equilibrium states does not involve the solution of the differential equations of the system, Equation (9–72), but only the solution of Equation (9–73).

Any isolated equilibrium state (that is, isolated from each other) can be shifted to the origin of the coordinates, or $\mathbf{f}(\mathbf{0}, t) = \mathbf{0}$, by a translation of coordinates. In this section and the next, we shall treat stability analysis of equilibrium states at the origin.

Stability in the sense of Liapunov.   In the following, we shall denote a spherical region of radius k about an equilibrium state $\mathbf{x}_e$ as

$$\|\mathbf{x} - \mathbf{x}_e\| \leq k$$

where $\|\mathbf{x} - \mathbf{x}_e\|$ is called the Euclidean norm and is defined by

$$\|\mathbf{x} - \mathbf{x}_e\| = [(x_1 - x_{1e})^2 + (x_2 - x_{2e})^2 + \cdots + (x_n - x_{ne})^2]^{1/2}$$

Let $S(\delta)$ consist of all points such that

$$\|\mathbf{x}_0 - \mathbf{x}_e\| \leq \delta$$

and let $S(\epsilon)$ consist of all points such that

$$\|\boldsymbol{\phi}(t; \mathbf{x}_0, t_0) - \mathbf{x}_e\| \leq \epsilon \qquad \text{for all } t \geq t_0$$

An equilibrium state $\mathbf{x}_e$ of the system of Equation (9–72) is said to be stable in the sense of Liapunov if, corresponding to each $S(\epsilon)$, there is an $S(\delta)$ such that trajectories starting in $S(\delta)$ do not leave $S(\epsilon)$ as t increases indefinitely. The real number δ depends on ϵ and, in general, also depends on t_0. If δ does not depend on t_0, the equilibrium state is said to be uniformly stable.

What we have stated here is that we first choose the region $S(\epsilon)$, and for each $S(\epsilon)$, there must be a region $S(\delta)$ such that trajectories starting within $S(\delta)$ do not leave $S(\epsilon)$ as t increases indefinitely.

Asymptotic stability.   An equilibrium state $\mathbf{x}_e$ of the system of Equation (9–72) is said to be asymptotically stable if it is stable in the sense of Liapunov and if every solution starting within $S(\delta)$ converges, without leaving $S(\epsilon)$, to $\mathbf{x}_e$ as t increases indefinitely.

In practice, asymptotic stability is more important than mere stability. Also, since asymptotic stability is a local concept, simply to establish asymptotic stability may not mean that the system will operate properly. Some knowledge of the size of the largest region of asymptotic stability is usually necessary. This region is called the *domain of attraction*. It is that part of the state space in which asymptotically stable trajectories originate. In other words, every trajectory originating in the domain of attraction is asymptotically stable.

Asymptotic stability in the large.   If asymptotic stability holds for all states (all points in the state space) from which trajectories originate, the equilibrium state is said to be asymptotically stable in the large. That is, the equilibrium state $\mathbf{x}_e$ of the system given by Equation (9–72) is said to be asymptotically stable in the large if it is stable and if every solution converges to $\mathbf{x}_e$ as t increases indefinitely. Obviously, a necessary condition for asymptotic stability in the large is that there be only one equilibrium state in the whole state space.

In control engineering problems, a desirable feature is asymptotic stability in the large. If the equilibrium state is not asymptotically stable in the large, then the problem becomes that of determining the largest region of asymptotic stability. This is usually very difficult.

For all practical purposes, however, it is sufficient to determine a region of asymptotic stability large enough so that no disturbance will exceed it.

Instability. An equilibrium state $\mathbf{x}_e$ is said to be unstable if for some real number $\epsilon > 0$ and any real number $\delta > 0$, no matter how small, there is always a state $\mathbf{x}_0$ in $S(\delta)$ such that the trajectory starting at this state leaves $S(\epsilon)$.

Graphical representation of stability, asymptotic stability, and instability.

A graphical representation of the foregoing definitions will clarify their notions.

Let us consider the two-dimensional case. Figures 9–11(a), (b), and (c) show equilibrium states and typical trajectories corresponding to stability, asymptotic stability, and instability, respectively. In Figure 9–11(a), (b), or (c), the region $S(\delta)$ bounds the initial state $\mathbf{x}_0$, and the region $S(\epsilon)$ corresponds to the boundary for the trajectory starting at $\mathbf{x}_0$.

Note that the foregoing definitions do not specify the exact region of allowable initial conditions. Thus the definitions apply to the neighborhood of the equilibrium state, unless $S(\epsilon)$ corresponds to the entire state plane.

Note that in Figure 9–11(c), the trajectory leaves $S(\epsilon)$ and implies that the equilibrium state is unstable. We cannot, however, say that the trajectory will go to infinity since it may approach a limit cycle outside the region $S(\epsilon)$. (If a linear time-invariant system is unstable, trajectories starting near the unstable equilibrium state go to infinity. But in the case of nonlinear systems, this is not necessarily true.)

Knowledge of the foregoing definitions is a minimum requirement for understanding the stability analysis of linear and nonlinear systems as presented in this section. Note that these definitions are not the only ones defining concepts of stability of an equilibrium state. In fact, various other ways are available in the literature. For example, in conventional or classical control theory, only systems that are asymptotically stable are called stable systems, and those systems that are stable in the sense of Liapunov, but are not asymptotically stable, are called unstable.

Positive definiteness of scalar functions.

A scalar function $V(\mathbf{x})$ is said to be *positive definite* in a region Ω (which includes the origin of the state space) if $V(\mathbf{x}) > 0$ for all nonzero states $\mathbf{x}$ in the region Ω and $V(\mathbf{0}) = 0$.

A time-varying function $V(\mathbf{x}, t)$ is said to be positive definite in a region Ω (which includes the origin of the state space) if it is bounded from below by a time-invariant positive-

Figure 9–11
(a) Stable equilibrium state and a representative trajectory; (b) asymptotically stable equilibrium state and a representative trajectory; (c) unstable equilibrium state and a representative trajectory.

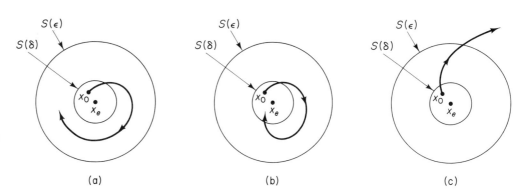

(a)　　　　(b)　　　　(c)

definite function, that is, if there exists a positive definite function $V(\mathbf{x})$ such that

$$V(\mathbf{x}, t) > V(\mathbf{x}) \qquad \text{for all } t \geq t_0$$
$$V(\mathbf{0}, t) = 0 \qquad \text{for all } t \geq t_0$$

Negative definiteness of scalar functions. A scalar function $V(\mathbf{x})$ is said to be *negative definite* if $-V(\mathbf{x})$ is positive definite.

Positive semidefiniteness of scalar functions. A scalar function $V(\mathbf{x})$ is said to be *positive semidefinite* if it is positive at all states in the region $\mathbf{\Omega}$ except at the origin and at certain other states, where it is zero.

Negative semidefiniteness of scalar functions. A scalar function $V(\mathbf{x})$ is said to be *negative semidefinite* if $-V(\mathbf{x})$ is positive semidefinite.

Indefiniteness of scalar functions. A scalar function $V(\mathbf{x})$ is said to be *indefinite* if in the region $\mathbf{\Omega}$ it assumes both positive and negative values, no matter how small the region $\mathbf{\Omega}$ is.

EXAMPLE 9–16

In this example, we give several scalar functions and their classifications according to the foregoing definitions. Here we assume $\mathbf{x}$ to be a two-dimensional vector.

1. $\qquad V(\mathbf{x}) = x_1^2 + 2x_2^2 \qquad\qquad$ positive definite
2. $\qquad V(\mathbf{x}) = (x_1 + x_2)^2 \qquad\qquad$ positive semidefinite
3. $\qquad V(\mathbf{x}) = -x_1^2 - (3x_1 + 2x_2)^2 \qquad$ negative definite
4. $\qquad V(\mathbf{x}) = x_1 x_2 + x_2^2 \qquad\qquad$ indefinite
5. $\qquad V(\mathbf{x}) = x_1^2 + \dfrac{2x_2^2}{1 + x_2^2} \qquad\qquad$ positive definite

Quadratic form. A class of scalar functions that plays an important role in the stability analysis based on the second method of Liapunov is the quadratic form. An example is

$$V(\mathbf{x}) = \mathbf{x}^T \mathbf{P} \mathbf{x} = [x_1 \quad x_2 \quad \cdots \quad x_n] \begin{bmatrix} p_{11} & p_{12} & \cdots & p_{1n} \\ p_{12} & p_{22} & \cdots & p_{2n} \\ \cdot & \cdot & & \cdot \\ \cdot & \cdot & & \cdot \\ \cdot & \cdot & & \cdot \\ p_{1n} & p_{2n} & \cdots & p_{nn} \end{bmatrix} \begin{bmatrix} x_1 \\ x_2 \\ \cdot \\ \cdot \\ \cdot \\ x_n \end{bmatrix}$$

Note that $\mathbf{x}$ is a real vector and $\mathbf{P}$ is a real symmetric matrix.

Hermitian form. If $\mathbf{x}$ is a complex n-vector and $\mathbf{P}$ is a Hermititan matrix, then the complex quadratic form is called the Hermitian form. An example is

$$V(\mathbf{x}) = \mathbf{x}^*\mathbf{P}\mathbf{x} = [\bar{x}_1 \quad \bar{x}_2 \quad \cdots \quad \bar{x}_n] \begin{bmatrix} p_{11} & p_{12} & \cdots & p_{1n} \\ \bar{p}_{12} & p_{22} & \cdots & p_{2n} \\ \cdot & \cdot & & \cdot \\ \cdot & \cdot & & \cdot \\ \cdot & \cdot & & \cdot \\ \bar{p}_{1n} & \bar{p}_{2n} & \cdots & p_{nn} \end{bmatrix} \begin{bmatrix} x_1 \\ x_2 \\ \cdot \\ \cdot \\ \cdot \\ x_n \end{bmatrix}$$

In the stability analysis in state space, we frequently use the Hermitian form rather than the quadratic form, since the former is more general than the latter. (For a real vector $\mathbf{x}$ and a real symmetric matrix $\mathbf{P}$, Hermitian form $\mathbf{x}^*\mathbf{P}\mathbf{x}$ becomes equal to quadratic form $\mathbf{x}^T\mathbf{P}\mathbf{x}$.)

The positive definiteness of the quadratic form or the Hermitian form $V(\mathbf{x})$ can be determined by Sylvester's criterion, which states that the necessary and sufficient conditions that the quadratic form or Hermitian form $V(\mathbf{x})$ be positive definite are that all the successive principal minors of $\mathbf{P}$ be positive; that is,

$$p_{11} > 0, \qquad \begin{vmatrix} p_{11} & p_{12} \\ \bar{p}_{12} & p_{22} \end{vmatrix} > 0, \qquad \cdots, \qquad \begin{vmatrix} p_{11} & p_{12} & \cdots & p_{1n} \\ \bar{p}_{12} & p_{22} & \cdots & p_{2n} \\ \cdot & \cdot & & \cdot \\ \cdot & \cdot & & \cdot \\ \cdot & \cdot & & \cdot \\ \bar{p}_{1n} & \bar{p}_{2n} & \cdots & p_{nn} \end{vmatrix} > 0$$

(Note that $\bar{p}_{ij}$ is the complex conjugate of p_{ij}. For the quadratic form, $\bar{p}_{ij} = p_{ij}$.)

$V(\mathbf{x}) = \mathbf{x}^*\mathbf{P}\mathbf{x}$ is positive semidefinite if $\mathbf{P}$ is singular and all the principal minors are nonnegative.

$V(\mathbf{x})$ is negative definite if $-V(\mathbf{x})$ is positive definite. Similarly, $V(\mathbf{x})$ is negative semidefinite if $-V(\mathbf{x})$ is positive semidefinite.

EXAMPLE 9–17   Show that the following quadratic form is positive definite:

$$V(\mathbf{x}) = 10x_1^2 + 4x_2^2 + x_3^2 + 2x_1x_2 - 2x_2x_3 - 4x_1x_3$$

The quadratic form $V(\mathbf{x})$ can be written

$$V(\mathbf{x}) = \mathbf{x}^T\mathbf{P}\mathbf{x} = [x_1 \quad x_2 \quad x_3] \begin{bmatrix} 10 & 1 & -2 \\ 1 & 4 & -1 \\ -2 & -1 & 1 \end{bmatrix} \begin{bmatrix} x_1 \\ x_2 \\ x_3 \end{bmatrix}$$

Applying Sylvester's criterion, we obtain

$$10 > 0, \qquad \begin{vmatrix} 10 & 1 \\ 1 & 4 \end{vmatrix} > 0, \qquad \begin{vmatrix} 10 & 1 & -2 \\ 1 & 4 & -1 \\ -2 & -1 & 1 \end{vmatrix} > 0$$

Since all the successive principal minors of the matrix $\mathbf{P}$ are positive, $V(\mathbf{x})$ is positive definite.

Second method of Liapunov.   From the classical theory of mechanics, we know that a vibratory system is stable if its total energy (a positive definite function) is continually decreasing (which means that the time derivative of the total energy must be negative definite) until an equilibrium state is reached.

The second method of Liapunov is based on a generalization of this fact: if the system has an asymptotically stable equilibrium state, then the stored energy of the system displaced within the domain of attraction decays with increasing time until it finally assumes its minimum value at the equilibrium state. For purely mathematical systems, however, there is no simple way of defining an "energy function." To circumvent this difficulty, Liapunov introduced the Liapunov function, a fictitious energy function. This idea is, however, more general than that of energy and is more widely applicable. In fact, any scalar function satisfying the hypotheses of Liapunov's stability theorems (see Theorems 9–1 and 9–2) can serve as Liapunov functions. (For simple systems, we may be able to guess suitable Liapunov functions; but, for a complicated system, finding a Liapunov function may be quite difficult.)

Liapunov functions depend on x_1, x_2, . . . , x_n, and t. We denote them by $V(x_1, x_2, . . . , x_n, t)$, or simply by $V(\mathbf{x}, t)$. If Liapunov functions do not include t explicitly, then we denote them by $V(x_1, x_2, . . . , x_n)$, or $V(\mathbf{x})$. In the second method of Liapunov, the sign behavior of $V(\mathbf{x}, t)$ and that of its time derivative $\dot{V}(\mathbf{x}, t) = dV(\mathbf{x}, t)/dt$ give us information as to the stability, asymptotic stability, or instability of an equilibrium state without requiring us to solve directly for the solution. (This applies to both linear and nonlinear systems.)

Liapunov's main stability theorem.   It can be shown that if a scalar function $V(\mathbf{x})$, where $\mathbf{x}$ is an n-vector, is positive definite, then the states $\mathbf{x}$ that satisfy

$$V(\mathbf{x}) = C$$

where C is a positive constant, lie on a closed hypersurface in the n-dimensional state space, at least in the neighborhood of the origin. If $V(\mathbf{x}) \to \infty$ as $\|\mathbf{x}\| \to \infty$, then such closed surfaces extend over the entire state space. The hypersurface $V(\mathbf{x}) = C_1$ lies entirely inside the hypersurface $V(\mathbf{x}) = C_2$ if $C_1 < C_2$.

For a given system, if a positive-definite scalar function $V(\mathbf{x})$ can be found such that its time derivative taken along a trajectory is always negative, then as time increases, $V(\mathbf{x})$ takes smaller and smaller values of C. As time increases, $V(\mathbf{x})$ finally shrinks to zero, and therefore $\mathbf{x}$ also shrinks to zero. This implies the asymptotic stability of the origin of the state space. Liapunov's main stability theorem, which is a generalization of the foregoing fact, provides a sufficient condition for asymptotic stability. This theorem may be stated as follows:

Theorem 9–1.   Suppose that a system is described by

$$\dot{\mathbf{x}} = \mathbf{f}(\mathbf{x}, t)$$

where

$$\mathbf{f}(\mathbf{0}, t) = \mathbf{0} \qquad \text{for all } t$$

If there exists a scalar function $V(\mathbf{x}, t)$ having continuous, first partial derivatives and satisfying the following conditions,

1. $V(\mathbf{x}, t)$ is positive definite
2. $\dot{V}(\mathbf{x}, t)$ is negative definite

then the equilibrium state at the origin is uniformly asymptotically stable.

If, in addition, $V(\mathbf{x}, t) \to \infty$ as $\|\mathbf{x}\| \to \infty$, then the equilibrium state at the origin is uniformly asymptotically stable in the large.

We shall not give details of the proof of this theorem here. (The proof follows directly from the definition of asymptotic stability. For details, see Problem A–9–15.)

EXAMPLE 9–18 Consider the system described by

$$\dot{x}_1 = x_2 - x_1(x_1^2 + x_2^2)$$

$$\dot{x}_2 = -x_1 - x_2(x_1^2 + x_2^2)$$

Clearly, the origin ($x_1 = 0, x_2 = 0$) is the only equilibrium state. Determine its stability.

If we define a scalar function $V(\mathbf{x})$ by

$$V(\mathbf{x}) = x_1^2 + x_2^2$$

which is positive definite, then the time derivative of $V(\mathbf{x})$ along any trajectory is

$$\dot{V}(\mathbf{x}) = 2x_1\dot{x}_1 + 2x_2\dot{x}_2$$
$$= -2(x_1^2 + x_2^2)^2$$

which is negative definite. This shows that $V(\mathbf{x})$ is continually decreasing along any trajectory; hence $V(\mathbf{x})$ is a Liapunov function. Since $V(\mathbf{x})$ becomes infinite with infinite deviation from the equilibrium state, by Theorem 9–1, the equilibrium state at the origin of the system is asymptotically stable in the large.

Note that if we let $V(\mathbf{x})$ take constant values $0, C_1, C_2, \ldots$ ($0 < C_1 < C_2 < \ldots$), then $V(\mathbf{x}) = 0$ corresponds to the origin of the state plane and $V(\mathbf{x}) = C_1, V(\mathbf{x}) = C_2, \ldots$ describe nonintersecting circles enclosing the origin of the state plane, as shown in Figure 9–12. Note also that since $V(\mathbf{x})$ is radially unbounded, or $V(\mathbf{x}) \to \infty$ as $\|\mathbf{x}\| \to \infty$, the circles extend over the entire state plane.

Since circle $V(\mathbf{x}) = C_k$ lies completely inside circle $V(\mathbf{x}) = C_{k+1}$, a representative trajectory crosses the boundary of V contours from outside toward the inside. From this, the geometric interpretation of a Liapunov function may be stated as follows: $V(\mathbf{x})$ is a measure of the distance of the state $\mathbf{x}$ from the origin of the state space. If the distance between the origin and the instantaneous state $\mathbf{x}(t)$ is continually decreasing as t increases [that is, $\dot{V}(\mathbf{x}(t)) < 0$], then $\mathbf{x}(t) \to \mathbf{0}$.

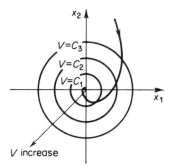

Figure 9–12
Constant-V contours and representative trajectory.

Comment. Although Theorem 9–1 is a basic theorem of the second method, it is somewhat restrictive because $\dot{V}(\mathbf{x}, t)$ must be negative definite. If, however, an additional restriction is imposed on $\dot{V}(\mathbf{x}, t)$ that it does not vanish identically along any trajectory except at the origin, then it is possible to replace the requirement of $\dot{V}(\mathbf{x}, t)$ being negative definite by stating that $\dot{V}(\mathbf{x}, t)$ be negative semidefinite.

Theorem 9–2. Suppose that a system is described by

$$\dot{\mathbf{x}} = \mathbf{f}(\mathbf{x}, t)$$

where

$$\mathbf{f}(\mathbf{0}, t) = \mathbf{0} \qquad \text{for all } t \geq t_0$$

If there exists a scalar function $V(\mathbf{x}, t)$ having continuous, first partial derivatives and satisfying the following conditions,

1. $V(\mathbf{x}, t)$ is positive definite
2. $\dot{V}(\mathbf{x}, t)$ is negative semidefinite
3. $\dot{V}(\boldsymbol{\phi}(t; \mathbf{x}_0, t_0), t)$ does not vanish identically in $t \geq t_0$ for any t_0 and any $\mathbf{x}_0 \neq \mathbf{0}$, where $\boldsymbol{\phi}(t; \mathbf{x}_0, t_0)$ denotes the trajectory or solution starting from $\mathbf{x}_0$ at t_0

then the equilibrium state at the origin of the system is uniformly asymptotically stable in the large.

Comment. Note that if $\dot{V}(\mathbf{x}, t)$ is not negative definite, but only negative semidefinite, then the trajectory of a representative point can become tangent to some particular surface $V(\mathbf{x}, t) = C$. However, since $\dot{V}(\boldsymbol{\phi}(t; \mathbf{x}_0, t_0), t)$ does not vanish identically in $t \geq t_0$ for any t_0 and any $\mathbf{x}_0 \neq \mathbf{0}$, the representative point cannot remain at the tangent point [the point that corresponds to $\dot{V}(\mathbf{x}, t) = 0$] and therefore must move toward the origin.

If, however, there exists a positive-definite scalar function $V(\mathbf{x}, t)$ such that $\dot{V}(\mathbf{x}, t)$ is identically zero, then the system can remain in a limit cycle. The equilibrium state at the origin, in this case, is said to be stable in the sense of Liapunov.

Instability. If an equilibrium state $\mathbf{x} = \mathbf{0}$ of a system is unstable, then there exists a scalar function $W(\mathbf{x}, t)$ which determines the instability of the equilibrium state. We shall present a theorem on instability in the following.

Theorem 9–3. Suppose a system is described by

$$\dot{\mathbf{x}} = \mathbf{f}(\mathbf{x}, t)$$

where

$$\mathbf{f}(\mathbf{0}, t) = \mathbf{0} \qquad \text{for all } t \geq t_0$$

If there exists a scalar function $W(\mathbf{x}, t)$ having continuous, first partial derivatives and satisfying the following conditions,

1. $W(\mathbf{x}, t)$ is positive definite in some region about the origin
2. $\dot{W}(\mathbf{x}, t)$ is positive definite in the same region

then the equilibrium state at the origin is unstable.

EXAMPLE 9–19   Consider the following system:

$$\begin{bmatrix} \dot{x}_1 \\ \dot{x}_2 \end{bmatrix} = \begin{bmatrix} 0 & 1 \\ -1 & -1 \end{bmatrix} \begin{bmatrix} x_1 \\ x_2 \end{bmatrix}$$

Clearly, the only equilibrium state is the origin, $\mathbf{x} = \mathbf{0}$. Determine the stability of this state.

Let us choose the following scalar function as a possible Liapunov function:

$$V(\mathbf{x}) = 2x_1^2 + x_2^2 = \text{positive definite}$$

Then $\dot{V}(\mathbf{x})$ becomes

$$\dot{V}(\mathbf{x}) = 4x_1\dot{x}_1 + 2x_2\dot{x}_2 = 2x_1x_2 - 2x_2^2$$

$\dot{V}(\mathbf{x})$ is indefinite. This implies that this particular $V(\mathbf{x})$ is not a Liapunov function, and therefore stability cannot be determined by its use. [Since the eigenvalues of the coefficient matrix are $(-1 + j\sqrt{3})/2$ and $(-1 - j\sqrt{3})/2$, clearly the origin of the system is stable. This means that we have not chosen a suitable Liapunov function.]

If we choose the following scalar function as a possible Liapunov function,

$$V(\mathbf{x}) = x_1^2 + x_2^2 = \text{positive definite}$$

then

$$\dot{V}(\mathbf{x}) = 2x_1\dot{x}_1 + 2x_2\dot{x}_2 = -2x_2^2$$

which is negative semidefinite. If $\dot{V}(\mathbf{x})$ is to vanish identically for $t \geq t_1$, then x_2 must be zero for all $t \geq t_1$. This requires that $\dot{x}_2 = 0$ for $t \geq t_1$. Since

$$\dot{x}_2 = -x_1 - x_2$$

x_1 must also be equal to zero for $t \geq t_1$. This means that $\dot{V}(x)$ vanishes identically only at the origin. Hence, by Theorem 9–2, the equilibrium state at the origin is asymptotically stable in the large.

To show that a different choice of a Liapunov function yields the same stability information, let us choose the following scalar function as another possible Liapunov function:

$$V(\mathbf{x}) = \tfrac{1}{2}[(x_1 + x_2)^2 + 2x_1^2 + x_2^2] = \text{positive definite}$$

Then $\dot{V}(\mathbf{x})$ becomes

$$\begin{aligned} \dot{V}(\mathbf{x}) &= (x_1 + x_2)(\dot{x}_1 + \dot{x}_2) + 2x_1\dot{x}_1 + x_2\dot{x}_2 \\ &= (x_1 + x_2)(x_2 - x_1 - x_2) + 2x_1x_2 + x_2(-x_1 - x_2) \\ &= -(x_1^2 + x_2^2) \end{aligned}$$

which is negative definite. Since $V(\mathbf{x}) \to \infty$ as $\|\mathbf{x}\| \to \infty$, by Theorem 9–1, the equilibrium state at the origin is asymptotically stable in the large.

Since the stability theorems of the second method require positive definiteness of $V(\mathbf{x})$, we often (but not always) choose $V(\mathbf{x})$ to be a quadratic form or Hermitian form in $\mathbf{x}$. (Note that the simplest positive-definite function is a quadratic form or Hermitian form.) Then we examine if $\dot{V}(\mathbf{x})$ is at least negative semidefinite.

9–8 LIAPUNOV STABILITY ANALYSIS OF LINEAR TIME-INVARIANT SYSTEMS

There are many approaches to the investigation of the asymptotic stability of linear time-invariant systems. For example, for a continuous-time system

$$\dot{\mathbf{x}} = \mathbf{A}\mathbf{x}$$

the necessary and sufficient condition for the asymptotic stability of the origin of the system can be stated that all eigenvalues of $\mathbf{A}$ have negative real parts, or the zeros of the characteristic polynomial

$$|s\mathbf{I} - \mathbf{A}| = s^n + a_1 s^{n-1} + \cdots + a_{n-1}s + a_n$$

have negative real parts.

Finding the eigenvalues becomes difficult or impossible in the case of higher-order systems or if some of the coefficients of the characteristic polynomial are nonnumerical. In such a case, Routh's stability criterion may conveniently be applied. One alternative to this approach is available; it is based on the second method of Liapunov. The Liapunov approach is algebraic and does not require factoring of the characteristic polynomial. In addition, this approach can be used to find solutions to certain optimal control problems. The purpose of this section is to present the Liapunov approach to the stability analysis of linear time-invariant systems. (Applications of the Liapunov approach to optimal control systems are presented in Sections 10–5 and 10–6.)

Liapunov stability analysis of linear time-invariant systems. Consider the following linear time-invariant system

$$\dot{\mathbf{x}} = \mathbf{A}\mathbf{x} \qquad (9\text{–}74)$$

where $\mathbf{x}$ is a state vector (n-vector) and $\mathbf{A}$ is an $n \times n$ constant matrix. We assume that $\mathbf{A}$ is nonsingular. Then the only equilibrium state is the origin $\mathbf{x} = \mathbf{0}$. The stability of the equilibrium state of the linear time-invariant system can be investigated easily by use of the second method of Liapunov.

For the system defined by Eq. (9–74), let us choose a possible Liapunov function as

$$V(\mathbf{x}) = \mathbf{x}^*\mathbf{P}\mathbf{x}$$

where $\mathbf{P}$ is a positive-definite Hermitian matrix. (If $\mathbf{x}$ is a real vector and $\mathbf{A}$ is a real matrix, then $\mathbf{P}$ can be chosen to be a positive-definite real symmetric matrix.) The time derivative of $V(\mathbf{x})$ along any trajectory is

$$\begin{aligned}
\dot{V}(\mathbf{x}) &= \dot{\mathbf{x}}^*\mathbf{P}\mathbf{x} + \mathbf{x}^*\mathbf{P}\dot{\mathbf{x}} \\
&= (\mathbf{A}\mathbf{x})^*\mathbf{P}\mathbf{x} + \mathbf{x}^*\mathbf{P}\mathbf{A}\mathbf{x} \\
&= \mathbf{x}^*\mathbf{A}^*\mathbf{P}\mathbf{x} + \mathbf{x}^*\mathbf{P}\mathbf{A}\mathbf{x} \\
&= \mathbf{x}^*(\mathbf{A}^*\mathbf{P} + \mathbf{P}\mathbf{A})\mathbf{x}
\end{aligned}$$

Since $V(\mathbf{x})$ was chosen to be positive definite, we require, for asymptotic stability, that $\dot{V}(\mathbf{x})$ be negative definite. Therefore, we require that

$$\dot{V}(\mathbf{x}) = -\mathbf{x}^*\mathbf{Q}\mathbf{x}$$

where

$$\mathbf{Q} = -(\mathbf{A}^*\mathbf{P} + \mathbf{P}\mathbf{A}) = \text{positive definite}$$

Hence, for the asymptotic stability of the system of Equation (9–74), it is sufficient that $\mathbf{Q}$ be positive definite. For a test of positive definiteness of an $n \times n$ matrix, we apply Sylvester's criterion, which states that a necessary and sufficient condition that the matrix be positive definite is that the determinants of all the successive principal minors of the matrix be positive.

Instead of first specifying a positive-definite matrix **P** and examining whether or not **Q** is positive definite, it is convenient to specify a positive-definite matrix **Q** first and then examine whether or not **P** determined from

$$\mathbf{A}^*\mathbf{P} + \mathbf{PA} = -\mathbf{Q}$$

is positive definite. Note that **P** being positive definite is a necessary and sufficient condition. We shall summarize what we have just stated in the form of a theorem.

Theorem 9–4. Consider the system described by

$$\dot{\mathbf{x}} = \mathbf{Ax}$$

where **x** is a state vector (*n*-vector) and **A** is an $n \times n$ constant nonsingular matrix. A necessary and sufficient condition that the equilibrium state **x** = **0** be asymptotically stable in the large is that, given any positive-definite Hermitian (or real symmetric) matrix **Q,** there exists a positive-definite Hermitian (or real symmetric) matrix **P** such that

$$\mathbf{A}^*\mathbf{P} + \mathbf{PA} = -\mathbf{Q}$$

The scalar function **x*****Px** is a Liapunov function for this system. [Note that in the linear system considered, if the equilibrium state (the origin) is asymptotically stable, then it is asymptotically stable in the large.]

Comments. In applying this theorem, several important remarks are in order.

1. If the system involves only real state vector **x** and real state matrix **A,** then the Liapunov function **x*****Px** becomes $\mathbf{x}^T\mathbf{Px}$ and the Liapunov equation becomes

$$\mathbf{A}^T\mathbf{P} + \mathbf{PA} = -\mathbf{Q}$$

2. If $\dot{V}(\mathbf{x}) = -\mathbf{x}^*\mathbf{Qx}$ does not vanish identically along any trajectory, then **Q** may be chosen to be positive semidefinite.

3. If we choose an arbitrary positive-definite matrix as **Q** or a positive-semidefinite matrix as **Q** if $\dot{V}(\mathbf{x})$ does not vanish identically along any trajectory and solve the matrix equation

$$\mathbf{A}^*\mathbf{P} + \mathbf{PA} = -\mathbf{Q}$$

to determine **P**, then the positive definitenes of **P** is a necessary and sufficient condition for the asymptotic stability of the equilibrium state **x** = **0**.

Note that $\dot{V}(\mathbf{x})$ does not vanish identically along any trajectory if a positive semidefinite matrix **Q** satisfies the following rank condition:

$$\text{rank} \begin{bmatrix} \mathbf{Q}^{1/2} \\ \mathbf{Q}^{1/2}\mathbf{A} \\ \cdot \\ \cdot \\ \cdot \\ \mathbf{Q}^{1/2}\mathbf{A}^{n-1} \end{bmatrix} = n$$

(See Problem A–9–17.)

4. The final result does not depend on a particular **Q** matrix chosen as long as it is positive definite (or positive semidefinite, as the case may be).

5. To determine the elements of the **P** matrix, we equate the matrices $\mathbf{A^*P + PA}$ and $-\mathbf{Q}$ element by element. This results in $n(n + 1)/2$ linear equations for the determination of the elements $p_{ij} = \bar{p}_{ji}$ of **P.** If we denote the eigenvalues of **A** by $\lambda_1, \lambda_2, \ldots, \lambda_n$, each repeated as often as its multiplicity as a root of the characteristic equation, and if for every sum of two roots

$$\lambda_j + \lambda_k \neq 0$$

then the elements of **P** are uniquely determined. Note that if the matrix **A** represents a stable system, then the sums $\lambda_j + \lambda_k$ are always nonzero.

6. In determining whether or not there exists a positive-definite Hermitian or a real symmetric matrix **P**, it is convenient to choose $\mathbf{Q = I}$, where **I** is the identity matrix. Then the elements of **P** are determined from

$$\mathbf{A^*P + PA = -I}$$

and the matrix **P** is tested for positive definiteness.

EXAMPLE 9–20. Consider the second-order system described by

$$\begin{bmatrix} \dot{x}_1 \\ \dot{x}_2 \end{bmatrix} = \begin{bmatrix} 0 & 1 \\ -1 & -1 \end{bmatrix} \begin{bmatrix} x_1 \\ x_2 \end{bmatrix}$$

Clearly, the equilibrium state is the origin. Determine the stability of this state.

Let us assume a tentative Liapunov function

$$V(\mathbf{x}) = \mathbf{x}^T \mathbf{P} \mathbf{x}$$

where **P** is to be determined from

$$\mathbf{A}^T \mathbf{P} + \mathbf{P} \mathbf{A} = -\mathbf{I}$$

This last equation can be written as

$$\begin{bmatrix} 0 & -1 \\ 1 & -1 \end{bmatrix} \begin{bmatrix} p_{11} & p_{12} \\ p_{12} & p_{22} \end{bmatrix} + \begin{bmatrix} p_{11} & p_{12} \\ p_{12} & p_{22} \end{bmatrix} \begin{bmatrix} 0 & 1 \\ -1 & -1 \end{bmatrix} = \begin{bmatrix} -1 & 0 \\ 0 & -1 \end{bmatrix}$$

By expanding this matrix equation, we obtain three simultaneous equations as follows:

$$-2p_{12} = -1$$

$$p_{11} - p_{12} - p_{22} = 0$$

$$2p_{12} - 2p_{22} = -1$$

Solving for p_{11}, p_{12}, p_{13}, we obtain

$$\begin{bmatrix} p_{11} & p_{12} \\ p_{12} & p_{22} \end{bmatrix} = \begin{bmatrix} \dfrac{3}{2} & \dfrac{1}{2} \\ \dfrac{1}{2} & 1 \end{bmatrix}$$

To test the positive definiteness of **P**, we check the determinants of the successive principal minors:

$$\frac{3}{2} > 0, \qquad \begin{vmatrix} \dfrac{3}{2} & \dfrac{1}{2} \\ \dfrac{1}{2} & 1 \end{vmatrix} > 0$$

Clearly, $\mathbf{P}$ is positive definite. Hence the equilibrium state at the origin is asymptotically stable in the large, and a Liapunov function is

$$V(\mathbf{x}) = \mathbf{x}^T\mathbf{P}\mathbf{x} = \tfrac{1}{2}(3x_1^2 + 2x_1x_2 + 2x_2^2)$$

and

$$\dot{V}(\mathbf{x}) = -(x_1^2 + x_2^2)$$

EXAMPLE 9–21.   Determine the stability range for the gain K of the system shown in Figure 9–13.

The state equation of the system is

$$\begin{bmatrix} \dot{x}_1 \\ \dot{x}_2 \\ \dot{x}_3 \end{bmatrix} = \begin{bmatrix} 0 & 1 & 0 \\ 0 & -2 & 1 \\ -K & 0 & -1 \end{bmatrix}\begin{bmatrix} x_1 \\ x_2 \\ x_3 \end{bmatrix} + \begin{bmatrix} 0 \\ 0 \\ K \end{bmatrix}u$$

In determining the stability range for K, we assume the input u to be zero. Then this last equation may be written as

$$\dot{x}_1 = x_2 \tag{9–75}$$

$$\dot{x}_2 = -2x_2 + x_3 \tag{9–76}$$

$$\dot{x}_3 = -Kx_1 - x_3 \tag{9–77}$$

From Equations (9–75) through (9–77), we find that the origin is the equilibrium state. Let us choose the positive-semidefinite real symmetric matrix $\mathbf{Q}$ to be

$$\mathbf{Q} = \begin{bmatrix} 0 & 0 & 0 \\ 0 & 0 & 0 \\ 0 & 0 & 1 \end{bmatrix} \tag{9–78}$$

This choice of $\mathbf{Q}$ is permissible since $\dot{V}(\mathbf{x}) = -\mathbf{x}^T\mathbf{Q}\mathbf{x}$ cannot be identically equal to zero except at the origin. To verify this, note that

$$\dot{V}(\mathbf{x}) = -\mathbf{x}^T\mathbf{Q}\mathbf{x} = -x_3^2$$

$\dot{V}(\mathbf{x})$ being identically zero implies that x_3 is identically zero. If x_3 is identically zero, then x_1 must be identically zero, since from Equation (9–77) we obtain

$$0 = -Kx_1 - 0$$

If x_1 is identically zero, then x_2 must also be identically zero, since from Equation (9–75)

$$0 = x_2$$

Thus, $\dot{V}(\mathbf{x})$ is identically zero only at the origin. Hence we may use the $\mathbf{Q}$ matrix defined by Equation (9–78) for stability analysis.

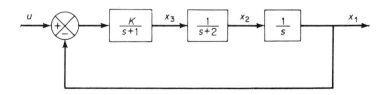

Figure 9–13
Control system.

Alternatively, we may check the rank of matrix

$$
\begin{bmatrix} \mathbf{Q}^{1/2} \\ \mathbf{Q}^{1/2}\mathbf{A} \\ \mathbf{Q}^{1/2}\mathbf{A}^2 \end{bmatrix} = \begin{bmatrix} 0 & 0 & 0 \\ 0 & 0 & 0 \\ 0 & 0 & 1 \\ 0 & 0 & 0 \\ 0 & 0 & 0 \\ -K & 0 & -1 \\ 0 & 0 & 0 \\ 0 & 0 & 0 \\ K & -K & 1 \end{bmatrix}
$$

Clearly, the rank is 3 for $K \neq 0$. Therefore, we may choose this $\mathbf{Q}$ for the Liapunov equation.

Now let us solve the Liapunov equation

$$
\mathbf{A}^T\mathbf{P} + \mathbf{PA} = -\mathbf{Q}
$$

which can be rewritten as

$$
\begin{bmatrix} 0 & 0 & -K \\ 1 & -2 & 0 \\ 0 & 1 & -1 \end{bmatrix}\begin{bmatrix} p_{11} & p_{12} & p_{13} \\ p_{12} & p_{22} & p_{23} \\ p_{13} & p_{23} & p_{33} \end{bmatrix} + \begin{bmatrix} p_{11} & p_{12} & p_{13} \\ p_{12} & p_{22} & p_{23} \\ p_{13} & p_{23} & p_{33} \end{bmatrix}\begin{bmatrix} 0 & 1 & 0 \\ 0 & -2 & 1 \\ -K & 0 & -1 \end{bmatrix} = \begin{bmatrix} 0 & 0 & 0 \\ 0 & 0 & 0 \\ 0 & 0 & -1 \end{bmatrix}
$$

Solving this equation for the elements of $\mathbf{P}$, we obtain

$$
\mathbf{P} = \begin{bmatrix} \dfrac{K^2 + 12K}{12 - 2K} & \dfrac{6K}{12 - 2K} & 0 \\[3mm] \dfrac{6K}{12 - 2K} & \dfrac{3K}{12 - 2K} & \dfrac{K}{12 - 2K} \\[3mm] 0 & \dfrac{K}{12 - 2K} & \dfrac{6}{12 - 2K} \end{bmatrix}
$$

For $\mathbf{P}$ to be positive definite, it is necessary and sufficient that

$$
12 - 2K > 0 \qquad \text{and} \qquad K > 0
$$

or

$$
0 < K < 6
$$

Thus, for $0 < K < 6$, the system is stable in the conventional sense: that is, the origin is asymptotically stable in the large.

9–9 LINEAR TIME-VARYING SYSTEMS*

An advantage of the state-space approach to control systems analysis is that it can easily be extended to linear time-varying systems. Most of the results obtained in this book so far carry over to linear time-varying systems by changing the state transition matrix $\mathbf{\Phi}(t)$ to $\mathbf{\Phi}(t, t_0)$. (This is because for time-varying systems the state transition matrix depends on

* The purpose of presenting this section is to provide a convenient reference to the reader to solve the state-space equation involving time-varying terms. This section may be omitted without losing continuity of the material.

both t and t_0 and not the difference $t - t_0$. Also, it is not always possible to set the initial time equal to zero.) It is important to point out, however, that there is an important difference between the time-invariant case and the time-varying case: The state transition matrix for the time-varying system cannot, in general, be given as a matrix exponential.

Solution of time-varying state equations.   For a scalar differential equation

$$\dot{x} = a(t)x$$

the solution can be given by

$$x(t) = e^{\int_{t_0}^{t} a(\tau)d\tau} x(t_0)$$

and the state-transition function is given by

$$\phi(t, t_0) = \exp\left[\int_{t_0}^{t} a(\tau)\, d\tau\right]$$

The same result, however, does not carry over to the vector-matrix differential equation.
Consider the state equation

$$\dot{\mathbf{x}} = \mathbf{A}(t)\mathbf{x} \tag{9–79}$$

where $\mathbf{x}(t)$ = state vector (n-vector)
$\mathbf{A}(t) = n \times n$ matrix whose elements are piecewise continuous functions of t in the interval $t_0 \leq t \leq t_1$

The solution to Equation (9–79) is given by

$$\mathbf{x}(t) = \mathbf{\Phi}(t, t_0)\mathbf{x}(t_0) \tag{9–80}$$

where $\mathbf{\Phi}(t, t_0)$ is the $n \times n$ nonsingular matrix satisfying the following matrix differential equation:

$$\dot{\mathbf{\Phi}}(t, t_0) = \mathbf{A}(t)\mathbf{\Phi}(t, t_0), \qquad \mathbf{\Phi}(t_0, t_0) = \mathbf{I} \tag{9–81}$$

The fact that Equation (9–80) is the solution of Equation (9–79) can be verified easily since

$$\mathbf{x}(t_0) = \mathbf{\Phi}(t_0, t_0)\mathbf{x}(t_0) = \mathbf{x}(t_0)$$

and from Equations (9–80) and (9–81) we have

$$\dot{\mathbf{x}}(t) = \frac{d}{dt}[\mathbf{\Phi}(t, t_0)\mathbf{x}(t_0)]$$

$$= \dot{\mathbf{\Phi}}(t, t_0)\mathbf{x}(t_0)$$

$$= \mathbf{A}(t)\mathbf{\Phi}(t, t_0)\mathbf{x}(t_0) = \mathbf{A}(t)\mathbf{x}(t)$$

We see that the solution of Equation (9–79) is simply a transformation of the initial state. The matrix $\mathbf{\Phi}(t, t_0)$ is the state-transition matrix for the time-varying system described by Equation (9–79).

State-transition matrix for the time-varying case.   It is important to note that the state-transition matrix $\mathbf{\Phi}(t, t_0)$ is given by a matrix exponential if and only if $\mathbf{A}(t)$ and $\int_{t_0}^{t}$

$\mathbf{A}(\tau) \, d\tau$ commute. That is,

$$\mathbf{\Phi}(t, t_0) = \exp\left[\int_{t_0}^{t} \mathbf{A}(\tau) \, d\tau \right] \qquad \text{(if and only if } \mathbf{A}(t) \text{ and } \int_{t_0}^{t} \mathbf{A}(\tau) \, d\tau \text{ commute)}$$

Note that, if $\mathbf{A}(t)$ is a constant matrix or diagonal matrix, $\mathbf{A}(t)$ and $\int_{t_0}^{t} \mathbf{A}(\tau) \, d\tau$ commute. If $\mathbf{A}(t)$ and $\int_{t_0}^{t} \mathbf{A}(\tau) \, d\tau$ do not commute, there is no simple way to compute the state-transition matrix.

To compute $\mathbf{\Phi}(t, t_0)$ numerically, we may use the following series expansion for $\mathbf{\Phi}(t, t_0)$:

$$\mathbf{\Phi}(t, t_0) = \mathbf{I} + \int_{t_0}^{t} \mathbf{A}(\tau) \, d\tau + \int_{t_0}^{t} \mathbf{A}(\tau_1) \left[\int_{t_0}^{\tau_1} \mathbf{A}(\tau_2) \, d\tau_2 \right] d\tau_1 + \cdots \qquad (9\text{--}82)$$

This, in general, will not give $\mathbf{\Phi}(t, t_0)$ in a closed form.

EXAMPLE 9–22. Obtain $\mathbf{\Phi}(t, 0)$ for the time-varying system

$$\begin{bmatrix} \dot{x}_1 \\ \dot{x}_2 \end{bmatrix} = \begin{bmatrix} 0 & 1 \\ 0 & t \end{bmatrix} \begin{bmatrix} x_1 \\ x_2 \end{bmatrix}$$

To compute $\mathbf{\Phi}(t, 0)$, let us use Equation (9–82). Since

$$\int_{0}^{t} \mathbf{A}(\tau) \, d\tau = \int_{0}^{t} \begin{bmatrix} 0 & 1 \\ 0 & \tau \end{bmatrix} d\tau = \begin{bmatrix} 0 & t \\ 0 & \dfrac{t^2}{2} \end{bmatrix}$$

$$\int_{0}^{t} \begin{bmatrix} 0 & 1 \\ 0 & \tau_1 \end{bmatrix} \left\{ \int_{0}^{\tau_1} \begin{bmatrix} 0 & 1 \\ 0 & \tau_2 \end{bmatrix} d\tau_2 \right\} d\tau_1 = \int_{0}^{t} \begin{bmatrix} 0 & 1 \\ 0 & \tau_1 \end{bmatrix} \begin{bmatrix} 0 & \tau_1 \\ 0 & \dfrac{\tau_1^2}{2} \end{bmatrix} d\tau_1 = \begin{bmatrix} 0 & \dfrac{t^3}{6} \\ 0 & \dfrac{t^4}{8} \end{bmatrix}$$

we obtain

$$\mathbf{\Phi}(t, 0) = \begin{bmatrix} 1 & 0 \\ 0 & 1 \end{bmatrix} + \begin{bmatrix} 0 & t \\ 0 & \dfrac{t^2}{2} \end{bmatrix} + \begin{bmatrix} 0 & \dfrac{t^3}{6} \\ 0 & \dfrac{t^4}{8} \end{bmatrix} + \cdots$$

$$= \begin{bmatrix} 1 & t + \dfrac{t^3}{6} + \cdots \\ 0 & 1 + \dfrac{t^2}{2} + \dfrac{t^4}{8} + \cdots \end{bmatrix}$$

Properties of the state-transition matrix $\mathbf{\Phi}(t, t_0)$. We shall list the properties of the state-transition matrix $\mathbf{\Phi}(t, t_0)$ in what follows.

1. $$\mathbf{\Phi}(t_2, t_1)\mathbf{\Phi}(t_1, t_0) = \mathbf{\Phi}(t_2, t_0)$$

To prove this, note that

$$\mathbf{x}(t_1) = \mathbf{\Phi}(t_1, t_0)\mathbf{x}(t_0)$$

$$\mathbf{x}(t_2) = \mathbf{\Phi}(t_2, t_0)\mathbf{x}(t_0)$$

Also

$$\mathbf{x}(t_2) = \mathbf{\Phi}(t_2, t_1)\mathbf{x}(t_1)$$

Hence

$$\mathbf{x}(t_2) = \mathbf{\Phi}(t_2, t_1)\mathbf{\Phi}(t_1, t_0)\mathbf{x}(t_0) = \mathbf{\Phi}(t_2, t_0)\mathbf{x}(t_0)$$

Thus

$$\mathbf{\Phi}(t_2, t_1)\mathbf{\Phi}(t_1, t_0) = \mathbf{\Phi}(t_2, t_0)$$

2.
$$\mathbf{\Phi}(t_1, t_0) = \mathbf{\Phi}^{-1}(t_0, t_1)$$

To prove this, note that

$$\mathbf{\Phi}(t_1, t_0) = \mathbf{\Phi}^{-1}(t_2, t_1)\mathbf{\Phi}(t_2, t_0)$$

If we let $t_2 = t_0$ in this last equation, then

$$\mathbf{\Phi}(t_1, t_0) = \mathbf{\Phi}^{-1}(t_0, t_1)\mathbf{\Phi}(t_0, t_0) = \mathbf{\Phi}^{-1}(t_0, t_1)$$

Solution of linear time-varying state equations.　Consider the following state equation:

$$\dot{\mathbf{x}} = \mathbf{A}(t)\mathbf{x} + \mathbf{B}(t)\mathbf{u} \tag{9–83}$$

where $\mathbf{x}$ = state vector (n-vector)
$\quad\mathbf{u}$ = control vector (r-vector)
$\mathbf{A}(t)$ = $n \times n$ matrix
$\mathbf{B}(t)$ = $n \times r$ matrix

The elements of $\mathbf{A}(t)$ and $\mathbf{B}(t)$ are assumed to be piecewise continuous functions of t in the interval $t_0 \leq t \leq t_1$.

To obtain the solution of Equation (9–83), let us put

$$\mathbf{x}(t) = \mathbf{\Phi}(t, t_0)\mathbf{\xi}(t)$$

where $\mathbf{\Phi}(t, t_0)$ is the unique matrix satisfying the following equation:

$$\dot{\mathbf{\Phi}}(t, t_0) = \mathbf{A}(t)\mathbf{\Phi}(t, t_0), \qquad \mathbf{\Phi}(t_0, t_0) = \mathbf{I}$$

Then

$$\begin{aligned}
\dot{\mathbf{x}}(t) &= \frac{d}{dt}[\mathbf{\Phi}(t, t_0)\mathbf{\xi}(t)] \\
&= \dot{\mathbf{\Phi}}(t, t_0)\mathbf{\xi}(t) + \mathbf{\Phi}(t, t_0)\dot{\mathbf{\xi}}(t) \\
&= \mathbf{A}(t)\mathbf{\Phi}(t, t_0)\mathbf{\xi}(t) + \mathbf{\Phi}(t, t_0)\dot{\mathbf{\xi}}(t)
\end{aligned}$$

If $\mathbf{x}(t) = \mathbf{\Phi}(t, t_0)\mathbf{\xi}(t)$ is substituted into Equation (9–83), we have

$$\dot{\mathbf{x}}(t) = \mathbf{A}(t)\mathbf{\Phi}(t, t_0)\mathbf{\xi}(t) + \mathbf{B}(t)\mathbf{u}(t)$$

Comparing the preceding two expressions for $\dot{\mathbf{x}}(t)$, we obtain

$$\boldsymbol{\Phi}(t, t_0)\dot{\boldsymbol{\xi}}(t) = \mathbf{B}(t)\mathbf{u}(t)$$

or

$$\dot{\boldsymbol{\xi}}(t) = \boldsymbol{\Phi}^{-1}(t, t_0)\mathbf{B}(t)\mathbf{u}(t)$$

Hence

$$\boldsymbol{\xi}(t) = \boldsymbol{\xi}(t_0) + \int_{t_0}^{t} \boldsymbol{\Phi}^{-1}(\tau, t_0)\mathbf{B}(\tau)\mathbf{u}(\tau)\, d\tau$$

Since

$$\boldsymbol{\xi}(t_0) = \boldsymbol{\Phi}^{-1}(t_0, t_0)\mathbf{x}(t_0) = \mathbf{x}(t_0)$$

the solution of Equation (9–83) is obtained as

$$\mathbf{x}(t) = \boldsymbol{\Phi}(t, t_0)\mathbf{x}(t_0) + \boldsymbol{\Phi}(t, t_0)\int_{t_0}^{t} \boldsymbol{\Phi}^{-1}(\tau, t_0)\mathbf{B}(\tau)\mathbf{u}(\tau)\, d\tau$$

$$= \boldsymbol{\Phi}(t, t_0)\mathbf{x}(t_0) + \int_{t_0}^{t} \boldsymbol{\Phi}(t, \tau)\mathbf{B}(\tau)\mathbf{u}(\tau)\, d\tau \tag{9–84}$$

Computing the right side of Equation (9–84) for practical cases will require a digital computer.

Example Problems and Solutions

A–9–1. Consider the system described by

$$\begin{bmatrix} \dot{x}_1 \\ \dot{x}_2 \end{bmatrix} = \begin{bmatrix} a_{11} & a_{12} \\ a_{21} & a_{22} \end{bmatrix} \begin{bmatrix} x_1 \\ x_2 \end{bmatrix}, \qquad a_{21} \neq 0$$

Assuming that the two eigenvalues λ_1 and λ_2 of the coefficient matrix are distinct, find a diagonalizing transformtion matrix $\mathbf{P}$ in terms of a_{ij}, λ_1, and λ_2.

Solution. If matrix $\mathbf{P}$ is a diagonalizing transformation matrix, then

$$\mathbf{P}^{-1}\mathbf{A}\mathbf{P} = \begin{bmatrix} \lambda_1 & 0 \\ 0 & \lambda_2 \end{bmatrix}$$

where

$$\mathbf{A} = \begin{bmatrix} a_{11} & a_{12} \\ a_{21} & a_{22} \end{bmatrix}, \qquad \mathbf{P} = \begin{bmatrix} p_{11} & p_{12} \\ p_{21} & p_{22} \end{bmatrix}$$

Matrix $\mathbf{P}$ must satisfy the following equation:

$$\begin{bmatrix} a_{11} & a_{12} \\ a_{21} & a_{22} \end{bmatrix}\begin{bmatrix} p_{11} & p_{12} \\ p_{21} & p_{22} \end{bmatrix} = \begin{bmatrix} p_{11} & p_{12} \\ p_{21} & p_{22} \end{bmatrix}\begin{bmatrix} \lambda_1 & 0 \\ 0 & \lambda_2 \end{bmatrix}$$

or

$$\begin{bmatrix} a_{11}p_{11} + a_{12}p_{21} & a_{11}p_{12} + a_{12}p_{22} \\ a_{21}p_{11} + a_{22}p_{21} & a_{21}p_{12} + a_{22}p_{22} \end{bmatrix} = \begin{bmatrix} p_{11}\lambda_1 & p_{12}\lambda_2 \\ p_{21}\lambda_1 & p_{22}\lambda_2 \end{bmatrix}$$

from which we obtain

$$p_{11}(\lambda_1 - a_{11}) = a_{12}p_{21}$$

$$p_{12}(\lambda_2 - a_{11}) = a_{12}p_{22}$$

$$p_{21}(\lambda_1 - a_{22}) = a_{21}p_{11}$$

$$p_{22}(\lambda_2 - a_{22}) = a_{21}p_{12}$$

These four equations can be satisfied in a number of ways. An example is

$$p_{11} = \lambda_1 - a_{22}, \qquad p_{12} = \lambda_2 - a_{22}, \qquad p_{21} = a_{21}, \qquad p_{22} = a_{21}$$

or

$$\mathbf{P} = \begin{bmatrix} \lambda_1 - a_{22} & \lambda_2 - a_{22} \\ a_{21} & a_{21} \end{bmatrix}$$

This matrix $\mathbf{P}$ is a diagonalizing transformation matrix. (There are infinitely many such diagonalizing transformation matrices for the given matrix $\mathbf{A}$.)

A–9–2. Consider the system described by

$$\dot{\mathbf{x}} = \mathbf{A}\mathbf{x} + \mathbf{B}\mathbf{u}$$

where $\mathbf{x}$ = state vector (n-vector)
$\quad\;\; \mathbf{u}$ = control vector (r-vector)
$\quad\;\; \mathbf{A}$ = $n \times n$ constant matrix
$\quad\;\; \mathbf{B}$ = $n \times r$ constant matrix

Obtain the response to each of the following inputs:

(a) The r components of $\mathbf{u}$ are impulse functions of various magnitudes.
(b) The r components of $\mathbf{u}$ are step functions of various magnitudes.
(c) The r components of $\mathbf{u}$ are ramp functions of various magnitudes.

Assume that each input is given at $t = 0$.

Solution. The solution to the given state equation is

$$\mathbf{x}(t) = e^{\mathbf{A}(t-t_0)}\mathbf{x}(t_0) + \int_{t_0}^{t} e^{\mathbf{A}(t-\tau)}\mathbf{B}\mathbf{u}(\tau)\,d\tau$$

[Refer to Equation (4–67).] Substituting $t_0 = 0-$ into this solution, we obtain

$$\mathbf{x}(t) = e^{\mathbf{A}t}\mathbf{x}(0-) + \int_{0-}^{t} e^{\mathbf{A}(t-\tau)}\mathbf{B}\mathbf{u}(\tau)\,d\tau$$

(a) *Impulse response*: Let us write the impulse input $\mathbf{u}(t)$ as

$$\mathbf{u}(t) = \delta(t)\mathbf{w}$$

where $\mathbf{w}$ is a vector whose components are the magnitudes of r impulse functions applied at $t = 0$. The solution to the impulse $\delta(t)\mathbf{w}$ given at $t = 0$ is

$$\mathbf{x}(t) = e^{\mathbf{A}t}\mathbf{x}(0-) + \int_{0-}^{t} e^{\mathbf{A}(t-\tau)}\mathbf{B}\delta(\tau)\mathbf{w}\,d\tau$$
$$= e^{\mathbf{A}t}\mathbf{x}(0-) + e^{\mathbf{A}t}\mathbf{B}\mathbf{w}$$

Example Problems and Solutions

741

(b) *Step response*: Let us write the step input $\mathbf{u}(t)$ as

$$\mathbf{u}(t) = \mathbf{k}$$

where $\mathbf{k}$ is a vector whose components are the magnitudes of r step functions applied at $t = 0$. The solution to the step input at $t = 0$ is given by

$$\mathbf{x}(t) = e^{\mathbf{A}t}\mathbf{x}(0) + \int_0^t e^{\mathbf{A}(t-\tau)}\mathbf{B}\mathbf{k}\,d\tau$$

$$= e^{\mathbf{A}t}\mathbf{x}(0) + e^{\mathbf{A}t}\left[\int_0^t \left(\mathbf{I} - \mathbf{A}\tau + \frac{\mathbf{A}^2\tau^2}{2!} - \cdots\right) d\tau\right]\mathbf{B}\mathbf{k}$$

$$= e^{\mathbf{A}t}\mathbf{x}(0) + e^{\mathbf{A}t}\left(\mathbf{I}t - \frac{\mathbf{A}t^2}{2!} + \frac{\mathbf{A}^2t^3}{3!} - \cdots\right)\mathbf{B}\mathbf{k}$$

If $\mathbf{A}$ is nonsingular, then this last equation can be simplified to give

$$\mathbf{x}(t) = e^{\mathbf{A}t}\mathbf{x}(0) + e^{\mathbf{A}t}[-(\mathbf{A}^{-1})(e^{-\mathbf{A}t} - \mathbf{I})]\mathbf{B}\mathbf{k}$$

$$= e^{\mathbf{A}t}\mathbf{x}(0) + \mathbf{A}^{-1}(e^{\mathbf{A}t} - \mathbf{I})\mathbf{B}\mathbf{k}$$

(c) *Ramp response*: Let us write the ramp input $\mathbf{u}(t)$ as

$$\mathbf{u}(t) = t\mathbf{v}$$

where $\mathbf{v}$ is a vector whose components are magnitudes of ramp functions applied at $t = 0$. The solution to the ramp input $t\mathbf{v}$ given at $t = 0$ is

$$\mathbf{x}(t) = e^{\mathbf{A}t}\mathbf{x}(0) + \int_0^t e^{\mathbf{A}(t-\tau)}\mathbf{B}\tau\mathbf{v}\,d\tau$$

$$= e^{\mathbf{A}t}\mathbf{x}(0) + e^{\mathbf{A}t}\int_0^t e^{-\mathbf{A}\tau}\tau\,d\tau\,\mathbf{B}\mathbf{v}$$

$$= e^{\mathbf{A}t}\mathbf{x}(0) + e^{\mathbf{A}t}\left(\frac{\mathbf{I}}{2}t^2 - \frac{2\mathbf{A}}{3!}t^3 + \frac{3\mathbf{A}^2}{4!}t^4 - \frac{4\mathbf{A}^3}{5!}t^5 + \cdots\right)\mathbf{B}\mathbf{v}$$

If $\mathbf{A}$ is nonsingular, then this last equation can be simplified to give

$$\mathbf{x}(t) = e^{\mathbf{A}t}\mathbf{x}(0) + (\mathbf{A}^{-2})(e^{\mathbf{A}t} - \mathbf{I} - \mathbf{A}t)\mathbf{B}\mathbf{v}$$

$$= e^{\mathbf{A}t}\mathbf{x}(0) + [\mathbf{A}^{-2}(e^{\mathbf{A}t} - \mathbf{I}) - \mathbf{A}^{-1}t]\mathbf{B}\mathbf{v}$$

A–9–3. Referring to the Cayley–Hamilton theorem, every $n \times n$ matrix $\mathbf{A}$ satisfies its own characteristic equation. The characteristic equation is not, however, necessarily the scalar equation of least degree that $\mathbf{A}$ satisfies. The least-degree polynomial having $\mathbf{A}$ as a root is called the *minimal polynomial*. That is, the minimal polynomial of an $n \times n$ matrix $\mathbf{A}$ is defined as the polynomial $\phi(\lambda)$ of least degree,

$$\phi(\lambda) = \lambda^m + a_1\lambda^{m-1} + \cdots + a_{m-1}\lambda + a_m \qquad m \le n$$

such that $\phi(\mathbf{A}) = \mathbf{0}$, or

$$\phi(\mathbf{A}) = \mathbf{A}^m + a_1\mathbf{A}^{m-1} + \cdots + a_{m-1}\mathbf{A} + a_m\mathbf{I} = \mathbf{0}$$

The minimal polynomial plays an important role in the computation of polynomials in an $n \times n$ matrix.

Let us suppose that $d(\lambda)$, a polynomial in λ, is the greatest common divisor of all the elements of adj $(\lambda\mathbf{I} - \mathbf{A})$. Show that if the coefficient of the highest-degree term in λ of $d(\lambda)$ is chosen as 1, then the minimal polynomial $\phi(\lambda)$ is given by

$$\phi(\lambda) = \frac{|\lambda\mathbf{I} - \mathbf{A}|}{d(\lambda)}$$

Solution. By assumption, the greatest common divisor of the matrix adj $(\lambda \mathbf{I} - \mathbf{A})$ is $d(\lambda)$. Therefore,

$$\text{adj } (\lambda \mathbf{I} - \mathbf{A}) = d(\lambda)\mathbf{B}(\lambda)$$

where the greatest common divisor of the n^2 elements (which are functions of λ) of $\mathbf{B}(\lambda)$ is unity. Since

$$(\lambda \mathbf{I} - \mathbf{A}) \text{ adj } (\lambda \mathbf{I} - \mathbf{A}) = |\lambda \mathbf{I} - \mathbf{A}|\mathbf{I}$$

we obtain

$$d(\lambda)(\lambda \mathbf{I} - \mathbf{A})\mathbf{B}(\lambda) = |\lambda \mathbf{I} - \mathbf{A}|\mathbf{I} \qquad (9\text{--}85)$$

from which we find that $|\lambda \mathbf{I} - \mathbf{A}|$ is divisible by $d(\lambda)$. Let us put

$$|\lambda \mathbf{I} - \mathbf{A}| = d(\lambda)\psi(\lambda) \qquad (9\text{--}86)$$

Because the coefficient of the highest-degree term in λ of $d(\lambda)$ has been chosen to be 1, the coefficient of the highest-degree term in λ of $\psi(\lambda)$ is also 1. From Equations (9–85) and (9–86), we have

$$(\lambda \mathbf{I} - \mathbf{A})\mathbf{B}(\lambda) = \psi(\lambda)\mathbf{I}$$

Hence

$$\psi(\mathbf{A}) = \mathbf{0}$$

Note that $\psi(\lambda)$ can be written as follows:

$$\psi(\lambda) = g(\lambda)\phi(\lambda) + \alpha(\lambda)$$

where $\alpha(\lambda)$ is of lower degree than $\phi(\lambda)$. Since $\psi(\mathbf{A}) = \mathbf{0}$ and $\phi(\mathbf{A}) = \mathbf{0}$, we must have $\alpha(\mathbf{A}) = \mathbf{0}$. Since $\phi(\lambda)$ is the minimal polynomial, $\alpha(\lambda)$ must be identically zero, or

$$\psi(\lambda) = g(\lambda)\phi(\lambda)$$

Note that because $\phi(\mathbf{A}) = \mathbf{0}$, we can write

$$\phi(\lambda)\mathbf{I} = (\lambda \mathbf{I} - \mathbf{A})\mathbf{C}(\lambda)$$

Hence

$$\psi(\lambda)\mathbf{I} = g(\lambda)\phi(\lambda)\mathbf{I} = g(\lambda)(\lambda \mathbf{I} - \mathbf{A})\mathbf{C}(\lambda)$$

and we obtain

$$\mathbf{B}(\lambda) = g(\lambda)\mathbf{C}(\lambda)$$

Note that the greatest common divisor of the n^2 elements of $\mathbf{B}(\lambda)$ is unity. Hence

$$g(\lambda) = 1$$

Therefore,

$$\psi(\lambda) = \phi(\lambda)$$

Then, from this last equation and Equation (9–86) we obtain

$$\phi(\lambda) = \frac{|\lambda \mathbf{I} - \mathbf{A}|}{d(\lambda)}$$

It is noted that the minimal polynomial $\phi(\lambda)$ of an $n \times n$ matrix $\mathbf{A}$ can be determined by the following procedure:

1. Form adj $(\lambda \mathbf{I} - \mathbf{A})$ and write the elements of adj $(\lambda \mathbf{I} - \mathbf{A})$ as factored polynomials in λ.

2. Determine $d(\lambda)$ as the greatest common divisor of all the elements of adj $(\lambda I - A)$. Choose the coefficient of the highest-degree term in λ of $d(\lambda)$ to be 1. If there is no common divisor, $d(\lambda) = 1$.

3. The minimal polynomial $\phi(\lambda)$ is then given as $|\lambda I - A|$ divided by $d(\lambda)$.

A–9–4. If an $n \times n$ matrix A has n distinct eigenvalues, then the minimal polynomial of A is identical with the characteristic polynomial. Also, if the multiple eigenvalues of A are linked in a Jordan chain, the minimal polynomial and the characteristic polynomial are identical. If, however, the multiple eigenvalues of A are not linked in a Jordan chain, the minimal polynomial is of lower degree than the characteristic polynomial.

Using the following matrices A and B as examples, verify the foregoing statements about the minimal polynomial when multiple eigenvalues are involved.

$$A = \begin{bmatrix} 2 & 1 & 4 \\ 0 & 2 & 0 \\ 0 & 3 & 1 \end{bmatrix}, \quad B = \begin{bmatrix} 2 & 0 & 0 \\ 0 & 2 & 0 \\ 0 & 3 & 1 \end{bmatrix}$$

Solution. First, consider the matrix A. The characteristic polynomial is given by

$$|\lambda I - A| = \begin{vmatrix} \lambda - 2 & -1 & -4 \\ 0 & \lambda - 2 & 0 \\ 0 & -3 & \lambda - 1 \end{vmatrix} = (\lambda - 2)^2(\lambda - 1)$$

Thus the eigenvalues of A are 2, 2, and 1. It can be shown that the Jordan canonical form of A is

$$\begin{bmatrix} 2 & 1 & 0 \\ 0 & 2 & 0 \\ 0 & 0 & 1 \end{bmatrix}$$

and the multiple eigenvalues are linked in the Jordan chain as shown. (For the procedure for deriving the Jordan canonical form of A, refer to the Appendix.)

To determine the minimal polynomial, let us first obtain adj $(\lambda I - A)$. It is given by

$$\text{adj}\,(\lambda I - A) = \begin{bmatrix} (\lambda - 2)(\lambda - 1) & (\lambda + 11) & 4(\lambda - 2) \\ 0 & (\lambda - 2)(\lambda - 1) & 0 \\ 0 & 3(\lambda - 2) & (\lambda - 2)^2 \end{bmatrix}$$

Notice that there is no common divisor of all the elements of adj $(\lambda I - A)$. Hence $d(\lambda) = 1$. Thus, the minimal polynomial $\phi(\lambda)$ is identical with the characteristic polynomial, or

$$\phi(\lambda) = |\lambda I - A| = (\lambda - 2)^2(\lambda - 1)$$
$$= \lambda^3 - 5\lambda^2 + 8\lambda - 4$$

A simple calculation proves that

$$A^3 - 5A^2 + 8A - 4I = 0$$

but

$$A^2 - 3A + 2I \neq 0$$

Thus, we have shown that the minimal polynomial and the characteristic polynomial of this matrix A are the same.

Next, consider the matrix B. The characteristic polynomial is given by

$$|\lambda I - B| = \begin{vmatrix} \lambda - 2 & 0 & 0 \\ 0 & \lambda - 2 & 0 \\ 0 & -3 & \lambda - 1 \end{vmatrix} = (\lambda - 2)^2(\lambda - 1)$$

A simple computation reveals that matrix $\mathbf{B}$ has three eigenvectors, and the Jordan canonical form of $\mathbf{B}$ is given by

$$\begin{bmatrix} 2 & 0 & 0 \\ 0 & 2 & 0 \\ 0 & 0 & 1 \end{bmatrix}$$

Thus, the multiple eigenvalues are not linked. To obtain the minimal polynomial, we first compute adj $(\lambda \mathbf{I} - \mathbf{B})$:

$$\text{adj}\,(\lambda\mathbf{I} - \mathbf{B}) = \begin{bmatrix} (\lambda - 2)(\lambda - 1) & 0 & 0 \\ 0 & (\lambda - 2)(\lambda - 1) & 0 \\ 0 & 3(\lambda - 2) & (\lambda - 2)^2 \end{bmatrix}$$

from which it is evident that

$$d(\lambda) = \lambda - 2$$

Hence

$$\phi(\lambda) = \frac{|\lambda\mathbf{I} - \mathbf{B}|}{d(\lambda)} = \frac{(\lambda - 2)^2(\lambda - 1)}{\lambda - 2} = \lambda^2 - 3\lambda + 2$$

As a check, let us compute $\phi(\mathbf{B})$:

$$\phi(\mathbf{B}) = \mathbf{B}^2 - 3\mathbf{B} + 2\mathbf{I} = \begin{bmatrix} 4 & 0 & 0 \\ 0 & 4 & 0 \\ 0 & 9 & 1 \end{bmatrix} - 3\begin{bmatrix} 2 & 0 & 0 \\ 0 & 2 & 0 \\ 0 & 3 & 1 \end{bmatrix} + 2\begin{bmatrix} 1 & 0 & 0 \\ 0 & 1 & 0 \\ 0 & 0 & 1 \end{bmatrix} = \begin{bmatrix} 0 & 0 & 0 \\ 0 & 0 & 0 \\ 0 & 0 & 0 \end{bmatrix}$$

For the given matrix $\mathbf{B}$, the degree of the minimal polynomial is lower by 1 than that of the characteristic polynomial. As shown here, if the multiple eigenvalues of an $n \times n$ matrix are not linked in a Jordan chain, the minimal polynomial is of lower degree than the characteristic polynomial.

A–9–5. Show that by use of the minimal polynomial, the inverse of a nonsingular matrix $\mathbf{A}$ can be expressed as a polynomial in $\mathbf{A}$ with scalar coefficients as follows:

$$\mathbf{A}^{-1} = -\frac{1}{a_m}(\mathbf{A}^{m-1} + a_1\mathbf{A}^{m-2} + \cdots + a_{m-2}\mathbf{A} + a_{m-1}\mathbf{I}) \qquad (9\text{--}87)$$

where $a_1, a_2, \ldots, a_m$ are coefficients of the minimal polynomial

$$\phi(\lambda) = \lambda^m + a_1\lambda^{m-1} + \cdots + a_{m-1}\lambda + a_m$$

Then obtain the inverse of the following matrix $\mathbf{A}$:

$$\mathbf{A} = \begin{bmatrix} 1 & 2 & 0 \\ 3 & -1 & -2 \\ 1 & 0 & -3 \end{bmatrix}$$

Solution. For a nonsingular matrix $\mathbf{A}$, its minimal polynomial $\phi(\mathbf{A})$ can be written as

$$\phi(\mathbf{A}) = \mathbf{A}^m + a_1\mathbf{A}^{m-1} + \cdots + a_{m-1}\mathbf{A} + a_m\mathbf{I} = \mathbf{0}$$

where $a_m \neq 0$. Hence,

$$\mathbf{I} = -\frac{1}{a_m}(\mathbf{A}^m + a_1\mathbf{A}^{m-1} + \cdots + a_{m-2}\mathbf{A}^2 + a_{m-1}\mathbf{A})$$

Premultiplying by A^{-1}, we obtain

$$A^{-1} = -\frac{1}{a_m}(A^{m-1} + a_1 A^{m-2} + \cdots + a_{m-2}A + a_{m-1}I)$$

which is Equation (9–87).

For the given matrix **A,** adj $(\lambda I - A)$ can be given as

$$\text{adj } (\lambda I - A) = \begin{bmatrix} \lambda^2 + 4\lambda + 3 & 2\lambda + 6 & -4 \\ 3\lambda + 7 & \lambda^2 + 2\lambda - 3 & -2\lambda + 2 \\ \lambda + 1 & 2 & \lambda^2 - 7 \end{bmatrix}$$

Clearly, there is no common divisor $d(\lambda)$ of all elements of adj $(\lambda I - A)$. Hence, $d(\lambda) = 1$. Consequently, the minimal polynomial $\phi(\lambda)$ is given by

$$\phi(\lambda) = \frac{|\lambda I - A|}{d(\lambda)} = |\lambda I - A|$$

Thus, the minimal polynomial $\phi(\lambda)$ is the same as the characteristic polynomial.

Since the characteristic polynomial is

$$|\lambda I - A| = \lambda^3 + 3\lambda^2 - 7\lambda - 17$$

we obtain

$$\phi(\lambda) = \lambda^3 + 3\lambda^2 - 7\lambda - 17$$

By identifying the coefficients a_i of the minimal polynomial (which is the same as the characteristic polynomial in this case), we have

$$a_1 = 3, \qquad a_2 = -7, \qquad a_3 = -17$$

The inverse of **A** can then be obtained from Equation (9–87) as follows:

$$A^{-1} = -\frac{1}{a_3}(A^2 + a_1 A + a_2 I) = \frac{1}{17}(A^2 + 3A - 7I)$$

$$= \frac{1}{17}\left\{\begin{bmatrix} 7 & 0 & -4 \\ -2 & 7 & 8 \\ -2 & 2 & 9 \end{bmatrix} + 3\begin{bmatrix} 1 & 2 & 0 \\ 3 & -1 & -2 \\ 1 & 0 & -3 \end{bmatrix} - 7\begin{bmatrix} 1 & 0 & 0 \\ 0 & 1 & 0 \\ 0 & 0 & 1 \end{bmatrix}\right\}$$

$$= \frac{1}{17}\begin{bmatrix} 3 & 6 & -4 \\ 7 & -3 & 2 \\ 1 & 2 & -7 \end{bmatrix}$$

$$= \begin{bmatrix} \frac{3}{17} & \frac{6}{17} & -\frac{4}{17} \\ \frac{7}{17} & -\frac{3}{17} & \frac{2}{17} \\ \frac{1}{17} & \frac{2}{17} & -\frac{7}{17} \end{bmatrix}$$

A–9–6 Consider the following polynomial in λ of degree $m - 1$, where we assume $\lambda_1, \lambda_2, \ldots, \lambda_m$ to be distinct:

$$p_k(\lambda) = \frac{(\lambda - \lambda_1) \cdots (\lambda - \lambda_{k-1})(\lambda - \lambda_{k+1}) \cdots (\lambda - \lambda_m)}{(\lambda_k - \lambda_1) \cdots (\lambda_k - \lambda_{k-1})(\lambda_k - \lambda_{k+1}) \cdots (\lambda_k - \lambda_m)}$$

where $k = 1, 2, \ldots, m$. Notice that

$$p_k(\lambda_i) = \begin{cases} 1 & \text{if } i = k \\ 0 & \text{if } i \neq k \end{cases}$$

Then the polynomial $f(\lambda)$ of degree $m - 1$

$$f(\lambda) = \sum_{k=1}^{m} f(\lambda_k) p_k(\lambda)$$

$$= \sum_{k=1}^{m} f(\lambda_k) \frac{(\lambda - \lambda_1) \cdots (\lambda - \lambda_{k-1})(\lambda - \lambda_{k+1}) \cdots (\lambda - \lambda_m)}{(\lambda_k - \lambda_1) \cdots (\lambda_k - \lambda_{k-1})(\lambda_k - \lambda_{k+1}) \cdots (\lambda_k - \lambda_m)}$$

takes on the values $f(\lambda_k)$ at the points λ_k. This last equation is commonly called *Lagrange's interpolation formula*. The polynomial $f(\lambda)$ of degree $m - 1$ is determined from m independent data $f(\lambda_1)$, $f(\lambda_2)$, . . . , $f(\lambda_m)$. That is, the polynomial $f(\lambda)$ passes through m points $f(\lambda_1)$, $f(\lambda_2)$, . . . , $f(\lambda_m)$. Since $f(\lambda)$ is a polynomial of degree $m - 1$, it is uniquely determined. Any other representations of the polynomial of degree $m - 1$ can be reduced to the Lagrange polynomial $f(\lambda)$.

Assuming that the eigenvalues of an $n \times n$ matrix $\mathbf{A}$ are distinct, substitute $\mathbf{A}$ for λ in the polynomial $p_k(\lambda)$. Then we get

$$p_k(\mathbf{A}) = \frac{(\mathbf{A} - \lambda_1 \mathbf{I}) \cdots (\mathbf{A} - \lambda_{k-1} \mathbf{I})(\mathbf{A} - \lambda_{k+1} \mathbf{I}) \cdots (\mathbf{A} - \lambda_m \mathbf{I})}{(\lambda_k - \lambda_1) \cdots (\lambda_k - \lambda_{k-1})(\lambda_k - \lambda_{k+1}) \cdots (\lambda_k - \lambda_m)}$$

Notice that $p_k(\mathbf{A})$ is a polynomial in $\mathbf{A}$ of degree $m - 1$. Notice also that

$$p_k(\lambda_i \mathbf{I}) = \begin{cases} \mathbf{I} & \text{if } i = k \\ \mathbf{0} & \text{if } i \neq k \end{cases}$$

Now define

$$f(\mathbf{A}) = \sum_{k=1}^{m} f(\lambda_k) p_k(\mathbf{A})$$

$$= \sum_{k=1}^{m} f(\lambda_k) \frac{(\mathbf{A} - \lambda_1 \mathbf{I}) \cdots (\mathbf{A} - \lambda_{k-1} \mathbf{I})(\mathbf{A} - \lambda_{k+1} \mathbf{I}) \cdots (\mathbf{A} - \lambda_m \mathbf{I})}{(\lambda_k - \lambda_1) \cdots (\lambda_k - \lambda_{k-1})(\lambda_k - \lambda_{k+1}) \cdots (\lambda_k - \lambda_m)} \tag{9–88}$$

Equation (9–88) is known as Sylvester's interpolation formula. Equation (9–88) is equivalent to the following equation:

$$\begin{vmatrix} 1 & 1 & \cdots & 1 & \mathbf{I} \\ \lambda_1 & \lambda_2 & \cdots & \lambda_m & \mathbf{A} \\ \lambda_1^2 & \lambda_2^2 & \cdots & \lambda_m^2 & \mathbf{A}^2 \\ \cdot & \cdot & & \cdot & \cdot \\ \cdot & \cdot & & \cdot & \cdot \\ \lambda_1^{m-1} & \lambda_2^{m-1} & \cdots & \lambda_m^{m-1} & \mathbf{A}^{m-1} \\ f(\lambda_1) & f(\lambda_2) & \cdots & f(\lambda_m) & f(\mathbf{A}) \end{vmatrix} = \mathbf{0} \tag{9–89}$$

Equations (9–88) and (9–89) are frequently used for evaluating functions $f(\mathbf{A})$ of matrix $\mathbf{A}$, for example, $(\lambda \mathbf{I} - \mathbf{A})^{-1}$, $e^{\mathbf{A}t}$, and so forth. Note that Equation (9–89) can also be written as

$$\begin{vmatrix} 1 & \lambda_1 & \lambda_1^2 & \cdots & \lambda_1^{m-1} & f(\lambda_1) \\ 1 & \lambda_2 & \lambda_2^2 & \cdots & \lambda_2^{m-1} & f(\lambda_2) \\ \cdot & \cdot & \cdot & & \cdot & \cdot \\ \cdot & \cdot & \cdot & & \cdot & \cdot \\ \cdot & \cdot & \cdot & & \cdot & \cdot \\ 1 & \lambda_m & \lambda_m^2 & \cdots & \lambda_m^{m-1} & f(\lambda_m) \\ \mathbf{I} & \mathbf{A} & \mathbf{A}^2 & \cdots & \mathbf{A}^{m-1} & f(\mathbf{A}) \end{vmatrix} = \mathbf{0} \tag{9–90}$$

Show that Equations (9–88) and (9–89) are equivalent. To simplify the arguments, assume that $m = 4$.

Solution. Equation (9–89), when $m = 4$, can be expanded as follows:

$$\mathbf{\Delta} = \begin{vmatrix} 1 & 1 & 1 & 1 & \mathbf{I} \\ \lambda_1 & \lambda_2 & \lambda_3 & \lambda_4 & \mathbf{A} \\ \lambda_1^2 & \lambda_2^2 & \lambda_3^2 & \lambda_4^2 & \mathbf{A}^2 \\ \lambda_1^3 & \lambda_2^3 & \lambda_3^3 & \lambda_4^3 & \mathbf{A}^3 \\ f(\lambda_1) & f(\lambda_2) & f(\lambda_3) & f(\lambda_4) & f(\mathbf{A}) \end{vmatrix}$$

$$= f(\mathbf{A}) \begin{vmatrix} 1 & 1 & 1 & 1 \\ \lambda_1 & \lambda_2 & \lambda_3 & \lambda_4 \\ \lambda_1^2 & \lambda_2^2 & \lambda_3^2 & \lambda_4^2 \\ \lambda_1^3 & \lambda_2^3 & \lambda_3^3 & \lambda_4^3 \end{vmatrix} - f(\lambda_4) \begin{vmatrix} 1 & 1 & 1 & \mathbf{I} \\ \lambda_1 & \lambda_2 & \lambda_3 & \mathbf{A} \\ \lambda_1^2 & \lambda_2^2 & \lambda_3^2 & \mathbf{A}^2 \\ \lambda_1^3 & \lambda_2^3 & \lambda_3^3 & \mathbf{A}^3 \end{vmatrix}$$

$$+ f(\lambda_3) \begin{vmatrix} 1 & 1 & 1 & \mathbf{I} \\ \lambda_1 & \lambda_2 & \lambda_4 & \mathbf{A} \\ \lambda_1^2 & \lambda_2^2 & \lambda_4^2 & \mathbf{A}^2 \\ \lambda_1^3 & \lambda_2^3 & \lambda_4^3 & \mathbf{A}^3 \end{vmatrix} - f(\lambda_2) \begin{vmatrix} 1 & 1 & 1 & \mathbf{I} \\ \lambda_1 & \lambda_3 & \lambda_4 & \mathbf{A} \\ \lambda_1^2 & \lambda_3^2 & \lambda_4^2 & \mathbf{A}^2 \\ \lambda_1^3 & \lambda_3^3 & \lambda_4^3 & \mathbf{A}^3 \end{vmatrix}$$

$$+ f(\lambda_1) \begin{vmatrix} 1 & 1 & 1 & \mathbf{I} \\ \lambda_2 & \lambda_3 & \lambda_4 & \mathbf{A} \\ \lambda_2^2 & \lambda_3^2 & \lambda_4^2 & \mathbf{A}^2 \\ \lambda_2^3 & \lambda_3^3 & \lambda_4^3 & \mathbf{A}^3 \end{vmatrix}$$

Since

$$\begin{vmatrix} 1 & 1 & 1 & 1 \\ \lambda_1 & \lambda_2 & \lambda_3 & \lambda_4 \\ \lambda_1^2 & \lambda_2^2 & \lambda_3^2 & \lambda_4^2 \\ \lambda_1^3 & \lambda_2^3 & \lambda_3^3 & \lambda_4^3 \end{vmatrix} = (\lambda_4 - \lambda_3)(\lambda_4 - \lambda_2)(\lambda_4 - \lambda_1)(\lambda_3 - \lambda_2)(\lambda_3 - \lambda_1)(\lambda_2 - \lambda_1)$$

and

$$\begin{vmatrix} 1 & 1 & 1 & \mathbf{I} \\ \lambda_i & \lambda_j & \lambda_k & \mathbf{A} \\ \lambda_i^2 & \lambda_j^2 & \lambda_k^2 & \mathbf{A}^2 \\ \lambda_i^3 & \lambda_j^3 & \lambda_k^3 & \mathbf{A}^3 \end{vmatrix} = (\mathbf{A} - \lambda_k \mathbf{I})(\mathbf{A} - \lambda_j \mathbf{I})(\mathbf{A} - \lambda_i \mathbf{I})(\lambda_k - \lambda_j)(\lambda_k - \lambda_i)(\lambda_j - \lambda_i)$$

we obtain

$$\begin{aligned}
\mathbf{\Delta} = & \; f(\mathbf{A})[(\lambda_4 - \lambda_3)(\lambda_4 - \lambda_2)(\lambda_4 - \lambda_1)(\lambda_3 - \lambda_2)(\lambda_3 - \lambda_1)(\lambda_2 - \lambda_1)] \\
& - f(\lambda_4)[(\mathbf{A} - \lambda_3 \mathbf{I})(\mathbf{A} - \lambda_2 \mathbf{I})(\mathbf{A} - \lambda_1 \mathbf{I})(\lambda_3 - \lambda_2)(\lambda_3 - \lambda_1)(\lambda_2 - \lambda_1)] \\
& + f(\lambda_3)[(\mathbf{A} - \lambda_4 \mathbf{I})(\mathbf{A} - \lambda_2 \mathbf{I})(\mathbf{A} - \lambda_1 \mathbf{I})(\lambda_4 - \lambda_2)(\lambda_4 - \lambda_1)(\lambda_2 - \lambda_1)] \\
& - f(\lambda_2)[(\mathbf{A} - \lambda_4 \mathbf{I})(\mathbf{A} - \lambda_3 \mathbf{I})(\mathbf{A} - \lambda_1 \mathbf{I})(\lambda_4 - \lambda_3)(\lambda_4 - \lambda_1)(\lambda_3 - \lambda_1)] \\
& + f(\lambda_1)[(\mathbf{A} - \lambda_4 \mathbf{I})(\mathbf{A} - \lambda_3 \mathbf{I})(\mathbf{A} - \lambda_2 \mathbf{I})(\lambda_4 - \lambda_3)(\lambda_4 - \lambda_2)(\lambda_3 - \lambda_2)] \\
= & \; \mathbf{0}
\end{aligned}$$

Solving this last equation for $\mathbf{f(A)}$, we obtain

$$\begin{aligned}
f(\mathbf{A}) = & \; f(\lambda_1) \frac{(\mathbf{A} - \lambda_2 \mathbf{I})(\mathbf{A} - \lambda_3 \mathbf{I})(\mathbf{A} - \lambda_4 \mathbf{I})}{(\lambda_1 - \lambda_2)(\lambda_1 - \lambda_3)(\lambda_1 - \lambda_4)} + f(\lambda_2) \frac{(\mathbf{A} - \lambda_1 \mathbf{I})(\mathbf{A} - \lambda_3 \mathbf{I})(\mathbf{A} - \lambda_4 \mathbf{I})}{(\lambda_2 - \lambda_1)(\lambda_2 - \lambda_3)(\lambda_2 - \lambda_4)} \\
& + f(\lambda_3) \frac{(\mathbf{A} - \lambda_1 \mathbf{I})(\mathbf{A} - \lambda_2 \mathbf{I})(\mathbf{A} - \lambda_4 \mathbf{I})}{(\lambda_3 - \lambda_1)(\lambda_3 - \lambda_2)(\lambda_3 - \lambda_4)} + f(\lambda_4) \frac{(\mathbf{A} - \lambda_1 \mathbf{I})(\mathbf{A} - \lambda_2 \mathbf{I})(\mathbf{A} - \lambda_3 \mathbf{I})}{(\lambda_4 - \lambda_1)(\lambda_4 - \lambda_2)(\lambda_4 - \lambda_3)} \\
= & \; \sum_{k=1}^{m} f(\lambda_k) \frac{(\mathbf{A} - \lambda_1 \mathbf{I}) \cdots (\mathbf{A} - \lambda_{k-1} \mathbf{I})(\mathbf{A} - \lambda_{k+1} \mathbf{I}) \cdots (\mathbf{A} - \lambda_m \mathbf{I})}{(\lambda_k - \lambda_1) \cdots (\lambda_k - \lambda_{k-1})(\lambda_k - \lambda_{k+1}) \cdots (\lambda_k - \lambda_m)}
\end{aligned}$$

where $m = 4$. Thus we have shown the equivalence of Equations (9–88) and (9–89). Although we assumed $m = 4$, the entire argument can be extended to an arbitrary positive integer m. (For the case where the matrix $\mathbf{A}$ involves multiple eigenvalues, refer to Problem A–9–7.)

A–9–7. Consider Sylvester's interpolation formula in the form given by Equation (9–90):

$$\begin{vmatrix} 1 & \lambda_1 & \lambda_1^2 & \cdots & \lambda_1^{m-1} & f(\lambda_1) \\ 1 & \lambda_2 & \lambda_2^2 & \cdots & \lambda_2^{m-1} & f(\lambda_2) \\ \cdot & \cdot & \cdot & & \cdot & \cdot \\ \cdot & \cdot & \cdot & & \cdot & \cdot \\ \cdot & \cdot & \cdot & & \cdot & \cdot \\ 1 & \lambda_m & \lambda_m^2 & \cdots & \lambda_m^{m-1} & f(\lambda_m) \\ \mathbf{I} & \mathbf{A} & \mathbf{A}^2 & \cdots & \mathbf{A}^{m-1} & f(\mathbf{A}) \end{vmatrix} = \mathbf{0}$$

This formula for the determination of $f(\mathbf{A})$ applies to the case where the minimal polynomial of $\mathbf{A}$ involves only distinct roots.

Suppose the minimal polynomial of $\mathbf{A}$ involves multiple roots. Then the rows in the determinant that correspond to the multiple roots become identical, and therefore modification of the determinant in Equation (9–90) becomes necessary.

Modify the form of Sylvester's interpolation formula given by Equation (9–90) when the minimal polynomial of $\mathbf{A}$ involves multiple roots. In deriving a modified determinant equation, assume that there are three equal roots ($\lambda_1 = \lambda_2 = \lambda_3$) in the minimal polynomial of $\mathbf{A}$ and that there are other roots ($\lambda_4, \lambda_5, \ldots, \lambda_m$) that are distinct.

Solution. Since the minimal polynomial of $\mathbf{A}$ involves three equal roots, the minimal polynomial $\phi(\lambda)$ can be written as

$$\begin{aligned} \phi(\lambda) &= \lambda^m + a_1 \lambda^{m-1} + \cdots + a_{m-1} \lambda + a_m \\ &= (\lambda - \lambda_1)^3 (\lambda - \lambda_4)(\lambda - \lambda_5) \ldots (\lambda - \lambda_m) \end{aligned}$$

An arbitrary function $f(\mathbf{A})$ of an $n \times n$ matrix $\mathbf{A}$ can be written as

$$f(\mathbf{A}) = g(\mathbf{A})\phi(\mathbf{A}) + \alpha(\mathbf{A})$$

where the minimal polynomial $\phi(\mathbf{A})$ is of degree m and $\alpha(\mathbf{A})$ is a polynomial in $\mathbf{A}$ of degree $m - 1$ or less. Hence we have

$$f(\lambda) = g(\lambda)\phi(\lambda) + \alpha(\lambda)$$

where $\alpha(\lambda)$ is a polynomial in λ of degree $m - 1$ or less, which can thus be written as

$$\alpha(\lambda) = \alpha_0 + \alpha_1 \lambda + \alpha_2 \lambda^2 + \cdots + \alpha_{m-1} \lambda^{m-1} \tag{9–91}$$

In the present case we have

$$\begin{aligned} f(\lambda) &= g(\lambda)\phi(\lambda) + \alpha(\lambda) \\ &= g(\lambda)[(\lambda - \lambda_1)^3 (\lambda - \lambda_4) \cdots (\lambda - \lambda_m)] + \alpha(\lambda) \end{aligned} \tag{9–92}$$

By substituting $\lambda_1, \lambda_4, \ldots, \lambda_m$ for λ in Equation (9–92), we obtain the following $m - 2$ equations:

$$f(\lambda_1) = \alpha(\lambda_1)$$

$$f(\lambda_4) = \alpha(\lambda_4)$$

$$\cdot$$

$$\cdot$$

$$\cdot$$

$$f(\lambda_m) = \alpha(\lambda_m)$$

$$\tag{9–93}$$

By differentiating Equation (9–92) with respct to λ, we obtain

$$\frac{d}{d\lambda}f(\lambda) = (\lambda - \lambda_1)^2 h(\lambda) + \frac{d}{d\lambda}\alpha(\lambda) \tag{9–94}$$

where

$$(\lambda - \lambda_1)^2 h(\lambda) = \frac{d}{d\lambda}[g(\lambda)(\lambda - \lambda_1)^3(\lambda - \lambda_4) \ldots (\lambda - \lambda_m)]$$

Substitution of λ_1 for λ in Equation (9–94) gives

$$\left.\frac{d}{d\lambda}f(\lambda)\right|_{\lambda=\lambda_1} = f'(\lambda_1) = \left.\frac{d}{d\lambda}\alpha(\lambda)\right|_{\lambda=\lambda_1}$$

Referring to Equation (9–91), this last equation becomes

$$f'(\lambda_1) = \alpha_1 + 2\alpha_2\lambda_1 + \cdots + (m-1)\alpha_{m-1}\lambda_1^{m-2} \tag{9–95}$$

Similarly, differentiating Equation (9–92) twice with respect to λ and substituting λ_1 for λ, we obtain

$$\left.\frac{d^2}{d\lambda^2}f(\lambda)\right|_{\lambda=\lambda_1} = f''(\lambda_1) = \left.\frac{d^2}{d\lambda^2}\alpha(\lambda)\right|_{\lambda=\lambda_1}$$

This last equation can be written as

$$f''(\lambda_1) = 2\alpha_2 + 6\alpha_3\lambda_1 + \cdots + (m-1)(m-2)\alpha_{m-1}\lambda_1^{m-3} \tag{9–96}$$

Rewriting Equations (9–96), (9–95), and (9–93), we get

$$\begin{aligned}
\alpha_2 + 3\alpha_3\lambda_1 + \cdots + \frac{(m-1)(m-2)}{2}\alpha_{m-1}\lambda_1^{m-3} &= \frac{f''(\lambda_1)}{2} \\
\alpha_1 + 2\alpha_2\lambda_1 + \cdots + (m-1)\alpha_{m-1}\lambda_1^{m-2} &= f'(\lambda_1) \\
\alpha_0 + \alpha_1\lambda_1 + \alpha_2\lambda_1^2 + \cdots + \alpha_{m-1}\lambda_1^{m-1} &= f(\lambda_1) \\
\alpha_0 + \alpha_1\lambda_4 + \alpha_2\lambda_4^2 + \cdots + \alpha_{m-1}\lambda_4^{m-1} &= f(\lambda_4) \\
&\quad\quad\vdots \\
\alpha_0 + \alpha_1\lambda_m + \alpha_2\lambda_m^2 + \cdots + \alpha_{m-1}\lambda_m^{m-1} &= f(\lambda_m)
\end{aligned} \tag{9–97}$$

These m simultaneous equations determine the α_k values (where $k = 0, 1, 2, \ldots, m-1$). Noting that $\phi(\mathbf{A}) = \mathbf{0}$ because it is a minimal polynomial, we have $f(\mathbf{A})$ as follows:

$$f(\mathbf{A}) = g(\mathbf{A})\phi(\mathbf{A}) + \alpha(\mathbf{A}) = \alpha(\mathbf{A})$$

Hence, referring to Equation (9–91), we have

$$f(\mathbf{A}) = \alpha(\mathbf{A}) = \alpha_0\mathbf{I} + \alpha_1\mathbf{A} + \alpha_2\mathbf{A}^2 + \cdots + \alpha_{m-1}\mathbf{A}^{m-1} \tag{9–98}$$

where the α_k values are given in terms of $f(\lambda_1), f'(\lambda_1), f''(\lambda_1), f(\lambda_4), f(\lambda_5), \ldots, f(\lambda_m)$. In terms of the determinant equation, $f(\mathbf{A})$ can be obtained by solving the following equation:

$$\begin{vmatrix} 0 & 0 & 1 & 3\lambda_1 & \cdots & \dfrac{(m-1)(m-2)}{2}\lambda_1^{m-3} & \dfrac{f''(\lambda_1)}{2} \\ 0 & 1 & 2\lambda_1 & 3\lambda_1^2 & \cdots & (m-1)\lambda_1^{m-2} & f'(\lambda_1) \\ 1 & \lambda_1 & \lambda_1^2 & \lambda_1^3 & \cdots & \lambda_1^{m-1} & f(\lambda_1) \\ 1 & \lambda_4 & \lambda_4^2 & \lambda_4^3 & \cdots & \lambda_4^{m-1} & f(\lambda_4) \\ \cdot & \cdot & \cdot & \cdot & & \cdot & \cdot \\ \cdot & \cdot & \cdot & \cdot & & \cdot & \cdot \\ \cdot & \cdot & \cdot & \cdot & & \cdot & \cdot \\ 1 & \lambda_m & \lambda_m^2 & \lambda_m^3 & \cdots & \lambda_m^{m-1} & f(\lambda_m) \\ \mathbf{I} & \mathbf{A} & \mathbf{A}^2 & \mathbf{A}^3 & \cdots & \mathbf{A}^{m-1} & f(\mathbf{A}) \end{vmatrix} = \mathbf{0} \qquad (9\text{–}99)$$

Equation (9–99) shows the desired modification in the form of the determinant. This equation gives the form of Sylvester's interpolation formula when the minimal polynomial of $\mathbf{A}$ involves three equal roots. (The necessary modification of the form of the determinant for other cases will be apparent.)

A–9–8. Using Sylvester's interpolation formula, compute $e^{\mathbf{A}t}$, where

$$\mathbf{A} = \begin{bmatrix} 2 & 1 & 4 \\ 0 & 2 & 0 \\ 0 & 3 & 1 \end{bmatrix}$$

Solution. Referring to Problem A–9–4, the characteristic polynomial and the minimal polynomial are the same for this $\mathbf{A}$. The minimal polynomial (characteristic polynomial) is given by

$$\phi(\lambda) = (\lambda - 2)^2(\lambda - 1)$$

Note that $\lambda_1 = \lambda_2 = 2$ and $\lambda_3 = 1$. Referring to Equation (9–98) and noting that $f(\mathbf{A})$ in this problem is $e^{\mathbf{A}t}$, we have

$$e^{\mathbf{A}t} = \alpha_0(t)\mathbf{I} + \alpha_1(t)\mathbf{A} + \alpha_2(t)\mathbf{A}^2$$

where $\alpha_0(t)$, $\alpha_1(t)$, and $\alpha_2(t)$ are determined from the equations

$$\alpha_1(t) + 2\alpha_2(t)\,\lambda_1 = te^{\lambda_1 t}$$

$$\alpha_0(t) + \alpha_1(t)\,\lambda_1 + \alpha_2(t)\,\lambda_1^2 = e^{\lambda_1 t}$$

$$\alpha_0(t) + \alpha_1(t)\,\lambda_3 + \alpha_2(t)\,\lambda_3^2 = e^{\lambda_3 t}$$

Substituting $\lambda_1 = 2$, and $\lambda_3 = 1$ into these three equations gives

$$\alpha_1(t) + 4\alpha_2(t) = te^{2t}$$

$$\alpha_0(t) + 2\alpha_1(t) + 4\alpha_2(t) = e^{2t}$$

$$\alpha_0(t) + \alpha_1(t) + \alpha_2(t) = e^t$$

Solving for $\alpha_0(t)$, $\alpha_1(t)$, and $\alpha_2(t)$, we obtain

$$\alpha_0(t) = 4e^t - 3e^{2t} + 2te^{2t}$$

$$\alpha_1(t) = -4e^t + 4e^{2t} - 3te^{2t}$$

$$\alpha_2(t) = e^t - e^{2t} + te^{2t}$$

Hence,

$$e^{\mathbf{A}t} = (4e^t - 3e^{2t} + 2te^{2t}) \begin{bmatrix} 1 & 0 & 0 \\ 0 & 1 & 0 \\ 0 & 0 & 1 \end{bmatrix} + (-4e^t + 4e^{2t} - 3te^{2t}) \begin{bmatrix} 2 & 1 & 4 \\ 0 & 2 & 0 \\ 0 & 3 & 1 \end{bmatrix}$$

$$+ (e^t - e^{2t} + te^{2t}) \begin{bmatrix} 4 & 16 & 12 \\ 0 & 4 & 0 \\ 0 & 9 & 1 \end{bmatrix}$$

$$= \begin{bmatrix} e^{2t} & 12e^t - 12e^{2t} + 13te^{2t} & -4e^t + 4e^{2t} \\ 0 & e^{2t} & 0 \\ 0 & -3e^t + 3e^{2t} & e^t \end{bmatrix}$$

A–9–9. Consider an $n \times n$ matrix $\mathbf{A}$. Show that

$$(s\mathbf{I} - \mathbf{A})^{-1} = \frac{\sum_{j=0}^{m-1} s^j \sum_{i=1+j}^{m} \alpha_i \mathbf{A}^{i-j-1}}{\sum_{i=0}^{m} \alpha_i s^i}$$

where the α_i's are coefficients of the minimal polynomial of $\mathbf{A}$:

$$\alpha_0 \mathbf{A}^m + \alpha_1 \mathbf{A}^{m-1} + \cdots + \alpha_{m-1} \mathbf{A} + \alpha_m \mathbf{I} = \mathbf{0}$$

where $\alpha_0 = 1$ and m is the degree of the minimal polynomial ($m \le n$).

Solution. Let us put

$$\mathbf{P} = (s\mathbf{I} - \mathbf{A})^{-1}$$

Then

$$s\mathbf{P} = \mathbf{A}\mathbf{P} + \mathbf{I}$$

By premultiplying $(s\mathbf{I} + \mathbf{A})$ to both sides of this equation, we obtain

$$s^2\mathbf{P} = \mathbf{A}^2\mathbf{P} + \mathbf{A} + s\mathbf{I}$$

Similarly, by premultiplying $(s\mathbf{I} + \mathbf{A})$ to both sides of this last equation, we obtain

$$s^3\mathbf{P} = \mathbf{A}^3\mathbf{P} + \mathbf{A}^2 + s\mathbf{A} + s^2\mathbf{I}$$

By repeating this process, we obtain the following set of equations

$$\mathbf{P} = \mathbf{P}$$

$$s\mathbf{P} = \mathbf{A}\mathbf{P} + \mathbf{I}$$

$$s^2\mathbf{P} = \mathbf{A}^2\mathbf{P} + \mathbf{A} + s\mathbf{I}$$

$$s^3\mathbf{P} = \mathbf{A}^3\mathbf{P} + \mathbf{A}^2 + s\mathbf{A} + s^2\mathbf{I}$$

$$\cdot$$
$$\cdot$$
$$\cdot$$

$$s^m\mathbf{P} = \mathbf{A}^m\mathbf{P} + \mathbf{A}^{m-1} + s\mathbf{A}^{m-2} + \cdots + s^{m-2}\mathbf{A} + s^{m-1}\mathbf{I}$$

where m is the degree of the minimal polynomial of $\mathbf{A}$. Then, by multiplying the $s^i\mathbf{P}$'s by α_{m-i} (where $i = 0, 1, 2, \ldots, m$) in the above $m + 1$ equations in this order and adding the product together, we get

$$\alpha_m \mathbf{P} + \alpha_{m-1} s \mathbf{P} + \alpha_{m-2} s^2 \mathbf{P} + \cdots + \alpha_0 s^m \mathbf{P}$$

$$= \sum_{i=0}^{m} \alpha_{m-i} \mathbf{A}^i \mathbf{P} + \sum_{i=1}^{m} \alpha_{m-i} \mathbf{A}^{i-1} + s \sum_{i=2}^{m} \alpha_{m-i} \mathbf{A}^{i-2} + \cdots$$

$$+ s^{m-2} \sum_{i=m-1}^{m} \alpha_{m-i} \mathbf{A}^{i-m+1} + s^{m-1} \alpha_0 \mathbf{I} \tag{9-100}$$

Noting that

$$\sum_{i=0}^{m} \alpha_{m-i} \mathbf{A}^i \mathbf{P} = (\alpha_0 \mathbf{A}^m + \alpha_1 \mathbf{A}^{m-1} + \cdots + \alpha_{m-1} \mathbf{A} + \alpha_m \mathbf{I}) \mathbf{P} = \mathbf{0}$$

we can simplify Equation (9–100) as follows:

$$\sum_{i=0}^{m} \alpha_{m-i} s^i \mathbf{P} = \sum_{j=0}^{m-1} s^j \sum_{i=1+j}^{m} \alpha_{m-i} \mathbf{A}^{i-j-1}$$

Therefore,

$$(s\mathbf{I} - \mathbf{A})^{-1} = \mathbf{P} = \frac{\displaystyle\sum_{j=0}^{m-1} s^j \sum_{i=1+j}^{m} \alpha_{m-i} \mathbf{A}^{i-j-1}}{\displaystyle\sum_{i=0}^{m} \alpha_{m-i} s^i}$$

If the minimal polynomial and characteristic polynomial of $\mathbf{A}$ are identical, then $m = n$. If $m = n$, then this last equation becomes

$$(s\mathbf{I} - \mathbf{A})^{-1} = \frac{\displaystyle\sum_{j=0}^{n-1} s^j \sum_{i=1+j}^{n} \alpha_{n-i} \mathbf{A}^{i-j-1}}{|s\mathbf{I} - \mathbf{A}|} \tag{9-101}$$

where

$$|s\mathbf{I} - \mathbf{A}| = \sum_{i=0}^{n} \alpha_{n-i} s^i, \qquad \alpha_0 = 1$$

A–9–10. A necessary and sufficient condition for complete state controllability is that no cancellation occurs in the transfer function or transfer matrix.

Consider the system defined by

$$\dot{\mathbf{x}} = \mathbf{A}\mathbf{x} + \mathbf{B}u, \qquad \mathbf{x}(0) = \mathbf{0} \tag{9-102}$$

where $\mathbf{x}$ = state vector (n-vector)

$\quad u$ = control signal (scalar)

$\quad \mathbf{A} = n \times n$ matrix

$\quad \mathbf{B} = n \times 1$ matrix

Laplace transforming Equation (9–102) and solving for $\mathbf{X}(s)$, we obtain

$$\mathbf{X}(s) = (s\mathbf{I} - \mathbf{A})^{-1} \mathbf{B} U(s) \tag{9-103}$$

where

$$(s\mathbf{I} - \mathbf{A})^{-1} \mathbf{B} = \frac{1}{|s\mathbf{I} - \mathbf{A}|} \begin{bmatrix} p_1(s) \\ p_2(s) \\ \cdot \\ \cdot \\ \cdot \\ p_n(s) \end{bmatrix}$$

and the $p_i(s)$ $(i = 1, 2, \ldots, n)$ are polynomials in s. We shall next define what we mean by *cancellations* in the transfer matrix $(s\mathbf{I} - \mathbf{A})^{-1}\mathbf{B}$. The matrix $(s\mathbf{I} - \mathbf{A})^{-1}\mathbf{B}$ is said to have no cancellation if and only if the polynomials $p_1(s), p_2(s), \ldots, p_n(s)$, and $|s\mathbf{I} - \mathbf{A}|$ have no common factor. If $(s\mathbf{I} - \mathbf{A})^{-1}\mathbf{B}$ has a cancellation, then the system cannot be controlled in the direction of the canceled mode.

Show that $(s\mathbf{I} - \mathbf{A})^{-1}\mathbf{B}$ has a cancellation if and only if the rank of

$$\mathbf{P} = [\mathbf{B} \mid \mathbf{AB} \mid \cdots \mid \mathbf{A}^{n-1}\mathbf{B}]$$

is less than n.

Solution. Let us define

$$\boldsymbol{\phi} \equiv (s\mathbf{I} - \mathbf{A})^{-1}\mathbf{B}$$

By the use of Equation (9–101), we can express $\boldsymbol{\phi}$ as

$$\boldsymbol{\phi} = \frac{\displaystyle\sum_{j=0}^{n-1} s^j \sum_{i=1+j}^{n} \alpha_{n-i}\mathbf{A}^{i-j-1}\mathbf{B}}{\displaystyle\sum_{i=0}^{n} \alpha_{n-i}s^i} \qquad (\alpha_0 = 1)$$

Define

$$\mathbf{v}_j = \sum_{i=1+j}^{n} \alpha_{n-i}\mathbf{A}^{i-j-1}\mathbf{B}$$

where $\mathbf{v}_j$ is an n-vector. Then

$$\boldsymbol{\phi} = \frac{\displaystyle\sum_{j=0}^{n-1} s^j\mathbf{v}_j}{\displaystyle\sum_{i=0}^{n} \alpha_{n-i}s^i} \qquad (9\text{--}104)$$

Suppose that $\boldsymbol{\phi}$ has a cancellation. Then the numerator of the right side of Equation (9–104) must have the following form:

$$\sum_{j=0}^{n-1} s^j\mathbf{v}_j = (s - s_k)\sum_{j=0}^{n-2} s^j\mathbf{w}_j \qquad (9\text{--}105)$$

where s_k is an eigenvalue of $\mathbf{A}$. By equating the coefficients of s^i $(i = 0, 1, 2, \ldots, n - 1)$ of both sides of Equation (9–105), we obtain

$$\mathbf{v}_0 = -s_k\mathbf{w}_0$$

$$\mathbf{v}_1 = \mathbf{w}_0 - s_k\mathbf{w}_1$$

$$\mathbf{v}_2 = \mathbf{w}_1 - s_k\mathbf{w}_2$$

$$\vdots$$

$$\mathbf{v}_{n-1} = \mathbf{w}_{n-2}$$

Then

$$\mathbf{v}_0 + s_k\mathbf{v}_1 + s_k^2\mathbf{v}_2 + \cdots + s_k^{n-1}\mathbf{v}_{n-1}$$
$$= (-s_k\mathbf{w}_0) + (s_k\mathbf{w}_0 - s_k^2\mathbf{w}_1) + (s_k^2\mathbf{w}_1 - s_k^3\mathbf{w}_2) + \cdots$$
$$+ (s_k^{n-2}\mathbf{w}_{n-3} - s_k^{n-1}\mathbf{w}_{n-2}) + s_k^{n-1}\mathbf{w}_{n-2}$$
$$= \mathbf{0} \qquad (9\text{--}106)$$

Equation (9–106) implies that

$$\sum_{i=1}^{n} \alpha_{n-i}\mathbf{A}^{i-1}\mathbf{B} + s_k \sum_{i=2}^{n} \alpha_{n-i}\mathbf{A}^{i-2}\mathbf{B} + \cdots + s_k^{n-1}\alpha_0\mathbf{B} = \mathbf{0}$$

which can be rewritten as

$$c_0\mathbf{B} + c_1\mathbf{AB} + c_2\mathbf{A}^2\mathbf{B} + \cdots + c_{n-1}\mathbf{A}^{n-1}\mathbf{B} = \mathbf{0} \qquad (9\text{–}107)$$

where

$$c_i = \sum_{j=0}^{n-i-1} \alpha_{n-i-j-1}s_k^j$$

Since the coefficient of $\mathbf{A}^{n-1}\mathbf{B}$ is $c_{n-1} = \alpha_0 = 1$, Equation (9–107) implies that the vectors $\mathbf{B}$, $\mathbf{AB}$, . . . , $\mathbf{A}^{n-1}\mathbf{B}$ are linearly dependent. Hence the rank of $\mathbf{P}$ is less than n. Thus we have proved that, if $\boldsymbol{\phi}$ has a cancellation, then the rank of $\mathbf{P}$ is less than n.

We shall next prove that, if the rank of $\mathbf{P}$ is less than n, then $\boldsymbol{\phi}$ has a cancellation. If the rank of $\mathbf{P}$ is less than n, then there exist constants γ_0, γ_1, . . . , γ_{n-1}, not all zero, such that

$$\gamma_0 \mathbf{B} + \gamma_1 \mathbf{AB} + \gamma_2 \mathbf{A}^2\mathbf{B} + \cdots + \gamma_{n-1}\mathbf{A}^{n-1}\mathbf{B} = \mathbf{0}$$

Since

$$\mathbf{B} = (s\mathbf{I} - \mathbf{A}) \, \boldsymbol{\phi}$$

we have

$$\sum_{i=0}^{n-1} \gamma_i\mathbf{A}^i(s\mathbf{I} - \mathbf{A})\boldsymbol{\phi} = (s\mathbf{I} - \mathbf{A}) \sum_{i=0}^{n-1} \gamma_i\mathbf{A}^i\boldsymbol{\phi} = \mathbf{0} \qquad (9\text{–}108)$$

For s such that $|s\mathbf{I} - \mathbf{A}| \neq 0$, Equation (9–108) implies that

$$\sum_{i=0}^{n-1} \gamma_i\mathbf{A}^i\boldsymbol{\phi} = \mathbf{0}$$

Using the identities

$$\boldsymbol{\phi} = \boldsymbol{\phi}$$

$$\mathbf{A}\boldsymbol{\phi} = s\boldsymbol{\phi} - \mathbf{B}$$

$$\mathbf{A}^2\boldsymbol{\phi} = s^2\boldsymbol{\phi} - \mathbf{AB} - s\mathbf{B}$$

$$.$$
$$.$$
$$.$$

$$\mathbf{A}^{n-1}\boldsymbol{\phi} = s^{n-1}\boldsymbol{\phi} - \mathbf{A}^{n-2}\mathbf{B} - s\mathbf{A}^{n-3}\mathbf{B} - \cdots - s^{n-2}\mathbf{B}$$

we obtain

$$\sum_{i=0}^{n-1} \gamma_i\mathbf{A}^i\boldsymbol{\phi} = \sum_{i=0}^{n-1} \gamma_i s^i\boldsymbol{\phi} - \sum_{i=1}^{n-1} \gamma_i\mathbf{A}^{i-1}\mathbf{B} - s \sum_{i=2}^{n-1} \gamma_i\mathbf{A}^{i-2}\mathbf{B}$$

$$- \cdots - s^{n-3} \sum_{i=n-2}^{n-1} \gamma_i\mathbf{A}^{i-n+2}\mathbf{B} - s^{n-2}\gamma_{n-1}\mathbf{B}$$

$$= \mathbf{0}$$

Hence,

$$\sum_{i=0}^{n-1} \gamma_i s^i \phi = \sum_{i=1}^{n-1} \gamma_i \mathbf{A}^{i-1}\mathbf{B} + s \sum_{i=2}^{n-1} \gamma_i \mathbf{A}^{i-2}\mathbf{B} + \cdots$$

$$+ s^{n-3} \sum_{i=n-2}^{n-1} \gamma_i \mathbf{A}^{i-n+2}\mathbf{B} + s^{n-2} \gamma_{n-1}\mathbf{B}$$

$$= \sum_{\substack{i=j+1 \\ j=0}}^{n-1} \gamma_i \mathbf{A}^{i-j-1}\mathbf{B} + s \sum_{\substack{i=j+1 \\ j=1}}^{n-1} \gamma_i \mathbf{A}^{i-j-1}\mathbf{B} + \cdots$$

$$+ s^{n-3} \sum_{\substack{i=j+1 \\ j=n-3}}^{n-1} \gamma_i \mathbf{A}^{i-j-1}\mathbf{B} + s^{n-2} \sum_{\substack{i=j+1 \\ j=n-2}}^{n-1} \gamma_i \mathbf{A}^{i-j-1}\mathbf{B}$$

$$= \sum_{j=0}^{n-2} s^j \sum_{i=j+1}^{n-1} \gamma_i \mathbf{A}^{i-j-1}\mathbf{B}$$

or

$$\phi = \frac{\displaystyle\sum_{j=0}^{n-2} s^j \sum_{i=j+1}^{n-1} \gamma_i \mathbf{A}^{i-j-1}\mathbf{B}}{\displaystyle\sum_{i=0}^{n-1} \gamma_i s^i} \tag{9-109}$$

The denominator of Equation (9–109) indicates that a cancellation occurred. This completes the proof. We have shown that the condition that the rank of the matrix

$$[\mathbf{B} \;\vdots\; \mathbf{A}\mathbf{B} \;\vdots\; \cdots \;\vdots\; \mathbf{A}^{n-1}\mathbf{B}]$$

is n is equivalent to the condition that no cancellation occurs in the transfer matrix $(s\mathbf{I} - \mathbf{A})^{-1}\mathbf{B}$.

A–9–11. Show that the system described by

$$\dot{\mathbf{x}} = \mathbf{A}\mathbf{x} + \mathbf{B}\mathbf{u} \tag{9-110}$$

$$\mathbf{y} = \mathbf{C}\mathbf{x} \tag{9-111}$$

where $\mathbf{x}$ = state vector (n-vector)
$\mathbf{u}$ = control vector (r-vector)
$\mathbf{y}$ = output vector (m-vector)　　$(m \leq n)$
$\mathbf{A}$ = $n \times n$ matrix
$\mathbf{B}$ = $n \times r$ matrix
$\mathbf{C}$ = $m \times n$ matrix

is completely output controllable if and only if the composite $m \times nr$ matrix $\mathbf{P}$ where

$$\mathbf{P} = [\mathbf{CB} \;\vdots\; \mathbf{CAB} \;\vdots\; \mathbf{CA}^2\mathbf{B} \;\vdots\; \cdots \;\vdots\; \mathbf{CA}^{n-1}\mathbf{B}]$$

is of rank m. (Notice that complete state controllability is neither necessary nor sufficient for complete output controllability.)

Solution. Suppose that the system is output controllable and the output $\mathbf{y}(t)$ starting from any $\mathbf{y}(0)$, the initial output, can be transferred to the origin of the output space in a finite time interval $0 \leq t \leq T$. That is,

$$\mathbf{y}(T) = \mathbf{C}\mathbf{x}(T) = \mathbf{0} \tag{9-112}$$

Since the solution of Equation (9–110) is

$$\mathbf{x}(t) = e^{\mathbf{A}t}\left[\mathbf{x}(0) + \int_0^t e^{-\mathbf{A}\tau}\mathbf{B}\mathbf{u}(\tau)\,d\tau\right]$$

at $t = T$, we have

$$\mathbf{x}(T) = e^{\mathbf{A}T}\left[\mathbf{x}(0) + \int_0^T e^{-\mathbf{A}\tau}\mathbf{B}\mathbf{u}(\tau)\,d\tau\right] \qquad (9\text{–}113)$$

Substituting Equation (9–113) into Equation (9–112), we obtain

$$\mathbf{y}(T) = \mathbf{C}\mathbf{x}(T)$$

$$= \mathbf{C}e^{\mathbf{A}T}\left[\mathbf{x}(0) + \int_0^T e^{-\mathbf{A}\tau}\mathbf{B}\mathbf{u}(\tau)\,d\tau\right] = \mathbf{0} \qquad (9\text{–}114)$$

On the other hand, $\mathbf{y}(0) = \mathbf{C}\mathbf{x}(0)$. Notice that the complete output controllability means that the vector $\mathbf{C}\mathbf{x}(0)$ spans the m-dimensional output space. Since $e^{\mathbf{A}T}$ is nonsingular, if $\mathbf{C}\mathbf{x}(0)$ spans the m-dimensional output space, so does $\mathbf{C}e^{\mathbf{A}T}\mathbf{x}(0)$, and vice versa. From Equation (9–114) we obtain

$$\mathbf{C}e^{\mathbf{A}T}\mathbf{x}(0) = -\mathbf{C}e^{\mathbf{A}T}\int_0^T e^{-\mathbf{A}\tau}\mathbf{B}\mathbf{u}(\tau)\,d\tau$$

$$= -\mathbf{C}\int_0^T e^{\mathbf{A}\tau}\mathbf{B}\mathbf{u}(T-\tau)\,d\tau$$

Note that $\int_0^T e^{\mathbf{A}\tau}\mathbf{B}\mathbf{u}(T-\tau)\,d\tau$ can be expressed as a sum of $\mathbf{A}^i\mathbf{B}_j$,

$$\int_0^T e^{\mathbf{A}\tau}\mathbf{B}\mathbf{u}(T-\tau)\,d\tau = \sum_{i=0}^{p-1}\sum_{j=1}^{r}\gamma_{ij}\mathbf{A}^i\mathbf{B}_j$$

where

$$\gamma_{ij} = \int_0^T \alpha_i(\tau)u_j(T-\tau)\,d\tau = \text{scalar}$$

and $\alpha_i(\tau)$ satisfies

$$e^{\mathbf{A}\tau} = \sum_{i=0}^{p-1}\alpha_i(\tau)\mathbf{A}^i \qquad (p\text{: degree of the minimal polynomial of }\mathbf{A})$$

and $\mathbf{B}_j$ is the jth column of $\mathbf{B}$. Therefore, we can write $\mathbf{C}e^{\mathbf{A}T}\mathbf{x}(0)$ as

$$\mathbf{C}e^{\mathbf{A}T}\mathbf{x}(0) = -\sum_{i=0}^{p-1}\sum_{j=1}^{r}\gamma_{ij}\mathbf{C}\mathbf{A}^i\mathbf{B}_j$$

From this last equation we see that $\mathbf{C}e^{\mathbf{A}T}\mathbf{x}(0)$ is a linear combination of $\mathbf{C}\mathbf{A}^i\mathbf{B}_j$ ($i = 0, 1, 2, \ldots, p - 1; j = 1, 2, \ldots, r$). Note that if the rank of $\mathbf{Q}$, where

$$\mathbf{Q} = [\mathbf{C}\mathbf{B} \mid \mathbf{C}\mathbf{A}\mathbf{B} \mid \mathbf{C}\mathbf{A}^2\mathbf{B} \mid \cdots \mid \mathbf{C}\mathbf{A}^{p-1}\mathbf{B}] \qquad (p \le n)$$

is m, then so is the rank of $\mathbf{P}$, and vice versa. [This is obvious if $p = n$. If $p < n$, then the $\mathbf{C}\mathbf{A}^h\mathbf{B}_j$ (where $p \le h \le n - 1$) are linearly dependent on $\mathbf{C}\mathbf{B}_j$, $\mathbf{C}\mathbf{A}\mathbf{B}_j$, $\ldots$, $\mathbf{C}\mathbf{A}^{p-1}\mathbf{B}_j$. Hence the rank of $\mathbf{P}$ is equal to that of $\mathbf{Q}$.] If the rank of $\mathbf{P}$ is m, then $\mathbf{C}e^{\mathbf{A}T}\mathbf{x}(0)$ spans the m-dimensional output space. This means that if the rank of $\mathbf{P}$ is m, then $\mathbf{C}\mathbf{x}(0)$ also spans the m-dimensional output space and the system is completely output controllable.

Conversely, suppose that the system is completely output controllable but the rank of $\mathbf{P}$ is k, where $k < m$. Then the set of all initial outputs that can be transferred to the origin is of k-dimensional space.

Hence the dimension of this set is less than m. This contradicts the assumption that the system is completely output controllable. This completes the proof.

Note that it can be immediately proved that, in the system of Equations (9–110) and (9–111), complete state controllability on $0 \le t \le T$ implies complete output controllability on $0 \le t \le T$ if and only if m rows of $\mathbf{C}$ are linearly independent.

A–9–12. Discuss the state controllability of the following system:

$$\begin{bmatrix} \dot{x}_1 \\ \dot{x}_2 \end{bmatrix} = \begin{bmatrix} -3 & 1 \\ -2 & 1.5 \end{bmatrix}\begin{bmatrix} x_1 \\ x_2 \end{bmatrix} + \begin{bmatrix} 1 \\ 4 \end{bmatrix}u \qquad (9\text{–}115)$$

Solution. For this system

$$\mathbf{A} = \begin{bmatrix} -3 & 1 \\ -2 & 1.5 \end{bmatrix}, \qquad \mathbf{B} = \begin{bmatrix} 1 \\ 4 \end{bmatrix}$$

Since

$$\mathbf{AB} = \begin{bmatrix} -3 & 1 \\ -2 & 1.5 \end{bmatrix}\begin{bmatrix} 1 \\ 4 \end{bmatrix} = \begin{bmatrix} 1 \\ 4 \end{bmatrix}$$

we see that vectors $\mathbf{B}$ and $\mathbf{AB}$ are not linearly independent and the rank of the matrix $[\mathbf{B} \ \vdots \ \mathbf{AB}]$ is 1. Therefore, the system is not completely state controllable. In fact, elimination of x_2 from Equation (9–115), or the following two simultaneous equations,

$$\dot{x}_1 = -3x_1 + x_2 + u$$

$$\dot{x}_2 = -2x_1 + 1.5\,x_2 + 4u$$

yields

$$\ddot{x}_1 + 1.5\dot{x}_1 - 2.5x_1 = \dot{u} + 2.5u$$

or, in the form of a transfer function,

$$\frac{X_1(s)}{U(s)} = \frac{s + 2.5}{(s + 2.5)(s - 1)}$$

Notice that cancellation of the factor $(s + 2.5)$ occurs in the numerator and denominator of the transfer function. Because of this cancellation, this system is not completely state controllable.

This is an unstable system. Remember that stability and controllability are quite different things. There are many systems that are unstable but are completely state controllable.

A–9–13. A state-space representation of a system in the controllable canonical form is given by

$$\begin{bmatrix} \dot{x}_1 \\ \dot{x}_2 \end{bmatrix} = \begin{bmatrix} 0 & 1 \\ -0.4 & -1.3 \end{bmatrix}\begin{bmatrix} x_1 \\ x_2 \end{bmatrix} + \begin{bmatrix} 0 \\ 1 \end{bmatrix}u \qquad (9\text{–}116)$$

$$y = \begin{bmatrix} 0.8 & 1 \end{bmatrix}\begin{bmatrix} x_1 \\ x_2 \end{bmatrix} \qquad (9\text{–}117)$$

The same system may be represented by the following state-space equation, which is in the observable canonical form:

$$\begin{bmatrix} \dot{x}_1 \\ \dot{x}_2 \end{bmatrix} = \begin{bmatrix} 0 & -0.4 \\ 1 & -1.3 \end{bmatrix}\begin{bmatrix} x_1 \\ x_2 \end{bmatrix} + \begin{bmatrix} 0.8 \\ 1 \end{bmatrix}u \qquad (9\text{–}118)$$

$$y = \begin{bmatrix} 0 & 1 \end{bmatrix}\begin{bmatrix} x_1 \\ x_2 \end{bmatrix} \qquad (9\text{–}119)$$

Show that the state-space representation given by Equations (9–116) and (9–117) gives a system that is state controllable but not observable. Show, on the other hand, that the state-space representation defined by Equations (9–118) and (9–119) gives a system that is not completely state controllable but is observable. Explain what causes the apparent difference in the controllability and observability of the same system.

Solution. Consider the system defined by Equations (9–116) and (9–117). The rank of the controllability matrix

$$[\mathbf{B} \,\vdots\, \mathbf{AB}] = \begin{bmatrix} 0 & 1 \\ 1 & -1.3 \end{bmatrix}$$

is 2. Hence the system is completely state controllable. The rank of the observability matrix

$$[\mathbf{C}^* \,\vdots\, \mathbf{A}^*\mathbf{C}^*] = \begin{bmatrix} 0.8 & -0.4 \\ 1 & -0.5 \end{bmatrix}$$

is 1. Hence the system is not observable.

Next consider the system defined by Equations (9–118) and (9–119). The rank of the controllability matrix

$$[\mathbf{B} \,\vdots\, \mathbf{AB}] = \begin{bmatrix} 0.8 & -0.4 \\ 1 & -0.5 \end{bmatrix}$$

is 1. Hence the system is not completely state controllable. The rank of the observability matrix

$$[\mathbf{C}^* \,\vdots\, \mathbf{A}^*\mathbf{C}^*] = \begin{bmatrix} 0 & 1 \\ 1 & -1.3 \end{bmatrix}$$

is 2. Hence the system is observable.

The apparent difference in the controllability and observability of the same system is caused by the fact that the original system has a pole–zero cancellation in the transfer function. Referring to Equation (9–34),

$$G(s) = \mathbf{C}(s\mathbf{I} - \mathbf{A})^{-1}\mathbf{B}$$

If we use Equations (9–116) and (9–117), then

$$G(s) = \begin{bmatrix} 0.8 & 1 \end{bmatrix} \begin{bmatrix} s & -1 \\ 0.4 & s + 1.3 \end{bmatrix}^{-1} \begin{bmatrix} 0 \\ 1 \end{bmatrix}$$

$$= \frac{1}{s^2 + 1.3s + 0.4} \begin{bmatrix} 0.8 & 1 \end{bmatrix} \begin{bmatrix} s + 1.3 & 1 \\ -0.4 & s \end{bmatrix} \begin{bmatrix} 0 \\ 1 \end{bmatrix}$$

$$= \frac{s + 0.8}{(s + 0.8)(s + 0.5)}$$

[Note that the same transfer function can be obtained by using Equations (9–118) and (9–119).] Clearly, cancellation occurs in this transfer function.

If a pole–zero cancellation occurs in the transfer function, then the controllability and observability vary, depending on how the state variables are chosen. Remember that to be completely state controllable and observable the transfer function must not have any pole–zero cancellations.

A–9–14. Prove that the system defined by

$$\dot{\mathbf{x}} = \mathbf{Ax}$$

$$\mathbf{y} = \mathbf{Cx}$$

Solution. To prove uniform asymptotic stability in the large, we need to prove the following. (*Note:* "uniform" implies "independent of time.")

1. The origin is uniformly stable.
2. Every solution is uniformly bounded.
3. Every solution converges to the origin when $t \to \infty$ uniformly in t_0 and $\|\mathbf{x}\| \leq \delta$, where δ is fixed but arbitrarily large. That is, given two real numbers $\delta > 0$ and $\mu > 0$, there is a real number $T(\mu, \delta)$ such that

$$\|\mathbf{x}_0\| \leq \delta$$

implies

$$\|\boldsymbol{\phi}(t); \mathbf{x}_0, t_0)\| \leq \mu \qquad \text{for all } t \geq t_0 + T(\mu, \delta)$$

where $\boldsymbol{\phi}(t; \mathbf{x}_0, t_0)$ is the solution to the given differential equation.

Since β is continuous and $\beta(0) = 0$, we can take a $\delta(\epsilon) > 0$ such that $\beta(\delta) < \alpha(\epsilon)$ for any $\epsilon > 0$. Figure 9–14 shows the curves $\alpha(\|\mathbf{x}\|)$, $\beta(\|\mathbf{x}\|)$, and $V(\mathbf{x}, t)$. Noting that

$$V(\boldsymbol{\phi}(t; \mathbf{x}_0, t_0), t) - V(\mathbf{x}_0, t_0) = \int_{t_0}^{t} \dot{V}(\boldsymbol{\phi}(\tau; \mathbf{x}_0, t_0), \tau)\, d\tau < 0 \qquad t > t_0$$

if $\|\mathbf{x}_0\| \leq \delta$, t_0 being arbitrary, we have

$$\alpha(\epsilon) > \beta(\delta) \geq V(\mathbf{x}_0, t_0) \geq V(\boldsymbol{\phi}(t; \mathbf{x}_0, t_0), t) \geq \alpha(\|\boldsymbol{\phi}(t; \mathbf{x}_0, t_0)\|)$$

for all $t \geq t_0$. Since α is nondecreasing and positive, this implies that

$$\|\boldsymbol{\phi}(t; \mathbf{x}_0, t_0)\| < \epsilon \qquad \text{for } t \geq t_0, \|\mathbf{x}_0\| \leq \delta$$

Hence, we have shown that, for each real number $\epsilon > 0$, there is a real number $\delta > 0$ such that $\|\mathbf{x}_0\| \leq \delta$ implies $\|\boldsymbol{\phi}(t; \mathbf{x}_0, t_0)\| \leq \epsilon$ for all $t \geq t_0$. Thus we have proved uniform stability.

Next we shall prove that $\|\boldsymbol{\phi}(t; \mathbf{x}_0, t_0)\| \to 0$ when $t \to \infty$ uniformly in t_0 and $\|\mathbf{x}_0\| \leq \delta$. Let us take any $0 < \mu < \|\mathbf{x}_0\|$ and find a $\nu(\mu) > 0$ such that $\beta(\nu) < \alpha(\mu)$. Let us denote by $\epsilon'(\mu, \delta) > 0$ the minimum of the continuous nondecreasing function $\gamma(\|\mathbf{x}\|)$ on the compact set $\nu(\mu) \leq \|\mathbf{x}\| \leq \epsilon(\delta)$. Let us define

$$T(\mu, \delta) = \frac{\beta(\delta)}{\epsilon'(\mu, \delta)} > 0$$

Suppose that $\|\boldsymbol{\phi}(t; \mathbf{x}_0, t_0)\| > \nu$ over the time interval $t_0 \leq t_1 = t_0 + T$. Then we have

$$0 < \alpha(\nu) \leq V(\boldsymbol{\phi}(t_1; \mathbf{x}_0, t_0), t_1) \leq V(\mathbf{x}_0, t_0) - (t_1 - t_0)\epsilon' \leq \beta(\delta) - T\epsilon' = 0$$

which is a contradiction. Hence, for some t in the interval $t_0 \leq t \leq t_1$, such as an arbitrary t_2, we have

$$\|\mathbf{x}_2\| = \|\boldsymbol{\phi}(t_2; \mathbf{x}_0, t_0)\| = \nu$$

Therefore,

$$\alpha(\|\boldsymbol{\phi}(t; \mathbf{x}_2, t_2)\|) < V(\boldsymbol{\phi}(t; \mathbf{x}_2, t_2), t) \leq V(\mathbf{x}_2, t_2) \leq \beta(\nu) < \alpha(\mu)$$

for all $t \geq t_2$. Hence,

$$\|\boldsymbol{\phi}(t; \mathbf{x}_0, t_0)\| < \mu$$

for all $t \geq t_0 + T(\mu, \delta) \geq t_2$, which proves uniform asymptotic stability. Since $\alpha(\|\mathbf{x}\|) \to \infty$ as $\|\mathbf{x}\| \to \infty$, there exists for arbitrarily large δ a constant $\epsilon(\delta)$ such that $\beta(\delta) < \alpha(\epsilon)$. Moreover, since $\epsilon(\delta)$ does not depend on t_0, the solution $\boldsymbol{\phi}(t; \mathbf{x}_0, t_0)$ is uniformly bounded. We thus have proved uniform asymptotic stability in the large.

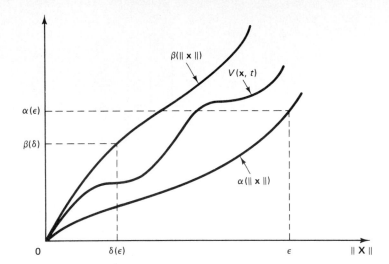

Figure 9–14
Curves $\alpha(\|\mathbf{x}\|)$, $\beta(\|\mathbf{x}\|)$, and $V(\mathbf{x}, t)$.

A–9–16. Consider the motion of a space vehicle about the principal axes of inertia. The Euler equations are

$$A\dot{\omega}_x - (B - C)\omega_y\omega_z = T_x$$

$$B\dot{\omega}_y - (C - A)\omega_z\omega_x = T_y$$

$$C\dot{\omega}_z - (A - B)\omega_x\omega_y = T_z$$

where A, B, and C denote the moments of inertia about the principal axes; ω_x, ω_y, and ω_z denote the angular velocities about the principal axes; and T_x, T_y, and T_z are the control torques.

Assume that the space vehicle is tumbling in orbit. It is desired to stop the tumbling by applying control torques, which are assumed to be

$$T_x = k_1 A\omega_x$$

$$T_y = k_2 B\omega_y$$

$$T_z = k_3 C\omega_z$$

Determine sufficient conditions for asymptotically stable operation of the system.

Solution. Let us choose the state variables as

$$x_1 = \omega_x, \qquad x_2 = \omega_y, \qquad x_3 = \omega_z$$

Then the system equations become

$$\dot{x}_1 - \left(\frac{B}{A} - \frac{C}{A}\right)x_2x_3 = k_1x_1$$

$$\dot{x}_2 - \left(\frac{C}{B} - \frac{A}{B}\right)x_3x_1 = k_2x_2$$

$$\dot{x}_3 - \left(\frac{A}{C} - \frac{B}{C}\right)x_1x_2 = k_3x_3$$

or

$$\begin{bmatrix} \dot{x}_1 \\ \dot{x}_2 \\ \dot{x}_3 \end{bmatrix} = \begin{bmatrix} k_1 & \dfrac{B}{A}x_3 & -\dfrac{C}{A}x_2 \\ -\dfrac{A}{B}x_3 & k_2 & \dfrac{C}{B}x_1 \\ \dfrac{A}{C}x_2 & -\dfrac{B}{C}x_1 & k_3 \end{bmatrix} \begin{bmatrix} x_1 \\ x_2 \\ x_3 \end{bmatrix}$$

The equilibrium state is the origin, or $\mathbf{x} = \mathbf{0}$. If we choose

$$V(\mathbf{x}) = \mathbf{x}^T\mathbf{P}\mathbf{x} = \mathbf{x}^T \begin{bmatrix} A^2 & 0 & 0 \\ 0 & B^2 & 0 \\ 0 & 0 & C^2 \end{bmatrix} \mathbf{x}$$

$$= A^2x_1^2 + B^2x_2^2 + C^2x_3^2$$

$$= \text{positive definite}$$

then the time derivative of $V(\mathbf{x})$ is

$$\dot{V}(\mathbf{x}) = \dot{\mathbf{x}}^T\mathbf{P}\mathbf{x} + \mathbf{x}^T\mathbf{P}\dot{\mathbf{x}}$$

$$= \mathbf{x}^T \begin{bmatrix} k_1 & -\dfrac{A}{B}x_3 & \dfrac{A}{C}x_2 \\ \dfrac{B}{A}x_2 & k_2 & -\dfrac{B}{C}x_1 \\ -\dfrac{C}{A}x_2 & \dfrac{C}{B}x_1 & k_3 \end{bmatrix} \begin{bmatrix} A^2 & 0 & 0 \\ 0 & B^2 & 0 \\ 0 & 0 & C^2 \end{bmatrix} \mathbf{x}$$

$$+ \mathbf{x}^T \begin{bmatrix} A^2 & 0 & 0 \\ 0 & B^2 & 0 \\ 0 & 0 & C^2 \end{bmatrix} \begin{bmatrix} k_1 & \dfrac{B}{A}x_3 & -\dfrac{C}{A}x_2 \\ -\dfrac{A}{B}x_3 & k_2 & \dfrac{C}{B}x_1 \\ \dfrac{A}{C}x_2 & -\dfrac{B}{C}x_1 & k_3 \end{bmatrix} \mathbf{x}$$

$$= \mathbf{x}^T \begin{bmatrix} 2k_1A^2 & 0 & 0 \\ 0 & 2k_2B^2 & 0 \\ 0 & 0 & 2k_3C^2 \end{bmatrix} \mathbf{x} = -\mathbf{x}^T\mathbf{Q}\mathbf{x}$$

For asymptotic stability, the sufficient condition is that $\mathbf{Q}$ be positive definite. Hence we require

$$k_1 < 0, \qquad k_2 < 0, \qquad k_3 < 0$$

If the k_i are negative, then noting that $V(\mathbf{x}) \to \infty$ as $\|\mathbf{x}\| \to \infty$, we see that the equilibrium state is asymptotically stable in the large.

A–9–17. Consider the system

$$\dot{\mathbf{x}} = \mathbf{A}\mathbf{x}$$

where $\mathbf{x}$ is a state vector (n-vector) and $\mathbf{A}$ is an $n \times n$ constant matrix. Assume that the origin, $\mathbf{x} = \mathbf{0}$, is the only equilibrium state. A possible Liapunov function for the system is

$$V(\mathbf{x}) = \mathbf{x}^*\mathbf{P}\mathbf{x}$$

The time derivative of $V(\mathbf{x})$ along any trajectory is

$$\dot{V}(\mathbf{x}) = \mathbf{x}^*(\mathbf{A}^*\mathbf{P} + \mathbf{P}\mathbf{A})\mathbf{x} = -\mathbf{x}^*\mathbf{Q}\mathbf{x}$$

where

$$\mathbf{Q} = -(\mathbf{A}^*\mathbf{P} + \mathbf{P}\mathbf{A})$$

Given any positive definite Hermitian (or real symmetric) matrix $\mathbf{Q}$, if this last equation is solved for matrix $\mathbf{P}$ and if it is found to be positive definite, then the origin of the system is asymptotically stable. If $\dot{V}(\mathbf{x}) = -\mathbf{x}^*\mathbf{Q}\mathbf{x}$ does not vanish identically along any trajectory, then $\mathbf{Q}$ may be chosen as positive semidefinite.

Show that a necessary and sufficient condition that $\dot{V}(\mathbf{x})$ does not vanish identically along any trajectory [meaning that $\dot{V}(\mathbf{x}) = 0$ only at $\mathbf{x} = \mathbf{0}$] is that the rank of the matrix

$$\begin{bmatrix} \mathbf{Q}^{1/2} \\ \mathbf{Q}^{1/2}\mathbf{A} \\ \cdot \\ \cdot \\ \cdot \\ \mathbf{Q}^{1/2}\mathbf{A}^{n-1} \end{bmatrix}$$

be n.

Solution. Suppose that, for a certain positive semidefinite matrix $\mathbf{Q}$, $\mathbf{P}$ is found to be positive definite. Since $\dot{V}(\mathbf{x})$ can be written as

$$\dot{V}(\mathbf{x}) = -\mathbf{x}^*\mathbf{Q}\mathbf{x} = -\mathbf{x}^*\mathbf{Q}^{1/2}\mathbf{Q}^{1/2}\mathbf{x}$$

$\dot{V}(\mathbf{x}) = 0$ means that

$$\mathbf{Q}^{1/2}\mathbf{x} = \mathbf{0} \tag{9–124}$$

Differentiating Equation (9–124) with respect to t gives

$$\mathbf{Q}^{1/2}\dot{\mathbf{x}} = \mathbf{Q}^{1/2}\mathbf{A}\mathbf{x} = \mathbf{0}$$

Differentiating this last equation once again, we get

$$\mathbf{Q}^{1/2}\mathbf{A}\dot{\mathbf{x}} = \mathbf{Q}^{1/2}\mathbf{A}^2\mathbf{x} = \mathbf{0}$$

Repeating this differentiation process, we obtain

$$\mathbf{Q}^{1/2}\mathbf{A}^3\mathbf{x} = \mathbf{0}$$
$$\cdot$$
$$\cdot$$
$$\cdot$$
$$\mathbf{Q}^{1/2}\mathbf{A}^{n-1}\mathbf{x} = \mathbf{0}$$

Combining the preceding equations, we obtain

$$\begin{bmatrix} \mathbf{Q}^{1/2} \\ \mathbf{Q}^{1/2}\mathbf{A} \\ \cdot \\ \cdot \\ \cdot \\ \mathbf{Q}^{1/2}\mathbf{A}^{n-1} \end{bmatrix}\mathbf{x} = \mathbf{0} \tag{9–125}$$

A necessary and sufficient condition that $\mathbf{x} = \mathbf{0}$ is the solution of Equation (9–125) is that

$$\text{rank}\begin{bmatrix} \mathbf{Q}^{1/2} \\ \mathbf{Q}^{1/2}\mathbf{A} \\ \cdot \\ \cdot \\ \cdot \\ \mathbf{Q}^{1/2}\mathbf{A}^{n-1} \end{bmatrix} = n \tag{9–126}$$

Hence, if Equation (9–126) is satisfied, then $\dot{V}(\mathbf{x}) = -\mathbf{x}^{*}\mathbf{Q}^{1/2}\mathbf{Q}^{1/2}\mathbf{x}$ becomes zero only at $\mathbf{x} = \mathbf{0}$. In other words, $\dot{V}(\mathbf{x})$ does not vanish identically along any trajectory, except at $\mathbf{x} = \mathbf{0}$.

A–9–18. Consider the system described by

$$\dot{\mathbf{x}} = \mathbf{A}\mathbf{x}$$

where

$$\mathbf{A} = \begin{bmatrix} 0 & 1 \\ -1 & -2 \end{bmatrix}$$

Determine the stability of the equilibrium state, $\mathbf{x} = \mathbf{0}$.

Solution. Instead of choosing $\mathbf{Q} = \mathbf{I}$, let us demonstrate the use of a positive-semidefinite matrix $\mathbf{Q}$ to solve the Liapunov equation

$$\mathbf{A}^{T}\mathbf{P} + \mathbf{P}\mathbf{A} = -\mathbf{Q} \tag{9–127}$$

For example, let us choose

$$\mathbf{Q} = \begin{bmatrix} 4 & 0 \\ 0 & 0 \end{bmatrix}$$

Then the rank of

$$\begin{bmatrix} \mathbf{Q}^{1/2} \\ \mathbf{Q}^{1/2}\mathbf{A} \end{bmatrix} = \begin{bmatrix} 2 & 0 \\ 0 & 0 \\ 0 & 2 \\ 0 & 0 \end{bmatrix}$$

is 2. Hence we may use this $\mathbf{Q}$ matrix and solve Equation (9–127), which can be rewritten as

$$\begin{bmatrix} 0 & -1 \\ 1 & -2 \end{bmatrix}\begin{bmatrix} p_{11} & p_{12} \\ p_{12} & p_{22} \end{bmatrix} + \begin{bmatrix} p_{11} & p_{12} \\ p_{12} & p_{22} \end{bmatrix}\begin{bmatrix} 0 & 1 \\ -1 & -2 \end{bmatrix} = \begin{bmatrix} -4 & 0 \\ 0 & 0 \end{bmatrix}$$

which can be simplified to

$$\begin{bmatrix} -2p_{12} & p_{11} - 2p_{12} - p_{22} \\ p_{11} - 2p_{12} - p_{22} & 2p_{12} - 4p_{22} \end{bmatrix} = \begin{bmatrix} -4 & 0 \\ 0 & 0 \end{bmatrix}$$

from which we obtain

$$p_{11} = 5, \qquad p_{12} = 2, \qquad p_{22} = 1$$

or

$$\mathbf{P} = \begin{bmatrix} 5 & 2 \\ 2 & 1 \end{bmatrix}$$

Matrix $\mathbf{P}$ is positive definite. Hence the equilibrium state, $\mathbf{x} = \mathbf{0}$, is asymptotically stable.

A–9–19.   Consider the second-order system

$$\dot{\mathbf{x}} = \mathbf{A}\mathbf{x}$$

where

$$\mathbf{x} = \begin{bmatrix} x_1 \\ x_2 \end{bmatrix}, \qquad \mathbf{A} = \begin{bmatrix} a_{11} & a_{12} \\ a_{21} & a_{22} \end{bmatrix} \qquad (a_{ij} = \text{real})$$

Find the real symmetric matrix $\mathbf{P}$ that satisfies

$$\mathbf{A}^T\mathbf{P} + \mathbf{P}\mathbf{A} = -\mathbf{I}$$

Then find the condition that $\mathbf{P}$ is positive definite. (Note that $\mathbf{P}$ being positive definite implies that the origin $\mathbf{x} = \mathbf{0}$ is asymptotically stable in the large.)

Solution.   The equation

$$\begin{bmatrix} a_{11} & a_{21} \\ a_{12} & a_{22} \end{bmatrix}\begin{bmatrix} p_{11} & p_{12} \\ p_{12} & p_{22} \end{bmatrix} + \begin{bmatrix} p_{11} & p_{12} \\ p_{12} & p_{22} \end{bmatrix}\begin{bmatrix} a_{11} & a_{12} \\ a_{21} & a_{22} \end{bmatrix} = \begin{bmatrix} -1 & 0 \\ 0 & -1 \end{bmatrix}$$

yields the following three simultaneous equations:

$$2(a_{11}p_{11} + a_{21}p_{12}) = -1$$

$$a_{11}p_{12} + a_{21}p_{22} + a_{12}p_{11} + a_{22}p_{12} = 0$$

$$2(a_{12}p_{12} + a_{22}p_{22}) = -1$$

Solving for the p_{ij}, we obtain

$$\mathbf{P} = \frac{1}{2(a_{11} + a_{22})|\mathbf{A}|}\begin{bmatrix} -(|\mathbf{A}| + a_{21}^2 + a_{22}^2) & a_{12}a_{22} + a_{21}a_{11} \\ a_{12}a_{22} + a_{21}a_{11} & -(|\mathbf{A}| + a_{11}^2 + a_{12}^2) \end{bmatrix}$$

$\mathbf{P}$ is positive definite if

$$P_{11} = -\frac{|\mathbf{A}| + a_{21}^2 + a_{22}^2}{2(a_{11} + a_{22})|\mathbf{A}|} > 0$$

$$|\mathbf{P}| = \frac{(a_{11} + a_{22})^2 + (a_{12} - a_{21})^2}{4(a_{11} + a_{22})^2\,|\mathbf{A}|}$$

from which we obtain

$$|\mathbf{A}| > 0, \qquad a_{11} + a_{22} < 0$$

as the conditions that $\mathbf{P}$ is positive definite.

A–9–20.   Determine the stability of the equilibrium state of the following system:

$$\begin{bmatrix} \dot{x}_1 \\ \dot{x}_2 \end{bmatrix} = \begin{bmatrix} -2 & -1 - j \\ -1 + j & -3 \end{bmatrix}\begin{bmatrix} x_1 \\ x_2 \end{bmatrix}$$

Solution.   In this problem both the state vector and the state matrix are complex. In determining the stability of the equilibrium state, the origin in this system, we solve the Liapunov equation $\mathbf{A}^*\mathbf{P} + \mathbf{P}\mathbf{A} = -\mathbf{Q}$ for $\mathbf{P}$, which is a Hermitian matrix:

$$\mathbf{P} = \begin{bmatrix} p_{11} & p_{12} \\ \bar{p}_{12} & p_{22} \end{bmatrix}$$

Let us choose $\mathbf{Q} = \mathbf{I}$. Then

$$\begin{bmatrix} -2 & -1-j \\ -1+j & -3 \end{bmatrix} \begin{bmatrix} p_{11} & p_{12} \\ \bar{p}_{12} & p_{22} \end{bmatrix} + \begin{bmatrix} p_{11} & p_{12} \\ \bar{p}_{12} & p_{22} \end{bmatrix} \begin{bmatrix} -2 & -1-j \\ -1+j & -3 \end{bmatrix} = \begin{bmatrix} -1 & 0 \\ 0 & -1 \end{bmatrix}$$

from which we obtain

$$4p_{11} + (1-j)p_{12} + (1+j)\bar{p}_{12} = 1$$

$$(1-j)p_{11} + 5\bar{p}_{12} + (1-j)p_{22} = 0$$

$$(1+j)p_{11} + 5p_{12} + (1+j)p_{22} = 0$$

$$(1-j)p_{12} + (1+j)\bar{p}_{12} + 6p_{22} = 1$$

Notice that since p_{11} and p_{22} are real, the second and the third of the above equations are equivalent. They are conjugate to each other. Solving the above equations for the p_{ij}'s, we obtain

$$p_{11} = \frac{3}{8}, \qquad p_{12} = -\frac{1}{8}(1+j), \qquad p_{22} = \frac{1}{4}$$

or

$$\mathbf{P} = \begin{bmatrix} \dfrac{3}{8} & -\dfrac{1}{8} - j\dfrac{1}{8} \\ -\dfrac{1}{8} + j\dfrac{1}{8} & \dfrac{1}{4} \end{bmatrix}$$

Applying Sylvester's criterion for positive definiteness of $\mathbf{P}$ matrix,

$$\frac{3}{8} > 0, \qquad \begin{vmatrix} \dfrac{3}{8} & -\dfrac{1}{8} - j\dfrac{1}{8} \\ -\dfrac{1}{8} + j\dfrac{1}{8} & \dfrac{1}{4} \end{vmatrix} = \frac{1}{16} > 0$$

we find $\mathbf{P}$ to be positive definite. Hence, we conclude that the origin of the system is asymptotically stable.

A–9–21. Obtain the response of the following time-varying system:

$$\dot{x} + tx = u$$

where u is an arbitrary input.

Solution. Referring to Equation (9–84), the solution of this system is given by

$$x(t) = \phi(t, t_0)x(t_0) + \int_{t_0}^{t} \phi(t, \tau)u(\tau)d\tau$$

In this system, since t and $\int_{t_0}^{t} \tau\, d\tau$ commute for all t, $\phi(t, t_0)$ can be expressed as

$$\phi(t, t_0) = \exp\left[-\int_{t_0}^{t} \tau d\tau \right] = e^{-(t^2 - t_0^2)/2}$$

Hence

$$\phi(t, \tau) = e^{-(t^2 - \tau^2)/2}$$

Therefore, the output $x(t)$ can be given by

$$x(t) = e^{-(t^2 - t_0^2)/2}x(t_0) + e^{-t^2/2}\int_{t_0}^{t} e^{\tau^2/2}u(\tau)\, d\tau$$

B-9-1. Consider the following matrix **A**:

$$\mathbf{A} = \begin{bmatrix} 0 & 1 & 0 & 0 \\ 0 & 0 & 1 & 0 \\ 0 & 0 & 0 & 1 \\ 1 & 0 & 0 & 0 \end{bmatrix}$$

Obtain the eigenvalues λ_1, λ_2, λ_3, and λ_4 of the matrix **A**. Then obtain a transformation matrix **P** such that

$$\mathbf{P}^{-1}\mathbf{A}\mathbf{P} = \text{diag}(\lambda_1, \lambda_2, \lambda_3, \lambda_4)$$

B-9-2. Consider the following matrix **A**:

$$\mathbf{A} = \begin{bmatrix} 0 & 1 \\ -2 & -3 \end{bmatrix}$$

Compute $e^{\mathbf{A}t}$ by three methods.

B-9-3. Given the system equation

$$\begin{bmatrix} \dot{x}_1 \\ \dot{x}_2 \\ \dot{x}_3 \end{bmatrix} = \begin{bmatrix} 2 & 1 & 0 \\ 0 & 2 & 1 \\ 0 & 0 & 2 \end{bmatrix} \begin{bmatrix} x_1 \\ x_2 \\ x_3 \end{bmatrix}$$

find the solution in terms of the initial conditions $x_1(0)$, $x_2(0)$, and $x_3(0)$.

B-9-4. Find $x_1(t)$ and $x_2(t)$ of the system described by

$$\begin{bmatrix} \dot{x}_1 \\ \dot{x}_2 \end{bmatrix} = \begin{bmatrix} 0 & 1 \\ -3 & -2 \end{bmatrix} \begin{bmatrix} x_1 \\ x_2 \end{bmatrix}$$

where the initial conditions are

$$\begin{bmatrix} x_1(0) \\ x_2(0) \end{bmatrix} = \begin{bmatrix} 1 \\ -1 \end{bmatrix}$$

B-9-5. Consider the following state equation and output equation:

$$\begin{bmatrix} \dot{x}_1 \\ \dot{x}_2 \\ \dot{x}_3 \end{bmatrix} = \begin{bmatrix} -6 & 1 & 0 \\ -11 & 0 & 1 \\ -6 & 0 & 0 \end{bmatrix} \begin{bmatrix} x_1 \\ x_2 \\ x_3 \end{bmatrix} + \begin{bmatrix} 2 \\ 6 \\ 2 \end{bmatrix} u$$

$$y = \begin{bmatrix} 1 & 0 & 0 \end{bmatrix} \begin{bmatrix} x_1 \\ x_2 \\ x_3 \end{bmatrix}$$

Show that the state equation can be transformed into the fol-

lowing form by use of a proper transformation matrix:

$$\begin{bmatrix} \dot{z}_1 \\ \dot{z}_2 \\ \dot{z}_3 \end{bmatrix} = \begin{bmatrix} 0 & 0 & -6 \\ 1 & 0 & -11 \\ 0 & 1 & -6 \end{bmatrix} \begin{bmatrix} z_1 \\ z_2 \\ z_3 \end{bmatrix} + \begin{bmatrix} 1 \\ 0 \\ 0 \end{bmatrix} u$$

Then obtain the output y in terms of z_1, z_2, and z_3.

B-9-6. The system shown in Figure 9–15 has two inputs, the reference input and the disturbance input, and one output. Obtain the transfer matrix between the output and the inputs.

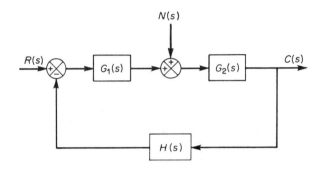

Figure 9–15 System having two inputs and one output.

B-9-7. Consider the mechanical system shown in Figure 9–16. Assume that the system is initially at rest. This system has two inputs $u_1(t)$ and $u_2(t)$ and two outputs $y_1(t)$ and $y_2(t)$. Obtain the transfer matrix between the outputs and the inputs.

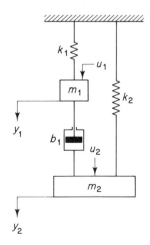

Figure 9–16 Mechanical system.

B-9-8. Obtain the transfer matrix of the system shown in Figure 9-17.

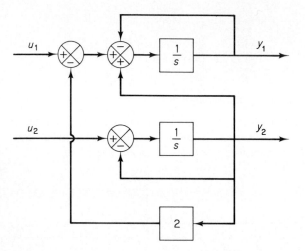

Figure 9-17 Multiple-input, multiple-output system.

B-9-9. Consider the system defined by

$$\begin{bmatrix} \dot{x}_1 \\ \dot{x}_2 \\ \dot{x}_3 \end{bmatrix} = \begin{bmatrix} -1 & -2 & -2 \\ 0 & -1 & 1 \\ 1 & 0 & -1 \end{bmatrix} \begin{bmatrix} x_1 \\ x_2 \\ x_3 \end{bmatrix} + \begin{bmatrix} 2 \\ 0 \\ 1 \end{bmatrix} u$$

$$y = \begin{bmatrix} 1 & 1 & 0 \end{bmatrix} \begin{bmatrix} x_1 \\ x_2 \\ x_3 \end{bmatrix}$$

Is the system completely state controllable and completely observable?

B-9-10. Consider the system given by

$$\begin{bmatrix} \dot{x}_1 \\ \dot{x}_2 \\ \dot{x}_3 \end{bmatrix} = \begin{bmatrix} 2 & 0 & 0 \\ 0 & 2 & 0 \\ 0 & 3 & 1 \end{bmatrix} \begin{bmatrix} x_1 \\ x_2 \\ x_3 \end{bmatrix} + \begin{bmatrix} 0 & 1 \\ 1 & 0 \\ 0 & 1 \end{bmatrix} \begin{bmatrix} u_1 \\ u_2 \end{bmatrix}$$

$$\begin{bmatrix} y_1 \\ y_2 \end{bmatrix} = \begin{bmatrix} 1 & 0 & 0 \\ 0 & 1 & 0 \end{bmatrix} \begin{bmatrix} x_1 \\ x_2 \\ x_3 \end{bmatrix}$$

Is the system completely state controllable and completely observable? Is the system completely output controllable?

B-9-11. Is the following system completely state controllable and completely observable?

$$\begin{bmatrix} \dot{x}_1 \\ \dot{x}_2 \\ \dot{x}_3 \end{bmatrix} = \begin{bmatrix} 0 & 1 & 0 \\ 0 & 0 & 1 \\ -6 & -11 & -6 \end{bmatrix} \begin{bmatrix} x_1 \\ x_2 \\ x_3 \end{bmatrix} + \begin{bmatrix} 0 \\ 0 \\ 1 \end{bmatrix} u$$

$$y = \begin{bmatrix} 20 & 9 & 1 \end{bmatrix} \begin{bmatrix} x_1 \\ x_2 \\ x_3 \end{bmatrix}$$

B-9-12. Consider the system defined by

$$\begin{bmatrix} \dot{x}_1 \\ \dot{x}_2 \\ \dot{x}_3 \end{bmatrix} = \begin{bmatrix} 0 & 1 & 0 \\ 0 & 0 & 1 \\ -6 & -11 & -6 \end{bmatrix} \begin{bmatrix} x_1 \\ x_2 \\ x_3 \end{bmatrix} + \begin{bmatrix} 0 \\ 0 \\ 1 \end{bmatrix} u$$

$$y = \begin{bmatrix} c_1 & c_2 & c_3 \end{bmatrix} \begin{bmatrix} x_1 \\ x_2 \\ x_3 \end{bmatrix}$$

Except for an obvious choice of $c_1 = c_2 = c_3 = 0$, find an example of a set of c_1, c_2, c_3 that will make the system unobservable.

B-9-13. Consider the system

$$\begin{bmatrix} \dot{x}_1 \\ \dot{x}_2 \\ \dot{x}_3 \end{bmatrix} = \begin{bmatrix} 2 & 0 & 0 \\ 0 & 2 & 0 \\ 0 & 3 & 1 \end{bmatrix} \begin{bmatrix} x_1 \\ x_2 \\ x_3 \end{bmatrix}$$

The output is given by

$$y = \begin{bmatrix} 1 & 1 & 1 \end{bmatrix} \begin{bmatrix} x_1 \\ x_2 \\ x_3 \end{bmatrix}$$

(a) Show that the system is not completely observable.
(b) Show that the system is completely observable if the output is given by

$$\begin{bmatrix} y_1 \\ y_2 \end{bmatrix} = \begin{bmatrix} 1 & 1 & 1 \\ 1 & 2 & 3 \end{bmatrix} \begin{bmatrix} x_1 \\ x_2 \\ x_3 \end{bmatrix}$$

B-9-14. Prove that a necessary and sufficient condition that the system

$$\dot{x} = Ax$$

$$y = Cx$$

where x = state vector (n-vector)
y = output signal (scalar)
A = $n \times n$ matrix
B = $1 \times n$ matrix

is completely observable is that $C(sI - A)^{-1}$ has no cancellation.

Note that for this system

$$\mathbf{X}(s) = (s\mathbf{I} - \mathbf{A})^{-1}\mathbf{x}(0)$$

and therefore

$$\mathbf{Y}(s) = \mathbf{C}(s\mathbf{I} - \mathbf{A})^{-1}\mathbf{x}(0)$$

Note also that $\mathbf{C}(s\mathbf{I} - \mathbf{A})^{-1}$ can be written as

$$\mathbf{C}(s\mathbf{I} - \mathbf{A})^{-1} = \frac{1}{|s\mathbf{I} - \mathbf{A}|}[q_1(s) \quad q_2(s) \quad \cdots \quad q_n(s)]$$

where the $q_i(s)$ are polynomials in s.

B–9–15. Determine whether or not the following quadratic form is positive definite.

$$Q = x_1^2 + 4x_2^2 + x_3^2 + 2x_1x_2 - 6x_2x_3 - 2x_1x_3$$

B–9–16. Determine whether or not the following quadratic form is negative definite.

$$Q = -x_1^2 - 3x_2^2 - 11x_3^2 + 2x_1x_2 - 4x_2x_3 - 2x_1x_3$$

B–9–17. Determine the stability of the origin of the following system:

$$\dot{x}_1 = -x_1 + x_2 + x_1(x_1^2 + x_2^2)$$

$$\dot{x}_2 = -x_1 - x_2 + x_2(x_1^2 + x_2^2)$$

Consider the following quadratic function as a possible Liapunov function:

$$V = x_1^2 + x_2^2$$

B–9–18. Write a few Liapunov functions for the system

$$\begin{bmatrix} \dot{x}_1 \\ \dot{x}_2 \end{bmatrix} = \begin{bmatrix} -1 & 1 \\ 2 & -3 \end{bmatrix} \begin{bmatrix} x_1 \\ x_2 \end{bmatrix}$$

Determine the stability of the origin of the system.

B–9–19. Determine the stability of the equilibrium state of the following system:

$$\dot{x}_1 = -x_1 - 2x_2 + 2$$

$$\dot{x}_2 = x_1 - 4x_2 - 1$$

B–9–20. Determine the stability of the equilibrium state of the following system:

$$\dot{x}_1 = x_1 + 3x_2$$

$$\dot{x}_2 = -3x_1 - 2x_2 - 3x_3$$

$$\dot{x}_3 = x_1$$

B–9–21. Consider the following time-varying system:

$$\dot{\mathbf{x}} = \mathbf{A}(t)\mathbf{x}$$

where

$$\mathbf{A}(t) = \begin{bmatrix} a & e^{-t} \\ -e^{-t} & b \end{bmatrix} \qquad (a \text{ and } b \text{ are constants})$$

Obtain the state transition matrix $\mathbf{\Phi}(t, 0)$ with $\mathbf{\Phi}(0, 0) = \mathbf{I}$.

CHAPTER 10

Design of Control Systems by State-Space Methods

10-1 INTRODUCTION

In Chapters 5 through 7 we presented the root-locus method and frequency-response methods that are quite useful for analyzing and designing single-input, single-output systems. For example, by means of open-loop frequency-response tests, we can predict the dynamic behavior of the closed-loop system. If necessary, the dynamic behavior of a complex system may be improved by inserting a simple lead or lag compensator. The techniques of conventional control theory are conceptually simple and require only a reasonable amount of computation.

In conventional control theory, only the input, output, and error signals are considered important; the analysis and design of control systems are carried out using transfer functions, together with a variety of graphical techniques, such as root-locus plots and Bode diagrams.

The main disadvantage of conventional control theory is that, generally, it is applicable only to linear time-invariant systems having a single input and a single output. It is powerless for time-varying systems, nonlinear systems (except simple ones), and multiple-input, multiple-output systems. Thus conventional techniques (the root-locus and frequency-response methods) do not apply to the design of optimal and adaptive control systems, which are mostly time-varying and/or nonlinear.

System design by conventional control theory is based on trial-and-error procedures that, in general, will not yield optimal control systems. System design by modern control theory via state-space methods, on the other hand, enables the engineer to design such systems having desired closed-loop poles (or desired characteristic equations) or optimal control systems with respect to given performance indexes. Also, modern control theory enables the designer to include initial conditions, if necessary, in the design. However, design by modern control theory (via state-space methods) requires accurate mathematical description of system dynamics. This is in contrast to the conventional methods, where, for example, experimental frequency-response curves that may not have sufficient accuracy can be incorporated in the design without their mathematical descriptions.

From the computational viewpoint, the state-space methods are particularly suited for digital-computer computations because of their time-domain approach. This relieves the engineer of the burden of tedious computations otherwise necessary and enables him to devote his efforts solely to the analytical aspects of the problem. This is one of the advantages of the state-space methods.

Finally, it is important to note that it is not necessary that the state variables represent physical quantities of the system. Variables that do not represent physical quantities and those that are neither measurable nor observable may be chosen as state variables. Such freedom in choosing state variables is another advantage of the state-space methods.

Performance indexes. In designing control systems, we may have the goal of finding a rule for determining the present control decision, subject to certain constraints that will minimize some measure of a deviation from ideal behavior. Such a measure is usually provided by a criterion of optimization, or performance index. The performance index defined in Chapter 4 is a function whose value indicates how well the actual performance of the system matches the desired performance. In most practical cases, system behavior is optimized by choosing the control vector in such a way that the performance index is minimized (or maximized as the case may be).

The performance index is important because it, to a large degree, determines the nature of the resulting control system. That is, the resulting control system may be linear, nonlinear, stationary or time varying, depending on the form of the performance index. The control engineer formulates this index based on the requirements of the problem. Thus, he influences the nature of the resulting system. The requirements of the problem usually include not only performance requirements but also restrictions on the form of the control to ensure physical realizability.

Choosing the most appropriate performance index for a given problem is very difficult, especially in complex systems. For example, consider the problem of the maximization of a payload of a space vehicle. Maximizing the payload will involve an optimization of both the thrust program and mission design for minimum propellant expenditures, as well as an optimal design of the components of the vehicle. In space-vehicle applications, other possible performance specifications may be minimum fuel expenditure, minimum target miss, and minimum time. In civilian, as differentiated from military, applications of control, the prime considerations are usually economic.

Major topics to be discussed in this chapter. In what follows we shall give an outline of each design problem to be discussed in this chapter.

Pole placement design and design of observers. In this chapter we shall present two approaches to the design of regulators. Regulator systems are feedback control systems that will bring nonzero states (caused by external disturbances) to the origin with a sufficient speed. One approach to design regulator systems is to construct an asymptotically stable closed-loop system by specifying the desired locations for the closed-loop poles. This may be accomplished by use of state feedback; that is, we assume the control vector to be $\mathbf{u} = -\mathbf{K}\mathbf{x}$ (where $\mathbf{u}$ is unconstrained) and determine the feedback gain matrix $\mathbf{K}$ such that the system will have a desired characteristic equation. This design scheme is referred to as pole placement.

In this case, the performance index can be written as follows:

$$\text{Performance index} = \sum_{i=1}^{n} (\mu_i - s_i)^2$$

where the μ_i's are the desired eigenvalues of the error dynamics of the system and the s_i's are the actual eigenvalues of the error dynamics of the designed system. In this case the performance index can be made zero by exactly matching the s_i's with the μ_i's, provided the system considered is completely state controllable.

The other approach to the design of regulator systems is to assume the state feedback control vector in the form $\mathbf{u} = -\mathbf{Kx}$ (where $\mathbf{u}$ is unconstrained) and determine the feedback gain matrix $\mathbf{K}$ such that a quadratic performance index is minimized. This formulation to determine the optimal control law is commonly called the quadratic optimal control problem.

Both the pole placement approach and the quadratic optimal control approach require the feedback of all state variables. Therefore, it becomes necessary that all state variables be available for feedback. However, some state variables may be unmeasurable and may not be available for feedback. Then we need to estimate such unmeasurable state variables by use of state observers. Design via pole placement is discussed in Section 10–2. We shall discuss an inverted pendulum system as an example to illustrate details of the pole placement technique. Design of state observers is presented in Section 10–3.

It is noted that it will be shown later that pole placement and quadratic optimal control are not possible if the system is not completely state controllable. Design of state observers (that are required in many state feedback schemes) is not possible if the system is not observable. Hence, controllability and observability play an important role in the design of control systems.

Design of servo systems.   We shall discuss design of type 1 servo systems based on the pole placement approach. We shall consider two cases: (1) the plant has one integrator; (2) the plant has no integrator. We shall present detailed procedures for design using example systems.

Optimal control systems based on quadratic performance indexes.   In many practical control systems, we desire to minimize some function of the error signal. For example, given the system

$$\dot{\mathbf{x}} = \mathbf{Ax} + \mathbf{Bu}$$

we may desire to minimize a generalized error function such as

$$J = \int_0^T [\boldsymbol{\xi}(t) - \mathbf{x}(t)]^* \mathbf{Q}[\boldsymbol{\xi}(t) - \mathbf{x}(t)] \, dt$$

where $\boldsymbol{\xi}(t)$ represents the desired state, $\mathbf{x}(t)$ the actual state [thus, $\boldsymbol{\xi}(t) - \mathbf{x}(t)$ is the error vector], and $\mathbf{Q}$ a positive-definite (or positive-semidefinite) Hermitian or real symmetric matrix, and the time interval $0 \leq t \leq T$ is either finite or infinite.

In addition to considering errors as a measure of system performance, however, we usually must pay attention to the energy required for control action. Since the control signal may have the dimension of force or torque, the control energy is proportional to the integral of such control signal squared. If the error function is minimized regardless of the energy required, then a design may result that calls for overly large values of the control signal. This is undesirable since all physical systems are subject to saturation. Large-amplitude control signals are ineffective outside the range determined by saturation. Thus, practical considerations place a constraint on the control vector, for example,

$$\int_0^T \mathbf{u}^*(t)\mathbf{R}\mathbf{u}(t) \, dt = K$$

where $\mathbf{R}$ is a positive-definite Hermitian or real symmetric matrix and K is a positive constant. The performance index of a control system over the time interval $0 \leq t \leq T$ may then be written, with the use of a Lagrange multiplier λ, as

$$J = \int_0^T [\boldsymbol{\xi}(t) - \mathbf{x}(t)]^*\mathbf{Q}[\boldsymbol{\xi}(t) - \mathbf{x}(t)] \, dt + \lambda \int_0^T \mathbf{u}^*(t)\mathbf{R}\mathbf{u}(t) \, dt \qquad (0 \leq t \leq T)$$

The Lagrange multiplier λ is a positive constant indicating the weight of control cost with respect to minimizing the error function. Note that in this formulation $\mathbf{u}(t)$ is unconstrained. Design based on this performance index has a practical significance that the resulting system compromises between minimizing the integral error squared and minimizing the control energy.

If $T = \infty$ and the desired state $\boldsymbol{\xi}$ is the origin, or $\boldsymbol{\xi} = \mathbf{0}$, then the preceding performance index can be expressed as

$$J = \int_0^\infty [\mathbf{x}^*(t)\mathbf{Q}\mathbf{x}(t) + \mathbf{u}^*(t)\mathbf{R}\mathbf{u}(t)] \, dt$$

where we have included λ in the positive definite matrix $\mathbf{R}$. In Section 10–5 we shall consider the design of regulator systems based on this performance index. It is called the quadratic performance index. Note that a choice of weighting matrices $\mathbf{Q}$ and $\mathbf{R}$ is in a sense arbitrary. Although minimizing an ''arbitrary'' quadratic performance index may not seem to have much significance, the advantage of the quadratic optimal control approach is that the resulting system is a stable system.

A regulator system designed by minimizing a quadratic performance index is called a quadratic optimal regulator system. This approach is alternative to the pole placement approach to the design of stable regulator systems.

Model-reference control systems and adaptive control systems. One useful method for specifying system performance is by means of a model that will produce the desired output for a given input. The model need not be an actual hardware. It can be only a mathematical model simulated on a computer. In a model reference control system, the output of the model and that of the plant are compared and the difference is used to generate the control signals. Using an example system, we shall discuss the design of a model-reference control system using the Liapunov approach. We also give general discussions of adaptive control systems.

Outline of the chapter. Section 10–1 has presented introductory materials for the chapter. Section 10–2 treats the design of regulator systems via pole placement. Section 10–3 discusses the design of state observers. Section 10–4 presents the design of servo systems.

We shall apply the pole placement approach to the design of type 1 servo systems. Section 10–5 treats the design of optimal control systems based on quadratic performance indexes. Here we shall use the Liapunov approach to the design. Section 10–6 presents model-reference control systems. Finally, Section 10–7 discusses adaptive control systems.

It is noted that the systems we deal with in this chapter, except in Sections 10–6 and 10–7, are linear time-invariant systems.

10–2 CONTROL SYSTEM DESIGN VIA POLE PLACEMENT

In this section we shall present a design method commonly called the *pole placement* or *pole assignment technique*. We assume that all state variables are measurable and are available for feedback. It will be shown that if the system considered is completely state controllable, then poles of the closed-loop system may be placed at any desired locations by means of state feedback through an appropriate state feedback gain matrix.

The present design technique begins with a determination of the desired closed-loop poles based on the transient-response and/or frequency-response requirements, such as speed, damping ratio, or bandwidth, as well as steady-state requirements.

Let us assume that we decide that the desired closed-loop poles are to be at $s = \mu_1$, $s = \mu_2, \ldots, s = \mu_n$. By choosing an appropriate gain matrix for state feedback, it is possible to force the system to have closed-loop poles at the desired locations, provided that the original system is completely state controllable.

In what follows, we shall treat the case where the control signal is a scalar and prove that a necessary and sufficient condition that the closed-loop poles can be placed at any arbitrary locations in the s plane is that the system be completely state controllable. Then we shall discuss three methods for determining the required state feedback gain matrix.

It is noted that when the control signal is a vector quantity, the mathematical aspects of the pole placement scheme become complicated. Therefore, we shall not discuss such a case in this book. (For the interested reader, refer to Ref. O–3 for the mathematical derivation in such a case.) It is also noted that when the control signal is a vector quantity, the state feedback gain matrix is not unique. It is possible to choose freely more than n parameters; that is, in addition to being able to place n closed-loop poles properly, we have the freedom to satisfy some or all of the other requirements, if any, of the closed-loop system.

Design via pole placement.   In the conventional approach to the design of a single-input, single-output control system, we design a controller (compensator) such that the dominant closed-loop poles have a desired damping ratio ζ and undamped natural frequency ω_n. In this approach, the order of the system may be raised by 1 or 2 unless pole–zero cancellation takes place. Note that in this approach we assume the effects on the responses of nondominant closed-loop poles to be negligible.

Different from specifying only dominant closed-loop poles (conventional design approach), the present pole placement approach specifies all closed-loop poles. (There is a cost associated with placing all closed-loop poles, however, because placing all closed-loop poles requires successful measurements of all state variables or else requires the inclusion of a state observer in the system.) There is also a requirement on the part of the system for the closed-loop poles to be placed at arbitrarily chosen locations. The requirement is that the system be completely state controllable. We shall prove this fact in this section.

Consider a control system

$$\dot{\mathbf{x}} = \mathbf{A}\mathbf{x} + \mathbf{B}u \qquad (10\text{--}1)$$

where $\mathbf{x}$ = state vector (n-vector)
$\quad u$ = control signal (scalar)
$\quad \mathbf{A}$ = $n \times n$ constant matrix
$\quad \mathbf{B}$ = $n \times 1$ constant matrix

We shall chooose the control signal to be

$$u = -\mathbf{K}\mathbf{x} \qquad (10\text{--}2)$$

This means that the control signal is determined by instantaneous state. Such a scheme is called *state feedback*. The $1 \times n$ matrix $\mathbf{K}$ is called the state feedback gain matrix. In the following analysis we assume that u is unconstrained.

Substituting Equation (10–2) into Equation (10–1) gives

$$\dot{\mathbf{x}}(t) = (\mathbf{A} - \mathbf{B}\mathbf{K})\mathbf{x}(t)$$

The solution of this equation is given by

$$\mathbf{x}(t) = e^{(\mathbf{A} - \mathbf{B}\mathbf{K})t}\mathbf{x}(0) \qquad (10\text{--}3)$$

where $\mathbf{x}(0)$ is the initial state caused by external disturbances. The stability and transient response characteristics are determined by the eigenvalues of matrix $\mathbf{A} - \mathbf{B}\mathbf{K}$. If matrix $\mathbf{K}$ is chosen properly, then matrix $\mathbf{A} - \mathbf{B}\mathbf{K}$ can be made an asymptotically stable matrix, and for all $\mathbf{x}(0) \neq \mathbf{0}$ it is possible to make $\mathbf{x}(t)$ approach $\mathbf{0}$ as t approaches infinity. The eigenvalues of matrix $\mathbf{A} - \mathbf{B}\mathbf{K}$ are called the regulator poles. If these regulator poles are located in the left-half s plane, then $\mathbf{x}(t)$ approaches $\mathbf{0}$ as t approaches infinity. The problem of placing the closed-loop poles at the desired location is called a pole placement problem.

Figure 10–1(a) shows the system defined by Equation (10–1). It is an open-loop control system, because the state $\mathbf{x}$ is not fed back to the control signal u. Figure 10–1(b) shows the system with state feedback. This is a closed-loop control system, because the state $\mathbf{x}$ is fed back to the control signal u.

In what follows, we shall prove that arbitrary pole placement for a given system is possible if and only if the system is completely state controllable.

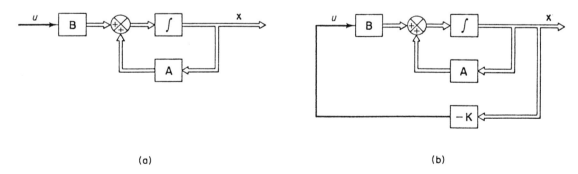

(a)   (b)

Figure 10–1
(a) Open-loop control system; (b) closed-loop control system with $u = -\mathbf{K}\mathbf{x}$.

Necessary and sufficient condition for arbitrary pole placement. Consider the control system defined by Equation (10–1). We assume that the magnitude of the control signal u is unbounded. If the control signal u is chosen as

$$u = -\mathbf{Kx}$$

where $\mathbf{K}$ is the state feedback gain matrix ($1 \times n$ matrix), then the system becomes a closed-loop control system as shown in Figure 10–1(b) and the solution to Equation (10–1) becomes as given by Equation (10–3), or

$$\mathbf{x}(t) = e^{(\mathbf{A} - \mathbf{BK})t}\mathbf{x}(0)$$

Note that the eigenvalues of matrix $\mathbf{A} - \mathbf{BK}$ (which we denote $\mu_1, \mu_2, \ldots, \mu_n$) are the desired closed-loop poles.

We shall now prove that a necessary and sufficient condition for arbitrary pole placement is that the system be completely state controllable. We shall first derive the necessary condition. We begin by proving that if the system is not completely state controllable, then there are eigenvalues of matrix $\mathbf{A} - \mathbf{BK}$ that cannot be controlled by state feedback.

Suppose the system of Equation (10–1) is not completely state controllable. Then the rank of the controllability matrix is less than n, or

$$\text{rank } [\mathbf{B} \mathbin{\vdots} \mathbf{AB} \mathbin{\vdots} \cdots \mathbin{\vdots} \mathbf{A}^{n-1}\mathbf{B}] = q < n$$

This means that there are q linearly independent column vectors in the controllability matrix. Let us define such q linearly independent column vectors as $\mathbf{f}_1, \mathbf{f}_2, \ldots, \mathbf{f}_q$. Also, let us choose $n - q$ additional n-vectors $\mathbf{v}_{q+1}, \mathbf{v}_{q+2}, \ldots, \mathbf{v}_n$ such that

$$\mathbf{P} = [\mathbf{f}_1 \mathbin{\vdots} \mathbf{f}_2 \mathbin{\vdots} \cdots \mathbin{\vdots} \mathbf{f}_q \mathbin{\vdots} \mathbf{v}_{q+1} \mathbin{\vdots} \mathbf{v}_{q+2} \mathbin{\vdots} \cdots \mathbin{\vdots} \mathbf{v}_n]$$

is of rank n. Then it can be shown that

$$\hat{\mathbf{A}} = \mathbf{P}^{-1}\mathbf{AP} = \begin{bmatrix} \mathbf{A}_{11} & \vdots & \mathbf{A}_{12} \\ \cdots & \cdots & \cdots \\ \mathbf{0} & \vdots & \mathbf{A}_{22} \end{bmatrix}, \qquad \hat{\mathbf{B}} = \mathbf{P}^{-1}\mathbf{B} = \begin{bmatrix} \mathbf{B}_{11} \\ \cdots \\ \mathbf{0} \end{bmatrix}$$

(See Problem A–10–1 for the derivation of the above equations.) Now define

$$\hat{\mathbf{K}} = \mathbf{KP} = [\mathbf{k}_1 \mathbin{\vdots} \mathbf{k}_2]$$

Then we have

$$|s\mathbf{I} - \mathbf{A} + \mathbf{BK}| = |\mathbf{P}^{-1}(s\mathbf{I} - \mathbf{A} + \mathbf{BK})\mathbf{P}|$$

$$= |s\mathbf{I} - \mathbf{P}^{-1}\mathbf{AP} + \mathbf{P}^{-1}\mathbf{BKP}|$$

$$= |s\mathbf{I} - \hat{\mathbf{A}} + \hat{\mathbf{B}}\hat{\mathbf{K}}|$$

$$= \left| s\mathbf{I} - \begin{bmatrix} \mathbf{A}_{11} & \vdots & \mathbf{A}_{12} \\ \cdots & \cdots & \cdots \\ \mathbf{0} & \vdots & \mathbf{A}_{22} \end{bmatrix} + \begin{bmatrix} \mathbf{B}_{11} \\ \cdots \\ \mathbf{0} \end{bmatrix} [\mathbf{k}_1 \mathbin{\vdots} \mathbf{k}_2] \right|$$

$$= \begin{vmatrix} s\mathbf{I}_q - \mathbf{A}_{11} + \mathbf{B}_{11}\mathbf{k}_1 & -\mathbf{A}_{12} + \mathbf{B}_{11}\mathbf{k}_2 \\ \mathbf{0} & s\mathbf{I}_{n-q} - \mathbf{A}_{22} \end{vmatrix}$$

$$= |s\mathbf{I}_q - \mathbf{A}_{11} + \mathbf{B}_{11}\mathbf{k}_1| \cdot |s\mathbf{I}_{n-q} - \mathbf{A}_{22}| = 0$$

where $\mathbf{I}_q$ is a q-dimensional identity matrix and $\mathbf{I}_{n-q}$ is an $(n - q)$-dimensional identity matrix.

Notice that the eigenvalues of $\mathbf{A}_{22}$ do not depend on $\mathbf{K}$. Thus, if the system is not completely state controllable, then there are eigenvalues of matrix $\mathbf{A}$ that cannot be arbitrarily placed. Therefore, to place the eigenvalues of matrix $\mathbf{A} - \mathbf{BK}$ arbitrarily, the system must be completely state controllable (necessary condition).

Next we shall prove a sufficient condition: that is, if the system is completely state controllable [meaning that matrix $\mathbf{M}$ given by Equation (10–5) below has the inverse], then all eigenvalues of matrix $\mathbf{A}$ can be arbitrarily placed.

In proving a sufficient condition, it is convenient to transform the state equation given by Equation (10–1) into the controllable canonical form.

Define a transformation matrix $\mathbf{T}$ by

$$\mathbf{T} = \mathbf{MW} \tag{10–4}$$

where $\mathbf{M}$ is the controllability matrix

$$\mathbf{M} = [\mathbf{B} \mid \mathbf{AB} \mid \cdots \mid \mathbf{A}^{n-1}\mathbf{B}] \tag{10–5}$$

and

$$\mathbf{W} = \begin{bmatrix} a_{n-1} & a_{n-2} & \cdots & a_1 & 1 \\ a_{n-2} & a_{n-3} & \cdots & 1 & 0 \\ \cdot & \cdot & & \cdot & \cdot \\ \cdot & \cdot & & \cdot & \cdot \\ \cdot & \cdot & & \cdot & \cdot \\ a_1 & 1 & \cdots & 0 & 0 \\ 1 & 0 & \cdots & 0 & 0 \end{bmatrix} \tag{10–6}$$

where the a_i's are coefficients of the characteristic polynomial

$$|s\mathbf{I} - \mathbf{A}| = s^n + a_1 s^{n-1} + \cdots + a_{n-1}s + a_n$$

Define a new state vector $\hat{\mathbf{x}}$ by

$$\mathbf{x} = \mathbf{T}\hat{\mathbf{x}}$$

If the rank of the controllability matrix $\mathbf{M}$ is n (meaning that the system is completely state controllable), then the inverse of matrix $\mathbf{T}$ exists and Equation (10–1) can be modified to

$$\dot{\hat{\mathbf{x}}} = \mathbf{T}^{-1}\mathbf{AT}\hat{\mathbf{x}} + \mathbf{T}^{-1}\mathbf{B}u \tag{10–7}$$

where

$$\mathbf{T}^{-1}\mathbf{AT} = \begin{bmatrix} 0 & 1 & 0 & \cdots & 0 \\ 0 & 0 & 1 & \cdots & 0 \\ \cdot & \cdot & \cdot & & \cdot \\ \cdot & \cdot & \cdot & & \cdot \\ \cdot & \cdot & \cdot & & \cdot \\ 0 & 0 & 0 & \cdots & 1 \\ -a_n & -a_{n-1} & -a_{n-2} & \cdots & -a_1 \end{bmatrix} \tag{10–8}$$

$$\mathbf{T}^{-1}\mathbf{B} = \begin{bmatrix} 0 \\ 0 \\ \cdot \\ \cdot \\ \cdot \\ 0 \\ 1 \end{bmatrix} \tag{10-9}$$

[See Problems A–10–2 and A–10–3 for the derivation of Equations (10–8) and (10–9).] Equation (10–7) is in the controllable canonical form. Thus, given a state equation, Equation (10–1), it can be transformed into the controllable canonical form if the system is completely state controllable and if we transform the state vector $\mathbf{x}$ into state vector $\hat{\mathbf{x}}$ by use of the transformation matrix $\mathbf{T}$ given by Equation (10–4).

Let us choose a set of the desired eigenvalues as $\mu_1, \mu_2, \ldots, \mu_n$. Then the desired characteristic equation becomes

$$(s - \mu_1)(s - \mu_2) \cdots (s - \mu_n) = s^n + \alpha_1 s^{n-1} + \cdots + \alpha_{n-1}s + \alpha_n = 0 \tag{10-10}$$

Let us write

$$\hat{\mathbf{K}} = \mathbf{KT} = [\delta_n \quad \delta_{n-1} \quad \cdots \quad \delta_1] \tag{10-11}$$

When $u = -\hat{\mathbf{K}}\hat{\mathbf{x}} = -\mathbf{KT}\hat{\mathbf{x}}$ is used to control the system given by Equation (10–7), the system equation becomes

$$\dot{\hat{\mathbf{x}}} = \mathbf{T}^{-1}\mathbf{AT}\hat{\mathbf{x}} - \mathbf{T}^{-1}\mathbf{BKT}\hat{\mathbf{x}}$$

The characteristic equation is

$$\left| s\mathbf{I} - \mathbf{T}^{-1}\mathbf{AT} + \mathbf{T}^{-1}\mathbf{BKT} \right| = 0$$

This characteristic equation is the same as the characteristic equation for the system, defined by Equation (10–1), when $u = -\mathbf{Kx}$ is used as the control signal. This can be seen as follows: Since

$$\dot{\mathbf{x}} = \mathbf{Ax} + \mathbf{Bu} = (\mathbf{A} - \mathbf{BK})\mathbf{x}$$

the characteristic equation for this system is

$$\left| s\mathbf{I} - \mathbf{A} + \mathbf{BK} \right| = \left| \mathbf{T}^{-1}(s\mathbf{I} - \mathbf{A} + \mathbf{BK})\mathbf{T} \right| = \left| s\mathbf{I} - \mathbf{T}^{-1}\mathbf{AT} + \mathbf{T}^{-1}\mathbf{BKT} \right| = 0$$

Now let us simplify the characteristic equation of the system in the controllable canonical form. Referring to Equations (10–8), (10–9), and (10–11), we have

$$\left| s\mathbf{I} - \mathbf{T}^{-1}\mathbf{AT} + \mathbf{T}^{-1}\mathbf{BKT} \right|$$

$$= \left| s\mathbf{I} - \begin{bmatrix} 0 & 1 & \cdots & 0 \\ \cdot & \cdot & & \cdot \\ \cdot & \cdot & & \cdot \\ 0 & 0 & \cdots & 1 \\ -a_n & -a_{n-1} & \cdots & -a_1 \end{bmatrix} + \begin{bmatrix} 0 \\ \cdot \\ \cdot \\ \cdot \\ 0 \\ 1 \end{bmatrix} [\delta_n \quad \delta_{n-1} \cdots \delta_1] \right|$$

$$
= \begin{vmatrix} s & -1 & \cdots & 0 \\ 0 & s & \cdots & 0 \\ \cdot & \cdot & & \cdot \\ \cdot & \cdot & & \cdot \\ \cdot & \cdot & & \cdot \\ a_n + \delta_n & a_{n-1} + \delta_{n-1} & \cdots & s + a_1 + \delta_1 \end{vmatrix}
$$

$$
= s^n + (a_1 + \delta_1)s^{n-1} + \cdots + (a_{n-1} + \delta_{n-1})s + (a_n + \delta_n) = 0 \qquad (10\text{--}12)
$$

This is the characteristic equation for the system with state feedback. Therefore, it must be equal to Equation (10–10), the desired characteristic equation. By equating the coefficients of like powers of s, we get

$$
a_1 + \delta_1 = \alpha_1
$$

$$
a_2 + \delta_2 = \alpha_2
$$

$$
\cdot
$$
$$
\cdot
$$
$$
\cdot
$$

$$
a_n + \delta_n = \alpha_n
$$

Solving the preceding equations for the δ_i's and substituting them into Equation (10–11), we obtain

$$
\mathbf{K} = \hat{\mathbf{K}}\mathbf{T}^{-1} = [\delta_n \quad \delta_{n-1} \quad \cdots \quad \delta_1]\mathbf{T}^{-1}
$$

$$
= [\alpha_n - a_n \ \vdots \ \alpha_{n-1} - a_{n-1} \ \vdots \ \cdots \ \vdots \ \alpha_2 - a_2 \ \vdots \ \alpha_1 - a_1]\mathbf{T}^{-1} \qquad (10\text{--}13)
$$

Thus, if the system is completely state controllable, all eigenvalues can be arbitrarily placed by choosing matrix $\mathbf{K}$ according to Equation (10–13) (sufficient condition).

We have thus proved that the necessary and sufficient condition for arbitrary pole placement is that the system be completely state controllable.

Design steps for pole placement.   Suppose that the system is defined by

$$
\dot{\mathbf{x}} = \mathbf{Ax} + \mathbf{B}u
$$

and the control signal is given by

$$
u = -\mathbf{Kx}
$$

The feedback gain matrix $\mathbf{K}$ that forces the eigenvalues of $\mathbf{A} - \mathbf{BK}$ to be $\mu_1, \mu_2, \ldots,$ μ_n (desired values) can be determined by the following steps. (If μ_i is a complex eigenvalue, then its conjugate must also be an eigenvalue of $\mathbf{A} - \mathbf{BK}$.)

Step 1: Check the controllability condition for the system. If the system is completely state controllable, then use the following steps.

Step 2: From the characteristic polynomial for matrix $\mathbf{A}$:

$$
|s\mathbf{I} - \mathbf{A}| = s^n + a_1 s^{n-1} + \cdots + a_{n-1}s + a_n
$$

determine the values of $a_1, a_2, \ldots, a_n$.

Step 3: Determine the transformation matrix **T** that transforms the system state equation into the controllable canonical form. (If the given system equation is already in the controllable canonical form, then **T** = **I**.) It is not necessary to write the state equation in the controllable canonical form. All we need here is to find the matrix **T**. The transformation matrix **T** is given by Equation (10–4), or

$$\mathbf{T} = \mathbf{MW}$$

where **M** is given by Equation (10–5) and **W** is given by Equation (10–6).

Step 4: Using the desired eigenvalues (desired closed-loop poles), write the desired characteristic polynomial:

$$(s - \mu_1)(s - \mu_2) \cdots (s - \mu_n) = s^n + \alpha_1 s^{n-1} + \cdots + \alpha_{n-1} s + \alpha_n$$

and determine the values of $\alpha_1, \alpha_2, \ldots, \alpha_n$.

Step 5: The required state feedback gain matrix **K** can be determined from Equation (10–13), rewritten thus:

$$\mathbf{K} = [\alpha_n - a_n \mid \alpha_{n-1} - a_{n-1} \mid \cdots \mid \alpha_2 - a_2 \mid \alpha_1 - a_1]\mathbf{T}^{-1}$$

Comments.   Note that if the system is of low order ($n \leq 3$), then direct substitution of matrix **K** into the desired characteristic polynomial may be simpler. For example, if $n = 3$, then write the state feedback gain matrix **K** as

$$\mathbf{K} = [k_1 \quad k_2 \quad k_3]$$

Substitute this **K** matrix into the desired characteristic polynomial $|s\mathbf{I} - \mathbf{A} + \mathbf{BK}|$ and equate it to $(s - \mu_1)(s - \mu_2)(s - \mu_3)$, or

$$|s\mathbf{I} - \mathbf{A} + \mathbf{BK}| = (s - \mu_1)(s - \mu_2)(s - \mu_3)$$

Since both sides of this characteristic equation are polynomials in s, by equating the coefficients of the like powers of s on both sides, it is possible to determine the values of k_1, k_2, and k_3. This approach is convenient if $n = 2$ or 3. (For $n = 4, 5, 6, \ldots$, this approach may become very tedious.)

There are other approaches for the determination of the state feedback gain matrix **K**. In what follows, we shall present a well-known formula, known as Ackermann's formula, for the determination of the state feedback gain matrix **K**.

Ackermann's formula.   Consider the system given by Equation (10–1), rewritten thus:

$$\dot{\mathbf{x}} = \mathbf{Ax} + \mathbf{B}u$$

We assume that the system is completely state controllable. We also assume that the desired closed-loop poles are at $s = \mu_1, s = \mu_2, \ldots, s = \mu_n$.

Use of the state feedback control

$$u = -\mathbf{Kx}$$

modifies the system equation to

$$\dot{\mathbf{x}} = (\mathbf{A} - \mathbf{BK})\mathbf{x} \tag{10–14}$$

Let us define

$$\tilde{\mathbf{A}} = \mathbf{A} - \mathbf{BK}$$

The desired characteristic equation is

$$|s\mathbf{I} - \mathbf{A} + \mathbf{BK}| = |s\mathbf{I} - \tilde{\mathbf{A}}| = (s - \mu_1)(s - \mu_2) \cdots (s - \mu_n)$$

$$= s^n + \alpha_1 s^{n-1} + \cdots + \alpha_{n-1} s + \alpha_n = 0$$

Since the Cayley–Hamilton theorem states that $\tilde{\mathbf{A}}$ satisfies its own characteristic equation, we have

$$\phi(\tilde{\mathbf{A}}) = \tilde{\mathbf{A}}^n + \alpha_1 \tilde{\mathbf{A}}^{n-1} + \cdots + \alpha_{n-1}\tilde{\mathbf{A}} + \alpha_n \mathbf{I} = \mathbf{0} \qquad (10\text{--}15)$$

We shall utilize Equation (10–15) to derive Ackermann's formula. To simplify the derivation, we consider the case where $n = 3$. (For any other positive integer n, the following derivation can be easily extended.)

Consider the following identities:

$$\mathbf{I} = \mathbf{I}$$

$$\tilde{\mathbf{A}} = \mathbf{A} - \mathbf{BK}$$

$$\tilde{\mathbf{A}}^2 = (\mathbf{A} - \mathbf{BK})^2 = \mathbf{A}^2 - \mathbf{ABK} - \mathbf{BK}\tilde{\mathbf{A}}$$

$$\tilde{\mathbf{A}}^3 = (\mathbf{A} - \mathbf{BK})^3 = \mathbf{A}^3 - \mathbf{A}^2\mathbf{BK} - \mathbf{ABK}\tilde{\mathbf{A}} - \mathbf{BK}\tilde{\mathbf{A}}^2$$

Multiplying the preceding equations in order by $\alpha_3, \alpha_2, \alpha_1, \alpha_0$ (where $\alpha_0 = 1$), respectively, and adding the results, we obtain

$$\alpha_3 \mathbf{I} + \alpha_2 \tilde{\mathbf{A}} + \alpha_1 \tilde{\mathbf{A}}^2 + \tilde{\mathbf{A}}^3$$

$$= \alpha_3 \mathbf{I} + \alpha_2 (\mathbf{A} - \mathbf{BK}) + \alpha_1 (\mathbf{A}^2 - \mathbf{ABK} - \mathbf{BK}\tilde{\mathbf{A}}) + \mathbf{A}^3 - \mathbf{A}^2\mathbf{BK}$$

$$- \mathbf{ABK}\tilde{\mathbf{A}} - \mathbf{BK}\tilde{\mathbf{A}}^2$$

$$= \alpha_3 \mathbf{I} + \alpha_2 \mathbf{A} + \alpha_1 \mathbf{A}^2 + \mathbf{A}^3 - \alpha_2 \mathbf{BK} - \alpha_1 \mathbf{ABK} - \alpha_1 \mathbf{BK}\tilde{\mathbf{A}} - \mathbf{A}^2\mathbf{BK}$$

$$- \mathbf{ABK}\tilde{\mathbf{A}} - \mathbf{BK}\tilde{\mathbf{A}}^2 \qquad (10\text{--}16)$$

Referring to Equation (10–15), we have

$$\alpha_3 \mathbf{I} + \alpha_2 \tilde{\mathbf{A}} + \alpha_1 \tilde{\mathbf{A}}^2 + \tilde{\mathbf{A}}^3 = \phi(\tilde{\mathbf{A}}) = \mathbf{0}$$

Also, we have

$$\alpha_3 \mathbf{I} + \alpha_2 \mathbf{A} + \alpha_1 \mathbf{A}^2 + \mathbf{A}^3 = \phi(\mathbf{A}) \neq \mathbf{0}$$

Substituting the last two equations into Equation (10–16), we have

$$\phi(\tilde{\mathbf{A}}) = \phi(\mathbf{A}) - \alpha_2 \mathbf{BK} - \alpha_1 \mathbf{BK}\tilde{\mathbf{A}} - \mathbf{BK}\tilde{\mathbf{A}}^2 - \alpha_1 \mathbf{ABK} - \mathbf{ABK}\tilde{\mathbf{A}} - \mathbf{A}^2\mathbf{BK}$$

Since $\phi(\tilde{\mathbf{A}}) = \mathbf{0}$, we obtain

$$\phi(\mathbf{A}) = \mathbf{B}(\alpha_2\mathbf{K} + \alpha_1\mathbf{K}\tilde{\mathbf{A}} + \mathbf{K}\tilde{\mathbf{A}}^2) + \mathbf{AB}(\alpha_1\mathbf{K} + \mathbf{K}\tilde{\mathbf{A}}) + \mathbf{A}^2\mathbf{BK}$$

$$= [\mathbf{B} \;\vdots\; \mathbf{AB} \;\vdots\; \mathbf{A}^2\mathbf{B}] \begin{bmatrix} \alpha_2\mathbf{K} + \alpha_1\mathbf{K}\tilde{\mathbf{A}} + \mathbf{K}\tilde{\mathbf{A}}^2 \\ \alpha_1\mathbf{K} + \mathbf{K}\tilde{\mathbf{A}} \\ \mathbf{K} \end{bmatrix} \qquad (10\text{--}17)$$

Since the system is completely state controllable, the inverse of the controllability matrix

$$[\mathbf{B} \;\vdots\; \mathbf{AB} \;\vdots\; \mathbf{A}^2\mathbf{B}]$$

exists. Premultiplying the inverse of the controllability matrix to both sides of Equation (10–17), we obtain

$$[\mathbf{B} \;\vdots\; \mathbf{AB} \;\vdots\; \mathbf{A}^2\mathbf{B}]^{-1}\phi(\mathbf{A}) = \begin{bmatrix} \alpha_2\mathbf{K} + \alpha_1\mathbf{K}\tilde{\mathbf{A}} + \mathbf{K}\tilde{\mathbf{A}}^2 \\ \alpha_1\mathbf{K} + \mathbf{K}\tilde{\mathbf{A}} \\ \mathbf{K} \end{bmatrix}$$

Premultiplying both sides of this last equation by $[0 \quad 0 \quad 1]$, we obtain

$$[0 \quad 0 \quad 1][\mathbf{B} \;\vdots\; \mathbf{AB} \;\vdots\; \mathbf{A}^2\mathbf{B}]^{-1}\phi(\mathbf{A}) = [0 \quad 0 \quad 1] \begin{bmatrix} \alpha_2\mathbf{K} + \alpha_1\mathbf{K}\tilde{\mathbf{A}} + \mathbf{K}\tilde{\mathbf{A}}^2 \\ \alpha_1\mathbf{K} + \mathbf{K}\tilde{\mathbf{A}} \\ \mathbf{K} \end{bmatrix} = \mathbf{K}$$

which can be rewritten as

$$\mathbf{K} = [0 \quad 0 \quad 1][\mathbf{B} \;\vdots\; \mathbf{AB} \;\vdots\; \mathbf{A}^2\mathbf{B}]^{-1}\phi(\mathbf{A})$$

This last equation gives the required state feedback gain matrix $\mathbf{K}$.

For an arbitrary positive integer n, we have

$$\mathbf{K} = [0 \quad 0 \quad \cdots \quad 0 \quad 1][\mathbf{B} \;\vdots\; \mathbf{AB} \;\vdots\; \cdots \;\vdots\; \mathbf{A}^{n-1}\mathbf{B}]^{-1}\phi(\mathbf{A}) \qquad (10\text{--}18)$$

Equation (10–18) is known as Ackermann's formula for the determination of the state feedback gain matrix $\mathbf{K}$.

EXAMPLE 10–1   Consider the system defined by

$$\dot{\mathbf{x}} = \mathbf{Ax} + \mathbf{B}u$$

where

$$\mathbf{A} = \begin{bmatrix} 0 & 1 \\ 20.6 & 0 \end{bmatrix}, \qquad \mathbf{B} = \begin{bmatrix} 0 \\ 1 \end{bmatrix}$$

The characteristic equation for the system is

$$|s\mathbf{I} - \mathbf{A}| = \begin{vmatrix} s & -1 \\ -20.6 & s \end{vmatrix} = s^2 - 20.6 = 0$$

Since the characteristic roots are $s = \pm 4.539$, the system is unstable. By using the state feedback control $u = -\mathbf{Kx}$, it is desired to have the closed-loop poles at $s = -1.8 \pm j2.4$ (that is, the eigenvalues of $\mathbf{A} - \mathbf{BK}$ to be $\mu_1 = -1.8 + j2.4$ and $\mu_2 = -1.8 - j2.4$). Determine the state feedback gain matrix $\mathbf{K}$.

We must first check the rank of the controllability matrix:

$$\mathbf{M} = [\mathbf{B} \mid \mathbf{AB}] = \begin{bmatrix} 0 & 1 \\ 1 & 0 \end{bmatrix}$$

Since the rank of matrix $\mathbf{M}$ is 2, arbitrary pole placement is possible.

We shall now solve this problem by using three methods.

Method 1: The first method is to use Equation (10–13). Noting that the given state equation is in the controllable canonical form, the transformation matrix $\mathbf{T}$ is a unity matrix, or $\mathbf{T} = \mathbf{I}$.

From the characteristic equation for the original system, we find

$$a_1 = 0, \qquad a_2 = -20.6$$

The desired characteristic polynomial is

$$(s - \mu_1)(s - \mu_2) = (s + 1.8 - j2.4)(s + 1.8 + j2.4)$$

$$= s^2 + 3.6s + 9 = s^2 + \alpha_1 s + \alpha_2$$

Hence

$$\alpha_1 = 3.6, \qquad \alpha_2 = 9$$

Referring to Equation (10–13) and noting that $\mathbf{T} = \mathbf{I}$, we have

$$\mathbf{K} = [\alpha_2 - a_2 \mid \alpha_1 - a_1]\mathbf{T}^{-1}$$

$$= [9 + 20.6 \mid 3.6 - 0]\mathbf{I}^{-1}$$

$$= [29.6 \quad 3.6]$$

Method 2: The second method is to use direct substitution of matrix $\mathbf{K} = [k_1 \quad k_2]$ into the desired characteristic polynomial. The characteristic polynomial for the desired system is

$$|s\mathbf{I} - \mathbf{A} + \mathbf{BK}| = \left| \begin{bmatrix} s & 0 \\ 0 & s \end{bmatrix} - \begin{bmatrix} 0 & 1 \\ 20.6 & 0 \end{bmatrix} + \begin{bmatrix} 0 \\ 1 \end{bmatrix} [k_1 \quad k_2] \right|$$

$$= \left| \begin{matrix} s & -1 \\ -20.6 + k_1 & s + k_2 \end{matrix} \right|$$

$$= s^2 + k_2 s - 20.6 + k_1$$

This characteristic polynomial must be equal to

$$(s + \mu_1)(s + \mu_2) = (s + 1.8 - j2.4)(s + 1.8 + j2.4)$$

$$= s^2 + 3.6s + 9$$

By equating the coefficients of the terms of the like powers of s, we obtain

$$k_1 = 29.6, \qquad k_2 = 3.6$$

or

$$\mathbf{K} = [k_1 \quad k_2] = [29.6 \quad 3.6]$$

Method 3: The third method is to use Ackermann's formula given by Equation (10–18). Since the desired characteristic polynomial is

$$|s\mathbf{I} - (\mathbf{A} - \mathbf{BK})| = |s\mathbf{I} - \tilde{\mathbf{A}}| = s^2 + 3.6s + 9 = \phi(s)$$

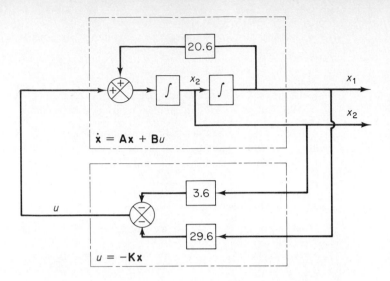

Figure 10–2
Block diagram of the system with state feedback.

we have

$$\phi(\mathbf{A}) = \mathbf{A}^2 + 3.6\mathbf{A} + 9\mathbf{I}$$

$$= \begin{bmatrix} 0 & 1 \\ 20.6 & 0 \end{bmatrix}\begin{bmatrix} 0 & 1 \\ 20.6 & 0 \end{bmatrix} + 3.6\begin{bmatrix} 0 & 1 \\ 20.6 & 0 \end{bmatrix} + 9\begin{bmatrix} 1 & 0 \\ 0 & 1 \end{bmatrix}$$

$$= \begin{bmatrix} 29.6 & 3.6 \\ 74.16 & 29.6 \end{bmatrix}$$

Thus

$$\mathbf{K} = [0 \quad 1][\mathbf{B} \ \vdots \ \mathbf{AB}]^{-1}\phi(\mathbf{A})$$

$$= [0 \quad 1]\begin{bmatrix} 0 & 1 \\ 1 & 0 \end{bmatrix}^{-1}\begin{bmatrix} 29.6 & 3.6 \\ 74.16 & 29.6 \end{bmatrix} = [29.6 \quad 3.6]$$

As a matter of course, the feedback gain matrices **K** obtained by the three methods are the same. With this state feedback, the closed-loop poles are located at $s = -1.8 \pm j2.4$, as desired. The damping ratio ζ of the closed-loop poles is 0.6, and the undamped natural frequency ω_n is 3 rad/sec. (The original unstable system is thus stabilized.) Figure 10–2 shows a block diagram of the system with state feedback.

It is noted that if the order n of the system were 4 or higher, methods 1 and 3 are recommended, since all matrix computations can be carried out by a computer. If method 2 is used, hand computations become necessary because a computer will not handle the characteristic equation with unknown parameters $k_1, k_2, \ldots, k_n$.

Comments. It is important to note that matrix **K** is not unique for a given system, but depends on the desired closed-loop pole locations (which determine the speed and damping of the response) selected. Note that the selection of the desired closed-loop poles or the desired characteristic equation is a compromise between the rapidity of the response of the error vector and the sensitivity to disturbances and measurement noises. That is, if we increase the speed of error response, then the adverse effects of disturbances and measurement noises generally increase. If the system is of second order, then the system dynamics (response characteristics) can be precisely correlated to the location of the desired closed-loop poles and the zero(s) of the plant. For higher-order systems, the location of the closed-loop poles and the system dynamics (response characteristics) are not easily correlated. Hence, in determining the state feedback gain matrix **K** for a given system, it is desirable to examine by computer simulations the response characteristics of the system for several different matrices **K** (based on several different desired characteristic equations) and to choose the one that gives the best overall system performance.

EXAMPLE 10–2   Consider the inverted pendulum system shown in Fig. 10–3, where an inverted pendulum is mounted on a motor-driven cart. Here we consider only the two-dimensional problem where the pendulum moves only in the plane of the paper. The inverted pendulum is unstable in that it may fall over any time unless a suitable control force is applied. Assume that the pendulum mass is concentrated at the end of the rod as shown in the figure. (We assume that the rod is massless.) The control force u is applied to the cart.

In the diagram, θ is the angle of the rod from the vertical line. We assume that angle θ is small so that we may approximate $\sin \theta$ by θ, $\cos \theta$ by 1, and also assume that $\dot{\theta}$ is small so that $\theta\dot{\theta}^2 \doteq 0$. (Under these conditions, the system's nonlinear equations can be linearized.)

It is desired to keep the pendulum upright in the presence of disturbances (such as a gust of wind acting on the mass m, an unexpected force applied to the cart). The slanted pendulum can be brought back to the vertical position when appropriate control force u is applied to the cart. At the end of each control process, it is desired to bring the cart back to $x = 0$, the reference position.

Design a control system such that given any initial conditions (caused by disturbances) the pendulum can be brought back to the vertical position and also the cart can be brought back to the reference position ($x = 0$) quickly (for example, with settling time of about 2 sec) with reasonable damping (for example, equivalent to $\zeta = 0.5$ in the standard second-order system). Assume the following numerical values for M, m, and l:

$$M = 2 \text{ kg}, \qquad m = 0.1 \text{ kg}, \qquad l = 0.5 \text{ m}$$

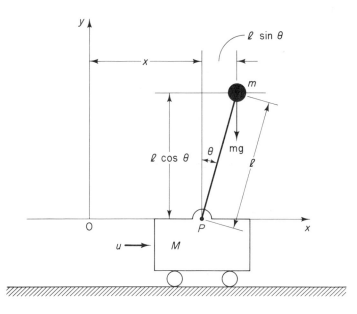

Figure 10–3
Inverted pendulum system.

The equations of motion for this system were derived in Section 2–3. For small angle θ, the equations of motion were given by Equations (2–17) and (2–18), rewritten thus:

$$(M + m)\ddot{x} + ml\ddot{\theta} = u \tag{10-19}$$

$$m\ddot{x} + ml\ddot{\theta} = mg\theta \tag{10-20}$$

By subtracting Equation (10–20) from Equation (10–19), we obtain

$$M\ddot{x} = u - mg\theta \tag{10-21}$$

By eliminating $\ddot{x}$ from Equations (10–19) and (10–21), we have

$$Ml\ddot{\theta} - (M + m)g\theta = -u$$

from which we obtain the plant transfer function

$$\frac{\Theta(s)}{-U(s)} = \frac{1}{Mls^2 - (M + m)g}$$

By substituting the given numerical values and noting that $g = 9.81$ m/sec^2, we have

$$\frac{\Theta(s)}{-U(s)} = \frac{1}{s^2 - 20.601} = \frac{1}{s^2 - (4.539)^2}$$

The inverted pendulum plant has one pole on the negative real axis ($s = -4.539$) and another on the positive real axis ($s = 4.539$). Hence, the plant is open-loop unstable.

In this example problem, we shall use the pole placement technique to stabilize the system and to have the desired dynamic characteristics.

The state-space equations for this system were derived in Example 2–3, rewritten thus:

$$\begin{bmatrix} \dot{x}_1 \\ \dot{x}_2 \\ \dot{x}_3 \\ \dot{x}_4 \end{bmatrix} = \begin{bmatrix} 0 & 1 & 0 & 0 \\ \dfrac{M+m}{Ml}g & 0 & 0 & 0 \\ 0 & 0 & 0 & 1 \\ -\dfrac{m}{M}g & 0 & 0 & 0 \end{bmatrix} \begin{bmatrix} x_1 \\ x_2 \\ x_3 \\ x_4 \end{bmatrix} + \begin{bmatrix} 0 \\ -\dfrac{1}{Ml} \\ 0 \\ \dfrac{1}{M} \end{bmatrix} u \tag{10-22}$$

$$\begin{bmatrix} y_1 \\ y_2 \end{bmatrix} = \begin{bmatrix} 1 & 0 & 0 & 0 \\ 0 & 0 & 1 & 0 \end{bmatrix} \begin{bmatrix} x_1 \\ x_2 \\ x_3 \\ x_4 \end{bmatrix} \tag{10-23}$$

where

$$x_1 = \theta$$

$$x_2 = \dot{\theta}$$

$$x_3 = x$$

$$x_4 = \dot{x}$$

By substituting the given numerical values for M, m, and l, we have

$$\frac{M+m}{Ml}g = 20.601, \qquad \frac{m}{M}g = 0.4905, \qquad \frac{1}{Ml} = 1, \qquad \frac{1}{M} = 0.5$$

Using these numerical values, Equations (10–22) and (10–23) can be rewritten as

$$\dot{\mathbf{x}} = \mathbf{A}\mathbf{x} + \mathbf{B}u$$

$$\mathbf{y} = \mathbf{C}\mathbf{x}$$

where

$$\mathbf{A} = \begin{bmatrix} 0 & 1 & 0 & 0 \\ 20.601 & 0 & 0 & 0 \\ 0 & 0 & 0 & 1 \\ -0.4905 & 0 & 0 & 0 \end{bmatrix}, \quad \mathbf{B} = \begin{bmatrix} 0 \\ -1 \\ 0 \\ 0.5 \end{bmatrix}, \quad \mathbf{C} = \begin{bmatrix} 1 & 0 & 0 & 0 \\ 0 & 0 & 1 & 0 \end{bmatrix}$$

We shall use the state feedback control scheme

$$u = -\mathbf{K}\mathbf{x}$$

Let us check if the system is completely state controllable. Since the rank of

$$\mathbf{M} = [\mathbf{B} \mid \mathbf{AB} \mid \mathbf{A^2B} \mid \mathbf{A^3B}] = \begin{bmatrix} 0 & -1 & 0 & -20.601 \\ -1 & 0 & -20.601 & 0 \\ 0 & 0.5 & 0 & 0.4905 \\ 0.5 & 0 & 0.4905 & 0 \end{bmatrix}$$

is 4, the system is completely state controllable.

The characteristic equation of the system is

$$|s\mathbf{I} - \mathbf{A}| = \begin{bmatrix} s & -1 & 0 & 0 \\ -20.601 & s & 0 & 0 \\ 0 & 0 & s & -1 \\ 0.4905 & 0 & 0 & s \end{bmatrix}$$

$$= s^4 - 20.601s^2$$
$$= s^4 + a_1 s^3 + a_2 s^2 + a_3 s + a_4 = 0$$

Therefore,

$$a_1 = 0, \quad a_2 = -20.601, \quad a_3 = 0, \quad a_4 = 0$$

Next we must choose the desired closed-loop pole locations. Since we require the system with reasonably small settling time (about 2 sec) and with reasonable damping (equivalent to $\zeta = 0.5$ in the standard second-order system), let us choose the desired closed-loop poles at $s = \mu_i$ $(i = 1, 2, 3, 4)$, where

$$\mu_1 = -2 + j3.464, \quad \mu_2 = -2 - j3.464, \quad \mu_3 = -10, \quad \mu_4 = -10$$

(In this case, μ_1 and μ_2 are a pair of dominant closed-loop poles with $\zeta = 0.5$ and $\omega_n = 4$. The remaining two closed-loop poles, μ_3 and μ_4, are located far to the left of the dominant pair of closed-loop poles and, therefore, the effect on the response of μ_3 and μ_4 is small. So the speed and damping requirements will be satisfied.) The desired characteristic equation becomes

$$(s - \mu_1)(s - \mu_2)(s - \mu_3)(s - \mu_4) = (s + 2 - j3.464)(s + 2 + j3.464)(s + 10)(s + 10)$$
$$= (s^2 + 4s + 16)(s^2 + 20s + 100)$$
$$= s^4 + 24s^3 + 196s^2 + 720s + 1600$$
$$= s^4 + \alpha_1 s^3 + \alpha_2 s^2 + \alpha_3 s + \alpha_4 = 0$$

Consequently, we have

$$\alpha_1 = 24, \quad \alpha_2 = 196, \quad \alpha_3 = 720, \quad \alpha_4 = 1600$$

Let us determine the state feedback gain matrix $\mathbf{K}$ by use of Equation (10–13), rewritten thus:

$$\mathbf{K} = [\alpha_4 - a_4 \ \vdots \ \alpha_3 - a_3 \ \vdots \ \alpha_2 - a_2 \ \vdots \ \alpha_1 - a_1] \, \mathbf{T}^{-1}$$

where matrix $\mathbf{T}$ is given by Equation (10–4), or

$$\mathbf{T} = \mathbf{MW}$$

and $\mathbf{M}$ and $\mathbf{W}$ are given by Equations (10–5) and (10–6), respectively. Thus,

$$\mathbf{M} = [\mathbf{B} \ \vdots \ \mathbf{AB} \ \vdots \ \mathbf{A^2B} \ \vdots \ \mathbf{A^3B}] = \begin{bmatrix} 0 & -1 & 0 & -20.601 \\ -1 & 0 & -20.601 & 0 \\ 0 & 0.5 & 0 & 0.4905 \\ 0.5 & 0 & 0.4905 & 0 \end{bmatrix}$$

$$\mathbf{W} = \begin{bmatrix} a_3 & a_2 & a_1 & 1 \\ a_2 & a_1 & 1 & 0 \\ a_1 & 1 & 0 & 0 \\ 1 & 0 & 0 & 0 \end{bmatrix} = \begin{bmatrix} 0 & -20.601 & 0 & 1 \\ -20.601 & 0 & 1 & 0 \\ 0 & 1 & 0 & 0 \\ 1 & 0 & 0 & 0 \end{bmatrix}$$

Then matrix $\mathbf{T}$ becomes

$$\mathbf{T} = \mathbf{MW} = \begin{bmatrix} 0 & 0 & -1 & 0 \\ 0 & 0 & 0 & -1 \\ -9.81 & 0 & 0.5 & 0 \\ 0 & -9.81 & 0 & 0.5 \end{bmatrix}$$

Hence

$$\mathbf{T}^{-1} = \begin{bmatrix} -\dfrac{0.5}{9.81} & 0 & -\dfrac{1}{9.81} & 0 \\ 0 & -\dfrac{0.5}{9.81} & 0 & -\dfrac{1}{9.81} \\ -1 & 0 & 0 & 0 \\ 0 & -1 & 0 & 0 \end{bmatrix}$$

The desired feedback gain matrix $\mathbf{K}$ becomes

$$\mathbf{K} = [\alpha_4 - a_4 \ \vdots \ \alpha_3 - a_3 \ \vdots \ \alpha_2 - a_2 \ \vdots \ \alpha_1 - a_1] \, \mathbf{T}^{-1}$$

$$= [1600 - 0 \ \vdots \ 720 - 0 \ \vdots \ 196 + 20.601 \ \vdots \ 24 - 0] \, \mathbf{T}^{-1}$$

$$= [1600 \quad 720 \quad 216.601 \quad 24] \begin{bmatrix} -\dfrac{0.5}{9.81} & 0 & -\dfrac{1}{9.81} & 0 \\ 0 & -\dfrac{0.5}{9.81} & 0 & -\dfrac{1}{9.81} \\ -1 & 0 & 0 & 0 \\ 0 & -1 & 0 & 0 \end{bmatrix}$$

$$= [-298.15 \quad -60.697 \quad -163.099 \quad -73.394]$$

The control signal u is given by

$$u = -\mathbf{Kx} = 298.15x_1 + 60.697x_2 + 163.099x_3 + 73.394x_4$$

Note that this system is a regulator system. The desired angle θ_d is always zero, and the desired location x_d of the cart is also always zero. Thus, the reference inputs are zeros. (We shall consider in Section 10–4 the case where the cart is to move according to the reference input.) Figure 10–4 shows

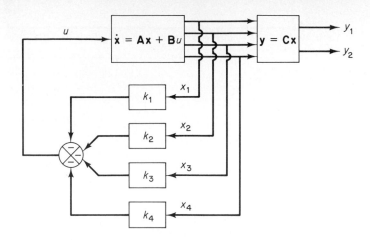

Figure 10–4
Inverted pendulum
system with state
feedback control.

the state feedback control scheme for the inverted-pendulum system. (Since the reference inputs are always zero in this system, they are not shown in the diagram.)

Once the state feedback gain matrix **K** is determined, the system performance must be examined by computer simulations. To simulate the system dynamics on the computer and to obtain responses to any initial conditions, we proceed as follows: The basic equations for the system are the state equation

$$\dot{\mathbf{x}} = \mathbf{Ax} + \mathbf{B}u$$

and the control equation

$$u = -\mathbf{Kx}$$

When the control equation is substituted into the state equation, we obtain

$$\dot{\mathbf{x}} = (\mathbf{A} - \mathbf{BK})\mathbf{x}$$

which, when the numerical values are substituted, can be given by

$$
\begin{bmatrix} \dot{x}_1 \\ \dot{x}_2 \\ \dot{x}_3 \\ \dot{x}_4 \end{bmatrix} =
\begin{bmatrix}
0 & 1 & 0 & 0 \\
-277.549 & -60.697 & -163.099 & -73.394 \\
0 & 0 & 0 & 1 \\
148.5845 & 30.3485 & 81.5495 & 36.697
\end{bmatrix}
\begin{bmatrix} x_1 \\ x_2 \\ x_3 \\ x_4 \end{bmatrix}
\tag{10-24}
$$

This state equation, which consists of four first-order differential equations, must be solved by a computer. A BASIC computer program for solving Equation (10–24) using the fourth-order Runge–Kutta method is given in Table 10–1. [In the program, the initial conditions are shown to be $x_1(0) = 0.1$ rad, $x_2(0) = 0$, $x_3(0) = 0$, and $x_4(0) = 0$.]

Figure 10–5 depicts response curves showing how the inverted-pendulum system comes back to the reference position ($\theta = 0$, $x = 0$), given arbitrary initial conditions. Specifically, Figure 10–5(a) shows the response curves when $\theta(0) = 0.1$ rad, $\dot{\theta}(0) = 0$, $x(0) = 0$, and $\dot{x}(0) = 0$. Figure 10–5(b) shows the response curves when $\theta(0) = 0.2$ rad, $\dot{\theta}(0) = 0$, $x(0) = 0.2$ m, and $\dot{x}(0) = 0$.

It is important to note that such response curves depend on the desired characteristic equation (that is, the desired closed-loop poles). For different desired characteristic equations, the response curves (for the same initial conditions) are different. To elaborate this point, let us consider the case where the desired closed-loop poles are at $s = \hat{\mu}_i$ ($i = 1, 2, 3, 4$):

$$\hat{\mu}_1 = -1 + j1.732, \qquad \hat{\mu}_2 = -1 - j1.732, \qquad \hat{\mu}_3 = -5, \qquad \hat{\mu}_4 = -5$$

Since the dominant closed-loop poles at $s = \hat{\mu}_1$ and $\hat{\mu}_2$ have $\zeta = 0.5$ and $\omega_n = 2$, and closed-loop poles at $s = \hat{\mu}_3$ and $\hat{\mu}_4$ are located five times farther to the left of the dominant closed-loop poles

Table 10–1 BASIC Computer Program for Solving Equation (10–24) with Initial Conditions $x_1(0) = 0.1$ rad, $x_2(0) = 0$, $x_3(0) = 0$, $x_4(0) = 0$

```
10   ORDER = 4
20   X(1)  = .1
30   X(2)  = 0
40   X(3)  = 0
50   X(4)  = 0
60   H = .01
70   T = 0
80   TK = 0
90   TF = 4
100  OPEN "O", #1, "ANS1.BAS"
110  OPEN "O", #2, "ANS2.BAS"
120  OPEN "O", #3, "ANS3.BAS"
130  OPEN "O", #4, "ANS4.BAS"
140  PRINT "      TIME          X(1)          X(2)          X(3)          X(4)          "
150  PRINT "----------------------------------------------------------------------------"
160  PRINT #1, USING "####.######"; X(1)
170  PRINT #2, USING "####.######"; X(2)
180  PRINT #3, USING "####.######"; X(3)
190  PRINT #4, USING "####.######"; X(4)
200  PRINT USING "####.######"; T, X(1), X(2), X(3), X(4)
210  IF T > TF THEN GOTO 5000
220  GOSUB 1000
230  GOTO 160
1000 TK = T
1010 GOSUB 2000
1020 FOR I = 1 TO ORDER
1030 XK(I) = X(I)
1040 K(1,I) = DX(I)
1050 T = TK + H/2
1060 X(I) = XK(I) + (H/2)*K(1,I)
1070 NEXT I
1080 GOSUB 2000
1090 FOR I = 1 TO ORDER
1100 K(2,I) = DX(I)
1110 X(I) = XK(I) + (H/2)*K(2,I)
1120 NEXT I
1130 GOSUB 2000
1140 FOR I = 1 TO ORDER
1150 K(3,I) = DX(I)
1160 T = TK + H
1170 X(I) = XK(I) + H*K(3,I)
1180 NEXT I
1190 GOSUB 2000
1200 FOR I = 1 TO ORDER
1210 K(4,I) = DX(I)
1220 X(I) = XK(I) + (H/6)*(K(1,I) + 2*K(2,I) + 2*K(3,I) + K(4,I))
1230 NEXT I
1240 RETURN
2000 DX(1) = X(2)
2010 DX(2) = - 277.549*X(1) - 60.697*X(2) - 163.099*X(3) - 73.394*X(4)
2020 DX(3) = X(4)
2030 DX(4) = 148.5845*X(1) + 30.3485*X(2) + 81.5495*X(3) + 36.697*X(4)
2040 RETURN
4900 CLOSE #1
4910 CLOSE #2
4920 CLOSE #3
4930 CLOSE #4
5000 END
```

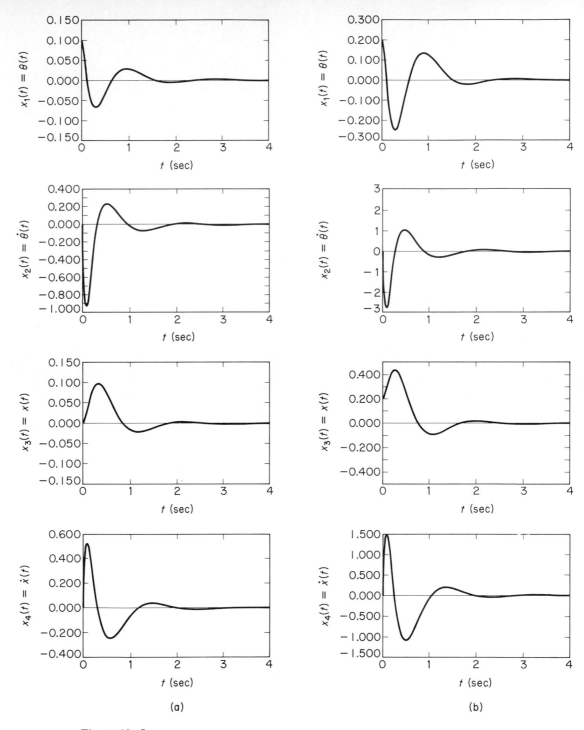

Figure 10–5

Response curves for the inverted pendulum system where the feedback gain matrix **K** is given by
K = [− 298.15   − 60.697   − 163.099   − 73.394]. (a) Initial conditions are $\theta(0)$ = 0.1 rad,
$\dot{\theta}(0)$ = 0, $x(0)$ = 0, $\dot{x}(0)$ = 0; (b) initial conditions are $\theta(0)$ = 0.2 rad, $\dot{\theta}(0)$ = 0, $x(0)$ = 0.2 m,
$\dot{x}(0)$ = 0.

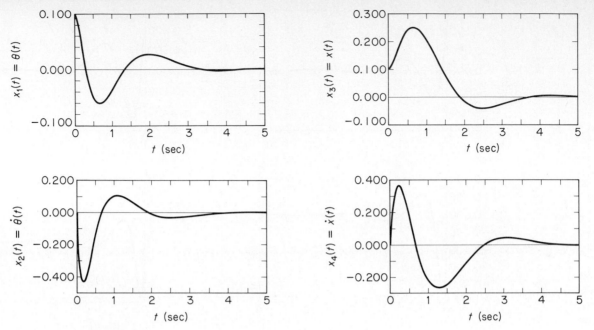

Figure 10–6
Response curves for the inverted pendulum system where the feedback gain matrix **K** is given by
K $= [-74.698 \quad -16.587 \quad -10.194 \quad -9.1743]$. The initial conditions are $\theta(0) = 0.1$ rad,
$\dot{\theta}(0) = 0$, $x(0) = 0.1$ m, $\dot{x}(0) = 0$.

measured from the $j\omega$ axis, this system will behave similarly to the standard second-order system with
$\zeta = 0.5$ and $\omega_n = 2$. For this set of the desired closed-loop poles, the corresponding desired characteristic
equation becomes

$$(s - \hat{\mu}_1)(s - \hat{\mu}_2)(s - \hat{\mu}_3)(s - \hat{\mu}_4) = (s + 1 - j1.732)(s + 1 + j1.732)(s + 5)(s + 5)$$

$$= (s^2 + 2s + 4)(s^2 + 10s + 25)$$

$$= s^4 + 12s^3 + 49s^2 + 90s + 100$$

$$= s^4 + \hat{\alpha}_1 s^3 + \hat{\alpha}_2 s^2 + \hat{\alpha}_3 s + \hat{\alpha}_4 = 0$$

Hence,

$$\hat{\alpha}_1 = 12, \qquad \hat{\alpha}_2 = 49, \qquad \hat{\alpha}_3 = 90, \qquad \hat{\alpha}_4 = 100$$

The state feedback gain matrix **K** for this desired characteristic equation is given by

$$\mathbf{K} = [\hat{\alpha}_4 - a_4 \; \vdots \; \hat{\alpha}_3 - a_3 \; \vdots \; \hat{\alpha}_2 - a_2 \; \vdots \; \hat{\alpha}_1 - a_1]\mathbf{T}^{-1}$$

$$= [100 - 0 \; \vdots \; 90 - 0 \; \vdots \; 49 + 20.601 \; \vdots \; 12 - 0]\mathbf{T}^{-1}$$

$$= [100 \quad 90 \quad 69.601 \quad 12] \begin{bmatrix} -\dfrac{0.5}{9.81} & 0 & -\dfrac{1}{9.81} & 0 \\ 0 & -\dfrac{0.5}{9.81} & 0 & -\dfrac{1}{9.81} \\ -1 & 0 & 0 & 0 \\ 0 & -1 & 0 & 0 \end{bmatrix}$$

$$= [-74.698 \quad -16.587 \quad -10.194 \quad -9.1743]$$

The control signal u in this case is

$$u = -\mathbf{K}\mathbf{x} = 74.698x_1 + 16.587x_2 + 10.194x_3 + 9.1743x_4$$

Figure 10–6 shows the response curves when $\theta(0) = 0.1$ rad, $\dot{\theta}(0) = 0$, $x(0) = 0.1$ m, and $\dot{x}(0) = 0$.

Notice that for the system with the first set of desired closed-loop poles ($\mu_1 = -2 + j3.464$, $\mu_2 = -2 - j3.464$, $\mu_3 = -10$, $\mu_4 = -10$) the speed of response is approximately twice as fast as that of the system with the second set of desired closed-loop poles ($\hat{\mu}_1 = -1 + j1.732$, $\hat{\mu}_2 = -1 - j1.732$, $\hat{\mu}_3 = -5$, $\hat{\mu}_4 = -5$). The damping is about the same for both systems.

However, the first system requires a larger control signal than the second one. In designing such a system as this, it is desired that the designer examine several different sets of desired closed-loop poles and determine the corresponding $\mathbf{K}$ matrices. After making computer simulations of the system and examining response curves, choose the matrix $\mathbf{K}$ that gives the "best overall" system performance. The criterion for the best overall system performance depends on the particular situation, including economic considerations.

10–3 DESIGN OF STATE OBSERVERS

In Section 10–2, where we presented the pole placement approach to the design of control systems, we assumed that all state variables are available for feedback. In practice, however, not all state variables are available for feedback. Then we need to estimate unavailable state variables. It is important to note that we should avoid differentiating a state variable to generate another one. Differentiation of a signal always decreases the signal-to-noise ratio because noise generally fluctuates more rapidly than the command signal. Sometimes the signal-to-noise ratio may be decreased by several times by a single differentiation process. There are methods available to estimate unmeasurable state variables without a differentiation process. Estimation of unmeasurable state variables is commonly called *observation*. A device (or a computer program) that estimates or observes the state variables is called a *state observer*, or simply an *observer*. If the state observer observes all state variables of the system, regardless of whether some state variables are available for direct measurement, it is called a *full-order state observer*. There are times when this will not be necessary, when we will need observation of only the unmeasurable state variables but not of those that are directly measurable as well. For example, since the output variables are observable and they are linearly related to the state variables, we need not observe all state variables, but observe only $n - m$ state variables, where n is the dimension of the state vector and m is the dimension of the output vector. The state observer that observes only the minimum number of state variables is called a *minimum-order state observer* or, simply, *minimum-order observer*. In this section, we shall discuss both the full-order state observer and the minimum-order state observer.

State observer. A state observer estimates the state variables based on the measurements of the output and control variables. Here the concept of observability discussed in Section 9–5 plays an important role. As we shall see later, state observers can be designed if and only if the observability condition is satisfied.

In the following discussions of state observers, we shall use the notation $\tilde{\mathbf{x}}$ to designate the observed state vector. In many practical cases, the observed state vector $\tilde{\mathbf{x}}$ is used in the state feedback to generate the desired control vector.

Consider the system defined by

$$\dot{\mathbf{x}} = \mathbf{Ax} + \mathbf{Bu} \qquad (10\text{–}25)$$

$$\mathbf{y} = \mathbf{Cx} \qquad (10\text{–}26)$$

Assume that the state $\mathbf{x}$ is to be approximated by the state $\tilde{\mathbf{x}}$ of the dynamic model

$$\dot{\tilde{\mathbf{x}}} = \mathbf{A\tilde{x}} + \mathbf{Bu} + \mathbf{K}_e(\mathbf{y} - \mathbf{C\tilde{x}}) \qquad (10\text{–}27)$$

which represents the state observer. Notice that the state observer has $\mathbf{y}$ and $\mathbf{u}$ as inputs and $\tilde{\mathbf{x}}$ as output. The last term on the right side of this model equation, Equation (10–27), is a correction term that involves the difference between the measured output $\mathbf{y}$ and the estimated output $\mathbf{C\tilde{x}}$. Matrix $\mathbf{K}_e$ serves as a weighting matrix. The correction term monitors the state $\tilde{\mathbf{x}}$. In the presence of discrepancies between the $\mathbf{A}$ and $\mathbf{B}$ matrices used in the model and those of the actual system, the addition of the correction term will help reduce the effects due to the difference between the dynamic model and the actual system. Figure 10–7 shows the block diagram of the system and the full-order state observer.

In what follows we shall discuss details of the state observer where dynamics are characterized by $\mathbf{A}$ and $\mathbf{B}$ matrices and by the additional correction term, which involves the difference between the measured output and the estimated output. In the current discussions we assume that the $\mathbf{A}$ and $\mathbf{B}$ matrices used in the model and those of the actual system are the same.

Full-order state observer. The order of the state observer that will be discussed here is the same as that of the system. Assume that the system is defined by Equations (10–25) and (10–26) and the observer model is defined by Equation (10–27).

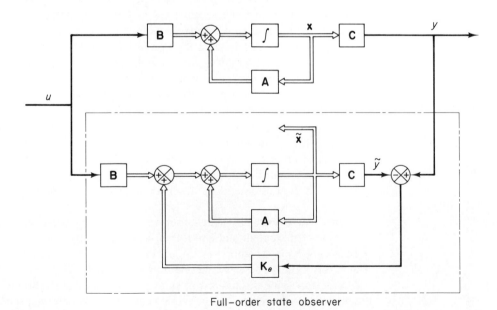

Full–order state observer

Figure 10–7
Block diagram of system and full-order state observer.

To obtain the observer error equation, let us subtract Equation (10–27) from Equation (10–25).

$$\dot{\mathbf{x}} - \dot{\tilde{\mathbf{x}}} = \mathbf{Ax} - \mathbf{A\tilde{x}} - \mathbf{K}_e(\mathbf{Cx} - \mathbf{C\tilde{x}})$$

$$= (\mathbf{A} - \mathbf{K}_e\mathbf{C})(\mathbf{x} - \tilde{\mathbf{x}}) \qquad (10\text{–}28)$$

Define the difference between $\mathbf{x}$ and $\tilde{\mathbf{x}}$ as the error vector $\mathbf{e}$, or

$$\mathbf{e} = \mathbf{x} - \tilde{\mathbf{x}}$$

Then, Equation (10–28) becomes

$$\dot{\mathbf{e}} = (\mathbf{A} - \mathbf{K}_e\mathbf{C})\mathbf{e} \qquad (10\text{–}29)$$

From Equation (10–29), we see that the dynamic behavior of the error vector is determined by the eigenvalues of matrix $\mathbf{A} - \mathbf{K}_e\mathbf{C}$. If matrix $\mathbf{A} - \mathbf{K}_e\mathbf{C}$ is a stable matrix, the error vector will converge to zero for any initial error vector $\mathbf{e}(0)$. That is, $\tilde{\mathbf{x}}(t)$ will converge to $\mathbf{x}(t)$ regardless of the values of $\mathbf{x}(0)$ and $\tilde{\mathbf{x}}(0)$. If the eigenvalues of matrix $\mathbf{A} - \mathbf{K}_e\mathbf{C}$ are chosen in such a way that the dynamic behavior of the error vector is asymptotically stable and is adequately fast, then any error vector will tend to zero (origin) with an adequate speed.

If the system is completely observable, then it can be proved that it is possible to choose matrix $\mathbf{K}_e$ such that $\mathbf{A} - \mathbf{K}_e\mathbf{C}$ has arbitrarily desired eigenvalues. That is, the observer gain matrix $\mathbf{K}_e$ can be determined to yield the desired matrix $\mathbf{A} - \mathbf{K}_e\mathbf{C}$. We shall discuss this matter in what follows.

Dual problem. The problem of designing a full-order observer becomes that of determining the observer gain matrix $\mathbf{K}_e$ such that the error dynamics defined by Equation (10–29) are asymptotically stable with sufficient speed of response. (The asymptotic stability and the speed of response of the error dynamics are determined by the eigenvalues of matrix $\mathbf{A} - \mathbf{K}_e\mathbf{C}$.) Hence the design of the full-order observer becomes that of determining an appropriate $\mathbf{K}_e$ such that $\mathbf{A} - \mathbf{K}_e\mathbf{C}$ has desired eigenvalues. Thus, the problem here becomes the same as the pole placement problem we discussed in Section 10–2.

Consider the system defined by

$$\dot{\mathbf{x}} = \mathbf{Ax} + \mathbf{B}u$$

$$y = \mathbf{Cx}$$

In designing the full-order state observer, we may solve the dual problem, that is, solve the pole placement problem for the dual system

$$\dot{\mathbf{z}} = \mathbf{A}^*\mathbf{z} + \mathbf{C}^*v$$

$$n = \mathbf{B}^*\mathbf{z}$$

assuming the control signal v to be

$$v = -\mathbf{Kz}$$

If the dual system is completely state controllable, then the state feedback gain matrix $\mathbf{K}$ can be determined such that matrix $\mathbf{A}^* - \mathbf{C}^*\mathbf{K}$ will yield a set of the desired eigenvalues.

If $\mu_1, \mu_2, \ldots, \mu_n$ are the desired eigenvalues of the state observer matrix, then by taking the same μ_i's as the desired eigenvalues of the state feedback gain matrix of the dual system we obtain

$$|s\mathbf{I} - (\mathbf{A}^* - \mathbf{C}^*\mathbf{K})| = (s - \mu_1)(s - \mu_2) \cdots (s - \mu_n)$$

Noting that the eigenvalues of $\mathbf{A}^* - \mathbf{C}^*\mathbf{K}$ and those of $\mathbf{A} - \mathbf{K}^*\mathbf{C}$ are the same, we have

$$|s\mathbf{I} - (\mathbf{A}^* - \mathbf{C}^*\mathbf{K})| = |s\mathbf{I} - (\mathbf{A} - \mathbf{K}^*\mathbf{C})|$$

Comparing the characteristic polynomial $|s\mathbf{I} - (\mathbf{A} - \mathbf{K}^*\mathbf{C})|$ and the characteristic polynomial $|s\mathbf{I} - (\mathbf{A} - \mathbf{K}_e\mathbf{C})|$ for the observer system, we find that $\mathbf{K}_e$ and $\mathbf{K}^*$ are related by

$$\mathbf{K}_e = \mathbf{K}^*$$

Thus, using the matrix $\mathbf{K}$ determined by the pole placement approach in the dual system, the observer gain matrix $\mathbf{K}_e$ for the original system can be determined by using the relationship $\mathbf{K}_e = \mathbf{K}^*$.

Necessary and sufficient condition for state observation.   As discussed above, a necessary and sufficient condition for the determination of the observer gain matrix $\mathbf{K}_e$ for the desired eigenvalues of $\mathbf{A} - \mathbf{K}_e\mathbf{C}$ is that the dual of the original system

$$\dot{\mathbf{z}} = \mathbf{A}^*\mathbf{z} + \mathbf{C}^*\mathbf{v}$$

be completely state controllable. The complete state controllability condition for this dual system is that the rank of

$$[\mathbf{C}^* \mid \mathbf{A}^*\mathbf{C}^* \mid \cdots \mid (\mathbf{A}^*)^{n-1}\mathbf{C}^*]$$

is n. This is the condition for complete observability of the original system defined by Equations (10–25) and (10–26). This means that a necessary and sufficient condition for the observation of the state of the system defined by Equations (10–25) and (10–26) is that the system be completely observable.

In what follows we shall present the direct approach (instead of the dual problem approach) to the solution of the state observer design problem. Later we shall use the dual problem approach to obtain Ackermann's formula for the determination of the observer gain matrix $\mathbf{K}_e$.

Design of full-order state observers.   Consider the system defined by

$$\dot{\mathbf{x}} = \mathbf{Ax} + \mathbf{B}u \tag{10–30}$$

$$y = \mathbf{Cx} \tag{10–31}$$

where $\mathbf{x}$ = state vector (n-vector)
u = control signal (scalar)
y = output signal (scalar)
$\mathbf{A} = n \times n$ constant matrix
$\mathbf{B} = n \times 1$ constant matrix
$\mathbf{C} = 1 \times n$ constant matrix

We assume that the system is completely observable. We assume further that the system configuration is the same as that shown in Figure 10–7.

In designing the full-order state observer, it is convenient if we transform the system equations given by Equations (10–30) and (10–31) into the observable canonical form. As presented earlier, this can be done as follows: Define a transformation matrix $\mathbf{Q}$ by

$$\mathbf{Q} = (\mathbf{WN}^*)^{-1} \qquad (10\text{–}32)$$

where $\mathbf{N}$ is the observability matrix

$$\mathbf{N} = [\mathbf{C}^* \mid \mathbf{A}^*\mathbf{C}^* \mid \cdots \mid (\mathbf{A}^*)^{n-1}\mathbf{C}^*] \qquad (10\text{–}33)$$

and $\mathbf{W}$ is defined by Equation (10–6), rewritten thus:

$$\mathbf{W} = \begin{bmatrix} a_{n-1} & a_{n-2} & \cdots & a_1 & 1 \\ a_{n-2} & a_{n-3} & \cdots & 1 & 0 \\ \cdot & \cdot & & \cdot & \cdot \\ \cdot & \cdot & & \cdot & \cdot \\ \cdot & \cdot & & \cdot & \cdot \\ a_1 & 1 & \cdots & 0 & 0 \\ 1 & 0 & \cdots & 0 & 0 \end{bmatrix}$$

where $a_1, a_2, \ldots, a_{n-1}$ are coefficients in the characteristic equation of the original state equation given by Equation (10–30):

$$|s\mathbf{I} - \mathbf{A}| = s^n + a_1 s^{n-1} + \cdots + a_{n-1}s + a_n = 0$$

(Since we assumed that the system is completely observable, the inverse of matrix $\mathbf{WN}^*$ exists.)

Define a new state vector (n-vector) $\boldsymbol{\xi}$ by

$$\mathbf{x} = \mathbf{Q}\boldsymbol{\xi} \qquad (10\text{–}34)$$

Then Equations (10–30) and (10–31) become

$$\dot{\boldsymbol{\xi}} = \mathbf{Q}^{-1}\mathbf{A}\mathbf{Q}\boldsymbol{\xi} + \mathbf{Q}^{-1}\mathbf{B}u \qquad (10\text{–}35)$$

$$y = \mathbf{C}\mathbf{Q}\boldsymbol{\xi} \qquad (10\text{–}36)$$

where

$$\mathbf{Q}^{-1}\mathbf{A}\mathbf{Q} = \begin{bmatrix} 0 & 0 & \cdots & 0 & -a_n \\ 1 & 0 & \cdots & 0 & -a_{n-1} \\ 0 & 1 & \cdots & 0 & -a_{n-2} \\ \cdot & \cdot & & \cdot & \cdot \\ \cdot & \cdot & & \cdot & \cdot \\ \cdot & \cdot & & \cdot & \cdot \\ 0 & 0 & \cdots & 1 & -a_1 \end{bmatrix} \qquad (10\text{–}37)$$

$$\mathbf{Q}^{-1}\mathbf{B} = \begin{bmatrix} b_n - a_n b_0 \\ b_{n-1} - a_{n-1} b_0 \\ \cdot \\ \cdot \\ \cdot \\ b_1 - a_1 b_0 \end{bmatrix} \qquad (10\text{–}38)$$

$$\mathbf{CQ} = [0 \quad 0 \quad \cdots \quad 0 \quad 1] \tag{10-39}$$

[See Problems A–10–6 and A–10–7 for the derivation of Equations (10–37) through (10–39).] Equations (10–35) and (10–36) are in the observable canonical form. Thus, given a state equation and output equation, they can be transformed into the observable canonical form if the system is completely observable and if the original state vector $\mathbf{x}$ is transformed into a new state vector $\boldsymbol{\xi}$ by use of the transformation given by Equation (10–34). Note that if matrix $\mathbf{A}$ is already in the observable canonical form, then $\mathbf{Q} = \mathbf{I}$.

As stated earlier, we choose the state observer dynamics to be given by

$$\dot{\tilde{\mathbf{x}}} = \mathbf{A}\tilde{\mathbf{x}} + \mathbf{B}u + \mathbf{K}_e(y - \mathbf{C}\tilde{\mathbf{x}})$$

$$= (\mathbf{A} - \mathbf{K}_e\mathbf{C})\tilde{\mathbf{x}} + \mathbf{B}u + \mathbf{K}_e\mathbf{C}\mathbf{x} \tag{10-40}$$

Now define

$$\tilde{\mathbf{x}} = \mathbf{Q}\tilde{\boldsymbol{\xi}} \tag{10-41}$$

By substituting Equation (10–41) into Equation (10–40), we have

$$\dot{\tilde{\boldsymbol{\xi}}} = \mathbf{Q}^{-1}(\mathbf{A} - \mathbf{K}_e\mathbf{C})\mathbf{Q}\tilde{\boldsymbol{\xi}} + \mathbf{Q}^{-1}\mathbf{B}u + \mathbf{Q}^{-1}\mathbf{K}_e\mathbf{C}\mathbf{Q}\boldsymbol{\xi} \tag{10-42}$$

Subtracting Equation (10–42) from Equation (10–35), we obtain

$$\dot{\boldsymbol{\xi}} - \dot{\tilde{\boldsymbol{\xi}}} = \mathbf{Q}^{-1}(\mathbf{A} - \mathbf{K}_e\mathbf{C})\mathbf{Q}(\boldsymbol{\xi} - \tilde{\boldsymbol{\xi}}) \tag{10-43}$$

Define

$$\boldsymbol{\epsilon} = \boldsymbol{\xi} - \tilde{\boldsymbol{\xi}}$$

Then Equation (10–43) becomes

$$\dot{\boldsymbol{\epsilon}} = \mathbf{Q}^{-1}(\mathbf{A} - \mathbf{K}_e\mathbf{C})\mathbf{Q}\boldsymbol{\epsilon} \tag{10-44}$$

We require the error dynamics to be asymptotically stable and $\boldsymbol{\epsilon}(t)$ to reach zero with sufficient speed. The procedure for determining matrix $\mathbf{K}_e$ is first to select the desired observer poles (the eigenvalues of $\mathbf{A} - \mathbf{K}_e\mathbf{C}$) and then to determine matrix $\mathbf{K}_e$ so that it will give the desired observer poles. Noting that $\mathbf{Q}^{-1} = \mathbf{W}\mathbf{N}^*$, we have

$$\mathbf{Q}^{-1}\mathbf{K}_e = \begin{bmatrix} a_{n-1} & a_{n-2} & \cdots & a_1 & 1 \\ a_{n-2} & a_{n-3} & \cdots & 1 & 0 \\ \cdot & \cdot & & \cdot & \cdot \\ \cdot & \cdot & & \cdot & \cdot \\ \cdot & \cdot & & \cdot & \cdot \\ a_1 & 1 & \cdots & 0 & 0 \\ 1 & 0 & \cdots & 0 & 0 \end{bmatrix} \begin{bmatrix} \mathbf{C} \\ \mathbf{CA} \\ \cdot \\ \cdot \\ \cdot \\ \mathbf{CA}^{n-2} \\ \mathbf{CA}^{n-1} \end{bmatrix} \begin{bmatrix} k_1 \\ k_2 \\ \cdot \\ \cdot \\ \cdot \\ k_{n-1} \\ k_n \end{bmatrix}$$

where

$$\mathbf{K}_e = \begin{bmatrix} k_1 \\ k_2 \\ \cdot \\ \cdot \\ \cdot \\ k_n \end{bmatrix}$$

Since $\mathbf{Q}^{-1}\mathbf{K}_e$ is an n-vector, let us write

$$\mathbf{Q}^{-1}\mathbf{K}_e = \begin{bmatrix} \delta_n \\ \delta_{n-1} \\ \cdot \\ \cdot \\ \cdot \\ \delta_1 \end{bmatrix} \tag{10–45}$$

Then, referring to Eq. (10–39), we have

$$\mathbf{Q}^{-1}\mathbf{K}_e\mathbf{CQ} = \begin{bmatrix} \delta_n \\ \delta_{n-1} \\ \cdot \\ \cdot \\ \cdot \\ \delta_1 \end{bmatrix} [0 \quad 0 \quad \cdots \quad 1] = \begin{bmatrix} 0 & 0 & \cdots & 0 & \delta_n \\ 0 & 0 & \cdots & 0 & \delta_{n-1} \\ \cdot & \cdot & & \cdot & \cdot \\ \cdot & \cdot & & \cdot & \cdot \\ \cdot & \cdot & & \cdot & \cdot \\ 0 & 0 & \cdots & 0 & \delta_1 \end{bmatrix}$$

and

$$\mathbf{Q}^{-1}(\mathbf{A} - \mathbf{K}_e\mathbf{C})\mathbf{Q} = \mathbf{Q}^{-1}\mathbf{AQ} - \mathbf{Q}^{-1}\mathbf{K}_e\mathbf{CQ}$$

$$= \begin{bmatrix} 0 & 0 & \cdots & 0 & -a_n - \delta_n \\ 1 & 0 & \cdots & 0 & -a_{n-1} - \delta_{n-1} \\ 0 & 1 & \cdots & 0 & -a_{n-2} - \delta_{n-2} \\ \cdot & \cdot & & \cdot & \cdot \\ \cdot & \cdot & & \cdot & \cdot \\ \cdot & \cdot & & \cdot & \cdot \\ 0 & 0 & \cdots & 1 & -a_1 - \delta_1 \end{bmatrix}$$

The characteristic equation

$$|s\mathbf{I} - \mathbf{Q}^{-1}(\mathbf{A} - \mathbf{K}_e\mathbf{C})\mathbf{Q}| = 0$$

becomes

$$\begin{vmatrix} s & 0 & 0 & \cdots & 0 & a_n + \delta_n \\ -1 & s & 0 & \cdots & 0 & a_{n-1} + \delta_{n-1} \\ 0 & -1 & s & \cdots & 0 & a_{n-2} + \delta_{n-2} \\ \cdot & \cdot & \cdot & & \cdot & \cdot \\ \cdot & \cdot & \cdot & & \cdot & \cdot \\ \cdot & \cdot & \cdot & & \cdot & \cdot \\ 0 & 0 & 0 & \cdots & -1 & s + a_1 + \delta_1 \end{vmatrix} = 0$$

or

$$s^n + (a_1 + \delta_1)s^{n-1} + (a_2 + \delta_2)s^{n-2} + \cdots + (a_n + \delta_n) = 0 \qquad (10\text{--}46)$$

It can be seen that each of $\delta_n, \delta_{n-1}, \ldots, \delta_1$ is associated with only one of the coefficients of the characteristic equation.

Suppose that the desired characteristic equation for the error dynamics is

$$(s - \mu_1)(s - \mu_2) \cdots (s - \mu_n)$$
$$= s^n + \alpha_1 s^{n-1} + \alpha_2 s^{n-2} + \cdots + \alpha_{n-1}s + \alpha_n = 0 \qquad (10\text{--}47)$$

(Note that the desired eigenvalues μ_i's determine how fast the observed state converges to the actual state of the plant.) Comparing the coefficients of terms of like powers of s in Equations (10–46) and (10–47), we obtain

$$a_1 + \delta_1 = \alpha_1$$
$$a_2 + \delta_2 = \alpha_2$$
$$\vdots$$
$$a_n + \delta_n = \alpha_n$$

from which we get

$$\delta_1 = \alpha_1 - a_1$$
$$\delta_2 = \alpha_2 - a_2$$
$$\vdots$$
$$\delta_n = \alpha_n - a_n$$

Then from Equation (10–45) we have

$$\mathbf{Q}^{-1}\mathbf{K}_e = \begin{bmatrix} \delta_n \\ \delta_{n-1} \\ \cdot \\ \cdot \\ \cdot \\ \delta_1 \end{bmatrix} = \begin{bmatrix} \alpha_n - a_n \\ \alpha_{n-1} - a_{n-1} \\ \cdot \\ \cdot \\ \cdot \\ \alpha_1 - a_1 \end{bmatrix}$$

Hence

$$\mathbf{K}_e = \mathbf{Q} \begin{bmatrix} \alpha_n - a_n \\ \alpha_{n-1} - a_{n-1} \\ \cdot \\ \cdot \\ \cdot \\ \alpha_1 - a_1 \end{bmatrix} = (\mathbf{WN}^*)^{-1} \begin{bmatrix} \alpha_n - a_n \\ \alpha_{n-1} - a_{n-1} \\ \cdot \\ \cdot \\ \cdot \\ \alpha_1 - a_1 \end{bmatrix} \qquad (10\text{--}48)$$

Equation (10–48) specifies the necessary state observer gain matrix $\mathbf{K}_e$.

As stated earlier, Equation (10–48) can also be obtained from Equation (10–13) by considering the dual problem. That is, consider the pole placement problem for the dual system and obtain the state feedback gain matrix **K** for the dual system. Then the state observer gain matrix $\mathbf{K}_e$ can be given by **K*** (see Problem A–10–9).

Once we select the desired eigenvalues (or desired characteristic equation), the full-order state observer can be designed, provided the system is completely observable. The desired eigenvalues or characteristic equation should be chosen so that the state observer responds at least two to five times faster than the closed-loop system considered. As stated earlier, the equation for the full-order state observer is

$$\dot{\tilde{\mathbf{x}}} = (\mathbf{A} - \mathbf{K}_e\mathbf{C})\tilde{\mathbf{x}} + \mathbf{B}u + \mathbf{K}_e y \qquad (10\text{–}49)$$

It is noted that thus far we assumed that the matrices **A** and **B** in the observer are exactly the same as those of the actual plant. In practice, this may not be true. Then the error dynamics may not be given by Equation (10–44). This means that the error may not approach zero. Hence, we should try to build an accurate mathematical model for the observer to make the error acceptably small.

Direct substitution approach to obtain state observer gain matrix $\mathbf{K}_e$.   Similar to the case of pole placement, if the system is of low order, then direct substitution of matrix $\mathbf{K}_e$ into the desired characteristic polynomial may be simpler. For example, if **x** is a 3-vector, then write the observer gain matrix $\mathbf{K}_e$ as

$$\mathbf{K}_e = \begin{bmatrix} k_{e1} \\ k_{e2} \\ k_{e3} \end{bmatrix}$$

Substitute this $\mathbf{K}_e$ matrix into the desired characteristic polynomial:

$$\left| s\mathbf{I} - (\mathbf{A} - \mathbf{K}_e\mathbf{C}) \right| = (s - \mu_1)(s - \mu_2)(s - \mu_3)$$

By equating the coefficients of the like powers of s on both sides of this last equation, we can determine the values of k_{e1}, k_{e2}, and k_{e3}. This approach is convenient if $n = 1$, 2, or 3, where n is the dimension of the state vector **x**. (Although this approach can be used when $n = 4, 5, 6, \ldots$, the computations involved may become very tedious.)

Another approach to the determination of the state observer gain matrix $\mathbf{K}_e$ is to use Ackermann's formula. This approach is presented in the following.

Ackermann's formula.   Consider the system defined by

$$\dot{\mathbf{x}} = \mathbf{A}\mathbf{x} + \mathbf{B}u \qquad (10\text{–}50)$$

$$y = \mathbf{C}\mathbf{x} \qquad (10\text{–}51)$$

In Section 10–2 we derived Ackermann's formula for pole placement for the system defined by Equation (10–50). The result was given by Equation (10–18), rewritten thus:

$$\mathbf{K} = \begin{bmatrix} 0 & 0 & \cdots & 0 & 1 \end{bmatrix} [\mathbf{B} \;\vdots\; \mathbf{AB} \;\vdots\; \cdots \;\vdots\; \mathbf{A}^{n-1}\mathbf{B}]^{-1}\phi(\mathbf{A})$$

For the dual of the system defined by Equations (10–50) and (10–51),

$$\dot{\mathbf{z}} = \mathbf{A}^*\mathbf{z} + \mathbf{C}^*v$$

$$n = \mathbf{B}^*\mathbf{z}$$

the preceding Ackermann's formula for pole placement is modified to

$$\mathbf{K} = [0 \quad 0 \quad \cdots \quad 0 \quad 1]\,[\mathbf{C}^* \;\vdots\; \mathbf{A}^*\mathbf{C}^* \;\vdots\; \cdots \;\vdots\; (\mathbf{A}^*)^{n-1}\mathbf{C}^*]^{-1}\phi(\mathbf{A}^*) \qquad (10\text{–}52)$$

As stated earlier, the state observer gain matrix $\mathbf{K}_e$ is given by $\mathbf{K}^*$, where $\mathbf{K}$ is given by Equation (10–52). Thus,

$$\mathbf{K}_e = \mathbf{K}^* = \phi(\mathbf{A}^*)^* \begin{bmatrix} \mathbf{C} \\ \mathbf{CA} \\ \cdot \\ \cdot \\ \cdot \\ \mathbf{CA}^{n-2} \\ \mathbf{CA}^{n-1} \end{bmatrix}^{-1} \begin{bmatrix} 0 \\ 0 \\ \cdot \\ \cdot \\ \cdot \\ 0 \\ 1 \end{bmatrix} = \phi(\mathbf{A}) \begin{bmatrix} \mathbf{C} \\ \mathbf{CA} \\ \cdot \\ \cdot \\ \cdot \\ \mathbf{CA}^{n-2} \\ \mathbf{CA}^{n-1} \end{bmatrix}^{-1} \begin{bmatrix} 0 \\ 0 \\ \cdot \\ \cdot \\ \cdot \\ 0 \\ 1 \end{bmatrix} \qquad (10\text{–}53)$$

where $\phi(s)$ is the desired characteristic polynomial for the state observer, or

$$\phi(s) = (s - \mu_1)(s - \mu_2) \cdots (s - \mu_n)$$

where $\mu_1, \mu_2, \ldots, \mu_n$ are the desired eigenvalues. Equation (10–53) is called Ackermann's formula for the determination of the observer gain matrix $\mathbf{K}_e$.

Comments on selecting the best $\mathbf{K}_e$.   Referring to Figure 10–7, notice that the feedback signal through the observer gain matrix $\mathbf{K}_e$ serves as a correction signal to the plant model to account for the unknowns in the plant. If significant unknowns are involved, the feedback signal through the matrix $\mathbf{K}_e$ should be relatively large. However, if the output signal is contaminated significantly by disturbances and measurement noises, then the output y is not reliable and the feedback signal through the matrix $\mathbf{K}_e$ should be relatively small. In determining the matrix $\mathbf{K}_e$, we should carefully examine the effects of disturbances and noises involved in the output y.

Remember that the observer gain matrix $\mathbf{K}_e$ depends on the desired characteristic equation

$$(s - \mu_1)(s - \mu_2) \cdots (s - \mu_n) = 0$$

The choice of a set of $\mu_1, \mu_2, \ldots, \mu_n$ is, in many instances, not unique. Hence, many different characteristic equations might be chosen as desired characteristic equations. For each desired characteristic equation, we have a different matrix $\mathbf{K}_e$.

In the design of the state observer, it is desirable to determine several observer gain matrices $\mathbf{K}_e$ based on several different desired characteristic equations. For each of the several different matrices $\mathbf{K}_e$, simulation tests must be run to evaluate the resulting system performance. Then we select the best $\mathbf{K}_e$ from the viewpoint of overall system performance. In many practical cases, the selection of the best matrix $\mathbf{K}_e$ boils down to a compromise between speedy response and sensitivity to disturbances and noises.

EXAMPLE 10–3   Consider the system

$$\dot{\mathbf{x}} = \mathbf{A}\mathbf{x} + \mathbf{B}u$$

$$y = \mathbf{C}\mathbf{x}$$

where

$$\mathbf{A} = \begin{bmatrix} 0 & 20.6 \\ 1 & 0 \end{bmatrix}, \qquad \mathbf{B} = \begin{bmatrix} 0 \\ 1 \end{bmatrix}, \qquad \mathbf{C} = [0 \quad 1]$$

Design a full-order state observer, assuming that the system configuration is identical to that shown in Figure 10–7. Assume that the desired eigenvalues of the observer matrix are

$$\mu_1 = -1.8 + j2.4, \qquad \mu_2 = -1.8 - j2.4$$

The design of the state observer reduces to the determination of an appropriate observer gain matrix $\mathbf{K}_e$.

Let us examine the observability matrix. The rank of

$$[\mathbf{C}^* \ \vdots \ \mathbf{A}^*\mathbf{C}^*] = \begin{bmatrix} 0 & 1 \\ 1 & 0 \end{bmatrix}$$

is 2. Hence the system is completely observable and the determination of the desired observer gain matrix is possible. We shall solve this problem by three methods.

Method 1: We shall determine the observer gain matrix by use of Equation (10–48). The given state matrix $\mathbf{A}$ is already in the observable canonical form. Hence the transformation matrix $\mathbf{Q} = (\mathbf{WN}^*)^{-1}$ is $\mathbf{I}$. Since the characteristic equation of the given system is

$$|s\mathbf{I} - \mathbf{A}| = \begin{vmatrix} s & -20.6 \\ -1 & s \end{vmatrix} = s^2 - 20.6 = s^2 + a_1 s + a_2 = 0$$

we have

$$a_1 = 0, \qquad a_2 = -20.6$$

The desired characteristic equation is

$$(s + 1.8 - j2.4)(s + 1.8 + j2.4) = s^2 + 3.6s + 9 = s^2 + \alpha_1 s + \alpha_2 = 0$$

Hence

$$\alpha_1 = 3.6, \qquad \alpha_2 = 9$$

Then the observer gain matrix $\mathbf{K}_e$ can be obtained from Equation (10–48) as follows:

$$\mathbf{K}_e = (\mathbf{WN}^*)^{-1} \begin{bmatrix} \alpha_2 - a_2 \\ \alpha_1 - a_1 \end{bmatrix} = \begin{bmatrix} 1 & 0 \\ 0 & 1 \end{bmatrix} \begin{bmatrix} 9 + 20.6 \\ 3.6 - 0 \end{bmatrix} = \begin{bmatrix} 29.6 \\ 3.6 \end{bmatrix}$$

Method 2: Referring to Equation (10–29),

$$\dot{\mathbf{e}} = (\mathbf{A} - \mathbf{K}_e\mathbf{C})\mathbf{e}$$

the characteristic equation for the observer becomes

$$|s\mathbf{I} - \mathbf{A} + \mathbf{K}_e\mathbf{C}| = 0$$

Define

$$\mathbf{K}_e = \begin{bmatrix} k_{e1} \\ k_{e2} \end{bmatrix}$$

Then the characteristic equation becomes

$$\begin{vmatrix} \begin{bmatrix} s & 0 \\ 0 & s \end{bmatrix} - \begin{bmatrix} 0 & 20.6 \\ 1 & 0 \end{bmatrix} + \begin{bmatrix} k_{e1} \\ k_{e2} \end{bmatrix} [0 \quad 1] \end{vmatrix} = \begin{vmatrix} s & -20.6 + k_{e1} \\ -1 & s + k_{e2} \end{vmatrix}$$

$$= s^2 + k_{e2}s - 20.6 + k_{e1} = 0 \qquad\qquad (10\text{–}54)$$

Since the desired characteristic equation is

$$s^2 + 3.6s + 9 = 0$$

by comparing Equation (10–54) with this last equation, we obtain

$$k_{e1} = 29.6, \qquad k_{e2} = 3.6$$

or

$$\mathbf{K}_e = \begin{bmatrix} 29.6 \\ 3.6 \end{bmatrix}$$

Method 3: We shall use Ackermann's formula given by Equation (10–53):

$$\mathbf{K}_e = \phi(\mathbf{A}) \begin{bmatrix} \mathbf{C} \\ \mathbf{CA} \end{bmatrix}^{-1} \begin{bmatrix} 0 \\ 1 \end{bmatrix}$$

where

$$\phi(s) = (s - \mu_1)(s - \mu_2) = s^2 + 3.6s + 9$$

Thus

$$\phi(\mathbf{A}) = \mathbf{A}^2 + 3.6\mathbf{A} + 9\mathbf{I}$$

and

$$\mathbf{K}_e = (\mathbf{A}^2 + 3.6\mathbf{A} + 9\mathbf{I}) \begin{bmatrix} 0 & 1 \\ 1 & 0 \end{bmatrix}^{-1} \begin{bmatrix} 0 \\ 1 \end{bmatrix}$$

$$= \begin{bmatrix} 29.6 & 74.16 \\ 3.6 & 29.6 \end{bmatrix} \begin{bmatrix} 0 & 1 \\ 1 & 0 \end{bmatrix} \begin{bmatrix} 0 \\ 1 \end{bmatrix} = \begin{bmatrix} 29.6 \\ 3.6 \end{bmatrix}$$

As a matter of course, we get the same $\mathbf{K}_e$ regardless of the method employed.

Notice that the system considered in Example 10–1 and the present system are dual to each other. The state feedback gain matrix obtained in Example 10–1 was $\mathbf{K} = [29.6 \quad 3.6]$. The observer gain matrix $\mathbf{K}_e$ obtained here is related to matrix $\mathbf{K}$ by the relationship $\mathbf{K}_e = \mathbf{K}^*$. (Since the present matrices $\mathbf{K}$ and $\mathbf{K}_e$ are real, we may write the relationship as $\mathbf{K}_e = \mathbf{K}^T$.)

The equation for the full-order state observer is given by Equation (10–49):

$$\dot{\tilde{\mathbf{x}}} = (\mathbf{A} - \mathbf{K}_e\mathbf{C})\tilde{\mathbf{x}} + \mathbf{B}u + \mathbf{K}_e y$$

or

$$\begin{bmatrix} \dot{\tilde{x}}_1 \\ \dot{\tilde{x}}_2 \end{bmatrix} = \begin{bmatrix} 0 & -9 \\ 1 & -3.6 \end{bmatrix} \begin{bmatrix} \tilde{x}_1 \\ \tilde{x}_2 \end{bmatrix} + \begin{bmatrix} 0 \\ 1 \end{bmatrix} u + \begin{bmatrix} 29.6 \\ 3.6 \end{bmatrix} y$$

Finally, it is noted that similar to the case of pole placement, if the system order n is 4 or higher, methods 1 and 3 are preferred, because all matrix computations can be carried out by a computer, while method 2 always requires hand computation of the characteristic equation involving unknown parameters $k_{e1}, k_{e2}, \ldots, k_{en}$.

EXAMPLE 10–4   Consider the system

$$\dot{\mathbf{x}} = \mathbf{Ax} + \mathbf{B}u$$

$$y = \mathbf{Cx}$$

where

$$\mathbf{A} = \begin{bmatrix} 0 & 1 & 0 \\ 0 & 0 & 1 \\ -6 & -11 & -6 \end{bmatrix}, \quad \mathbf{B} = \begin{bmatrix} 0 \\ 0 \\ 1 \end{bmatrix}, \quad \mathbf{C} = [1 \quad 0 \quad 0]$$

Design a full-order state observer, assuming that the system configuration is identical to that shown in Figure 10–7. Assume that the desired eigenvalues of the observer matrix are

$$\mu_1 = -2 + j3.464, \qquad \mu_2 = -2 - j3.464, \qquad \mu_3 = -5$$

Let us examine the observability matrix. The rank of

$$\mathbf{N} = [\mathbf{C}^* \; \vdots \; \mathbf{A}^*\mathbf{C}^* \; \vdots \; (\mathbf{A}^*)^2\mathbf{C}^*] = \begin{bmatrix} 1 & 0 & 0 \\ 0 & 1 & 0 \\ 0 & 0 & 1 \end{bmatrix}$$

is 3. Hence, the system is completely observable and the determination of the observer gain matrix $\mathbf{K}_e$ is possible.

Since the characteristic equation of the given system is

$$|s\mathbf{I} - \mathbf{A}| = \begin{vmatrix} s & -1 & 0 \\ 0 & s & -1 \\ 6 & 11 & s + 6 \end{vmatrix}$$

$$= s^3 + 6s^2 + 11s + 6$$

$$= s^3 + a_1 s^2 + a_2 s + a_3 = 0$$

we have

$$a_1 = 6, \qquad a_2 = 11, \qquad a_3 = 6$$

The desired characteristic equation is

$$(s - \mu_1)(s - \mu_2)(s - \mu_3) = (s + 2 - j3.464)(s + 2 + j3.464)(s + 5)$$

$$= s^3 + 9s^2 + 36s + 80$$

$$= s^3 + \alpha_1 s^2 + \alpha_2 s + \alpha_3 = 0$$

Hence

$$\alpha_1 = 9, \qquad \alpha_2 = 36, \qquad \alpha_3 = 80$$

We shall solve this problem by using Equation (10–48):

$$\mathbf{K}_e = (\mathbf{W}\mathbf{N}^*)^{-1} \begin{bmatrix} \alpha_3 - a_3 \\ \alpha_2 - a_2 \\ \alpha_1 - a_1 \end{bmatrix}$$

Noting that

$$\mathbf{N}^* = \begin{bmatrix} 1 & 0 & 0 \\ 0 & 1 & 0 \\ 0 & 0 & 1 \end{bmatrix}, \quad \mathbf{W} = \begin{bmatrix} 11 & 6 & 1 \\ 6 & 1 & 0 \\ 1 & 0 & 0 \end{bmatrix}$$

we have

$$(\mathbf{W}\mathbf{N}^*)^{-1} = \left\{ \begin{bmatrix} 11 & 6 & 1 \\ 6 & 1 & 0 \\ 1 & 0 & 0 \end{bmatrix} \begin{bmatrix} 1 & 0 & 0 \\ 0 & 1 & 0 \\ 0 & 0 & 1 \end{bmatrix} \right\}^{-1} = \begin{bmatrix} 0 & 0 & 1 \\ 0 & 1 & -6 \\ 1 & -6 & 25 \end{bmatrix}$$

Hence

$$\mathbf{K}_e = \begin{bmatrix} 0 & 0 & 1 \\ 0 & 1 & -6 \\ 1 & -6 & 25 \end{bmatrix} \begin{bmatrix} 80 - 6 \\ 36 - 11 \\ 9 - 6 \end{bmatrix} = \begin{bmatrix} 3 \\ 7 \\ -1 \end{bmatrix}$$

Referring to Equation (10–49), the full-order state observer is given by

$$\dot{\tilde{\mathbf{x}}} = (\mathbf{A} - \mathbf{K}_e\mathbf{C})\tilde{\mathbf{x}} + \mathbf{B}u + \mathbf{K}_e y$$

or

$$\begin{bmatrix} \dot{\tilde{x}}_1 \\ \dot{\tilde{x}}_2 \\ \dot{\tilde{x}}_3 \end{bmatrix} = \begin{bmatrix} -3 & 1 & 0 \\ -7 & 0 & 1 \\ -5 & -11 & -6 \end{bmatrix} \begin{bmatrix} \tilde{x}_1 \\ \tilde{x}_2 \\ \tilde{x}_3 \end{bmatrix} + \begin{bmatrix} 0 \\ 0 \\ 1 \end{bmatrix} u + \begin{bmatrix} 3 \\ 7 \\ -1 \end{bmatrix} y$$

(The solutions to this example problem by use of the direct substitution method and by use of Ackermann's formula are given in Problem A–10–11.)

Effects of the addition of the observer on a closed-loop system. In the pole placement design process, we assumed that the actual state $\mathbf{x}(t)$ was available for feedback. In practice, however, the actual state $\mathbf{x}(t)$ may not be measurable, so we will need to design an observer and use the observed state $\tilde{\mathbf{x}}(t)$ for feedback as shown in Figure 10–8. The design process, therefore, becomes a two-stage process, the first stage being the determination of the feedback gain matrix $\mathbf{K}$ to yield the desired characteristic equation and the second stage being the determination of the observer gain matrix $\mathbf{K}_e$ to yield the desired observer characteristic equation.

Let us now investigate the effects of the use of the observed state $\tilde{\mathbf{x}}(t)$, rather than the actual state $\mathbf{x}(t)$, on the characteristic equation of a closed-loop control system.

Consider the completely state controllable and completely observable system defined by the equations

$$\dot{\mathbf{x}} = \mathbf{A}\mathbf{x} + \mathbf{B}u$$

$$y = \mathbf{C}\mathbf{x}$$

For the state feedback control based on the observed state $\tilde{\mathbf{x}}$,

$$u = -\mathbf{K}\tilde{\mathbf{x}}$$

With this control, the state equation becomes

$$\dot{\mathbf{x}} = \mathbf{A}\mathbf{x} - \mathbf{B}\mathbf{K}\tilde{\mathbf{x}} = (\mathbf{A} - \mathbf{B}\mathbf{K})\mathbf{x} + \mathbf{B}\mathbf{K}(\mathbf{x} - \tilde{\mathbf{x}}) \qquad (10\text{–}55)$$

The difference between the actual state $\mathbf{x}(t)$ and the observed state $\tilde{\mathbf{x}}(t)$ has been defined as the error $\mathbf{e}(t)$:

$$\mathbf{e}(t) = \mathbf{x}(t) - \tilde{\mathbf{x}}(t)$$

Substitution of the error vector $\mathbf{e}(t)$ into Equation (10–55) gives

$$\dot{\mathbf{x}} = (\mathbf{A} - \mathbf{B}\mathbf{K})\mathbf{x} + \mathbf{B}\mathbf{K}\mathbf{e} \qquad (10\text{–}56)$$

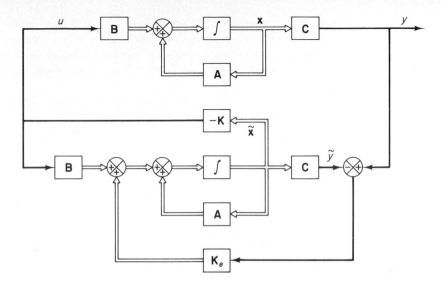

Figure 10–8
Observed-state
feedback control
system.

Note that the observer error equation was given by Equation (10–29), repeated here:

$$\dot{\mathbf{e}} = (\mathbf{A} - \mathbf{K}_e\mathbf{C})\mathbf{e} \qquad (10\text{–}57)$$

Combining Equations (10–56) and (10–57), we obtain

$$\begin{bmatrix} \dot{\mathbf{x}} \\ \dot{\mathbf{e}} \end{bmatrix} = \begin{bmatrix} \mathbf{A} - \mathbf{BK} & \mathbf{BK} \\ \mathbf{0} & \mathbf{A} - \mathbf{K}_e\mathbf{C} \end{bmatrix} \begin{bmatrix} \mathbf{x} \\ \mathbf{e} \end{bmatrix} \qquad (10\text{–}58)$$

Equation (10–58) describes the dynamics of the observed-state feedback control system. The characteristic equation for the system is

$$\begin{vmatrix} s\mathbf{I} - \mathbf{A} + \mathbf{BK} & -\mathbf{BK} \\ \mathbf{0} & s\mathbf{I} - \mathbf{A} + \mathbf{K}_e\mathbf{C} \end{vmatrix} = 0$$

or

$$|s\mathbf{I} - \mathbf{A} + \mathbf{BK}\|s\mathbf{I} - \mathbf{A} + \mathbf{K}_e\mathbf{C}| = 0$$

Notice that the closed-loop poles of the observed-state feedback control system consist of the poles due to the pole placement design alone plus the poles due to the observer design alone. This means that the pole placement design and the observer design are independent of each other. They can be designed separately and combined together to form the observed-state feedback control system. Note that, if the order of the plant is n, then the observer is also of nth order (if the full-order state observer is used), and the resulting characteristic equation for the entire closed-loop system becomes of order $2n$.

The desired closed-loop poles to be generated by state feedback (pole placement) are chosen in such a way that the system satisfies the performance requirements. The poles of the observer are usually chosen so that the observer response is much faster than the system response. A rule of thumb is to choose an observer response at least two to five times faster than the system response. Since the observer is, in general, not a hardware structure but is programmed on the computer, it is possible to increase the response speed so that the observed state quickly converges to the actual state. The maximum response speed of the observer is generally limited only by noise and sensitivity problem involved in the control system. It is noted that since the observer poles are located to the left of the desired closed-loop poles in the pole placement process, the latter will dominate in the response.

Transfer function for the controller–observer.   Consider the system defined by

$$\dot{\mathbf{x}} = \mathbf{A}\mathbf{x} + \mathbf{B}u$$

$$y = \mathbf{C}\mathbf{x}$$

Assume that the system is completely observable but $\mathbf{x}$ is not available for direct measurement. Assume that we employ the observed-state feedback control

$$u = -\mathbf{K}\tilde{\mathbf{x}} \tag{10–59}$$

In the observed-state feedback control system as shown in Figure 10–8, the observer equation is

$$\dot{\tilde{\mathbf{x}}} = (\mathbf{A} - \mathbf{K}_e\mathbf{C})\tilde{\mathbf{x}} + \mathbf{B}u + \mathbf{K}_e y \tag{10–60}$$

Let us take the Laplace transform of Equation (10–59).

$$U(s) = -\mathbf{K}\tilde{\mathbf{X}}(s) \tag{10–61}$$

The Laplace transform of the observer equation given by Equation (10–60) is

$$s\tilde{\mathbf{X}}(s) = (\mathbf{A} - \mathbf{K}_e\mathbf{C})\tilde{\mathbf{X}}(s) + \mathbf{B}U(s) + \mathbf{K}_e Y(s) \tag{10–62}$$

where we assumed the initial observed state to be zero, or $\tilde{\mathbf{x}}(0) = \mathbf{0}$. By substituting Equation (10–61) into Equation (10–62) and solving the resulting equation for $\tilde{\mathbf{X}}(s)$, we obtain

$$\tilde{\mathbf{X}}(s) = (s\mathbf{I} - \mathbf{A} + \mathbf{K}_e\mathbf{C} + \mathbf{B}\mathbf{K})^{-1}\mathbf{K}_e Y(s)$$

By substituting this last equation into Equation (10–61), we get

$$U(s) = -\mathbf{K}(s\mathbf{I} - \mathbf{A} + \mathbf{K}_e\mathbf{C} + \mathbf{B}\mathbf{K})^{-1}\mathbf{K}_e Y(s) \tag{10–63}$$

In the present discussions, both u and y are scalars. Equation (10–63) gives the transfer function between $U(s)$ and $-Y(s)$.

Figure 10–9 shows the block diagram representation for the system. Notice that the transfer function

$$\mathbf{K}(s\mathbf{I} - \mathbf{A} + \mathbf{K}_e\mathbf{C} + \mathbf{B}\mathbf{K})^{-1}\mathbf{K}_e$$

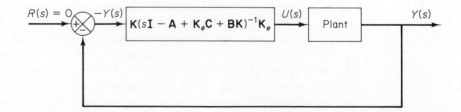

Figure 10–9
Block diagram representation of system with a controller-observer.

acts as a controller for the system. Hence we call the following transfer function:

$$\frac{U(s)}{-Y(s)} = \mathbf{K}(s\mathbf{I} - \mathbf{A} + \mathbf{K}_e\mathbf{C} + \mathbf{BK})^{-1}\mathbf{K}_e \qquad (10\text{-}64)$$

the controller–observer transfer function.

EXAMPLE 10–5. Consider the design of a regulator system for the following plant:

$$\dot{\mathbf{x}} = \mathbf{Ax} + \mathbf{B}u \qquad (10\text{-}65)$$

$$y = \mathbf{Cx} \qquad (10\text{-}66)$$

where

$$\mathbf{A} = \begin{bmatrix} 0 & 1 \\ 20.6 & 0 \end{bmatrix}, \qquad \mathbf{B} = \begin{bmatrix} 0 \\ 1 \end{bmatrix}, \qquad \mathbf{C} = [1 \quad 0]$$

Suppose that we use the pole placement approach to the design of the system and that the desired closed-loop poles for this sytem are at $s = \mu_i$ ($i = 1, 2$), where $\mu_1 = -1.8 + j2.4$ and $\mu_2 = -1.8 - j2.4$. The state feedback gain matrix $\mathbf{K}$ for this case was obtained in Example 10–1 as follows:

$$\mathbf{K} = [29.6 \quad 3.6]$$

Using this state feedback gain matrix $\mathbf{K}$, the control signal u is given by

$$u = -\mathbf{Kx} = -[29.6 \quad 3.6]\begin{bmatrix} x_1 \\ x_2 \end{bmatrix}$$

Suppose that we use the observed-state feedback control instead of actual-state feedback control, or

$$u = -\mathbf{K}\tilde{\mathbf{x}} = -[29.6 \quad 3.6]\begin{bmatrix} \tilde{x}_1 \\ \tilde{x}_2 \end{bmatrix}$$

where we choose the eigenvalues of the observer gain matrix to be

$$\mu_1 = \mu_2 = -8$$

Obtain the observer gain matrix $\mathbf{K}_e$ and draw a block diagram for the observed-state feedback control system. Then obtain the transfer function $U(s)/[-Y(s)]$ for the controller–observer and draw a block diagram for the system.

For the system defined by Equation (10–65), the characteristic polynomial is

$$|s\mathbf{I} - \mathbf{A}| = \begin{vmatrix} s & -1 \\ -20.6 & s \end{vmatrix} = s^2 - 20.6 = s^2 + a_1s + a_2$$

Thus

$$a_1 = 0, \qquad a_2 = -20.6$$

The desired characteristic polynomial for the observer is

$$(s - \mu_1)(s - \mu_2) = (s + 8)(s + 8) = s^2 + 16s + 64$$
$$= s^2 + \alpha_1s + \alpha_2$$

Hence

$$\alpha_1 = 16, \qquad \alpha_2 = 64$$

For the determination of the observer gain matrix, we use Equation (10–48), or

$$\mathbf{K}_e = (\mathbf{WN*})^{-1} \begin{bmatrix} \alpha_2 - a_2 \\ \alpha_1 - a_1 \end{bmatrix}$$

where

$$\mathbf{N} = [\mathbf{C*} \mid \mathbf{A*C*}] = \begin{bmatrix} 1 & 0 \\ 0 & 1 \end{bmatrix}$$

$$\mathbf{W} = \begin{bmatrix} a_1 & 1 \\ 1 & 0 \end{bmatrix} = \begin{bmatrix} 0 & 1 \\ 1 & 0 \end{bmatrix}$$

Hence

$$\mathbf{K}_e = \left\{ \begin{bmatrix} 0 & 1 \\ 1 & 0 \end{bmatrix} \begin{bmatrix} 1 & 0 \\ 0 & 1 \end{bmatrix} \right\}^{-1} \begin{bmatrix} 64 + 20.6 \\ 16 - 0 \end{bmatrix}$$

$$= \begin{bmatrix} 0 & 1 \\ 1 & 0 \end{bmatrix} \begin{bmatrix} 84.6 \\ 16 \end{bmatrix} = \begin{bmatrix} 16 \\ 84.6 \end{bmatrix} \tag{10–67}$$

Equation (10–67) gives the observer gain matrix $\mathbf{K}_e$. The observer equation is given by Equation (10–49):

$$\dot{\tilde{\mathbf{x}}} = (\mathbf{A} - \mathbf{K}_e\mathbf{C})\tilde{\mathbf{x}} + \mathbf{B}u + \mathbf{K}_e y \tag{10–68}$$

Since

$$u = -\mathbf{K}\tilde{\mathbf{x}}$$

Equation (10–68) becomes

$$\dot{\tilde{\mathbf{x}}} = (\mathbf{A} - \mathbf{K}_e\mathbf{C} - \mathbf{BK})\tilde{\mathbf{x}} + \mathbf{K}_e y$$

or

$$\begin{bmatrix} \dot{\tilde{x}}_1 \\ \dot{\tilde{x}}_2 \end{bmatrix} = \left\{ \begin{bmatrix} 0 & 1 \\ 20.6 & 0 \end{bmatrix} - \begin{bmatrix} 16 \\ 84.6 \end{bmatrix} [1 \quad 0] - \begin{bmatrix} 0 \\ 1 \end{bmatrix} [29.6 \quad 3.6] \right\} \begin{bmatrix} \tilde{x}_1 \\ \tilde{x}_2 \end{bmatrix} + \begin{bmatrix} 16 \\ 84.6 \end{bmatrix} y$$

$$= \begin{bmatrix} -16 & 1 \\ -93.6 & -3.6 \end{bmatrix} \begin{bmatrix} \tilde{x}_1 \\ \tilde{x}_2 \end{bmatrix} + \begin{bmatrix} 16 \\ 84.6 \end{bmatrix} y$$

The block diagram of the system with observed-state feedback is shown in Figure 10–10.
Referring to Equation (10–64), the transfer function of the controller–observer is

$$\frac{U(s)}{-Y(s)} = \mathbf{K}(s\mathbf{I} - \mathbf{A} + \mathbf{K}_e\mathbf{C} + \mathbf{BK})^{-1}\mathbf{K}_e$$

$$= [29.6 \quad 3.6] \begin{bmatrix} s + 16 & -1 \\ 93.6 & s + 3.6 \end{bmatrix}^{-1} \begin{bmatrix} 16 \\ 84.6 \end{bmatrix}$$

$$= \frac{778.16s + 3690.72}{s^2 + 19.6s + 151.2}$$

Figure 10–11 shows a block diagram of the system.
The dynamics of the observed-state feedback control system just designed can be described by the

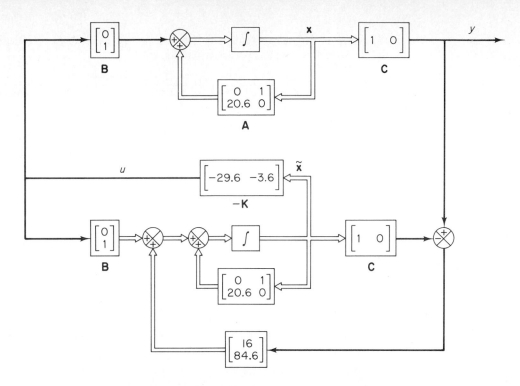

Figure 10–10
Block diagram of system with observed-state feedback. (Example 10–5.)

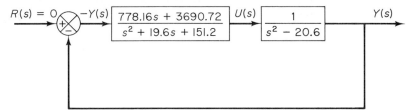

Figure 10–11
Block diagram of transfer function system. (Example 10–5.)

following equations: For the plant,

$$\begin{bmatrix} \dot{x}_1 \\ \dot{x}_2 \end{bmatrix} = \begin{bmatrix} 0 & 1 \\ 20.6 & 0 \end{bmatrix} \begin{bmatrix} x_1 \\ x_2 \end{bmatrix} + \begin{bmatrix} 0 \\ 1 \end{bmatrix} u$$

$$y = [1 \quad 0] \begin{bmatrix} x_1 \\ x_2 \end{bmatrix}$$

$$u = -[29.6 \quad 3.6] \begin{bmatrix} \tilde{x}_1 \\ \tilde{x}_2 \end{bmatrix}$$

For the observer,

$$\begin{bmatrix} \dot{\tilde{x}}_1 \\ \dot{\tilde{x}}_2 \end{bmatrix} = \begin{bmatrix} -16 & 1 \\ -93.6 & -3.6 \end{bmatrix} \begin{bmatrix} \tilde{x}_1 \\ \tilde{x}_2 \end{bmatrix} + \begin{bmatrix} 16 \\ 84.6 \end{bmatrix} y$$

The system, as a whole, is of fourth order. The characteristic equation for the system is

$$|s\mathbf{I} - \mathbf{A} + \mathbf{BK}||s\mathbf{I} - \mathbf{A} + \mathbf{K}_e\mathbf{C}| = (s^2 + 3.6s + 9)(s^2 + 16s + 64)$$
$$= s^4 + 19.6s^3 + 130.6s^2 + 374.4s + 576 = 0$$

The characteristic equation can also be obtained from the block diagram for the sytem shown in Figure 10–11. Since the closed-loop transfer function is

$$\frac{Y(s)}{R(s)} = \frac{778.16s + 3690.72}{(s^2 + 19.6s + 151.2)(s^2 - 20.6) + 778.16s + 3690.72}$$

the characteristic equation is

$$(s^2 + 19.6s + 151.2)(s^2 - 20.6) + 778.16s + 3690.72$$
$$= s^4 + 19.6s^3 + 130.6s^2 + 374.4s + 576 = 0$$

As a matter of course, the characteristic equation is the same for the system in state-space representation and that in transfer-function representation.

Minimum-order observer.   The observers discussed thus far are designed to reconstruct all the state variables. In practice, some of the state variables may be accurately measured. Such accurately measurable state variables need not be estimated. An observer that estimates fewer than n state variables, where n is the dimension of the state vector, is called a *reduced-order observer*. If the order of the reduced-order observer is the minimum possible, the observer is called a *minimum-order observer*.

Suppose the state vector $\mathbf{x}$ is an n-vector and the output vector $\mathbf{y}$ is an m-vector that can be measured. Since m output variables are linear combinations of the state variables, m state variables need not be estimated. We need to estimate only $n - m$ state variables. Then the reduced-order observer becomes an $(n - m)$th-order observer. Such an $(n - m)$th-order observer is the minimum-order observer. Figure 10–12 shows the block diagram of a system with a minimum-order observer.

It is important to note, however, that if the measurement of output variables involves significant noises and is relatively inaccurate, then the use of the full-order observer may result in a better system performance.

To present the basic idea of the minimum-order observer, without undue mathematical

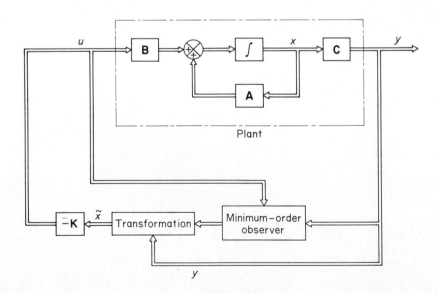

Figure 10–12
Observed-state feedback control system with a minimum-order observer.

complications, we shall present the case where the output is a scalar (that is, $m = 1$) and derive the state equation for the minimum-order observer. Consider the system

$$\dot{\mathbf{x}} = \mathbf{Ax} + \mathbf{B}u$$

$$y = \mathbf{Cx}$$

where the state vector $\mathbf{x}$ can be partitioned into two parts x_a (a scalar) and $\mathbf{x}_b$ [an $(n - 1)$-vector]. Here the state variable x_a is equal to the output y and thus can be directly measured and $\mathbf{x}_b$ is the unmeasurable portion of the state vector. Then the partitioned state and output equations become as

$$\begin{bmatrix} \dot{x}_a \\ \hline \dot{\mathbf{x}}_b \end{bmatrix} = \begin{bmatrix} A_{aa} & \vdots & \mathbf{A}_{ab} \\ \hline \mathbf{A}_{ba} & \vdots & \mathbf{A}_{bb} \end{bmatrix} \begin{bmatrix} x_a \\ \hline \mathbf{x}_b \end{bmatrix} + \begin{bmatrix} B_a \\ \hline \mathbf{B}_b \end{bmatrix} u \tag{10-69}$$

$$y = [1 \; \vdots \; \mathbf{0}] \begin{bmatrix} x_a \\ \hline \mathbf{x}_b \end{bmatrix} \tag{10-70}$$

where A_{aa} = scalar

$\quad \mathbf{A}_{ab} = 1 \times (n - 1)$ matrix

$\quad \mathbf{A}_{ba} = (n - 1) \times 1$ matrix

$\quad \mathbf{A}_{bb} = (n - 1) \times (n - 1)$ matrix

$\quad B_a$ = scalar

$\quad \mathbf{B}_b = (n - 1) \times 1$ matrix

From Equation (10–69), the equation for the measured portion of the state becomes

$$\dot{x}_a = A_{aa}x_a + \mathbf{A}_{ab}\mathbf{x}_b + B_a u$$

or

$$\dot{x}_a - A_{aa}x_a - B_a u = \mathbf{A}_{ab}\mathbf{x}_b \tag{10-71}$$

The terms on the left side of Equation (10–71) can be measured. Equation (10–71) acts as the output equation. In designing the minimum-order observer, we consider the left side of Equation (10–71) to be known quantities. Thus, Equation (10–71) relates the measurable quantities and unmeasurable quantities of the state.

From Equation (10–69), the equation for the unmeasured portion of the state becomes

$$\dot{\mathbf{x}}_b = \mathbf{A}_{ba}x_a + \mathbf{A}_{bb}\mathbf{x}_b + \mathbf{B}_b u \tag{10-72}$$

Noting that terms $\mathbf{A}_{ba}x_a$ and $\mathbf{B}_b u$ are known quantities, Equation (10–72) describes the dynamics of the unmeasured portion of the state.

In what follows we shall present a method for designing a minimum-order observer. The design procedure can be simplified if we utilize the design technique developed for the full-order state observer.

Let us compare the state equation for the full-order observer with that for the minimum-order observer. The state equation for the full-order observer is

$$\dot{\mathbf{x}} = \mathbf{Ax} + \mathbf{B}u$$

and the "state equation" for the minimum-order observer is

$$\dot{\mathbf{x}}_b = \mathbf{A}_{bb}\mathbf{x}_b + \mathbf{A}_{ba}x_a + \mathbf{B}_b u$$

The output equation for the full-order observer is

$$y = \mathbf{Cx}$$

and the "output equation" for the minimum-order observer is

$$\dot{x}_a - A_{aa}x_a - B_a u = \mathbf{A}_{ab}\mathbf{x}_b$$

The design of the minimum-order observer can be carried out as follows: First, note that the observer equation for the full-order observer was given by Equation (10–49), which we repeat here:

$$\dot{\tilde{\mathbf{x}}} = (\mathbf{A} - \mathbf{K}_e\mathbf{C})\tilde{\mathbf{x}} + \mathbf{B}u + \mathbf{K}_e y \qquad (10\text{–}73)$$

Then, making the substitutions of Table 10–2 into Equation (10–73), we obtain

$$\dot{\tilde{\mathbf{x}}}_b = (\mathbf{A}_{bb} - \mathbf{K}_e\mathbf{A}_{ab})\tilde{\mathbf{x}}_b + \mathbf{A}_{ba}x_a + \mathbf{B}_b u + \mathbf{K}_e(\dot{x}_a - A_{aa}x_a - B_a u) \qquad (10\text{–}74)$$

where the state observer gain matrix $\mathbf{K}_e$ is an $(n - 1) \times 1$ matrix. In Equation (10–74) notice that in order to estimate $\tilde{\mathbf{x}}_b$ we need the derivative of x_a. This is undesirable and so we need to modify Equation (10–74).

Let us rewrite Equation (10–74) as follows. Noting that $x_a = y$, we have

$$\dot{\tilde{\mathbf{x}}}_b - \mathbf{K}_e\dot{x}_a = (\mathbf{A}_{bb} - \mathbf{K}_e\mathbf{A}_{ab})\tilde{\mathbf{x}}_b + (\mathbf{A}_{ba} - \mathbf{K}_e A_{aa})y + (\mathbf{B}_b - \mathbf{K}_e B_a)u$$

$$= (\mathbf{A}_{bb} - \mathbf{K}_e\mathbf{A}_{ab})(\tilde{\mathbf{x}}_b - \mathbf{K}_e y)$$

$$+ [(\mathbf{A}_{bb} - \mathbf{K}_e\mathbf{A}_{ab})\mathbf{K}_e + \mathbf{A}_{ba} - \mathbf{K}_e A_{aa}]y$$

$$+ (\mathbf{B}_b - \mathbf{K}_e B_a)u \qquad (10\text{–}75)$$

Define

$$\mathbf{x}_b - \mathbf{K}_e y = \mathbf{x}_b - \mathbf{K}_e x_a = \boldsymbol{\eta}$$

and

$$\tilde{\mathbf{x}}_b - \mathbf{K}_e y = \tilde{\mathbf{x}}_b - \mathbf{K}_e x_a = \tilde{\boldsymbol{\eta}} \qquad (10\text{–}76)$$

Table 10–2 List of Necessary Substitutions for Writing the Observer Equation for the Minimum-Order State Observer

Full-Order State Observer	Minimum-Order State Observer
$\tilde{\mathbf{x}}$	$\tilde{\mathbf{x}}_b$
$\mathbf{A}$	$\mathbf{A}_{bb}$
$\mathbf{B}u$	$\mathbf{A}_{ba}x_a + \mathbf{B}_b u$
y	$\dot{x}_a - A_{aa}x_a - B_a u$
$\mathbf{C}$	$\mathbf{A}_{ab}$
$\mathbf{K}_e$ ($n \times 1$ matrix)	$\mathbf{K}_e$ [$(n - 1) \times 1$ matrix]

Then, Equation (10–75) becomes

$$\dot{\boldsymbol{\eta}} = (\mathbf{A}_{bb} - \mathbf{K}_e\mathbf{A}_{ab})\tilde{\boldsymbol{\eta}} + [(\mathbf{A}_{bb} - \mathbf{K}_e\mathbf{A}_{ab})\mathbf{K}_e$$

$$+ \mathbf{A}_{ba} - \mathbf{K}_e A_{aa}]y + (\mathbf{B}_b - \mathbf{K}_e B_a)u \qquad (10\text{–}77)$$

Equation (10–77) together with Equation (10–76) define the minimum-order observer.

Next we shall derive the observer error equation. Using Equation (10–71), Equation (10–74) can be modified to

$$\dot{\tilde{\mathbf{x}}}_b = (\mathbf{A}_{bb} - \mathbf{K}_e\mathbf{A}_{ab})\tilde{\mathbf{x}}_b + \mathbf{A}_{ba}x_a + \mathbf{B}_b u + \mathbf{K}_e\mathbf{A}_{ab}\mathbf{x}_b \qquad (10\text{–}78)$$

By subtracting Equation (10–78) from Equation (10–72), we obtain

$$\dot{\mathbf{x}}_b - \dot{\tilde{\mathbf{x}}}_b = (\mathbf{A}_{bb} - \mathbf{K}_e\mathbf{A}_{ab})(\mathbf{x}_b - \tilde{\mathbf{x}}_b) \qquad (10\text{–}79)$$

Define

$$\mathbf{e} = \mathbf{x}_b - \tilde{\mathbf{x}}_b = \boldsymbol{\eta} - \tilde{\boldsymbol{\eta}}$$

Then Equation (10–79) becomes

$$\dot{\mathbf{e}} = (\mathbf{A}_{bb} - \mathbf{K}_e\mathbf{A}_{ab})\mathbf{e} \qquad (10\text{–}80)$$

This is the error equation for the minimum-order observer. Note that $\mathbf{e}$ is an $(n - 1)$-vector.

The error dynamics can be chosen as desired by following the technique developed for the full-order observer, provided that the rank of matrix:

$$\begin{bmatrix} \mathbf{A}_{ab} \\ \mathbf{A}_{ab}\mathbf{A}_{bb} \\ \cdot \\ \cdot \\ \cdot \\ \mathbf{A}_{ab}\mathbf{A}_{bb}^{n-2} \end{bmatrix}$$

is $n - 1$. (This is the complete observability condition applicable to the minimum-order observer.)

The characteristic equation for the minimum-order observer is obtained from Equation (10–80) as follows:

$$|s\mathbf{I} - \mathbf{A}_{bb} + \mathbf{K}_e\mathbf{A}_{ab}| = (s - \mu_1)(s - \mu_2) \cdots (s - \mu_{n-1})$$

$$= s^{n-1} + \hat{\alpha}_1 s^{n-2} + \cdots + \hat{\alpha}_{n-2}s + \hat{\alpha}_{n-1} = 0 \qquad (10\text{–}81)$$

where $\mu_1, \mu_2, \ldots, \mu_{n-1}$ are desired eigenvalues for the minimum-order observer. The observer gain matrix $\mathbf{K}_e$ can be determined by first choosing the desired eigenvalues for the minimum-order observer [that is, by placing the roots of the characteristic equation, Equation (10–81), at the desired locations] and then using the procedure developed for the full-order observer with appropriate modifications. For example, if the formula for determining matrix $\mathbf{K}_e$ given by Equation (10–48) is to be used, it should be modified to

$$\mathbf{K}_e = \hat{\mathbf{Q}} \begin{bmatrix} \hat{\alpha}_{n-1} - \hat{a}_{n-1} \\ \hat{\alpha}_{n-2} - \hat{a}_{n-2} \\ \cdot \\ \cdot \\ \cdot \\ \hat{\alpha}_1 - \hat{a}_1 \end{bmatrix} = (\hat{\mathbf{W}}\hat{\mathbf{N}}^*)^{-1} \begin{bmatrix} \hat{\alpha}_{n-1} - \hat{a}_{n-1} \\ \hat{\alpha}_{n-2} - \hat{a}_{n-2} \\ \cdot \\ \cdot \\ \cdot \\ \hat{\alpha}_1 - \hat{a}_1 \end{bmatrix} \tag{10-82}$$

where $\mathbf{K}_e$ is an $(n-1) \times 1$ matrix and

$$\hat{\mathbf{N}} = [\mathbf{A}_{ab}^* \mid \mathbf{A}_{bb}^*\mathbf{A}_{ab}^* \mid \cdots \mid (\mathbf{A}_{bb}^*)^{n-2}\mathbf{A}_{ab}^*] = (n-1) \times (n-1) \text{ matrix}$$

$$\hat{\mathbf{W}} = \begin{bmatrix} \hat{a}_{n-2} & \hat{a}_{n-3} & \cdots & \hat{a}_1 & 1 \\ \hat{a}_{n-3} & \hat{a}_{n-4} & \cdots & 1 & 0 \\ \cdot & \cdot & & \cdot & \cdot \\ \cdot & \cdot & & \cdot & \cdot \\ \cdot & \cdot & & \cdot & \cdot \\ \hat{a}_1 & 1 & \cdots & 0 & 0 \\ 1 & 0 & \cdots & 0 & 0 \end{bmatrix} = (n-1) \times (n-1) \text{ matrix}$$

Note that $\hat{a}_1, \hat{a}_2, \ldots, \hat{a}_{n-2}$ are coefficients in the charcteristic equation for the state equation

$$|s\mathbf{I} - \mathbf{A}_{bb}| = s^{n-1} + \hat{a}_1 s^{n-2} + \cdots + \hat{a}_{n-2}s + \hat{a}_{n-1} = 0$$

Also, if Ackermann's formula given by Equation (10–53) is to be used, then it should be modified to

$$\mathbf{K}_e = \phi(\mathbf{A}_{bb}) \begin{bmatrix} \mathbf{A}_{ab} \\ \mathbf{A}_{ab}\mathbf{A}_{bb} \\ \cdot \\ \cdot \\ \cdot \\ \mathbf{A}_{ab}\mathbf{A}_{bb}^{n-3} \\ \mathbf{A}_{ab}\mathbf{A}_{bb}^{n-2} \end{bmatrix}^{-1} \begin{bmatrix} 0 \\ 0 \\ \cdot \\ \cdot \\ \cdot \\ 0 \\ 1 \end{bmatrix} \tag{10-83}$$

where

$$\phi(\mathbf{A}_{bb}) = \mathbf{A}_{bb}^{n-1} + \hat{\alpha}_1 \mathbf{A}_{bb}^{n-2} + \cdots + \hat{\alpha}_{n-2}\mathbf{A}_{bb} + \hat{\alpha}_{n-1}\mathbf{I}$$

EXAMPLE 10–6 Consider the same system as discussed in Example 10–4. Assume that the output y can be accurately measured. Thus, state variable x_1 (which is equal to y) need not be estimated. Design a minimum-order observer. (The minimum-order observer is of second order.) Assume that the desired eigenvalues for the minimum-order observer are

$$\mu_1 = -2 + j3.464, \qquad \mu_2 = -2 - j3.464$$

Referring to Equation (10–81), the characteristic equation for the minimum-order observer is

$$\begin{aligned} |s\mathbf{I} - \mathbf{A}_{bb} + \mathbf{K}_e\mathbf{A}_{ab}| &= (s - \mu_1)(s - \mu_2) \\ &= (s + 2 - j3.464)(s + 2 + j3.464) \\ &= s^2 + 4s + 16 = 0 \end{aligned}$$

We shall use Ackermann's formula given by Equation (10–83). [Problem A–10–13 presents the determination of $\mathbf{K}_e$ by use of Equation (10–82).]

$$\mathbf{K}_e = \phi(\mathbf{A}_{bb}) \left[\begin{array}{c} \mathbf{A}_{ab} \\ \hline \mathbf{A}_{ab}\mathbf{A}_{bb} \end{array} \right]^{-1} \left[\begin{array}{c} 0 \\ 1 \end{array} \right] \qquad (10\text{–}84)$$

where

$$\phi(\mathbf{A}_{bb}) = \mathbf{A}_{bb}^2 + \hat{\alpha}_1\mathbf{A}_{bb} + \hat{\alpha}_2\mathbf{I} = \mathbf{A}_{bb}^2 + 4\mathbf{A}_{bb} + 16\mathbf{I}$$

Since

$$\mathbf{x} = \left[\begin{array}{c} x_a \\ \hline \mathbf{x}_b \end{array} \right] = \left[\begin{array}{c} x_1 \\ \hline x_2 \\ x_3 \end{array} \right], \qquad \mathbf{A} = \left[\begin{array}{c|cc} 0 & 1 & 0 \\ \hline 0 & 0 & 1 \\ -6 & -11 & -6 \end{array} \right], \qquad \mathbf{B} = \left[\begin{array}{c} 0 \\ \hline 0 \\ 1 \end{array} \right]$$

we have

$$A_{aa} = 0, \qquad \mathbf{A}_{ab} = [1 \quad 0], \qquad \mathbf{A}_{ba} = \left[\begin{array}{c} 0 \\ -6 \end{array} \right]$$

$$\mathbf{A}_{bb} = \left[\begin{array}{cc} 0 & 1 \\ -11 & -6 \end{array} \right], \qquad B_a = 0, \qquad \mathbf{B}_b = \left[\begin{array}{c} 0 \\ 1 \end{array} \right]$$

Equation (10–84) now becomes

$$\mathbf{K}_e = \left\{ \left[\begin{array}{cc} 0 & 1 \\ -11 & -6 \end{array} \right]^2 + 4\left[\begin{array}{cc} 0 & 1 \\ -11 & -6 \end{array} \right] + 16\left[\begin{array}{cc} 1 & 0 \\ 0 & 1 \end{array} \right] \right\} \left[\begin{array}{cc} 1 & 0 \\ 0 & 1 \end{array} \right]^{-1} \left[\begin{array}{c} 0 \\ 1 \end{array} \right]$$

$$= \left[\begin{array}{cc} 5 & -2 \\ 22 & 17 \end{array} \right] \left[\begin{array}{c} 0 \\ 1 \end{array} \right] = \left[\begin{array}{c} -2 \\ 17 \end{array} \right]$$

Referring to Equations (10–76) and (10–77), the equation for the minimum-order observer can be given by

$$\dot{\tilde{\boldsymbol{\eta}}} = (\mathbf{A}_{bb} - \mathbf{K}_e\mathbf{A}_{ab})\tilde{\boldsymbol{\eta}} + [(\mathbf{A}_{bb} - \mathbf{K}_e\mathbf{A}_{ab})\mathbf{K}_e + \mathbf{A}_{ba} - \mathbf{K}_e A_{aa}]y + (\mathbf{B}_b - \mathbf{K}_e B_a)u \qquad (10\text{–}85)$$

where

$$\tilde{\boldsymbol{\eta}} = \tilde{\mathbf{x}}_b - \mathbf{K}_e y = \tilde{\mathbf{x}}_b - \mathbf{K}_e x_1$$

Noting that

$$\mathbf{A}_{bb} - \mathbf{K}_e\mathbf{A}_{ab} = \left[\begin{array}{cc} 0 & 1 \\ -11 & -6 \end{array} \right] - \left[\begin{array}{c} -2 \\ 17 \end{array} \right][1 \quad 0] = \left[\begin{array}{cc} 2 & 1 \\ -28 & -6 \end{array} \right]$$

the equation for the minimum-order observer, Equation (10–85), becomes

$$\left[\begin{array}{c} \dot{\tilde{\eta}}_2 \\ \dot{\tilde{\eta}}_3 \end{array} \right] = \left[\begin{array}{cc} 2 & 1 \\ -28 & -6 \end{array} \right] \left[\begin{array}{c} \tilde{\eta}_2 \\ \tilde{\eta}_3 \end{array} \right] + \left\{ \left[\begin{array}{cc} 2 & 1 \\ -28 & -6 \end{array} \right] \left[\begin{array}{c} -2 \\ 17 \end{array} \right] \right.$$

$$+ \left[\begin{array}{c} 0 \\ -6 \end{array} \right] - \left[\begin{array}{c} -2 \\ 17 \end{array} \right]0 \right\} y + \left\{ \left[\begin{array}{c} 0 \\ 1 \end{array} \right] - \left[\begin{array}{c} -2 \\ 17 \end{array} \right]0 \right\} u$$

or

$$\left[\begin{array}{c} \dot{\tilde{\eta}}_2 \\ \dot{\tilde{\eta}}_3 \end{array} \right] = \left[\begin{array}{cc} 2 & 1 \\ -28 & -6 \end{array} \right] \left[\begin{array}{c} \tilde{\eta}_2 \\ \tilde{\eta}_3 \end{array} \right] + \left[\begin{array}{c} 13 \\ -52 \end{array} \right] y + \left[\begin{array}{c} 0 \\ 1 \end{array} \right] u$$

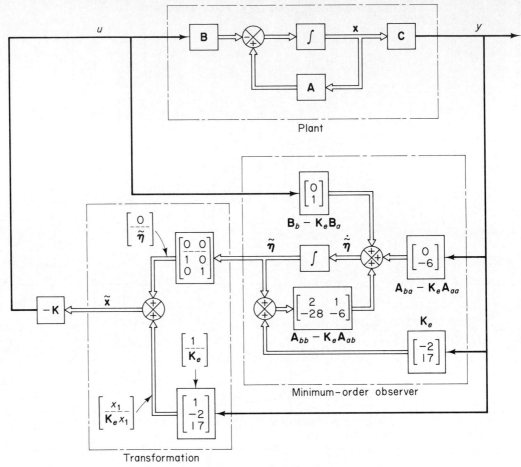

Figure 10–13
System with observed-state feedback where the observer is the minimum order observer designed in Example 10–6.

where

$$\begin{bmatrix} \tilde{\eta}_2 \\ \tilde{\eta}_3 \end{bmatrix} = \begin{bmatrix} \tilde{x}_2 \\ \tilde{x}_3 \end{bmatrix} - \mathbf{K}_e y$$

or

$$\begin{bmatrix} \tilde{x}_2 \\ \tilde{x}_3 \end{bmatrix} = \begin{bmatrix} \tilde{\eta}_2 \\ \tilde{\eta}_3 \end{bmatrix} + \mathbf{K}_e x_1$$

If the observed-state feedback is used, then the control signal u becomes

$$u = -\mathbf{K}\tilde{\mathbf{x}} = -\mathbf{K} \begin{bmatrix} x_1 \\ \tilde{x}_2 \\ \tilde{x}_3 \end{bmatrix}$$

where $\mathbf{K}$ is the state feedback gain matrix. (The matrix $\mathbf{K}$ is not determined in this example.) Figure 10–13 is a block diagram showing the configuration of the system with observed-state feedback, where the observer is the minimum-order observer.

Observed-state feedback control system with minimum-order observer. For the case of the observed-state feedback control system with full-order state observer, we have shown that the closed-loop poles of the observed-state feedback control system consist of the poles due to the pole placement design alone, plus the poles due to the observer design alone. Hence the pole placement design and the full-order observer design are independent of each other.

For the observed-state feedback control sytem with minimum-order observer, the same conclusion applies. The system characteristic equation can be derived as

$$|s\mathbf{I} - \mathbf{A} + \mathbf{BK}\,\|\,s\mathbf{I} - \mathbf{A}_{bb} + \mathbf{K}_e\mathbf{A}_{ab}| = 0$$

(See Problem A–10–12 for the detail.) The closed-loop poles of the observed-state feedback control system with a minimum-order observer comprise the closed-loop poles due to pole placement [the eigenvalues of matrix $(\mathbf{A} - \mathbf{BK})$] and the closed-loop poles due to the minimum-order observer [the eigenvalues of matrix $(\mathbf{A}_{bb} - \mathbf{K}_e\mathbf{A}_{ab})$]. Therefore, the pole placement design and the design of the minimum-order observer are independent of each other.

10–4 DESIGN OF SERVO SYSTEMS

In Section 4–7, we discussed the system types according to the number of the integrators in the feedforward transfer function. The type 1 system has one integrator in the feedforward path and the system will exhibit no steady-state error in the step response. In this section we shall discuss the pole placement approach to the design of type 1 servo systems. Here we shall limit our systems each to have a scalar control signal u and a scalar output y.

In Chapter 7 we discussed I–PD control systems. In the I–PD control system, an integrator is placed in the feedforward path to integrate the error signal, and the proportional and derivative controls were inserted in the minor loop. Figure 10–14 shows a block diagram of an I–PD control of a plant $G_p(s)$, where we assume the plant has no integrator. Figure 10–15 shows a block diagram of a type 1 servo system. It is a more general case of the

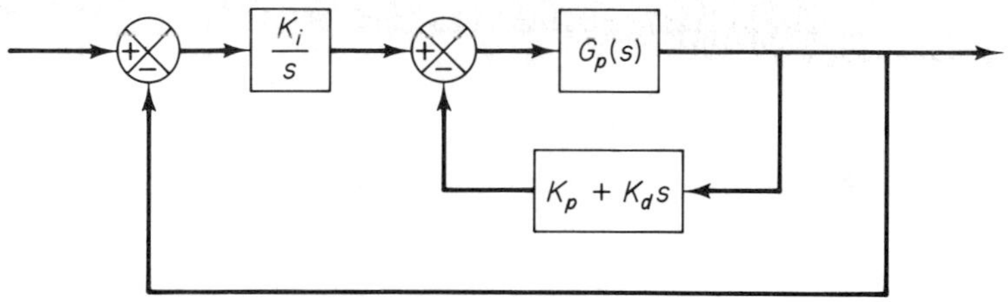

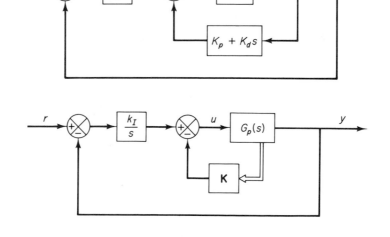

Figure 10–14
I-PD control of plant $G_p(s)$.

Figure 10–15
Type 1 servo system.

I–PD control of the plant $G_p(s)$. Such a configuration of the type 1 servo system is commonly encountered in practice. (Other configurations are also used in practice.) In the servo system shown in Figure 10–15, the integral control action together with state feedback scheme is used to properly stabilize the system. This system will exhibit no steady-state error in the response to the step input.

In what follows we shall first discuss a problem of designing a type 1 servo system where the plant involves an integrator. Then we shall discuss the design of type 1 servo systems where the plant has no integrator.

Type 1 servo system when plant has an integrator.

Assume that the plant is defined by

$$\dot{\mathbf{x}} = \mathbf{A}\mathbf{x} + \mathbf{B}u \tag{10–86}$$

$$y = \mathbf{C}\mathbf{x} \tag{10–87}$$

where $\mathbf{x}$ = state vector for the plant (n-vector)
  u = control signal (scalar)
  y = output signal (scalar)
  $\mathbf{A} = n \times n$ constant matrix
  $\mathbf{B} = n \times 1$ constant matrix
  $\mathbf{C} = 1 \times n$ constant matrix

As stated earlier, we assume that both the control signal u and the output signal y are scalars. By a proper choice of a set of state variables, it is possible to choose the output to be equal to one of the state variables. (See the method presented in Section 2–2 for obtaining a state-space representation of the transfer function system in which the output y becomes equal to x_1.)

Figure 10–16 shows a general configuration of the type 1 servo system where the plant has an integrator. Here we assumed that $y = x_1$. In the present analysis we assume that the reference input r is a step function. In this system we use the following state feedback control scheme:

$$u = -\begin{bmatrix} 0 & k_2 & k_3 & \cdots & k_n \end{bmatrix} \begin{bmatrix} x_1 \\ x_2 \\ \cdot \\ \cdot \\ \cdot \\ x_n \end{bmatrix} + k_1(r - x_1)$$

$$= -\mathbf{K}\mathbf{x} + k_1 r \tag{10–88}$$

where

$$\mathbf{K} = \begin{bmatrix} k_1 & k_2 & \cdots & k_n \end{bmatrix} \tag{10–89}$$

Assume that the reference input (step function) is applied at $t = 0$. Then, for $t > 0$, the system dynamics can be described by Equations (10–86) and (10–88), or

$$\dot{\mathbf{x}} = \mathbf{A}\mathbf{x} + \mathbf{B}u = (\mathbf{A} - \mathbf{B}\mathbf{K})\mathbf{x} + \mathbf{B}k_1 r \tag{10–90}$$

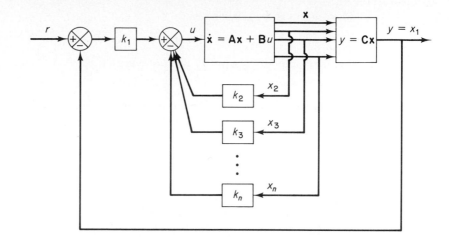

Figure 10–16
Type 1 servo system
where the plant has an
integrator.

We shall design the type 1 servo system such that the closed-loop poles are located at desired positions. The designed system will be an asymptotically stable system, and $y(\infty)$ will approach constant value r and $u(\infty)$ will approach zero.

Notice that at steady state we have

$$\dot{\mathbf{x}}(\infty) = (\mathbf{A} - \mathbf{BK})\mathbf{x}(\infty) + \mathbf{B}k_1 r(\infty) \qquad (10\text{--}91)$$

Noting that $r(t)$ is a step input, we have $r(\infty) = r(t) = r$ (constant) for $t > 0$. By subtracting Equation (10–91) from Equation (10–90), we obtain

$$\dot{\mathbf{x}}(t) - \dot{\mathbf{x}}(\infty) = (\mathbf{A} - \mathbf{BK})[\mathbf{x}(t) - \mathbf{x}(\infty)] \qquad (10\text{--}92)$$

Define

$$\mathbf{x}(t) - \mathbf{x}(\infty) = \mathbf{e}(t)$$

Then Equation (10–92) becomes

$$\dot{\mathbf{e}} = (\mathbf{A} - \mathbf{BK})\mathbf{e} \qquad (10\text{--}93)$$

Equation (10–93) describes the error dynamics.

The design of the type 1 servo system here becomes that of a design of an asymptotically stable regulator system such that $\mathbf{e}(t)$ approaches zero, given any initial condition $\mathbf{e}(0)$. If the system defined by Equation (10–86) is completely state controllable, then by specifying the desired eigenvalues $\mu_1, \mu_2, \ldots, \mu_n$ for the matrix $\mathbf{A} - \mathbf{BK}$, matrix $\mathbf{K}$ can be determined by the pole placement technique presented in Section 10–2.

The steady-state values of $\mathbf{x}(t)$ and $u(t)$ can be found as follows: At steady state ($t = \infty$), we have, from Equation (10–90),

$$\dot{\mathbf{x}}(\infty) = \mathbf{0} = (\mathbf{A} - \mathbf{BK})\mathbf{x}(\infty) + \mathbf{B}k_1 r$$

Since the desired eigenvalues of $\mathbf{A} - \mathbf{BK}$ are all in the left-half s plane, the inverse of matrix $\mathbf{A} - \mathbf{BK}$ exists. Consequently, $\mathbf{x}(\infty)$ can be determined as

$$\mathbf{x}(\infty) = -(\mathbf{A} - \mathbf{BK})^{-1}\mathbf{B}k_1 r$$

Also, $u(\infty)$ can be obtained as

$$u(\infty) = -\mathbf{K}\mathbf{x}(\infty) + k_1 r = 0$$

(See Example 10–7 to verify this last equation.)

EXAMPLE 10–7   Consider the design of a type 1 servo system where the plant transfer function has the integrator

$$\frac{Y(s)}{U(s)} = \frac{1}{s(s + 1)(s + 2)}$$

It is desired to design a type 1 servo system such that the closed-loop poles are at $-2 \pm j3.464$ and -10. Assume that the system configuration is the same as that shown in Figure 10–16 and the reference input r is a step function.

Define state variables x_1, x_2, and x_3 as follows:

$$x_1 = y$$

$$x_2 = \dot{x}_1$$

$$x_3 = \dot{x}_2$$

Then the state space representation of the system becomes

$$\dot{\mathbf{x}} = \mathbf{A}\mathbf{x} + \mathbf{B}u \qquad (10\text{–}94)$$
$$y = \mathbf{C}\mathbf{x} \qquad (10\text{–}95)$$

where

$$\mathbf{A} = \begin{bmatrix} 0 & 1 & 0 \\ 0 & 0 & 1 \\ 0 & -2 & -3 \end{bmatrix}, \qquad \mathbf{B} = \begin{bmatrix} 0 \\ 0 \\ 1 \end{bmatrix}, \qquad \mathbf{C} = \begin{bmatrix} 1 & 0 & 0 \end{bmatrix}$$

Referring to Figure 10–16 and noting that $n = 3$, the control signal u is given by

$$u = -(k_2 x_2 + k_3 x_3) + k_1(r - x_1) = -\mathbf{K}\mathbf{x} + k_1 r \qquad (10\text{–}96)$$

where

$$\mathbf{K} = \begin{bmatrix} k_1 & k_2 & k_3 \end{bmatrix}$$

Our problem here is to determine the state feedback gain matrix $\mathbf{K}$ by the pole placement approach.

Let us examine the controllability matrix for the system. The rank of

$$\mathbf{M} = \begin{bmatrix} \mathbf{B} & \vdots & \mathbf{A}\mathbf{B} & \vdots & \mathbf{A}^2\mathbf{B} \end{bmatrix} = \begin{bmatrix} 0 & 0 & 1 \\ 0 & 1 & -3 \\ 1 & -3 & 7 \end{bmatrix}$$

is 3. Hence the plant is completely state controllable. The characteristic equation for the system is

$$|s\mathbf{I} - \mathbf{A}| = \begin{vmatrix} s & -1 & 0 \\ 0 & s & -1 \\ 0 & 2 & s + 3 \end{vmatrix}$$

$$= s^3 + 3s^2 + 2s$$

$$= s^3 + a_1 s^2 + a_2 s + a_3 = 0$$

Hence

$$a_1 = 3, \qquad a_2 = 2, \qquad a_3 = 0$$

For the determination of the matrix $\mathbf{K}$ by the pole placement approach, we shall use Equation (10–13), rewritten as

$$\mathbf{K} = [\alpha_3 - a_3 \; \vdots \; \alpha_2 - a_2 \; \vdots \; \alpha_1 - a_1]\mathbf{T}^{-1} \qquad (10\text{–}97)$$

Since the state equation for the system, Equation (10–94), is already in the controllable canonical form, we have $\mathbf{T} = \mathbf{I}$.

By substituting Equation (10–96) into Equation (10–94), we obtain

$$\dot{\mathbf{x}} = \mathbf{A}\mathbf{x} + \mathbf{B}(-\mathbf{K}\mathbf{x} + k_1 r) = (\mathbf{A} - \mathbf{B}\mathbf{K})\mathbf{x} + \mathbf{B}k_1 r \qquad (10\text{–}98)$$

where the input r is a step function. Then, as t approaches infinity, $\mathbf{x}(t)$ approaches $\mathbf{x}(\infty)$, a constant vector. At steady state, we have

$$\dot{\mathbf{x}}(\infty) = (\mathbf{A} - \mathbf{B}\mathbf{K})\mathbf{x}(\infty) + \mathbf{B}k_1 r \qquad (10\text{–}99)$$

Subtracting Equation (10–99) from Equation (10–98), we have

$$\dot{\mathbf{x}}(t) - \dot{\mathbf{x}}(\infty) = (\mathbf{A} - \mathbf{B}\mathbf{K})[\mathbf{x}(t) - \mathbf{x}(\infty)]$$

Define

$$\mathbf{x}(t) - \mathbf{x}(\infty) = \mathbf{e}(t)$$

Then

$$\dot{\mathbf{e}}(t) = (\mathbf{A} - \mathbf{B}\mathbf{K})\mathbf{e}(t) \qquad (10\text{–}100)$$

Equation (10–100) defines the error dynamics.

Since the desired eigenvalues of $\mathbf{A} - \mathbf{B}\mathbf{K}$ are

$$\mu_1 = -2 + j3.464, \qquad \mu_2 = -2 - j3.464, \qquad \mu_3 = -10$$

we have the desired characteristic equation as follows:

$$\begin{aligned}
(s - \mu_1)(s - \mu_2)(s - \mu_3) &= (s + 2 - j3.464)(s + 2 + j3.464)(s + 10) \\
&= s^3 + 14s^2 + 56s + 160 \\
&= s^3 + \alpha_1 s^2 + \alpha_2 s + \alpha_3 = 0
\end{aligned}$$

Hence

$$\alpha_1 = 14, \qquad \alpha_2 = 56, \qquad \alpha_3 = 160$$

The state feedback gain matrix $\mathbf{K}$ is given by Equation (10–97), or

$$\begin{aligned}
\mathbf{K} &= [\alpha_3 - a_3 \; \vdots \; \alpha_2 - a_2 \; \vdots \; \alpha_1 - a_1]\mathbf{T}^{-1} \\
&= [160 - 0 \; \vdots \; 56 - 2 \; \vdots \; 14 - 3]\mathbf{I} \\
&= [160 \quad 54 \quad 11]
\end{aligned}$$

The step response of this system can be obtained easily by a computer simulation. Since

$$\mathbf{A} - \mathbf{B}\mathbf{K} = \begin{bmatrix} 0 & 1 & 0 \\ 0 & 0 & 1 \\ 0 & -2 & -3 \end{bmatrix} - \begin{bmatrix} 0 \\ 0 \\ 1 \end{bmatrix}[160 \quad 54 \quad 11] = \begin{bmatrix} 0 & 1 & 0 \\ 0 & 0 & 1 \\ -160 & -56 & -14 \end{bmatrix}$$

the state equation for the designed system is

$$\begin{bmatrix} \dot{x}_1 \\ \dot{x}_2 \\ \dot{x}_3 \end{bmatrix} = \begin{bmatrix} 0 & 1 & 0 \\ 0 & 0 & 1 \\ -160 & -56 & -14 \end{bmatrix} \begin{bmatrix} x_1 \\ x_2 \\ x_3 \end{bmatrix} + \begin{bmatrix} 0 \\ 0 \\ 160 \end{bmatrix} r$$

and the output equation is

$$y = \begin{bmatrix} 1 & 0 & 0 \end{bmatrix} \begin{bmatrix} x_1 \\ x_2 \\ x_3 \end{bmatrix}$$

The unit-step response curve $y(t)$ versus t obtained in the computer simulation is shown in Figure 10–17.

Notice that $\dot{\mathbf{x}}(\infty) = \mathbf{0}$. Hence we have, from Equation (10–99),

$$(\mathbf{A} - \mathbf{BK})\mathbf{x}(\infty) = -\mathbf{B}k_1 r$$

Since

$$(\mathbf{A} - \mathbf{BK})^{-1} = \begin{bmatrix} 0 & 1 & 0 \\ 0 & 0 & 1 \\ -160 & -56 & -14 \end{bmatrix}^{-1} = \begin{bmatrix} -\dfrac{7}{20} & -\dfrac{7}{80} & -\dfrac{1}{160} \\ 1 & 0 & 0 \\ 0 & 1 & 0 \end{bmatrix}$$

we have

$$\mathbf{x}(\infty) = -(\mathbf{A} - \mathbf{BK})^{-1}\mathbf{B}k_1 r = -\begin{bmatrix} -\dfrac{7}{20} & -\dfrac{7}{80} & -\dfrac{1}{160} \\ 1 & 0 & 0 \\ 0 & 1 & 0 \end{bmatrix} \begin{bmatrix} 0 \\ 0 \\ 1 \end{bmatrix}(160)r$$

$$= \begin{bmatrix} \dfrac{1}{160} \\ 0 \\ 0 \end{bmatrix}(160)r = \begin{bmatrix} 1 \\ 0 \\ 0 \end{bmatrix} r = \begin{bmatrix} r \\ 0 \\ 0 \end{bmatrix}$$

Clearly, $x_1(\infty) = y(\infty) = r$. There is no steady-state error in the step response.

Note that since

$$u(\infty) = -\mathbf{K}\mathbf{x}(\infty) + k_1 r(\infty) = -\mathbf{K}\mathbf{x}(\infty) + k_1 r$$

we have

$$u(\infty) = -\begin{bmatrix} 160 & 54 & 11 \end{bmatrix} \begin{bmatrix} x_1(\infty) \\ x_2(\infty) \\ x_3(\infty) \end{bmatrix} + 160r$$

$$= -\begin{bmatrix} 160 & 54 & 11 \end{bmatrix} \begin{bmatrix} r \\ 0 \\ 0 \end{bmatrix} + 160r$$

$$= -160r + 160r = 0$$

At steady state the control signal u becomes zero.

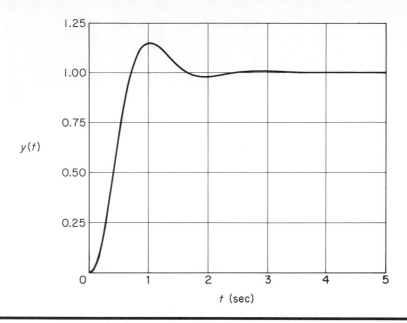

Figure 10–17
Unit-step response curve $y(t)$ versus t for the system designed in Example 10–7.

Design of type 1 servo system where the plant has no integrator. Since the plant has no integrator (type 0 plant), the basic principle of the design of a type 1 servo system is to insert an integrator in the feedforward path between the error comparator and the plant as shown in Figure 10–18. (The block diagram of Figure 10–18 is a basic form of the type 1 servo system where the plant has no integrator.) From the diagram we obtain

$$\dot{\mathbf{x}} = \mathbf{A}\mathbf{x} + \mathbf{B}u \tag{10–101}$$

$$y = \mathbf{C}\mathbf{x} \tag{10–102}$$

$$u = -\mathbf{K}\mathbf{x} + k_1\xi \tag{10–103}$$

$$\dot{\xi} = r - y = r - \mathbf{C}\mathbf{x} \tag{10–104}$$

where $\mathbf{x} = $ state vector of the plant (n-vector)
   $u = $ control signal (scalar)
   $y = $ output signal (scalar)
   $\xi = $ output of the integrator (state variable of the system, scalar)
   $r = $ reference input signal (step function, scalar)
   $\mathbf{A} = n \times n$ constant matrix
   $\mathbf{B} = n \times 1$ constant matrix
   $\mathbf{C} = 1 \times n$ constant matrix

We assume that the plant given by Equation (10–101) is completely state controllable. The transfer function of the plant can be given by

$$G_p(s) = \mathbf{C}(s\mathbf{I} - \mathbf{A})^{-1}\mathbf{B}$$

To avoid the possibility of the inserted integrator being canceled by the zero at the origin of the plant, we assume that $G_p(s)$ has no zero at the origin.

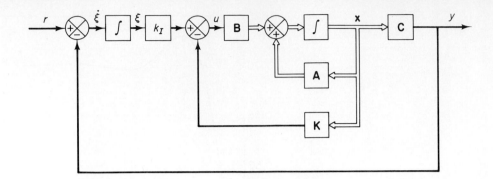

Figure 10–18
Type 1 servo system.

Assume that the reference input (step function) is applied at $t = 0$. Then, for $t > 0$, the system dynamics can be described by an equation that is a combination of Equations (10–101) and (10–104):

$$\begin{bmatrix} \dot{\mathbf{x}}(t) \\ \dot{\xi}(t) \end{bmatrix} = \begin{bmatrix} \mathbf{A} & \mathbf{0} \\ -\mathbf{C} & 0 \end{bmatrix} \begin{bmatrix} \mathbf{x}(t) \\ \xi(t) \end{bmatrix} + \begin{bmatrix} \mathbf{B} \\ 0 \end{bmatrix} u(t) + \begin{bmatrix} \mathbf{0} \\ 1 \end{bmatrix} r(t) \qquad (10\text{–}105)$$

We shall design an asymptotically stable system such that $\mathbf{x}(\infty)$, $\xi(\infty)$, and $u(\infty)$ approach constant values, respectively. Then, at steady state $\dot{\xi}(t) = 0$, and we get $y(\infty) = r$.

Notice that at steady state we have

$$\begin{bmatrix} \dot{\mathbf{x}}(\infty) \\ \dot{\xi}(\infty) \end{bmatrix} = \begin{bmatrix} \mathbf{A} & \mathbf{0} \\ -\mathbf{C} & 0 \end{bmatrix} \begin{bmatrix} \mathbf{x}(\infty) \\ \xi(\infty) \end{bmatrix} + \begin{bmatrix} \mathbf{B} \\ 0 \end{bmatrix} u(\infty) + \begin{bmatrix} \mathbf{0} \\ 1 \end{bmatrix} r(\infty) \qquad (10\text{–}106)$$

Noting that $r(t)$ is a step input, we have $r(\infty) = r(t) = r$ (constant) for $t > 0$. By subtracting Equation (10–106) from Equation (10–105), we obtain

$$\begin{bmatrix} \dot{\mathbf{x}}(t) - \dot{\mathbf{x}}(\infty) \\ \dot{\xi}(t) - \dot{\xi}(\infty) \end{bmatrix} = \begin{bmatrix} \mathbf{A} & \mathbf{0} \\ -\mathbf{C} & 0 \end{bmatrix} \begin{bmatrix} \mathbf{x}(t) - \mathbf{x}(\infty) \\ \xi(t) - \xi(\infty) \end{bmatrix} + \begin{bmatrix} \mathbf{B} \\ 0 \end{bmatrix} [u(t) - u(\infty)] \qquad (10\text{–}107)$$

Define

$$\mathbf{x}(t) - \mathbf{x}(\infty) = \mathbf{x}_e(t)$$

$$\xi(t) - \xi(\infty) = \xi_e(t)$$

$$u(t) - u(\infty) = u_e(t)$$

Then Equation (10–107) can be written as

$$\begin{bmatrix} \dot{\mathbf{x}}_e(t) \\ \dot{\xi}_e(t) \end{bmatrix} = \begin{bmatrix} \mathbf{A} & \mathbf{0} \\ -\mathbf{C} & 0 \end{bmatrix} \begin{bmatrix} \mathbf{x}_e(t) \\ \xi_e(t) \end{bmatrix} + \begin{bmatrix} \mathbf{B} \\ 0 \end{bmatrix} u_e(t) \qquad (10\text{–}108)$$

where

$$u_e(t) = -\mathbf{K}\mathbf{x}_e(t) + k_I \xi_e(t) \qquad (10\text{–}109)$$

Define a new $(n + 1)$th-order error vector $\mathbf{e}(t)$ by

$$\mathbf{e}(t) = \begin{bmatrix} \mathbf{x}_e(t) \\ \xi_e(t) \end{bmatrix} = (n + 1)\text{-vector}$$

Then Equation (10–108) becomes

$$\dot{\mathbf{e}} = \hat{\mathbf{A}}\mathbf{e} + \hat{\mathbf{B}}u_e \qquad (10–110)$$

where

$$\hat{\mathbf{A}} = \begin{bmatrix} \mathbf{A} & \mathbf{0} \\ -\mathbf{C} & 0 \end{bmatrix}, \qquad \hat{\mathbf{B}} = \begin{bmatrix} \mathbf{B} \\ 0 \end{bmatrix}$$

and Equation (10–109) becomes

$$u_e = -\hat{\mathbf{K}}\mathbf{e} \qquad (10–111)$$

where

$$\hat{\mathbf{K}} = [\mathbf{K} \mid -k_I]$$

The basic idea of designing the type 1 servo system here is to design a stable $(n + 1)$th-order regulator system that will bring the new error vector $\mathbf{e}(t)$ to zero, given any initial condition $\mathbf{e}(0)$.

Equations (10–110) and (10–111) describe the dynamics of the $(n + 1)$th-order regulator system. If the system defined by Equation (10–110) is completely state controllable, then, by specifying the desired characteristic equation for the system, matrix $\hat{\mathbf{K}}$ can be determined by the pole placement technique presented in Section 10–2.

The steady-state values of $\mathbf{x}(t)$, $\xi(t)$, and $u(t)$ can be found as follows: At steady state $(t = \infty)$, from Equations (10–101) and (10–104), we have

$$\dot{\mathbf{x}}(\infty) = \mathbf{0} = \mathbf{A}\mathbf{x}(\infty) + \mathbf{B}u(\infty)$$

$$\dot{\xi}(\infty) = 0 = r - \mathbf{C}\mathbf{x}(\infty)$$

which can be combined into one vector-matrix equation:

$$\begin{bmatrix} \mathbf{0} \\ 0 \end{bmatrix} = \begin{bmatrix} \mathbf{A} & \mathbf{B} \\ -\mathbf{C} & 0 \end{bmatrix} \begin{bmatrix} \mathbf{x}(\infty) \\ u(\infty) \end{bmatrix} + \begin{bmatrix} \mathbf{0} \\ r \end{bmatrix}$$

If matrix $\mathbf{P}$, defined by

$$\mathbf{P} = \begin{bmatrix} \mathbf{A} & \mathbf{B} \\ -\mathbf{C} & 0 \end{bmatrix} \qquad (10–112)$$

is of rank $n + 1$, then its inverse exists and

$$\begin{bmatrix} \mathbf{x}(\infty) \\ u(\infty) \end{bmatrix} = \begin{bmatrix} \mathbf{A} & \mathbf{B} \\ -\mathbf{C} & 0 \end{bmatrix}^{-1} \begin{bmatrix} \mathbf{0} \\ -r \end{bmatrix}$$

Also, from Equation (10–103) we have

$$u(\infty) = -\mathbf{K}\mathbf{x}(\infty) + k_I\xi(\infty)$$

and therefore we have

$$\xi(\infty) = \frac{1}{k_I}[u(\infty) + \mathbf{K}\mathbf{x}(\infty)]$$

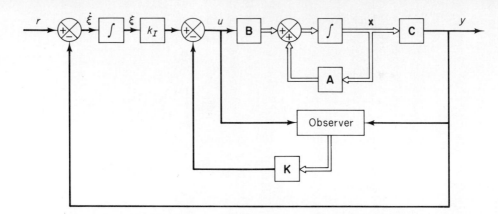

Figure 10–19
Type 1 servo system
with state observer.

It is noted that, if matrix **P** given by Equation (10–112) has rank $n + 1$, then the system defined by Equation (10–110) becomes completely state controllable (see Problem A–10–14). Therefore, if the rank of matrix **P** given by Equation (10–112) is $n + 1$, then the solution to this problem can be obtained by the pole placement approach.

The state error equation can be obtained by substituting Equation (10–111) into Equation (10–110).

$$\dot{\mathbf{e}} = (\hat{\mathbf{A}} - \hat{\mathbf{B}}\hat{\mathbf{K}})\mathbf{e} \qquad (10\text{–}113)$$

If the desired eigenvalues of matrix $\hat{\mathbf{A}} - \hat{\mathbf{B}}\hat{\mathbf{K}}$ (that is, the desired closed-loop poles) are specified as $\mu_1, \mu_2, \ldots, \mu_{n+1}$, then the state feedback gain matrix **K** and the integral gain constant k_I can be determined. In the actual design, it is necessary to consider several different matrices $\hat{\mathbf{K}}$ (which correspond to several different sets of desired eigenvalues) and carry out computer simulations to find the one that yields the best overall system performance. Then choose the best one as the matrix $\hat{\mathbf{K}}$.

As is usually the case, not all state variables can be directly measurable. If this is the case, we need to use a state observer. Figure 10–19 shows a block diagram of a type 1 servo system with a state observer.

EXAMPLE 10–8

Referring to Example 10–2, consider the inverted pendulum system shown in Figure 10–3. In this example, we are concerned only with the motion of the pendulum and motion of the cart on the plane of the page. We assume that the pendulum angle θ and the angular velocity $\dot{\theta}$ are small so that sin $\theta \doteq \theta$, cos $\theta \doteq 1$, and $\theta\dot{\theta}^2 \doteq 0$. We also assume the same numerical values for M, m, and l as we used in Example 10–2.

It is desired to keep the inverted pendulum upright as much as possible and yet control the position of the cart, for instance, move the cart in a step fashion. To control the position of the cart, we need to build a type 1 servo system. The inverted-pendulum system mounted on a cart does not have an integrator. Therefore, we feed the position signal y (which indicates the position of the cart) back to the input and insert an integrator in the feedforward path, as shown in Figure 10–20.

As in the case of Example 10–2, we define state variables x_1, x_2, x_3, and x_4 by

$$x_1 = \theta$$

$$x_2 = \dot{\theta}$$

$$x_3 = x$$

$$x_4 = \dot{x}$$

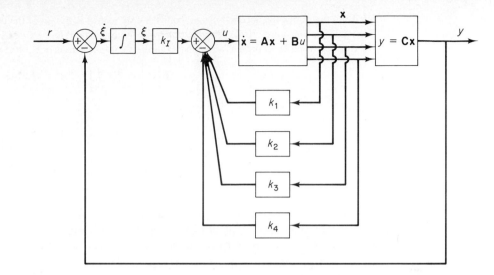

Figure 10–20
Inverted-pendulum control system. (Type 1 servo system where the plant has no integrator.)

Then, referring to Example 10–2 and Figure 10–20, the equations for the inverted pendulum system are

$$\dot{\mathbf{x}} = \mathbf{A}\mathbf{x} + \mathbf{B}u \tag{10–114}$$

$$y = \mathbf{C}\mathbf{x} \tag{10–115}$$

$$u = -\mathbf{K}\mathbf{x} + k_I\xi \tag{10–116}$$

$$\dot{\xi} = r - y = r - \mathbf{C}\mathbf{x} \tag{10–117}$$

where

$$\mathbf{A} = \begin{bmatrix} 0 & 1 & 0 & 0 \\ 20.601 & 0 & 0 & 0 \\ 0 & 0 & 0 & 1 \\ -0.4905 & 0 & 0 & 0 \end{bmatrix}, \quad \mathbf{B} = \begin{bmatrix} 0 \\ -1 \\ 0 \\ 0.5 \end{bmatrix}, \quad \mathbf{C} = [0 \quad 0 \quad 1 \quad 0]$$

For the type 1 servo system, we have the state error equation as given by Equation (10–110):

$$\dot{\mathbf{e}} = \hat{\mathbf{A}}\mathbf{e} + \hat{\mathbf{B}}u_e \tag{10–118}$$

where

$$\hat{\mathbf{A}} = \begin{bmatrix} \mathbf{A} & \mathbf{0} \\ -\mathbf{C} & 0 \end{bmatrix} = \begin{bmatrix} 0 & 1 & 0 & 0 & 0 \\ 20.601 & 0 & 0 & 0 & 0 \\ 0 & 0 & 0 & 1 & 0 \\ -0.4905 & 0 & 0 & 0 & 0 \\ 0 & 0 & -1 & 0 & 0 \end{bmatrix}, \quad \hat{\mathbf{B}} = \begin{bmatrix} \mathbf{B} \\ 0 \end{bmatrix} = \begin{bmatrix} 0 \\ -1 \\ 0 \\ 0.5 \\ 0 \end{bmatrix}$$

and the control signal is given by Equation (10–111):

$$u_e = -\hat{\mathbf{K}}\mathbf{e}$$

where

$$\hat{\mathbf{K}} = [\mathbf{K} \;\vdots\; -k_I] = [k_1 \quad k_2 \quad k_3 \quad k_4 \;\vdots\; -k_I]$$

We shall determine the necessary state feedback gain matrix $\hat{\mathbf{K}}$ by use of the pole placement technique. We shall use Equation (10–13) for the determination of matrix $\hat{\mathbf{K}}$.

Before we proceed further, we must examine the rank of matrix $\mathbf{P}$, where

$$\mathbf{P} = \begin{bmatrix} \mathbf{A} & \mathbf{B} \\ -\mathbf{C} & 0 \end{bmatrix}$$

Matrix $\mathbf{P}$ is given by

$$\mathbf{P} = \begin{bmatrix} \mathbf{A} & \mathbf{B} \\ -\mathbf{C} & 0 \end{bmatrix} = \begin{bmatrix} 0 & 1 & 0 & 0 & 0 \\ 20.601 & 0 & 0 & 0 & -1 \\ 0 & 0 & 0 & 1 & 0 \\ -0.4905 & 0 & 0 & 0 & 0.5 \\ 0 & 0 & -1 & 0 & 0 \end{bmatrix} \tag{10–119}$$

The rank of this matrix is 5. (See Problem A–10–16 for the rank test.) Therefore, the system defined by Equation (10–118) is completely state controllable and arbitrary pole placement is possible. (Refer to Problem A–10–14.) We shall next obtain the characteristic equation for the system given by Equation (10–118).

$$|s\mathbf{I} - \hat{\mathbf{A}}| = \begin{vmatrix} s & -1 & 0 & 0 & 0 \\ -20.601 & s & 0 & 0 & 0 \\ 0 & 0 & s & -1 & 0 \\ 0.4905 & 0 & 0 & s & 0 \\ 0 & 0 & 1 & 0 & s \end{vmatrix}$$

$$= s^3(s^2 - 20.601)$$

$$= s^5 - 20.601s^3$$

$$= s^5 + a_1 s^4 + a_2 s^3 + a_3 s^2 + a_4 s + a_5 = 0$$

Hence

$$a_1 = 0, \qquad a_2 = -20.601, \qquad a_3 = 0, \qquad a_4 = 0, \qquad a_5 = 0$$

To obtain a reasonable speed and damping in the response of the designed system (for example, the settling time of approximately 4 $\sim$ 5 sec and the maximum overshoot of 15% $\sim$ 16% in the step response of the cart), let us choose the desired closed-loop poles at $s = \mu_i$ ($i = 1, 2, 3, 4, 5$), where

$$\mu_1 = -1 + j1.732, \qquad \mu_2 = -1 - j1.732, \qquad \mu_3 = -5, \qquad \mu_4 = -5, \qquad \mu_5 = -5$$

(This is a possible set of desired closed-loop poles. Other sets can be chosen.) Then the desired characteristic equation becomes

$$(s - \mu_1)(s - \mu_2)(s - \mu_3)(s - \mu_4)(s - \mu_5)$$
$$= (s + 1 - j1.732)(s + 1 + j1.732)(s + 5)(s + 5)(s + 5)$$
$$= s^5 + 17s^4 + 109s^3 + 335s^2 + 550s + 500$$
$$= s^5 + \alpha_1 s^4 + \alpha_2 s^3 + \alpha_3 s^2 + \alpha_4 s + \alpha_5 = 0$$

Hence

$$\alpha_1 = 17, \qquad \alpha_2 = 109, \qquad \alpha_3 = 335, \qquad \alpha_4 = 550, \qquad \alpha_5 = 500$$

The next step is to obtain the transformation matrix $\mathbf{T}$ given by Equation (10–4):

$$\mathbf{T} = \mathbf{MW}$$

where **M** and **W** are given by Equations (10–5) and (10–6), respectively:

$$\mathbf{M} = [\hat{\mathbf{B}} \mid \hat{\mathbf{A}}\hat{\mathbf{B}} \mid \hat{\mathbf{A}}^2\hat{\mathbf{B}} \mid \hat{\mathbf{A}}^3\hat{\mathbf{B}} \mid \hat{\mathbf{A}}^4\hat{\mathbf{B}}]$$

$$= \begin{bmatrix} 0 & -1 & 0 & -20.601 & 0 \\ -1 & 0 & -20.601 & 0 & -(20.601)^2 \\ 0 & 0.5 & 0 & 0.4905 & 0 \\ 0.5 & 0 & 0.4905 & 0 & 10.1048 \\ 0 & 0 & -0.5 & 0 & -0.4905 \end{bmatrix}$$

$$\mathbf{W} = \begin{bmatrix} a_4 & a_3 & a_2 & a_1 & 1 \\ a_3 & a_2 & a_1 & 1 & 0 \\ a_2 & a_1 & 1 & 0 & 0 \\ a_1 & 1 & 0 & 0 & 0 \\ 1 & 0 & 0 & 0 & 0 \end{bmatrix} = \begin{bmatrix} 0 & 0 & -20.601 & 0 & 1 \\ 0 & -20.601 & 0 & 1 & 0 \\ -20.601 & 0 & 1 & 0 & 0 \\ 0 & 1 & 0 & 0 & 0 \\ 1 & 0 & 0 & 0 & 0 \end{bmatrix}$$

Then

$$\mathbf{T} = \mathbf{MW} = \begin{bmatrix} 0 & 0 & 0 & -1 & 0 \\ 0 & 0 & 0 & 0 & -1 \\ 0 & -9.81 & 0 & 0.5 & 0 \\ 0 & 0 & -9.81 & 0 & 0.5 \\ 9.81 & 0 & -0.5 & 0 & 0 \end{bmatrix}$$

The inverse of matrix **T** is

$$\mathbf{T}^{-1} = \begin{bmatrix} 0 & -\dfrac{0.25}{(9.81)^2} & 0 & -\dfrac{0.5}{(9.81)^2} & \dfrac{1}{9.81} \\[2mm] -\dfrac{0.5}{9.81} & 0 & -\dfrac{1}{9.81} & 0 & 0 \\[2mm] 0 & -\dfrac{0.5}{9.81} & 0 & -\dfrac{1}{9.81} & 0 \\[2mm] -1 & 0 & 0 & 0 & 0 \\[2mm] 0 & -1 & 0 & 0 & 0 \end{bmatrix}$$

Referring to Equation (10–13), matrix $\hat{\mathbf{K}}$ is given by

$$\hat{\mathbf{K}} = [\alpha_5 - a_5 \mid \alpha_4 - a_4 \mid \alpha_3 - a_3 \mid \alpha_2 - a_2 \mid \alpha_1 - a_1]\mathbf{T}^{-1}$$
$$= [500 - 0 \mid 550 - 0 \mid 335 - 0 \mid 109 + 20.601 \mid 17 - 0]\mathbf{T}^{-1}$$
$$= [500 \mid 550 \mid 335 \mid 129.601 \mid 17]\mathbf{T}^{-1}$$
$$= [-157.6336 \quad -35.3733 \quad -56.0652 \quad -36.7466 \quad 50.9684]$$
$$= [k_1 \quad k_2 \quad k_3 \quad k_4 \quad -k_I]$$

Thus we get

$$\mathbf{K} = [k_1 \quad k_2 \quad k_3 \quad k_4] = [-157.6336 \quad -35.3733 \quad -56.0652 \quad -36.7466]$$

and

$$k_I = -50.9684$$

Once we determine the feedback gain matrix **K** and the integral gain constant k_I, the step response in the cart position can be obtained by solving the following equation:

$$\begin{bmatrix} \dot{\mathbf{x}} \\ \dot{\xi} \end{bmatrix} = \begin{bmatrix} \mathbf{A} & 0 \\ -\mathbf{C} & 0 \end{bmatrix} \begin{bmatrix} \mathbf{x} \\ \zeta \end{bmatrix} + \begin{bmatrix} \mathbf{B} \\ 0 \end{bmatrix} u + \begin{bmatrix} 0 \\ 1 \end{bmatrix} r \qquad (10\text{--}120)$$

Since

$$u = -\mathbf{K}\mathbf{x} + k_I \xi$$

Equation (10–120) can be written as follows:

$$\begin{bmatrix} \dot{\mathbf{x}} \\ \dot{\xi} \end{bmatrix} = \begin{bmatrix} \mathbf{A} - \mathbf{B}\mathbf{K} & \mathbf{B}k_I \\ -\mathbf{C} & 0 \end{bmatrix} \begin{bmatrix} \mathbf{x} \\ \xi \end{bmatrix} + \begin{bmatrix} 0 \\ 1 \end{bmatrix} r$$

or

$$\begin{bmatrix} \dot{x}_1 \\ \dot{x}_2 \\ \dot{x}_3 \\ \dot{x}_4 \\ \dot{\xi} \end{bmatrix} = \begin{bmatrix} 0 & 1 & 0 & 0 & 0 \\ -137.0326 & -35.3733 & -56.0652 & -36.7466 & 50.9684 \\ 0 & 0 & 0 & 1 & 0 \\ 78.3263 & 17.6867 & 28.0326 & 18.3733 & -25.4842 \\ 0 & 0 & -1 & 0 & 0 \end{bmatrix} \begin{bmatrix} x_1 \\ x_2 \\ x_3 \\ x_4 \\ \xi \end{bmatrix} + \begin{bmatrix} 0 \\ 0 \\ 0 \\ 0 \\ 1 \end{bmatrix} r \qquad (10\text{--}121)$$

Figure 10–21 shows the response curves $x_1(t)$ versus t, $x_2(t)$ versus t, $x_3(t)$ versus t, $x_4(t)$ versus t, $\xi(t)$ versus t, and $u(t)$ versus t, where the input $r(t)$ to the cart is a step function of magnitude 0.5 [that is, $r(t) = 0.5$ m]. Note that $x_1 = \theta$, $x_2 = \dot{\theta}$, $x_3 = x$, and $x_4 = \dot{x}$. All initial conditions are set equal to zero.

The step response in $x_3(t)$ [$= x(t)$] shows the settling time of approximately 4.5 sec and the maximum overshoot of approximately 14.8%, as desired. An interesting point in the position curve [$x_3(t)$ versus t curve] is that the cart moves backward for the first 0.6 sec or so to make the pendulum fall forward. Then the cart accelerates to move in the positive direction.

The response curve $x_3(t)$ versus t clearly shows that $x_3(\infty)$ approaches r. Also, $x_1(\infty) = 0$, $x_2(\infty) = 0$, $x_4(\infty) = 0$, and $\xi(\infty) = 0.55$. This result can be verified by the analytical approach as given below. At steady state, from Equations (10–114) and (10–117) we have

$$\dot{\mathbf{x}}(\infty) = \mathbf{0} = \mathbf{A}\mathbf{x}(\infty) + \mathbf{B}u(\infty)$$

$$\dot{\xi}(\infty) = 0 = r - \mathbf{C}\mathbf{x}(\infty)$$

which can be combined into

$$\begin{bmatrix} 0 \\ 0 \end{bmatrix} = \begin{bmatrix} \mathbf{A} & \mathbf{B} \\ -\mathbf{C} & 0 \end{bmatrix} \begin{bmatrix} \mathbf{x}(\infty) \\ u(\infty) \end{bmatrix} + \begin{bmatrix} 0 \\ r \end{bmatrix}$$

Since we earlier found that the rank of matrix

$$\begin{bmatrix} \mathbf{A} & \mathbf{B} \\ -\mathbf{C} & 0 \end{bmatrix}$$

is 5, it has the inverse. Hence

$$\begin{bmatrix} \mathbf{x}(\infty) \\ u(\infty) \end{bmatrix} = \begin{bmatrix} \mathbf{A} & \mathbf{B} \\ -\mathbf{C} & 0 \end{bmatrix}^{-1} \begin{bmatrix} 0 \\ -r \end{bmatrix}$$

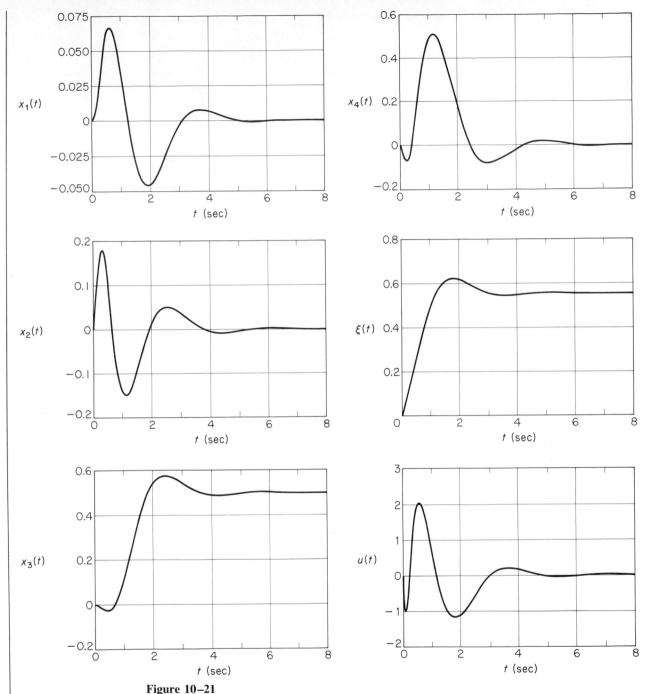

Figure 10–21
Response curves $x_1(t)$ versus t, $x_2(t)$ versus t, $x_3(t)$ versus t, $x_4(t)$ versus t, $\xi(t)$ versus t, and $u(t)$ versus t for the system defined by Eqs. (10–121) and (10–116). (The input to the cart is a step input of magnitude 0.5.)

Referring to Equation (10–119), we have

$$\begin{bmatrix} \mathbf{A} & \mathbf{B} \\ -\mathbf{C} & 0 \end{bmatrix}^{-1} = \begin{bmatrix} 0 & \dfrac{0.5}{9.81} & 0 & \dfrac{1}{9.81} & 0 \\ 1 & 0 & 0 & 0 & 0 \\ 0 & 0 & 0 & 0 & -1 \\ 0 & 0 & 1 & 0 & 0 \\ 0 & 0.05 & 0 & 2.1 & 0 \end{bmatrix}$$

Hence

$$\begin{bmatrix} x_1(\infty) \\ x_2(\infty) \\ x_3(\infty) \\ x_4(\infty) \\ u(\infty) \end{bmatrix} = \begin{bmatrix} 0 & \dfrac{0.5}{9.81} & 0 & \dfrac{1}{9.81} & 0 \\ 1 & 0 & 0 & 0 & 0 \\ 0 & 0 & 0 & 0 & -1 \\ 0 & 0 & 1 & 0 & 0 \\ 0 & 0.05 & 0 & 2.1 & 0 \end{bmatrix} \begin{bmatrix} 0 \\ 0 \\ 0 \\ 0 \\ -r \end{bmatrix} = \begin{bmatrix} 0 \\ 0 \\ r \\ 0 \\ 0 \end{bmatrix}$$

Consequently,

$$y(\infty) = \mathbf{C}\mathbf{x}(\infty) = \begin{bmatrix} 0 & 0 & 1 & 0 \end{bmatrix} \begin{bmatrix} x_1(\infty) \\ x_2(\infty) \\ x_3(\infty) \\ x_4(\infty) \end{bmatrix} = x_3(\infty) = r$$

Since

$$\dot{\mathbf{x}}(\infty) = \mathbf{0} = \mathbf{A}\mathbf{x}(\infty) + \mathbf{B}u(\infty)$$

or

$$\begin{bmatrix} 0 \\ 0 \\ 0 \\ 0 \end{bmatrix} = \begin{bmatrix} 0 & 1 & 0 & 0 \\ 20.601 & 0 & 0 & 0 \\ 0 & 0 & 0 & 1 \\ -0.4905 & 0 & 0 & 0 \end{bmatrix} \begin{bmatrix} 0 \\ 0 \\ r \\ 0 \end{bmatrix} + \begin{bmatrix} 0 \\ -1 \\ 0 \\ 0.5 \end{bmatrix} u(\infty)$$

we get

$$u(\infty) = 0$$

Since $u(\infty) = 0$, we have, from Equation (10–116),

$$u(\infty) = 0 = -\mathbf{K}\mathbf{x}(\infty) + k_I \xi(\infty)$$

and so

$$\xi(\infty) = \frac{1}{k_I}[\mathbf{K}\mathbf{x}(\infty)] = \frac{1}{k_I} k_3 x_3(\infty) = \frac{-56.0652}{-50.9684} r = 1.1r$$

Hence, for $r = 0.5$, we have

$$\xi(\infty) = 0.55$$

as shown in Figure 10–21.

It is noted that as in any design problem, if the speed and damping are not quite satisfactory, then we must modify the desired characteristic equation and determine a new matrix $\hat{\mathbf{K}}$. Computer simulations must be repeated until a satisfactory result is obtained.

10–5 QUADRATIC OPTIMAL CONTROL SYSTEMS

In this section we shall consider the design of stable control systems based on quadratic performance indexes. The control system that we shall consider here may be defined by

$$\dot{\mathbf{x}} = \mathbf{A}\mathbf{x} + \mathbf{B}\mathbf{u} \tag{10-122}$$

where $\mathbf{x}$ = state vector (n-vector)
$\mathbf{u}$ = control vector (r-vector)
$\mathbf{A}$ = $n \times n$ constant matrix
$\mathbf{B}$ = $n \times r$ constant matrix

Control systems considered in this section are mostly regulator systems.

In designing control systems, we are often interested in choosing the control vector $\mathbf{u}(t)$ such that a given performance index is minimized. It can be proved that a quadratic performance index, where the limits of integration are 0 and ∞, such as

$$J = \int_0^\infty L(\mathbf{x}, \mathbf{u}) \, dt$$

where $L(\mathbf{x}, \mathbf{u})$ is a quadratic function or Hermitian function of $\mathbf{x}$ and $\mathbf{u}$, will yield linear control laws, that is

$$\mathbf{u}(t) = -\mathbf{K}\mathbf{x}(t)$$

where $\mathbf{K}$ is an $r \times n$ matrix, or

$$
\begin{bmatrix} u_1 \\ u_2 \\ \cdot \\ \cdot \\ \cdot \\ u_r \end{bmatrix} = -
\begin{bmatrix}
k_{11} & k_{12} & \cdots & k_{1n} \\
k_{21} & k_{22} & \cdots & k_{2n} \\
\cdot & \cdot & & \cdot \\
\cdot & \cdot & & \cdot \\
\cdot & \cdot & & \cdot \\
k_{r1} & k_{r2} & \cdots & k_{rn}
\end{bmatrix}
\begin{bmatrix} x_1 \\ x_2 \\ \cdot \\ \cdot \\ \cdot \\ x_n \end{bmatrix}
$$

Therefore, the design of optimal control systems and optimal regulator systems based on such quadratic performance indexes boils down to the determination of the elements of the matrix $\mathbf{K}$.

In the following, we consider the problem of determining the optimal control vector $\mathbf{u}(t)$ for the system described by Equation (10–122) and the performance index given by

$$J = \int_0^\infty (\mathbf{x}^*\mathbf{Q}\mathbf{x} + \mathbf{u}^*\mathbf{R}\mathbf{u}) \, dt \tag{10-123}$$

where $\mathbf{Q}$ is a positive-definite (or positive-semidefinite) Hermitian or real symmetric matrix, $\mathbf{R}$ is a positive-definite Hermitian or real symmetric matrix, and $\mathbf{u}$ is unconstrained. The optimal control system is to minimize the performance index. Such a system is stable. Among many different approaches to the solution of this type of problem, we shall present here one approach that is based on the second method of Liapunov.

It is noted that in the discussions of quadratic optimal control problems that follow we use the complex quadratic performance indexes (Hermitian performance indexes), rather than the real quadratic performance indexes, since the former include the latter as a special case.

For systems with real vectors and real matrices, $\int_0^\infty$ ($\mathbf{x}^*\mathbf{Qx}$ + $\mathbf{u}^*\mathbf{Ru}$) dt is the same as $\int_0^\infty$ ($\mathbf{x}^T\mathbf{Qx}$ + $\mathbf{u}^T\mathbf{Ru}$) dt.

Control system optimization via the second method of Liapunov. Classically, control systems are first designed and then their stability is examined. An approach different from this is where the conditions for stability are formulated first and then the system is designed within these limitations. (See Section 10–6 for the design of nonlinear control systems based on the Liapunov aproach.) If the second method of Liapunov is utilized to form the basis for the design of an optimal controller, then we are assured that the system will work; that is, the system output will be continually driven toward its desired value. Thus the designed system has a configuration with inherent stability characteristics.

For a large class of control systems, a direct relationship can be shown between Liapunov functions and quadratic performance indexes used in the synthesis of optimal control systems. We shall begin the Liapunov approach to the solution of optimization problems by considering a simple case, known as the parameter-optimization problem.

Parameter-optimization problem solved by the second method of Liapunov. In the following, we shall discuss a direct relationship between Liapunov functions and quadratic performance indexes and solve the parameter-optimization problem using this relationship. Let us consider the system:

$$\dot{\mathbf{x}} = \mathbf{Ax}$$

where all eigenvalues of $\mathbf{A}$ have negative real parts, or the origin $\mathbf{x} = \mathbf{0}$ is asymptotically stable. (We shall call such a matrix $\mathbf{A}$ a stable matrix.) We assume that matrix $\mathbf{A}$ involves an adjustable parameter (or parameters). It is desired to minimize the following performance index:

$$J = \int_0^\infty \mathbf{x}^*\mathbf{Qx}\, dt$$

where $\mathbf{Q}$ is a positive-definite (or positive-semidefinite) Hermitian or real symmetric matrix. The problem thus becomes that of determining the value(s) of the adjustable parameter(s) so as to minimize the performance index.

We shall show that a Liapunov function can effectively be used in the solution of this problem. Let us assume that

$$\mathbf{x}^*\mathbf{Qx} = -\frac{d}{dt}(\mathbf{x}^*\mathbf{Px})$$

where $\mathbf{P}$ is a positive-definite Hermitian or real symmetric matrix. Then we obtain

$$\mathbf{x}^*\mathbf{Qx} = -\dot{\mathbf{x}}^*\mathbf{Px} - \mathbf{x}^*\mathbf{P}\dot{\mathbf{x}} = -\mathbf{x}^*\mathbf{A}^*\mathbf{Px} - \mathbf{x}^*\mathbf{PAx} = -\mathbf{x}^*(\mathbf{A}^*\mathbf{P} + \mathbf{PA})\mathbf{x}$$

By the second method of Liapunov, we know that for a given $\mathbf{Q}$ there exists $\mathbf{P}$, if $\mathbf{A}$ is stable, such that

$$\mathbf{A}^*\mathbf{P} + \mathbf{PA} = -\mathbf{Q} \tag{10–124}$$

Hence we can determine the elements of $\mathbf{P}$ from this equation.

The performance index J can be evaluated as

$$J = \int_0^\infty \mathbf{x}^*\mathbf{Q}\mathbf{x}\,dt = -\mathbf{x}^*\mathbf{P}\mathbf{x}\,\Big|_0^\infty = -\mathbf{x}^*(\infty)\mathbf{P}\mathbf{x}(\infty) + \mathbf{x}^*(0)\mathbf{P}\mathbf{x}(0)$$

Since all eigenvalues of $\mathbf{A}$ have negative real parts, we have $\mathbf{x}(\infty) \rightarrow \mathbf{0}$. Therefore, we obtain

$$J = \mathbf{x}^*(0)\mathbf{P}\mathbf{x}(0) \tag{10–125}$$

Thus the performance index J can be obtained in terms of the initial condition $\mathbf{x}(0)$ and $\mathbf{P}$, which is related to $\mathbf{A}$ and $\mathbf{Q}$ by Equation (10–124). If, for example, a system parameter is to be adjusted so as to minimize the performance index J, then it can be accomplished by minimizing $\mathbf{x}^*(0)\mathbf{P}\mathbf{x}(0)$ with respect to the parameter in question. Since $\mathbf{x}(0)$ is the given initial condition and $\mathbf{Q}$ is also given, $\mathbf{P}$ is a function of the elements of $\mathbf{A}$. Hence this minimization process will result in the optimal value of the adjustable parameter.

It is important to note that the optimal value of this parameter depends, in general, on the initial condition $\mathbf{x}(0)$. However, if $\mathbf{x}(0)$ invovles only one nonzero component, for example, $x_1(0) \neq 0$, and other initial conditions are zero, then the optimal value of the parameter does not depend on the numerical value of $x_1(0)$. (See the following example.)

EXAMPLE 10–9

Consider the system shown in Figure 10–22. Determine the value of the damping ratio $\zeta > 0$ so that when the system is subjected to a unit-step input $r(t) = 1(t)$, the following performance index is minimized:

$$J = \int_{0+}^\infty (e^2 + \mu\dot{e}^2)\,dt \qquad (\mu > 0)$$

where e is the error signal and is given by $e = r - c$. The system is assumed to be at rest initially.

From Figure 10–22, we find

$$\frac{C(s)}{R(s)} = \frac{1}{s^2 + 2\zeta s + 1}$$

or

$$\ddot{c} + 2\zeta\dot{c} + c = r$$

In terms of the error signal e, we obtain

$$\ddot{e} + 2\zeta\dot{e} + e = \ddot{r} + 2\zeta\dot{r}$$

Since the input $r(t)$ is a unit-step input, we have $\dot{r}(0+) = 0$, $\ddot{r}(0+) = 0$. Hence, for $t \geq 0+$, we have

$$\ddot{e} + 2\zeta\dot{e} + e = 0, \qquad e(0+) = 1, \qquad \dot{e}(0+) = 0$$

Now define the state variables as follows:

$$x_1 = e$$
$$x_2 = \dot{e}$$

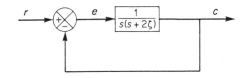

Figure 10–22
Control system.

Then the state equation becomes

$$\dot{\mathbf{x}} = \mathbf{A}\mathbf{x}$$

where

$$\mathbf{A} = \begin{bmatrix} 0 & 1 \\ -1 & -2\zeta \end{bmatrix}$$

The performance index J can be rewritten as

$$J = \int_{0+}^{\infty} (e^2 + \mu \dot{e}^2)\, dt = \int_{0+}^{\infty} (x_1^2 + \mu x_2^2)\, dt$$

$$= \int_{0+}^{\infty} \begin{bmatrix} x_1 & x_2 \end{bmatrix} \begin{bmatrix} 1 & 0 \\ 0 & \mu \end{bmatrix} \begin{bmatrix} x_1 \\ x_2 \end{bmatrix} dt$$

$$= \int_{0+}^{\infty} \mathbf{x}^T \mathbf{Q} \mathbf{x}\, dt$$

where

$$\mathbf{x} = \begin{bmatrix} x_1 \\ x_2 \end{bmatrix} = \begin{bmatrix} e \\ \dot{e} \end{bmatrix}, \qquad \mathbf{Q} = \begin{bmatrix} 1 & 0 \\ 0 & \mu \end{bmatrix}$$

Since $\mathbf{A}$ is a stable matrix, referring to Equation (10–125), the value of J can be given by

$$J = \mathbf{x}^T(0+)\mathbf{P}\mathbf{x}(0+)$$

where $\mathbf{P}$ is determined from

$$\mathbf{A}^T\mathbf{P} + \mathbf{P}\mathbf{A} = -\mathbf{Q} \qquad\qquad (10\text{–}126)$$

Equation (10–126) can be rewriten as

$$\begin{bmatrix} 0 & -1 \\ 1 & -2\zeta \end{bmatrix} \begin{bmatrix} p_{11} & p_{12} \\ p_{12} & p_{22} \end{bmatrix} + \begin{bmatrix} p_{11} & p_{12} \\ p_{12} & p_{22} \end{bmatrix} \begin{bmatrix} 0 & 1 \\ -1 & -2\zeta \end{bmatrix} = \begin{bmatrix} -1 & 0 \\ 0 & -\mu \end{bmatrix}$$

This equation results in the following three equations:

$$-2p_{12} = -1$$

$$p_{11} - 2\zeta p_{12} - p_{22} = 0$$

$$2p_{12} - 4\zeta p_{22} = -\mu$$

Solving these three equations for the p_{ij}, we obtain

$$\mathbf{P} = \begin{bmatrix} p_{11} & p_{12} \\ p_{12} & p_{22} \end{bmatrix} = \begin{bmatrix} \zeta + \dfrac{1+\mu}{4\zeta} & \dfrac{1}{2} \\ \dfrac{1}{2} & \dfrac{1+\mu}{4\zeta} \end{bmatrix}$$

The performance index J can be given by

$$J = \mathbf{x}^T(0+)\mathbf{P}\mathbf{x}(0+)$$

$$= \left(\zeta + \frac{1+\mu}{4\zeta} \right) x_1^2(0+) + x_1(0+)x_2(0+) + \frac{1+\mu}{4\zeta} x_2^2(0+)$$

Substituting the initial conditions $x_1(0+) = 1$, $x_2(0+) = 0$ into this last equation, we obtain

$$J = \zeta + \frac{1+\mu}{4\zeta}$$

To minimize J with respect to ζ, we set $\partial J/\partial \zeta = 0$, or

$$\frac{\partial J}{\partial \zeta} = 1 - \frac{1 + \mu}{4\zeta^2} = 0$$

This yields

$$\zeta = \frac{\sqrt{1 + \mu}}{2}$$

Thus the optimal value of ζ is $\sqrt{1 + \mu}/2$. For example, if $\mu = 1$, then the optimal value of ζ is $\sqrt{2}/2$ or 0.707.

Quadratic optimal control problems.   We shall now consider the optimal control problem that, given the system equation

$$\dot{\mathbf{x}} = \mathbf{Ax} + \mathbf{Bu} \qquad (10\text{–}127)$$

determine the matrix $\mathbf{K}$ of the optimal control vector

$$\mathbf{u}(t) = -\mathbf{Kx}(t) \qquad (10\text{–}128)$$

so as to minimize the performance index

$$J = \int_0^\infty (\mathbf{x}^*\mathbf{Qx} + \mathbf{u}^*\mathbf{Ru})\, dt \qquad (10\text{–}129)$$

where $\mathbf{Q}$ is a positive-definite (or positive-semidefinite) Hermitian or real symmetric matrix and $\mathbf{R}$ is a positive-definite Hermitian or real symmetric matrix. Note that the second term on the right side of Equation (10–129) accounts for the expenditure of the energy of the control signals. The matrices $\mathbf{Q}$ and $\mathbf{R}$ determine the relative importance of the error and the expenditure of this energy. In this problem, we assume that the control vector $\mathbf{u}(t)$ is unconstrained.

As will be seen later, the linear control law given by Equation (10–128) is the optimal control law. Therefore, if the unknown elements of the matrix $\mathbf{K}$ are determined so as to minimize the performance index, then $\mathbf{u}(t) = -\mathbf{Kx}(t)$ is optimal for any initial state $\mathbf{x}(0)$. The block diagram showing the optimal configuration is shown in Figure 10–23.

Now let us solve the optimization problem. Substituting Equation (10–128) into Equation (10–127), we obtain

$$\dot{\mathbf{x}} = \mathbf{Ax} - \mathbf{BKx} = (\mathbf{A} - \mathbf{BK})\mathbf{x}$$

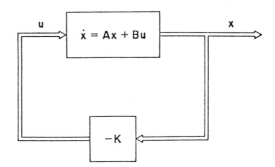

Figure 10–23
Optimal control system.

In the following derivations, we assume that the matrix $\mathbf{A} - \mathbf{BK}$ is stable, or that the eigenvalues of $\mathbf{A} - \mathbf{BK}$ have negative real parts.

Substituting Equation (10–128) into Equation (10–129) yields

$$J = \int_0^\infty (\mathbf{x}^*\mathbf{Qx} + \mathbf{x}^*\mathbf{K}^*\mathbf{RKx})\, dt$$

$$= \int_0^\infty \mathbf{x}^*(\mathbf{Q} + \mathbf{K}^*\mathbf{RK})\mathbf{x}\, dt$$

Following the discussion given in solving the parameter-optimization problem, we set

$$\mathbf{x}^*(\mathbf{Q} + \mathbf{K}^*\mathbf{RK})\mathbf{x} = -\frac{d}{dt}(\mathbf{x}^*\mathbf{Px})$$

Then we obtain

$$\mathbf{x}^*(\mathbf{Q} + \mathbf{K}^*\mathbf{RK})\mathbf{x} = -\dot{\mathbf{x}}^*\mathbf{Px} - \mathbf{x}^*\mathbf{P}\dot{\mathbf{x}} = -\mathbf{x}^*[(\mathbf{A} - \mathbf{BK})^*\mathbf{P} + \mathbf{P}(\mathbf{A} - \mathbf{BK})]\,\mathbf{x}$$

Comparing both sides of this last equation and noting that this equation must hold true for any $\mathbf{x}$, we require that

$$(\mathbf{A} - \mathbf{BK})^*\mathbf{P} + \mathbf{P}(\mathbf{A} - \mathbf{BK}) = -(\mathbf{Q} + \mathbf{K}^*\mathbf{RK}) \qquad (10\text{–}130)$$

By the second method of Liapunov, if $\mathbf{A} - \mathbf{BK}$ is a stable matrix, there exists a positive-definite matrix $\mathbf{P}$ that satisfies Equation (10–130). Then, using the same approach as we used in deriving Equation (10–125) and noting that $\mathbf{x}(\infty) = \mathbf{0}$, the performance index can be written as

$$J = \mathbf{x}^*(0)\mathbf{Px}(0) \qquad (10\text{–}131)$$

To obtain the solution to the quadratic optimal control problem, we proceed as follows: Since $\mathbf{R}$ has been assumed to be a positive-definite Hermitian or real symmetric matrix, we can write

$$\mathbf{R} = \mathbf{T}^*\mathbf{T}$$

where $\mathbf{T}$ is a nonsingular matrix. Then Equation (10–130) can be written as

$$(\mathbf{A}^* - \mathbf{K}^*\mathbf{B}^*)\mathbf{P} + \mathbf{P}(\mathbf{A} - \mathbf{BK}) + \mathbf{Q} + \mathbf{K}^*\mathbf{T}^*\mathbf{TK} = \mathbf{0}$$

which can be rewritten as

$$\mathbf{A}^*\mathbf{P} + \mathbf{PA} + [\mathbf{TK} - (\mathbf{T}^*)^{-1}\mathbf{B}^*\mathbf{P}]^*[\mathbf{TK} - (\mathbf{T}^*)^{-1}\mathbf{B}^*\mathbf{P}] - \mathbf{PBR}^{-1}\mathbf{B}^*\mathbf{P} + \mathbf{Q} = \mathbf{0}$$

The minimization of J with respect to $\mathbf{K}$ requires the minimization of

$$\mathbf{x}^*[\mathbf{TK} - (\mathbf{T}^*)^{-1}\mathbf{B}^*\mathbf{P}]^*[\mathbf{TK} - (\mathbf{T}^*)^{-1}\mathbf{B}^*\mathbf{P}]\mathbf{x}$$

with respect to $\mathbf{K}$. (See problem A–10–17.) Since this last expression is nonnegative, the minimum occurs when it is zero, or when

$$\mathbf{TK} = (\mathbf{T}^*)^{-1}\mathbf{B}^*\mathbf{P}$$

Hence

$$\mathbf{K} = \mathbf{T}^{-1}(\mathbf{T}^*)^{-1}\mathbf{B}^*\mathbf{P} = \mathbf{R}^{-1}\mathbf{B}^*\mathbf{P} \qquad (10\text{–}132)$$

Equation (10–132) gives the optimal matrix $\mathbf{K}$. Thus, the optimal control law to the quadratic optimal control problem where the performance index is given by Equation (10–129) is linear and is given by

$$\mathbf{u}(t) = -\mathbf{K}\mathbf{x}(t) = -\mathbf{R}^{-1}\mathbf{B}^*\mathbf{P}\mathbf{x}(t)$$

The matrix $\mathbf{P}$ in Equation (10–132) must satisfy Equation (10–130) or the following reduced equation:

$$\mathbf{A}^*\mathbf{P} + \mathbf{P}\mathbf{A} - \mathbf{P}\mathbf{B}\mathbf{R}^{-1}\mathbf{B}^*\mathbf{P} + \mathbf{Q} = 0 \qquad (10\text{–}133)$$

Equation (10–133) is called the reduced-matrix Riccati equation. The design steps may be stated as follows:

1. Solve Equation (10–133), the reduced-matrix Riccati equation, for the matrix $\mathbf{P}$.
2. Substitute this matrix $\mathbf{P}$ into Equation (10–132). The resulting matrix $\mathbf{K}$ is the optimal one.

A design example based on this approach is given in Example 10–10.

If the matrix $\mathbf{A} - \mathbf{B}\mathbf{K}$ is stable, the present method always gives the correct result. It can be shown that the requirement of $\mathbf{A} - \mathbf{B}\mathbf{K}$ being a stable matrix is equivalent to that of the rank of

$$\begin{bmatrix} \mathbf{Q}^{1/2} \\ \mathbf{Q}^{1/2}\mathbf{A} \\ \cdot \\ \cdot \\ \cdot \\ \mathbf{Q}^{1/2}\mathbf{A}^{n-1} \end{bmatrix} \qquad (10\text{–}134)$$

being n. (See Problem B–10–16.) This rank condition may conveniently be applied to check if the matrix $\mathbf{A} - \mathbf{B}\mathbf{K}$ is stable.

An alternate approach to the determination of the optimal feedback gain matrix $\mathbf{K}$ is available. The design steps based on the alternate approach may be stated as follows:

1. Determine the matrix $\mathbf{P}$ that satisfies Equation (10–130) as a function of $\mathbf{K}$.
2. Substitute the matrix $\mathbf{P}$ into Equation (10–131). Then the performance index becomes a function of $\mathbf{K}$.
3. Determine the elements of $\mathbf{K}$ so that the performance index J is minimized. The minimization of J with respect to the elements k_{ij} of $\mathbf{K}$ can be accomplished by setting $\partial J/\partial k_{ij}$ equal to zero and solving for the optimal values of k_{ij}.

For details of this design approach, see Problems A–10–18 and A–10–19. When the number of the elements k_{ij} is not small, this approach is not convenient.

Finally, note that if the performance index is given in terms of the output vector rather than the state vector, that is,

$$J = \int_0^\infty (\mathbf{y}^*\mathbf{Q}\mathbf{y} + \mathbf{u}^*\mathbf{R}\mathbf{u})\, dt$$

then the index can be modified by using the output equation

$$\mathbf{y} = \mathbf{Cx}$$

to

$$J = \int_0^\infty (\mathbf{x}^*\mathbf{C}^*\mathbf{QCx} + \mathbf{u}^*\mathbf{Ru})\, dt$$

and the design steps presented in this section can be applied to obtain the optimal matrix **K**.

EXAMPLE 10–10 Consider the system shown in Figure 10–24. Assuming the control signal to be

$$u(t) = -\mathbf{Kx}(t)$$

determine the optimal feedback gain matrix **K** such that the following performance index is minimized:

$$J = \int_0^\infty (\mathbf{x}^T\mathbf{Qx} + u^2)\, dt$$

where

$$\mathbf{Q} = \begin{bmatrix} 1 & 0 \\ 0 & \mu \end{bmatrix} \qquad (\mu \geq 0)$$

From Figure 10–24, we find that the state equation for the plant is

$$\dot{\mathbf{x}} = \mathbf{Ax} + \mathbf{B}u$$

where

$$\mathbf{A} = \begin{bmatrix} 0 & 1 \\ 0 & 0 \end{bmatrix}, \qquad \mathbf{B} = \begin{bmatrix} 0 \\ 1 \end{bmatrix}$$

Noting that

$$\mathbf{Q} = \begin{bmatrix} 1 & 0 \\ 0 & \sqrt{\mu} \end{bmatrix}\begin{bmatrix} 1 & 0 \\ 0 & \sqrt{\mu} \end{bmatrix}$$

we find the rank of the matrix given by (10–134), or

$$\begin{bmatrix} \mathbf{Q}^{1/2} \\ \mathbf{Q}^{1/2}\mathbf{A} \end{bmatrix} = \begin{bmatrix} 1 & 0 \\ 0 & \sqrt{\mu} \\ 0 & 1 \\ 0 & 0 \end{bmatrix}$$

to be 2. Thus, $\mathbf{A} - \mathbf{BK}$ is a stable matrix, and the Liapunov aproach presented in this section yields the correct result.

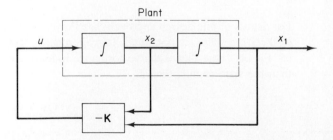

Figure 10–24
Control system.

We shall demonstrate the use of the reduced-matrix Riccati equation in the design of the optimal control system. Let us solve Equation (10–133), rewritten as

$$\mathbf{A}^*\mathbf{P} + \mathbf{P}\mathbf{A} - \mathbf{P}\mathbf{B}\mathbf{R}^{-1}\mathbf{B}^*\mathbf{P} + \mathbf{Q} = \mathbf{0}$$

Noting that matrix $\mathbf{A}$ is real and matrix $\mathbf{Q}$ is real symmetric, matrix $\mathbf{P}$ is a real symmetric matrix. Hence this last equation can be written as

$$\begin{bmatrix} 0 & 0 \\ 1 & 0 \end{bmatrix}\begin{bmatrix} p_{11} & p_{12} \\ p_{12} & p_{22} \end{bmatrix} + \begin{bmatrix} p_{11} & p_{12} \\ p_{12} & p_{22} \end{bmatrix}\begin{bmatrix} 0 & 1 \\ 0 & 0 \end{bmatrix}$$

$$- \begin{bmatrix} p_{11} & p_{12} \\ p_{12} & p_{22} \end{bmatrix}\begin{bmatrix} 0 \\ 1 \end{bmatrix}[1][0 \quad 1]\begin{bmatrix} p_{11} & p_{12} \\ p_{12} & p_{22} \end{bmatrix} + \begin{bmatrix} 1 & 0 \\ 0 & \mu \end{bmatrix} = \begin{bmatrix} 0 & 0 \\ 0 & 0 \end{bmatrix}$$

This equation can be simplified to

$$\begin{bmatrix} 0 & 0 \\ p_{11} & p_{12} \end{bmatrix} + \begin{bmatrix} 0 & p_{11} \\ 0 & p_{12} \end{bmatrix} - \begin{bmatrix} p_{12}^2 & p_{12}p_{22} \\ p_{12}p_{22} & p_{22}^2 \end{bmatrix} + \begin{bmatrix} 1 & 0 \\ 0 & \mu \end{bmatrix} = \begin{bmatrix} 0 & 0 \\ 0 & 0 \end{bmatrix}$$

from which we obtain the following three equations:

$$1 - p_{12}^2 = 0$$

$$p_{11} - p_{12}p_{22} = 0$$

$$\mu + 2p_{12} - p_{22}^2 = 0$$

Solving these three simultaneous equations for p_{11}, p_{12}, and p_{22}, requiring $\mathbf{P}$ to be positive definite, we obtain

$$\mathbf{P} = \begin{bmatrix} p_{11} & p_{12} \\ p_{12} & p_{22} \end{bmatrix} = \begin{bmatrix} \sqrt{\mu + 2} & 1 \\ 1 & \sqrt{\mu + 2} \end{bmatrix}$$

Referring to Equation (10–132), the optimal feedback gain matrix $\mathbf{K}$ is obtained as

$$\mathbf{K} = \mathbf{R}^{-1}\mathbf{B}^*\mathbf{P}$$

$$= [1]\,[0 \quad 1]\begin{bmatrix} p_{11} & p_{12} \\ p_{12} & p_{22} \end{bmatrix}$$

$$= [p_{12} \quad p_{22}]$$

$$= [1 \quad \sqrt{\mu + 2}]$$

Thus the optimal control signal is

$$u = -\mathbf{K}\mathbf{x} = -x_1 - \sqrt{\mu + 2}\,x_2 \tag{10–135}$$

Note that the control law given by Equation (10–135) yields an optimal result for any initial state under the given performance index. Figure 10–25 is the block diagram for this system.

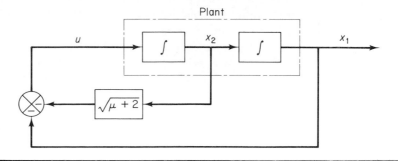

Figure 10–25
Optimal control of the plant shown in Fig. 10–24.

Concluding comments

1. Given any initial state $\mathbf{x}(t_0)$, the optimal control problem is to find an allowable control vector $\mathbf{u}(t)$ that transfers the state to the desired region of the state space and for which the performance index is minimized. For the existence of an optimal control vector $\mathbf{u}(t)$, the system must be completely state controllable.

2. The system that minimizes (or maximizes as the case may be) the selected performance index is, by definition, optimal. Although the controller may have nothing to do with "optimality" in many practical applications, the important point is that the design based on the quadratic performance index yields a stable control system.

3. The characteristic of an optimal control law based on a quadratic performance index is that it is a linear function of the state variables, which implies that we need to feed back all state variables. This requires that all such variables be available for feedback. If not all state variables are available for feedback, then we need to employ a state observer to estimate unmeasurable state variables and use estimated values to generate optimal control signals.

4. When the optimal control system is designed in the time domain, it is desirable to investigate the frequency-response characteristics to compensate for noise effects. The system frequency-response characteristics must be such that the system attenuates highly in the frequency range where noise and resonance of components are expected. (To compensate for noise effects, we must in some cases either modify the optimal configuration and accept suboptimal performance or modify the performance index.)

5. If the upper limit of integration in the performance index J given by Equation (10–129) is finite, then it can be shown that the optimal control vector is still a linear function of the state variables, but with time-varying coefficients. (Therefore, the determination of the optimal control vector involves that of optimal time-varying matrices.)

10–6 MODEL-REFERENCE CONTROL SYSTEMS

In this chapter thus far we have presented design techniques of linear time-invariant control systems. Since all physical plants are nonlinear to some degree, the designed system will behave satisfactorily only over a limited range of operation. If the assumption of linearity in the plant equation is removed, the design techniques of this chapter presented thus far are not applicable. In such a case, the model-reference approach to the system design presented in this section may be useful.

Model-reference control systems. One useful method for specifying system performance is by means of a model that will produce the desired output for a given input. The model need not be actual hardware. It can be only a mathematical model simulated on a computer. In a model-reference control system, the output of the model and that of the plant are compared and the difference is used to generate the control signals.

Model-reference control has been used to obtain acceptable performance in some very difficult control situations involving nonlinearity and/or time-varying parameters.

Design of a controller.* We shall assume that the plant is characterized by the following state equation:

$$\dot{\mathbf{x}} = \mathbf{f}(\mathbf{x}, \mathbf{u}, t) \qquad (10\text{–}136)$$

* References M-6 and V-2.

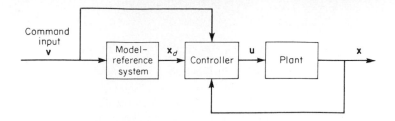

Figure 10–26
Model-reference
control system.

where $\mathbf{x}$ = state vector (n-vector)

$\quad\mathbf{u}$ = control vector (r-vector)

$\quad\mathbf{f}$ = vector-valued function

It is desired that the control system follow closely some model system. Our design problem here is to synthesize a controller that always generates a signal that forces the plant state toward the model state. Figure 10–26 is the block diagram showing the system configuration.

We shall assume that the model-reference system is linear and described by

$$\dot{\mathbf{x}}_d = \mathbf{A}\mathbf{x}_d + \mathbf{B}\mathbf{v} \qquad (10\text{–}137)$$

where $\mathbf{x}_d$ = state vector of the model (n-vector)

$\quad\mathbf{v}$ = input vector (r-vector)

$\quad\mathbf{A}$ = $n \times n$ constant matrix

$\quad\mathbf{B}$ = $n \times r$ constant matrix

We assume that the eigenvalues of $\mathbf{A}$ have negative real parts so that the model-reference system has an asymptotically stable equilibrium state.

Let us define the error vector $\mathbf{e}$ by

$$\mathbf{e} = \mathbf{x}_d - \mathbf{x} \qquad (10\text{–}138)$$

In the present problem, we wish to reduce the error vector to zero by a suitable control vector $\mathbf{u}$. From Equations (10–136), (10–137), and (10–138), we obtain

$$\begin{aligned}\dot{\mathbf{e}} = \dot{\mathbf{x}}_d - \dot{\mathbf{x}} &= \mathbf{A}\mathbf{x}_d + \mathbf{B}\mathbf{v} - \mathbf{f}(\mathbf{x}, \mathbf{u}, t) \\ &= \mathbf{A}\mathbf{e} + \mathbf{A}\mathbf{x} - \mathbf{f}(\mathbf{x}, \mathbf{u}, t) + \mathbf{B}\mathbf{v} \end{aligned} \qquad (10\text{–}139)$$

Equation (10–139) is a differential equation for the error vector.

We now design a controller such that at steady state $\mathbf{x} = \mathbf{x}_d$ and $\dot{\mathbf{x}} = \dot{\mathbf{x}}_d$, or $\mathbf{e} = \dot{\mathbf{e}} = \mathbf{0}$. Thus the origin $\mathbf{e} = \mathbf{0}$ will be an equilibrium state.

A convenient starting point in the synthesis of the control vector $\mathbf{u}$ is the construction of a Liapunv function for the system given by Equation (10–139).

Let us assume that the form of the Liapunov function is

$$V(\mathbf{e}) = \mathbf{e}^*\mathbf{P}\mathbf{e}$$

where $\mathbf{P}$ is a positive-definite Hermitian or real symmetric matrix. Taking the derivative of $V(\mathbf{e})$ with respect to time gives

$$\begin{aligned}\dot{V}(\mathbf{e}) &= \dot{\mathbf{e}}^*\mathbf{P}\mathbf{e} + \mathbf{e}^*\mathbf{P}\dot{\mathbf{e}} \\ &= [\mathbf{e}^*\mathbf{A}^* + \mathbf{x}^*\mathbf{A}^* - \mathbf{f}^*(\mathbf{x}, \mathbf{u}, t) + \mathbf{v}^*\mathbf{B}^*]\mathbf{P}\mathbf{e} \\ &\quad + \mathbf{e}^*\mathbf{P}[\mathbf{A}\mathbf{e} + \mathbf{A}\mathbf{x} - \mathbf{f}(\mathbf{x}, \mathbf{u}, t) + \mathbf{B}\mathbf{v}] \\ &= \mathbf{e}^*(\mathbf{A}^*\mathbf{P} + \mathbf{P}\mathbf{A})\mathbf{e} + 2M \end{aligned} \qquad (10\text{–}140)$$

where

$$M = \mathbf{e}^*\mathbf{P}[\mathbf{Ax} - \mathbf{f}(\mathbf{x}, \mathbf{u}, t) + \mathbf{Bv}] = \text{scalar quantity}$$

The assumed $V(\mathbf{e})$ function is a Liapunov function if:

1. $\mathbf{A}^*\mathbf{P} + \mathbf{PA} = -\mathbf{Q}$ is a negative-definite matrix.
2. The control vector $\mathbf{u}$ can be chosen to make the scalar quantity M nonpositive.

Then, noting that $V(\mathbf{e}) \to \infty$ as $\|\mathbf{e}\| \to \infty$, we see that the equilibrium state $\mathbf{e} = \mathbf{0}$ is asymptotically stable in the large. Condition 1 can always be met by a proper choice of $\mathbf{P}$, since the eigenvalues of $\mathbf{A}$ are assumed to have negative real parts. The problem here is to choose an appropriate control vector $\mathbf{u}$ so that M is either zero or negative.

We shall illustrate the application of the present approach to the design of a nonlinear controller by means of an example.

EXAMPLE 10–11 Consider a nonlinear time-varying plant described by

$$\begin{bmatrix} \dot{x}_1 \\ \dot{x}_2 \end{bmatrix} = \begin{bmatrix} 0 & 1 \\ -b & -a(t)x_2 \end{bmatrix} \begin{bmatrix} x_1 \\ x_2 \end{bmatrix} + \begin{bmatrix} 0 \\ 1 \end{bmatrix} u$$

where $a(t)$ is time-varying and b is a positive constant. Assuming the reference model equation to be

$$\begin{bmatrix} \dot{x}_{d1} \\ \dot{x}_{d2} \end{bmatrix} = \begin{bmatrix} 0 & 1 \\ -\omega_n^2 & -2\zeta\omega_n \end{bmatrix} \begin{bmatrix} x_{d1} \\ x_{d2} \end{bmatrix} + \begin{bmatrix} 0 \\ \omega_n^2 \end{bmatrix} v \tag{10–141}$$

design a nonlinear controller that will give stable operation of the system.

Define the error vector by

$$\mathbf{e} = \mathbf{x}_d - \mathbf{x}$$

and a Liapunov function by

$$V(\mathbf{e}) = \mathbf{e}^*\mathbf{Pe}$$

where $\mathbf{P}$ is a positive-definite real symmetric matrix. Then, referring to Equation. (10–140), we obtain $\dot{V}(\mathbf{e})$ as

$$\dot{V}(\mathbf{e}) = \mathbf{e}^*(\mathbf{A}^*\mathbf{P} + \mathbf{PA})\mathbf{e} + 2M$$

where

$$M = \mathbf{e}^*\mathbf{P}[\mathbf{Ax} - \mathbf{f}(\mathbf{x}, \mathbf{u}, t) + \mathbf{Bv}]$$

By identifying the matrices $\mathbf{A}$ and $\mathbf{B}$ from Equation (10–141) and choosing the matrix $\mathbf{Q}$ to be

$$\mathbf{Q} = \begin{bmatrix} q_{11} & 0 \\ 0 & q_{22} \end{bmatrix} = \text{positive definite}$$

we obtain

$$\dot{V}(\mathbf{e}) = -(q_{11}e_1^2 + q_{22}e_2^2) + 2M$$

where

$$M = \begin{bmatrix} e_1 & e_2 \end{bmatrix} \begin{bmatrix} p_{11} & p_{12} \\ p_{12} & p_{22} \end{bmatrix} \left\{ \begin{bmatrix} 0 & 1 \\ -\omega_n^2 & -2\zeta\omega_n \end{bmatrix} \begin{bmatrix} x_1 \\ x_2 \end{bmatrix} - \begin{bmatrix} 0 & 1 \\ -b & -a(t)x_2 \end{bmatrix} \begin{bmatrix} x_1 \\ x_2 \end{bmatrix} - \begin{bmatrix} 0 \\ u \end{bmatrix} + \begin{bmatrix} 0 \\ \omega_n^2 v \end{bmatrix} \right\}$$

$$= (e_1 p_{12} + e_2 p_{22})[-(\omega_n^2 - b)x_1 - 2\zeta\omega_n x_2 + a(t)x_2^2 + \omega_n^2 v - u]$$

If we choose u so that

$$u = -(\omega_n^2 - b)x_1 - 2\zeta\omega_n x_2 + \omega_n^2 v + a_m x_2^2 \, \text{sign} \, (e_1 p_{12} + e_2 p_{22}) \qquad (10\text{–}142)$$

where

$$a_m = \max |a(t)|$$

then

$$M = (e_1 p_{12} + e_2 p_{22})[a(t) - a_m \, \text{sign} \, (e_1 p_{12} + e_2 p_{22})]x_2^2$$
$$= \text{nonpositive}$$

With the control function u given by Equation (10–142), the equilibrium state $\mathbf{e} = \mathbf{0}$ is asymptotically stable in the large. Thus Equation (10–142) defines a nonlinear control law that will yield an asymptotically stable operation. The block diagram for the present control system is shown in Figure 10–27.

Note that the rate of convergence of the transient response depends on the matrix $\mathbf{P}$, which in turn depends on the matrix $\mathbf{Q}$ chosen at the start of the design.

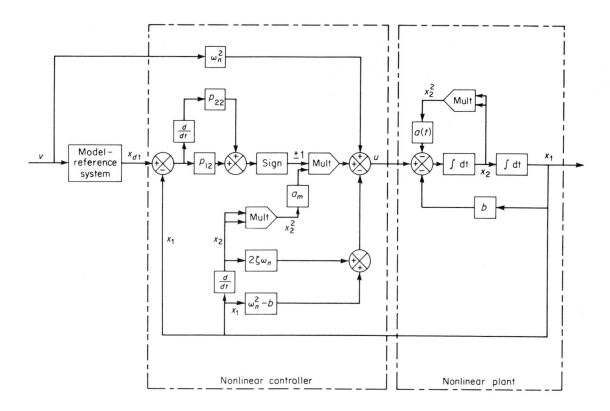

Figure 10–27
Model-reference control of a nonlinear plant.

In recent years, interest in adaptive control systems has increased rapidly along with interest and progress in robotics and other control fields. The term adaptive system has a variety of specific meanings, but it usually implies that the system is capable of accommodating unpredictable environmental changes, whether these changes arise within the system or external to it. This concept has a great deal of appeal to the systems designer since a highly adaptive system, besides accommodating environmental changes, would also accommodate moderate engineering design errors or uncertainties and would compensate for the failure of minor system components, thereby increasing system reliability.

We shall first present some basic concepts of adaptive control systems and explain what such systems are. Then we shall discuss the necessary functions a controller must perform in order that it be called adaptive. Finally, we shall introduce some concepts of learning systems.

Introduction.   In most feedback control systems, small deviations in parameter values from their design values will not cause any problem in the normal operation of the system, provided these parameters are inside the loop. If plant parameters vary widely according to environmental changes, however, then the control system may exhibit satisfactory response for one environmental condition but may fail to provide satisfactory performance under other conditions. In certain cases, large variations of plant parameters may even cause instability.

In the simplest analysis, one may consider different sets of values of the plant parameters. It is then desirable to design a control system that works well for all sets. As soon as this demand is formulated, the strict optimal control problem loses its importance. By asking for good performance over a range, we have to abandon the best performance for one parameter set.

If the plant transfer function or plant state equation can be identified continuously, then we can compensate for variations in the transfer function or state equation of the plant simply by varying adjustable parameters of the controller and thereby obtain satisfactory system performance continuously under various environmental conditions. Such an adaptive approach is quite useful to cope with a problem where the plant is normally exposed to varying environments so that plant parameters change from time to time. (Since changes are not predictable in most practical cases, a fixed-parameter or preprogrammed time-varying controller cannot provide an answer.)

Definition of adaptive control systems.   Adaptation is a fundamental characteristic of living organisms since they attempt to maintain physiological equilibrium in the midst of changing environmental conditions. An approach to the design of adaptive systems is then to consider the adaptive aspects of human or animal behavior and to develop systems that behave somewhat analogously.

There are different definitions of adaptive control systems now in use in the literature. The vagueness surrounding most definitions and classifications of adaptive systems is due to the large variety of mechanisms by which adaptation may be achieved and to a failure to differentiate between the external manifestations of adaptive behavior and the internal mechanisms used to achieve it. Primarily, different definitions arise because of the various classifications and delineations that divide control systems into those that are adaptive and those that are not. (The small degree of adaptivity required by most system specifications could be achieved by the use of familiar feedback techniques with fixed gains, compensators, and, in some cases, nonlinearities.)

We shall find it necessary to define adaptive system characteristics that are fundamentally different from those of conventional feedback systems so that we may restrict our attention to only those unique aspects of adaptive system behavior and design. In this book we define adaptive control systems as follows:

Definition. An adaptive control system is one that continuously and automatically measures the dynamic characteristics (such as the transfer function or state equation) of the plant, compares them with the desired dynamic characteristics, and uses the difference to vary adjustable system parameters (usually controller characteristics) or to generate an actuating signal so that the optimal performance can be maintained regardless of the environmental changes; alternatively, such a system may continuously measure its own performance according to a given performance index and modify, if necessary, its own parameters so as to maintain optimal performance regardless of the environmental changes.

To be called an adaptive system, self-organizing features must exist. If the adjustment of the system parameters is done only by direct measurement of the environment, the system is not adaptive.

A seemingly adaptive system is exemplified by the aircraft autopilot, which is designed to adjust its loop gains as a function of altitude to compensate for corresponding changes in aircraft parameters. The adjustment is based on direct information about the environment (in this case, atmospheric pressure) and not on a self-organizing scheme. These systems do not possess any self-organizing features and therefore are essentially conventional closed-loop systems.

Another example of systems that may seem to be adaptive but really are not exists in the field of model-reference control systems. Some of these systems (such as the one considered in Example 10–11) merely use the difference between model response and plant response as an input signal to the plant, as shown in Figure 10–28(a). These systems cannot be considered truly adaptive since block-diagram manipulation will, in this case, reduce the configuration to that of Figure 10–28(b), which is simply a basic feedback loop with a prefilter. (Note that some authors call this type of control *model adaptation*. The model is either a physical model or a simulated system on the computer. The model has no variable parameters.)

Performance indexes. The very basis of adaptive control rests in the premise that there is some condition of operation or performance for the system that is better than any

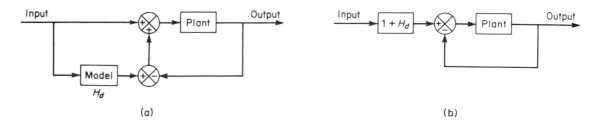

(a) (b)

Figure 10–28
(a) Model-reference control system; (b) simplified block diagram.

other. Thus it becomes necesary to define what constitutes optimal performance. In adaptive control systems, such performance is defined in terms of the performance index, which we must decide upon after stating our goals. These goals will be as diverse as the systems to which they are applied, but it is usually possible to generalize the object of optimization to that of minimizing the cost of operation, or maximizing the profit.

Some general characteristics that are usually considered desirable are:

1. Reliability
2. Selectivity
3. Applicability

Hence, the performance index should be reliable, or it should be a uniform measure of "goodness" for systems of all orders. It should be selective, or it should involve a sharply defined optimum as a function of system parameters. It should not have local optima or saddle points. The performance index should be easily applicable to practical systems and should be easily measurable.

If the performance index assumes a value of zero at the optimal operating condition, rather than a maximum or a minimum, then it can be used as the adaptive-loop error signal and can be used for feedback directly in some systems.

Note that, in general, all the mathematically tractable performance indexes (such as quadratic performance indexes) have one serious drawback in common; although they specify the cost of system operation in terms of error and energy, they do not give us information about the transient-response characteristics of the system. Thus a system that is designed to operate optimally from the standpoint of maximum "profit" may have undesirable transient characteristics or may even be unstable. Therefore, to assure satisfactory response characteristics, we may need secondary criteria relating to response characteristics in order to influence the choice of cost weighting elements.

Finally, remember that the performance index used in an adaptive control system defines an optimal performance for that system. This means that the performance index essentially gives the upper limit of the performance of the system. Therefore, the selection of a proper performance index is most important.

Adaptive controllers. An adaptive controller may consist of the following three functions:

1. Identification of dynamic characteristics of the plant
2. Decision making based on the identification of the plant
3. Modification or actuation based on the decision made

If the plant is imperfectly known, perhaps because of random time-varying parameters or because of the effects of environmental changes on the plant dynamic characteristics, then the initial identification, decision, and modification procedures will not be sufficient to minimize (or maximize) the performance index. It then becomes necessary to carry out these procedures continuously or at intervals of time, depending on how fast the plant parameters are changing. This constant "self-redesign" or self-organization of the system to compensate for unpredictable changes in the plant is the aspect of performance that is usually considered in defining an adaptive control system.

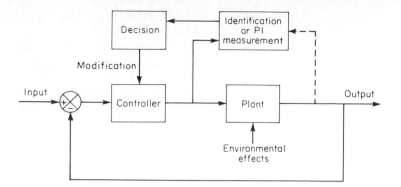

Figure 10–29
Block diagram representation of an adaptive control system.

A block diagram representation of an adaptive control system is shown in Figure 10–29. In this system, the plant is identified and the performance index measured continuously or periodically. Once these have been accomplished, the performance index is compared with the optimal and a decision is made based on the findings as to how to modify the actuating signal. Since the plant is identified within the system itself, adjustment of the parameters is a closed-loop operation. Note that in such closed-loop adaptation the question of stability may arise.

In the following, we shall explain in some detail the three functions: identification, decision, and modification.

Identification of the dynamic characteristics of the plant.

The dynamic characteristics of the plant must be measured and identified continuously or at least very frequently. This should be accomplished without affecting the normal operation of the system. To identify the characteristics of a system, we must perform a test and analyze the results. (For a control system, this entails imposing a control signal on the plant and analyzing system response.) Identification may be made from normal operating data of the plant or by use of test signals, such as sinusoidal ones of small amplitude or various stochastic signals of small amplitude. In practice, no direct application of step or impulse inputs can be made. (Except in certain special cases, the plant will be in normal operation during the test so that the test signals imposed should not unduly disturb normal outputs; furthermore, normal inputs and system noise should not disturb and confuse the test.) Normal inputs are ideal as test signals since no difficulties with undesired outputs or confusing inputs will arise. However, identification with normal inputs is only possible when they have adequate signal characteristics (bandwidth, amplitude, and so on) for proper identification.

Stochastic test signals are quite convenient in certain applications. By using cross-correlation techniques, we may analyze the output as a function of the stochastic input to determine the response characteristics. With a stochastic input, the excitation energy can be spread over a band of frequencies making the effect tolerable. Also, since the crosscorrelator can be designed to closely correlate inputs and outputs, the level of the test signal may be kept low.

Identification must not take too long since if it does, further variations of the plant parameter may occur. Identification time should be sufficiently short compared with the rate of environmental changes. With time for identification limited, it is usually impossible to identify the plant completely; the best one can expect is only partial identification.

It is important to note that not all adaptive systems require identification explicitly. Some systems have already been identified to the extent that the measurement of the value of the performance index can indicate what controller parameters must be changed. That is, the system is fairly well known so that a measurement of the performance index completes the identification.

On the other hand, if plant identification is very difficult, we need to measure directly the performance index and build an adaptive controller based on it. If identification is not called for and adaptivity is based on measurements of the performance index alone, the control system is called an optimizing control system. Since self-organization is achieved with this approach, we shall consider such a system adaptive.

The difficulty of making a realistic identification will depend on how much information about the plant is required and on the amount of prior knowledge of the plant. In general, these are also the factors that will determine whether to use an identification approach or a direct search of the controller parameter space as a function of the performance index, which is discussed later under optimizing control systems.

Decision making based on the identification of the plant.   Decision here refers to one made on the basis of plant characteristics that have been identified and on the performance index computed.

Once the plant has been identified, it is compared with the optimal characteristics (or optimal performance), and then a decision must be made as to how the adjustable parameters (controller characteristics) should be varied to maintain optimal performance. The decision is accomplished by a computer.

Modification based on the decision made.   Modification refers to the change of control signals according to the results of the identification and decision. In most schemes, the decision and modification are conceptually a single operation with the modification consisting of a means of mechanizing the transformation of a decision output signal into the control signal (the input to the plant).

This control signal, or the input signal to the plant, can be modified in two ways. The first approach is to adjust the controller parameters to compensate for changes in the plant dynamics. This is called controller parameter modification. The second approach is to synthesize the optimal control signal, based on the plant transfer function or plant state equation, performance index, and desired transient response. This is called control-signal synthesis.

The choice between controller parameter modification and control-signal synthesis is primarily a hardware decision since the two approaches are conceptually equivalent. Where reliability is very important, as in aerospace applications, the use of parameter change adaptation is often favored over the use of control-signal synthesis. (This is because the system can operate even after the failure of the adaptive loop if the control signal is not entirely dependent on the adaptive portion of the system.)

Optimalizing control systems.   Optimalizing control systems heavily rely on optimization techniques. In general, optimization consists of searching the space of variable controller parameters as a function of some performance index to determine where the performance index is maximized or minimized. Implicit in the previous statement is the fact that a scalar performance index that is a function of system outputs can be defined so that its extremum represents the best possible system performance. This is generally possible and necessary for any adaptive control system.

The methods of finding the optimal operating point are basically trial-and-error procedures. In the steepest-descent approach, the gradient of the performance index surface is measured by noting the effects of small changes in the variable parameters. (This may be called the derivative sensing method.) The parameter vector is then moved in the direction of maximum slope, either a fixed amount or else an amount determined by the surface gradient. Where the parameters are slowly varying, the gradient may be calculated relatively infrequently. Under conditions of more rapid parameter changes, however, a procedure kown as alternate biasing is superior to the derivative-sensing method. With alternate biasing, the system is never operated at the optimal condition but is alternately operated a fixed distance on either side of the calculated optimum, and a new optimum is calculated from the difference in the values of the performance index.

Probably the greatest advantage of the optimalizing control approach is that no restrictions have been placed on the plant. It may be nonlinear; multiple-input, multiple-output; time-varying; and so on. A major difficulty of this optimization approach is that no satisfactory method has been found for discriminating against local extremum. Therefore, this approach is useful for any physical process whose performance surface has a single optimum and whose variations are slow enough for the control system to accommodate them.

Learning systems. One significant difference between a skilled human operator and the adaptive controller discussed above is that the human operator recognizes familiar inputs and can use past learned experiences to react in an optimal manner. Adaptive control systems are designed to modify the control signal as the system environment changes so that performance is always optimal.

A system that is capable of recognizing the familiar feature and patterns of a situation and that uses its past learned experiences in behaving in an optimal fashion is called a *learning system*.

A learning system is a higher-level system than an adaptive system. The space of all control systems may be divided into four basic hierarchal levels:

1. Open loop
2. Closed loop
3. Adaptive loop
4. Learning loop

where each level is responsive to a performance index or control error measured at the next lower level and where levels higher than the fourth will exist for more complex environments.

A learning system responding to a familiar situation will not require identification of the system. The approach to the design of such a system is to "teach" the system the best choice for each situation. Once the system has learned the optimal control law for each possible situation, it may operate near the optimal condition regardless of environmental changes. (Learning capability is very important in robotic systems.)

A learning system, when subjected to a new situation, learns how to behave by an adaptive approach. If the system experiences the same situation it has learned before, it will recognize it and behave optimally without going through the same adaptive approach. Models of human behavior that are being developed by many researchers will undoubtedly yield useful results for application to learning systems.

Concluding comments. Recent developments of high-performance aerospace vehicles, robotic systems, and high-efficiency manufacturing plants impose more and more stringent requirements on their associated control systems. In designing such control systems, we are concerned with the development of systems that will meet the specifications imposed by the users under the anticipated operating conditions. Most control systems that require exacting performance over a wide range of operating conditions will necessarily be adaptive. When high adaptivity is clearly called for, most present-day requirements will be met by an identification–decision–modification system with either sequential or continuous modification, depending on the rate of change of the varying parameters.

The most fascinating developments in adaptive control systems lie in the areas of pattern recognition and learning systems. Pattern-recognition techniques may answer the need for a general identification procedure. When tied in with learning approaches, they will accomplish adaptive-learning control.

This section has presented only an outline of adaptive control systems. The reader interested in such systems should refer to recent research results available in the literature.

Example Problems and Solutions

A–10–1. Consider the system defined by

$$\dot{\mathbf{x}} = \mathbf{A}\mathbf{x} + \mathbf{B}u$$

Suppose that this system is not completely state controllable. Then the rank of the controllability matrix is less than n, or

$$\text{rank } [\mathbf{B} \ \vdots \ \mathbf{AB} \ \vdots \ \cdots \ \vdots \ \mathbf{A}^{n-1}\mathbf{B}] = q < n \tag{10–143}$$

This means that there are q linearly independent column vectors in the controllability matrix. Let us define such q linearly independent column vectors as $\mathbf{f}_1, \mathbf{f}_2, \ldots, \mathbf{f}_q$. Also, let us choose $n - q$ additional n-vectors $\mathbf{v}_{q+1}, \mathbf{v}_{q+2}, \ldots, \mathbf{v}_n$ such that

$$\mathbf{P} = [\mathbf{f}_1 \ \vdots \ \mathbf{f}_2 \ \vdots \ \cdots \ \vdots \ \mathbf{f}_q \ \vdots \ \mathbf{v}_{q+1} \ \vdots \ \mathbf{v}_{q+2} \ \vdots \ \cdots \ \vdots \ \mathbf{v}_n]$$

is of rank n. By using matrix $\mathbf{P}$ as the transformation matrix, define

$$\mathbf{P}^{-1}\mathbf{A}\mathbf{P} = \hat{\mathbf{A}}, \qquad \mathbf{P}^{-1}\mathbf{B} = \hat{\mathbf{B}}$$

Show that $\hat{\mathbf{A}}$ can be given by

$$\hat{\mathbf{A}} = \begin{bmatrix} \mathbf{A}_{11} & \vdots & \mathbf{A}_{12} \\ \cdots & \cdots & \cdots \\ \mathbf{0} & \vdots & \mathbf{A}_{22} \end{bmatrix}$$

where $\mathbf{A}_{11}$ is a $q \times q$ matrix, $\mathbf{A}_{12}$ is a $q \times (n - q)$ matrix, $\mathbf{A}_{22}$ is an $(n - q) \times (n - q)$ matrix, and $\mathbf{0}$ is an $(n - q) \times q$ matrix. Show also that matrix $\hat{\mathbf{B}}$ can be given by

$$\hat{\mathbf{B}} = \begin{bmatrix} \mathbf{B}_{11} \\ \cdots \\ \mathbf{0} \end{bmatrix}$$

where $\mathbf{B}_{11}$ is a $q \times 1$ matrix and $\mathbf{0}$ is an $(n - q) \times 1$ matrix.

Solution. Notice that

$$\mathbf{AP} = \mathbf{P}\hat{\mathbf{A}}$$

or

$$[\mathbf{Af}_1 \ \vdots \ \mathbf{Af}_2 \ \vdots \ \cdots \ \vdots \ \mathbf{Af}_q \ \vdots \ \mathbf{Av}_{q+1} \ \vdots \ \cdots \ \vdots \ \mathbf{Av}_n]$$
$$= [\mathbf{f}_1 \ \vdots \ \mathbf{f}_2 \ \vdots \ \cdots] \ \mathbf{f}_q \ \vdots \ \mathbf{v}_{q+1} \ \vdots \ \cdots \ \vdots \ \mathbf{v}_n]\hat{\mathbf{A}} \qquad (10\text{–}144)$$

Also,

$$\mathbf{B} = \mathbf{P}\hat{\mathbf{B}} \qquad (10\text{–}145)$$

Since we have q linearly independent column vectors $\mathbf{f}_1, \mathbf{f}_2, \ldots, \mathbf{f}_q$, we can use the Cayley–Hamilton theorem to express vectors $\mathbf{Af}_1, \mathbf{Af}_2, \ldots, \mathbf{Af}_q$ in terms of these q vectors. That is,

$$\mathbf{Af}_1 = a_{11}\mathbf{f}_1 + a_{21}\mathbf{f}_2 + \cdots + a_{q1}\mathbf{f}_q$$
$$\mathbf{Af}_2 = a_{12}\mathbf{f}_1 + a_{22}\mathbf{f}_2 + \cdots + a_{q2}\mathbf{f}_q$$
$$\vdots$$
$$\mathbf{Af}_q = a_{1q}\mathbf{f}_1 + a_{2q}\mathbf{f}_2 + \cdots + a_{qq}\mathbf{f}_q$$

Hence Equation (10–144) may be written as follows:

$$[\mathbf{Af}_1 \ \vdots \ \mathbf{Af}_2 \ \vdots \ \cdots \ \vdots \ \mathbf{Af}_q \ \vdots \ \mathbf{Av}_{q+1} \ \vdots \ \cdots \ \vdots \ \mathbf{Av}_n]$$

$$= [\mathbf{f}_1 \ \vdots \ \mathbf{f}_2 \ \vdots \ \cdots \ \vdots \ \mathbf{f}_q \ \vdots \ \mathbf{v}_{q+1} \ \vdots \ \cdots \ \vdots \ \mathbf{v}_n]
\begin{bmatrix}
a_{11} & \cdots & a_{1q} & a_{1\,q+1} & \cdots & a_{1n} \\
a_{21} & \cdots & a_{2q} & a_{2\,q+1} & \cdots & a_{2n} \\
\vdots & & \vdots & \vdots & & \vdots \\
a_{q1} & \cdots & a_{qq} & a_{q\,q+1} & \cdots & a_{qn} \\
\hline
0 & \cdots & 0 & a_{q+1\,q+1} & \cdots & a_{q+1\,n} \\
\vdots & & \vdots & \vdots & & \vdots \\
0 & \cdots & 0 & a_{n\,q+1} & \cdots & a_{nn}
\end{bmatrix}$$

Define

$$\begin{bmatrix}
a_{11} & \cdots & a_{1q} \\
a_{21} & \cdots & a_{2q} \\
\vdots & & \vdots \\
\vdots & & \vdots \\
a_{q1} & \cdots & a_{qq}
\end{bmatrix} = \mathbf{A}_{11}$$

$$\begin{bmatrix}
a_{1\,q+1} & \cdots & a_{1n} \\
a_{2\,q+1} & \cdots & a_{2n} \\
\vdots & & \vdots \\
\vdots & & \vdots \\
a_{q\,q+1} & \cdots & a_{qn}
\end{bmatrix} = \mathbf{A}_{12}$$

$$\begin{bmatrix} 0 & \cdots & 0 \\ \cdot & & \cdot \\ \cdot & & \cdot \\ \cdot & & \cdot \\ 0 & \cdots & 0 \end{bmatrix} = \mathbf{A}_{21} = (n - q) \times q \text{ zero matrix}$$

$$\begin{bmatrix} a_{q+1\,q+1} & \cdots & a_{q+1\,n} \\ \cdot & & \cdot \\ \cdot & & \cdot \\ \cdot & & \cdot \\ a_{n\,q+1} & \cdots & a_{nn} \end{bmatrix} = \mathbf{A}_{22}$$

Then Equation (10–144) can be written as

$$[\mathbf{Af}_1 \ \vdots \ \mathbf{Af}_2 \ \vdots \ \cdots \ \vdots \ \mathbf{Af}_q \ \vdots \ \mathbf{Av}_{q+1} \ \vdots \ \cdots \ \vdots \ \mathbf{Av}_n]$$

$$= [\mathbf{f}_1 \ \vdots \ \mathbf{f}_2 \ \vdots \ \cdots \ \vdots \ \mathbf{f}_q \ \vdots \ \mathbf{v}_{q+1} \ \vdots \ \cdots \ \vdots \ \mathbf{v}_n] \begin{bmatrix} \mathbf{A}_{11} & \vdots & \mathbf{A}_{12} \\ \cdots & + & \cdots \\ \mathbf{0} & \vdots & \mathbf{A}_{22} \end{bmatrix}$$

Thus

$$\mathbf{AP} = \mathbf{P} \begin{bmatrix} \mathbf{A}_{11} & \vdots & \mathbf{A}_{12} \\ \cdots & + & \cdots \\ \mathbf{0} & \vdots & \mathbf{A}_{22} \end{bmatrix}$$

Hence

$$\mathbf{P}^{-1}\mathbf{AP} = \hat{\mathbf{A}} = \begin{bmatrix} \mathbf{A}_{11} & \vdots & \mathbf{A}_{12} \\ \cdots & + & \cdots \\ \mathbf{0} & \vdots & \mathbf{A}_{22} \end{bmatrix}$$

Next, referring to Equation (10–145), we have

$$\mathbf{B} = [\mathbf{f}_1 \ \vdots \ \mathbf{f}_2 \ \vdots \ \cdots \ \vdots \ \mathbf{f}_q \ \vdots \ \mathbf{v}_{q+1} \ \vdots \ \cdots \ \vdots \ \mathbf{v}_n]\, \hat{\mathbf{B}} \tag{10–146}$$

Referring to Equation (10–143), notice that vector $\mathbf{B}$ can be written in terms of q linearly independent column vectors $\mathbf{f}_1, \mathbf{f}_2, \ldots, \mathbf{f}_q$. Thus, we have

$$\mathbf{B} = b_{11}\mathbf{f}_1 + b_{21}\mathbf{f}_2 + \cdots + b_{q1}\mathbf{f}_q$$

Consequently, Equation (10–146) may be written as follows:

$$b_{11}\mathbf{f}_1 + b_{21}\mathbf{f}_2 + \cdots + b_{q1}\mathbf{f}_q = [\mathbf{f}_1 \ \vdots \ \mathbf{f}_2 \ \vdots \ \cdots \ \vdots \ \mathbf{f}_q \ \vdots \ \mathbf{v}_{q+1} \ \vdots \ \cdots \ \vdots \ \mathbf{v}_n] \begin{bmatrix} b_{11} \\ b_{21} \\ \cdot \\ \cdot \\ \cdot \\ b_{q1} \\ 0 \\ \cdot \\ \cdot \\ \cdot \\ 0 \end{bmatrix}$$

Thus

$$\hat{\mathbf{B}} = \left[\begin{array}{c} \mathbf{B}_{11} \\ \hline \mathbf{0} \end{array}\right]$$

where

$$\mathbf{B}_{11} = \begin{bmatrix} b_{11} \\ b_{21} \\ \cdot \\ \cdot \\ \cdot \\ b_{q1} \end{bmatrix}$$

A–10–2. Consider a completely state controllable system

$$\dot{\mathbf{x}} = \mathbf{A}\mathbf{x} + \mathbf{B}u$$

Define the controllability matrix as $\mathbf{M}$:

$$\mathbf{M} = [\mathbf{B} \;\vdots\; \mathbf{AB} \;\vdots\; \cdots \;\vdots\; \mathbf{A}^{n-1}\mathbf{B}]$$

Show that

$$\mathbf{M}^{-1}\mathbf{A}\mathbf{M} = \begin{bmatrix} 0 & 0 & \cdots & 0 & -a_n \\ 1 & 0 & \cdots & 0 & -a_{n-1} \\ 0 & 1 & \cdots & 0 & -a_{n-2} \\ \cdot & \cdot & & \cdot & \cdot \\ \cdot & \cdot & & \cdot & \cdot \\ \cdot & \cdot & & \cdot & \cdot \\ 0 & 0 & \cdots & 1 & -a_1 \end{bmatrix}$$

where $a_1, a_2, \ldots, a_n$ are the coefficients of the characteristic polynomial

$$|s\mathbf{I} - \mathbf{A}| = s^n + a_1 s^{n-1} + \cdots + a_{n-1}s + a_n$$

Solution. Let us consider the case where $n = 3$. We shall show that

$$\mathbf{A}\mathbf{M} = \mathbf{M}\begin{bmatrix} 0 & 0 & -a_3 \\ 1 & 0 & -a_2 \\ 0 & 1 & -a_1 \end{bmatrix} \tag{10–147}$$

The left side of Equation (10–147) is

$$\mathbf{A}\mathbf{M} = \mathbf{A}[\mathbf{B} \;\vdots\; \mathbf{AB} \;\vdots\; \mathbf{A}^2\mathbf{B}] = [\mathbf{AB} \;\vdots\; \mathbf{A}^2\mathbf{B} \;\vdots\; \mathbf{A}^3\mathbf{B}]$$

The right side of Equation (10–147) is

$$[\mathbf{B} \;\vdots\; \mathbf{AB} \;\vdots\; \mathbf{A}^2\mathbf{B}]\begin{bmatrix} 0 & 0 & -a_3 \\ 1 & 0 & -a_2 \\ 0 & 1 & -a_1 \end{bmatrix} = [\mathbf{AB} \;\vdots\; \mathbf{A}^2\mathbf{B} \;\vdots\; -a_3\mathbf{B} - a_2\mathbf{AB} - a_1\mathbf{A}^2\mathbf{B}] \tag{10–148}$$

The Cayley–Hamilton theorem states that matrix $\mathbf{A}$ satisfies its own characteristic equation, or in the case of $n = 3$,

$$\mathbf{A}^3 + a_1\mathbf{A}^2 + a_2\mathbf{A} + a_3\mathbf{I} = \mathbf{0} \tag{10–149}$$

Using Equation (10–149), the third column of the right side of Equation (10–148) becomes

$$-a_3\mathbf{B} - a_2\mathbf{AB} - a_1\mathbf{A}^2\mathbf{B} = (-a_3\mathbf{I} - a_2\mathbf{A} - a_1\mathbf{A}^2)\mathbf{B} = \mathbf{A}^3\mathbf{B}$$

Thus, Equation (10–148), the right side of Equation (10–147), becomes

$$[\mathbf{B} \;\vdots\; \mathbf{AB} \;\vdots\; \mathbf{A}^2\mathbf{B}] \begin{bmatrix} 0 & 0 & -a_3 \\ 1 & 0 & -a_2 \\ 0 & 1 & -a_1 \end{bmatrix} = [\mathbf{AB} \;\vdots\; \mathbf{A}^2\mathbf{B} \;\vdots\; \mathbf{A}^3\mathbf{B}]$$

Hence, the left side and the right side of Equation (10–147) are the same. We have thus shown that Equation (10–147) is true. Consequently,

$$\mathbf{M}^{-1}\mathbf{AM} = \begin{bmatrix} 0 & 0 & -a_3 \\ 1 & 0 & -a_2 \\ 0 & 1 & -a_1 \end{bmatrix}$$

The preceding derivation can be easily extended to the general case of any positive integer n.

A–10–3. Consider a completely state controllable system

$$\dot{\mathbf{x}} = \mathbf{Ax} + \mathbf{B}u$$

Define

$$\mathbf{M} = [\mathbf{B} \;\vdots\; \mathbf{AB} \;\vdots\; \cdots \;\vdots\; \mathbf{A}^{n-1}\mathbf{B}]$$

and

$$\mathbf{W} = \begin{bmatrix} a_{n-1} & a_{n-2} & \cdots & a_1 & 1 \\ a_{n-2} & a_{n-3} & \cdots & 1 & 0 \\ \cdot & \cdot & & \cdot & \cdot \\ \cdot & \cdot & & \cdot & \cdot \\ \cdot & \cdot & & \cdot & \cdot \\ a_1 & 1 & \cdots & 0 & 0 \\ 1 & 0 & \cdots & 0 & 0 \end{bmatrix}$$

where the a_i's are coefficients of the characteristic polynomial

$$|s\mathbf{I} - \mathbf{A}| = s^n + a_1 s^{n-1} + \cdots + a_{n-1}s + a_n$$

Define also

$$\mathbf{T} = \mathbf{MW}$$

Show that

$$\mathbf{T}^{-1}\mathbf{AT} = \begin{bmatrix} 0 & 1 & 0 & \cdots & 0 \\ 0 & 0 & 1 & \cdots & 0 \\ \cdot & \cdot & \cdot & & \cdot \\ \cdot & \cdot & \cdot & & \cdot \\ \cdot & \cdot & \cdot & & \cdot \\ 0 & 0 & 0 & \cdots & 1 \\ -a_n & -a_{n-1} & -a_{n-2} & \cdots & -a_1 \end{bmatrix}, \quad \mathbf{T}^{-1}\mathbf{B} = \begin{bmatrix} 0 \\ 0 \\ \cdot \\ \cdot \\ \cdot \\ 0 \\ 1 \end{bmatrix}$$

Solution. Let us consider the case where $n = 3$. We shall show that

$$\mathbf{T}^{-1}\mathbf{A}\mathbf{T} = (\mathbf{MW})^{-1}\mathbf{A}(\mathbf{MW}) = \mathbf{W}^{-1}(\mathbf{M}^{-1}\mathbf{A}\mathbf{M})\mathbf{W} = \begin{bmatrix} 0 & 1 & 0 \\ 0 & 0 & 1 \\ -a_3 & -a_2 & -a_1 \end{bmatrix} \tag{10–150}$$

Referring to Problem A–10–2, we have

$$\mathbf{M}^{-1}\mathbf{A}\mathbf{M} = \begin{bmatrix} 0 & 0 & -a_3 \\ 1 & 0 & -a_2 \\ 0 & 1 & -a_1 \end{bmatrix}$$

Hence, Equation (10–150) can be rewritten as

$$\mathbf{W}^{-1} \begin{bmatrix} 0 & 0 & -a_3 \\ 1 & 0 & -a_2 \\ 0 & 1 & -a_1 \end{bmatrix} \mathbf{W} = \begin{bmatrix} 0 & 1 & 0 \\ 0 & 0 & 1 \\ -a_3 & -a_2 & -a_1 \end{bmatrix}$$

Therefore, we need to show that

$$\begin{bmatrix} 0 & 0 & -a_3 \\ 1 & 0 & -a_2 \\ 0 & 1 & -a_1 \end{bmatrix} \mathbf{W} = \mathbf{W} \begin{bmatrix} 0 & 1 & 0 \\ 0 & 0 & 1 \\ -a_3 & -a_2 & -a_1 \end{bmatrix} \tag{10–151}$$

The left side of Equation (10–151) is

$$\begin{bmatrix} 0 & 0 & -a_3 \\ 1 & 0 & -a_2 \\ 0 & 1 & -a_1 \end{bmatrix} \begin{bmatrix} a_2 & a_1 & 1 \\ a_1 & 1 & 0 \\ 1 & 0 & 0 \end{bmatrix} = \begin{bmatrix} -a_3 & 0 & 0 \\ 0 & a_1 & 1 \\ 0 & 1 & 0 \end{bmatrix}$$

The right side of Equation (10–151) is

$$\begin{bmatrix} a_2 & a_1 & 1 \\ a_1 & 1 & 0 \\ 1 & 0 & 0 \end{bmatrix} \begin{bmatrix} 0 & 1 & 0 \\ 0 & 0 & 1 \\ -a_3 & -a_2 & -a_1 \end{bmatrix} = \begin{bmatrix} -a_3 & 0 & 0 \\ 0 & a_1 & 1 \\ 0 & 1 & 0 \end{bmatrix}$$

Clearly, Equation (10–151) holds true. Thus, we have shown that

$$\mathbf{T}^{-1}\mathbf{A}\mathbf{T} = \begin{bmatrix} 0 & 1 & 0 \\ 0 & 0 & 1 \\ -a_3 & -a_2 & -a_1 \end{bmatrix}$$

Next, we shall show that

$$\mathbf{T}^{-1}\mathbf{B} = \begin{bmatrix} 0 \\ 0 \\ 1 \end{bmatrix} \tag{10–152}$$

Note that Equation (10–152) can be written as

$$\mathbf{B} = \mathbf{T} \begin{bmatrix} 0 \\ 0 \\ 1 \end{bmatrix} = \mathbf{MW} \begin{bmatrix} 0 \\ 0 \\ 1 \end{bmatrix}$$

Example Problems and Solutions

861

Noting that

$$\mathbf{T}\begin{bmatrix} 0 \\ 0 \\ 1 \end{bmatrix} = [\mathbf{B} \ \vdots \ \mathbf{AB} \ \vdots \ \mathbf{A^2B}]\begin{bmatrix} a_2 & a_1 & 1 \\ a_1 & 1 & 0 \\ 1 & 0 & 0 \end{bmatrix}\begin{bmatrix} 0 \\ 0 \\ 1 \end{bmatrix} = [\mathbf{B} \ \vdots \ \mathbf{AB} \ \vdots \ \mathbf{A^2B}]\begin{bmatrix} 1 \\ 0 \\ 0 \end{bmatrix} = \mathbf{B}$$

we have

$$\mathbf{T^{-1}B} = \begin{bmatrix} 0 \\ 0 \\ 1 \end{bmatrix}$$

The derivation shown here can be easily extended to the general case of any positive integer n.

A–10–4. Consider the state equation

$$\dot{\mathbf{x}} = \mathbf{Ax} + \mathbf{B}u$$

where

$$\mathbf{A} = \begin{bmatrix} 1 & 1 \\ -4 & -3 \end{bmatrix}, \qquad \mathbf{B} = \begin{bmatrix} 0 \\ 2 \end{bmatrix}$$

The rank of the controllability matrix $\mathbf{M}$,

$$\mathbf{M} = [\mathbf{B} \ \vdots \ \mathbf{AB}] = \begin{bmatrix} 0 & 2 \\ 2 & -6 \end{bmatrix}$$

is 2. Thus, the system is completely state controllable. Transform the given state equation into the controllable canonical form.

Solution. Since

$$|s\mathbf{I} - \mathbf{A}| = \begin{vmatrix} s - 1 & -1 \\ 4 & s + 3 \end{vmatrix} = (s - 1)(s + 3) + 4$$
$$= s^2 + 2s + 1 = s^2 + a_1s + a_2$$

we have

$$a_1 = 2, \qquad a_2 = 1$$

Define

$$\mathbf{T} = \mathbf{MW}$$

where

$$\mathbf{M} = \begin{bmatrix} 0 & 2 \\ 2 & -6 \end{bmatrix}, \qquad \mathbf{W} = \begin{bmatrix} 2 & 1 \\ 1 & 0 \end{bmatrix}$$

Then

$$\mathbf{T} = \begin{bmatrix} 0 & 2 \\ 2 & -6 \end{bmatrix}\begin{bmatrix} 2 & 1 \\ 1 & 0 \end{bmatrix} = \begin{bmatrix} 2 & 0 \\ -2 & 2 \end{bmatrix}$$

and

$$\mathbf{T^{-1}} = \begin{bmatrix} 0.5 & 0 \\ 0.5 & 0.5 \end{bmatrix}$$

Define

$$\mathbf{x} = \mathbf{T}\hat{\mathbf{x}}$$

Then the state equation becomes

$$\dot{\hat{\mathbf{x}}} = \mathbf{T}^{-1}\mathbf{A}\mathbf{T}\hat{\mathbf{x}} + \mathbf{T}^{-1}\mathbf{B}u$$

Since

$$\mathbf{T}^{-1}\mathbf{A}\mathbf{T} = \begin{bmatrix} 0.5 & 0 \\ 0.5 & 0.5 \end{bmatrix} \begin{bmatrix} 1 & 1 \\ -4 & -3 \end{bmatrix} \begin{bmatrix} 2 & 0 \\ -2 & 2 \end{bmatrix} = \begin{bmatrix} 0 & 1 \\ -1 & -2 \end{bmatrix}$$

and

$$\mathbf{T}^{-1}\mathbf{B} = \begin{bmatrix} 0.5 & 0 \\ 0.5 & 0.5 \end{bmatrix} \begin{bmatrix} 0 \\ 2 \end{bmatrix} = \begin{bmatrix} 0 \\ 1 \end{bmatrix}$$

we have

$$\begin{bmatrix} \dot{\hat{x}}_1 \\ \dot{\hat{x}}_2 \end{bmatrix} = \begin{bmatrix} 0 & 1 \\ -1 & -2 \end{bmatrix} \begin{bmatrix} \hat{x}_1 \\ \hat{x}_2 \end{bmatrix} + \begin{bmatrix} 0 \\ 1 \end{bmatrix} u$$

which is in the controllable canonical form.

A–10–5. Consider a system defined by

$$\dot{\mathbf{x}} = \mathbf{A}\mathbf{x} + \mathbf{B}u$$

$$y = \mathbf{C}\mathbf{x}$$

where

$$\mathbf{A} = \begin{bmatrix} 0 & 1 \\ -2 & -3 \end{bmatrix}, \qquad \mathbf{B} = \begin{bmatrix} 0 \\ 2 \end{bmatrix}, \qquad \mathbf{C} = [1 \quad 0]$$

The characteristic equation of the system is

$$|s\mathbf{I} - \mathbf{A}| = \begin{vmatrix} s & -1 \\ 2 & s+3 \end{vmatrix} = s^2 + 3s + 2 = (s+1)(s+2) = 0$$

The eigenvalues of matrix $\mathbf{A}$ are -1 and -2.

It is desired to have eigenvalues at -3 and -5 by using a state feedback control $u = -\mathbf{K}\mathbf{x}$. Determine the necessary feedback gain matrix $\mathbf{K}$ and the control signal u.

Solution. The given system is completely state controllable, since the rank of

$$\mathbf{M} = [\mathbf{B} \ \vdots \ \mathbf{A}\mathbf{B}] = \begin{bmatrix} 0 & 2 \\ 2 & -6 \end{bmatrix}$$

is 2. Hence arbitrary pole placement is possible.

Since the characteristic equation of the original system is

$$s^2 + 3s + 2 = s^2 + a_1 s + a_2 = 0$$

we have

$$a_1 = 3, \qquad a_2 = 2$$

The desired characteristic equation is

$$(s + 3)(s + 5) = s^2 + 8s + 15 = s^2 + \alpha_1 s + \alpha_2 = 0$$

Hence

$$\alpha_1 = 8, \qquad \alpha_2 = 15$$

It is important to point out that the original state equation is not in the controllable canonical form, because matrix **B** is not

$$\begin{bmatrix} 0 \\ 1 \end{bmatrix}$$

Hence the transformation matrix **T** must be determined.

$$\mathbf{T} = \mathbf{MW} = [\mathbf{B} \; \vdots \; \mathbf{AB}]\begin{bmatrix} a_1 & 1 \\ 1 & 0 \end{bmatrix} = \begin{bmatrix} 0 & 2 \\ 2 & -6 \end{bmatrix}\begin{bmatrix} 3 & 1 \\ 1 & 0 \end{bmatrix} = \begin{bmatrix} 2 & 0 \\ 0 & 2 \end{bmatrix}$$

Hence

$$\mathbf{T}^{-1} = \begin{bmatrix} 0.5 & 0 \\ 0 & 0.5 \end{bmatrix}$$

Referring to Equation (10–13), the necessary feedback gain matrix is given by

$$\mathbf{K} = [\alpha_2 - a_2 \; \vdots \; \alpha_1 - a_1]\mathbf{T}^{-1}$$

$$= [15 - 2 \; \vdots \; 8 - 3]\begin{bmatrix} 0.5 & 0 \\ 0 & 0.5 \end{bmatrix} = [6.5 \quad 2.5]$$

Thus the control signal u becomes

$$u = -\mathbf{Kx} = -[6.5 \quad 2.5]\begin{bmatrix} x_1 \\ x_2 \end{bmatrix}$$

A–10–6. Consider a completely observable system

$$\dot{\mathbf{x}} = \mathbf{Ax}$$

$$y = \mathbf{Cx}$$

Define the observability matrix as **N**:

$$\mathbf{N} = [\mathbf{C}^* \; \vdots \; \mathbf{A}^*\mathbf{C}^* \; \vdots \; \cdots \; \vdots \; (\mathbf{A}^*)^{n-1}\mathbf{C}^*]$$

Show that

$$\mathbf{N}^*\mathbf{A}(\mathbf{N}^*)^{-1} = \begin{bmatrix} 0 & 1 & 0 & \cdots & 0 \\ 0 & 0 & 1 & \cdots & 0 \\ \cdot & \cdot & \cdot & & \cdot \\ \cdot & \cdot & \cdot & & \cdot \\ \cdot & \cdot & \cdot & & \cdot \\ 0 & 0 & 0 & \cdots & 1 \\ -a_n & -a_{n-1} & -a_{n-2} & \cdots & -a_1 \end{bmatrix} \qquad (10\text{–}153)$$

where $a_1, a_2, \ldots, a_n$ are the coefficients of the characteristic polynomial

$$|s\mathbf{I} - \mathbf{A}| = s^n + a_1 s^{n-1} + \cdots + a_{n-1}s + a_n$$

Solution. Let us consider the case where $n = 3$. Then Equation (10–153) can be written as

$$\mathbf{N}^*\mathbf{A}(\mathbf{N}^*)^{-1} = \begin{bmatrix} 0 & 1 & 0 \\ 0 & 0 & 1 \\ -a_3 & -a_2 & -a_1 \end{bmatrix} \qquad (10\text{–}154)$$

Equation (10–154) may be rewritten as

$$\mathbf{N}^*\mathbf{A} = \begin{bmatrix} 0 & 1 & 0 \\ 0 & 0 & 1 \\ -a_3 & -a_2 & -a_1 \end{bmatrix} \mathbf{N}^* \qquad (10\text{–}155)$$

We shall show that Equation (10–155) holds true. The left side of Equation (10–155) is

$$\mathbf{N}^*\mathbf{A} = \begin{bmatrix} \mathbf{C} \\ \mathbf{CA} \\ \mathbf{CA}^2 \end{bmatrix} \mathbf{A} = \begin{bmatrix} \mathbf{CA} \\ \mathbf{CA}^2 \\ \mathbf{CA}^3 \end{bmatrix} \qquad (10\text{–}156)$$

The right side of Equation (10–155) is

$$\begin{bmatrix} 0 & 1 & 0 \\ 0 & 0 & 1 \\ -a_3 & -a_2 & -a_1 \end{bmatrix} \mathbf{N}^* = \begin{bmatrix} 0 & 1 & 0 \\ 0 & 0 & 1 \\ -a_3 & -a_2 & -a_1 \end{bmatrix} \begin{bmatrix} \mathbf{C} \\ \mathbf{CA} \\ \mathbf{CA}^2 \end{bmatrix}$$

$$= \begin{bmatrix} \mathbf{CA} \\ \mathbf{CA}^2 \\ -a_3\mathbf{C} - a_2\mathbf{CA} - a_1\mathbf{CA}^2 \end{bmatrix} \qquad (10\text{–}157)$$

The Cayley–Hamilton theorem states that matrix $\mathbf{A}$ satisfies its own characteristic equation, or

$$\mathbf{A}^3 + a_1\mathbf{A}^2 + a_2\mathbf{A} + a_3\mathbf{I} = \mathbf{0}$$

Hence

$$-a_1\mathbf{CA}^2 - a_2\mathbf{CA} - a_3\mathbf{C} = \mathbf{CA}^3$$

Thus, the right side of Equation (10–157) becomes the same as the right side of Equation (10–156). Consequently,

$$\mathbf{N}^*\mathbf{A} = \begin{bmatrix} 0 & 1 & 0 \\ 0 & 0 & 1 \\ -a_3 & -a_2 & -a_1 \end{bmatrix} \mathbf{N}^*$$

which is Equation (10–155). This last equation can be modified to

$$\mathbf{N}^*\mathbf{A}(\mathbf{N}^*)^{-1} = \begin{bmatrix} 0 & 1 & 0 \\ 0 & 0 & 1 \\ -a_3 & -a_2 & -a_1 \end{bmatrix}$$

The derivation presented here can be extended to the general case of any positive integer n.

A–10–7. Consider a completely observable system defined by

$$\dot{\mathbf{x}} = \mathbf{Ax} + \mathbf{B}u \qquad (10\text{–}158)$$

$$y = \mathbf{Cx} + Du \qquad (10\text{–}159)$$

Example Problems and Solutions

865

Define

$$\mathbf{N} = [\mathbf{C}^* \,\vdots\, \mathbf{A}^*\mathbf{C}^* \,\vdots\, \cdots \,\vdots\, (\mathbf{A}^*)^{n-1}\mathbf{C}^*]$$

and

$$\mathbf{W} = \begin{bmatrix} a_{n-1} & a_{n-2} & \cdots & a_1 & 1 \\ a_{n-2} & a_{n-3} & \cdots & 1 & 0 \\ \cdot & \cdot & & \cdot & \cdot \\ \cdot & \cdot & & \cdot & \cdot \\ \cdot & \cdot & & \cdot & \cdot \\ a_1 & 1 & \cdots & 0 & 0 \\ 1 & 0 & \cdots & 0 & 0 \end{bmatrix}$$

where the a's are coefficients of the characteristic polynomial

$$|s\mathbf{I} - \mathbf{A}| = s^n + a_1 s^{n-1} + \cdots + a_{n-1}s + a_n$$

Define also

$$\mathbf{Q} = (\mathbf{WN}^*)^{-1}$$

Show that

$$\mathbf{Q}^{-1}\mathbf{A}\mathbf{Q} = \begin{bmatrix} 0 & 0 & \cdots & 0 & -a_n \\ 1 & 0 & \cdots & 0 & -a_{n-1} \\ 0 & 1 & \cdots & 0 & -a_{n-2} \\ \cdot & \cdot & & \cdot & \cdot \\ \cdot & \cdot & & \cdot & \cdot \\ \cdot & \cdot & & \cdot & \cdot \\ 0 & 0 & \cdots & 1 & -a_1 \end{bmatrix}$$

$$\mathbf{CQ} = \begin{bmatrix} 0 & 0 & \cdots & 0 & 1 \end{bmatrix}$$

$$\mathbf{Q}^{-1}\mathbf{B} = \begin{bmatrix} b_n - a_n b_0 \\ b_{n-1} - a_{n-1} b_0 \\ \cdot \\ \cdot \\ \cdot \\ b_1 - a_1 b_0 \end{bmatrix}$$

where the b_k's ($k = 0, 1, 2, \ldots, n$) are those coefficients appearing in the numerator of the transfer function when $\mathbf{C}(s\mathbf{I} - \mathbf{A})^{-1}\mathbf{B} + D$ is written as follows:

$$\mathbf{C}(s\mathbf{I} - \mathbf{A})^{-1}\mathbf{B} + D = \frac{b_0 s^n + b_1 s^{n-1} + \cdots + b_{n-1}s + b_n}{s^n + a_1 s^{n-1} + \cdots + a_{n-1}s + a_n}$$

where $D = b_0$.

Solution. Let us consider the case where $n = 3$. We shall show that

$$\mathbf{Q}^{-1}\mathbf{A}\mathbf{Q} = (\mathbf{WN}^*)\mathbf{A}(\mathbf{WN}^*)^{-1} = \begin{bmatrix} 0 & 0 & -a_3 \\ 1 & 0 & -a_2 \\ 0 & 1 & -a_1 \end{bmatrix} \qquad (10\text{--}160)$$

Note that, by referring to Problem A–10–6, we have

$$(\mathbf{WN^*})\mathbf{A}(\mathbf{WN^*})^{-1} = \mathbf{W}[\mathbf{N^*A}(\mathbf{N^*})^{-1}]\mathbf{W}^{-1} = \mathbf{W}\begin{bmatrix} 0 & 1 & 0 \\ 0 & 0 & 1 \\ -a_3 & -a_2 & -a_1 \end{bmatrix}\mathbf{W}^{-1}$$

Hence we need to show that

$$\mathbf{W}\begin{bmatrix} 0 & 1 & 0 \\ 0 & 0 & 1 \\ -a_3 & -a_2 & -a_1 \end{bmatrix}\mathbf{W}^{-1} = \begin{bmatrix} 0 & 0 & -a_3 \\ 1 & 0 & -a_2 \\ 0 & 1 & -a_1 \end{bmatrix}$$

or

$$\mathbf{W}\begin{bmatrix} 0 & 1 & 0 \\ 0 & 0 & 1 \\ -a_3 & -a_2 & -a_1 \end{bmatrix} = \begin{bmatrix} 0 & 0 & -a_3 \\ 1 & 0 & -a_2 \\ 0 & 1 & -a_1 \end{bmatrix}\mathbf{W} \tag{10–161}$$

The left side of Equation (10–161) is

$$\mathbf{W}\begin{bmatrix} 0 & 1 & 0 \\ 0 & 0 & 1 \\ -a_3 & -a_2 & -a_1 \end{bmatrix} = \begin{bmatrix} a_2 & a_1 & 1 \\ a_1 & 1 & 0 \\ 1 & 0 & 0 \end{bmatrix}\begin{bmatrix} 0 & 1 & 0 \\ 0 & 0 & 1 \\ -a_3 & -a_2 & -a_1 \end{bmatrix}$$

$$= \begin{bmatrix} -a_3 & 0 & 0 \\ 0 & a_1 & 1 \\ 0 & 1 & 0 \end{bmatrix}$$

The right side of Equation (10–161) is

$$\begin{bmatrix} 0 & 0 & -a_3 \\ 1 & 0 & -a_2 \\ 0 & 1 & -a_1 \end{bmatrix}\mathbf{W} = \begin{bmatrix} 0 & 0 & -a_3 \\ 1 & 0 & -a_2 \\ 0 & 1 & -a_1 \end{bmatrix}\begin{bmatrix} a_2 & a_1 & 1 \\ a_1 & 1 & 0 \\ 1 & 0 & 0 \end{bmatrix}$$

$$= \begin{bmatrix} -a_3 & 0 & 0 \\ 0 & a_1 & 1 \\ 0 & 1 & 0 \end{bmatrix}$$

Thus, we see that Equation (10–161) holds true. Hence, we have proved Equation (10–160).

Next we shall show that

$$\mathbf{CQ} = [0 \quad 0 \quad 1]$$

or

$$\mathbf{C}(\mathbf{WN^*})^{-1} = [0 \quad 0 \quad 1]$$

Notice that

$$[0 \quad 0 \quad 1](\mathbf{WN^*}) = [0 \quad 0 \quad 1]\begin{bmatrix} a_2 & a_1 & 1 \\ a_1 & 1 & 0 \\ 1 & 0 & 0 \end{bmatrix}\begin{bmatrix} \mathbf{C} \\ \mathbf{CA} \\ \mathbf{CA}^2 \end{bmatrix}$$

$$= [1 \quad 0 \quad 0]\begin{bmatrix} \mathbf{C} \\ \mathbf{CA} \\ \mathbf{CA}^2 \end{bmatrix} = \mathbf{C}$$

Hence, we have shown that

$$[0 \quad 0 \quad 1] = \mathbf{C}(\mathbf{WN}^*)^{-1} = \mathbf{CQ}$$

Next define

$$\mathbf{x} = \mathbf{Q}\hat{\mathbf{x}}$$

Then Equation (10–158) becomes

$$\dot{\hat{\mathbf{x}}} = \mathbf{Q}^{-1}\mathbf{AQ}\hat{\mathbf{x}} + \mathbf{Q}^{-1}\mathbf{B}u \tag{10–162}$$

and Equation (10–159) becomes

$$y = \mathbf{CQ}\hat{\mathbf{x}} + Du \tag{10–163}$$

For the case of $n = 3$, Equation (10–162) becomes

$$\begin{bmatrix} \dot{\hat{x}}_1 \\ \dot{\hat{x}}_2 \\ \dot{\hat{x}}_3 \end{bmatrix} = \begin{bmatrix} 0 & 0 & -a_3 \\ 1 & 0 & -a_2 \\ 0 & 1 & -a_1 \end{bmatrix} \begin{bmatrix} \hat{x}_1 \\ \hat{x}_2 \\ \hat{x}_3 \end{bmatrix} + \begin{bmatrix} \gamma_3 \\ \gamma_2 \\ \gamma_1 \end{bmatrix} u$$

where

$$\begin{bmatrix} \gamma_3 \\ \gamma_2 \\ \gamma_1 \end{bmatrix} = \mathbf{Q}^{-1}\mathbf{B}$$

The transfer function $G(s)$ for the system defined by Equations (10–162) and (10–163) is

$$G(s) = \mathbf{CQ}(s\mathbf{I} - \mathbf{Q}^{-1}\mathbf{AQ})^{-1}\mathbf{Q}^{-1}\mathbf{B} + D$$

Noting that

$$\mathbf{CQ} = [0 \quad 0 \quad 1]$$

we have

$$G(s) = [0 \quad 0 \quad 1] \begin{bmatrix} s & 0 & a_3 \\ -1 & s & a_2 \\ 0 & -1 & s + a_1 \end{bmatrix}^{-1} \begin{bmatrix} \gamma_3 \\ \gamma_2 \\ \gamma_1 \end{bmatrix} + D$$

Note that $D = b_0$. Since

$$\begin{bmatrix} s & 0 & a_3 \\ -1 & s & a_2 \\ 0 & -1 & s + a_1 \end{bmatrix}^{-1} = \frac{1}{s^3 + a_1 s^2 + a_2 s + a_3} \begin{bmatrix} s^2 + a_1 s + a_2 & -a_3 & -a_3 s \\ s + a_1 & s^2 + a_1 s & -a_2 s - a_3 \\ 1 & s & s^2 \end{bmatrix}$$

we have

$$G(s) = \frac{1}{s^3 + a_1 s^2 + a_2 s + a_3} [1 \quad s \quad s^2] \begin{bmatrix} \gamma_3 \\ \gamma_2 \\ \gamma_1 \end{bmatrix} + D$$

$$= \frac{\gamma_1 s^2 + \gamma_2 s + \gamma_3}{s^3 + a_1 s^2 + a_2 s + a_3} + b_0$$

$$= \frac{b_0 s^3 + (\gamma_1 + a_1 b_0)s^2 + (\gamma_2 + a_2 b_0)s + \gamma_3 + a_3 b_0}{s^3 + a_1 s^2 + a_2 s + a_3}$$

$$= \frac{b_0 s^3 + b_1 s^2 + b_2 s + b_3}{s^3 + a_1 s^2 + a_2 s + a_3}$$

Hence

$$\gamma_1 = b_1 - a_1 b_0, \qquad \gamma_2 = b_2 - a_2 b_0, \qquad \gamma_3 = b_3 - a_3 b_0$$

Thus, we have shown that

$$\mathbf{Q}^{-1}\mathbf{B} = \begin{bmatrix} \gamma_3 \\ \gamma_2 \\ \gamma_1 \end{bmatrix} = \begin{bmatrix} b_3 - a_3 b_0 \\ b_2 - a_2 b_0 \\ b_1 - a_1 b_0 \end{bmatrix}$$

Note that what we have derived here can be easily extended to the case where n is any positive integer.

A–10–8. Consider a system defined by

$$\dot{\mathbf{x}} = \mathbf{Ax} + \mathbf{B}u$$

$$y = \mathbf{Cx}$$

where

$$\mathbf{A} = \begin{bmatrix} 1 & 1 \\ -4 & -3 \end{bmatrix}, \qquad \mathbf{B} = \begin{bmatrix} 0 \\ 2 \end{bmatrix}, \qquad \mathbf{C} = \begin{bmatrix} 1 & 1 \end{bmatrix}$$

The rank of the observability matrix $\mathbf{N}$,

$$\mathbf{N} = [\mathbf{C}^* \;\vdots\; \mathbf{A}^*\mathbf{C}^*] = \begin{bmatrix} 1 & -3 \\ 1 & -2 \end{bmatrix}$$

is 2. Hence the system is completely observable. Transform the system equations into the observable canonical form.

Solution. Since

$$|s\mathbf{I} - \mathbf{A}| = s^2 + 2s + 1 = s^2 + a_1 s + a_2$$

we have

$$a_1 = 2, \qquad a_2 = 1$$

Define

$$\mathbf{Q} = (\mathbf{WN}^*)^{-1}$$

where

$$\mathbf{N} = \begin{bmatrix} 1 & -3 \\ 1 & -2 \end{bmatrix}, \qquad \mathbf{W} = \begin{bmatrix} a_1 & 1 \\ 1 & 0 \end{bmatrix} = \begin{bmatrix} 2 & 1 \\ 1 & 0 \end{bmatrix}$$

Then

$$\mathbf{Q} = \left\{ \begin{bmatrix} 2 & 1 \\ 1 & 0 \end{bmatrix} \begin{bmatrix} 1 & 1 \\ -3 & -2 \end{bmatrix} \right\}^{-1} = \begin{bmatrix} -1 & 0 \\ 1 & 1 \end{bmatrix}^{-1} = \begin{bmatrix} -1 & 0 \\ 1 & 1 \end{bmatrix}$$

and

$$\mathbf{Q}^{-1} = \begin{bmatrix} -1 & 0 \\ 1 & 1 \end{bmatrix}$$

Define

$$\mathbf{x} = \mathbf{Q}\hat{\mathbf{x}}$$

Then the state equation becomes

$$\dot{\hat{\mathbf{x}}} = \mathbf{Q}^{-1}\mathbf{A}\mathbf{Q}\hat{\mathbf{x}} + \mathbf{Q}^{-1}\mathbf{B}u$$

or

$$\begin{bmatrix} \dot{\hat{x}}_1 \\ \dot{\hat{x}}_2 \end{bmatrix} = \begin{bmatrix} -1 & 0 \\ 1 & 1 \end{bmatrix}\begin{bmatrix} 1 & 1 \\ -4 & -3 \end{bmatrix}\begin{bmatrix} -1 & 0 \\ 1 & 1 \end{bmatrix}\begin{bmatrix} \hat{x}_1 \\ \hat{x}_2 \end{bmatrix} + \begin{bmatrix} -1 & 0 \\ 1 & 1 \end{bmatrix}\begin{bmatrix} 0 \\ 2 \end{bmatrix}u$$

$$= \begin{bmatrix} 0 & -1 \\ 1 & -2 \end{bmatrix}\begin{bmatrix} \hat{x}_1 \\ \hat{x}_2 \end{bmatrix} + \begin{bmatrix} 0 \\ 2 \end{bmatrix}u \qquad (10\text{--}164)$$

The output equation becomes

$$y = \mathbf{C}\mathbf{Q}\hat{\mathbf{x}}$$

or

$$y = \begin{bmatrix} 1 & 1 \end{bmatrix}\begin{bmatrix} -1 & 0 \\ 1 & 1 \end{bmatrix}\begin{bmatrix} \hat{x}_1 \\ \hat{x}_2 \end{bmatrix} = \begin{bmatrix} 0 & 1 \end{bmatrix}\begin{bmatrix} \hat{x}_1 \\ \hat{x}_2 \end{bmatrix} \qquad (10\text{--}165)$$

Equations (10–164) and (10–165) are in the observable canonical form.

A–10–9. For the system defined by

$$\dot{\mathbf{x}} = \mathbf{A}\mathbf{x} + \mathbf{B}u$$

$$y = \mathbf{C}\mathbf{x}$$

consider the problem of designing a state observer such that the desired eigenvalues for the observer gain matrix are $\mu_1, \mu_2, \ldots, \mu_n$.

Show that the observer gain matrix given by Equation (10–48), rewritten as

$$\mathbf{K}_e = (\mathbf{W}\mathbf{N}^*)^{-1}\begin{bmatrix} \alpha_n - a_n \\ \alpha_{n-1} - a_{n-1} \\ \cdot \\ \cdot \\ \cdot \\ \alpha_1 - a_1 \end{bmatrix} \qquad (10\text{--}166)$$

can be obtained from Equation (10–13) by considering the dual problem. That is, the matrix $\mathbf{K}_e$ can be determined by considering the pole placement problem for the dual system, obtaining the state feedback gain matrix $\mathbf{K}$, and taking its conjugate transpose, or $\mathbf{K}_e = \mathbf{K}^*$.

Solution. The dual of the given system is

$$\dot{\mathbf{z}} = \mathbf{A}^*\mathbf{z} + \mathbf{C}^*v \qquad (10\text{--}167)$$

$$n = \mathbf{B}^*\mathbf{z}$$

Using the state feedback control

$$v = -\mathbf{K}\mathbf{z}$$

Equation (10–167) becomes

$$\dot{z} = (A^* - C^*K)z$$

Equation (10–13), which is rewritten here, is

$$K = [\alpha_n - a_n \ \vdots \ \alpha_{n-1} - a_{n-1} \ \vdots \ \cdots \ \vdots \ \alpha_2 - a_2 \ \vdots \ \alpha_1 - a_1]T^{-1} \qquad (10\text{–}168)$$

where

$$T = MW = [C^* \ \vdots \ A^*C^* \ \vdots \ \cdots \ \vdots \ (A^*)^{n-1}C^*]W$$

Referring to Equation (10–33),

$$[C^* \ \vdots \ A^*C^* \ \vdots \ \cdots \ \vdots \ (A^*)^{n-1}C^*] = N$$

Hence

$$T = NW$$

Since $W = W^*$, we have

$$T^* = W^*N^* = WN^*$$

and

$$(T^*)^{-1} = (WN^*)^{-1}$$

Taking the conjugate transpose of both sides of Equation (10–168), we have

$$K^* = (T^{-1})^* \begin{bmatrix} \alpha_n - a_n \\ \alpha_{n-1} - a_{n-1} \\ \cdot \\ \cdot \\ \cdot \\ \alpha_1 - a_1 \end{bmatrix} = (T^*)^{-1} \begin{bmatrix} \alpha_n - a_n \\ \alpha_{n-1} - a_{n-1} \\ \cdot \\ \cdot \\ \cdot \\ \alpha_1 - a_1 \end{bmatrix} = (WN^*)^{-1} \begin{bmatrix} \alpha_n - a_n \\ \alpha_{n-1} - a_{n-1} \\ \cdot \\ \cdot \\ \cdot \\ \alpha_1 - a_1 \end{bmatrix}$$

The right side of this last equation is the same as the right side of Equation (10–166). Thus, we have $K_e = K^*$.

A–10–10. Consider the double-integrator system

$$\ddot{y} = u$$

Define $x_1 = y$, $x_2 = \dot{y}$. Then the state and output equations become

$$\begin{bmatrix} \dot{x}_1 \\ \dot{x}_2 \end{bmatrix} = \begin{bmatrix} 0 & 1 \\ 0 & 0 \end{bmatrix} \begin{bmatrix} x_1 \\ x_2 \end{bmatrix} + \begin{bmatrix} 0 \\ 1 \end{bmatrix} u$$

$$y = [1 \quad 0] \begin{bmatrix} x_1 \\ x_2 \end{bmatrix}$$

Design a state observer such that the eigenvalues of the observer gain matrix are

$$\mu_1 = -2 + j3.464, \qquad \mu_2 = -2 - j3.464$$

Solution. The rank of the observability matrix for this system

$$N = [C^* \ \vdots \ A^*C^*] = \begin{bmatrix} 1 & 0 \\ 0 & 1 \end{bmatrix}$$

is 2. Hence the design of a state observer with an arbitrary observer gain matrix is possible. The characteristic equation for the system is

$$\left| s\mathbf{I} - \begin{bmatrix} 0 & 1 \\ 0 & 0 \end{bmatrix} \right| = \begin{vmatrix} s & -1 \\ 0 & s \end{vmatrix} = s^2 = s^2 + a_1 s + a_2 = 0$$

Hence

$$a_1 = 0, \qquad a_2 = 0$$

The desired characteristic equation for the observer is

$$\begin{aligned} (s - \mu_1)(s - \mu_2) &= (s + 2 - j3.464)(s + 2 + j3.464) \\ &= s^2 + 4s + 16 \\ &= s^2 + \alpha_1 s + \alpha_2 = 0 \end{aligned}$$

Thus,

$$\alpha_1 = 4, \qquad \alpha_2 = 16$$

Referring to Equation (10–48), we have

$$\mathbf{K}_e = (\mathbf{WN}^*)^{-1} \begin{bmatrix} \alpha_2 - a_2 \\ \alpha_1 - a_1 \end{bmatrix}$$

where $\mathbf{N}$ was obtained earlier as

$$\mathbf{N} = \begin{bmatrix} 1 & 0 \\ 0 & 1 \end{bmatrix}$$

and

$$\mathbf{W} = \begin{bmatrix} a_1 & 1 \\ 1 & 0 \end{bmatrix} = \begin{bmatrix} 0 & 1 \\ 1 & 0 \end{bmatrix}$$

Thus

$$\begin{aligned} \mathbf{K}_e &= \left\{ \begin{bmatrix} 0 & 1 \\ 1 & 0 \end{bmatrix} \begin{bmatrix} 1 & 0 \\ 0 & 1 \end{bmatrix} \right\}^{-1} \begin{bmatrix} 16 - 0 \\ 4 - 0 \end{bmatrix} \\ &= \begin{bmatrix} 0 & 1 \\ 1 & 0 \end{bmatrix} \begin{bmatrix} 16 \\ 4 \end{bmatrix} = \begin{bmatrix} 4 \\ 16 \end{bmatrix} \end{aligned}$$

The state observer is given by

$$\dot{\tilde{\mathbf{x}}} = (\mathbf{A} - \mathbf{K}_e \mathbf{C})\tilde{\mathbf{x}} + \mathbf{B}u + \mathbf{K}_e y$$

or

$$\begin{bmatrix} \dot{\tilde{x}}_1 \\ \dot{\tilde{x}}_2 \end{bmatrix} = \begin{bmatrix} -4 & 1 \\ -16 & 0 \end{bmatrix} \begin{bmatrix} \tilde{x}_1 \\ \tilde{x}_2 \end{bmatrix} + \begin{bmatrix} 0 \\ 1 \end{bmatrix} u + \begin{bmatrix} 4 \\ 16 \end{bmatrix} y$$

A–10–11. Consider the system

$$\dot{\mathbf{x}} = \mathbf{A}\mathbf{x} + \mathbf{B}u$$

$$y = \mathbf{C}\mathbf{x}$$

where

$$
\mathbf{A} = \begin{bmatrix} 0 & 1 & 0 \\ 0 & 0 & 1 \\ -6 & -11 & -6 \end{bmatrix}, \quad \mathbf{B} = \begin{bmatrix} 0 \\ 0 \\ 1 \end{bmatrix}, \quad \mathbf{C} = [1 \quad 0 \quad 0]
$$

It is desired to design a full-order state observer. Determine the observer gain matrix $\mathbf{K}_e$ by use of (a) the direct substitution method and (b) Ackermann's formula. Assume that the desired eigenvalues of the observer gain matrix are

$$
\mu_1 = -2 + j3.464, \qquad \mu_2 = -2 - j3.464, \qquad \mu_3 = -5
$$

Solution. Since the rank of

$$
\mathbf{N} = [\mathbf{C}^* \; \vdots \; \mathbf{A}^*\mathbf{C}^* \; \vdots \; (\mathbf{A}^*)^2\mathbf{C}^*] = \begin{bmatrix} 1 & 0 & 0 \\ 0 & 1 & 0 \\ 0 & 0 & 1 \end{bmatrix}
$$

is 3, the system is completely observable. Hence the design of a state observer is possible.

(a) Define the observer gain matrix as $\mathbf{K}_e$. Then $\mathbf{K}_e$ can be written as

$$
\mathbf{K}_e = \begin{bmatrix} k_{e1} \\ k_{e2} \\ k_{e3} \end{bmatrix}
$$

The characteristic polynomial for the observer system is

$$
\begin{aligned}
| s\mathbf{I} - \mathbf{A} + \mathbf{K}_e\mathbf{C} | &= \left| \begin{bmatrix} s & 0 & 0 \\ 0 & s & 0 \\ 0 & 0 & s \end{bmatrix} - \begin{bmatrix} 0 & 1 & 0 \\ 0 & 0 & 1 \\ -6 & -11 & -6 \end{bmatrix} + \begin{bmatrix} k_{e1} \\ k_{e2} \\ k_{e3} \end{bmatrix} [1 \quad 0 \quad 0] \right| \\
&= \begin{vmatrix} s + k_{e1} & -1 & 0 \\ k_{e2} & s & -1 \\ k_{e3} + 6 & 11 & s + 6 \end{vmatrix} \\
&= s^3 + (k_{e1} + 6)s^2 + (6k_{e1} + k_{e2} + 11)s + 11k_{e1} \\
&\quad + 6k_{e2} + k_{e3} + 6
\end{aligned} \tag{10–169}
$$

The desired characteristic polynomial is

$$
\begin{aligned}
(s + \mu_1)(s + \mu_2)(s + \mu_3) &= (s + 2 - j3.464)(s + 2 + j3.464)(s + 5) \\
&= s^3 + 9s^2 + 36s + 80
\end{aligned} \tag{10–170}
$$

By equating coefficients of like powers of s of Equations (10–169) and (10–170), we obtain

$$
k_{e1} + 6 = 9
$$

$$
6k_{e1} + k_{e2} + 11 = 36
$$

$$
11k_{e1} + 6k_{e2} + k_{e3} + 6 = 80
$$

from which we get

$$
k_{e1} = 3, \qquad k_{e2} = 7, \qquad k_{e3} = -1
$$

Hence

$$\mathbf{K}_e = \begin{bmatrix} 3 \\ 7 \\ -1 \end{bmatrix}$$

(b) Next we shall obtain the observer gain matrix $\mathbf{K}_e$ by use of Ackermann's formula. Referring to Equation (10–53), we have

$$\mathbf{K}_e = \phi(\mathbf{A}) \begin{bmatrix} \mathbf{C} \\ \mathbf{CA} \\ \mathbf{CA}^2 \end{bmatrix}^{-1} \begin{bmatrix} 0 \\ 0 \\ 1 \end{bmatrix}$$

Since the desired characteristic polynomial is

$$\phi(s) = s^3 + 9s^2 + 36s + 80$$

we have

$$\phi(\mathbf{A}) = \mathbf{A}^3 + 9\mathbf{A}^2 + 36\mathbf{A} + 80\mathbf{I}$$

$$= \begin{bmatrix} 0 & 1 & 0 \\ 0 & 0 & 1 \\ -6 & -11 & -6 \end{bmatrix}^3 + 9\begin{bmatrix} 0 & 1 & 0 \\ 0 & 0 & 1 \\ -6 & -11 & -6 \end{bmatrix}^2$$

$$+ 36\begin{bmatrix} 0 & 1 & 0 \\ 0 & 0 & 1 \\ -6 & -11 & -6 \end{bmatrix} + 80\begin{bmatrix} 1 & 0 & 0 \\ 0 & 1 & 0 \\ 0 & 0 & 1 \end{bmatrix}$$

$$= \begin{bmatrix} 74 & 25 & 3 \\ -18 & 41 & 7 \\ -42 & -95 & -1 \end{bmatrix}$$

Also

$$\begin{bmatrix} \mathbf{C} \\ \mathbf{CA} \\ \mathbf{CA}^2 \end{bmatrix}^{-1} = \begin{bmatrix} 1 & 0 & 0 \\ 0 & 1 & 0 \\ 0 & 0 & 1 \end{bmatrix}^{-1} = \begin{bmatrix} 1 & 0 & 0 \\ 0 & 1 & 0 \\ 0 & 0 & 1 \end{bmatrix}$$

Hence

$$\mathbf{K}_e = \begin{bmatrix} 74 & 25 & 3 \\ -18 & 41 & 7 \\ -42 & -95 & -1 \end{bmatrix}\begin{bmatrix} 1 & 0 & 0 \\ 0 & 1 & 0 \\ 0 & 0 & 1 \end{bmatrix}\begin{bmatrix} 0 \\ 0 \\ 1 \end{bmatrix} = \begin{bmatrix} 3 \\ 7 \\ -1 \end{bmatrix}$$

A–10–12. Consider an observed-state feedback control system with a minimum-order observer described by the following equations:

$$\dot{\mathbf{x}} = \mathbf{Ax} + \mathbf{B}u \qquad (10\text{–}171)$$

$$y = \mathbf{Cx}$$

$$u = -\mathbf{K}\tilde{\mathbf{x}} \qquad (10\text{–}172)$$

where

$$\mathbf{x} = \begin{bmatrix} x_a \\ \hline \mathbf{x}_b \end{bmatrix}, \qquad \tilde{\mathbf{x}} = \begin{bmatrix} x_a \\ \hline \tilde{\mathbf{x}}_b \end{bmatrix}$$

(x_a is the state variable that can be directly measured and $\tilde{x}_b$ corresponds to the observed state variables.)

Show that the closed-loop poles of the system comprise the closed-loop poles due to pole placement [the eigenvalues of matrix $(\mathbf{A} - \mathbf{BK})$] and the closed-loop poles due to the minimum-order observer [the eigenvalues of matrix $(\mathbf{A}_{bb} - \mathbf{K}_e\mathbf{A}_{ab})$].

Solution. The error equation for the minimum-order observer may be derived as given by Equation (10–80), rewritten thus:

$$\dot{\mathbf{e}} = (\mathbf{A}_{bb} - \mathbf{K}_e\mathbf{A}_{ab})\mathbf{e} \qquad (10\text{–}173)$$

where

$$\mathbf{e} = \mathbf{x}_b - \tilde{\mathbf{x}}_b$$

From Equations (10–171) and (10–172), we obtain

$$\dot{\mathbf{x}} = \mathbf{A}\mathbf{x} - \mathbf{BK}\tilde{\mathbf{x}} = \mathbf{A}\mathbf{x} - \mathbf{BK}\begin{bmatrix} x_a \\ \hline \tilde{x}_b \end{bmatrix} = \mathbf{A}\mathbf{x} - \mathbf{BK}\begin{bmatrix} x_a \\ \hline x_b - \mathbf{e} \end{bmatrix}$$

$$= \mathbf{A}\mathbf{x} - \mathbf{BK}\left\{ \mathbf{x} - \begin{bmatrix} 0 \\ \hline \mathbf{e} \end{bmatrix} \right\} = (\mathbf{A} - \mathbf{BK})\mathbf{x} + \mathbf{BK}\begin{bmatrix} 0 \\ \hline \mathbf{e} \end{bmatrix} \qquad (10\text{–}174)$$

Combining Equations (10–173) and (10–174) and writing

$$\mathbf{K} = [\mathbf{K}_a \mid \mathbf{K}_b]$$

we obtain

$$\begin{bmatrix} \dot{\mathbf{x}} \\ \dot{\mathbf{e}} \end{bmatrix} = \begin{bmatrix} \mathbf{A} - \mathbf{BK} & \mathbf{BK}_b \\ \mathbf{0} & \mathbf{A}_{bb} - \mathbf{K}_e\mathbf{A}_{ab} \end{bmatrix}\begin{bmatrix} \mathbf{x} \\ \mathbf{e} \end{bmatrix} \qquad (10\text{–}175)$$

Equation (10–175) describes the dynamics of the observed-state feedback control system with a minimum-order observer. The characteristic equation for this system is

$$\begin{vmatrix} s\mathbf{I} - \mathbf{A} + \mathbf{BK} & -\mathbf{BK}_b \\ \mathbf{0} & s\mathbf{I} - \mathbf{A}_{bb} + \mathbf{K}_e\mathbf{A}_{ab} \end{vmatrix} = 0$$

or

$$|s\mathbf{I} - \mathbf{A} + \mathbf{BK}| \, \| \, s\mathbf{I} - \mathbf{A}_{bb} + \mathbf{K}_e\mathbf{A}_{ab}| = 0$$

The closed-loop poles of the observed-state feedback control system with a minimum order-observer consist of the closed-loop poles due to pole placement and the closed-loop poles due to the minimum-order observer. (Therefore, the pole placement design and the design of the minimum-order observer are independent of each other.)

A–10–13. Consider the system defined by

$$\dot{\mathbf{x}} = \mathbf{A}\mathbf{x} + \mathbf{B}u$$

$$y = \mathbf{C}\mathbf{x}$$

where

$$\mathbf{A} = \begin{bmatrix} 0 & 1 & 0 \\ \hline 0 & 0 & 1 \\ -6 & -11 & -6 \end{bmatrix}, \qquad \mathbf{B} = \begin{bmatrix} 0 \\ 0 \\ 1 \end{bmatrix}, \qquad \mathbf{C} = [1 \mid 0 \quad 0]$$

Suppose that the state variable x_1 (which is equal to y) is measurable and need not be observed. Determine the observer gain matrix $\mathbf{K}_e$ for the minimum-order observer. The desired eigenvalues are

$$\mu_1 = -2 + j3.464, \qquad \mu_2 = -2 - j3.464$$

Solution. From the partitioned matrix, we have

$$A_{aa} = 0, \qquad \mathbf{A}_{ab} = [1 \quad 0]$$

$$\mathbf{A}_{ba} = \begin{bmatrix} 0 \\ -6 \end{bmatrix}, \qquad \mathbf{A}_{bb} = \begin{bmatrix} 0 & 1 \\ -11 & -6 \end{bmatrix}$$

$$B_a = 0, \qquad \mathbf{B}_b = \begin{bmatrix} 0 \\ 1 \end{bmatrix}$$

The characteristic polynomial for the unobserved portion of the system is

$$|s\mathbf{I} - \mathbf{A}_{bb}| = \begin{vmatrix} s & -1 \\ 11 & s+6 \end{vmatrix} = s^2 + 6s + 11 = s^2 + \hat{a}_1 s + \hat{a}_2$$

Hence

$$\hat{a}_1 = 6, \qquad \hat{a}_2 = 11$$

Notice that

$$\hat{\mathbf{N}} = [\mathbf{A}_{ab}^* \ \vdots \ \mathbf{A}_{bb}^* \mathbf{A}_{ab}^*] = \begin{bmatrix} 1 & 0 \\ 0 & 1 \end{bmatrix}$$

$$\hat{\mathbf{W}} = \begin{bmatrix} \hat{a}_1 & 1 \\ 1 & 0 \end{bmatrix} = \begin{bmatrix} 6 & 1 \\ 1 & 0 \end{bmatrix}$$

Therefore,

$$\hat{\mathbf{Q}} = (\hat{\mathbf{W}}\hat{\mathbf{N}}^*)^{-1} = \begin{bmatrix} 6 & 1 \\ 1 & 0 \end{bmatrix}^{-1} = \begin{bmatrix} 0 & 1 \\ 1 & -6 \end{bmatrix}$$

The desired characteristic polynomial for the minimum-order observer is

$$(s + \mu_1)(s + \mu_2) = (s + 2 - j3.464)(s + 2 + j3.464)$$
$$= s^2 + 4s + 16 = s^2 + \hat{\alpha}_1 s + \hat{\alpha}_2$$

Hence

$$\hat{\alpha}_1 = 4, \qquad \hat{\alpha}_2 = 16$$

Referring to Equation (10–82), we have

$$\mathbf{K}_e = \hat{\mathbf{Q}} \begin{bmatrix} \hat{\alpha}_2 - \hat{a}_2 \\ \hat{\alpha}_1 - \hat{a}_1 \end{bmatrix} = \begin{bmatrix} 0 & 1 \\ 1 & -6 \end{bmatrix} \begin{bmatrix} 16 - 11 \\ 4 - 6 \end{bmatrix} = \begin{bmatrix} 0 & 1 \\ 1 & -6 \end{bmatrix} \begin{bmatrix} 5 \\ -2 \end{bmatrix} = \begin{bmatrix} -2 \\ 17 \end{bmatrix}$$

As a matter of course, the matrix $\mathbf{K}_e$ determined here is the same as that determined by use of Ackermann's formula (see Example 10–6).

A–10–14. Consider a completely state controllable system defined by

$$\dot{\mathbf{x}} = \mathbf{A}\mathbf{x} + \mathbf{B}u \qquad\qquad (10\text{–}176)$$

$$y = \mathbf{C}\mathbf{x}$$

where $\mathbf{x} =$ state vector (n-vector)
$u =$ control signal (scalar)

y = output signal (scalar)
$\mathbf{A}$ = $n \times n$ constant matrix
$\mathbf{B}$ = $n \times 1$ constant matrix
$\mathbf{C}$ = $1 \times n$ constant matrix

Suppose that the rank of the following $(n + 1) \times (n + 1)$ matrix

$$\begin{bmatrix} \mathbf{A} & \mathbf{B} \\ -\mathbf{C} & 0 \end{bmatrix}$$

is $n + 1$. Show that the system defined by

$$\dot{\mathbf{e}} = \hat{\mathbf{A}}\mathbf{e} + \hat{\mathbf{B}}u_e \tag{10–177}$$

where

$$\hat{\mathbf{A}} = \begin{bmatrix} \mathbf{A} & 0 \\ -\mathbf{C} & 0 \end{bmatrix}, \qquad \hat{\mathbf{B}} = \begin{bmatrix} \mathbf{B} \\ 0 \end{bmatrix}, \qquad u_e = u(t) - u(\infty)$$

is completely state controllable.

Solution. Define

$$\mathbf{M} = [\mathbf{B} \ \vdots \ \mathbf{AB} \ \vdots \ \cdots \ \mathbf{A}^{n-1}\mathbf{B}]$$

Because the system given by Equation (10–176) is completely state controllable, the rank of matrix $\mathbf{M}$ is n. Then the rank of

$$\begin{bmatrix} \mathbf{M} & 0 \\ 0 & 1 \end{bmatrix}$$

is $n + 1$. Consider the following equation:

$$\begin{bmatrix} \mathbf{A} & \mathbf{B} \\ -\mathbf{C} & 0 \end{bmatrix} \begin{bmatrix} \mathbf{M} & 0 \\ 0 & 1 \end{bmatrix} = \begin{bmatrix} \mathbf{AM} & \mathbf{B} \\ -\mathbf{CM} & 0 \end{bmatrix} \tag{10–178}$$

Since matrix

$$\begin{bmatrix} \mathbf{A} & \mathbf{B} \\ -\mathbf{C} & 0 \end{bmatrix}$$

is of rank $n + 1$, the left side of Equation (10–178) is of rank $n + 1$. Therefore, the right side of Equation (10–178) is also of rank $n + 1$. Since

$$\begin{bmatrix} \mathbf{AM} & \mathbf{B} \\ -\mathbf{CM} & 0 \end{bmatrix} = \begin{bmatrix} \mathbf{A}\,[\mathbf{B} \ \vdots \ \mathbf{AB} \ \vdots \ \cdots \ \vdots \ \mathbf{A}^{n-1}\mathbf{B}] & \mathbf{B} \\ -\mathbf{C}\,[\mathbf{B} \ \vdots \ \mathbf{AB} \ \vdots \ \cdots \ \vdots \ \mathbf{A}^{n-1}\mathbf{B}] & 0 \end{bmatrix}$$

$$= \begin{bmatrix} \mathbf{AB} \ \vdots & \mathbf{A}^2\mathbf{B} \ \vdots & \cdots & \vdots & \mathbf{A}^n\mathbf{B} & \vdots & \mathbf{B} \\ -\mathbf{CB} \ \vdots & -\mathbf{CAB} \ \vdots & & \vdots & -\mathbf{CA}^{n-1}\mathbf{B} & \vdots & 0 \end{bmatrix}$$

$$= [\hat{\mathbf{A}}\hat{\mathbf{B}} \ \vdots \ \hat{\mathbf{A}}^2\hat{\mathbf{B}} \ \vdots \ \cdots \ \vdots \ \hat{\mathbf{A}}^n\hat{\mathbf{B}} \ \vdots \ \hat{\mathbf{B}}]$$

we find that the rank of

$$[\hat{\mathbf{B}} \ \vdots \ \hat{\mathbf{A}}\hat{\mathbf{B}} \ \vdots \ \hat{\mathbf{A}}^2\hat{\mathbf{B}} \ \vdots \ \cdots \ \vdots \ \hat{\mathbf{A}}^n\hat{\mathbf{B}}]$$

is $n + 1$. Thus, the system defined by Equation (10–177) is completely state controllable.

A–10–15. Show that the maximum number of linearly independent rows of a matrix $\mathbf{A}$ is equal to the maximum number of linearly independent columns.

Solution. Let $\mathbf{A}$ be any $n \times m$ matrix. Let $\mathbf{P}$ be an $n \times n$ nonsingular matrix and $\mathbf{Q}$ be an $m \times m$ nonsingular matrix such that

$$\mathbf{PAQ} = \begin{bmatrix} \mathbf{I}_r & \mathbf{0} \\ \mathbf{0} & \mathbf{0} \end{bmatrix}$$

where $\mathbf{I}_r$ is the $r \times r$ identity matrix ($r \leqslant n$, $r \leqslant m$). The maximum number of linearly independent rows of $\mathbf{PAQ}$ is r, which is the dimension of the $r \times r$ identity matrix.

The maximum number of linearly independent rows of $\mathbf{PA}$ is the same as that of $\mathbf{A}$, since $\mathbf{P}$ is a nonsingular matrix. Since

$$\mathbf{PA} = \begin{bmatrix} \mathbf{I}_r & \mathbf{0} \\ \mathbf{0} & \mathbf{0} \end{bmatrix} \mathbf{Q}^{-1} = \begin{bmatrix} \mathbf{I}_r & \mathbf{0} \\ \mathbf{0} & \mathbf{0} \end{bmatrix} \begin{bmatrix} \mathbf{Q}_{11} & \mathbf{Q}_{12} \\ \mathbf{Q}_{21} & \mathbf{Q}_{22} \end{bmatrix} = \begin{bmatrix} \mathbf{Q}_{11} & \mathbf{Q}_{12} \\ \mathbf{0} & \mathbf{0} \end{bmatrix}$$

where $\mathbf{Q}_{11}$ and $\mathbf{Q}_{12}$ are r-rowed matrix and since $\mathbf{Q}^{-1}$ is nonsingular, the m rows are linearly independent. Hence, the maximum number of linearly independent rows of $\mathbf{PA}$ is r. Thus, the maximum number of linearly independent rows of $\mathbf{A}$ is also r.

Consider next

$$\mathbf{Q}^*\mathbf{A}^*\mathbf{P}^* = \begin{bmatrix} \mathbf{I}_r & \mathbf{0} \\ \mathbf{0} & \mathbf{0} \end{bmatrix}$$

By a discussion similar to the preceding, we find that the maximum number of linearly independent rows of $\mathbf{Q}^*\mathbf{A}^*$ to be r. Hence, the maximum number of linearly independent rows of $\mathbf{A}^*$ is r, which means that the maximum number of linearly independent columns of $\mathbf{A}$ is r. Thus, we have proved that the maximum number of linearly independent rows of a matrix is equal to that of linearly independent columns.

A–10–16. The computation of determinants for the determination of the rank of a matrix may be time consuming. The computational effort can be reduced greatly by means of elementary row or column operations, since elementary operations do not alter the rank of a matrix.

The following row (or column) operations on a matrix $\mathbf{A}$ are called the elementary row (or column) operations:

1. The interchange of any two rows (or columns).
2. The multiplication of a row (or column) by a constant c, where $c \neq 0$.
3. The addition of one row (or column) multiplied by a constant to another row (or column).

An $n \times n$ matrix obtained from the $n \times n$ identity matrix $\mathbf{I}$ by means of one elementary row (or column) operation is called an elementary matrix. The elementary matrices are nonsingular. (The inverse of an elementary matrix is also an elementary matrix.) A few examples of elementary matrices are

$$\begin{bmatrix} 0 & 0 & 1 \\ 0 & 1 & 0 \\ 1 & 0 & 0 \end{bmatrix}, \quad \begin{bmatrix} c & 0 & 0 \\ 0 & 1 & 0 \\ 0 & 0 & 1 \end{bmatrix}, \quad \begin{bmatrix} 1 & 0 & 0 \\ c & 1 & 0 \\ 0 & 0 & 1 \end{bmatrix}$$

The invariance of the rank under elementary operations can be seen as follows: Since the rank of a matrix $\mathbf{A}$ is the maximum number of linearly independent rows (or columns), elementary operations of types 1 and 2 clearly will not alter the rank of $\mathbf{A}$. Since the maximum number of linearly independent rows (or columns) is not altered by the addition of one row (or column) multiplied by a constant to another row (or column), the elementary operations of type 3 also will not alter the rank of $\mathbf{A}$. Hence, the rank of a matrix $\mathbf{A}$ is invariant under elementary operations.

To determine the rank of a given matrix $\mathbf{A}$, we may therefore premultiply and/or postmultiply $\mathbf{A}$ by a series of elementary matrices and transform matrix $\mathbf{A}$ into a simpler form with many zeros, that

is, a triangular matrix, identity matrix, and so on, so that the rank can be determined immediately.
Determine the rank of the following matrix $\mathbf{P}$:

$$\mathbf{P} = \begin{bmatrix} 0 & 1 & 0 & 0 & 0 \\ 20.601 & 0 & 0 & 0 & -1 \\ 0 & 0 & 0 & 1 & 0 \\ -0.4905 & 0 & 0 & 0 & 0.5 \\ 0 & 0 & -1 & 0 & 0 \end{bmatrix}$$

Solution. Interchanging columns 1 and 2 and interchanging columns 3 and 5 give

$$\begin{bmatrix} 1 & 0 & 0 & 0 & 0 \\ 0 & 20.601 & -1 & 0 & 0 \\ 0 & 0 & 0 & 1 & 0 \\ 0 & -0.4905 & 0.5 & 0 & 0 \\ 0 & 0 & 0 & 0 & -1 \end{bmatrix}$$

Interchanging columns 3 and 4 yields

$$\begin{bmatrix} 1 & 0 & 0 & 0 & 0 \\ 0 & 20.601 & 0 & -1 & 0 \\ 0 & 0 & 1 & 0 & 0 \\ 0 & -0.4905 & 0 & 0.5 & 0 \\ 0 & 0 & 0 & 0 & -1 \end{bmatrix}$$

Adding (0.4905/20.601) times row 2 to row 4 simplifies the matrix to a triangular form.

$$\begin{bmatrix} 1 & 0 & 0 & 0 & 0 \\ 0 & 20.601 & 0 & -1 & 0 \\ 0 & 0 & 1 & 0 & 0 \\ 0 & 0 & 0 & 0.4762 & 0 \\ 0 & 0 & 0 & 0 & -1 \end{bmatrix}$$

The determinant of this triangular matrix is nonzero. Hence, the rank of matrix $\mathbf{P}$ is 5.

A–10–17. Consider the following scalar system:

$$\dot{x} = ax + bu \tag{10–179}$$

where $a < 0$ and the performance index is given by

$$J = \int_0^\infty (qx^2 + ru^2)\, dt \tag{10–180}$$

where $q > 0$ and $r > 0$. The optimal control law that will minimize the performance index J can be given by

$$u = -Kx \tag{10–181}$$

Substituting Equation (10–181) into Equation (10–179) gives

$$\dot{x} = (a - bK)x \tag{10–182}$$

Also, substituting Equation (10–181) into Equation (10–180) gives

$$J = \int_0^\infty (q + rK^2)x^2\, dt \tag{10–183}$$

Example Problems and Solutions

Using the Liapunov approach, we set

$$(q + rK^2)x^2 = -\frac{d}{dt}(px^2)$$

or

$$(q + rK^2)x^2 = -2px\dot{x} = -2p(a - bK)x^2$$

which can be simplified to

$$[q + rK^2 + 2p(a - bK)]x^2 = 0$$

This last equation must hold true for any $x(t)$. Hence, we require

$$q + rK^2 + 2p(a - bK) = 0 \qquad (10\text{--}184)$$

Note that by the second method of Liapunov we know that for a given $q + rK^2$ there exists a p such that

$$(a - bK)p + p(a - bK) = -q - rK^2$$

which is the same as Equation (10–184). Hence, there exists a p that satisfies Equation (10–184).

Show that the optimal control law can be given by

$$u = -Kx = -\frac{pb}{r}x$$

and p can be determined as a positive root of the following equation:

$$q + 2ap - \frac{p^2b^2}{r} = 0 \qquad (10\text{--}185)$$

Solution. For a stable system, we have $x(\infty) = 0$. Hence the performance index can be evaluated as follows:

$$J = \int_0^\infty (q + rK^2)x^2 \, dt = -\int_0^\infty \frac{d}{dt}(px^2) \, dt$$

$$= -[px^2(\infty) - px^2(0)] = px^2(0)$$

To minimize the value of J [for a given $x(0)$] with respect to K, we set

$$\frac{\partial p}{\partial K} = 0 \qquad (10\text{--}186)$$

where, referring to Equation (10–184),

$$p = -\frac{q + rK^2}{2(a - bK)} \qquad (10\text{--}187)$$

Thus

$$\frac{\partial p}{\partial K} = -\frac{2rK(a - bK) - (q + rK^2)(-b)}{2(a - bK)^2} = 0$$

which yields

$$2rK(a - bK) + b(q + rK^2) = 0$$

Hence we have

$$\frac{q + rK^2}{2(a - bK)} = -\frac{rK}{b} \tag{10-188}$$

From Equations (10–187) and (10–188), we obtain

$$p = \frac{rK}{b}$$

or

$$K = \frac{pb}{r} \tag{10-189}$$

By substituting Equation (10–189) into Equation (10–184), we obtain

$$q + 2pa - \frac{p^2b^2}{r} = 0 \tag{10-190}$$

which is Equation (10–185). The value of p can be determined as a positive root of the quadratic equation given by Equation (10–190).

The same results can be obtained in a different way. First note that Equation (10–184) can be modified as follows:

$$q + 2pa + \left(\sqrt{r}\,K - \frac{pb}{\sqrt{r}} \right)^2 - \frac{p^2b^2}{r} = 0 \tag{10-191}$$

Then, considerig this last equation as a function of K, the minimum of the left side of this last equation with respect to K occurs when

$$\sqrt{r}\,K - \frac{pb}{\sqrt{r}} = 0$$

or

$$K = \frac{pb}{r} \tag{10-192}$$

which is Equation (10–189). Thus, the minimization of J with respect to K is the same as the minimization of the left side of Equation (10–184) with respect to K. By substituting Equation (10–192) into Equation (10–191), we obtain

$$q + 2pa - \frac{p^2b^2}{r} = 0$$

which is Equation (10–185).

A–10–18. Consider the control system described by

$$\dot{\mathbf{x}} = \mathbf{A}\mathbf{x} + \mathbf{B}u \tag{10-193}$$

where

$$\mathbf{A} = \begin{bmatrix} 0 & 1 \\ 0 & 0 \end{bmatrix}, \qquad \mathbf{B} = \begin{bmatrix} 0 \\ 1 \end{bmatrix}$$

Assuming the linear control law

$$u = -\mathbf{K}\mathbf{x} = -k_1 x_1 - k_2 x_2 \tag{10-194}$$

determine the constants k_1 and k_2 so that the following performance index is minimized:

$$J = \int_0^\infty \mathbf{x}^T\mathbf{x}\, dt$$

Consider only the case where the initial condition is

$$\mathbf{x}(0) = \begin{bmatrix} c \\ 0 \end{bmatrix}$$

Choose the undamped natural frequency to be 2 rad/sec.

Solution. Substituting Equation (10–194) into Equation (10–193), we obtain

$$\dot{\mathbf{x}} = \mathbf{Ax} - \mathbf{BKx}$$

or

$$\begin{bmatrix} \dot{x}_1 \\ \dot{x}_2 \end{bmatrix} = \begin{bmatrix} 0 & 1 \\ 0 & 0 \end{bmatrix}\begin{bmatrix} x_1 \\ x_2 \end{bmatrix} + \begin{bmatrix} 0 \\ 1 \end{bmatrix}[-k_1x_1 - k_2x_2]$$

$$= \begin{bmatrix} 0 & 1 \\ -k_1 & -k_2 \end{bmatrix}\begin{bmatrix} x_1 \\ x_2 \end{bmatrix} \qquad (10\text{--}195)$$

Thus

$$\mathbf{A} - \mathbf{BK} = \begin{bmatrix} 0 & 1 \\ -k_1 & -k_2 \end{bmatrix}$$

Elimination of x_2 from Equation (10–195) yields

$$\ddot{x}_1 + k_2\dot{x}_1 + k_1x_1 = 0$$

Since the undamped natural frequency is specified as 2 rad/sec, we obtain

$$k_1 = 4$$

Therefore,

$$\mathbf{A} - \mathbf{BK} = \begin{bmatrix} 0 & 1 \\ -4 & -k_2 \end{bmatrix}$$

$\mathbf{A} - \mathbf{BK}$ is a stable matrix if $k_2 > 0$. Our problem now is to determine the value of k_2 so that the performance index

$$J = \int_0^\infty \mathbf{x}^T\mathbf{x}\, dt = \mathbf{x}^T(0)\mathbf{P}(0)\mathbf{x}(0)$$

is minimized, where the matrix $\mathbf{P}$ is determined from Equation (10–130), rewritten

$$(\mathbf{A} - \mathbf{BK})^*\mathbf{P} + \mathbf{P}(\mathbf{A} - \mathbf{BK}) = -(\mathbf{Q} + \mathbf{K}^*\mathbf{RK})$$

Since in this system $\mathbf{Q} = \mathbf{I}$ and $\mathbf{R} = \mathbf{0}$, this last equation can be simplified to

$$(\mathbf{A} - \mathbf{BK})^*\mathbf{P} + \mathbf{P}(\mathbf{A} - \mathbf{BK}) = -\mathbf{I} \qquad (10\text{--}196)$$

Since the system involves only real vectors and real matrices, $\mathbf{P}$ becomes a real symmetric matrix. Then Equation (10–196) can be written as

$$\begin{bmatrix} 0 & -4 \\ 1 & -k_2 \end{bmatrix}\begin{bmatrix} p_{11} & p_{12} \\ p_{12} & p_{22} \end{bmatrix} + \begin{bmatrix} p_{11} & p_{12} \\ p_{12} & p_{22} \end{bmatrix}\begin{bmatrix} 0 & 1 \\ -4 & -k_2 \end{bmatrix} = \begin{bmatrix} -1 & 0 \\ 0 & -1 \end{bmatrix}$$

Solving for the matrix $\mathbf{P}$, we obtain

$$\mathbf{P} = \begin{bmatrix} p_{11} & p_{12} \\ p_{12} & p_{22} \end{bmatrix} = \begin{bmatrix} \dfrac{5}{2k_2} + \dfrac{k_2}{8} & \dfrac{1}{8} \\ \dfrac{1}{8} & \dfrac{5}{8k_2} \end{bmatrix}$$

The performance index is then

$$J = \mathbf{x}^T(0)\mathbf{P}\mathbf{x}(0)$$

$$= [c \quad 0] \begin{bmatrix} p_{11} & p_{12} \\ p_{12} & p_{22} \end{bmatrix} \begin{bmatrix} c \\ 0 \end{bmatrix} = p_{11}c^2$$

$$= \left(\frac{5}{2k_2} + \frac{k_2}{8} \right) c^2 \tag{10--197}$$

To minimize J, we differentiate J with respect to k_2 and set $\partial J/\partial k_2$ equal to zero as follows:

$$\frac{\partial J}{\partial k_2} = \left(\frac{-5}{2k_2^2} + \frac{1}{8} \right) c^2 = 0$$

Hence

$$k_2 = \sqrt{20}$$

With this value of k_2, we have $\partial^2 J/\partial k_2^2 > 0$. Thus the minimum value of J is obtained by substituting $k_2 = \sqrt{20}$ into Equation (10--197), or

$$J_{\min} = \frac{\sqrt{5}}{2} c^2$$

The designed system has the control law

$$u = -4x_1 - \sqrt{20}x_2$$

The designed system is optimal in that it results in a minimum value for the performance index J under the assumed initial condition.

A–10–19. Consider the control system shown in Figure 10–24. The equation for the plant is

$$\dot{\mathbf{x}} = \mathbf{A}\mathbf{x} + \mathbf{B}u \tag{10--198}$$

where

$$\mathbf{A} = \begin{bmatrix} 0 & 1 \\ 0 & 0 \end{bmatrix}, \qquad \mathbf{B} = \begin{bmatrix} 0 \\ 1 \end{bmatrix}$$

Assuming the linear control law

$$u = -\mathbf{K}\mathbf{x} = -k_1x_1 - k_2x_2 \tag{10--199}$$

determine the constants k_1 and k_2 so that the following performance index is minimized:

$$J = \int_0^\infty (\mathbf{x}^T\mathbf{x} + u^2)\, dt$$

Solution. By substituting Equation (10--199) into Equation (10--198), we obtain

$$\dot{\mathbf{x}} = \mathbf{A}\mathbf{x} - \mathbf{B}\mathbf{K}\mathbf{x} = (\mathbf{A} - \mathbf{B}\mathbf{K})\mathbf{x} = \begin{bmatrix} 0 & 1 \\ -k_1 & -k_2 \end{bmatrix} \mathbf{x}$$

If we assume k_1 and k_2 to be positive constants, then $\mathbf{A} - \mathbf{BK}$ becomes a stable matrix and $\mathbf{x}(\infty) = \mathbf{0}$. Hence the performance index can be written

$$J = \int_0^\infty (\mathbf{x}^T\mathbf{x} + \mathbf{x}^T\mathbf{K}^T\mathbf{K}\mathbf{x})\, dt$$

$$= \int_0^\infty \mathbf{x}^T(\mathbf{I} + \mathbf{K}^T\mathbf{K})\mathbf{x}\, dt$$

$$= \mathbf{x}^T(0)\mathbf{P}\mathbf{x}(0)$$

where $\mathbf{P}$ is determined from Equation (10–130):

$$(\mathbf{A} - \mathbf{BK})^*\mathbf{P} + \mathbf{P}(\mathbf{A} - \mathbf{BK}) = -(\mathbf{Q} + \mathbf{K}^*\mathbf{RK}) = -(\mathbf{I} + \mathbf{K}^*\mathbf{K})$$

where we substituted $\mathbf{Q} = \mathbf{I}$ and $\mathbf{R} = I_1 = 1$. For this system, since matrix $\mathbf{P}$ is real symmetric, this last equation can be rewritten as

$$\begin{bmatrix} 0 & -k_1 \\ 1 & -k_2 \end{bmatrix}\begin{bmatrix} p_{11} & p_{12} \\ p_{12} & p_{22} \end{bmatrix} + \begin{bmatrix} p_{11} & p_{12} \\ p_{12} & p_{22} \end{bmatrix}\begin{bmatrix} 0 & 1 \\ -k_1 & -k_2 \end{bmatrix} = -\begin{bmatrix} 1 & 0 \\ 0 & 1 \end{bmatrix} - \begin{bmatrix} k_1^2 & k_1k_2 \\ k_1k_2 & k_2^2 \end{bmatrix}$$

This matrix equation results in the following three equations in p_{ij}:

$$-2k_1p_{12} = -1 - k_1^2$$

$$p_{11} - k_2p_{12} - k_1p_{22} = -k_1k_2$$

$$2p_{12} - 2k_2p_{22} = -1 - k_2^2$$

Solving these three equations for the p_{ij}, we obtain

$$\mathbf{P} = \begin{bmatrix} p_{11} & p_{12} \\ p_{12} & p_{22} \end{bmatrix} = \begin{bmatrix} \dfrac{1}{2}\left(\dfrac{k_2}{k_1} + \dfrac{k_1}{k_2}\right) + \dfrac{k_1}{2k_2}\left(\dfrac{1}{k_1} + k_1\right) & \dfrac{1}{2}\left(\dfrac{1}{k_1} + k_1\right) \\ \dfrac{1}{2}\left(\dfrac{1}{k_1} + k_1\right) & \dfrac{1}{2}\left(\dfrac{1}{k_2} + k_2\right) + \dfrac{1}{2k_2}\left(\dfrac{1}{k_1} + k_1\right) \end{bmatrix}$$

Now

$$J = \mathbf{x}^T(0)\mathbf{P}\mathbf{x}(0)$$

$$= \left[\frac{1}{2}\left(\frac{k_2}{k_1} + \frac{k_1}{k_2}\right) + \frac{k_1}{2k_2}\left(\frac{1}{k_1} + k_1\right)\right]x_1^2(0) + \left(\frac{1}{k_1} + k_1\right)x_1(0)x_2(0)$$

$$+ \left[\frac{1}{2}\left(\frac{1}{k_2} + k_2\right) + \frac{1}{2k_2}\left(\frac{1}{k_1} + k_1\right)\right]x_2^2(0)$$

To minimize J, we set $\partial J/\partial k_1 = 0$ and $\partial J/\partial k_2 = 0$, or

$$\frac{\partial J}{\partial k_1} = \left[\frac{1}{2}\left(\frac{-k_2}{k_1^2} + \frac{1}{k_2}\right) + \frac{k_1}{k_2}\right]x_1^2(0) + \left(\frac{-1}{k_1^2} + 1\right)x_1(0)x_2(0) + \left[\frac{1}{2k_2}\left(\frac{-1}{k_1^2} + 1\right)\right]x_2^2(0) = 0$$

$$\frac{\partial J}{\partial k_2} = \left[\frac{1}{2}\left(\frac{1}{k_1} - \frac{k_1}{k_2^2}\right) + \frac{-k_1}{2k_2^2}\left(\frac{1}{k_1} + k_1\right)\right]x_1^2(0) + \left[\frac{1}{2}\left(\frac{-1}{k_2^2} + 1\right) - \frac{1}{2k_2^2}\left(\frac{1}{k_1} + k_1\right)\right]x_2^2(0) = 0$$

For any given initial conditions $x_1(0)$ and $x_2(0)$, the value of J becomes minimum when

$$k_1 = 1, \qquad k_2 = \sqrt{3}$$

Note that k_1 and k_2 are positive constants as we assumed in the solution. Hence for the optimal control law

$$\mathbf{K} = [k_1 \quad k_2] = [1 \quad \sqrt{3}]$$

The block diagram of this optimal control system becomes the same as that shown in Figure 10–25 if $\mu = 1$ is substituted.

A–10–20. Consider the system

$$\dot{\mathbf{x}} = \mathbf{f}(\mathbf{x}, \mathbf{u})$$

which may be linear or nonlinear. It is desired to determine the optimal control law $\mathbf{u} = \mathbf{g}(\mathbf{x})$ such that the performance index

$$J = \int_0^\infty L(\mathbf{x}, \mathbf{u}) \, dt$$

is minimized, where $\mathbf{u}$ is unconstrained.
   If the origin of the system described by

$$\dot{\mathbf{x}} = \mathbf{f}(\mathbf{x}, \mathbf{g}(\mathbf{x}))$$

is asymptotically stable, and thus a Liapunov function $V(\mathbf{x})$ exists such that $\dot{V}(\mathbf{x})$ is negative definite, then show that a sufficient condition for a control vector $\mathbf{u}_1$ to be optimal is that $H(\mathbf{x}, \mathbf{u})$, where

$$H(\mathbf{x}, \mathbf{u}) = \frac{dV}{dt} + L(\mathbf{x}, \mathbf{u}) \tag{10–200}$$

is minimum with $\mathbf{u} = \mathbf{u}_1$, or

$$\min_{\mathbf{u}} H(\mathbf{x}, \mathbf{u}) = \min_{\mathbf{u}} \left[\frac{dV}{dt} + L(\mathbf{x}, \mathbf{u}) \right]$$

$$= \frac{dV}{dt} \bigg|_{\mathbf{u}=\mathbf{u}_1} + L(\mathbf{x}, \mathbf{u}_1) \tag{10–201}$$

and

$$\frac{dV}{dt} \bigg|_{\mathbf{u}=\mathbf{u}_1} = -L(\mathbf{x}, \mathbf{u}_1) \tag{10–202}$$

Solution. Let us integrate both sides of Equation (10–202). Then

$$V(\mathbf{x}(\infty)) - V(\mathbf{x}(0)) = -\int_0^\infty L(\mathbf{x}(t), \mathbf{u}_1(t)) \, dt \tag{10–203}$$

Since the origin of the system is asymptotically stable, $\mathbf{x}(\infty) = \mathbf{0}$ and $V(\mathbf{x}(\infty)) = 0$. Then Equation (10–203) becomes

$$V(\mathbf{x}(0)) = \int_0^\infty L(\mathbf{x}(t), \mathbf{u}_1(t)) \, dt \tag{10–204}$$

   To prove that $\mathbf{u}_1(t)$ is optimal, assume that $\mathbf{u}_1(t)$ is not optimal and that the control vector $\mathbf{u}_2(t)$ will yield a smaller value of J. Then

$$\int_0^\infty L(\mathbf{x}(t), \mathbf{u}_2(t)) \, dt < \int_0^\infty L(\mathbf{x}(t), \mathbf{u}_1(t)) \, dt$$

Note that, from Equation (10–201), the minimum value of $H(\mathbf{x}, \mathbf{u})$ occurs at $\mathbf{u} = \mathbf{u}_1$. Note also that, from Equation (10–202), this minimum value is equal to zero. Hence

$$H(\mathbf{x}, \mathbf{u}) \geq 0$$

Example Problems and Solutions

for all **u.** Therefore,

$$H(\mathbf{x}, \mathbf{u}_2) = \left.\frac{dV}{dt}\right|_{\mathbf{u}=\mathbf{u}_2} + L(\mathbf{x}, \mathbf{u}_2) \geq 0$$

Integrating both sides of this inequality from 0 to ∞, we obtain

$$V(\mathbf{x}(\infty)) - V(\mathbf{x}(0)) \geq -\int_0^\infty L(\mathbf{x}(t), \mathbf{u}_2(t))\, dt$$

Since $V(\mathbf{x}(\infty)) = 0$, we have

$$V(\mathbf{x}(0)) \leq \int_0^\infty L(\mathbf{x}(t), \mathbf{u}_2(t))\, dt \qquad (10\text{--}205)$$

Then from Equations (10–204) and (10–205)

$$\int_0^\infty L(\mathbf{x}(t), \mathbf{u}_1(t))\, dt \leq \int_0^\infty L(\mathbf{x}(t), \mathbf{u}_2(t))\, dt$$

This is a contradiction. Hence $\mathbf{u}_1(t)$ is the optimal control vector.

PROBLEMS

B–10–1. Consider the system defined by

$$\dot{\mathbf{x}} = \mathbf{Ax} + \mathbf{B}u$$

$$y = \mathbf{Cx}$$

where

$$\mathbf{A} = \begin{bmatrix} 1 & 2 \\ -4 & -3 \end{bmatrix}, \quad \mathbf{B} = \begin{bmatrix} 1 \\ 2 \end{bmatrix}, \quad \mathbf{C} = [1 \quad 1]$$

Transform the system equations into the controllable canonical form.

B–10–2. Consider the system defined by

$$\dot{\mathbf{x}} = \mathbf{Ax} + \mathbf{B}u$$

$$y = \mathbf{Cx}$$

where

$$\mathbf{A} = \begin{bmatrix} -1 & 0 & 1 \\ 1 & -2 & 0 \\ 0 & 0 & -3 \end{bmatrix}, \quad \mathbf{B} = \begin{bmatrix} 0 \\ 0 \\ 1 \end{bmatrix}, \quad \mathbf{C} = [1 \quad 1 \quad 0]$$

Transform the system equations into the controllable canonical form.

B–10–3. Consider the system defined by

$$\dot{\mathbf{x}} = \mathbf{Ax} + \mathbf{B}u$$

where

$$\mathbf{A} = \begin{bmatrix} 0 & 1 & 0 \\ 0 & 0 & 1 \\ -1 & -5 & -6 \end{bmatrix}, \quad \mathbf{B} = \begin{bmatrix} 0 \\ 0 \\ 1 \end{bmatrix}$$

By using the state feedback control $u = -\mathbf{Kx}$, it is desired to have the closed-loop poles at $s = -2 \pm j4$, $s = -10$. Determine the state feedback gain matrix **K.**

B–10–4. Referring to Example 10–2, consider the inverted pendulum system with the same numerical values for M, m, and l as used in that example. Suppose that the desired locations for the closed-loop poles $s = \mu_i$ ($i = 1, 2, 3, 4$) are changed to

$$\mu_1 = -2, \qquad \mu_2 = -2, \qquad \mu_3 = -10, \qquad \mu_4 = -10$$

Determine the state feedback gain matrix **K.** (Use the same definition of the state variables as used in Example 10–2; that is, $x_1 = \theta$, $x_2 = \dot{\theta}$, $x_3 = x$, and $x_4 = \dot{x}$.)

Write a computer program to obtain the response to arbitrary initial conditions. In particular, for the initial conditions

$$x_1(0) = 0.1 \text{ rad}, \qquad x_2(0) = 0, \qquad x_3(0) = 0, \qquad x_4(0) = 0$$

obtain a computer solution and plot curves $x_1(t)$ versus t, $x_2(t)$ versus t, $x_3(t)$ versus t, and $x_4(t)$ versus t. Compare these curves with those shown in Figure 10–5(a). In Example 10–2, the desired closed-loop poles included a pair of dominant complex-conjugate poles. In this problem, the desired closed-loop poles are all real and located in the left half-plane. Which set of the desired closed-loop poles may be preferred?

B–10–5. Consider the inverted pendulum system discussed in Example 10–2. Referring to the schematic diagram shown in Figure 10–3, assume that

$$M = 2 \text{ kg}, \quad m = 0.5 \text{ kg}, \quad l = 1 \text{ m}$$

Define state variables as follows,

$$x_1 = \theta, \quad x_2 = \dot{\theta}, \quad x_3 = x, \quad x_4 = \dot{x}$$

and output variables as follows:

$$y_1 = \theta = x_1, \quad y_2 = x = x_3$$

Derive the state-space equations for this system.

It is desired to have closed-loop poles at $s = \mu_i$ ($i = 1, 2, 3, 4$), where

$$\mu_1 = -4 + j4, \quad \mu_2 = -4 - j4, \quad \mu_3 = -20, \quad \mu_4 = -20$$

Determine the state feedback gain matrix $\mathbf{K}$.

Using the state feedback gain matrix $\mathbf{K}$ thus determined, examine the performance of the system by computer simulation. Write a computer program to obtain the response of the system to an arbitrary initial condition. Obtain the response curves $x_1(t)$ versus t, $x_2(t)$ versus t, $x_3(t)$ versus t, and $x_4(t)$ versus t for the following set of initial conditions:

$$x_1(0) = 0, \quad x_2(0) = 0, \quad x_3(0) = 0, \quad x_4(0) = 1 \text{ m/s}$$

B–10–6. A regulator system has a plant

$$\frac{Y(s)}{U(s)} = \frac{10}{(s + 1)(s + 2)(s + 3)}$$

Define state variables as

$$x_1 = y$$

$$x_2 = \dot{x}_1$$

$$x_3 = \dot{x}_2$$

By use of the state feedback control $u = -\mathbf{Kx}$, it is desired to place the closed-loop poles at $s = \mu_i$ ($i = 1, 2, 3$), where

$$\mu_1 = -2 + j3.464, \quad \mu_2 = -2 - j3.464, \quad \mu_3 = -10$$

Determine the necessary state feedback gain matrix $\mathbf{K}$.

B–10–7. Consider the system defined by

$$\begin{bmatrix} \dot{x}_1 \\ \dot{x}_2 \end{bmatrix} = \begin{bmatrix} -1 & 1 \\ 0 & 2 \end{bmatrix} \begin{bmatrix} x_1 \\ x_2 \end{bmatrix} + \begin{bmatrix} 1 \\ 0 \end{bmatrix} u$$

Show that the system cannot be stabilized by the state feedback control scheme $u = -\mathbf{Kx}$ whatever matrix $\mathbf{K}$ is chosen.

B–10–8. Consider the system defined by

$$\dot{\mathbf{x}} = \mathbf{Ax} + \mathbf{Bu}$$

$$y = \mathbf{Cx}$$

where

$$\mathbf{A} = \begin{bmatrix} -1 & 0 & 1 \\ 1 & -2 & 0 \\ 0 & 0 & -3 \end{bmatrix}, \quad \mathbf{B} = \begin{bmatrix} 0 \\ 1 \\ 1 \end{bmatrix}, \quad \mathbf{C} = [1 \quad 1 \quad 1]$$

Transform the system equations into the observable canonical form.

B–10–9. Consider the system defined by

$$\dot{\mathbf{x}} = \mathbf{Ax}$$

$$y = \mathbf{Cx}$$

where

$$\mathbf{A} = \begin{bmatrix} -1 & 1 \\ 1 & -2 \end{bmatrix}, \quad \mathbf{C} = [1 \quad 0]$$

Design a full-order state observer. The desired eigenvalues for the observer matrix are $\mu_1 = -5$, $\mu_2 = -5$.

B–10–10. Consider the system defined in Problem B–10–9. Assuming that the output y is accurately measurable, design a minimum-order observer. The desired eigenvalue for the observer matrix is $\mu = -5$; that is, the desired characteristic equation for the minimum-order observer is

$$s + 5 = 0$$

B–10–11. Consider a system defined by

$$\begin{bmatrix} \dot{x}_1 \\ \dot{x}_2 \\ \dot{x}_3 \end{bmatrix} = \begin{bmatrix} 0 & 1 & 0 \\ 0 & 0 & 1 \\ 1.244 & 0.3956 & -3.145 \end{bmatrix} \begin{bmatrix} x_1 \\ x_2 \\ x_3 \end{bmatrix} + \begin{bmatrix} 0 \\ 0 \\ 1.244 \end{bmatrix} u$$

$$y = [1 \quad 0 \quad 0] \begin{bmatrix} x_1 \\ x_2 \\ x_3 \end{bmatrix}$$

Given the following set of desired eigenvalues of the observer gain matrix,

$$\mu_1 = -5 + j8.66, \quad \mu_2 = -5 - j8.66, \quad \mu_3 = -10$$

design a full-order state observer.

Design also a minimum-order observer assuming that the output y is accurately measurable. Choose the desired eigenvalues of the minimum-order observer matrix to be

$$\mu_1 = -5 + j8.66, \qquad \mu_2 = -5 - j8.66$$

B–10–12. Consider the system designed in Example 10–7. Using the same **K** matrix determined in the example, it is desired to obtain the response of the system by computer simulation. Write a computer program to obtain the response to a ramp input.

B–10–13. Consider the type 1 servo system shown in Figure 10–30. Matrices **A**, **B**, and **C** in Figure 10–30 are given by

$$\mathbf{A} = \begin{bmatrix} 0 & 1 & 0 \\ 0 & 0 & 1 \\ 0 & -5 & -6 \end{bmatrix}, \qquad \mathbf{B} = \begin{bmatrix} 0 \\ 0 \\ 1 \end{bmatrix}, \qquad \mathbf{C} = [1 \quad 0 \quad 0]$$

Determine the feedback gain constants k_1, k_2, and k_3 such that the closed-loop poles are located at $s = -2 \pm j4$, $s = -10$.

Simulate the designed system on the computer. Obtain a computer solution to the unit-step response and plot the curve $y(t)$ versus t.

B–10–14. Consider the inverted-pendulum system discussed in Example 10–8. (The system diagram is shown in Figure 10–20.) In this problem, use the same numerical values for matrices **A**, **B**, and **C** as given in Example 10–8, including the locations of the desired closed-loop poles (dominant closed-loop poles at $s = -1 \pm j1.732$ and a triple pole at $s = -5$).

Define the state variables as follows:

$$x_1 = \theta$$

$$x_2 = \dot{\theta}$$

$$x_3 = x$$

$$x_4 = \dot{x}$$

$$x_5 = \xi$$

Simulate the system on the computer. The input $r(t)$ to the cart is a step function of magnitude 0.5 [that is, $r(t) = 0.5$ m]. Obtain the response curves $x_1(t)$ versus t, $x_2(t)$ versus t, $x_3(t)$ versus t, $x_4(t)$ versus t, and $x_5(t)$ versus t for the following two sets of initial conditions:

(a)
$$x_1(0) = 0.1 \text{ rad}$$
$$x_2(0) = 0$$
$$x_3(0) = 0$$
$$x_4(0) = 0$$
$$x_5(0) = 0$$

(b)
$$x_1(0) = 0$$
$$x_2(0) = 0.1 \text{ rad/sec}$$
$$x_3(0) = 0$$
$$x_4(0) = 0$$
$$x_5(0) = 0$$

B–10–15. Consider the system

$$\dot{\mathbf{x}} = \mathbf{Ax} + \mathbf{Bu}$$

Show that if the control vector **u** is given by

$$\mathbf{u} = -\mathbf{B}^*\mathbf{Px}$$

where matrix **P** is a positive-definite Hermitian matrix satisfying the condition that

$$\mathbf{A}^*\mathbf{P} + \mathbf{PA} = -\mathbf{I}$$

then the origin of the system is asymptotically stable in the large.

B–10–16. Consider a system defined by

$$\dot{\mathbf{x}} = \mathbf{Ax} + \mathbf{B}u$$

and the performance index to be minimized is given by

$$J = \int_0^\infty \mathbf{x}^*\mathbf{Qx} \, dt$$

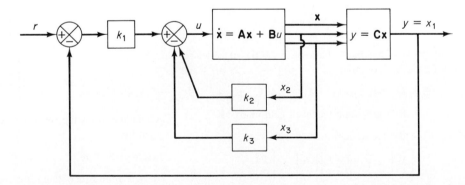

Figure 10–30

Assume that a state feedback control $u = -\mathbf{Kx}$ is used.

Show that the requirement of $\mathbf{A} - \mathbf{BK}$ being a stable matrix is equivalent to that of the rank of

$$
\begin{bmatrix}
\mathbf{Q}^{1/2} \\
\mathbf{Q}^{1/2}\mathbf{A} \\
\cdot \\
\cdot \\
\cdot \\
\mathbf{Q}^{1/2}\mathbf{A}^{n-1}
\end{bmatrix}
$$

being n. (See Problem A–9–17 for a hint to the solution of this problem.)

B–10–17. Consider the system defined by

$$\dot{\mathbf{x}} = \mathbf{Ax}$$

where

$$
\mathbf{A} = \begin{bmatrix}
0 & 1 & 0 \\
0 & 0 & 1 \\
-1 & -2 & -a
\end{bmatrix}
$$

a = adjustable parameter > 0

Determine the value of the parameter a so as to minimize the following performance index:

$$J = \int_0^{\infty} \mathbf{x}^T\mathbf{x}\, dt$$

Assume that the initial state $\mathbf{x}(0)$ is given by

$$
\mathbf{x}(0) = \begin{bmatrix}
c_1 \\
0 \\
0
\end{bmatrix}
$$

B–10–18. Consider the system shown in Figure 10–31. Determine the value of the gain K so that the damping ratio ζ of the closed-loop system is equal to 0.5. Then determine also the undamped natural frequency ω_n of the closed-loop system. Assuming that $e(0) = 1$ and $\dot{e}(0) = 0$, evaluate

$$\int_0^{\infty} e^2\,(t)\, dt$$

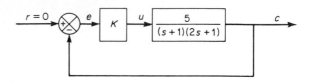

Figure 10–31 Control system.

B–10–19. Determine the optimal control signal u for the system defined by

$$\dot{\mathbf{x}} = \mathbf{Ax} + \mathbf{B}u$$

where

$$
\mathbf{A} = \begin{bmatrix}
0 & 1 \\
0 & -1
\end{bmatrix}, \qquad
\mathbf{B} = \begin{bmatrix}
0 \\
1
\end{bmatrix}
$$

such that the following performance index is minimized:

$$J = \int_0^{\infty} (\mathbf{x}^T\mathbf{x} + u^2)\, dt$$

B–10–20. Consider the system

$$
\begin{bmatrix}
\dot{x}_1 \\
\dot{x}_2
\end{bmatrix} =
\begin{bmatrix}
0 & 1 \\
0 & 0
\end{bmatrix}
\begin{bmatrix}
x_1 \\
x_2
\end{bmatrix} +
\begin{bmatrix}
0 \\
1
\end{bmatrix} u
$$

It is desired to find the optimal control signal u such that the following performance index:

$$
J = \int_0^{\infty} (\mathbf{x}^T\mathbf{Q}\mathbf{x} + u^2)\, dt, \qquad
\mathbf{Q} = \begin{bmatrix}
1 & 0 \\
0 & \mu
\end{bmatrix}
$$

is minimized.

Referring to the sufficient condition for the optimal control vector presented in Problem A–10–20, determine the optimal control signal $u(t)$.

APPENDIX

Vector-Matrix Analysis

A–1 DEFINITIONS

Conjugate matrix. The conjugate of a matrix $\mathbf{A}$ is the matrix in which each element is the complex conjugate of the corresponding element of $\mathbf{A}$. The conjugate of $\mathbf{A}$ is denoted by $\bar{\mathbf{A}} = [\bar{a}_{ij}]$, where $\bar{a}_{ij}$ is the complex conjugate of a_{ij}. For example, if $\mathbf{A}$ is given by

$$\mathbf{A} = \begin{bmatrix} 0 & 1 & 0 \\ -1+j & -3-j3 & -1+j4 \\ -1+j & -1 & -2+j3 \end{bmatrix} \tag{A–1}$$

then

$$\bar{\mathbf{A}} = \begin{bmatrix} 0 & 1 & 0 \\ -1-j & -3+j3 & -1-j4 \\ -1-j & -1 & -2-j3 \end{bmatrix}$$

Transpose. If the rows and columns of an $n \times m$ matrix $\mathbf{A}$ are interchanged, then the resulting $m \times n$ matrix is called the transpose of $\mathbf{A}$. The transpose of $\mathbf{A}$ is denoted by $\mathbf{A}^T$. That is, if $\mathbf{A}$ is given by

$$\mathbf{A} = \begin{bmatrix} a_{11} & a_{12} & \cdots & a_{1m} \\ a_{21} & a_{22} & \cdots & a_{2m} \\ \cdot & \cdot & & \cdot \\ \cdot & \cdot & & \cdot \\ \cdot & \cdot & & \cdot \\ a_{n1} & a_{n2} & \cdots & a_{nm} \end{bmatrix}$$

then $\mathbf{A}^T$ is given by

$$\mathbf{A}^T = \begin{bmatrix} a_{11} & a_{21} & \cdots & a_{n1} \\ a_{12} & a_{22} & \cdots & a_{n2} \\ \cdot & \cdot & & \cdot \\ \cdot & \cdot & & \cdot \\ \cdot & \cdot & & \cdot \\ a_{1m} & a_{2m} & \cdots & a_{nm} \end{bmatrix}$$

Note that $(\mathbf{A}^T)^T = \mathbf{A}.$ It is easy to verify that if $\mathbf{A} + \mathbf{B}$ and $\mathbf{AB}$ can be defined, then

$$(\mathbf{A} + \mathbf{B})^T = \mathbf{A}^T + \mathbf{B}^T, \qquad (\mathbf{AB})^T = \mathbf{B}^T\mathbf{A}^T$$

Conjugate transpose. The conjugate transpose is the conjugate of the transpose of a matrix. Given a matrix $\mathbf{A} = [a_{ij}]$, the conjugate transpose is denoted by $\bar{\mathbf{A}}^T$ or $\mathbf{A}^*$; that is,

$$\bar{\mathbf{A}}^T = \mathbf{A}^* = [\bar{a}_{ji}]$$

For example, if $\mathbf{A}$ is as given by Equation (A–1), then

$$\bar{\mathbf{A}}^T = \mathbf{A}^* = \begin{bmatrix} 0 & -1 - j & -1 - j \\ 1 & -3 + j3 & -1 \\ 0 & -1 - j4 & -2 - j3 \end{bmatrix}$$

Clearly, the conjugate of $\mathbf{A}^T$ is the same as the transpose of $\bar{\mathbf{A}}$. Note that $(\mathbf{A}^*)^* = \mathbf{A}.$ It can easily be shown that if $\mathbf{A} + \mathbf{B}$ and $\mathbf{AB}$ can be defined, then

$$(\mathbf{A} + \mathbf{B})^* = \mathbf{A}^* + \mathbf{B}^*, \qquad (\mathbf{AB})^* = \mathbf{B}^*\mathbf{A}^*$$

Note also that if c is a complex number, then

$$(c\,\mathbf{A})^* = \bar{c}\mathbf{A}^*$$

If $\mathbf{A}$ is a real matrix (a matrix whose elements are all real), the conjugate transpose $\mathbf{A}^*$ is the same as the transpose $\mathbf{A}^T$.

Symmetric matrix and skew-symmetric matrix. A *symmetric* matrix is a matrix that is equal to its transpose. That is, for a symmetric matrix $\mathbf{A}$,

$$\mathbf{A}^T = \mathbf{A} \qquad \text{or} \qquad a_{ji} = a_{ij}$$

If a matrix $\mathbf{A}$ is equal to the negative of its transpose, or

$$\mathbf{A}^T = -\mathbf{A} \qquad \text{or} \qquad a_{ji} = -a_{ij}$$

then it is called a *skew-symmetric* matrix.

If $\mathbf{A}$ is any square matrix, then $\mathbf{A} + \mathbf{A}^T$ is a symmetric matrix and $\mathbf{A} - \mathbf{A}^T$ is a skew-symmetric matrix. For example, if $\mathbf{A}$ is given by

$$\mathbf{A} = \begin{bmatrix} 1 & 2 & 3 \\ 4 & 5 & 6 \\ 7 & 8 & 9 \end{bmatrix}$$

then

$$\mathbf{A} + \mathbf{A}^T = \begin{bmatrix} 2 & 6 & 10 \\ 6 & 10 & 14 \\ 10 & 14 & 18 \end{bmatrix} = \text{symmetric matrix}$$

and

$$\mathbf{A} - \mathbf{A}^T = \begin{bmatrix} 0 & -2 & -4 \\ 2 & 0 & -2 \\ 4 & 2 & 0 \end{bmatrix} = \text{skew-symmetric matrix}$$

Notice that if $\mathbf{A}$ is a rectangular matrix, then $\mathbf{A}^T\mathbf{A} = \mathbf{B}$ is a symmetric matrix. Notice also that the inverse of a symmetric matrix, if the inverse exists, is symmetric. To establish this fact, take the transpose of $\mathbf{BB}^{-1} = \mathbf{I}$. We have $(\mathbf{B}^{-1})^T\mathbf{B}^T = \mathbf{I}^T = \mathbf{I}$. Noting that $\mathbf{B} = \mathbf{B}^T$, we have $(\mathbf{B}^{-1})^T\mathbf{B}^T = (\mathbf{B}^{-1})^T\mathbf{B} = \mathbf{I} = \mathbf{B}^{-1}\mathbf{B}$. Hence, $\mathbf{B}^{-1} = (\mathbf{B}^{-1})^T$. Thus, the inverse of a symmetric matrix is symmetric.

Orthogonal matrix.   A matrix $\mathbf{A}$ is called an orthogonal matrix if it is real and satisfies the relationship $\mathbf{A}^T\mathbf{A} = \mathbf{A}\mathbf{A}^T = \mathbf{I}$. (This implies that $|\mathbf{A}| = \pm 1$ and thus $\mathbf{A}$ is nonsingular.) Examples of orthogonal matrices are

$$\mathbf{A} = \begin{bmatrix} \cos\theta & -\sin\theta \\ \sin\theta & \cos\theta \end{bmatrix}, \qquad \mathbf{B} = \begin{bmatrix} 0.6 & 0.8 & 0 \\ -0.8 & 0.6 & 0 \\ 0 & 0 & 1 \end{bmatrix}$$

In an orthogonal matrix the inverse is exactly equal to the transpose.

$$\mathbf{A}^{-1} = \mathbf{A}^T$$

To demonstrate, let us obtain $\mathbf{A}^{-1}$ and $\mathbf{B}^{-1}$ for the matrices $\mathbf{A}$ and $\mathbf{B}$ that were just given:

$$\mathbf{A}^{-1} = \begin{bmatrix} \cos\theta & \sin\theta \\ -\sin\theta & \cos\theta \end{bmatrix} = \mathbf{A}^T, \qquad \mathbf{B}^{-1} = \begin{bmatrix} 0.6 & -0.8 & 0 \\ 0.8 & 0.6 & 0 \\ 0 & 0 & 1 \end{bmatrix} = \mathbf{B}^T$$

If $\mathbf{A}$ and $\mathbf{B}$ are $n \times n$ orthogonal matrices, then so are $\mathbf{A}^{-1}$, $\mathbf{A}^T$, and $\mathbf{AB}$. This may be seen as follows. Since $\mathbf{A}$ is orthogonal, $\mathbf{AA}^T = \mathbf{I}$ and $(\mathbf{A}^T)^T\mathbf{A}^T = \mathbf{I}$. Hence, $\mathbf{A}^T$ is orthogonal. Since $\mathbf{A}^{-1} = \mathbf{A}^T$, $\mathbf{A}^{-1}$ is also orthogonal. Since $\mathbf{B}^T = \mathbf{B}^{-1}$, $\mathbf{A}^T = \mathbf{A}^{-1}$, $(\mathbf{AB})^T = \mathbf{B}^T\mathbf{A}^T$, and $(\mathbf{AB})^{-1} = \mathbf{B}^{-1}\mathbf{A}^{-1}$, we have $(\mathbf{AB})^T = (\mathbf{AB})^{-1}$, and so $\mathbf{AB}$ is orthogonal.

Hermitian matrix and skew-Hermitian matrix.   A matrix whose elements are complex quantities is called a *complex* matrix. If a complex matrix $\mathbf{A}$ satisfies the relationship

$$\mathbf{A}^* = \mathbf{A} \qquad \text{or} \qquad a_{ij} = \bar{a}_{ji}$$

where $\bar{a}_{ji}$ is the complex conjugate of a_{ji}, then $\mathbf{A}$ is called a *Hermitian* matrix. Note that a Hermitian matrix must be square and that the main diagonal elements must be real. An example is

$$\mathbf{A} = \begin{bmatrix} 1 & 4+j3 & j5 \\ 4-j3 & 2 & 2+j \\ -j5 & 2-j & 0 \end{bmatrix}$$

If a Hermitian matrix $\mathbf{A}$ is written as $\mathbf{A} = \mathbf{B} + j\mathbf{C}$, where $\mathbf{B}$ and $\mathbf{C}$ are real matrices, then

$$\mathbf{B} = \mathbf{B}^T, \qquad \mathbf{C} = -\mathbf{C}^T$$

In the foregoing example,

$$\mathbf{A} = \mathbf{B} + j\mathbf{C} = \begin{bmatrix} 1 & 4 & 0 \\ 4 & 2 & 2 \\ 0 & 2 & 0 \end{bmatrix} + j \begin{bmatrix} 0 & 3 & 5 \\ -3 & 0 & 1 \\ -5 & -1 & 0 \end{bmatrix}$$

Notice that the inverse of a Hermitian matrix $\mathbf{A}$ is Hermitian, or $\mathbf{A}^{-1} = (\mathbf{A}^{-1})^*$. Notice also that every square matrix can be expressed uniquely as $\mathbf{A} = \mathbf{G} + j\mathbf{H}$, where $\mathbf{G}$ and $\mathbf{H}$ are Hermitian and are given by

$$\mathbf{G} = \frac{1}{2}(\mathbf{A} + \mathbf{A}^*), \qquad \mathbf{H} = \frac{1}{2j}(\mathbf{A} - \mathbf{A}^*)$$

The fact that $\mathbf{G}$ and $\mathbf{H}$ are Hermitian is seen as follows:

$$\mathbf{G}^* = \frac{1}{2}(\mathbf{A}^* + \mathbf{A}) = \mathbf{G}, \qquad \mathbf{H}^* = -\frac{1}{2j}(\mathbf{A}^* - \mathbf{A}) = \mathbf{H}$$

It can be easily verified that for $n \times n$ Hermitian matrices $\mathbf{A}$ and $\mathbf{B}$, the matrices $\mathbf{A} + \mathbf{B}$, $\mathbf{A} - \mathbf{B}$, and $\mathbf{AB} + \mathbf{BA}$ are also Hermitian. The product $\mathbf{AB}$ is Hermitian, however, if and only if $\mathbf{A}$ and $\mathbf{B}$ commute, since $\mathbf{AB} = \mathbf{A}^*\mathbf{B}^* = (\mathbf{BA})^*$. The determinant of a Hermitian matrix is always real, since

$$|\mathbf{A}| = |\mathbf{A}^*| = |\bar{\mathbf{A}}^T| = |\bar{\mathbf{A}}|$$

If a matrix $\mathbf{A}$ satisfies the relationship $\mathbf{A}^* = -\mathbf{A}$, then $\mathbf{A}$ is called a *skew-Hermitian* matrix. An example is

$$\mathbf{A} = \begin{bmatrix} j5 & -2 + j3 & -4 + j6 \\ 2 + j3 & j4 & -2 + j2 \\ 4 + j6 & 2 + j2 & j \end{bmatrix}$$

Note that a skew-Hermitian matrix must be square and that the main diagonal elements must be imaginary or zero.

If a skew-Hermitian matrix $\mathbf{A}$ is written as $\mathbf{A} = \mathbf{B} + j\mathbf{C}$, where $\mathbf{B}$ and $\mathbf{C}$ are real matrices, then

$$\mathbf{B} = -\mathbf{B}^T, \qquad \mathbf{C} = \mathbf{C}^T$$

In the present example,

$$\mathbf{A} = \mathbf{B} + j\mathbf{C} = \begin{bmatrix} 0 & -2 & -4 \\ 2 & 0 & -2 \\ 4 & 2 & 0 \end{bmatrix} + j \begin{bmatrix} 5 & 3 & 6 \\ 3 & 4 & 2 \\ 6 & 2 & 1 \end{bmatrix}$$

Unitary matrix.   A unitary matrix is a complex matrix in which the inverse is equal to the conjugate of the transpose. That is,

$$\mathbf{A}^{-1} = \mathbf{A}^* \qquad \text{or} \qquad \mathbf{AA}^* = \mathbf{A}^*\mathbf{A} = \mathbf{I}$$

An example of a unitary matrix is

$$\mathbf{A} = \begin{bmatrix} \dfrac{1}{\sqrt{15}}(2+j) & \dfrac{1}{\sqrt{15}}(3+j) \\ \dfrac{1}{\sqrt{15}}(-3+j) & \dfrac{1}{\sqrt{15}}(2-j) \end{bmatrix}$$

To demonstrate, let us compute $\mathbf{A}^{-1}$ and $\mathbf{A}^*$. Since for a unitary matrix $\mathbf{A}$ the determinant of $\mathbf{A}$ is equal to unity, or $|\mathbf{A}| = 1$, we obtain

$$\mathbf{A}^{-1} = \begin{bmatrix} \dfrac{1}{\sqrt{15}}(2-j) & -\dfrac{1}{\sqrt{15}}(3+j) \\ \dfrac{1}{\sqrt{15}}(3-j) & \dfrac{1}{\sqrt{15}}(2+j) \end{bmatrix}$$

The conjugate transpose $\mathbf{A}^*$ is given by

$$\mathbf{A}^* = \begin{bmatrix} \dfrac{1}{\sqrt{15}}(2-j) & -\dfrac{1}{\sqrt{15}}(3+j) \\ \dfrac{1}{\sqrt{15}}(3-j) & \dfrac{1}{\sqrt{15}}(2+j) \end{bmatrix}$$

Hence we have verified that $\mathbf{A}^{-1} = \mathbf{A}^*$.

Orthogonal matrices satisfy the relationship $\mathbf{AA}^* = \mathbf{A}^*\mathbf{A} = \mathbf{I}$; hence they are unitary.

Notice that if $\mathbf{A}$ is unitary, then so is the inverse $\mathbf{A}^{-1}$. To see this, notice that since $\mathbf{AA}^* = \mathbf{A}^*\mathbf{A} = \mathbf{I}$, we obtain

$$(\mathbf{A}^*)^{-1}\mathbf{A}^{-1} = (\mathbf{A}^{-1})^*(\mathbf{A}^{-1}) = \mathbf{I}$$

$$\mathbf{A}^{-1}(\mathbf{A}^*)^{-1} = (\mathbf{A}^{-1})(\mathbf{A}^{-1})^* = \mathbf{I}$$

If $n \times n$ matrices $\mathbf{A}$ and $\mathbf{B}$ are unitary, then so is matrix $\mathbf{AB}$. To prove this, notice that since $\mathbf{AA}^* = \mathbf{A}^*\mathbf{A} = \mathbf{I}$ and $\mathbf{BB}^* = \mathbf{B}^*\mathbf{B} = \mathbf{I}$, we have

$$(\mathbf{AB})(\mathbf{AB})^* = \mathbf{ABB}^*\mathbf{A}^* = \mathbf{AA}^* = \mathbf{I}$$

$$(\mathbf{AB})^*(\mathbf{AB}) = \mathbf{B}^*\mathbf{A}^*\mathbf{AB} = \mathbf{B}^*\mathbf{B} = \mathbf{I}$$

Hence, it is shown that $\mathbf{AB}$ is unitary.

Normal matrix. A matrix that commutes with its conjugate transpose is called a *normal* matrix. Specifically, for a normal matrix $\mathbf{A}$,

$$\mathbf{AA}^* = \mathbf{A}^*\mathbf{A} \qquad \text{if } \mathbf{A} \text{ is a complex matrix}$$

$$\mathbf{AA}^T = \mathbf{A}^T\mathbf{A} \qquad \text{if } \mathbf{A} \text{ is a real matrix}$$

Notice that if $\mathbf{A}$ is normal and $\mathbf{U}$ is unitary, then $\mathbf{U}^{-1}\mathbf{AU}$ is also normal, since

$$(\mathbf{U}^{-1}\mathbf{AU})(\mathbf{U}^{-1}\mathbf{AU})^* = \mathbf{U}^{-1}\mathbf{AUU}^*\mathbf{A}^*(\mathbf{U}^{-1})^* = \mathbf{U}^{-1}\mathbf{AA}^*(\mathbf{U}^{-1})^* = \mathbf{U}^*\mathbf{A}^*\mathbf{AU}$$
$$= \mathbf{U}^*\mathbf{A}^*(\mathbf{U}^{-1})^*\mathbf{U}^{-1}\mathbf{AU} = (\mathbf{U}^{-1}\mathbf{AU})^*(\mathbf{U}^{-1}\mathbf{AU})$$

A matrix is normal if it is a real symmetric, a Hermitian, a real skew-symmetric, a skew-Hermitian, a unitary, or an orthogonal matrix.

Summary. In summarizing the definitions of various matrices, the reader may find the following list useful:

$$\mathbf{A}^T = \mathbf{A} \qquad\qquad \mathbf{A} \text{ is symmetric}$$

$$\mathbf{A}^T = -\mathbf{A} \qquad\qquad \mathbf{A} \text{ is skew-symmetric}$$

$$\mathbf{A}\mathbf{A}^T = \mathbf{A}^T\mathbf{A} = \mathbf{I} \qquad\qquad \mathbf{A} \text{ is orthogonal}$$

$$\mathbf{A}^* = \mathbf{A} \qquad\qquad \mathbf{A} \text{ is Hermitian}$$

$$\mathbf{A}^* = -\mathbf{A} \qquad\qquad \mathbf{A} \text{ is skew-Hermitian}$$

$$\mathbf{A}\mathbf{A}^* = \mathbf{A}^*\mathbf{A} = \mathbf{I} \qquad\qquad \mathbf{A} \text{ is unitary}$$

$$\mathbf{A}\mathbf{A}^* = \mathbf{A}^*\mathbf{A} \quad \text{or} \quad \mathbf{A}\mathbf{A}^T = \mathbf{A}^T\mathbf{A} \qquad \mathbf{A} \text{ is normal}$$

A–2 DETERMINANTS

Determinants of a 2 × 2 matrix, a 3 × 3 matrix, and a 4 × 4 matrix. For a 2 × 2 matrix $\mathbf{A}$, we have

$$|\mathbf{A}| = \begin{vmatrix} a_1 & a_2 \\ b_1 & b_2 \end{vmatrix} = a_1 b_2 - b_1 a_2$$

For a 3 × 3 matrix $\mathbf{A}$,

$$|\mathbf{A}| = \begin{vmatrix} a_1 & a_2 & a_3 \\ b_1 & b_2 & b_3 \\ c_1 & c_2 & c_3 \end{vmatrix} = a_1 b_2 c_3 + b_1 c_2 a_3 + c_1 a_2 b_3 - c_1 b_2 a_3 - b_1 a_2 c_3 - a_1 b_3 c_2$$

For a 4 × 4 matrix $\mathbf{A}$,

$$|\mathbf{A}| = \begin{vmatrix} a_1 & a_2 & a_3 & a_4 \\ b_1 & b_2 & b_3 & b_4 \\ c_1 & c_2 & c_3 & c_4 \\ d_1 & d_2 & d_3 & d_4 \end{vmatrix}$$

$$= \begin{vmatrix} a_1 & a_2 \\ b_1 & b_2 \end{vmatrix} \begin{vmatrix} c_3 & c_4 \\ d_3 & d_4 \end{vmatrix} - \begin{vmatrix} a_1 & a_2 \\ c_1 & c_2 \end{vmatrix} \begin{vmatrix} b_3 & b_4 \\ d_3 & d_4 \end{vmatrix}$$

$$+ \begin{vmatrix} a_1 & a_2 \\ d_1 & d_2 \end{vmatrix} \begin{vmatrix} b_3 & b_4 \\ c_3 & c_4 \end{vmatrix} + \begin{vmatrix} b_1 & b_2 \\ c_1 & c_2 \end{vmatrix} \begin{vmatrix} a_3 & a_4 \\ d_3 & d_4 \end{vmatrix}$$

$$- \begin{vmatrix} b_1 & b_2 \\ d_1 & d_2 \end{vmatrix} \begin{vmatrix} a_3 & a_4 \\ c_3 & c_4 \end{vmatrix} + \begin{vmatrix} c_1 & c_2 \\ d_1 & d_2 \end{vmatrix} \begin{vmatrix} a_3 & a_4 \\ b_3 & b_4 \end{vmatrix}$$

(This expansion is called Laplace's expansion by the minors.)

Properties of the determinant. The determinant of an $n \times n$ matrix has the following properties:

1. If two rows (or two columns) of the determinant are interchanged, only the sign of the determinant is changed.

2. The determinant is invariant under the addition of a scalar multiple of a row (or a column) to another row (or column).

3. If an $n \times n$ matrix has two identical rows (or columns), then the determinant is zero.

4. For an $n \times n$ matrix $\mathbf{A}$,

$$|\mathbf{A}^T| = |\mathbf{A}|, \qquad |\mathbf{A}*| = |\bar{\mathbf{A}}|$$

5. The determinant of a product of two square matrices $\mathbf{A}$ and $\mathbf{B}$ is the product of their determinants:

$$|\mathbf{AB}| = |\mathbf{A}|\,|\mathbf{B}| = |\mathbf{BA}|$$

6. If a row (or a column) is multiplied by a scalar k, then the determinant is multiplied by k.

7. If all elements of an $n \times n$ matrix are multiplied by k, then the determinant is multiplied by k^n; that is,

$$|k\mathbf{A}| = k^n|\mathbf{A}|$$

8. If the eigenvalues of $\mathbf{A}$ are λ_i ($i = 1, 2, \ldots, n$), then

$$|\mathbf{A}| = \lambda_1\lambda_2 \ldots \lambda_n$$

Hence $|\mathbf{A}| \neq 0$ implies $\lambda_i \neq 0$ for $i = 1, 2, \ldots, n$. (For details of the eigenvalue, see Section A–6.)

9. If matrices $\mathbf{A}$, $\mathbf{B}$, $\mathbf{C}$, and $\mathbf{D}$ are an $n \times n$, an $n \times m$, an $m \times n$, and an $m \times m$ matrix, respectively, then

$$\begin{vmatrix} \mathbf{A} & \mathbf{B} \\ \mathbf{0} & \mathbf{D} \end{vmatrix} = \begin{vmatrix} \mathbf{A} & \mathbf{0} \\ \mathbf{C} & \mathbf{D} \end{vmatrix} = |\mathbf{A}|\,|\mathbf{D}| \qquad \text{if } |\mathbf{A}| \neq 0 \text{ and } |\mathbf{D}| \neq 0 \tag{A–2}$$

$$\begin{vmatrix} \mathbf{A} & \mathbf{B} \\ \mathbf{0} & \mathbf{D} \end{vmatrix} = \begin{vmatrix} \mathbf{A} & \mathbf{0} \\ \mathbf{C} & \mathbf{D} \end{vmatrix} = 0 \qquad \text{if } |\mathbf{A}| = 0 \text{ or } |\mathbf{D}| = 0 \text{ or } |\mathbf{A}| = |\mathbf{D}| = 0$$

Also,

$$\begin{vmatrix} \mathbf{A} & \mathbf{B} \\ \mathbf{C} & \mathbf{D} \end{vmatrix} = \begin{cases} |\mathbf{A}|\,|\mathbf{D} - \mathbf{CA}^{-1}\mathbf{B}| & \text{if } |\mathbf{A}| \neq 0 & \text{(A–3)} \\ |\mathbf{D}|\,|\mathbf{A} - \mathbf{BD}^{-1}\mathbf{C}| & \text{if } |\mathbf{D}| \neq 0 & \text{(A–4)} \end{cases}$$

[For the derivation of Equation (A–2), see Problem A–1. For derivations of Equations (A–3) and (A–4), refer to Problem A–2.]

10. For an $n \times m$ matrix $\mathbf{A}$ and an $m \times n$ matrix $\mathbf{B}$,

$$|\mathbf{I}_n + \mathbf{AB}| = |\mathbf{I}_m + \mathbf{BA}| \tag{A–5}$$

(For the proof, see Problem A–3.) In particular, for $m = 1$, that is, for an $n \times 1$ matrix $\mathbf{A}$ and a $1 \times n$ matrix $\mathbf{B}$, we have

$$|\mathbf{I}_n + \mathbf{AB}| = 1 + \mathbf{BA} \tag{A–6}$$

Equations (A–2) through (A–6) are useful in computing the determinants of matrices of large order.

A–3 INVERSION OF MATRICES

Nonsingular matrix and singular matrix. A square matrix $\mathbf{A}$ is called a nonsingular matrix if a matrix $\mathbf{B}$ exists such that $\mathbf{BA} = \mathbf{AB} = \mathbf{I}$. If such a matrix $\mathbf{B}$ exists, then it is denoted by $\mathbf{A}^{-1}$. $\mathbf{A}^{-1}$ is called the *inverse* of $\mathbf{A}$. The inverse matrix $\mathbf{A}^{-1}$ exists if $|\mathbf{A}|$ is nonzero. If $\mathbf{A}^{-1}$ does not exist, $\mathbf{A}$ is said to be *singular*.

If $\mathbf{A}$ and $\mathbf{B}$ are nonsingular matrices, then the product $\mathbf{AB}$ is a nonsingular matrix and

$$(\mathbf{AB})^{-1} = \mathbf{B}^{-1}\mathbf{A}^{-1}$$

Also,

$$(\mathbf{A}^T)^{-1} = (\mathbf{A}^{-1})^T$$

and

$$(\mathbf{A}^*)^{-1} = (\mathbf{A}^{-1})^*$$

Properties of the inverse matrix. The inverse of a matrix has the following properties.

1. If k is a nonzero scalar and $\mathbf{A}$ is an $n \times n$ nonsingular matrix, then

$$(k\mathbf{A})^{-1} = \frac{1}{k}\mathbf{A}^{-1}$$

2. The determinant of $\mathbf{A}^{-1}$ is the inverse of the determinant of $\mathbf{A}$, or

$$|\mathbf{A}^{-1}| = \frac{1}{|\mathbf{A}|}$$

This can be verified easily as follows:

$$|\mathbf{AA}^{-1}| = |\mathbf{A}|\,|\mathbf{A}^{-1}| = 1$$

Useful formulas for finding the inverse of a matrix.

1. For a 2×2 matrix $\mathbf{A}$, where

$$\mathbf{A} = \begin{bmatrix} a & b \\ c & d \end{bmatrix} \qquad ad - bc \neq 0$$

the inverse matrix is given by

$$\mathbf{A}^{-1} = \frac{1}{ad - bc} \begin{bmatrix} d & -b \\ -c & a \end{bmatrix}$$

2. For a 3×3 matrix $\mathbf{A}$, where

$$\mathbf{A} = \begin{bmatrix} a & b & c \\ d & e & f \\ g & h & i \end{bmatrix} \qquad |\mathbf{A}| \neq 0$$

the inverse matrix is given by

$$\mathbf{A}^{-1} = \frac{1}{|\mathbf{A}|} \begin{bmatrix} \begin{vmatrix} e & f \\ h & i \end{vmatrix} & -\begin{vmatrix} b & c \\ h & i \end{vmatrix} & \begin{vmatrix} b & c \\ e & f \end{vmatrix} \\[2mm] -\begin{vmatrix} d & f \\ g & i \end{vmatrix} & \begin{vmatrix} a & c \\ g & i \end{vmatrix} & -\begin{vmatrix} a & c \\ d & f \end{vmatrix} \\[2mm] \begin{vmatrix} d & e \\ g & h \end{vmatrix} & -\begin{vmatrix} a & b \\ g & h \end{vmatrix} & \begin{vmatrix} a & b \\ d & e \end{vmatrix} \end{bmatrix}$$

3. If $\mathbf{A}$, $\mathbf{B}$, $\mathbf{C}$, and $\mathbf{D}$ are, respectively, an $n \times n$, an $n \times m$, an $m \times n$, and an $m \times m$ matrix, then

$$(\mathbf{A} + \mathbf{BDC})^{-1} = \mathbf{A}^{-1} - \mathbf{A}^{-1}\mathbf{B}(\mathbf{D}^{-1} + \mathbf{CA}^{-1}\mathbf{B})^{-1}\mathbf{CA}^{-1} \tag{A–7}$$

provided the indicated inverses exist. Equation (A–7) is commonly referred to as the *matrix inversion lemma*. (For the proof, see Problem A–4.)

If $\mathbf{D} = \mathbf{I}_m$, then Equation (A–7) simplifies to

$$(\mathbf{A} + \mathbf{BC})^{-1} = \mathbf{A}^{-1} - \mathbf{A}^{-1}\mathbf{B}(\mathbf{I}_m + \mathbf{CA}^{-1}\mathbf{B})^{-1}\mathbf{CA}^{-1}$$

In this last equation, if $\mathbf{B}$ and $\mathbf{C}$ are an $n \times 1$ matrix and a $1 \times n$ matrix, respectively, then

$$(\mathbf{A} + \mathbf{BC})^{-1} = \mathbf{A}^{-1} - \frac{\mathbf{A}^{-1}\mathbf{BCA}^{-1}}{1 + \mathbf{CA}^{-1}\mathbf{B}} \tag{A–8}$$

Equation (A–8) is useful in that if an $n \times n$ matrix $\mathbf{X}$ can be written as $\mathbf{A} + \mathbf{BC}$, where $\mathbf{A}$ is an $n \times n$ matrix whose inverse is known and $\mathbf{BC}$ is a product of a column vector and a row vector, then $\mathbf{X}^{-1}$ can be obtained easily in terms of the known $\mathbf{A}^{-1}$, $\mathbf{B}$, and $\mathbf{C}$.

4. If $\mathbf{A}$, $\mathbf{B}$, $\mathbf{C}$, and $\mathbf{D}$ are, respectively, an $n \times n$, an $n \times m$, an $m \times n$, and an $m \times m$ matrix, then

$$\begin{bmatrix} \mathbf{A} & \mathbf{B} \\ \mathbf{C} & \mathbf{D} \end{bmatrix}^{-1} = \begin{bmatrix} \mathbf{A}^{-1} + \mathbf{A}^{-1}\mathbf{B}(\mathbf{D} - \mathbf{CA}^{-1}\mathbf{B})^{-1}\mathbf{CA}^{-1} & -\mathbf{A}^{-1}\mathbf{B}(\mathbf{D} - \mathbf{CA}^{-1}\mathbf{B})^{-1} \\ -(\mathbf{D} - \mathbf{CA}^{-1}\mathbf{B})^{-1}\mathbf{CA}^{-1} & (\mathbf{D} - \mathbf{CA}^{-1}\mathbf{B})^{-1} \end{bmatrix} \tag{A–9}$$

provided $|\mathbf{A}| \neq 0$ and $|\mathbf{D} - \mathbf{CA}^{-1}\mathbf{B}| \neq 0$, or

$$\begin{bmatrix} \mathbf{A} & \mathbf{B} \\ \mathbf{C} & \mathbf{D} \end{bmatrix}^{-1} = \begin{bmatrix} (\mathbf{A} - \mathbf{BD}^{-1}\mathbf{C})^{-1} & -(\mathbf{A} - \mathbf{BD}^{-1}\mathbf{C})^{-1}\mathbf{BD}^{-1} \\ -\mathbf{D}^{-1}\mathbf{C}(\mathbf{A} - \mathbf{BD}^{-1}\mathbf{C})^{-1} & \mathbf{D}^{-1}\mathbf{C}(\mathbf{A} - \mathbf{BD}^{-1}\mathbf{C})^{-1}\mathbf{BD}^{-1} + \mathbf{D}^{-1} \end{bmatrix} \tag{A–10}$$

provided $|\mathbf{D}| \neq 0$ and $|\mathbf{A} - \mathbf{BD}^{-1}\mathbf{C}| \neq 0$. In particular, if $\mathbf{C} = \mathbf{0}$ or $\mathbf{B} = \mathbf{0}$, then Equations (A–9) and (A–10) can be simplified as follows:

$$\begin{bmatrix} \mathbf{A} & \mathbf{B} \\ \mathbf{0} & \mathbf{D} \end{bmatrix}^{-1} = \begin{bmatrix} \mathbf{A}^{-1} & -\mathbf{A}^{-1}\mathbf{BD}^{-1} \\ \mathbf{0} & \mathbf{D}^{-1} \end{bmatrix} \tag{A–11}$$

or

$$\begin{bmatrix} \mathbf{A} & \mathbf{0} \\ \mathbf{C} & \mathbf{D} \end{bmatrix}^{-1} = \begin{bmatrix} \mathbf{A}^{-1} & \mathbf{0} \\ -\mathbf{D}^{-1}\mathbf{CA}^{-1} & \mathbf{D}^{-1} \end{bmatrix} \tag{A–12}$$

[For the derivation of Equations (A–8) through (A–12), refer to Problems A–5 and A–6.]

A–4 RULES OF MATRIX OPERATIONS

In this section we shall review some of the rules of algebraic operations with matrices and then give definitions of the derivative and the integral of matrices. Then the rules of differentiation of matrices will be presented.

Note that matrix algebra differs from ordinary number algebra in that matrix multiplication is not commutative and cancellation of matrices is not valid.

Multiplication of a matrix by a scalar.   The product of a matrix and a scalar is a matrix in which each element is multiplied by the scalar. That is,

$$
k\mathbf{A} = \begin{bmatrix} ka_{11} & ka_{12} & \cdots & ka_{1m} \\ ka_{21} & ka_{22} & \cdots & ka_{2m} \\ \cdot & \cdot & & \cdot \\ \cdot & \cdot & & \cdot \\ \cdot & \cdot & & \cdot \\ ka_{n1} & ka_{n2} & \cdots & ka_{nm} \end{bmatrix}
$$

Multiplication of a matrix by a matrix.   Multiplication of a matrix by a matrix is possible between matrices in which the number of columns in the first matrix is equal to the number of rows in the second. Otherwise, multiplication is not defined.

Consider the product of an $n \times m$ matrix $\mathbf{A}$ and an $m \times r$ matrix $\mathbf{B}$:

$$
\mathbf{AB} = \begin{bmatrix} a_{11} & a_{12} & \cdots & a_{1m} \\ a_{21} & a_{22} & \cdots & a_{2m} \\ \cdot & \cdot & & \cdot \\ \cdot & \cdot & & \cdot \\ \cdot & \cdot & & \cdot \\ a_{n1} & a_{n2} & \cdots & a_{nm} \end{bmatrix} \begin{bmatrix} b_{11} & b_{12} & \cdots & b_{1r} \\ b_{21} & b_{22} & \cdots & b_{2r} \\ \cdot & \cdot & & \cdot \\ \cdot & \cdot & & \cdot \\ \cdot & \cdot & & \cdot \\ b_{m1} & b_{m2} & \cdots & b_{mr} \end{bmatrix}
$$

$$
= \begin{bmatrix} c_{11} & c_{12} & \cdots & c_{1r} \\ c_{21} & c_{22} & \cdots & c_{2r} \\ \cdot & \cdot & & \cdot \\ \cdot & \cdot & & \cdot \\ \cdot & \cdot & & \cdot \\ c_{n1} & c_{n2} & \cdots & c_{nr} \end{bmatrix}
$$

where

$$
c_{ik} = \sum_{j=1}^{m} a_{ij} b_{jk}
$$

Thus multiplication of an $n \times m$ matrix by an $m \times r$ matrix yields an $n \times r$ matrix. It should be noted that, in general, matrix multiplication is not commutative; that is

$$
\mathbf{AB} \neq \mathbf{BA} \qquad \text{in general}
$$

For example,

$$\mathbf{AB} = \begin{bmatrix} a_{11} & a_{12} \\ a_{21} & a_{22} \end{bmatrix} \begin{bmatrix} b_{11} & b_{12} \\ b_{21} & b_{22} \end{bmatrix} = \begin{bmatrix} a_{11}b_{11} + a_{12}b_{21} & a_{11}b_{12} + a_{12}b_{22} \\ a_{21}b_{11} + a_{22}b_{21} & a_{21}b_{12} + a_{22}b_{22} \end{bmatrix}$$

and

$$\mathbf{BA} = \begin{bmatrix} b_{11} & b_{12} \\ b_{21} & b_{22} \end{bmatrix} \begin{bmatrix} a_{11} & a_{12} \\ a_{21} & a_{22} \end{bmatrix} = \begin{bmatrix} b_{11}a_{11} + b_{12}a_{21} & b_{11}a_{12} + b_{12}a_{22} \\ b_{21}a_{11} + b_{22}a_{21} & b_{21}a_{12} + b_{22}a_{22} \end{bmatrix}$$

Thus, in general, $\mathbf{AB} \neq \mathbf{BA}$. Hence, the order of multiplication is significant and must be preserved. If $\mathbf{AB} = \mathbf{BA}$, matrices $\mathbf{A}$ and $\mathbf{B}$ are said to commute. In the preceding matrices $\mathbf{A}$ and $\mathbf{B}$, if, for example, $a_{12} = a_{21} = b_{12} = b_{21} = 0$, then $\mathbf{A}$ and $\mathbf{B}$ commute.

For $n \times n$ diagonal matrices $\mathbf{A}$ and $\mathbf{B}$,

$$\mathbf{AB} = [a_{ij}\delta_{ij}][b_{ij}\delta_{ij}] = \begin{bmatrix} a_{11}b_{11} & & & & 0 \\ & a_{22}b_{22} & & & \\ & & \cdot & & \\ & & & \cdot & \\ 0 & & & & a_{nn}b_{nn} \end{bmatrix}$$

If $\mathbf{A}$, $\mathbf{B}$, and $\mathbf{C}$ are an $n \times m$ matrix, an $m \times r$ matrix, and an $r \times p$ matrix, respectively, then the following associativity law holds true:

$$(\mathbf{AB})\mathbf{C} = \mathbf{A}(\mathbf{BC})$$

This may be proved as follows:

$$(i, k)\text{th element of } \mathbf{AB} = \sum_{j=1}^{m} a_{ij}b_{jk}$$

$$(j, h)\text{th element of } \mathbf{BC} = \sum_{k=1}^{r} b_{jk}c_{kh}$$

$$(i, h)\text{th element of } (\mathbf{AB})\mathbf{C} = \sum_{k=1}^{r} \left(\sum_{j=1}^{m} a_{ij}b_{jk} \right) c_{kh} = \sum_{j=1}^{m} \sum_{k=1}^{r} (a_{ij}b_{jk})c_{kh}$$

$$= \sum_{j=1}^{m} \sum_{k=1}^{r} a_{ij}(b_{jk}c_{kh}) = \sum_{j=1}^{m} a_{ij} \left[\sum_{k=1}^{r} b_{jk}c_{kh} \right]$$

$$= (i, h)\text{th element of } \mathbf{A}(\mathbf{BC})$$

Since the associativity of multiplication of matrices holds true, we have

$$\mathbf{ABCD} = (\mathbf{AB})(\mathbf{CD}) = \mathbf{A}(\mathbf{BCD}) = (\mathbf{ABC})\mathbf{D}$$

$$\mathbf{A}^{m+n} = \mathbf{A}^m\mathbf{A}^n \qquad m, n = 1, 2, 3, \ldots$$

If $\mathbf{A}$ and $\mathbf{B}$ are $n \times m$ matrices and $\mathbf{C}$ and $\mathbf{D}$ are $m \times r$ matrices, then the following distributivity law holds true:

$$(\mathbf{A} + \mathbf{B})(\mathbf{C} + \mathbf{D}) = \mathbf{AC} + \mathbf{AD} + \mathbf{BC} + \mathbf{BD}$$

This can be proved by comparing the (i, j)th element of $(\mathbf{A} + \mathbf{B})(\mathbf{C} + \mathbf{D})$ and the (i, j)th element of $(\mathbf{AC} + \mathbf{AD} + \mathbf{BC} + \mathbf{BD})$.

Remarks on cancellation of matrices. Cancellation of matrices is not valid in matrix algebra. Consider the product of two singular matrices $\mathbf{A}$ and $\mathbf{B}$. Take, for example,

$$\mathbf{A} = \begin{bmatrix} 2 & 1 \\ 6 & 3 \end{bmatrix} \neq \mathbf{0}, \qquad \mathbf{B} = \begin{bmatrix} 1 & -2 \\ -2 & 4 \end{bmatrix} \neq \mathbf{0}$$

Then

$$\mathbf{AB} = \begin{bmatrix} 2 & 1 \\ 6 & 3 \end{bmatrix} \begin{bmatrix} 1 & -2 \\ -2 & 4 \end{bmatrix} = \begin{bmatrix} 0 & 0 \\ 0 & 0 \end{bmatrix} = \mathbf{0}$$

Clearly, $\mathbf{AB} = \mathbf{0}$ implies neither $\mathbf{A} = \mathbf{0}$ nor $\mathbf{B} = \mathbf{0}$. In fact, $\mathbf{AB} = \mathbf{0}$ implies one of the following three:

1. $\mathbf{A} = \mathbf{0}$.
2. $\mathbf{B} = \mathbf{0}$.
3. Both $\mathbf{A}$ and $\mathbf{B}$ are singular.

It can easily be proved that if both $\mathbf{A}$ and $\mathbf{B}$ are nonzero matrices and $\mathbf{AB} = \mathbf{0}$, then both $\mathbf{A}$ and $\mathbf{B}$ must be singular. Assume that $\mathbf{B}$ is nonzero and $\mathbf{A}$ is not singular. Then $|\mathbf{A}| \neq \mathbf{0}$ and $\mathbf{A}^{-1}$ exists. Then we obtain

$$\mathbf{A}^{-1}\mathbf{AB} = \mathbf{B} = \mathbf{0}$$

which contradicts the assumption that $\mathbf{B}$ is nonzero. In this way we can prove that both $\mathbf{A}$ and $\mathbf{B}$ must be singular if $\mathbf{A} \neq \mathbf{0}$ and $\mathbf{B} \neq \mathbf{0}$.

Similarly, notice that if $\mathbf{A}$ is singular, then neither $\mathbf{AB} = \mathbf{AC}$ nor $\mathbf{BA} = \mathbf{CA}$ implies $\mathbf{B} = \mathbf{C}$. If, however, $\mathbf{A}$ is a nonsingular matrix, then $\mathbf{AB} = \mathbf{AC}$ implies $\mathbf{B} = \mathbf{C}$ and $\mathbf{BA} = \mathbf{CA}$ also implies $\mathbf{B} = \mathbf{C}$.

Derivative and integral of a matrix. The derivative of an $n \times m$ matrix $\mathbf{A}(t)$ is defined by the matrix whose (i, j)th element is the derivative of the (i, j)th element of the original matrix, provided that all the elements $a_{ij}(t)$ have derivatives with respect to t:

$$\frac{d}{dt}\mathbf{A}(t) = \begin{bmatrix} \dfrac{d}{dt}a_{11}(t) & \cdots & \dfrac{d}{dt}a_{1m}(t) \\ \vdots & & \vdots \\ \dfrac{d}{dt}a_{n1}(t) & \cdots & \dfrac{d}{dt}a_{nm}(t) \end{bmatrix}$$

In the case of an n-dimensional vector $\mathbf{x}(t)$,

$$\frac{d}{dt}\mathbf{x}(t) = \begin{bmatrix} \dfrac{d}{dt}x_1(t) \\ \cdot \\ \cdot \\ \cdot \\ \dfrac{d}{dt}x_n(t) \end{bmatrix}$$

Similarly, the integral of an $n \times m$ matrix $\mathbf{A}(t)$ with respect to t is defined by the matrix whose (i, j)th element is the integral of the (i, j)th element of the original matrix, or

$$\int \mathbf{A}(t)\, dt = \begin{bmatrix} \displaystyle\int a_{11}(t)\, dt & \cdots & \displaystyle\int a_{1m}(t)\, dt \\ \cdot & & \cdot \\ \cdot & & \cdot \\ \cdot & & \cdot \\ \displaystyle\int a_{n1}(t)\, dt & \cdots & \displaystyle\int a_{nm}(t)\, dt \end{bmatrix}$$

provided that the $a_{ij}(t)$'s are integrable as functions of t.

Differentiation of a matrix. If the elements of matrices $\mathbf{A}$ and $\mathbf{B}$ are functions of t, then

$$\frac{d}{dt}(\mathbf{A} + \mathbf{B}) = \frac{d}{dt}\mathbf{A} + \frac{d}{dt}\mathbf{B} \tag{A–13}$$

$$\frac{d}{dt}(\mathbf{AB}) = \frac{d\mathbf{A}}{dt}\mathbf{B} + \mathbf{A}\frac{d\mathbf{B}}{dt} \tag{A–14}$$

If $k(t)$ is a scalar and is a function of t, then

$$\frac{d}{dt}[\mathbf{A}k(t)] = \frac{d\mathbf{A}}{dt}k(t) + \mathbf{A}\frac{dk(t)}{dt} \tag{A–15}$$

Also,

$$\int_a^b \frac{d\mathbf{A}}{dt}\mathbf{B}\, dt = \mathbf{AB}\,\Bigg|_a^b - \int_a^b \mathbf{A}\frac{d\mathbf{B}}{dt}\, dt \tag{A–16}$$

It is important to note that the derivative of $\mathbf{A}^{-1}$ is given by

$$\frac{d}{dt}\mathbf{A}^{-1} = -\mathbf{A}^{-1}\frac{d\mathbf{A}}{dt}\mathbf{A}^{-1} \tag{A–17}$$

Equation (A–17) can be derived easily by differentiating $\mathbf{AA}^{-1}$ with respect to t. Since

$$\frac{d}{dt}\mathbf{AA}^{-1} = \frac{d\mathbf{A}}{dt}\mathbf{A}^{-1} + \mathbf{A}\frac{d\mathbf{A}^{-1}}{dt}$$

and also

$$\frac{d}{dt} \mathbf{A} \mathbf{A}^{-1} = \frac{d}{dt} \mathbf{I} = \mathbf{0}$$

we obtain

$$\mathbf{A} \frac{d\mathbf{A}^{-1}}{dt} = - \frac{d\mathbf{A}}{dt} \mathbf{A}^{-1}$$

or

$$\mathbf{A}^{-1} \mathbf{A} \frac{d\mathbf{A}^{-1}}{dt} = \frac{d\mathbf{A}^{-1}}{dt} = - \mathbf{A}^{-1} \frac{d\mathbf{A}}{dt} \mathbf{A}^{-1}$$

which is the desired result.

Derivatives of a scalar function with respect to a vector.   If $J(\mathbf{x})$ is a scalar function of a vector $\mathbf{x}$, then

$$\frac{\partial J}{\partial \mathbf{x}} = \begin{bmatrix} \dfrac{\partial J}{\partial x_1} \\ \cdot \\ \cdot \\ \cdot \\ \dfrac{\partial J}{\partial x_n} \end{bmatrix}, \quad \frac{\partial^2 J}{\partial \mathbf{x}^2} = \begin{bmatrix} \dfrac{\partial^2 J}{\partial^2 x_1} & \dfrac{\partial^2 J}{\partial x_1 \, \partial x_2} & \cdots & \dfrac{\partial^2 J}{\partial x_1 \, \partial x_n} \\ \cdot & \cdot & & \cdot \\ \cdot & \cdot & & \cdot \\ \cdot & \cdot & & \cdot \\ \dfrac{\partial^2 J}{\partial x_n \, \partial x_1} & \dfrac{\partial^2 J}{\partial x_n \, \partial x_2} & \cdots & \dfrac{\partial^2 J}{\partial x_n^2} \end{bmatrix}$$

Also, for a scalar function $V(\mathbf{x}(t))$, we have

$$\frac{d}{dt} V(\mathbf{x}(t)) = \left(\frac{\partial V}{\partial \mathbf{x}} \right)^T \frac{d\mathbf{x}}{dt}$$

Jacobian.   If an $m \times 1$ matrix $\mathbf{f}(\mathbf{x})$ is a vector function of an n-vector $\mathbf{x}$ (*note:* an n-vector is meant as an n-dimensional vector), then

$$\frac{\partial \mathbf{f}}{\partial \mathbf{x}} = \begin{bmatrix} \dfrac{\partial f_1}{\partial x_1} & \dfrac{\partial f_2}{\partial x_1} & \cdots & \dfrac{\partial f_m}{\partial x_1} \\ \dfrac{\partial f_1}{\partial x_2} & \dfrac{\partial f_2}{\partial x_2} & \cdots & \dfrac{\partial f_m}{\partial x_2} \\ \cdot & \cdot & & \cdot \\ \cdot & \cdot & & \cdot \\ \cdot & \cdot & & \cdot \\ \dfrac{\partial f_1}{\partial f_n} & \dfrac{\partial f_2}{\partial x_n} & \cdots & \dfrac{\partial f_m}{\partial x_n} \end{bmatrix} \tag{A–18}$$

Such an $n \times m$ matrix is called a *Jacobian*.

Notice that by using this definition of the Jacobian, we have

$$\frac{\partial}{\partial \mathbf{x}} \mathbf{A} \mathbf{x} = \mathbf{A}^T \tag{A–19}$$

The fact that Equation (A–19) holds true can be easily seen from the following example. If $\mathbf{A}$ and $\mathbf{x}$ are given by

$$\mathbf{A} = \begin{bmatrix} a_{11} & a_{12} & a_{13} \\ a_{21} & a_{22} & a_{23} \end{bmatrix}, \qquad \mathbf{x} = \begin{bmatrix} x_1 \\ x_2 \\ x_3 \end{bmatrix}$$

then

$$\mathbf{A}\mathbf{x} = \begin{bmatrix} a_{11} & a_{12} & a_{13} \\ a_{21} & a_{22} & a_{23} \end{bmatrix} \begin{bmatrix} x_1 \\ x_2 \\ x_3 \end{bmatrix} = \begin{bmatrix} a_{11}x_1 + a_{12}x_2 + a_{13}x_3 \\ a_{21}x_1 + a_{22}x_2 + a_{23}x_3 \end{bmatrix} = \begin{bmatrix} f_1 \\ f_2 \end{bmatrix}$$

and

$$\frac{\partial}{\partial \mathbf{x}}\mathbf{A}\mathbf{x} = \begin{bmatrix} \dfrac{\partial f_1}{\partial x_1} & \dfrac{\partial f_2}{\partial x_1} \\[2mm] \dfrac{\partial f_1}{\partial x_2} & \dfrac{\partial f_2}{\partial x_2} \\[2mm] \dfrac{\partial f_1}{\partial x_3} & \dfrac{\partial f_2}{\partial x_3} \end{bmatrix} = \begin{bmatrix} a_{11} & a_{21} \\ a_{12} & a_{22} \\ a_{13} & a_{23} \end{bmatrix} = \mathbf{A}^T$$

Also, we have the following useful formula. For an $n \times n$ real matrix $\mathbf{A}$ and a real n-vector $\mathbf{x}$,

$$\frac{\partial}{\partial \mathbf{x}}\mathbf{x}^T\mathbf{A}\mathbf{x} = \mathbf{A}\mathbf{x} + \mathbf{A}^T\mathbf{x} \qquad\qquad \text{(A–20)}$$

In addition, if matrix $\mathbf{A}$ is a real symmetric matrix, then

$$\frac{\partial}{\partial \mathbf{x}}\mathbf{x}^T\mathbf{A}\mathbf{x} = 2\mathbf{A}\mathbf{x}$$

Note that if $\mathbf{A}$ is an $n \times n$ Hermitian matrix and $\mathbf{x}$ is a complex n-vector, then

$$\frac{\partial}{\partial \bar{\mathbf{x}}}\mathbf{x}^*\mathbf{A}\mathbf{x} = \mathbf{A}\mathbf{x} \qquad\qquad \text{(A–21)}$$

[For derivations of Equations (A–20) and (A–21), see Problem A–7.]

For an $n \times m$ real matrix $\mathbf{A}$, a real n-vector $\mathbf{x}$, and a real m-vector $\mathbf{y}$, we have

$$\frac{\partial}{\partial \mathbf{x}}\mathbf{x}^T\mathbf{A}\mathbf{y} = \mathbf{A}\mathbf{y} \qquad\qquad \text{(A–22)}$$

$$\frac{\partial}{\partial \mathbf{y}}\mathbf{x}^T\mathbf{A}\mathbf{y} = \mathbf{A}^T\mathbf{x} \qquad\qquad \text{(A–23)}$$

Similarly, for an $n \times m$ complex matrix $\mathbf{A}$, a complex n-vector $\mathbf{x}$, and a complex m-vector $\mathbf{y}$, we have

$$\frac{\partial}{\partial \bar{\mathbf{x}}}\mathbf{x}^*\mathbf{A}\mathbf{y} = \mathbf{A}\mathbf{y} \qquad\qquad \text{(A–24)}$$

$$\frac{\partial}{\partial \mathbf{y}} \mathbf{x}^*\mathbf{A}\mathbf{y} = \mathbf{A}^T\bar{\mathbf{x}} \qquad\qquad (A-25)$$

[For derivations of Equations (A–22) through (A–25), refer to Problem A–8.] Note that Equation (A–25) is equivalent to the following equation:

$$\overline{\frac{\partial}{\partial \mathbf{y}} \mathbf{x}^*\mathbf{A}\mathbf{y}} = \mathbf{A}^*\mathbf{x}$$

A–5 VECTORS AND VECTOR ANALYSIS

Linear dependence and independence of vectors. Vectors $\mathbf{x}_1, \mathbf{x}_2, \ldots, \mathbf{x}_n$ are said to be *linearly independent* if the equation

$$c_1\mathbf{x}_1 + c_2\mathbf{x}_2 + \cdots + c_n\mathbf{x}_n = \mathbf{0}$$

where $c_1, c_2, \ldots, c_n$ are constants, implies that $c_1 = c_2 = \cdots = c_n = 0$. Conversely, vectors $\mathbf{x}_1, \mathbf{x}_2, \ldots, \mathbf{x}_n$ are said to be *linearly dependent* if and only if $\mathbf{x}_i$ can be expressed as a linear combination of $\mathbf{x}_j$ ($j = 1, 2, \ldots, n; j \neq i$).

It is important to note that if vectors $\mathbf{x}_1, \mathbf{x}_2, \ldots, \mathbf{x}_n$ are linearly independent and vectors $\mathbf{x}_1, \mathbf{x}_2, \ldots, \mathbf{x}_n, \mathbf{x}_{n+1}$ are linearly dependent, then $\mathbf{x}_{n+1}$ can be expressed as a unique linear combination of $\mathbf{x}_1, \mathbf{x}_2, \ldots, \mathbf{x}_n$.

Necessary and sufficient conditions for linear independence of vectors. It can be proved that the necessary and sufficient conditions for n-vectors $\mathbf{x}_i$ ($i = 1, 2, \ldots, m$) to be linearly independent are that

1. $m \leq n$.

2. There exists at least one nonzero m-column determinant of the $n \times m$ matrix whose columns cosist of $\mathbf{x}_1, \mathbf{x}_2, \ldots, \mathbf{x}_m$.

Hence, for n vectors $\mathbf{x}_1, \mathbf{x}_2, \ldots, \mathbf{x}_n$ the necessary and sufficient condition for linear independence is

$$|\mathbf{A}| \neq 0$$

where $\mathbf{A}$ is the $n \times n$ matrix whose ith column is made up of the components of $\mathbf{x}_i$ ($i = 1, 2, \ldots, n$).

Inner product. Any rule that assigns to each pair of vectors $\mathbf{x}$ and $\mathbf{y}$ in a vector space a scalar quantity is called an *inner product* or *scalar product* and is given the symbol $\langle \mathbf{x}, \mathbf{y} \rangle$, provided that the following four axioms are satisfied:

1.
$$\langle \mathbf{y}, \mathbf{x} \rangle = \overline{\langle \mathbf{x}, \mathbf{y} \rangle}$$

where the bar denotes the conjugate of a complex number

2.
$$\langle c\mathbf{x}, \mathbf{y} \rangle = \bar{c}\langle \mathbf{x}, \mathbf{y} \rangle = \langle \mathbf{x}, \bar{c}\mathbf{y} \rangle$$

where c is a complex number

3.
$$\langle \mathbf{x} + \mathbf{y}, \mathbf{z} + \mathbf{w} \rangle = \langle \mathbf{x}, \mathbf{z} \rangle + \langle \mathbf{y}, \mathbf{z} \rangle + \langle \mathbf{x}, \mathbf{w} \rangle + \langle \mathbf{y}, \mathbf{w} \rangle$$

4.
$$\langle \mathbf{x}, \mathbf{x} \rangle > 0 \qquad \text{for } \mathbf{x} \neq \mathbf{0}$$

In any finite dimensional vector space, there are many different definitions of the inner product, all having the properties required by the definition.

In this book, unless the contrary is stated, we shall adopt the following definition of the inner product: The inner product of a pair of n-vectors $\mathbf{x}$ and $\mathbf{y}$ in a vector space V is given by

$$\langle \mathbf{x}, \mathbf{y} \rangle = \bar{x}_1 y_1 + \bar{x}_2 y_2 + \cdots + \bar{x}_n y_n = \sum_{i=1}^{n} \bar{x}_i y_i \tag{A-26}$$

where the summation is a complex number and where the $\bar{x}_i$'s are the complex conjugates of the x_i's. This definition clearly satisfies the four axioms. The inner product can then be expressed as follows:

$$\langle \mathbf{x}, \mathbf{y} \rangle = \mathbf{x}^*\mathbf{y}$$

where $\mathbf{x}^*$ denotes the conjugate transpose of $\mathbf{x}$. Also,

$$\langle \mathbf{x}, \mathbf{y} \rangle = \overline{\langle \mathbf{y}, \mathbf{x} \rangle} = \overline{\mathbf{y}^*\mathbf{x}} = \mathbf{y}^T\bar{\mathbf{x}} = \mathbf{x}^*\mathbf{y} \tag{A-27}$$

The inner product of two n-vectors $\mathbf{x}$ and $\mathbf{y}$ with real components is therefore given by

$$\langle \mathbf{x}, \mathbf{y} \rangle = x_1 y_1 + x_2 y_2 + \cdots + x_n y_n = \sum_{i=1}^{n} x_i y_i \tag{A-28}$$

In this case, clearly we have

$$\langle \mathbf{x}, \mathbf{y} \rangle = \mathbf{x}^T\mathbf{y} = \mathbf{y}^T\mathbf{x} \qquad \text{for real vectors } \mathbf{x} \text{ and } \mathbf{y}$$

It is noted that the real or complex vector $\mathbf{x}$ is said to be *normalized* if $\langle \mathbf{x}, \mathbf{x} \rangle = 1$.

It is also noted that for an n-vector $\mathbf{x}$, $\mathbf{x}^*\mathbf{x}$ is a nonnegative scalar but $\mathbf{x}\mathbf{x}^*$ is an $n \times n$ matrix. That is,

$$\mathbf{x}^*\mathbf{x} = \langle \mathbf{x}, \mathbf{x} \rangle = \bar{x}_1 x_1 + \bar{x}_2 x_2 + \cdots + \bar{x}_n x_n$$
$$= |x_1|^2 + |x_2|^2 + \cdots + |x_n|^2$$

and

$$\mathbf{x}\mathbf{x}^* = \begin{bmatrix} x_1\bar{x}_1 & x_1\bar{x}_2 & \cdots & x_1\bar{x}_n \\ x_2\bar{x}_1 & x_2\bar{x}_2 & \cdots & x_2\bar{x}_n \\ \cdot & \cdot & & \cdot \\ \cdot & \cdot & & \cdot \\ \cdot & \cdot & & \cdot \\ x_n\bar{x}_1 & x_n\bar{x}_2 & \cdots & x_n\bar{x}_n \end{bmatrix}$$

Notice that for an $n \times n$ complex matrix $\mathbf{A}$ and complex n-vectors $\mathbf{x}$ and $\mathbf{y}$, the inner product of $\mathbf{x}$ and $\mathbf{Ay}$ and that of $\mathbf{A}^*\mathbf{x}$ and $\mathbf{y}$ are the same, or

$$\langle \mathbf{x}, \mathbf{Ay} \rangle = \mathbf{x}^*\mathbf{Ay}, \qquad \langle \mathbf{A}^*\mathbf{x}, \mathbf{y} \rangle = \mathbf{x}^*\mathbf{Ay}$$

Similarly, for an $n \times n$ real matrix $\mathbf{A}$ and real n-vectors $\mathbf{x}$ and $\mathbf{y}$, the inner product of $\mathbf{x}$ and $\mathbf{Ay}$ and that of $\mathbf{A}^T\mathbf{x}$ and $\mathbf{y}$ are the same, or

$$\langle \mathbf{x}, \mathbf{Ay} \rangle = \mathbf{x}^T\mathbf{Ay}, \qquad \langle \mathbf{A}^T\mathbf{x}, \mathbf{y} \rangle = \mathbf{x}^T\mathbf{Ay}$$

Unitary transformation. If $\mathbf{A}$ is a unitary matrix (that is, if $\mathbf{A}^{-1} = \mathbf{A}^*$), then the inner product $\langle \mathbf{x}, \mathbf{x} \rangle$ is invariant under the linear transformation $\mathbf{x} = \mathbf{A}\mathbf{y}$, because

$$\langle \mathbf{x}, \mathbf{x} \rangle = \langle \mathbf{A}\mathbf{y}, \mathbf{A}\mathbf{y} \rangle = \langle \mathbf{y}, \mathbf{A}^*\mathbf{A}\mathbf{y} \rangle = \langle \mathbf{y}, \mathbf{A}^{-1}\mathbf{A}\mathbf{y} \rangle = \langle \mathbf{y}, \mathbf{y} \rangle$$

Such a transformation $\mathbf{x} = \mathbf{A}\mathbf{y}$, where $\mathbf{A}$ is a unitary matrix, which transforms $\sum_{i=1}^{n} \bar{x}_i x_i$ into $\sum_{i=1}^{n} \bar{y}_i y_i$, is called a *unitary transformation*.

Orthogonal transformation. If $\mathbf{A}$ is an orthogonal matrix (that is, if $\mathbf{A}^{-1} = \mathbf{A}^T$), then the inner product $\langle \mathbf{x}, \mathbf{x} \rangle$ is invariant under the linear transformation $\mathbf{x} = \mathbf{A}\mathbf{y}$, because

$$\langle \mathbf{x}, \mathbf{x} \rangle = \langle \mathbf{A}\mathbf{y}, \mathbf{A}\mathbf{y} \rangle = \langle \mathbf{y}, \mathbf{A}^T\mathbf{A}\mathbf{y} \rangle = \langle \mathbf{y}, \mathbf{A}^{-1}\mathbf{A}\mathbf{y} \rangle = \langle \mathbf{y}, \mathbf{y} \rangle$$

Such a transformation $\mathbf{x} = \mathbf{A}\mathbf{y}$, which transforms $\sum_{i=1}^{n} x_i^2$ into $\sum_{i=1}^{n} y_i^2$, is called an *orthogonal transformation*.

Norms of a vector. Once we define the inner product, we can use this inner product to define norms of a vector $\mathbf{x}$. The concept of a norm is somewhat similar to that of the absolute value. A norm is a function that assigns to every vector $\mathbf{x}$ in a given vector space a real number denoted by $\|\mathbf{x}\|$ such that

1. $$\|\mathbf{x}\| > 0 \qquad \text{for } \mathbf{x} \neq \mathbf{0}$$
2. $$\|\mathbf{x}\| = 0 \qquad \text{if and only if } \mathbf{x} = \mathbf{0}$$
3. $$\|k\mathbf{x}\| = |k|\,\|\mathbf{x}\|$$
   where k is a scalar and $|k|$ is the absolute value of k
4. $$\|\mathbf{x} + \mathbf{y}\| \leq \|\mathbf{x}\| + \|\mathbf{y}\| \qquad \text{for all } \mathbf{x} \text{ and } \mathbf{y}$$
5. $$|\langle \mathbf{x}, \mathbf{y} \rangle| \leq \|\mathbf{x}\|\,\|\mathbf{y}\| \qquad \text{(Schwarz inequality)}$$

Several different definitions of norms are commonly used in the literature. However, the following definition is widely used. A norm of a vector is defined as the nonnegative square root of $\langle \mathbf{x}, \mathbf{x} \rangle$:

$$\|\mathbf{x}\| = \langle \mathbf{x}, \mathbf{x} \rangle^{1/2} = (\mathbf{x}^*\mathbf{x})^{1/2} = \sqrt{|x_1|^2 + |x_2|^2 + \cdots + |x_n|^2} \qquad \text{(A–29)}$$

If $\mathbf{x}$ is a real vector, the quantity $\|\mathbf{x}\|^2$ can be interpreted geometrically as the square of the distance from the origin to the point represented by the vector $\mathbf{x}$. Note that

$$\|\mathbf{x} - \mathbf{y}\| = \langle \mathbf{x} - \mathbf{y}, \mathbf{x} - \mathbf{y} \rangle^{1/2} = \sqrt{(x_1 - y_1)^2 + (x_2 - y_2)^2 + \cdots + (x_n - y_n)^2}$$

The five properties of norms listed earlier may be obvious, except perhaps the last two inequalities. These two inequalities may be proved as follows. From the definitions of the inner product and the norm, we have

$$\|\lambda\mathbf{x} + \mathbf{y}\|^2 = \langle \lambda\mathbf{x} + \mathbf{y}, \lambda\mathbf{x} + \mathbf{y} \rangle = \langle \lambda\mathbf{x}, \lambda\mathbf{x} \rangle + \langle \mathbf{y}, \lambda\mathbf{x} \rangle + \langle \lambda\mathbf{x}, \mathbf{y} \rangle + \langle \mathbf{y}, \mathbf{y} \rangle$$
$$= \bar{\lambda}\lambda\|\mathbf{x}\|^2 + \lambda\langle \mathbf{y}, \mathbf{x} \rangle + \bar{\lambda}\langle \mathbf{x}, \mathbf{y} \rangle + \|\mathbf{y}\|^2$$
$$= \bar{\lambda}(\lambda\|\mathbf{x}\|^2 + \langle \mathbf{x}, \mathbf{y} \rangle) + \lambda\overline{\langle \mathbf{x}, \mathbf{y} \rangle} + \|\mathbf{y}\|^2 \geq 0$$

If we choose

$$\lambda = -\frac{\langle \mathbf{x}, \mathbf{y} \rangle}{\|\mathbf{x}\|^2} \qquad \text{for } \mathbf{x} \neq \mathbf{0}$$

then

$$\lambda \overline{\langle \mathbf{x}, \mathbf{y} \rangle} + \|\mathbf{y}\|^2 = -\frac{\langle \mathbf{x}, \mathbf{y} \rangle \overline{\langle \mathbf{x}, \mathbf{y} \rangle}}{\|\mathbf{x}\|^2} + \|\mathbf{y}\|^2 \geq 0$$

and

$$\|\mathbf{x}\|^2 \|\mathbf{y}\|^2 \geq \langle \mathbf{x}, \mathbf{y} \rangle \overline{\langle \mathbf{x}, \mathbf{y} \rangle} = |\langle \mathbf{x}, \mathbf{y} \rangle|^2 \qquad \text{for } \mathbf{x} \neq \mathbf{0}$$

For $\mathbf{x} = \mathbf{0}$, clearly,

$$\|\mathbf{x}\|^2 \|\mathbf{y}\|^2 = |\langle \mathbf{x}, \mathbf{y} \rangle|^2$$

Therefore, we obtain the Schwarz inequality,

$$|\langle \mathbf{x}, \mathbf{y} \rangle| \leq \|\mathbf{x}\| \, \|\mathbf{y}\| \qquad\qquad\qquad (A\text{--}30)$$

By use of the Schwarz inequality we obtain the following inequality:

$$\|\mathbf{x} + \mathbf{y}\| \leq \|\mathbf{x}\| + \|\mathbf{y}\| \qquad\qquad\qquad (A\text{--}31)$$

This can be proved easily, since

$$
\begin{aligned}
\|\mathbf{x} + \mathbf{y}\|^2 &= \langle \mathbf{x} + \mathbf{y}, \mathbf{x} + \mathbf{y} \rangle \\
&= \langle \mathbf{x}, \mathbf{x} \rangle + \langle \mathbf{x}, \mathbf{y} \rangle + \langle \mathbf{y}, \mathbf{x} \rangle + \langle \mathbf{y}, \mathbf{y} \rangle \\
&= \|\mathbf{x}\|^2 + \langle \mathbf{x}, \mathbf{y} \rangle + \overline{\langle \mathbf{x}, \mathbf{y} \rangle} + \|\mathbf{y}\|^2 \\
&= \|\mathbf{x}\|^2 + \|\mathbf{y}\|^2 + 2 \operatorname{Re} \langle \mathbf{x}, \mathbf{y} \rangle \\
&\leq \|\mathbf{x}\|^2 + \|\mathbf{y}\|^2 + 2 |\langle \mathbf{x}, \mathbf{y} \rangle| \\
&\leq \|\mathbf{x}\|^2 + \|\mathbf{y}\|^2 + 2\|\mathbf{x}\| \, \|\mathbf{y}\| \\
&= (\|\mathbf{x}\| + \|\mathbf{y}\|)^2
\end{aligned}
$$

Equations (A–26) through (A–31) are useful in modern control theory.

As stated earlier, there are different definitions of norms used in the literature. Three such definitions of norms follow.

1. A norm $\|\mathbf{x}\|$ may be defined as follows:

$$\|\mathbf{x}\| = [(\mathbf{Tx})^*(\mathbf{Tx})]^{1/2} = (\mathbf{x^*T^*Tx})^{1/2} = (\mathbf{x^*Qx})^{1/2}$$

$$= \left[\sum_{i=1}^{n} \sum_{j=1}^{n} q_{ij} \bar{x}_i x_j \right]^{1/2} \geq 0$$

The matrix $\mathbf{Q} = \mathbf{T^*T}$ is Hermitian, since $\mathbf{Q^*} = \mathbf{T^*T} = \mathbf{Q}$. The norm $\|\mathbf{x}\| = (\mathbf{x^*Qx})^{1/2}$ is a generalized form of $(\mathbf{x^*x})^{1/2}$, which can be written as $(\mathbf{x^*Ix})^{1/2}$.

2. A norm may be defined as the sum of the magnitudes of all the components x_i:

$$\|\mathbf{x}\| = \sum_{i=1}^{n} |x_i|$$

3. A norm may be defined as the maximum of the magnitudes of all the components x_i:

$$\|\mathbf{x}\| = \max_i \{|x_i|\}$$

It can be shown that the various norms just defined are equivalent. Among these definitions of norms, norm $(\mathbf{x^*x})^{1/2}$ is most commonly used in explicit calculations.

Norms of a matrix. The concept of norms of a vector can be extended to matrices. There are several different definitions of norms of a matrix. Some of them follow.

1. A norm $\|\mathbf{A}\|$ of an $n \times n$ matrix $\mathbf{A}$ may be defined by

$$\|\mathbf{A}\| = \min k$$

such that

$$\|\mathbf{A}\mathbf{x}\| \leq k\|\mathbf{x}\|$$

For the norm $(\mathbf{x}^*\mathbf{x})^{1/2}$, this definition is equivalent to

$$\|\mathbf{A}\|^2 = \max_{\mathbf{x}} \{\mathbf{x}^*\mathbf{A}^*\mathbf{A}\mathbf{x}; \mathbf{x}^*\mathbf{x} = 1\}$$

which means that $\|\mathbf{A}\|^2$ is the maximum of the ''absolute value'' of the vector $\mathbf{A}\mathbf{x}$ when $\mathbf{x}^*\mathbf{x} = 1$.

2. A norm of an $n \times n$ matrix $\mathbf{A}$ may be defined by

$$\|\mathbf{A}\| = \sum_{i=1}^{n} \sum_{j=1}^{n} |a_{ij}|$$

where $|a_{ij}|$ is the absolute value of a_{ij}.

3. A norm may be defined by

$$\|\mathbf{A}\| = \left(\sum_{i=1}^{n} \sum_{j=1}^{n} |a_{ij}|^2 \right)^{1/2}$$

4. Another definition of a norm is given by

$$\|\mathbf{A}\| = \max_{i} \left(\sum_{j=1}^{n} |a_{ij}| \right)$$

Note that all definitions of norms of an $n \times n$ matrix $\mathbf{A}$ have the following properties:

1. $$\|\mathbf{A}\| = \|\mathbf{A}^*\| \quad \text{or} \quad \|\mathbf{A}\| = \|\mathbf{A}^T\|$$
2. $$\|\mathbf{A} + \mathbf{B}\| \leq \|\mathbf{A}\| + \|\mathbf{B}\|$$
3. $$\|\mathbf{A}\mathbf{B}\| \leq \|\mathbf{A}\| \, \|\mathbf{B}\|$$
4. $$\|\mathbf{A}\mathbf{x}\| \leq \|\mathbf{A}\| \, \|\mathbf{x}\|$$

Orthogonality of vectors. If the inner product of two vectors $\mathbf{x}$ and $\mathbf{y}$ is zero, or $\langle \mathbf{x}, \mathbf{y} \rangle = 0$, then vectors $\mathbf{x}$ and $\mathbf{y}$ are said to be *orthogonal to each other*. For example, vectors

$$\mathbf{x}_1 = \begin{bmatrix} 1 \\ 1 \\ 0 \end{bmatrix}, \quad \mathbf{x}_2 = \begin{bmatrix} 0 \\ 0 \\ 1 \end{bmatrix}, \quad \mathbf{x}_3 = \begin{bmatrix} 1 \\ -1 \\ 0 \end{bmatrix}$$

are orthogonal in pairs and thus form an orthogonal set.

In an n-dimensional vector space, vectors $\mathbf{x}_1, \mathbf{x}_2, \ldots, \mathbf{x}_n$ defined by

$$\mathbf{x}_1 = \begin{bmatrix} 1 \\ 0 \\ \cdot \\ \cdot \\ \cdot \\ 0 \end{bmatrix}, \qquad \mathbf{x}_2 = \begin{bmatrix} 0 \\ 1 \\ \cdot \\ \cdot \\ \cdot \\ 0 \end{bmatrix}, \qquad \ldots, \qquad \mathbf{x}_n = \begin{bmatrix} 0 \\ 0 \\ \cdot \\ \cdot \\ \cdot \\ 1 \end{bmatrix}$$

satisfy the conditions $\langle \mathbf{x}_i, \mathbf{x}_j \rangle = \delta_{ij}$, or

$$\langle \mathbf{x}_i, \mathbf{x}_i \rangle = 1$$

$$\langle \mathbf{x}_i, \mathbf{x}_j \rangle = 0 \qquad i \neq j$$

where $i, j = 1, 2, \ldots, n$. Such a set of vectors is said to be *orthonormal*, since the vectors are orthogonal to each other and each vector is normalized.

A nonzero vector $\mathbf{x}$ can be normalized by dividing $\mathbf{x}$ by $\|\mathbf{x}\|$. The normalized vector $\mathbf{x}/\|\mathbf{x}\|$ is a unit vector. Unit vectors $\mathbf{x}_1, \mathbf{x}_2, \ldots, \mathbf{x}_n$ form an orthonormal set if they are orthogonal in pairs.

Consider a unitary matrix $\mathbf{A}$. By partitioning $\mathbf{A}$ into column vectors $\mathbf{A}_1, \mathbf{A}_2, \ldots, \mathbf{A}_n$, we have

$$\mathbf{A}^*\mathbf{A} = \begin{bmatrix} \mathbf{A}_1^* \\ \hline \mathbf{A}_2^* \\ \hline \cdot \\ \cdot \\ \cdot \\ \hline \mathbf{A}_n^* \end{bmatrix} [\mathbf{A}_1 \mid \mathbf{A}_2 \mid \cdots \mid \mathbf{A}_n]$$

$$= \begin{bmatrix} \mathbf{A}_1^*\mathbf{A}_1 & \mathbf{A}_1^*\mathbf{A}_2 & \cdots & \mathbf{A}_1^*\mathbf{A}_n \\ \mathbf{A}_2^*\mathbf{A}_1 & \mathbf{A}_2^*\mathbf{A}_2 & \cdots & \mathbf{A}_2^*\mathbf{A}_n \\ \cdot & \cdot & & \cdot \\ \cdot & \cdot & & \cdot \\ \cdot & \cdot & & \cdot \\ \mathbf{A}_n^*\mathbf{A}_1 & \mathbf{A}_n^*\mathbf{A}_2 & \cdots & \mathbf{A}_n^*\mathbf{A}_n \end{bmatrix}$$

$$= \begin{bmatrix} 1 & 0 & \cdots & 0 \\ 0 & 1 & \cdots & 0 \\ \cdot & \cdot & & \cdot \\ \cdot & \cdot & & \cdot \\ \cdot & \cdot & & \cdot \\ 0 & 0 & \cdots & 1 \end{bmatrix}$$

it follows that

$$\mathbf{A}_i^*\mathbf{A}_i = \langle \mathbf{A}_i, \mathbf{A}_i \rangle = 1$$

$$\mathbf{A}_i^*\mathbf{A}_j = \langle \mathbf{A}_i, \mathbf{A}_j \rangle = 0 \qquad i \neq j$$

Thus we see that the column vectors (or row vectors) of a unitary matrix $\mathbf{A}$ are orthonormal. The same is true for orthogonal matrices, since they are unitary.

A–6 EIGENVALUES, EIGENVECTORS, AND SIMILARITY TRANSFORMATION

In this section we shall first review important properties of the rank of a matrix and then give definitions of eigenvalues and eigenvectors. Finally, we shall discuss Jordan canonical forms, similarity transformation, and the trace of an $n \times n$ matrix.

Rank of a matrix. A matrix $\mathbf{A}$ is called of rank m if the maximum number of linearly independent rows (or columns) is m. Hence, if there exists an $m \times m$ submatrix $\mathbf{M}$ of $\mathbf{A}$ such that $|\mathbf{M}| \neq 0$ and the determinant of every $r \times r$ submatrix (where $r \geq m + 1$) of $\mathbf{A}$ is zero, then the rank of $\mathbf{A}$ is m. [Note that, if the determinant of every $(m + 1) \times (m + 1)$ submatrix of $\mathbf{A}$ is zero, then any determinant of order s (where $s > m + 1$) is zero, since any determinant of order $s > m + 1$ can be expressed as a linear sum of determinants of order $m + 1$.]

Properties of rank of a matrix. We shall list important properties of the rank of a matrix in the following.

1. The rank of a matrix is invariant under the interchange of two rows (or columns), or the addition of a scalar multiple of a row (or column) to another row (or column), or the multiplication of any row (or column) by a nonzero scalar.
2. For an $n \times m$ matrix $\mathbf{A}$

$$\text{rank } \mathbf{A} \leq \min{(n, m)}$$

3. For an $n \times n$ matrix $\mathbf{A}$, a necessary and sufficient condition for rank $\mathbf{A} = n$ is that $|\mathbf{A}| \neq 0$.
4. For an $n \times m$ matrix $\mathbf{A}$,

$$\text{rank } \mathbf{A}^* = \text{rank } \mathbf{A} \qquad \text{or} \qquad \text{rank } \mathbf{A}^T = \text{rank } \mathbf{A}$$

5. The rank of a product of two matrices $\mathbf{AB}$ cannot exceed the rank of $\mathbf{A}$ or the rank of $\mathbf{B}$; that is,

$$\text{rank } \mathbf{AB} \leq \min{(\text{rank } \mathbf{A}, \text{rank } \mathbf{B})}$$

Hence, if $\mathbf{A}$ is an $n \times 1$ matrix and $\mathbf{B}$ is a $1 \times m$ matrix, then rank $\mathbf{AB} = 1$ unless $\mathbf{AB} = \mathbf{0}$. If a matrix has rank 1, then this matrix can be expressed as a product of a column vector and a row vector.

6. For an $n \times n$ matrix $\mathbf{A}$ (where $|\mathbf{A}| \neq 0$) and an $n \times m$ matrix $\mathbf{B}$,

$$\text{rank } \mathbf{AB} = \text{rank } \mathbf{B}$$

Similarly, for an $m \times m$ matrix $\mathbf{A}$ (where $|\mathbf{A}| \neq 0$) and an $n \times m$ matrix $\mathbf{B}$,

$$\text{rank } \mathbf{BA} = \text{rank } \mathbf{B}$$

Eigenvalues of a square matrix. For an $n \times n$ matrix $\mathbf{A}$, the determinant

$$|\lambda\mathbf{I} - \mathbf{A}|$$

is called the *characteristic polynomial* of **A**. It is an *n*th-degree polynomial in λ. The characteristic equation is given by

$$|\lambda\mathbf{I} - \mathbf{A}| = 0$$

If the determinant $|\lambda\mathbf{I} - \mathbf{A}|$ is expanded, the characteristic equation becomes

$$|\lambda\mathbf{I} - \mathbf{A}| = \begin{vmatrix} \lambda - a_{11} & -a_{12} & \cdots & -a_{1n} \\ -a_{21} & \lambda - a_{22} & \cdots & -a_{2n} \\ \cdot & \cdot & & \cdot \\ \cdot & \cdot & & \cdot \\ \cdot & \cdot & & \cdot \\ -a_{n1} & -a_{n2} & \cdots & \lambda - a_{nn} \end{vmatrix}$$

$$= \lambda^n + a_1\lambda^{n-1} + \cdots + a_{n-1}\lambda + a_n = 0$$

The *n* roots of the characteristic equation are called the *eigenvalues* of **A**. They are also called the *characteristic roots*.

It is noted that an $n \times n$ real matrix **A** does not necessarily possess real eigenvalues. However, for an $n \times n$ real matrix **A**, the characteristic equation $|\lambda\mathbf{I} - \mathbf{A}| = 0$ is a polynomial with real coefficients, and therefore any complex eigenvalues must occur in conjugate pairs; that is, if $\alpha + j\beta$ is an eigenvalue of **A**, then $\alpha - j\beta$ is also an eigenvalue of **A**.

There is an important relationship between the eigenvalues of an $n \times n$ matrix **A** and those of $\mathbf{A}^{-1}$. If we assume the eigenvalues of **A** to be λ_i and those of $\mathbf{A}^{-1}$ to be μ_i, then

$$\mu_i = \lambda_i^{-1} \qquad i = 1, 2, \ldots, n$$

That is, if λ_i is an eigenvalue of **A**, then λ_i^{-1} is an eigenvalue of $\mathbf{A}^{-1}$. To prove this, notice that the characteristic equation for matrix **A** can be written as

$$|\lambda\mathbf{I} - \mathbf{A}| = |\lambda\mathbf{A}^{-1} - \mathbf{I}|\,|\mathbf{A}| = |\lambda|\,|\mathbf{A}^{-1} - \lambda^{-1}\mathbf{I}|\,|\mathbf{A}| = 0$$

or

$$|\lambda^{-1}\mathbf{I} - \mathbf{A}^{-1}| = 0$$

By assumption, the characteristic equation for the inverse matrix $\mathbf{A}^{-1}$ is

$$|\mu\mathbf{I} - \mathbf{A}^{-1}| = 0$$

By comparing the last two equations, we see that

$$\mu = \lambda^{-1}$$

Hence, if λ is an eigenvalue of **A**, then $\mu = \lambda^{-1}$ is an eigenvalue of $\mathbf{A}^{-1}$.

Finally, note that it is possible to prove that, for two square matrices **A** and **B**,

$$|\lambda\mathbf{I} - \mathbf{AB}| = |\lambda\mathbf{I} - \mathbf{BA}|$$

(For the proof, see Problem A–9.)

Eigenvectors of an $n \times n$ matrix. Any nonzero vector $\mathbf{x}_i$ such that

$$\mathbf{A}\mathbf{x}_i = \lambda_i\mathbf{x}_i$$

is said to be an *eigenvector* associated with an eigenvalue λ_i of **A**, where **A** is an $n \times n$ matrix. Since the components of $\mathbf{x}_i$ are determined from *n* linear homogeneous algebraic equations within a constant factor, if $\mathbf{x}_i$ is an eigenvector, then for any scalar $\alpha \neq 0$, $\alpha\mathbf{x}_i$ is also an eigenvector. The eigenvector is said to be a *normalized* eigenvector if its length or absolute value is unity.

Similar matrices.   The $n \times n$ matrices $\mathbf{A}$ and $\mathbf{B}$ are said to be *similar* if a nonsingular matrix $\mathbf{P}$ exists such that

$$\mathbf{P}^{-1}\mathbf{A}\mathbf{P} = \mathbf{B}$$

The matrix $\mathbf{B}$ is said to be obtained from $\mathbf{A}$ by a *similarity transformation*, in which $\mathbf{P}$ is the transformation matrix. Notice that $\mathbf{A}$ can be obtained from $\mathbf{B}$ by a similarity transformation with a transformation matrix $\mathbf{P}^{-1}$, since

$$\mathbf{A} = \mathbf{P}\mathbf{B}\mathbf{P}^{-1} = (\mathbf{P}^{-1})^{-1}\mathbf{B}(\mathbf{P}^{-1})$$

Diagonalization of matrices.   If an $n \times n$ matrix $\mathbf{A}$ has n distinct eigenvalues, then there are n linearly independent eigenvectors. If matrix $\mathbf{A}$ has a multiple eigenvalue of multiplicity k, then there are at least one and not more than k linearly independent eigenvectors associated with this eigenvalue.

If an $n \times n$ matrix has n linearly independent eigenvectors, it can be diagonalized by a similarity transformation. However, a matrix that does not have a complete set of n linearly independent eigenvectors cannot be diagonalized. Such a matrix can be transformed into a Jordan canonical form.

Jordan canonical form.   A $k \times k$ matrix $\mathbf{J}$ is said to be in the Jordan canonical form if

$$\mathbf{J} = \begin{bmatrix} \mathbf{J}_{p_1} & & & & \mathbf{0} \\ & \mathbf{J}_{p_2} & & & \\ & & \cdot & & \\ & & & \cdot & \\ & & & & \cdot \\ \mathbf{0} & & & & \mathbf{J}_{p_s} \end{bmatrix}$$

where the $\mathbf{J}_{p_i}$'s are $p_i \times p_i$ matrices of the form

$$\mathbf{J}_{p_i} = \begin{bmatrix} \lambda & 1 & 0 & \cdots & 0 & 0 \\ 0 & \lambda & 1 & \cdots & 0 & 0 \\ \cdot & \cdot & \cdot & & \cdot & \cdot \\ \cdot & \cdot & \cdot & & \cdot & \cdot \\ \cdot & \cdot & \cdot & & \cdot & \cdot \\ 0 & 0 & 0 & \cdots & \lambda & 1 \\ 0 & 0 & 0 & \cdots & 0 & \lambda \end{bmatrix}$$

The matrices $\mathbf{J}_{p_i}$ are called p_ith order Jordan blocks. Note that the λ in $\mathbf{J}_{p_i}$ and that in $\mathbf{J}_{p_j}$ may or may not be the same, and that

$$p_1 + p_2 + \cdots + p_s = k$$

For example, in a 7×7 matrix $\mathbf{J}$, if $p_1 = 3$, $p_2 = 2$, $p_3 = 1$, $p_4 = 1$, and the eigenvalues of $\mathbf{J}$ are $\lambda_1, \lambda_1, \lambda_1, \lambda_1, \lambda_1, \lambda_6, \lambda_7$, then the Jordan canonical form may be given by

$$
\mathbf{J} = \begin{bmatrix} \mathbf{J}_3(\lambda_1) & & & & \mathbf{0} \\ & \mathbf{J}_2(\lambda_1) & & & \\ & & \mathbf{J}_1(\lambda_6) & \\ \mathbf{0} & & & \mathbf{J}_1(\lambda_7) \end{bmatrix} = \begin{bmatrix} \lambda_1 & 1 & 0 & & & & \mathbf{0} \\ 0 & \lambda_1 & 1 & & & & \\ 0 & 0 & \lambda_1 & & & & \\ \hline & & & \lambda_1 & 1 & & \\ & & & 0 & \lambda_1 & & \\ \hline & & & & & \lambda_6 & \\ \hline \mathbf{0} & & & & & & \lambda_7 \end{bmatrix}
$$

Notice that a diagonal matrix is a special case of the Jordan canonical form.

Jordan canonical forms have the properties that the elements on the main diagonal of the matrix are the eigenvalues of $\mathbf{A}$ and that the elements immediately above (or below) the main diagonal are either 1 or 0 and all other elements are zeros.

The determination of the exact form of the Jordan block may not be simple. To illustrate some possible structures, consider a 3×3 matrix having a triple eigenvalue of λ_1. Then any one of the following Jordan canonical forms is possible:

$$
\begin{bmatrix} \lambda_1 & 1 & 0 \\ 0 & \lambda_1 & 1 \\ 0 & 0 & \lambda_1 \end{bmatrix}, \quad \begin{bmatrix} \lambda_1 & 1 & 0 \\ 0 & \lambda_1 & 0 \\ \hline 0 & 0 & \lambda_1 \end{bmatrix}, \quad \begin{bmatrix} \lambda_1 & 0 & 0 \\ \hline 0 & \lambda_1 & 0 \\ \hline 0 & 0 & \lambda_1 \end{bmatrix}
$$

Each of the three preceding matrices has the same characteristic equation $(\lambda - \lambda_1)^3 = 0$. The first one corresponds to the case where there exists only one linearly independent eigenvector, since by denoting the first matrix by $\mathbf{A}$ and solving the following equation for $\mathbf{x}$,

$$
(\mathbf{A} - \lambda_1 \mathbf{I})\mathbf{x} = \mathbf{0}
$$

we obtain only one eigenvector:

$$
\mathbf{x} = \begin{bmatrix} a \\ 0 \\ 0 \end{bmatrix} \qquad a = \text{nonzero constant}
$$

The second and third of these matrices have, respectively, two and three linearly independent eigenvectors. (Notice that only the diagonal matrix has three linearly independent eigenvectors.)

As we have seen, if a $k \times k$ matrix $\mathbf{A}$ has a k-multiple eigenvalue, then the following can be shown:

1. If the rank of $\lambda \mathbf{I} - \mathbf{A}$ is $k - s$ (where $1 \le s \le k$), then there exist s linearly independent eigenvectors associated with λ.
2. There are s Jordan blocks corresponding to the s eigenvectors.
3. The sum of the orders p_i of the Jordan blocks equals the multiplicity k.

Therefore, as demonstrated in the preceding three 3×3 matrices, even if the multiplicity of eigenvalue is the same, the number of Jordan blocks and their orders may be different depending on the structure of matrix $\mathbf{A}$.

Similarity transformation when an $n \times n$ matrix has distinct eigenvalues.
If n eigenvalues of $\mathbf{A}$ are distinct, there exists one eigenvector associated with each eigenvalue λ_i. It can be proved that such n eigenvectors $\mathbf{x}_1, \mathbf{x}_2, \ldots, \mathbf{x}_n$ are linearly independent.

Let us define an $n \times n$ matrix $\mathbf{P}$ such that

$$\mathbf{P} = [\mathbf{P}_1 \mid \mathbf{P}_2 \mid \cdots \mid \mathbf{P}_n] = [\mathbf{x}_1 \mid \mathbf{x}_2 \mid \cdots \mid \mathbf{x}_n]$$

where column vector $\mathbf{P}_i$ is equal to column vector $\mathbf{x}_i$, or

$$\mathbf{P}_i = \mathbf{x}_i \qquad i = 1, 2, \ldots, n$$

Matrix $\mathbf{P}$ defined in this way is nonsingular, and $\mathbf{P}^{-1}$ exists. Noting that eigenvectors $\mathbf{x}_1, \mathbf{x}_2, \ldots, \mathbf{x}_n$ satisfy the equations

$$\mathbf{A}\mathbf{x}_1 = \lambda_1 \mathbf{x}_1$$

$$\mathbf{A}\mathbf{x}_2 = \lambda_2 \mathbf{x}_2$$

$$\vdots$$

$$\mathbf{A}\mathbf{x}_n = \lambda_n \mathbf{x}_n$$

we may combine these n equations into one, as follows:

$$\mathbf{A}[\mathbf{x}_1 \mid \mathbf{x}_2 \mid \cdots \mid \mathbf{x}_n] = [\mathbf{x}_1 \mid \mathbf{x}_2 \mid \cdots \mid \mathbf{x}_n] \begin{bmatrix} \lambda_1 & & & & 0 \\ & \lambda_2 & & & \\ & & \ddots & & \\ & & & \ddots & \\ 0 & & & & \lambda_n \end{bmatrix}$$

or, in terms of matrix $\mathbf{P}$,

$$\mathbf{A}\mathbf{P} = \mathbf{P} \begin{bmatrix} \lambda_1 & & & & 0 \\ & \lambda_2 & & & \\ & & \ddots & & \\ & & & \ddots & \\ 0 & & & & \lambda_n \end{bmatrix}$$

By premultiplying this last equation by $\mathbf{P}^{-1}$, we obtain

$$\mathbf{P}^{-1}\mathbf{A}\mathbf{P} = \begin{bmatrix} \lambda_1 & & & & 0 \\ & \lambda_2 & & & \\ & & \ddots & & \\ & & & \ddots & \\ 0 & & & & \lambda_n \end{bmatrix} = \text{diag}\,(\lambda_1, \lambda_2, \ldots, \lambda_n)$$

Thus matrix $\mathbf{A}$ is transformed into a diagonal matrix by a similarity transformation.

The process that transforms matrix $\mathbf{A}$ into a diagonal matrix is called the *diagonalization* of matrix $\mathbf{A}$.

As noted earlier, a scalar multiple of eigenvector $\mathbf{x}_i$ is also an eigenvector, since $\alpha\mathbf{x}_i$ satisfies the following equation:

$$\mathbf{A}(\alpha\mathbf{x}_i) = \lambda_i(\alpha\mathbf{x}_i)$$

Consequently, we may choose an α such that the transformation matrix $\mathbf{P}$ becomes as simple as possible.

To summarize, if the eigenvalues of an $n \times n$ matrix $\mathbf{A}$ are distinct, then there are exactly n eigenvectors and they are linearly independent. A transformation matrix $\mathbf{P}$ that transforms $\mathbf{A}$ into a diagonal matrix can be constructed from such n linearly independent eigenvectors.

Similarity transformation when an $n \times n$ matrix has multiple eigenvalues.

Let us assume that an $n \times n$ matrix $\mathbf{A}$ involves a k-multiple eigenvalue λ_1 and other eigenvalues $\lambda_{k+1}, \lambda_{k+2}, \ldots, \lambda_n$ that are all distinct and different from λ_1. That is, the eigenvalues of $\mathbf{A}$ are

$$\lambda_1, \lambda_1, \ldots, \lambda_1, \lambda_{k+1}, \lambda_{k+2}, \ldots, \lambda_n$$

We shall first consider the case where the rank of $\lambda_1 \mathbf{I} - \mathbf{A}$ is $n - 1$. For such a case there exists only one Jordan block for the multiple eigenvalue λ_1 and there is only one eigenvector associated with this multiple eigenvalue. The order of the Jordan block is k, which is the same as the order of multiplicity of the eigenvalue λ_1.

Note that, when an $n \times n$ matrix $\mathbf{A}$ does not possess n linearly independent eigenvectors, it cannot be diagonalized, but can be reduced to a Jordan canonical form.

In the present case, only one linearly independent eigenvector exists for λ_1. We shall now investigate whether it is possible to find $k - 1$ vectors that are somehow associated with this eigenvalue and that are linearly independent of the eigenvectors. Without proof, we shall show that this is possible. First, note that the eigenvector $\mathbf{x}_1$ is a vector that satisfies the equation

$$(\mathbf{A} - \lambda_1\mathbf{I})\mathbf{x}_1 = \mathbf{0}$$

so that $\mathbf{x}_1$ is annihilated by $\mathbf{A} - \lambda_1\mathbf{I}$. Since we do not have enough vectors that are annihilated by $\mathbf{A} - \lambda_1\mathbf{I}$, we seek vectors that are annihilated by $(\mathbf{A} - \lambda_1\mathbf{I})^2$, $(\mathbf{A} - \lambda_1\mathbf{I})^3$, and so on, until we obtain $k - 1$ vectors. The $k - 1$ vectors determined in this way are called *generalized eigenvectors*.

Let us define the desired $k - 1$ generalized eigenvectors as $\mathbf{x}_2, \mathbf{x}_3, \ldots, \mathbf{x}_k$. Then these $k - 1$ generalized eigenvectors can be determined from the equations

$$(\mathbf{A} - \lambda_1\mathbf{I})\mathbf{x}_1 = \mathbf{0}$$

$$(\mathbf{A} - \lambda_1\mathbf{I})^2\mathbf{x}_2 = \mathbf{0}$$

$$\vdots$$

$$(\mathbf{A} - \lambda_1\mathbf{I})^k\mathbf{x}_k = \mathbf{0} \tag{A–32}$$

which can be rewritten as

$$(\mathbf{A} - \lambda_1\mathbf{I})\mathbf{x}_1 = \mathbf{0}$$

$$(\mathbf{A} - \lambda_1\mathbf{I})\mathbf{x}_2 = \mathbf{x}_1$$

.

.

.

$$(\mathbf{A} - \lambda_1\mathbf{I})\mathbf{x}_k = \mathbf{x}_{k-1}$$

Notice that

$$(\mathbf{A} - \lambda_1\mathbf{I})^{k-1}\mathbf{x}_k = (\mathbf{A} - \lambda_1\mathbf{I})^{k-2}\mathbf{x}_{k-1} = \cdots = (\mathbf{A} - \lambda_1\mathbf{I})\mathbf{x}_2 = \mathbf{x}_1$$

or

$$(\mathbf{A} - \lambda_1\mathbf{I})^{k-1}\mathbf{x}_k = \mathbf{x}_1 \qquad\qquad (\text{A--33})$$

The eigenvector $\mathbf{x}_1$ and the $k - 1$ generalized eigenvectors $\mathbf{x}_2, \mathbf{x}_3, \ldots, \mathbf{x}_k$ determined in this way form a set of k linearly independent vectors.

A proper way to determine the generalized eigenvectors is to start with $\mathbf{x}_k$. That is, we first determine the $\mathbf{x}_k$ that will satisfy Equation (A–32) and at the same time will yield a nonzero vector $(\mathbf{A} - \lambda_1\mathbf{I})^{k-1}\mathbf{x}_k$. Any such nonzero vector can be considered as a possible eigenvector $\mathbf{x}_1$. Therefore, to find eigenvector $\mathbf{x}_1$, we apply a row reduction process to $(\mathbf{A} - \lambda_1\mathbf{I})^k$ and find k linearly independent vectors satisfying Equation (A–32). Then these vectors are tested to find one that yields a nonzero vector on the right side of Equation (A–33). (Note that if we start with $\mathbf{x}_1$, then we must make arbitrary choices at each step along the way to determine $\mathbf{x}_2, \mathbf{x}_3, \ldots, \mathbf{x}_k$. This is time consuming and inconvenient. For this reason, this approach is not recommended.)

To summarize what we have discussed so far, the eigenvector $\mathbf{x}_1$ and the generalized eigenvectors $\mathbf{x}_2, \mathbf{x}_3, \ldots, \mathbf{x}_k$ satisfy the following equations:

$$\mathbf{A}\mathbf{x}_1 = \lambda_1\mathbf{x}_1$$

$$\mathbf{A}\mathbf{x}_2 = \mathbf{x}_1 + \lambda_1\mathbf{x}_2$$

.

.

.

$$\mathbf{A}\mathbf{x}_k = \mathbf{x}_{k-1} + \lambda_1\mathbf{x}_k$$

The eigenvectors $\mathbf{x}_{k+1}, \mathbf{x}_{k+2}, \ldots, \mathbf{x}_n$ associated with distinct eigenvalues $\lambda_{k+1}, \lambda_{k+2}, \ldots, \lambda_n$, respectively, can be determined from

$$\mathbf{A}\mathbf{x}_{k+1} = \lambda_{k+1}\mathbf{x}_{k+1}$$

$$\mathbf{A}\mathbf{x}_{k+2} = \lambda_{k+2}\mathbf{x}_{k+2}$$

.

.

.

$$\mathbf{A}\mathbf{x}_n = \lambda_n\mathbf{x}_n$$

Now define

$$\mathbf{S} = [\mathbf{S}_1 \mid \mathbf{S}_2 \mid \cdots \mid \mathbf{S}_n] = [\mathbf{x}_1 \mid \mathbf{x}_2 \mid \cdots \mid \mathbf{x}_n]$$

where the n column vectors of $\mathbf{S}$ are linearly independent. Thus, matrix $\mathbf{S}$ is nonsingular. Then, combining the preceding eigenvector equations and generalized eigenvector equations into one, we obtain

$$\mathbf{A}[\mathbf{x}_1 \mid \mathbf{x}_2 \mid \cdots \mid \mathbf{x}_k \mid \mathbf{x}_{k+1} \mid \cdots \mid \mathbf{x}_n]$$

$$= [\mathbf{x}_1 \mid \mathbf{x}_2 \mid \cdots \mid \mathbf{x}_k \mid \mathbf{x}_{k+1} \mid \cdots \mid \mathbf{x}_n] \begin{bmatrix} \lambda_1 & 1 & & & & & 0 & & & 0 \\ & \lambda_1 & 1 & & & & & & & \\ & & \cdot & & & & & & & \\ & & & \cdot & & & & & & \\ & & & & \cdot & 1 & & & & \\ 0 & & & & & \lambda_1 & 0 & & & \\ & & & & & 0 & \lambda_{k+1} & & & 0 \\ & & & & & & & \cdot & & \\ & & & & & & & & \cdot & \\ 0 & & & & & & 0 & & & \lambda_n \end{bmatrix}$$

Hence

$$\mathbf{AS} = \mathbf{S} \begin{bmatrix} \mathbf{J}_k(\lambda_1) & & & & 0 \\ & \lambda_{k+1} & & & \\ & & \cdot & & \\ & & & \cdot & \\ 0 & & & & \lambda_n \end{bmatrix}$$

By premultiplying this last equation by $\mathbf{S}^{-1}$, we obtain

$$\mathbf{S}^{-1}\mathbf{AS} = \begin{bmatrix} \mathbf{J}_k(\lambda_1) & & & & 0 \\ & \lambda_{k+1} & & & \\ & & \cdot & & \\ & & & \cdot & \\ 0 & & & & \lambda_n \end{bmatrix}$$

In the preceding discussion we considered the case where the rank of $\lambda_1\mathbf{I} - \mathbf{A}$ was $n - 1$. Next we shall consider the case where the rank of $\lambda_1\mathbf{I} - \mathbf{A}$ is $n - s$ (where $2 \le s \le n$). Since we assumed that matrix $\mathbf{A}$ involves the k-multiple eigenvalue λ_1 and other eigenvalues $\lambda_{k+1}, \lambda_{k+2}, \ldots, \lambda_n$ that are all distinct and different from λ_1, we have s linearly independent eigenvectors associated with eigenvalue λ_1. Hence, there are s Jordan blocks corresponding to eigenvalue λ_1.

For notational convenience, let us define the s linearly independent eigenvectors associated with eigenvalue λ_1 as $\mathbf{v}_{11}, \mathbf{v}_{21}, \ldots, \mathbf{v}_{s1}$. We shall define the generalized eigenvectors associated with $\mathbf{v}_{i1}$ as $\mathbf{v}_{i2}, \mathbf{v}_{i3}, \ldots, \mathbf{v}_{ip_i}$, where $i = 1, 2, \ldots, s$. Then there are altogether k such vectors (eigenvectors and generalized eigenvectors), which are

$$\mathbf{v}_{11}, \mathbf{v}_{12}, \ldots, \mathbf{v}_{1p_1}, \mathbf{v}_{21}, \mathbf{v}_{22}, \ldots, \mathbf{v}_{2p_2}, \ldots, \mathbf{v}_{s1}, \mathbf{v}_{s2}, \ldots, \mathbf{v}_{sp_s}$$

The generalized eigenvectors are determined from

$$(\mathbf{A} - \lambda_1\mathbf{I})\mathbf{v}_{11} = \mathbf{0}, \qquad \cdots \qquad (\mathbf{A} - \lambda_1\mathbf{I})\mathbf{v}_{s1} = \mathbf{0}$$

$$(\mathbf{A} - \lambda_1\mathbf{I})\mathbf{v}_{12} = \mathbf{v}_{11}, \qquad \cdots \qquad (\mathbf{A} - \lambda_1\mathbf{I})\mathbf{v}_{s2} = \mathbf{v}_{s1}$$

$$\vdots \qquad\qquad\qquad\qquad\qquad \vdots$$

$$(\mathbf{A} - \lambda_1\mathbf{I})\mathbf{v}_{1p_1} = \mathbf{v}_{1p_1-1}, \qquad \cdots \qquad (\mathbf{A} - \lambda_1\mathbf{I})\mathbf{v}_{sp_s} = \mathbf{v}_{sp_s-1}$$

where the s eigenvectors $\mathbf{v}_{11}, \mathbf{v}_{21}, \ldots, \mathbf{v}_{s1}$ are linearly independent and

$$p_1 + p_2 + \cdots + p_s = k$$

Note that $p_1, p_2, \ldots, p_s$ represent the order of each of the s Jordan blocks. (For the determination of the generalized eigenvectors, we follow the method discussed earlier. For an example showing the details of such a determination, see Problem A–11.)

Let us define an $n \times k$ matrix consisting of $\mathbf{v}_{11}, \mathbf{v}_{12}, \ldots, \mathbf{v}_{sp_s}$ as

$$\mathbf{S}(\lambda_1) = [\mathbf{v}_{11} \vdots \mathbf{v}_{12} \vdots \cdots \vdots \mathbf{v}_{1p_1} \vdots \cdots \vdots \mathbf{v}_{s1} \vdots \mathbf{v}_{s2} \vdots \cdots \vdots \mathbf{v}_{sp_s}]$$

$$= [\mathbf{x}_1 \vdots \mathbf{x}_2 \vdots \cdots \vdots \mathbf{x}_{p_1} \vdots \cdots \vdots \mathbf{x}_k]$$

$$= [\mathbf{S}_1 \vdots \mathbf{S}_2 \vdots \cdots \vdots \mathbf{S}_k]$$

and define

$$\mathbf{S} = [\mathbf{S}(\lambda_1) \vdots \mathbf{S}_{k+1} \vdots \mathbf{S}_{k+2} \vdots \cdots \vdots \mathbf{S}_n]$$

$$= [\mathbf{S}_1 \vdots \mathbf{S}_2 \vdots \cdots \vdots \mathbf{S}_n]$$

where

$$\mathbf{S}_{k+1} = \mathbf{x}_{k+1}, \qquad \mathbf{S}_{k+2} = \mathbf{x}_{k+2}, \ldots, \mathbf{S}_n = \mathbf{x}_n$$

Note that $\mathbf{x}_{k+1}, \mathbf{x}_{k+2}, \ldots, \mathbf{x}_n$ are eigenvectors associated with eigenvalues $\lambda_{k+1}, \lambda_{k+2}, \ldots, \lambda_n$, respectively. Matrix $\mathbf{S}$ defined in this way is nonsingular. Now we obtain

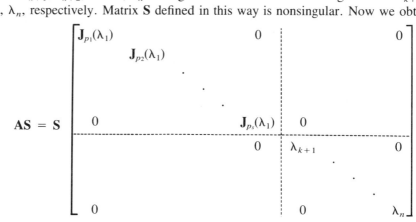

where $\mathbf{J}_{p_i}(\lambda_1)$ is in the form

$$\mathbf{J}_{p_i}(\lambda_1) = \begin{bmatrix} \lambda_1 & 1 & & & & 0 \\ & \lambda_1 & 1 & & & \\ & & & \cdot & & \\ & & & & \cdot & \cdot \\ & & & & & \cdot \quad 1 \\ 0 & & & & & \lambda_1 \end{bmatrix}$$

which is a $p_i \times p_i$ matrix. Hence,

$$\mathbf{S}^{-1}\mathbf{AS} = \left[\begin{array}{cccc:ccc} \mathbf{J}_{p_1}(\lambda_1) & & & 0 & & & 0 \\ & \mathbf{J}_{p_2}(\lambda_1) & & & & & \\ & & \cdot & & & & \\ 0 & & & \mathbf{J}_{p_s}(\lambda_1) & 0 & & \\ \hdashline & & & 0 & \lambda_{k+1} & & 0 \\ & & & & & \cdot & \\ 0 & & & & 0 & & \lambda_n \end{array}\right]$$

Thus, as we have shown, by using a set of n linearly independent vectors (eigenvectors and generalized eigenvectors), any $n \times n$ matrix can be reduced to a Jordan canonical form by a similarity transformation.

Similarity transformation when an $n \times n$ matrix is normal. First, recall that a matrix is normal if it is a real symmetric, a Hermitian, a real skew-symmetric, a skew-Hermitian, an orthogonal, or a unitary matrix.

Assume that an $n \times n$ normal matrix has a k-multiple eigenvalue λ_1 and that its other $n - k$ eigenvalues are distinct and different from λ_1. Then the rank of $\mathbf{A} - \lambda_1\mathbf{I}$ becomes $n - k$. (Refer to Problem A–12 for the proof.) If the rank of $\mathbf{A} - \lambda_1\mathbf{I}$ is $n - k$, there are k linerly independent eigenvectors $\mathbf{x}_1, \mathbf{x}_2, \dots, \mathbf{x}_k$ that satisfy the equation

$$(\mathbf{A} - \lambda_1\mathbf{I})\mathbf{x}_i = \mathbf{0} \qquad i = 1, 2, \dots, k$$

Therefore, there exist k Jordan blocks for eigenvalue λ_1. Since the number of Jordan blocks is the same as the multiplicity number of eigenvalue λ_1, all k Jordan blocks become first order. Since the remaining $n - k$ eigenvalues are distinct, the eigenvectors associated with these eigenvalues are linearly independent. Hence the $n \times n$ normal matrix possesses altogether n linearly independent eigenvectors and the Jordan canonical form of the normal matrix becomes a diagonal matrix.

It can be proved that if $\mathbf{A}$ is an $n \times n$ normal matrix, then regardless of whether or not the eigenvalues include multiple eigenvalues, there exists an $n \times n$ unitary matrix $\mathbf{U}$ such that

$$\mathbf{U}^{-1}\mathbf{AU} = \mathbf{U}^*\mathbf{AU} = \mathbf{D} = \text{diag}\,(\lambda_1, \lambda_2, \dots, \lambda_n)$$

where $\mathbf{D}$ is a diagonal matrix with n eigenvalues as diagonal elements.

Trace of an $n \times n$ matrix. The trace of an $n \times n$ matrix $\mathbf{A}$ is defined as follows:

$$\text{trace of } \mathbf{A} = \text{tr } \mathbf{A} = \sum_{i=1}^{n} a_{ii}$$

The trace of an $n \times n$ matrix $\mathbf{A}$ has the following properties:

1.
$$\text{tr } \mathbf{A}^T = \text{tr } \mathbf{A}$$

2. For $n \times n$ matrices $\mathbf{A}$ and $\mathbf{B}$,

$$\text{tr } (\mathbf{A} + \mathbf{B}) = \text{tr } \mathbf{A} + \text{tr } \mathbf{B}$$

3. If the eigenvalues of $\mathbf{A}$ are denoted by $\lambda_1, \lambda_2, \ldots, \lambda_n$, then

$$\text{tr } \mathbf{A} = \lambda_1 + \lambda_2 + \cdots + \lambda_n \tag{A–34}$$

4. For an $n \times m$ matrix $\mathbf{A}$ and and $m \times n$ matrix $\mathbf{B}$, regardless of whether $\mathbf{AB} = \mathbf{BA}$ or $\mathbf{AB} \neq \mathbf{BA}$, we have

$$\text{tr } \mathbf{AB} = \text{tr } \mathbf{BA} = \sum_{i=1}^{n} \sum_{j=1}^{m} a_{ij} b_{ji}$$

If $m = 1$, then by writing $\mathbf{A}$ and $\mathbf{B}$ as $\mathbf{a}$ and $\mathbf{b}$, respectively, we have

$$\text{tr } \mathbf{ab} = \mathbf{ba}$$

Hence, for an $n \times m$ matrix $\mathbf{C}$, we have

$$\mathbf{a}^T \mathbf{C} \mathbf{a} = \text{tr } \mathbf{aa}^T \mathbf{C}$$

Note that Equation (A–34) may be proved as follows. By use of a similarity transformation we have

$$\mathbf{P}^{-1} \mathbf{A} \mathbf{P} = \mathbf{D} = \text{diagonal matrix}$$

or

$$\mathbf{S}^{-1} \mathbf{A} \mathbf{S} = \mathbf{J} = \text{Jordan canonical form}$$

That is,

$$\mathbf{A} = \mathbf{PDP}^{-1} \quad \text{or} \quad \mathbf{A} = \mathbf{SJS}^{-1}$$

Hence, by using property 4 listed here, we have

$$\text{tr } \mathbf{A} = \text{tr } \mathbf{PDP}^{-1} = \text{tr } \mathbf{P}^{-1} \mathbf{PD} = \text{tr } \mathbf{D} = \lambda_1 + \lambda_2 + \cdots + \lambda_n$$

Similarly,

$$\text{tr } \mathbf{A} = \text{tr } \mathbf{SJS}^{-1} = \text{tr } \mathbf{S}^{-1} \mathbf{SJ} = \text{tr } \mathbf{J} = \lambda_1 + \lambda_2 + \cdots + \lambda_n$$

Invariant properties under similarity transformation. If an $n \times n$ matrix $\mathbf{A}$ can be reduced to a similar matrix that has a simple form, then important properties of $\mathbf{A}$ can be readily observed. A property of a matrix is said to be invariant if it is possessed by all similar matrices. For example, the determinant and the characteristic polynomial are invariant under a similarity transformation, as shown in the following. Suppose that $\mathbf{P}^{-1} \mathbf{A} \mathbf{P} = \mathbf{B}$.

Then

$$|\mathbf{B}| = |\mathbf{P}^{-1}\mathbf{AP}| = |\mathbf{P}^{-1}|\,|\mathbf{A}|\,|\mathbf{P}| = |\mathbf{A}|\,|\mathbf{P}^{-1}|\,|\mathbf{P}| = |\mathbf{A}|\,|\mathbf{P}^{-1}\mathbf{P}|$$
$$= |\mathbf{A}|\,|\mathbf{I}| = |\mathbf{A}|$$

and

$$|\lambda\mathbf{I} - \mathbf{B}| = |\lambda\mathbf{I} - \mathbf{P}^{-1}\mathbf{AP}| = |\mathbf{P}^{-1}(\lambda\mathbf{I})\mathbf{P} - \mathbf{P}^{-1}\mathbf{AP}|$$
$$= |\mathbf{P}^{-1}(\lambda\mathbf{I} - \mathbf{A})\mathbf{P}| = |\mathbf{P}^{-1}|\,|\lambda\mathbf{I} - \mathbf{A}|\,|\mathbf{P}|$$
$$= |\lambda\mathbf{I} - \mathbf{A}|\,|\mathbf{P}^{-1}|\,|\mathbf{P}| = |\lambda\mathbf{I} - \mathbf{A}|$$

Notice that the trace of a matrix is also invariant under similarity transformation, as was shown earlier:

$$\text{tr } \mathbf{A} = \text{tr } \mathbf{P}^{-1}\mathbf{AP}$$

The property of symmetry of a matrix, however, is not invariant.

Notice that only invariant properties of matrices present intrinsic characteristics of the class of similar matrices. To determine the invariant properties of a matrix $\mathbf{A}$, we examine the Jordan canonical form of $\mathbf{A}$, since the similarity of two matrices can be defined in terms of the Jordan canonical form: The necessary and sufficient condition for $n \times n$ matrices $\mathbf{A}$ and $\mathbf{B}$ to be similar is that the Jordan canonical form of $\mathbf{A}$ and that of $\mathbf{B}$ be identical.

A–7 QUADRATIC FORMS

Quadratic forms. For an $n \times n$ real symmetric matrix $\mathbf{A}$ and a real n-vector $\mathbf{x}$, the form

$$\mathbf{x}^T\mathbf{Ax} = \sum_{i=1}^{n}\sum_{j=1}^{n} a_{ij}x_ix_j \qquad a_{ji} = a_{ij}$$

is called a *real quadratic form* in x_i. Frequently, a real quadratic form is called simply a *quadratic form*. Note that $\mathbf{x}^T\mathbf{Ax}$ is a real scalar quantity.

Any real quadratic form can always be written as $\mathbf{x}^T\mathbf{Ax}$. For example,

$$x_1^2 - 2x_1x_2 + 4x_1x_3 + x_2^2 + 8x_3^2 = \begin{bmatrix} x_1 & x_2 & x_3 \end{bmatrix} \begin{bmatrix} 1 & -1 & 2 \\ -1 & 1 & 0 \\ 2 & 0 & 8 \end{bmatrix} \begin{bmatrix} x_1 \\ x_2 \\ x_3 \end{bmatrix}$$

It is worthwhile to mention that for an $n \times n$ real matrix $\mathbf{A}$, if we define

$$\mathbf{B} = \tfrac{1}{2}(\mathbf{A} + \mathbf{A}^T) \qquad \text{and} \qquad \mathbf{C} = \tfrac{1}{2}(\mathbf{A} - \mathbf{A}^T)$$

then

$$\mathbf{A} = \mathbf{B} + \mathbf{C}$$

Notice that

$$\mathbf{B}^T = \mathbf{B} \qquad \text{and} \qquad \mathbf{C}^T = -\mathbf{C}$$

Hence an $n \times n$ real matrix $\mathbf{A}$ can be expressed as a sum of a real symmetric and a real skew-symmetric matrix. Noting that, since $\mathbf{x}^T\mathbf{C}\mathbf{x}$ is a real scalar quantity, we have

$$\mathbf{x}^T\mathbf{C}\mathbf{x} = (\mathbf{x}^T\mathbf{C}\mathbf{x})^T = \mathbf{x}^T\mathbf{C}^T\mathbf{x} = -\mathbf{x}^T\mathbf{C}\mathbf{x}$$

Consequently, we have

$$\mathbf{x}^T\mathbf{C}\mathbf{x} = 0$$

This means that a quadratic form for a real skew-symmetric matrix is zero. Hence

$$\mathbf{x}^T\mathbf{A}\mathbf{x} = \mathbf{x}^T(\mathbf{B} + \mathbf{C})\mathbf{x} = \mathbf{x}^T\mathbf{B}\mathbf{x}$$

and we see that the real quadratic form $\mathbf{x}^T\mathbf{A}\mathbf{x}$ involves only the symmetric component $\mathbf{x}^T\mathbf{B}\mathbf{x}.$ This is the reason why the real quadratic form is defined only for a real symmetric matrix.

For a Hermitian matrix $\mathbf{A}$ and a complex n-vector $\mathbf{x},$ the form

$$\mathbf{x}^*\mathbf{A}\mathbf{x} = \sum_{i=1}^{n}\sum_{j=1}^{n} a_{ij}\bar{x}_i x_j \qquad a_{ji} = \bar{a}_{ij}$$

is called a *complex quadratic form*, or Hermitian form. Notice that the scalar quantity $\mathbf{x}^*\mathbf{A}\mathbf{x}$ is real, because

$$\overline{\mathbf{x}^*\mathbf{A}\mathbf{x}} = \mathbf{x}^T\bar{\mathbf{A}}\bar{\mathbf{x}} = (\mathbf{x}^T\bar{\mathbf{A}}\bar{\mathbf{x}})^T = \bar{\mathbf{x}}^T\bar{\mathbf{A}}^T\mathbf{x} = \mathbf{x}^*\mathbf{A}\mathbf{x}$$

Bilinear forms. For an $n \times m$ real matrix $\mathbf{A},$ a real n-vector $\mathbf{x},$ and a real m-vector $\mathbf{y},$ the form

$$\mathbf{x}^T\mathbf{A}\mathbf{y} = \sum_{i=1}^{n}\sum_{j=1}^{m} a_{ij}x_i y_j$$

is called a *real bilinear form* in x_i and $y_j.$ $\mathbf{x}^T\mathbf{A}\mathbf{y}$ is a real scalar quantity.

For an $n \times m$ complex matrix $\mathbf{A},$ a complex n-vector $\mathbf{x},$ and a complex m-vector $\mathbf{y},$ the form

$$\mathbf{x}^*\mathbf{A}\mathbf{y} = \sum_{i=1}^{n}\sum_{j=1}^{m} a_{ij}\bar{x}_i y_j$$

is called a *complex bilinear form.* $\mathbf{x}^*\mathbf{A}\mathbf{y}$ is a complex scalar quantity.

Definiteness and semidefiniteness. A quadratic form $\mathbf{x}^T\mathbf{A}\mathbf{x},$ where $\mathbf{A}$ is a real symmetric matrix (or a Hermitian form $\mathbf{x}^*\mathbf{A}\mathbf{x}$ where $\mathbf{A}$ is a Hermitian matrix), is said to be positive definite if

$$\mathbf{x}^T\mathbf{A}\mathbf{x} > 0 \qquad (\text{or } \mathbf{x}^*\mathbf{A}\mathbf{x} > 0) \qquad \text{for } \mathbf{x} \neq \mathbf{0}$$

$$\mathbf{x}^T\mathbf{A}\mathbf{x} = 0 \qquad (\text{or } \mathbf{x}^*\mathbf{A}\mathbf{x} = 0) \qquad \text{for } \mathbf{x} = \mathbf{0}$$

$\mathbf{x}^T\mathbf{A}\mathbf{x}$ (or $\mathbf{x}^*\mathbf{A}\mathbf{x}$) is said to be positive semidefinite if

$$\mathbf{x}^T\mathbf{A}\mathbf{x} \geq 0 \qquad (\text{or } \mathbf{x}^*\mathbf{A}\mathbf{x} \geq 0) \qquad \text{for } \mathbf{x} \neq \mathbf{0}$$

$$\mathbf{x}^T\mathbf{A}\mathbf{x} = 0 \qquad (\text{or } \mathbf{x}^*\mathbf{A}\mathbf{x} = 0) \qquad \text{for } \mathbf{x} = \mathbf{0}$$

$\mathbf{x}^T\mathbf{A}\mathbf{x}$ (or $\mathbf{x}*\mathbf{A}\mathbf{x}$) is said to be negative definite if

$$\mathbf{x}^T\mathbf{A}\mathbf{x} < 0 \qquad (\text{or } \mathbf{x}*\mathbf{A}\mathbf{x} < 0) \qquad \text{for } \mathbf{x} \neq \mathbf{0}$$

$$\mathbf{x}^T\mathbf{A}\mathbf{x} = 0 \qquad (\text{or } \mathbf{x}*\mathbf{A}\mathbf{x} = 0) \qquad \text{for } \mathbf{x} = \mathbf{0}$$

$\mathbf{x}^T\mathbf{A}\mathbf{x}$ (or $\mathbf{x}*\mathbf{A}\mathbf{x}$) is said to be negative semidefinite if

$$\mathbf{x}^T\mathbf{A}\mathbf{x} \leq 0 \qquad (\text{or } \mathbf{x}*\mathbf{A}\mathbf{x} \leq 0) \qquad \text{for } \mathbf{x} \neq \mathbf{0}$$

$$\mathbf{x}^T\mathbf{A}\mathbf{x} = 0 \qquad (\text{or } \mathbf{x}*\mathbf{A}\mathbf{x} = 0) \qquad \text{for } \mathbf{x} = \mathbf{0}$$

If $\mathbf{x}^T\mathbf{A}\mathbf{x}$ (or $\mathbf{x}*\mathbf{A}\mathbf{x}$) can be of either sign, then $\mathbf{x}^T\mathbf{A}\mathbf{x}$ (or $\mathbf{x}*\mathbf{A}\mathbf{x}$) is said to be indefinite.

Note that if $\mathbf{x}^T\mathbf{A}\mathbf{x}$ or $\mathbf{x}*\mathbf{A}\mathbf{x}$ is positive (or negative) definite, then we say that $\mathbf{A}$ is a positive (or negative) definite matrix. Similarly, matrix $\mathbf{A}$ is called a positive (or negative) semidefinite matrix if $\mathbf{x}^T\mathbf{A}\mathbf{x}$ or $\mathbf{x}*\mathbf{A}\mathbf{x}$ is positive (or negative) semidefinite; matrix $\mathbf{A}$ is called an indefinite matrix if $\mathbf{x}^T\mathbf{A}\mathbf{x}$ or $\mathbf{x}*\mathbf{A}\mathbf{x}$ is indefinite.

Note also that the eigenvalues of an $n \times n$ real symmetric or Hermitian matrix are real. (For the proof, see Problem A–13.) It can be shown that an $n \times n$ real symmetric or Hermitian matrix $\mathbf{A}$ is a positive-definite matrix if all eigenvalues λ_i ($i = 1, 2, \ldots, n$) are positive. Matrix $\mathbf{A}$ is positive semidefinite if all eigenvalues are nonnegative, or $\lambda_i \geq 0$ ($i = 1, 2, \ldots, n$), and at least one of them is zero.

Notice that if $\mathbf{A}$ is a positive definite matrix, then $|\mathbf{A}| \neq 0$, because all eigenvalues are positive. Hence, the inverse matrix always exists for a positive definite matrix.

In the process of determining the stability of an equilibrium state, we frequently encounter a scalar function $V(\mathbf{x})$. A scalar function $V(\mathbf{x})$, which is a function of $x_1, x_2, \ldots, x_n$, is said to be positive definite if

$$V(\mathbf{x}) > 0 \qquad \text{for } \mathbf{x} \neq \mathbf{0}$$

$$V(\mathbf{0}) = 0$$

$V(\mathbf{x})$ is said to be positive semidefinite if

$$V(\mathbf{x}) \geq 0 \qquad \text{for } \mathbf{x} \neq \mathbf{0}$$

$$V(\mathbf{0}) = 0$$

If $-V(\mathbf{x})$ is positive definite (or positive semidefinite), then $V(\mathbf{x})$ is said to be negative definite (or negative semidefinite).

Necessary and sufficient conditions for the quadratic form $\mathbf{x}^T\mathbf{A}\mathbf{x}$ (or the Hermitian form $\mathbf{x}*\mathbf{A}\mathbf{x}$) to be positive definite, negative definite, positive semidefinite, or negative semidefinite have been given by J. J. Sylvester. Sylvester's criteria follow.

Sylvester's criterion for positive definiteness of a quadratic form or Hermitian form.
A necessary and sufficient condition for a quadratic form $\mathbf{x}^T\mathbf{A}\mathbf{x}$ (or a Hermitian form $\mathbf{x}*\mathbf{A}\mathbf{x}$), where $\mathbf{A}$ is an $n \times n$ real symmetric matrix (or Hermitian matrix), to be positive definite is that the determinant of $\mathbf{A}$ be positive and the successive principal minors of the determinant of $\mathbf{A}$ (the determinants of the $k \times k$ matrices in the top-left corner of matrix $\mathbf{A}$, where $k = 1, 2, \ldots, n - 1$) be positive; that is, we must have

$$a_{11} > 0, \qquad \begin{vmatrix} a_{11} & a_{12} \\ a_{21} & a_{22} \end{vmatrix} > 0, \qquad \begin{vmatrix} a_{11} & a_{12} & a_{13} \\ a_{21} & a_{22} & a_{23} \\ a_{31} & a_{32} & a_{33} \end{vmatrix} > 0, \quad \ldots, \quad |\mathbf{A}| > 0$$

where

$$a_{ij} = a_{ji} \qquad \text{for real symmetric matrix } \mathbf{A}$$

$$a_{ij} = \bar{a}_{ji} \qquad \text{for Hermitian matrix } \mathbf{A}$$

Sylvester's criterion for negative definiteness of a quadratic form or Hermitian form.   A necessary and sufficient condition for a quadratic form $\mathbf{x}^T\mathbf{A}\mathbf{x}$ (or a Hermitian form $\mathbf{x}^*\mathbf{A}\mathbf{x}$), where $\mathbf{A}$ is an $n \times n$ real symmetric matrix (or Hermitian matrix), to be negative definite is that the determinant of $\mathbf{A}$ be positive if n is even and negative if n is odd, and that the successive principal minors of even order be positive and the successive principal minors of odd order be negative; that is, we must have

$$a_{11} < 0, \qquad \begin{vmatrix} a_{11} & a_{12} \\ a_{21} & a_{22} \end{vmatrix} > 0, \qquad \begin{vmatrix} a_{11} & a_{12} & a_{13} \\ a_{21} & a_{22} & a_{23} \\ a_{31} & a_{32} & a_{33} \end{vmatrix} < 0, \; \ldots$$

$$|\mathbf{A}| > 0 \qquad (n \text{ even})$$

$$|\mathbf{A}| < 0 \qquad (n \text{ odd})$$

where

$$a_{ij} = a_{ji} \qquad \text{for real symmetric matrix } \mathbf{A}$$

$$a_{ij} = \bar{a}_{ji} \qquad \text{for Hermitian matrix } \mathbf{A}$$

[This condition can be derived by requiring that $\mathbf{x}^T(-\mathbf{A})\mathbf{x}$ be positive definite.]

Sylvester's criterion for positive semidefiniteness of a quadratic form or Hermitian form.   A necessary and sufficient condition for a quadratic form $\mathbf{x}^T\mathbf{A}\mathbf{x}$ (or a Hermitian form $\mathbf{x}^*\mathbf{A}\mathbf{x}$), where $\mathbf{A}$ is a real symmetric matrix (or a Hermitian matrix), to be positive semidefinite is that $\mathbf{A}$ be singular ($|\mathbf{A}| = 0$) and all the principal minors be nonnegative:

$$a_{ii} \geq 0, \qquad \begin{vmatrix} a_{ii} & a_{ij} \\ a_{ji} & a_{jj} \end{vmatrix} \geq 0, \qquad \begin{vmatrix} a_{ii} & a_{ij} & a_{ik} \\ a_{ji} & a_{jj} & a_{jk} \\ a_{ki} & a_{kj} & a_{kk} \end{vmatrix} \geq 0, \quad \ldots, \quad |\mathbf{A}| = 0$$

where $i < j < k$ and

$$a_{ij} = a_{ji} \qquad \text{for real symmetric matrix } \mathbf{A}$$

$$a_{ij} = \bar{a}_{ji} \qquad \text{for Hermitian matrix } \mathbf{A}$$

(It is important to point out that in the positive semidefiniteness test or negative semidefiniteness test we must check the signs of all the principal minors, not just successive principal minors. See Problem A–15.)

Sylvester's criterion for negative semidefiniteness of a quadratic form or a Hermitian form.   A necessary and sufficient condition for a quadratic form $\mathbf{x}^T\mathbf{A}\mathbf{x}$ (or a Hermitian form $\mathbf{x}^*\mathbf{A}\mathbf{x}$), where $\mathbf{A}$ is an $n \times n$ real symmetric matrix (or Hermitian matrix), to be negative semidefinite is that $\mathbf{A}$ be singular ($|\mathbf{A}| = 0$) and that all the principal minors of even order be nonnegative and those of odd order be nonpositive:

$$a_{ii} \leq 0, \qquad \begin{vmatrix} a_{ii} & a_{ij} \\ a_{ji} & a_{jj} \end{vmatrix} \geq 0, \qquad \begin{vmatrix} a_{ii} & a_{ij} & a_{ik} \\ a_{ji} & a_{jj} & a_{jk} \\ a_{ki} & a_{kj} & a_{kk} \end{vmatrix} \leq 0, \quad \ldots \, , \quad |\mathbf{A}| = 0$$

where $i < j < k$ and

$$a_{ij} = a_{ji} \qquad \text{for real symmetric matrix } \mathbf{A}$$

$$a_{ij} = \bar{a}_{ji} \qquad \text{for Hermitian matrix } \mathbf{A}$$

A–8 PSEUDOINVERSES

The concept of pseudoinverses of a matrix is a generalization of the notion of an inverse. It is useful for finding a "solution" to a set of algebraic equations in which the number of unknown variables and the number of independent linear equations are not equal.

In what follows, we shall consider pseudoinverses that enable us to determine minimum norm solutions.

Minimum norm solution that minimizes $\|\mathbf{x}\|$. Consider a linear algebraic equation

$$x_1 + 5x_2 = 1$$

Since we have two variables and only one equation, no unique solution exists. Instead, there exist an infinite number of solutions. Graphically, any point on line $x_1 + 5x_2 = 1$, as shown in Figure A–1, is a possible solution. However, if we decide to pick the point that is closest to the origin, the solution becomes unique.

Consider the vector-matrix equation

$$\mathbf{A}\mathbf{x} = \mathbf{b} \qquad\qquad (A–35)$$

where $\mathbf{A}$ is an $n \times m$ matrix, $\mathbf{x}$ is an m-vector, and $\mathbf{b}$ is an n-vector. We assume that $m > n$ (that is, that the number of unknown variables is greater than the number of equations) and that the equation has an infinite number of solutions. Let us find the unique solution $\mathbf{x}$ that is located closest to the origin or that has the minimum norm $\|\mathbf{x}\|$.

Let us define the minimum norm solution as $\mathbf{x}^\circ$. That is, $\mathbf{x}^\circ$ satisfies the condition that $\mathbf{A}\mathbf{x}^\circ = \mathbf{b}$ and $\|\mathbf{x}^\circ\| \leq \|\mathbf{x}\|$ for all $\mathbf{x}$ that satisfy $\mathbf{A}\mathbf{x} = \mathbf{b}$. This means that the solution point $\mathbf{x}^\circ$ is nearest to the origin of the m-dimensional space among all possible solutions of Equation (A–35). We shall obtain such a minimum norm solution in the following.

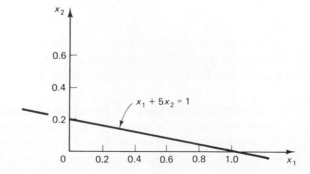

Figure A–1
Line $x_1 + 5x_2 = 1$
on the $x_1 x_2$ plane.

Right pseudoinverse matrix.   For a vector-matrix equation

$$\mathbf{Ax} = \mathbf{b}$$

where $\mathbf{A}$ is an $n \times m$ matrix having rank n, $\mathbf{x}$ is an m-vector, and $\mathbf{b}$ is an n-vector, the solution that minimizes the norm $\|\mathbf{x}\|$ is given by

$$\mathbf{x}^\circ = \mathbf{A}^{RM}\mathbf{b}$$

where $\mathbf{A}^{RM} = \mathbf{A}^T(\mathbf{AA}^T)^{-1}$.

This can be proved as follows. First, note that norm $\|\mathbf{x}\|$ can be written as follows:

$$\|\mathbf{x}\| = \|\mathbf{x} - \mathbf{x}^\circ + \mathbf{x}^\circ\| = \|\mathbf{x}^\circ\| + \|\mathbf{x} - \mathbf{x}^\circ\| + 2(\mathbf{x}^\circ)^T(\mathbf{x} - \mathbf{x}^\circ)$$

The last term, $2(\mathbf{x}^\circ)^T(\mathbf{x} - \mathbf{x}^\circ)$, can be shown to be zero, since

$$
\begin{aligned}
(\mathbf{x}^\circ)^T(\mathbf{x} - \mathbf{x}^\circ) &= [\mathbf{A}^T(\mathbf{AA}^T)^{-1}\mathbf{b}]^T [\mathbf{x} - \mathbf{A}^T(\mathbf{AA}^T)^{-1}\mathbf{b}] \\
&= \mathbf{b}^T(\mathbf{AA}^T)^{-1}\mathbf{A}[\mathbf{x} - \mathbf{A}^T(\mathbf{AA}^T)^{-1}\mathbf{b}] \\
&= \mathbf{b}^T(\mathbf{AA}^T)^{-1} [\mathbf{Ax} - (\mathbf{AA}^T)(\mathbf{AA}^T)^{-1}\mathbf{b}] \\
&= \mathbf{b}^T(\mathbf{AA}^T)^{-1}(\mathbf{b} - \mathbf{b}) \\
&= 0
\end{aligned}
$$

Hence

$$\|\mathbf{x}\| = \|\mathbf{x}^\circ\| + \|\mathbf{x} - \mathbf{x}^\circ\|$$

which can be rewritten as

$$\|\mathbf{x}\| - \|\mathbf{x}^\circ\| = \|\mathbf{x} - \mathbf{x}^\circ\|$$

Since $\|\mathbf{x} - \mathbf{x}^\circ\| \geq 0$, we obtain

$$\|\mathbf{x}\| \geq \|\mathbf{x}^\circ\|$$

Thus, we have shown that $\mathbf{x}^\circ$ is the solution that gives the minimum norm $\|\mathbf{x}\|$.

The matrix $\mathbf{A}^{RM} = \mathbf{A}^T(\mathbf{AA}^T)^{-1}$ that yields the minimum norm solution ($\|\mathbf{x}^\circ\| = \text{minimum}$) is called the *right pseudoinverse* or *minimal right inverse* of $\mathbf{A}$.

Summary on the right pseudoinverse matrix.   The right pseudoinverse $\mathbf{A}^{RM}$ gives the solution $\mathbf{x}^\circ = \mathbf{A}^{RM}\mathbf{b}$ that minimizes the norm, or gives $\|\mathbf{x}^\circ\| = \text{minimum}$. Note that the right pseudoinverse $\mathbf{A}^{RM}$ is an $m \times n$ matrix, since $\mathbf{A}$ is an $n \times m$ matrix and

$$
\begin{aligned}
\mathbf{A}^{RM} &= \mathbf{A}^T(\mathbf{AA}^T)^{-1} \\
&= (m \times n \text{ matrix})(n \times n \text{ matrix})^{-1} \\
&= m \times n \text{ matrix} \qquad m > n
\end{aligned}
$$

Notice that the dimension of $\mathbf{AA}^T$ is smaller than the dimension of vector $\mathbf{x}$, which is m. Notice also that the right pseudoinverse $\mathbf{A}^{RM}$ possesses the property that it is indeed an ''inverse'' matrix if premultiplied by $\mathbf{A}$:

$$\mathbf{AA}^{RM} = \mathbf{A}[\mathbf{A}^T(\mathbf{AA}^T)^{-1}] = \mathbf{AA}^T(\mathbf{AA}^T)^{-1} = \mathbf{I}_n$$

Solution that minimizes $\|\mathbf{Ax} - \mathbf{b}\|$.   Consider a vector-matrix equation

$$\mathbf{Ax} = \mathbf{b} \tag{A-36}$$

where $\mathbf{A}$ is an $n \times m$ matrix, $\mathbf{x}$ is an m-vector, and $\mathbf{b}$ is an n-vector. Here we assume that

$n > m$. That is, the number of unknown variables is smaller than the number of equations. In the classical sense, there may or may not exist any solution.

If no solution exists, we may wish to find a unique "solution" that minimizes the norm $\|\mathbf{Ax} - \mathbf{b}\|$. Let us define a "solution" to Equation (A–36) that will minimize $\|\mathbf{Ax} - \mathbf{b}\|$ as $\mathbf{x}^\circ$. In other words, $\mathbf{x}^\circ$ satisfies the condition

$$\|\mathbf{Ax} - \mathbf{b}\| \geq \|\mathbf{Ax}^\circ - \mathbf{b}\| \qquad \text{for all } \mathbf{x}$$

Note that $\mathbf{x}^\circ$ is not a solution in the classical sense, since it does not satisfy the original vector-matrix equation $\mathbf{Ax} = \mathbf{b}$. Therefore, we may call $\mathbf{x}^\circ$ an "approximate solution," in that it minimizes norm $\|\mathbf{Ax} - \mathbf{b}\|$. We shall obtain such an approximate solution in the following.

Left pseudoinverse matrix. For a vector-matrix equation

$$\mathbf{Ax} = \mathbf{b}$$

where $\mathbf{A}$ is an $n \times m$ matrix having rank m, $\mathbf{x}$ is an m-vector, and $\mathbf{b}$ is an n-vector, the vector $\mathbf{x}^\circ$ that minimizes the norm $\|\mathbf{Ax} - \mathbf{b}\|$ is given by

$$\mathbf{x}^\circ = \mathbf{A}^{LM}\mathbf{b} = (\mathbf{A}^T\mathbf{A})^{-1}\mathbf{A}^T\mathbf{b}$$

where $\mathbf{A}^{LM} = (\mathbf{A}^T\mathbf{A})^{-1}\mathbf{A}^T$.

To verify this, first note that

$$\begin{aligned}
\|\mathbf{Ax} - \mathbf{b}\| &= \|\mathbf{A}(\mathbf{x} - \mathbf{x}^\circ) + \mathbf{Ax}^\circ - \mathbf{b}\| \\
&= \|\mathbf{A}(\mathbf{x} - \mathbf{x}^\circ)\| + \|\mathbf{Ax}^\circ - \mathbf{b}\| + 2[\mathbf{A}(\mathbf{x} - \mathbf{x}^\circ)]^T(\mathbf{Ax}^\circ - \mathbf{b})
\end{aligned}$$

The last term can be shown to be zero as follows:

$$\begin{aligned}
[\mathbf{A}(\mathbf{x} - \mathbf{x}^\circ)]^T(\mathbf{Ax}^\circ - \mathbf{b}) &= (\mathbf{x} - \mathbf{x}^\circ)^T\mathbf{A}^T[\mathbf{A}(\mathbf{A}^T\mathbf{A})^{-1}\mathbf{A}^T - \mathbf{I}_n]\mathbf{b} \\
&= (\mathbf{x} - \mathbf{x}^\circ)^T[(\mathbf{A}^T\mathbf{A})(\mathbf{A}^T\mathbf{A})^{-1}\mathbf{A}^T - \mathbf{A}^T]\mathbf{b} \\
&= (\mathbf{x} - \mathbf{x}^\circ)^T(\mathbf{A}^T - \mathbf{A}^T)\mathbf{b} \\
&= 0
\end{aligned}$$

Hence

$$\|\mathbf{Ax} - \mathbf{b}\| = \|\mathbf{A}(\mathbf{x} - \mathbf{x}^\circ)\| + \|\mathbf{Ax}^\circ - \mathbf{b}\|$$

Noting that $\|\mathbf{A}(\mathbf{x} - \mathbf{x}^\circ)\| \geq 0$, we obtain

$$\|\mathbf{Ax} - \mathbf{b}\| - \|\mathbf{Ax}^\circ - \mathbf{b}\| = \|\mathbf{A}(\mathbf{x} - \mathbf{x}^\circ)\| \geq 0$$

or

$$\|\mathbf{Ax} - \mathbf{b}\| \geq \|\mathbf{Ax}^\circ - \mathbf{b}\|$$

Thus

$$\mathbf{x}^\circ = \mathbf{A}^{LM}\mathbf{b} = (\mathbf{A}^T\mathbf{A})^{-1}\mathbf{A}^T\mathbf{b}$$

minimizes $\|\mathbf{Ax} - \mathbf{b}\|$.

The matrix $\mathbf{A}^{LM} = (\mathbf{A}^T\mathbf{A})^{-1}\mathbf{A}^T$ is called the *left pseudoinverse* or *minimal left inverse* of matrix $\mathbf{A}$. Note that $\mathbf{A}^{LM}$ is indeed the inverse matrix of $\mathbf{A}$, in that if postmultiplied by $\mathbf{A}$ it will give an identity matrix $\mathbf{I}_m$:

$$\mathbf{A}^{LM}\mathbf{A} = (\mathbf{A}^T\mathbf{A})^{-1}\mathbf{A}^T\mathbf{A} = (\mathbf{A}^T\mathbf{A})^{-1}(\mathbf{A}^T\mathbf{A}) = \mathbf{I}_m$$

For illustrative applications of the use of the right and left pseudoinverses for obtaining minimum norm solutions of vector-matrix equations, refer to Problems A–16 and A–17.

Example Problems and Solutions

A–1. Show that if matrices **A**, **B**, **C**, and **D** are an $n \times n$, an $n \times m$, an $m \times n$, and an $m \times m$ matrix, respectively, and if $|\mathbf{A}| \neq 0$ and $|\mathbf{D}| \neq 0$, then

$$\begin{vmatrix} \mathbf{A} & \mathbf{B} \\ \mathbf{0} & \mathbf{D} \end{vmatrix} = \begin{vmatrix} \mathbf{A} & \mathbf{0} \\ \mathbf{C} & \mathbf{D} \end{vmatrix} = |\mathbf{A}|\,|\mathbf{D}| \neq 0 \qquad \text{if } |\mathbf{A}| \neq 0 \text{ and } |\mathbf{D}| \neq 0$$

Solution. Since matrix **A** is nonsingular, we have

$$\begin{bmatrix} \mathbf{A} & \mathbf{B} \\ \mathbf{0} & \mathbf{D} \end{bmatrix} = \begin{bmatrix} \mathbf{A} & \mathbf{0} \\ \mathbf{0} & \mathbf{I} \end{bmatrix}\begin{bmatrix} \mathbf{I} & \mathbf{0} \\ \mathbf{0} & \mathbf{D} \end{bmatrix}\begin{bmatrix} \mathbf{I} & \mathbf{A}^{-1}\mathbf{B} \\ \mathbf{0} & \mathbf{I} \end{bmatrix}$$

Hence,

$$\begin{vmatrix} \mathbf{A} & \mathbf{B} \\ \mathbf{0} & \mathbf{D} \end{vmatrix} = \begin{vmatrix} \mathbf{A} & \mathbf{0} \\ \mathbf{0} & \mathbf{I} \end{vmatrix}\begin{vmatrix} \mathbf{I} & \mathbf{0} \\ \mathbf{0} & \mathbf{D} \end{vmatrix}\begin{vmatrix} \mathbf{I} & \mathbf{A}^{-1}\mathbf{B} \\ \mathbf{0} & \mathbf{I} \end{vmatrix} = |\mathbf{A}|\,|\mathbf{D}|$$

Similarly, since **D** is nonsingular, we get

$$\begin{vmatrix} \mathbf{A} & \mathbf{0} \\ \mathbf{C} & \mathbf{D} \end{vmatrix} = \begin{vmatrix} \mathbf{A} & \mathbf{0} \\ \mathbf{0} & \mathbf{I} \end{vmatrix}\begin{vmatrix} \mathbf{I} & \mathbf{0} \\ \mathbf{0} & \mathbf{D} \end{vmatrix}\begin{vmatrix} \mathbf{I} & \mathbf{0} \\ \mathbf{D}^{-1}\mathbf{C} & \mathbf{I} \end{vmatrix} = |\mathbf{A}|\,|\mathbf{D}|$$

A–2. Show that if matrices **A**, **B**, **C**, and **D** are an $n \times n$, an $n \times m$, an $m \times n$, and an $m \times m$ matrix, respectively, then

$$\begin{vmatrix} \mathbf{A} & \mathbf{B} \\ \mathbf{C} & \mathbf{D} \end{vmatrix} = \begin{cases} |\mathbf{A}|\,|\mathbf{D} - \mathbf{C}\mathbf{A}^{-1}\mathbf{B}| & \text{if } |\mathbf{A}| \neq 0 \\ |\mathbf{D}|\,|\mathbf{A} - \mathbf{B}\mathbf{D}^{-1}\mathbf{C}| & \text{if } |\mathbf{D}| \neq 0 \end{cases}$$

Solution. If $|\mathbf{A}| \neq 0$, the matrix

$$\begin{bmatrix} \mathbf{A} & \mathbf{B} \\ \mathbf{C} & \mathbf{D} \end{bmatrix}$$

can be written as a product of two matrices:

$$\begin{bmatrix} \mathbf{A} & \mathbf{0} \\ \mathbf{C} & \mathbf{I}_m \end{bmatrix} \qquad \text{and} \qquad \begin{bmatrix} \mathbf{I}_n & \mathbf{A}^{-1}\mathbf{B} \\ \mathbf{0} & \mathbf{D} - \mathbf{C}\mathbf{A}^{-1}\mathbf{B} \end{bmatrix}$$

or

$$\begin{bmatrix} \mathbf{A} & \mathbf{B} \\ \mathbf{C} & \mathbf{D} \end{bmatrix} = \begin{bmatrix} \mathbf{A} & \mathbf{0} \\ \mathbf{C} & \mathbf{I}_m \end{bmatrix}\begin{bmatrix} \mathbf{I}_n & \mathbf{A}^{-1}\mathbf{B} \\ \mathbf{0} & \mathbf{D} - \mathbf{C}\mathbf{A}^{-1}\mathbf{B} \end{bmatrix}$$

Hence

$$\begin{vmatrix} \mathbf{A} & \mathbf{B} \\ \mathbf{C} & \mathbf{D} \end{vmatrix} = \begin{vmatrix} \mathbf{A} & \mathbf{0} \\ \mathbf{C} & \mathbf{I}_m \end{vmatrix}\begin{vmatrix} \mathbf{I}_n & \mathbf{A}^{-1}\mathbf{B} \\ \mathbf{0} & \mathbf{D} - \mathbf{C}\mathbf{A}^{-1}\mathbf{B} \end{vmatrix}$$

$$= |\mathbf{A}|\,|\mathbf{I}_m|\,|\mathbf{I}_n|\,|\mathbf{D} - \mathbf{C}\mathbf{A}^{-1}\mathbf{B}|$$

$$= |\mathbf{A}|\,|\mathbf{D} - \mathbf{C}\mathbf{A}^{-1}\mathbf{B}|$$

Similarly, if $|\mathbf{D}| \neq 0$, then

$$\begin{bmatrix} \mathbf{A} & \mathbf{B} \\ \mathbf{C} & \mathbf{D} \end{bmatrix} = \begin{bmatrix} \mathbf{I}_n & \mathbf{B} \\ \mathbf{0} & \mathbf{D} \end{bmatrix}\begin{bmatrix} \mathbf{A} - \mathbf{B}\mathbf{D}^{-1}\mathbf{C} & \mathbf{0} \\ \mathbf{D}^{-1}\mathbf{C} & \mathbf{I}_m \end{bmatrix}$$

and therefore

$$\begin{vmatrix} \mathbf{A} & \mathbf{B} \\ \mathbf{C} & \mathbf{D} \end{vmatrix} = \begin{vmatrix} \mathbf{I}_n & \mathbf{B} \\ \mathbf{0} & \mathbf{D} \end{vmatrix} \begin{vmatrix} \mathbf{A} - \mathbf{BD}^{-1}\mathbf{C} & \mathbf{0} \\ \mathbf{D}^{-1}\mathbf{C} & \mathbf{I}_m \end{vmatrix}$$

$$= |\mathbf{I}_n|\,|\mathbf{D}|\,|\mathbf{A} - \mathbf{BD}^{-1}\mathbf{C}|\,|\mathbf{I}_m|$$

$$= |\mathbf{D}|\,|\mathbf{A} - \mathbf{BD}^{-1}\mathbf{C}|$$

A–3. For an $n \times m$ matrix **A** and an $m \times n$ matrix **B,** show that

$$|\mathbf{I}_n + \mathbf{AB}| = |\mathbf{I}_m + \mathbf{BA}|$$

Solution. Consider the following matrix:

$$\begin{bmatrix} \mathbf{I}_n & -\mathbf{A} \\ \mathbf{B} & \mathbf{I}_m \end{bmatrix}$$

Referring to Problem A–2,

$$\begin{vmatrix} \mathbf{A} & \mathbf{B} \\ \mathbf{C} & \mathbf{D} \end{vmatrix} = \begin{cases} |\mathbf{A}|\,|\mathbf{D} - \mathbf{CA}^{-1}\mathbf{B}| & \text{if } |\mathbf{A}| \neq 0 \\ |\mathbf{D}|\,|\mathbf{A} - \mathbf{BD}^{-1}\mathbf{C}| & \text{if } |\mathbf{D}| \neq 0 \end{cases}$$

Hence

$$\begin{vmatrix} \mathbf{I}_n & -\mathbf{A} \\ \mathbf{B} & \mathbf{I}_m \end{vmatrix} = \begin{cases} |\mathbf{I}_n|\,|\mathbf{I}_m + \mathbf{BA}| = |\mathbf{I}_m + \mathbf{BA}| \\ |\mathbf{I}_m|\,|\mathbf{I}_n + \mathbf{AB}| = |\mathbf{I}_n + \mathbf{AB}| \end{cases}$$

and we have

$$|\mathbf{I}_n + \mathbf{AB}| = |\mathbf{I}_m + \mathbf{BA}|$$

A–4. If **A**, **B**, **C**, and **D** are, respectively, an $n \times n$, an $n \times m$, an $m \times n$, and an $m \times m$ matrix, then we have the following matrix inversion lemma:

$$(\mathbf{A} + \mathbf{BDC})^{-1} = \mathbf{A}^{-1} - \mathbf{A}^{-1}\mathbf{B}(\mathbf{D}^{-1} + \mathbf{CA}^{-1}\mathbf{B})^{-1}\mathbf{CA}^{-1}$$

where we assume the indicated inverses to exist. Prove this matrix inversion lemma.

Solution. Let us premultiply both sides of the equation by $(\mathbf{A} + \mathbf{BDC})$:

$$(\mathbf{A} + \mathbf{BDC})(\mathbf{A} + \mathbf{BDC})^{-1} = (\mathbf{A} + \mathbf{BDC})[\mathbf{A}^{-1} - \mathbf{A}^{-1}\mathbf{B}(\mathbf{D}^{-1} + \mathbf{CA}^{-1}\mathbf{B})^{-1}\mathbf{CA}^{-1}]$$

or

$$\begin{aligned} \mathbf{I} &= \mathbf{I} + \mathbf{BDCA}^{-1} - \mathbf{B}(\mathbf{D}^{-1} + \mathbf{CA}^{-1}\mathbf{B})^{-1}\mathbf{CA}^{-1} - \mathbf{BDCA}^{-1}\mathbf{B}(\mathbf{D}^{-1} + \mathbf{CA}^{-1}\mathbf{B})^{-1}\mathbf{CA}^{-1} \\ &= \mathbf{I} + \mathbf{BDCA}^{-1} - (\mathbf{B} + \mathbf{BDCA}^{-1}\mathbf{B})(\mathbf{D}^{-1} + \mathbf{CA}^{-1}\mathbf{B})^{-1}\mathbf{CA}^{-1} \\ &= \mathbf{I} + \mathbf{BDCA}^{-1} - \mathbf{BD}(\mathbf{D}^{-1} + \mathbf{CA}^{-1}\mathbf{B})(\mathbf{D}^{-1} + \mathbf{CA}^{-1}\mathbf{B})^{-1}\mathbf{CA}^{-1} \\ &= \mathbf{I} + \mathbf{BDCA}^{-1} - \mathbf{BDCA}^{-1} \\ &= \mathbf{I} \end{aligned}$$

Hence we have proved the matrix inversion lemma.

A–5. Prove that if **A**, **B**, **C**, and **D** are, respectively, an $n \times n$, an $n \times m$, an $m \times n$, and an $m \times m$ matrix, then

$$\begin{bmatrix} \mathbf{A} & \mathbf{B} \\ \mathbf{0} & \mathbf{D} \end{bmatrix}^{-1} = \begin{bmatrix} \mathbf{A}^{-1} & -\mathbf{A}^{-1}\mathbf{BD}^{-1} \\ \mathbf{0} & \mathbf{D}^{-1} \end{bmatrix} \tag{A–37}$$

provided $|\mathbf{A}| \neq 0$ and $|\mathbf{D}| \neq 0$.

Prove also that

$$\begin{bmatrix} \mathbf{A} & \mathbf{0} \\ \mathbf{C} & \mathbf{D} \end{bmatrix}^{-1} = \begin{bmatrix} \mathbf{A}^{-1} & \mathbf{0} \\ -\mathbf{D}^{-1}\mathbf{C}\mathbf{A}^{-1} & \mathbf{D}^{-1} \end{bmatrix} \tag{A–38}$$

provided $|\mathbf{A}| \neq 0$ and $|\mathbf{D}| \neq 0$.

Solution. Note that

$$\begin{bmatrix} \mathbf{A}^{-1} & -\mathbf{A}^{-1}\mathbf{B}\mathbf{D}^{-1} \\ \mathbf{0} & \mathbf{D}^{-1} \end{bmatrix} \begin{bmatrix} \mathbf{A} & \mathbf{B} \\ \mathbf{0} & \mathbf{D} \end{bmatrix} = \begin{bmatrix} \mathbf{I}_n & \mathbf{A}^{-1}\mathbf{B} - \mathbf{A}^{-1}\mathbf{B} \\ \mathbf{0} & \mathbf{I}_m \end{bmatrix} = \begin{bmatrix} \mathbf{I}_n & \mathbf{0} \\ \mathbf{0} & \mathbf{I}_m \end{bmatrix}$$

Hence, Equation (A–37) is proved. Similarly,

$$\begin{bmatrix} \mathbf{A}^{-1} & \mathbf{0} \\ -\mathbf{D}^{-1}\mathbf{C}\mathbf{A}^{-1} & \mathbf{D}^{-1} \end{bmatrix} \begin{bmatrix} \mathbf{A} & \mathbf{0} \\ \mathbf{C} & \mathbf{D} \end{bmatrix} = \begin{bmatrix} \mathbf{I}_n & \mathbf{0} \\ -\mathbf{D}^{-1}\mathbf{C} + \mathbf{D}^{-1}\mathbf{C} & \mathbf{I}_m \end{bmatrix} = \begin{bmatrix} \mathbf{I}_n & \mathbf{0} \\ \mathbf{0} & \mathbf{I}_m \end{bmatrix}$$

Hence, we have proved Equation (A–38).

A–6. Prove that if $\mathbf{A}$, $\mathbf{B}$, $\mathbf{C}$, and $\mathbf{D}$ are, respectively, an $n \times n$, an $n \times m$, an $m \times n$, and an $m \times m$ matrix, then

$$\begin{bmatrix} \mathbf{A} & \mathbf{B} \\ \mathbf{C} & \mathbf{D} \end{bmatrix}^{-1} = \begin{bmatrix} \mathbf{A}^{-1} + \mathbf{A}^{-1}\mathbf{B}(\mathbf{D} - \mathbf{C}\mathbf{A}^{-1}\mathbf{B})^{-1}\mathbf{C}\mathbf{A}^{-1} & -\mathbf{A}^{-1}\mathbf{B}(\mathbf{D} - \mathbf{C}\mathbf{A}^{-1}\mathbf{B})^{-1} \\ -(\mathbf{D} - \mathbf{C}\mathbf{A}^{-1}\mathbf{B})^{-1}\mathbf{C}\mathbf{A}^{-1} & (\mathbf{D} - \mathbf{C}\mathbf{A}^{-1}\mathbf{B})^{-1} \end{bmatrix}$$

provided $|\mathbf{A}| \neq 0$ and $|\mathbf{D} - \mathbf{C}\mathbf{A}^{-1}\mathbf{B}| \neq 0$.

Prove also that

$$\begin{bmatrix} \mathbf{A} & \mathbf{B} \\ \mathbf{C} & \mathbf{D} \end{bmatrix}^{-1} = \begin{bmatrix} (\mathbf{A} - \mathbf{B}\mathbf{D}^{-1}\mathbf{C})^{-1} & -(\mathbf{A} - \mathbf{B}\mathbf{D}^{-1}\mathbf{C})^{-1}\mathbf{B}\mathbf{D}^{-1} \\ -\mathbf{D}^{-1}\mathbf{C}(\mathbf{A} - \mathbf{B}\mathbf{D}^{-1}\mathbf{C})^{-1} & \mathbf{D}^{-1}\mathbf{C}(\mathbf{A} - \mathbf{B}\mathbf{D}^{-1}\mathbf{C})^{-1}\mathbf{B}\mathbf{D}^{-1} + \mathbf{D}^{-1} \end{bmatrix}$$

provided $|\mathbf{D}| \neq 0$ and $|\mathbf{A} - \mathbf{B}\mathbf{D}^{-1}\mathbf{C}| \neq 0$.

Solution. First, note that

$$\begin{bmatrix} \mathbf{A} & \mathbf{B} \\ \mathbf{C} & \mathbf{D} \end{bmatrix} = \begin{bmatrix} \mathbf{A} & \mathbf{0} \\ \mathbf{C} & \mathbf{I}_m \end{bmatrix} \begin{bmatrix} \mathbf{I}_n & \mathbf{A}^{-1}\mathbf{B} \\ \mathbf{0} & \mathbf{D} - \mathbf{C}\mathbf{A}^{-1}\mathbf{B} \end{bmatrix} \tag{A–39}$$

By taking the inverse of both sides of Equation (A–39), we obtain

$$\begin{bmatrix} \mathbf{A} & \mathbf{B} \\ \mathbf{C} & \mathbf{D} \end{bmatrix}^{-1} = \begin{bmatrix} \mathbf{I}_n & \mathbf{A}^{-1}\mathbf{B} \\ \mathbf{0} & \mathbf{D} - \mathbf{C}\mathbf{A}^{-1}\mathbf{B} \end{bmatrix}^{-1} \begin{bmatrix} \mathbf{A} & \mathbf{0} \\ \mathbf{C} & \mathbf{I}_m \end{bmatrix}^{-1}$$

By referring to Problem A–5, we find

$$\begin{bmatrix} \mathbf{I}_n & \mathbf{A}^{-1}\mathbf{B} \\ \mathbf{0} & \mathbf{D} - \mathbf{C}\mathbf{A}^{-1}\mathbf{B} \end{bmatrix}^{-1} = \begin{bmatrix} \mathbf{I}_n & -\mathbf{A}^{-1}\mathbf{B}(\mathbf{D} - \mathbf{C}\mathbf{A}^{-1}\mathbf{B})^{-1} \\ \mathbf{0} & (\mathbf{D} - \mathbf{C}\mathbf{A}^{-1}\mathbf{B})^{-1} \end{bmatrix}$$

and

$$\begin{bmatrix} \mathbf{A} & \mathbf{0} \\ \mathbf{C} & \mathbf{I}_m \end{bmatrix}^{-1} = \begin{bmatrix} \mathbf{A}^{-1} & \mathbf{0} \\ -\mathbf{C}\mathbf{A}^{-1} & \mathbf{I}_m \end{bmatrix}$$

Hence

$$\begin{bmatrix} \mathbf{A} & \mathbf{B} \\ \mathbf{C} & \mathbf{D} \end{bmatrix}^{-1} = \begin{bmatrix} \mathbf{I}_n & \mathbf{A}^{-1}\mathbf{B} \\ \mathbf{0} & \mathbf{D} - \mathbf{C}\mathbf{A}^{-1}\mathbf{B} \end{bmatrix}^{-1} \begin{bmatrix} \mathbf{A} & \mathbf{0} \\ \mathbf{C} & \mathbf{I}_m \end{bmatrix}^{-1}$$

$$= \begin{bmatrix} \mathbf{I}_n & -\mathbf{A}^{-1}\mathbf{B}(\mathbf{D} - \mathbf{C}\mathbf{A}^{-1}\mathbf{B})^{-1} \\ \mathbf{0} & (\mathbf{D} - \mathbf{C}\mathbf{A}^{-1}\mathbf{B})^{-1} \end{bmatrix} \begin{bmatrix} \mathbf{A}^{-1} & \mathbf{0} \\ -\mathbf{C}\mathbf{A}^{-1} & \mathbf{I}_m \end{bmatrix}$$

$$= \begin{bmatrix} A^{-1} + A^{-1}B(D - CA^{-1}B)^{-1}CA^{-1} & -A^{-1}B(D - CA^{-1}B)^{-1} \\ -(D - CA^{-1}B)^{-1}CA^{-1} & (D - CA^{-1}B)^{-1} \end{bmatrix}$$

provided $|A| \neq 0$ and $|D - CA^{-1}B| \neq 0$.

Similarly, notice that

$$\begin{bmatrix} A & B \\ C & D \end{bmatrix} = \begin{bmatrix} I_n & B \\ 0 & D \end{bmatrix} \begin{bmatrix} A - BD^{-1}C & 0 \\ D^{-1}C & I_m \end{bmatrix} \qquad (A\text{--}40)$$

By taking the inverse of both sides of Equation (A–40) and referring to Problem A–5, we obtain

$$\begin{bmatrix} A & B \\ C & D \end{bmatrix}^{-1} = \begin{bmatrix} A - BD^{-1}C & 0 \\ D^{-1}C & I_m \end{bmatrix}^{-1} \begin{bmatrix} I_n & B \\ 0 & D \end{bmatrix}^{-1}$$

$$= \begin{bmatrix} (A - BD^{-1}C)^{-1} & 0 \\ -D^{-1}C(A - BD^{-1}C)^{-1} & I_m \end{bmatrix} \begin{bmatrix} I_n & -BD^{-1} \\ 0 & D^{-1} \end{bmatrix}$$

$$= \begin{bmatrix} (A - BD^{-1}C)^{-1} & -(A - BD^{-1}C)^{-1}BD^{-1} \\ -D^{-1}C(A - BD^{-1}C)^{-1} & D^{-1}C(A - BD^{-1}C)^{-1}BD^{-1} + D^{-1} \end{bmatrix}$$

provided $|D| \neq 0$ and $|A - BD^{-1}C| \neq 0$.

A–7.   For an $n \times n$ real matrix A and real n-vectors $\mathbf{x}$ and $\mathbf{y}$, show that

(a)
$$\frac{\partial}{\partial \mathbf{x}} \mathbf{y}^T \mathbf{x} = \mathbf{y}$$

(b)
$$\frac{\partial}{\partial \mathbf{x}} \mathbf{x}^T A \mathbf{x} = A\mathbf{x} + A^T\mathbf{x}$$

For an $n \times n$ Hermitian matrix A and a complex n-vector $\mathbf{x}$, show that

(c)
$$\frac{\partial}{\partial \bar{\mathbf{x}}} \mathbf{x}^* A \mathbf{x} = A\mathbf{x}$$

Solution.

(a) Note that

$$\mathbf{y}^T \mathbf{x} = y_1 x_1 + y_2 x_2 + \cdots + y_n x_n$$

which is a scalar quantity. Hence

$$\frac{\partial}{\partial \mathbf{x}} \mathbf{y}^T \mathbf{x} = \begin{bmatrix} \dfrac{\partial}{\partial x_1} \mathbf{y}^T \mathbf{x} \\ \vdots \\ \dfrac{\partial}{\partial x_n} \mathbf{y}^T \mathbf{x} \end{bmatrix} = \begin{bmatrix} y_1 \\ \vdots \\ y_n \end{bmatrix} = \mathbf{y}$$

(b) Notice that

$$\mathbf{x}^T A \mathbf{x} = \sum_{i=1}^{n} \sum_{j=1}^{n} a_{ij} x_i x_j$$

which is a scalar quantity. Hence

$$\frac{\partial}{\partial \mathbf{x}} \mathbf{x}^T \mathbf{A} \mathbf{x} = \begin{bmatrix} \dfrac{\partial}{\partial x_1}\left(\displaystyle\sum_{i=1}^{n} \sum_{j=1}^{n} a_{ij} x_i x_j \right) \\ \cdot \\ \cdot \\ \cdot \\ \dfrac{\partial}{\partial x_n}\left(\displaystyle\sum_{i=1}^{n} \sum_{j=1}^{n} a_{ij} x_i x_j \right) \end{bmatrix} = \begin{bmatrix} \displaystyle\sum_{j=1}^{n} a_{1j} x_j + \sum_{i=1}^{n} a_{i1} x_i \\ \cdot \\ \cdot \\ \cdot \\ \displaystyle\sum_{j=1}^{n} a_{nj} x_j + \sum_{i=1}^{n} a_{in} x_i \end{bmatrix}$$

$$= \mathbf{A}\mathbf{x} + \mathbf{A}^T \mathbf{x}$$

which is Equation (A–20)

If matrix $\mathbf{A}$ is a real symmetric matrix, then

$$\frac{\partial}{\partial \mathbf{x}} \mathbf{x}^T \mathbf{A} \mathbf{x} = 2\mathbf{A}\mathbf{x} \qquad \text{if } \mathbf{A} = \mathbf{A}^T$$

(c) For a Hermitian matrix $\mathbf{A}$, we have

$$\mathbf{x}^* \mathbf{A} \mathbf{x} = \sum_{i=1}^{n} \sum_{j=1}^{n} a_{ij} \bar{x}_i x_j$$

and

$$\frac{\partial}{\partial \bar{\mathbf{x}}} \mathbf{x}^* \mathbf{A} \mathbf{x} = \begin{bmatrix} \dfrac{\partial}{\partial \bar{x}_1}\left(\displaystyle\sum_{i=1}^{n} \sum_{j=1}^{n} a_{ij} \bar{x}_i x_j \right) \\ \cdot \\ \cdot \\ \cdot \\ \dfrac{\partial}{\partial \bar{x}_n}\left(\displaystyle\sum_{i=1}^{n} \sum_{j=1}^{n} a_{ij} \bar{x}_i x_j \right) \end{bmatrix} = \begin{bmatrix} \displaystyle\sum_{j=1}^{n} a_{1j} x_j \\ \cdot \\ \cdot \\ \cdot \\ \displaystyle\sum_{j=1}^{n} a_{nj} x_j \end{bmatrix} = \mathbf{A}\mathbf{x}$$

which is Equation (A–21).

Note that

$$\frac{\partial}{\partial \mathbf{x}} \mathbf{x}^* \mathbf{A} \mathbf{x} = \begin{bmatrix} \dfrac{\partial}{\partial x_1}\left(\displaystyle\sum_{i=1}^{n} \sum_{j=1}^{n} a_{ij} \bar{x}_i x_j \right) \\ \cdot \\ \cdot \\ \cdot \\ \dfrac{\partial}{\partial x_n}\left(\displaystyle\sum_{i=1}^{n} \sum_{j=1}^{n} a_{ij} \bar{x}_i x_j \right) \end{bmatrix} = \begin{bmatrix} \displaystyle\sum_{i=1}^{n} a_{i1} \bar{x}_i \\ \cdot \\ \cdot \\ \cdot \\ \displaystyle\sum_{i=1}^{n} a_{in} \bar{x}_i \end{bmatrix} = \mathbf{A}^T \bar{\mathbf{x}}$$

Therefore,

$$\overline{\frac{\partial}{\partial \mathbf{x}} \mathbf{x}^* \mathbf{A} \mathbf{x}} = \mathbf{A}^* \mathbf{x} = \mathbf{A}\mathbf{x}$$

A–8. For an $n \times m$ complex matrix $\mathbf{A}$, a complex n-vector $\mathbf{x}$, and a complex m-vector $\mathbf{y}$, show that

(a)
$$\frac{\partial}{\partial \bar{\mathbf{x}}} \mathbf{x}^* \mathbf{A} \mathbf{y} = \mathbf{A}\mathbf{y}$$

(b)
$$\frac{\partial}{\partial \mathbf{y}} \mathbf{x}^* \mathbf{A} \mathbf{y} = \mathbf{A}^T \bar{\mathbf{x}}$$

Example Problems and Solutions

Solution.

(a) Notice that

$$\mathbf{x}^*\mathbf{A}\mathbf{y} = \sum_{i=1}^{n} \sum_{j=1}^{m} a_{ij} \bar{x}_i y_j$$

Hence

$$\frac{\partial}{\partial \bar{\mathbf{x}}} \mathbf{x}^*\mathbf{A}\mathbf{y} = \begin{bmatrix} \dfrac{\partial}{\partial \bar{x}_1}\left(\displaystyle\sum_{i=1}^{n}\sum_{j=1}^{m} a_{ij}\bar{x}_i y_j\right) \\ \cdot \\ \cdot \\ \cdot \\ \dfrac{\partial}{\partial \bar{x}_n}\left(\displaystyle\sum_{i=1}^{n}\sum_{j=1}^{m} a_{ij}\bar{x}_i y_j\right) \end{bmatrix} = \begin{bmatrix} \displaystyle\sum_{j=1}^{m} a_{1j} y_j \\ \cdot \\ \cdot \\ \cdot \\ \displaystyle\sum_{j=1}^{m} a_{nj} y_j \end{bmatrix} = \mathbf{A}\mathbf{y}$$

which is Equation (A–24).

(b) Notice that

$$\frac{\partial}{\partial \mathbf{y}} \mathbf{x}^*\mathbf{A}\mathbf{y} = \begin{bmatrix} \dfrac{\partial}{\partial y_1}\left(\displaystyle\sum_{i=1}^{n}\sum_{j=1}^{m} a_{ij}\bar{x}_i y_j\right) \\ \cdot \\ \cdot \\ \cdot \\ \dfrac{\partial}{\partial y_m}\left(\displaystyle\sum_{i=1}^{n}\sum_{j=1}^{m} a_{ij}\bar{x}_i y_j\right) \end{bmatrix} = \begin{bmatrix} \displaystyle\sum_{i=1}^{n} a_{i1}\bar{x}_i \\ \cdot \\ \cdot \\ \cdot \\ \displaystyle\sum_{i=1}^{n} a_{im}\bar{x}_i \end{bmatrix} = \mathbf{A}^T\bar{\mathbf{x}}$$

which is Equation (A–25).

Similarly, for an $n \times m$ real matrix $\mathbf{A}$, a real n-vector $\mathbf{x}$, and a real m-vector $\mathbf{y}$, we have

$$\frac{\partial}{\partial \mathbf{x}} \mathbf{x}^T\mathbf{A}\mathbf{y} = \mathbf{A}\mathbf{y}, \qquad \frac{\partial}{\partial \mathbf{y}} \mathbf{x}^T\mathbf{A}\mathbf{y} = \mathbf{A}^T\mathbf{x}$$

which are Equations (A–22) and (A–23), respectively.

A–9. Given two $n \times n$ matrices $\mathbf{A}$ and $\mathbf{B}$, prove that the eigenvalues of $\mathbf{AB}$ and those of $\mathbf{BA}$ are the same, even if $\mathbf{AB} \neq \mathbf{BA}$.

Solution. First, we shall consider the case where $\mathbf{A}$ (or $\mathbf{B}$) is nonsingular. In this case,

$$|\lambda\mathbf{I} - \mathbf{BA}| = |\lambda\mathbf{I} - \mathbf{A}^{-1}(\mathbf{AB})\mathbf{A}| = |\mathbf{A}^{-1}(\lambda\mathbf{I} - \mathbf{AB})\mathbf{A}| = |\mathbf{A}^{-1}||\lambda\mathbf{I} - \mathbf{AB}||\mathbf{A}| = |\lambda\mathbf{I} - \mathbf{AB}|$$

Next we shall consider the case where both $\mathbf{A}$ and $\mathbf{B}$ are singular. There exist $n \times n$ nonsingular matrices $\mathbf{P}$ and $\mathbf{Q}$ such that

$$\mathbf{PAQ} = \begin{bmatrix} \mathbf{I}_r & \mathbf{0} \\ \mathbf{0} & \mathbf{0} \end{bmatrix}$$

where $\mathbf{I}_r$ is the $r \times r$ identity matrix and r is the rank of $\mathbf{A}$, $r < n$. We have

$$|\lambda\mathbf{I} - \mathbf{BA}| = |\lambda\mathbf{I} - \mathbf{Q}^{-1}\mathbf{BAQ}| = |\lambda\mathbf{I} - \mathbf{Q}^{-1}\mathbf{BP}^{-1}\mathbf{PAQ}|$$

$$= \left|\lambda\mathbf{I} - \begin{bmatrix} \mathbf{G}_{11} & \mathbf{G}_{12} \\ \mathbf{G}_{21} & \mathbf{G}_{22} \end{bmatrix}\begin{bmatrix} \mathbf{I}_r & \mathbf{0} \\ \mathbf{0} & \mathbf{0} \end{bmatrix}\right|$$

where

$$Q^{-1}BP^{-1} = \begin{bmatrix} G_{11} & G_{12} \\ G_{21} & G_{22} \end{bmatrix}$$

Then

$$|\lambda I - BA| = \left| \lambda I - \begin{bmatrix} G_{11} & 0 \\ G_{21} & 0 \end{bmatrix} \right| = \begin{vmatrix} \lambda I_r - G_{11} & 0 \\ -G_{21} & \lambda I_{n-r} \end{vmatrix}$$

$$= |\lambda I_r - G_{11}| \, |\lambda I_{n-r}|$$

Also,

$$|\lambda I - AB| = |\lambda I - PABP^{-1}| = |\lambda I - PAQQ^{-1}BP^{-1}|$$

$$= \left| \lambda I - \begin{bmatrix} I_r & 0 \\ 0 & 0 \end{bmatrix} \begin{bmatrix} G_{11} & G_{12} \\ G_{21} & G_{22} \end{bmatrix} \right|$$

$$= \left| \lambda I - \begin{bmatrix} G_{11} & G_{12} \\ 0 & 0 \end{bmatrix} \right|$$

$$= \begin{vmatrix} \lambda I_r - G_{11} & -G_{12} \\ 0 & \lambda I_{n-r} \end{vmatrix}$$

$$= |\lambda I_r - G_{11}| \, |\lambda I_{n-r}|$$

Hence, we have proved that

$$|\lambda I - BA| = |\lambda I - AB|$$

or that the eigenvalues of AB and BA are the same regardless of whether $AB = BA$ or $AB \neq BA$.

A–10. Show that the following 2×2 matrix A has two distinct eigenvalues and that the eigenvectors are linearly independent of each other:

$$A = \begin{bmatrix} 1 & 1 \\ 0 & 2 \end{bmatrix}$$

Then normalize the eigenvectors.

Solution. The eigenvalues are obtained from

$$|\lambda I - A| = \begin{vmatrix} \lambda - 1 & -1 \\ 0 & \lambda - 2 \end{vmatrix} = (\lambda - 1)(\lambda - 2) = 0$$

as

$$\lambda_1 = 1 \quad \text{and} \quad \lambda_2 = 2$$

Thus matrix A has two distinct eigenvalues.

There are two eigenvectors x_1 and x_2, associated with λ_1 and λ_2, respectively. If we define

$$x_1 = \begin{bmatrix} x_{11} \\ x_{21} \end{bmatrix}, \qquad x_2 = \begin{bmatrix} x_{12} \\ x_{22} \end{bmatrix}$$

then the eigenvector x_1 can be found from

$$Ax_1 = \lambda_1 x_1$$

or

$$(\lambda_1 \mathbf{I} - \mathbf{A})\mathbf{x}_1 = \mathbf{0}$$

Noting that $\lambda_1 = 1$, we have

$$\begin{bmatrix} 1-1 & -1 \\ 0 & 1-2 \end{bmatrix} \begin{bmatrix} x_{11} \\ x_{21} \end{bmatrix} = \begin{bmatrix} 0 \\ 0 \end{bmatrix}$$

which gives

$$x_{11} = \text{arbitrary constant} \quad \text{and} \quad x_{21} = 0$$

Hence, eigenvector $\mathbf{x}_1$ may be written as

$$\mathbf{x}_1 = \begin{bmatrix} x_{11} \\ x_{21} \end{bmatrix} = \begin{bmatrix} c_1 \\ 0 \end{bmatrix}$$

where $c_1 \neq 0$ is an arbitrary constant.

Similarly, for the eigenvector $\mathbf{x}_2$, we have

$$\mathbf{A}\mathbf{x}_2 = \lambda_2 \mathbf{x}_2$$

or

$$(\lambda_2 \mathbf{I} - \mathbf{A})\mathbf{x}_2 = \mathbf{0}$$

Noting that $\lambda_2 = 2$, we obtain

$$\begin{bmatrix} 2-1 & -1 \\ 0 & 2-2 \end{bmatrix} \begin{bmatrix} x_{12} \\ x_{22} \end{bmatrix} = \begin{bmatrix} 0 \\ 0 \end{bmatrix}$$

from which we get

$$x_{12} - x_{22} = 0$$

Hence the eigenvector associated with $\lambda_2 = 2$ may be selected as

$$\mathbf{x}_2 = \begin{bmatrix} x_{12} \\ x_{22} \end{bmatrix} = \begin{bmatrix} c_2 \\ c_2 \end{bmatrix}$$

where $c_2 \neq 0$ is an arbitrary constant.

The two eigenvectors are therefore given by

$$\mathbf{x}_1 = \begin{bmatrix} c_1 \\ 0 \end{bmatrix} \quad \text{and} \quad \mathbf{x}_2 = \begin{bmatrix} c_2 \\ c_2 \end{bmatrix}$$

The fact that eigenvectors $\mathbf{x}_1$ and $\mathbf{x}_2$ are linearly independent can be seen from the fact that the determinant of the matrix $[\mathbf{x}_1 \ \mathbf{x}_2]$ is nonzero:

$$\begin{vmatrix} c_1 & c_2 \\ 0 & c_2 \end{vmatrix} \neq 0$$

To normalize the eigenvectors, we choose $c_1 = 1$ and $c_2 = 1/\sqrt{2}$, or

$$\mathbf{x}_1 = \begin{bmatrix} 1 \\ 0 \end{bmatrix}, \quad \mathbf{x}_2 = \begin{bmatrix} \dfrac{1}{\sqrt{2}} \\ \dfrac{1}{\sqrt{2}} \end{bmatrix}$$

Clearly, the absolute value of each eigenvector becomes unity and therefore the eigenvectors are normalized.

A–11. Obtain a transformation matrix **T** that transforms the matrix

$$\mathbf{A} = \begin{bmatrix} 0 & 1 & 0 & 3 \\ 0 & -1 & 1 & 1 \\ 0 & 0 & 0 & 1 \\ 0 & 0 & -1 & -2 \end{bmatrix}$$

into a Jordan canonical form.

Solution. The characteristic equation is

$$|\lambda\mathbf{I} - \mathbf{A}| = \left|\begin{array}{cc|cc} \lambda & -1 & 0 & -3 \\ 0 & \lambda+1 & -1 & -1 \\ \hline 0 & 0 & \lambda & -1 \\ 0 & 0 & 1 & \lambda+2 \end{array}\right| = \begin{vmatrix} \lambda & -1 \\ 0 & \lambda+1 \end{vmatrix} \begin{vmatrix} \lambda & -1 \\ 1 & \lambda+2 \end{vmatrix}$$

$$= (\lambda + 1)^3\lambda = 0$$

Hence matrix **A** involves eigenvalues

$$\lambda_1 = -1, \qquad \lambda_2 = -1, \qquad \lambda_3 = -1, \qquad \lambda_4 = 0$$

For the multiple eigenvalue -1, we have

$$\lambda_1\mathbf{I} - \mathbf{A} = \begin{bmatrix} -1 & -1 & 0 & -3 \\ 0 & 0 & -1 & -1 \\ 0 & 0 & -1 & -1 \\ 0 & 0 & 1 & 1 \end{bmatrix}$$

which is of rank 2, or rank $(4 - 2)$. From the rank condition we see that there must be two Jordan blocks for eigenvalue -1, that is, one $p_1 \times p_1$ Jordan block and one $p_2 \times p_2$ Jordan block, where $p_1 + p_2 = 3$. Notice that for $p_1 + p_2 = 3$, there is only one combination (2 and 1) for the orders of p_1 and p_2. Let us choose

$$p_1 = 2 \qquad \text{and} \qquad p_2 = 1$$

Then there are one eigenvector and one generalized eigenvector for Jordan block $\mathbf{J}_{p_1}$ and one eigenvector for Jordan block $\mathbf{J}_{p_2}$.

Let us define an eigenvector and a generalized eigenvector for Jordan block $\mathbf{J}_{p_1}$ as $\mathbf{v}_{11}$ and $\mathbf{v}_{12}$, respectively, and an eigenvector for Jordan block $\mathbf{J}_{p_2}$ as $\mathbf{v}_{21}$. Then there must be vectors $\mathbf{v}_{11}$, $\mathbf{v}_{12}$, and $\mathbf{v}_{21}$ that satisfy the following equations:

$$(\mathbf{A} - \lambda_1\mathbf{I})\mathbf{v}_{11} = \mathbf{0}, \qquad (\mathbf{A} - \lambda_1\mathbf{I})\mathbf{v}_{21} = \mathbf{0}$$

$$(\mathbf{A} - \lambda_1\mathbf{I})\mathbf{v}_{12} = \mathbf{v}_{11}$$

For $\lambda_1 = -1$, $\mathbf{A} - \lambda_1\mathbf{I}$ can be given as follows:

$$\mathbf{A} - \lambda_1\mathbf{I} = \begin{bmatrix} 1 & 1 & 0 & 3 \\ 0 & 0 & 1 & 1 \\ 0 & 0 & 1 & 1 \\ 0 & 0 & -1 & -1 \end{bmatrix}$$

Noting that

$$(\mathbf{A} - \lambda_1\mathbf{I})^2 = \begin{bmatrix} 1 & 1 & -2 & 1 \\ 0 & 0 & 0 & 0 \\ 0 & 0 & 0 & 0 \\ 0 & 0 & 0 & 0 \end{bmatrix}$$

we determine vector $\mathbf{v}_{12}$ to be such that it will satisfy the equation

$$(\mathbf{A} - \lambda_1\mathbf{I})^2\mathbf{v}_{12} = \mathbf{0}$$

and at the same time will make $(\mathbf{A} - \lambda_1\mathbf{I})\mathbf{v}_{12}$ nonzero. An example of such a generalized eigenvector $\mathbf{v}_{12}$ can be found to be

$$\mathbf{v}_{12} = \begin{bmatrix} -a \\ 0 \\ 0 \\ a \end{bmatrix} \qquad a = \text{arbitrary nonzero constant}$$

The eigenvector $\mathbf{v}_{11}$ is then found to be a nonzero vector $(\mathbf{A} - \lambda_1\mathbf{I})\mathbf{v}_{12}$:

$$\mathbf{v}_{11} = (\mathbf{A} - \lambda_1\mathbf{I})\mathbf{v}_{12} = \begin{bmatrix} 2a \\ a \\ a \\ -a \end{bmatrix}$$

Since a is an arbitrary nonzero constant, let us choose $a = 1$. Then we have

$$\mathbf{v}_{11} = \begin{bmatrix} 2 \\ 1 \\ 1 \\ -1 \end{bmatrix} \quad \text{and} \quad \mathbf{v}_{12} = \begin{bmatrix} -1 \\ 0 \\ 0 \\ 1 \end{bmatrix}$$

Next, we determine $\mathbf{v}_{21}$ so that $\mathbf{v}_{21}$ and $\mathbf{v}_{11}$ are linearly independent. For $\mathbf{v}_{21}$ we may choose

$$\mathbf{v}_{21} = \begin{bmatrix} b + 3c \\ -b \\ c \\ -c \end{bmatrix}$$

where b and c are arbitrary constants. Let us choose, for example, $b = 1$ and $c = 0$. Then

$$\mathbf{v}_{21} = \begin{bmatrix} 1 \\ -1 \\ 0 \\ 0 \end{bmatrix}$$

Clearly, $\mathbf{v}_{11}$, $\mathbf{v}_{12}$, and $\mathbf{v}_{21}$ are linearly independent. Let us define

$$\mathbf{v}_{11} = \mathbf{x}_1, \qquad \mathbf{v}_{12} = \mathbf{x}_2, \qquad \mathbf{v}_{21} = \mathbf{x}_3$$

and

$$\mathbf{T}(\lambda_1) = [\mathbf{v}_{11} \mid \mathbf{v}_{12} \mid \mathbf{v}_{21}] = [\mathbf{x}_1 \mid \mathbf{x}_2 \mid \mathbf{x}_3] = \begin{bmatrix} 2 & -1 & 1 \\ 1 & 0 & -1 \\ 1 & 0 & 0 \\ -1 & 1 & 0 \end{bmatrix}$$

For the distinct eigenvalue $\lambda_4 = 0$, the eigenvector $\mathbf{x}_4$ can be determined from

$$(\mathbf{A} - \lambda_4\mathbf{I})\mathbf{x}_4 = \mathbf{0}$$

Noting that

$$\mathbf{A} - \lambda_4\mathbf{I} = \mathbf{A} = \begin{bmatrix} 0 & 1 & 0 & 3 \\ 0 & -1 & 1 & 1 \\ 0 & 0 & 0 & 1 \\ 0 & 0 & -1 & -2 \end{bmatrix}$$

we find

$$\mathbf{x}_4 = \begin{bmatrix} d \\ 0 \\ 0 \\ 0 \end{bmatrix}$$

where $d \neq 0$ is an arbitrary constant. By choosing $d = 1$, we have

$$\mathbf{T}(\lambda_4) = \mathbf{x}_4 = \begin{bmatrix} 1 \\ 0 \\ 0 \\ 0 \end{bmatrix}$$

Thus the transformation matrix $\mathbf{T}$ can be written as

$$\mathbf{T} = [\mathbf{T}(\lambda_1) \mathbin{\vdots} \mathbf{T}(\lambda_4)] = \begin{bmatrix} 2 & -1 & 1 & 1 \\ 1 & 0 & -1 & 0 \\ 1 & 0 & 0 & 0 \\ -1 & 1 & 0 & 0 \end{bmatrix}$$

Then

$$\mathbf{T}^{-1}\mathbf{A}\mathbf{T} = \begin{bmatrix} 0 & 0 & 1 & 0 \\ 0 & 0 & 1 & 1 \\ 0 & -1 & 1 & 0 \\ 1 & 1 & -2 & 1 \end{bmatrix} \begin{bmatrix} 0 & 1 & 0 & 3 \\ 0 & -1 & 1 & 1 \\ 0 & 0 & 0 & 1 \\ 0 & 0 & -1 & -2 \end{bmatrix} \begin{bmatrix} 2 & -1 & 1 & 1 \\ 1 & 0 & -1 & 0 \\ 1 & 0 & 0 & 0 \\ -1 & 1 & 0 & 0 \end{bmatrix}$$

$$= \begin{bmatrix} -1 & 1 & 0 & 0 \\ 0 & -1 & 0 & 0 \\ \hline 0 & 0 & -1 & 0 \\ 0 & 0 & 0 & 0 \end{bmatrix} = \operatorname{diag}\,[\mathbf{J}_2(-1), \mathbf{J}_1(-1), \mathbf{J}_1(0)]$$

A–12. Assume that an $n \times n$ normal matrix $\mathbf{A}$ has a k-multiple eigenvalue λ_1. Prove that the rank of $\mathbf{A} - \lambda_1\mathbf{I}$ is $n - k$.

Solution. Suppose that the rank of $\mathbf{A} - \lambda_1\mathbf{I}$ is $n - m$. Then the equation

$$(\mathbf{A} - \lambda_1\mathbf{I})\mathbf{x} = \mathbf{0} \tag{A–41}$$

will have m linearly independent vector solutions. Let us choose m such vectors so that they are orthogonal to each other and normalized. That is, vectors $\mathbf{x}_1, \mathbf{x}_2, \ldots, \mathbf{x}_m$ will satisfy Equation (A–41) and will be orthonormal.

Let us consider $n - m$ vectors $\mathbf{x}_{m+1}, \mathbf{x}_{m+2}, \ldots, \mathbf{x}_n$ such that all n vectors

$$\mathbf{x}_1, \mathbf{x}_2, \ldots, \mathbf{x}_n$$

will be orthonormal to each other. Then matrix $\mathbf{U}$, defined by

$$\mathbf{U} = [\mathbf{x}_1 \mathbin{\vdots} \mathbf{x}_2 \mathbin{\vdots} \cdots \mathbin{\vdots} \mathbf{x}_n]$$

is a unitary matrix.

Since for $1 \leq i \leq m$, we have

$$\mathbf{A}\mathbf{x}_i = \lambda_1\mathbf{x}_i$$

and therefore we can write

$$\mathbf{AU} = \mathbf{U} \begin{bmatrix} \lambda_1 \mathbf{I}_m & \mathbf{B} \\ \mathbf{0} & \mathbf{C} \end{bmatrix}$$

or

$$\mathbf{U*AU} = \begin{bmatrix} \lambda_1 \mathbf{I}_m & \mathbf{B} \\ \mathbf{0} & \mathbf{C} \end{bmatrix}$$

Noting that

$$
\begin{aligned}
\|\mathbf{Ax}_i - \lambda \mathbf{x}_i\|^2 &= \langle (\mathbf{A} - \lambda \mathbf{I})\mathbf{x}_i, (\mathbf{A} - \lambda \mathbf{I})\mathbf{x}_i \rangle \\
&= \langle (\mathbf{A*} - \bar{\lambda} \mathbf{I})(\mathbf{A} - \lambda \mathbf{I})\mathbf{x}_i, \mathbf{x}_i \rangle \\
&= \langle (\mathbf{A} - \lambda \mathbf{I})(\mathbf{A*} - \bar{\lambda} \mathbf{I})\mathbf{x}_i, \mathbf{x}_i \rangle \\
&= \langle (\mathbf{A*} - \bar{\lambda} \mathbf{I})\mathbf{x}_i, (\mathbf{A*} - \bar{\lambda} \mathbf{I})\mathbf{x}_i \rangle \\
&= \|\mathbf{A*x}_i - \bar{\lambda} \mathbf{x}_i\|^2 \\
&= 0
\end{aligned}
$$

we have

$$\mathbf{A*x}_i = \bar{\lambda} \mathbf{x}_i$$

Therefore, we can write

$$\mathbf{A*U} = \mathbf{U} \begin{bmatrix} \bar{\lambda}_1 \mathbf{I}_m & \mathbf{B}_1 \\ \mathbf{0} & \mathbf{C}_1 \end{bmatrix}$$

or

$$\mathbf{U*A*U} = \begin{bmatrix} \bar{\lambda}_1 \mathbf{I}_m & \mathbf{B}_1 \\ \mathbf{0} & \mathbf{C}_1 \end{bmatrix}$$

Hence

$$\begin{bmatrix} \lambda_1 \mathbf{I}_m & \mathbf{B} \\ \mathbf{0} & \mathbf{C} \end{bmatrix} = \mathbf{U*AU} = (\mathbf{U*A*U})* = \begin{bmatrix} \bar{\lambda}_1 \mathbf{I}_m & \mathbf{B}_1 \\ \mathbf{0} & \mathbf{C}_1 \end{bmatrix}^* = \begin{bmatrix} \lambda_1 \mathbf{I}_m & \mathbf{0} \\ \mathbf{B}_1^* & \mathbf{C}_1^* \end{bmatrix}$$

Comparing the left and right sides of this last equation, we obtain

$$\mathbf{B} = \mathbf{0}$$

Hence we get

$$\mathbf{A} = \mathbf{U} \begin{bmatrix} \lambda_1 \mathbf{I}_m & \mathbf{0} \\ \mathbf{0} & \mathbf{C} \end{bmatrix} \mathbf{U*}$$

Then

$$\mathbf{A} - \lambda \mathbf{I} = \mathbf{U} \begin{bmatrix} (\lambda_1 - \lambda)\mathbf{I}_m & \mathbf{0} \\ \mathbf{0} & \mathbf{C} - \lambda \mathbf{I}_{n-m} \end{bmatrix} \mathbf{U*}$$

The determinant of this last equation is

$$|\mathbf{A} - \lambda \mathbf{I}| = (\lambda_1 - \lambda)^m |\mathbf{C} - \lambda \mathbf{I}_{n-m}| \tag{A-42}$$

On the other hand, we have

$$\text{rank}\,(\mathbf{A} - \lambda_1\mathbf{I}) = n - m = \text{rank}\left\{\mathbf{U}\begin{bmatrix} \mathbf{0} & \mathbf{0} \\ \mathbf{0} & \mathbf{C} - \lambda_1\mathbf{I}_{n-m} \end{bmatrix}\mathbf{U}^*\right\}$$

$$= \text{rank}\begin{bmatrix} \mathbf{0} & \mathbf{0} \\ \mathbf{0} & \mathbf{C} - \lambda_1\mathbf{I}_{n-m} \end{bmatrix} = \text{rank}\,(\mathbf{C} - \lambda_1\mathbf{I}_{n-m})$$

Hence, we conclude that the rank of $\mathbf{C} - \lambda_1\mathbf{I}_{n-m}$ is $n - m$. Consequently,

$$|\mathbf{C} - \lambda_1\mathbf{I}_{n-m}| \neq 0$$

and from Equation (A–42), λ_1 is shown to be the m-multiple eigenvalue of $|\mathbf{A} - \lambda\mathbf{I}| = 0$. Since λ_1 is the k-multiple eigenvalue of $\mathbf{A}$, we must have $m = k$. Therefore, the rank of $\mathbf{A} - \lambda_1\mathbf{I}$ is $n - k$.

Note that since the rank of $\mathbf{A} - \lambda_1\mathbf{I}$ is $n - k$, the equation

$$(\mathbf{A} - \lambda_1\mathbf{I})\mathbf{x}_i = \mathbf{0}$$

will have k linearly independent eigenvectors $\mathbf{x}_1, \mathbf{x}_2, \ldots, \mathbf{x}_k$.

A–13. Prove that the eigenvalues of an $n \times n$ Hermitian matrix and of an $n \times n$ real symmetric matrix are real. Prove also that the eigenvalues of a skew-Hermitian matrix and of a real skew-symmetric matrix are either zero or purely imaginary.

Solution. Let us define any eigenvalue of an $n \times n$ Hermitian matrix $\mathbf{A}$ by $\lambda = \alpha + j\beta$. There exists a vector $\mathbf{x} \neq \mathbf{0}$ such that

$$\mathbf{A}\mathbf{x} = (\alpha + j\beta)\mathbf{x}$$

The conjugate transpose of this last equation is

$$\mathbf{x}^*\mathbf{A}^* = (\alpha - j\beta)\mathbf{x}^*$$

Since $\mathbf{A}$ is Hermitian, $\mathbf{A}^* = \mathbf{A}$. Therefore, we obtain

$$\mathbf{x}^*\mathbf{A}\mathbf{x} = (\alpha - j\beta)\mathbf{x}^*\mathbf{x}$$

On the other hand, since $\mathbf{A}\mathbf{x} = (\alpha + j\beta)\mathbf{x}$, we have

$$\mathbf{x}^*\mathbf{A}\mathbf{x} = (\alpha + j\beta)\mathbf{x}^*\mathbf{x}$$

Hence we obtain

$$[(\alpha - j\beta) - (\alpha + j\beta)]\mathbf{x}^*\mathbf{x} = 0$$

or

$$-2j\beta\mathbf{x}^*\mathbf{x} = 0$$

Since $\mathbf{x}^*\mathbf{x} \neq 0$ (for $\mathbf{x} \neq \mathbf{0}$), we conclude that

$$\beta = 0$$

This proves that any eigenvalue of an $n \times n$ Hermitian matrix $\mathbf{A}$ is real. It follows that the eigenvalues of a real symmetric matrix are also real, since it is Hermitian.

To prove the second half of the problem, notice that if $\mathbf{B}$ is skew-Hermitian, then $j\mathbf{B}$ is Hermitian. Hence, the eigenvalues of $j\mathbf{B}$ are real, which implies that the eigenvalues of $\mathbf{B}$ are either zero or purely imaginary.

The eigenvalues of a real skew-symmetric matrix are also either zero or purely imaginary, since a real skew-symmetric matrix is skew-Hermitian.

Note that, in the real skew-symmetric matrix, purely imaginary eigenvalues always occur in conjugate pairs, since the coefficients of the characteristic equation are real. Note also that an $n \times n$ real skew-symmetric matrix is singular if n is odd, since such a matrix must include at least one zero eigenvalue.

A–14. Examine whether or not the following 3×3 matrix $\mathbf{A}$ is positive definite:

$$\mathbf{A} = \begin{bmatrix} 2 & 2 & -1 \\ 2 & 6 & 0 \\ -1 & 0 & 1 \end{bmatrix}$$

Solution. We shall demonstrate three different ways to test the positive definiteness of matrix $\mathbf{A}$.

1. We may first apply Sylvester's criterion for positive definiteness of a quadratic form $\mathbf{x}^T\mathbf{A}\mathbf{x}$. For the given matrix $\mathbf{A}$, we have

$$2 > 0, \qquad \begin{vmatrix} 2 & 2 \\ 2 & 6 \end{vmatrix} > 0, \qquad \begin{vmatrix} 2 & 2 & -1 \\ 2 & 6 & 0 \\ -1 & 0 & 1 \end{vmatrix} > 0$$

Thus, the successive principal minors are all positive. Hence matrix $\mathbf{A}$ is positive definite.

2. We may examine the positive definiteness of $\mathbf{x}^T\mathbf{A}\mathbf{x}$. Since

$$\mathbf{x}^T\mathbf{A}\mathbf{x} = [x_1 \, x_2 \, x_3] \begin{bmatrix} 2 & 2 & -1 \\ 2 & 6 & 0 \\ -1 & 0 & 1 \end{bmatrix} \begin{bmatrix} x_1 \\ x_2 \\ x_2 \end{bmatrix}$$

$$= 2x_1^2 + 4x_1x_2 - 2x_1x_3 + 6x_2^2 + x_3^2$$

$$= (x_1 - x_3)^2 + (x_1 + 2x_2)^2 + 2x_2^2$$

we find that $\mathbf{x}^T\mathbf{A}\mathbf{x}$ is positive except at the origin ($\mathbf{x} = \mathbf{0}$). Hence, we conclude that matrix $\mathbf{A}$ is positive definite.

3. We may examine the eigenvalues of matrix $\mathbf{A}$. Note that

$$|\lambda\mathbf{I} - \mathbf{A}| = \lambda^3 - 9\lambda^2 + 15\lambda - 2$$
$$= (\lambda - 2)(\lambda - 0.1459)(\lambda - 6.8541)$$

Hence

$$\lambda_1 = 2, \qquad \lambda_2 = 0.1459, \qquad \lambda_3 = 6.8541$$

Since all eigenvalues are positive, we conclude that $\mathbf{A}$ is a positive definite matrix.

A–15. Examine whether the following matrix $\mathbf{A}$ is positive semidefinite:

$$\mathbf{A} = \begin{bmatrix} 1 & 2 & 1 \\ 2 & 4 & 2 \\ 1 & 2 & 0 \end{bmatrix}$$

Solution. In the positive semidefiniteness test, we need to examine the signs of all principal minors in addition to the sign of the determinant of the given matrix, which must be zero; that is, $|\mathbf{A}|$ must be equal to 0.

For the 3×3 matrix

$$\begin{bmatrix} a_{11} & a_{12} & a_{13} \\ a_{21} & a_{22} & a_{23} \\ a_{31} & a_{32} & a_{33} \end{bmatrix}$$

there are six principal minors:

$$a_{11}, \qquad a_{22}, \qquad a_{33}, \qquad \begin{vmatrix} a_{11} & a_{12} \\ a_{21} & a_{22} \end{vmatrix}, \qquad \begin{vmatrix} a_{22} & a_{23} \\ a_{32} & a_{33} \end{vmatrix}, \qquad \begin{vmatrix} a_{11} & a_{13} \\ a_{31} & a_{33} \end{vmatrix}$$

We need to examine the signs of all six principal minors and the sign of $|\mathbf{A}|$.

For the given matrix $\mathbf{A}$,

$$a_{11} = 1 > 0$$

$$a_{22} = 4 > 0$$

$$a_{33} = 0$$

$$\begin{vmatrix} a_{11} & a_{12} \\ a_{21} & a_{22} \end{vmatrix} = \begin{vmatrix} 1 & 2 \\ 2 & 4 \end{vmatrix} = 0$$

$$\begin{vmatrix} a_{22} & a_{23} \\ a_{32} & a_{33} \end{vmatrix} = \begin{vmatrix} 4 & 2 \\ 2 & 0 \end{vmatrix} = -4 < 0$$

$$\begin{vmatrix} a_{11} & a_{13} \\ a_{31} & a_{33} \end{vmatrix} = \begin{vmatrix} 1 & 1 \\ 1 & 0 \end{vmatrix} = -1 < 0$$

$$\begin{vmatrix} a_{11} & a_{12} & a_{13} \\ a_{21} & a_{22} & a_{23} \\ a_{31} & a_{32} & a_{33} \end{vmatrix} = \begin{vmatrix} 1 & 2 & 1 \\ 2 & 4 & 2 \\ 1 & 2 & 0 \end{vmatrix} = 0$$

Clearly, two of the principal minors are negative. Hence we conclude that matrix $\mathbf{A}$ is not positive semidefinite.

It is important to note that had we tested the signs of only the successive principal minors and the determinant of $\mathbf{A}$,

$$1 > 0, \qquad \begin{vmatrix} 1 & 2 \\ 2 & 4 \end{vmatrix} = 0, \qquad |\mathbf{A}| = \begin{vmatrix} 1 & 2 & 1 \\ 2 & 4 & 2 \\ 1 & 2 & 0 \end{vmatrix} = 0$$

we would have reached the wrong conclusion that matrix $\mathbf{A}$ is positive semidefinite.

In fact, for the given matrix $\mathbf{A}$,

$$|\lambda \mathbf{I} - \mathbf{A}| = \begin{vmatrix} \lambda - 1 & -2 & -1 \\ -2 & \lambda - 4 & -2 \\ -1 & -2 & \lambda \end{vmatrix} = (\lambda^2 - 5\lambda - 5)\lambda$$

$$= (\lambda - 5.8541)\lambda(\lambda + 0.8541)$$

and so the eigenvalues are

$$\lambda_1 = 5.8541, \qquad \lambda_2 = 0, \qquad \lambda_3 = -0.8541$$

For matrix $\mathbf{A}$ to be positive semidefinite, all eigenvalues must be nonnegative and at least one of them must be zero. Clearly, matrix $\mathbf{A}$ is an indefinite matrix.

A–16. Consider the equation

$$x_1 + 5x_2 = 1 \tag{A–43}$$

or, in vector-matrix form,

$$\mathbf{A}\mathbf{x} = b$$

where

$$\mathbf{A} = [1 \quad 5], \qquad \mathbf{x} = \begin{bmatrix} x_1 \\ x_2 \end{bmatrix}, \qquad b = 1$$

Find the solution that will give the minimum norm $\|\mathbf{x}\|$, that is, the solution closest to the origin.

Solution. The minimum norm solution is given by

$$\mathbf{x}^\circ = \mathbf{A}^{RM}b$$

where the right pseudoinverse $\mathbf{A}^{RM}$ is given by

$$\mathbf{A}^{RM} = \mathbf{A}^T(\mathbf{A}\mathbf{A}^T)^{-1}$$

In the present example, the right pseudoinverse matrix becomes

$$\mathbf{A}^{RM} = \begin{bmatrix} 1 \\ 5 \end{bmatrix} \left\{ [1 \quad 5] \begin{bmatrix} 1 \\ 5 \end{bmatrix} \right\}^{-1} = \begin{bmatrix} 1 \\ 5 \end{bmatrix} (26)^{-1} = \begin{bmatrix} \frac{1}{26} \\ \frac{5}{26} \end{bmatrix}$$

Hence the minimum norm solution becomes

$$\mathbf{x}^\circ = \begin{bmatrix} x_1^\circ \\ x_2^\circ \end{bmatrix} = \begin{bmatrix} \frac{1}{26} \\ \frac{5}{26} \end{bmatrix}$$

It is instructive to examine the minimum norm solution graphically. In Figure A–2, the minimum norm solution is located at point P. This point is the intersection of line $x_1 + 5x_2 = 1$ and the line that is perpendicular to it and passes through the origin. This perpendicular line can be written as

$$x_1 - \tfrac{1}{5}x_2 = 0 \tag{A–44}$$

The solution of the simultaneous Equations (A–43) and (A–44) gives point P:

$$x_1 = \tfrac{1}{26}, \qquad x_2 = \tfrac{5}{26}$$

which is the same as the result obtained by use of the right pseudoinverse matrix.

A–17. Consider a vector-matrix equation

$$\mathbf{A}\mathbf{x} = \mathbf{b}$$

where

$$\mathbf{A} = \begin{bmatrix} 1 & 1 \\ 1 & 2 \\ 1 & 4 \end{bmatrix}, \qquad \mathbf{x} = \begin{bmatrix} x_1 \\ x_2 \end{bmatrix}, \qquad \mathbf{b} = \begin{bmatrix} 1 \\ 2 \\ 2 \end{bmatrix}$$

Clearly, no solution exists in the classical sense.

Figure A–2
Graphical representation of the minimum norm solution for Eq. (A–43) in Problem A–16.

Appendix / Vector-Matrix Analysis

Find the minimum norm solution such that norm $\|\mathbf{Ax} - \mathbf{b}\|$ is minimum.

Solution. The desired minimum norm solution is given by

$$\mathbf{x}^\circ = \mathbf{A}^{LM}\mathbf{b} = (\mathbf{A}^T\mathbf{A})^{-1}\mathbf{A}^T\mathbf{b}$$

Hence

$$\mathbf{x}^\circ = \left\{ \begin{bmatrix} 1 & 1 & 1 \\ 1 & 2 & 4 \end{bmatrix} \begin{bmatrix} 1 & 1 \\ 1 & 2 \\ 1 & 4 \end{bmatrix} \right\}^{-1} \begin{bmatrix} 1 & 1 & 1 \\ 1 & 2 & 4 \end{bmatrix} \begin{bmatrix} 1 \\ 2 \\ 2 \end{bmatrix} = \begin{bmatrix} 1 \\ \frac{2}{7} \end{bmatrix}$$

This problem can, of course, be solved in several different ways. The use of the left pseudoinverse matrix $\mathbf{A}^{LM}$ is one approach, as just demonstrated. An additional method, based on an ordinary minimization method, follows.

Noting that minimizing $\|\mathbf{Ax} - \mathbf{b}\|$ is the same as minimizing $\|\mathbf{Ax} - \mathbf{b}\|^2$, let us minimize $\|\mathbf{Ax} - \mathbf{b}\|^2$. We shall begin the solution by writing $\|\mathbf{Ax} - \mathbf{b}\|^2$ as follows:

$$\|\mathbf{Ax} - \mathbf{b}\|^2 = (x_1 + x_2 - 1)^2 + (x_1 + 2x_2 - 2)^2 + (x_1 + 4x_2 - 2)^2$$

Let

$$L = (x_1 + x_2 - 1)^2 + (x_1 + 2x_2 - 2)^2 + (x_1 + 4x_2 - 2)^2$$

Then, by differentiating L with respect to x_1 and x_2, respectively, and equating each of the resulting equations to zero, we obtain

$$\frac{\partial L}{\partial x_1} = 2(x_1 + x_2 - 1) + 2(x_1 + 2x_2 - 2) + 2(x_1 + 4x_2 - 2) = 0$$

$$\frac{\partial L}{\partial x_2} = 2(x_1 + x_2 - 1) + 4(x_1 + 2x_2 - 2) + 8(x_1 + 4x_2 - 2) = 0$$

which can be simplified to read

$$3x_1 + 7x_2 - 5 = 0$$

$$7x_1 + 21x_2 - 13 = 0$$

The solution to these two simultaneous equations is

$$x_1 = 1, \qquad x_2 = \tfrac{2}{7}$$

or

$$\mathbf{x}^\circ = \begin{bmatrix} 1 \\ \frac{2}{7} \end{bmatrix}$$

which is the same as the solution obtained by use of the left pseudoinverse of $\mathbf{A}$.

REFERENCES

A–1   Ackermann, J.E., "Der Entwulf Linearer Regelungs Systeme im Zustandstraum," *Regelungstechnik und Prozessdatenverarbeitung*, **7**(1972), pp. 297–300.

A–2   Athans, M., and P. L. Falb, *Optimal Control: An Introduction to the Theory and Its Applications*. New York: McGraw-Hill Book Company, 1965.

B–1   Barnet, S., "Matrices, Polynomials, and Linear Time-Invariant Systems," *IEEE Trans. Automatic Control*, **AC-18** (1973), pp. 1–10.

B–2   Bayliss, L. E., *Living Control Systems*. London: English Universities Press Limited, 1966.

B–3   Bellman, R., *Introduction to Matrix Analysis*. New York: McGraw-Hill Book Company, 1960.

B–4   Bode, H. W., *Network Analysis and Feedback Design*. New York: Van Nostrand Reinhold, 1945.

B–5   Brogan, W. L., *Modern Control Theory*. Englewood Cliffs, N.J.: Prentice-Hall, 1985.

B–6   Butman, S., and R. Sivan (Sussman), "On Cancellations, Controllability and Observability," *IEEE Trans. Automatic Control*, **AC-9** (1964), pp. 317–8.

C–1   Campbell, D. P., *Process Dynamics*. New York: John Wiley & Sons, Inc., 1958.

C–2   Cannon, R., *Dynamics of Physical Systems*. New York: McGraw-Hill Book Company, 1967.

C–3   Cheng, D. K., *Analysis of Linear Systems*. Reading, Mass.: Addison-Wesley Publishing Company, Inc., 1959.

C–4   Churchill, R. V., *Operational Mathematics*, 3rd ed. New York: McGraw-Hill Book Company, 1972.

C–5   Coddington, E. A., and N. Levinson, *Theory of Ordinary Differential Equations*. New York: McGraw-Hill Book Company, 1955.

C–6   Craig, J. J., *Introduction to Robotics, Mechanics and Control*. Reading, Mass.: Addison-Wesley Publishing Company, Inc., 1986.

C–7 Cunningham, W. J., *Introduction to Nonlinear Analysis*. New York: McGraw-Hill Book Company, 1958.

E–1 Enns, M., J. R. Greenwood, III, J. E. Matheson, and F. T. Thompson, "Practical Aspects of State-Space Methods Part I: System Formulation and Reduction," *IEEE Trans. Military Electronics*, **MIL-8** (1964), pp. 81–93.

E–2 Evans, W. R., "Graphical Analysis of Control Systems," *AIEE Trans. Part II*, **67** (1948), pp. 547–51.

E–3 Evans, W. R., "Control System Synthesis by Root Locus Method," *AIEE Trans. Part II*, **69** (1950), pp. 66–9.

E–4 Evans, W. R., "The Use of Zeros and Poles for Frequency Response or Transient Response," *ASME Trans.*, **76** (1954), pp. 1335–44.

F–1 Franklin, G. F., J. D. Powell, and A. Emami-Naeini, *Feedback Control of Dynamic Systems*. Reading, Mass.: Addison-Wesley Publishing Company, Inc., 1986.

F–2 Friedland, B., *Control System Design*. New York: McGraw-Hill Book Company, 1986.

F–3 Fu, K. S., R. C. Gonzalez, and C. S. G. Lee, *Robotics*: *Control, Sensing, Vision, and Intelligence*. New York: McGraw-Hill Book Company, 1987.

G–1 Gantmacher, F. R., *Theory of Matrices*, Vols. I and II. New York: Chelsea Publishing Co., Inc., 1959.

G–2 Gibson, J. E., *Nonlinear Automatic Control*. New York: McGraw-Hill Book Company, 1963.

G–3 Gilbert, E. G., "Controllability and Observability in Multivariable Control Systems," *J.SIAM Control*, ser. A, **1** (1963), pp. 128–51.

G–4 Graham, D., and R. C. Lathrop, "The Synthesis of Optimum Response: Criteria and Standard Forms," *AIEE Trans. Part II*, **72** (1953), pp. 273–88.

H–1 Hahn, W., *Theory and Application of Liapunov's Direct Method*. Englewood Cliffs, N.J.: Prentice-Hall, Inc., 1963.

H–2 Halmos, P. R., *Finite Dimensional Vector Spaces*. New York: Van Nostrand Reinhold, 1958.

H–3 Higdon, D. T., and R. H. Cannon, Jr., "On the Control of Unstable Multiple-Output Mechanical Systems," *ASME Paper no.* **63**-*WA-148*, 1963.

I–1 Irwin, J. D., *Basic Engineering Circuit Analysis*. New York: Macmillan Inc., 1984.

K–1 Kailath, T., *Linear Systems*. Englewood Cliffs, N.J.: Prentice-Hall, Inc., 1980.

K–2 Kalman, R. E., "Contributions to the Theory of Optimal Control," *Bol. Soc. Mat. Mex.*, **5** (1960), pp. 102–19.

K–3 Kalman, R. E., "On the General Theory of Control Systems," *Proc. First Intern, Cong. IFAC*, Moscow, 1960; *Automatic and Remote Control*. London: Butterworths & Company, Ltd., 1961, pp. 481–92.

K–4 Kalman, R. E., "Canonical Structure of Linear Dynamical Systems," *Proc. Natl. Acad. Sci.*, *USA*, **48** (1962), pp. 596–600.

K–5 Kalman, R. E., "When Is a Linear Control System Optimal?" *ASME J. Basic Engineering*, ser. D, **86** (1964), pp. 51–60.

K–6 Kalman, R. E., and J. E. Bertram, "Control System Analysis and Design via the Second Method of Lyapunov: I Continuous-Time Systems," *ASME J. Basic Engineering*, ser. D, **82** (1960), pp. 371–93.

K–7 Kalman, R. E., Y. C. Ho, and K. S. Narendra, "Controllability of Linear Dynamic Systems," in *Contributions to Differential Equations*, Vol. 1. New York: Wiley-Interscience Publishers, Inc., 1962.

K–8 Kochenburger, R. J., "A Frequency Response Method for Analyzing and Synthesizing Contactor Servomechanisms," *AIEE Trans.*, **69** (1950), pp. 270–83.

K–9 Kreindler, E., and P. E. Sarachick, "On the Concepts of Controllability and Observability of Linear Systems," *IEEE Trans. Automatic Control*, **AC-9** (1964), pp. 129–36.

K–10 Kuo, B. C., *Automatic Control Systems*, 5th ed. Englewood Cliffs, N.J.: Prentice-Hall, Inc., 1987.

L–1 LaSalle, J. P., and S. Lefschetz, *Stability by Liapunov's Direct Method with Applications*. New York: Academic Press, Inc., 1961.

L–2 Luenberger, D. G., "Observing the State of a Linear System," *IEEE Trans. Military Electr.*, **MIL-8** (1964), pp. 74–80.

M–1 Mason, S. J., "Feedback Theory: Some Properties of Signal Flow Graphs," *Proc. IRE*, **41** (1953), pp. 1144–56.

M–2 Mason, S. J., "Feedback Theory: Further Properties of Signal Flow Graphs," *Proc. IRE*, **44** (1956), pp. 920–6.

M–3 Melbourne, W. G., "Three Dimensional Optimum Thrust Trajectories for Power-Limited Propulsion Systems," *ARS J.*, **31** (1961), pp. 1723–8.

M–4 Melbourne, W. G., and C. G. Sauer, Jr., "Optimum Interplanetary Rendezvous with Power-Limited Vehicles, *AIAA J.*, **1** (1963), pp. 54–60.

M–5 Minorsky, N., *Nonlinear Oscillations*. New York: Van Nostrand Reinhold, 1962.

M–6 Monopoli, R. V., "Controller Design for Nonlinear and Time-Varying Plants," *NASA* **CR-152,** Jan., 1965.

N–1 Noble, B., and J. Daniel, *Applied Linear Algebra*, 2nd ed. Englewood Cliffs, N.J.: Prentice-Hall, Inc., 1977.

N–2 Nyquist, H., "Regeneration Theory," *Bell System Tech. J.*, **11** (1932), pp. 126–47.

O–1 Ogata, K., *State Space Analysis of Control Systems*. Englewood Cliffs, N.J.: Prentice-Hall, Inc., 1967.

O–2 Ogata, K., *System Dynamics*. Englewood Cliffs, N.J.: Prentice-Hall, Inc., 1978.

O–3 Ogata, K., *Discrete-Time Control Systems*. Englewood Cliffs, N.J.: Prentice Hall, Inc., 1987.

P–1 Phillips, C. L., and R. D. Harbor, *Feedback Control Systems*. Englewood Cliffs, N.J.: Prentice-Hall, Inc., 1988.

R–1 Rekasius, Z. V., "A General Performance Index for Analytical Design of Control Systems," *IRE Trans. Automatic Control*, **AC-6** (1961), pp. 217–22.

S–1 Schultz, W. C., and V. C. Rideout, "Control System Performance Measures: Past, Present, and Future," *IRE Trans. Automatic Control*, **AC-6** (1961), pp. 22–35.

S–2 Smith, R. J., *Electronics*: *Circuits and Devices*, 2d ed. New York: John Wiley & Sons, Inc., 1980.

S–3 Staats, P. F., "A Survey of Adaptive Control Topics'" *Plan B paper*, Dept. of Mech. Eng., University of Minnesota, Mar., 1966.

S–4 Strang, G., *Linear Algebra and Its Applications*. New York: Academic Press, Inc., 1976.

T–1 Truxal, J. G., *Automatic Feedback Systems Synthesis*. New York: McGraw-Hill Book Company, 1955.

V–1 Valkenburg, M. E., *Network Analysis*. Englewood Cliffs, N.J.: Prentice-Hall, Inc., 1974.

V–2 Van Landingham, H. F., and W. A. Blackwell, "Controller Design for Nonlinear and Time-Varying Plants," *Educational Monograph*, College of Engineering, Oklahoma State University, 1967.

W–1 Wadel, L. B., "Describing Function as Power Series," *IRE Trans. Automatic Control*, **AC-7** (1962), p. 50.

W–2 Waltz, M. D., and K. S. Fu, "A Learning Control System," *Proc. Joint Automatic Control Conference*, 1964, pp. 1–5.

W–3 Wilcox, R. B., "Analysis and Synthesis of Dynamic Performance of Industrial Organizations— The Application of Feedback Control Techniques to Organizational Systems," *IRE Trans. Automatic Control*, **AC-7** (1962), pp. 55–67.

W–4 Willems, J. C., and S. K. Mitter, "Controllability, Observability, Pole Allocation, and State Reconstruction," *IEEE Trans. Automatic Control*, **AC-16** (1971), pp. 582–95.

W–5 Wojcik, C. K., "Analytical Representation of the Root Locus," *ASME J. Basic Engineering*, ser. D, **86** (1964), pp. 37–43.

W–6 Wonham, W. M., "On Pole Assignment in Multi-Input Controllable Linear Systems," *IEEE Trans. Automatic Control*, **AC-12** (1967), pp. 660–65.

Z–1 Ziegler, J. G., and N. B. Nichols, "Optimum Settings for Automatic Controllers," *ASME Trans.* **64** (1942), pp. 759–68.

Z–2 Ziegler, J. G., and N. B. Nichols, "Process Lags in Automatic Control Circuits," *ASME Trans.* **65** (1943), pp. 433–44.

Index

Block diagram, 43–45
    procedures for drawing, 47–51
    realization of state equation and output
        equation, 97, 99
    reduction, 48–52
Block diagram algebra:
    rules of, 48–49
Bode diagrams, 432–51
    error in asymptotic expression of, 437
    of first-order factors, 435
    of gain K, 433
    of integral and derivative factors, 434–
        35
    of quadratic factors, 439–41
Branch, 52
    point, 44
Break frequency, 436
Breakaway point, 355, 357, 370
Break-in point, 357, 361, 370
Bridged-T networks, 607–8
Business systems, 12

C

Cancellation:
    in transfer matrix, 754–56
    of undesirable complex-conjugate poles,
        607
    of undesirable poles, 606–8
Capacitance, 191–93
    of liquid level system, 126
    of pressure system, 191–93
    of thermal system, 133–37
Cauchy–Riemann conditions, 14
Cauchy's theorem, 516
Cayley–Hamilton theorem, 686–87, 742
Change of time scale, 24
Characteristic equation, 330
Characteristic polynomial, 912
Characteristic roots, 680, 912
Circular root locus, 362
Classical control theory:
    versus modern control theory, 67
Closed-loop control system, 4
Closed-loop frequency response, 495–505
Closed-loop transfer function, 45–46
Compensation, 537
    feedback, 538
    lag, 544
    lag–lead, 544
    lead, 544
    parallel, 538
    series, 538
Compensator, 537
    lag, 561–64

    lag–lead, 576–80
    lead, 544–50
Complete observability, 707–8
Complete state controllability, 701–2, 753
    alternate form of, 703–5
    in the s plane, 705
Complex bilinear form, 923
Complex differentiation theorem, 31
Complex function, 13
Complex impedance, 108–9
Complex matrix, 892
Complex quadratic form, 726–27, 923
Complex variable, 13
Conditionally stable system, 379–80, 474–
    75
Conduction heat transfer, 133
Conformal mapping, 464, 482–83, 516–17,
    524
    for relative stability analysis, 482–83
Conjugate matrix, 890
Conjugate transpose, 891
Constant magnitude loci (M circles), 496–
    97
Constant phase-angle loci (N circles), 497–
    99
Control, 2
    feedback, 3
    playback mode of, 9
    programmed, 4
    of temperature, 11
Control action, 182
    integral, 186
    on–off, 184–86
    proportional, 186
    proportional-plus-derivative, 187–88
    proportional-plus-integral, 187
    proportional-plus-integral-plus-derivative,
        188–89
    two-position, 184–86
Control law, 69
Control-signal synthesis, 854
Control system:
    adaptive, 5
    classification of, 6, 289
    closed-loop, 4
    deterministic, 6
    distributed-parameter, 6
    feedback, 3
    inventory, 12
    learning, 5
    lumped-parameter, 6
    numerical, 9–10
    open-loop, 4
    process, 4
    robot, 7–9
    robot arm, 8–9

robot hand grasping, 9
satellite attitude, 42–43
speed, 6–7
stochastic, 6
temperature, 10–11
time-invariant, 6
time-varying, 6
traffic, 11
Controllability, 699–706
complete state, 703–5, 753
matrix, 702
output, 706
Controllable canonical form, 715
Controlled variable, 2
Controller–observer, 810–12
Convection heat transfer, 133
Conventional control theory:
versus modern control theory, 60, 67
Convolution, 31
integral, 31–33, 251–53
Coordinate transformation, 139–43
Corner frequency, 436, 440
Critical damping, 260
Cutoff frequency, 493–94
Cutoff rate, 494

D

Damped natural frequency, 261
Damper, 101
Damping ratio, 260
lines of constant, 552
Dashpot, 101, 212
DC servomotors, 118–22
armature control of, 118–21
permanent magnet, 118
speed control of, 121–22
DC tachometer, 274–75
Dead space, 94
Dead time, 381–82, 385, 507
Dead-zone nonlinearity, 94
Decade, 434
Decibel, 433
Degrees of freedom, 153–55
Delay time, 265
Derivative control, 376
Derivative control action, 223
Derivative time, 188
Describing function, 652–60
analysis, 661–65
for backlash nonlinearity, 674–75
for dead zone nonlinearity, 657–58
for hysteresis nonlinearity, 674–75
for on–off nonlinearity, 653–55

for on–off nonlinearity with dead zone, 668–69
for on–off nonlinearity with dead zone and hysteresis, 670–72
for on–off nonlinearity with hysteresis, 655–56
for saturation nonlinearity, 658–59
for spring preload nonlinearity, 672–73
Design, 68
basic approach to, 68–69
steps, 69
Determinants, 895
properties of, 895–96
Deterministic control system, 6
Diagonal canonical form, 719
Diagonal matrix, 682
Diagonalization:
of $n \times n$ matrix, 682, 913
Diagonalizing transformation matrix, 740
Differential amplifier, 112
Differential gap, 185–86, 667
Dirac delta function, 23
Direct programming method, 713–15
Distributed-parameter control system, 6
Distributive law, 901
Disturbance, 3
external, 3
internal, 3
measurable, 608
unmeasurable, 608
Disturbance-feedforward scheme, 634
Disturbance feedforward transfer function, 609–10
Dither signal, 146
Domain of attraction, 724
Dominant closed-loop poles, 282
Duffing's equation, 646

E

e^{At}:
computation of, 688–93
machine computation of, 687–88
Edge-position control system, 246–47
Eigenvalue, 680, 911–12
invariance of, 681–82
Eigenvector, 912, 935
generalized, 916–20
normalized, 913
Electric-pneumatic transducer, 246
Electric solenoid-operated valve, 185
Electrical lag network, 562–63
Electrical lag-lead network, 577–78
Electrical lead network, 547–48
Electromagnetic valve, 185

Input node, 52
Instability, 725, 730
Integral absolute-error criterion, 297
Integral control:
    action, 186, 219
    of load element, 223
    with observed-state feedback, 613
    with state feedback, 613
Integral-of-time-multiplied absolute-error
        criterion, 297–98
Integral-of-time-multiplied square-error cri-
        terion, 297
Integral square-error criterion, 296
Integral time, 187
Inventory control system, 12
Inverse Laplace transformation, 34–35
Inverse matrix, 897
    properties of, 897
Inverse polar plots, 477–78
Inversion integral, 35
Inverted pendulum system, 104–7, 787–95,
        830–36
Inverting amplifier, 112–13
I-PD control, 611–12, 821
I-PD controlled system:
    with feedforward control, 611–12
ISE performance index, 300–302
ITAE criterion:
    applied to nth-order systems, 299–300
ITSE performance index, 302–3

J

Jacobian, 903
Jordan blocks, 704, 913–15, 920, 937
Jordan canonical form, 683, 704, 721–22,
        913–14, 920, 937
Jump resonance, 647

K

Kalman, R. E., 699, 711
Kirchhoff's laws:
    current (node), 107
    voltage (loop), 107

L

Lag compensation, 574
    frequency-response approach to, 570–71
    root-locus approach to, 566–67
Lag compensator:
    Bode diagram of, 565

characteristics of, 564
electronic, 561–62
polar plot of, 564
Lag network, 538–39
    electrical, 562–63
    mechanical, 563–64
    polar plot of, 614
Lag–lead compensation, 575–92
    frequency-response approach to, 589–90
    root-locus approach to, 580
Lag–lead compensator, 589–92
    Bode diagram of, 580–81
    characteristics of, 579–80
    electronic, 576
    polar plot of, 580
Lag–lead networks, 538–39
    electrical, 577
    mechanical, 578–79
Lagrange's interpolation formula, 747
Laminar flow, 126
    resistance, 127
Laplace transform:
    existence of, 16
    inverse, 34–35
    properties of, 33–34
    table, 20–21
    of translated function, 21–22
Laplace transform method:
    for solving differential equations, 39–41
Laplace transformation:
    definition of, 16
Lead compensation:
    frequency-response approach to, 555–56
    root-locus approach to, 550–51
Lead compensator, 544, 555
    Bode diagram of, 550
    characteristics of, 549–50
    polar plot of, 549
Lead network, 538–39
    electrical, 547–48
    electronic, 544–47
    mechanical, 548
    polar plot of, 614
Learning control system, 5
Learning system, 855
Left pseudoinverse matrix, 928–29, 945
Liapunov, A. M., 723
Liapunov function, 728
Liapunov stability analysis, 722–36
    of linear time-invariant systems, 732–34
Liapunov's main stability theorem, 728–29,
        761–63
Limit cycle, 648–49
    stability of, 662–63
Linear dependence:
    of vectors, 905

N

O

P

Relative stability, 250, 264, 482–83
Relative stability analysis, 288
  via conformal mapping, 482–83
Repeats per minute, 187
Reset:
  control, 186
  rate, 187
Residue, 36
  theorem, 516–18
Resistance, 191–92
  of liquid-level system, 126
  of pressure system, 191–92
  of thermal system, 133–37
Resonant (peak) frequency, 442, 454–55,
    488–89
Resonant peak magnitude, 454, 488–89
Resonant peak value, 442
Reynolds number, 126
Right pseudoinverse matrix, 927, 944
Rise time, 265–67
Robot:
  arm control system, 8–9
  arm simulator, 139–43
  arm system, 137–39, 164–66
  control system, 7–9
  hand grasping control system, 9
  industrial, 7–9
Robot systems:
  circular coordinate, 138
  multiple-joint-type, 138
  polar coordinate, 138
  rectangular coordinate, 138
Root contour, 348, 385
  plots, 385, 387
Root loci:
  asymptotes of, 355
  general rules for constructing, 367–73
Root-locus method, 347–48
Root-locus plot:
  for positive feedback system, 395–98
Routh's stability criterion, 283–88
Runge–Kutta equations:
  fourth-order, 314, 336–39
  third-order, 314
Runge–Kutta method, 312–14

S

Satellite attitude control system, 42–43
Saturation nonlinearity, 94
Scalar product, 905
Schwarz inequality, 908
Second method of Liapunov, 723, 727–31
  optimization via, 838–45

Second-order system:
  critically damped, 262
  impulse response of, 270–72
  overdamped, 262–64
  proportional-plus-derivative control of,
    273–74
  step response of, 260–78
  underdamped, 261–62
Self-excited oscillation, 648–49
Self-operated controller, 184
Sensitivity, 225
Sensor, 183, 189
  first-order, 189
  overdamped second-order, 189
  underdamped second-order, 189
Series compensation, 538
Servodriver, 121–22
Servomechanism, 3
Servo system, 3, 257–60, 674
  with velocity feedback, 275–76
Set point, 183
Settling time, 265, 268–69
  5% criterion, 268–69
  2% criterion, 268–69
Sign inverter, 113
Signal flow graph, 52
Signal stabilization, 649
Similar matrices, 913
Similarity transformation, 913, 915–16,
    920
  invariant properties under, 921–22
Singular matrix, 897
Singular point, 15
Sink, 52
Sinusoidal describing function, 652
Sinusoidal function, 19
Sinusoidal-signal generators, 506
Sinusoidal transfer function, 427–29
Skew-Hermitian matrix, 893
Skew-symmetric matrix, 891–92
Soft spring, 646
Source, 52
Space vehicle attitude control system, 642
Space vehicle control system, 642
Speed control system, 6–7, 238–39, 241
Spring–mass–dashpot system, 101
Spring preload nonlinearity, 673
Square-law nonlinearity, 94
Stable matrix, 838
Stability:
  in the sense of Liapunov, 730
Stability analysis:
  in the complex plane, 282–83
Stack controller, 199
State, 60